Handbook of
Thin Film Materials

Handbook of Thin Film Materials

Volume 5

Nanomaterials and Magnetic Thin Films

Edited by

Hari Singh Nalwa, M.Sc., Ph.D.
Stanford Scientific Corporation
Los Angeles, California, USA

Formerly at
Hitachi Research Laboratory
Hitachi Ltd., Ibaraki, Japan

ACADEMIC PRESS

A Division of Harcourt, Inc.

San Diego   San Francisco   New York   Boston   London   Sydney   Tokyo

ACADEMIC PRESS
A division of Harcourt, Inc.
525 B Street, Suite 1900, San Diego, CA 92101-4495, USA
http://www.academicpress.com

Academic Press
Harcourt Place, 32 Jamestown Road, London, NW1 7BY, UK
http://www.academicpress.com

Library of Congress Catalog Card Number: 00-2001090614
International Standard Book Number, Set: 0-12-512908-4
International Standard Book Number, Volume 5: 0-12-512913-0

Printed in the United States of America
01   02   03   04   05   06   07   MB   9   8   7   6   5   4   3   2   1

1002667002

To my children
Surya, Ravina, and Eric

Preface

Thin film materials are the key elements of continued technological advances made in the fields of electronic, photonic, and magnetic devices. The processing of materials into thin-films allows easy integration into various types of devices. The thin film materials discussed in this handbook include semiconductors, superconductors, ferroelectrics, nanostructured materials, magnetic materials, etc. Thin film materials have already been used in semiconductor devices, wireless communication, telecommunications, integrated circuits, solar cells, light-emitting diodes, liquid crystal displays, magneto-optic memories, audio and video systems, compact discs, electro-optic coatings, memories, multilayer capacitors, flat-panel displays, smart windows, computer chips, magneto-optic disks, lithography, microelectromechanical systems (MEMS) and multifunctional protective coatings, as well as other emerging cutting edge technologies. The vast variety of thin film materials, their deposition, processing and fabrication techniques, spectroscopic characterization, optical characterization probes, physical properties, and structure-property relationships compiled in this handbook are the key features of such devices and basis of thin film technology.

Many of these thin film applications have been covered in the five volumes of the *Handbook of Thin Film Devices* edited by M. H. Francombe (Academic Press, 2000). The *Handbook of Thin Film Materials* is complementary to that handbook on devices. The publication of these two handbooks, selectively focused on thin film materials and devices, covers almost every conceivable topic on thin films in the fields of science and engineering.

This is the first handbook ever published on thin film materials. The 5-volume set summarizes the advances in thin film materials made over past decades. This handbook is a unique source of the in-depth knowledge of deposition, processing, spectroscopy, physical properties, and structure–property relationship of thin film materials. This handbook contains 65 state-of-the-art review chapters written by more than 125 world-leading experts from 22 countries. The most renowned scientists write over 16,000 bibliographic citations and thousands of figures, tables, photographs, chemical structures, and equations. It has been divided into 5 parts based on thematic topics:

Volume 1: Deposition and Processing of Thin Films
Volume 2: Characterization and Spectroscopy of Thin Films
Volume 3: Ferroelectric and Dielectric Thin Films
Volume 4: Semiconductor and Superconductor Thin Films
Volume 5: Nanomaterials and Magnetic Thin Films

Volume 1 has 14 chapters on different aspects of thin film deposition and processing techniques. Thin films and coatings are deposited with chemical vapor deposition (CVD), physical vapor deposition (PVD), plasma and ion beam techniques for developing materials for electronics, optics, microelectronic packaging, surface science, catalytic, and biomedical technological applications. The various chapters include: methods of deposition of hydrogenated amorphous silicon for device applications, atomic layer deposition, laser applications in transparent conducting oxide thin film processing, cold plasma processing in surface science and technology, electrochemical formation of thin films of binary III–V compounds, nucleation, growth and crystallization of thin films, ion implant doping and isolation of GaN and related materials, plasma etching of GaN and related materials, residual stresses in physically vapor deposited thin films, Langmuir–Blodgett films of biological molecules, structure formation during electrocrystallization of metal films, epitaxial thin films of intermetallic compounds, pulsed laser deposition of thin films: expectations and reality and b″-alumina single-crystal films. This volume is a good reference source of information for those individuals who are interested in the thin film deposition and processing techniques.

Volume 2 has 15 chapters focused on the spectroscopic characterization of thin films. The characterization of thin films using spectroscopic, optical, mechanical, X-ray, and electron microscopy techniques. The various topics in this volume include: classification of cluster morphologies, the band structure and orientations of molecular adsorbates on surfaces by angle-resolved electron spectroscopies, electronic states in GaAs-AlAs short-period superlattices: energy levels and symmetry, ion beam characterization in superlattices, *in situ* real time spectroscopic ellipsometry studies: carbon-based materials and metallic TiNx thin films growth, *in situ* Faraday-modulated fast-nulling single-wavelength ellipsometry of the growth of semiconductor, dielectric and metal thin films, photocurrent spectroscopy of thin passive films, low frequency noise spectroscopy for characterization of polycrystalline semiconducting thin films and polysilicon thin film transistors, electron energy loss spectroscopy for surface study, theory of low-energy electron diffraction and photoelectron spectroscopy from ultra-thin films, *in situ* synchrotron structural studies of the growth of oxides and metals, operator formalism in polarization nonlinear optics and spectroscopy of polarization inhomogeneous media, secondary ion mass spectrometry (SIMS) and its application to thin films characterization, and a solid state approach to Langmuir monolayers, their phases, phase transitions and design.

Volume 3 focuses on dielectric and ferroelectric thin films which have applications in microelectronics packaging, ferroelectric random access memories (FeRAMs), microelectromechanical systems (MEMS), metal–ferroelectric–semiconductor field-effect transistors (MFSFETs), broad band wireless communication, etc. For example, the ferroelectric materials such as barium strontium titanate discussed in this handbook have applications in a number of tunable circuits. On the other hand, high-permittivity thin film materials are used in capacitors and for integration with MEMS devices. Volume 5 of the *Handbook of Thin Film Devices* summarizes applications of ferroelectrics thin films in industrial devices. The 12 chapters on ferroelectrics thin films in this volume are complimentary to Volume 5 as they are the key components of such ferroelectrics devices. The various topics include electrical properties of high dielectric constant and ferroelectrics thin films for very large scale integration (VLSI) integrated circuits, high permittivity (Ba, Sr)TiO$_3$ thin films, ultrathin gate dielectric films for Si-based microelectronic devices, piezoelectric thin films: processing and properties, fabrication and characterization of ferroelectric oxide thin films, ferroelectric thin films of modified lead titanate, point defects in thin insulating films of lithium fluoride for optical microsystems, polarization switching of ferroelecric crystals, high temperature superconductor and ferroelectrics thin films for microwave applications, twinning in ferroelectrics thin films: theory and structural analysis, and ferroelectrics polymers Langmuir–Blodgett films.

Volume 4 has 13 chapters dealing with semiconductor and superconductor thin film materials. Volumes 1, 2, and 3 of the *Handbook of Thin Film Devices* summarize applications of semiconductor and superconductors thin films in various types of electronic, photonic and electro-optics devices such as infrared detectors, quantum well infrared photodetectors (QWIPs), semiconductor lasers, quantum cascade lasers, light emitting diodes, liquid crystal and plasma displays, solar cells, field effect transistors, integrated circuits, microwave devices, SQUID magnetometers, etc. The semiconductor and superconductor thin film materials discussed in this volume are the key components of such above mentioned devices fabricated by many industries around the world. Therefore this volume is in coordination to Volumes 1, 2, and 3 of the *Handbook of Thin Film Devices*. The various topics in this volume include; electrochemical passivation of Si and SiGe surfaces, optical properties of highly excited (Al, In)GaN epilayers and heterostructures, electical conduction properties of thin films of cadmium compounds, carbon containing heteroepitaxial silicon and silicon/germanium thin films on Si(001), germanium thin films on silicon for detection of near-infrared light, physical properties of amorphous gallium arsenide, amorphous carbon thin films, high-T_c superconducting thin films, electronic and optical properties of strained semiconductor films of group V and III-V materials, growth, structure and properties of plasma-deposited amorphous hydrogenated carbon–nitrogen films, conductive metal oxide thin films, and optical properties of dielectric and semiconductor thin films.

Volume 5 has 12 chapters on different aspects of nanostructured materials and magnetic thin films. Volume 5 of the *Handbook of Thin Film Devices* summarizes device applications of magnetic thin films in permanent magnets, magneto-optical recording, microwave, magnetic MEMS, etc. Volume 5 of this handbook on magnetic thin film materials is complimentary to Volume 5 as they are the key components of above-mentioned magnetic devices. The various topics covered in this volume are; nanoimprinting techniques, the energy gap of clusters, nanoparticles and quantum dots, spin waves in thin films, multi-layers and superlattices, quantum well interference in double quantum wells, electro-optical and transport properties of quasi-two-dimensional nanostrutured materials, magnetism of nanoscale composite films, thin magnetic films, magnetotransport effects in semiconductors, thin films for high density magnetic recording, nuclear resonance in magnetic thin films, and multilayers, and magnetic characterization of superconducting thin films.

I hope these volumes will be very useful for the libraries in universities and industrial institutions, governments and independent institutes, upper-level undergraduate and graduate students, individual research groups and scientists working in the field of thin films technology, materials science, solid-state physics, electrical and electronics engineering, spectroscopy, superconductivity, optical engineering, device engineering nanotechnology, and information technology, everyone who is involved in science and engineering of thin film materials.

I appreciate splendid cooperation of many distinguished experts who devoted their valuable time and effort to write excellent state-of-the-art review chapters for this handbook. Finally, I have great appreciation to my wife Dr. Beena Singh Nalwa for her wonderful cooperation and patience in enduring this work, great support of my parents Sri Kadam Singh and Srimati Sukh Devi and love of my children, Surya, Ravina and Eric in this exciting project.

Hari Singh Nalwa
Los Angeles, CA, USA

Contents

Chapter 1. NANOIMPRINT TECHNIQUES

Hella-C. Scheer, Hubert Schulz, Thomas Hoffmann,
Clivia M. Sotomayor Torres

Chapter 2. THE ENERGY GAP OF CLUSTERS, NANOPARTICLES, AND QUANTUM DOTS

Klaus Sattler

Chapter 3. ELECTRONIC STATES IN GaAs-AlAs SHORT-PERIOD SUPERLATTICES: ENERGY LEVELS AND SYMMETRY

Jian-Bai Xia, Weikun Ge

Chapter 4. SPIN WAVES IN THIN FILMS, SUPERLATTICES AND MULTILAYERS

Zhang Zhi-Dong

Chapter 5. QUANTUM WELL INTERFERENCE IN DOUBLE QUANTUM WELLS

Zhang Zhi-Dong

Chapter 6. ELECTRO-OPTICAL AND TRANSPORT PROPERTIES OF QUASI-TWO-DIMENSIONAL NANOSTRUCTURED MATERIALS

Rodrigo A. Rosas, Raúl Riera, José L. Marín, Germán Campoy

Chapter 7.   MAGNETISM OF NANOPHASE COMPOSITE FILMS

D. J. Sellmyer, C. P. Luo, Y. Qiang, J. P. Liu

Chapter 8.   THIN MAGNETIC FILMS

Hans Hauser, Rupert Chabicovsky, Karl Riedling

Chapter 9.   MAGNETOTRANSPORT EFFECTS IN SEMICONDUCTORS

Nicola Pinto, Roberto Murri, Marian Nowak

Chapter 10.   THIN FILMS FOR HIGH-DENSITY MAGNETIC RECORDING

Genhua Pan

Chapter 11. NUCLEAR RESONANCE IN MAGNETIC THIN FILMS AND MULTILAYERS

Mircea Serban Rogalski

Chapter 12. MAGNETIC CHARACTERIZATION OF SUPERCONDUCTING THIN FILMS

M. R. Koblischka

About the Editor

Dr. Hari Singh Nalwa is the Managing Director of the Stanford Scientific Corporation in Los Angeles, California. Previously, he was Head of Department and R&D Manager at the Ciba Specialty Chemicals Corporation in Los Angeles (1999–2000) and a staff scientist at the Hitachi Research Laboratory, Hitachi Ltd., Japan (1990–1999). He has authored over 150 scientific articles in journals and books. He has 18 patents, either issued or applied for, on electronic and photonic materials and devices based on them.

He has published 43 books including *Ferroelectric Polymers* (Marcel Dekker, 1995), *Nonlinear Optics of Organic Molecules and Polymers* (CRC Press, 1997), *Organic Electroluminescent Materials and Devices* (Gordon & Breach, 1997), *Handbook of Organic Conductive Molecules and Polymers*, Vols. 1–4 (John Wiley & Sons, 1997), *Handbook of Low and High Dielectric Constant Materials and Their Applications*, Vols. 1–2 (Academic Press, 1999), *Handbook of Nanostructured Materials and Nanotechnology*, Vols. 1–5 (Academic Press, 2000), *Handbook of Advanced Electronic and Photonic Materials and Devices*, Vols. 1–10 (Academic Press, 2001), *Advanced Functional Molecules and Polymers*, Vols. 1–4 (Gordon & Breach, 2001), *Photodetectors and Fiber Optics* (Academic Press, 2001), *Silicon-Based Materials and Devices*, Vols. 1–2 (Academic Press, 2001), *Supramolecular Photosensitive and Electroactive Materials* (Academic Press, 2001), *Nanostructured Materials and Nanotechnology*–Condensed Edition (Academic Press, 2001), and *Handbook of Thin Film Materials*, Vols. 1–5 (Academic Press, 2002). The *Handbook of Nanostructured Materials and Nanotechnology* edited by him received the 1999 Award of Excellence in Engineering Handbooks from the Association of American Publishers.

Dr. Nalwa is the founder and Editor-in-Chief of the *Journal of Nanoscience and Nanotechnology* (2001–). He also was the founder and Editor-in-Chief of the *Journal of Porphyrins and Phthalocyanines* published by John Wiley & Sons (1997–2000) and serves or has served on the editorial boards of *Journal of Macromolecular Science-Physics* (1994–), *Applied Organometallic Chemistry* (1993–1999), *International Journal of Photoenergy* (1998–) and *Photonics Science News* (1995–). He has been a referee for many international journals including *Journal of American Chemical Society, Journal of Physical Chemistry, Applied Physics Letters, Journal of Applied Physics, Chemistry of Materials, Journal of Materials Science, Coordination Chemistry Reviews, Applied Organometallic Chemistry, Journal of Porphyrins and Phthalocyanines, Journal of Macromolecular Science-Physics, Applied Physics, Materials Research Bulletin*, and *Optical Communications*.

Dr. Nalwa helped organize the First International Symposium on the Crystal Growth of Organic Materials (Tokyo, 1989) and the Second International Symposium on Phthalocyanines (Edinburgh, 1998) under the auspices of the Royal Society of Chemistry. He also proposed a conference on porphyrins and phthalocyanies to the scientific community that, in part, was intended to promote public awareness of the *Journal of Porphyrins and Phthalocyanines*, which he founded in 1996. As a member of the organizing committee, he helped effectuate the First International Conference on Porphyrins and Phthalocyanines, which was held in Dijon, France in 2000. Currently he is on the organizing committee of the BioMEMS and Smart Nanostructures, (December 17–19, 2001, Adelaide, Australia) and the World Congress on Biomimetics and Artificial Muscles (December 9–11, 2002, Albuquerque, USA).

Dr. Nalwa has been cited in the *Dictionary of International Biography, Who's Who in Science and Engineering, Who's Who in America,* and *Who's Who in the World.* He is a member of the American Chemical Society (ACS), the American Physical Society (APS), the Materials Research Society (MRS), the Electrochemical Society and the American Association for the Advancement of Science (AAAS). He has been awarded a number of prestigious fellowships including a National Merit Scholarship, an Indian Space Research Organization (ISRO) Fellowship, a Council of Scientific and Industrial Research (CSIR) Senior fellowship, a NEC fellowship, and Japanese Government Science & Technology Agency (STA) Fellowship. He was an Honorary Visiting Professor at the Indian Institute of Technology in New Delhi.

Dr. Nalwa received a B.Sc. degree in biosciences from Meerut University in 1974, a M.Sc. degree in organic chemistry from University of Roorkee in 1977, and a Ph.D. degree in polymer science from Indian Institute of Technology in New Delhi in 1983. His thesis research focused on the electrical properties of macromolecules. Since then, his research activities and professional career have been devoted to studies of electronic and photonic organic and polymeric materials. His endeavors include molecular design, chemical synthesis, spectroscopic characterization, structure-property relationships, and evaluation of novel high performance materials for electronic and photonic applications. He was a guest scientist at Hahn-Meitner Institute in Berlin, Germany (1983) and research associate at University of Southern California in Los Angeles (1984–1987) and State University of New York at Buffalo (1987–1988). In 1988 he moved to the Tokyo University of Agriculture and Technology, Japan as a lecturer (1988–1990), where he taught and conducted research on electronic and photonic materials. His research activities include studies of ferroelectric polymers, nonlinear optical materials for integrated optics, low and high dielectric constant materials for microelectronics packaging, electrically conducting polymers, electroluminescent materials, nanocrystalline and nanostructured materials, photocuring polymers, polymer electrets, organic semiconductors, Langmuir-Blodgett films, high temperature-resistant polymer composites, water-soluble polymers, rapid modeling, and stereolithography.

List of Contributors

Numbers in parenthesis indicate the pages on which the author's contribution begins.

GERMÁN CAMPOY (207)
Departamento de Investigatión en Física, Universidad de Sonora, Hermosillo, Sonora, Mexico

RUPERT CHABICOVSKY (375)
Institute of Industrial Electronics and Material Science, Vienna University of Technology,
A-1040 Vienna, Austria

WEIKUN GE (99)
Department of Physics, Hong Kong University of Science and Technology, Clearwater Bay,
Kowloon, Hong Kong, China

HANS HAUSER (375)
Institute of Industrial Electronics and Material Science, Vienna University of Technology,
A-1040 Vienna, Austria

THOMAS HOFFMANN (1)
IMEC vzw, SPT Optical Lithography, Heverlee (Leuven), Belgium

M. R. KOBLISCHKA (589)
Experimentalphysik, Universität des Saarlandes, D-66041 Saarbrücken, Germany

J. P. LIU (337)
Department of Physics and Institute for Micromanufacturing, Louisiana Tech University,
Ruston, Louisiana, USA

C. P. LUO (337)
Behlen Laboratory of Physics and Center for Materials Research and Analysis,
University of Nebraska, Lincoln, Nebraska, USA

JOSÉ L. MARÍN (207)
Departamento de Investigatión en Física, Universidad de Sonora, Hermosillo, Sonora, Mexico

ROBERTO MURRI (439)
INFM, Dipartimento di Matematica e Fisica, Università di Camerino, 62032 Camerino, Italy

MARIAN NOWAK (439)
Silesian Technical University, Institute of Physics, 40-019 Katowice, Poland

GENHUA PAN (495)
Centre for Research in Information Storage Technology, Department of Communication and
Electronic Engineering, University of Plymouth, Plymouth, Devon PL4 5AA, United Kingdom

NICOLA PINTO (439)
INFM, Dipartimento di Matematica e Fisica, Università di Camerino, 62032 Camerino, Italy

Y. QIANG (337)
Behlen Laboratory of Physics and Center for Materials Research and Analysis,
University of Nebraska, Lincoln, Nebraska, USA

KARL RIEDLING (375)
Institute of Industrial Electronics and Material Science, Vienna University of Technology,
A-1040 Vienna, Austria

RAÚL RIERA (207)
Departamento de Investigatión en Física, Universidad de Sonora, Hermosillo, Sonora, Mexico

MIRCEA SERBAN ROGALSKI (555)
Faculdade de Ciências e Tecnologia, Universidade do Algarve, Gambelas, 8000 Faro, Portugal

RODRIGO A. ROSAS (207)
Departamento de Física, Universidad de Sonora, Hermosillo, Sonora, México

KLAUS SATTLER (61)
Department of Physics and Astronomy, University of Hawaii at Manoa,
Honolulu, Hawaii, USA

HELLA-C. SCHEER (1)
Microstructure Engineering, Department of Electrical and Information Engineering,
University of Wuppertal, 42097 Wuppertal, Germany

HUBERT SCHULZ (1)
Microstructure Engineering, Department of Electrical and Information Engineering,
University of Wuppertal, 42097 Wuppertal, Germany

D. J. SELLMYER (337)
Behlen Laboratory of Physics and Center for Materials Research and Analysis,
University of Nebraska, Lincoln, Nebraska, USA

CLIVIA M. SOTOMAYOR TORRES (1)
Institute of Materials Science, Department of Electrical and Information Engineering,
University of Wuppertal, 42097 Wuppertal, Germany

JIAN-BAI XIA (99)
Department of Physics, Hong Kong University of Science and Technology, Clearwater Bay,
Kowloon, Hong Kong, China and
Institute of Semiconductors, Chinese Academy of Sciences, Beijing 100083, China

ZHANG ZHI-DONG (141, 169)
Shenyang National Laboratory for Materials Science and International Centre for
Material Physics, Institute of Metal Research, Academia Sinica,
Shenyang 110015, People's Republic of China

Handbook of Thin Film Materials

Edited by H.S. Nalwa

Volume 1. DEPOSITION AND PROCESSING OF THIN FILMS

Volume 2. CHARACTERIZATION AND SPECTROSCOPY OF THIN FILMS

Volume 3. FERROELECTRIC AND DIELECTRIC THIN FILMS

Volume 4. SEMICONDUCTOR AND SUPERCONDUCTING THIN FILMS

Volume 5. NANOMATERIALS AND MAGNETIC THIN FILMS

Chapter 1

Nanoimprint Techniques

Hella-C. Scheer, Hubert Schulz
*Microstructure Engineering, Department of Electrical and Information Engineering,
University of Wuppertal, 42097 Wuppertal, Germany*

Thomas Hoffmann
IMEC vzw, SPT Optical Lithography, Heverlee (Leuven), Belgium

Clivia M. Sotomayor Torres
*Institute of Materials Science, Department of Electrical and Information Engineering,
University of Wuppertal, 42097 Wuppertal, Germany*

Contents

Handbook of Thin Film Materials, edited by H.S. Nalwa
Volume 5: Nanomaterials and Magnetic Thin Films

ISBN 0-12-512913-0/$35.00

1. INTRODUCTION

The modification of thin-film surfaces by means of patterning nanometer-sized features obeys a number of requirements, ranging from those of the microelectronics industry to biotechnology. In addition to industrial-driven demands, the quest to understand matter and its interactions at the nanometer scale needs laboratory techniques that open the door to the study of single nanostructures, be it a small cluster, a quantum dot, or a nanovalve, by means of, preferentially, scanning probe techniques. The response of a nanostructure array such as those used in biosensors, information storage devices, diffractive optics, and photonic crystals, to name but a few, is also becoming increasingly important. This follows the research trend toward high-density information acquisition and processing systems, which are based on arrays containing the smallest possible controllable and reproducible features, with well-defined properties. Among these research fields, one finds quantum electronics, quantum computing, bioelectronics, and neuroinformatics.

An added flexibility is offered by the combination of nanoimprint lithography, as a lateral patterning technique, with multilayer thin-film technology to unleash the potential of three-dimensional nanofabrication. In particular, combinations of materials, such as polymer–semiconductor, organic semiconductor–dielectric, polymer–superconductor, ceramic–metal, biological material–oxide, and colloid–glass, bring about a whole range of new science and, above all, potential applications.

The different approaches to nanometer-scaled printing range from molding of a thin polymer layer by a stamp under controlled temperature and pressure, usually referred to as hot embossing or nanoimprinting, to transfer of a monolayer of self-assembled molecules from an elastomeric stamp to a substrate, usually known as microcontact printing. Classification schemes may be constructed depending on whether processes are primarily physical or chemical, whether they require pressure or only ultraviolet curing, and so on. In this chapter, several classification criteria are explored.

In this context, nanoimprint techniques are seen as lithography. In this section, the current status of lithography is presented, followed by leading to the basic ideas of printing-based techniques.

1.1. Current Lithography Situation

The actual status of lithography results from 3 decades of ongoing miniaturization for the production of silicon integrated circuits where the minimum active feature sizes were reduced by a factor of 1.4 every 3 years. Despite the fact that patterns down to 180 nm are actually used in the mass fabrication of DRAMs (dynamic random-access memories), however, the lithographic definition of such pattern sizes is far from being trivial.

Over the years, optical lithography concepts have been improved to meet these challenges. Exposure wavelengths have reached the deep ultraviolet (DUV) [1], requiring development of new machine concepts for beam forming and specific optical materials. Resolution enhancement methods such as phase-shifting masks [2] were developed to improve the contrast of optical images by ruling out diffraction effects. Resolution enhancement enables the reproduction of features at approximately half the wavelength of the incident light [3]. Specific antireflection coatings were implemented in the layer sequence and multilayer resist systems were used, where optical lithography is combined with dry development of an underlying resist. As a consequence, optical lithography as used in the fabrication of devices has become a very intricate and cost-intensive technique.

On the other hand, as the limit of optical lithography had been expected to impose an ultimate boundary on the minimum feature size, currently estimated at 70 nm [1, 3], a number of alternative techniques have been developed to define patterns in the deep submicrometer range [3]. One of them is X-ray lithography, [4] offering a high focal depth compared to all other lithography techniques due to the nearly parallel radiation supplied by a synchrotron source. Major challenges are the realization of a compact electron storage ring as a lithography tool, as well as the defect-free fabrication and in-process handling of masks having the same scale as the features exposed, because X-ray lithography is a 1 : 1 technique. Soft X-ray reflection techniques have been presented as extreme ultraviolet (EUV) [3].

Ion beam projection lithography (IPL) has been developed as well, where a beam of, for example gallium ions, exposes the resist in a demagnifying system. Lithography systems require high electric fields to realize the electrostatic lenses for formation of the beam of heavy ions, and pattern definition has to take into account the use of masks with physical holes. Both techniques, X-ray and ion beam as well, have demonstrated pattern definition in the sub-100-nm range, but are not yet widely used in device fabrication. The main reasons are the high cost of these systems and techniques as well as the changes required in a production line upon replacement of optical lithography.

Furthermore, such lithography concepts are not convenient for definition of nanometer scaled patterns for research. Here electron beam writing is the technique of choice, and low-cost systems are in use, where a writing unit is combined with a high-resolution scanning electron microscope (SEM). In addition, surface probe methods (SPMs) have emerged, where an atomic force microscope (AFM) tip develops a resist in liquid environment via current, a technique for smallest volume fabrication of nanometer structures due to its sequential nature [5–7].

Electron beam lithography is not only the workhorse in research when pattern sizes in the sub-100-nm range are envisaged, it is also the current technique for optical mask fabrication for device production of pattern sizes of 1–2 μm and below. Moreover, it is the only production-proven technique for pattern sizes beyond optical lithography. Shaped beam techniques have been developed in order to increase productivity, where writing speed and writing fields of production systems have been improved substantially. Furthermore, projection systems such as SCALPEL (scattering with angular limitation projection electron beam lithography) have been developed where contrast is obtained by scattering of electrons out of the optical system [3]. Nevertheless, electron beam lithography is a technique with limited throughput, leading to high costs in device production.

The contest of lithography techniques for reliable fabrication of future integrated nanometer-scaled devices is not yet settled.

1.2. Emerging Imprint-Based Lithography Concepts: Basic Ideas and Expected Advantages

Stimulated by this open situation, a number of imprint-based alternative lithography concepts have emerged during the last few years. They borrow concepts from techniques that had not been recognized in connection with lithography in its classical form until recently, at least not by scientists and engineers working in the technology of submicrometer- and nanometer-scaled devices. They are derived from replication techniques [8] well known from the fabrication of diffractive optical elements (DOEs) [9] and integrated optics [10], from micromachining techniques such as LIGA (lithography with galvanic replication) [11], and from printing.

Common to all these new techniques is the use of a template with the patterns envisaged, the master. The master's features are replicated in a thin layer of material on a substrate. Replication can be done by molding, embossing, or printing as well as by combinations thereof, and the replica used can be rigid or soft.

The fact that the basic ideas are borrowed from well-established fields of research does not mean that the techniques have reached production maturity. Replication techniques have not yet been developed into a lithographic technology [12, 13] to pattern semiconductors, metals, and dielectrics used in the fabrication of ultra large scale integrated (ULSI) devices. It is not only a question of feature sizes or of patterned areas, as replication of, for example, DOEs, requires patterns down to several 100 nm covering the surface of an optically suitable material. State-of-the-art replication is a surface-oriented technique where just the uppermost part of a film, either deposited on a substrate or freestanding, is patterned. Often the pattern depth is small compared to the film thickness and the unpatterned base acts as a support. It is not a layer-oriented technique as is required for lithography, where a mostly thin layer of a polymer has to be patterned in its full thickness to form a mask for a subsequent process (e.g., dry etching). As will be discussed in detail in Sections 2 and 3, full-layer patterning imposes a number of new challenges to replication due to the thin layers and the rigid substrate involved. One of these challenges is the mainly lateral flow of material in planes parallel to the surface. Additional ones include the need for a high and uniform thickness contrast and the need for "masking" a subsequent process. The same argument holds for the thermoplastic or reactive injection molding techniques used with high throughput for large-area patterning in the fabrication of CDs (compact discs) or in LIGA parts and tools. Overlaps between these classical surface-oriented production techniques and the lithography concepts exist and will increase with (i) the development of these techniques toward nanometer-scaled features [14], (ii) the use of masters made from Si wafers [15], and (iii) the improvement of stability of the plastic components when a rigid substrate is used as a support [16].

In the following subsections, three basically different approaches and their expected advantages will be introduced. These three approaches designate the main ideas for alternative imprint-based lithography concepts. For a critical assessment of these, the reader is referred to the detailed discussion in Sections 2–6. Moreover, a survey of the wide range of process modifications and their nomenclature will be given.

1.2.1. Nanoimprint Lithography/Hot-Embossing Lithography

The term "nanoimprint lithography" was introduced by Chou in 1995 [13, 17] when he filed a patent under this name. It makes use of the mechanical deformation of a polymer layer under pressure and elevated temperature. The idea of the process is explained in Fig. 1. The master with its patterned surface and a sample, prepared with a layer of poly(methyl methacrylate) (PMMA), are heated up to about 200°C. Then the stamp is pressed against the sample with a pressure of about 130 bar. After cooldown to near 100°C, the pressure is released and the mold is separated from the sample. The resulting patterned polymer forms a mask on the substrate.

The main advantages expected in comparison to electron beam writing are the parallel nature of the process, the high throughput, and the low system costs, because no energetic particle beam generator is needed [17, 18]. In addition, nanoimprinting eliminates a number of adverse effects of conventional lithography techniques [12, 18] such as wave diffraction, scattering and interference in the resist, electron backscattering from the substrate, and chemical issues of resists and developers. Large-area processing over more than 50×50 mm^2 is expected to work without the defect problems of state-of-the-art lithography [12]. Low-cost mass production of nanometer-scaled features can be achieved by multiple imprint, where only the mold has to be fabricated in a high-resolution, low-throughput process [18].

This technique is also known as hot-embossing lithography [19] and is expected to be a powerful nanomanufacturing technique for the future, in particular, when resist structures with high aspect ratio are needed [20].

1.2.2. Mold-Assisted Lithography

Mold-assisted lithography was introduced by Haisma and co-workers in 1996 [21] and is based on the molding of a monomer in a vacuum contact printer and subsequent curing by ultraviolet (UV) radiation. The idea is illustrated in Fig. 2. For a lithography process, UV curing has to be performed through the mold, which is therefore made from fused silica. The polymer layer features a thickness contrast after the

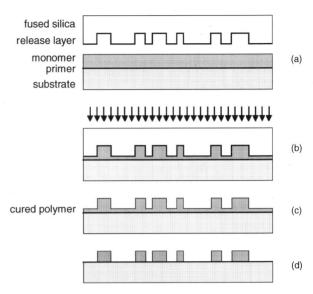

Fig. 2. Principle of mold-assisted lithography. A fused silica mold is prepared with a release layer, the substrate with a primer and a monomer (a). Under vacuum contact, the monomer fills the mold cavities and is then cured by UV radiation (b). The thin residual layer of polymer remaining between the patterns after separation (c) is removed by etching, resulting in the polymer mask (d).

process, and a thin residual layer has to be removed in a reactive ion etching (RIE) step to form the local polymer mask. To prevent sticking during polymerization, the master and the substrate are treated with a mold release layer and a primer, respectively.

A major advantage of this process is its direct compatibility with current Si technology, as a commercially available vacuum contact printer is employed, enabling wafer-scale processing. Commonly used fused silica photomasks are the ideal choice for the master and UV radiation of 300–400 nm is used for curing. The alignment and parallelism adjustments of the printer can be used. Thus, mold-assisted lithography should be a low-cost technique because no specific new tool development is required.

1.2.3. Microcontact Printing

The concept of microcontact printing was presented by Whitesides and co-workers in 1993/1994 [22, 23]. The idea is to print a self-assembling monolayer (SAM) on a substrate using an elastomeric stamp. The principle is shown in Fig. 3, starting with the preparation of the stamp. A precursor of poly(dimethylsiloxane) (PDMS), an elastomeric material, is poured over the template, cured, and peeled off. The surface of this elastomeric stamp is inked with alkanethiol and printed onto a Si substrate prepared with a thin Au layer. Upon contact, an SAM is formed within a few seconds which protects the Au surface so that a selective wet etch can be performed.

The main advantages of this technique result from the conformal contact to nonflat surfaces, which is due to the elasticity of the stamp. As a consequence, flatness of the surface over large areas is not a problem, and even three-dimensional

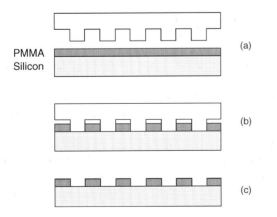

Fig. 1. Principle of nanoimprint lithography or hot-embossing lithography. A patterned stamp and a Si wafer prepared with PMMA are heated to a temperature above the glass transition of the polymer (a); under pressure application, the polymer layer is patterned mechanically (b), resulting in a polymer mask on top of the wafer (c).

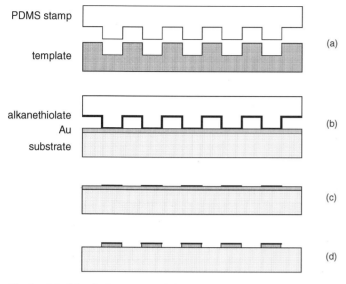

PDMS stamp

template

alkanethiolate
Au
substrate

(a)

(b)

(c)

(d)

Fig. 3. Principle of microcontact printing. A template is replicated in PDMS via molding (a). After peel-off, the PDMS stamp is inked over its whole surface and prints the alkanethiol on Au (b). After separation, a patterned SAM remains on the substrate (c). It is used as a mask in a wet-etch process to pattern the Au layer (d).

surfaces can be printed. A further advantage is that the template may consist of a patterned polymer layer on a substrate. Thus, in contrast to hot-embossing and mold-assisted lithography, the master fabrication for microcontact printing needs only lithography.

1.3. Overview of Imprint-Based Techniques and Nomenclature

The example of microcontact printing makes clear the many links among the three techniques. The stamp for microcontact printing is made in a curing process resembling the mold lithography procedure presented before. Curing processes for crosslinking of polymers can also be performed by a temperature treatment, not only by radiation. The former is also employed for the fabrication of embossing stamps, linking nanoimprinting and molding.

The situation becomes less clear as various basic replication lithography techniques have been proposed under a number of different names. Some of these techniques are surface oriented than layer oriented. Most of them claim nanometer patterning potential, whereas others have implicitly developed in this direction.

To introduce a structured nomenclature, the nanoimprint techniques are listed in Tables I to III according to technological aspects (Table I), to stamp materials (Table II), and to their suitability for a lithography process (Table III).

In particular, the term "nanoimprint lithography" is often used in a more general way for all these new emerging lithography techniques and does not only refer to hot-embossing lithography. As a consequence, we will use the somewhat more precise term "hot-embossing lithography" for the physical/mechanical replication process. The terms "imprinting"

and "embossing" are used as synonyms in the remainder of this chapter.

1.4. This Chapter

The results and challenges of the different techniques are presented. Hot-embossing lithography is discussed in Section 2, emphasizing on the thermomechanical behavior of polymers as one main issue of this technique. Some aspects of mold-assisted lithography and microcontact printing are summarized in Sections 3 and 4 in order to complement the section on hot embossing. Results on achieved pattern sizes, pattern fidelity, uniformity, and large-scale approaches will be presented as far as they are available. In Sections 5 and 6, general topics relevant to all imprint techniques are addressed, in particular, stamp fabrication and wear as well as the issue of sticking and antisticking layers. Section 7 presents some applications of these techniques, and concluding remarks and prospects comprise Section 8.

2. HOT-EMBOSSING LITHOGRAPHY

This section discusses details on nanoimprint lithography under pressure and elevated temperature, namely, hot-embossing lithography (HEL). The organization is as follows. After a short historical overview differentiating hot-embossing lithography and classical embossing, as used, for example, for optical elements, the principle behind the technique is discussed in detail, followed by a summary of the achievements concerning pattern sizes, fidelity, and embossed materials. Then the typical hardware used for imprinting is presented, in particular, large-area approaches. Processing details correlated to the hardware used are also given.

Before entering a discussion on fundamental process challenges, namely, the polymer flow behavior, a detailed résumé on the viscoelastic properties of polymers is given, explaining the main issues of polymer behavior during hot embossing.

Section 2 ends with multilayer techniques, ideas for room temperature embossing, and combinations with optical lithography.

2.1. History of Classical Hot Embossing

Embossing of micro- and nanostructures has a long history—not as a lithography process, but as a surface-oriented technique. As many of the embossing lithography ideas are borrowed from there, a short survey is given.

One of the first applications still in use is the embossing of metals to make coins, the surfaces of which show micrometer-scaled patterns. A little bit closer to the hot-embossing lithography lies the use of sealing wax patterned in its viscous state with a signet tool (e.g. a ring). In both cases, the embossing procedure attributes meaning or value to the material embossed, which exceeds the value of the pure material many times. This is also the case in modern embossing techniques, both surface-oriented and layer-oriented ones. The low-cost

Table I. Novel Patterning Techniques for the Nanometer and Micrometer Range Based on Replication (Nanoimprint Techniques)

Physical					Chemical		Self-Assembly
					Curing	Reactive/thermal	Reactive
Self-assembly	Elevated temperature	Pressure	Room temperature	Solvent	UV radiation		
LISA Lithographically induced self-assembly [109]	**NIL** Nanoimprint lithography [13]	DNP Direct nanoprinting [411]	***SAMIM*** Solvent-assisted microcontact molding [148]	***MIMIC*** Micromolding in capillaries [142]	**MAL** Mold-assisted lithography [21] UV-based imprint [116] Nanoimprint lithography [156]	**CAS** Casting [23] (preparation of elastomeric stamps for μCP, μTM, SAMIN, MMIC, REM, LIM)	**μCP** Microcontact printing [22]
LISC Lithographically induced self-construction [112]	**HEL** Hot-embossing lithography [19]		***RTNIL*** Room temperature nanoimprint lithography [149]		**SFIL** Step-and-flash imprint lithography [36]		**LIM** Lithographic molding [141]
	SSIL Step-and-stamp imprint lithography [36]				***μTM*** Microtransfer molding [158]		
	DNI Direct nanoimprint [42]				REM Replica molding [125]		
	NSE Nanoscale embossing [114]				PIL Photoimprint lithography [191]		
	NSP Nanoscale printing [190]						
	Thermoplastic polymers Thermosets Functionalized polymers [42]	Metals [41] Sol-gels [45]	Thermoplastic polymers		Monomers, oligomers, prepolymers, and mixtures thereof		SAMs

The different techniques are ordered with respect to the type of patterning (physical, chemical, self-assembly) and further characteristic process parameters like the temperature or the means of crosslinking. The name of the techniques according to the nomenclature of the authors is given (boldface, layer-related lithography process, i.e., full patterning of a thin layer on a substrate in order to define a local mask; normal, surface-related process, i.e., patterning of the surface of a thicker film without mask definition purpose; boldface italics, suitable for both types of procedures, surface and layer patterning).

Table II.   Nanoimprint Techniques Listed in Table I, Ordered According to the Type of Stamp Used (Soft, Hard)

| Soft stamp/elastomer | | Hard stamp/metal, SiO₂, plastic | | | |
| | | Curing | | Pressure | |
Printing	Molding	Radiation	Reactive/ thermal	Elevated temperature	Room temparature
μCP	*MIMIC*	**MAL**	CAS	**NIL/HEL**	DNP
LIM	μTM	**SFIL**		**SSIL**	RTNIL
	REM	PIL		NSE/NSP	
	SAMIM			DNI	

material "polymer" obtains added value in the embossing process, serving as a mold for a high-end micromachined tool or as a mask for fabrication of a nanometer-scaled device.

Embossing is the main technique for the fabrication of large-area holograms [24], in particular, relief phase holograms. The resolution required is 100 nm in depth and several 100 nm laterally. Replication is accomplished with a metallic master stamp, which is heated and then embossed into a transparent material, typically a vinyl tape or sheet. The metallic master, usually made of Ni, is the galvanic replication of a photoresist pattern where the hologram has been initially inscribed [9].

Embossing is also widely used for volume production of optical components [8]. Once again, a Ni copy of the original microrelief is obtained in a galvanic process, which is then used to emboss a thermoplastic polymer, a sheet of poly(vinyl chloride) (PVC) or a layer of PVC on a rigid substrate. Typical process parameters are temperatures of 140°C and pressures of 30 bar, and typical patterned areas can reach up to 1000 cm².

Another early example is the embossing of optical waveguides in PMMA [25] using a glass fiber as the embossing tool. After filling of the embossed groove with a monomer and subsequent curing, the optical waveguide is completed.

2.2. Principle of Hot-Embossing Lithography

Reviewing the literature and collecting the main ideas, the principle of hot embossing as a lithography process can be summarized according to Fig. 4.

A master with a patterned surface—the "stamp"—is prepared by state-of-the-art Si technology, for example, electron

Table III.   Nanoimprint Techniques Listed in Table I, Ordered According to
the Type of Patterning
(Lithography Type, Surface Patterning Type)

Layer patterning technique suitable for lithography		Surface patterning technique
MAL	*MIMIC*	NSE/NSP
SFIL	μTM	DNP
NIL/HEL	*SAMIM*	CAS
SSIL		REM
μCP		PIL
LIM		LISA
RTNIL		

beam writing of small patterns into a resist followed by dry-etch pattern transfer into the substrate. This stamp is pressed into a thin polymer layer spun onto a substrate. The embossing is carried out at a temperature well above the glass transition temperature T_g of the polymer, where it has a relatively low viscosity and can flow under force. Suitable polymers should be thermoplastic ones moldable at elevated temperatures, with a T_g around 100°C. Typical process temperatures are about 200°C and typical pressures are about 100 bar. For specific pattern types, lower pressure and temperature values may be sufficient (see Section 2.7). To separate the stamp from the sample, both are cooled down under pressure until the temperature is below T_g.

For easy separation of stamp and sample after imprinting, the polymer should feature good adherence to the substrate and substantially lower adherence to the stamp. This can be achieved by suitable choice of the materials involved or with the help of an antisticking layer on the stamp (see Section-6.2).

The initial thickness of the polymer layer has to be tuned to the pattern sizes, their fill factor, and the depth of the pattern relief in the master, as will be discussed in Section 2.5. As soon as the stamp is in full area contact with the polymer, which is the case when the polymer fills the whole stamp relief pattern conformally, the effective pressure decreases and the imprint "stops." In fact, it slows down as a consequence of polymer transport phenomena as discussed in Section 2.7.

A residual layer of polymer remains within the compressed regions. This residual layer protects the rigid master from contact with the hard substrate and thus prolongs the lifetime of the stamp [18]. When a lithography process is envisaged, this residual layer has to be removed in order to result in a polymer mask on top of the substrate. This is generally done in an anisotropic RIE step in oxygen. To minimize the demands for this dry-etch step, the residual layer thickness after embossing should be small in comparison to the pattern height achieved in the imprinted polymer. After RIE, the lithography is completed and the polymer mask can be used for further processing of the surface. In most cases, this is either pattern transfer or lift-off. Pattern transfer of the mask into the underlying layer or in the substrate itself is performed by etching, preferably by RIE, as this is the only technique suitable for pattern transfer in the nanometer range. When layer patterning

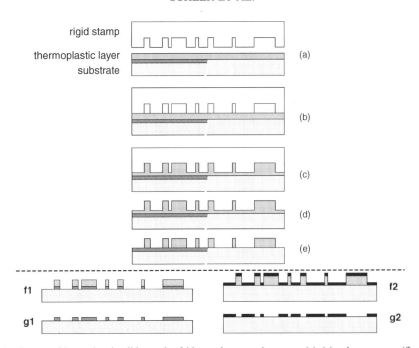

Fig. 4. Schematics of patterning by use of hot-embossing lithography. Lithography comprises steps (a)–(e), whereas steps (f) and (g) refer to the subsequent processing with the HEL mask. (a) Sample (substrate with or without additional layer, covered with a thermoplastic polymer) is heated to $T > T_g$ together with a patterned rigid stamp; (b) stamp and sample are brought into contact; (c) polymer imprint via pressure application; (d) imprinted sample after cooldown and separation; (e) sample with polymer mask after residual layer removal. Left branch (with additional layer), dry etching: (f1) dry etch of additional layer with the polymer mask; (g1) patterned layer after mask stripping. Right branch (no additional layer), lift-off: (f2) evaporation of metal over masked sample; (g2) patterned metal on top of substrate after polymer dissolution in solvent. The lift-off pattern is negative compared to the patterned additional layer of (g1).

is envisaged, the substrate has to be covered with the respective layer before spinning the polymer (left branch of Fig. 4). In case of lift-off (right branch of Fig. 4), the surface of the patterned polymer is covered with a thin layer (e.g. a metal) preferably by evaporation. Due to the low pressure regularly involved in this step, the metal atoms travel straight to the sample without scattering with the background gas. As a consequence, the evaporated layer will not cover the pattern slopes as long as they are sufficiently steep and the aspect ratio is high enough. Dissolution of the polymer will then remove the metal deposited on the polymer and a patterned metal film remains on the substrate surface.

In addition to dry etching and lift-off, galvanic filling of HEL mask patterns has also been reported [26].

2.3. Fundamental Achievements

2.3.1. Pattern Size and Resolution

The first reports claimed to be able to fabricate sub-100-nm feature sizes by hot-embossing lithography. Dots of 25-nm diameter and 120-nm pitch having a height of 100 nm, as well as 60-nm trenches, were reported [13]. The template was written by electron beam (e-beam) lithography in PMMA and a dry-etch mask for pattern transfer into silicon oxide was prepared by lift-off with Cr. The stamp patterns had a height of 250 nm. To demonstrate the successful imprint, holes and trenches in the polymer were subjected to a lift-off process so that the small imprinted features became visible as a metal pattern.

In later publications [12, 17, 18, 27–29] these feature sizes were improved, reaching 15-nm trenches with a 60-nm period [27] as well as 6-nm-diameter holes with a 60-nm period [28]. The stamps for the latter had 10-nm dots produced by RIE and were thinned to a diameter of only 6 nm by dipping in HF. An example of pattern definition on gold, which is a high-atomic-number substrate, is given in the same paper. 20-nm dots and lines were replicated in Ni by lift-off after imprint on Au. This is noteworthy as electron beam lithography would not be suitable for such a patterning process due to the high level of electron backscattering from the substrate.

Imprinted holes and trenches are smaller than lines. The smallest lines reported in the polymer were 70 nm wide and 200 nm high with an estimated period of 220 nm [12, 18]. Lines of 30 nm with a 70-nm period could only be documented using lift-off as the polymer melted during SEM inspection.

The finest structures reported were imprinted over areas up to about 6 cm^2 [12, 28], where the stamp was not fully patterned. Typical minimum pattern sizes of 50–200 nm were imprinted over fully patterned areas of up to 10 cm^2, with a very low residual layer thickness of only 20–50 nm [19]. Typical large-area imprinted features are shown in Fig. 5.

2.3.2. Fidelity of Pattern Definition

To the best of our knowledge, there is no publication where imprinted features are quantitatively compared to those of the stamp.

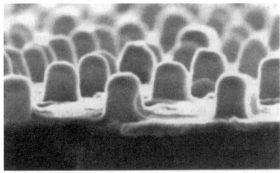

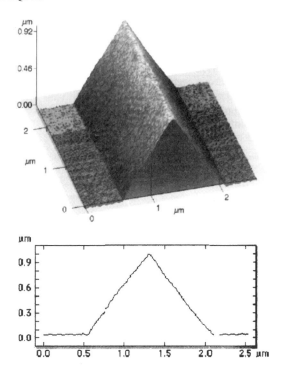

Fig. 5. "Tombstones" and lines of 125 nm prepared by hot embossing in PMMA. Reprinted with permission from R. W. Jaszewski, H. Schift, J. Gobrecht, and P. Smith, *Microelectron. Eng.* 41/42, 575 (1988). © 1998 by Elsevier Science.

Fig. 6. AFM measurement of pyramidal lines prepared by hot embossing in PMMA. The stamp was fabricated by anisotropic wet etching in Si(100). Reprinted with permission from H. Schift, R. W. Jaszewski, C. David, and J. Gobrecht, *Microelectron. Eng.* 46, 121 (1999). © 1999 by Elsevier Science.

Chou et al. compares the size of the lift-off patterns with the size of imprinted trenches and holes in the polymer and use this as a qualitative measure of pattern definition fidelity [28]. This makes sense as it also includes the O_2 RIE used for removal of the residual polymer layer in the imprinted regions and thus characterizes the polymer mask fabrication process as a whole. However, it is specific to lift-off. Stamps with sloped walls might result in a polymer pattern unfavorable for dry etching, as demonstrated in [30], where 15-nm trenches after imprint resulted in 25-nm lines after lift-off due to broadening in the dry-etch step.

An indirect assessment is given in [31], where infrared-selective surfaces prepared by HEL and by e-beam lithography are compared. A slightly shifted response of the filters is attributed to dimensional differences obtained with the two methods.

To measure the achievable quality of shape transfer, three-dimensional sloped patterns were embossed [20, 32]. A stamp with V grooves prepared by anisotropic wet etching of Si(100) was used to emboss a PMMA layer under vacuum and pressure control. The resulting polymer structures were investigated with AFM in intermittent contact mode and both the volume and the shape of the imprinted features were found to conform precisely to the stamp as demonstrated in Fig. 6. The end radius of the polymer ridge was around 10 nm, which is on the order of the AFM tip size itself, so that no deviations were measurable.

2.3.3. Embossed Materials

Polymers for Lithography-Type Applications. Most of the hot-embossing experiments have been performed in PMMA, a well-known e-beam resist. This material is available in a broad range of molecular weights. It is a thermoplastic polymer, which softens and flows with increasing temperature. Its glass transition temperature is 105°C. A detailed discussion of the physical behavior of thermoplastic polymers is given in Section 2.6.

PMMA is reported to be a good choice in combination with a mold from SiO_2 because the polymer has a hydrophilic surface [17] and will not stick to the stamp. Its thermal expansion coefficient is $5 \times 10^{-5} K^{-1}$, and its pressure shrinkage coefficient is 5.5×10^{-6} bar^{-1}, so that the thermal shrinkage is below 0.8% and the pressure shrinkage below 0.07% within a typical cooldown step [18, 27].

Excellent polymer flow is reported for 15 k PMMA (PMMA with a mean molecular weight of 15 kg/mol) at quite low temperature and pressure (175°C, 45 bar) [33], and a low-molecular-weight polymer (50 k PMMA) displays much better flow than one with a high-molecular-weight (950 k) [34]. This supports theoretical expectations with respect to changes of viscosity for different molecular weights (see Section 2.6).

PMMA also offers preparation advantages. Solved in chlorobenzene, it lends itself to spin coating free from surface nonuniformities, which could be caused by local density fluctuations or a highly volatile solvent [19].

Imprinting in AZ and Shipley novolack resin-based resists is mentioned to be suitable but no details have been given [27]. Furthermore, commercial polymers such as polystyrene (PS) were tested successfully [35] (see also Section 2.5.2).

New polymers optimized for nanoimprinting have been reported, featuring better mask selectivity than PMMA in a fluorine-based dry-etch process [36–40]. This is a result of including aromatic groups in the resist formulation. Their imprint behavior is excellent [40] and they show even less sticking to SiO_2 than PMMA. One of these polymers, commercially available as mrt-8000, is a thermoplast with a glass temperature of about 115°C. Another, mrt-9000, is a thermoplastic prepolymer for a duroplast that can be crosslinked in a temperature treatment during or after embossing to form a patterned polymer layer with high thermal and chemical stability.

In addition, highly polar copolymers were tested for imprint, featuring a light network due to H bonding, which is degradable at elevated temperatures. These polymers had to be imprinted using a specific procedure schematically shown in Fig. 18.

Embossing beyond Lithography. Other interesting materials have been imprinted that are worth mentioning. As in this case the embossed material itself provides the functionality envisaged, the process is referred to as "direct nanoprinting" (DNP) [41] (see also Section 2.9.1) or "direct nanoimprint" (DNI) [42].

Functionalized polymers have been imprinted, in particular, active organic light-emitting materials (DCMII-doped Alq_3), in either their pure form or embedded in a PMMA matrix [42]. Gratings of 200-nm period have been imprinted in vacuum as this material degrades in an atmosphere containing water and oxygen. The optical properties of the material were not degraded by the imprint process at 55 bar and 150°C. Spontaneous molecular alignment of the chromophores in the plane of the film during the patterning process was found [43].

Besides being a technique for polymer pattern definition, there are many other applications for embossing. For example, direct nanoprinting of Al substrates has been reported using a SiC mold [41], a technique known to work for a number of metals for structures in the micrometer range [44]. This process works at room temperature (see also Section 2.9.1) and arrays of 800-nm-deep dots of 100-nm width were reported. The pressure applied is substantially higher than in a hot-embossing process. Taking into account the patterned area, we estimate it to be about 2400 bar.

Another interesting material class is sol–gels, precursors for oxide materials. Gels have a quite high free volume, which can be compressed by an applied pressure without raising the temperature. Imprint experiments with 50 bar at room temperature into a wet spin-coated gel film of 600-nm thickness resulted in sub-100-nm SiO_2-TiO_2 gratings after a 400°C anneal following the imprint for oxide formation [45].

These last three examples are not lithography processes in the original meaning, but direct patterning processes of a functionalized material. According to our definition in Section 1.3,

they are surface-oriented processes and the patterned surface can be used directly, for example, for an optical application. This is independent of layer thicknesses and demonstrates again that there is a gray area where hot-embossing lithography and hot embossing of nanometer patterns can hardly be distinguished.

2.4. Imprint Systems Used for Hot Embossing

2.4.1. Systems for Feasibility Test

Often simple commercially available hydraulic presses are used, equipped with two heated and cooled stages or hotplates for the sample and the stamp as shown in Fig. 7. Sometimes a ball joint in the moving part is added for mechanical wedge error compensation [34] or a piezo drive for precision imprint feed. For such hydraulic systems, the pressure is given as a process parameter without further information about whether the system features a pressure control or only a manual pressure application. The latter may have adverse effects on the imprint results, because maintaining the pressure during the cooling phase is important (see Section 2.5.1). In such simple systems, no control of the imprint depth during the process

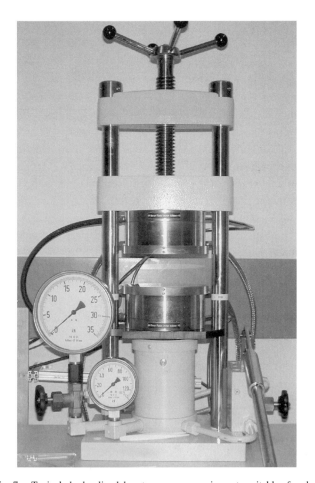

Fig. 7. Typical hydraulic laboratory press equipment suitable for hot embossing. Contact and pressure application between the upper and the lower sample stages is performed manually. The stages are supplied with electrical heating lines and water cooling. Courtesy of D. Echterhoff.

is available and the thickness contrast resulting from a certain pressure is measured after the process in an SEM or AFM. Typical maximum pressures supplied in such systems amount to forces of up to 20 tons (200 kN), maximum temperatures up to 300°C and stage diameters of 8–15 cm. In general, heating is electrically generated and cooling works under ambient atmosphere or with water and the units are not optimized for short processing times. Heating and cooling times between the room temperature and the process temperature will lie in the range of 5 to 30 min. This fact might enhance embossing yields because in such a system the polymer, a highly viscous liquid, will have enough time to flow and to conform with the mold. Shorter processing times are desirable; however, detailed investigations of which processing times can be reached under specific temperature and pressure conditions remain to be carried out. This would only be possible with a quickly reacting system.

Some groups also use simple clamp arrangements for their first experiments, where pressure is applied via clamps or screws without control or measurement of the force and where the whole assembly is put into an oven for heating [30]. The results demonstrate that such simple systems work well, although they will not be suitable for process development.

An improved version of the latter uses a dual-plate assembly, where parallelism is maintained by a sliding-rod guidance. Force is applied by loading the top plate with weights [46]. The whole assembly is heated in an oven for processing, and shock cooldown is performed by flushing with liquid N₂.

Commercially available hot-embossing machines used for micromechanics and LIGA production are fully controllable and allow for vertical feed control [9, 44, 47] and evacuation. Cycle times for such "stamper hot embossing" systems (see Fig. 8) may be as low as 5–7 min when loading and unloading is performed slightly below the glass temperature and embossing is done at moderate temperatures. Such systems have to be modified to accept flat samples instead of volume tools. An automated mold release may be a critical development [47].

In analogy to the widely used "roller hot embossing" approach [9] for the shaping of foils, a roller tool for hot-embossing lithography has been reported [48]. The roller press consists of a roller, a movable platform, and a hinge. The roller is a hollow tube with a halogen bulb mounted inside, constituting a small thermal mass. The platform holding the sample is heated by a strip heater and moves back and forth in a direction vertical to the fixed roller. The hinge allows the application of a constant force between roller and sample using a weight. During embossing, the roller rotates around its fixed axis while the platform slides underneath. Two different techniques have been investigated so far. Either a cylinder mold or a flat mold is used. The effective length of the line contact between roller and platform was estimated to be 2 mm; the sample used in the reported experiments was up to 2 cm wide. With loads of up to 100 N, pressures of 20–330 bar were achieved. For the cylinder mold, a Ni compact disc master was used as a flexible stamp produced by Ni electroplating, a process well known from state-of-the-art roller embossing

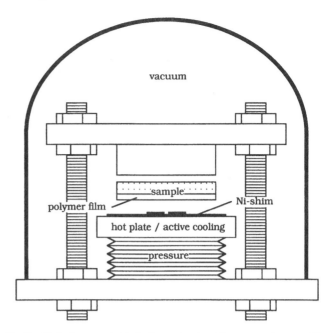

Fig. 8. Flatbed press suitable for hot embossing. The system shown enables evacuation of the procesing volume. Only the lower stage is heated. Reprinted with permission from F. Nikolajeff, S. Hard, and B. Curtis, *Appl. Opt.* 36, 8481 (1997). © 1997 by the Optical Society of America.

techniques [9]. At first glance, such a system suggests large-area processing, although this has not been demonstrated.

2.4.2. Dedicated Large-Area Approaches

Although simple systems have demonstrated wafer-scale embossing with nice results [49], systems developed or modified and tested for large-area processing are of more geneal interest for future applications of HEL. They rely either on full wafer-scale parallel processing or on a sequential approach.

Wafer-Scale Parallel Processing. One approach is based on a modified injection molding system. A first prototype was built for 2-in.-diameter samples [50–52], followed by a 6-in. imprint machine [53, 54] both of which are now commercially available. The 2-in. system as shown in Fig. 9 has been described in detail. The maximum temperatures and hydraulic pressures are 300°C and 200 bar, respectively. As common for injection molding, it is a horizontal system. Stamp and sample are fixed to the respective stages by vacuum, and the parallelism of both surfaces is obtained by an oil balance concept. At the beginning of the imprint sequence, stamp and sample are brought into "gentle" contact so that the surfaces can assume parallel alignment.

The small 2-in. system is placed in a laminar flow bench to avoid particle contamination. The substrate is mounted to the fixed part of the system with a graphite heater and a water cooling unit below, where all parts are electrically isolated. The stamp is neither heated nor cooled. This heating–cooling arrangement allows for fast temperature regulation of the system.

Fig. 9. Two-inch-diameter horizontal imprint system for hot embossing, with the substrate holder at the right and the stamp at the left part of the equipment. The system is operated automatically and uses an oil balance principle for parallel adjustment of stamp and sample. Reprinted with permission from L. Montelius, B. Heidari, M. Graczyk, T. Ling, I. Maximov, and E.-L. Sarwe, *Proc. SPIE* 3997, 442 (2000). © 2000 by SPIE.

Another 4-in. wafer-scale system has been reported [55], which resembles the simple hydraulic presses mentioned earlier. Here only a schematic diagram is shown (see Fig. 10). The system consists of two heated press plates equipped with a heating–cooling line, where the pressure is applied hydraulically via a ball joint for automatic parallelism adjustment. An elastic cushion layer is placed between the plates and the stamp or the sample in order to compensate for thickness variations of the backside of the wafers, which are typically a few micrometers.

An additional large-area instrument currently under development is the modification of a wafer-bonding system commercially available for microsystem technologies. The system can handle 4- or 6-in. wafers. The maximum force is 40 kN and maximum temperature is 400°C. Initial experiments with such a system demonstrate that hot embossing should be feasible over whole wafer areas. Fully patterned 4-in. wafers with pattern sizes ranging from 400 nm to 100 μm were used for these tests. Specific modifications of the system with respect to hot-embossing lithography processes are under way [56]. The advantage of this type of system is that standard adjustment as used for UV lithography is available and that such systems, in their wafer-bonding function, are already in use within Si production lines.

Sequential Processing. A totally different large-area approach is a "step-and-stamp" procedure [36] imitating an optical stepper. To test this approach, a commercially available flip chip bonder is used. Figure 11 shows the basic idea of stamping. The stamp is fixed to the SiC plate of the bonding arm with a thermally stable silicon adhesive. The stamp is heated by radiation. The sample is mounted on the wafer chuck, also made from SiC, which is, in turn, mounted on a motorized xy stage with a spatial resolution of 0.1 μm. Preleveling to 20 μrad and lateral adjustment are performed with the respective bonder systems before stamp and sample are brought into contact. Maximum temperatures of chuck and bonder arm are 450°C, the maximum force of the bonder arm is 50 kg. To exert pressures of up to 100 bar for the embossing process, the stamp size had to be limited to 5×5 mm^2.

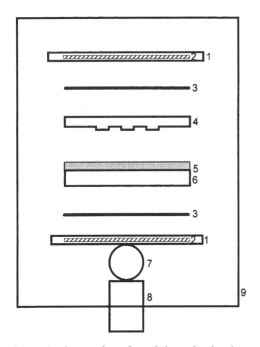

Fig. 10. Schematic of a press for wafer-scale hot embossing: 1, press platen; 2, heating–cooling line; 3, cushion layer; 4, mold; 5 and 6, substrate with polymer layer; 7, ball joint; 8, hydraulic unit. Reprinted with permission from D.-Y. Khang and H. H. Lee, *Appl. Phys. Lett.* 75, 2599 (1999). © 1999 by the American Institute of Physics.

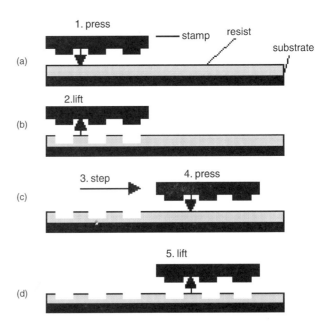

Fig. 11. Principle of step-and-stamp imprint lithography which was implemented in a commercial flip-chip bonder system. Reprinted with permission from T. Haatainen, J. Ahopelto, G. Gruetzner, M. Fink, and K. Pfeiffer, *Proc. SPIE* 3997, 874 (2000). © 2000 by SPIE.

2.5. Processing Details

2.5.1. Hot-Embossing Process

Basic Parameters and Process Sequences. Typical process data for hot-embossing lithography include a pressure of 40–100 bar and a temperature 50–100°C above the glass transition temperature T_g of the polymer.

Whereas the first publications reported quite high values of 200°C and 130 bar [12, 17, 18, 27], values of 175°C and 45 bar became a standard for the imprint of PMMA [19, 33, 34, 57]. Depending on the type of features to be imprinted, higher or lower pressure and temperature values could be required. This is discussed in detail in Section 2.7.

In most cases, both the stamp and the sample are heated to the imprint temperature, then pressure is applied and held during the cooldown step until a temperature below T_g is reached. Figure 12 (top) shows this standard procedure. Maintaining the pressure during the cooling phase is a critical point for successful pattern definition. The parameter "embossing time" t_{emb} is not clearly defined. Often the time at full pressure and highest temperature is meant (full line arrows in Fig. 12), but a certain percentage of the cooldown time (or heating time) under pressure (a certain minimum level above T_g) will be efficient, too, for local polymer transport (dashed arrows in Fig. 12). This is of concern, in particular, in cases where heating and/or cooldown times under pressure are long compared to the holding time (time at highest temperature).

Demolding, the separation of stamp and sample, is recommended around T_g where polymers react soft-elastically to mechanical stresses by conformational changes (see Section 2.6) in order to prevent damage to the embossed pattern during separation. This is only possible in systems where the sample and the stamp are fixed to the stages and in which pressure release and demolding are coupled. In simple systems, this is often not the case and demolding has to be performed manually [58]. This might require application of shear forces (e.g., via a scalpel). The demolding step is critical and can lead to extended damage of the stamp [59] and of the polymer film, in particular, for structures with high aspect ratio. Demolding problems are also correlated with sticking effects, discussed in detail in Section 6.

When heating and cooling lines are integrated into the stamp and sample stages, both of which represent a high thermal load, the cooldown to room temperature is a relatively slow process. Therefore, cooling is initiated as soon as the imprint temperature is reached [only dashed part of t_{emb} in Fig. 12 (top)]. In this way, the stack will remain at elevated temperatures for a period of some 10 min before T_g is reached. This appears to be sufficient time for the highly viscous polymer to flow and to fill the mold patterns conformally [20].

Sometimes the pressure is applied before heating [30, 60]. This is the case when using a simple clamp mechanism to apply pressure. The clamp is then put into an oven for heating. High-temperature steps of about 20-min duration are reported in accordance with the processing scheme described in Fig. 12 (bottom). These two different methods do not seem to lead to different results and are just a consequence of the equipment used.

Pressure Maintenance. In a clamp mechanism, the process parameters are not well defined. When a pressure is applied to the cold stack, this will increase during heating due to the thermal expansion of the clamping materials. Pressures occurring in the heated state may be as high as twice the pressure in the cold state [61]. There have been no schemes proposed to control or measure this parameter. Therefore, the pressures reported in such configurations are likely to have a large uncertainty. The same happens in a hydraulic system without pressure control, when the pressure is applied before the maximum temperature is reached.

Thermal expansion is also the reason that in a hydraulic press with heated stages the pressure has to be controlled during cooldown. A pressure decrease due to material contraction could cause adverse effects, in particular, when three-dimensional patterns are imprinted [59]. This is analogous to injection molding where the polymer feed pressure is even increased during sample cooling [62] to ensure dimensional accuracy.

As will be discussed in Section 2.7, the actual pressure in the embossing process is a question of the actual contact area between stamp and polymer surface. Thus, as long as the contact area is small, the effective pressure will be high and its height will depend on the specific pattern embossed. As a consequence, the pressures reported can be misleading and have to be seen in context with the patterns and the overall patterned area. It would be more straightforward to report the force applied and the overall patterned area or even better the elevated stamp area.

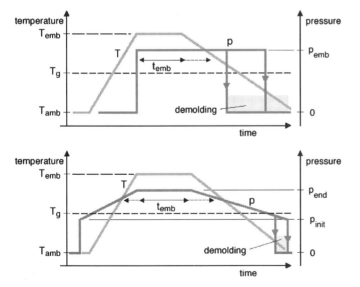

Fig. 12. Schematics of typical processing sequences during hot embossing used in conjunction with hydraulic systems with heating–cooling lines (top) and simple clamp devices put into an oven (bottom). Note that a constant pressure is only achieved under pressure control as long as the temperature of the system changes.

Vacuum. The need for a vacuum in order to prevent air bubbles has been mentioned [18]. Nevertheless, most of the simple embossing systems do not provide a vacuum and without a need for evacuation future production-grade imprint systems will remain simple [58]. Only one paper mentions explicitly that a vacuum is not required for hot embossing lithography [19]. These authors used a stamper hot-embossing system (similar to Fig. 8) where evacuation is provided. To investigate this question, experiments have been performed using stamps with inverted pyramids made by anisotropic wet etching of Si(100). During embossing of these pyramids, entrapped air, compressed to 1% of the pyramid volume under 100 bar during imprint, should be concentrated in the tip region, leading to bubble-like defects at the apex on the order of 20% of the pyramid height. Alternatively, when due to specific flow effects, as discussed in Section 2.7, the filling of stamp cavities is irregular (compare, e.g., Figs. 34–36), enclosed air bubbles should be found in the patterned polymer when investigating cross sections. Neither of these defects could be detected during investigation of pyramids or V-shaped lines. Even pyramids of large base length are replicated defect-free [59] (see Fig. 13a), suggesting that during the embossing process the air will dissolve in the polymer. Amorphous polymers feature a free volume between the disordered polymer chains. This free volume is filled by small gaseous molecules when the polymer is held in atmosphere. These gases are compressible, and air stemming from enclosed volumes between stamp and sample will enter the polymer free volume under pressure and may even form bonds with the PMMA molecules [33]. The solubility of air in PMMA increases with pressure and temperature so that air bubbles can be prevented by the choice of embossing conditions and, if necessary, by increasing the thickness of the polymer [33]. The latter should be avoided for embossing lithography where small residual layer thicknesses are required.

When large bubble-like defects occur (see Fig. 30), they are often a result of the characteristic polymer flow [58], an issue discussed in Section 2.7.

Real bubbles (see Fig. 13b) do not result from trapped air, but from solvent remaining in the polymer layer [33] and entering the gaseous state at elevated temperatures. This is the case when the prebaking after spin-on was too short or at too low a temperature for efficient solvent removal. Prebake temperatures should be chosen above T_g so that the polymer chains become mobile and help the evaporation of the solvent.

Procedures Adapted to Specific Systems. For specific types of systems, such as the roller type one or the large area approaches shown in Figs. 9–11, the processes have to be adjusted accordingly.

In the case of the roller-type approach [48], it is not possible to maintain the pressure during cooldown. As a consequence, the embossed polymer has to be cooled efficiently directly after imprinting in order to remain stable. This is achieved by having the sample platform at a temperature below the T_g of the polymer (e.g., at 50–70°C in the case

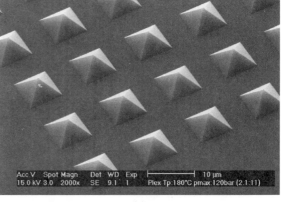

(a)

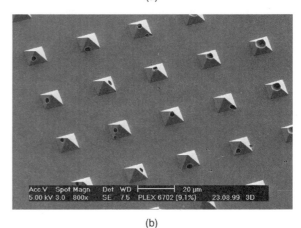

(b)

Fig. 13. Pyramids replicated by hot embossing without evacuation of the imprint volume. (a) Large size pyramids (7-μm base length) with perfect apex; (b) pyramids with random bubbles presumably due to remaining solvent in the polymer.

of PMMA), while the roller remains heated to 170–200°C. Thus, only during contact between roller and sample the temperature rises in the contact zone and the polymer can flow and conform to the mold. At higher platform temperatures the polymer patterns deform after embossing. Pressure (20–330 bar) and rolling speed (0.5–1.5 cm/min) are critical parameters, too. When the pressure is too low or the speed too fast, the polymer does not have enough time to be shaped in the embossing process. The results reported in the only paper on roller nanoimprint reflect these problems. The pattern is not as clearly defined as in a parallel process when using the cylinder molds. Moreover, the imprint depth is only 40 nm for a mold with 180-nm-high patterns when using the flat mold roller process reported in [48].

For the large-scale approaches, even more different processing sequences were followed. For the wafer-scale modified injection molding system for 2- and 6-in. diameter (see Fig. 9), the temperatures and pressures are within the usual range, namely, 150–180°C and 40–100 bar, but the imprint time is much shorter. Cooling starts 3–5 s after contact, and the cooldown takes only 1 min, resulting in a total processing time as low as 2 min [50, 51], as shown in Fig. 14a.

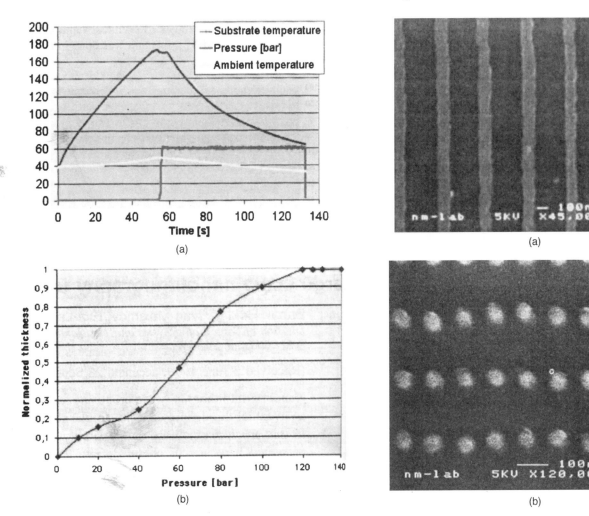

Fig. 14. Imprint sequence measured in the 2-in. horizontal hot-embossing system of Fig. 9. (a) Substrate temperature (topmost line), pressure (middle), and ambient temperature (bottom line); (b) normalized pattern depth resulting from a hot-embossing process with different imprint pressures. Reprinted with permission from L. Montelius, B. Heidari, M. Graczyk, T. Ling, and E.-L. Sarwe, *Proc. SPIE* 3997, 442 (2000) and from B. Heidari, I. Maximov, E.-L. Sarwe, and L. Montelius, *J. Vac. Sci. Technol., B* 17, 2691 (1999). © 2000 and 1999 by SPIE and the American Vacuum Society.

Fig. 15. Cr features prepared by hot embossing in the horizontal imprint system of Fig. 9 and subsequent lift-off. (a) 100-nm-wide lines; (b) 50-nm-diameter dots. Reprinted with permission from B. Heidari, I. Maximov, E.-L. Sarwe, and L. Montelius, *J. Vac. Sci. Technol., B* 17, 2961 (1999). © 1999 by the American Vacuum Society.

Within this short time, polymer patterns were achieved with about 160-nm thickness contrast and a residual layer thickness of 50 nm, using a stamp with 300- to 400-nm pattern height and 950 k PMMA. It has to be mentioned that the patterned area of the 2-in. wafer was only about 0.16 mm² arranged over the wafer in four blocks and the patterns were positive e-beam-written line structures (see Section 2.7.) down to 100 nm [52] and dots of 50 nm [50], as shown in Fig. 15.

In the 4-in. wafer-scale system (see Fig. 10), an "asymmetric heating and quenching method" is used and is reported to be necessary for clean mold release [55]. In this method, the stamp is heated to 150–180°C above the T_g of the polymer used—in this case a 200- to 500-nm-thick polystyrene (PS) layer with a T_g of 95°C—while the polymer-coated sample is maintained below T_g. After pressing, which lasts typically 5–30 min, the heated mold is quenched to room temperature

by water cooling, resulting in a cooling time of only a few minutes, despite the high initial temperature of the stamp. It has been reported that with this procedure the mold detaches automatically from the polymer without damage to the resist layer. Pressures used in these imprints range from 300 to 2500 bar. This seems to be high at first glance, but in these experiments about 10% of the wafer area was patterned with 49 dies distributed over the wafer with different pattern sizes down to 100-nm lines. There is no result concerning the imprint depth or residual layer thickness achieved, but details of the mold pattern are replicated to size in the PS layer (see Fig. 16).

In the step-and-stamp apparatus (see Fig. 11), the process used has some similarity to the roller approach. Being also a sequential procedure, the polymer on the sample that has just been imprinted has to be hindered from flowing and so a substrate temperature below T_g is chosen. For stamping of a 340-nm-thick layer (mrt-8000, $T_g = 115°C$), the best results were obtained with a sample temperature of 80°C, a stamp temperature of 180°C and a force of 25 kg applied to a stamp

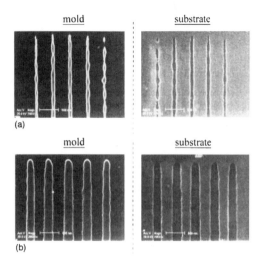

mold   substrate

(a)

mold   substrate

(b)

Fig. 16. Various patterns imprinted in polystyrene in a system according to the schematics of Fig. 10. The figure compares mold patterns (left) and the respective imprinted features (right). Reprinted with permission from D.-Y. Khang and H. H. Lee, *Appl. Phys. Lett.* 75, 2599 (1999). © 1999 by the American Institute of Physics.

of 5×5 mm², fully patterned with mixed structures down to 50 nm. The printing time was 5 min, and cooldown of the sample required an additional 3 min resulting in a cycle time of about 8 min; the procedure follows the schematic of Fig. 12 (top). A result for subsequently imprinted step fields is given in Fig. 17.

Crosslinked and Crosslinking Polymers. Two types of crosslinked or crosslinking polymers have been investigated. The first type refers to MAA/MA (methyl methacrylate/methacrylic acid) copolymers where the highly polar acid groups form a weakly bound network via H bonding, thus tailoring T_g [61]. At about 180°C, adjacent acid groups may form anhydrides, leading to a loss of polarity, a decrease in H bonding, and thus a decrease in T_g, which should lower the

viscosity and thus improve the embossing quality at a temperature above 180°C. This could not be confirmed in the experiments, most probably because the effect is too weak during typical imprint times. However, it could be shown that for such lightly crosslinked polymers an improvement in the embossing quality can be achieved when pressure is applied from T_g on. Using a modified procedure as shown in Fig. 18 (top), the definition of patterns from 400 nm up to 10 μm was improved substantially (Fig. 19). In addition, the gradual increase in pressure also resulted in a reduction in the observed sticking between stamp and sample.

The second type of investigated polymers are thermosets where a crosslink is initiated by thermal treatment. This causes a behavior opposite to the one described for the first polymer type. Above the threshold, the viscosity increases with increasing temperature.

Thermosets have shown high etch resistance and thus good mask selectivity in a dry-etch process subsequent to hot-embossing lithography [38, 39], and they have high thermal stability in their crosslinked duroplastic state [37]. The polymers developed and investigated for hot embossing (mrt-9000; see Section 2.3.3) feature a crosslink threshold of about 120°C. They are embossed in their thermoplastic prepolymer form. Typical glass transition temperatures are around 70°C. Up to 120°C, they can be imprinted using the standard process of Fig. 12 (top). When higher temperatures are applied for embossing, the full time the polymer remains viscous has to be used in order to achieve sufficient polymer flow before the crosslinking process increases the polymer T_g and decreases viscosity, thus inhibiting pattern transfer. In this case, pressure should again be applied from T_g on according to Fig. 18 (bottom) [40, 63], and in the case of high heating rates, it should be sustained for some time at a temperature slightly below the crosslinking threshold to enable flow.

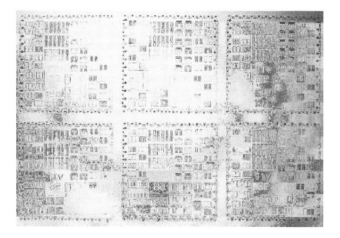

Fig. 17. Optical micrograph of six step fields imprinted with the step-and-stamp system according to the schematics of Fig. 11. Reprinted with permission from T. Haatainen, J. Anopelto, G. Gruetzner, M. Fink, and K. Pfeiffer, *Proc. SPIE* 3997, 874 (2000). © 2000 by SPIE.

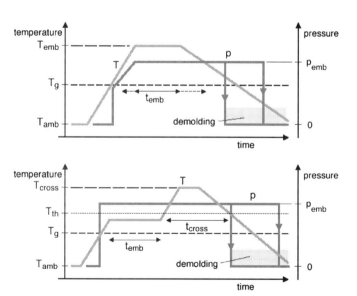

Fig. 18. Schematics of processing sequences used for hot embossing of thermally modified (MAA/MA) copolymers (top) and crosslinkable polymers (bottom).

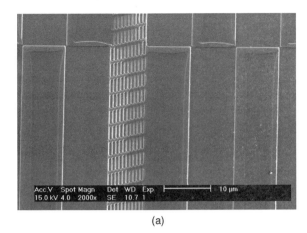

(a)

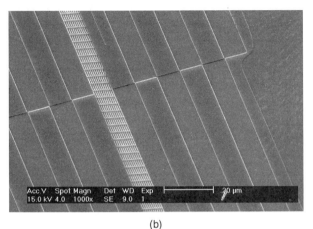

(b)

Fig. 19. Imprint results obtained with thermally modified (MAA/MA) copolymers. (a) Imprint result according to the standard process as shown in Fig. 12 (top); (b) imprint result according to the process of Fig. 18 (top). The latter results in a complete pattern transfer visible, in particular, for the 10-μm-wide lines. Reprinted with permission from K. Pfeiffer, G. Bleidiessel, G. Gruetzner, H. Schulz, T. Hoffmann, H.-C. Scheer, C. M. Sotomayor-Torres, and J. Ahopelto, *Microelectron. Eng.* 46, 431 (1999). © 1999 by Elsevier Science.

When embossing thermoset polymers, measures for the prevention of sticking have to be taken (see Section 6).

Results for the imprint quality of thermosets are found in Section 5.

Process Uniformity. As the large-area approaches are still in their infancy, only a few results on uniformity have been reported and they are only of a qualitative nature.

A first experiment tested the uniformity over a total embossed area of 15×18 mm^2. Within this area, five patterned fields with 30-nm lines and 150-nm pitch were distributed. The imprinted lines were reproduced with a lift-off process and thus visualized, and no significant differences between the patterns in the different fields were visible.

An additional uniformity test was reported for replication over a 4-in. wafer area, where the stamp was manufactured by optical lithography [34]. Five fields of 24×24 mm^2 each were arranged side by side in a crosslike geometry. Uniformity was good when low-molecular-weight PMMA was used (50

k), whereas for high-molecular-weight (PMMA) the replication was often nonhomogeneous due to the polymer's higher viscosity.

A third test was performed in the 4-in. system with asymmetric heating and quenching according to Fig. 10. As mentioned there, 49 dies of 0.5×0.3 cm^2 were distributed evenly over a 4-in. wafer, featuring e-beam-written patterns down to 100 nm. In [55], the patterns on the stamp and the respective imprint were compared for five different locations in the center and at the periphery of the wafer. In every location, the imprints conformed ideally with the stamp. Two of the examples are given in Fig. 16. Another uniformity test over 4-in. has been reported, where a 100-nm linewidth grating was embossed. Details are given in [64].

2.5.2. Residual Layer Removal

The removal of the residual layer completes the lithography step, as it transforms the patterned polymer into a mask. An anisotropic dry-etch step is needed here, which reduces the height of the whole resist layer by at least the residual layer's thickness ideally without affecting the slope of the patterns. This is done in general by means of an RIE process in oxygen or an O$_2$/Ar mixture. The Ar is helpful in enhancing the physical dry-etch effect, sputtering away nonvolatile etch products and potential residues. To render this dry-etch step uncritical, the residual layer thickness should be small in comparison to the thickness contrast of the resist layer. As uniformity over a whole wafer can never be assured, an overetch time has to be included in this step in order to open all the mask windows. Residual layers reported are between 10 [13] and 60 nm [39], and process parameters are 90–125 mTorr O$_2$ with a radiofrequency (rf) power of 150–400 W, resulting in etching times of about 10 s [18, 27, 36].

Only two groups showed SEM pattern profiles before and after residual layer removal. One reported on the etch behavior of newly developed polymers for imprinting (mrt-8000; see Section 2.3.3) in an inductively coupled reactor with sample cooling to 20°C [39]. No distinct pattern loss was observed. However, since the focus is on pattern transfer, process data on residual layer removal were not mentioned. The other group [35] used PS as a masking polymer, and the residual layer of about 100 nm was removed in a capacitively coupled O$_2$-RIE process at 5 mTorr, 50 sccm, and 450 V_{dc} (see Fig. 20). The slope of the PS lines was affected, revealing that the process does not work fully anisotropically. In a different investigation, a decrease in the pattern aspect ratio from $3:2$ down to $1:1$ was reported during residual layer removal [19] measured by scanning force microscopy (SFM). These results indicate that residual polymer layer removal by anisotropic RIE in O$_2$ is not trivial and this maskless process still requires optimization.

2.6. Viscoelastic Properties of Polymers

Optimum pattern transfer requires the complete filling of the mold cavities by displacement of the polymer material. Moreover, it is desirable that these deformations of the polymer

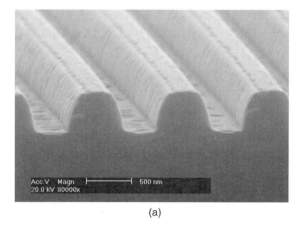

(a)

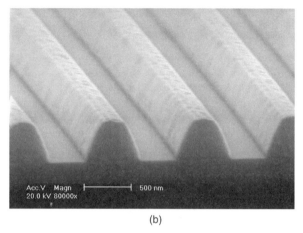

(b)

Fig. 20.   Cross sections of PS lines. (a) After imprint; pattern height is 450 nm, residual layer thickness is 100 nm. (b) After residual layer removal in O_2-RIE (5 mTorr, 50 sccm, 450 V_{dc}). Reprinted with permission from D. Lyebyedyev, H. Schulz, and H.-C. Scheer, *Proc. SPIE* (2001). To appear. © 2001 by SPIE.

layer by irreversible flow take place within a reasonable processing time. This time depends on (i) the transport distances, determined by the size and arrangement of the patterns (cf. Section 2.7.1); (ii) the applied pressure; and (iii) the viscoelastic properties of the polymer at the processing temperature. The material parameters, which describe the linear viscoelastic behavior of polymer melts, are the zero shear rate viscosity η_0 and the steady-state shear compliance J_e^0, both defined in the limit of small shear stress or shear rates. They are related to the irreversible and reversible contributions to the overall response of a polymer melt, respectively. Both parameters define a characteristic flow or stress relaxation time, which can be related to the typical times in the embossing process.

A brief phenomenological description of the linear viscoelastic behavior of polymeric materials will be given. The linear response to small stresses and strains can be explained in a microscopic picture that considers (i) the conformational flexibility of single polymer chains and (ii) the presence of entanglements in polymer melts. A quantitative description is possible by the material parameters η_0 and J_e^0, which depend on temperature and molecular weight. Their relation to the embossing process will be discussed for the case of PMMA.

The discussion is restricted to amorphous polymers because their homogeneity is of major interest for lithographic processes. Semicrystalline polymers consist of small crystallites that are typically some tens of nanometers thick and extend laterally a few micrometers [65], thus exhibiting inhomogeneity on the scale of the desired lithographic resolution.

2.6.1. Creep and Recovery

To illustrate the linear viscoelastic behavior of an amorphous polymer, a creep experiment is considered. At a time t_0, a constant small shear stress σ_{zx}^0 is acting instantaneously on the polymer. The first index refers to the normal of the plane the force acts on, the second to the direction of the force. The time-dependent deformation or shear strain $e_{zx}(t)$ is displayed in Fig. 21. To distinguish between reversible and irreversible contributions, the deformation can be monitored after unloading the sample at time t_1 (recovery experiment). The instantaneous deformation of the polymer corresponds to the elastic response (Fig. 21) as described by Hooke's law:

$$e_{zx} = J\sigma_{zx}^0 \tag{1}$$

with shear compliance J. The elastic contribution to the overall deformation is fully recovered upon unloading. A second contribution (Fig. 21) is given by the viscous response of the polymer and according to Newton's law increases linearly with time:

$$e_{zx}(t) = (\sigma_{zx}^0/\eta_o) \times t \tag{2}$$

The viscous behavior is irreversible and its contribution to the overall deformation can be determined from the recovery curve. Furthermore, a retarded deformation is observed (Fig. 21), which requires some time to establish itself. Upon unloading, this part of the deformation recovers entirely after some time. This retarded response is sometimes addressed as *anelastic* behavior, but frequently both reversible parts are taken together as elastic contributions. The combination of the

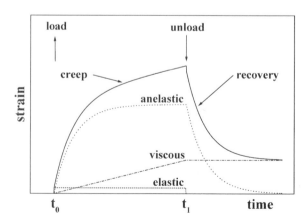

Fig. 21.   Deformation of a polymer under constant stress applied at time t_0 (creep experiment). The different contributions to the overall deformation are plotted separately. The reversible and irreversible contributions can be distinguished by recording the recovery after unloading the polymer at time t_1.

elastic, anelastic, and viscous response is generally addressed as *viscoelastic behavior*. The relative weights of the three contributions to the overall deformation depend on the type of polymer and, in particular, on the temperature [65].

Because the deformation of a viscoelastic material changes with time, it is convenient to define the *time-dependent shear compliance* $J(t)$ as

$$J(t) = e_{zx}(t)/\sigma_{zx}^0 \qquad (3)$$

An instantaneously applied small deformation leads to a stress that relaxes with time. Correspondingly, the *relaxation shear modulus* $G(t)$ is defined as

$$G(t) = \sigma_{zx}(t)/e_{zx}^0 \qquad (4)$$

Similar definitions apply to the time-dependent tensile modulus $E(t)$ and creep compliance $D(t)$. In addition, dynamic experiments are frequently performed to characterize the mechanical behavior of polymers. For instance, a periodically varying external stress yields both the storage modulus (in-phase response) and the loss modulus (out-of-phase response) [65].

2.6.2. Conformational Flexibility and Entanglements in Polymer Melts

The viscoelastic response of a polymer as illustrated in Fig. 21 can be explained by the large conformational flexibility of the polymer chain and the formation of a topological network of entangled chains. A typical polymer molecule is composed of a large number of covalently bonded monomers. For linear homopolymers, all the monomers that form one long chain without branching are identical. Usually, this chain is made up of carbon–carbon single bonds with each bond allowing rotation about its bond axis (Fig. 22). The rotation potential $U(\phi)$ of each bond possesses three distinct minima, which correspond to the trans, gauche$^+$, and gauche$^-$ rotational isomeric states. The projection of a polymer chain along one C—C bond axis indicates that the minima are related to staggered positions of the atoms that are attached to both central C atoms (Fig. 22). If the continuations of the chain are in opposite positions ($\phi = 0°$), they closely approach each other and repulsive interactions between the electron wave functions of the atoms contribute significantly to the potential energy. If, on the other hand, the continuations are separated by the maximum possible distance, they make up an angle of 180°, corresponding to the rotational isomeric state of lowest energy (trans state). The energy difference between the trans and both gauche states is about 2–3 kJ mol^{-1}. This value corresponds to the average thermal energy at room temperature, whereas the barriers separating the minima have energies on the order of 15 kJ mol^{-1} [65, 66]. In a simplified picture, it can be assumed that at room temperature the different minima are independently populated with probabilities according to their energy differences. From time, to time a bond will gain sufficient thermal

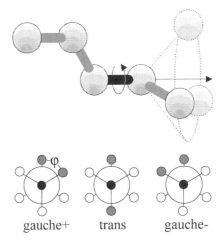

Fig. 22. Rotation of a polymer chain along one of the backbone bonds (black). For simplicity, only the carbon backbone atoms are displayed (top). Three different states of minimum energy (gauche$^+$, trans, gauche$^-$) are seen in a projection along the bond axis. The black circles correspond to the carbon atoms that form the bond. Gray circles indicate the continuation of the polymer chain and white circles indicate atoms or side groups attached to the central atoms (bottom).

energy to cross the barrier and rotate into a different rotational isomeric state. The number of distinct conformations that a chain of N bonds can take is 3^N, corresponding to possible sequences of trans, gauche$^+$, and gauche$^-$ states (Fig. 22). Only a few of these sequences lead to chains with large extension in space; in particular, only the all-trans sequence leads to an entirely stretched-out chain. Therefore, the most likely conformations of a flexible macromolecule are those of a random coil (Fig. 23).

Although many specific aspects are ignored, this picture provides a useful view of the large degree of conformational flexibility that is common to most polymers. Clearly, the specific chemical structure determines the interactions between chain atoms and alters the form of the bond and rotational potential. They lead to interdependence between the rotational isomeric states of adjacent bonds. This correlation persists on length scales of a few bond distances, resulting in stiffer chains. Furthermore, excluded volume effects must be considered to account correctly for the statistics of chain conformations. Strong repulsive forces, which result if a chain folds back onto itself, reduce the number of likely conformations [65–68].

Conformational flexibility can be related to the mechanical behavior by considering a single chain subjected to a small external force applied at two arbitrarily chosen chain atoms. The chain segment between these two points will be stretched out (Fig. 23) by passing through a sequence of conformational changes. The external force is counterbalanced by the attempt of the segment to return to a more probable conformation. Thus, upon unloading, the segment will recoil into an equilibrium conformation by subsequent jumps across rotational barriers. The counteracting force and the relaxation are driven by entropy. The rate at which the segment is stretched or relaxes into its equilibrium conformation increases with increasing

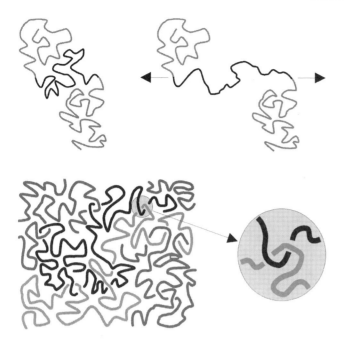

Fig. 23. Conformational flexibility and response to an external force (arrows) of a single polymer chain are schematically depicted. The segment that is stretched by the force is displayed in black. On unloading, this segment recoils or relaxes into an equilibrium conformation (top). Schematic picture of a temporary network of entanglements, which are displayed as knots. The topological obstacles due to the presence of neighboring chains set constraints to the motion of a polymer chain. Motion takes place on time scales longer than those typical for the stretching and the relaxation of the segments between entanglements (bottom).

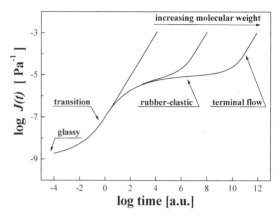

Fig. 24. Time dependence of the shear compliance for an amorphous polymer of different molecular weight. The extension of the rubber-elastic plateau in time increases with increasing molecular weight.

temperature. Thus, the response can be retarded as described previously for anelastic behavior. For sufficiently high temperatures, the polymer responds by immediate large deformation corresponding to the *rubber-elastic* behavior.

In multichain systems, such as concentrated solutions or melts, the interactions between the macromolecules impose constraints on the motion of entire chains. These interactions are generally called entanglements and can be pictured either as particular sites where two chains meet or as an effective field of long-range interaction [68]. Such particular sites can be pictured as knots (Fig. 23). These entanglements persist over times longer than those typical for the relaxation of segments between them. If an external force acts over time scales shorter than the average lifetime of entanglements but comparable to the segment relaxation times, the polymer responds by large reversible deformations. The entanglements act as a temporary network. In this picture, flow sets in when chains pass one another by slippage of entanglements [65, 69–71].

The characteristic time of both relaxation processes shows up in the time-dependent response, leading to the appearance of regions with distinct behavior. For instance, the time-dependent shear compliance $J(t)$ can be monitored at constant temperature as the response to a constant shear stress. The general form of this behavior is depicted in Fig. 24. At the shortest times, both relaxation processes are too slow to contribute significantly to the deformation. In this time range, the

deformation is small and reversible as expected from elastic behavior. In this *glassy region*, the response is almost exclusively determined by the bond interactions within the chains (intramolecular) or between the chains (intermolecular). The glassy region is followed by a *transition region*, which is characterized by an increase in the deformation by 3–4 orders of magnitude. At the corresponding times, the response to the external force is gradually taken over by conformational changes of the segments. Correspondingly, the obtained deformations are larger than those in the glassy states. The process levels off into a constant deformation where the external force is counterbalanced by stretched segments. The observed plateau of $J(t)$ corresponds to the *rubber-elastic region*, which is characterized by large reversible deformations. At the beginning of the plateau, flow is still negligible. With increasing time, it gradually contributes more to the response and a significant contribution of flow is seen in the *terminal region*. Finally, $J(t)$ increases linearly with time as expected for linear viscous flow [65, 70].

A particular dependence of the plateau on the molecular weight M is observed (Fig. 24). Below a certain molecular weight, the transition and the terminal flow region merge and a plateau does not appear. Above this critical molecular weight M_c, the plateau extends increasingly in time with increasing M. In the aforementioned microscopic picture, the dependence on M can be interpreted as the presence of entanglements and their average number per chain. Below M_c, the average number of entanglements per chain is not sufficient to build a temporary network that extends effectively over the entire polymer melt. Therefore, the rubber-elastic plateau does not appear for polymers of low molecular weight. The difference in the extension of the plateau indicates that the average number of entanglements per chain increases with molecular weight and the onset of flow is shifted to longer times. Furthermore, the approximately constant value of $J(t)$ in the rubber-elastic region can be understood by assuming that the average segment length between entanglements is mainly independent of M [65, 69, 70].

Flow can be completely suppressed by chemical crosslinks, which form a permanent network. A weakly crosslinked polymer still exhibits rubber-elastic behavior, because chain segments can be extended by external forces. These weakly crosslinked polymers are known as *elastomers* because they are characterized by an extended rubber-elastic plateau beyond the glass transition. The value of the shear compliance at the plateau depends on the degree of crosslinking. Highly crosslinked polymers do not show a transition to rubber-like behavior. They are called *duromers* because they show deformations typical of glassy polymers over the whole range of temperature. Chemical crosslinks may be obtained by either thermally controlled or photochemical reactions (cf. Section 3).

2.6.3. Time–Temperature Superposition

In practical terms, it is almost impossible to record the entire response over 15–20 time decades with a single experimental setup. A convenient way to overcome this limitation is to record the response at various temperatures. Depending on the temperature, different responses will appear within the experimentally accessible time range. For example, the transition region appears at shorter time scales as the temperature is increased because the involved segmental relaxation processes are faster. This *time–temperature superposition* can be effectively used to construct one single master curve from all experimentally obtained curves [65, 70–73]. The curves have to be shifted along the time axis by an amount according to the temperature at which they were recorded. An example of the time dependent shear relaxation modulus $G(t)$ is depicted in Fig. 25. At temperatures below the *glass transition temperature* T_g, the conformational changes are very slow and the polymer acquires essentially the properties of an elastic body. Correspondingly, the modulus is typically on the order of gigapascals for most amorphous polymers [74–81]. On approaching the glass transition temperature T_g from below, conformational changes become faster and the modulus

begins to decrease. In the transition zone, this segment motion is still retarded and does not follow immediately the external deformation, leading to anelastic behavior. Increasing the temperature further allows conformational changes to take place immediately. The modulus of the polymer is now 3–4 orders lower than those observed in the glassy region. Typical for most amorphous polymers are values of 0.1–1 MPa [65, 74–76, 78–81]. Finally, the terminal flow region is reached with increasing temperature. In this region, the motion of entire chains takes place on time scales that are comparable to the experimentally accessible time ranges.

The amount by which each observed curve has to be shifted to obtain the master curve at a single reference temperature T_0 depends on the difference between the experimental temperature and T_0. An empirical equation for this temperature-dependent shift factor a_T was first proposed by Williams and co-workers [70, 73]. The shift factor is usually expressed with the aid of two parameters, C_1 and C_2, as

$$\log a_T = C_1(T - T_0)/(C_2 + T - T_0) \qquad (5)$$

The shift factor is related to the temperature dependence of the viscosity η_0, which can be described by a Vogel–Fulcher law, as will be discussed later. Usually this equation holds for a range between T_g and 100 K above it. Dynamic measurements on PMMA showed good agreement up to almost 140 K above T_g, which corresponds to a maximum temperature of 240°C [75, 76]. Although other theoretical models for the description of the glass transition have been proposed and the validity of the empirical Vogel–Fulcher law has been questioned [82, 83], the method of Williams, and co-workers has proved to be a powerful technique for analyzing the mechanical properties of amorphous polymers.

2.6.4. Terminal Flow Region

The region of major interest for hot-embossing lithography is certainly the terminal flow region because here deformations are mainly determined by irreversible flow. Even in this range, however, the recoverable contributions have to be considered. The linear viscous behavior of a Newtonian liquid is completely characterized by its viscosity η_0, and the shear compliance is given by

$$J(t) = t/\eta_0 \qquad (6)$$

In contrast, a polymer melt requires an additional term J_e^0, and the shear compliance may be expressed in its simplest form as

$$J(t) = J_e^0 + (t/\eta_0) \qquad (7)$$

As mentioned previously, J_e^0 is a measure of the reversible part of the deformation that recovers when the shear stress ceases. Both quantities determine the characteristic flow time given by

$$\tau_{\text{flow}} = J_e^0 \eta_0 \qquad (8)$$

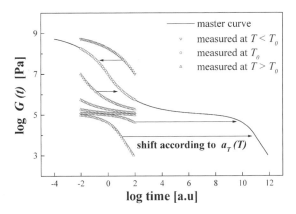

Fig. 25. Construction of a single master curve $G(t)$ at a reference temperature T_0 from experimental data obtained at different temperatures and within a limited time range. Each data set is displaced along the time axis by an amount according to the temperature-dependent shift factor $a_T(T)$ (cf [74]).

This flow time can also be interpreted as the characteristic time of stress relaxation after the deformation of the melt has ceased [65]. This has implications for the embossing process, as discussed later. However, a brief summary of the temperature and molecular-weight dependence of η_0 and J_e^0 is given first.

The main contribution to the temperature dependence of flow comes from the viscosity. Its temperature dependence is described by an empirical Vogel–Fulcher law, which has the form

$$\eta_0(T) = B \exp(-T_A/(T - T_V)) \qquad (9)$$

where T_A is the activation temperature and T_V is the *Vogel temperature*. This equation is formally equivalent to that for the shift factor a_T [Eq. (5)], and therefore both quantities are related by

$$a_T = \log \eta_0(T)/\log \eta_0(T_0) \qquad (10)$$

which relates the viscosity at a temperature T to that at a reference temperature T_0. The steady-state shear compliance increases linearly with increasing temperature due to the entropy-driven rubber-elastic forces.

The zero shear rate viscosity exhibits a strong dependence on molecular weight M, which shows up in two distinct regions separated by the aforementioned critical molecular weight M_c. For low molecular weights, the viscosity increases linearly as

$$\eta_0 \propto M \qquad M < M_c \qquad (11)$$

whereas above M_c the relation

$$\eta_0 \propto M^{3.4 \pm 0.2} \qquad M > M_c \qquad (12)$$

is found to hold for a large variety of polymers [65, 69, 84]. The recoverable shear compliance increases linearly with increasing M up to M_c and remains constant above M_c. The scaling behavior for the flow time can then be obtained by combining Eqs. (8), (11), and (12). It follows that

$$\tau \propto M^2 \qquad M < M_c \qquad (13)$$

and

$$\tau \propto M^{3.4} \qquad M > M_c \qquad (14)$$

Interestingly, there is experimental evidence that PMMA shows a distinct behavior with respect to J_e^0. It was found that J_e^0 increases linearly with M above M_c. This discrepancy has been attributed to the presence of polar groups in PMMA, although the effect of branching could not be disregarded with the particular set of PMMA samples [76]. To the best of our knowledge, this point has not been further clarified. Values of the critical molecular weight in terms of the number of chain backbone atoms for many polymers have been summarized [85]. The value of M_c for PMMA is approximately 3×10^3 g mol^{-1} [86]. The previously discussed scaling laws have been explained in terms of topological obstacles caused by neighboring chains by different theoretical models. The various approaches have been reviewed to some extent in [69].

2.6.5. Implications of Flow Behavior on Hot-Embossing Lithography

The temperature and molecular-weight dependence of η_0, J_e^0, and τ has a number of implications for the embossing process. The most important consequences are as follows:

(i) For a polymer of a given molecular weight, the flow time depends strongly on the temperature due to the strong temperature dependence of the viscosity. Therefore, it is desirable to choose higher temperatures during embossing where the maximum temperature is set by the chemical stability of the polymer in question. Under this condition, the filling of the mold cavities may be completed within a reasonable time.

(ii) A low-molecular-weight polymer may be embossed either at lower pressures or within shorter times. The limit is set by the entanglement limit as polymers with a molecular weight lower than M_c are more brittle. This brittleness in connection with nanometer-sized features may lead to fracture of the features when the mold is separated from the polymer. The latter problem can be avoided by embossing into low-molecular-weight compounds (monomers, prepolymers), which can be subsequently crosslinked to achieve mechanical stability (cf. Section 3.2).

(iii) When the deformation of the polymer slows down during the embossing step, the reversible part of the deformation attempts to recover its original shape. As a result of this attempt, the polymer exerts forces on the mold structures, which may additionally contribute to sticking when the mold is separated from the imprinted polymer pattern. Whether these mechanical forces contribute to sticking depends on both the amount of stress and the characteristic times at which the induced stress in the polymer layer relaxes.

Clear experimental evidence for the latter implication is lacking. However, a comparison of flow times with typical embossing times demonstrates that stress relaxation has to be considered as a contributing factor to sticking. Table IV

Table IV. Summary of Viscosity and Flow Times for PMMA of Moderate Molecular Weight

PMMA		$M_w \approx 1.1 \times 10^5$ g mol^{-1}
$T(^\circ$ C)	η_0(Pa s)	τ(s)
140	$4.6 \times 10^{9(a)}$	$9.6 \times 10^{4(a)}$
160	$7.1 \times 10^{7(a)}$	$1.5 \times 10^{3(b)}$
200	$1.5 \times 10^{6(a)}$	$3.1 \times 10^{1(b)}$
220	$1.4 \times 10^{6(c)}$	$2.2 \times 10^{1(d)}$

[a] Value from [67]: $M_w = 1.11 \times 10^5$ g mol^{-1}, $M_w/M_n = 1.66$.
[b] Value computed from $\tau = J_e^0 \eta_0$: $J_e^0 = 2.08 \times 10^{-5}$Pa^{-1} [67].
[c] Value from Fig. 2 in [65]: $M_w = 1.14 \times 10^5$ g mol, $M_w/M_n = 1.15$.
[d] Value computed from $\tau = J_e^0 \eta_0$: $J_e^0 \approx 1.50 \times 10^{-5}$ Pa^{-1} from Fig. 7 in [66].

summarizes values of τ computed from Eq. (8) with experimentally determined values of η_0 and J_e^0. The data were obtained from different sets of PMMA samples of similar mean molecular weight but with slightly different distributions and a T_g of approximately 105° [75–77]. The embossing time, which is of interest for the comparison, is determined by cooling the polymer from the imprint temperature down to T_g. Typical values are in the range of $3 - 6 \times 10^2$ s. At embossing temperatures of 200°C, the stress relaxation times are much shorter; thus, any stress induced during the embossing process ceases sufficiently fast. At temperatures below 160°C the relaxation times are at least 1 order of magnitude larger than the embossing times (Table IV) and stress may be frozen in when cooling below T_g. In practical terms, it is difficult to address and separate the different possible contributions to sticking (cf. Section 6.1.2).

The flow time agree values fairly well with experimental observation for a number of different polymer resists. The best quality of pattern transfer characterized by complete filling of the mold cavities and no destruction of the imprinted features is achieved at about 90° C above the glass transition temperature. In contrast, incomplete filling and damage of imprinted features are observed with imprints carried out at 50°C above T_g (cf. Section 2.7.1).

So far, the flow properties of polymers have been discussed under the assumption that the polymer is linear and possesses a narrow molecular-weight distribution. Both the amount of branching and the dispersity of the molecular-weight distribution have an influence on the rheological behavior and, in particular, on the values of η_0 and J_e^0 [87–91]. Therefore, it may be necessary to measure these values for a given polymer to determine its optimum imprint conditions and minimum processing times. Furthermore, it remains unclear whether the nonlinear behaviour has to be considered for a full understanding of the embossing process. For instance, nonlinear behaviour shows up in the dependence of η_0 on the shear rate, leading to a decrease in the viscosity by 3–5 orders of magnitude for sufficiently high shear rates [92]. It appears that this cannot be disregarded because at the very beginning of the embossing process shear rates are sufficiently high to produce nonlinear behavior where penetration of the mold into the polymer proceeds fast (cf. Section 2.7.1). These questions have to be addressed in future work, for example, by means of force–displacement measurements during the embossing process. These types of measurements will also provide insight into the complex flow patterns and thus lead to better process control as required for automated mass fabrication.

2.7. Fundamental Process Challenges

2.7.1. Polymer Flow Characteristics

Pattern Size. During hot embossing, the polymer, as a highly viscous liquid, has to be displaced from elevated stamp regions to recessed ones. As discussed in Section 2.6, the higher the temperature and the higher the pressure applied the lower

the viscosity of the polymer and thus the shorter the time required for it to conform to the mold (see Table IV). When the stamp has small patterns and, in particular, when the stamp pattern is periodic, polymer transport is required over short distances corresponding to the separation between adjacent patterns. When the stamp patterns are large, however, the polymer underneath these patterns has to be displaced over larger distances (Fig. 26). Therefore, in a hot-embossing process, it should be easy to replicate small and periodic patterns, whereas the replication of larger patterns requires a higher temperature, a higher pressure, or more time.

This is confirmed by experiments [58] where fully patterned stamps with mixed structures ranging from 400 nm to 100 μm, prepared by UV lithography and dry etching, were embossed. When low temperatures and pressures are used, the small patterns are replicated, but the recessed stamp areas around large patterns are not filled with polymer. Instead, a three-level configuration is found in the polymer where the front of polymer transport is visible as a step around larger patterns in the SEM image (see Fig. 27). With increasing pressure and temperature, this polymer front moves and the recessed stamp cavities fill up. At temperatures of 90–100°C and pressures of about 100 bar, a conformal replication of patterns up to 100 μm is achieved.

In Fig. 27, a critical pattern region is shown where the smallest patterns, a field of 400-nm lines and spaces, and the largest patterns, $100 \times 100 \ \mu m^2$ bond pads, are in close proximity. It is clearly seen that the small periodic patterns are replicated for the three parameter sets tested. Thus, as a gift of nature, hot embossing is a technique per se ideally suited for the replication of small patterns.

As a consequence, hot-embossing lithography of nanometer patterns is relatively easy and requires only moderate temperatures and pressures. Pressure of 100 bar and temperatures 100°C above T_g are only required when large pattern sizes are involved. In Fig. 27, the $100 \times 100 \ \mu m^2$ bond pads are probably a good example of maximum feature sizes for an integrated device.

Isolated and Periodic Patterns. With respect to the considerations concerning pattern size mentioned previously, some definitions are required. Isolated patterns consist either of small elevated regions surrounded by large recessed regions or the opposite. In this case, the size of the larger areas dominates the overall flow quality. Pressure and temperature have to be chosen to fit to these, and thus isolated lines at the stamp can be imprinted at much lower pressure and temperature than

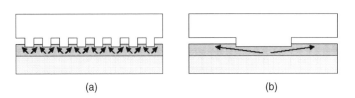

(a) (b)

Fig. 26. Local range of polymer transfer during hot embossing in case of small patterns (a) and in case of large patterns (b).

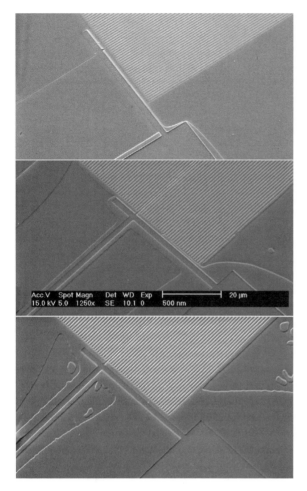

Fig. 27. Typical pattern-size-dependent imprint result for three different embossing conditions: (top) 60 bar, $T_g + 50°$ C; (middle) 100 bar, $T_g + 50°$ C; (bottom) 100 bar, $T_g + 70°$ C. With increasing pressure and temperature (top to bottom), the polymer fronts around larger elevated stamp patterns (visible in the bottom part of the SEMs) move and the cavities become filled. The fine lines are replicated in good quality in all three cases. Reprinted with permission from H.-C. Scheer, H. Schulz, T. Hoffmann, and C. M. Sotomayor Torres, *J. Vac. Sci. Technol., B* 16, 3917 (1998). © 1998 by the American Vacuum Society.

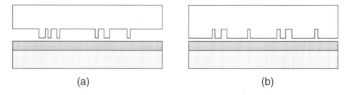

(a)   (b)

Fig. 28. Positive (left) and negative (right) stamps. Note the different contact area between stamp and sample during the initial stage of hot embossing.

imprint [94] and this will be discussed next in terms of effective pressure considerations and thermodynamics.

Effective Pressure. We define the "effective pressure" considering the contact areas of the stamp and the polymer surface normal to the applied force. Let us assume that stamp and sample are brought into contact, that the stamp patterns have nearly vertical walls, and that a certain pressure or force is applied. Then this force or pressure acts between the elevated stamp features and the polymer surface. When the elevated stamp surface is large, which is the case for a negative stamp, the effective pressure is only slightly increased with respect to the overall system pressure. However, when the elevated stamp area is small, as is the case for a positive stamp, the effective pressure is high because it is acting directly on a smaller area. As a consequence, the imprint depth per unit of time at a given temperature and force is higher for a positive stamp than for a negative stamp. Therefore, a positive stamp seems to imprint more easily than a negative one. On the other hand, when examining conformal filling of the large spaces between the positive stamp features, this takes time and, as long as no full contact is obtained over the whole stamp area, the result is the three-level situation discussed in Fig. 27.

In the more realistic case of a stamp with sloped walls, the effective pressure will decrease with increasing imprint depth, because increasing the relative contact area slows down the imprint. These considerations suggest that "pressure" is not an adequate parameter to characterize a hot-embossing process, unless detailed information on the stamp used is provided.

Recovery. When large stamp patterns are imprinted, complete filling of the stamp cavities takes time. Even when full contact is reached between stamp and sample, it cannot be assumed that polymer from all the area in between the recessed stamp structures has been transferred to these gaps. This is only the case if after full contact the pressure is maintained for a time long enough to enable complete flow at the elevated temperature. Otherwise the recessed stamp patterns are filled with polymer near the boundary of the cavities, whereas the polymer in the middle of elevated stamp patterns is not transferred but compressed. This compressed polymer reacts elastically and when the pressure is removed it recovers. As a consequence, in between recessed stamp patterns the residual polymer layer bows as the elastic deformation is restored. For negative stamps with large elevated stamp areas, this bowing may be as high as the initial layer thickness. This mechanical

isolated stamp trenches. Periodic patterns are governed by the larger of the two dimensions, and when this periodicity exists over the whole area the processing parameters can be chosen accordingly. Small periodic patterns are imprinted most easily.

Positive and Negative Patterns. The different imprint characteristics of "positive" and "negative" stamp patterns (Fig. 28) have been discussed first in [19] and later in [93] and are directly correlated to the preceding discussion. In a positive stamp, the percentage of elevated areas is small, whereas in a negative stamp it is high. In [19], it is stated that the negative profiles proved to be less difficult to emboss than the positive ones; however, there is no common reference to degrees of difficulty. In any case, for the reasons given in the preceding section, a difference is expected. Another investigation concludes that negative stamp patterns are more difficult to

recovery of the polymer will result in an increased nonuniformity of the residual layer thickness and, simultaneously, in a reduction the in imprint depth. These effects are most obvious when imprinted regions of large elevated stamp features are investigated, in particular, over hours and weeks (Fig. 29). Such a delayed recovery will not hamper the process as long as the time scale of processing after imprint is quick compared to the polymer's time constant. However, recovery is a hint that polymer transfer is not yet complete. Figure 29b shows this effect taking the example of a 100-μm square pad structure.

Nevertheless, recovery is not only a long-term effect. Under unsatisfactory imprint conditions, such as too low pressure, temperature, or time, it is evident directly after imprint [95].

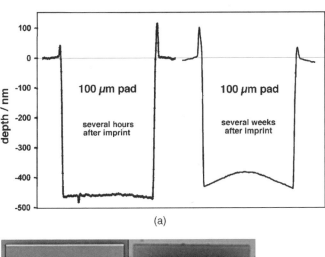

(a)

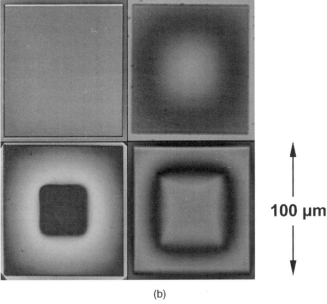

100 μm

(b)

Fig. 29. Elastic recovery effects of a polymer after imprint, taking the example of a 100-μm square pad geometry imprinted at low temperature. (a) Pad profiles measured directly after imprint and several weeks after. (b) Different types of bowing within the pad observed several weeks after imprint in a laser microscope. The four examples refer to different types of polymers and demonstrate their different flow behavior. The situation at the right of (a) resembles the top right laser micrograph of (b).

Although the polymer's mechanical recovery may seem, in general, to be a negative effect, it has been used for specific device formation, namely, magnetic ring structures [96].

Imprint "Quality." When in the case of positive stamp patterns the polymer is displaced over a short distance only, the three distinct polymer thicknesses as discussed in Fig. 27 remain within the stamp cavities. However, the imprint depth of the positive stamp patterns is high compared to the upper level and compared to the initial polymer thickness. Thus, removal of the residual layer poses no major problem. The fact that the remaining polymer mask has an inhomogeneous thickness does not disturb a subsequent dry-etch or lift-off process. On the contrary, high steps improve lift-off and high mask walls may counteract the elevated etch rates at the mask edges in a low-pressure ion-dominated dry-etch process known as "faceting."

To imprint a negative stamp pattern, compared to a positive one, a higher pressure is needed for the same values of temperature and printing time. Thus, comparing identical processing conditions, the imprint quality of positive stamps is better than that of negative stamps. This is particularly the case when a negative stamp contains only a small patterned area, as is the case for the imprint in Fig. 30. Periodic patterns are easier to

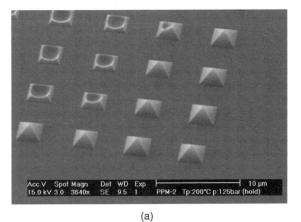

(a)

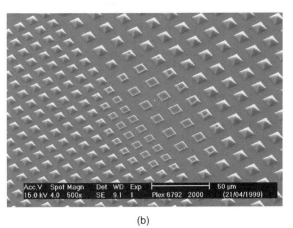

(b)

Fig. 30. Embossing result for negative stamps with only small patterned area [94]. (a) Just the boundary of the patterned field is replicated. (b) Nearly perfect pattern transfer over the whole field at optimized processing conditions, excluding only a small center region.

imprint than isolated ones and the smaller the patterns are the better the imprint quality obtained.

Nevertheless, imprint is not always as simple as this picture suggests. Considering stamps with positive and negative patterns in different areas, for example, positive stamp patterns within a certain region surrounded by a negative stamp area, the overall imprint situation may become unpredictable. Because warping of the Si stamp may occur when some areas are in contact with the polymer and others are not, the imprint of patterns of different type and size may be difficult to control. Moreover, it becomes rather complex when effects associated with "self-assembly" are involved, as discussed in Section 2.10, affecting the mask functionality of the embossed polymer for pattern transfer.

Residual Layer Thickness. As mentioned in the discussion of the hot-embossing process in Section 2.5.1, as soon as the imprint depth equals the height of the stamp features the effective pressure, which had been increased to some extent due to the ratio of elevated to recessed stamp area before, drops to the externally applied pressure. This is the moment when the polymer has filled the recessed stamp features and the maximum achievable thickness contrast in the polymer is obtained. When this situation is met, the imprint slows down substantially and appears to "stop."

To imprint deeper to reduce the residual layer thickness, polymer would have to flow within distances corresponding to the overall stamp size. In fact, the polymer would have to be transported from the center of the overall imprinted area to the periphery. As polymers are highly viscous liquids, this would require enormous pressures and very long times and is therefore not practical in the context of lithography [97, 98].

It appears that the only way to tailor the residual layer thickness is by an exact choice of the initially spun polymer thickness. The residual layer thickness then results from the polymer volume when the volume corresponding to the imprinted depth specific to each stamp pattern is subtracted.

Thermodynamics. The following approximation [94], which is, in fact, a differential equation [97, 98] may help to illustrate the situation:

$$v(t) \approx (p_{\mathrm{eff}} \cdot h(t)^3)/(\eta R^2) \qquad (15)$$

where $v(t)$ is the vertical imprint velocity, p_{eff} is the effective pressure referring to the elevated stamp area in contact with the polymer, $h(t)$ is the actual polymer layer height below the elevated stamp features, η is the polymer viscosity at the imprint temperature, and R is the radius of the largest stamp feature in contact with the polymer, that is, the wafer radius for an unpatterned stamp, or the maximum elevated feature size. To achieve comparable imprint conditions, indicated by a certain value of $v(t)$, a large elevated area R (negative stamp) has to be compensated for either by an increased pressure p_{eff}, or, alternatively, by an increased polymer layer thickness $h(t)$. The latter is not suitable for lithography purposes where the residual layer thickness should be minimized. Equation (15)

shows that embossing for nanoimprint lithography is different from and much harder to accomplish than embossing of thick volume materials, which is the classical embossing technique (see Section 1.3). When small thicknesses are involved, the imprint velocity slows down according to the third power of h. This is the case toward the end of the imprint when the residual layer thickness is reached.

The positive/negative stamp discussion becomes clearer when interpreting Eq. (15). The effective pressure depends linearly on the imprint velocity, whereas the feature size R has a quadratic dependence. For a positive stamp, R is small and p_{eff} is high. For a negative stamp, R may assume huge values, comparable to the overall stamp size. Thus, the high R value will efficiently enhance the effect of the low p_{eff} for a negative stamp, and this is why it is suggested that negative stamps are much harder to imprint than positive ones. As Fig. 31 shows, the imprint of negative stamps was only successful with a very high layer thickness h to counteract the high value of R and the low p_{eff} according to Eq. (15).

Imprint Depth. In a number of experiments, the imprint depth was chosen to be less than the pattern height of the stamp [12, 13, 17, 18, 27, 36, 48, 50–52]. In most of these publications, small positive patterns were imprinted where an adequate thickness contrast and residual layer thickness were obtainable without having full contact between stamp and sample. Alternatively, full imprint depth was not achieved because embossing times were too short compared to the polymer time constants at the applied temperature and pressure. Imprint depth can then be correlated with pressure (see Fig. 14b). One group reports increased sticking upon full contact between stamp and polymer [50, 52]. In most papers in the field, no statement is made concerning imprint depth.

Imprint depth becomes particularly important when patterns with specific three-dimensional shape have to be imprinted [20, 59], as would be the case for "gray-tone lithography." Moreover, to estimate the potential of hot-embossing

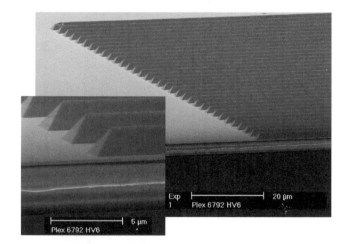

Fig. 31. Imprinted negative stamp with a small amount of patterned area, similar to Fig. 30b. The cross section shows that successful imprint is due to the high layer thickness used (about 4μm), a situation incompatible with a lithography process.

lithography at the wafer scale for future production application, the consequences of polymer displacement phenomena have to be investigated further [40, 58, 63, 99]. To overcome such polymer transport problems in micromechanical applications, retaining structures are foreseen to provide some volume for the excess polymer. This is unlikely to be a solution for a lithography process. For reliable hot-embossing lithography, the processing parameters have to be chosen in accordance with the largest recessed stamp areas involved and the procedure should provide for complete intrusion of the stamp.

2.7.2. Embossing-Induced Damage

During hot-embossing lithography, damage could be caused to the substrate due to the applied high pressure, for example, in the case of highly crystalline semiconductors susceptible to uniaxial strain. Furthermore, the imprinted material itself could suffer from the temperature or pressure applied, as could be the case during imprint of a functionalized polymer. The latter was addressed in Section 2.3.3, so that we can concentrate here on substrate damage.

Only few investigations where potential drawbacks associated with the high pressure applied in the embossing lithography process for device fabrication have been reported. A wire channel Si field effect transistor was investigated [29] where the channel as well as the source and drain areas were imprinted at 175°C and 45 bar with a 30-nm-thick sacrificial oxide layer between the polymer and the substrate for protection of the active Si channel. Similar devices were fabricated using electron beam writing and no difference in the electrical properties of the two types of devices was found. This is explained in terms of the yield strength of Si of about 3×10^9 Pa, which is 3 orders of magnitude higher than the applied pressure so that no structural damage of the Si substrate could be expected.

Quantum point contacts were investigated in [60, 100, 101] where the substrate is a modulation-doped GaAs/GaAlAs high-mobility structure grown by MBE (molecular beam epitaxy). Once again, no differences were found between devices fabricated by electron beam lithography and by hot-embossing lithography.

A further investigation covered GaAs MSM (metal–semiconductor–metal) photodetectors [102] and compared samples prepared by electron beam lithography to others prepared by hot embossing. It was found that there was no degradation of the device characteristics if the imprint pressure were kept at 600 psi (40 bar) or below. At higher pressure levels, an increased dark current was observed. The imprint had been performed in PMMA at 175°C.

2.8. Hot Embossing in a Multilevel, Multilayer, or Multistep Lithography Sequence

2.8.1. Embossing over Topography

So far, only hot-embossing lithography on flat surfaces has been discussed. However, it is clear that a lithography process aimed at realistic device production has to work over topography. Only a few publications address this.

It is often argued that in a number of processing sequences for device fabrication only one critical layer with really small dimensions is needed. In this case, this lithography step can be the first one where the wafer is still flat. This holds for simple devices such as sensors [50–52] but not for complex microelectronic devices, in which, for example, high integration levels and vertical transistors are needed.

Profiled Stamps. Hot embossing has been used as the second lithography step in the fabrication of quantum point contacts, with a profiled stamp where the pattern to be imprinted is located on a pedestal [60] and the stamp fits within the sample topography. This solution is viable when few devices have to be fabricated and when the aspect ratio of the wafer topography is low, but it cannot be a general solution of the topography problem.

Single-Layer Schemes. In principle, it is possible to imprint into a thick polymer layer over a topography. The resulting residual layer, which has to level the surface topography, would then be quite high and of different thickness. This would require a highly anisotropic and highly selective dry-etch step. Needless to say, both requirements are hardly met simultaneously. Moreover, the local topography will affect the local flow behavior of the polymer. Consequently, the single-layer scheme is likely to be successful only in a few cases where the underlying topography is very regular. An example is given in [18] where a 75-nm oxide step was covered with a 300-nm-thick polymer layer. With embossing, a 200-nm-thick contrast was achieved and 150 nm of residual layer had to be removed by dry etching. Due to the long etching time required, the imprinted linewidth was reduced from 60 to 40 nm. This corresponds to the discussion in Section 2.5.2 and confirms that, ideally, residual layers should be thin compared to the imprint depth.

Multilayer Schemes. Another approach follows the ideas used in microelectronics to overcome the topography problem. It is based on the well-known concept of "planarization" and uses a "trilevel" resist scheme, which is state of the art with classical lithography when the depth of focus is limited.

A thick base layer of polymer is used to planarize the topography. On top of this polymer a thin inorganic layer, the "hard mask" (oxide, nitride, or metal), is deposited followed by an additional thin polymer layer. The uppermost polymer layer is imprinted, then the thin residual layer is removed and the hard mask is patterned by RIE using the top polymer as a mask. In a last step, the bottom resist layer is etched with the hard mask, the "dry development."

Five planarizing methods potentially suitable for hot embossing are given in [103]. (i) The first one follows the classical "trilevel" scheme as explained previously using 30-nm SiO_2 as a hard mask. It was found that for successful hot embossing of the top layer, without deformation of the bottom layer, the glass temperatures of the two polymers have to differ by at least 100°C with the top layer featuring the lower T_g. 250 k PMMA was used as the bottom layer and 2 k polystyrene (PS) with a glass temperature as low as 0°C [103] as the top layer. (ii) The second method starts with a two-level scheme using a thermoset polymer as the bottom layer and PMMA as the top layer. After spin-on, the bottom layer is crosslinked in a temperature treatment and will not dissolve during application of the second polymer layer. As in the first method, the top layer is imprinted without deformation of the bottom layer. In a lift-off process, Cr is deposited on the imprinted resist, leading to the tone reversal. This Cr pattern is then used as a hard mask for dry development of the bottom resist, resulting in a negative tone pattern. (iii) The third method is a two-level scheme with PMMA as the bottom layer and the low-T_g PS as a top layer, but this time without intermediate oxide. The top layer is imprinted and used directly as a mask for dry development. Mask selectivity is critical in this procedure. A value of 1 : 2.4 is reported in a low-pressure O_2 RIE process (150 W, 3 mTorr). (iv) The fourth method uses only one thick polymer layer, which is embossed on its surface. Then the elevated features of the imprinted polymer are printed in a roller process to build the hard mask. Material choice is extremely critical here. In a different approach, evaporation of Cr is reported at a glancing angle so that only the top of the patterns is covered, thus providing the hard mask [104]. (v) The last approach also follows a microelectronics idea, the "etchback." A single thick resist layer (polymer 1) that is imprinted on its upper part is deposited. Then a second polymer layer (polymer 2) is applied and the stack is etched back in an anisotropic dry-etch process with equal etch rates of the two polymers (selectivity 1) until a situation is reached where a polymer 2 pattern is embedded in a polymer 1 matrix. Then the process parameters have to be changed to a highly selective anisotropic process in order to use the patterned polymer 2 as a mask for polymer 1. Once again, the material choice is very critical as is the choice of dry-etch parameters as two totally different polymer dry-etch processes are needed. Whereas the latter techniques were not illustrated with examples, these are not easily implemented because of the respective material problems and polymer flow challenges over topography. However, it is easy to see that the first two approaches will work.

Because planarization should always be successful in hot-embossing lithography over topography and as this is the state-of-the-art strategy in microelectronics integrated device fabrication without additional steps in a production process, it may be argued that topography is not a bottleneck for a wide range of applications of hot embossing as a nanometer-scaled lithography process.

2.8.2. Trilevel for Mask Selectivity Improvement

The trilevel technique may also be used as a means to improve the mask selectivity of nanoimprinting during pattern transfer. When very small pattern sizes are involved, the thickness contrast obtained in the polymer is often of the same order of magnitude as the pattern size. Reasons may be a limited stamp aspect ratio due to conventional processing as well as the fact that sticking effects may limit the performance at high aspect ratios for dense, nanometer-scaled patterns [46]. In such cases, trilevel techniques increase the effective mask selectivity for pattern transfer.

Trilayer systems of PGMI/Ge/PMMA (50 k), PGMI/Ge/S1805, and PGMI/Ge/hybrane have been reported [105, 106], where PGMI, mechanically stable up to 200°C, represents the bottom layer and 10-nm Ge (evaporated) represents the "hard mask." Top layers of PMMA and S1805 could be imprinted at 160°C/8 bar and 100–165°C/12–30 bar, respectively, where S1805 featured a smaller thickness contrast after embossing than PMMA due to its higher viscosity. Successful dry development of the PGMI and use of this mask for lift-off, reactive ion etching, and electroplating were demonstrated for 100-nm-pitch dot arrays [105].

2.8.3. Embossing of Layer Stacks

As an alternative to the trilevel approach, stacks of layers can also be imprinted directly in one embossing step. In [30], a stack of 80-nm PMMA (bottom resist), 10-nm Ni (hard mask) and 30-nm PMMA (top resist) was imprinted with 25-nm lines through the metal layer. The metal in the trenches was removed by wet chemical etching and the metal not punctuated by the imprint was used as a mask for the dry development step. This simplification of the trilevel scheme, proven to work well when fine positive patterns are imprinted where the thin metal layer is smeared over the trench sidewalls, may not be suitable for larger patterns or negative stamps.

Layer stack embossing can also help to improve the lift-off capability of nanoimprint. Removal of the thin residual layer remaining after the embossing step often tends to decrease the slope of the patterns, resulting in problems when used as a lift-off mask. A sound approach to optimize hot embossing for lift-off is to imprint a bilayer of two different polymers [46], namely, 495 k PMMA as the first layer and 73 k PMMA/MAA (a copolymer of PMMA with methacrylic acid) as the second layer. The two polymers need different solvents so that spinning of the second layer on top of the first dried layer is not a problem. Imprint is performed at 180°C and the residual layer is removed either by O_2 RIE or by soaking the sample in methanol, which removes the copolymer isotropically. In both cases, the PMMA bottom layer is then treated in chlorobenzene in order to underetch the top resist. Dense patterns down to 50 nm were prepared in this way with polymer layers of 50 nm each and successful lift-off was demonstrated.

2.8.4. Multilayer Imprint and Alignment

The imprint over topography (see Section 2.8.1) is directly correlated to the question of adjustment of a hot-embossing step to a previous lithography.

Whereas the multilevel examples in [103] represent a feasibility study where adjustment is not a topic, the example given in [60] relies on alignment of the gate structure with respect to the source and drain regions. As in most of the examples, an alignment tool is used, in this case a mask aligner with an accuracy of $\pm 2\,\mu$m, using a backside infrared (IR) illumination alignment principle. It is stated that transport of the aligned stack from the mask aligner to the imprint equipment may be the most critical step.

Submicrometer alignment over a 2-in. area has been demonstrated for 4-in. wafers. Once again, alignment and imprint were carried out in separate equipment. The alignment was performed in a modified commercial aligner before the stack of stamp and sample was placed in the imprint system [64].

For a more general investigation of the accuracy of pattern placement, taking into account the dimensional changes of the imprinted material, the high-performance alignment system of an X-ray stepper was employed [34]. The alignment procedure uses different alignment marks separated from each other by 24 mm and performs an alignment between marks at the sample and marks at the stamp by use of three independent lamps. Marks were prepared on the sample by imprint and lift-off and fine alignment to the respective stamp marks could be performed with a standard deviation of only 30 nm at a mask to wafer gap of 40 nm. An example of the alignment is given in Fig. 32 where a second set of marks was embossed after alignment into a planarizing resist layer.

This quite complex investigation reveals one of the major challenges of nanoimprinting the alignment, a challenge these alternative lithography concepts share with the classical ones. It is far from trivial to align nanometer-scaled features with nanometer accuracy, in particular, when larger areas are involved. This situation is further aggravated for a hot-embossing process, as identical coefficients of thermal expansion for all materials involved are hardly obtained. As a consequence, a multilevel alignment may be one of the most complex parts of an imprint lithography system and may

increase costs substantially. The preceding casts doubts over the large-area and low-cost lithography tool for production as suggested in the first publications. Fortunately, alignment systems of other lithography concepts can be adapted for imprint.

2.9. Room Temperature Embossing of Polymers

For temperature cycles typical of hot-embossing lithography, from room temperature up to 150–220°C, the thermal expansion and shrinkage of the materials involved may lead to mechanical stresses and may limit the pattern definition accuracy. Thus, room temperature embossing is of general interest.

2.9.1. Direct Nanoprinting

Under the name "direct nanoprinting," the surface of substrates (600-μm-thick polycarbonate (PC) foil [107], 350-μm-thick Al foil [41]) were embossed with a hard SiC stamp featuring patterns in the 100-nm range and below. Densely packed dots of 100-, 70-, and 63-nm pitch were embossed in PC within 5 min at a pressure of 300 bar. A temperature increase to about the polymer's T_g (145°C) resulted in a slight increase in the imprint depth. For the 100-nm-pitch dots, an imprint depth of 70% of the stamp height was already achieved at room temperature. This success may be attributed to the fact that a thick material was embossed and that the stamp pattern was a periodic positive pattern with very small feature size (compare Section 2.7). The results were evaluated by AFM and limited by the tip used and thus hardly assessed.

Results obtained for Al are summarized in Section 2.3.3 together with embossing results for sol–gel materials [45] and functionalized polymers [42]. Direct nanoprinting is a volume technique and even when performed onto a polymer substrate it is hardly suited for most lithography purposes.

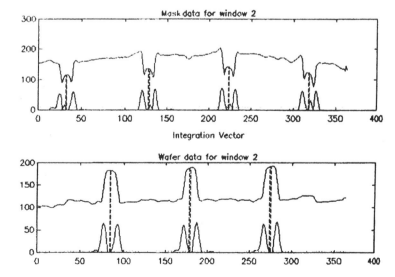

Fig. 32.   Aligned hot embossing using alignment marks compatible with an X-ray stepper. (Left) Alignment marks (thicker lines, wafer; thinner lines, stamp); (Right) fine alignment signals of stamp (top) and wafer (bottom). Reprinted with permission from A. Lebib, Y. Chen, J. Bourneix, F. Carcenac, E. Cambril, L. Couraud, and H. Launois, *Microelectron. Eng.* 46, 319 (1999). © 1999 by Elsevier Science.

2.9.2. *Solvent Vapor Treatment*

An idea coming straight from polymer preparation makes use of solvent vapor rather than temperature to soften the polymer. It corresponds with the experience that polymer layers with insufficient prebake are easy to emboss as the incorporated solvent reduces the glass transition temperature of the polymer [108]. In addition, the entangling limit is increased by the dilution effect and both effects together result in a decrease in viscosity. The method is a modified version of the solvent-assisted microcontact molding (SAMIM) introduced in Section 4.5.5.

Starting with a well-dried spin-coated layer of polymer, viscosity is decreased by adding solvent in a controlled amount in an oven held between 25 and 60°C for a time ranging from 5 to 40 min, resulting in 5–15 wt% of solvent incorporated in to the polymer. This layer is then embossed at pressures ranging from 300 to 1500 bar within periods of 5–40 min. It is subsequently separated from the mold and the solvent is allowed to reevaporate. The principle is schematically shown in Fig. 33.

Features of 100- to 500-nm size could be replicated with a Si mold. Neither size differences between mold and imprint nor pattern distortions due to reevaporation of the solvent were found. Residual layer thicknesses and usable polymer layer thicknesses were not reported and may depend on the time of solvent treatment because, due to diffusion, the solvent exhibits a concentration gradient in the polymer. Investigation of this topic is essential for use of solvent vapor treatment as a room temperature embossing lithography technique.

Solvent vapor treatment

Room-temperature imprinting

Pattern transfer to polymer
(removal of the mold)

Fig. 33. Schematic of room temperature embossing after solvent vapor treatment. Reprinted with permission from D.-Y. Khang and H. H. Lee, *Appl. Phys. Lett.* 76, 870, (2000). © 2000 by the American Institute of Physics.

2.10. **Self-Assembly and Wafer-Scale Embossing**

In a hot-embossing assembly with the polymer layer at imprint temperature, holding the stamp with spacers some 165 nm above the polymer surface, a regular pillar formation was observed for thin layers (≈95 nm) of low-viscosity polymers (2 k, 15 k PMMA). When using a patterned stamp, the polymer fills the gap in elevated stamp regions only. This effect is referred to as "lithography induced self-assembly" (LISA) [109]. The gap fills up with pillars starting from the periphery and propagating to the center of the stamp feature. Regular arrays are formed where the pillar diameter and pillar arrangement depend on the stamp geometries. Pillar size and pillar distances are in the micrometer range; pillar height equals the gap size where the gap width may amount to 2–7 times the initial polymer layer thickness. When stamp feature areas were larger than 50 μm, multidomain regimes were seen to evolve. Besides pillar structures [109, 110], concentric ring structures were also found [111]. Potential reasons for pillar formation include electrostatic forces, that is, polymer charges and image charges on the stamp, and hydrodynamic forces. Charging effects could explain the start of pillar formation at the pattern edges [109].

It was also observed that the gap may fill completely without forming pillars. This is referred to as "lithography induced self-construction" (LISC) [112]. Whereas for a high difference in surface energy between polymer and stamp, which refers to limited wetting and adhesion (see Section 6.1), the pillars are formed, a decreased difference in surface energies leads to the LISC situation, which is the result of a lateral spreading and merging of pillars, building up a mesa structure finally.

LISA and LISC effects play an important role in the development of hot embossing into a large-area processing technique. Even flatness of bare Si wafers is in the micrometer range and becomes even worse during processing. Thus, in a wafer-scale embossing process, it is likely that in certain regions of the wafer the full contact between polymer and stamp is not obtained in recessed stamp areas. This is in particular the case for larger pattern sizes and mixed patterns (see Section 6.2.1). In such a region, we have exactly the LISA situation: The polymer surface under processing temperature is a submicrometer distance from the stamp surface within larger recessed stamp regions. Dots, mounds, and trenches are observed in such areas and may have different shapes and periodicities, depending on the specific process conditions, the local imprint parallelism, and the polymers used. Examples are shown in Figs. 34 and 35. Details are published in [93, 94].

It has been found that when trenches remain, due to incomplete embossing without full stamp intrusion, these go down to the substrate (see Fig. 36) and will then represent serious mask defects. The conditions for large-area processing have to take into account such effects, which might require higher temperatures or longer imprint times compared to processing of small areas.

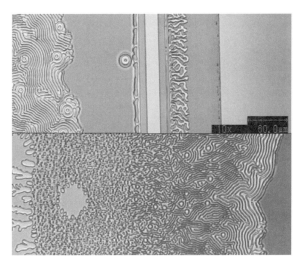

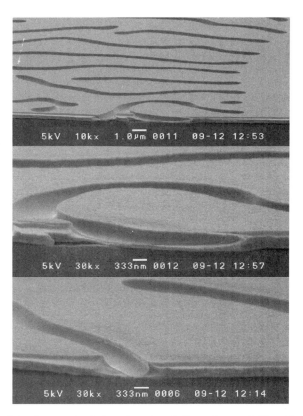

Fig. 34. Mounds and trenches of different types formed in front of a polymer front propagating from the white line (top) to the left (bottom) in the case of an uncomplete imprint [94]. In the laser micrograph, the dark color characterizes the upper polymer level, the light color the lower one.

2.11. Combining Hot Embossing with Other Lithography Concepts

Hot-embossing lithography may be combined with any other lithography where the different methods can be combined in such a way that either one single polymer layer is patterned by different lithography methods to form a mask, referred to as multistep lithography, or different lithography types are

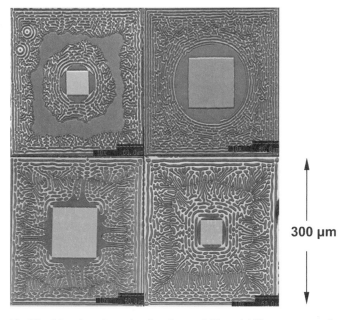

Fig. 35. Mounds and trenches found around 50- and 100-μm square pads enclosed in a 300-μm square box [94]. The mound patterns seem to propagate from the inside (pad) and from the outside (box boundary) as well. The four different examples refer to different locations in the sample. In the situation at the top left, a medium level of polymer is discernible between pad and boundary, which lies between the upper level (dark) and the lower level (light). This is the initial polymer thickness after spinning.

Fig. 36. Cross-sectional view of trenches similar to the ones in Figs. 34 and 35. (Top) Survey; (middle and bottom) detailed view. The trenches go down to the substrate [94].

used for patterning of different layers. Whereas the second approach is the classical mix-and-match one, corresponding to, for example, combining e-beam lithography for definition of fine patterns with UV lithography for definition of coarse patterns, the multistep approach is typical for hot-embossing lithography as embossing only makes use of the mechanical properties of the polymer whereas other properties such as radiation sensitivity may be addressed by e-beam or UV exposure of the same layer additionally. Most resists for optical or e-beam lithography are thermoplastic, suggesting combinations with hot embossing.

2.11.1. Mix-and-Match Approaches

Most of the mix-and-match approaches rely on the combination of hot-embossing lithography with optical or UV lithography [29, 36, 51, 60]. The UV lithography defines the bond pads, whereas hot-embossing lithography defines the small patterns and is thus used as a replacement for electron beam lithography. In most cases, the hot-embossing process is the first-level lithography and UV lithography is performed aligned to the first-level pattern as the second step.

Only in [60], as mentioned in Section 2.8.1, the procedure is different: For the fabrication of the split-gate quantum point contact device, the Hall bar and the contacts as well as alignment marks are first defined by optical lithography and in the second step the split gate is defined by hot-embossing.

2.11.2. Multistep Lithography

For multistep processing, hot embossing was successfully combined with optical lithography [113] where a commercial negative-tone resist was used. In a first step, the radiation-sensitive resist was imprinted and afterwards subjected to a lithography step where a different pattern was exposed on the imprinted resist. After development, a three-level profile was obtained, featuring the initial layer thickness and the residual layer thickness in the exposed areas (see Fig. 37). After residual layer removal, the mask with mixed patterns was ready (Fig. 38).

Because imprint is problematic for larger pattern sizes, the combination with UV lithography for easy mixed-pattern definition is favorable as then imprint can be performed at relatively low pressures and temperatures. Alternatively, when the residual layer is chosen in the thickness range of the imprint depth, a multistep resist pattern can be achieved, which can be of interest for specific devices or for profiled stamp fabrication as well as for nonlithography applications of hot embossing. One example for the latter is the fabrication of multiple-layered plastic devices in cellulose acetate for biological purposes [114].

An additional approach refers to a combination of hot embossing and scanning probe lithography (SPL) [115] using

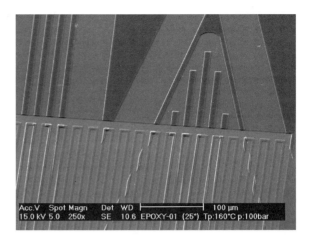

Fig. 38. Resist pattern obtained by the combination of hot embossing with UV lithography according to Fig. 37 (feasibility test). In the bottom part of the view graph the UV pattern is superimposed on the imprinted pattern. In contrast to the basic idea of using imprint for the small patterns and UV lithography for the larger ones, the respective pattern sizes are reversed in this feasibility study [113].

a commercial P(MMA/MAA) copolymer, the e-beam resist ARP-610. Hot embossing was used for large-area patterning, whereas SPL was performed locally after imprint, either to remove excess polymer or to induce controlled additional slopes in the imprinted patterns by adjusting the local strength of the load force during SPL.

3. MOLD-ASSISTED LITHOGRAPHY

3.1. Principle of Mold-Assisted Lithography

Surprisingly, the literature on mold-assisted lithography (MAL) is limited compared to the other imprint techniques. The basic reference is still the first one [21], where the principle is given along with a number of important processing details. In contrast to the other imprint techniques this first publication starts with a 4-in.-diameter study of MAL in which a piece of semiconductor production equipment is used, a vacuum contact printer (see Section 1.2.2). This first evaluation of MAL is thus an operational, production-oriented one, not a small-scale feasibility study under laboratory conditions.

Nevertheless, this study was not performed under realistic MAL conditions. Differing from the principle given (Figs. 2 and 39), Haisma et al. used a nontransparent mold, a patterned Si wafer, but a transparent substrate (fused silica) spin-coated with a low-viscosity monomer. The Si mold prepared with a mold release agent by immersion was aligned and pressed into the monomer with a copy pressure of only 0.1 bar. A compression time of 25 s was sufficient to allow the monomer to fill the stamp pattern. Polymerization was accomplished by UV radiation (mercury lamp, 1000 mJ/cm^2, 300–400 nm wavelength range). Due to the mold release agent used and due to the inherent shrinkage upon polymerisation of 5–15%, the separation of sample and mold was easy. The whole replication process took 5 min.

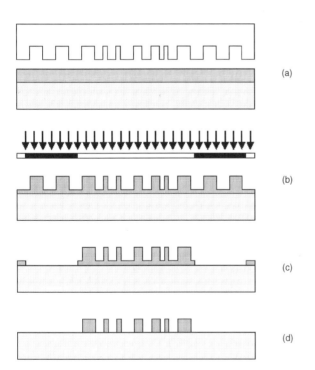

Fig. 37. Principle for combining hot embossing with UV lithography. Hot embossing of a thermoplastic photocurable polymer (a), followed by exposure of parts of the area to UV radiation through a mask (b). The radiation crosslinks the polymer in exposed areas, whereas it remains soluble in unexposed ones and can be removed (c). After residual layer removal, the mask pattern is complete (d). The schematic makes best use of both techniques. The whole area is patterned by imprint with features of smaller size. These features are dummies in regions removed during the UV-lithography step and are used to facilitate the imprint step only.

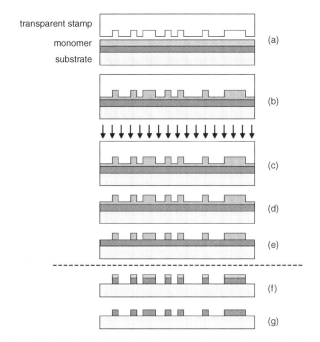

transparent stamp
monomer
substrate

(a)

(b)

(c)

(d)

(e)

(f)

(g)

Fig. 39. Schematic of patterning by use of mold-assisted lithography. Lithography comprises steps (a)–(e); steps (f) and (g) refer to pattern transfer with the MAL mask. (a) Sample (substrate with an additional layer, covered with a monomer or other curable precursor of low viscosity) and transparent mold (prepared with a release layer) before contact; (b) conformal filling of the mold cavities by the monomer upon contact at approximately 1 bar; (c) curing by flood UV radiation; (d) crosslinked patterned polymer layer on substrate after mold release; (e) polymer mask on substrate after etching (wet/dry); (f) pattern transfer with polymer mask into the underlying layer by RIE; (g) patterned layer after polymer stripping. Note that after UV curing the polymer is crosslinked and thus no longer soluble. Therefore, MAL is not suitable for a subsequent lift-off process.

Residual layer removal was performed by H_2SO_4-based wet etching, and transfer of the polymer pattern into the silica substrate was accomplished by dry etching without loss of resolution.

The uniformity of the residual layer is problematic. Nonuniformities of more than 500 nm were found over 4-in. wafers, whereas values smaller than 50 nm were typical for smaller molds of 30×30 mm^2. The high residual layer nonuniformity is clearly related to wafer-scale processing and results from the leveling mechanism of the system (wedge error compensation) together with the limited flatness of the processed wafer. To overcome this limitation, a stepping procedure was proposed [21].

Residual layer thickness and uniformity are of central importance for molding as a lithography process. As discussed in [21], no really anisotropic and thus vertical etch process is available for residual layer removal (see Section 2.5.2). To preserve the molded pattern during etching, residual layers smaller than the structure height and smaller than twice the structure width are required. Thus, a high residual layer thickness limits the pattern resolution. In addition, without making use of a multilevel technique (see Section 2.8.2), MAL can only work over a limited topography as the minimum residual layer thickness corresponds to the topography height.

A recent publication [116] reported successful MAL over areas of up to 4 cm^2 with a Si substrate and curing through a quartz mold. The mold had 25-mm-diameter and 4-μm line patterns. With initial layers of 300-nm thickness, residual layers down to 110 nm were achieved at a contact pressure of 1 bar. The residual layer thickness was found to be determined by the initial layer thickness for a limited pressure [117], a behavior similar to hot embossing as discussed in Section 2.7.1. The overall process time was 10 min, including alignment. For high-precision printing, it is intended to explore a step-and-repeat process as proposed in [21]. Such a step-and-repeat concept has been developed and tested [118]. This large-area approach is discussed in detail in Section 3.4.

3.2. Material Issues

The central material issues in MAL are (i) the photocuring ability of the material to be patterned, (ii) its viscosity before curing, (iii) its wetting of the mold, and (iv) its masking selectivity after curing. The latter is not specific to MAL but common to all the discussed imprint lithography processes.

Photocurable liquids are solvent-free, UV-reactive systems formulated from monomers, oligomers, or low-molecular-weight polymers together with a photoinitiator (PI). UV light cracks the photoinitiator, and its radicals start a chain reaction, leading to a crosslinked three-dimensional network in the presence of multifunctional components. Typical curing times range from milliseconds to a few seconds. The principle of curing as sketched in Fig. 40 is well known for a number of technical systems and commercial applications [119] such as, for example, photolithography with negative-tone resists. It is even more convenient for many nonlithography techniques known as casting and molding. In particular, early applications exist for optical elements [10, 25] demonstrating even submicrometer resolution.

The features of the resulting polymer may be tuned by the specific formulation of the components so that different characteristics are obtained when using acrylate-, ionic-, thiolene-, or vinyl ether–based systems. Among others, their mechanical properties may be adjusted from hard and brittle to elastic. As a mask material, the hard ones are used mainly for other purposes such as, for example, stamp replication (see Section 5.2.2), elastic materials are also of interest. In the latter case, the curing process is often driven by temperature rather than by radiation.

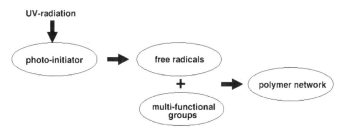

UV-radiation

photo-initiator → free radicals
+
multi-functional groups → polymer network

Fig. 40. Illustration of a radiation curing process: UV-initiated free-radical polymerization leads to a crosslinked polymer network when multifunctional groups are present.

The viscosity of the polymer precursor has to match two conflicting demands. On the one hand, it has to be low enough to fill the mold cavities conformally under relatively low contact pressure [see Eq. (15)]. On the other hand, it has to be high enough to enable the formation of a homogeneous layer of adequate thickness on top of the substrate, for example, in a spin-coating process. The need to optimize materials is obvious. Oligomers, prepolymers, or mixtures thereof may be better suited for spinning than pure monomers. But then a contact pressure of 1 bar may be too low for mold replication. As a consequence, the mold size has to be reduced and a step-and-repeat process invoked for wafer-scale processing [116].

Although obvious, the third point, namely, the need for wetting of the mold, is first mentioned in [118] for a modified MAL procedure as discussed in Section 3.4. Capillary forces have to be utilized for stamp cavity filling, making wetting of central importance.

In [21], two different compounds were used, a mixture of either HDDA (hexanedioldiacrylate) or HEBDM [bis(hydroxyethyl)bisphenol-A-diacetate] as the monomer, with DMPA (dimethoxyphenylacetophenone) as the photoinitiator. The monomer viscosity is very low, 7 mPa, enabling conformation to a patterned 10-cm-diameter mold at lowest contact pressure (0.1 bar). After curing, the rigid polymer is reported to have a dry-etch selectivity higher than 1 with respect to SiO_2. In [116], modified compounds based on acrylate and epoxide were investigated with a viscosity of 600–800 mPa and a dry-etch selectivity of 2. Due to the higher viscosity, a contact pressure of 0.8 bar was used for replication of areas of 4 cm^2.

3.3. Achieved Patterns

In [21], the smallest patterns reported were dots of about 35-nm diameter with a 75-nm pitch. The process truly replicated the mold as mold defects demonstrated. Residual layer removal and dry-etch pattern transfer using the MAL pattern as a mask were demonstrated for lines as small as 70 nm wide and a 200-nm pitch with a stamp 250 nm high (i.e., an aspect ratio of about 3). At higher aspect ratios, the mold release became difficult. Resolution limits of less than 25 nm were postulated. 80-nm dot arrays 130 nm high were shown in [116], successfully replicated in polymer layers of 300-nm to 4-μm thickness using molds with an overall e-beam-written area of 1.2 × 1.2 mm^2.

3.4. Step-and-Flash Imprint Lithography

A dedicated large-area approach, step-and-flash imprint lithography (SFIL) [118, 120–122] implements the ideas of mold-assisted lithography for large-area wafer-scale application. The basic concept of MAL is modified in order to overcome the limitations detected [21] by implementing some of the ideas of the micromolding concept as discussed in Section 4.5.2.

3.4.1. Basic Step-and-Flash Imprint Lithography Process

The SFIL process differs somewhat from the basic MAL procedure as Fig. 41 illustrates. Molds with dry-etched submicrometer patterns made from quartz are cut, cleaned, and prepared with a fluorinated alkylsiloxane SAM (see Section 6.2.3) as a mold release layer. Then a mold is closely aligned over the substrate and a drop of a low-viscosity liquid photopolymer precursor is dispensed to the gap, filling it by capillary action. After filling, the gap is "closed" and the stack is illuminated by UV radiation through the transparent mold, resulting in a polymer replica of the mold on top of the substrate. Thus, single fields of the substrate are processed and the whole area is filled in a stepping procedure as illustrated in Fig. 41. As in hot-embossing lithography, the residual layer is removed in an anisotropic dry-etch process to open the mask windows (see Section 2.5.2).

3.4.2. Step-and-Flash Imprint Lithography Modifications

As wafer-scale MAL suffered from residual layer uniformity problems [118], SFIL uses the step-and-repeat approach, where templates of 1×1 in. are used as a mold, resulting in less than 80-nm nonuniformity of the residual layer thickness over 1 in.^{2} [121] in a self-aligning procedure for local parallelism of mold and sample, which makes use of flexure orientation stages [120, 122].

To achieve a very thin residual layer, the gap is closed after filling with the liquid prepolymer. As the pressure applicable in such a system is limited, the viscosity of the prepolymer has to be very low to enable compression [see Eq. (15)]. At these low viscosities, layer formation by spincoating is hardly feasible, so the prepolymer liquid is dispensed to the gap between substrate and mold. Gap filling takes place by capillary action, which is aided by the low viscosity of the liquid. This is exactly the basic idea of micromolding in capillaries (MIMIC) (see Section 4.5.2 [123, 124] though a rigid mold is used instead of an elastic one. Even at very low viscosities capillary filling requires capillary dimensions of several micrometers when reasonable filling times for 1-in. channel lengths are envisaged [123, 125]. As the stamp patterns are in the submicrometer range, this can only be accomplished with the gap between mold and substrate playing the role of the capillary, resulting in large amounts of prepolymer being expelled into the kerf upon closing of the gap. To work with smaller initial gaps, a lateral translation of the template has been proposed [126].

The improvement of the photopolymerizable prepolymer is a good example of how single components influence the features of the final polymer (see Section 3.2). Here the optimization result is a Si-rich material of high etch resistance and low surface energy. As a consequence of the latter, the interface between this layer and the SAM-treated mold is the one with the lowest value of adhesion energy in the system (see Section 6.1.1), enabling defect-free detachment of mold and sample [118]. The high etch resistance due to its silicon

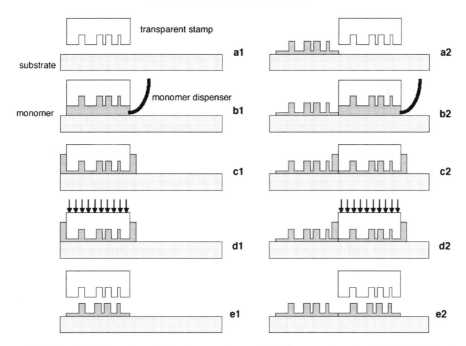

Fig. 41. Schematic of step-and-flash imprint lithography, referring to two subsequent stepping procedures 1 and 2. Note that for simplicity the exact substrate configuration (Si wafer, additional layer to be patterned, and bottom resist as transfer layer) is not shown. (a) Positioning of transparent stamp in close vicinity to the substrate; (b) filling of the cavity with monomer from a dispenser; (c) closing of the gap; (d) local curing by UV radiation (the uncured monomer, which was ejected into the kerf, remains soluble and can be removed); (e) sample with patterned polymer region after stamp removal.

content makes it an etch barrier useful as a hard mask in a multilayer process (see Section 2.8.2).

Because in MAL an increase in sticking was observed for high-aspect-ratio molds [21], the SFIL process relies on patterning of a thin layer only with a low-aspect-ratio mold and makes use of contrast enhancement by dry development utilizing the etch barrier quality. Thus, the samples are prepared with a 100-nm-thick bottom resist (not shown in Fig. 41), the "transfer layer," usually a conventional photoresist, PMMA, or PS before the stepping procedure.

3.4.3. Achievements

Material optimization resulted in low imprint and release forces in SFIL. Imprint forces of only 20 N were sufficient to replicate patterns down to 60 nm with 50-nm height in the barrier layer. They were successfully transferred into a 350-nm-thick transfer layer, resulting in a mask with an aspect ratio of up to 6 : 1 [118, 121]. Moreover, due to this two-layer concept, SFIL is intrinsically suitable to pattern over topography as the transfer layer acts as a planarizing layer (see Section 2.8.1).

An automated stepper for SFIL is available [120–122], allowing for multiple imprints over a 200-mm-diameter wafer, and has been employed for defect investigations. SFIL turned out to be a self-cleaning process as well because after about eight imprints the stamp was cleaned from particles, probably by successive entrainment of the particles in the etch barrier layer [122, 127]. SFIL has also been tested for lithography on curved surfaces [126].

3.5. Optical Lithography and Mold-Assisted Lithography

A combination of mold-assisted lithography with optical lithography concepts, in particular, the near-field approach, was introduced in [128], the mold-assisted near-field optical lithography. It is an analog of the near-field attempts with elastomeric stamps [129] (see Section 4.6). Photocuring was performed in the near field of a partially transparent patterned mold and enabled reproduction of features down to 70 nm.

Figure 42 shows the different types of molds discussed and the principle of the concept. Whereas mold I (the typical MAL

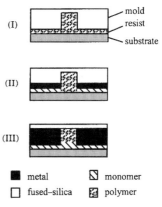

Fig. 42. Three different stamp configurations for lithography with transparent stamps as used in MAL. Type I leads to a pure thickness contrast, type III results in a pure chemical contrast, and type II is able to create both a chemical and a thickness contrast. Reprinted with permission from Y. Chen, F. Carcenac, C. Ecoffet, D. J. Lougnot, and H. Launois, *Microelectron. Eng.* 46, 69, (1999). © 1999 by Elsevier Science.

configuration) will result in a pure thickness contrast and mold III, with its thick nanometer-patterned absorbing layer, in a purely chemical contrast, mold II will give both chemical and thickness contrast. Calculations have shown that thin residual layers are required in order to obtain a high contrast in material composition. The residual layer is easily removed by a solvent dip and does not require RIE, thus preserving the cured pattern slopes.

4. MICROCONTACT PRINTING

4.1. Principle of Microcontact Printing

As illustrated in Fig. 43, the idea of microcontact printing (μCP) is to imprint a surface with an inked elastic stamp under soft contact. This is one of the classical imprint techniques, stamping or contact printing. The new elements implemented include (i) the use of an elastomeric stamp made from poly(dimethylsiloxane) (PDMS), a material that is capable of replicating features in the submicrometer range, and (ii) the use of a solution of molecules capable of forming a SAM as an ink.

The rubber stamp is not the original but just a replication (see Fig. 3). Often the master used is simply a patterned resist on a substrate prepared by conventional lithography techniques [22, 23, 125, 130]. As an alternative, commercially available templates such as transmission electron microscopy (TEM) grids or patterned metallized semiconductor wafers are used. Typical pattern heights in the photoresist template are 1–2 μm for micrometer-sized patterns [22, 23] and 0.4–1.5 μm for submicrometer patterns [130]. Some details on rubber stamp preparation are given in Section 5.2.2.

The procedure of μCP is as follows (see Fig. 43). A stamp is prepared from PDMS and "inked" with the SAM solution, 0.1–10 mM in ethanol or diethyl ether, by using a moistened lint-free paper as an ink pad or a soaked Q-tip or simply by pouring the solution over the stamp [23]. Often the whole stamp surface is covered with the ink [22]. Then the stamp is dried in nitrogen flow. The substrate is prepared with a thin layer of Au, typically 200 nm, on, for example, a 5-nm-thick Ti base layer. Stamp and substrate are brought into soft contact without applying pressure. In contact areas, the ink forms a dense monolayer on Au with a typical thickness of 1–3 nm. Upon removal of the stamp after a few seconds, the SAM patterns formed represent a surface mask for the gold and the "lithography" step is completed.

This mask may then be used in two ways. It can protect the Au surface during a wet-etching step, thus resulting in a patterned gold layer [22]. Alternatively, it can be used as a mask for selective deposition of metals by CVD (chemical vapor deposition), electroplating, and electroless processes [130]. SAMs are not suitable for RIE or lift-off [131].

For lithography purposes, the main procedure is to use a Au surface layer and wet etching to transfer the SAM mask. However, in a classic Si semiconductor process, etching is performed via dry processes only and metals such as Au are avoided as they may result in deep traps in the semiconductor. Thus, μCP is mainly suited as a soft lithography process for the large number of non- (Si-) microelectronic applications.

To improve the compatibility of μCP with state-of-the-art semiconductor microfabrication, a trilayer approach was demonstrated [132] (see Section 2.8.2). Substrates were prepared with gold on top of a bottom resist (e.g., PMMA). SAMs

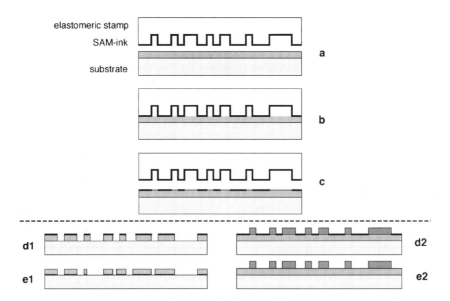

Fig. 43. Schematic of patterning by use of microcontact printing. Steps (a)–(c) refer to the lithography process; steps (d) and (e) refer to subsequent processing using the SAM mask. (a) Preparation of the substrate with a thin Au layer and of an elastomeric stamp. The stamp surface may be inked as a whole (as shown) or in the elevated regions only. The latter is the case when an ink pad is used. (b) Soft contact of inked stamp with the sample. (c) Patterned SAM layer formed at the substrate after separation from the stamp. Left branch, etching: (d1) wet etching of the Au layer with the SAM acting as a mask; (e1) patterned Au layer after SAM removal. Right branch, selective deposition: (d2) selective deposition of material in the SAM windows by CVD or wet processes. Depending on the deposited materials used, the windows are filled laterally to a more or less amount; (e2) deposited pattern after SAM removal.

were applied by μCP and the pattern was transferred into the Au layer by wet etching. The resulting Au mask was then used for dry development of the bottom resist in an O_2RIE process with parameters of 10 W, 30 mTorr, and 200 nm/min. The patterned bottom resist thus represented a mask with good aspect ratio for further dry-etch pattern transfer into the substrate.

4.2. Material Issues

The characteristic material in μCP is the SAM ink. A very good summary on these inks for μCP is given in [130] and we will only summarize some important topics of general interest here.

Molecules forming self-assembled monolayers are long alkyl chains $(CH_2)_n$ of regular configuration. They feature a head group at one end, providing the functionality of the final SAM surface, and a base group at the other end, chosen to chemisorb spontaneously on specific surfaces. The systems best characterized are the alkanethiolates [sulfur base group, $CH_3(CH_2)_nS$] on Au, Ag, and Cu (see, e.g., Fig. 44). Application of an SAM to a surface is classically done by immersion of the substrate for several minutes into the thiol solution. The molecules assemble on the substrate in an ordered form, building a two-dimensional organic crystal. The layer thickness refers to the chain length of the molecules and may thus be tuned. Minimum chain lengths of $n = 11$ are required to result in a low-defect and dense monolayer [133].

μCP has been performed primarily using the previously mentioned alkanethiolates on Au, and, in particular, HDT [hexadecanethiolate, $CH_3(CH_2)_{15}SH$]. They efficiently protect the Au underneath in a chemical wet-etch process when solved in KOH/KCN saturated with oxygen [23]. The SAM patterns are stable for at least 9 months without degradation even when washed repeatedly in polar and nonpolar liquids [23].

Some experience also exists using alkylsiloxanes ($SiCl_3$ base group) on OH-terminated surfaces such as Si/SiO_2 or Al/Al_2O_3 [130, 134] (see also Section 6.2.3). In comparison to the thiols, these layers are less ordered and in general need more time to form. Their protection capability for SiO_2 in a

wet-etching process in aqueous HF/NH_4F is insufficient so that they are not directly suitable as an etch mask. Nevertheless, it was possible to assemble PMMA in the hydrophilic surface regions between the SAM patterns and to use these complementary PMMA patterns as an etch mask. However, the PMMA patterns reported were smaller than the SAM clearance [134]. A more complex approach [135] based on catalyst-activated monolayers between the printed SAMs, enabling patterned growth of a polymer there, showed better accuracy of the negative pattern. The polymer thickness may be up to 100 nm and is thus more suited to serve as a mask in a dry-etch process.

Stamp material issues are treated in Section 5.2.2.

4.3. Achieved Patterns

In its basic process, that is, commercially available PDMS as the stamp and HDT as the ink, patterns in the micrometer range were demonstrated over areas of up to 50 cm^2 [136]. Smaller feature sizes down to 300 nm were produced routinely over smaller areas and features down to 100 nm were reported to be feasible although not reproducible [23]. Accuracy and resolution were said to be limited due to soaking of the ink, potential swelling of the stamp [23], reactive spreading of the ink on the Au substrate [130], and, in particular, limited stability of the elastomeric stamp material [133, 137] (for stamp stability, see Section 5.3). An investigation of the distortions during μCP resulted in relatively large values. For printing by hand, smallest relative distortions of 500 nm were found for pairs of patterns over 0.25-cm^2 areas when stiff, thick 8-mm stamps were used. Absolute distortions in the same range were found in an automated process over 1 cm^2 with thin elastomers casted against rigid substrates [138].

Extensive work was carried out at IBM to improve the critical components of the process, in particular, the stamp (see Section 5.3). With an optimized "hybrid" stamp, high-density patterns of 100-nm size have been printed reproducibly over 5×5 cm^2 areas [139]. Long-distance accuracy is below 500 nm and overlay accuracies on the order of 1 μm were achieved over the same areas.

4.4. Curved Substrates and Stamps and Large Area Printing

The use of an elastomeric stamp is essential for the μCP technique as patterning of the surface means an ink transfer from the stamp to the substrate and thus relies on conformal contact between both. In addition to the fact that surface roughness is not a problem, which is important for the polycrystalline Au substrate layer, this offers additional flexibility. An elastomeric stamp is capable of printing on curved substrates, such as lens structures. Curvature radii down to 25 μm were successfully printed [140], which correspond to typical glass fiber geometries. In the case of a larger curvature radius, the stamp itself was mechanically deformed to contact the nonflat surface. In the case of very small radii, the substrate, a small cylinder, was rolled on top of the patterned flat elastomeric stamp. A

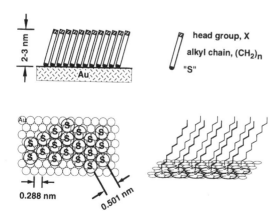

Fig. 44. Characteristics of a highly ordered monolayer of alkanethiolate on Au(111). Reprinted with permission from Y. Xia, X.-M. Zhao, and G. M. Whitesides, *Microelectron. Eng.* 32, 255, (1996). © 1996 by Elsevier Science.

similar pattern transfer fidelity is reported for flat and nonflat surfaces [140].

The flexibility of the stamp also offers a specific large-area approach, namely, the μCP with a cylindrical rolling stamp [136] (Fig. 45). For "roller imprint" the stamp was inked and dried, then mounted at the roller and used for patterning surfaces of up to 3-in. diameter (50 cm^2) within only 20 s with patterns down to 300 nm. The rolling speed was about 2 cm/s and only minor differences were detected when rolling along or across line patterns. Such a technique could be developed into a continuous printing technique resembling the rotary printing scheme for patterning of very large areas, such as for displays.

4.5. Related Soft-Contact Techniques

There are a number of techniques related to μCP in that they also make use of an elastomeric stamp and soft contact to the sample. They can be partly classified as lithography processes as they result in a patterned masking layer on a substrate and partly as techniques for surface patterning of volume materials (see Tables I and III). Most of them do not stamp SAMs but are rather of the UV-molding type.

The examples found in the literature show clearly that nanometer-scaled patterns are not the primary goal of all these μCP-related techniques. In most cases, feature sizes are in the 1-to 100-μm range. The goal is to establish a number of chemically oriented techniques capable of modifying surfaces laterally and to use these techniques in a flexible way to prepare patterned surfaces for different applications without the need for a classical photolithography process, thus providing low-cost micro- and nanopatterning techniques to chemists, biologists, and material scientists [133]. In this context, layer-oriented techniques (lithography) and volume techniques (casting, molding) are of equal importance.

4.5.1. Lithographic Molding

In lithographic molding [141], a normal μCP step is followed by an anisotropic etch into the Si substrate with the patterned Au as a mask. The resulting V grooves and inverted pyramids are molded again into PDMS, and this three-dimensional stamp is then used for μCP to define much smaller patterns than the original ones. Additionally, application of some pressure can tailor the pattern sizes obtained.

4.5.2. Micromolding in Capillaries

Another technique is micromolding in capillaries [124, 142]. The elastomeric stamp is used as a mold and brought into contact with a substrate, thus forming micrometer-sized channels. When a continuous network of channels is built in this way, these channels are open at least at one end. When a drop of liquid is placed at the channel opening, the channels may be filled with the liquid driven by capillary forces (see Fig. 46).

To enable filling of the capillaries, the surface energies and surface tensions of the materials involved have to ensure that the liquid wets the substrate at least partially. PDMS with its low surface energy of 21.6 mJ/m^2 features good adhesion to

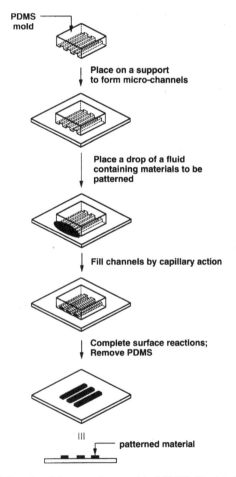

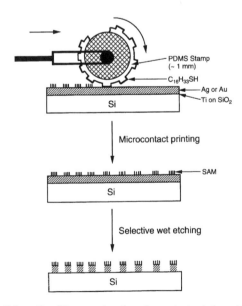

Fig. 45. Schematic of the procedure for microcontact printing with a rolling stamp. Reprinted with permission from Y. Xia, D. Qin, and G. M. Whitesides, *Adv. Mater.* 8, 1015 (1996). © 1996 by Wiley–VCH.

Fig. 46. Schematic of the procedure used in MIMIC. Reprinted with permission from E. Kim, Y. Xia, and G. M. Whitesides, *J. Am. Chem. Soc.* 118, 5722 (1996). © 1996 by the American Chemical Society.

any higher surface energy substrate so that closed channels are formed easily. The physical background of this technique is explained in detail in [124, 142]. If the liquid filling the capillaries is a monomer or a prepolymer, then it can be cured thermally or photochemically through the UV-transparent PDMS mold. As a result, a patterned mask remains on the substrate after mold removal.

MIMIC works with liquids of very low viscosity [< 400 cP (40 Pa)]. Filled channel lengths are in the centimeter range and processed areas cover a few square centimeters [124]. An interesting technical implementation is the SFIL approach [118] discussed in Section 3.4.

Beyond lithography, MIMIC is also capable of producing freestanding patterned films by peeling off the crosslinked network from the substrate [124]. In addition, the liquid chosen can be used as a carrier for metal or polymer particles (e.g., latex microspheres), resulting in a laterally patterned regular sphere array after evaporation of the liquid [123].

4.5.3. Micro- (Transfer) Molding, Channel Stamping

To overcome the small-area limitation of MIMIC, the idea of microtransfer molding (μTM) was born [143], also referred to as channel stamping [144]. The PDMS mold is directly filled with the liquid prepolymer and afterwards brought into contact with the substrate [143]. The prepolymer is subsequently cured. With this technique, a thin continuous polymer layer connecting the thicker polymer patterns, a residual layer, cannot be avoided. It has to be removed in an RIE process, similar to the removal of the thin residual layer after hot embossing. In [144], the technique is discussed in detail, dealing with capillarity, channel filling, debonding, and the wetting angles of the liquid required for successful processing.

μTM was demonstrated to work over a lift-off topography when preparing arrays of Schottky diodes [143]. In addition, the preparation of two- and three-layer networks by consecutive μTM steps with partial crosslink in between the molding steps was shown [145].

The prepolymer may also be applied to the substrate instead of to the mold before making contact. The process is then referred to as micro-molding [146] and was successfully used to pattern sol–gel precursors during preparation of glassy microstructures [146]. In this configuration, the only difference to MAL (see Section 3) is the use of an elastomeric stamp instead of a rigid one.

4.5.4. Replica Molding

Replica molding represents the basic UV-molding technique (see Section 3), making use of an elastomeric mold that is easily peeled off from the rigid replica after curing. It is a surface-patterning process unsuitable as a lithography technique such as MIMIC or μTM. Due to the low surface energy of PDMS (21.6 mJ/m^2), preparation of the mold surface with an anti-sticking layer is not required. Complex optical structures were prepared by replica molding in PU (polyurethane) [125] with a deformed mold. Under lateral compression, a reduction of pattern geometries (e.g., from 50 to 30 nm) could be shown [125, 147].

4.5.5. Solvent-Assisted Microcontact Molding

Another idea in this context is solvent-assisted microcontact molding (SAMIM) [148], which combines characteristics of replica molding and embossing. A polymer is spun onto a substrate and a PDMS mold is filled with a solvent for this polymer. After contact, the solvent dissolves the polymer and the dissolved or gel-like polymer conforms to the mold under minimum pressure and without raising the temperature. By evaporation of the solvent, the pattern resolidifies, conforming to the mold. Areas to be patterned are limited to several square centimeters as evaporation of the solvent takes place from the uncovered substrate surface. A processing time of only 5 min has been reported due to the quick solvent diffusion [148].

This idea is not limited to the use of an elastomeric stamp. It was implemented in an interesting modification where solvent vapor was allowed to soften a prebaked polymer layer from its top, followed by a room temperature embossing step [149] (see Section 2.9.2).

SAMIM was also used to prepare surface patterns for optical lithography, as introduced at the end of the next section.

4.6. Optical Lithography with PDMS Stamps and Patterned Polymers

The fact that the PDMS stamp is transparent down to 300 nm gave rise to a number of new ideas where the PDMS stamp is used as a photomask in direct contact with a substrate.

As the elasticity of the stamp material allows conformal contact, near-field effects could be exploited [129] (see also Section 3.5 for use with rigid stamps). Flood exposure of such a system, namely, PDMS stamp on a photoresist covered substrate, with micrometer-sized patterns could "image" the diffraction intensity profiles in the exposed and developed resist layer when the irradiance level fluctuations met the almost linear middle part of the sensitivity characteristics of the resist [150], which is otherwise avoided during optical lithography. With a stamp pattern height equivalent to a phase shift of π for the exposure light in the medium, the stamp was operated as a binary phase-shifting mask featuring a sharp minimum of intensity near the slopes of the stamp walls. This effect generated 90-nm-wide resist features with micrometer-sized stamp patterns and a broadband flood exposure (330–460 nm) [129].

In addition, three-dimensional PDMS stamps were tested as contact masks. When the stamps were replicated from anisotropically etched Si(100) masters, their slope angle of 54.7° was higher than the critical angle for total reflection in PDMS (47°). Thus, the sloped parts of the stamp reflected the light, leading once again to a doubling of patterns when "underetched" Si masters were used or when pressure was applied during exposure [151].

Preparation of stable elastomeric stamps from high-modulus PDMS material (10–15 N/mm^2; see Section 5.2.2) with pattern sizes down to 100 nm enabled lensless subwavelength optical lithography [152]. Again, the PDMS acted as a contact mask during flood exposure (248 nm). Due to the good matching of the refractive indices of PDMS and resist, the stamp acts as a light guide, directing the intensity into the contact regions. These light-coupling masks (LCMs) demonstrated definition of patterns down to 100 nm as the resolution limit of the vacuum wavelength is reduced by the refractive index of the stamp ($n \approx 1.6$). To apply this technique to larger patterns or mixed patterns, the noncontact surface of the stamp had to be covered with a reflecting or absorbing layer.

A modified approach makes use of an embossed polymer on the substrate itself rather than the PDMS contact mask [153] where embossing was done by SAMIM (see Section 4.5.5). In this case, the patterned photoresist acts as its own optical element during flood exposure: Either the previously mentioned line doubling occurred with rectangular resist geometries or, alternatively, focusing effects could be observed with three-dimensional sloped resist patterns. This technique is called topographically directed photolithography (TOP).

5. MASTERS, STAMPS, AND MOLDS

It is clear that all the imprint techniques introduced are not self-sufficient. They rely on a master, a template, carrying the pattern to be replicated. As all the techniques feature direct contact of the master and the sample during imprint, the patterns on the master have to have, in general, the same scale as the patterns envisaged.

At first glance, this dependence on a master seems to be a major drawback of the imprint techniques. However, this is a situation well known and commonly accepted in classical lithography: The master, stamp, or mold is the analog of the photomask in a broad-beam lithography process.

Classical lithography can only work successfully when three prerequisites are fulfilled. There has to be (i) a mask technology providing the original patterns, (ii) a polymer sensitive to exposure at the respective wavelength, and (iii) the exposure system with the lithography process itself [3]. These three prerequisites have to fit together and the mask technology is often a critical part [4]. Mask requirements differ to a large extent and include the phase-shifting masks for advanced optical lithography, the transmission masks for X-ray lithography, the reflection masks for EUV, the stencil masks for IPL, and the scattering masks for electron beam projection (SCALPEL). Direct writing techniques are the only ones scoping the mask problem, but as serial techniques they have to face the throughput issue.

Thus, the need for a master with the envisaged pattern is a general aspect of broad-beam lithography techniques and not specific to nanoimprinting. Moreover, as in any lithography process, the achieved quality is directly related to the mask quality where distortion, defects, and particle contamination are crucial. In cases of demagnifying exposure systems, the demands for pattern accuracy are slightly relaxed as is the case for SCALPEL, EUV, or IPL, where 4 : 1 masks are supposed [3] similar to state-of-the-art optical lithography. X-ray lithography with its 1 : 1 masks is comparable to nanoimprint lithography as far as pattern accuracy is concerned. We should bear in mind, however, that pattern accuracy is only one aspect of mask technology besides material and processing issues, which may be even more challenging in complex mask fabrication.

5.1. Master Fabrication

Masters for nanoimprint techniques are fabricated in most cases using state-of-the-art Si technology. Nanometer-scaled patterns are e-beam written into a resist layer on a substrate as is common for mask fabrication in today's advanced lithography. When, in addition, larger pattern sizes are required over large areas, a mix and match with UV lithography may be the choice.

In many cases, the patterned polymer layer on a quartz substrate or on a Si wafer is not robust enough to serve as a master during the imprint process. In these cases, transfer of the polymer pattern into the underlying layer or substrate is required. This is done by dry etching (RIE). As typical electron beam resists such as PMMA offer only moderate mask selectivity in an RIE process [104], "hard" dry-etch masks have to be provided indirectly from metal by lift-off (see Section 2.2) or from oxide or nitride by trilevel techniques (see Section 2.8.2). Often the pattern is transferred into quartz or into a SiO_2 layer on Si [13, 28, 50, 51] or, alternatively, into poly-Si over oxide on Si [36, 40, 58, 61] or into bulk Si [19, 59]. Standard dry-etch processes used for contact hole etching or gate etching can be used in this step. Different types of stamps are shown in Fig. 47.

Two-Dimensional and Three-Dimensional Masters. In general, the information represented by such a master is a two-dimensional (2-D) one where local grooves are defined on the master surface so that at least two different but constant height levels exist. We will refer to such a "digital" geometry as a 2-D master.

Due to sidewall passivation effects during etching and due to a basic roughness stemming from the lithography process, the sidewalls of the masters are often rough and sloped, giving rise to nonideal behavior. Roughness enlarges the contact area between stamp and sample during HEL and MAL. It may also increase sticking (see Section 6) and thus affect the imprint quality. On the other hand, sloped walls may facilitate imprint as they facilitate the separation of stamp and sample after the process by inhibiting a potential clamping effect due to unrelaxed strain (see Section 6). Then separation over the whole surface area is an instantaneous process without sliding effects.

Anisotropic wet-etching processes could yield vertical sidewalls in Si(110) with minimum roughness but are restricted to test purposes [19] because of the very limited pattern design flexibility. In Si(100), the well-known 54.7° sloped walls arise

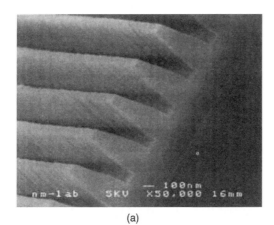

(a)

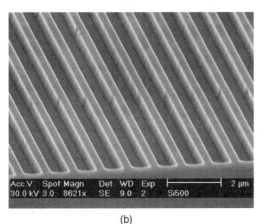

(b)

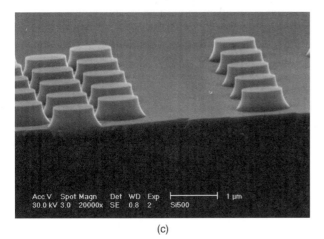

(c)

Fig. 47. Different stamps used for hot-embossing lithography. (a) SiO_2 stamp fabricated by e-beam lithography and RIE as used for embossing according to Figs. 14 and 9. Reprinted with permission from B. Heidari, I. Maximov, E.-L. Sarwe, and L. Montelius, *J. Vac. Sci. Technol., B* 17, 2961, (1999). © 2000 by SPIE. (b) and (c) Stamps from poly-Si over SiO_2 on Si fabricated by UV lithography and RIE. The line pattern shows relatively high roughness. Both examples feature somewhat bowed sidewall geometries and a sharp edge at the bottom due to selectivity because the dry-etch process stops when the SiO_2 layer is reached.

and such stamps are also used for testing the potential of nanoimprinting [20, 59] as the smooth walls, the sharp edges and tips, and the definite slope angles are easily assessed for measurement in the imprinted polymer. We will refer to

such samples as 3D masters as they represent fully three-dimensional, "analog" information.

Polymer Masters. An alternative for low-cost master fabrication is direct e-beam writing into a solid polymer sheet of, for example, perspex (PMMA) [154]. To minimize charging effects during exposure, the plastic is covered with a 30-nm NiCr layer in a low-temperature-load evaporation process before writing. It can serve as a stamp for hot embossing up to temperatures around its T_g where the polymer is still glassy. Such rigid stamps from plastic offer elastic response without viscous flow in the time regime of the imprint process and could thus reduce the danger of damage in case of contact between stamp and sample substrate toward the end of the imprint process which is, in general, carefully avoided [27]. In combination with high-aspect-ratio stamp patterns, hot-embossing lithography of thin low-T_g polymer layers could thus be feasible with minimum residual layer thickness over large areas.

In a different approach, a commercial crosslinkable resist (SU-8) is used for e-beam direct writing of a master [155]. In this case, a thin layer of the resist on 10-nm Cr on Si is used. The polymer is cured in the ebeam, and the patterns so formed result in a temperature-stable stamp for hot embossing.

5.2. Replicated Stamps

In nanoimprint lithography, cost issues have led to consideration of working with copies of the relatively expensive e-beam-written masters. In the case of hot-embossing lithography, the main reason for this is the potential wear of the stamp [28, 51, 57, 63], which would result in a cost-efficient master technology when e-beam-written originals are used. In the case of microcontact printing, the reason is inherent to the process as a "soft" elastic stamp is required, thus making replication of the master into an elastomeric material an essential part of the process. In mold-assisted lithography, a quartz master is often used [116, 156], as the stamp has to be transparent and thus the material choice is somewhat limited. Thus, master replication is not as straightforward as for the other techniques although it offers interesting advantages such as sticking elimination without using an additional layer, by a suitable material choice (see Section 6). In SFIL (see Section 3.4), the wetting characteristics are also optimized using a PC mold.

As far as cost and wear aspects are concerned, the strategy to replicate the masters is analogous to the concept of working masks elaborated in the past for optical lithography and X-ray lithography.

Replicated stamps are fabricated from rigid materials, such as metal or plastic, or from elastic ones. In general, hot-embossing lithography uses rigid stamps, microcontact printing uses elastomeric stamps, and mold-assisted lithography and other molding techniques may use both types as long as their optical properties are adequate.

5.2.1. Rigid Replicas

Metals. Metallic replicas are exclusively used for hot-embossing lithography [51, 53]. The idea follows the concept of classical embossing and injection molding for microfabrication [9]. The replication of the master is performed via electroforming in a galvanic process. The metal widely used is Ni with an electroplating process optimized, in particular, for CD production [14] and where low-stress processing over large areas is feasible. Typical Ni shim tools have a thickness of 100–300 nm and are flexible.

The fabrication process is as follows. The original microrelief, in general, a polymer layer patterned by e-beam or laser exposure, is covered with a thin metal layer (≤ 50 nm Au or Ag), for example, by sputtering, which acts as a conductive starting layer, the cathode, during the electroplating process. When the micropattern is filled and the shim has its desired thickness, it is removed from the plating bath and has to be separated from the original. During demolding, the original is lost, as the patterned photoresist of the master is dissolved to free the shim. Therefore, a second replication of this Ni shim by molding or embossing into a polymer and a second electroforming procedure are advised in order to multiply the number of replicas available from the first Ni copy. These Ni stamps are negative replicas of the master microrelief [53]. In most cases, they are positive stamps (see Section 2.7.1) because the exposed pattern is removed in the polymer layer.

Plastics. Rigid replicas can also be provided from plastics where duromers with their high temperature and solvent stability, and their relatively high hardness offer some advantages compared to thermoplastic materials. The latter are applicable as long as the difference in glass temperature or respective relaxation time of the stamp material and the embossed material is high enough [154].

Duroplastic polymers feature a high level of crosslinking and can be prepared from monomers or prepolymers by thermal treatment or irradiation, depending on the chemical formulation. As a consequence, mold-assisted lithography or hot embossing are suggested replication techniques. Obviously, injection molding would be applicable, too [14], but it is often convenient to replicate the master with the same technique used for nanoimprinting.

Replication of the master into polycarbonate (PC) is performed in some cases for step-and-flash imprint lithography [121]. Thus, a master from Si or GaAs can be fabricated, whereas the PS copy is used as the "compliant template" for the UV-curing process.

Replication into duroplastic prepolymers by hot embossing [37, 40] has been proposed as a single and double copying process [63]. Before embossing of the crosslinkable prepolymer, the master had to be prepared with an anti-sticking layer [157] (see Section 6). A prepolymer of the duroplastic material, a "thermoset," featuring thermoplastic behavior, was spin-coated onto a Si wafer and imprinted where processing had to allow for sufficient viscous flow before the temperature-induced crosslinking process became predominant (see Section 2.5 1). Unlike a real lithography process, no removal of the residual layer is required and thus the choice of prepolymer thickness is not critical. Because of the branched structure of the prepolymer and the low degree of crosslinking due to temperature, the inherent shrink during crosslinking is negligible. After separation, a duroplastic stamp is available for embossing lithography, which is a positive copy of the master (left branch of Fig. 48). In case a positive copy is preferred, the replication process can be repeated (right branch of Fig. 48). Then a 1-h temperature treatment of the stamp at about 180°C is advisable in order to finalize the crosslinking procedure and thus avoid sticking (see Section 6) during the second replication.

Duroplastic stamps prepared in this way are temperature stable up to more than 200°C. They have demonstrated excellent imprint results in a hot-embossing lithography process with thermoplastic polymers at temperatures up to 200°C and pressures of 100 bar (see Figs. 49 and 50) without any further temperature or surface treatment [63].

Alternatively, UV-curable materials may be used for stamp replication. Processes such as mold-assisted lithography [21, 116] (Section 3) are applicable without removal of the residual layer. By use of adequate prepolymers instead of monomers, the shrink should be reduced to about 1% or less so that pattern transfer fidelity is not affected substantially. In addition, molding techniques (see Section 4.5) such as replica molding [125, 147] or microtransfer molding [143] can be used which demonstrated excellent replication results with a UV-curable polystyrene prepolymer. In these cases, an intermediate elastomeric mold is used in order to facilitate the demolding step. Thus, the procedure is a double replication and the PS sample is a positive copy of the master.

5.2.2. Elastomeric Replicas

Elastomeric stamps are in use primarily for microcontact printing (see Section 4) and derived techniques thereof [22, 23, 123–125, 130, 134, 136, 140–142, 147, 148, 153, 158] where the use in a replica molding process (see Section 4.5.4) to fabricate a rigid stamp from duromers is of specific importance for rigid stamp fabrication as mentioned previously. The elastomeric material of choice is PDMS and the stamp is fabricated in a casting and curing process.

Preparation. A liquid prepolymer of PDMS is mixed 10 : 1 with the respective curing agent and poured over the master. Curing is done in two steps, in laboratory ambient at room temperature for about 1 h, followed by a 1-h treatment at 65°C. No need for vacuum is reported during this preparation [22]. After cooling to room temperature, the PDMS is carefully peeled from the master, cut to isolate the patterned area, and used as an elastomeric stamp [22, 23, 141] or an elastomeric mold [125, 147]. Typical PDMS stamps have a thickness of a few millimeters.

The replication can be performed from any patterned surface, as well as a patterned photoresist layer. Thus, as in the

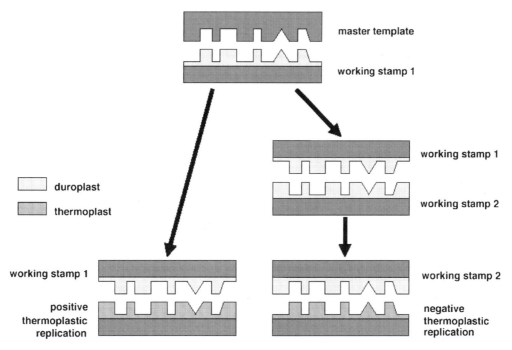

Fig. 48. Preparation of duroplastic working stamps from a patterned master template, either by single (left) or by double (right) replication, and their use for hot embossing into a thermoplastic polymer (bottom).

galvanic replication process, dry etching is not essential for master fabrication [159]. Here the master is not lost during the demolding process as the crosslinked PDMS can be peeled off from the master. Due to its very low surface energy, PDMS has antisticking qualities and the master can be used several times [133].

Advantages and Limitations. The elasticity of these stamps enables conformal contact to rough, warped, or curved substrates [125] in a roller imprint process [136] and even reduction of pattern sizes from 1–2 μm to 100–200 nm by compression of the stamp during microcontact printing [125, 130] and from 50 to 30 nm by bending the mold in a replica molding process [147].

Use of an elastomeric stamp not only results in the advantages just mentioned but also in some limitations with respect to the aspect ratios processable. It was shown in detail [137] (see Fig. 51) that for a defect-free and sharp pattern definition by μCP with a PDMS stamp made from standard material the aspect ratio of the stamp patterns had to lie within a small range, namely, 0.4–0.6. For aspect ratios greater than 2.0, the lines collapsed under gravity. For aspect ratios down to 0.6, the patterns were very sensitive to self-adhesion and capillary forces involved in the inking, drying, and printing process and coalesced in pairs. These stamp defects may be repaired to a large degree by adequate washing but remain blemished [137]. Stamp patterns with aspect ratios below 0.4 showed an increased tendency toward rounded and blurred lines due to line tension. In addition, when features are widely separated a sagging of the stamp is observed [133]. Optimized stamps from such material should therefore feature a pattern height well adapted to the envisaged lateral pattern size and thus should provide locally different aspect ratios of the stamp features when mixed pattern sizes are involved.

A thorough study was carried out in order to optimize the stamps for μCP [139, 160]. The investigations in [137] showed that the goal of printing smaller features in accurate layouts is best accomplished with a stamp polymer having a higher modulus and a greater surface hardness than the standard PDMS [Sylgard 184: compression modulus, 2 N/mm^2; shear modulus, 1 N/mm^2; surface hardness, 65 Shore A ($\approx$ 0.7% of glass); strain at break, 50%]. Modifications tested included the ratio of crosslinker to resin, the curing times and temperatures, the filling with nanoparticles as well as copolymerization; two optimized materials were prepared. In these materials, patterns as low as 80 nm with a stamp aspect ratio of 1.25 were possible. Finally, different types of "hybrid" stamps were prepared [139] using a quartz backplane as shown in Fig. 52. The trilayer stamps (a–c) feature stiff patterns on top of a soft cushion layer molded to a hard backplane. These stamps fit to mask aligner tools made for hard-contact lithography. Due to the soft cushion, they conform well to a surface topography. Separation after printing may be problematic, however, as the hard backplane hampers sequential release. As an alternative, flexible stamps were prepared (d) using a thin backplane (100 nm). Details on these "hybrid" stamps and their materials are given in [139].

5.3. Stamp Wear

As nanoimprinting is a contact process, one of the major questions in view of the potential production application of these techniques is the question of stamp wear during the process.

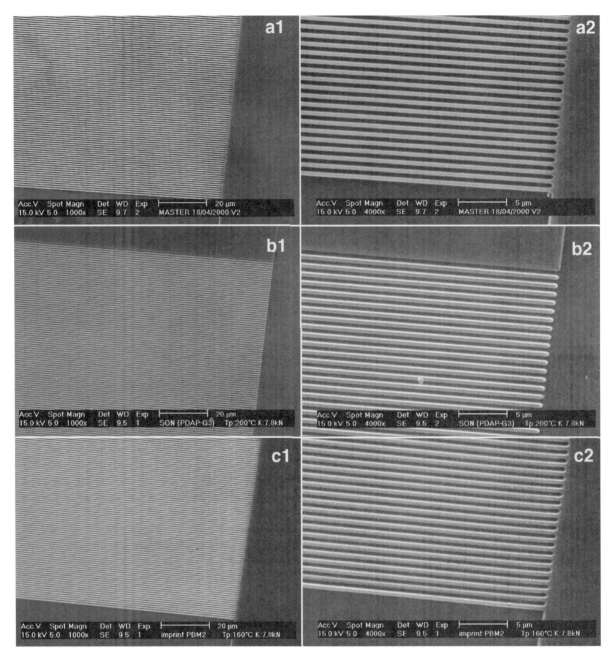

Fig. 49.   Examples of hot embossing with duroplastic working stamps according to the left branch of the scheme in Fig. 48. 400-nm line patterns in two different magnifications (1) and (2) are shown, giving an idea of the large-area patterning quality obtained. (a) Master (patterned poly-Si on top of SiO$_2$ and Si); (b) working stamp prepared by replication of the master into a thermoset polymer layer on Si; (c) imprint with the working stamp into a thermoplastic polymer layer.

How often can a stamp or mold be used before wear will affect the imprint quality, and how often can a stamp or mold be used before cleaning is required?

As the imprint techniques are still in their infancy and as automated imprint systems are rare, results with respect to wear are scarce. In the case of hot-embossing lithography, mechanical wear effects may be dominant because of the high pressures used. In the case of microcontact printing, irreversible merging of lines during inking and cleaning [137] may be critical, and in the case of mold-assisted lithography wear will be correlated, to a large extent, to the lifetime of the antisticking layers used (see Section 6.3) or, when polymeric molds are used, to degradation of the optical transparency.

Investigations have started in the case of hot embossing, which is supposed to be the most critical with respect to stamp wear. Mold inspection after 20 [30], 30 [18, 27], and 40 imprints [19] did not show detectable stamp defects where in some cases [18, 27] mold release agents had been used. Even critical stamps such as, for example, 10-nm-diameter

Fig. 50. Examples of hot embossing with duroplastic working stamps according to the right branch of the scheme in Fig. 48. Positive and negative line and dot patterns of 1-μm lateral size are shown. (a) Master (patterned poly-Si on top of SiO$_2$ and Si); (b) first replication into thermoset material, working stamp 1; (c) second replication into thermoset material, working stamp 2; (d) imprint into a thermoplastic polymer layer using working stamp 2.

SiO$_2$ dots were without damage after 12 imprints at 175°C and 45 bar.

More than 50 imprints without the need for an intermediate cleaning have been reported, but dust particles are said to result in degradation of the stamp [19]. The latter is in contrast to findings where hot embossing is claimed to be a "self-cleaning process" [57]. In these experiments, a mold release agent was used, making dust particles only weakly bound to the stamp. The molten polymer on the substrate acts as a glue for the particles so that after a few imprints the mold is cleaned from the particles.

In the case of microcontact printing with elastomeric stamps, more than 100 imprints without damage to the stamp have been reported when a thorough clean between single imprints is performed (rinsing in hexane and ethanol) [134].

It is quite clear that with respect to the wear of stamps and molds a large amount of work is still required to provide numbers suitable for production evaluation of the imprint techniques, and such numbers are hardly obtained in a research environment.

6. STICKING CHALLENGE

As all imprint techniques rely on contact between stamp and sample, the wetting and adhesion characteristics of the polymer materials to the substrate, on the one hand, and to the stamp or mold, on the other hand, are critical issues.

In microcontact printing, the transfer of the ink from the stamp to the substrate requires complete wetting of both surfaces and higher adhesion of the ink to the substrate in order to get a successful transfer. The latter is driven by controlled chemical bonding of the SAM "ink" to the sample surface. In mold-assisted lithography, the monomer or prepolymer has to wet the substrate but should not adhere to the mold after curing. In this case, adhesion is strongly related to the buildup of chemical bonds between the crosslinking polymer and the mold, and, in general, mold release agents are used on the stamp surface to reduce this effect. In hot-embossing lithography, the polymer should adhere to the substrate but not to the stamp. As this process is based on mechanical deformation of the thermoplastic material, relaxation mechanisms may additionally contribute to the resulting adhesion behavior of the polymer.

Whenever in HEL or MAL the adhesion balance between the different surfaces in contact with the polymer is disadvantageous, "sticking" is observed. In this case, more or less polymer remains in between the stamp or mold patterns and is peeled off from the substrate during separation. Thus, not only does the lithography process fail, but the stamp or mold has to

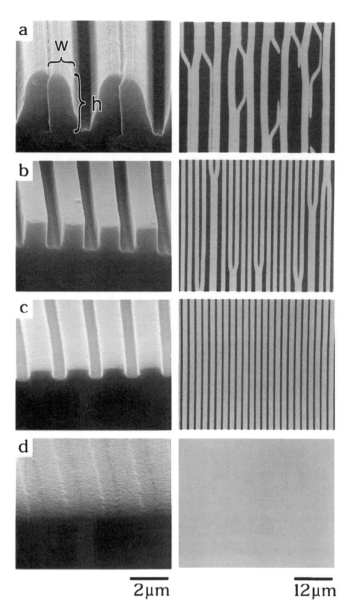

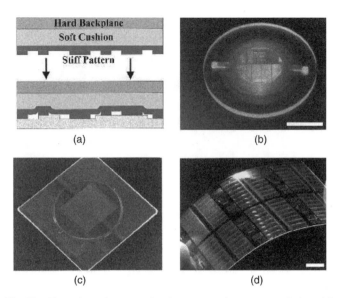

Fig. 52. Examples of stamps for large-area microcontact printing. (a) Scheme of trilayer stamp; (b) trilayer stamp with 270-nm features, bar = 10 mm; (c) stamp 1 mm thick having > 5-μm patterns molded in a soft siloxane polymer on a 125-mm glass plate; (d) example of two-layer thin-film stamp composed of a 100-nm glass backplane and a 30-μm-thick hard siloxane material with 270-nm features, bar = 10 mm. Reprinted with permission from P. M. Moran and F. F. Lange, *Appl. Phys. Lett.* 74, 1332 (1999). © 1999 by the American Chemical Society.

Fig. 51. Stability of a mold of commercial PDMS at different pattern aspect ratios. Line width w, 1.2 μm; space, 0.8 μm; height of patterns from top to bottom, 2.3 μm, 1.2 μm, 0.5 μm, 0.2 μm. (Left) Cross sectional view; (right) top view. Reprinted with permission from E. Delamarche, H. Schmid, B. Michel, and H. Biebuyck, *Adv. Mater.* 9, 741 (1997). © 1997 by Wiley–VCH.

undergo a thorough cleaning process to remove the polymer residues that otherwise could affect the next imprint step.

In summary, sticking effects will reduce the lifetime of stamps and molds, limit the quality of the replication, and complicate the processing procedures. For reliable processing, sticking has to be eliminated by choice of appropriate preventive measures. This may require the help of antiadhesive layers. However, as such layers result in further increase in process complexity and thus may be a further source of defects and failure, antiadhesive layers should be the ultimate attempt. "Natural" methods are preferred, based on the right choice of materials under contact or the right choice of pro-

cessing parameters and sequences. A prerequisite for this is an understanding of adhesion and of the basic effects becoming apparent as "sticking." The next section addresses this topic.

6.1. Physics of Adhesion

6.1.1. Surface Energy and Adhesion Forces

The strength of adhesion between two materials, for example, a polymer P and a substrate or stamp S, is characterized by the amount of energy required to separate the two materials. A number of methods exists to measure adhesion energies [161], including direct methods, where a fairly well-defined shear stress or normal stress is applied, or indirect methods, where a complex stress field is applied via an indenter or tip or where a crack is introduced and its growth resistance is monitored. In a number of cases, in particular, when very thin films are involved, measurements are difficult or even impossible. As a consequence, the respective data for a typical imprint situation may not be available from direct measurements.

Physical Background. During separation of two materials, the "internal" surface of their interface is transformed into two "external" surfaces interfacing with air. Assuming γ_P and γ_S are the surface energies of the polymer and the substrate in air and γ_{PS} the surface energy of the intact interface, then the adhesion energy W_{adh} is simply given by the surface energy difference between the two states [118], the intact one W_{int} and the separated one W_{sep} as

$$W_{adh} = W_{sep} - W_{int} = (\gamma_P + \gamma_S) - \gamma_{PS} \qquad (16)$$

Thus, knowledge of the strength of adhesion requires knowledge of the surface energies of the involved materials and the respective interface.

For a liquid, the surface energy is identical to the well-known parameter surface tension. In the case of solid materials, surface energies are determined by contact-angle measurement (discussed later) and numerous values are available from the literature (e.g., [108] for polymers). Typical values for standard polymers lie in the range of 30–40 mJ/m^2 [PMMA, 41.1 mJ/m^2; polystyrene (PS), 40.7 mJ/m^2]. Commercial Teflon (PTFE) and PDMS are low-surface-energy materials with values of 20 and 21.6 mJ/m^2, respectively. Surface energies of oxides (silica) depend on the specific preparation and amount to about 70 mJ/m2 [118] (wetting angle 20°).

Values of the surface energy of interfaces are not easily obtained. They are often derived from measurements of the adhesion energy W_{adh} according to Eq. (16). When no measurements of W_{adh} exist but when the surface energies of the materials in contact are known, an estimate following [108] is possible. The authors assume for the interfacial surface energy:

$$\gamma_{PS} = \gamma_P + \gamma_S - 2C\sqrt{(\gamma_P\gamma_S)} \approx (\sqrt{\gamma_P} - \sqrt{\gamma_S})^2 \quad \text{for} \quad C = 1 \tag{17}$$

with C as a constant, leading to an expression for the adhesion energy, which requires the surface energies only [162]:

$$W_{adh} \approx \gamma_P + \gamma_S - (\sqrt{\gamma_P} - \sqrt{\gamma_S})^2 \tag{18}$$

This approximation is often used for polymers.

Surface Energy Measurement. As mentioned previously, surface energies of solid materials are measured via the contact-angle method. Definite droplets of at least two liquids of different known surface tension (surface energy) are placed on the solid surface and the developing "contact angles" δ of the liquids are measured (see Fig. 53). Values $\delta < 90°$ refer to a wetting situation, whereas values $\delta > 90°$ refer to a dewetting situation, where δ is measured from the solid surface inside the droplet. Often the measurement systems are designed for specific test liquids. In general, one of the liquids is water; the others may be diiodomethane (CH_2I_2), paraffin oil [163], or

glycerol [118, 122]. The basic measurement relies on an equilibrium between solid, liquid, and air, which can be described by Young's equation from 1805 [108, 164]:

$$\gamma_L \cos\delta = (\gamma_S - \gamma_{SL}) - (\gamma_S - \gamma_{SV}) \approx (\gamma_S - \gamma_{SL}) \tag{19}$$

where γ_L is the known surface tension of the test liquid, γ_S is the unknown surface energy of the solid, and γ_{SL} and γ_{SV} are the interfacial energies of the solid surface to the liquid and to air. In general, the second term in parentheses is near zero and may be omitted, resulting in the previous approximation.

Contact-angle measurements have to supply at least two data points ($\delta_1, \gamma_{L1}; \delta_2, \gamma_{L2}$), which are used for an extrapolation toward ideal (complete) wetting. In the case of complete wetting, δ will tend to zero (thus, $\cos\delta = 1$) and the interfacial energy between solid and liquid γ_{SL} tends to zero. In this case, Eq. (19) is reduced to

$$\gamma_L = \gamma_S \tag{20}$$

so that γ_S can be deduced from extrapolation of the $\cos\delta$ data to complete wetting. The extrapolation idea is illustrated in Fig. 53. The curve $\cos\delta$ against γ_L follows approximately a hyperbolic law as indicated by Eq. (19). Historically, plots of $\gamma_L \cos\delta$ vs. γ_L were also used for extrapolation [162].

6.1.2. Adhesion Mechanisms and Their Impact for Imprint Techniques

Adhesion mechanisms have been studied in detail for functionalized surface coatings of metals, oxides, and plastics in tribology. The major mechanisms can be classified into three categories: mechanical, physical, and chemical [161] (see Fig. 54). Adhesion is promoted mechanically by a large surface area in contact, that is, surface roughness, frictioninterlocking, and dovetail interlocking. Physical adhesion is due to weak bonds at the interface with interface energy around 50 kJ/mol, such as van der Waals bonds or H bridges effective over relatively long distances of 0.2–0.5 nm. Chemical adhesion is caused by chemical bonds (ionic, atomic, or metal) in between the surfaces with interfacial energies of 100–1000 kJ/mol and an even lower reach than the physical ones. Typical adhesion forces are 500 N/mm^2 for hydrogen bonds and more than 5000 N/mm^2 for chemical bonds.

In the case of imprint techniques, the adhesion forces with respect to the imprinted monomer or polymer are of interest. Thermoplastic polymers are expected to undergo physi-

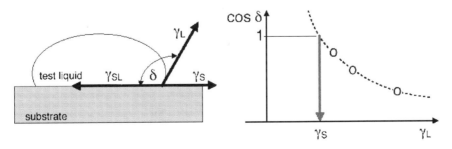

Fig. 53. Contact-angle definition and principle of determination of surface energy from measurements with test liquids of different surface tension.

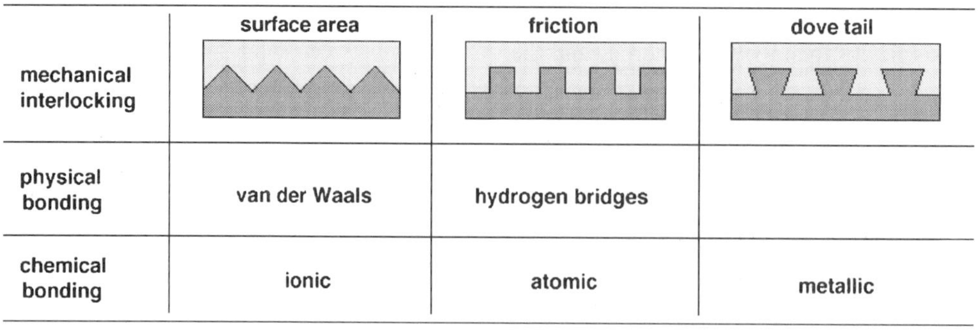

	surface area	friction	dove tail
mechanical interlocking			
physical bonding	van der Waals	hydrogen bridges	
chemical bonding	ionic	atomic	metallic

Fig. 54. Schematic of different mechanisms resulting in adhesion between two surfaces, that is, mechanical interlocking, physical bonding, and chemical bonding.

cal bonding at surfaces in contact where the surface area is responsible for the resulting force level. This may be critical as in many cases the substrate is flat (bare wafer or planarized sample), whereas the stamp or mold is patterned and, due to lithography and etching steps, its surface is rough, resulting in a much higher contact area between stamp and polymer than between stamp and sample. This has consequences for the imprint techniques.

Mold-Assisted Lithography. Molding techniques working with monomers, oligomers, or prepolymers rely on solidification of the patterns by, for example, UV radiation. The radiation induces or continues a chemical crosslinking process where new bonds are built up within the polymer and to the surfaces in contact as well, the latter resulting in a severe sticking due to the chemical or physical bonds at the polymer–mold interface. There is no chance to overcome this challenge other than by using a "release layer," which renders the stamp surface hydrophilic and chemically inert (see Section 6). Such layers are essential for any curing-assisted molding technique [21, 118]. Dovetailed stamp features have to be avoided to enable a defect-free demolding process. The release layers are not stressed mechanically as any crosslinking process is accompanied by a certain shrink so that when chemical or physical bonding to the mold is avoided a natural detachment of the surfaces will take place. Thus, chemical functionality in particular is demanded for MAL and SFIL and mechanical wear of the release layers should be of minor impact.

Hot-Embossing Lithography. During hot embossing where the polymer is under viscous flow at imprint temperature and pressure, the polymer will perfectly conform to the stamp surface. As a consequence, stamp roughness should be as low as possible and dovetailed stamp features have to be definitely avoided. Potential bonds developing between polymer and stamp are in general physical ones and thus only weak, even when large contact areas are involved. If necessary, physical bonding may be reduced by an appropriate hydrophobic stamp surface layer.

Nevertheless, physical bonding at the interface is not a crucial point for hot embossing, the key point is friction interlocking. When a stamp with approximately vertical sidewalls (in total or in parts) is used, the polymer may clamp in between the stamp patterns due to internal stresses as long as these have not been relaxed. As relaxation effects in polymers require time (see Section 2.6), and the typical time constants are strongly temperature dependent, an appropriate processing temperature is required in order to obtain practically interesting processing times in the minute range. These temperatures are also needed for filling of larger recessed stamp areas as discussed in Section 2.6 and have been evaluated to be on the order of 90–100°C above the respective T_g of the polymer [157]. In cases where short processing times as well as low temperatures are desired, a stamp coating may be used to reduce sticking due to friction effects. In this case, the coating has to facilitate gliding under forces normal to the pattern walls, which often relies on abrasion of a small amount of molecules or clusters from the coating serving as a dry lubricant of the interface. This effect has been visualized experimentally when nonideal processing conditions were used [157] (Fig. 55).

As previously explained, high enough imprint temperatures, long enough imprint times, or both are effective in minimizing sticking in a hot-embossing process. In addition, adequate processing sequences may contribute to low-sticking procedures without additional stamp coatings. Specific processing conditions can be utilized to support "natural detachment" of stamp and sample during separation, as, for example, the "asymmetric heating and quenching method" introduced in Section 2.5.1 [55]. Furthermore, the separation temperature may play a role. Separation around T_g can use the still relatively high elasticity of the polymer in order to minimize polymer pattern damage during separation.

When thermoset polymers are imprinted at temperatures where crosslinking is taking place, an antisticking layer must be used [157] as is the case in a radiation curing process.

One has to be aware that antisticking layers are mechanically stressed during hot embossing by the applied pressure and during separation of stamp and sample. Thus, mechanical stability, lubrication properties, and wear will probably override the chemical functionality of this layer in importance for the process. These strong additional demands, which are not easily fulfilled, may favor a strategy of sticking optimization

Fig. 55. Embossed polymer patterns with feature sizes from 400 nm to 10 μm under nonideal processing conditions (imprint temperature only 70° C above T_g). (Top) Without ASL; (bottom) with ASL.

by process optimization rather than by introducing an auxiliary layer.

6.2. Antisticking Layers

The use of intermediate relief layers is a common technique in macroscopic replication techniques where details of the form and surface of a body have to be copied true to scale. In general, such layers should be thin compared to the patterns to be replicated, they should cover the surface conformally, perhaps requiring wetting of the surface, and they should not react with the molded material. In some cases, the layer is destroyed upon separation of the copy from the mold and has to be removed and renewed before every replication, an undesirable situation for nanoimprinting.

As an antisticking layer for hot embossing and molding, the search is for a few monolayers of a chemically inert material that is easy to apply uniformly and conformally over the patterned surface of the stamp or mold. It should feature minimum wear during imprint so that a large number of replications can be performed without cleaning and, in particular, without renewal of the layer. In view of the application as a lithography process, the layer deposition technique should, in addition, fit with Si technology and its materials and it should be cleanroom compatible.

As such, layers should have (i) antiadhesive properties to prevent bonding in general and (ii) lubricating properties to facilitate gliding and to reduce friction effects in the case of HEL. As both properties together reduce the assessed sticking, we will refer to such layers as antisticking layers (ASLs).

6.2.1. Metallic Antisticking Layers

It is well known from Si technology and from micromachining that the adhesion of polymers to metal surfaces is weak. Therefore, metallic layers of Cr, Ni, and Al have been tested as antiadhesion layers in a hot-embossing process [50–52]. For these experiments, the stamp was made from SiO_2 over Si, and so SiO_2 makes up most of the stamp surface (see Fig. 47b and c). PMMA was the polymer embossed, which is known for its good adhesion to Si and SiO_2. As a stamp protection, a 20-nm thin layer of the metal was deposited on the stamp in a thermal evaporation process. Ni is reported to show the best protection properties. Results on wear or thickness optimization have not yet been published, but a patent has been filed with respect to this surface passivation technique [165].

6.2.2. Spin Coated Antisticking Layers

The use of branched perfluorinated polymers as a spin-coatable release agent for SiO_2 molds with high aspect ratio is reported in [46]. The solution is applied to the mold so that its meniscus covers the whole surface, followed by immediate spinning at 6000 rpm before the solvent evaporates significantly. Solvent removal after spinning is performed in vacuum (30 mTorr). Film thicknesses achieved are as low as 5–10 nm as measured by AFM. Despite the high aspect ratio (500 nm height, 50-nm patterns), the ASL works effectively at imprint temperatures of 180°C.

6.2.3. Self-Assembled Antisticking Layers from Alkylsiloxanes

Self-assembly is a widely used phenomenon for realizing very thin and homogeneous layers on top of a substrate. In general, amphiphilic linear aliphatic chain molecules are used, featuring a head group at one end, which provides the functionality of the final surface layer, such as wettability, adhesion, or lubrication [130], and a base group at the other end, which provides a chemical link to the surface. Because of the chemical bonds to the surface, they are stable compared to physisorbed LB (Langmuir–Blodgett) films. The chain length determines the thickness of the self-assembled monolayer. The chains have to feature a regular conformation to enable defect-free two-dimensional assembly at the surface. This prerequisite is fulfilled for the commonly used $-(CH_2)_n-$ chains.

Such self-assembling molecules are used as an ink in microcontact printing as introduced in Section 4 where the base group is thiol ($-S$) for bonding to a Au surface (see Section 4). When self-assembled layers are used for antisticking purposes, silica and silica-like materials (Si, metals, and their oxides) are the substrates of interest. As mentioned in Section 5, stamps and molds are often fabricated from Si/SiO_2. SAM ASLs have been tested for mold assisted lithography [21, 118] as well as for hot-embossing lithography [56, 60].

Typically, the layers are prepared by immersion of the substrate at room temperature into a water-free solution of the SAM material [60]. Alternatively, deposition from the gas phase at elevated temperatures (>300°C) is possible but only short-chain SAMs have a high enough vapor pressure for this preparation [166, 167].

Self-Assembled Monolayer Layer Formation. For chemical anchoring of the chains to a silica surface, chlorosilane base groups ($—SiH_nCl_{3-n}$) are used. The most popular alkylsiloxane is OTS (octadecyl-tri-chlorosilane, $CH_3—(CH_2)_{17}—SiCl_3$), consisting of a chain of 18 C atoms with a triclorosilane group at the end [60, 168]. Due to its chain length, a dense self-assembled layer of OTS has a thickness of 22 Å [169].

Silica surfaces are, in general, hydroxyl ($—OH$) terminated, unless prepared under specific high-temperature and low-pressure conditions where the $—OH$ groups condense and are removed from the surface in a process called dehydration. Due to their polarity, the $—OH$ groups tend to adsorb water; thus, they are hydrophilic. This water film can comprise one or more monolayers and is crucial for anchoring of the chlorosilane group to the surface [167, 169–171].

Layer formation is suggested to proceed in the following three steps [171] illustrated in Fig. 56: (i) physical adsorption of the SAM to the previously formed water film at the solid surface, (ii) hydration of the chlorosilane groups where the Cl atoms are exchanged by OH groups under reaction with water to HCl, and (iii) polymerization of the SAM at the surface where the siloxane molecules become linked together laterally via $—O$ atoms under formation of water (dehydration) and where also a few links to the surface are built up.

This model is supported by the finding that the SAM layer has a lower roughness than the substrate surface. In addition, it can be proven by Fourier transform infrared spectroscopy (FTIR) measurements that the hydrolyzed silica surface remains nearly unchanged during SAM formation and that only a few surface bonds are formed [168, 170]. An OTS layer is thus an internally polymerized dense layer, anchored occasionally to the substrate.

The SAM coverage depends on the number of $—OH$ groups at the silica surface but complete hydration is not necessary for complete coverage of the surface [170]. On the contrary, a thick water layer gives rise to a polymerized SAM layer without surface bonds sitting on top of the surface water so that it can easily be floated off [167]. In the extreme case, bulk polymerization occurs in the solvent, manifesting itself as particulates at the surface [172]. As the hydration of the surface is a critical factor, it might be advisable to dehydrate the samples as mentioned previously. At temperatures from 170 to 400°C, the silanol groups ($—OH$) at the surface condense and eliminate water. Within this temperature range more than 50% of the $—OH$ groups will remain. Under such conditions, the process is reversible, and rehydration starting from these $—OH$ groups is easily performed in room temperature water within several minutes [170].

A study comparing different chlorosilanes ($CH_3SiH_nCl_{3-n}$ with $n = 0-2$) reveals clearly that when polymerization of the SAM layer occurs, as is, in particular, the case for $n = 0$, self-condensation is favored over surface reaction and a curing step is needed to complete the self-condensation and to undergo

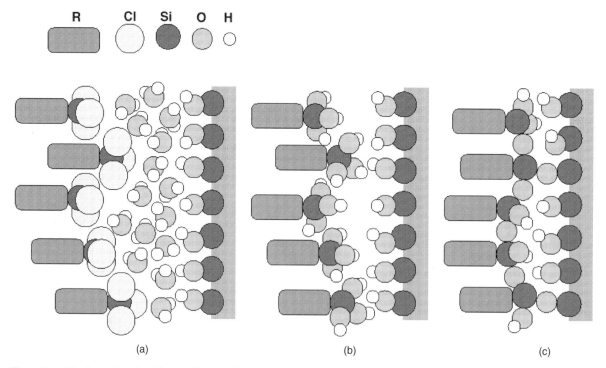

Fig. 56.   Illustration of the formation of a self-assembling crosslinked siloxane layer on Si as a three-stage process as supported by [171]: physical adsorption to the OH-terminated Si covered with surface water (a), hydration of the siloxane under HCL formation (not shown) and water consumption (b) and polymerization of the layer (c) under water release. The final layer is crosslinked but is not necessarily linked regularly to the Si surface.

some surface reactions [167]. Without polymerization ($n = 2$), the SAM molecules attach to the surface as no competing process exists, but they are not crosslinked internally.

Preparation. The preparation of an OTS layer on Si may proceed in the following steps [172, 173]: (i) HF dip to remove oxide layers from the Si surface; (ii) controlled chemical reoxidation in a 5 : 1 : 1 mixture of H_2O : H_2O_2 : NH_4OH at 70°C; (iii) rinsing with methanol and then the OTS solvent, for example, anhydrous toluene; (iv) immersion into a 1% [171] or even less concentrated [118, 170, 174] solution of OTS in toluene, because the previously used mixtures of hexadecane and CCl_4 have been replaced for environmental reasons; (v) rinsing in pure toluene to remove excess OTS molecules and adsorbed layers from the surface; (vi) rinsing of the solvent from the surface with methanol; and (vii) drying/anneal at about 100–120°C for a few hours in nitrogen or vacuum. Alternatively, aging in air for a few days has been proposed [172]. When drying is performed in an oven under N_2 or vacuum, the OTS films are stable up to temperatures of about 400°C [173], which is not the case for samples annealed in air or oxygen. Ultrasonic waves are reported to assist in anchored monolayer formation [169]. For oxide or silica surfaces, the first two steps may be skipped.

The temperature during the self-assembly process should be below 28°C for OTS in order to form defect-free and dense monolayers at relatively low layer formation speed within a treatment time of about 15–90 min [174] at temperatures of 26 and 18°C, respectively. At higher temperatures and thus increased film formation speed, the film is disordered and not closely packed, resulting in a layer thickness down to about 1.5 nm.

Layer Functionality. We have not yet discussed the head group. In the case of OTS, it is —CH_3, which is hydrophobic in nature and thus provides good antiwetting properties and has a low surface energy. Water contact angles of such films exhibit values up to 110°, corresponding to a surface energy of 20 mJ/m^2 [173], which is comparable to commercial PTFE. Improvement is even possible with F-containing head groups or chains. Different fluorinated trichlorosilanes have been tested [118, 172, 173] that were developed mainly for antiadhesive coatings in micromechanical devices [175]. In the case of —CF_3, a F-terminated surface of the SAM exists with antiwetting properties better than PTFE characterized by surface energies of 5–12 mJ/m^2 [122,172].

Somewhat more complex are the Ormocers [163]. They consist of an organic–inorganic network linked to the surface and feature among other properties increased abrasion resistance. They are applicable in immersion to glass or polymers. Their basic antiwetting properties are somewhat lower than those of OTS with contact angles of 55° in water. For these coatings, an admixture as low as 1% of a fluorinated compound was successful in increasing the contact angles by a factor of about 2, meeting the PTFE features. This improvement in functionality did not increase with increasing admixtures and thus the abrasion resistance was not affected.

Imprint Application. The discussion shows that reliable formation of an SAM for antisticking purposes requires care in preparation and is not a quick process. To our knowledge, layer formation and layer defects have not yet been studied in the case of patterned surfaces. Pinholes and grain boundaries should not affect the dewetting properties in an imprint process as long as they are small. The temperature used in a hot-embossing process should not be a problem as most siloxanes on SiO_2 are stable up to 400°C when prepared appropriately (see previous discussion) [130, 173]. However, the suitability of SAM layers for applications to surfaces under repeated contact or even under sliding contact, as may be the case in imprinting, is doubtful [176]. Direct investigations only exist for micromechanical applications [172] without high pressure and temperature. Frictional forces may break the upper part of the chains as they are polymerized only at the surface and may also tear off the film when anchorage to the surface is too scarce. Thus, SAM-based ASLs may be more suited to a curing process than to hot embossing. Fluorinated trichlorosilanes have been tested successfully for SFIL under UV curing [122, 127].

An additional problem with self-assembled ASLs may be the lack of reproducibility caused by the large number of process variables, as up to now there is neither process automation for layer deposition nor real-time monitoring of the layer formation process [172]. Cl residues may cause corrosion effects [173]. Furthermore, scaling up to wafer sizes may be problematic. Thus, manufacturability at wafer scale may limit actual SAM use to the laboratory scale [172].

6.2.4. Plasma-Deposited Teflon-like Antisticking Layers

Teflon is a polymer well known for its chemical inertness and low surface energy, resulting in its hydrophobic properties and making it an excellent antiwetting material. Casted volume material consists of long almost linear chains of —$(CF_2)_n$—. Because of the very tight C—F bond, its chemical reactivity is low. In the form of a powder, it can serve as a gliding agent or lubricant as it is not wetted by a large number of materials.

Teflon layers can be deposited in different processes where low-pressure deposition plays an important role when very thin layers are involved. In plasma polymerization processes, a precursor gas supplying the C and F components is used and C_nF_m components are deposited. Alternatively, C_nF_m molecules or clusters may be sputtered from a Teflon target. The film composition will be more or less Teflon-like as the deposited layers will build up a somewhat disordered and partly crosslinked material rather than long chains. The percentage of CF_2 is commonly used as a measure for Teflon-likeliness. In addition, good adhesion of such films to the stamp substrate is important for a long lifetime, which is likely supported by such a morphology.

Two different types of Teflon-like antisticking layers have been tested for hot-embossing lithography with a Ni stamp in PC and PMMA, namely, sputtered films and plasma-deposited films [177]. For sputtering, CHF_3 was used as a sputtering gas in an rf discharge at 5×10^{-2} mbar and 50 W, resulting in a deposition rate of 0.1 nm/s. Plasma polymerization was performed in a microwave electron cyclotron resonance plasma system with CF_4/H_2 as a precursor gas at 10^{-1} mbar and 200 W. Both processes resulted in about 5-nm-thick layers where the plasma deposition process was self-thickness limited [177–179]: After an initial growth phase, the deposited CH_n/CF_m functionalities were transformed during a treatment phase to a large amount into $—CF_2—$ due to ion bombardment where finally a balance between deposition and etching was obtained, limiting the layer thickness. This change in chemical composition was detected by *in situ* XPS (X-ray photoelectron spectroscopy) analysis [179]. The transition from growth to treatment was self-induced by negative biasing of the isolating deposited layer. Contact-angle measurements revealed a very low surface energy of the plasma-polymerized films of 4 mJ/m^2 [179].

Comparison of the two types of films by XPS analysis (Fig. 57) revealed a high content of CF_2 in the plasma-polymerized films, indicating relatively long chains with only minor contents of CF, CCF, and CCC, indicating a certain amount of crosslinking. This result corresponds to the measured low surface energy. In contrast, the sputtered film is chemically more heterogeneous and less Teflon-like. In addition, C—Ni bonds were detected at the interface to the Ni stamp, proving good adherence of the film on the stamp [178].

In a different approach, Teflon-like layers in the same thickness range were deposited on poly–Si stamps in a capacitively coupled hollow cathode discharge with C_4F_8 as a precursor gas at 0.3 mbar and 50 W with the stamp at floating potential. XPS measurements showed a large amount of CF_2 with a relative amount of the other functionalities lying in between the two curves of Fig. 57. CF_2 bonds were also verified by FTIR analysis [157].

These ASLs were used during imprint of thermosetting polymers where sticking due to the chemical crosslinking at elevated temperature during embossing is observed. Excellent embossing results were obtained over areas of 4 cm^2 when ASLs on the stamp were used (see Fig. 58), whereas sticking dominated the imprint without them.

6.3. Wear and Lifetime

Spin-coated ASLs from perfluorinated polymers are found to attach to the imprinted sample [46] after hot embossing at 180°C. However, no indication is given on the amount transferred or on the number of imprints that can be done before renewal of the ASL is required. For metallic ASLs, wear investigations have not yet been reported.

Self-assembled ASLs developed for micromechanical devices are only tested for this specific purpose. Often only contact under low pressure (26 kPa [157]) is investigated and 40 million operation cycles in touch mode are reported without wear featuring low static friction coefficients of 0.08. Study of tribological properties use sidewall friction test structures and report kinetic friction coefficients of 0.02. Wear manifests itself by increase in this value to 0.27 after 2000 cycles under 84 MPa [172]. Conclusions from such results for applicability in an imprint process have not yet been drawn. Only one paper gave qualitative indications for the stability of an OTS ASL in a hot-embossing process [60]. It was mentioned that OTS forms a stable monolayer, that the protection lasts many imprint cycles, and that it is resistant to mold cleaning in pyrrholidon. In SFIL applications [122], durability was proved by an unchanged contact angle of the SAM-coated surface after tens of imprints under UV curing, storage in the open laboratory, and a number of aggressive cleaning cycles.

A quantitative study on wear exists for the Teflon-like plasma-polymerized ASLs introduced in Fig. 57. It comprises the compositional change of the layers themselves due to a hot-embossing process as well as the investigation of the embossed polymer [177, 178]. The study was performed with flat Ni stamps and embossing was done into PC and PMMA at 205°C and 80 bar. No sticking was observed with both types

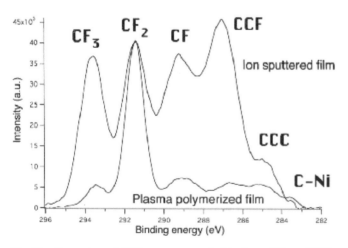

Fig. 57. Comparison of XPS measurements (bulk) of two different PTFE-like ASLs (plasma-polymerized, ion sputtered) normalized to the CF_2 peak. Reprinted with permission from R. W. Jaszewski, H. Schift, B. Schnyder, A. Schneuwly, and P. Gröning, *Appl. Surf. Sci.* 143, 301 (1999). © 1999 by Elsevier Science.

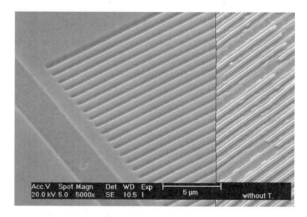

Fig. 58. Imprinted thermoset polymer (array of 250-nm lines) with (left) and without (right) a plasma-deposited PTFE-like ASL. Reprinted with permission from H. Schulz, H.-C. Scheer, F. Osenberg, and J. Engemann, *Proc. SPIE* 3996, 244 (2000). © 2000 by the American Vacuum Society.

of ASLs even after 50 embossings. However, it was found by XPS that the composition of the ASLs changed. As shown in Fig. 59a and b, the films lose F. The relative amount of CF_2 decreases, whereas the relative amount of CF and CC increases. The film thickness decrease is estimated to be less than 0.1 nm/embossing.

The lost F could be monitored at the embossed polymer (see Fig. 59c). The amount of F transferred to the replica decreased first exponentially and reached a minimum constant value of about 2 at% after 10 embossings. The authors assumed two different transfer processes. Within the first phase, weakly bound fluorinated molecules diffuse out of the film and to the embossed polymer. This process depends on the contact time and the amount of transfer is higher for the chemically more homogeneous (less Teflon-like) sputtered films. In a later phase, abrasion of a few submonolayers from the top of the film is responsible for the low constant and time independent amount of F transfer. Abrasion may be chemically induced, for example, by oxidation processes at the polymer–ASL film interface. This abrasion effect was also found with casted Teflon sheets containing only CF_2 chains and with dispersion-deposited Teflon layers commercially available for tribological purposes.

The results obtained for flat stamps are comparable to those found with low-aspect-ratio patterned stamps [178].

The observed abrasion effect supports the assumption that ASLs also act as a dry lubricant, providing a sacrificial interface that can be easily separated [177]. In this way, an ASL can decrease the "brake" effects due to friction discussed in Section 6.1.2 when high aspect ratio patterns and vertical walls are involved [157].

The investigations show that plasma-polymerized layers should be thin enough and durable enough to serve as an ASL for more than 100 embossings without renewal [177]. Because (i) the deposition process is fast, (ii) the film lifetime is high, (iii) processes developed for tribological applications are available [179], and (iv) PECVD (plasma-enhanced chemical vapor deposition) techniques fit into Si production technology, the reliable use of ASLs should not be a bottleneck for nanoimprinting.

7. APPLICATIONS

Applications already demonstrated include diffractive optics, magnetic arrays, biosensors, quantum point contacts, single-electron transistors, (MESFETS), and interdigitated fingers.

All benefit first from the capability to pattern large areas with small features and second the use of a polymer on a given substrate, thereby adding several benefits by choosing a polymer the properties of which may be enhanced by nanostructuring. For example, a polymer suitable for printing in the 10- to 100-nm scale spun on glass or plastic can be used as a low-cost biosensor, which can be disposed of after one use, thus minimizing contamination. Moreover, low-cost planar approaches are bound to become a growth area for the

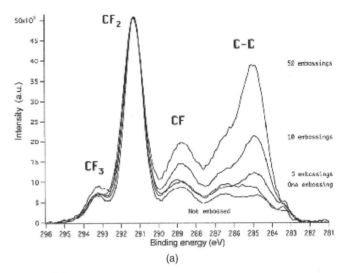

(a)

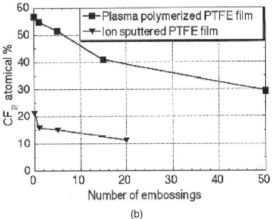

(b)

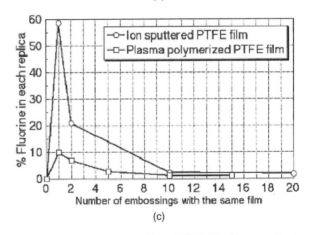

(c)

Fig. 59. Evolution of the composition of PTFE-like films as a function of the number of embossings in PC for the films introduced in Fig. 56. (a) Film composition of plasma-polymerized layer, normalized to the CF_2 peak; (b) CF_2 content of the films; (c) amount of F detected at the PC replica after embossing. Reprinted with permission from R. W. Jaszewski, H. Schift, B. Schnyder, A. Schneuwly, and P. Gröning, *Appl. Surf. Sci.* 143, 301 (1999). © 1999 by Elsevier Science.

fabrication of lab-on-a-chip-based sensors in medical technologies. The flexibility offered by polymers is especially relevant because these can be made to react to different environments by building chemical, optical, mechanical, or electrical responses. In this section, we describe applications in data storage, electronic devices, photodetectors and light emitters, gratings and integrated optics, and biosensors.

7.1. Data Storage

For magnetic arrays, the holy grail is a large area with features down to 10 nm lateral size separated also by about 10 nm as long as the magnetic domains of, for example, TbFeCo, do not interact with their neighbors. 10-nm features are essential for terabit magnetic memories. Depending on the technology to make it to the market for read-only devices, a strong requirement on surface roughness may pose a nonnegligible challenge to nanoimprint lithography. Early work at IBM [180] used silicon oxide masters written by electron beam lithography. The pattern is transferred to PMMA by means of photopolymerization and subsequently further crosslinked. The smallest features obtained were 50×100 nm marks, confirmed by atomic force microscopy, albeit with slight broadening. This feature size translates into 10 Gb/cm^2 in CD-like patterns. Concerning roughness, the authors measured 0.2- to 0.5-nm rms for both the SiO$_2$ master and the replica and argued that this was below that of polycarbonates, which are known to have an acceptable signal-to-noise level. In a similar approach, Chou et al. [28] and Krauss and Chou [181] showed nanoimprint data, which would lead to 400 Gb/in^2.

The preparation of stamps or molds for high-density data storage devices is one of the main challenges. Carcenac et al. [182] reported fabrication details to produce a SiO$_2$ master by electron beam lithography with pillars down to 30-nm diameter, 60-nm period, and 100 nm high. Further attempts resulted in lines of 20-nm width and 20-nm spacing.

One way to fabricate Cr stamps and imprint tests on PMMA has been reported by Zankovych et al. [183]. The 2 $\times$ 2 cm^2 silicon master contains areas of 200×200 μm^2 Cr pillars and is fabricated by e-beam lithography and lift-off. These Cr pillars have 50-nm diameter, and 100-nm period and are 60 nm high (Fig. 60, left). An image using atomic force microscopy of the imprint on PMMA ($M_W = 200,000$) shows the high fidelity of the imprint on a 100-nm-thick film (Fig. 60, right). The roughness remains to be evaluated over large areas; however, preliminary results show roughness well below 1 nm.

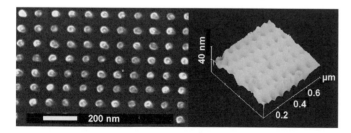

Fig. 60. SEM micrographs of a Cr stamp consisting of 60-nm-high columns with 50-nm diameter and 100-nm period (left) and an AFM image of 50-nm features imprinted into a 100-nm-thick PMMA layer (right). The stamp is 200×200 μm^2. Courtesy of S. Zankovych.

7.2. Electronic Devices

Probably the first set silicon-based electronic devices fabricated by nanoimprint lithography were those of Chou et al. [28]. The authors reported silicon quantum dots, wires, and rings on a SOI substrate. The stamps were written by e-beam lithography and the alignment marks by optical lithography. The latter was needed for the definition of larger areas containing the source and drain. The patterns were reactive ion etched in CHF$_4$, resulting in 110-nm-tall features. The imprints were carried out on top of the (100) p-type Si layer of the SOI substrate. A sacrificial layer was a purposely-grown oxide used to protect the upper Si layer during nanoimprinting. The stamp was used to pattern PMMA, which subsequently was subjected to Cr lift-off. The Cr features were then used as a mask to transfer the pattern into the SOI in a two-step etching process. The devices were subjected to further reduction to fabricate the gate oxide. A layer of polysilicon was deposited to define the gate over the transistor channel. The self-aligned source and drain are defined by ion implantation, which is also used to dope the gate. The 100-nm wire channel metal–oxide–semiconductor field effect transistor (MOSFET) with 2-μm channel length was found to behave well with a threshold of 0.2 V and a slope of 90 mV/decade.

One example of a quantum point contact structure fabricated by nanoimprinting on a two-dimensional electron gas of GaAs-GaAlAs was reported by Martini et al. [60]. The stamp was made of Si and had a split-gate pattern. The pattern was imprinted on PMMA spun over the GaAs–GaAlAs substrate. Metal evaporation and lift-off were subsequently used to define the Schottky gate. An optical mask aligner was used to align mold and substrate. This suggests that no measurable damage is incurred by using nanoimprint lithography to fabricate highly sensitive quantum devices.

7.3. Photodetectors and Light Emitters

The fact that printing can be carried out on a polymer spun on a silicon substrate, followed by standard lift-off techniques, leads to low-cost metal–semiconductor–metal (MSM) detectors. The pressure and temperature used in imprint lithography can be kept low enough to avoid degradation of the underlying CMOS [complementary MOSFET (metal-oxide-semiconductor field effect transistor)] platform. Yu et al. [102] have demonstrated MSM photodetectors with lines and spacing between 200 and 600 nm. The silicon mold had features between 193 and 330 nm high and patterned areas of 14×14 μm^2, which were printed on a PMMA coating the GaAs substrate. Control MSM devices were fabricated by photolithography. As long as the pressure during nanoimprinting was kept below 600 psi, the I–V curves revealed no detectable device degradation. Small differences were ascribed to surface recombination. In principle, there is nothing on the way toward larger areas patterned with 60-nm features for even higher frequency operation.

Patterning of optically active organic material has been demonstrated in two types of light-emitting structures. One structure contains an electron-conducting small molecule (Alq$_3$ doped with 2 wt% DCMII dye) and the other contains PMMA loaded with semiconducting small molecules usually incorporated in organic light-emitting diodes and optically pumped organic lasers [184]. The polymer films were typically 200 nm thick and prepared by spin coating. The SiO$_2$ stamps had gratings of period 200 and 300 nm and were 180 nm high. The imprint parameters were 800 psi, 150°C under a vacuum of about 1 Torr or in a nitrogen atmosphere. The broad luminescence spectra of both types of patterned organic light-emitting diodes showed a negligible decrease in peak intensity compared to spectra recorded on the films before patterning. The luminescence data recorded on samples with either a glass or a Si substrate were indistinguishable. Degradation was only observed in samples printed in an air atmosphere. The remaining test is to study the longer term behavior of these light-emitting devices and smaller features of organic light-emitting molecules, the ordered extent of which becomes comparable with the feature size.

Distributed feedback (DFB) resonators and Bragg reflector resonators (DBRs) have been fabricated by μCP and mold-assisted replication [185]. Comparing the performance of nanoimprinted spin-coated gain media to photolithographically defined SiO$_2$ gratings, the author reported these to be similar. The emission was either single mode or multimode. The latter suggests that the uniformity, or spatial coherence, of the DFB resonator over the excited area was insufficient. Some of the reasons include distortions in the stamp, uneven polymer displacement during patterning, and some form of peeling during separation of stamp and mold. On the other hand, these features could be used in a careful design to tune the resonators.

7.4. Gratings and Integrated Optics

One early attempt to replicate patterns on the surface of optical elements with submicrometer resolution was that of Aumiller et al. [10], who used a hard stamp (a metallic die) to pattern dielectric material for integrated optics. Wang et al. [186] later used nanoimprint lithography to fabricate a double-layer metal grating to act as a broadband waveguide polarizer with 190-nm period. The idea was to make a waveguide polarizer, which is transverse magnetic (TM)-pass, which lets through light with TM polarization. The double-layer metal grating is positioned at the interface between the core and the cladding of the waveguide. The fabrication requires SiN as top layer, SiO$_2$, and silicon. The process combines nanoimprint lithography, lift-off, and dry etching and yields waveguide stripes over the double-layer metal grating. The extinction ratio of TM to transverse electric (TE) polarization between 720 and 820 nm was measured to be larger than 50 dB/mm and the loss of the TM mode 2 dB/mm. The wavelength range of operation depends on the transparency of the materials in the wavelength of interest.

The ease promised by a single-step process to pattern a thin polymer film as a grating motivated the experiments of

Seekamp et al. [187] in which a 5×5 cm^2 silicon grating with 1250 lines/mm (i.e., 800-nm period and 300 nm high) was printed into PMMA. Figure 61 shows images of this grating recorded with an optical microscope at different angles. The uniformity seen in the lower panels suggests a high grating quality for different wavelengths. A more qualitative result is shown in Fig. 62, where the full width at half-maximum of each peak is about 20 nm from the UV to the near infrared. The efficiency of this first grating is 4% compared to 20% of a commercially available equivalent tested in the same monochromator setup as a reference.

A nanoimprinted Y branch printed into polystyrene on oxidized silicon using a Si stamp is shown in Fig. 63. The stamp was produced by UV lithography. Coupling light of 514 nm into the waveguide via a monomode fiber shows the scattered light at the entrance, the split, and the exits of the Y branch.

Seekamp et al. [187] also investigated the prospects of printing polymer films in polymer waveguides, extending the fabrication to photonic crystal elements. The stamp was made by combining photolithography to write the Y branch and electron beam lithography to write a small two-dimensional photonic crystal in one of the branches (Fig. 64). A detail of the stamp and the imprint is shown in Fig. 65, left and right panels, respectively, where a two-dimensional photonic crystal with hexagonal lattice is clearly seen to have been transferred with remarkable fidelity. Tests are in progress for waveguide parameters suitable for the telecommunication range at 1.3 μm, according to photonic crystal waveguide models. In any case, the feasibility of printing a polymer waveguide with micrometer features combined with photonic crystal elements has been demonstrated.

The feature sizes demonstrated previously suggest that nanoimprint lithography will be a strong candidate for the fab-

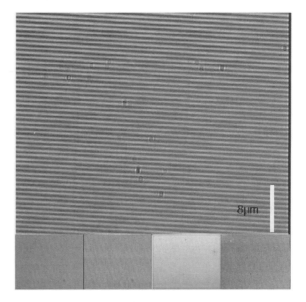

Fig. 61. Light microscope image of an 800-nm-period structure printed in PMMA on silicon. The images in the bottom panels show the homogeneity of the dispersion over the area of roughly 0.1×0.1 mm^2. Courtesy of J. Seekamp.

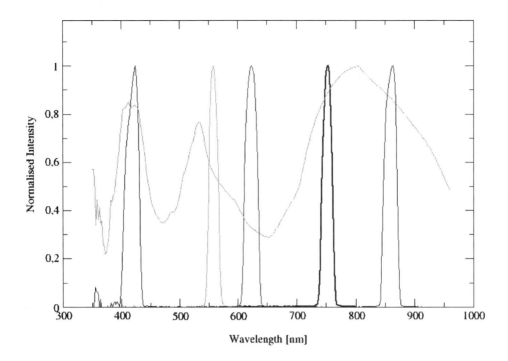

Fig. 62. Transmission and zeroth-order characteristics of a printed 5×5 mm^2 grating showing a set of transmission lines from the UV to the near infrared. The light-gray line in the background gives the characteristics for the zeroth-order reflection. Courtesy of J. Seekamp.

rication of compact optical sharp bends (right angles) and of devices sporting multimode interference in, for example, SOI. The moderate refractive index contrast of about 2 favors compact designs. Likewise, the fabrication of frequency-selective surfaces for the near- and midinfrared should be viable again using SOI and nanoimprint techniques.

7.5. Biosensors

One key aspect when considering the fabrication of biosensors is volume production. Montelius et al. [51] have shown that nanoimprint lithography is highly suited to fabricate interdigitated metal electrodes and contact pads with feature size of 100 nm and spacings varying from 100 to 500 nm. The

authors tested the capacitance of about 17 pF and found that it did not vary significantly between 1 kHz and 1 MHz. Likewise, the 1-GΩ resistance between the fingers suggested there was no unwanted current on the surface. Future work includes the fabrication of biocompatible materials and site selectivity.

Other representative work in this area is that of the Glasgow University team, who has explored patterning of glass and plastic surfaces for cell cultivation [114, 154, 188].

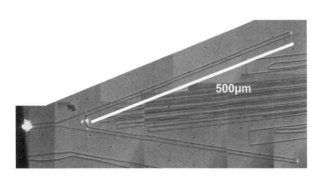

Fig. 63. Y branch printed in polystyrene on oxidized silicon. The SiO$_2$ stamp was pressed 400 nm into the polymer, leaving a 100- to 150-nm residual layer of polystyrene. Light of 514-nm wavelength is coupled into the structures from a monomode fiber with a mode field diameter of about 4 μm on the left. The light propagates in the waveguide, reaching both ends of the branch despite losses at the joint. Courtesy of J. Seekamp.

Fig. 64. Stamp of a Y branch with a photonic crystal in one arm of the branch fabricated by combined UV lithography and electron beam lithography. The top image shows a Y branch after UV lithography. A Cr mask is applied on one arm as depicted in the lower left image of a triangular lattice. This structure is subsequently dry etched to achieve the stamp shown in the lower right image for a triangular structure. Courtesy of J. Seekamp.

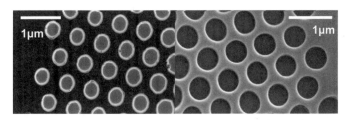

Fig. 65. Pattern transfer by nanoimprinting. The 300-nm columns etched into SiO$_2$ on Si (left) are accurately transferred into PMMA on SOI (right). Courtesy of J. Seekamp.

8. PERSPECTIVES

The prospects for nanoimprint-based techniques are wide open. These are based not only on the low cost of their nanometer-scaled fabrication processes but also on the potentially huge range of applications, mainly at the end of less demanding applications. However, to make it as an industrial process in, for example, the electronics industry, the proposers of nanoimprint lithography and its variations have to address multilevel and throughput issues, which, while being key determinants, are not the only aspects needing closer scrutiny. Concerning throughput, a rough estimate considering 50-nm features covering a 6-in. wafer (182 cm^2) printed in 20 min, including alignment and heating and cooling cycles, gives an "exposure rate" of 0.152 cm^2/s for 50-nm resolution. This exposure rate is close to that of SCALPEL, which, for 100-nm resolution, reaches an exposure rate of 0.25 cm^2/s [189].

Among some of the issues requiring further investigation and discussion, one finds fidelity, and validation as well as alignment and parallelism. Different degrees of complexity are expected for the different techniques. Concerning fidelity, there is a strong need to report work using similar qualitative comparison parameters and or procedures, beyond the examination by scanning electron microscopy. Validation, or the quality control issue, needs agreement as to what counts as tolerances for a good print. Optical diffraction techniques lend themselves to monitoring the quality of periodic structures. For meaningful comparisons, a reference system concerning process, instrumentation, and materials will be essential. It is clear that much of the preceding will strongly depend on the type of applications in mind, having relative importance in some cases. For example, in the case of hot-embossing lithography for applications in advanced electronics, the alignment and the range of features that can be placed together in one stamp will determine if printing-based techniques will find an application in industry.

On the science front, the relative ease of these techniques is bound to make them a common laboratory tool to study surface and interface phenomena, beyond the solid state, by putting within the reach of scientists the use of patterned substrates down to 10 nm for their research.

Acknowledgments

The authors acknowledge the support of the European Union ESPRIT Project NANOTECH and the EU IST Project CHANIL. H-CS, HS, and CMST are also grateful to the German Research Council and the Deutsche Forschungsgemeinschaft for their support. It is a pleasure to thank the members of the CHANIL team for numerous discussions and fruitful collaborations over the recent years. In particular, HS and H-CS thank J. Ahopelto, VTT Electronics Espoo; L. Montelius, NSC University Lund; and K. Zimmer, IOM Leipzig, for the gift of stamps; and K. Pfeiffer, microresist technology, for polymers and interesting discussions of polymer behavior. Without their support, our work would not have been successful. HS and H-CS are also indebted to N. Roos and D. Lyebyedyev for careful reading of the manuscript and critical comments, and to M. Knippschild for administration of the database and its implementation for the references, as well as to D. Echterhoff for technical support. We are grateful to Suzanne Broch for her assistance with the preparation of this chapter.

REFERENCES

1. W. H. Arnold, *Microelectron. Eng.* 46, 7 (1999).
2. M. Yomazzo and D. Van Den Brocke, *Microelectron. Eng.* 41/42, 53 (1998).
3. L. R. Harriott, *Mater. Sci. Semicond. Process.* 1, 93 (1998).
4. S. Hector, *Microelectron. Eng.* 41/42, 25 (1998).
5. M. Calleja, J. Anguita, K. Birkelund, F. Perez-Murano, and J. Dagata, *Nanotechnology* 10, 34 (1999).
6. L. J. Balk, P. M. Koschinski, G. B. M. Fiege, and F. J. Reineke, *Proc. SPIE* 2723, 402 (1996).
7. J. Jersch, F. Demming, and K. Dickmann, *Appl. Phys. A* 64, 29 (1997).
8. C. Puech, *Opt. Commun.* 7, 135 (1973).
9. M. T. Gale, *Microelectron. Eng.* 34, 321 (1997).
10. G. D. Aumiller, E. A. Chandross, W. J. Tomlinson, and H. P. Weber, *J. Appl. Phys.* 45, 4557 (1974).
11. W. Ehrfeld, V. Hessel, H. Löwe, C. Schulz, and L. Weber, *Microsystem Technol.* 5, 105 (1999).
12. S. Y. Chou, P. R. Krauss, and P. J. Renstrom, *Science* 272, 85 (1996).
13. S. Y. Chou, P. R. Krauss, and P. J. Renstrom, *Appl. Phys. Lett.* 67, 3114 (1995).
14. A. Macintyre and S. Thomas, *Microelectron. Eng.* 41/42, 211 (1998).
15. H. Becker and U. Heim, *Proc. MEMS* 99, 228 (1999).
16. F. Nikolajeff, S. Hard, and B. Curtis, *Appl. Opt.* 36, 8481 (1997).
17. S. Y. Chou, U.S. Patent 5,772,905, 1998.
18. S. Y. Chou, P. R. Krauss, and P. J. Renstrom, *J. Vac. Sci. Technol., B* 14, 4129 (1996).
19. R. W. Jaszewski, H. Schift, J. Gobrecht, and P. Smith, *Microelectron. Eng.* 41/42, 575 (1998).
20. H. Schift, R. W. Jaszewski, C. David, and J. Gobrecht, *Microelectron. Eng.* 46, 121 (1999).
21. J. Haisma, M. Verheijen, K. van den Heuvel, and J. van den Berg, *J. Vac. Sci. Technol., B* 14, 4124 (1996).
22. A. Kumar and G. M. Whitesides, *Appl. Phys. Lett.* 63, 2002 (1993).
23. A. Kumar, H. A. Biebuyck, and G. M. Whitesides, *Langmuir* 10, 1498 (1994).
24. R. Bartolini, W. Hannan, D. Karlsons, and M. Lurie, *Appl. Opt.* 9, 2283 (1970).
25. R. Ulrich, H. P. Weber, E. A. Chandross, W. J. Tomlinson, and E. A. Franke, *Appl. Phys. Lett.* 20, 213 (1972).

26. L. J. Heyderman, H. Schift, C. David, B. Ketterer, M. Auf der Maur and J. Gobrecht, Proc. Micro- and Nano-Engineering Conf. 2000, Jena, Germany.

27. S. Y. Chou and P. R. Krauss, *Microelectron. Eng.* 35, 237 (1997).

28. S. Y. Chou, P. R. Krauss, W. Zhang, L. Guo, and L. Zhuang, *J. Vac. Sci. Technol., B.* 15, 2897 (1997).

29. L. Guo, P. R. Krauss, and S. Y. Chou, *Appl. Phys. Lett.* 71, 1881 (1997).

30. D. Eisert, W. Braun, S. Kuhn, J. Koeth, and A. Forchel, *Microelectron. Eng.* 46, 179 (1999).

31. I. Puscasu, G. Boreman, R. C. Tiberio, D. Spencer, and R. R. Krchnavek, *J. Vac. Sci. Technol., B* 18, 3578 (2000).

32. H. Schift, C. David, J. Gobrecht, A. D'Amore, D. Simoneta, and W. Kaiser, *J. Vac. Sci. Technol., B* 18, 3564 (2000).

33. M. Li, J. Wang, L. Zhuang, and S. Y. Chou, *Appl. Phys. Lett.* 76, 673 (2000).

34. A. Lebib, Y. Chen, J. Bourneix, F. Carcenac, E. Cambril, L. Couraud, and H. Launois, *Microelectron. Eng.* 46, 319 (1999).

35. D. Lyebyedyev, H. Schulz, and H.-C. Scheer, *Proc. SPIE* (2001). To appear.

36. T. Haatainen, J. Ahopelto, G. Gruetzner, M. Fink, and K. Pfeiffer, *Proc. SPIE* 3997, 874 (2000).

37. K. Pfeiffer, M. Fink, G. Bleidiessel, G. Gruetzner, H. Schulz, H.-C. Scheer, T. Hoffmann, C. M. Sotomayor Torres, F. Gaboriau, and C. Cardinaud, *Microelectron. Eng.* 53, 411 (2000).

38. F. Gaboriau, M. C. Peignon, G. Turban, C. Cardinaud, K. Pfeiffer, G. Bleidießel, and G. Grützner, *Le Vide: Sci. Tech. Appl.* 291, 245 (1999).

39. F. Gaboriau, M. C. Peignon, A. Barreau, G. Turban, C. Cardinaud, K. Pfeiffer, G. Bleidießel, and G. Grützner, *Microelectron. Eng.* 53, 501 (2000).

40. H. Schulz, H.-C. Scheer, T. Hoffmann, C. M. Sotomayor Torres, K. Pfeiffer, G. Bleidiessel, G. Grützner, C. Cardinaud, F. Gaboriau, M.-C. Peignon, J. Ahopelto, and B. Heidari, *J. Vac. Sci. Technol., B* 18, 3582 (2000).

41. S. W. Pang, T. Tamamura, M. Nakao, A. Ozawa, and H. Masuda, *J. Vac. Sci. Technol., B* 16, 1145 (1998).

42. J. Wang, X. Sun, L. Chen, and S. Y. Chou, *Appl. Phys. Lett.* 75, 2767 (1999).

43. J. Wang, X. Sun, L. Chen, L. Zhuang, and S. Y. Chou, *Appl. Phys. Lett.* 77, 166 (2000).

44. B. Hillerich and M. Rode, *Proc. MEMS* 98, 545 (1998).

45. H. Tan, L. Chen, M. Li, J. Wang, W. Wu, and S. Y. Chou, Electron, Ion and Photon Beam Technology and Nanofabrication "EIPBN 2000," 2000.

46. B. Faircloth, H. Rohrs, R. Tiberio, R. Ruoff, and R. R. Krchnavek, *J. Vac. Sci. Technol., B* 18, 1866 (2000).

47. H. Becker and C. Gärtner, *Electrophoresis* 21, 12 (2000).

48. H. Tan, A. Gilbertson, and S. Y. Chou, *J. Vac. Sci. Technol., B* 16, 3926 (1998).

49. H.-C. Scheer, H. Schulz, and D. Lyebyedyev, *Proc. SPIE* (2001). To appear.

50. B. Heidari, I. Maximov, E.-L. Sarwe, and L. Montelius, *J. Vac. Sci. Technol., B* 17, 2961 (1999).

51. L. Montelius, B. Heidari, M. Graczyk, I. Maximov, E.-L. Sarwe, and T. G. I. Ling, "Micro- and Nano-Engineering '99 Conference," 2000, Vol. 53, p. 521.

52. L. Montelius, B. Heidari, M. Graczyk, T. Ling, I. Maximov, and E.-L. Sarwe, *Proc. SPIE* 3997, 442 (2000).

53. B. Heidari, I. Maximov, E.-L. Sarwe, and L. Montelius, *J. Vac. Sci. Technol., B* 18, 3575 (2000).

54. L. Montelius, B. Heidari, and I. Maximov, Micro- and Nano-Engineering (2001). To appear.

55. D.-Y. Khang and H. H. Lee, *Appl. Phys. Lett.* 75, 2599 (1999).

56. N. Roos, T. Luxbacher, T. Glinsner, K. Pfeiffer, H. Schulz, and H.-C. Scheer, *Proc. SPIE* 4343 (2001). To appear.

57. W. Wu, B. Cui, X.-Y. Sun, W. Zhang, L. Zhuang, L. Kong, and S. Y. Chou, *J. Vac. Sci. Technol., B* 16, 3825 (1998).

58. H.-C. Scheer, H. Schulz, T. Hoffmann, and C. M. Sotomayor Torres, *J. Vac. Sci. Technol., B* 16, 3917 (1998).

59. K. Zimmer, K. Otte, A. Braun, S. Rudschuck, H. Friedrich, H. Schulz, H.-C. Scheer, T. Hoffmann, C. M. Sotomayor Torres, R. Mehnert, and F. Bigl, *Proc. EUSPEN* 1, 534 (1999).

60. I. Martini, D. Eisert, S. Kuhn, M. Kamp, L. Worschech, J. Koeth, and A. Forchel, *J. Vac. Sci. Technol., B* 18, 3561 (2000).

61. K. Pfeiffer, G. Bleidiessel, G. Gruetzner, H. Schulz, T. Hoffmann, H.-C. Scheer, C. M. Sotomayor-Torres, and J. Ahopelto, *Microelectron. Eng.* 46, 431 (1999).

62. W. Menz and P. Bley, "Mikrosystemtechnik für Ingenieure." VCH, Weinheim, 1993.

63. H. Schulz, D. Lebedev, H.-C. Scheer, K. Pfeiffer, G. Bleidiessel, G. Grützner, and J. Ahopelto, *J. Vac. Sci. Technol., B* 18, 3582 (2000).

64. W. Zhang and S. Y. Chou, Electron, Ion and Photon Beam Technology and Nanofabrication "EIPBN 2000," 2000.

65. G. R. Strobl, "The Physics of Polymers." Springer-Verlag, Berlin, 1997.

66. P. J. Flory, "Statistical Mechanics of Chain Molecules." Interscience, New York, 1969.

67. M. Doi and F. Edwards, "The Theory of Polymer Dynamics." Clarendon, Oxford, 1986.

68. M. Doi, "Introduction to Polymer Physics." Clarendon, Oxford, 1996.

69. T. P. Lodge, N. A. Rotstein, and S. Prager, *Adv. Chem. Phys.* 79, 1 (1990).

70. J. D. Ferry, "Viscoelastic Properties of Polymers." Wiley, New York, 1980.

71. N. C. Robinson, in "Polymers for Electronic and Photonic Applications." Academic Press, Boston, (1983).

72. H. L. Williams, "Polymer Engineering." Elsevier, Amsterdam, 1975.

73. M. L. Williams, R. F. Landel, and J. D. Ferry, *J. Am. Chem. Soc.* 77, 3701 (1955).

74. J. R. McLoughlin and A. V. Tobolsky, *J. Colloid Sci.* 7, 555 (1952).

75. S. Onogi, T. Masuda, T. Ibaragi, and Z. Kolloid, *Z. Polym.* 222, 110 (1967).

76. T. Masuda, K. Kitagawa, and S. Onogi, *Polym. J.* 1, 418 (1970).

77. J. L. Halary, A. K. Oultache, J. F. Louyot, B. Jasse, T. Sarraf, and R. Müller, *J. Polym. Sci., Part B: Polym. Phys.* 29, 933 (1991).

78. A. Dufresne, S. Etienne, J. Perez, P. Demont, M. Diffalah, C. Lacabanne, and J. J. Martinez, *Polym. J.* 37, 2359 (1996).

79. A. V. Tobolsky, J. J. Aklonis, and G. Akovali, *J. Chem. Phys.* 42, 723 (1965).

80. S. Onogi, T. Masuda, and K. Kitagawa, *Macromolecules* 3, 109 (1970).

81. E. Catsiff and A. V. Tobolsky, *J. Colloid Sci.* 10, 375 (1955).

82. S. S. N. Murthy, *J. Polym. Sci., Part C: Polym. Lett.* 26, 361 (1988).

83. S. S. N. Murthy, *J. Polym. Sci., Part B: Polym. Phys.* 31, 475 (1993).

84. G. C. Berry and T. G. Fox, *Adv. Polym. Sci.* 5, 261 (1968).

85. S. M. Aharoni, *Macromolecules* 16, 1722 (1983).

86. V. P. Privalko and Y. S. Lipatov, *Makromol. Chem.* 175, 641 (1974).

87. T. Masuda, K. Kitagawa, T. Inoue, and S. Onogi, *Macromolecules* 3, 116 (1970).

88. W. M. Prest and R. S. Porter, *Polym. J.* 4, 154 (1973).

89. W. M. J. Prest, *Polym. J.* 4, 163 (1973).

90. J. P. Montfort, G. Marin, and P. Monge, *Macromolecules* 19, 1979 (1986).

91. A. Gourari, M. Bendaoud, C. Labanne, and R. F. Boyer, *J. Polym. Sci. Part B: Polym. Phys.* 23, 889 (1985).

92. R. B. Bird, R. C. Armstrong, and O. Hassager, "Dynamics of Polymeric Liquids," Fluid Mechanics, Vol. 1. Wiley, New York, 1977.

93. L. J. Heyderman, H. Schift, C. David, J. Gobrecht, and T. Schweizer, *Microelectron. Eng.* 54, 229 (2000).

94. H.-C. Scheer and H. Schulz, *Microelectron. Eng.* (2001). to appear.

95. D. Lyebyedyev, H. Schulz, and H.-C. Scheer, Micro- and Nano-Engineering. To appear.

96. Y. Chen, A. Lebib, S. P. Li, M. Natali, D. Peyrade, and E. Cambril, Micro- and Nano-Engineering. To appear.

97. V. Sirotkin, A. Svintsov, and S. Zaitsev, "Fourth MEL-ARI/NID Workshop," 1999.

98. L. G. Baraldi, Heissprägen in Polymeren für die Herstellung integriert-optischer Systemkomponenten, Dissertation, Eidgenössische Technische Hochschule, Zurich, 1994.

99. T. Hoffmann, F. Gottschalch, and C. M. Sotomayor Torres, unpublished manuscript.

100. I. Martini, J. Koeth, M. Kamp, and A. Forchel, Micro- and Nano-Engineering. To appear.

101. I. Martini, D. Eisert, M. Kamp, L. Worschech, and A. Forchel, *Appl. Phys. Lett.* 77, 2237 (2000).

102. Z. Yu, J. Steven, J. Schablitsky, and S. Y. Chou, *Appl. Phys. Lett.* 74, 2381 (1999).

103. X. Sun, L. Zhuang, W. Zhang, and S. Y. Chou, *J. Vac. Sci. Technol., B* 16, 3922 (1998).

104. B. Cui, W. Wu, L. Kong, X. Sun, and S. Y. Chou, *J. Appl. Phys.* 85, 5534 (1999).

105. A. Lebib, Y. Chen, F. Carcenac, E. Cambril, L. Manin, L. Couraud, and H. Launois, *Microelectron. Eng.* 53, 175 (2000).

106. A. Lebib, S. Li, M. Natali, E. Cambril, L. Manin, and Y. Chen, Micro- and Nano-Engineering. To appear.

107. A. Yokoo, M. Nakao, H. Yoshikawa, H. Masuda, and T. Tamamura, *Jpn. J. Appl. Phys.* 38, 7268 (1999).

108. D. W. van Krevelen, "Properties of Polymers," Elsevier, Amsterdam, 1996.

109. S. Y. Chou and L. Zhuang, *J. Vac. Sci. Technol., B* 17, 3197 (1999).

110. S. Y. Chou, P. Deshpande, L. Zhuang, and L. Chen, Electron, Ion and Photon Beam Technology and Nanofabrication "EIPBN 2000," 2000.

111. P. Deshpande, L. Chen, L. He, and S. Y. Chou, Electron, Ion and Photon Beam Technology and Nanofabrication "EIPBN 2000," 2000.

112. S. Y. Chou, L. Zhuang, and L. Guo, *Appl. Phys. Lett.* 75, 1004 (1999).

113. K. Pfeiffer, M. Fink, G. Gruetzner, G. Bleidiessel, H. Schulz, and H. Scheer, *Microelectron. Eng.* (2001). To appear.

114. B. G. Casey, W. Monaghain, and C. D. W. Wilkinson, *Microelectron. Eng.* 35, 393 (1997).

115. H. Schulz, A. S. Körbes, H.-C. Scheer, and L. J. Balk, *Microelectron. Eng.* 53, 221 (2000).

116. M. Bender, M. Otto, B. Hadam, B. Vratzow, B. Spangenberg, and H. Kurz, *Microelectron. Eng.* 53, 233 (2000).

117. M. Otto, M. Bender, B. Hadam, B. Spangenberg, and H. Kurz, Micro- and Nano-Engineering. To appear.

118. M. Colburn, S. Johnson, M. Stewart, S. Damle, T. Bailey, B. Choi, M. Wedlake, T. Michaelson, S. V. Sreenivasan, J. Ekerdt, and C. G. Willson, *Proc. SPIE* 3676, 379 (1999).

119. R. Mehnert, A. Pincus, I. Janorsky, R. Stowe, and A. Berejka, "UV & EB Curing Technology & Equipment, Vol. 1. Wiley, in association with SITA Technology, New York/London, 1998.

120. B. J. Choi, S. Johnson, S. V. Sreenivasan, M. Colburn, T. Bailey, and C. G. Willson, "Proceedings of DETC2000: 2000 ASME Design Engineering Technical Conference," 2000.

121. M. Colburn, A. Grot, M. Amistoso, B. J. Choi, T. Bailey, J. Ekerdt, S. V. Sreenivasan, J. Hollenhorst, and C. G. Willson, *Proc. SPIE* 3997, 453 (2000).

122. M. Colburn, B. J. Choi, T. Bailey, A. Grot, S. V. Sreenivasan, J. G. Ekerdt, and C. G. Willson, Electron, Ion and Photon Beam Technology and Nanofabrication "EIPBN 2000," 2000.

123. E. Kim, Y. Xia, and G. M. Whitesides, *Adv. Mater.* 8, 245 (1996).

124. E. Kim, Y. Xia, and G. M. Whitesides, *Nature* 376, 581 (1995).

125. Y. Xia, E. Kim, X.-M. Zhao, J. A. Rogers, M. Prentiss, and G. M. Whitesides, *Science* 273, 347 (1996).

126. P. Ruchhoeft, M. Colburn, B. Choi, H. Nounu, S. Johnson, T. Bailey, S. Damle, M. Stewart, J. Ekerdt, S. V. Sreenivasan, J. C. Wolfe, and C. G. Willson, *J. Vac. Sci. Technol., B* 17, 2965 (1999).

127. T. Bailey, J. Choi, M. Colburn, A. Grot, M. Meissl, S. Shaya, J. G. Ekerdt, S. V. Sreenivasan, and C. G. Willson, *J. Vac. Sci. Technol., B* 18, 3572 (2000).

128. Y. Chen, F. Carcenac, C. Ecoffet, D. J. Lougnot, and H. Launois, *Microelectron. Eng.* 46, 69 (1999).

129. J. A. Rogers, K. E. Paul, R. J. Jackman, and G. M. Whitesides, *Appl. Phys. Lett.* 70, 2658 (1997).

130. Y. Xia, X.-M. Zhao, and G. M. Whitesides, *Microelectron. Eng.* 32, 255 (1996).

131. G. Schmidt, M. Tormen, G. Müller, L. W. Molenkamp, Y. Chen, A. Lebib, and H. Launois, *Electron. Lett.* 35, 1731 (1999).

132. Y. Chen, A. Lebib, F. Carcenac, H. Launois, G. Schmidt, M. Tormen, G. Müller, L. W. Molenkamp, M. Liebau, J. Huskens, and S. N. Reinhoudl, *Microelectron. Eng.* 53, 253 (2000).

133. Y. Xia and G. M. Whitesides, *Angew. Chem. Int. Ed.* 37, 550 (1998).

134. Y. Xia, M. Mrksich, E. Kim, and G. M. Whitesides, *J. Am. Chem. Soc.* 117, 9576 (1995).

135. N. J. Jeon, I. S. Choi, and G. M. Whitesides, *Appl. Phys. Lett.* 75, 4201 (1999).

136. Y. Xia, D. Qin, and G. M. Whitesides, *Adv. Mater.* 8, 1015 (1996)

137. E. Delamarche, H. Schmid, B. Michel, and H. Biebuyck, *Adv. Materials* 9, 741 (1997).

138. J. A. Rogers, K. E. Paul, and G. M. Whitesides, *J. Vac. Sci. Technol., B* 16, 88 (1998).

139. H. Schmid and B. Michel, *Macromolecules* 33, 3042 (2000).

140. R. J. Jackmann, J. L. Wilbur, and G. M. Whitesides, *Science* 269, 664 (1995).

141. J. L. Wilbur, E. Kim, Y. Xia, and G. M. Whitesides, *Adv. Mater.* 7, 649 (1995).

142. E. Kim, Y. Xia, and G. M. Whitesides, *J. Am. Chem. Soc.* 118, 5722 (1996).

143. J. Hu, T. Deng, R. G. Beck, R. M. Westervelt, and G. M. Whitesides, *Sens. Actuators* 75, 65 (1999).

144. P. M. Moran and F. F. Lange, *Appl. Phys. Lett.* 74, 1332 (1999).

145. X.-M. Zhao, Y. Xia, and G. M. Whitesides, *Adv. Mater.* 8, 837 (1996).

146. C. Marzolin, S. P. Smith, M. Prentiss, and G. M. Whitesides, *Adv. Mater.* 10, 571 (1998).

147. Y. Xia, J. J. McClelland, R. Gupta, D. Qin, X.-M. Zhao, L. J. Sohn, R. J. Celotta, and G. M. Whitesides, *Adv. Mater.* 9, 147 (1997).

148. E. Kim, Y. Xia, X.-M. Zhao, and G. M. Whitesides, *Adv. Mater.* 9, 651 (1997).

149. D.-Y. Khang and H. H. Lee, *Appl. Phys. Lett.* 76, 870 (2000).

150. J. Aizenberg, J. A. Rogers, K. E. Paul, and G. M. Whitesides, *Appl. Phys. Lett.* 71, 3773 (1997).

151. D. Qin, Y. Xia, A. J. Black, and G. M. Whitesides, *J. Vac. Sci. Technol., B* 16, 98 (1998).

152. H. Schmid, H. Biebuyck, B. Michel, and O. J. F. Martin, *Appl. Phys. Lett.* 72, 2379 (1998).

153. K. E. Paul, T. L. Breen, J. Aizenberg, and G. M. Whitesides, *Appl. Phys. Lett.* 73, 2893 (1998).

154. B. G. Casey, D. R. S. Cumming, I. I. Khandaker, A. S. G. Curtis, and C. D. W. Wilkinson, *Microelectron. Eng.* 46, 125 (1999).

155. A. L. Bogdanov, B. Heidari, and L. Montelius, Micro- and Nano-Engineering. To appear.

156. D. L. White and O. R. Wood II, *J. Vac. Sci. Technol., B,* 18(6), 3552, (2000), p. 277.

157. H. Schulz, H.-C. Scheer, F. Osenberg, and J. Engemann, *Proc. SPIE* 3996, 244 (2000).

158. X.-M. Zhao, S. P. Smith, M. Prentiss, and G. M. Whitesides, "Cleo '97," 1997.

159. K. W. Rhee, L. M. Shirey, P. I. Isaacson, W. J. Dressick, and S. L. Brandow, *J. Vac. Sci. Technol., B* 18, 3569 (2000).

160. A. Bietsch, E. Delamarche, B. Michel, H. Rothuizen, and H. Schmid, "Fourth MEL-ARI/NID Workshop," 1999, p. 27.

161. H. Weiss, *Surf. Coat. Technol.* 71, 201 (1995).

162. K. N. G. Fuller and F. R. S. Tabor, *Proc. R. Soc. London, Ser. A* 345, 327 (1975).

163. K.-H. Haas, S. Amberg-Schwab, and K. Rose, *Thin Solid Films* 351, 198 (1999).

164. D. Y. Kwok and A. W. Neumann, *Adv. Colloid Interface Sci.* 81, 167 (1999).

165. L. Montelius, B. Heidari, and T. G. Ling, Swedish Patent 00 00173-5, 2000.

166. C. P. Tripp, R. P. N. Veregin, and M. L. Hair, *Langmuir* 9, 3518 (1993).

167. C. P. Tripp and M. L. Hair, *Langmuir* 11, 149 (1995).

168. C. P. Tripp and M. L. Hair, *Langmuir* 11, 1215 (1995).

169. P. Silberzan, L. Léger, D. Ausserré, and J. J. Benattar, *Langmuir* 7, 1647 (1991).

170. J. D. Le Grange and J. L. Markham, *Langmuir* 9, 1749 (1993).

171. X. Zhao and R. Kopelman, *J. Phys. Chem.* 100, 11014 (1996).

172. R. Maboudian, W. R. Ashurst, and C. Carraro, *Sens. Actuators* 82, 219 (2000).

173. R. Maboudian and R. T. How, *J. Vac. Sci. Technol., B* 15, 1 (1997).

174. M. Goldmann, J. V. Davidovits, and P. Silberzan, *Thin Solid Films* 327–329, 166 (1998).

175. M. C. B. A. Michielsen, V. B. Marriott, J. J. Ponjée, H. van der Wel, F. J. Touwslager, and J. A. H. M. Moonen, *Microelectron. Eng.* 11, 475 (1990).

176. B. K. Smith, J. J. Sniegowski, G. La Vigne, and C. Brown, *Sens. Actuators, A* 70, 159 (1998).

177. R. W. Jaszewski, H. Schift, B. Schnyder, A. Schneuwly, and P. Gröning, *Appl. Surf. Sci.* 143, 301 (1999).

178. R. W. Jaszewski, H. Schift, P. Gröning, and G. Margaritondo, *Microelectron. Eng.* 35, 381 (1997).

179. P. Gröning, A. Schneuwly, L. Schlapbach, and M. T. Gale, *J. Vac. Sci. Technol., B* 14, 3043 (1996).

180. B. D. Terris, H. J. Mamin, M. E. Best, J. A. Logan, and D. Rugar, *Appl. Phys. Lett.* 69, 4262 (1996).

181. P. R. Krauss and S. Y. Chou, *Appl. Phys. Lett.* 71, 3174 (1997).

182. F. Carcenac, C. Vieu, A. Lebib, Y. Chen, L. Manin-Ferlazzo, and H. Launois, *Microelectron. Eng.* 53, 163 (2000).

183. S. Zankovych, T. Hoffmann, J. Seekamp, J.-U. Bruch, and C. M. Sotomayor Torres. Nanotechnology 12, 91–95 (2001).

184. J. Wang, X. Sun, L. Chen, and S. Y. Chou, *Appl. Phys. Lett.* 75, 2767 (1999).

185. J. A. Rogers, M. Meier, and A. Dodabalapur, *Appl. Phys. Lett.* 73, 1766 (1998).

186. J. Wang, S. Schablitsky, Z. Yu, W. Wu, and S. Y. Chou, *J. Vac. Sci. Technol., B* 17, 2957 (1999).

187. J. Seekamp, S. Zankovich, T. Maka, R. Slaby, and C. M. Sotomayor Torres. unpublished manuscript.

188. C. D. W. Wilkinson, A. S. G. Curtis, and J. Crossan, *J. Vac. Sci. Technol., B* 16, 3132 (1998).

189. R. Companó (Ed.), "Technology Road Map for Nanoelectronics," 2nd ed. European Commission, 2000.

190. J. A. Rogers, M. Meier, A. Dodabalapur, E. J. Laskowski, and M. A. Cappuzzo, *Appl. Phys. Lett.* 74, 3257 (1999).

191. T. H. Dam and P. Pantano, *Rev. Sci. Instrum.* 70, 3982 (1999).

Chapter 2

THE ENERGY GAP OF CLUSTERS, NANOPARTICLES, AND QUANTUM DOTS

Klaus Sattler

Department of Physics and Astronomy, University of Hawaii at Manoa, Honolulu, Hawaii, USA

Contents

Handbook of Thin Film Materials, edited by H.S. Nalwa
Volume 5: Nanomaterials and Magnetic Thin Films
Copyright © 2002 by Academic Press
All rights of reproduction in any form reserved.

ISBN 0-12-512913-0/$35.00

1. INTRODUCTION

The energy gap between valence and conduction band is of fundamental importance for the properties of a solid. Most of a material's behavior, such as intrinsic conductivity, optical transitions, or electronic transitions, depend on it. Any change of the gap may significantly alter the material's physics and chemistry. This occurs when the size of a solid is reduced to the nanometer length scale. Therefore, the science and the technology of nanomaterials needs to take into account a bandgap, which is different from that of the bulk.

The aim of the research in this field is to determine experimentally and theoretically the dependence of the energy gap for particles, which are reduced in size to the nanometer range. In addition, effects such as structural changes, lattice contraction, atomic relaxation, surface reconstruction, surface passivation, or strain induced by a host material, can change the gap. We consider nanoparticles of metals, semiconductors, and carbon, with sizes typically smaller than 10 nm, which is the range where size effects become observable. The quantum size effect (QSE) predicts the formation of a bandgap with decreasing particle size for metals and widening of the intrinsic gap for semiconductors.

For clusters of simple metals, two models are used for their description: the tight-binding (molecular orbit) model and the electron-shell model. They can lead to different bandgaps for small clusters and they can lead to different critical sizes for the cluster-to-bulk transition. For divalent metals, a bandgap is expected to open with decreasing particle size due to narrowing and shift of energy bands. The critical size where this transition occurs has been determined by various groups.

For transition and noble metals, a bandgap opens for small cluster sizes and varies strongly with the cluster size. Both, tight-binding and electron-shell approaches have been used theoretically.

The bandgap for semiconductor quantum dots is usually quite well described by an extended effective mass approximation (EMA). This describes a bandgap, which gradually increases for smaller sized particles. It is illustrated in Figure 1, which shows calculated optical band energies for silicon crystallites with respect to their diameter. Experimental bandgap data from photoemission and photoluminescence studies as well as pseudopotential calculations [1] are given in Figure 2.

For very small clusters of semiconductors, the EMA does not apply. The basic approach to cluster properties is to start from the atom, and calculate the gap with increasing cluster size. For clusters containing just a few atoms, surface passivation would change their intrinsic properties significantly. Therefore, the properties of a pristine cluster have to be studied with the assumption of a bare surface. Most often, the clusters do not have the bulk atomic structure. One finds for instance that not the diamond structure but rather the close-packed structure gives the global minimum of small silicon clusters. A covalent-metallic transition is predicted which leads to a bandgap for silicon clusters much smaller than the 1.1-eV bulk gap.

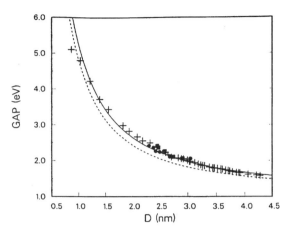

Fig. 1. Calculated optical bandgap energies for silicon crystallites dependent on their diameter [122]. The continuous line is an extrapolation by a $d^{-1.39}$ power law. The dashed curve includes the Coulomb energy between the electron and the hole. The black dots and the squares are experimental results [391]. Reproduced with permission from [122], copyright 1992, American Institute of Physics. Reproduced with permission from [391], copyright 1988, American Physical Society.

Nanoparticles can be produced in many ways and environments. Free clusters and particles are usually generated by cooling of a vapor in contact with an inert gas or during supersonic expansion. Clusters and particles can also be grown on solid supports after atomic vapor has been deposited. A scanning tunneling microscopy (STM) image of a supported silicon cluster is shown in Figure 3. It has a diameter of 2.5 Å and is surrounded by the clean graphite surface, imaged atom by atom.

Morphology and crystalline structure of nanoparticles can be studied by electron microscopy, diffraction methods,

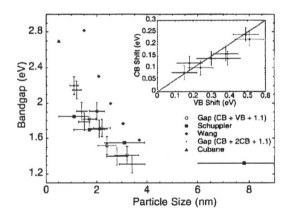

Fig. 2. Bandgap of silicon particles as a function of particle size [1]. Photoelectron data are given by the open and filled circles; the photoluminescence data of Schuppler et al. [586] for oxidized silicon particles is given with the filled squares; the pseudopotential calculation of Wang and Zunger [587] is given with the filled diamonds and the cubane data are given by the filled triangle. The inset shows the conduction band shift versus the valence band shift. Reproduced with permission from [1], copyright 1998, American Physical Society. Reproduced with permission from [586], copyright 1995, American Physical Society. Reproduced with permission from [587], copyright 1996, American Physical Society.

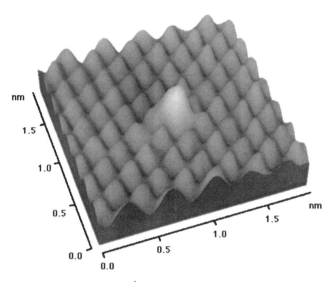

nm

1.5

1.0

0.5

nm

1.5

1.0

0.5

0.0

0.0

Fig. 3. STM image of a 2.5-Å silicon cluster supported on the substrate of highly oriented pyrolytic graphite (HOPG) [48]. Reproduced with permission from [48], copyright 2000, American Physical Society.

surface reactivity, and Raman spectroscopy. The question is if the particles are fragments of the crystalline bulk or if they have their own atomic structure. Silicon and germanium nanoparticles formed as deposits or composite materials typically show nearly spherical geometries in high-resolution transmission electron microscopy (HRTEM) for particles having diameters as large as 200 nm [2–4]. Frequently, micrographs reveal lattice fringes corresponding to the crystalline diamond phases, such as the {111} interplanar spacings of 0.31 nm for Si and 0.33 nm for Ge. The atomic spacing can critically influence the electronic structure of the particles. A change to reduced lattice constant was observed for Ge particles of diameter less than 4 nm [3]. A nanoparticle with a diamond crystal structure is shown in Figure 4.

For silicon particles, a structural transformation is predicted for smaller sizes. In molecular dynamic studies, Si_n cluster growth follows nondiamond structure up to $n = 2000$ (ca. 4 nm

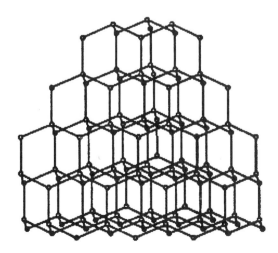

Fig. 4. Model of a 120-atom silicon cluster with a diamond-type crystal structure.

diameter) at which point crossover to a diamond growth pattern occurs [5]. Large crystalline Si and Ge particles, with $d > 50$ nm have their Raman peaks unchanged when the size is varied [2]. Therefore, the critical size for structural transformation of silicon particles is probably around 4 nm.

Small carbon clusters and nanoparticles can have amorphous, graphitic, diamond, or cage structures. In addition, mono- and polycyclic rings, chains, and other low-density structures are possible. We will report bandgap measurements on vapor-grown (amorphous) carbon particles as a function of size.

There is no strict distinction in literature between the terms "clusters, nanoparticles, and quantum dots." However, often "clusters" are used for agglomerates of very few atoms, "nanoparticles" are used for larger agglomerates (usually of metals or carbon), and "quantum dots" are used for semiconductor particles and islands where quantum confinement of charge carriers or excitons determines their properties.

In the following sections, we give the status of research of the bandgap and we consider possible applications of the particles for optoelectronic and other uses.

2. EXPERIMENTAL TECHNIQUES

2.1. Optical Spectroscopy

A method for probing the band structure of semiconductors is to measure the optical absorption or luminescence spectrum. In the absorption process, a photon of known energy excites an electron from a lower to a higher energy state. For determination of the fundamental gap, band-to-band transitions are probed. However, because the transitions are subject to selection rules, the determination of the energy gap from the "absorption edge" of the luminescence peak is not a straightforward process.

Because the momentum of a photon is very small compared to the crystal momentum, the optical process should conserve the momentum of the electron. In a direct-gap semiconductor, momentum-conserving transition connects states having the same k-values. In an indirect bandgap semiconductor momentum is conserved via a phonon interaction. Although a broad spectrum of phonons is available, only those with the required momentum change are usable. These are usually the longitudinal- and the transverse-acoustic phonons. In addition, the indirect process is a two-step event. Therefore, indirect optical processes have very low transition probabilities.

Clusters and nanoparticles of a semiconductor may be optically active even though the bulk material is not. This can have various reasons: The energy band structure and the phonon distribution can be entirely different for clusters compared to the crystalline bulk. Clusters have localized states that are not present in the bulk. The transition from an upper to a lower energy state may proceed via one or more localized intermediate states. The intermediate steps may or may not be radiative. Yet, they have a profound effect on the actual efficiency of the radiative transition.

Nanoparticles have many of their atoms at surface positions. A surface is a strong perturbation to any lattice, creating many dangling bonds. These unsaturated bonds are energetically unfavorable. The particles can lower their free energy by side- and back-bonding of these bonds. For silicon and germanium particles, this can be at the cost of giving up the sp^3-hybridization. Therefore, the sp^3-characteristics of the diamond-like bulk may not be present for these small particles.

A high concentration of deep and shallow levels can occur at the surface of nanoparticles, and these may act as electron-hole recombination centers. The distribution of surface states is discontinuous for small clusters, but can be continuous for nanoparticles, where the surface consists of several facets. When electrons or holes are within a diffusion length of the surface, they will recombine, with transitions through a continuum of states being nonradiative.

Optical spectroscopy studies of nanoparticles demonstrate their atomlike discrete level structure by showing very narrow transition line widths [6–11]. Optical techniques probe the allowed transitions between valence band and conduction band states for nanoparticles which do not have defect or impurity states in the energy gap. Interpretation of optical spectra often is not straightforward and needs correlation with theoretical models [12–14].

Photoluminescence peak energies were found to increase with decreasing particle size, for instance from ∼1.3 to 1.6 eV for Si nanocrystals that decrease in size from 5 to 2 nm [15]. Nanocrystalline silicon films showed optical bandgaps of 1.9 to 2.4 eV [16]. The photoluminescence results from various studies are plotted in Figure 5 [17]. Even blue luminescence was observed, both for Si and Ge nanocrystals [18].

The bandgap of semiconductor clusters depends strongly on their atomic structure. Fullerenes, for instance, have bandgaps strongly varying with size and pentagon–hexagon arrangements. Amorphous carbon (a-C) has a bandgap of typically between 0.4 and 0.7 eV and can behave like a semiconductor. In some samples of a-C, energy gaps up to 3 eV were observed [19]. There is a significant bandgap variation for differently prepared a-C samples. This is related to varying microscopic mass densities and described theoretically by microcrystalline and cluster models [20]. Tetrahedral amorphous carbon (ta-C) contains more than 80% sp^3-bonded carbon [21]. Clusters and rings with sp^2-coordination in the otherwise amorphous material have been suggested to model this material. A gap of ∼2 eV is found for ta-C [22].

Thin films have been prepared that consist of 5- to 50-nm-sized Si crystallites [23]. Silicon crystallites are embedded in films of SiO_2 glasses [24, 25]. Semiconductor-doped glasses (SDG) have large nonlinear susceptibilities for optical transitions near their optical bandgap and a fast electron-hole recombination time of psec order. Bandgaps of Si particles in SiO_2 in the range 1.2 to 1.5 eV have been observed and depend on the film preparation [24].

In early studies of photoluminescence (PL), surface defects were often considered to control the optical properties of quantum dots [26–32]. More recent work [7, 33–38] on CdSe QDs

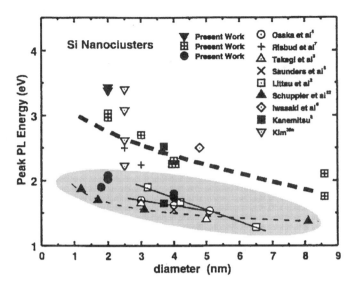

Fig. 5. Summary of data on peak luminescence versus silicon particle size, after Wilcoxon, Samara, and Provencio [17] with data from Osaka et al. [588], Risbud, Liu, and Shackelford [589], Takagi et al. [267], Saunders et al. [590], Littau et al. [269], Schuppler et al. [402], Iwasaki, Ida, and Kimura [591], Kanemitsu [592], and Kim [367].

however suggests that the near-edge emission (which leads to the determination of the bandgap) is determined by their core, and not by the surface structure.

2.2. Scanning Tunneling Spectroscopy

Historically, tunneling spectroscopy with metal-insulator-metal (MIM) tunnel junctions was first demonstrated by Giaever [39]. His MIM tunneling experiment provided a direct measurement of the energy gap of superconductors, which was a critical evidence for the Bardeen-Cooper-Schrieffer (BCS) theory of superconductivity.

A systematic method of obtaining local tunneling spectra with STM was developed by Feenstra, Thompson, and Fein [40]. It is an extension of the tunneling junction experiment of Giaever. Scanning tunneling spectroscopy (STS) is complementary to topographic imaging by STM. It can be accomplished in a number of ways. Most STS studies today are performed with constant tip-sample separation at a fixed location on the surface. This is accomplished by momentarily interrupting the feedback controller and then ramping the applied voltage over the desired interval while simultaneously measuring the tunneling current. If no spatial resolution is required and a large flat area is investigated, the method is straightforward [41]. However, if one wants to correlate the tunneling spectra with the local surface structure, the I–V measurement must be performed together with the topography measurement. While the tip is scanning over the sample surface, a map of tunneling spectra is generated.

Scanning tunneling spectroscopy has been applied extensively to determine the electronic structure of semiconductors [42–44], superconductors [45], and metals [46]. STS is used to study solid surfaces and thin films [47] as well as adsorbed

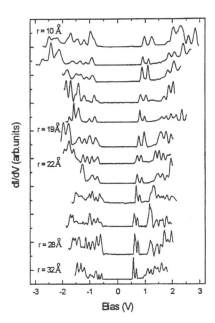

Fig. 6. Size evolution of dI/dV versus V characteristics of InAs quantum dots, measured by scanning tunneling spectroscopy [49]. Reproduced with permission from [49], copyright 2000, American Physical Society.

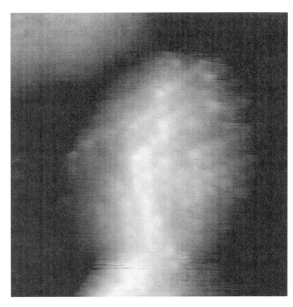

Fig. 7. An 11×11-nm STM image of a palladium particle, with atomic resolution [475]. Reproduced with permission from [475].

clusters [48, 49], fullerenes [50, 51], and molecules [52–54]. STS studies were reported for semiconductor quantum dots such as Si [48], CdS [55], CdSe [56, 57], InAs [49, 58] and others [59–61]. In addition, tunneling spectra of metal quantum dots have been obtained [62, 63]. A sequence of STS spectra is shown in Figure 6 for quantum dots of InAs [49].

STS theory has its focus mostly on the relationship between I–V curves and the local density of states (LDOS) [64–71]. Most calculations used the assumption of imaging at low bias voltage and were following the theory of Tersoff and Hamann [64]. In some theoretical treatments, this has been extended to finite bias voltages [72–74]. An STS theory has been developed specifically for semiconductors [75], since they need a treatment different from the one for metals [76, 77].

Scanning tunneling microscopy can be used to obtain the atomic structure of particles. This is illustrated in Figure 7, which shows an STM image of a palladium particle with atomic resolution. The combined use of STM and atomic force microscope (AFM) [78, 79] can also give information about the interplay between atomic and electronic structure. The STM probes the density of states (DOS) of a sample at the Fermi level E_F (if a small bias voltage of a few millivolts is applied). The AFM probes the total DOS of the valence electrons, over a relatively wide energy range. This can lead to differences in the images obtained from the two methods. For instance, when assembled nanoparticles are imaged, and there are low-energy grain boundaries between the particles, these boundaries usually have different DOS at E_F, well seen in STM imaging. Therefore, the particles are individually imaged by STM. The AFM however is not sensitive to small DOS variations at E_F and therefore does not well resolve particles with low-energy grain boundaries. This has been seen in experimental studies of nanophase materials [79].

As expected from the Bardeen formula (derived for MIM tunnel junctions) [80], the tip density of states (DOS) plays an equal role as the sample DOS in determining the tunneling spectra. In an STS experiment, the goal is to obtain the DOS of the sample, and one requires a tip with a constant, structureless DOS. When the tip is a metal, there is a sharp onset of the tunneling current when the bias voltage exceeds half the gap energy and this is independent of the tip DOS. This argument has been used by Giaever and many others for the determination of bandgaps in tunnel junctions and it applies to gap studies of nanoparticles by STS as well.

For nanoparticles, a gap may appear in the I–V curve which is not related to the electronic structure. If the particle charges up, a Coulomb blockade gap may be the dominant I–V feature around zero bias [81–83]. This gap is size dependent, increasing with smaller particle diameter [84]. Single-electron charging has also been studied for molecules in STS experiments [85].

The effect of the Coulomb blockade can be seen from the appearance of supported clusters in STM images. On a gray scale image, a supported cluster usually appears brighter than the substrate. If the substrate in an STM image is set to a gray level, a cluster appears white. If however the cluster is charged, the tunneling current is blocked and the cluster appears black.

STS has been used to study the interplay between single-electron tunneling (SET) and quantum-size effects. This can experimentally be observed when the charging energy of the particle by a single electron is comparable to the electronic level separation, and both are larger than $k_B T$ [50].

In optical spectroscopy, the bandgap is determined from allowed transitions between valence and conduction band states in the particle. In tunneling experiments, on the other hand, the discrete levels in a particle's valence and conduction band are probed separately. Therefore, STS provides complimentary information on the electronic structure of the particles.

3. THEORY

3.1. Methods and Results for Metal Clusters

The electronic structure of metal clusters is calculated within two theoretical ab initio approaches: density functional calculations in the local-density approximation (LDA) [86] and Hartree-Fock configuration interaction (HFCI) calculations [87]. The LDA gives accurate ground state properties for neutral and charged clusters, but does not well describe excited states. Simulated annealing [86] can be used together with LDA to find the atomic structure of a cluster in the ground state. The calculations can yield the vertical detachment energies (VDE) of metal clusters which are related to the electron affinity [88]. The HFCI method yields electronic ground and excited states of neutral and charged metal clusters. The optimum geometry of a cluster can only be determined for very small sizes because the calculations for larger clusters require too much computer time.

For metal clusters with highly delocalized valence electrons, the electronic-shell model can be applied [89]. It assumes that the valence electrons of each atom in the cluster become free electrons confined by the boundaries of the cluster. For spherical clusters, the eigenstates are electronic shells with defined angular momentum. Shell closings occur at $n = 2, 8, 18, 20, 34, 40, \ldots$ electrons which lead to magic numbers for alkali metal clusters with these n-values. With increasing cluster size, the 1s, 1p, 1d, 2s, 1f,...shells become successively occupied.

The shell model neglects the core potentials of the atoms in the cluster. It uses the jellium model, which considers the positive charges of the atoms being smeared out to a homogeneous background. Therefore, the model is not used to determine properties which are related to the atomic structure of the clusters.

While metal clusters with closed electron shells are spherical, the clusters with open electronic shells are deformed from the spherical shape. In theory, this deformation has been studied within the jellium approach [90–95], with ab initio calculations [96–98], and tight-binding models [99, 100]. The clusters are deformed due to the Jahn–Teller effect [101, 102]. The deformations have effects on the electron-shell energies [101].

Ab initio molecular dynamics calculations were used to study the deformations of Na clusters at high temperatures (500–1100 K) [103]. A large highest occupied molecular orbital–lowest unoccupied molecular orbital (HOMO–LUMO) gap of 1.1 eV was found for the closed-shell cluster Na_8. This gap is unchanged at 550 K, where the cluster is liquid-like. In the open-shell Na_{14} the deformation opens a gap which remains in the liquid state. The liquid Na_{14} cluster favors axially deformed shapes, with two isomers, prolate and oblate.

The deformation is driven by the opening of the HOMO–LUMO gap. Among three isomers for Na_{14}, prolate, oblate, and spherical, the prolate isomer has the lowest energy and the highest HOMO–LUMO gap (0.4 eV). When the Na_{14} cluster is heated to 1100 K, the cluster structure oscillates between different isomers and the time-averaged level density no longer shows a pronounced HOMO–LUMO gap.

The shell model is not restricted to alkali metal clusters. Electron detachment energies of Pb_n^- ($n = 24$–204) [104] for instance are well described by local-density approximation in the spherical jellium model.

The simplest atomic structure is the linear chain. For linear arrangement of metal atoms, one usually obtains HOMO–LUMO gaps which oscillate between odd and even numbers of atoms per cluster. This is illustrated in Figure 8, which gives the orbital energy distributions of linear Cu_n clusters [105] (up to $n = 10$).

Transition metal clusters are particularly interesting for studies of the formation of the bulk band structure with increasing particle size. Transition metal atoms include both localized 3d-electrons and delocalized 4s-electrons. It can be expected that these states have a very different cluster size dependence. Cu, for example, has the atomic structure $3d^{10}4s$. Like the alkali atoms, Cu has a single s-electron outside a closed shell. The 3d-states however lie just below the 4s-states in energy and are expected to influence the cluster properties. In bulk Cu, the 3d-states contribute significantly to the density of states at the Fermi level, which leads to the very high electrical conductivity of Cu. In Cu clusters, the 4s-states are delocalized and therefore show strong discontinuous size-dependent variations, whereas the 3d-band evolves monotonically with cluster size. The calculated density of states for Cu_8 and Cu_{20} clusters is shown in Figure 9.

In photoemission experiments, the electron affinities for Cu clusters have been studied as a function of cluster size [106–108]. A multitude of distinct features was observed in the spectra [108] and attempts were made to explain these features theoretically [109–111].

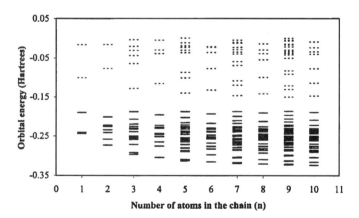

Fig. 8. Occupied (continuous line) and unoccupied (dashed line) energy levels for linear Cu atomic chain [105]. Energies in the neighborhood of the HOMO and LUMO are shown. Reproduced with permission from [105], copyright 1999, American Chemical Society.

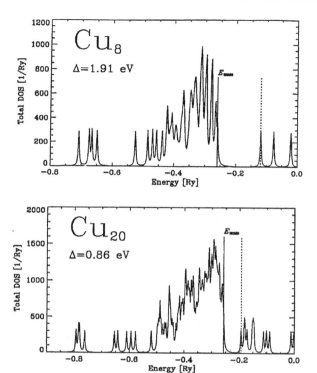

Fig. 9. Total electronic density of states calculated for the two closed-shell clusters Cu$_8$ and Cu$_{20}$ in their lowest energy structures [102]. Reproduced with permission from [102], copyright 1994, American Physical Society.

Clusters from noble metals exhibit similarities to simple metal clusters with respect to their electronic structures. This is well illustrated with tight-binding calculations for Cu clusters ($n = 2$–95) [102] (Figure 10). The clusters have large HOMO–LUMO gaps with the magic (electron) numbers $n = 2, 8, 18, 20, 34$, which are the numbers known from the spherical shell model. Shell closures result in particularly large HOMO–LUMO gaps for these clusters. The theoretical and experimental values for Cu$_8$ coincide very well, 1.91 [102] and 1.93 eV [112], respectively. A lower value of $E_g = 1.33$ eV for Cu$_8$ was also calculated [113]. For the closed-shell Cu$_{20}$ cluster, $E_g = 0.83$ eV is obtained. The typical band features of bulk Cu are not yet evolved for Cu$_{20}$.

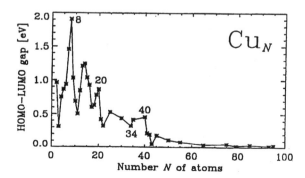

Fig. 10. HOMO–LUMO gap calculated for Cu clusters versus size [102]. Reproduced with permission from [102], copyright 1994, American Physical Society.

The same general characteristics for the electron level structure of Cu clusters has been obtained from several studies [102, 114, 115]. In contrast to the bulk, Cu clusters exhibit an almost complete s–d-band separation. Low lying occupied valence states have s-character, whereas the higher occupied levels have predominantly d-character. The states directly at the HOMO and the unoccupied states have mostly s–p-character with minor d-contributions. This shows that the atomic d-electrons form a narrow d-band in the clusters with the extended s–p-states being located in separate energy regions, below and above the d-band. Hybridization between these two band characters is found to be small. The molecular orbitals of the extended states are associated with the states of the electron-shell model. An s- and d-band separation similar to that for Cu clusters has also been obtained in calculations of surface slab model clusters Ni$_{20}$ and Ni$_{51}$, where the HOMO was found to have s-character [116]. We note that bulk Cu and Ni is characterized by completely overlapping s- and d-bands.

3.2. Methods and Results for Semiconductor Clusters

Early theoretical studies of semiconductor particles used the effective mass approximation (EMA) [117–119]. This approach can sometimes be successful in fitting luminescence data [120]. The EMA assumes that the semiconductor bands are parabolic and that the kinetic energies of electrons and holes can be expressed in terms of their effective masses, m_e and m_h. Direct bandgap semiconductors have parabolic bands at the top of the valence band and the bottom of the conduction band. Therefore, the EMA gives a good estimate of the bandgap in large clusters of direct gap materials. The approximation is not accurate for states away from the Γ point, or for indirect bandgap semiconductors.

Early calculations using EMA assumed an infinite potential step at the cluster surface, and applied perturbation theory to solve the Schroedinger equation with the effective mass (EM) Hamiltonian. The calculated exciton energies agree qualitatively with the experiment for larger clusters, but are usually too large for small clusters [118, 119]. Calculations with a more realistic finite potential step at the surface have given results in much better agreement with the experiment [121].

First-principles and semiempirical theories show that the EMA-derived bandgap of semiconductor particles is overestimated. The gap-to-diameter dependence is still described by an inverse power law, but with a smaller exponent [122]. Another applied method is an empirical pseudopotential theory, which has been used for nanoparticles by imposing a boundary condition on the pseudopotential solution for an infinite lattice [123–125]. There is better agreement with experiments compared to EMA, but the method does not include surface effects, as an infinite potential step is required at the surface. Yet, the surface of nanoparticles may influence or may even dominate the particle properties. Many experiments, as, for example, time-resolved fluorescence measurements [29], or electrooptical Stark effect measurements [126] show the surface trapping of excited charge carriers. Such results cannot be explained by models, which account for size quantization alone but ignore the surface. Green's function recursion within the tight-binding approximation was also used to estimate bandgaps and exciton energies of nanoparticles [127]. The method models the atomic structure of clusters realistically and can be used to study surface effects.

Theoretically, small silicon clusters were first proposed to be fragments of the crystalline bulk [128, 129]. The clusters were assumed to have a diamond structure with non-relaxed surface atoms. Such tetrahedral-bond-network (TBN) clusters have many dangling orbitals and very low average coordination numbers. In further studies, crystalline structures were shown either to correspond to high-energy local minima or to be highly unstable [130–134]. Ab initio electronic structure calculations have been used to predict lowest energy structures for Si clusters in the size range 2–14 atoms [130, 131, 135–144]. Raghavachari [130, 139], Raghavachari and Logovinsky [140], and Raghavachari and Rohlfing [141] have used (uncorrelated) Hartree–Fock (HF) wave functions to optimize the cluster structures. Tomanek and Schlüter [142, 143] have used a local-density functional (LDF) method, Pacchioni and Koutecky [131] and Balasubramanian [135, 136] performed configuration interaction (CI) calculations, and Ballone et al. [137] used simulated annealing techniques for geometry optimization. All of these calculations are largely in agreement as to the equilibrium structures for very small silicon clusters.

4. METALS

4.1. Alkali Metal Clusters

Clusters of metal atoms encompass the evolution of collective properties from an atom to a solid. In particular, the development of metallicity with increasing size is a fascinating field to study [145]. The free atom has sharp and well-separated electronic energy levels, which undergo major changes once the atoms are assembled to clusters. With increasing particle size, the HOMO–LUMO gap becomes gradually narrower and finally vanishes. A nonmetal–metal transition occurs as a function of size. Experimentally, the closure of the gap is observed in the shrinking difference between ionization potential and electron affinity. The Kubo effect (the quantum size effect, QSE) describes the discrete levels in small systems. Yet, the QSE is a continuum approach, which breaks down at very small particle sizes.

A number of experimental [146] and theoretical [147–151] studies of alkali clusters have been performed. The alkali metals have simple electronic structures as they have energy bands very similar to those of free electrons. For alkali metal clusters, two theoretical models are used for describing their electronic properties. The jellium model has been successful for the calculation of mass spectra and optical spectra [152, 153]. The valence electrons are treated as completely free, only confined to a spherical box with a constant positive background. The model leads to a shell structure for the electrons

(electron-shell model), For other properties, quantum chemical calculations [147–149] were successfully used (molecular orbit model). They are expensive in theoretical input and computing time, but lead to the best results for very small clusters with less than about 50 atoms. The two models yield different bandgap behavior as a function of particle size. An overlap of the two theories has been achieved by including pseudopotentials to the jellium calculation [150].

4.2. Noble and Transition Metal Clusters

For metal particles, the quantum-size effect (QSE) predicts the opening of a gap with decreasing size. Experiments usually show HOMO–LUMO gaps much smaller than predicted by the QSE. In addition, the nonmetal to metal transition seems to occur at quite small cluster sizes. For example, while Pt clusters with up to six atoms revealed nonzero energy gaps [154], 60-atom Pt clusters showed zero-energy gaps [155]. For Ni clusters with $n < 20$, ab initio molecular orbit calculations yield HOMO–LUMO gaps up to 0.25 eV, approaching zero at about $n = 16$ [156]. Studies of photodetachment spectroscopy [157] suggest that the electronic structure of Ni clusters approaches the bulk limit for $n > 14$, while measurements of vertical ionization potentials [158] put this limit beyond $n = 100$. Structural, electronic, and magnetic properties are strongly correlated for transition metal clusters such as Ni_n [159].

In a study of vanadium clusters V_n ($n < 65$), three size regions of spectral evolution were observed: molecular-like behavior for $n = 3–12$, transition from molecular to bulklike for $n = 13–17$, and gradual convergence to bulk for $n > 17$ with bulk behavior at about $n = 60$ [157]. For Mn clusters, the strong dependence of the gap on the cluster structure was demonstrated. For Mn_5, there are several isomers whose energies lie very close to the energy of the ground state but with the HOMO–LUMO gap varying between zero and 2.63 eV [160]. For Cu_{13}, a gap of 0.36 eV was calculated [161]. This value is smaller than the experimental results of about 0.7 eV [109]. In the range $4 \leq n \geq 30$, the gaps for Cu clusters vary between 0.25 and 1.4 eV but do not approach zero at $n = 30$ [109] (Fig. 11). The density of states for Cu clusters has almost fully converged to the bulk DOS for $n \approx 500$ [102]. Theoretical studies of Al clusters with $n \leq 15$ show strong discontinuous size-dependent gap variations (between 0.2 and 1.2 eV) [162] (Fig. 12). They roughly agree with earlier theoretical studies where HOMO–LUMO gaps in the range 0.4 to 2 eV were found for Al clusters up to $n = 13$ [163]. The gap was calculated to be close to zero for Al clusters with $n \geq 54$ [163] (Fig. 13). For small Ag_n ($n = 2–6$), the HOMO–LUMO gaps were experimentally determined: 6.5 eV (Ag_2), 3.8 eV (Ag_3), 5.0 eV (Ag_4), 4.2 eV (Ag_5), 5.1 eV (Ag_6) [164]. An odd–even effect in these values is visible.

Aluminum clusters show a strong magic number tendency for $n = 13$ [165, 166]. Therefore, the gap for Al_{13} has typically a much greater value compared to the other clusters in this size range [163]. A gap of 1.89 eV was obtained by photoemission for anions of Al_{13}, dramatically exceeding the gaps of

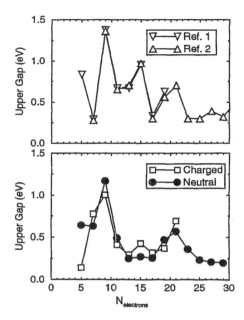

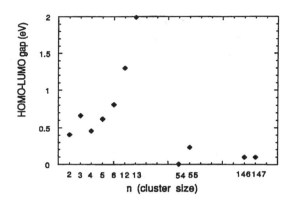

Fig. 13. HOMO–LUMO gap calculated for neutral Al clusters [163]. Reproduced with permission from [163], copyright 1993, American Physical Society.

Fig. 11. Gap between the two upper electronic levels for odd-electron clusters. Upper panel: experimental results on Cu_n^- [593, 594]. Lower panel: calculations for neutral and negatively charged Cu_n^- clusters. The data are plotted as a function of the odd-electron number [109].

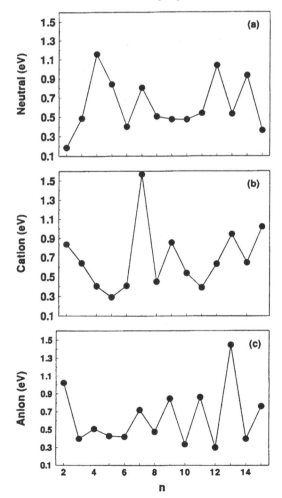

Fig. 12. HOMO–LUMO gap calculated for (a) neutral, (b) positively charged, and negatively charged Al clusters [162]. Reproduced with permission from [162], copyright 1999, American Institute of Physics.

0.25 and 0.32 eV of Al_{12} and Al_{14}, respectively [167]. Angle-resolved photoemission studies indicate that the nonmetal to metal transition for Pd particles is at about 25 Å [168]. Photoelectron spectroscopy studies of Ti_n ($n = 3 - 65$) yield electron affinities (EA) which increase monotonically toward the bulk work function starting from Ti_8 [169]. The EAs of the larger clusters are well described by the spherical drop model. This suggests that titanium clusters become metallic at $n \approx 8$. A number of first-principles studies were performed for small Ni clusters but with large discrepancies in the calculated ionization potentials [156]. Such discrepancies arise from the particular choice of geometries, interatomic spacing, atomic basis functions, approximations in the exchange-correlation potentials, and the treatment of the core electrons.

Scanning tunneling spectroscopy has been applied to determine the HOMO–LUMO gap of metal clusters. The tunneling conductance through the clusters decreases markedly for a range of bias voltages around zero and for small enough clusters. Pd, Ag, Cd, and Au clusters with diameters greater than 1 nm show the zero-gap characteristics of the bulk metals in STS experiments [170]. Below 1 nm, an energy gap opens, and its value increases with smaller cluster sizes. Energy gaps up to 70 meV were measured. These are much smaller than the gap values obtained from calculations.

Metal clusters usually have an atomic structure which is different from that of the bulk. Studies by scanning transmission electron diffraction (STED) showed that Pd clusters with diameters less than 20 Å prefer the non-fcc (face-centered cubic) icosahedral over the standard fcc bulk structure [171]. On the other hand, studies of Cu clusters with extended X-ray absorption fine structure (EXAFS) indicated fcc structure even for clusters with about 13 atoms [172]. For Au, Pt, Rh, Ni, and Ag clusters, the fcc bulk structure seems to be favored over the icosahedral structure for a diameter bigger than about 20 Å [173, 174].

The reactivity of small clusters with simple molecules can vary significantly with cluster size [175–184]. Features in the electronic structure of the bare clusters can be obtained from measurements of vertical detachment energies (approximately giving the electron affinities) and the ionization potentials [177, 185–188]. The reactivities of neutral Nb and Fe clusters with H_2 were found to correlate with their ionization potential [185, 189]. This shows that the reactivity of a cluster is linked to its electronic structure. Various models have been proposed for the size-dependent reactivity. The most obvious explanation is the promotion of an electron from the cluster to the antibonding state of the attached molecule [190]. Other models have emphasized the details of the electronic structure of the clusters [191]. In studies of photoelectron spectra of Fe, Co, and Ni clusters, it was found that the size dependence of reactivity with H_2 molecules correlates well with the HOMO–LUMO gap variation [186].

The reactivity of metal clusters with inert gases such as argon and krypton can be very similar to those with hydrogen [192]. In addition, the measured reactivities can be almost independent of the charge state of the clusters (neutral, positive, negative) [176, 178, 182, 189]. This suggests that the electron transfer model, based on the reactivity correlation with the electronic structure needs to be replaced by an alternative model in some cases. The geometry of the clusters may instead determine the adsorption characteristics. It has been considered in some cases to be responsible for the reactivity of metal clusters [193, 194].

Magic number clusters are highly stable due to electronic or geometric shell closings. They have high binding energies per atom, large ionization potentials, and wide HOMO–LUMO gaps. This is well seen with Nb clusters, which show shell closings for Nb_8, Nb_{10}, and Nb_{16}. These magic clusters are relatively unreactive for hydrogen absorption [195]. The reactivity pattern for Nb clusters are roughly independent on the charge state of the clusters, with Nb_n^+, Nb_n^-, and Nb_n^0 giving similar results. This indicates that the geometrical structure of the clusters might be important for their chemical activity. The dependence of HOMO–LUMO gaps for Nb clusters is shown in Figure 14.

For a comparison of theory with experiments, one needs to take into account that neutral, positively and negatively charged clusters may have different atomic and electronic structures. This has been illustrated with calculations for Al_n [162], Cu_n [196], and others. For example, while the bandgap for the anion Al_{13} of 1.5 eV is much higher than the gaps of 0.3 and 0.4 eV gaps for Al_{12} and Al_{14}, respectively, $n = 13$ is

much less magic for the neutral and positively charged clusters. Another magic Al cluster is Al_7, but only if positively charged. Gaps of 1.6, 0.40, and 0.45 eV were calculated for Al_7^+, Al_6^+ and Al_8^+, respectively.

Structural isomers may also differ considerably in their electronic properties. This has been shown by a number of theoretical studies, for instance for Co_n ($n = 1$–8) [197], Cu_n ($n = 2$–5) [196], Cu_n ($n \leq 9$) [111], Al_5 [198], and others. Various isomers of Na_{19}^+ lead to very different photoabsorption spectra [199].

Bond-length contraction is another effect, which has an influence on cluster properties. Metal clusters usually have reduced interatomic distances compared to the bulk metals. The bond-length contraction $\Delta R / R$ becomes very pronounced for clusters with just a few atoms. This is illustrated in calculations of rhodium clusters with $\Delta R/R = 17\%$ (Rh_2), 10.9% (Rh_3), 7.4% (Rh_4), 5.8% (Rh_6) and 4.0% (Rh_{12}) [200]. Average interatomic distances and coordination numbers for Ni_n ($n = 3$–23) have been calculated [201] and are shown in Figure 15. For Pd clusters, several theoretical studies predict a

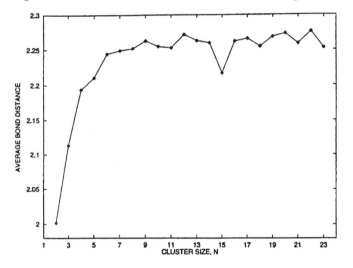

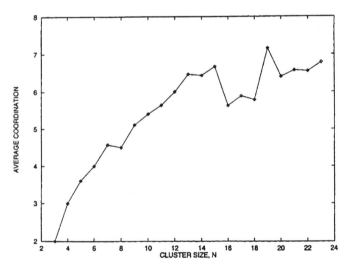

Fig. 15. Average interatomic distances and coordination numbers calculated for Ni clusters [201]. Reproduced with permission from [201], copyright 1997, American Chemical Society.

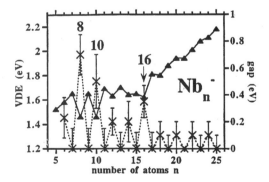

Fig. 14. Vertical detachment energies (triangles, left scale) and HOMO–LUMO gaps (crosses, right scale), extracted from photoelectron spectra of Nb_n^- clusters [195]. Reproduced with permission from [195], copyright 1998, American Institute of Physics.

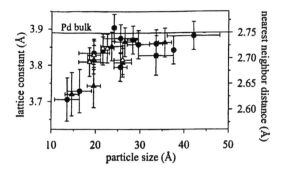

Fig. 16. Atomic spacings in supported Pd clusters as a function of cluster size evaluated from electron diffraction patterns [206]. Reproduced with permission from [206], copyright 1999, American Chemical Society.

decrease of the lattice constant with a decreasing cluster size [202–205]. An experimental study on Pd particles in the size range 15–45 Å showed a lattice constant reduction of about 5% [206] (Fig. 16). A reduction in lattice constant has also been observed for Ta and Pt clusters [207, 208]. In general, a cluster may be viewed as bulk matter under pressure.

Another consideration is the interaction with the substrate for supported clusters. It has been studied for several metal clusters and substrates: Cu_4, Ag_4, Ni_4, and Pd_4 on MgO [209], Cu_n ($n \leq 13$) on MgO [210], Cr_n on Ru(001) [211], Cu_n and Au_n on Cu(001) [212], Al_n on highly oriented pyrolytic graphite (HOPG) [213] and Pd_n on HOPG [214]. Some of these substrates may strongly influence a cluster's atomic and electronic structure [211, 212]. Size-selected clusters can completely change their free atomic structure, after softly landing on the support. This has been illustrated for Cu and Au clusters (with $n = 13$ and 55), which become flat after deposition on a $(1 \times 1)O/Ru(001)$ substrate and stay in a metastable configuration when the substrate is kept at low temperature [212]. The clusters then show a high degree of epitaxy on the substrate lattice.

Platinum clusters adsorbed on HOPG affect the area around the adsorption site by inducing periodic charge density modulations (PCDM) [215]. This occurs due to relatively strong particle-substrate interaction. The particle is surrounded by superstructures, which decay exponentially away from the adsorption site. These structures result from a local distortion of the periodic charge density of the substrate.

Nanoparticles of metals may emit light when stimulated by electron injection. For semiconductor particles, photon emission is mostly due to interband transitions and can be used to determine bandgaps. For metal particles, however the luminescence is originated by the collective excitation of the electron gas leading to a Mie plasmon resonance [216]. The peak position of the plasmon resonance for Ag particles shows a pronounced blueshift with decreasing particle size with a $1/d$ behavior (d-cluster diameter). This is caused by intrinsic size effects of the particles.

Metal clusters in beams can be used for scientific studies but their amounts are usually to small for industrial applications. In addition, supported nanoparticles [214, 217–219]

are limited to scientific studies. Other techniques, such as solution-phase synthesis [220] have been applied for metal particles with high (macroscopic) quantities. Additionally, metal nanoparticles (Pt, Pd, Ag, Au) were synthesized in polymer matrices [221]. The large scale production of nanopowders allows their application in nanotechnology.

4.3. Divalent Metal Clusters

A transition from van der Waals to metallic bonding is expected for divalent metal clusters (e.g., Be_n, Mg_n, Hg_n) as a function of size. The atoms of these elements have a closed-shell s^2-atomic configuration and a fairly large sp-promotion energy E_{sp} (e.g., $E_{sp}(Hg) \sim 6$ eV, $E_{sp}(Be) \sim 4$ eV, where $E_{sp} = E(s_1 p_1) - E(s_2)$. Therefore, one expects the smaller clusters to be insulating and bonded through weak, mainly van der Waals-like forces. In contrast, the corresponding bulk materials are strongly bound (e.g., $E_{coh}(Hg) = 0.67$ eV, $E_{coh}(Be) = 3.32$ eV) and have metallic properties which result from the overlap between the s- and p-bands. Consequently, a strong qualitative change in the nature of the chemical bonding (from van der Waals to metallic) should take place with increasing size. For Hg_n a transition from van der Waals to metallic bonding has been estimated to occur in the range $20 < n < 50$ [222].

The first experiments which confirmed qualitatively the expectations were ionization energy studies, performed on beams of Hg clusters [223]. Metallic behavior was reached at $n = 80$–100. Similar results have been obtained from autoionizaton resonance energy measurements [224]. Studies of photoelectron spectra on Hg clusters revealed a much bigger cluster size of $n = 400 \pm 30$ for bandgap closure [225] (Fig. 17).

Electron-shell and supershell effects, originally observed in mass spectra of monovalent-metal sodium clusters [226, 227], were later also found for divalent metal clusters [228]. The electron-shell structure was identified for Hg clusters containing up to 1500 valence electrons.

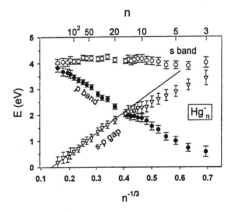

Fig. 17. The size dependence of the binding energies of the 6s- (open circles) and the 6p- (full circles) electrons in the photoelectron spectra of Hg anion clusters. The s–p-bandgap is the difference between these values (open triangles). The solid line is a linear fit to the bandgap in the range $n = 50$–250, and extrapolates to zero at about $n = 400$ [225]. Reproduced with permission from [225], copyright 1998, American Chemical Society.

5. SEMICONDUCTORS

5.1. Binary Semiconductor Nanocrystals

The properties of binary semiconductor nanocrystals and quantum dots (QDs) have received much attention in the past, following the pioneering work of Efros and Efros [117] and Ekimov, Efros, and Onushchenko [229], Brus [119], and Weller et al. [230]. Many size-dependent phenomena have then been observed in the following years [231–233]. These include light emission from silicon nanocrystals and porous silicon [234], and bandgap tunability in the nanoparticles [235].

The systematic investigation of II–VI quantum dots (ZnS, CdS, CdSe, etc.) began with Brus' work using solution-phase synthesis [236]. It also was inspired by a theoretical work which showed that the linear and resonant nonlinear optical properties exhibit the greatest enhancement when the nanoparticle radius R is much smaller than the Bohr radius of the exciton (a_B) in the corresponding bulk material [237]. Deviation of properties are expected for semiconductor QDs in the strong-confinement limit, where $R/a_B \ll 1$. The Bohr radii of excitons vary strongly between different semiconductors: 10 Å (CuCl), 60 Å (CdSe), 200 Å (PbS), 340 Å (InAs), 460 Å (PbSe), 540 Å (InSb).

The electronic spectra of PbS, PbSe, and PbTe QDs are simple, with energy gaps that can be much larger than the gaps of the parent bulk materials (which are 0.2–0.4 eV for the lead salts). In contrast, the energy level structure of II–VI and III–V QDs are much more complicated, with closely spaced hole levels and valence band mixing [6] which complicates the study of these materials.

Particles of Cd_3P_2 are found to be strongly size quantized [238]. This is seen from a 1.5-eV blueshift of the first electronic transition for 27-Å particles compared to the bulk bandgap of 0.5 eV. The shift is explained by quantum confinement of the exciton. The radius of the exciton is 180 Å, as calculated from the effective masses of the electron ($0.05m_0$) and the hole ($0.4m_0$) and the high frequency dielectric constant (~15). The large increase in the gap is due to the restriction of the large exciton to the 27-Å sized particle.

The electronic structure and the luminescence of II–VI nanopowders and quantum dots has been studied by several groups [11, 29, 239–244]. Many articles concerning the distribution of electronic states have appeared in literature [127, 231, 232, 245–248]. In addition to optical spectroscopy, conductance spectroscopy was used to study the bandgap [57].

Photoelectron spectroscopy of cluster beams is now an established method to study the electronic structure. The electron affinity of cluster anions as a function of size can be studied via photodetachment [249]. Such beam experiments are restricted to small sizes (for instance $Ga_nP_n^-$ ($n = 1–9$) [249] or $In_nP_n^-$ ($n = 1–13$) [250]) but allow studying both stoichiometric and nonstoichiometric clusters.

The physical properties of the larger nanoparticles are dominated by the confinement of electrons [251]. For CdSe particles, a widening HOMO–LUMO gap was observed in optical

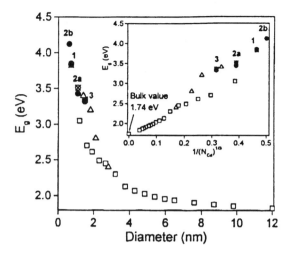

Fig. 18. Size dependence of the bandgap of CdSe particles [252]. Filled circles are taken from Soloviev et al. [252], squares and triangles are taken from Murray, Noms, and Bawendi [235] and Rogach et al. [595]. The inset displays for the same data the dependence of the gap on $1/(N_{Cd})^{1/3}$, with N_{Cd} being the number of Cd atoms.

absorption and band-edge emission (photoluminescence) studies [235, 241]. From another study, the size dependence of the gap for CdSe particles is given in Figure 18 [252]. Absorption and emission spectra of InP quantum dots show similar dependence [253] illustrated in Figure 19 [254]. Although the experiment and the QSE theory agree reasonably well at large sizes, theory diverges from the experimental values for small sizes ($d < 7$ nm). This may be due to the nonparabolicity of the bands at higher wave vectors and the finite potential barrier at the surface of the particles. Tight-binding calculations can yield better agreement for the smaller sizes [127, 245].

The effective mass approximation gives $1/R^2$ for the dependence of the bandgap on the radius of nanoparticles. However, it usually overestimates E_g. For PbS particles, for instance, the

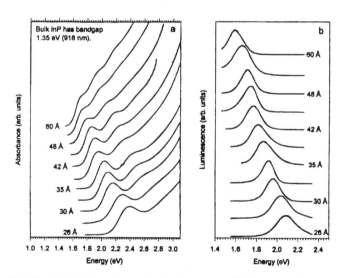

Fig. 19. Optical absorption and emission spectra of InP quantum dots as a function of diameter [254]. Reproduced with permission from [254], copyright 1998, American Chemical Society.

energy gap increases as $\sim 1/R^{1.3}$ for $R \geq 2.5$ nm [255]. For the very small sizes, the bandgap is usually not described by a simple R-dependence but rather changes discontinuously. This is shown in the gaps of stoichiometric Ga_nAs_n clusters ($n = 2$–6) which were determined using first-principle pseudopotential calculations: 5.8 eV (Ga_2As_2), 5.7 eV (Ga_3As_3), 4.3 eV (Ga_4As_4), 5.1 eV (Ga_5As_5), and 4.8 eV (Ga_6As_6) [256].

Williamson et al. [257] studied the conditions under which the bandgaps of free standing and embedded semiconductor quantum dots are direct or indirect. The results are obtained from experimental and theoretical investigations of quantum dots and quantum wells. It is found that free dots of CdSe, InP, and InAs have direct bandgaps for all sizes, while the direct bulk gap of GaAs and InSb dots becomes indirect below a critical size. Dots embedded in a direct gap host matrix either stay direct (InAs/GaAs at zero pressure) or become indirect at a critical size. Direct gap dots in an indirect gap host have a transition to indirect gap for sufficiently small particles (GaAs/AlAs and InP/GaP quantum well) or are always indirect (InP/GaP dots, InAs/GaAs above 43-kbar pressure and GeSi/Si dots). Theoretical studies predict a change from an indirect to a direct semiconductor for particles smaller than ~ 2 nm [258, 259] for GaP.

The bandgap in a bulk semiconductor depends on the temperature. Studies generally report that this dependency is similar for nanoparticles and the bulk [260]. The variation may be slightly weaker [261] in nanoparticles of varying size. The response of the bandgap to temperature can be measured in optical experiments by the temperature coefficients dE_g/dT of lowest electron-hole pair energies. For PbS and PbSe quantum dots, it was found that dE_g/dT decreases with decreasing dot size by more than an order of magnitude from the bulk value and even becomes negative (below ~ 3 nm) [262]. This shows that nanoparticles of semiconductors can have their electron levels frozen, i.e., with no or little response to a change of temperature. Such weak temperature dependence is expected for atomic-like levels. The observation is explained with the characteristics of quantum dots and nanoparticles: With a decrease in size, the continuum states of the bulk semiconductor transform gradually to the discrete states of the particle. This has effects on the thermal expansion of the lattice, the thermal expansion of the wave function envelope, the mechanical strain, and the electron-phonon coupling. The localized states respond little to variations of interatomic distances and angles.

Semiconductor particles are starting to find applications due to their delta-function-like density of states and their photoluminescence. However, the luminescence often is quenched above 300 K, at temperatures where the devices still have to function. This illustrates that besides having the bandgap well adjusted to a particular application, the matrix elements for the optical transitions may play a decisive role. Thermal escape of carriers to nonradiative recombination centers such as surface and interface states or other electron acceptors can drastically reduce carrier recombination rates [263]. The effect of temperature on the photoluminescence of semiconductor particles has well been studied [264].

Particles in solid host materials are protected from ambient air exposure. The embedded particles can show effects very similar to those of the free particles. For example, GaAs particles in Vycor glass show the typical nonlinear optical properties due to quantum confinement [265].

5.2. Silicon and Germanium Clusters

The early work of silicon clusters was inspired by studies of cluster beams. Such clusters are produced under high vacuum conditions and consequently do not have their surface bonds saturated. In the early 1990s, there was a new development for nanostructured silicon. Bulk silicon, when electrochemically etched, was made porous and was found to show photoluminescence in the visible range [266]. Soon the new effect was attributed to silicon nanowires or particles present in the porous structure. Indeed, STM studies of the porous silicon surface show the presence of nanoparticles (Fig. 20). It was supported by the observation that individual silicon particles showed photoluminescence [267–274]. The Si particles have been synthesized by liquid-solution-phase growth [275, 276], from silane via slow combustion [277], microwave plasma [267], chemical vapor deposition [25], gas evaporation [278–281], sputtering or ablation [24], ultrasonification of porous silicon [282, 283] and various other techniques.

This development led to an intense scientific activity to understand both the increase of the bandgap (made responsible for the luminescence in the visible range due to interband transitions), and the observed high transition probability. The silicon particles, responsible for the photoluminescence, were considered to have their surface bonds passivated, in accordance with the experimental conditions during the production of porous silicon. Indeed, unpassivated silicon clusters did not emit any light after UV excitation [274]. Therefore, interest in

Fig. 20.   AFM image (656×656 nm) of light-emitting porous silicon [475]. Reproduced with permission from [475].

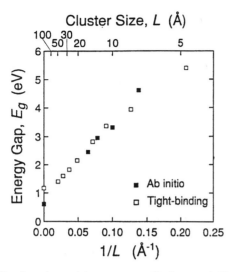

Fig. 21. Size dependence of the energy gap of hydrogenated silicon clusters; results from ab initio [287] and tight-binding calculations [596].

coming years moved away from pristine to passivated silicon clusters. Most of the work concentrated on hydrogen passivated [1, 284–290] (Fig. 21) and on oxidized silicon clusters [291–294].

Since the discovery of room-temperature visible photoluminescence in silicon nanocrystals [267] and porous silicon [266], the size dependence of the energy gap of Si nanostructures has been discussed extensively [287, 295–297]. The quantum-size effect, resulting in a blueshift of the energy gap with decreasing size, was widely believed to be at the heart of the novel optical properties of porous silicon [1, 267, 298–300]. The energy gap was found to increase significantly with decreasing size, for instance between 1.3 and 2.5 eV for particles of 5 to 1 nm in size [1]. The experimental data often were described by the effective mass approximation (EMA). An inverse power law for the bandgap behavior as a function of size is predicted by EMA and leads to very large bandgaps for particles of small size. Different exponents in the power law have been discussed [301].

Obviously, simple models like the effective mass approximation are not applicable to pristine silicon clusters. EMA is a continuum approach, which does not consider the complexity of the atomic structure, the unsaturated surface, and the significant changes in hybridization, which can occur when the size of the pristine particles is reduced.

$(Si)_n$ with just a few atoms ($n < 10$) were theoretically found to have compact cubooctahedral or icosahedral structure. The average coordination number for silicon clusters with $n = 12–24$ tends to lie above the coordination number 4 of the bulk diamond structure [302] (Fig. 22). Theoretical results [143, 303–307] predicting the compact structure for small Si clusters have been confirmed experimentally by Raman spectroscopy [308] and anion photoelectron spectroscopy [309]. The close-packed structure is typical for metallic rather than covalent systems. As bulk silicon is a semiconductor, a major change in the electronic properties can therefore be expected

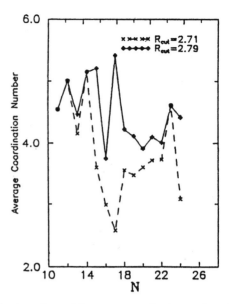

Fig. 22. Size dependence of the average coordination number calculated for small silicon clusters (with different distance cutoffs) [302]. Reproduced with permission from [302], copyright 1995.

with size reduction. The critical size n^* for a transition from covalent to metallic bonding was estimated, but with very different values: $n^* = 100–1000$ [143] and $n^* = 50$ [303].

Electronic structure and bandgap are reasonably well known for porous silicon and large passivated Si particles [287, 288, 310–314]. However, the HOMO–LUMO gap of small, unpassivated Si clusters is much less understood.

Gaps between zero and several electron volts were predicted for small unpassivated silicon clusters. For example, a zero gap or close to zero gap was theoretically obtained for Si_{28} [315] (Fig. 23), Si_{29} to Si_{45} [316] (Fig. 24), while gaps of 1.2 eV for Si_{20} [317] and Si_{46} [318] and 0.3–3 eV for Si_3-Si_{11} [143] were calculated (Fig. 25).

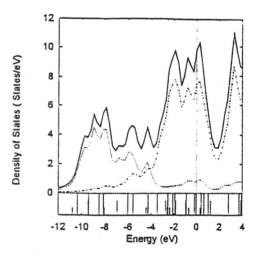

Fig. 23. Electronic density of states and eigenvalue distribution for an Si_{28} cluster. Dotted and dashed lines denote the partial densities of states for 3s and 3p, respectively, [315]. Reproduced with permission from [315], copyright 1995, American Physical Society.

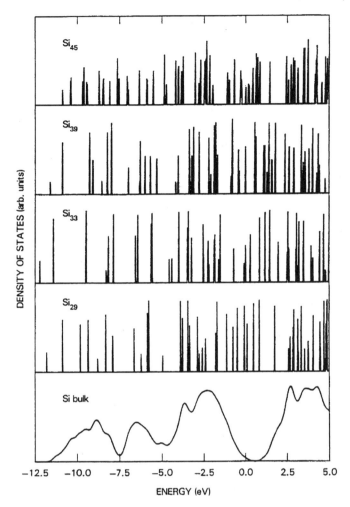

Fig. 24. Electron level distribution calculated for Si_{29}, Si_{33}, Si_{39}, and Si_{45}, in comparison with the density of states of bulk silicon [316]. Reproduced with permission from [316], copyright 1996, Trans Tech Publications.

For medium-sized silicon clusters, with up to several hundred atoms per cluster various atomic structures have been proposed [303, 315, 319–322] and the HOMO–LUMO gap was calculated to depend strongly on the assumed atomic structure [320, 323]. Due to the relatively large number of

atoms, semiempirical techniques such as the interatomic force field method [324] were used instead of ab initio techniques. In this size range, there seems to be a structural transformation. Jarrold and Constant [325] found that clusters up to ~27 atoms have a prolate shape while larger clusters have more spherical shapes. Other abrupt changes in properties of medium-sized Si clusters have been observed in photoionization measurements, at $n^* = 20$–30 [326]. A structural transformation at n^* would imply a sudden change of the electric polarizabilities of the clusters. In a theoretical study of Si_n with 10–20 atoms [327] the polarizability is found to be a slowly varying function of n. This indicates that n^* might be greater than 20. Photoelectron spectroscopy of Si cluster anions ($n = 10$–20) [328] yielded HOMO–LUMO gaps to scatter discontinuously between 0.6 and 2.1 eV (Fig. 26).

Clusters may have disordered structures if produced by fast cooling of a molecular beam or vapor [2]. The atoms do not have the time to find low-energy positions. The atomic structure of such particles therefore can be significantly different from both the bulk and the nanocrystal structures [329]. The electronic structure and the bandgap of amorphous silicon clusters (a-Si_n) have been calculated with a tight-binding approximation [284]. A comparison of the size dependence of the HOMO-LUMO gap for amorphous bare (a-Si_n), H-passivated (a-Si_n), and crystalline (c-Si) particles is shown in Figure 27. A strong blueshift with size reduction is predicted for all three types of particles.

The size-dependent bandgap of Ge nanoparticles has been calculated over a wide size range, from 12 to 1 nm, and the bandgap was found to change gradually from 1 to 4 eV, respectively, [330]. We note that 12-nm Ge particles would have a gap much larger than the bulk gap of 0.6 eV. Indeed, there are many reports of strong visible photoluminescence in germanium nanocrystals [3, 18, 331–340]. If interband transitions are responsible for the luminescence, the bandgaps of the Ge particles have to be several electron volts wide.

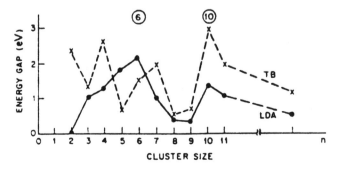

Fig. 25. HOMO–LUMO gaps calculated for Si clusters; results from local-density functional (solid line) and tight-binding (dashed line) results) [143]. Reproduced with permission from [143], copyright 1987, American Physical Society.

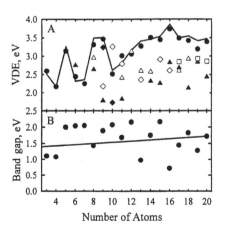

Fig. 26. (A) Vertical detachment energies (VDE) of Si_n anions; the solid line is measured with photoelectron spectroscopy. The symbols mark values calculated for silicon clusters of various structures. (B) Bandgaps calculated for lowest energy neutral Si clusters [328]. Reproduced with permission from [328], copyright 2000, American Physical Society.

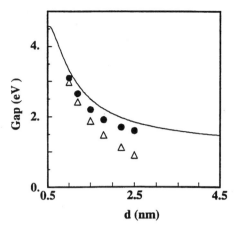

Fig. 27. HOMO–LUMO gap of amorphous clusters compared to crystallites: amorphous bare Si_n (empty triangle), hydrogen passivated Si_n (filled circle), and crystallites (full line) [284]. Reproduced with permission from [284], copyright 1997, American Physical Society.

Gap calculations within the effective mass model yield such wide gaps but quantitatively disagree with the optical results [335, 341]. Additionally, the tight-binding calculations for Ge particle gaps by Niquet et al. [330] are inconsistent with the experiments.

The blue–green photoluminescence is observed with little change in color for particles from about 2 to 15 nm [335, 341]. This would imply a bandgap of about 2 eV, independent of the particle size. Photoelectron spectroscopy studies have revealed HOMO–LUMO gaps for Ge_n clusters ($n = 4$–34) which do not show a blueshift in this size range but rather scatter around the bulk value of 0.6 eV [342] (Fig. 28). It indicates that the observed photoluminescence radiation is not due to the size-dependent quantum confinement but rather to surface or defect state transitions.

A wide spread of energy gaps is predicted for structural isomers of clusters with the same size. For example, calculations of germanium clusters with different atomic structures

give gap values of 0.1–1.35 eV for Ge_8, 0–1.84 eV for Ge_9, and 1.13–2.64 eV for Ge_{10} [288].

6. UNPASSIVATED SILICON PARTICLES

6.1. Particle Preparation

In the following, we show tunneling spectroscopy studies [48] determining the energy gap of pristine silicon particles over a size range where major size-dependent changes are expected. Using STM and STS as local probes allows studying individual clusters. STM is used to image the clusters and the surrounding support and to determine cluster sizes and shapes. STS is used to measure the energy gap.

The clusters were grown on highly oriented pyrolytic graphite (HOPG) upon submonolayer deposition of silicon vapor. The deposition was done by direct current (dc) magnetron sputtering of silicon in ultra-high vacuum (UHV) base pressure. Clusters form after surface diffusion by quasi-free growth on the inert substrate. After the deposition, the samples were transferred to the STM chamber (1×10^{-10} Torr) without braking vacuum. Since local STS of clusters is not feasible at room temperature due to the thermal drift, voltage-dependent STM [343] was the STS method of choice. It has been shown that local STS and voltage-dependent STM give comparable results for spectra of an adsorbate [343].

At submonolayer coverage, small clusters are formed on the support, well separated from each other. An area of pure graphite typically surrounds a particle (Fig. 29). Once a Si cluster was selected, a series of STM images at different bias voltages was recorded. If the cluster has a gap and the bias voltage is tuned to be within the gap, there are no states available for tunneling. Accordingly, the cluster is "invisible" until the bias voltage is high enough to allow tunneling into the "conduction band" or out of the "valence band" of the cluster. In the constant-current mode, this is reflected by the apparent height Δz of the clusters as it varies with the bias voltage. The equivalent measurement in the constant-height mode is the difference ΔI in tunneling current on the substrate and on the cluster. Each series of STM images yields a plot of ΔI as a function of V. Clusters in the size range from a few angstroms to a few nanometers were analyzed. For long exposure and high currents, nanoscale silicon wires can be grown [344].

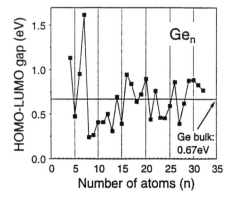

Fig. 28. HOMO–LUMO gaps of Ge clusters obtained from photoelectron spectroscopy (PES). The line at 0.67 eV indicates the energy gap of bulk Ge [342]. Reproduced with permission from [342], copyright 1998, Elsevier Science.

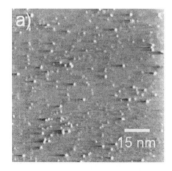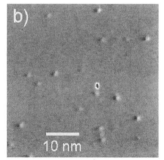

Fig. 29. Pristine silicon particles on HOPG, generated by vapor deposition in UHV [389]. Reproduced with permission from [389].

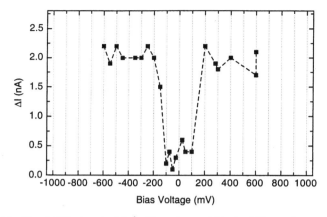

Fig. 30. $\Delta I(V)$ plot of a 2.5-Å silicon cluster. ΔI is the difference between the tunneling currents measured with the tip located on top of the cluster or on the substrate [48]. Reproduced with permission from [48], copyright 2000, American Physical Society.

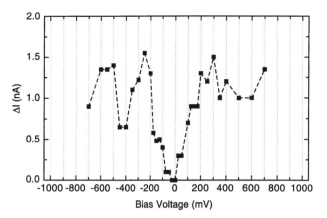

Fig. 32. $\Delta I(V)$ plot of a 9-Å silicon cluster [48]. Reproduced with permission from [48], copyright 2000, American Physical Society.

The well-known nonreactive nature of HOPG and the observed low sticking coefficients for many atoms and molecules (at room temperature) shows that chemisorption is not a process to occur for silicon clusters adsorbed on HOPG. This is supported by studies [345] reporting high scattering yields of Si_n^+ and Si_n^- ($n = 5$–24) impinging on HOPG. Impinging Si clusters have very low sticking probability and are easily reemitted. If the clusters are formed at the substrate, they are kept on the support by physisorption due to small electrostatic dipole forces.

6.2. Gap Measurement of Si Particles by STS

In Figures 30–33 $\Delta I(V)$ curves for several silicon clusters are displayed. They have in common that a bandgap is visible around zero bias. Figure 31, for example, shows the $\Delta I(V)$ curve for a 10-Å cluster. A bandgap with steep edges can be seen. The plot is taken with data points typically 25 mV apart. ΔI is zero in the gap region since the cluster completely vanishes in the STM image. There are no accessible states in the cluster and the tunneling electrons pass through the cluster

without interaction. Accordingly, only the substrate is visible in the STM image. The gap is nearly symmetric around zero bias. Outside of the gap, the ΔI values fluctuate due to quality differences of the STM images. For positive bias voltages, electrons tunnel from the tip to the sample and vice versa for negative bias voltages. ΔI values are observed to be higher in the positive than in the negative bias range. This shows that the cluster has a higher state density at the conduction band edge compared to the valence band edge. The states are difficult to identify, as the electronic features of small silicon clusters are little known.

In Figure 34 the energy gap of the analyzed clusters is plotted versus the cluster size. In the range between 15 and 40 Å, only clusters with zero gap are observed. For smaller clusters, zero gaps are found as well but nonzero gaps predominate. Below 15 Å, the gaps tend to increase with decreasing cluster size. The largest gap recorded is 450 meV, for clusters with 5 and 8.5 Å. There is a significant scatter of the data points in the small size range. Clusters of similar size can have very different energy gaps. For example, at 8 ± 1 Å a zero-gap cluster and one with a 250-meV gap is found.

The size of a cluster was determined with an error bar of typically $\pm$ 10%. For a spherical cluster, one can relate a

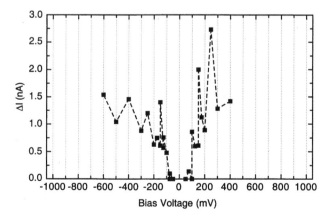

Fig. 31. $\Delta I(V)$ plot of a 10-Å silicon cluster [48]. Reproduced with permission from [48], copyright 2000, American Physical Society.

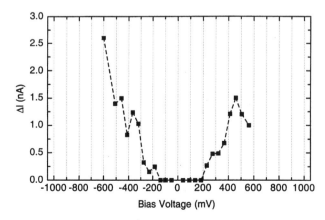

Fig. 33. $\Delta I(V)$ plot of a 8.5-Å silicon cluster [48]. Reproduced with permission from [48], copyright 2000, American Physical Society.

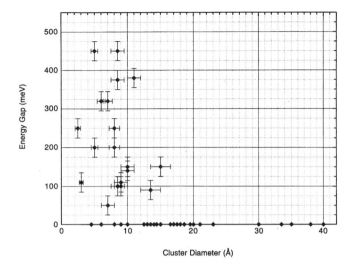

Fig. 34. Energy gap values from STS measurements as a function of silicon cluster size [48]. Reproduced with permission from [48], copyright 2000, American Physical Society.

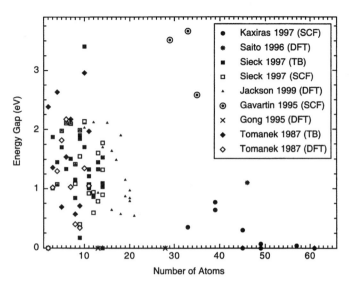

Fig. 35. Calculated energy gaps for pristine silicon clusters using density functional theory (DFT) [143, 302, 315, 317, 327], tight-binding (TB) methods [143, 349], or self-consistent-field (SCF) algorithms [349–351].

given diameter to the number of atoms using the expression $d(n) = (3/4\pi\rho)^{1/3}n^{1/3}$, where ρ is the mass density and n is the number of atoms in the cluster. Using the mass density of bulk silicon, one obtains $d(n) = 3.3685n^{1/3}$ (Å). It follows that 26 atoms are in a spherical 10-Å Si cluster. 9- and 11-Å Si clusters contain $n = 19$ and 35 atoms, respectively. These small-sized clusters have properties depending strongly on their size. One atom more or less may have a pronounced effect on the cluster's electronic structure. Therefore, the energy gap can strongly differ for clusters of similar size. In the STS experiment, one does not determine the number of atoms in a cluster but rather the diameter. In the uncertainty window of ±10%, there are clusters with slightly different numbers of atoms. Such a spread in the number of atoms per cluster explains the observed scatter of gap data.

The gaps for clusters smaller than 15 Å lie far below the value of silicon bulk or the enhanced values due to the quantum-size effect. To check whether clusters with very large gaps are present, we applied bias voltages up to ± 2.5 V. Therefore, clusters with gaps up to 5 eV could have been detected.

The results are surprising at first. Bulk silicon has a bandgap of 1.1 eV and one may expect it to be much higher for small particle sizes due to the quantum-size effect. For surface passivated silicon clusters, this has in fact been confirmed both experimentally [1, 269, 273, 274, 291–294, 296, 346, 347] and theoretically [122, 284, 285, 287, 288, 290, 295, 301, 348] with gaps up to 4 eV. The STS results show that the gaps of unpassivated and passivated clusters are entirely different.

Theoretical results about the HOMO–LUMO gaps of silicon clusters show strong variations, depending on calculation methods, assumptions, and types of approaches. A variety of calculated gaps is shown in Figure 35. They have been obtained by density functional theory (DFT) [143, 302,

315, 317, 327], tight-binding methods [143, 349], or self-consistent-field (SCF) algorithms [349–351]. Over the range from Si_2 to Si_{61}, clusters with zero bandgaps are found, as well as clusters with gaps up to 3.5 eV.

On the extended surface of bulk silicon, the density of dangling bonds is known to be high. It is a consequence of the strong directional properties of the sp^3-hybrids and the missing neighbors at the surface. The same can be expected for unpassivated silicon clusters. Each dangling bond contributes a partially filled surface state.

Since they are partially filled, the surface states associated with the dangling bonds are located in the energy gap around the Fermi level. This has been well demonstrated for the surface of bulk silicon using a number of methods including tunneling spectroscopy [352–354]. These experiments show that surface states mostly fill the bandgap of the unpassivated silicon surface. For clusters of silicon, with unpassivated surfaces, a similar situation can be expected. In fact, a calculation of Si_{29}, Si_{87}, and Si_{357} shows density of states with zero-energy gap between HOMO and LUMO [355]. The passivated clusters, $Si_{29}H_{36}$, $Si_{87}H_{76}$, and $Si_{357}H_{204}$, however show the broad bandgaps of 3.44, 2.77, and 1.99 eV, respectively, just as expected for particles in this size range. Evidently, the zero gap of the pristine particles originates from the surface states.

The surface of bulk silicon has a high density of extra states which are located in the fundamental bandgap [356] and tend to form surface state bands [357, 358]. The ideal bulk surface of silicon reconstructs to a variety of complicated atomic arrangements, as, for example, the 7×7 reconstruction of the (111) crystal facet [359]. This reduces some of the dangling orbitals but still leaves many extra states in the gap. Local tunneling spectroscopy measurements show substantial tunneling currents even at very low bias, indicating a high density of states near the Fermi energy [353]. Zero bandgap, due to the metallic surface state density, has also been observed in combined photoemission and inverse photoemission studies [360].

Two transport mechanisms explain the conducting properties of the cluster surface. For high density, the surface states overlap forming a conduction band. The cluster surface then shows the transport properties of a metal. For lower state density, thermally activated hopping between the localized surface states leads to the observed conducting behavior. At ~15 Å, corresponding to $n^* \sim 90$ atoms per cluster, a major change in the cluster properties occurs. The energy gap suddenly opens up and subsequently widens for smaller clusters. This change to nonzero energy gaps shows a transformation of the electronic structure of the clusters. We associate it with the covalent-metallic transformation suggested previously for silicon clusters by several groups. Accordingly, large silicon clusters are sp^3-hybridized with strong covalent bonds and rigid bond angles giving rise to high surface state density. Around a critical size n^*, the sp^3-nature of the bonds gradually changes to hybridization other than sp^3 or even to the atomic s- and p-configurations. In fact, ab initio calculations show that sp^3 is not the favored hybridization for small silicon particles. The change of the electronic structure at the critical size also means a change in the atomic structure of the clusters. Below n^*, the clusters are not fragments of the bulk anymore but rather have their cluster-specific structures. These are close-packed structures resembling those of metals. In this sense, one may say that at n^* a covalent-metallic transition occurs.

Some of the studied silicon clusters exhibit zero-energy gaps even below n^*. These clusters have the surface state configurations similar to those above n^*. It shows that the transition at $N^* = 90$ does not occur for all silicon clusters. Some clusters have sp^3-coordinated bonds even at very small size. They coexist with the compact clusters as structural isomers. This is not surprising because silicon bonds are very strong and many isomers are stable at room temperature. The observation of various bandgaps for similarly sized clusters is also a consequence of the analysis of individual clusters using STM and STS.

A silicon cluster of 15-Å diameter contains about 90 atoms, with the bulk structure of silicon being assumed for the volume density. This critical size of $n^* = 90$ can be compared to the widely different theoretical estimates ($n^* = 30 - 4000$) [132, 142, 303, 361, 362]. A covalent-metallic transition has been postulated but it has previously not been observed. Yet, there are some indications for such a transition. Chemical reactions of Si_n^+ show significant changes in the chemisorption probabilities at $n = 29 - 36$ (for O_2 adsorption) [363]. The dissociation energy of Si_n^+ starts to deviate from the smooth size behavior below about $n = 40$ [364]. The photoionization threshold suddenly has a drop at $n \sim 20 - 30$ [326]. We note that these experiments have been performed on charged clusters or by transfer from neutral to charged. The stability and structure of neutral and charged silicon clusters however may be different.

The highest observed gap (~450 meV) in the STS experiment is less than half the bulk value. Interestingly, the gap of

surface states of Si (111) (2 × 1) was found to be in this range (450 meV) [357]. A similar result has been obtained from studies of silicon clusters in beams, where the bandgaps were found to be far below the 1.1-eV gap of bulk silicon [365].

6.3. Coulomb Blockade of Si Particles

Quantum dots of silicon and germanium can show Coulomb blockade and single-electron tunneling effects. These have been studied in particular with respect to the fabrication of a quantum dot transistor [366–370]. Coulomb blockade oscillations have also been studied for (10-nm diameter) silicon nanowires within a metal-oxide-semiconductor field-effect transistor (MOSFET) [371].

Single-electron tunneling (SET) was observed in conductance spectroscopy for metal particles [372, 373–375]. The SET studies on metal particles were done at low temperatures, at 4 K [84, 376] and 77 K [61, 377]. A Coulomb staircase in the I–V characteristics was also observed for CdSe particles at 77 K [61]. In some tunneling studies of quantum dots, sub-Kelvin temperatures were applied [378–380]. In these systems, the single-electron charging energy is substantially larger than the level spacing. Low temperatures are required for these particles to show the characteristic Coulomb gaps.

Electrochemically it was demonstrated that gold particles in solution show Coulomb-staircase charging behavior [381]. Incremental charging was also found for a molecule using an STM at room temperature [85]. Silicon particles on HOPG can also show Coulomb blockade at room temperature [371, 382–388].

In Figure 36 an STM image is shown with both, "white" and "black" dots on the "gray" background of the substrate.

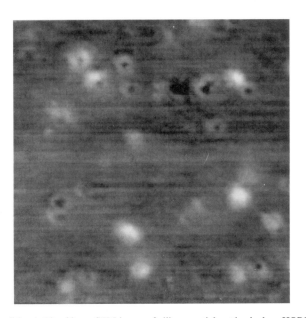

Fig. 36. A 29 × 29-nm STM image of silicon particles adsorbed on HOPG. Some of the particles appear black due to Coulomb blockade of the tunneling current [487]. Reproduced with permission from [487], Fig. 37. Reproduced with permission from [50], copyright 1996, American Institute of Physics.

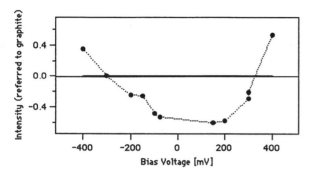

Fig. 37. $\Delta I(V)$ plot for a silicon particle which shows Coulomb blockade. The intensity of the cluster referred to graphite is below zero in a bias voltage range of about 600 mV due to Coulomb blockade [597]. Reproduced with permission from [597].

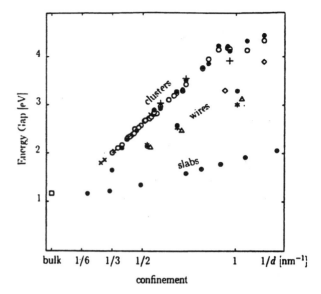

Fig. 38. Energy gap calculated for H-terminated slabs (two-dimensional), wires (one-dimensional) and clusters (zero-dimensional) (after Delley and Stelgmeier [290]). Data are plotted for studies by Delley and Steigmeier [290, 301], Delley, Steigmeir, and Auderset [598], Wang and Zunger [295], Hirao, Uda, and Murayama [599], Mintmire [600], Read et al. [601], and Buda, Kohanoff, and Parrinello [409].

The white dots are due to silicon particles, which are conductive for the tunneling electrons. The black dots however, block the tunneling electrons because they are insulating and charged by single electrons. The white halo around the black dots indicates that the surface of the Si particles is conducting. This is expected for particles with unpassivated surface bonds. Electron transfer between tip and substrate can occur through the conducting surface of the particles. One finds that the core of these particles is insulating while the surface is conducting.

In Figure 37 the STS plot for a black cluster is given. For values above and below the zero line, the cluster appears white or black, respectively. The voltage range where the particle appears black gives the value of the Coulomb gap. The plot was taken from a 13-Å silicon particle. It shows a Coulomb energy gap of 640 meV [389].

7. PASSIVATED SILICON PARTICLES

7.1. Pristine versus Passivated Silicon Clusters

While reconstruction leads to a reduction of surface dangling bonds, it does not eliminate them completely. Therefore, the reconstructed silicon surface still contains a large number of dangling bonds and shows strong chemical reactivity. The bulk silicon surface is oxidized rapidly under ambient conditions. The reactivity of silicon clusters is reduced [363] compared to the bulk and the reduction depends on the size of the particles. In fact, magic Si clusters where found with weak interaction strength [390]. For such clusters, most (or even all) of the dangling bonds have disappeared by backbonding and relaxation.

What a difference the surface passivation makes was illustrated in a calculation of pristine silicon clusters and their hydrogenated counterparts [355]. The pristine clusters show a density of states with zero-energy gaps between HOMO and LUMO. The passivated clusters show wide bandgaps up to 3.4 eV.

Passivated silicon particles have been investigated experimentally and theoretically [289, 391–398]. The diamond structure is stabilized by the hydrogen passivation and therefore the bulk crystal structure has been applied to the smallest silicon clusters [289].

H-passivated silicon wires were also studied [399]. Clusters, wires, and slabs of silicon show the same qualitative behavior with a blueshift of the bandgap with decreasing size [290] (Fig. 38). Large energy gaps have been reported, such as 4.6–2.45 eV for 7.3–15.5 Å [287], 5.3–1.7 eV for 5 to 50 Å [400] and ~5 to ~2 eV for 10 to 27 Å [288] hydrogenated silicon particles. Often, the purpose of the research was the interest in understanding the visible photoluminescence of porous silicon (π-Si) [287, 297, 311, 401–412].

The quantum efficiency and the stability of the luminescence depend on the passivation of the grains and the grain boundaries in porous silicon (π-Si). The π-Si material can significantly be improved by passivation with oxygen through rapid thermal processing [31, 413]. The porous silicon structure can be envisioned as small units remaining in the lattice after electrochemical edging. These units where first considered to be silicon nanowires [266] and later to be rather nanoparticles. In both cases, their surface is passivated as it was exposed to the ions in the acid during the electrochemical process. The atomic structure of both particles and wires is usually assumed to be the diamond structure of the bulk. This structure is stabilized by the surrounding atoms of the host and by the passivation of the dangling bonds. It is usually not the equilibrium structure of the corresponding free nanoparticle or nanowire.

The passivation of the surface is often debated with respect to the luminescence. Bulk silicon is an indirect bandgap semiconductor with no optical activity. In the crystalline bulk material, the momentum conservation rules have to be fulfiled. In a nanoparticle however, the strict rules may be lifted due to disorder and surface effects.

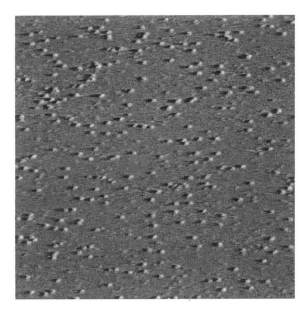

Fig. 39. STM image (152 × 152 nm) of hydrogenated silicon particles on HOPG [389]. Reproduced with permission from [389].

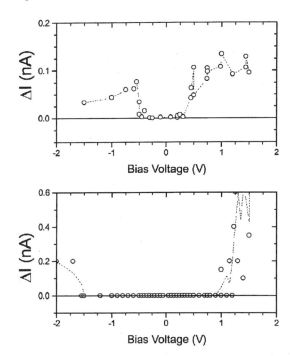

Fig. 40. $\Delta I(V)$ plots of two hydrogenated silicon clusters (a) 6.8 Å, (b) 9.8 Å [389]. Reproduced with permission from [389].

Passivation with hydrogen removes dangling bonds at the surface of the particles. Passivation with oxygen has further consequences for the particles. For instance, the silicon atoms at the surface may become inactive for luminescence once a layer of silicon dioxide has formed. Then, after a particle is oxidized it can appear with smaller diameter compared to the original pristine particle [272].

7.2. Energy Gap Studies of H-Passivated Si Particles

Dangling bond states are localized states in the bulk bandgap. If they are removed by passivation, the gap measured in the STS study should drastically increase. Passivation of surface states is achieved after particle formation by exposing the surface to foreign atoms. In studies of H-passivated silicon particles by STM–STS [389], atomic hydrogen gas was used for surface passivation of silicon particles. H atoms are obtained when igniting a plasma by applying a 1-kV dc voltage to very sharp tips in close proximity to the grounded sample holder. Figure 39 shows an STM image of silicon particles which have been grown on the graphite support in UHV and subsequently passivated with atomic hydrogen. Figure 40 shows $\Delta I(V)$ curves for two hydrogenated clusters. The 6.8- and 9.8 Å diameter particles have energy gaps of 0.9 and 2.4 eV, respectively. These values are far greater than the gap values, which were measured by STS for the pristine particles. It shows that dangling surface states were removed by passivation with hydrogen.

8. NANOWIRES OF SILICON

8.1. Nanowires with Various Geometries

Semiconductor wires thinner than 100 nm are attracting much attention because of their fascinating quantum properties. They

may be used for developing one-dimensional quantum wire high speed field-effect transistors and light-emitting devices with extremely low power consumption. Sakaki [414] calculated, that for one-dimensional GaAs channels the electron mobility exceeds 10^6 cm^2/V at low temperature, which is more than 1 order of magnitude larger than the calculated electron mobility of a two-dimensional electron gas [415]. The effect of quantum confinement in quantum wires (QWRs) has been evidenced in the luminescence [416–418], two-photon optical absorption [419], inelastic light scattering [420], and various other studies. It is obvious that impurity scattering and boundary effects become increasingly important when the width of the QWRs is reduced. Additionally, the atomic structure of the QWRs is fundamentally important for their overall properties. In many studies, the atomic structure of QWRs has been assumed to be the same as in the crystalline bulk.

In addition to theoretical studies of electronic and optical properties [399], a number of experimental studies have been reported for Si QWRs: transmission electron microscopy [421], electron transport [422], photoluminescence [423, 424], infrared-induced emission [425], Raman spectroscopy [423, 426]. The atomic structure of the wires in these studies however was not known and was assumed to be either amorphous or the diamond structure of the bulk. This may be justified when the wires are formed by methods like lithography and orientation dependent etching [427]. It may not apply to self-forming quantum wires, which were grown freely by vapor condensation.

Crystalline silicon does not have the tendency to grow in one dimension, as there is no preferential direction associated with the diamond-type lattice. The formation of sp^3-bonds in silicon leads to fourfold coordination with four equivalent directions for growth. This is in contrast to carbon, which can also occur in sp^1- and sp^2-configurations and therefore has various forms of one-dimensional structures. As small clusters, carbon grows in the form of linear chains or monocyclic rings. It also grows in the form of nanotubes [428]. Such quasi-one-dimensional structures are of great interest for scientists and engineers due to their exceptional quantum properties not found in the three-dimensional bulk.

Due to the technological importance, efforts have been made to produce nanometer scale silicon wires in a controlled manner, using common semiconductor processing steps [429–432]. Columns with small diameters have been found after electrochemical treatment of Si wafers with hydrofluoric acid [266, 433]. There has been much discussion about whether quantum confinement in these wires explains the visible photoluminescence of porous silicon.

Individual nanowires of silicon have been produced by a number of research groups [427, 430, 431, 434–445], for example, by natural masking [435], lithography [427, 441], wet-chemical etching [431, 437] and vapor–liquid–solid growth [438]. These methods use growth techniques, which lead to natural surface passivation of the wires, usually by oxidation. Silicon is very easily oxidized, because the diamond-type crystal structure leads to a high density of dangling bonds at the surface.

8.2. Fullerene-Structured Si Nanowires

In an article [344], the formation of fullerene-structured silicon nanowires was reported. The results of this work are shown in this chapter.

The nanowires were grown from the atomic vapor in ultra-high vacuum and analyzed by scanning tunneling microscopy (STM). Such an STM image (114×114 nm) is displayed in Figure 41. The picture shows several bundles with 20–30 wires per bundle. The nanowires are more than 100-nm long, with diameters from 3 to 7 nm. The width of the wires is uniform in each bundle.

For structural consideration of vapor grown silicon nanowires, one may start from the requirement of a distinct wire axis, and keep bond angles close to the bulk ones. Additionally one may apply the topological restrictions that only five- and six-membered rings occur. This leads to the construction of several linear polyhedric networks:

(a) Si_{12}-cage polymer structure (12 atoms per unit cell).
(b) Si_{15}-cage polymer structure (10 atoms per unit cell).
(c) Si_{20}-cage polymer structure (based on the I_h dodecahedron, 30 atoms per unit cell).
(d) Si_{24}-cage polymer structure (based on the $D6d$ icosahedron, 36 atoms per unit cell).

The suggested structures are shown in Figure 42. All models have a stacking of Si cages in common, in the center of which lies the wire axis. While these lattices deviate from the diamond-structured bulk, tetrahedral configuration of the Si atoms is maintained.

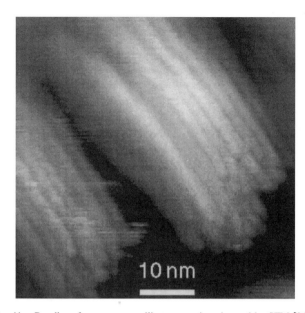

Fig. 41.   Bundles of vapor-grown silicon nanowires, imaged by STM [344]. Reproduced with permission from [344], copyright 1999, American Physical Society.

In structure "a" (C_{3v} symmetry), the axis of the wire passes through the centers of buckled Si_6-rings. Two adjacent rings are connected by three bonds and form an Si_{12} cage. This cage represents the unit cell, which is repeated every 6.31 Å. The surface of the wire consists of buckled hexagons.

Structure "b" (C_{5v} symmetry) consists of planar pentagons, joined through five outward oriented interstitial atoms. Two pentagons together with the interstitial atoms form a cage. The

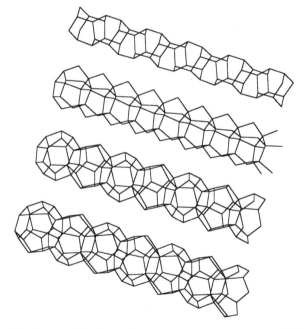

Fig. 42.   Four models of nanowire core structures, as polymers of Si_{12-}, Si_{15-}, Si_{20-}, and Si_{24-} cages [344]. Reproduced with permission from [344], copyright 1999, American Physical Society.

unit cell of this structure contains 10 atoms and the repeat distance is 3.84 Å. The surface of the wire consists of buckled hexagons.

Structure "c" (C_{5v} symmetry) is built from Si_{20}-cages. These dodecahedra (I_h symmetry) are the smallest possible fullerene structures, consisting of 12 pentagons. In the wire structure, two adjacent cages share one pentagon. The 30 atoms of one and a half such cages build the unit cell, which is repeated every 9.89 Å. Pentagons make up the surface net of this wire.

Structure "d" (C_{6v} symmetry) is similar to structure c, only that Si_{24}-cages (D_{6d} symmetry) are used as building blocks. These units contain 12 pentagons and 2 hexagons. The hexagons are shared by two adjacent cages and are concentric to the wire axis. The unit cell consists of 36 atoms and the lattice parameter is 10.03 Å. The surface of the wire consists of pentagons.

The Si_{20} and Si_{24} cages correspond to the smallest fullerenes. Si_{20} is a 12-hedron consisting entirely of pentagons. It has six equivalent fivefold symmetry axes. The fullerenic 13-hedron (Si_{22}) does not exist. The fullerenic 14-hedron (Si_{24}) has twofold linear (30°-twisted) coordination.

To find the most stable of the proposed structures, their binding energies and HOMO–LUMO gaps were calculated. The PM3 self-consistent-field molecular orbit (SCFMO) theory derived by Stewart [446] was applied. The calculations show energy gaps between 1.8 and 3.2 eV (Fig. 43) for nanowires containing up to 60 atoms. The wires are between ~4 Å (structure a) and 10 Å (structure d) thick. The gap

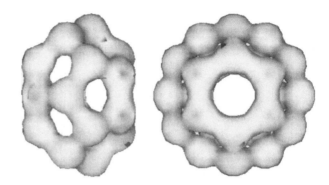

Fig. 44. Charge density isosurface of the fullerenic Si_{24} cluster [344]. Reproduced with permission from [344], copyright 1999, American Institute of Physics.

becomes rapidly smaller with increasing length of the wire. For a given size, the energy gap of structure d is largest.

For small length, structure d has the highest binding energy. The Si_{24} fullerene cage is thermodynamically favorable over the other structures, and also has a symmetry axis specified for preferential addition of further cages. Therefore, the Si_{24}-based wire may have the best chance to form.

Among the four considered configurations, structure d has the highest binding energy per atom and the largest energy gap. If energetics is responsible for the wire formation in its early stage, then structure d should grow preferentially. Once the Si_{24}-based polymer has formed, the wire may continue to add layer by layer, further increasing its diameter.

Dodecahedral Si_{20} clusters were predicted to be stable, using a model potential for an sp^3-hybridized atom [129]. It has been argued that these clusters may not be found isolated in experiments because the dangling bonds (on the cages' outer surfaces) would make them "unextractable" [317] from other silicon material. Fullerenic Si_{20} units were synthesized as bulk body-centered cubic (bcc) solids, with additional Si atoms at half the interstitial sites (three sites per Si_{20}), leading to a silicon lattice with $(Si_3Si_{20})_2 = Si_{46}$ units [54]. Such hollow silicon materials may have novel properties. In fact, superconductivity was found in the $Na_2Ba_6Si_{46}$ phase [447].

In Figure 44, the charge density distribution of Si_{24} is shown, as obtained from the PM3 analysis. The left side of the figure gives a side view with the pentagonal-type surface. The axis for wire growth goes through the two hexagons on the opposite sides of the cluster. The hexagons are rotated by 30° relative to each other. On the right side of the figure, the cluster is viewed from the wire axis.

9. CARBON PARTICLES

9.1. Electronic Structure of Carbon Particles

The fast growing technological importance of carbon-based nanostructures is well known [448]. From the backbones of organic molecules to particulates in air pollution [449, 450] or diamond-like films, carbon is crucial to the stability and the properties of many natural and artificial structures. In fact,

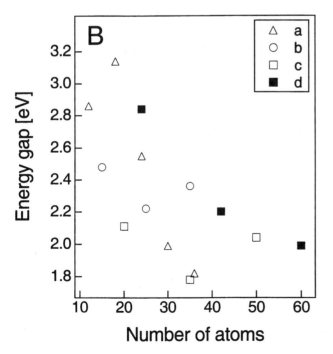

Fig. 43. PM3-calculated energy gaps as a function of size for the four wire structures [344]. Reproduced with permission from [344], copyright 1999, American Physical Society.

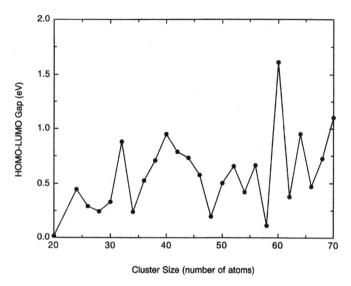

Fig. 45. HOMO–LUMO gap, calculated for fullerene-structured Si_{20} to Si_{70} clusters [461]. Reproduced with permission from [461], copyright 1992, American Institute of Physics.

ever, this flexibility in bond type becomes a difficult issue if small clusters of elemental carbon are considered. Many metastable isomeric structures are possible, each having a pronounced minimum in the potential-energy hypersurface [451, 452]. Fullerenes are a well-studied class of such isomers, but they form only a small fraction of the overall possibilities. While carbon clusters with up to about 30 atoms in the ground state are believed to be linear chains, monocyclic or polycyclic rings [453], the atomic structures and properties of the larger (nonfullerene) clusters are little known.

Graphite and some forms of amorphous carbon [454] are semimetallic with zero bandgaps. If carbon particles are reduced in size to a level where quantum effects become important, large energy gaps may appear [145, 455]. Indeed, energy gaps of up to 8.5 eV have been predicted for carbon clusters with fewer than 10 atoms [456]. Using high-resolution electron energy loss spectroscopy (EELS) a semimetal to semiconductor transition has been found for noncrystalline, graphite-like carbon particles with diameters of about 1 nm [457].

Simulation studies obtained the value of 2.2 eV [458] for C_{60} and gaps below 2 eV for cages composed of 24 to 80 atoms [459–461]. In another theoretical study of fullerenes from C_{20} to C_{70} [461] the HOMO–LUMO gaps vary between 0.015 (C_{20}) to 1.61 eV (C_{60}). This variation is discontinuous

a very large number of covalent carbon structures may exist with either diamond-like (sp^3), graphite-like (sp^2), linear (sp), or mixed (sp^3/sp^2, sp^2/sp, sp^3/sp) bonding. The "flexibility" in the s–p-hybridization makes carbon unique leading to the enormously large variety of hydrocarbon molecules. How-

Table I. Energy Gaps for Carbon Clusters

Reference	Year	Clusters	Energy Gap	Methods
Pitzer and Clementi [580]	1959	$C_2, \ldots, C_{10}$	Increasing with decreasing size	LCAO calculations
Liang and Schaefer [456]	May 1990	C_4, C_6, C_8, C_{10}	C_4: 8.14 eV C_6: 7.2 eV C_8: 6.6 eV C_{10}: 6.19 eV	SCF and CISD (single and double configuration interaction)
Liang and Schaefer [456]	September 1990	C_{10}, isomers	C_{10}: 8.47 eV	ab initio quantum mechanical method with SCF, CISD
Feng, Wang, and Lerner [582]	1990	C_{60}, isomers	5.4 eV	Intermediate neglect of differential overlap (INDO), INDO/CI
Zhang et al. [461]	1992	$C_{20}, C_{24}, \ldots, C_{70}$	Refer to graph	Tight-binding molecular dynamics and geometry optimization
Woo, Kim, and Lee [460]	1993	$C_{50}, C_{60}, C_{70}, C_{80}$	C_{50}: 0.5 eV C_{60}: 2.2 eV C_{70}: 1.55 eV C_{80}: 0.3–0.4 eV	Empirical tight-binding energy calculations
Wang et al. [583]	1993	C_{60}, C_{70}	Values around 2–3 eV	Analytical quasi-particle energy calculations
Endredi and Ladik [584]	1993	C_{60}	8.31 eV	Density functional theory
Tomanek [585]	1995	C_{60}	2.2 eV	Linear combination of atomic orbitals (LCAO) method
Yao et al. [51]	1996	C_{60}	1.8 eV	STM experiment

CLUSTERS, NANOPARTICLES, AND QUANTUM DOTS

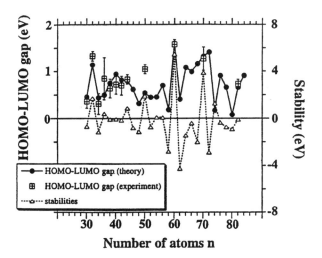

Fig. 46. HOMO–LUMO gaps for fullerenes derived from photoelectron spectra, as well as theoretical gap and stability data [464]. Reproduced with permission from [464], copyright 1998, American Physical Society.

with respect to the cluster size (Fig. 45). A comparison of theoretical results for energy gaps of carbon particles is given in Table I.

Several groups have measured the HOMO–LUMO gap experimentally by scanning tunneling spectroscopy. For C_{60}, a gap of 1.8 eV was determined [51]. The importance of charging effects has been used to explain a small observed gap value of 0.7 eV for STS of C_{60} [50]. Various gap values (0.8, 1.3, and 1.4 eV) were determined for C_{60} clusters at different adsorption sites of the Si(111) 7×7 substrate [462]. For C_{36}, a gap of 0.8 eV was obtained by STS [463]. Photoelectron spectroscopy has been used to study free fullerenes. HOMO–LUMO gaps between zero and 1.6 eV were determined for C_{30} to C_{82}, and the size dependence was found to change discontinuously with cluster size [464] (Fig. 46).

9.2. Bandgap Studies of Carbon Particles

In this chapter, we show STS results for carbon clusters [465] with diameters between 4 and 1100 Å. Voltage-dependent STM was used as the STS method.

Figure 47 shows STM images of a 5.5 Å carbon cluster at different bias voltages. For this cluster, the complete sequence consists of 36 images in the bias range from −1.0 to +1.0 V. Six of these images, with bias voltages of −550, −275, −225, +225, +275, and +550 mV are displayed in the upper panel of Figure 47. While the cluster appears bright at −550 mV, it is less bright at −275 mV and it is no longer visible at −225 and at +225 mV. It reappears again at +275 mV. The lower panel of Figure 47 shows cross sections of the images. The difference ΔI between the currents with the tip on the cluster and on the bare substrate is a measure of the density of states of the cluster.

ΔI values as a function of the bias voltage are shown for three different clusters in Figure 48. Each data point corresponds to an STM image. Distinct electronic structure features are observed, varying with size. For some of the clusters, a small asymmetry of the current-voltage plots with respect to zero bias was observed. Such shifts may be due to the presence of a contact potential between the cluster and the substrate, which is superimposed to the applied bias voltage [466]. The interaction of carbon clusters with the HOPG substrate is negligible [467, 468]. This is seen from the fact that the clusters are easily moved by the STM tip and often are even picked up.

Energy gaps of 650, 450, and 150 meV are found for clusters with diameters of 4, 5.5, and 7.5 Å, respectively. The cluster diameters and energy gaps are plotted in Figure 49. The results show close to zero-energy gaps for large clusters. At about 15 Å, an energy gap opens.

To understand the observed size behavior of the energy gap, the effective mass approximation (EMA) is used. For a spherical well, the solutions to the Schrödinger equation are given by spherical Bessel functions. If the potential well is infinite, the

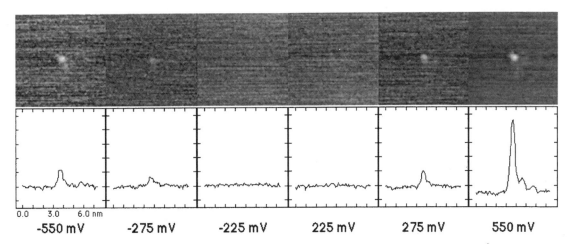

Fig. 47. STM images (upper panel) and profiles (lower panel) of a carbon cluster at different bias voltages. The cluster is 5.5 Å in diameter [465]. Reproduced with permission from [465], copyright 1999, Elsevier Science.

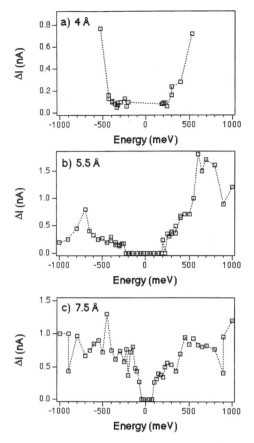

Fig. 48. $\Delta I(V)$ plots for three different clusters. (a) A 4-Å cluster with an energy gap of 300 meV, (b) a 5.5-Å cluster with an energy gap of 450 meV, (c) a 7.5-Å cluster with a 150-meV gap [465]. Reproduced with permission from [465], copyright 1999, Elsevier Science.

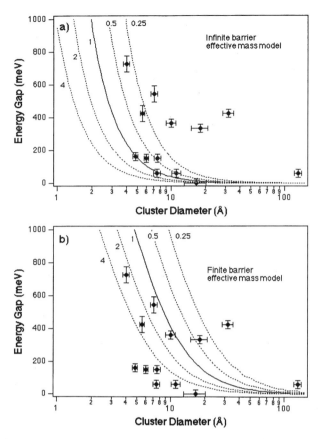

Fig. 49. Energy gaps for carbon clusters of different size; (a) experimental data and infinite barrier effective mass model fit (solid line is for $m^*/m = 1$), (b) experimental data and finite barrier effective mass model fit [465]. Reproduced with permission from [465], copyright 1999, Elsevier Science.

wave function vanishes at this barrier and the energy eigenvalues are determined from the zeros of the Bessel functions. This leads to a size-dependent shift of the energy gap, which scales with the inverse square of the diameter d.

In Figure 49a, this result is plotted for different effective masses. The curves show a significant opening of the energy gap for particle sizes below about 100 Å. Only for very high effective masses, do the curves get close to the observed gap values. Therefore, it appears that the "infinite barrier effective mass model" does not well describe the particle properties.

The infinite barrier approach does not take into account the finite ionization potential of the clusters. It can be introduced by considering a finite potential well. In this case, electrons have a nonzero probability to move out into an area surrounding the particle. For clusters, due to their small sizes, the spill-out of the electron density significantly affects their overall properties [290, 469]. The radial wave function exhibits an oscillatory behavior in the internal region and a decreasing exponential behavior in the external region. If V_0 is the depth of the well, the energies E of bound states are within $-V_0$ and zero. For convenience, we introduce the parameters $k = \{2m^*(V_0 + E)\}^{1/2}$ (inside) and $\kappa = \{2m^*|E|\}^{1/2}$ (outside). The internal and external solutions join smoothly at the boundary only for those k and κ at which the transcendental equation

$ctg(kd/2) = -\kappa/k$ holds. The resulting energy gap values are smaller than in the infinite wall approximation, as can be seen in Figure 49b. Again, the curves for different effective masses are plotted. It is found that in the size regime of interest the energy gaps are not sensitive to the depth of the potential well. Essentially the same results are obtained for $V_0 = 5$ eV and for $V_0 = 8$ eV.

The experimental gap data do not follow a single curve in Figure 49. There is a spread of gaps even for the same cluster size. This is not surprising as in this small size range (below about 15 Å) the individual atomic structures and bond conditions become crucial. For small clusters, size-dependent properties are discontinuous. In addition, as small covalent clusters are stable in a large number of isomeric structures, a wide range of energy gaps is expected for the same clusters size.

10. THIN FILMS OF PARTICLES

10.1. Formation and Properties

Nanoparticles, produced in vacuum, by an aerosol or liquid phase techniques, can form thin compact films after deposition on a support material. Narrow particle size distributions

($<\sim5\%$) can be obtained. Several groups have studied ordered superlattices of crystalline metal [470, 471] and semiconductor particles [472–474].

Another technique is the deposition of atomic vapor on an inert substrate. The particles form on the substrate by quasi-free growth, diffuse at the substrate, and form a granular solid film. By using appropriate generation conditions, one can achieve quite narrow size distributions. Depending on the substrate temperature and on the cluster-substrate interaction, the film thickness can be more or less uniform. Layer-by-layer growth for particle films can be achieved.

Nanoparticle films are interesting from many points of view. They still may have many properties of the free nanoparticles. The coupling between the particles however leads to interface formation, which can result in changes of electronic and optical properties. The particles in the films can form ordered lattices with translational symmetry. The particles in gold particle films on mica, for instance, are ordered in a zigzag pattern [475]. The type of order can be due to the coupling to the substrate or due to the interaction between the particles. For spherical symmetric particles of metal atoms, one can expect close-packed structures. The type of structure however can differ over macroscopic regions. Often, the nanoparticle films have ordered areas in the nano-to-micrometer size range.

Silicon nanoparticle films have been studied by several groups concerning their structural [16], electrical [476], electronic [477–479], and optical properties [480–482]. The films were produced from particles deposited after laser ablation of silicon [476], magnetron sputtering [483, 484], laser induced decomposition of SiH_4 [480] and a few other techniques [355, 485, 486].

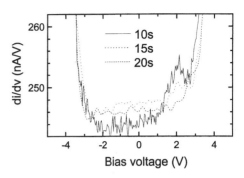

10.2. Bandgap Studies of Si Particle Films

An image of a silicon particle film, grown by vapor condensation on highly oriented pyrolytic graphite (HOPG), is shown in Figure 50 [487]. The silicon nanoparticles were grown upon deposition of several monolayers of atomic silicon using magnetron sputtering. By removing some of the particles in small areas, using high tunneling currents, it was determined that the films are between one and three layers thick.

In Figure 51 di/dV curves are shown taken on several of these films produced under different conditions. All di/dV curves show a peak at a positive bias of about 2 V. Very likely, this peak is not a band structure feature. Therefore, one obtains a bandgap of about 3.5 eV. This value is relatively large compared to the gap expected for free 5-nm silicon particles. It suggests that the coupling between the particles can alter their electronic structure and that the interface has an important influence on the film properties. These films have been grown from pristine silicon particles, which are expected to be highly reactive after formation. They strongly couple with their neighbors once they are consolidated. The dangling bonds, originally present at the particle surface, can be passivated by the interaction with their neighbors. Cross-linking can change the particle's bandgaps and other features in the electron density of states. This is seen in the peak at 2 V, which seems to be related to interparticle interaction, as it has not been observed for free silicon particles.

11. APPLICATIONS

11.1. Photonic Devices

The potential value of quantum-confined particles for nonlinear optical and electrooptical applications has long been anticipated [237, 488, 489]. In fact, there are plenty of possible applications for quantum dots in nanoelectronics [490–494] and for photonic devices [495–502]. Nonlinear optical response has for instance been found for gold particles with 25-, 90-, and 150-Å radius [503].

Nanomaterials with nonlinear properties find use as optical limiters, for eye protection and switching applications. Carbon black (particle) suspensions (CBSs) have been extensively investigated as optical limiters [504, 505]. A strong optical

Fig. 50.   An 820 × 820-nm STM image of a thin film of silicon particles [487]. Reproduced with permission from [487], to be published.

limiting effect was found for 56-Å Ag particles, while Ni (58 Å), CdS (50 Å), and PbS (66 Å) show poor limiting performance [506]. Gold particles show a size-dependent optical limiting effect in the range 25–150 Å [503].

The field of nanoelectronics is often considered as the key technology in the twenty-first century [507–509]. The required nanometer-sized structures however are often too small for the currently used industrial methods [510]. Technology such as new patterning methods have to be developed [511].

The tunability of the energy gap can be used to make light sources and detectors for specific use. In addition, when discrete energy level are formed with decreasing particle size, third-order optical nonlinearities are observed, which can be used in optical switches, waveguides, shutters, or information processors [265, 488]. Nonlinear optical response occurs for metal particles [512, 513] and semiconductor particles. For CdSe nanocrystal quantum dots, for example, the second harmonic generation (SHG) response shows a pronounced size dependence [514]. Additionally, third-harmonic linearities were found for nanocrystals and were discussed in relation to quantum confinement [8, 515, 516].

Further use of the particles has been demonstrated for metal-insulator-semiconductor field-effect-transistors (MISFET) [371, 517], photodetectors [518, 519], quantum dot lasers [520–523], solar cells [524], single-electron transistors [525], infrared detectors [518], or optical memory devices [526].

Silicon single-electron transistors (SETs) [368] have the potential of very low power consumption devices [527]. Often, SETs need to be operated at 4.2 K [366]. To make the devices however practical for circuits application, the room temperature operation and the integration into silicon are essential. Additionally, a Si dot size less than 10 nm is required. This is the size range where both the electronic and Coulomb energy gaps become large and strongly dependent on the particle sizes. The carrier transport properties through the dots [382, 528–530] needs to be known for such applications. It critically depends on the bandgap.

A silicon nanocrystal-based single-electron memory [531, 532] could well be embedded in integrated circuits. It would allow using techniques of the silicon industry to incorporate the memory function in silicon chips.

Semiconductor nanoparticles were also used as probes in biological diagnostics [533, 534]. Nanobiology and nanochemistry are large fields where clusters and nanoparticles can be used for novel processes and materials [495].

Metal particles find many applications in fields such as photocatalysis, ferrofluids, chemisorption, aerosols, and powder metallurgy. The heterogeneous catalysis by metal particles has been studied extensively [535] and is widely used in industry. The linear and nonlinear optical properties [536–538] of metal clusters have attracted a great deal of interest and may be used in future applications.

Carbon clusters are important species in astrophysics, materials science, and combustion processes. They play a key role in the preparation of thin diamond films via chemical vapor deposition or cold plasma techniques [539–544].

A type of material with potential applications is composed of organic ligand stabilized metal clusters. The metal core of a particle is encapsulated by an insulating organic monolayer. The insulating layer is a conductivity barrier through which electron tunneling or hopping occurs. Both components, the core and the encapsulant, can exert strong influences on the macroscopic electrical conductivity. A size-induced metal to semiconductor transition was reported for dodecanethiol-stabilized gold clusters [545]. Thiol-protected Au particles have been found to assemble in highly ordered two- and three-dimensional superlattices [546].

11.2. Cluster-Assembled Materials

Assembling clusters and nanoparticles can form a new class of materials and structures. Such so-called cluster-assembled materials [355, 470, 547–549], nanophase [79], nanocomposite materials [550] have novel mechanical, electrical, and optical properties. A hypothetical cluster assembled crystal with six atomic gold cluster units is shown in Figure 52.

The particles can be assembled in the form of mono- or multilayers on solid supports [551]. Such nanoparticle films have many uses for mechanical protection as well as thermal and electrical shielding.

Cross-linking of clusters may be an important effect for future cluster assembled materials. For carbon, crystals with C_{20} and C_{22} fullerene units have been postulated [552]. The clusters can be arranged in a face-centered cubic lattice. Electronic structure calculations predict for these hypothetical

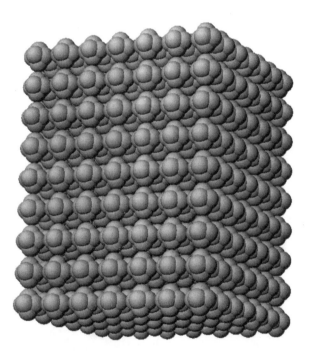

Fig. 52. Model for a hypothetical cluster assembled material consisting of Au_6 clusters in a simple cubic lattice [475]. Reproduced with permission from [475], unpublished.

crystals large bandgaps of up to 6 eV. Porous carbon structures such as hollow diamonds [553] could be very strong and light materials.

From the work on porous silicon [265, 488] and silicon nanosize films [518], one can conclude that silicon nanoparticles have the potential for use in silicon-based photonic and optoelectronic devices [554–556]. While III–V semiconductor quantum dots show interesting optical properties [520], they may eventually be replaced by silicon dots. One of the greatest challenges is the integration of such dots into serviceable architectures such as electroluminescent (EL) devices [557]. Their efficiency not only depends on the optical but also on the electrical (carrier transport) properties.

Light-emitting silicon particles may be integrated into microelectronic circuits [558]. Charge transport in nanostructures [559] is becoming an important issue as miniaturization of devices proceeds. One needs to know the detailed atomic structure of the sample at the nanoscale to fully understand the type of transport. The energy gap dependence on size and structure needs to be known for crystallites and grain boundaries. A silicon-based single electron tunneling transistor has been realized but at a temperature of 4.2 K [560], much to low for technical applications. Room-temperature single-electron tunneling can be achieved for the transport of carriers through smaller nanostructures such as a few nanometer-sized quantum dots.

A size-dependent opening of the bandgap for nanoparticles gives numerous new opportunities for their use in industrial applications. Powders of such particles can be assembled and compacted leading to nanogranular solids with new functions. If GaAs, for example, can be replaced by Si in light-emitting devices, it is an advantage as the silicon-based junction is easier to prepare and is environmentally much friendlier. Most two-component semiconductor materials are poisonous and can lead to severe health problems for exposed humans. Nanoparticle films of silicon may therefore replace intermediate and large bandgap bulk semiconductors.

11.3. Nanolithography

Optical nanolithography is limited to structures whose sizes are comparable to the wavelength of the used radiation. For nanoscale structures, X-rays can be used, but the techniques require extensive setups of special lenses and mirror systems. Alternative ways to produce small nanosized features are neutral atom lithography [561] or nanoimprint lithography [562].

A different approach is the use of STM and AFM [563]. The decomposition of organic gases in the gap between the tip and the sample was used to fabricate nanosized structures [564–568]. For instance, aluminum features were produced on silicon surfaces, using tunneling and field emission modes [569]. Using the STM, nanowires of silicon were generated by the decomposition of silane [570].

Electroluminescent devices, using silicon nanowires [441] are based on the intergration of quantum optics and silicon. Electric-field assisted AFM can also be used for nanofabrication [571]. Pits have been formed with the STM tip on

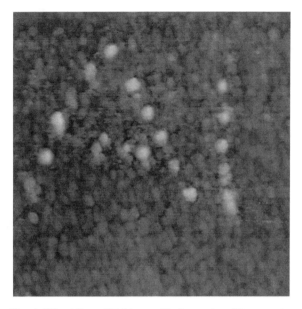

Fig. 53. A 250 × 250-nm STM image: The letters A and I were written on a thin film of silicon particles using the STM tip and high tunneling currents. For obtaining a white dot, two to three particles of the substrate where fused together [487]. Reproduced with permission from [487], unpublished.

gold [572], graphite [573], sputtered carbon [574], and metal oxide layers [575]. Nanometer scale rings on a thin Si-oxide layer were fabricated [576]. Another approach is the nanooxidation of hydrogen-passivated Si surfaces using an STM [577–579].

In Figure 53 an STM image is shown where the letters A I are written on a thin film of silicon particles [487]. The indi-

Fig. 54. A 600 × 600-nm STM image: A three level terrace structure was produced on a silicon particle film by fusion of clusters in a small square with subsequent vaporization of a surrounding square. The white square in the center is stabilized in the first process and does not vaporize in the second [487]. Reproduced with permission from [487], unpublished.

vidual points, making up the letters, are produced by scanning over a few particles with an enhanced tunneling current. This leads to the fusion of two to three silicon particles, which then appear white with respect to the gray background.

The STM can also be used to vaporize particles. This occurs when a high bias voltage is temporarily applied. In this case, nanostructures can be formed which appear black on a gray background of the silicon particle film. The combination of fusion and vaporization can be used to shape a silicon particle film [487]. This is demonstrated in Figure 54. A square-shaped terrace, higher than the substrate level, is surrounded by a trench, separating the inner terrace from the particle film.

Both the fusion and the vaporization are possible due to the large bandgap of the silicon particles forming the thin films. For the low-conducting layer, the separation between the tip and silicon surface is very small and the electric field is very high. A combination of both, very high tunneling current and very high bias voltage, and their variation, allows the formation of shapes as shown in the preceding figures.

Acknowledgments

The author thanks his coworkers who have contributed to experiments and discussions which made this review possible: Manuel Lonfat, Hugues-Vincent Roy, Bjorn Marsen, and Paul Scheier. Also, helpful discussions with David Harwell and John Head (from the Hawaii Advanced Nanotechnology Alliance, HANA) are acknowledged.

REFERENCES

1. T. van Buuren, L. N. Dinh, L. L. Chase, W. J. Siekhaus, and L. J. Terminello, *Phys. Rev. Lett.* 80, 3803 (1998).
2. S. Hayashi and H. Abe, *Jpn. J. Appl. Phys.* 23, L824 (1984).
3. Y. Kanemitsu, H. Uto, Y. Masumoto, and Y. Maeda, *Appl. Phys. Lett.* 61, 2187 (1992).
4. V. Cracium, I. W. Boyd, A. H. Reader, and D. E. W. Vandenhoudt, *Appl. Phys. Lett.* 65, 3233 (1994).
5. M. R. Zachariah, M. J. Carrier, and E. Blaisten-Barijas, *J. Phys. Chem.* 100, 14,856 (1996).
6. D. J. Norris, A. Sacra, C. B. Murray, and M. G. Bawendi, *Phys. Rev. Lett.* 72, 2612 (1994).
7. D. J. Norris and M. G. Bawendi, *Phys. Rev. B* 53, 16338 (1996).
8. U. Banin, J. C. Lee, A. A. Guzelian, A. V. Kadavanich, A. P. Alivisatos, W. Jaskolski, G. W. Bryant, A. L. Efros, and M. Rosen, *J. Chem. Phys.* 109, 2306 (1998).
9. U. Banin, J. C. Lee, A. Guzelian, A. V. Kadavanich, and A. P. Alivisatos, *Superlattices Microstruct.* 22, 559 (1997).
10. M. Leon, P. M. Petroff, D. Leonard, and S. Fafard, *Science* 267, 1966 (1995).
11. S. A. Empedocles, D. J. Norris, and M. G. Bawendi, *Phys. Rev. Lett.* 77, 3873 (1996).
12. A. I. Ekimov, F. Hache, M. C. Schanne-Klein, D. Ricard, C. Flytzanis, I. A. Kudryavtsev, T. V. Yazeva, A. V. Rodina, and A. L. Efros, *J. Opt. Soc. Am. B* 10, 100 (1993).
13. H. Fu and A. Zunger, *Phys. Rev. B* 57, R15064 (1998).
14. H. Fu, L. W. Wang, and A. Zunger, *Phys. Rev. B* 57, 9971 (1998).
15. L.-W. Wang and A. Zunger, *J. Phys. Chem.* 98, 2158 (1994).
16. B. Garrido, A. Perez-Rodrigues, J. R. Morante, A. Achiq, F. Gourbilleau, R. Madelon, and R. Rizk, *J. Vac. Sci. Technol. B* 16, 1851 (1998).

17. J. P. Wilcoxon, G. A. Samara, and P. N. Provencio, *Phys. Rev. B* 60, 2704 (1999).
18. A. K. Dutta, *Appl. Phys. Lett.* 68, 1189 (1996).
19. V. S. Veerasamy, G. A. J. Amaratunga, W. I. Milne, P. Hewitt, P. J. Fallon, D. R. McKenzie, and C. A. Davis, *Diamond Relat. Mater.* 2, 782 (1993).
20. U. Stephan, T. Frauenheim, P. Blaudeck, and G. Jingnickel, *Phys. Rev. B* 49, 1489 (1994).
21. D. R. McKenzie, D. A. Muller, and B. A. Pailthorpe, *Phys. Rev. Lett.* 67, 773 (1991).
22. D. A. Drabold, P. A. Fedders, and M. P. Grumbach, *Phys. Rev. B* 54, 5480 (1996).
23. D. Kwon, C.-C. Chen, J. D. Cohen, H.-C. Jin, E. Hollar, I. Robertson, and J. R. Abelson, *Phys. Rev. B* 60, 4442 (1999).
24. M. Yamamoto, R. Hayashi, K. Tsunemoto, K. Kohno, and Y. Osaka, *Jpn. J. Appl. Phys.* 30, 136 (1991).
25. D. J. DiMaria, J. R. Kirtley, E. J. Pakulis, D. W. Dong, T. S. Kuan, F. L. Pesavento, T. N. Theis, and J. A. Cutro, *J. Appl. Phys.* 56, 401 (1984).
26. M. O'Neil, J. Marohn, and G. McLendon, *J. Phys. Chem.* 94, 4356 (1990).
27. J. A. Eychmuller, A. Hasselbrath, L. Katsikas, and H. Weller, *Ber. Bunsen-Ges. Phys. Chem.* 95, 79 (1991).
28. A. Hasselbarth, A. Eychmuller, and H. Weller, *Chem. Phys. Lett.* 203, 271 (1993).
29. M. G. Bawendi, P. J. Carroll, and W. L. Wilson, *J. Chem. Phys.* 96, 946 (1992).
30. M. Nirmal, C. B. Murray, and M. G. Bawendi, *Phys. Rev. B* 50, 2293 (1994).
31. V. Petrova-Koch, T. Muschik, A. Kux, B. K. Meyer, and F. Koch, *Appl. Phys. Lett.* 61, 943 (1992).
32. W. Hoheisel, Y. L. Colvin, C. S. Johnson, and A. P. Alivisatos, *J. Chem. Phys.* 101, 845 (1994).
33. M. Nirmal, D. J. Norris, M. Kuno, M. G. Bawendi, A. L. Efros, and M. Rosen, *Phys. Rev. Lett.* 75, 3728 (1995).
34. A. L. Efros, M. Rosen, M. Kuno, M. Nirmal, D. J. Norris, and M. Bawendi, *Phys. Rev. B* 54, 4843 (1996).
35. D. J. Norris, A. L. Efros, M. Rosen, and M. G. Bawendi, *Phys. Rev. B* 53, 16,347 (1996).
36. M. Chamarro, C. Gourdon, P. Lavallard, O. Lublinskaya, and A. I. Ekimov, *Phys. Rev. B* 53, 1336 (1996).
37. M. Chamarro, C. Gourdon, and P. Lavallard, *J. Lumin.* 70, 222 (1996).
38. U. Woggon, F. Gindele, O. Wind, and C. Klingshirn, *Phys. Rev. B* 54, 1506 (1996).
39. I. Giaever, *Phys. Rev. Lett.* 5, 147 (1960).
40. R. M. Feenstra, W. A. Thompson, and A. P. Fein, *Phys. Rev. Lett.* 56, 608 (1986).
41. R. M. Feenstra, J. A. Stroscio, and A. P. Fein, *Surf. Sci.* 181, 295 (1987).
42. R. M. Feenstra, *J. Vac. Sci. Technol. B* 7, 925 (1989).
43. P. Avouris and I. Lyo, *Surf. Sci.* 242, 1 (1991).
44. R. J. Hamers, R. M. Tromp, and J. E. Demuth, *Phys. Rev. Lett.* 56, 1972 (1986).
45. H. F. Hess, R. B. Robinson, R. C. Dynes, J. Valles, J. M., and J. V. Waszczak, *J. Vac. Sci. Technol. A* 8, 450 (1990).
46. L. C. Davis, M. P. Everson, R. C. Jaklevic, and W. Shen, *Phys. Rev. B* 43, 3821 (1991).
47. R. M. Feenstra, *Phys. Rev. B* 60, 4478 (1999-I).
48. B. Marsen, M. Lonfat, P. Scheier, and K. Sattler, *Phys. Rev. B* 62, 6892 (2000).
49. O. Millo, D. Katz, Y. W. Cao, and U. Banin, *Phys. Rev. B* 61, 16,773 (2000-II).
50. D. Porath, Y. Levi, M. Tarabiah, and O. Millo, *Phys. Rev. B* 56, 9829 (1997-I).
51. X. Yao, T. G. Ruskell, R. K. Workman, D. Sarid, and D. Chen, *Surf. Sci.* 366, L743 (1996).
52. W. Mizutani, M. Shigeno, and Y. Sakakibara, *J. Vac. Sci. Technol.* 8, 675 (1990).

53. A. Okumura, K. Miyamura, and Y. Gohshi, *J. Vac. Sci. Technol.* 8, 625 (1990).
54. A. Okumura, H. Nakagawa, and Y. Gohshi, *Jpn. J. Appl. Phys.* 34, 2055 (1995).
55. S. Ogawa, F.-R. F. Fan, and A. J. Bard, *J. Phys. Chem.* 99, 11,182 (1995).
56. B. Alperson, I. Rubinstein, G. Hodes, D. Porath, and O. Millo, *Appl. Phys. Lett.* 75, 1751 (1999).
57. B. Alperson, S. Cohen, I. Rubinstein, and G. Hodes, *Phys. Rev. B* 52, R17,017 (1995).
58. L. I. Halaoui, R. L. Wells, and J. Coury, L. A., *Chem. Mater.* 12, 1205 (2000).
59. U. Banin, Y. W. Cao, D. Katz, and O. Millo, *Nature (London)* 400, 542 (1999).
60. D. L. Klein, R. Roth, A. K. L. Lim, A. P. Alivisatos, and P. L. McEuen, *Nature (London)* 389, 699 (1997).
61. D. L. Klein, P. L. McEuen, J. E. Bowen Katari, R. Roth, and A. P. Alivisatos, *Appl. Phys. Lett.* 68, 2574 (1996).
62. U. Sivan, R. Berkovits, Y. Aloni, O. Prus, A. Auerbach, and G. BenYoseph, *Phys. Rev. Lett.* 77, 1123 (1996).
63. S. R. Patel, S. M. Cronenwett, D. R. Stewart, A. G. Hubers, C. M. Marcus, C. I. Duruoz, J. S. Harris, K. Campman, and A. C. Gossard, *Phys. Rev. Lett.* 80, 4522 (1998).
64. J. Tersoff and D. R. Hamann, *Phys. Rev. B* 31, 805 (1985).
65. N. D. Lang, *Phys. Rev. Lett.* 55, 230 (1985).
66. N. D. Lang, *Phys. Rev. Lett.* 55, 2925 (1985).
67. C. J. Chen, *J. Vac. Sci. Technol. A* 6, 319 (1988).
68. C. R. Leavens and G. C. Aers, *Phys. Rev. B* 38, 7357 (1988).
69. E. Teckman and S. Ciraci, *Phys. Rev. B* 40, 10,386 (1989).
70. C. J. Chen, *Phys. Rev. Lett.* 65, 448 (1990).
71. G. Doyen, D. Drakova, and M. Scheffler, *Phys. Rev. B* 47, 9778 (1993).
72. N. D. Lang, *Phys. Rev. B* 34, 5947 (1986).
73. A. Selloni, P. Carnevalli, E. Tosati, and C. D. Chen, *Phys. Rev. B* 31, 2602 (1985).
74. M. Tsukada, K. Kobayashi, N. Isshiki, and H. Kageshima, *Surf. Sci. Rep.* 13, 2 (1991).
75. F. Zypman, L. F. Fonseca, and Y. Goldstein, *Phys. Rev. B* 49, 1981 (1994-I).
76. R. Tromp, *J. Phys. Condens. Matter* 1, 10,211 (1989).
77. R. J. Hamers, *Annu. Rev. Phys. Chem.* 40, 531 (1989).
78. S. Morita, Y. Sugawara, and Y. Fukano, *Jpn. J. Appl. Phys., Part 1*, R32, 2983 (1993).
79. K. Sattler, G. Raina, M. Ge, J. Xhie, N. Venkateswaran, Y. X. Liao, and R. W. Siegel, *J. Appl. Phys.* 76, 546 (1994).
80. J. Bardeen, *Phys. Rev. Lett.* 6, 57 (1960).
81. K. Makoshi, N. Shima, and T. Mii, *Surf. Sci.* 386, 335 (1997).
82. H. Matsuoka and S. Kimura, *Jpn. J. Appl. Phys.* 34, 1326 (1995).
83. K. A. Matveev, L. I. Glazman, and H. U. Baranger, *Phys. Rev. B* 54, 5637 (1996).
84. B. Wang, X. Xiao, X. Huang, P. Sheng, and J. G. Hou, *Appl. Phys. Lett.* 77, 1179 (2000).
85. H. Nejo, M. Aono, and V. A. Tkachenko, *J. Vac. Sci. Technol.* 14, 2399 (1996).
86. R. O. Jones, *J. Chem. Phys.* 99, 1194 (1993).
87. V. Bonacic-Koutecky, L. Cespiva, P. Fantucci, J. Pittner, and J. Koutecky, *J. Chem. Phys.* 100, 490 (1994).
88. C.-Y. Cha, G. Ganteför, and W. Eberhardt, *J. Chem. Phys.* 100, 995 (1994).
89. W. A. de Heer, *Rev. Mod. Phys.* 65, 611 (1993).
90. K. Clemenger, *Phys. Rev. B* 32, 1359 (1985).
91. W. Ekardt and Z. Penzar, *Phys. Rev. B* 43, 1322 (1991).
92. S. Frauendorf and V. V. Pashkevich, *Z. Phys. D* 26, S98 (1993).
93. A. Bulgac and C. Lewenkopf, *Phys. Rev. Lett.* 71, 4130 (1993).
94. C. Yannouleas and U. Landman, *Phys. Rev. B* 51, 1902 (1995).
95. G. Lauritsch, P.-G. Reinhard, J. Meyer, and M. Brack, *Phys. Lett. A* 160, 179 (1991).
96. J. Martins, J. Buttet, and R. Car, *Phys. Rev. B* 31, 1804 (1985).
97. U. Roetlisberger and W. Andreoni, *J. Chem. Phys.* 94, 8129 (1991).
98. V. Bonacic-Koutecky, P. Fantucci, and J. Koutecky, *Chem. Rev.* 91, 1035 (1991).
99. D. M. Lindsay, Y. Wang, and T. F. George, *J. Chem. Phys.* 86, 3500 (1987).
100. A. Yoshida, T. Dossing, and M. Manninen, *J. Chem. Phys.* 101, 3041 (1994).
101. F. Chandezon, S. Bjornholm, J. Borggreen, and K. Hansen, *Phys. Rev. B* 55, 5485 (1997-II).
102. U. Lammers and G. Borstel, *Phys. Rev. B* 49, 17,360 (1994-II).
103. H. Häkkinen and M. Manninen, *Phys. Rev. B* 52, 1540 (1995).
104. C. Lueder and K. H. Meiwes-Broer, *Chem. Phys. Lett.* 294, 391 (1998).
105. P. B. Balbuena, P. A. Derosa, and M. J. Seminario, *J. Phys. Chem.* 103, 2830 (1999).
106. O. Cheshnovsky, K. J. Taylor, J. Conceicao, and R. E. Smalley, *Phys. Rev. Lett.* 64, 1785 (1990).
107. K. J. Taylor, C. L. Pettiette-Hall, O. Cheshnovsky, and R. E. Smalley, *J. Chem. Phys.* 96, 3319 (1993).
108. C. Y. Cha, G. Ganteför, and W. Eberhardt, *J. Chem. Phys.* 99, 6308 (1993).
109. O. B. Christensen, *Phys. Rev. B* 50, 1844 (1994-I).
110. C. Massobrio, A. Pasquarello, and R. Car, *Phys. Rev. B* 54, 8913 (1996-II).
111. C. Massobrio, A. Pasquarello, and R. Car, *Phys. Rev. Lett.* 75, 2104 (1995).
112. H. Akeby, I. Panas, L. G. M. Pettersson, P. Siegbahn, and U. Wahlgren, *J. Chem. Phys.* 94, 5471 (1990).
113. O. B. Christensen, K. W. Jacobsen, J. K. Norskov, and M. Manninen, *Phys. Rev. Lett.* 66, 2219 (1991).
114. J. Demuynck, M.-M. Rohmer, A. Strich, and A. Veillard, *J. Chem. Phys.* 75, 3443 (1981).
115. H. Tekewaki, E. Miyoshi, and T. Nakamura, *J. Chem. Phys.* 76, 5073 (1982).
116. G. L. Estiu, M. G. Cory, and M. C. Zerner, *J. Phys. Chem.* 104, 233 (2000).
117. A. L. Efros and A. L. Efros, *Sov. Phys. Semicond.* 16, 772 (1982).
118. L. E. Brus, *J. Chem. Phys.* 79, 5566 (1983).
119. L. E. Brus, *J. Chem. Phys.* 80, 4403 (1984).
120. P. F. Trwoga, A. J. Kenyon, and C. W. Pitt, *J. Appl. Phys.* 83, 3789 (1998).
121. D. B. T. Thoai, Y. Z. Hu, and S. W. Koch, *Phys. Rev. B* 42, 11,261 (1990).
122. J. P. Proot, C. Delerue, and G. Allan, *Appl. Phys. Lett.* 61, 1948 (1992).
123. M. V. Rama Krishna and R. A. Friesner, *Phys. Rev. Lett.* 67, 629 (1991).
124. M. V. Rama Krishna and R. A. Friesner, *J. Chem. Phys.* 96, 873 (1992).
125. M. V. Rama Krishna and R. A. Friesner, *J. Chem. Phys.* 95, 8309 (1992).
126. V. L. Colvin and A. P. Alivisatos, *J. Chem. Phys.* 97, 730 (1992).
127. P. E. Lippens and M. Lannoo, *Phys. Rev. B* 39, 10,935 (1989).
128. J. C. Phillips, *J. Chem. Phys.* 83, 3330 (1985).
129. S. Saito, S. Ohnishi, and S. Sugano, *Phys. Rev. B* 33, 7036 (1986).
130. K. Raghavachari, *J. Chem. Phys.* 84, 5672 (1986).
131. G. Pacchioni and J. Koutecky, *J. Chem. Phys.* 84, 3301 (1986).
132. R. Biswas and D. R. Hamann, *Phys. Rev. B.* 36, 6434 (1987).
133. U. Roethlisberger, W. Andreoni, and P. Giannozzi, *J. Chem. Phys.* 96, 1248 (1992).
134. U. Roethlisberger, W. Andreoni, and M. Parrinello, *Phys. Rev. Lett.* 72, 665 (1994).
135. K. Balasubramanian, *Chem. Phys. Lett* 125, 400 (1986).
136. K. Balasubramanian, *Chem. Phys. Lett.* 135, 283 (1987).
137. P. Ballone, W. Andreoni, R. Car, and M. Parrinello, *Phys. Rev. Lett.* 60, 271 (1988).
138. C. H. Patterson and R. P. Messmer, *Phys. Rev. B* 42, 7530 (1990).
139. K. Raghavachari, *J. Chem. Phys.* 83, 3520 (1985).
140. K. Raghavachari and V. Logovinsky, *Phys. Rev. Lett.* 55, 2853 (1985).
141. K. Raghavachari and C. M. Rohlfing, *J. Chem. Phys.* 89, 2219 (1988).
142. D. Tomanek and M. A. Schlueter, *Phys. Rev. Lett.* 56, 1055 (1986).
143. D. Tomanek and M. A. Schlüter, *Phys. Rev. B* 36, 1208 (1987).

144. N. Binggeli and J. R. Chelikowsky, *Phys. Rev. B* 50, 11,764 (1994).

145. M. R. Hunt and R. E. Palmer, *Philos. Trans. R. Soc. London, Ser. A* 356, 231 (1998).

146. C. Brechignac, P. Cahuzac, F. Carlier, M. de Frutos, and J. Leygnier, *Chem. Phys. Lett.* 189, 28 (1992).

147. V. Bonacic-Koutecky, J. Pittner, C. Fuchs, P. Fantucci, M. F. Guest, and J. Koutecky, *J. Chem. Phys.* 104, 1427 (1996).

148. V. Bonacic-Koutecky, J. Jellinek, M. Wiechert, and P. Fantucci, *J. Chem. Phys.* 107, 6321 (1997).

149. U. Roethlisberger and W. Andreoni, *J. Chem. Phys.* 94, 8129 (1991).

150. J. M. Pacheco and W. D. Schoene, *Phys. Rev. Lett.* 79, 4989 (1997).

151. R. A. Broglia and J. M. Pacheco, *Phys. Rev. B* 44, 5901 (1991).

152. Z. Penzar and W. Ekardt, *Z. Phys. D* 19, 109 (1991).

153. O. Genzken, M. Brack, and E. Chabanat, *Ber. Bunsen-Ges. Phys. Chem.* 96, 1217 (1992).

154. W. Eberhardt, P. Fayet, D. M. Cox, Z. Fu, A. Kaldor, R. Sherwood, and D. Sondericker, *Phys. Rev. Lett.* 64, 780 (1990).

155. A. Bettac, L. Koeler, V. Rank, and K. H. Meiwes-Broer, *Surf. Sci.* 402, 475 (1998).

156. B. V. Reddy, S. K. Nayak, S. N. Khanna, B. K. Rao, and P. Jena, *J. Phys. Chem.* 102, 1748 (1998).

157. H. Wu, S. R. Desai, and L.-S. Wang, *Phys. Rev. Lett.* 77, 2436 (1996).

158. L. Lian, C.-X. Su, and P. B. Armentrout, *J. Chem. Phys.* 96, 7542 (1992).

159. G. L. Estiu and M. C. Zerner, *J. Phys. Chem.* 100, 16,874 (1996).

160. S. K. Nayak, B. K. Rao, and P. Jena, *J. Phys.: Condens. Matter* 10, 10,863 (1998).

161. Q. Wang, X. G. Gong, Q. Q. Zheng, D. Y. Sun, and G. H. Wang, *Phys. Rev. B* 54, 10,896 (1996-I).

162. B. K. Rao and P. Jena, *J. Chem. Phys.* 111, 1890 (1999).

163. S. H. Yang, D. A. Drabold, J. B. Adams, and A. Sachdev, *Phys. Rev. B* 47, 1567 (1993-I).

164. R. C. Baetzold, *J. Phys. Chem.* 101, 8180 (1997).

165. J. E. Fowler and J. M. Ugalde, *Phys. Rev. A* 58, 383 (1998).

166. V. Kumar, *Phys. Rev. B* 57, 8827 (1998).

167. J. Akola, M. Manninen, H. Hakkinen, U. Landman, X. Li, and L.-S. Wank, *Phys. Rev. B* 60, R11,297 (1999-II).

168. Y. Q. Cai, A. M. Bradshaw, Q. Guo, and D. W. Goodman, *Surf. Sci.* 399, L357 (1998).

169. H. Wu, S. R. Desai, and L.-S. Wang, *Phys. Rev. Lett.* 76, 212 (1996).

170. C. P. Vinod, G. U. Kulkarni, and C. N. R. Rao, *Chem. Phys. Lett.* 289, 329 (1998).

171. H. Poppa, R. D. Moorhead, and M. Avalos-Borja, *J. Vac. Sci. Technol. A* 7, 2882 (1989).

172. P. A. Montano, G. K. Shenoy, E. E. Alp, W. Schulze, and J. Urban, *Phys. Rev. Lett.* 56, 1285 (1983).

173. S. Ijima and T. Ichihashi, *Phys. Rev. B* 43, 10,647 (1991).

174. H.-P. Cheng and R. S. Berry, *Phys. Rev. B* 43, 10,647 (1991).

175. M. E. Geusic, M. D. Morse, and R. E. Smalley, *J. Chem. Phys.* 82, 590 (1985).

176. M. D. Morse, M. E. Geusic, J. R. Heathm, and R. E. Smalley, *J. Chem. Phys.* 83, 2293 (1985).

177. R. L. Whetten, M. R. Zakin, D. M. Cox, D. J. Trevor, and A. Kaldor, *J. Chem. Phys.* 85, 1697 (1986).

178. J. L. Elkind, F. D. Weiss, J. M. Alford, R. T. Laaksonen, and R. E. Smalley, *J. Chem. Phys.* 88, 5215 (1988).

179. P. Fayet, A. Kaldor, and D. M. Cox, *J. Chem. Phys.* 92, 254 (1990).

180. J. L. Persson, M. Andersson, and A. Rosen, *Z. Phys. D* 23, 1 (1992).

181. J. Ho, L. Zhu, E. K. Parks, and S. J. Riley, *J. Chem. Phys.* 99, 140 (1993).

182. Y. M. Hamrick and M. D. Morse, *J. Chem. Phys.* 93, 6494 (1989).

183. C. Q. Jiao and B. S. Freiser, *J. Phys. Chem.* 99, 10,723 (1995).

184. C. Berg, T. Schindler, G. Niedern-Schatteburg, and V. E. Bondybey, *J. Chem. Phys.* 104, 4770 (1995).

185. R. L. Whetten, D. M. Cox, D. J. Trevor, and A. Kaldor, *Phys. Rev. Lett.* 54, 1494 (1985).

186. J. Conceicao, R. T. Laaksonen, L.-S. Wang, T. Guo, P. Nordlander, and R. E. Smalley, *Phys. Rev. B* 51, 4668 (1995).

187. M. B. Knickelbein and S. Yang, *J. Chem. Phys.* 93, 1476 (1990).

188. K. Athanassenas, D. Kreisle, B. A. Collings, D. M. Rayner, and P. A. Hackett, *Chem. Phys. Lett.* 213, 105 (1993).

189. M. R. Zakin, R. O. Brickman, D. M. Cox, and A. Kaldor, *J. Chem. Phys.* 88, 6605 (1988).

190. J. Y. Saillard and R. Hoffman, *J. Am. Chem. Soc.* 106, 2006 (1984).

191. M. A. Nygren, P. Siegbahn, C. Jin, T. Guo, and R. E. Smalley, *J. Chem. Phys.* 95, 6181 (1991).

192. M. B. Knickelbein and W. J. C. Menezes, *J. Phys. Chem.* 96, 1992 (1992).

193. S. K. Nayak, B. K. Rao, S. N. Khanna, and P. Jena, *Chem. Phys. Lett.* 259, 588 (1996).

194. J. Zhao, X. Chen, and G. Wang, *Phys. Lett. A* 214, 211 (1996).

195. H. Kietzmann, J. Morenzin, P. S. Bechthold, G. Ganteför, and W. Eberhardt, *J. Chem. Phys.* 109, 2275 (1998).

196. P. Calaminici, A. M. Koster, N. Russo, and D. R. Salahub, *J. Chem. Phys.* 105, 9546 (1996).

197. H.-J. Fan, C.-W. Liu, and M.-S. Liao, *Chem. Phys. Lett.* 273, 353 (1997).

198. G. D. Geske, A. I. Boldyrev, X. Li, and L.-S. Wang, *J. Chem. Phys.* 113, 5130 (2000).

199. S. Kümmel, M. Brack, and P.-G. Reinhard, *Phys. Rev. B* 62, 7602 (2000-I).

200. C. Barreteau, D. Spanjaard, and M. C. Desjonquères, *Phys. Rev. B* 58, 9721 (1998-I).

201. S. K. Nayak, S. N. Khanna, B. K. Rao, and P. Jena, *J. Phys. Chem. A* 101, 1072 (1997).

202. E. Z. D. Silva and A. Antonelli, *Phys. Rev. B* 54, 17,057 (1996).

203. H.-G. Fritsche, *Z. Phys. Chem.* 191, 219 (1995).

204. H.-G. Fritsche, *Z. Phys. Chem.* 192, 21 (1995).

205. S. Krüger, S. Vent, and N. Rösch, *Ber. Bunsen-Ges. Phys. Chem. Int. Ed.* 101, 1640 (1997).

206. S. A. Nepijko, M. Klimenkov, M. Adelt, H. Kuhlenbeck, R. Schlögl, and H.-J. Freund, *Langmuir* 15, 5309 (1999).

207. M. Klimenkov, S. Nepijko, H. Kuhlenbeck, M. Bäumer, R. Schlögel, and H.-J. Freund, *Surf. Sci.* 391, 27 (1997).

208. S. A. Nepijko, M. Klimenkov, H. Kuhlenbeck, D. Zemlyanov, D. Herein, R. Schögel, and H.-J. Freund, *Surf. Sci.* 413, 192 (1998).

209. A. V. Matveev, K. M. Neyman, G. Pacchioni, and N. Rösch, *Chem. Phys. Lett.* 299, 603 (1999).

210. V. Musolino, A. Selloni, and R. Car, *Surf. Sci.* 402–404, 413 (1998).

211. J. T. Lau, A. Achleitner, and W. Wurth, *Chem. Phys. Lett.* 317, 269 (2000).

212. F. J. Palacios, M. P. Iniguez, M. J. López, and J. A. Alonso, *Phys. Rev. B* 60, 2908 (1999-II).

213. Q. Ma and R. A. Rosenberg, *Surf. Sci.* 391, L1224 (1997).

214. A. Bifone, L. Casalis, and R. Riva, *Phys. Rev. B* 51, 11,043 (1995).

215. J. Xhie, K. Sattler, U. Mueller, G. Raina, and N. Venkateswaran, *Phys. Rev. B* 43, 8917 (1991).

216. N. Nilius, N. Ernst, and H.-J. Freund, *Phys. Rev. Lett.* 84, 2000 (2000).

217. Y. S. Gordeev, M. V. Gomoyunova, A. K. Grigorev, V. M. Mikushin, I. I. Pronin, S. E. Sysoev, and N. S. Faradzhev, *Fiz. Tverd. Tela* 36, 8 (1994).

218. C. Goyhenex, M. Croci, C. Claeys, and C. R. Henry, *Surf. Sci.* 352, 5 (1996).

219. J. G. A. Dubois, J. W. Gerritsen, and H. van Kempen, *Physica B* 218, 262 (1996).

220. D. A. Handley, in "Colloidal Gold: Principles, Methods, and Applications" (M. A. Hayat, Ed.), Vol. 1, pp. 13–32. Academic Press, San Diego, CA, 1989.

221. Y. N. C. Chan, G. S. W. Craig, R. R. Schrock, and R. E. Cohen, *Chem. Mater.* 4, 885 (1992).

222. D. Tomanek, S. Mukherjee, and K. H. Bennemann, *Phys. Rev. B* 28, 665 (1983).

223. K. Rademann, B. Kaiser, U. Even, and F. Hensel, *Phys. Rev. Lett.* 59, 2319 (1987).

224. C. Brechignac, M. Broyer, P. Cahuzac, G. Delacretaz, P. Labastie, and L. Woeste, *Chem. Phys. Lett.* 120, 55 (1985).

225. R. Busani, M. Folkers, and O. Cheshnovsky, *Phys. Rev. Lett.* 81, 3836 (1998).

226. J. Pedersen, S. Bjornholm, J. Borggreen, K. Hansen, T. P. Martin, and H. D. Rasmussen, *Nature* 353, 733 (1991).

227. T. P. Martin, S. Bjornholm, J. Borggreen, C. Brechignac, P. Cahuzac, K. Hansen, and J. Pedersen, *Chem. Phys. Lett.* 186, 53 (1991).

228. H. Ito, T. Sakurai, T. Matsuo, T. Ichihara, and I. Katakuse, *Phys. Rev. B* 48, 4741 (1993-I).

229. A. I. Ekimov, A. L. Efros, and A. A. Onushchenko, *Solid State Commun.* 56, 921 (1985).

230. H. Weller, U. Koch, M. Gutierez, and A. Henglein, *Ber. Bunsen-Ges. Phys. Chem.* 88, 649 (1984).

231. A. P. Alivisatos, *J. Phys. Chem.* 100, 13,226 (1996).

232. N. A. Hill and K. B. Whaley, *Chem. Phys.* 210, 117 (1996).

233. S. V. Gaponenko, "Optical Properties of Semiconductor Nanocrystals," Cambridge Univ. Press, New York, 1998.

234. L. Brus, *J. Phys. Chem.* 98, 3575 (1994).

235. C. B. Murray, D. J. Norris, and M. G. Bawendi, *J. Am. Chem. Soc.* 115, 8706 (1993).

236. N. Chestnoy, T. D. Harris, R. Hull, and L. E. Brus, *J. Phys. Chem.* 90, 3393 (1986).

237. S. Schmitt-Rink, D. S. Miller, and D. S. Chemla, *Phys. Rev. B* 35, 8113 (1987).

238. A. Kornowski, R. Eichberger, M. Giersig, H. Weller, and A. Eychmuller, *J. Phys. Chem.* 100, 12,467 (1996).

239. K. W. Berryman, S. A. Lyon, and M. Segev, *J. Vac. Sci. Technol.* 15, 1045 (1997).

240. H. Fu and A. Zunger, *Phys. Rev. B* 56, 1496 (1997).

241. M. Kuno, J. K. Lee, B. O. Dabbousi, F. V. Mikulec, and M. G. Bawendi, *J. Chem. Phys.* 106, 9869 (1997).

242. O. I. Micic, J. R. Sprague, C. J. Curtis, K. M. Jones, J. L. Machol, A. J. Notik, H. Giessen, B. Fluegel, G. Mohs, and N. Peyghambarian, *J. Phys. Chem.* 99, 7754 (1995).

243. L. M. Ramaniah and S. V. Nair, *Physica B* 212, 245 (1995).

244. J.-Y. Marzin, J.-M. Gerard, A. Izrael, D. Barrier, and G. Bastard, *Phys. Rev. Lett.* 73, 716 (1994).

245. M. Lannoo, *Mater. Sci. Eng. B* 9, 485 (1991).

246. C. Sikorski and U. Merkt, *Phys. Rev. Lett.* 62, 2164 (1989).

247. P. N. Brounkov, A. Polimeni, S. T. Stoddart, M. Henini, L. Eaves, P. C. Main, A. R. Kovsh, Y. G. Musikhin, and S. G. Konnikov, *Appl. Phys. Lett.* 73, 1092 (1998).

248. D. Gammon, E. S. Snow, B. V. Shanabrook, D. S. Katzer, and D. Park, *Phys. Rev. Lett.* 76, 3005 (1996).

249. T. R. Taylor, K. R. Asmis, C. Xu, and D. M. Neumark, *Chem. Phys. Lett.* 20, 133 (1998).

250. K. R. Asmis, T. R. Taylor, and D. M. Neumark, *Chem. Phys. Lett.* 308 (1999).

251. I. Kang and F. W. Wise, *J. Opt. Soc. Am.* 14, 1632 (1997).

252. V. N. Soloviev, A. Eichhöfer, D. Fenske, and U. Banin, *J. Am. Chem. Soc.* 122, 2673 (2000).

253. O. I. Micic, H. M. Cheong, H. Fu, A. Zunger, J. R. Sprague, A. Mascarenhas, and A. J. Nozik, *J. Phys. Chem. B* 101, 4904 (1997).

254. O. I. Micic, K. M. Jones, A. Cahill, and A. J. Nozik, *J. Phys. Chem. B* 102, 9791 (1998).

255. I. Kang and F. W. Wise, *J. Opt. Soc. Am. B* 14, 1632 (1997).

256. J.-Y. Yi, *Chem. Phys. Lett.* 325, 269 (2000).

257. A. J. Williamson, A. Franceschetti, H. Fu, L. W. Wang, and A. Zunger, *J. Electron. Mater.* 28, 414 (1999).

258. A. Tomasulo and M. V. Ramakrishna, *J. Chem. Phys.* 105, 3612 (1996).

259. M. V. Ramakrishna and R. A. Friesner, *J. Chem. Phys.* 95, 8309 (1991).

260. J. P. Zheng and H. S. Kwok, *J. Opt. Soc. Am. B* 9, 2047 (1992).

261. S. Nomura and T. Kobayashi, *Phys. Rev. B* 45, 1305 (1992).

262. A. Olkhovets, R.-C. Hsu, A. Lopovskii, and F. W. Wise, *Phys. Rev. Lett.* 81, 3539 (1998).

263. M. S. El-Shall, S. Li, and I. Germanenko, "Abstracts of Papers of the American Chemical Society," 1999, Vol. 217, p. 1.

264. W. H. Jiang, X. L. Ye, B. Xu, H. Z. Xu, D. Ding, J. B. Liang, and Z. G. Wang, *J. Appl. Phys.* 88, 2529 (2000).

265. B. L. Justus, R. J. Tonucci, and A. D. Berry, *Appl. Phys. Lett.* 61, 3152 (1992).

266. L. T. Canham, *Appl. Phys. Lett.* 57, 1046 (1990).

267. H. Takagi, H. Ogawa, Y. Yamazaki, A. Ishizaki, and T. Nakagiri, *Appl. Phys. Lett.* 56, 2379 (1990).

268. V. G. Baru, S. Bayliss, L. Zaharov, and Yu, *Microelectron. Eng.* 36, 111 (1997).

269. K. A. Littau, P. J. Szajowski, A. J. Muller, A. R. Kortan, and L. E. Brus, *J. Phys. Chem.* 97, 1224 (1993).

270. X.-N. Liu, X.-W. Wu, X.-M. Bao, and Y.-L. He, *Appl. Phys. Lett.* 64, 220 (1994).

271. P. D. Milewski, D. J. Lichtenwalner, P. Mehta, A. I. Kingon, D. Zhang, and R. M. Kolbas, *J. Electron Mater.* 23, 57 (1994).

272. H. Morizaki, F. W. Ping, H. Ono, and K. Yazawa, *J. Appl. Phys.* 70, 1869 (1991).

273. S. Nozaki, S. Sato, H. Ono, and H. Morisaki, *Mater. Res. Soc. Symp. Proc.* 351, 399 (1994).

274. A. A. Seraphin, S.-T. Ngiam, and K. D. Kolenbrander, *J. Appl. Phys.* 80, 6429 (1996).

275. J. R. Heath, *Science* 258, 1131 (1992).

276. R. Bley, A. Kauzlarich, and M. Susan, *J. Am. Chem. Soc.* 118, 12,461 (1996).

277. A. Frojtik, H. Weller, S. Fiechter, and A. Henglein, *Chem. Phys. Lett.* 134, 477 (1987).

278. M. F. Jarrold, *Science* 252, 1085 (1991).

279. J. M. Hunter, J. L. Fye, M. F. Jarold, and J. E. Bower, *Phys. Rev. Lett.* 73, 2063 (1994).

280. S. Iijima, *Jpn. J. Appl. Phys.* 26, 365 (1987).

281. S. Iijima, *Jpn. J. Appl. Phys.* 26, 357 (1987).

282. J. L. Heinrich, C. L. Curtis, G. M. Credo, K. L. Kavamagh, and M. J. Sailor, *Science* 255, 66 (1992).

283. R. A. Bley, S. M. Kauzlarich, and H. W. H. Lee, *Chem. Mater.* 8, 1881 (1996).

284. G. Allan, C. Delerue, and M. Lannoo, *Phys. Rev. Lett.* 78, 3161 (1997).

285. L.-W. Wang and A. Zunger, *J. Chem. Phys.* 100, 2394 (1994).

286. M. O. Watanabe, T. Miyazaki, and T. Kanayama, *Phys. Rev. Lett.* 81, 5362 (1998).

287. M. Hirao and T. Uda, *Surf. Sci.* 306, 87 (1994).

288. S. Oeguet, J. R. Chelikowsky, and S. G. Louie, *Phys. Rev. Lett.* 79 (1997).

289. G. Onida and W. Andreoni, *Chem. Phys. Lett.* 243 (1995).

290. B. Delley and E. F. Steigmeier, *Appl. Phys. Lett.* 67, 2370 (1995).

291. Y. Kanemitsu, S. Okamoto, M. Otobe, and S. Oda, *Phys. Rev. B* 55, R7375 (1997).

292. A. B. Filonov, A. N. Kholod, F. d'Avitaya, and X. Arnaud, *Phys. Rev. B* 57, 1394 (1998).

293. W. L. Wilson, P. F. Szajowski, and L. E. Brus, *Science* 262, 1242 (1993).

294. D. Zhang, R. M. Kolbas, and J. M. Zavada, *Appl. Phys. Lett.* 65, 2684 (1994).

295. L.-W. Wang and A. Zunger, *J. Phys. Chem.* 98, 2158 (1994).

296. H. Yorikawa, H. Uchida, and S. Muramatsu, *J. Appl. Phys.* 79, 3619 (1996).

297. A. Kux and M. B. Chorin, *Phys. Rev. B* 51, 17,535 (1995).

298. L. E. Brus, P. F. Szajowski, W. L. Wilson, T. D. Harris, S. Schuppler, and P. H. Citrin, *J. Am. Chem. Soc.* 117, 2915 (1995).

299. L. N. Dinh, L. L. Chase, M. Balooch, W. J. Siekhaus, and F. Wooten, *Phys. Rev. B* 54, 5029 (1996).

300. S. S. Iyer and Y. H. Xie, *Science* 260, 40 (1993).

301. B. Delley and E. F. Steigmeier, *Phys. Rev. B* 47, 1397 (1993).

302. X. G. Gong, Q. Q. Zheng, and Y.-Z. He, *J. Phys. Condens. Matter* 7, 577 (1995).

303. J. R. Chelikowsky, *Phys. Rev. Lett.* 60, 2669 (1988).

304. C. M. Rohlfing and K. Raghavachari, *Chem. Phys. Lett.* 167, 559 (1990).

305. N. Binggeli, J. L. Martins, and J. R. Chelikowsky, *Phys. Rev. Lett.* 68, 2956 (1992).

306. X. Jing, N. Troullier, and Y. Saad, *Phys. Rev. B* 50, 12,234 (1994).

307. M. Menon and K. R. Subbaswamy, *Phys. Rev. B* 47, 12,754 (1993).

308. E. C. Honea, A. Ogura, C. A. Murray, K. Raghavachari, W. O. Sprenger, M. F. Jarrold, and W. L. Brown, *Nature* 366, 42 (1993).

309. C. Xu, T. R. Taylor, and D. M. Neumark, *J. Chem. Phys* 108, 1395 (1998).

310. T. van Buuren, T. Tiedje, and J. R. Dahn, *Appl. Phys. Lett.* 63, 2911 (1993).

311. K. Murayama, H. Komasu, S. Miyazaki, and M. Hirose, *Solid State Commun.* 103, 155 (1997).

312. D. J. Wales, *Phys. Rev. A* 49, 2195 (1994).

313. H. Kimura, S. Imanaga, Y. Hayafuji, and H. Adachi, *J. Phys. Soc. Jpn.* 62, 2663 (1993).

314. B.-L. Gu, Z.-Q. Li, and J.-L. Zhu, *J. Phys.: Condens. Matter* 5, 5255 (1993).

315. X. G. Gong, *Phys. Rev. B* 52, 14,677 (1995).

316. E. Kaxiras, in "Cluster Assembled Materials" (K. Sattler, Ed.), Vol. 232, p. 67. Trans Tech Publications, Zurich, 1996.

317. S. Saito, in "Cluster Assembled Materials" (K. Sattler, Ed.), Vol. 232, p. 233. Trans Tech Publications, Zurich, 1996.

318. S. Saito, *Phys. Rev. B* 51, 51 (1995).

319. K. M. Ho, A. A. Shvartsburg, B. Pan, Z.-Y. Lu, C.-Z. Wang, J. G. Wacker, J. L. Fye, and M. F. Jarrold, *Nature* 392, 582 (1998).

320. D. A. Jelski, B. L. Swift, and T. T. Rantala, *J. Chem. Phys.* 95, 8552 (1991).

321. E. Kaxiras and K. Jackson, *Phys. Rev. Lett.* 71, 727 (1993).

322. A. M. Mazzone, *Phys. Rev. B* 54, 5970 (1996).

323. R. Fournier, S. B. Sinnott, and A. E. DePristo, *J. Chem. Phys.* 97, 414 (1992).

324. J. R. Chelikowsky, K. Glassford, and J. C. Phillips, *Phys. Rev. B* 44, 1538 (1991).

325. M. F. Jarrold and V. A. Constant, *Phys. Rev. Lett.* 67, 2994 (1991).

326. K. Fuke, K. Tsukamoto, and F. Misaizu, *J. Chem. Phys.* 99, 7807 (1993).

327. K. Jackson, M. Pederson, and K.-M. Ho, *Phys. Rev. A* 59, 3685 (1999).

328. J. Müller, B. Liu, A. A. Shvartsburg, S. Ögut, J. R. Chelikowsky, K. W. M. Siu, K.-M. Ho, and G. Ganteför, *Phys. Rev. Lett.* 85, 1666 (2000).

329. J. A. Cogordan, L. E. Sansores, and A. A. Valladares, *J. Non-Cryst. Solids* 181, 135 (1995).

330. Y. M. Niquet, G. Allan, C. Delerue, and M. Lannoo, *Appl. Phys. Lett.* 77, 1182 (2000).

331. Y. Maeda, N. Tsukamoto, Y. Yazawa, Y. Kanemitsu, and Y. Masumoto, *Appl. Phys. Lett.* 59, 3168 (1991).

332. D. C. Paine, C. Caragiania, T. Y. Kim, Y. Shigesto, and T. Ishahara, *Appl. Phys. Lett.* 62, 2842 (1993).

333. M. H. Ludwig, R. E. Hummel, and S.-S. Chank, *J. Vac. Sci. Technol. B* 12, 3023 (1994).

334. M. Nogami and Y. Abe, *Appl. Phys. Lett.* 65, 2545 (1994).

335. Y. Maeda, *Phys. Rev. B* 51, 1658 (1995).

336. S. Okamoto and Y. Kanemitsu, *Phys. Rev. B* 54, 16,421 (1996).

337. V. Cracium, C. Boulmer-Leborgne, and I. W. Boyd, *Appl. Phys. Lett.* 69, 1506 (1996).

338. A. Saito and T. Suemoto, *Phys. Rev. B* 56, R1688 (1997).

339. M. Zacharias and P. M. Fauchet, *Appl. Phys. Lett.* 71, 380 (1997).

340. W. K. Choi, V. Ng, S. P. Ng, H. H. Thio, Z. X. Shen, and W. S. Li, *J. Appl. Phys.* 86, 1398 (1999).

341. S. Takeoka, M. Fujii, S. Hayashi, and K. Yamamoto, *Phys. Rev. B* 58, 7921 (1998).

342. Y. Negishi, H. Kawamata, F. Hayakawa, A. Nakajima, and K. Kaya, *Chem. Phys. Lett.* 294, 370 (1998).

343. M. F. Crommie, C. P. Lutz, and D. M. Eigler, *Phys. Rev. B* 48, 2851 (1993).

344. B. Marsen and K. Sattler, *Phys. Rev. B* 60, 11,593 (1999).

345. P. M. St. John and R. L. Whetten, *Chem. Phys. Lett.* 196, 330 (1992).

346. P. E. Batson and J. R. Heath, *Phys. Rev. Lett.* 71, 911 (1993).

347. Y. Kanemitsu, H. Uto, Y. Masumoto, T. Matsumoto, T. Futagi, and H. Mimura, *Phys. Rev. B* 48, 2827 (1993).

348. C. Delerue, E. Martin, J.-F. Lampin, G. Allan, and M. Lannoo, *J. Phys. IV* 3, 359 (1993).

349. A. Sieck, D. Porezag, T. Frauenheim, M. R. Pederson, and K. Jackson, *Phys. Rev. A* 56, 4890 (1997).

350. E. Kaxiras, *Phys. Rev. B* 56, 13,455 (1997).

351. J. L. Gavartin and C. C. Matthai, *Mater. Sci. Eng.* 35, 459 (1995).

352. J. A. Stroscio, R. M. Feenstra, and A. P. Fein, *Phys. Rev. Lett.* 57, 2579 (1986).

353. R. J. Hamers, P. Avouris, and F. Bozso, *Phys. Rev. Lett.* 59, 2071 (1987).

354. R. J. Hamers, *Annu. Rev. Phys. Chem.* 40, 531 (1989).

355. P. Melinon, P. Keghelian, B. Prevel, A. Perez, G. Guiraud, J. LeBrusq, J. Lerme, M. Pellarin, and M. Broyer, *J. Chem. Phys.* 107, 10,278 (1997).

356. S. G. Louis and M. L. Cohen, *Phys. Rev. B* 13, 2461 (1976).

357. R. Matz, H. Lueth, and A. Ritz, *Solid State Commun.* 46, 343 (1983).

358. K. C. Pandey, *Phys. Rev. Lett.* 47, 1913 (1981).

359. R. S. Becker, J. A. Golovchenko, D. R. Hamann, and B. S. Swartzentruber, *Phys. Rev. Lett.* 55, 2032 (1985).

360. F. J. Himpsel and T. Fauster, *J. Vac. Sci. Technol. A* 2, 815 (1984).

361. E. Kaxiras, *Chem. Phys. Lett.* 163, 323 (1989).

362. E. Kaxiras, *Phys. Rev. Lett.* 64, 551 (1990).

363. M. F. Jarrold, U. Ray, and M. Creegan, *J. Chem. Phys.* 93, 224 (1990).

364. M. F. Jarrold and E. C. Honea, *J. Phys. Chem.* 95, 9181 (1991).

365. M. Maus, G. Ganteför, and W. Eberhardt, *Appl. Phys. A* 70, 535 (2000).

366. A. Nakajama, T. Futatsugi, K. Kosemura, T. Fukano, and N. Yokoyama, *J. Vac. Sci. Technol.* 17, 2163 (1999).

367. K. Kim, *Phys. Rev. B* 57, 13,072 (1998).

368. E. Leobandung, L. Guo, Y. Wang, and S. Y. Chou, *Appl. Phys. Lett.* 67, 938 (1995).

369. F. G. Pikus and K. K. Likharev, *Appl. Phys. Lett.* 71, 3661 (1997).

370. S. K. Zhang, H. J. Zhu, F. Lu, Z. M. Jiang, and X. Wang, *Phys. Rev. Lett.* 80, 3340 (1998).

371. H. Ishikuro, T. Fujii, T. Saraya, G. Hashiguchi, T. Hiramoto, and T. Ikoma, *Appl. Phys. Lett.* 68, 3585 (1996).

372. A. E. Hanna and M. Tinkham, *Phys. Rev. B* 44, 5919 (1991).

373. "Single Charge Tunneling" (H. Grabert and M. H. Devoret, Eds.), Vol. X. Plenum, New York, 1992.

374. E. Bar-Sadeh, Y. Goldstein, B. Abeles, and O. Millo, *Phys. Rev. B* 50, R8961 (1994).

375. C. Schonenberger, H. van Houten, and H. C. Donkersloot, *Europhys. Lett.* 20, 249 (1992).

376. R. Wilkins, E. Ben-Jacob, and R. C. Jaklevic, *Phys. Rev. Lett.* 63, 801 (1989).

377. G. Markovich, D. V. Leff, S. W. Chung, H. M. Soyez, B. Dunn, and J. R. Heath, *Appl. Phys. Lett.* 70, 3107 (1997).

378. D. C. Ralph, C. T. Black, and M. Tinkham, *Phys. Rev. Lett.* 74, 3241 (1995).

379. O. Agam, N. S. Wingreen, B. L. Altshuler, D. C. Ralph, and M. Tinkham, *Phys. Rev. Lett.* 78, 1956 (1997).

380. J. G. A. Dubois, J. W. Geritsen, S. E. Shafranjuk, E. J. G. Boon, G. Schmid, and H. van Kempen, *Europhys. Lett.* 33, 279 (1995).

381. R. S. Ingram, J. M. Hostetler, and R. W. Murray, *J. Am. Chem. Soc.* 119, 9175 (1997).

382. R. Tsu, X.-L. Li, and E. H. Nicollian, *Appl. Phys. Lett.* 65, 842 (1994).

383. K. Yano, T. Ishii, T. Hashimoto, T. Kobayashi, F. Murai, and K. Seki, *IEEE Trans. Electron Devices* ED-41, 1628 (1994).

384. Y. Takahashi, M. Nagase, H. Namatsu, K. Kurihara, K. Iwadate, Y. Nakajima, S. Horiguchi, K. Murase, and M. Tabe, *Electron. Lett.* 31, 136 (1995).

385. T. Hiramoto, H. Ishikuro, T. Fujii, T. Saraya, G. Hashiguchi, and T. Ikoma, *Physica B* 227, 95 (1996).

386. H. Ishikuro and T. Hiramoto, *Appl. Phys. Lett.* 71, 3691 (1997).

387. L. Zhuang, G. L., and S. Y. Chou, *Appl. Phys. Lett.* 72, 1205 (1998).

388. N. Takahashi, H. Ishikuro, and T. Hiramoto, (2000). *Appl. Phys. Lett.* 76, 209 (2000).

389. B. Marsen, Ph.D. thesis, University of Hawaii, Honolulu, 2000.

390. M.-S. Ho, I.-S. Hwang, and T. T. Tsong, *Phys. Rev. Lett.* 94, 5792 (2000).

391. S. Furukawa and T. Miyasato, *Phys. Rev. B* 38, 5726 (1988).
392. G. R. Gupte and R. Prasad, *Int. J. Mod. Phys.* 12, 1737 (1998).
393. G. R. Gupte and R. Prasad, *Int. J. Mod. Phys.* 12, 1607 (1998).
394. V. Meleshko, X. Xu, and Q. Zhang, *Chem. Phys. Lett.* 300, 118 (1999).
395. T. Miyazaki, T. Uda, and K. Terakura, *Chem. Phys. Lett.* 261, 346 (1996).
396. H. K. T. Murakami, *Appl. Phys. Lett.* 67, 2341 (1995).
397. M. T. Swihart and S. L. Girshick, *Chem. Phys. Lett.* 307, 527 (1999).
398. T. Uda, *Surf. Rev. Lett.* 3, 127 (1996).
399. A. M. Saitta, F. Buda, G. Fiumara, and P. V. Giaquinta, *Phys. Rev. B* 53, 1446 (1996).
400. S. Y. Ren, *Phys. Rev. B* 55, 4665 (1997).
401. S. V. Gaponenko, I. N. Germanenko, E. P. Petrov, A. P. Stupak, V. P. Bondarenko, and A. M. Dorofeev, *Appl. Phys. Lett.* 64, 85 (1994).
402. S. Schuppler, S. L. Friedmann, M. A. Marcus, D. L. Adler, Y.-H. Xie, F. M. Ross, T. D. Harris, W. L. Brown, Y. J. Chabal, L. E. Brus, and P. H. Citrin, *Phys. Rev. Lett.* 72, 2648 (1994).
403. J. L. Gole and D. A. Dixon, *J. Appl. Phys.* 83, 5985 (1998).
404. J. L. Gole and S. M. Prokes, *Phys. Rev. B* 58, 4761 (1998).
405. J. L. Gole and D. A. Dixon, *Phys. Rev. B* 57, 12,002 (1998).
406. J. L. Gole and D. A. Dixon, *J. Phys. Chem. B* 102, 33 (1998).
407. G. B. Amisola, R. Behrensmeier, J. M. Galligan, and F. A. Otter, *Appl. Phys. Lett.* 61, 2595 (1992).
408. L. Tsybeskov, S. P. Duttagupta, and P. M. Fauchet, *Solid State Commun.* 95, 429 (1995).
409. Buda, F., J. Kohanoff, and M. Parrinello, *Phys. Rev. Lett.* 69, 1272 (1992).
410. Z. C. Feng and R. Tsu, "Porous Silicon," World Scientific, Singapore, 1994.
411. K. Ito, S. Ohyama, Y. Uehara, and S. Ushioda, *Appl. Phys. Lett.* 67, 2536 (1995).
412. R. Laiho, A. Pavlov, and Y. Pavlova, *Thin Solid Films* 297, 138 (1997).
413. Y. Xiao, M. J. Heben, J. M. McCullough, Y. S. Tsuo, J. I. Pankove, and S. K. Deb, *Appl. Phys. Lett.* 62, 1152 (1993).
414. H. Sakaki, *Jpn. J. Appl. Phys.* 19, L735 (1980).
415. S. Mori and T. Ando, *J. Phys. Soc. Jpn.* 48, 865 (1980).
416. J. Bloch, U. Bockelmann, and F. Laruelle, *Europhys. Lett.* 28, 501 (1994).
417. H. Akiyama, T. Someya, and H. Sakaki, *Phys. Rev. B* 53, R4229 (1996).
418. F. Vouilloz, D. Y. Oberli, M.-A. Dupertuis, A. Gustafsson, F. Reinhardt, and E. Kapon, *Phys. Rev. Lett.* 78, 1580 (1997).
419. R. Cingolani, M. Lepore, R. Tommasi, I. M. Catalano, H. Lage, D. Heitmann, K. Ploog, A. Shimizu, H. Sakaki, and T. Ogawa, *Phys. Rev. Lett.* 69, 1276 (1992).
420. A. R. Goni, A. Pinczuk, J. S. Weiner, B. S. Dennis, L. N. Pfeiffer, and K. W. West, *Phys. Rev. Lett.* 70, 1151 (1993).
421. G. W. Zhou, Z. Zhang, and D. P. Yu, *Appl. Phys. Lett.* 73, 677 (1998).
422. V. Ng, H. Ahmed, and T. Shimada, *Appl. Phys. Lett.* 73, 972 (1998).
423. D. Papadimitriou and A. G. Nassiopoulos, *J. Appl. Phys.* 84, 1059 (1998).
424. D. P. Yu, Z. G. Bai, and S. Q. Geng, *Phys. Rev. B* 59, R2498 (1999).
425. N. T. Bagraev, E. T. Chaikina, and A. M. Malyarenko, *Solid-State Electron.* 42, 1199 (1998).
426. B. Li, D. Yu, and S.-L. Zhang, *Phys. Rev. B* 59, 1645 (1999).
427. H. Namatsu, K. Kurihara, and T. Makino, *Appl. Phys. Lett.* 70, 619 (1997).
428. M. Ge and K. Sattler, *Science* 260, 515 (1993).
429. H. I. Liu, D. K. Biegelsen, F. A. Ponce, N. M. Johnson, and R. F. W. Pease, *Appl. Phys. Lett.* 64, 1383 (1994).
430. H. Namatsu, Y. Takahashi, M. Nagase, and K. Murase, *J. Vac. Sci. Technol. B* 13, 2166 (1995).
431. Y. Shi, J. L. Liu, F. Wang, Y. Lu, R. Zhang, S. L. Gu, P. Han, L. Q. Hu, Y. D. Zheng, C. Y. Lin, and D. A. Du, *J. Vac. Sci. Technol. A* 14, 1194 (1996).
432. G. S. Chen, C. B. Boothroyd, and C. J. Humphreys, *Appl. Phys. Lett.* 62, 1949 (1993).
433. V. Lehmann and U. Goesele, *Appl. Phys. Lett.* 58, 856 (1991).
434. J. L. Liu, Y. Shi, and Y. D. Zheng, *Appl. Phys. Lett.* 68, 352 (1996).
435. M. Gotza, B. Saint-Cricq, and P.-H. Jouneau, *Microelectron. Eng.* 27, 129 (1995).
436. G. Hashiguchi and H. Mimura, *Jpn. J. Appl. Phys. Part 2* 33, L1649 (1994).
437. J. L. Liu, Y. Shi, and Y. D. Zheng, *J. Vac. Sci. Technol. B* 13, 2137 (1995).
438. A. M. Morales and C. M. Lieber, *Science* 279, 208 (1998).
439. T. Ono, H. Saitoh, and M. Esashi, *Appl. Phys. Lett.* 70, 1852 (1997).
440. G. D. Sanders and Y.-C. Chang, *Appl. Phys. Lett.* 60, 2525 (1992).
441. A. G. Nassiopoulos, S. Grigoropoulos, and D. Papadimitriou, *Thin Solid Films* 297, 176 (1997).
442. J. Westwater, D. P. Gosain, and H. Ruda, *J. Vac. Sci. Technol. B* 15, 554 (1997).
443. J. Westwater, D. P. Gosain, and S. Usui, *Phys. Status Solidi A* 165, 37 (1998).
444. S. Yanagiya, S. Kamimura, and H. Koinuma, *Appl. Phys. Lett.* 71, 1409 (1997).
445. D. P. Yu, Z. G. Bai, Y. Ding, Q. L. Hang, H. Z. Zhang, J. J. Wang, Y. H. Zou, W. Qian, G. C. Xiong, H. T. Zhou, and S. Q. Feng, *Appl. Phys. Lett.* 72, 3458 (1998).
446. J. J. P. Stewart, *J. Comput. Chem.* 10, 221 (1989).
447. H. Kawaji, H. Horie, S. Yamanaka, and M. Ishikawa, *Phys. Rev. Lett.* 74, 1427 (1995).
448. P. L. McEuen, *Nature* 393, 15 (1998).
449. M. Kasper, K. Siegmann, and K. Sattler, *J. Aerosol Sci.* 28, 1569 (1997).
450. M. Kasper, K. Sattler, K. Siegmann, U. Matter, and H. C. Siegmann, *J. Aerosol Sci.* 30, 217 (1999).
451. P. Dugourd, R. R. Hudgins, and M. F. Jarrold, *Phys. Rev. Lett.* 80, 4197 (1998).
452. F. Jensen, *Chem. Phys. Lett.* 209, 417 (1993).
453. M. T. Bowers, P. R. Kemper, G. von Helden, and P. A. M. van Koppen, *Science* 260, 1446 (1993).
454. T. Frauenheim, G. Jungnickel, T. Koehler, and U. Stephan, *J. Non-Cryst. Solids* 182, 186 (1995).
455. R. Kubo, *J. Phys. Soc. Jpn.* 17, 975 (1962).
456. C. Liang and H. F. Schaefer III, *Chem. Phys. Lett.* 169, 150 (1990).
457. G. P. Lopinski, V. I. Merkulov, and J. S. Lannin, *Phys. Rev. Lett.* 80, 4241 (1998).
458. S. Saito, S. Okada, S. Sawada, and N. Hamada, *Phys. Rev. Lett.* 75, 685 (1995).
459. S. Saito and A. Oshiyama, *Phys. Rev. B* 44, 11,532 (1991).
460. S. J. Woo, E. Kim, and Y. H. Lee, *Phys. Rev. B* 47, 6721 (1993).
461. B. L. Zhang, C. Z. Wang, K. M. Ho, C. H. Xu, and J. Chan, *J. Chem. Phys.* 97, 5007 (1992).
462. H. Wang, C. Zeng, Q. Li, B. Wang, J. Yang, J. G. Hou, and Q. Zhu, *Surf. Sci.* 442, L1024 (1999).
463. P. G. Collins, J. C. Grossman, M. Cote, M. Ishigami, C. Piskoti, S. G. Louie, M. L. Cohen, and A. Zettl, *Phys. Rev. Lett.* 82, 165 (1999).
464. H. Kietzmann, R. Rochow, G. Gantefor, W. Eberhardt, K. Vietze, G. Seifert, and P. W. Fowler, *Phys. Rev. Lett.* 81, 1998 (1998).
465. M. Lonfat, B. Marsen, and K. Sattler, *Chem. Phys. Lett.* 313, 539 (1999).
466. J. W. G. Wildoer, L. C. Venema, A. G. Rinzler, R. E. Smalley, and C. Dekker, *Nature* 391, 59 (1998).
467. M. R. C. Hunt, P. J. Durston, and R. E. Palmer, *Surf. Sci.* 364, 266 (1996).
468. Z. Y. Li, K. M. Hock, and R. E. Palmer, *Phys. Rev. Lett.* 67, 1562 (1991).
469. W. A. de Heer, P. Milani, and A. Chatelain, *Phys. Rev. Lett.* 63, 2834 (1989).
470. P. C. Ohara, D. V. Leff, J. R. Heath, and W. M. Gelhart, *Phys. Rev. Lett.* 75, 3466 (1995).
471. M. J. Hostetler, S. J. Green, J. J. Stokes, and R. W. Murray, *J. Am. Chem. Soc.* 118, 4212 (1996).
472. N. Herron, J. C. Calabrese, W. Farneth, and Y. Wang, *Science* 259, 1426 (1993).

473. C. B. Murray, C. R. Kagan, and M. G. Bawendi, *Science* 270, 1335 (1995).

474. J. J. Shiang, J. R. Heath, C. P. Collier, and R. J. Saykally, *J. Phys. Chem.* 102, 3425 (1998).

475. K. Sattler, unpublished results.

476. T. A. Burr, A. A. Seraphin, and K. D. Kolenbrander, *Phys. Rev. B* 56, 4818 (1997).

477. A. B. Filinov, A. N. Kholod, V. A. Novikov, V. E. Borisenko, L. Vervoort, F. Bassani, A. Saul, and F. A. d'Avitaya, *Appl. Phys. Lett.* 70, 744 (1997).

478. M. Higo, K. Nishino, and S. Kamata, *Appl. Surf. Sci.* 51, 61 (1991).

479. M. Higo, M. Isobata, and S. Kamata, *J. Phys. Chem.* 97, 4491 (1993).

480. F. Huisken, B. Kohn, and V. Paillard, *Appl. Phys. Lett.* 74, 3776 (1999).

481. M. Ehbrecht, B. Kohn, and V. Paillard, *Phys. Rev. B* 56, 6958 (1997).

482. M. A. Laguna, V. Paillard, and H. Hofmeister, *J. Lumin.* 80, 223 (1998).

483. S. Kerdiles, R. Rizk, and J. R. Morante, *Solid-State Electron* 42, 2315 (1998).

484. M. Tzolov, Y. Jeliazova, and N. Tzenov, *Solid state Phenomena* 67–68, 107 (1999).

485. J. R. Heath, S. M. Gates, and C. A. Chess, *Appl. Phys. Lett.* 64, 3569 (1994).

486. R. Cherfi, G. Farhi, and M. Aoucher, *Solid state Phenomena* 67–68, 113 (1999).

487. P. Scheier, B. Marsen, and K. Sattler, to be published.

488. E. Hanamura, *Phys. Rev.* 37, 1273 (1988).

489. L. Banyai, Y. Z. Hu, M. Lindberg, and S. W. Koch, *Phys. Rev. B* 38, 8142 (1988).

490. S. Bandyopadhyay and V. P. Roychowdhury, *IEEE Potentials* 15, 8 (1996).

491. R. T. Bate, *Solid State Technol.* 32, 101 (1989).

492. F. A. Buot, *Phys. Rep.* 234, 1 (1993).

493. A. Schenk, U. Krumbein, and W. Fichtner, *IEICE Trans. Electron.* 77, 148 (1994).

494. B. Sweryda-Krawiec, T. Cassagneau, and J. H. Fendler, *Adv. Mater.* 11, 659 (1999).

495. G. D. Stucky, *Naval Res. Rev.* 43, 28 (1991).

496. J. H. Sinfelt and G. D. Meitzner, *Acc. Chem. Res.* 26, 1 (1993).

497. M. L. Steigerwald and L. E. Brus, *Acc. Chem. Rev.* 23, 183 (1990).

498. H. Weller, *Angew. Chem. Int. Ed. Engl.* 32, 41 (1993).

499. H. Weller, *Adv. Mater.* 5, 88 (1993).

500. L. Brus, *IEEE J. Quantum Electron.* QE-22, 1909 (1986).

501. V. Dneprovskii, A. Eev, and E. Dovidenko, *Phys. Status Solidi* 188, 297 (1995).

502. L. Pavesi and R. Guardini, *Braz. J. Phys.* 26, 151 (1996).

503. L. François, M. Mostafavi, J. Belloni, J.-F. Delouis, J. Delaire, and P. Feneyrou, *J. Phys. Chem. B* 104, 6133 (2000).

504. K. Mansour, M. J. Soileau, and E. W. Van Stryland, *J. Opt. Soc. Am. B* 9, 1100 (1992).

505. K. M. Nashold and D. P. Walter, *J. Opt. Soc. Am. B* 12, 1228 (1995).

506. Y. P. Sun, J. E. Riggs, H. W. Rollins, and R. J. Guduru, *J. Phys. Chem.* 103, 77 (1999).

507. J. N. Randall, M. A. Reed, and G. A. Frazier, *J. Vac. Sci. Technol.* 7, 1398 (1989).

508. W. M. Toiles, *Nanotechnology* 7, 59 (1996).

509. H. Sakaki, T. Matsusue, and M. Tsuchiya, *IEEE J. Quantum Electron.* 25, 2498 (1989).

510. M. Van Rossum, *Mater. Sci. Eng.* 20, 128 (1993).

511. J. R. Tucker, C. Wang, and T. C. Shen, *Nanotechnology* 7, 275 (1996).

512. K. Puech, W. Blau, A. Grund, C. Bubeck, and G. Cardenas, *Opt. Lett.* 20, 1613 (1995).

513. H. H. Huang, F. Q. Yan, Y. M. Kek, C. H. Chew, G. Q. Xu, W. Ji, P. S. Oh, and S. H. Tang, *Langmuir* 13, 172 (1997).

514. M. Jacobsohn and U. Banin, *J. Phys. Chem.* 104, 1 (2000).

515. L. E. Brus, *Appl. Phys. A* 53, 465 (1991).

516. D. J. Norris, M. G. Bawendi, and L. E. Brus, in "Molecular Electronics," (J. Jortner and M. Ratner, Eds.), Chap. 9. Blackwell Sci. Oxford, U.K., 1997.

517. K. H. Schmidt, G. Medeiros-Ribeiro, and P. M. Petroff, *Phys. Rev. B* 58, 3597 (1998).

518. S. Maimon, E. Finkman, G. Bahir, S. E. Achacham, J. M. Garcia, and P. M. Petroff, *Appl. Phys. Lett.* 73, 203 (1998).

519. D. Pan and E. Towe, *Electron. Lett.* 34, 1883 (1998).

520. D. L. Fuffaker, G. Park, Z. Zou, O. B. Shchekin, and D. G. Deppe, *Appl. Phys. Lett.* 73, 2564 (1998).

521. Y. Arakawa and H. Sakaki, *Appl. Phys. Lett.* 40, 939 (1982).

522. F. Schafer, J. P. Reithaier, and A. Forchel, *Appl. Phys. Lett.* 74, 2915 (1999).

523. F. Heinrichsdorff, C. Ribbat, M. Grundmann, and D. Bimberg, *Appl. Phys. Lett.* 76, 556 (2000).

524. M. Tzolov, F. Finger, R. Carius, and P. Hapke, *J. Appl. Phys.* 81, 7378 (1997).

525. A. Fujuwara, Y. Takahashi, and K. Murase, *Phys. Rev. Lett.* 78, 1532 (1997).

526. K. Tamamura, Y. Sugiyama, Y. Nakata, S. Muto, and N. Yokoyama, *Jpn. J. Appl. Phys.* 34, L1445 (1995).

527. J. R. Tucker, *J. Appl. Phys.* 72, 4399 (1992).

528. R. Tsu, *Physica B* 189, 235 (1993).

529. M. Tsukada, *Z. Phys. D* 19, 283 (1991).

530. H. Grabert, G.-L. Ingord, M. H. Devoret, D. Esteve, H. Pothier, and C. Urbina, *Z. Phys. B* 84, 143 (1991).

531. S. Tiwari, F. Rana, and K. Chan, *Appl. Phys. Lett.* 68, 1377 (1996).

532. A. Nakajima, T. Futatsugi, K. Kosemura, T. Fukano, and N. Yokoyama, *Appl. Phys. Lett.* 70, 1742 (1997).

533. C. W. Chan and S. Nie, *Science* 281, 2016 (1998).

534. M. Bruchez Jr., M. Moronne, P. Gin, S. Weiss, and A. P. Alivisatos, *Science* 281, 2013 (1998).

535. H. Hirai, H. Wakabayashi, and M. Komiyama, *Bull. Chem. Soc. Jpn.* 59, 367 (1986).

536. M. J. Bloemer, J. W. Haus, and P. R. Ashley, *J. Opt. Soc. Am. B* 7, 790 (1990).

537. T. W. Roberti, B. A. Smith, and J. Z. J. Zhang, *J. Chem. Phys.* 102, 3860 (1995).

538. E. J. Heilweil and R. M. Hochstrasser, *J. Chem. Phys.* 82, 4762 (1985).

539. E. A. Bollick and D. A. Ramsey, *J. Chem. Phys.* 29, 1418 (1958).

540. E. A. Bollick and D. A. Ramsey, *Astrophys. J.* 137, 84 (1963).

541. K. H. Hinkle, J. J. Keady, and P. F. Bernath, *Science* 241, 1319 (1988).

542. F. G. Celii and J. E. Butler, *Annu. Rev. Phys. Chem.* 42, 643 (1991).

543. R. Levy Gyyer and D. E. Koshland, *Science* 250, 1640 (1990).

544. R. F. Curl, *Rev. Mod. Phys.* 69, 691 (1997).

545. A. W. Snow and H. Wohltjen, *Chem. Mater.* 10, 947 (1998).

546. R. L. Whetten, J. T. Khoury, M. M. Alvarez, S. Murthy, I. Vezmar, Z. L. Wang, P. W. Stephens, C. L. Cleveland, W. D. Luedtke, and U. Landman, *Adv. Mater.* 8, 428 (1996).

547. "Cluster Assembled Materials" (K. Sattler, Ed.), Vol. 232. Trans Tech Publications, Zurich, 1997.

548. P. Melinon, V. Paillard, and J. Lerme, *Int. J. Mod. Phys.* 9, 339 (1995).

549. L. Motte, F. Billoudet, and M. P. Pileni, *Adv. Mater.* 8, 1018 (1996).

550. S. Komarneni, "Nanophase and Nanocomposite Materials," p. 459. Materials Research Soc., Boston, MA, 1992.

551. B. Michel, E. Delamarche, and C. Gerber, *Adv. Mater.* 8, 719 (1996).

552. G. Benedek, I. Colombo, and S. Serra, *J. Chem. Phys.* 106, 2311 (1997).

553. G. Benedek, E. Galvani, and S. Serra, *Chem. Phys. Lett.* 244, 339 (1995).

554. "Light Emission in Silicon: From Physics to Devices" (D. J. Lockwood, Ed.), Vol. 49. Academic Press, San Diego, 1998.

555. L. Tsybeskov, S. P. Duttagupta, K. D. Hirschmann, and P. M. Fauchet, *Appl. Phys. Lett.* 68, 2058 (1996).

556. L. Tsybeskov, K. L. Moore, and P. M. Fauchet, *Appl. Phys. Lett.* 69, 3411 (1996).

557. N. Lalic and J. Linnros, *J. Lumin.* 80, 263 (1998).

558. K. D. Hirschman, L. Tsybeskov, S. P. Duttagupta, and P. M. Fauchet, *Nature* 384, 338 (1996).

559. D. K. Ferry and S. M. Goodnick, "Transport in Nanostructures," Cambridge Univ. Press, New York, 1997.

560. A. Ohata, H. Niiyama, and A. Toriumi, *Jpn. J. Appl. Phys.* 34, 4485 (1995).

561. A. S. Bell, B. Brezger, U. Drodofsky, S. Nowak, T. Pfau, J. Stuhler, T. Schulze, and J. Mlynek, *Surf. Sci.* 433, 40 (1999).

562. B. Faircloth, H. Rohrs, R. Tiberio, R. Ruoff, and R. R. Krchnavek, *J. Vac. Sci. Technol. B* 18, 1866 (2000).

563. K. Kragler, E. Gunther, and G. Saemann-Ischenko, *Appl. Phys. Lett.* 67, 1163 (1995).

564. S. Dong and F. Zhu, *Vac. Sci. Technol.* 10, 379 (1990).

565. F. Thibaudau, J. R. Roche, and F. Salvan, *Appl. Phys. Lett.* 64, 523 (1994).

566. E. E. Ehrichs, S. Yoon, and A. L. de-Lozanne, *Appl. Phys. Lett.* 53, 2287 (1988).

567. D. P. Adams, T. M. Mayer, and B. S. Schwartzentruber, *Appl. Phys. Lett.* 68, 2210 (1996).

568. F. Marchi, D. Tonneau, H. Dallaporta, R. Pierrisnard, V. Bouchiat, V. I. Saforov, P. Doppelt, and R. Even, *Microelectron. Eng.* 50, 59 (2000).

569. A. Laracuente, M. J. Bronikowski, and A. Gallangher, *Appl. Surf. Sci.* 107, 11 (1996).

570. T. M. H. Wong, S. J. O'Shea, A. W. McKinnon, and M. E. Welland, *Appl. Phys. Lett.* 67, 786 (1995).

571. T. Muhl, H. Bruckl, and G. Reiss, *J. Appl. Phys.* 82, 5255 (1997).

572. C. Lebreton and Z. Z. Wang, *Surf. Sci.* 382, 193 (1997).

573. T. Abe, K. Hane, and S. Okuma, *J. Appl. Phys.* 75, 1228 (1994).

574. S. C. Eagle and G. K. Feder, *Appl. Phys. Lett.* 74, 3902 (1999).

575. J. H. Chen, G. Jianian, and J. A. Switzer, *Appl. Phys. Lett.* 71, 1637 (1997).

576. L. Nan, T. Yoshinobu, and H. Iwasaki, *Appl. Phys. Lett.* 74, 1621 (1999).

577. J. A. Dagata, J. Schneier, H. H. Harary, C. J. Evans, M. T. Postek, and J. Bennet, *Appl. Phys. Lett.* 56, 2001 (1990).

578. H. C. Day and D. R. Allee, *Appl. Phys. Lett.* 62, 2691 (1993).

579. E. S. Snow and P. M. Campbell, *Appl. Phys. Lett.* 64, 1932 (1994).

580. K. S. Pitzer and E. Clementi, *J. Am. Chem. Soc.* 81, 4477 (1959).

581. C. Liang and H. F. Schaefer III, *J. Chem. Phys.* 93, 8844 (1990).

582. J. Feng, Z. Wang, and M. C. Zerner, *Int. J. Quantum Chem.* 37, 599 (1990).

583. L. Wang, P. S. Davids, A. Saxena, and A. R. Bishop, *J. Phys. Chem. Solids* 54, 1493 (1993).

584. G. Endredi and J. Ladik, *J. Mol. Struct.* 300, 405 (1993).

585. D. Tomanek, *Comments At. Mol. Phys.* 31, 337 (1995).

586. S. Schuppler, S. L. Friedmann, M. A. Marcus, D. L. Adler, Y.-H. Xie, F. M. Ross, Y. J. Chabal, T. D. Harris, L. E. Brus, W. L. Brown, E. E. Chaban, P. F. Szajowski, S. B. Christman, and P. H. Citrin, *Phys. Rev. B* 52, 4910 (1995).

587. L.-W. Wang and A. Zunger, in "Nanocrystalline Semiconductor Materials," (P. V. Kamat and D. Meisel, Eds.), Elsevier, New York, 1996.

588. Y. Osaka, K. Tsunemoto, F. Toyomura, H. Myoren, and K. Kohno, *Jpn. J. Appl. Phys.* 31, L565 (1992).

589. S. H. Risbud, L. C. Liu, and J. F. Shackelford, *Appl. Phys. Lett.* 63, 1648 (1993).

590. W. A. Saunders, P. C. Sercel, R. B. Lee, H. A. Atwater, K. J. Vahala, R. C. Flagan, and E. J. Escorcia-Aparcio, *Appl. Phys. Lett.* 63, 1549 (1993).

591. S. Iwasaki, T. Ida, and K. Kimura, *Jpn. J. Appl. Phys.* 35, L551 (1996).

592. Y. Kanemitsu, *Phys. Rev. B* 49, 16,845 (1994).

593. C. Y. Cha, G. Ganteför, and W. Eberhardt, *Z. Phys. D* 26, 307 (1993).

594. C. L. Pettiette, S. H. Yang, M. J. Craycraft, J. Conceicao, R. T. Laaksonen, O. Cheshnovsky, and R. E. Smalley, *J. Chem. Phys.* 88, 5377 (1988).

595. A. L. Rogach, A. Koronowski, M. Gao, A. Eichmüller, and H. Weller, *J. Phys. Chem.* 103, 3065 (1999).

596. S. Y. Ren and J. D. Dow, *Phys. Rev. B* 45, 6492 (1992).

597. M. Lonfat, B. Marsen, and K. Sattler, to be published.

598. B. Delley, E. F. Steigmeier, and H. Auderset, *Bull. Am. Phys. Soc.* 37, 719 (1992).

599. M. Hirao, T. Uda, and Y. Murayama, *Mater. Res. Soc. Symp. Proc.* 283, 425 (1993).

600. J. W. Mintmire, *J. Vac. Sci. Technol. A* 11, 1733 (1993).

601. A. J. Read, R. J. Needs, K. J. Nash, L. T. Canham, P. D. J. Calcott, and A. Qteish, *Phys. Rev. Lett.* 69, 1232 (1992).

Chapter 3

ELECTRONIC STATES IN GaAs-AlAs SHORT-PERIOD SUPERLATTICES: ENERGY LEVELS AND SYMMETRY

Jian-Bai Xia

Department of Physics, Hong Kong University of Science and Technology,
Clearwater Bay, Kowloon, Hong Kong, China
and
Institute of Semiconductors, Chinese Academy of Sciences, Beijing 100083, China

Weikun Ge

Department of Physics, Hong Kong University of Science and Technology,
Clearwater Bay, Kowloon, Hong Kong, China

Contents

1. INTRODUCTION

The short-period superlattices (SPSLs) $(GaAs)_n/(AlAs)_m$ have been studied for a long time. Experimentally, following the pioneering work of Gossard et al. [1] on molecular beam epitaxy (MBE)–grown samples, SPSLs have been successfully grown by MBE [2, 3] and metal-organic chemical vapor deposition (MOCVD). [4] In experimental and theoretical works the fundamental band gap and its properties are a common focus of concern. Figure 1a [5] schematically shows the real space energy band alignment of a SPSL. The electron at the Γ valley of the conduction band is confined in the GaAs

well, and the electron at the X valley is confined in the AlAs well. The electron at the X valley along the z direction has a much heavier effective mass than the electron at the Γ valley. Therefore, in some cases (for example, very short periods) the energy of the lowest confined X state may be lower than that of the lowest confined Γ state. This results in a staggered band alignment in which the lowest conduction band state of the superlattice is located in the AlAs barrier and has the symmetry properties of the AlAs X-point, and the top of the valence band is still located at the GaAs Γ-point. The conduction band bottom and the valence band top are at different points in both the real space and the $\boldsymbol{k}$ space; this kind of superlattice is

Handbook of Thin Film Materials, edited by H.S. Nalwa
Volume 5: Nanomaterials and Magnetic Thin Films

ISBN 0-12-512913-0/$35.00

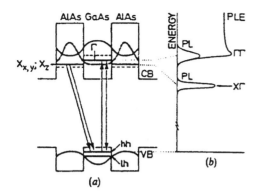

Fig. 1. (a) Schematic illustration of a real space energy band diagram of a type II GaAs/AlAs SPSL. (b) Impact of the staggered band alignment on the optical properties of the superlattice. Reproduced with permission from [5], copyright 1990.

called a type II superlattice, in which the transition is pseudo-direct. If the lowest conduction band state is a Γ state located in the GaAs well, the superlattice is called a type I super-lattice, in which the transition is direct. The effects of these peculiar electronic properties on the emission and absorption processes of SPSLs are shown in Figure 1b. The luminescence arises from the lowest energy levels of the superlattice, for the type II superlattice it is from the X–Γ transition, whereas the absorption has a considerable strength only above the direct Γ–Γ edge. Therefore, the comparison of photoluminescence (PL) and photoluminescence excitation (PLE) provides complementary information on the type I and type II ga ps in SPSLs. When the PL and PLE spectra were measured for a set of symmetric (i.e., $n = m$) $(GaAs)_n/(AlAs)_m$ SPSLs with an increasing number of monolayers, it was found that the PL peak (type II gap) and the PLE onset (type I gap) approach each other in energy and finally merge for a certain number of monolayers. Above this critical thickness the $(GaAs)_n/(AlAs)_m$ SPSLs recover the direct character and the energy splitting between PL and PLE reduces to the usual Stokes shift. The thickness-induced type I to type II cross-over has been determined in this way for different SPSL configurations by many groups. Additional information on this size-induced change of the band alignment also comes from luminescence measurements in electric fields, under hydrostatic pressure, and by photoacoustic and high excitation intensity spectroscopy, etc.

First we discuss the symmetry of the SPSLs (Section 2), then introduce the experimental researches, including PL and PLE spectra (Section 3), time decay and the temperature dependence of PL and PLE spectra (Section 4), PL spectra under hydrostatic pressure (Section 5), PL spectra in applied fields (Section 6), very short period superlattices (Section 7), and other spectra (Section 8). Finally we discuss theoretical research (Section 9) and applications (Section 10).

2. SYMMETRY OF SHORT-PERIOD SUPERLATTICES

The symmetry of SPSLs was first discussed by Lu and Sham [6] and Sham and Lu [7]. Assume that an idealized superlat-

tice is composed of n layers of an BA compound and m layers of an DA compound with a period along the z axis (here we specify (GaAs)/(AlAs)-like superlattices with a common atom in two compounds). The coordinate axes are chosen along the principal symmetry directions (100), (010), and (001). The Bravais lattice of the zinc blende semiconductors is fcc. We refer to a pair of adjacent cation and anion planes as a "mono-layer," or simply a "layer." Viewed along the z axis, atoms of the same element in alternating planes do not lie on top of each other. A period of the $(BA)_n/(DA)_m$ superlattice con-sists of $n + m$ layers. The basis vector of the unit cell points along the grown direction if $n + m$ is even and points along (0, 1, $n + m$) if $n + m$ is odd. Therefore, the Bravais lattices are different in the two cases. When $n + m$ is even, the Bravais lattice is simple tetragonal, with the basis vectors

$$(1, 1, 0)a/2, \qquad (-1, 1, 0)a/2, \qquad (0, 0, n+m)a/2, \quad (1)$$

where a is the bulk lattice constant. When $n + m$ is odd, the Bravais lattice is body-centered tetragonal, with the basis vec-tors

$$(1, 1, 0)a/2, \qquad (-1, 1, 0)a/2, \qquad (0, 1, n+m)a/2 \quad (2)$$

The corresponding Brillouin zones (BZ) are shown in Figure 2b and c. Different parts of the three-dimensional fcc Brillouin zone are folded into a superlattice zone as shown in Figure 2.

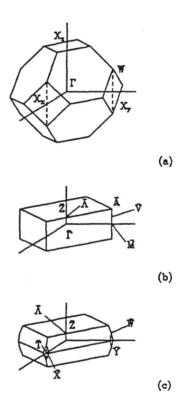

Fig. 2. (a) Brillouin zone of the bulk fcc lattice. (b) Brillouin zone of a superlattice with even $n + m$. (c) Brillouin zone of a superlattice with odd $n + m$. Reproduced with permission from [6], copyright 1989, American Physical Society.

Table I. Folding Relationships between Some Symmetry Points of the Bulk BZ and Mini-BZ

$m + n$ In bulk BZ	Even In mini BZ	Odd In mini BZ
Γ	Γ	Γ
X_z	Γ	Z
X_x	M	X
X_z	M	Y
L	R, X	Σ

Reproduced with permission from [8], copyright 1994, Elsevier Science.

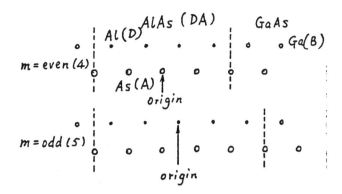

Fig. 3. Schematic figure for choosing the origin of the J operator for SL with even and odd m in the DA region. The dashed lines represent interfaces.

Figure 2b is the simple tetragonal mini-BZ for even $n + m$, and Figure 2c is the body-centered tetragonal mini-BZ for odd $n + m$. The symmetry points and lines in the mini-BZ are marked by a bar over the symbol to distinguish them from points in the bulk zone. Table I relates the symmetry points in the bulk BZ to those in the mini-BZ. Usually electronic states are referred to the symmetry points in the bulk BZ from which the superlattice states are derived, because the perturbation due to the superlattice potential is usually rather small, and each minizone state retains most of the characteristics of its parent.

The point group for the bulk zinc-blende semiconductors is T_d ($\bar{4}3m$). For the superlattice only those operators of the T_d group, which do not change the z coordinate, leave the superlattice invariant. They are C_{2z} (twofold rotation about the z axis) and σ_v and σ_v' (reflections in the $x = \pm y$ planes), and form the D_{2d} ($\bar{4}2m$) point group. The space group is symmorphic and is D_{2d}^5 ($P\bar{4}m2$) when $n + m$ is even and D_{2d}^9 ($I\bar{4}m2$) when $n + m$ is odd [9].

Of particular interest are the subbands along the k_z direction through the bulk conduction valleys at Γ, X, L, i.e., with parallel wave vector $\mathbf{k}_{//} = (0, 0)$, $(1, 0)$, and $(0.5, 0.5)$ $(2\pi/a)$, respectively. The k_z varies in the Brillouin zone of the superlattice: $-c^* < k_z < c^*$, where $c^* = 2\pi/(n + m)a$. The point groups (k group) associated with the various symmetry points for the superlattices are given in Table II.

The space group operators can be divided into those not changing the z coordinate (denoted by E) and those changing z to $-z$ (denoted by J). An A plane between a B and a D plane

may be regarded as an interface plane between a BA region and a DA region. For an operator of J, the origin is taken at an atom equidistant from two nearest interface planes of A atoms. For definiteness, if we consider the symmetry (parity) of states confined in the DA region, the origin will be chosen to be the midpoint in a DA region. Thus, it is at an A atom if m is even and at a D atom if m is odd (see Fig. 3). At a number of symmetry points, such as $\bar{\Gamma}$ and $\bar{M}$ in Figure 2b (derived from Γ and X points of the bulk Brillouin zone), we are interested in states with irreducible representations of unity under operators in E and ± 1 under J, which will be referred to, respectively, as even and odd parity states. The point group for the more general superlattice with neither common anion nor common cation depends on the interfaces (see ref. 7).

In calculating conduction bands of the (001) (GaAs)/(Al$_x$Ga$_{1-x}$As) and (Al$_x$Ga$_{1-x}$As)/(AlAs) superlattices Ting et al. [10] found that the symmetry of the confined X-valley states is critically dependent on whether the slabs with higher Al concentration contain an even or odd number of monolayers. As a result, the amount of Γ–X mixing is extremely sensitive to the layer thickness. The energy levels of the $\mathbf{k}_{//} = 0$, $k_z = 0$ states of the (001) (Al$_x$Ga$_{1-x}$As)/(AlAs) superlattice are given in Figure 4 as functions of x near the crossover point [10]. When x increases, the bottom of the Γ well rises, resulting in an increase in the energies of the Γ states confined in the Al$_x$Ga$_{1-x}$As well. Meanwhile, the top of the barrier of the X well shifts down, and the energies of the X states confined in the AlAs well decrease. These two kinds of states can be clearly distinguished in Figure 4, where the solid lines stand for the even-parity states, and the dashed lines stand for the odd-parity states. The superlattices are (28,8) and (28,7) for Figure 4a and 4b, respectively. Note that in Figure 4a, the crossover composition x_c is well defined. In Figure 4b, however, the curves for the lowest X-well state and the lowest Γ-well state anticross, leaving no clearly defined crossover point. The difference is due to the parity properties of both states, they cross if they have different parities and anticross if they have the same parity. What is puzzling is this: in going from Figure 4a to Figure 4b, the only change made is to reduce the thickness of the AlAs layers from eight to seven. Notice that the Γ-well states remain at the same parity, and all of the X-well states have switched parity. For

Table II. Symmetry of k Group G_k, $a^* = 2\pi/a$, $c^* = 2\pi/(n + m)a$ [6]

$n + m$ Bravais lattice Space group	Even Simple tetragonal $D_{2d}^5(P\bar{4}m2)$		Odd Body-centered tetragonal $D_{2d}^9(I\bar{4}m2)$	
$k (0 \le k_z \le c^*)$	Notation	G_k	Notation	G_k
$(0, 0, 0)$	$\bar{\Gamma}$	D_{2d}	$\bar{\Gamma}$	D_{2d}
$(0, 0, k_z)$	$\bar{\Lambda}$	C_{2v}	$\bar{\Lambda}$	C_{2v}
$(0, 0, c^*)$	$\bar{Z}$	D_{2d}	$\bar{Z}$	D_{2d}
$(a^*, 0, 0)$	$\bar{M}$	D_{2d}	$\bar{X}$	D_2
$(a^*, 0, k_z)$	$\bar{V}$	C_{2v}	$\bar{T}$	C_2
$(a^*, 0, c^*/2)$	$\bar{V}$	C_{2v}	$\bar{W}$	S_4
$(a^*, 0, c^*)$	$\bar{A}$	D_{2d}	$\bar{Y}$	D_2

Reproduced with permission from [6], copyright 1989, American Physical Society.

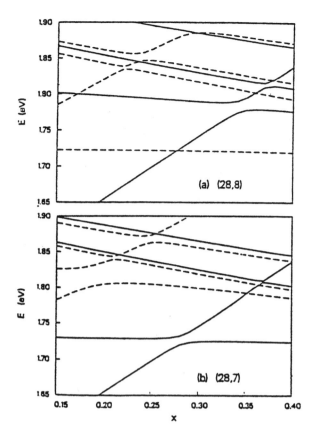

Fig. 4. Energy levels of the (001) $(Al_xGa_{1-x}As)/(AlAs)$ superlattice as functions of x near the crossover point. The layer thicknesses are (28,8) in (a) and (28,7) in (b). The solid lines are for the even-parity states and dashed lines are for the odd-parity states. Reproduced with permission from [10], copyright 1987, American Physical Society.

instance, the lowest X-well state in the (28,7) superlattice has even parity, but the lowest X-well state in the (28,8) superlattice has odd parity. To understand this we need to look at the wave functions.

The coordinate space envelope functions of the lowest two states for each of the two superlattices at the crossover composition $x_c = 0.28$ are shown in Figure 5. For the case of a (28,8) superlattice the X-well state ($n = 1$) has odd parity, and the Γ-well state has even parity around the origin of the AlAs region ($28 < \lambda < 36$), so the energy levels of two states cross. For the case of a (28,7) superlattice both states have even parity, and the energy levels anticross, resulting in hybridization of wave functions of the two states (mixing of states), as shown in the lower two graphs of Figure 5.

Later on Lu and Sham [6] and Sham and Lu [7] also calculated the conduction band structure of $(GaAs)_N/(AlAs)_M$ (001) superlattices with a second-neighbor tight-binding method. The parameters are fitted to the conduction band valleys of the bulk materials. In the same way they found the striking general feature of the valley mixing to depend sensitively on n and m, which is explained in terms of the symmetry of the superlattice. The parity behavior may be understood with the help of the effective mass approximation, in which an eigen function is the product of the bulk Bloch wave function at a band edge and the envelope function. The parity of the superlattice state

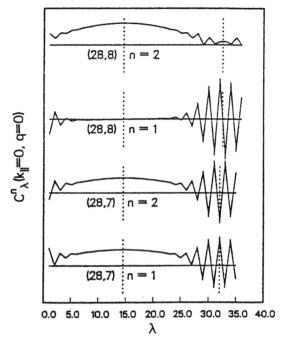

Fig. 5. Coordinate space envelope functions of the lowest two $k_{//} = 0$, $k_z = 0$ states of the (28,8) and (28,7) (001) $(Al_xGa_{1-x}As)/(AlAs)$ superlattices. λ is the layer number. Reproduced with permission from [10], copyright 1987, American Physical Society.

is the product of the parity of the bulk band edge state and the parity of the envelope function. The bulk Γ state has a zero wave vector along the z axis, and, therefore, the Bloch function has a constant coefficient for each layer wave function $u(r)$ (periodic part of the Bloch function) over the GaAs region of a supercell. The superlattice Γ state thus has the same parity as the envelope wave function. On the other hand, the bulk X Bloch state has an exponential factor $\exp(ik_X \cdot r)$, where $k_X = (0, 0, 1)2\pi/a$, thus has an alternating ± 1 as coefficients of the layer wave function $u(r)$ in the AlAs region. Lu and Sham [6] found that the bulk X state $u(r)$ with wave vector $(0,0,a^*)$ has zero coefficients for (Z_a, S_c) and alternating ± 1 coefficients of (S_a, Z_c) in the AlAs region, where S_a represents the S state of the anion (As atom), Z_a represents the P_z state of the anion, etc. In the AlAs region with M layers, the ratio of the coefficients of the layer function on the two interface As planes is $(-1)^M$ (see Fig. 3). Therefore, for the even parity envelope functions of the $(2n)X$ levels, the parity of the state is the same as that of M and, for odd parity envelope functions of the $(2n+1)X$ levels, the parity of the state is opposite that of M. This conclusion is contrary to that of Ting et al. [10] shown in Figure 4.

The conduction energy levels as functions of N for the $(GaAs)_N/(AlAs)_N$ superlattices calculated by Lu et al. are shown in Figure 6. At $N = 12$ (even) the 0Γ level and $0X$ level cross with the same even parity. At $N = 16$ the 1Γ level (odd parity) and the $4X$ level (even parity) cross. But the anticrossing of the 0Γ and $0X$ levels has not been seen from the figure.

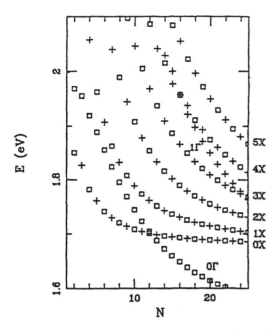

Fig. 6.   Conduction energy levels as functions of N for the $(GaAs)_N/(AlAs)_N$ superlattice. Labels $n\Gamma$ and nX denote the number of a level and its origin in the bulk conduction valley. An open square denotes even parity, and a cross, odd parity. Reproduced with permission from [6], copyright 1989, American Physical Society.

The reason for the contradiction between Lu and Sham [6] and Ting and Chang [10] can be explained by the values of the function $u(r)$ at atoms. The former declared that the coefficients of Z_a and S_c equal zero, and those of S_a and Z_c do not equal zero. From Figure 5 we see that the latter obtained the opposite results: zero coefficient of S_a and nonzero coefficient of S_c. Because they used the one-band Wannier orbital model, there is only the S state for each atom. If it is the latter case the parity of the $u(r)$ of the X state will be $(-1)^{m+1}$. What is the actual situation? We calculated the energy band and wave functions of AlAs by the empirical pseudopotential method and verified that the former is correct. In using the empirical tight-binding model or the one-band Wannier model, to obtain the correct conduction band and effective mass, one has to fit many interaction parameters empirically and artificially. As a result, on the one hand we can obtain a more precise conduction band structure; on the other hand it may bring about unphysical wave function symmetry; for example, the lowest conduction X state in AlAs is X_3 rather than X_1.

Ting and Chang [10] also found that the X_x and X_y states have properties similar to those of Γ and X_z states: the symmetry of the confined X-valley states is critically dependent on whether the slabs with higher Al concentration contain an even or an odd number of monolayers. As a result, the amount of X_x-X_y mixing is extremely sensitive to the layer thickness. The X_x state originates from the bulk X state at $k = (1, 0, 0)2\pi/a$, and the X_y state originates from the X state at $k = (0, 1, 0)2\pi/a$. Because they have the same effective mass in the z direction (m_t), they have the same confined energies at $k = (1, 0, 0)2\pi/a$ in the X well. Similarly, with the help of the effective mass approximation, in which an eigenfunction

is the product of the bulk Bloch wave function at a band edge and the envelope function, the parity of the superlattice state is the product of the parity of the bulk band edge state and the parity of the envelope function. Related to $k_{//} = (1, 0)2\pi/a$, the bulk X_x state has a zero wave vector along the z axis, and, therefore, the Bloch function has a constant coefficient for each layer wave function $u(r)$ (periodic part of the Bloch function) over the AlAs region of a supercell. The superlattice X_x state thus has the same parity as the envelope wave function. On the other hand, the bulk X_y Bloch state has an exponential factor $\exp(ik_X \cdot r)$, where $k_X = (1, 0, 1)2\pi/a$ and thus has an alternating ± 1 as coefficients of the layer wave function $u(r)$ in the AlAs region. [Note: the (010) point is equivalent to the (101) point in k space because they differ by a reciprocal-lattice vector $(1\bar{1}1)$.]

The most striking feature of the symmetry for SPSLs is the qualitative change in the conduction subbands as m and n vary by unit. All such changes can be understood by a symmetry-group analysis, together with a consideration of the effective-mass envelope wave functions. The reciprocal Bravais lattice changes from simple to body-centered tetragonal as $n + m$ change from even to odd. The symmetry operators that change the sign of the z coordinate preserve the anion plane as the origin when m is even and shift to the cation plane when m is odd. Thus, the symmetry properties of the states change depending on whether n or m is even or odd. Such sensitive dependence of the electronic properties on n or m deduced from the theoretical analysis of the perfect superlattices forms a good basis for inferring the properties of the interface. For example, the Γ-X_z mixing in the vicinity of the level crossing is stronger for even m and weaker for odd m. Therefore, if the decay time of the pseudodirect transition is due to the valley mixing in a perfect superlattice, it is expected that the decay time will vary by about an order of magnitude as m goes from even to odd. Such strong dependence on m has not been observed experimentally because of interface disorder scattering [11,12].

3. PHOTOLUMINESCENCE AND PHOTOLUMINESCENCE EXCITATION SPECTRA OF SHORT-PERIOD SUPERLATTICES

Ishibashi et al. [4] first investigated the optical properties of the superlattices $(GaAs)_n/(AlAs)_m$ grown by MOCVD. They concluded that the superlattices $(GaAs)_n/(AlAs)_m$ ($n = 1$–24) have direct energy gaps. Finkman et al. [13], Nagle et al. [14], and Moore et al. [15] measured the PL and PLE spectra of SPSLs and found that the samples of SPSLs ($n < 10$ or the well width < 35 Å) exhibit a large shift between the PL and the PLE threshold. They attributed the luminescence to X states and the PLE threshold to the onset of the Γ transition; hence the SPSLs have an indirect or pseudodirect energy gap.

Moore et al. [15] systematically investigated the effects of decreasing the AlAs layer thickness from 41 to 5 Å on the alignment of (GaAs)/(AlAs) superlattices in which the GaAs thickness was kept constant nominally at 25 Å. The low-temperature (6 K) PL and PLE spectra for superlattices with AlAs thickness 5, 8, 19, 28, and 41 Å are shown in Figures 7 and 8, respectively. First consider the low-temperature PL spectra (Fig. 7) from the superlattice in which the thicknesses of the AlAs layers are 41, 28, and 19 Å. The main PL transition (labeled $E1X$-$HH1$) in these three samples is zero-phonon recombination of type II excitons involving electrons confined at the lowest X point of the AlAs and holes at the Γ point in the GaAs. The PLE peaks correspond to type I exciton transition $E1\Gamma$-$HH1$ and $E1\Gamma$-$LH1$, involving $n = 1$ GaAs Γ electron ($E1\Gamma$) and $n = 1$ heavy ($HH1$) and light hole ($LH1$), respectively, as shown in Figure 8. These transitions occur at a much higher energy than the PL peak. When the thickness of AlAs decreases from 41 Å to 19 Å, the type II PL peaks shift to higher energy, but the type I PLE peaks shift to lower energy. The former reflects the increase in the electron confinement at the X minimum in the AlAs. The latter is a result of coupling between adjacent GaAs layers through thin AlAs layers. This reduces the energy of all of the confined states and thus decreases the Γ transition energies.

When the AlAs thickness is reduced still further to 8 and 5 Å the Γ states become extensively coupled and a dramatic shift of the PL peaks to lower energy is observed, simultaneously with a dramatic modification of their spectral shape. In these samples the $E1\Gamma$-$HH1$ exciton peaks in the PLE spectra (Fig. 8) coincide with the energy of the emission lines (Fig. 7), which means that the GaAs $n = 1$ Γ state now has

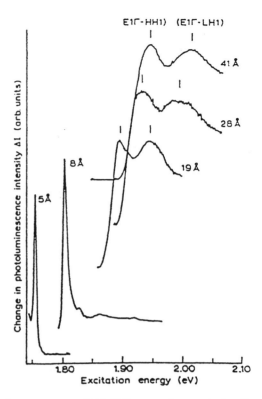

Fig. 8.   Low-temperature (6 K) PLE spectra for superlattices described in Figure 7. Reproduced with permission from [7], copyright 1989, American Physical Society.

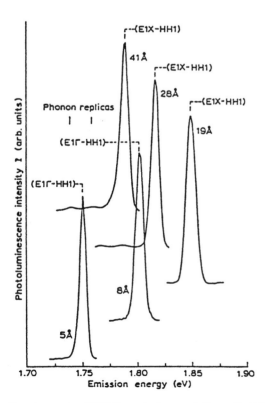

Fig. 7.   Low-temperature (6 K) PL spectra for superlattices with AlAs layers of 5, 8, 19, 28, and 41 Å. Reproduced with permission from [7], copyright 1989, American Physical Society.

been lower than the AlAs $n = 1$ X state and represents the lowest-energy electron state of the superlattice. Therefore, in these samples the nature of the emission has changed to be type I exciton recombination of GaAs Γ-state electrons and holes. This clearly displays the transition from a type II to a type I superlattice. For the superlattices with a GaAs constant thickness of 25 Å this type II–type I crossover occurs between 10 and 15 Å of AlAs.

To explain the experimental results Moore et al. [15] calculated the $E1X$-$HH1$ and $E1\Gamma$-$HH1$ transition energies as functions of AlAs width within the effective-mass envelope function approximation. It is found that the parameter, which most strongly influences the size of the type II energy gap, is the value of the fractional valence-band offset Q_v. The Q_v value is determined to be 0.33 by comparing the experimental and theoretical results.

The type I–type II crossover in the symmetric superlattices $(GaAs)_n/(AlAs)_n$ is an interesting topic because they represent a kind of prototype for experimental and theoretical studies. There have been many experimental and theoretical works, but at this early stage the conclusions about the number of monolayers n at which the type I–type II crossover occurs (denoted as n_c) are divergent, especially for the theoretical works. This is not surprising, because the key parameters for deciding n_c is the fractional valence-band offset Q_v. If we take a smaller Q_v value, for example, 0.15, then the bottom of the X valley in AlAs will rise relative to the bottom of the Γ valley in GaAs. Then the lowest conduction state will always be the

Γ state in GaAs, it would yield no type II alignment for the X valley and no valley mixing effects. This has been pointed out by Schulman et al. [16] and Xia [17]. Actually, as we have seen above, the accurate valence-band offset is determined by comparing the experimental and theoretical results, not by theoretical calculation alone. Even the self-consistent pseudopotential method, or the first principle calculation, cannot give an accurate valence-band offset. And if an accurate valence-band offset is used, all of the theoretical calculations, including empirical pseudopotential, tight binding, even the effective-mass envelope function without coupling of Γ and X valleys, will give basically the same results for n_c.

Table III shows the n_c values of $(GaAs)_n/(AlAs)_n$ superlattices, determined from experimental and theoretical works since 1986. From the table one can see that the results are convergent.

The commonly used method for determining n_c is PL and PLE spectra, which can simultaneously determine the energies of type I and type II emissions. Typical low-temperature PL and PLE spectra for a $(GaAs)_6/(AlAs)_6$ superlattice are shown in Figure 9, where the PL peak at 1.889 eV [in (a) of the figure] is assigned to the recombination of localized excitons built from electron states at X_z in AlAs and holes at Γ in GaAs. The sharply rising peak at 2.06 eV of the PLE spectrum [in (c) of the figure] is caused by the transition from the electron state at Γ in GaAs to the hole state at Γ in GaAs. Setting the detection energy at 1.902 eV and performing PLE over the energy range spanned by (b) of Figure 9 reveals a distinct feature at 1.9265 eV. At this energy excitons are being created also via states at the X_z minima in AlAs. Therefore the X_z state in AlAs is lower than the Γ state in GaAs, the sample is a type II superlattice.

When n decreases, both the energies of the PL peak (X_z exciton) and the PLE peak (Γ exciton) increase, though there is a strong coupling between X wells or Γ wells. It is not

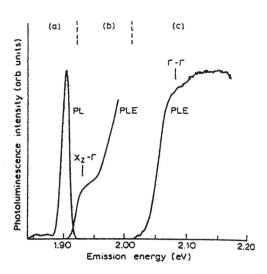

Fig. 9. Low-temperature (5 K) PL and PLE spectra for a $(GaAs)_6/(AlAs)_6$ superlattice. Reproduced with permission from [19], copyright 1988, American Physical Society.

like the case of superlattices with a constant thickness of GaAs [15] as in Figure 8, where the energy of type I PLE peak decreases with decreasing AlAs thickness. Figure 10 shows the low-temperature PL and PLE spectra in a set of

Table III. n_c Values of $(GaAs)_n/(AlAs)_n$ Superlattices, Determined from Experimental and Theoretical Works

Year	First author	Ref.	n_c	Experiment	Temperature	Theory
1986	Gell	18	8			Pseudo
1986	Finkman	13	> 8	PL & PLE	1.7K	
1987	Nagle	14	11			Effective
1988	Xia	17	> 10			Pseudo
1988	Moore	19	> 8	PL & PLE	5K	
1988	Jiang	20	< 15	PL & PLE	2K	
1989	Muñoz	21	8–9			Tight-bind.
1989	Lu	6	12			Tight-bind.
1989	Cingolani	22	12	PAS, HEI	300, 10K	
1989	Li	23	11	PL in HP	300, 77K	
1989	Kato	24	14	PL & AB	20K	
1989	Gapalan	25	> 7			First-princ.
1990	Fujimoto	26	10	PL & PR	300K	
1990	Holtz	27	14	PL in HP	10K	
1990	Xia	28	12			Tight-bind.
1991	Matsuoka	29	14, 12	PL & PR	< 150, 250K	
1992	Nakayama	30	13	PL & AB	2K	
1993	Nakayama	31	13	PL & PLE	10K	

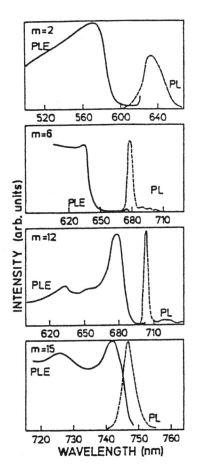

Fig. 10. Low-temperature PL and PLE spectra in a set of $(GaAs)_m/(AlAs)_m$ superlattices with increasing m. Reproduced with permission from [5], copyright 1990.

$(GaAs)_m/(AlAs)_m$ superlattices with increasing m. When the well width increases, the confinement energies of the GaAs Γ states and AlAs X states decrease and both the PLE and PL spectra shift to the red. However, because of the larger electron mass at the AlAs X point, the red shift of the PLE spectra is more pronounced. As a consequence, the PL peak (type II gap) and the PLE onset (type I gap) approach each other in energy when the number of monolayers is increased and finally merge for $(m = n) > 12$ monolayers. Above this critical thickness the $(GaAs)_m/(AlAs)_m$ superlattices recover the direct character, and the energy splitting between PL and PLE reduces to the usual Stokes shift, which is due to weak localization of the exciton resulting from random potentials at interfaces.

Later people also used other methods instead of the PLE spectrum to determine the type I Γ transition energies, for example, absorption, [24] high-excitation intensity luminescence (HEI-PL), photoreflectance, etc. The HEI-PL and normal continuous wave luminescence (here denoted as LEI-PL for comparison) spectra for a set of symmetric $(GaAs)_m/(AlAs)_m$ superlattices with $m = 2, 10$, and 15 are shown in Figure 11. For fully type II SPSL ($m = 2$) the LEI-PL and HEI-PL spectra peak at the same energy, corresponding to the type II emission. When the number of monolayers is increased, the SPSL approaches the type I–type II crossover. For the $m = 10$ SPSL the HEI-PL spectra exhibit a type II emission, coincident with the LEI-PL. At higher photogeneration rates the higher Γ states in GaAs are also occupied by electrons, resulting in a sharp emission line (Γ) arising on the high-energy side of the type II luminescence, at energy coincident with the type I transition measured in PLE. Above the direct–indirect crossover ($m = 15$) the LEI-PL and HEI-PL spectra coincide at the energy of the type I transition, as expected for direct-gap SPSLs. Nevertheless, at a large enough carrier density a shoulder on the high-energy side of the direct Γ emission is observed, indicating the population of the higher energy X states in AlAs. These phenomena have been successfully used to study size-induced type I–type II transitions in

these SPSLs, giving a critical monolayer thickness n_c, in perfect agreement with the one deduced from PLE measurements like those in Table III.

Analysis of the photoreflectance (PR) spectra gives several critical-point energies, which coincide with the PL peak energies in superlattices with high accuracy. The experimental data of PR are analyzed by the third-derivative formula derived by Aspnes et al. [32, 33]. The expression for the modulated reflectance in the presence of the electric field perturbation involves both the real and imaginary parts of the complex dielectric function ε. For normal incidence, the change in the reflectance, ΔR, has the form

$$\Delta R/R = \alpha \Delta \varepsilon_1 + \beta \Delta \varepsilon_2 \tag{3}$$

where $\Delta \varepsilon = \Delta \varepsilon_1 + i \Delta \varepsilon_2$ is the perturbation-induced change in ε, and α and β are the Seraphin coefficients, which are a function of ε_1 and ε_2. The modulated reflectance spectrum in the region of a weak electric field is well expressed by a line-shape formula derived by Aspnes [33],

$$\Delta R/R = \sum_{j}^{p} \mathrm{Re}[C_j \exp(i\theta_j)(E - E_{gj} + i\Gamma_j)^{-m_j}] \tag{4}$$

where p is the number of critical points; E is the phonon energy; and C_j, θ_j, E_{gj}, and Γ_j are the amplitude, phase, energy gap, and broadening parameter, respectively. The value of m_j depends on the dimension of the critical point, which is taken to be 2.5 for the three-dimensional critical point in SPSLs. Figure 12 shows room-temperature PR (open circles) and PL (dot-dashed curve) experimental spectra together with the best-fit PR spectra (solid curve) determined with the use of the fitting formula for $(GaAs)_n/(AlAs)_n$ superlattices with $n = 3, 5, 8, 10$, and 12. The vertical arrow indicates the lowest interband transition energy obtained by the fitting procedure. It is clearly seen in Figure 12a ($n = 12$) and Figure 12b ($n = 10$) that the lowest critical-point energies obtained from the PR are very close to the photon energies corresponding to the PL peaks. Therefore Fujimoto et al. [26] concluded that the SPSLs with $n > 10$ have direct energy gaps. On the other hand, optical properties of the SPSLs with $n < 10$ exhibit quite different features compared with those for $n > 10$. In Figure 12c for the $n = 8$ SPSL the lowest transition energy, 1.881 eV, obtained by the best-fit procedure of the PR spectrum is higher than the PL peak energy, 1.85 eV. Similar results are obtained for the SPSLs with $n = 5$ and 3, which are shown in Figure 12d and e, respectively. In the cases of Figure 12d and e there are two peaks in the PL spectra. The high-energy peak of the PL coincides with the lowest direct transition energy determined from the PR analysis, indicating that the PL peak arises from the direct recombination. The low-energy peak of the PL is caused by the type II transition, the intensities of the two PL peaks are very weak compared with those of the SPSLs with $n > 10$. These results indicate that the SPSLs have a pseudodirect band gap for $n < 10$ and a direct band gap for $n \geq 10$ at room temperature.

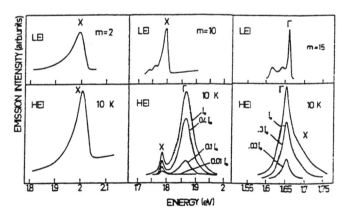

Fig. 11. HEI-PL and LEI-PL spectra measured in a set of $(GaAs)_m/(AlAs)_m$ superlattices with $m = 2, 10$, and 15 ($I_0 \approx 300$ kW cm^{-2}). Reproduced with permission from [22], copyright 1989, American Physical Society.

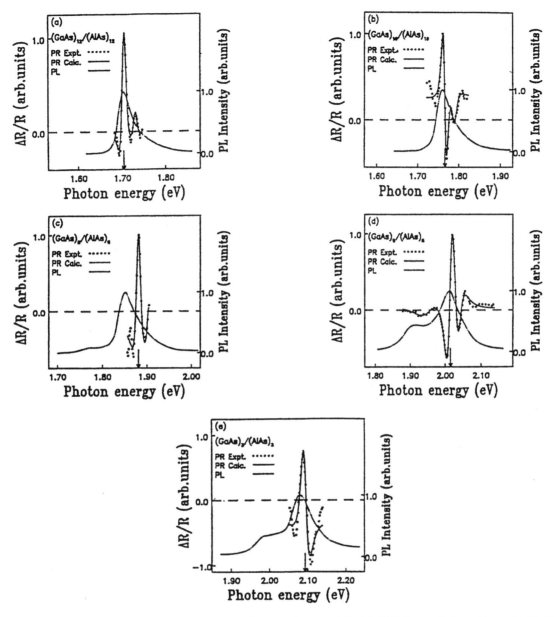

Fig. 12. Room-temperature PR (open circles) and PL (dot-dashed curve) spectra for a set of $(GaAs)_n/(AlAs)_n$ superlattices with $n =$ (a) 12, (b) 10, (c) 8, (d) 5, and (e) 3. The solid curve is determined by fitting the line shape formula to the PR spectrum. The vertical arrow indicates the lowest interband transition energy obtained by the fitting procedure. Reproduced with permission from [26], copyright 1990, American Physical Society.

4. TIME DECAY AND TEMPERATURE DEPENDENCE OF PHOTOLUMINESCENCE AND PHOTOLUMINESCENCE EXCITATION SPECTRA

Finkman et al. [13] first investigated the time-resolved PL and PLE spectra of superlattices with approximately equal thicknesses of AlAs and GaAs and with periods of 18–60 Å. The strongest PL feature shows nonexponential decay after pulse excitation, as shown in Figure 13. The decay has been successfully fit with the use of a model [34] derived to describe disorder-induced direct recombination in bulk indirect-bandgap AlGaAs alloys. For localized excitons at zone boundary, this model predicts the decay of the zero-phonon and phonon-

sideband emissions to take, respectively, the forms,

$$I_{zp}(t) = \exp(-w_n t)(1 + 2w_r t)^{-3/2} \qquad (5)$$

$$I_{ps}(t) = (w_n/w_r)\exp(-w_n t)(1 + 2w_r t)^{-1/2} \qquad (6)$$

The parameter w_r is the mean recombination rate due to random mixing of the direct and indirect exciton states by disorder scattering, and w_n is the net rate due to all nonrandom processes. The full line is fitted by $I_{zp}(t)$ with $w_n \sim 0$, and $w_r = 1.5 \times 10^6$ s^{-1}. The fact that the time decay can be fitted without any nonrandom component led Finkman and Sturge et al. [11, 13] to the conclusion that the X exciton is not mixed with Γ by the superlattice potential. It must therefore be associated with X_x and X_y minima, which have wave vectors parallel to the layer, rather than with the X_z minimum. Thus, the lowest conduction minimum of these superlattices is indirect X_x and X_y, not the pseudodirect X_z. This conclusion is contrary to most experimental and theoretical results, but is supported by Ihm's theoretical calculation [35]. According to the calculation the degeneracy of X_x and X_y will be removed because of the superlattice potential, and the symmetric combination X_{xy} can be pushed below X_z. Comparison of the mean decay rates in a series of SPSLs with different period [13], with the use of the theory developed for indirect excitons in bulk $Al_xGa_{1-x}As$ [36], shows that the density of the scattering centers responsible for the Γ-X mixing varies approximately inversely as the period, confirming that the scattering is due to disorder at the interfaces.

Nakayama et al. [30] studied the CW and time-resolved PL spectra of $(GaAs)_n/(AlAs)_n$ ($n = 8$–15) superlattices. For the type II superlattices ($n \leq 12$) they obtained decay curves similar to these in Figures 13 and 14 and fitted them to Eq. (5). The fitted values of w_r and w_n are given in Table IV. The values of w_n are in all cases very small, about three orders of magnitude smaller than w_r. Furthermore, the magnitude of w_r does not scale monotonously with the energy separation between the X and Γ peaks; that is, it oscillates as the layer number n changes from even to odd, and it occurs dominantly for an odd number of AlAs monolayers. Because w_r is the recombination rate due to random disorder scattering, this result is similar to that of Finkman and Sturge et al. [11, 13].

The conclusion that the lowest conduction state is X_{xy}, not X_z in type II superlattices caused arguments by, for example, Moore et al. [12, 19]. Because of the anisotropic nature of the X minima, the X minima in the growth direction X_z have a much larger confinement mass ($1.1m_0$) than the X minima in the layer plane X_x and X_y ($0.19m_0$). This means that the folded X_z state is always at a lower confinement energy than the unfolded X_{xy} states in type II superlattices. This has been verified by a series of theoretical calculations, including

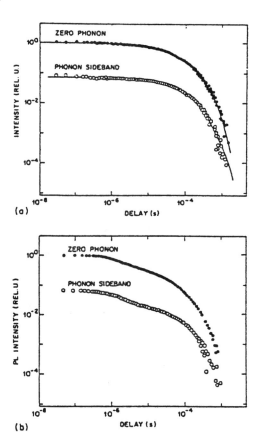

Fig. 14. Decay of the PL intensity in the zero-phonon line and in the lowest-energy phonon replica at 6 K for the sample $(GaAs)_{11}/(AlAs)_{24}$ (a) at low excitation intensity (3×10^3 W/cm^2), and (b) at higher excitation intensity (3×10^4 W/cm^2). The solid curves in (a) are fitted by Eqs. (5) and (6). Reproduced with permission from [12], copyright 1988, American Vacuum Society.

the effective-mass envelope function and tight-binding methods [6, 7, 10, 28]. In Ihm's tight-binding model the complete neglect of the next-nearest-neighbor interaction gives an infinite transverse effective mass, m_t, in the X valley of the bulk bands, resulting in a situation in which the $X_{x,y}$ levels always stay at the bottom of the bulk X valley, lower than the X_z level. But the effective-mass envelope function and tight-binding methods fail in the case of very short-period superlattices (for example, $n < 4$). Ge et al. [37] investigated the energy bands of very short-period $(GaAs)_n/(AlAs)_n$ superlattices with $n \leq 4$ by PL and showed that the lowest conduction bands are

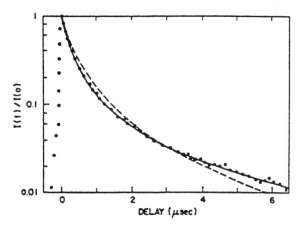

Fig. 13. Decay of the PL peak after short pulse excitation (points). The solid line represents the calculated decay of zone-boundary excitons due to purely random Γ-X mixing, and the dashed line represents that of indirect excitons not at the zone boundary. Reproduced with permission from [13], copyright 1986, American Physical Society.

Table IV.   Time Decay Fitting Parameters

(n, n)	$w_r (10^6 \text{s}^{-1})$	$w_n (10^6 \text{s}^{-1})$
(8, 8)	0.75	< 0.03
(9, 9)	2.5	< 0.002
(10, 10)	1.2	< 0.001
(11, 11)	8.2	< 0.01
(12, 12)	3.2	< 0.01

Reproduced with permission from [30], copyright 1992, Elsevier Science.

X_{xy}-like for $n \leq 3$ and X_z-like for $n \geq 4$. This problem is discussed in Section 7 for very short period superlattices.

Wilson et al. [12] studied further the time-resolved PL spectra for a series of type II (GaAs)/(AlAs) superlattices with different GaAs layer thicknesses. They found that at 6 K the decay curves of the type II emission contain power-law and exponential components with a rate of $\sim 10^3$ s^{-1}, indicating significant Γ-X mixing, both by random and nonrandom processes, contrary to the results of Finkman et al. [13] and Sturge et al. [11]. The decay of the PL intensity in the zero-phonon line and in the lowest-energy phonon replica at 6 K for the sample $(GaAs)_{11}/(AlAs)_{24}$ is shown in Figure 14. At a low excitation intensity of $\leq 3 \times 10^3$ W/cm^2, corresponding to 10^{14} photons/ cm^2/ pulse, the decay curves of both the zero-phonon and phonon-sideband lines can be fitted with Eqs. (5) and (6) over five decades in time, as shown in Figure 14a. At a higher intensity of 3×10^4 W/cm^2, as seen in Figure 14b, the decays contain additional components at early times, which are manifested spectrally as broad, rapidly shifting PL lines. The nature of these additional recombination processes in the type II energy region is not yet fully understood. Although the model gives a good account of the observed decay curves, the fits are relatively insensitive to the precise form of the power-law component arising from the random processes. Comparing the relative intensities in the zero-phonon and phonon-sideband emissions as a function of decay, the model predicts

$$I_{zp}(t)/I_{ps}(t) = (w_n/w_r)(1 + 2w_r t) \qquad (7)$$

In fact, the observed ratio is clearly sublinear, as would be the case if there were additional nonrandom radiative contributions to the zero-phonon line. Such contributions are expected from the symmetry breaking associated with individual interfaces, as well as from the mixing by the periodic potential of the multiplayer structure. A comparison of the low-temperature decay statistics for three samples, $(GaAs)_5/(AlAs)_{24}$, $(GaAs)_8/(AlAs)_{24}$, and $(GaAs)_{11}/(AlAs)_{24}$, reveals that both w_r and w_n scale with the GaAs layer thickness; they are weakest in the five-monolayer (ML) sample and strongest in the 11-ML sample. This behavior is expected, because any Γ-X mixing will scale as $1/\Delta^2$, where Δ is the energy difference between the occupied X states and the nearest-energy Γ state confined in the GaAs layer. The exponential component of the decay is observed to increase exponentially with temperature. These facts indicate that the lowest conduction state would be the X_z state, and the PL is associated with the significant Γ-X mixing.

Another Nakayama, M. Nakayama et al. [38], also studied CW and time-resolved PL and PLE spectra for type II (GaAs)$_n$/(AlAs)$_n$ ($n = 8$–13) superlattices. At the low excitation intensity as in [11, 13] the decay profiles shown in Figure 15 are remarkably different from that given by Eq. (5). The decay has two components, fast and slow. The fast component of the PL decay decreases as the detection energy is lowered, whereas the slow component, the decay time of which is independent of the detection energy, becomes dominant. Similar decay features have been observed in all samples. It is

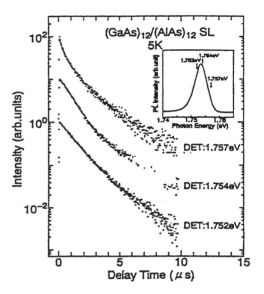

Fig. 15. Time decay profiles of the $11HX_z$ exciton detected at different energies in the (12,12) SL at 5 K, where the inset shows the integrated PL spectrum, and the arrows indicate the detection energies. Reproduced with permission from [38], copyright 1994, American Physical Society.

assumed that the fast component of the PL decay is related to relaxation processes of exciton localization, and the slow component corresponds to the intrinsic PL decay. The random scattering process in the radiative recombination is negligible in this case. The time decay profiles on the low-energy side for all samples can be least-square fitted based on a double-exponential model:

$$I(t) = A_f \exp(-w_f t) + A_s \exp(-w_s t) \qquad (8)$$

where the subscripts f and s mean the fast and slow components, respectively. The PL decay profiles are successfully explained by Eq. (8). The values of w_f and w_s are listed in Table V, with the values of relative PL intensity $I(11HX_z)/I(11H\Gamma)$ at $1/T = 0$. From the table it can be seen that w_s is almost independent of the layer thickness. Because w_s is attributed to the radiative decay rate of the $11HX_z$ exciton corresponding to the oscillation strength, this suggests that the oscillation strength is hardly affected by the layer thickness in the range of 8–13 monolayers. This result verified further that the lowest conduction state is X_z and the transition is pseudodirect, not indirect. The obvious difference between this work and previous works suggests that the decay rate depends remarkably on the interface quality.

The temperature dependence of the relative PL intensity provides a means of determining the magnitude of the Γ-X mixing [29]. The PL intensity is proportional to the product of the density of states at the conduction band and the transition probability, and the ratio of PL emission between the direct and the pseudodirect band gaps is therefore written as

$$\frac{I(\Gamma)}{I(X)} = \left(\frac{m_\Gamma}{m_X}\right)^{3/2} \frac{W_\Gamma}{W_X} \exp\left(\frac{E_X - E_\Gamma}{k_B T}\right) \qquad (9)$$

Table V. Decay Rates of the Fast and Slow Components (w_f and w_s) of the $11HX_z$ Exciton at 5 K and the Relative PL Intensity of $I(11HX_z)/I(11H\Gamma)$ at $1/T = 0$

(n, n)	$w_f(10^6 \text{s}^{-1})$	$w_s(10^6 \text{s}^{-1})$	$I(X_z)/I(\Gamma)$ at $1/T = 0(10^{-3})$
(8, 8)	0.95	0.28	0.97
(9, 9)	1.1	0.46	1.2
(10, 10)	1.2	0.46	1.1
(12, 12)	1.2	0.50	1.0
(13, 13)	1.1	0.46	1.1

Reproduced with permission from [38], copyright 1994, American Physical Society.

where $I(\Gamma)$, E_Γ, and W_Γ are the PL intensity, transition energy, and transition probability for the direct allowed transition; $I(X)$, E_X, and W_X for designate these variables for the pseudodirect transition; and m_Γ and m_X are the density-of-states effective masses in the Γ-like and X-like conduction bands. The factor $(m_\Gamma/m_X)^{3/2}$ is the ratio of the density of states in two bands. Figure 16 shows the intensity ratio $I(\Gamma)/I(X)$ as a function of inverse temperature with the least-square-fit lines for $(GaAs)_n/(AlAs)_n$ superlattices with $n = 5, 7, 8, 10$, and 12. The energy separation $E_\Gamma - E_X$ obtained from the slope of the line is in reasonable agreement with the PL peak energy separation. From the intercept of the line at zero the ratio of the transition probabilities W_X/W_Γ is estimated to be 6.7, 1.6, 2.5, 2.7, and 8.3 ($\times 10^{-4}$) for $n = 5, 7, 8, 10$, and 12, respectively, taking $m_\Gamma = 0.068m_0$ and $m_X = 0.34m_0$. Therefore, these results strongly suggest that the lower-energy weak transition is ascribed to a pseudodirect band gap associated with the X_z-like state, and the higher-energy transition is to a direct band gap associated with the Γ state.

From the ratio of the transition probabilities W_X/W_Γ obtained from the ratio of the PL peak intensity of the pseudodirect transition to the direct transition $I(X)/I(\Gamma)$ at $1/T =$

0 M, Nakayama et al. [31] evaluated the Γ-X mixing factor as a function of the layer thickness by using a first-order perturbation theory. The PL intensity ratio $I(X)/I(\Gamma)$ at $1/T = 0$ obtained by Nakayama et al. [38] is listed in Table V. According to the first-order perturbation theory the relative oscillator strength, W_X/W_Γ, is given by

$$W_X/W_\Gamma = |\langle \Psi_\Gamma | V | \Psi_X \rangle|^2 / \Delta E^2 \qquad (10)$$

where Ψ_Γ and Ψ_X are the total wave functions of the Γ and X_z states, V is a Γ-X mixing potential, and ΔE is the energy difference of the Γ and X_z states. Figure 17 shows the estimated values of the Γ-X mixing factor, $\langle \Psi_\Gamma | V | \Psi_X \rangle$ (closed circles), and the overlap integral of the envelope functions of the Γ and X_z states, $\langle \phi_\Gamma | \phi_X \rangle$ (solid line), as a function of the layer thickness. As shown in Figure 17, the value of the Γ-X mixing factor changes from ~ 0.8 meV in the (13,13) SL to ~ 1.9 meV in the (8,8) SL with decreasing layer thickness. The values are consistent with those estimated from other experimental and theoretical works. Meynadier et al. [39] reported evidence of the Γ-X mixing from the electric-field-induced anticrossing of the PL band of the $11HX_z$ exciton and $11H\Gamma$ exciton in a $(GaAs)_{12}/(AlAs)_{28}$ superlattice. The appearance of the anticrossing originates from the Γ-X mixing, and the estimated mixing potential is very small, about 1 meV. The tight-binding calculation for the same superlattice in the electric field [28] also obtained a similar anticrossing behavior of the Γ and X_z states at an electric field strength of 4.5×10^4 V/cm. The energy splitting at the crossing field is 2 meV, the same as the experimental value. From Figure 17 it is evident that the layer-thickness dependence of the Γ-X mixing factor almost agrees with that of the overlap integral of the envelope functions. As the envelope function varies slowly on the scale of the Bloch functions, the mixing factor in Eq. (39) can be approximately rewritten as

$$\langle \Psi_\Gamma | V | \Psi_X \rangle = \langle \psi_\Gamma | V | \psi_X \rangle \langle \phi_\Gamma | \phi_X \rangle \qquad (11)$$

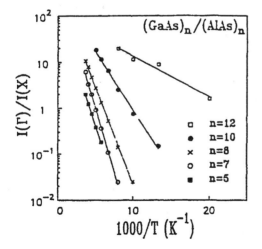

Fig. 16. Ratio of the PL peak intensity of the direct transition to the pseudodirect transition as a function of inverse temperature for $(GaAs)_n/(AlAs)_n$ superlattices with $n = 5, 7, 8, 10$, and 12. Reproduced with permission from [29], copyright 1994, American Physical Society.

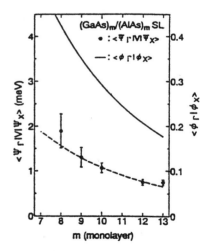

Fig. 17. Values of the Γ-X mixing factors, $\langle \psi_\Gamma | V | \psi_X \rangle$, (closed circles), estimated from Eq. (10) and the overlap integral of the envelope functions, $\langle \phi_\Gamma | \phi_X \rangle$ (solid line), as a function of the layer thickness. Reproduced with permission from [31], copyright 1993, Elsevier Science.

where ψ and ϕ are the Bloch function and the envelope function, respectively. The estimated values of $\langle \psi_\Gamma \mid V \mid \psi_X \rangle$ are 4.2, 3.6, 3.6, 3.8, and 4.5 meV for $n = 13$, 12, 10, 9, and 8, respectively. Theoretical studies [6, 7, 10] predict that the Γ-X mixing in type II SLs with perfect interfaces drastically changes as a function of the AlAs layer thickness in monolayer units because of the symmetry change of the wave function of the X state. The present results suggest that the Bloch-function term of the Γ-X mixing is almost independent of the parity of the AlAs monolayers. It seems that the inconsistency between the theoretical predictions and the experimental results originates from the imperfection of interfaces, the so-called monolayer fluctuation, in real SLs.

5. PHOTOLUMINESCENCE SPECTRUM UNDER HYDROSTATIC PRESSURE

The measurements of PL spectra under hydrostatic pressure provide a powerful tool for the investigation of the type I–type II transition in the GaAs/AlAs SPSLs. Under hydrostatic pressure the conduction Γ, X, and L valleys of bulk GaAs and other III–V compounds move at different rates. For example, the pressure coefficients of Γ, X, and L valleys of GaAs are 10.7, -1.3, and 4.5 meV/kbar, respectively [40]. Therefore, it is easy to identify the character of the lowest conduction subband of a GaAs/AlAs superlattice by measuring the pressure coefficients of the PL peak energies. The lowest conduction subband is Γ-like for type I superlattices, so that the PL peak moves to higher energy under pressure with a pressure coefficient similar to that of a Γ valley in bulk GaAs. On the other hand, the lowest state in type II superlattices is X-like and the PL peak moves to lower energy at a rate similar to that of a bulk GaAs X valley.

Wolford et al. [41] first studied the pressure dependence of quantum-well confined states formed in the GaAs/Al$_x$Ga$_{1-x}$As heterostructure system. For a type I superlattice at the critical pressure, a crossing occurs between the Γ-like state and the X-like state. Confirmed by wave function calculations, the new transition occurs across the heterointerface, between X-confined electrons within the Al$_x$Ga$_{1-x}$As and Γ-confined holes within the GaAs. Comparing the experimental and theoretical results, the valence band offset is determined directly with potentially meV resolution, which is $\Delta E_v \approx (0.32 \pm 0.02)\Delta E_g^\Gamma$ for $x \approx 0.28$ and 0.70. Figure 18 shows PL spectra for a GaAs/Al$_{0.28}$Ga$_{0.72}$As superlattice with both well and barrier widths of 70 Å from 21.7 to 50.4 kbar. The narrow (7.4 meV) line at 21.7 kbar results from the $\Gamma_{1e} - \Gamma_{1hh}$ transition, so at atmospheric pressure the superlattice is type I. The $\Gamma_{1e} - \Gamma_{1hh}$ peak shifts to higher energy with increasing pressure, and the PL intensity decreases rapidly. On the other hand, at 31.3 kbar there appears a new peak labeled $X_e - \Gamma_{1hh}$, which shifts to lower energy with increasing pressure. The other peaks, the weak peak at 21.7 kbar and those labeled D_Γ and BA, result from substrate emission. They also shift to higher energy with increasing pressure, which represents the shift of the direct GaAs band edge.

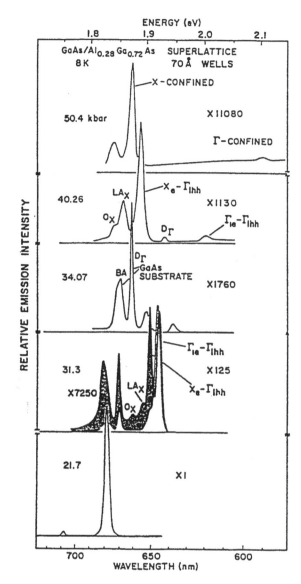

Fig. 18. PL spectra (8 K) of a GaAs/Al$_{0.28}$Ga$_{0.72}$As superlattice with well and barrier widths of 70 Å under the indicated pressures. Reproduced with permission from [41], copyright 1986, American Vacuum Society.

Figure 19 compares the collected pressure results from the two superlattices of different barrier composition ($x \approx 0.28$ and 0.70). For reference the Γ and X band-gap dependences on pressure of bulk GaAs are shown, all referenced to the top of the bulk GaAs valence band. It is apparent that the $\Gamma_{1e} - \Gamma_{1hh}$ transitions track the Γ gaps of bulk GaAs, and the $X_{1e} - \Gamma_{1hh}$ transitions follow closely the pressure dependence of the X conduction band edges. This result, together with the accompanying near-zone-boundary replicas (LA$_X$ and O$_X$ in Figs. 18 and 19), identifies the bound electron as being formed from X-conduction states. The crossings between these new X-confined states and the Γ_{1e} states occur at different threshold pressures in the two superlattices. For $x \approx 0.28$ this crossing occurs near 29.5 kbar, whereas for $x \approx 0.70$ it occurs near 18 kbar. This is due to the fact that the X band edge is lower for the superlattice of larger x, so that the crossings of the X-confined states and the Γ_{1e} states occur at smaller pressure.

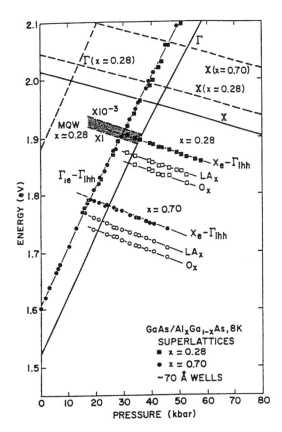

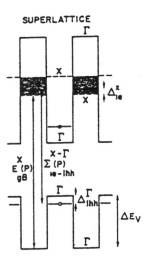

Fig. 20. Schematic band diagram showing a GaAs/Al$_x$Ga$_{1-x}$As superlattice and those energies necessary for obtaining valence-band offsets. Reproduced with permission from [41], copyright 1986, American Vacuum Society.

Fig. 19. PL transition energies (8 K) of a GaAs/Al$_x$Ga$_{1-x}$As superlattice with well and barrier widths of 70 Å, $x \approx 0.28$ and 0.70, as functions of pressure. Reproduced with permission from [41], copyright 1986, American Vacuum Society.

With the use of Figure 19 it is easy to obtain directly the valence band offset ΔE_v for the GaAs/Ga$_x$Al$_{1-x}$As superlattices. In Figure 19 the dashed line labeled $X(x = 0.28)$ is the band gap between the X conduction band bottom and the Γ valence band top for the bulk alloy Al$_{0.28}$Ga$_{0.72}$As $E_{gB}^X(P)$. At the critical pressure the Γ_{1e}–Γ_{1hh} transition and the X_{1e}–Γ_{1hh} transition cross. From Figure 20, the schematic band diagram of the GaAs/Al$_x$Ga$_{1-x}$As superlattice, the valence band offset can be written as

$$\Delta E_v = E_{gB}^X(P) - \sum_{1e-1hh}^{X-\Gamma}(P) + \Delta_{1hh}^\Gamma + \Delta_{1e}^X \qquad (12)$$

where $\sum_{1e-1hh}^{X-\Gamma}(P)$, Δ_{1hh}^Γ and Δ_{1e}^X are energy spacing of the X_1 confined electron state and the Γ_1 confined heavy hole state and the confined energies of the X_1 state and the Γ_1 state, respectively (see Fig. 20). The PL peak energy is

$$PL_{1e-1hh}^{X-\Gamma}(P) = \sum_{1e-1hh}^{X-\Gamma}(P) - E_x^{X-\Gamma} - \Delta_{ss}^{X-\Gamma} \qquad (13)$$

where $E_x^{X-\Gamma}$ is the binding energy of the indirect X–Γ exciton, because of minimal electron-hole spatial overlap, which will be neglected. $\Delta_{ss}^{X-\Gamma}$ is the Stokes shift between peak emission and excitation, which is assumed to be 4 meV. From the

effective-mass theoretical calculation for a 70-Å well, $\Delta_{1e}^X \approx$ 4 meV and $\Delta_{1hh}^\Gamma \approx 10$ meV. Thus, with the data of Figure 19 surrounding the X_{1e}–Γ_{1e} crossing and the $E_{gB}^X(P)$ at the same pressures, the valence-band offset is found to be 0.11 and 0.32 meV for $x = 0.28$ and 0.70, respectively [41]. The fractional band offset is approximately 0.68:0.32, which differs considerably from the original 0.85:0.15 rule, and somewhat from the other once suggested ratio, 0.60:0.40.

At the same time Gell et al. [42] used the empirical pseudopotential method to calculate the electronic and optical properties of GaAs/Al$_x$Ga$_{1-x}$As superlattices and the effects of hydrostatic pressure. For consideration of the pressure effects, the symmetric and antisymmetric form factors for GaAs and AlAs are fitted to the experimental energy band gaps at 0, 10, 20, $\ldots$, 50 kbar. The virtual crystal approximation and the form factors for GaAs and AlAs are used to model the band structure of the alloy Al$_x$Ga$_{1-x}$As. The calculated pressure coefficients for various states for a GaAs(50 Å)/Al$_x$Ga$_{1-x}$As (102 Å) superlattice are shown in Table VI, together with those for bulk GaAs and AlAs. From the table it can be seen that the pressure coefficients of bulk GaAs are in good agreement with the experimental values [40], and the corresponding states in superlattices EΓ and EX shift with nearly the same rates as Γ and X band edges in bulk materials, which is also consistent with experimental results [41]. Figure 21 shows the energy variation of EΓ1-HH1 and EX1-HH1 PL peaks with pressure for a GaAs(68 Å)/Al$_{0.28}$Ga$_{0.72}$As(68 Å) superlattice, where the dashed lines are the calculation results [42] and the dots are the experimental results [41]. The calculations have been performed with a 0.71:0.29 offset, and assuming the binding energy of indirect exciton to be 20 meV. The calculated wave function of the EX1 state shows that it is localized mainly in the Al$_{0.28}$Ga$_{0.72}$As layer with the k components around the X_z points in the bulk Brillouin zone. The oscillator strength for EΓ1-HH1 and EX1-HH1 transitions are about 5 and 5×10^{-3}, respectively, and are nearly independent of pressure.

Table VI. Pressure Coefficients for Various States for a
GaAs(50 Å)/Al$_x$Ga$_{1-x}$As (102 Å) Superlattice, Together with
Those for Bulk GaAs and AlAs

System	Level or state	Pressure coefficient (meV/kbar)
Bulk GaAs	Γ_c	10.6
	X_c	−1.3
Bulk AlAs	Γ_c	10.2
	X_c	−1.7
Superlattice; $x = 0.25$	EΓ1	10.5
	EΓ2	10.2
	EΓR1	10.4
	EX1	−1.4
Superlattice; $x = 1.0$	EΓ1	10.4
	EX1	−1.7
	EX3	−1.6
	EX5	−1.5
	EX7	−1.3

Reproduced with permission from [42], copyright 1987, American Physical Society.

Danan et al. [43] performed PL experiments at 1.7 K for three series of (GaAs)/(AlAs) SPSL samples, characterized by two parameters: the period $P = L_{GaAs} + L_{AlAs}$ and the average aluminum concentration, $x = L_{AlAs}/P$. The PL measurements under hydrostatic pressure are used to distinguish type I and type II transitions according to their energy shifts under pressure. All investigated samples are displayed on a P versus x graph in Figure 22, and are categorized into direct (crosses),

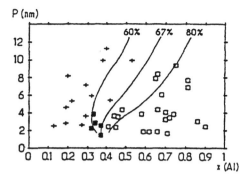

Fig. 22. Samples distinguished as direct (+), pseudodirect (□), and ambiguous (■) are displayed on the P versus x graph. The theoretical crossover curves for some different values of Δ are plotted. Reproduced with permission from [43], copyright 1987, American Physical Society.

pseudodirect (open squares), and ambiguous (filled squares). The calculated crossover curves with the effective-mass envelope function theory for different fractional conduction band offsets Δ, which separates the two domains in the P versus x graph, are plotted in Figure 22. The first domain includes direct type I samples; it corresponds roughly to $x \leq 0.35$ or $L_{GaAs} \geq 4$ nm. The other domain includes pseudodirect type II superlattices. From Figure 22 the best fitted fractional conduction band offset Δ = 0.67 is determined, in agreement with the above result [41].

Li et al. [23] systematically studied a series of (GaAs)$_m$/(AlAs)$_n$ SPSLs with $(6,6) \leq (m, n) \leq (17,17)$ from PL spectra measured at room temperature under hydrostatic pressure in the range of 0–50 kbar. At room temperature, because of the Boltzmann distribution of electrons in the conduction band, it is easy to observe the luminescence related to both Γ-like and X-like states simultaneously under a wide pressure range and to obtain important information about the different luminescence lines and their relative intensi-

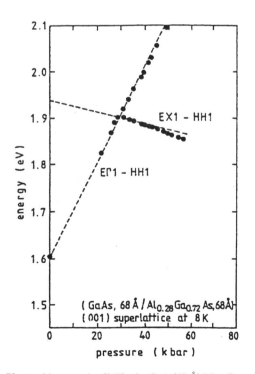

Fig. 21. PL transition energies (8 K) of a GaAs(68 Å)/Al$_{0.28}$Ga$_{0.72}$As(68 Å) superlattice as functions of pressure. Dashed lines and dots are calculated and experimental results, respectively. Reproduced with permission from [42], copyright 1987, American Physical Society.

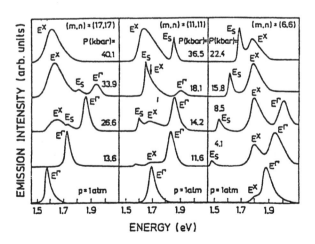

Fig. 23. Room-temperature PL spectra of three representative (GaAs)$_n$/(AlAs)$_n$ superlattices under different hydrostatic pressures indicated in the figure. The intensities have been normalized according to the corresponding strongest peak. Reproduced with permission from [23], copyright 1989, American Physical Society.

ties. The room-temperature PL spectra of three representative $(GaAs)_n/(AlAs)_n$ superlattices obtained under different hydrostatic pressures indicated are shown in Figure 23, where E^Γ, E^X, and E_S represent PL peaks from the conduction Γ-like state, the X-like state transitions, and from the GaAs substrate, respectively. The intensities have been normalized according to the corresponding strongest peak. The pressure dependence of the room-temperature PL peak energies obtained from Figure 23 for three representative samples is shown in Figure 24. For type I superlattices the lowest conduction state is Γ-like, the PLE threshold energy will only have a small Stokes shift away from the main PL peak, and then it is impossible to determine the X-Γ separation, i.e., the energy of the lowest X state. Therefore, there have been no data on the X-Γ separation energy in type I superlattices before. From Figure 24 it can be seen that the pressure dependence of PL energies for E^Γ, E^X, and E_S can be well fitted by a linear relation $E^i(P) = E_0^i + a^i P$, where $i = \Gamma, X, S, E_0^i$ represents the peak energy at atmosphere pressure, and a^i is the pressure coefficient. Through an extrapolation of the pressure dependence of the X-like or Γ-like states to atmospheric pressure, the X-Γ energy separation can be obtained for both type I and type II superlattices, as shown in Figure 24. The values obtained for E_0^i and a^i, and the energy separation $E_0^\Gamma - E_0^X$ are listed in Table VII. Inspection of these data in Table VII reveals that the pressure coefficients for E^Γ and E^X are close to their counterparts for bulk GaAs (10.7 and –1.3 meV/kbar, respectively). The type I–type II transition at atmospheric pressure takes place in the $(GaAs)_{11}/(AlAs)_{11}$ superlattice ($E_0^\Gamma - E_0^X \sim 0$).

It is found that the intensities of both Γ and X PL peaks decrease with increasing pressure, and that of the Γ PL peak decreases faster than that of the X PL peak. The dependence of the room-temperature PL intensity ratio $I(\Gamma)/I(X)$ on the relative pressure $(P - P_0)$ for a series of samples is shown in Figure 25 [23], in which the solid lines are the least-squares fits to the experimental data by the exponential expression

$$\frac{I(\Gamma)}{I(X)} = \left(\frac{I(\Gamma)}{I(X)}\right)_0 \exp\left(-\frac{a'(P - P_0)}{k_B T}\right) \quad (14)$$

The fact that the PL intensity ratio follows the exponential rule means that more electrons transfer from the Γ-like state to

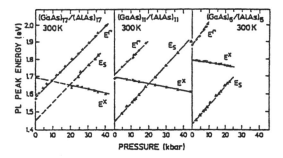

Fig. 24. Pressure dependence of the room-temperature PL peak energies obtained from Fig. 23 for three representative samples. Reproduced with permission from [23], copyright 1989, American Physical Society.

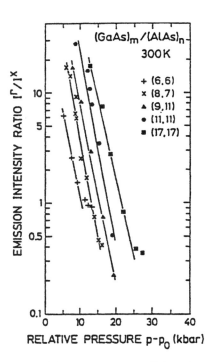

Fig. 25. Dependence of the room-temperature PL intensity ratio $I(\Gamma)/I(X)$ on the relative pressure $(P - P_0)$. Reproduced with permission from [23], copyright 1989, American Physical Society.

the X-like state when the energy separation $E_0^\Gamma - E_0^X$ increases with increasing $P - P_0$. In Eq. (14) a' nearly equals the experimental values of the difference in pressure coefficients $a^\Gamma - a^X$. The values of $(I(\Gamma)/I(X))_0$ obtained by the extrapolated solid lines intercepting the coordinate axis at $(P - P_0) = 0$ are given in Table VII. Assuming that the occupation probabilities of electrons on the two levels at P_0 are comparable, the intensity ratio $(I(\Gamma)/I(X))_0$ can be approximately expressed as

$$\left(\frac{I(\Gamma)}{I(X)}\right)_0 = \frac{N_\Gamma W_\Gamma}{2N_X W_X} = \frac{m_\Gamma W_\Gamma}{2m_{tX} W_X} \quad (15)$$

where N_Γ, N_X, m_Γ, and m_{tX} represent the two-dimensional density of states of Γ-like and X-like states and effective masses of Γ- and X-valley electrons in the xy plane, respectively. W_Γ and W_X are the probabilities for the transition of Γ and X electrons to the heavy-hole states. The factor 2 is due to the existence of two X valleys in the (001) direction. Taking $m_\Gamma = 0.0665m_0$ and $m_{tX} = 0.19m_0$, the $(W)_X/W_\Gamma$ values for different superlattices are calculated and listed in Table VII. From these data it is found that $W_X/W_\Gamma \sim 10^{-4}$ to 10^{-3}, which is the same order of magnitude as other experimental values from the temperature dependence [26] and theoretical predictions by Gell et al. [42] and Xia [17]. The data of Table VII show that the value of W_X/W_Γ increases from 1.4×10^{-4} [for (17,17)] to 4.6×10^{-3} [for (6,6)], indicating that the state mixing is relatively more pronounced for superlattices with shorter period lengths.

Skolnick et al. [44] observed the phonon satellites (PSs) of X pseudodirect excitonic recombination in a $(GaAs)_{12}(AlAs)_8$ superlattice under hydrostatic pressure up to 50 kbar. The 2 K

Table VII. Experimentally Determined Parameters for $(GaAs)_m/(AlAs)_n$ Superlattices[a]

(m, n)	$E_0^\Gamma (eV)$	a^Γ (meV/kbar)	E_0^X (eV)	a^X (meV/kbar)	$E_0^\Gamma - E_0^X$ (meV)	P_0 (kbar)	(I^Γ/I^X) at $P = P_0$	$W_X/W_\Gamma (10^{-3})$
(6, 6)	1.874	14.7	1.795	−1.6	79	−4.9	38.6	4.6
(8, 7)	1.765	11.1	1.748	−2.0	17	−1.3	118	1.5
(9, 11)	1.774	10.9	1.728	−2.2	46	−3.5	261	0.68
(11, 11)	1.708	10.0	1.699	−2.0	9	−0.8	516	0.34
(17, 17)	1.588	10.2	1.691	−2.2	−103	8.3	2736	0.14

[a]Some symbols are explained in the text. P_0 is the pressure at which the energies E^Γ and E^X are equal. $(I^\Gamma/I^X)_0$ denotes the extrapolated intensity ratio of I^Γ and I^X at $P = P_0$. W_Γ and W_X are transition probabilities for two electron states to the Γ heavy-hole state. Reproduced with permission from [23], copyright 1989, American Physical Society.

PL spectra from $P = 0$ to 36.1 kbar are shown in Figure 26. At $P = 0$ (Fig. 26a), the spectrum is composed of a direct exciton zero-phonon line (ZPL) $[E_{1h}(\Gamma)]$ and a PS 36.7 meV lower in energy, very close to the Γ-point LO phonon energy of GaAs. The Stokes shift between PL and PLE at $P = 0$ is 7.6 meV. At the lowest pressure (2.4 kbar, Fig. 26b), the form of the PS is very different. Three PSs are observed (labeled Y_1, Y_2, and Y_3) below the ZPL $[E_{1h}(X)]$. The overall character of the PS remains the same up to 50 kbar. At 3.3 kbar, the Y_1, Y_2, and Y_3 energies are 27.8, 35.1, and 48.5 meV, respectively. The Y_1 satellite is ascribed to a zone boundary, momentum con-

serving (MC) LA(X) phonon because it is close in energy to the LA(X) energies of AlAs and GaAs. The Y_2 and Y_3 satellites fall in the range of the optic-phonon branches of GaAs and AlAs, respectively, but because of the small optic-phonon dispersion they cannot be attributed unambiguously to Γ or X point phonons on this evidence alone. The X and Γ transition energies as a function of P are similar to those reported in other papers, and type I–type II crossover occurs between 1 and 2.4 kbar. The intensity ratio R_3 of the PS (Y_3) to the ZPL and its dependence on P are shown in Figure 27. Similar results are obtained for $R_1(P)$ and $R_2(P)$. From the figure it can be seen that $R_3(P)$ increases rapidly up to 15 kbar, then nearly saturates at a value of ~ 1.4–1.5% for $P > 20$ kbar. This result can be understood by the $\Gamma - X$ mixing mechanism, which gives oscillator strength to the ZPL and the MC PS satellites, proceeding via the same Γ intermediate state. The electric dipole matrix element for the X pseudodirect transition between conduction and valence bands is given by $M_{c,v} = \langle \psi_c | p | \psi_v^\Gamma \rangle$, where ψ_c and ψ_v^Γ are wave functions at the conduction and valence-band extreme and p is the dipole operator. $M_{c,v}$ is only nonzero if ψ_c has a Γ-wave-function character. Within first-order perturbation theory for $\Gamma - X$ mixing by a superlattice potential V, the matrix element becomes

$$M_{c.v} = \frac{\langle \psi_c^X | V | \psi_c^\Gamma \rangle \langle \psi_c^\Gamma | p | \psi_v^\Gamma \rangle}{E_\Gamma - E_X} \quad (16)$$

For MC phonon-assisted PL transitions, the analysis is similar, with

$$M_{c,v}^{ph} = \frac{\langle \psi_c^X | H_{e-ph} | \psi_c^\Gamma \rangle \langle \psi_c^\Gamma | p | \psi_v^\Gamma \rangle}{(E_\Gamma + \hbar\omega_{ph}) - E_X}, \quad (17)$$

where H_{e-ph} is the electron–phonon interaction Hamiltonian, $\hbar\omega_{ph}$ the phonon energy. The term $E_\Gamma + \hbar\omega_{ph}$ is the energy of the virtual state. The oscillator strength is proportional to $(M_{c,v}^{ph})^2$ and hence to $1/(\Delta + \hbar\omega_{ph})^2$, and the PS fractional intensity R_n is proportional to $[\Delta/(\Delta + \hbar\omega_{ph})]^2$, where Δ is the $\Gamma - X$ splitting ($E_\Gamma - E_X$). The pressure dependence of $[\Delta/(\Delta + \hbar\omega_{ph})]^2$ is plotted as the solid line in Figure 27, using the variation of Δ with P obtained experimentally. The initial rapid increase of R_3 with P, as well as the near saturation above 20 kbar when $\Delta \gg \hbar\omega_{ph}$, are well accounted for, showing that Y_3 is a symmetry-break MC phonon satellite. The variations in the relative intensities of all three PSs with P are

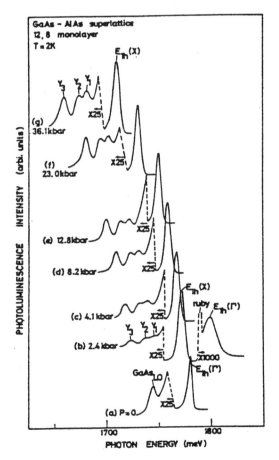

Fig. 26. 2K PL spectra for a $(GaAs)_{12}(AlAs)_8$ superlattice from $P = 0$ to 36.1 kbar [44]. Reproduced with permission from [44], copyright 1989, American Physical Society.

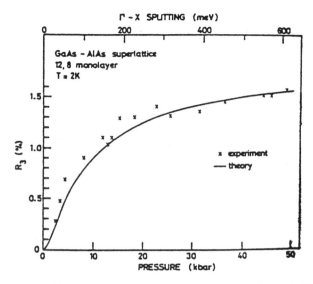

Fig. 27. $R_3(P)$, the ratio of the intensity of the Y_3 phonon satellite to the ZPL as a function of P. ×, experimental points; solid line, fit to $[\Delta/(\Delta + \hbar\omega_{ph})]^2$. Reproduced with permission from [44], copyright 1989, American Physical Society.

very similar, showing that they are all momentum conserving and that they are all scattered through the same Γ_1 intermediate state.

6. SHORT-PERIOD SUPERLATTICES IN EXTERNALLY APPLIED FIELDS

The characters of type I and type II superlattices and type I–type II transition can also be investigated by the application of an external field, for example, an electric field or a magnetic field. Under an electric or magnetic field the subbands of electron Γ-like and X-like states will shift with different rates so that at some critical field strength a type I to type II transition will occur, which can be detected by PL spectra. The signals of optically detected magnetic resonance (ODMR) or cyclotron resonance (CR) depend on the effective masses of electron, which are different for Γ and X electrons. Therefore, magnetic resonance can be used directly to distinguish Γ and X electrons in the lowest conduction state.

Meynadier et al. [39] found dramatic electric field effects in type II (GaAs)/(AlAs) superlattices designed with layer thicknesses so that the X state in the AlAs layer is just lower than the Γ state in the GaAs layer. Under the application of an electric field these electrons at the X state can be made to undergo a transition to the GaAs Γ state by making use of energy shifts of the quantum states. The material is therefore switched from type II to type I by the electric field, which is an extremely fast process. In a type II superlattice, as a result of the spatial separation of electrons and holes, the indirect band-gap energy is shifted toward higher energy under the application of an electric field perpendicular to the layers. The quantum confined Stark effect on the spatially direct Γ band gap remains a red shift. It is these opposing blue and red shifts

under an electric field that lead to the switch of a superlattice from type II to type I. The PL spectra of a p-i-n (GaAs)(35 Å)/(AlAs)(80 Å) superlattice, obtained under various applied voltages, are shown in Figure 28. At very low fields, they consist of two weak peaks at 1.735 and 1.720 eV, labeled Γ and X, respectively. As voltage is applied, from 2.25 to 3.45 V, the X peak intensity increases and moves to higher energies. The blue shift of the X peak results from the existence of a voltage drop between adjacent layers. When the applied voltage is above 4.5 V, Γ becomes the lowest-lying line. At 5 V the X peak has been driven to an energy of 1.749 eV, a shift of ≈ 30 meV from its original position. The energies of the direct and indirect transitions at high fields as functions of the applied field are shown in Figure 29, where the linear blue shift of the X transition follows the drop of voltage across a half-period $eFd/2$, and the red quantum confined Stark shift is saturated in this field regime. Crossing of the Γ and X transition energies occurs at about 4.5×10^4 V cm^{-1}, and the transitions are anticrossing, with a splitting at the crossover of about 2.5 meV, manifesting the existence of mixing. The potential responsible for the mixing of the (000) and (001) Bloch wave functions, which is twice as small, is therefore on the order of 1 meV. The time decay measurements are also performed on a series of superlattices with periods ranging from 38 to 88 Å, corresponding to $\Gamma - X$ spacing of 100 to 200 meV. Decay times ranged from 5 μs for the 200-meV $\Gamma - X$ spacing sample to 220 ns for the 100-meV $\Gamma - X$ spacing sample. The resulting potential V_{mix} responsible for the mixing is then given by

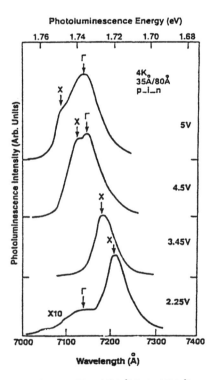

Fig. 28. PL spectra of a p-i-n (GaAs)(35 Å)/(AlAs)(80 Å) superlattice, taken for various applied voltages. Reproduced with permission from [39], copyright 1988, American Physical Society.

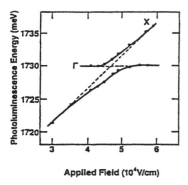

Fig. 29. Energies of the direct and indirect transitions at high fields, showing the anticrossing between the X and Γ transitions. Reproduced with permission from [39], copyright 1988, American Physical Society.

perturbation theory as

$$\frac{\omega(X)}{\omega(\Gamma)} = \left(\frac{V_{\mathrm{mix}}}{E_X - E_\Gamma}\right)^2 \qquad (18)$$

where $\omega(X)$ and $\omega(\Gamma)$ are the radiative recombination rates of the X and Γ transitions, respectively. Assuming that the $\omega(\Gamma)$ is constant and equal to 5×10^9 s^{-1} for all samples, the potential V_{mix} is found to be 1 to 3 meV for all eight samples investigated, in agreement with the result of splitting energy at anticrossing.

Xia and Chans [28] calculated the electronic and optical properties of SPSLs with an empirical tight-binding model, including second-neighbor interactions. Figure 30 shows the lowest two conduction-band energy levels (measured with

respect to the highest valence subband level) as functions of the applied electric field F for a $(GaAs)_{12}/(AlAs)_{28}$ superlattice (as in ref. 39). The corresponding experimental results of (39) are also shown (solid circles) for comparison. From Figure 30 it can be seen that the Γ level does not change appreciably with the applied electric field; that is, the Stark effect is not apparent for the SPSLs. The X level rises as the applied field increases because of the potential difference between the centers of two adjacent layers. The variation in the Γ and X levels with the applied electric field is in good agreement with the experimental results. The Γ and X energy levels anticross at $F = 4.5 \times 10^4$ V/cm, as does the experimental value. The energy splitting at the crossing field is about 2 meV, also the same as the experimental value of 2 meV. The numbers in parentheses along the energy axis denote the experimental transition energies. The difference between theoretical and experimental transition energies may partly be due to the exciton binding energy, which is ignored in the calculation.

For SPSLs, in which the barrier widths of Γ wells and X wells are so small that the confined states in neighbor wells will be coupled, the discrete energy levels form a series of minibands with definite widths W. Under an electric field F the energy of states in the nth well will be added to an electrostatic energy $nedF$, where d is the superlattice period. When the electrostatic energy between neighbor wells is larger than the bandwidth, $eFd > W$, the miniband will not exist, and the state in each well becomes a localized state with the energy $E(n) = E_0 + nedF$, where E_0 is the confined energy in a single well. This is known as the Wannier–Stark effect, and the localized states are called a Stark ladder [45]. Transition energies between Stark ladders are approximately expressed as

$$E = \Delta E + \nu eFd \qquad (19)$$

where

$$\nu = 0, \pm 1, \pm 2, \ldots \qquad \text{for type I transitions}$$
$$\nu = \pm\tfrac{1}{2}, \pm\tfrac{3}{2}, \ldots \qquad \text{for type II transitions}$$

ΔE is the interband transition energy in the same well. In type II superlattices, the shifts of transition energy correspond to separations of $1/2, 2/3, \ldots$ of the superlattice period, because the electrons and holes are confined in adjacent wells.

Morifuji et al. [46] studied electronic structures of (GaAs)/(AlAs) SPSLs in an applied electric field with a second-neighbor tight-binding model. The electrostatic energy eFz_i is added to the diagonal matrix elements of the Hamiltonian for the atomic site at z_i, so the secular equation has an infinite number of superlattice periods. To solve the secular equation they considered 10 periods of $(GaAs)_n/(AlAs)_n$ superlattices. A typical calculated charge density of a $(GaAs)_8/(AlAs)_8$ superlattice near the band edge at 60 kV/cm is shown in Figure 31. The vertical axis is not exactly to scale but arbitrary in the figure. All states ($\Gamma_{hh}, \Gamma_c, X_c^1$, and X_c^2) are well localized in each GaAs and AlAs layer, respectively. The states Γ_c and X_c^2 mix with each

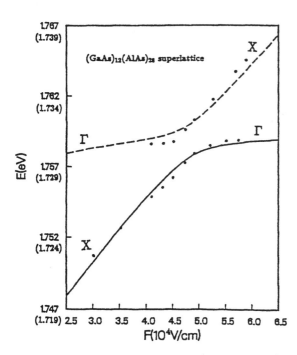

Fig. 30. Lowest two conduction-band energy levels of a $(GaAs)_{12}/(AlAs)_{28}$ superlattice as functions of the applied electric field F. Solid and dashed lines: theory. Solid circles: data from [39]. Reproduced with permission from [28], copyright 1990, American Physical Society.

other and are delocalized because they are close in energy. Calculated transition energies and transition probabilities for a $(GaAs)_8/(AlAs)_8$ superlattice are shown in Figure 32. There are three kinds of transitions: $\Gamma_{hh} - \Gamma_c$, which includes $\nu = 0, \pm 1$, with three branches; and $\Gamma_{hh} - X_c^1$ and $\Gamma_{hh} - X_c^2$, which include $\nu = \pm 1/2$, with two branches. The energy of the $\Gamma_{hh} - \Gamma_c(0)$ transition basically does not change with the electric field, the electric field dependence of other transitions is well approximated as νeFd. $\Gamma_{hh} - \Gamma_c(0)$ and $\Gamma_{hh} - X_c^2(-1/2)$ show anticrossing at $F = 90\text{--}100$ kV/cm. Anticrossing between $\Gamma_{hh} - \Gamma_c(+1)$ and $\Gamma_{hh} - X_c^2(+1/2)$ and kinking in $\Gamma_{hh} - \Gamma_c(-1)$ take place at the same electric field. These are due to strong mixing between the Γ_{hh} and X_c^2 states, as shown in Figure 31. Behavior of the transition probability in Figure 32 is quite complicated. The direct transition in real space, $\Gamma_{hh} - \Gamma_c(0)$, becomes stronger, and other oblique transitions become weak with an increasing electric field. This is due to the fact that the wave functions are strongly confined in the layers at larger electric fields. At the anticrossing electric field $F = 90\text{--}100$ kV/cm there is a large difference in transition probabilities between the plus index state and the minus index state due to the mixing effect.

On the theoretical basis of [46] Yamaguchi et al. [47] studied the Stark-ladder transitions in a type II $(GaAs)_8/(AlAs)_8$ superlattice under an electric field with the use of electroreflectance (ER) spectroscopy. A typical ER spectrum in a $(GaAs)_8/(AlAs)_8$ superlattice at an applied voltage of –3 V and a comparison with the theoretical curve are shown in Figure 33. The theoretical curve is best fitted to the data with the use of the third-derivative formula of Eq. (4). The vertical arrows indicate the transition energies estimated from the best fitting. The fitted transition energies from the ER data as functions of the electric field are shown in Figure 34 (denoted by full and open circles). The solid and dashed lines are theoretical curves as in Figure 32, corresponding to $\Gamma_{hh} -$

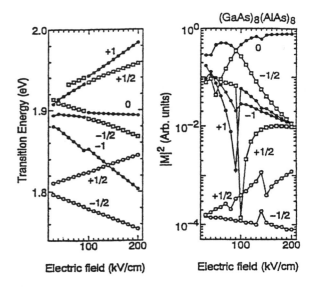

Fig. 32. Transition energies (left-hand side) and optical matrix elements (right-hand side) calculated for a $(GaAs)_8/(AlAs)_8$ superlattice. Filled circles, open circles, and open squares denote the transitions $\Gamma_{hh} - \Gamma_c$, $\Gamma_{hh} - X_c^1$, and $\Gamma_{hh} - X_c^2$, respectively. Indices 0, ± 1, and $\pm 1/2$ denote the relative positions of localized states. Reproduced with permission from [46], copyright 1994, American Physical Society.

$\Gamma_c(0, \pm 1)$ transitions (here denoted as E_1H_10 and ± 1), and $\Gamma_{hh} - X_c^1(\pm 1/2)$ and $\Gamma_{hh} - X_c^2(\pm 1/2)$ transitions (here denoted as $X_1H_1 \pm 1/2$ and $X_2H_1 \pm 1/2$), respectively. For each electric field there are five transition energies that are well resolved. From Figure 34 it is found that the experimental data agree well with the calculated results. The observed ER signals are attributed to the Stark-ladder transitions of E_1H_10, $E_1H_1 + 1$, and $X_2H_1 + 1/2$, an indirect transition from the Γ-point heavy-hole localized state to the X-point electron state. The open circles may be interpreted in terms of the transitions associated with the light hole, E_1L_10 and $E_1L_1 - 1$, which have not been calculated. It may be noticed in Figure 34 that all tran-

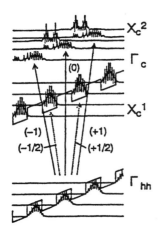

Fig. 31. Typical calculated charge densities of a $(GaAs)_8/(AlAs)_8$ superlattice near the band edge at 60 kV/cm. Heavy-hole valence states ((Γ_{hh}), conduction states of the GaAs Γ edge (Γ_c), and first and second conduction states of the AlAs X edge (X_c^1 and X_c^2) are shown with oblique steps, indicating the layers of GaAs and AlAs. Reproduced with permission from [46], copyright 1994, American Physical Society.

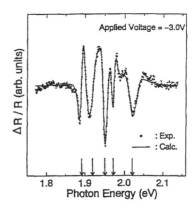

Fig. 33. A typical electroreflectance spectrum in a $(GaAs)_8/(AlAs)_8$ superlattice at an applied voltage of -3 V and a comparison with the theoretical curve, where the full circles represent experimental data and the full curve is best fitted to the data with Eq. (4). The vertical arrows indicate the transition energies estimated from the best fitting. Reproduced with permission from [47], copyright 1994, Institute of Physics.

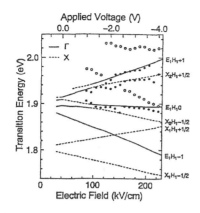

Fig. 34. Fitted transition energies from the ER data as functions of the electric field. The open and full circles are the transition energies determined from the ER measurements. The full curves correspond to the transitions E_1H_10 and ± 1, and the dashed lines to the transitions $X_1H_1 \pm 1/2$ and $X_2H_1 \pm 1/2$. Reproduced with permission from [47], copyright 1994, Institute of Physics.

sitions with lower energies have not been observed, including $E_1H_1 - 1$, $X_2H_1 - 1/2$, $X_1H_1 + 1/2$, and $X_1H_1 - 1/2$. Theoretically, the transition probabilities of the $E_1H_1 - 1$, $X_2H_1 - 1/2$ transitions are comparable to that of $E_1H_1 + 1$, as shown in Figure 32. This discrepancy has not yet been explained.

Van Kesteren et al. [48] studied a type II superlattice with an ODMR technique. The type II transition lifetimes are typically in the microsecond range. In an ODMR experiment the microwave-induced transition rate should be higher than or comparable to the optical transition rate to make detection of magnetic resonances feasible. For the microwave powers available for the measurements at liquid-helium temperature, this requirement implies lifetimes of the spin system studied of microseconds or longer. Therefore, in contrast to type I systems, type II systems can, in principle, be studied by the ODMR technique. For a (GaAs)(25 Å)/(AlAs)(25 Å) superlattice the ODMR experiments were carried out at 9.68 GHz with a sample temperature of 1.6 K. The ODMR spectrum is shown in Figure 35, obtained at a chopping frequency of 1 kHz. Three resonance lines are observed with equal intensity but opposite polarity for the σ^+ and σ^- detections. It was shown that the outer two ODMR lines correspond to electron-

spin transitions in the heavy-hole exciton and that they split apart by the exchange interaction of the electron and the hole. The resonance line in between is ascribed to the unbound electrons in the AlAs layer. To assign the resonance lines the Hamiltonian for excitons in an applied magnetic field along the z axis is considered,

$$H_{ex} = H_e + H_h + H_{ex} \qquad (20)$$

where H_e, H_h, and H_{ex} are Hamiltonians for electrons, holes, and the exchange coupling of the electron and hole forming the exciton, respectively.

$$H_e = \mu_B g_{ez} S_z B$$
$$H_h = -D\left[J_z^2 - \tfrac{1}{3}J(J+1)\right] - 2\mu_B B\left(\kappa J_z + q J_z^3\right)$$
$$H_{ex} = -a\overline{J} \cdot \overline{S} - b\left(J_x^3 S_x + J_y^3 S_y + J_z^3 S_z\right)$$

In these expressions, S denotes the electron spin, J is the effective hole spin including angular momentum, μ_B is the Bohr magneton, and B is the magnetic field strength. The parameter g_{ez} is the g factor for the electrons, κ and q are the Luttinger parameters for the Zeeman energy splitting of the hole, and a and b are the exchange-coupling constants. For the magnetic field used in the ODMR experiments the Zeeman energy splittings are much smaller than the difference in confinement energy for the light and heavy holes, which amounts to 61 meV in the quantum well studied. Therefore, the light-hole exciton ($J_z = \pm 1/2$, $S_z = \pm 1/2$) and the heavy-hole exciton ($J_z = \pm 3/2$, $S_z = \pm 1/2$) can be considered separately. For the heavy-hole exciton the microwave-induced transitions with $\Delta S_z = \pm 1$, $\Delta J_z = 0$ are allowed as shown in Figure 36. Because of the exchange coupling, these two transitions occur at different magnetic fields. The experimental result in Figure 35 is in agreement with a heavy-hole

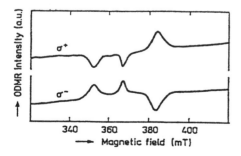

Fig. 35. ODMR signals of the heavy-hole excitons and the unbound electrons for a (GaAs)(25 Å)/(AlAs)(25 Å) superlattice obtained at a microwave chopping frequency of 1 kHz and detected as intensity changes in the σ^+ and σ^- circularly polarized components of the emission. Reproduced with permission from [48], copyright 1988, American Physical Society.

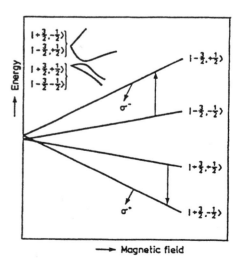

Fig. 36. Energy-level diagram for the heavy-hole excitons. The microwave-induced transitions and the σ^+ and σ^- emissions are indicated by arrows. Reproduced with permission from [48], copyright 1988, American Physical Society.

exciton assignment. From the Hamiltonian (20) the two resonances with $\Delta S_z = \pm 1$, $\Delta J_z = 0$ occur for $h\nu = \mu_B g_{ez} B \pm (1.5a + 3.375b)$. From the exciton resonance fields the g factor of the electron $g_{ez} = 1.89$ and the exciton exchange splitting $1.5a + 3.375b = 1.7\mu eV$ are obtained. The value of the g factor is far from that of 0.4 for Γ-conduction-band electrons in GaAs and is consistent with other experimental estimates for X electrons in AlAs. This verifies directly that the lowest conduction state in the (GaAs)(25 Å)/(AlAs)(25 Å) superlattice is not the Γ state, but the X state.

In a subsequent paper van Kesteren et al. [49] studied the ODMR of electrons and excitons in type II (GaAs)/(AlAs) superlattices as a function of the orientation of the quantum well with respect to the direction of the magnetic field. For an arbitrarily oriented magnetic field the spin Hamiltonian for electrons with spin $S = 1/2$ is given by

$$H_e = \mu_B \left(g_{xx} B_x S_x + g_{yy} B_y S_y + g_{zz} B_z S_z \right) \qquad (21)$$

where the g value is written as a tensor with diagonal elements g_{xx}, g_{yy}, and g_{zz}, with x, y, and z corresponding to the [100], [010], and [001] crystal axes, respectively. The X electron has an anisotropic g factor with g_l in the principal axis direction, and g_t perpendicular to the principal axis direction. For the D_{2d} symmetry of the quantum well, the $X_{xy}(A_1)$ and $X_{xy}(B_1)$ states have $g_{xx} = g_{yy} = (g_l + g_t)/2$, $g_{zz} = g_t$, and the $X_z(A_1)$ state has $g_{xx} = g_{yy} = g_t$, $g_{zz} = g_l$. So both the X_{xy} and X_z states show a uniaxial symmetry along the [001] axis. For the D_2 symmetry the Zeeman splitting shows a uniaxial symmetry along an axis, which corresponds to the principal axis of the valley involved. Therefore, from the valley symmetry axis, the number of spin resonances observed, and the difference in magnitude of the anisotropy for the X_{xy} and X_z states, the ordering of the X valleys in various samples can be determined. This method reveals that for (GaAs)(25 Å)/(AlAs) superlattices with AlAs layers thinner than ~ 55 Å the X_z conduction-band valley has the lowest energy, whereas for superlattices with thicker AlAs layers the X_x and X_y valleys are lowest in energy. This is a consequence of the lattice mismatch strain splitting of various X valleys in AlAs.

With a very large magnetic field, (a pulsed magnetic field up to 500 T), Miura et al. [50] reported the study of the type I–type II transition in (GaAs)/(AlAs) SPSLs. The magnetic field is so large that the cyclotron energy $\hbar\omega_c$ exceeds various energy spacings in superlattices. Thus, various types of magnetic-field-induced level crossover take place in very high fields. Pulsed higher magnetic fields can be produced by three different techniques. First, ultrahigh magnetic fields up to 550 T can be produced by electromagnetic flux compression. Second, a field of up to about 200 T can be produced by the single-turn coil technique. Both of these techniques for fields exceeding 100 T are destructive, and the rise time of the field to the top is a few microseconds. For high-precision measurements in the lower field range, conventional nondestructive longpulse magnets are employed to produce fields of up to about 50 T. The rise time of the field in this case is about 10 ms. Figure 37 shows a magnetic photoluminescence (MPL)

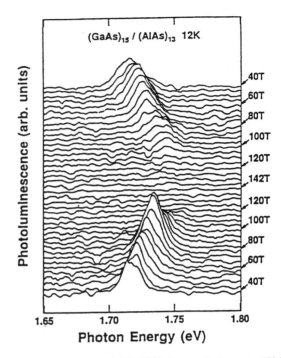

Fig. 37. MPL spectra in a (GaAs)$_{15}$/(AlAs)$_{13}$ superlattice up to 150 T. The data were obtained as a streak photograph in a one-shot pulse field. The time proceeds from top to bottom. Reproduced with permission from [50], copyright 1996, Institute of Physics.

spectra for a (GaAs)$_{15}$/(AlAs)$_{13}$ superlattice in a shot of pulse. It can be seen that the MPL peak from the exciton in the Γ minimum diminishes with increasing field, and it reappears when the field is decreased again. Because the effective mass of the Γ electron is much smaller than that of the X electron, and the lowest magnetic energy level rises as $\hbar eB/2m^*c$, at a critical magnetic field the lowest Γ magnetic energy level will exceed the lowest X level, and the type I–type II transition then occurs. The simultaneous application of high magnetic fields and high pressures can cause the transition to be easily realized in a wide range of samples. Figure 38 shows the phase diagram of the type I–type II transition point on

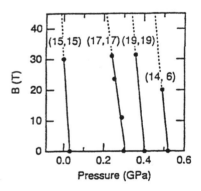

Fig. 38. Phase diagram of the type I to type II transition in (GaAs)$_m$/(AlAs)$_n$ superlattices with various (m, n) on a pressure-magnetic field plane. Reproduced with permission from [50], copyright 1996, Institute of Physics.

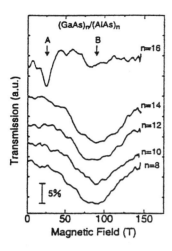

Fig. 39. CR in $(GaAs)_n/(AlAs)_n$ superlattices with $n = 8$–16 in the transmission of 23-μm radiation. Reproduced with permission from [51], copyright 1996, Elsevier Science.

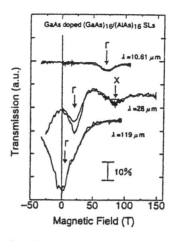

Fig. 40. CR for a $(GaAs)_{16}/(AlAs)_{16}$ superlattice in various radiation transmissions (λ) = 10.61, 28, and 119 μm). Reproduced with permission from [51], copyright 1996, Elsevier Science.

a field-pressure plane. The gradients of the lines are almost parallel with each other, indicating nearly the same effective mass of excitons and the same pressure coefficient between the samples.

CR was measured in doped samples with $n = 8$–16 in fields of up to 150 T by the single-coil technique [51]. Figure 39 shows the CR traces for these samples at a wavelength of 23μm ($\hbar\omega = 54$ meV). Two resonance lines were observed. The peaks labeled A are associated with the Γ point, and those labeled B are associated with the X point, corresponding to the effective masses $m_\Gamma = 0.053m_0$ and $m_X = 0.193m_0$, respectively. The Γ peak is observed for $n > 15$, and the X peak is observed for $n < 14$. This indicates that with varying n, the type I–type II transition takes place at around $n = 14$ at $\sim$30 K. It is also found that when the magnetic field is tilted from the normal direction, the peak from the X exciton shifts to a higher field because of the decrease in the effective mass that deviates from that along the principal axis.

Actually in the CR experiment the high magnetic field also increases the lowest Γ level relative to the lowest X level, and in some cases causes $\Gamma - X$ crossover, as shown in Figure 37. Figure 40 shows the CR of a $(GaAs)_{16}/(AlAs)_{16}$ superlattice at 30 K in various microwave wavelengths of (10.61, 28, and 119 μm) [51]. For the transmission of 119-μm radiation only one resonance was observed, which corresponds to Γ-like states. For the transmission of 28-μm radiation two resonances were observed: one corresponds to Γ-like states ($0.055m_0$) and the other to X-like states ($0.22m_0$). The CR peak due to X-like states was not observed for the transmission of 10.6-μm radiation because of the limited range of the magnetic field applied. This behavior indicates that the $\Gamma - X$ crossover occurs at ~ 60 T for $n = 16$ superlattices, and that the CR peak of X_z states becomes visible at a higher magnetic field.

7. ULTRA-SHORT-PERIOD SUPERLATTICES

The term *ultra-short-period superlattices* (USPSLs) generally refers to those $(GaAs)_m/(AlAs)_n$ superlattices with $m, n \leq 4$. They have some special properties different from those of general SPSLs. Their energy gaps of $\Gamma - \Gamma$ and $X - \Gamma$ do not vary with the monolayer number n and m as the effective-mass theory or the tight-binding method predicts. Determining whether X_z or X_{xy} is the lowest conduction state has long been an open problem attracting many arguments since the first paper of Finkman et al. [13]. For the identification of the symmetry properties X_z or X_{xy} states the PL spectrum under uniaxial stress is specially suited, because under uniaxial stress in different directions the electron X_z and X_{xy} levels shift in opposite directions, then the PL peaks from $X_z - \Gamma$ and $X_{xy} - \Gamma$ transitions can easily be distinguished. Furthermore, the $X_{xy} - \Gamma$ transition is indirect, and the $X_z - \Gamma$ transition is pseudodirect because of mixing of X_z and Γ states caused by the superlattice potential. Therefore, the time decay of the $X_{xy} - \Gamma$ PL peak can be fitted by a nonexponential function of t, and has a smaller decay rate, whereas that of the $X_z - \Gamma$ PL peak has a larger decay rate. For the $X_{xy} - \Gamma$ PL peak the ratio of the phonon satellites (PSs) to the zero-phonon peak (ZP) is large so that the PS peaks can be observed clearly, but for the $X_z - \Gamma$ PL peak a strong ZP peak with very weak PS peaks is generally observed. All of these experimental results together with results of uniaxial stress can be used to unambiguously assign the lowest conduction state in USPSLs. Ge et al. [8, 37] made a comprehensive PL study of the low-lying conduction band states of $(GaAs)_m/(AlAs)_n$ USPSLs to determine their energy and symmetry. As for the theoretical calculations, they gave results not as good as those of SPSLs with larger period ($m, n > 4$) because of the limit of methods or the ideal approximation about the interface. The theoretical results were divergent depending on various models taken, but they promoted corresponding experimental studies. Only after the definite experimental results came out could reasonable theoretical results be obtained while taking into account actual modification of the ideal model.

Lefebvre et al. [52, 53] first studied the symmetry of the conduction-band states for type II superlattices under on-axis and in-plane uniaxial stress. From their stress dependence, the symmetry of the electron states X_z or X_{xy} is determined. In some structures X_z is the conduction ground state, and the ordering between X_z and X_{xy} can be reversed by in-plane stress. In others, the situation is reversed and the ordering between X_{xy} and X_z is reversed by on-axis stress. The shear tetragonal deformation potential of X states in AlAs is obtained.

Under a uniaxial stress the energy shift of the X state can be written as

$$\Delta E(X) = E_2 \vec{n} \cdot \left[\overleftrightarrow{e} - \tfrac{1}{3}(e_{xx} + e_{yy} + e_{zz}) \overleftrightarrow{I} \right] \cdot \vec{n} \quad (22)$$

where $\vec{n}$ is the unit vector in the direction of the band extreme in k space, $\overleftrightarrow{I}$ is the unit dyadic, $\overleftrightarrow{e}$ is the strain tensor, and E_2 is the corresponding shear deformation potential. For the cases of [001]- and [110]-oriented stress, the energy shifts,

$$\Delta E(X_z) = -\tfrac{2}{3}\delta, \quad \Delta E(X_x) = \Delta E(X_y) = \tfrac{1}{3}\delta \quad \text{[001] stress}$$

$$\Delta E(X_z) = \tfrac{1}{3}\delta, \quad \Delta E(X_x) = \Delta E(X_y)$$
$$= -\tfrac{1}{6}\delta, \quad \text{[110] stress} \quad (23)$$

are obtained, where
$$\delta = -E_2(S_{11} - S_{12})\sigma$$

S_{ij} is the elastic compliance coefficient, and σ represents the magnitude of stress in the corresponding direction and is negative for a compressive stress (these are the same for the following). The variation of energy levels of X_x, X_y, and X_z states with stress is schematically shown in Figure 41 for a compressive stress. The energy shift of the valence band top Γ state is

$$\Delta E(HH_1) = b(S_{11} - S_{12})\sigma \quad \text{[001] stress}$$

$$\Delta E(HH_1) = -0.5b(S_{11} - S_{12})\sigma \quad \text{[110] stress} \quad (24)$$

and those of the band edges are

$$\Delta E_i = a_i(S_{11} + 2S_{12})\sigma \quad (25)$$

which are related to the hydrostatic pressure effect, and thus are independent of the orientation of stress. In Eqs. (24) and (25) a_i and b are shear deformation potentials. Then the energy shift of the PL peak under a uniaxial stress is

$$\Delta E_{PL} = \Delta E(X) + \Delta E_g - \Delta E(HH_1)$$
$$= \Delta E(X) + (a_X - a_v)(S_{11} + 2S_{12})\sigma - \Delta E(HH_1) \quad (26)$$

ΔE_g is the energy shift of the energy gap. From the experimental energy shift of the PL peak as a function of σ the energy shift of the X state can be derived from Eq. (26). The deformation potential parameters for the X-related state in AlAs and the Γ-related state in GaAs are $a_\Gamma - a_v = -8.49$ eV, $a_X - a_v = 1.43$ eV, $b = 1.76$ eV, and $E_2 = 5.10$ eV [53]. The elastic compliance coefficients for GaAs are $S_{11} = 1.16$ Mbar^{-1}, $S_{12} = -0.37$ Mbar^{-1}, and those for AlAs are $S_{11} = 1.20$ Mbar^{-1}, $S_{12} = -0.39$ Mbar^{-1}.

Figure 42 shows PL spectra of a $(GaAs)_{12}/(AlAs)_{22}$ superlattice under [110] and [001] stress. In the case of zero stress the direct luminescence band labeled e is seen at 1800 meV. Lower in energy, a strong PL peak a appears, and some weak peaks b, c, and d are also observed on its lower-energy side, which may be attributed to phonon-assisted transition from X states, involving three different phonon energies. When stress is applied, the intensity of transition (a) collapses, whereas the intensity of the phonon-assisted transitions is only weakly dependent on the stress. The stress shift of line a is small (~ -0.4 meV kbar^{-1}), whereas the lower band shifts faster (~ -4.5 meV kbar^{-1}). The [001] stress dependence is rather

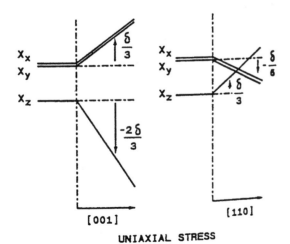

UNIAXIAL STRESS

Fig. 41. Schematic behavior of the of X_x, X_y, and X_z states under both directions of compressive stress. Reproduced with permission from [52], copyright 1989, American Physical Society.

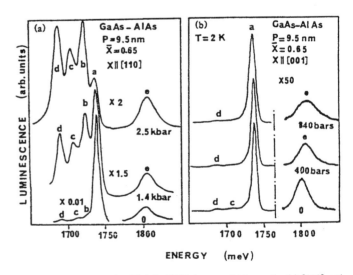

Fig. 42. PL spectra of a $(GaAs)_{12}/(AlAs)_{22}$ superlattice under (a) [110] and (b) [001] stress. P is the period of the superlattice, and x represents the thickness of AlAs measured in units of the period P. Reproduced with permission from [52], copyright 1989, American Physical Society.

different: line a shifts to lower energy, but no drastic decrease in intensity is observed. Lines b, c, and d have a smaller slope than line a and still maintain weak intensity. To extract the conduction energy shifts from these experimental data, the influence of the stress on valence band states and the hydrostatic contribution should be subtracted. Using the deformation potential and elastic compliance parameters, we obtain

$$\Delta E(X) = \Delta E_{PL} - \Delta E_g + \Delta E(HH_1)$$
$$\Delta E(HH_1) = 1.4256\sigma[001], \qquad -0.7128\sigma[110]$$
$$\Delta E_g = 0.0932\sigma \qquad (27)$$

The unit of σ is kbar, and that of the energy is meV. The electron shifts thus obtained for line a on the one hand and for lines b, c, and d on the other hand match the scheme of Figure 41, with $\delta \sim 7.8$ meV kbar^{-1}. As a consequence, line a is assigned to the zero-phonon transition associated with the X_z levels, and lines b, c, and d to phonon-assisted transitions from the $X_{x,y}$ levels, lying at a slightly higher energy than that of X_z by about 1 meV at zero stress. The decrease in the intensity of the line a under [110] stress then results in the change of ordering of the conduction levels from $E(X_z) < E(X_{x,y})$ to the opposite in a range of stress of less than 200 bars. On the other hand, no such sensitivity is expected if X_z remains in the ground state under stress, which is actually the case with [001] stress, in agreement with Figure 41.

The conduction state shifts derived from the PL spectra under uniaxial stresses for a $(GaAs)_6/(AlAs)_9$ superlattice are shown in Figure 43, after subtraction of, the energy shifts of

the valence band state and the energy gap. In the case of [110] stress the conduction level rises with a slope $\delta/3$ as the stress increases up to ~ 1.2 kbar, then it descends with a slope of $-\delta/6$. In the case of [001] stress the PL peak energy decreases as the stress increases and reaches a slope of $-2\delta/3$ in the limit of large stress. Comparing with Figure 41, it is deduced that the lowest conduction state is the X_z state, and the zero-stress splitting between the X_z and $X_{x,y}$ states is ~ 6.5 meV. The X_z and $X_{x,y}$ crossover occurs at ~ 1.2 kbar. For a USPSL $(GaAs)_{1.5}/(AlAs)_2$ no noticeable change of intensities under stress is observed. The shifts of all of the transition lines are the same. This identifies the symmetry of the lowest conduction state as $X_{x,y}$. When a lattice mismatch of AlAs is included in the effective-mass calculation, which decreases the energy of the $X_{x,y}$ state relative to the X_z state for ~ 23 meV, a result of $E(X_{x,y}) = E(X_z) - 5$ meV is obtained, in qualitative agreement with the data.

Ge et al. [37] investigated the energy levels of $(GaAs)_n/(AlAs)_n (n \leq 4)$ USPSLs by PL spectrum, combined with PL decay, PL excitation, and PL under uniaxial stress. Experimental results show that these superlattices are type II, but the lowest conduction states are $X_{x,y}$ for $n \leq 3$ and X_z for $n = 4$, respectively. Figure 44 shows the low-temperature PL spectra for four superlattice samples ($n = 1$–4) and one $Al_{0.5}Ga_{0.5}Al$ alloy sample. As n varies from 4 to 1, the zero-phonon (ZP) line (peak A) gradually shifts to higher energy. There is, however, an abrupt change in the phonon sideband structure between $n = 4$ and $n = 3$. The former has two weak phonon satellite (PS) peaks labeled C and D, involving $LO(X)_{GaAs}$ (31 meV) and $LO(X, \Gamma)_{AlAs}$ (48 meV) phonons, respectively. The integrated intensity of C and D for $n = 4$ is only 20% of that of A: the transition is known to be the pseudodirect, $X_z \rightarrow \Gamma$. For $n \leq 3$ relative to the ZP line, the PSs

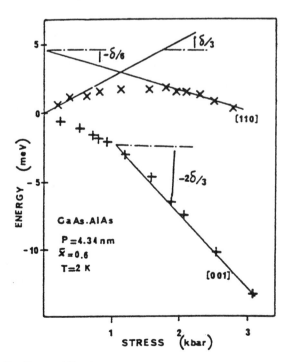

Fig. 43. Energy shifts of the conduction state for a $(GaAs)_6/(AlAs)_9$ superlattice as functions of stress. $\times$, [110] stress; $+$, [001] stress. Reproduced with permission from [52], copyright 1989, American Physical Society.

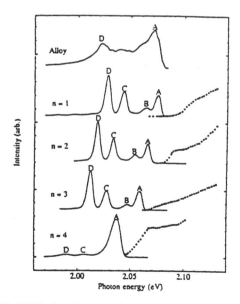

Fig. 44. PL (full line) and PLE (dots) spectra of $(GaAs)_n/(AlAs)_n$ superlattices with $n = 1$, 2, 3, and 4, and an $Ga_{0.5}Al_{0.5}As$ sample at 2.2 K. Reproduced with permission from [37], copyright 1989, American Institue of Physics.

are much stronger: under weak excitation and at low temperature, C and D together have an integrated intensity four times that of A. Peak B, 11 meV below A, is a defect line. Strong PS and low quantum efficiency are characteristic of emission from the $X_{x,y}$ state, which is brought below X_z by lattice mismatch strain. The PS is strong relative to the ZP line because this is an indirect transition requiring simultaneous emissions of momentum-conserving (MC) phonons. The ZP line, while weak, is still present, presumably because of disorder.

Figure 45a shows the PL spectra of a (4,4) sample under various [110] stresses, clearly showing the crossover of the conduction band minimum (CBM) from X_z to $X_{x,y}$. Figure 45b shows the PL peak shift of the (4,4) sample as a function of [110] and [001] stress. It clearly shows that the CBM varies continuously from X_z (under low [110] stress) to $X_{x,y}$ (under high [110] stress). Before and after crossover the change of the ratio of the PS intensity to the ZP intensity is clearly seen. Figure 46a and b shows the PL spectra of a (1,1) sample under [110] and [100] (not [001]) stress, respectively.

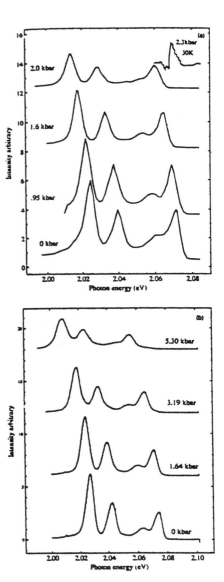

Fig. 46. PL spectra of a (1,1) sample at 2 K and excited at 514.5 nm, under (a) [110] stress and (b) [100] stress. Reproduced with permission from [8], copyright 1994, Elsevier science.

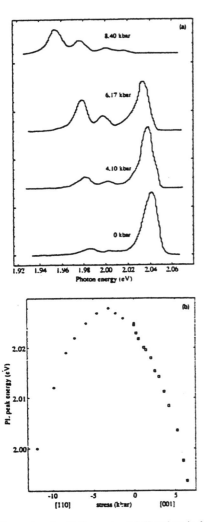

Fig. 45. (a) PL spectra of a (4,4) sample at 2 K and excited at 514.5 nm, under various [110] stresses. (b) PL peak shift of the (4,4) sample under [001] and [110] stresses. Reproduced with permission from [8], copyright 1994, Elsevier Science.

For the case of [110] stress the PL spectra are obviously different from those of the (4,4) sample (Fig. 45a). The PL peaks have similar structures and continuously shift to lower energy, showing that the CBM is the $X_{x,y}$ state. The PL spectra under [100] stress have a shape similar to the shapes those under [110] stress. The data for [001] stress are not shown; it has a strong ZP peak with high quantum efficiency and a short lifetime, showing that the X_z state is pushed down and becomes the CBM. The same spectral shape, the same ZP-PS separation, similar PS/ZP intensity ratios, and similar lifetimes are observed for the (2,2) and (3,3) samples, suggesting that the CBM is the same for (1,1), (2,2), and (3,3), being derived from $X_{x,y}$.

The time decays of PL spectra and PL spectra under uniaxial stress are measured [8]. The intensity $I(t)$ after excitation by a short pulse can be fitted to Eq. (5), which describes transitions induced by nonrandom (w_n) and random (w_r) scattering.

For the (4,4) sample at zero stress case the fitting parameters are $w_n = 5 \times 10^4 s^{-1}$ and $w_r = 10^6 s^{-1}$, so the nonrandom decay rate w_n is smaller than the random decay rate w_r by one order of magnitude, and the superlattice potential is random if the interfaces are not perfectly flat. Figure 47 shows decay curves under different stresses for (4,4) and (1,1) samples. For the (4,4) and (1,1) samples the stress causes X_z to $X_{x,y}$ crossover and $X_{x,y}$ to X_z crossover, respectively. For the former the time decay at zero stress (full line) is faster than that at [110] stress. For the latter it is the opposite. The decay rates (w_r) of the X_z and $X_{x,y}$ states fitted by Eq. (5) for different superlattices are listed in Table VIII. Table VIII shows that the ratio $w_z/w_{x,y}$ is in the range of 1.9 to 7.5, showing that the superlattice potential is contributing substantially to w_z. For the forbidden $X_{x,y} \rightarrow \Gamma$ transition, which is predominantly due to phonon-

Table VIII. Decay Rates of the X_z and $X_{x,y}$ States for Different Superlattices

(m, n)	(4, 4)	(3, 3)	(2, 2)	(1, 1)	(2, 3)	(1, 2)
w_z (ms^{-1})	1000	780	340	1200	42±10	15±5
$w_{x,y}$ (ms^{-1})	180	200	180	160	57±10	20±5
ΔE (meV)	159	120	118	121	203	356
w^*	455	288	251	234	235	253

Reproduced with permission from [8], copyright 1994, Elsevier Science.

assisted processes, the decay rate is considerably slower than the average decay rate of the pseudodirect $X_z \rightarrow \Gamma$ transition.

Figure 48 compares the zero stress PL transition energies as functions of n for $(GaAs)_n/(AlAs)_n$ superlattices with the theoretical predictions of Lu and Sham [6]. It can be seen that the theoretical predictions are not in agreement with the experimental results, even taking into account the effect of mismatch strain, which will lower $X_{x,y}$ by about 20 meV relative to X_z. The experimental results are that the $X_{x,y}$ states are lower than the X_z state for $n \leq 3$, and the energy difference is about 20 meV, which is the opposite of the theoretical prediction. The variation of transition energies with n is not as large as the theoretical prediction. As we said at beginning of this section, this discrepancy is due to the limit of methods and the ideal approximation about the interface.

Γ-Γ PL is very weak in type II SPSLs because it competes with fast nonradiative transfer from Γ to X. At high excitation (≈ 500 W/cm^2) and high gain Γ-Γ PL spectra for $n = 1, 2, 3$, and 4 $(GaAs)_n/(AlAs)_n$ superlattices have been observed [54]. Figure 49 shows the PL spectrum of a (2,2) superlattice at 2.3 K. The integrated intensity ratio of X to Γ

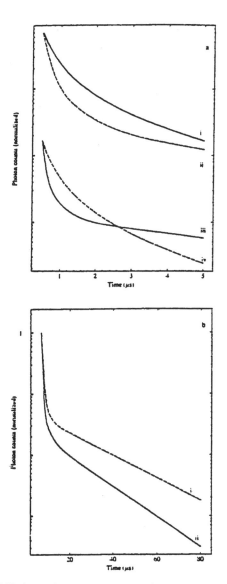

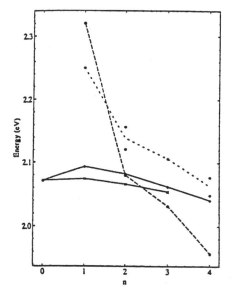

Fig. 47. ZP PL decay after pulse excitation (514.5 nm) at 2 K. (i) (4,4) sample at zero stress. (ii) (4,4) under a [110] stress of 5.96 kbar. (iii) (1,1) at zero stress. (iv) (1,1) at a [001] stress of 1.43 kbar. Reproduced with permission from [8], copyright 1994, Elsevier Science.

Fig. 48. Comparison of experimental transition energies as functions of n for $(GaAs)_n/(AlAs)_n$ superlattices with the theoretical predictions of ref. 6. Pseudodirect X_z-Γ transitions: +, expt; ○, theory. Indirect $X_{x,y}$-Γ transition: ×, expt; ●, theory. $n = 0$ refers to the $Al_{0.5}Ga_{0.5}As$ alloy. Reproduced with permission from [37], copyright 1990, American Institute of Physics.

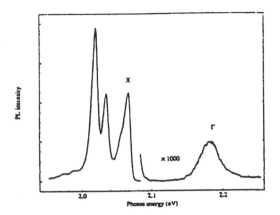

Fig. 49. PL spectrum of (2,2) SPSL sample at 2.3 K. The excitation level is approximately 500 W/cm² at 514.5 nm. Reproduced with permission from [54], copyright 1991, Elsevier Science.

emission is 10^3 to 10^4 for all SPSL samples studied. The Γ-Γ transition energies were obtained from the high-excitation PL spectra for (1,1), (2,2), (3,3), and (4,4) SPSLs. It is found that the direct Γ gap of SPSLs does not change substantially from $n = 1$ to $n = 4$ (≈ 2.18 eV), contrary to the predictions of the tight-binding and effective-mass theoretical calculations. Using the samples grown on the vicinal (001) GaAs substrates, Schwabe et al. [55] observed strong Γ-Γ luminescence in a series of type II superlattices. The samples consisted of 50 periods of 2, 4, 5, 6, or 7 monolayers (MLs) of GaAs embedded in 28 ML of AlAs. The most intense Γ-Γ emission has been observed for samples grown on GaAs substrates misoriented 6° toward the nearest (111) plane of group V atoms. Figure 50 shows the low-temperature luminescence spectra of a set of $(\text{GaAs})_n/(\text{AlAs})_{28}$ superlattices for $n = 7(A)$, $6(B)$, $5(C)$, $4(D)$, and $2(E)$ [55]. The peaks at the low-energy side are due to the X-Γ transition (type II) and its phonon replica, and the peak at the high-energy side is due to the Γ-Γ transition (type I). The experimental optical type I and

type II transition energies are in agreement with the calculation results of the effective-mass approach and the empirical tight-binding Green's function scheme. The intensive type I emission is probably connected with the interface peculiarities: the interface structure prevents the loss of photoexcited carriers from GaAs layers to the surrounding AlAs materials.

Laks and Zunger [56] calculated the $(\text{GaAs})_1/(\text{AlAs})_1$ superlattice by the first-principles pseudopotential method and found that the resulting energy levels are very different for the case of partial intermixing of the Al and Ga atoms relative to the abrupt case. The formation energy is lowered by the partial intermixing, making this superlattice even more stable at low temperatures than the fully randomized $\text{Al}_{0.5}\text{Ga}_{0.5}\text{Al}$ alloy. Consequently, the CBM reverts from the GaAs L-derived state to the $X_{x,y}$-derived AlAs state. The previously noted discrepancy between theory (assuming an abrupt interface) and experiment is therefore attributed to insufficient interfacial abruptness in the samples used in experimental studies. Figure 51 shows calculated conduction energy levels of the abrupt and the 1/3 intermixed $(\text{GaAs})_1/(\text{AlAs})_1$ superlattices compared with those of the bulk $\text{Al}_{0.5}\text{Ga}_{0.5}\text{Al}$ alloy. With superlattice states denoted by an overbar, followed by the zinc-blende states from which they originate in parentheses, the basic symmetry compatibility relations between the alloy and the superlattice are

$$\left\langle X_{1c}^{x,y,z}\right\rangle \rightarrow \overline{M}_{5c}\left(X_{1c}^{x,y}\right)+\overline{\Gamma}_{4c}\left(X_{1c}^{z}\right)$$
$$\left\langle L_{1c}\right\rangle \rightarrow \overline{R}_{1c}(L_{1c})+\overline{R}_{4c}(L_{1c}) \quad\quad (28)$$
$$\left\langle X_{3c}^{z}\right\rangle + \left\langle \Gamma_{1c}\right\rangle \rightarrow \overline{\Gamma}_{1c}^{(1)}\left(\Gamma_{1c}+X_{3c}^{z}\right)+\overline{\Gamma}_{1c}^{(2)}\left(\Gamma_{1c}+X_{3c}^{z}\right)$$

Three degenerate x, y, z zinc-blende X_{1c} valleys fold in the superlattice into the double degenerate M_{5c} state and Γ_{1c} state. Both states have cation-pd and anion-s character, with zero cation-s character in AlAs. The four zinc-blende L_{1c} valleys at $k = \pi/a(1, \pm 1, \pm 1)$ fold into the superlattice R_{1c} and R_{4c} states. R_{1c} is a $(\text{Ga}-s)+(\text{As}-s)$ state with zero s character on the Al site, and the complementary R_{4c} state has $(\text{Al}-s)+(\text{As}-s)$ character with zero s character on the Ga site. Hence, the R_{1c}-R_{4c} energy splitting reflects the potential difference $V_s(\text{Ga})$-$V_s(\text{Al})$, an atomic energy difference of about 1 eV between Al, s and Ga, s. This large R_{1c}-R_{4c} energy splitting makes R_{1c} the CBM, despite the fact that in the alloy the L_{1c} state, from which R_{1c} is derived, is about 0.3 eV above X_{1c}, as shown in Figure 51. This result was found in a number of first-principles calculations. The far smaller splitting between the atomic p energy levels (as opposed to the s levels) of Al and Ga results in a small (0.06 ± 0.05 eV) M_{5c}-Γ_{4c} splitting in the superlattice. The potential difference dominates over the kinetic energy effects, so that the X_z-derived Γ_{1c} state is above the $X_{x,y}$-derived M_{5c} state, despite the fact that kinetic energy arguments would place the heavier (longitudinal) mass X_z state below the lighter-mass $X_{x,y}$ state. The higher zinc-blende X_{3c}^z state folds into the superlattice $\Gamma_{1c}^{(2)}$ state, with the same symmetry as the original zinc-blende $\Gamma_{1c}^{(1)}$ state. The interaction between the two

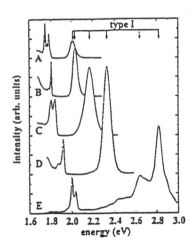

Fig. 50. Low-temperature luminescence spectra of a set of $(\text{GaAs})_n/(\text{AlAs})_{28}$ superlattices for $n = 7$ (A), 6 (B), 5 (C), 4 (D), and 2 (E). Reproduced with permission from [55], copyright 1997, American Physical Society.

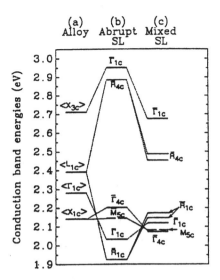

Fig. 51. Calculated (local density approximation and spin-orbit corrected) energy levels of (a) the bulk $Al_{0.5}Ga_{0.5}As$ alloy, (b) the abrupt $(GaAs)_1/(AlAs)_1$ superlattice, and (c) the 1/3 intermixed superlattice with a (3×1) interfacial unit cell. Reproduced with permission from [56], copyright 1992, American Physical Society.

states repels the $\Gamma_{1c}^{(1)}$ state down. To explain the discrepancy between theory and experimental results, Laks and Zunger [56] replaced the abrupt (1×1) interface in the xy plane with larger (2×1), (3×1), and (4×1) interfacial unit cells, and a fraction (1/2, 1/3, or 1/4) of the Al atoms on the AlAs side is exchanged with Ga atoms on the GaAs side. They first calculated the changes in total energy due to this local atomic intermixing. The pseudopotential calculated excess enthalpy of an abrupt $(GaAs)_1/(AlAs)_1$ superlattice taken with respect to bulk GaAs+AlAs is 13.7 meV/$AlGaAs_2$-units, and the mixing enthalpy of the random $Al_{0.5}Ga_{0.5}As$ alloy is calculated to be 10.6 meV/$AlGaAs_2$-units. However, total-energy calculations for the 1/2, 1/3, and 1/4 fraction intermixed (4×1) superlattices show that the excess enthalpy is lowered to 5.5, 7.3, and 8.8 meV/$AlGaAs_2$-units, respectively. Hence, local atomic mixing at the interface stabilizes the superlattice with respect to the random alloy. The calculated energy levels for the (3×1) intermixed superlattice are shown in Figure 51c. Comparison with Figure 51a shows that the intermixed superlattice is indeed not an alloy. Reconstruction removes the symmetry constraint, so that the energy splitting of R_{1c} and R_{4c} states decreases and the R_{1c} state moves up by 0.23 meV. The s-like $\Gamma_{1c}^{(1)}$ level also moves up in energy and is 40 meV above the M_{5c} level. The CBM now is either M_{5c} or Γ_{4c} (the energy splitting is only 5 meV). The position of the CBM is now at 2.08 eV, close to the experimentally determined (zero-phonon) CBM at 2.07 eV [8, 37]. Therefore, partial intermixing of the Al and Ga atoms relative to the abrupt case removes the conflicts between experiment and the theoretically calculated levels of the abrupt superlattice.

8. OTHER EXPERIMENTAL METHODS AND OTHER ORIENTED SHORT-PERIOD SUPERLATTICES

The SPSLs discussed in previous sections are all grown on [001]-oriented substrates, because of the high structural quality that can be achieved for this orientation. Improvements in molecular beam epitaxy (MBE) have made possible the preparation of quantum wells and superlattices grown along a variety of other crystallographic directions, including [110], [111], and [012]. Calle et al. [57] studied [111]-oriented $(GaAs)_m/(AlAs)_n$ SPSLs by resonant Raman scattering. It is found that samples with a moderately large period $[(m, n) > 9]$ show a direct band gap, whereas for a (6,6) sample the lowest-energy transition appears to be indirect in k space (but direct in real space), occurring between the GaAs conduction L and valence Γ states. Confined optical phonons are clearly observed in resonant conditions, for both GaAs and AlAs; they yield the bulk optical-phonon dispersions of these materials along the $\Gamma - L$ direction. The polarization selection rules for Raman backscattering are

$$Z'(X'X')\bar{Z}' : \qquad a^2 + \tfrac{1}{3}d^2(\text{LO}) + \frac{2}{3}d^2(\text{TO})$$

$$Z'(Y'X')\bar{Z}' : \qquad \tfrac{2}{3}d^2(\text{TO}) \tag{29}$$

where X', Y', and Z' represent the $[1\bar{1}0]$, $[11\bar{2}]$, and $[111]$ directions, respectively. a is the intraband Fröhlich interaction and d is the deformation potential term. When the incident light and the scattered light have the same polarization, LO and TO phonons are allowed, whereas only TO modes are active for the cross-polarized configuration. This is illustrated in Figure 52, where Raman spectra of a (6,6) [111] sample, recorded in parallel and crossed polarization and nonresonant conditions, are shown. Two groups of Raman

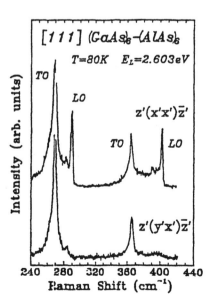

Fig. 52. Raman spectra of a (6,6) [111] (GaAs)/(AlAs) superlattice, recorded in parallel and crossed polarizations for out-of-resonance excitation. Reproduced with permission from [57], copyright 1991, American Physical Society.

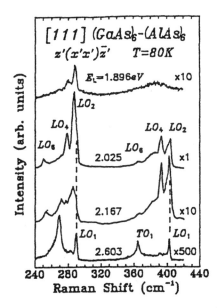

Fig. 53.   Raman spectra of the sample of Figure. 52, in the $Z'(X'X')\overline{Z}'$ configuration, for several laser energies close to resonance with the lowest direct energy gap. Reproduced with permission from [57], copyright 1991, American Physical Society.

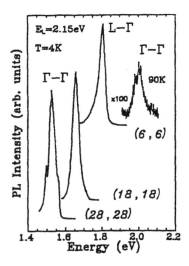

Fig. 54.   PL spectra of three [111] (m, m) (GaAs)/(AlAs) superlattices, recorded at 4 K with 2.15-eV laser excitation. The emission at 2.0 eV in the (6,6) sample can only be observed at 90 K. Reproduced with permission from [57], copyright 1991, American Physical Society.

peaks correspond to optical phonon modes in GaAs and AlAs, respectively.

When the energy of the incident photons approaches one of the superlattice electronic transition energies, the phonon intensities are resonantly enhanced. The Raman spectra in the $Z'(X'X')\overline{Z}'$ polarization geometry are displayed in Figure 53 for the same (6,6) sample at several laser energies. The LO-to-TO intensity ratio increases rapidly as the laser energy is reduced from 2.6 eV (far from resonance) to 2.0 eV, where only the even-parity confined LO modes are seen. This clearly shows that the Fröhlich interaction is the dominant scattering mechanism under resonance conditions. The PL spectra of three (m, m) samples are shown in Figure 54. For the (18,18) and (28,28) samples only a strong Γ-Γ peak is observed, which means that they are type I superlattices. But for the (6,6) sample as the sample temperature is raised (90 K), a weaker peak becomes visible at a higher energy of 2.00 eV, close to the maximum in the resonant profile. This behavior leads to the conclusion that the lowest energy gap in the [111] (6,6) superlattice (at 1.81 eV) is indirect in k space (though direct in real space), occurring between the GaAs conduction L and valence Γ states. Therefore, this is a new type of superlattice that is different from type II superlattices, in which the energy gap is indirect in both k space and real space. The higher-energy luminescence at 2.00 eV represents the direct $\Gamma - \Gamma$ transition. This conclusion is in agreement with the theoretical prediction by Xia and Chang [28] The optical-phonon dispersion relations of GaAs and AlAs in the [111] direction are obtained from the higher-order confined modes in the resonant Raman spectra of [111] $(GaAs)_m/(AlAs)_m$ superlattices with different m [57].

Before the above work Xia and Chang [28] calculated the electronic energy levels as functions of the monolayer number

of GaAs for [111] $(GaAs)_m/(AlAs)_6$ superlattices, as shown in Figure 55. Instead of Γ-X crossover in the [001] case, a Γ-L crossover in the [111] case is found. The Γ-L crossover is due to the different effective masses of the two band edges. It differs from the Γ-X crossover in that both Γ and L states are localized in the same GaAs layer (i.e., no spatial separation). Therefore the energy splitting caused by the Γ-L mixing is apparently larger than that caused by the Γ-X mixing, as seen in Figure 55. The Γ-L crossover occurs near $m = 6$, where the mixing between Γ- and L-like states is strongest, resulting in comparable optical matrix elements for transitions from the first heavy-hole state (HH1) to the two lowest conduction-band states.

Most of the optical investigations of SPSLs have been restricted to the region near the fundamental absorption edge. It is also desirable to know the optical properties well above the fundamental absorption edge, which are of basic physi-

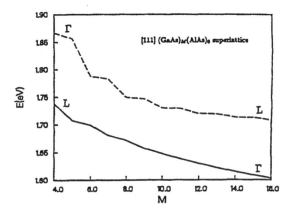

Fig. 55.   Energies of lowest two conduction-band states (relative to the valence band top of bulk GaAs) of [111] $(GaAs)_m/(AlAs)_6$ superlattices as functions of m. Reproduced with permission from [28], copyright 1990, American Physical Society.

cal interest and should become important for optoelectronic devices, such as optical modulators. Spectroscopic ellipsometry is a well-established method for obtaining the dielectric function of solids and heterostructures, in a wide range of photon energies [58]. Ellipsometry measures the amplitude ratio Ψ of and the phase shift Δ between the parallel and perpendicular polarized parts of the reflected light beam. Applying the simplest model with one surface and Fresnel's boundary conditions, the dielectric function can be derived from Ψ and Δ [59] as

$$\varepsilon = \varepsilon_1 + i\varepsilon_2 = \varepsilon_a \sin^2 \varphi_0 \left\{ 1 + \tan^2 \varphi_0 \left[\frac{1 - \tan \Psi e^{i\Delta}}{1 + \tan \Psi e^{i\Delta}} \right]^2 \right\} \quad (30)$$

where φ_0 is the angle of incidence, ε_a is the dielectric constant of the ambient. The dielectric function can be calculated by [28]

$$\varepsilon_1(\hbar\omega) = 1 + \frac{8\pi e^2 \hbar^2}{mV} \sum_{\vec{k}, ij} \frac{|\langle \vec{k}, i|\vec{p}|\vec{k}, j\rangle|^2}{(E_j - E_i)[(E_j - E_i)^2 - \hbar^2\omega^2]}$$

$$\varepsilon_2(\hbar\omega) = \frac{4\pi^2 e^2}{mV\omega^2} \sum_{\vec{k}, ij} |\langle \vec{k}, i|\vec{p}|\vec{k}, j\rangle|^2 \delta(E_j - E_i - \hbar\omega)$$

$$(31)$$

where the summation over $\vec{k}$ means over the whole Brillouin zone, and over j and i means over all of the conduction- and valence-band states, respectively. $\vec{p}$ is the momentum operator. For the special need of the ellipsometry measurement the samples after growth were transported without breaking the ultra-high-vacuum (UHV) conditions from the MBE equipment to the UHV ellipsometer at the synchrotron radiation source (BESSY I, Berlin). The measured imaginary parts of the dielectric functions, ε_2, for the bulk GaAs and AlAs samples and for a $(GaAs)_9/(AlAs)_7$ superlattice are shown in Figure 56. The three characteristic structures E_1, E_2, and E_1' in each spectrum correspond to transitions at critical points, as shown in Figure 57. Around the transition at the critical point the joint density of states has a singularity, resulting in a peak in the ε_2, as expected from Eq. (31). The E_1 structure is known to result from transition from the uppermost spin-orbit-split valence band to the lowest conduction band near the L point along the Λ line in the Brillouin zone. The energy splitting of the E_1 structure is due to spin-orbit splitting of the uppermost valence band. The position of E_1 structure in the superlattice is close to that of GaAs but is shifted by more than 200 meV to a higher energy compared with the GaAs E_1 structure. Hence, the E_1 transition in the superlattice takes place mainly in the GaAs component, and the observed blueshift is due to a confinement effect of the valence band state. The E_2 structure in GaAs is assigned to the transition at the X point overlapped with the Γ-related E_0' transition (see Fig. 57), whereas the E_2 structure in AlAs is attributed to transitions from the spin-orbit-split valence band to the lowest conduction band at X. The second-order derivative spectrum

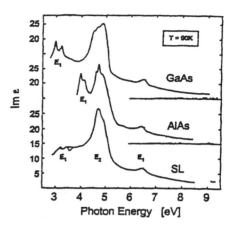

Fig. 56. Imaginary part of the dielectric functions for the bulk GaAs, AlAs samples and a $(GaAs)_9/(AlAs)_7$ superlattice. Reproduced with permission from [58], copyright 1995, American Physical Society.

of the superlattice in the vicinity of the E_2 structure is AlAs-like and very different from that of GaAs. Therefore the corresponding transitions in the superlattice are assigned mainly to the AlAs component. The E_1' structure is assigned to the transition from the spin-orbit-split valence band to the second conduction band at L. The E_1' peak for the superlattice is an AlAs-like transition with no confinement effect; therefore the transition should take place mainly in the AlAs component.

Xia and Chang [28] calculated the dielectric functions of bulk materials and superlattices by using the tight-binding method. From the calculation the structures in ε_2 can be clearly assigned. Figure 58 shows the imaginary part of the dielectric function $\varepsilon_2(E)$ for a $(GaAs)_6/(AlAs)_6$ superlattice, compared with the average $\varepsilon_2(E)$ of bulk GaAs and AlAs. Curves 1–4 are various contributions for the superlattice from regions 1–4 defined in Figure 59. The lower-energy tail comes from region 1 around the Γ point (i.e., the transition near the Γ band edge). By comparing the peak position with that in the bulk components one can assign the component, from which the peak in the superlattice is mainly derived. The E_1 peak

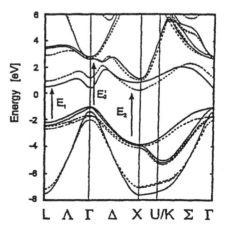

Fig. 57. Calculated band structure within LDA for GaAs (solid lines) and AlAs (dashed lines), and corresponding transitions in ε_2. Reproduced with permission from [58], copyright 1995, American Physical Society.

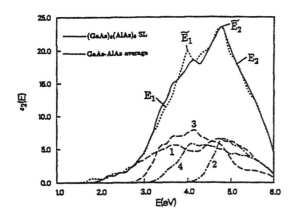

Fig. 58. Imaginary part of the dielectric function $\varepsilon_2(E)$ for a $(GaAs)_6/(AlAs)_6$ superlattice (solid curve), compared with the average $\varepsilon_2(E)$ of bulk GaAs and AlAs (dotted curve). Curves 1–4 are various contributions for the superlattice from regions 1–4 defined in Figure. 59. Reproduced with permission from [28], copyright 1990, American Physical Society.

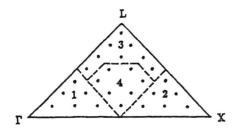

Fig. 59. Two-dimensional Brillouin zone for (001) superlattices divided into four regions (labeled 1–4). Reproduced with permission from [28], copyright 1990, American Physical Society.

is derived mainly from region 3 near the L point. It is contributed from the E_1 peak in GaAs. A hump $\widetilde{E}_1$ appearing in the superlattice ε_2 spectrum near 4.0 eV, which is mainly contributed from regions 3 and 4, is related to the AlAs E_1 peak. Another strong peak $\widetilde{E}_2$ appears at 4.7 eV, which is mainly derived from region 2 (near the X point). The peak is derived from the E_2 peak of AlAs. The shoulder on the higher-energy side of E_2 is related to the GaAs E_2 peak.

9. THEORETICAL RESEARCH

Accompanying the experimental research on SPSLs in the past 15 years, theoretical research has also been developed to bring to light the physics of SPSLs. Here we will not give a comprehensive summing up of the theoretical, research, but only outline some important works on the nature of SPSLs from the authors' points of view.

Caruthers and Lin-Chung [60] were the first to study the (GaAs)/(AlAs) superlattice by the empirical pseudopotential method. They used a table of atomic form factors $v_\alpha(q)$ for $\alpha =$ Ga, Al, and As (with a single As potential), adjusting the Ga and Al form factors to fit the GaAs and AlAs bulk energy bands, respectively. They found that the energy band structure of the $(GaAs)_1/(AlAs)_1$ superlattice is obviously different

from that of the $Al_{0.5}Ga_{0.5}As$ alloy and represents a new type of material. But they did not notice the band offset problem. Hence, the resulting $(AlAs)_1/(GaAs)_1$ [001] superlattice had a quasi-direct gap $\Gamma_{4c}(X_z)$ of 1.585 eV (i.e., ~0.5 eV below the experimental values), which was probably due to a too large valence band offset.

Gell et al. [18] and Xia and Baldereschi [17, 61] independently proposed a new empirical pseudopotential method to study the electron states of superlattices. They did not use the direct plane wave expansion method as in [58], instead they used the pseudopotential Hamiltonian of a bulk crystal (e.g., GaAs) or an alloy in the virtual crystal approximation as the zero-order Hamiltonian H_0, and the difference in the potentials between the superlattice and GaAs or alloy as a perturbation V. The wave function of the superlattice is expanded by the eigenfunctions ϕ_{nk} of H_0,

$$\Psi = \frac{1}{\sqrt{\Omega}} \sum_{nk} A_{nk} \phi_{nk} \tag{32}$$

where $\vec{k}$ is the wave vector in the bulk Brillouin zone determined by the period condition of the superlattice, n is the order of energy band, Ω is the volume, and

$$\begin{aligned} \vec{k} &= \vec{k}' + \vec{k} \\ \vec{K} &= l\frac{2\pi}{Na}\hat{z} \end{aligned} \tag{33}$$

where $\vec{k}'$ is the wave vector of the superlattice, $-\pi/Na < k'_z < \pi/Na$, Na is the superlattice period, and $\hat{z}$ is the unit vector in the z direction,

$$\begin{aligned} N &= \frac{m+n}{2} \\ l &= -(N-1), -(N-2), \ldots, -1, 0, 1, \ldots, (N-1), N \end{aligned} \tag{34}$$

Several bands close to the energy gap (for example, five bands, VB2, VB3, VB4, CB1, and CB2) are included in the expansion [18]. The superlattice Schrödinger equation is written as

$$(H_0 + V)\Psi = E\Psi \tag{35}$$

Inserting Ψ [Eq. (32)] into Eq. (35) and multiplying Eq. (35) from the left by $(1/\sqrt{\Omega})\phi^*_{n'k'}$ and integrating over the volume of the crystal, we are left with a set of linear equations,

$$A_{n'k'}(E_{n'k'} - E) + \frac{1}{\Omega} \sum_{nk} A_{nk} \langle \phi_{n'k'} | V | \phi_{nk} \rangle = 0 \tag{36}$$

the solutions of which are obtained by direct diagonalization.

This method has some advantages over the usual direct plane wave expansion method. First, because the superlattice potential is of long range, the superlattice states are composed of bulk states of energy bands close to the energy gap, unlike the deep level center. Hence, at the beginning we define a subspace including all basic functions related to the superlattice states, and drop all unrelated states, for example, those of the lowest valence band and high-excited conduction bands. In this way the computation effort can be decreased greatly, especially for the long-period superlattices. Second, the wave function of the superlattice is expanded in terms of the bulk or alloy eigenfunctions [Eq. (32)], the calculated results will give a clear physical picture of the composition of superlattice wave functions, Γ-X mixing, etc. This method has been applied to calculate other semiconductor microstructures, for example, quantum wires, quantum dots, clusters, etc.

Gell et al. [18] constructed the potentials, using the original discrete bulk form factors of GaAs and AlAs and taking the 60:40 band offset. They found that the pseudodirect-to-direct transition of the [001] $(\text{GaAs})_n/(\text{AlAs})_n$ superlattices occurs at $n_c \approx 8$ monolayers. It was the first reasonable theoretical result recorded by experiment.

Xia [17] studied the electronic structures of $(\text{GaAs})_m/(\text{AlAs})_n$ SPSLs and corresponding alloys, $\text{Al}_{n/(m+n)}\text{Ga}_{m/(m+n)}\text{As}$. Continuous, parameterized functions for the pseudopotential form factors are fitted to the bulk band structures and to the valence-band offset. In determining form factor functions it was found that the band offset has a critical effect on the relative positions of states with different symmetries. When ΔE_v increases by 0.1 eV, $E_{C\Gamma 1} - E_{CX1}$ increases by 0.044 eV for the $(\text{GaAs})_4/(\text{AlAs})_4$ superlattice. Though the form factor $v(0)$ is meaningless in calculating the energy band of bulk semiconductors, in superlattices the relative amplitude of $v(0)$ for two composed materials determines the band offset. Hence, caution should be exercised in fitting form factors of the two materials of superlattices. The calculation found that for $(\text{GaAs})_m/(\text{AlAs})_n$ superlattices with $n \le 10$ the $C\Gamma 1$ state is higher than the $CX1$ state, so the superlattices are of type II. The optical transition, matrix elements for $CX1 - H\Gamma 1$ are not zero because of Γ-X mixing, but their amplitudes are smaller than those for $C\Gamma 1 - H\Gamma 1$ for (1,1) and (4,4) superlattices by about two orders of magnitude. The transition is pseudodirect. The electronic structures of SPSLs and corresponding alloys are similar in many aspects. The direct and indirect energy gaps vary with $x = n/(m + n)$ in the same way; the folding of energy bands does not change the properties of the Γ and X states essentially. On the other hand, except for $n = 1$, the energy gaps of other superlattices are all systematically lower than those of alloys.

Ting and Chang [10] studied the conduction bands of [001] $(\text{GaAs})/(\text{Al}_x\text{Ga}_{1-x}\text{As})$ and $(\text{Al}_x\text{Ga}_{1-x}\text{As})/(\text{AlAs})$ superlattices with a one-band Wannier orbital model. From the dependence of the superlattice conduction-band energy levels on layer thickness, alloy composition, etc., they first found that the symmetry of the confined X-valley states is critically dependent on whether the slabs with higher Al concentration contain an even or odd number of monolayers. As a result, the amount of Γ-X mixing is extremely sensitive to the layer thickness. These results have been introduced in Section 2, though the one-band Wannier orbital model was not found to be good for describing the electronic states of superlattices.

Lu and Sham [6] and Sham and Lu [7] calculated the subband structure of [001] $(\text{GaAs})_n/(\text{AlAs})_m$ superlattices with the use of a second-neighbor tight-binding method with the parameters fitted to the conduction valleys of the bulk materials. They made a comprehensive symmetry analysis of [001] superlattices, giving the unit cell, Brillouin zone, space group, and point group, etc., for the two cases of $(n + m)$ being even and odd. They pointed out correctly the symmetry of the conduction X states and explained the parity bahavior with the help of the effective-mass approximation. All of these are introduced in detail in Section 2. They also determined that for $(\text{GaAs})_n/(\text{AlAs})_n$ superlattices the lowest-level anticrossing occurs at $n = 12$ (34Å).

Xia and Chang [28] investigated the electronic and optical properties of (GaAs)/(AlAs) SPSLs with an empirical tight-binding model, which includes second-neighbor interactions. The Γ- and X-like electronic energy levels are obtained as functions of the number of monolayers, the applied electric field, and the parallel wave vector. The Γ-X crossover occurs at $n = 12$ for $(\text{GaAs})_n/(\text{AlAs})_n$ superlattices. The conduction energy levels of [111] superlattices were also calculated, suggesting that the Γ-L crossover occurs at $n = 6$. Dielectric functions of superlattices over the full energy range were calculated with the use of a newly developed empirical method to obtain optical matrix elements. These results are all introduced in corresponding sections. The authors also pointed out the reason for the discrepancy about the parity of the X states as a function of monolayer number m of the AlAs layer: $(-1)^m$ [6, 7] or $(-1)^{m+1}$ [10, 28]. It was said that because the tight-binding parameters are not unique, one can obtain an "equally" good fit to the empirical pseudo-potential method (EPM) band structure, but with different characters in the states at X. Thus, in a different tight-binding model the lowest conduction-band state at X may consist of an anion s-like component and a cation z-like component (due to a switch of the roles of the fifth and sixth conduction bands at X—X_1 and X_3), and the prediction regarding even or odd n for the Γ-X mixing to occur would be reversed. The lowest conduction X_1 states have an anion s-like component and a cation z-like component, and the X_3 state is the opposite. In [10] and [25] the fitted lowest conduction state is X_3-like, though the energy band is fitted well.

Wei and Zunger [62] performed self-consistent band structure and total energy calculations for GaAs, AlAs, and the [001] $(\text{GaAs})_n/(\text{AlAs})_n$ superlattice ($n = 1, 2$) within the local density approximation (LDA), using the first principles, all electron, general potential linearized augmented plane wave (LAPW) method. Because the excited-state properties, such as band gaps, are systematically underestimated by LDA, for each superlattice level they identified the proportion of GaAs and AlAs character in its wave function and shifted the calculated superlattice energy level by the weighted average of the errors of the corresponding states in bulk GaAs and AlAs. The energy levels of the $(\text{GaAs})_n/(\text{AlAs})_n$ superlattice for $n = 1$ and $n = 2$ were calculated, and the evolution of the energy levels from $n = 1$ to $n = 2$ is analyzed. Table IX relates the symmetry points in the bulk BZ to those in the mini-BZ for $n = 1$ (odd) and $n = 2$ (even) superlattices (see Fig. 2b). It

Table IX.   Folding Relationships between Bulk States and Superlattice
States for $n = 1$ and 2 at Symmetry Points in the Bulk BZ
and Mini-BZ, Respectively

Bulk	(1, 1)	(2, 2)
Γ_{1c}	$\overline{\Gamma}_{1c}$	$\overline{\Gamma}_{1c}$
$X_{1c}^{x,y}$	$\overline{M}_{5c}$	$\overline{M}_{1c}, \overline{M}_{2c}$
X_{1c}^{z}	$\overline{\Gamma}_{4c}$	$\overline{\Gamma}_{1c}$
$X_{3c}^{x,y}$	$\overline{M}_{1c}, \overline{M}_{2c}$	$\overline{M}_{5c}$
X_{3c}^{z}	$\overline{\Gamma}_{1c}$	$\overline{\Gamma}_{4c}$
L_{1c}	$\overline{R}_{1c}, \overline{R}_{4c}$	$\overline{X}_{1c}, \overline{X}_{4c}$

was found that there are two classes of states. The first class
is those states that have the same symmetry of the superlattice. For example, for $n = 1$ the bulk states Γ_{1c} and X_{3c}
each yield a $\overline{\Gamma}_{1c}$ state in the superlattice. Such states of identical symmetry will repel each other, lowering the energy of
one of the two states. Hence, even if this state were mass-
delocalized in the absence of level repulsion, it would become
localized because of this energy lowering. This class of states
is referred as "repulsion localized states." The second class of
states is those that have different symmetries of the superlattice, but will experience a symmetry-enforced splitting caused
by the superlattice potential, for example, for $n = 1$ the pair of
states $\overline{R}_{1c}(L_{1c})$ and $\overline{R}_{4c}(L_{1c})$ originating from GaAs and AlAs
states L_{1c}, respectively. The level splitting $E(\overline{R}_{4c}) - E(\overline{R}_{1c}) \approx$
$V_s(\text{Al}) - V_s(\text{Ga}) \sim 1$ eV reflects the atomic energy difference
between Al, s and Ga, s, as shown in Figure 51b. Such a pair
of states is referred to as "segregating states." The energy splitting of segregating states depends strongly on the superlattice
period. For $n = 2$, the two L_{1c}-derived states are forced to
equally experience the superlattice potential on the Ga and Al
site. The splitting is caused by the potential difference between
the interfacial As atom and the interior As atom. This results
in a near degeneracy of the two L_{1c}-derived states for $n = 2$.
This will help the analysis of the variation of energy levels
with the superlattice period. For the abrupt interface the low-
est conduction states are $\overline{R}_1(L_1)$ and $\Gamma_1(X_z)$, and the transition
energies are 1.92 and 2.0 eV, for $n = 1$ and 2, respectively.

The first-principles calculation result of the lowest conduction states $\overline{R}_1(L_1)$ for the $n = 1$ superlattice is not in agreement with the experimental result [37]. Morrison et al. [63]
calculated the subband structure of (GaAs)/(AlAs) superlattices with microscopically imperfect interfaces by the empirical pseudopotential method. It is shown that the imperfect
interface has a significant effect on the order of the conduction
states in superlattices. To model a disordered interface (i.e.,
an interface that includes monolayer fluctuations at the interface plane) they chose a larger unit cell in the x-y plane, eight
times larger than the unit cell associated with the ordered interface. The imperfect nature of this interface is then included
by exchanging one Ga atom in this unit cell adjacent to an
interface with an Al atom. The irregularity at the interface
is repeated with the periodicity of the large unit cell. In the

$(\text{GaAs})_3(\text{AlAs})_3$ superlattice the unit cell was twice as long
as that of the superlattice with the ordered interfaces. This
exchange of atomic species is only made at one of the four
interfaces in the large unit cell, so the exchange ratio of the
Ga and Al atoms at the interface is 1/16. The band structure
is significantly different from that obtained from the perfect
interface calculation. The splitting between X_z and $X_{x,y}$ states
is reduced by 50 meV from the perfect to the imperfect case,
and the Γ state is significantly reduced in energy. With the
inclusion of strain the X_z state will be pushed up in energy to
within 10 meV of the $X_{x,y}$ state. It can be imagined that if the
exchange ratio increases from 1/16 to 1/8 or 1/4, then the X_z
state will become higher than the $X_{x,y}$ state.

Laks and Zunger [56] calculated the [001] $(\text{GaAs})_1/(\text{AlAs})_1$
superlattice with the first-principles pseudopotential method. It
shows that the partial intermixing of the Al and Ga atoms relative to the abrupt case lowers its formation energy, making this superlattice even more stable than the randomized
$\text{Al}_{0.5}\text{Ga}_{0.5}\text{As}$ alloy. For the 1/3 intermixed superlattice the
CBM reverts from the GaAs L-derived state to the $X_{x,y}$-
derived AlAs state, in agreement with the experimental results
[37]. The details were introduced in Section 7.

Up to then (1992) the basic problems associated with
SPSLs had seemed to be clear. Since then some meticulous theories have been developed, which have yielded results
in better and quantitative agreement with the experimental
results. Mäder and Zunger [64] fitted the form factors of
atomic pseudopotential with a linear combination of four
Gaussians, multiplied by a smooth function that allows adjustment of the small q components,

$$v_\alpha(q) = \Omega_\alpha \sum_{i=1}^{4} a_{i\alpha} \exp[-c_{i\alpha}(q - b_{i\alpha})^2]$$
$$\times \{1 + f_{0\alpha} \exp(-\beta_\alpha q^2)\}, \qquad (37)$$

where Ω_α is the atomic normalization volume and a, b, c, f,
and β are all fitting parameters. In superlattice the As pseudopotentials in AlAs and GaAs are not constrained to be
identical, and the pseudopotential of As at the interface coor-
dinated by $(4 - n)$ Ga atoms and n Al atoms is taken as the
weighted average,

$$v_{\text{As}}(\text{Ga}_{4-n}\text{Al}_n\text{As}) = \frac{4-n}{4}v_{\text{As}}(\text{GaAs}) + \frac{n}{4}v_{\text{As}}(\text{AlAs}). \quad (38)$$

To demonstrate the quality of the present EPM, they calculated
the electron energy levels for [001] $(\text{GaAs})_n/(\text{AlAs})_n$ superlattices and obtained the following electronic properties of ideal-
structure SPSLs: (i) an L-like CBM of the $n = 1$ superlattice,
(ii) an X-like indirect (at $X_{x,y}$) or pseudodirect (at X_z) CBM
for $1 < n \leq 4$, (iii) a type II to type I transition at $n \approx 9$, and
(iv) wave functions yielded by the pseudopotential that closely
resemble the LDA results even for SPSLs. All of these "bench-
mark" results are in good agreement with *ab initio* theoretical
results or experimental results.

The empirical tight-binding method has been proposed for
the approximately 45 years since Slater and Koster [65].

Within a minimal sp^3 basis and interactions only between nearest-neighbor atoms, the empirical Slater–Koster model can describe the valence-band energy dispersion satisfactorily but fails to reproduce the conduction band structure. To overcome this difficulty a series of improved models have been proposed, such as the sp^3s^* model, the sp^3d^2 model, models including next-neighbor interaction, etc. But the calculated conduction band properties are still not satisfactory, for example, the transverse masses at X and L points and the second conduction band are in poor agreement with experiment. Jancu et al. [66] made a breakthrough which this problem. Except for the sp^3s^* basis they included five d states as the basis and developed a 10 atom-like orbitals $sp^3d^5s^*$ tight-binding model. All tight-binding parameters, on-site energies, and interaction energies between nearest neighbors are fitted to measured energies, pseudopotential results, and the free-electron band structure. The calculated energy bands of element and compound semiconductors are in good agreement with pseudopotential calculations up to about 6 eV above the valence-band maximum. The transverse effective masses at conduction band valleys and deformation potentials are correctly reproduced.

Scholz et al. [67] applied the $sp^3d^5s^*$ tight-binding model to calculate the electronic states of $(GaAs)_n/(AlAs)_n$ SPSLs. The deformation of the AlAs region is treated as usual with classical elasticity, resulting in homogeneous tetragonal strain, and the modifications of the AlAs TB parameters are accounted for by including the distance dependences derived from the bulk deformation potentials. The only further input required is the valence band offset, chosen in accordance with experimental data as $\Delta E_v = 0.55$ eV. For the $n = 1$ superlattice the CBM is the L-derived state with nearly half of the wave function localized on the Ga sublattice, in agreement with first-principle calculations. Consistent with the *ab initio* and empirical pseudopotential investigations, this energy level is found to oscillate strongly with the superlattice period n. For $n = 2$, the TB CBM derives from the bulk AlAs $X_{x,y}$ states because tetragonal strain shifts the X_z level above $X_{x,y}$ by about 20 meV. A crossover $X_{x,y} \rightarrow X_z$ for the CBM occurs between $n = 2$ and 3, whereas experimentally it occurs at $n = 3$ and 4. For $3 \leq n \leq 13$ the CBM derives from the bulk AlAs X_z state, and the type II–type I transition occurs between $n = 13$ and 14. This calculated value is in good agreement with the experimental results for the crossover thickness of $n_c \approx 11$ at room temperature and of $n_c \approx 14$ at low temperature. The overall agreement between theory and experiment clearly demonstrates the transferability of the TB parameters to quantum structures with energy offset at the interface region.

10. APPLICATION OF SHORT-PERIOD SUPERLATTICES

An important property of the SPSLs not mentioned before is the miniband. In superlattices, if the barrier width is small enough, the energy levels of adjacent wells are strongly coupled, and their wave functions are delocalized, which leads to the formation of minibands. The miniband of the superlattice is formed by the energy band folding. The miniband is the energy dispersion $E(k_z)$ in the growth direction as a function of k_z with $-\pi/d < k_z < \pi/d$ (one-dimensional Brillouin zone of superlattices), where d is the superlattice period. The width of the miniband Δ depends on the confinement potential and the barrier width. Generally when the confinement potential is weaker, or the barrier width is smaller, the miniband width Δ is larger. Electron transport in superlattice minibands was first considered by Esaki and Tsu [68], who reported that a region of negative differential conductivity (NDC) should occur when electrons are accelerated to the miniband zone boundary. They also pointed out that the Bloch oscillation of electrons is probably observable in superlattices because of their much smaller Brillouin zone relative to the bulk Brillouin zone. Electrons can move to the zone boundary under a modest field before being scattered. The NDC can be used to design devices working at very high frequencies up to 200 GHz, and the Bloch oscillation, if it is realized in superlattices, will become an ideal solid-state oscillation source of THz. People have made great effort in this direction.

Following Esaki and Tsu [68], the classical equation of electron motion in a one-dimensional solid, to which an electric field F is applied, is expressed as

$$\hbar \frac{dk}{dt} = eF, \qquad v = \frac{1}{\hbar} \frac{\partial E(k)}{\partial k} \tag{39}$$

where F, k, and v are all in the z direction, and v is the electron velocity. From the first equation of Eq. (39) we obtain

$$k(t) = k(0) + \frac{eFt}{\hbar} \tag{40}$$

Unlike a free electron, the $k(t)$ would not increase with t to infinity. When $k = \pi/d$ (zone boundary) it will transfer to $-\pi/d$ because of the superlattice periodicity. The periodic movement of electrons in the k space is called Bloch oscillation; its frequency equals

$$\nu = \frac{1}{T} = \frac{eFd}{h}. \tag{41}$$

Actually, an electron in the superlattice will be scattered by impurities, phonons, etc., losing its original phase of k. If the mean scattering time τ is larger than the period of the Bloch oscillation T, then the Bloch oscillation is possible, so the condition for a Bloch oscillation is

$$\frac{eFd\tau}{h} > 1 \tag{42}$$

Because d is much larger than a, the lattice constant, condition Eq. (42) is more easily satisfied for superlattices than for the bulk material.

Assuming an exponential temporal decay of the probability of collisionless (ballistic) transport at time t, the average drift velocity

$$V = \int_0^\infty \exp(-t/\tau) dv = \frac{eF}{\hbar^2} \int_0^\infty \frac{\partial^2 E(k)}{\partial k^2} \exp(-t/\tau) dt$$

$$= \frac{\mu F}{1 + (F/F_c)^2}, \tag{43}$$

where the miniband dispersion is assumed to be

$$E(k) = \frac{\Delta}{2}[1 - \cos(kd)] \qquad (44)$$

and

$$\mu = \frac{e\Delta\tau d^2}{2\hbar^2}, \qquad F_c = \frac{\hbar}{e\tau d} \qquad (45)$$

where Δ is the miniband width, which is determined by the superlattice structure, especially the width of the barrier. From Eq. (43) it can be seen that at low field $F \ll F_c$ the drift velocity is proportional to the field strength, and μ is the low-field mobility. When the field increases and becomes larger than F_c the drift velocity reaches a maximum and then decreases. The negative differential velocity (NDV) appears. F_c is called the critical field. By comparing with Eq. (42), it is found that F_c is just the field strength, which satisfies the condition of Bloch oscillation.

Sibille et al. [69, 70] measured the I–V characteristics of a series of GaAs/AlAs superlattices and verified the NDV predicted by Esaki and Tsu. Figure 60 presents the experimental and simulated I–V characteristics of a $(GaAs)_{13}/(AlAs)_7$ superlattice [69]. They used Eq. (43) (curve b in the inset of Fig. 58) and another model with the power of (F/F_c) equal to 1 (curve d) to fit the I–V characteristics and found that an excellent agreement could be obtained only with Eq. (43) at $F_c = 16.5$ kV/cm, which verifies the existence of NDV.

To make a comparison with the Esaki–Tsu model, the experimental peak velocity $V_p = \mu F_c/2$ divided by the period d and the product of the critical field F_c with d vs. miniband width in a series of GaAs/AlAs superlattices are shown at the top and bottom of Figure 61, respectively. [70] According to the Esaki–Tsu model, $V_p/d = \Delta/4\hbar$ and $edF_c = \hbar/\tau$.

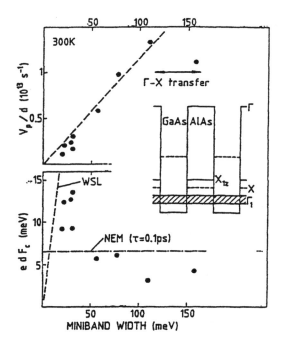

Fig. 61. (Top) Dependence on Δ of the peak velocity divided by the superlattice period. The dashed line is only a guide to the eye. (Bottom) Dependence on Δ of the voltage drop per period at the critical field, including predicted dependences for the NDV and WSL mechanisms. (Inset) Conduction-band diagram of the relevant energy levels. Reproduced with permission from [70], copyright 1989, American Physical Society.

It is obvious from the top of Figure 61 that the experimental V_p/d appears to vary linearly with Δ, at least up to $\Delta \sim 100$ meV. The deviation from the straight line for the point at $\Delta \sim 150$ meV is due to the Γ-X transfer, as shown in the inset. The NDV can also be caused by another mechanism: the Wannier–Stark quantized level (WSL) model proposed by Tsu and Döhler [71], which predicts $edF_c \sim \Delta$ and a superlinear dependence of V_p/d on Δ. At the bottom of Figure 61 the experimental values of edF_c rapidly deviate from the line predicted by the WSL mechanism and approach a constant predicted by the NDV model. These two features verify the NDV model for the superlattices studied. Generally the NDV model is suitable for superlattices with large Δ, whereas the WSL model is for superlattices with small Δ.

The transport of electrons and holes through the miniband was directly observed through time-resolved photoluminescence measurements [72]. The "stepwise graded-gap superlattice" (GGSL) samples were designed in such a way as to produce a well-defined miniband in a variable gap structure as shown in the inset of Figure 62 [72]. In the GaAs/$Al_x Ga_{1-x}$As superlattice sample the well width and barrier width are kept constant. The Al composition x in the barrier is changed in steps of 2% every 800 Å, from 35% at the surface to 17% ~ 1 μm away. One enlarged well (EW) is introduced in the last step (typically 2000 Å) to collect the carries. Thus, the measurement of luminescence spectra at different decays directly determines the positions of photoexcited carries as a function of time. Figure 62 shows the luminescence spectra of a 20 Å/20 Å graded-gap superlattice at var-

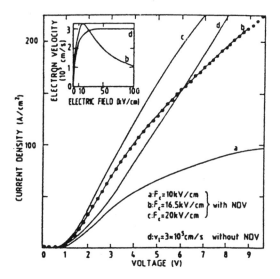

Fig. 60. Experimental (dots) and simulated (full lines) I–V characteristics of a $(GaAs)_{13}/(AlAs)_7$ superlattice. Curves b and d in the inset correspond to Eq. (43) and a model without NDV, respectively. Reproduced with permission from [69], copyright 1989, American Physical Society.

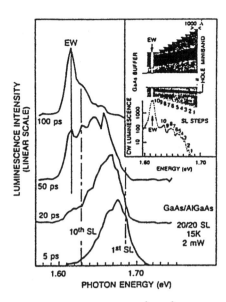

Fig. 62. Luminescence spectra of the 20 Å/20 Å GGSL sample. (Inset) Schematic energy band structure of the GGSL sample: 10 steps of 1000 Å each with superlattice layers and the EW in the last step. Also displayed is the cw luminescence spectrum (logarithmic scale) showing the luminescence peaks in 10 steps and the EW. Reproduced with permission from [72], copyright 1987, American Physical Society.

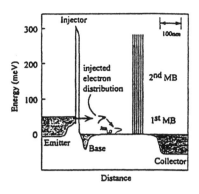

Fig. 63. Schematic conduction band diagram of a three-terminal device. The miniband positions are indicated by the two shaded areas. The emitter is biased at 50 meV; the base and collector are grounded. Reproduced with permission from [73], copyright 1997, American Institute of Physics.

ious delays [72]. The 5-ps spectrum shows that the carriers are created near the surface of the sample as the luminescence peaking at ~1.68 eV mainly originates in the first step of superlattices. For a decay time of 100 ps, most of the carriers have been transferred into the EW with a luminescence peak energy of ~1.615 eV. This shift exactly reflects the motion of the packet of photoexcited electrons and holes in the superlattice minibands. The diffusion coefficient and the mobility are deduced by fitting the measured motion with a diffusion model including the step boundary. For the 20 Å/20 Å superlattice, the diffusion coefficient is 2.2 cm²/s and the mobility is 1800 cm²/V · s. Because any separation of the two types of carriers would give rise to a large electric field between the two distributions at these densities, transport is expected to be ambipolar with a mobility approximately twice the hole mobility. So the hole mobility for the 20 Å/20 Å superlattice is 900 cm²/V · s. But for the 30 Å/30 Å superlattice the mobility is deduced to be 50 cm²/V · s by a factor of 20. It is partly due to an order-of-magnitude reduction in the hole miniband width. The heavy-hole miniband widths are 23 meV and 2.5 meV for the 20 Å/20 Å and 30 Å/30 Å superlattices, respectively. Graded-gap superlattices are widely used in quantum cascade lasers.

Rauch et al. [73] designed a three-terminal device with the use of a hot electron transistor structure as an electron energy spectrometer to study the ballistic transport in the mininband of superlattices. The structure is shown in Figure 63 [73]: a highly doped n^+-GaAs collector contact layer ($n = 1 \times 10^{18}$ cm⁻³), a superlattice, and a drift region of 200 nm, followed by a highly doped (2×10^{18} cm⁻³) n^+-GaAs base of 13 nm width. On top of the base layer a 13-nm undoped $Al_{0.3}Ga_{0.7}As$

barrier, a spacer, and a n^+-GaAs layer are successively grown. Finally, a n^+-GaAs contact layer is grown on top of the heterostructure to form the emitter. Three different superlattices with six $Al_{0.3}Ga_{0.7}As$ barriers (2.5 nm) and five GaAs wells of varying widths, 6.5 nm, 8.5 nm, and 15 nm, were studied. In the case of an applied free field (flat band condition) the transfer ratios $\alpha = I_C/I_E$ as functions of injection energy ($\approx eU_{EB}$) at lower injection energies for three samples are shown in Figure 64 [73], where I_C and I_E are collector and emitter currents, respectively. The first peak positions are in agreement with the calculated first miniband positions shown by the solid lines in the figure, so the first peak is due to ballistic transport through the lowest miniband. The dashed line indicates the broadening due to the energy distribution of the injected electron beam of about 20 meV. The second peak is the phonon replica of the first miniband shifted 36 meV to higher injection energies. At the higher energy side there is another peak

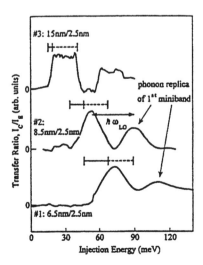

Fig. 64. Transfer ratio α vs. injection energy at lower injection energies for three samples. The solid line indicates the calculated miniband position, and the dashed line indicates the broadening due to the energy distribution of the injected electron beam. A double arrow represents the energy of a longitudinal optical phonon (36 meV). Reproduced with permission from [73], copyright 1997, American Institute of Physics.

corresponding to the transport through the second mininband, which is also in agreement with the calculated results.

Rauch et al. [74] studied further the coherent (ballistic) and incoherent (diffusive) transport in the similar three terminal structures under an electric field bias. For a few-period superlattice the transmittance is found to be independent of the electric field; and hence the transport is coherent. For a superlattice longer than the coherence length, the transmission becomes asymmetric and dependent on the electric field direction, and the transport is incoherent. The total miniband transmission T_α is defined as twice the area of the lower energy side of the first transfer ratio peak (Fig. 64) as a measure of the average current through the first miniband at given bias conditions. The analysis of T_α for the 5-, 10-, 20-, and 30-period superlattices is shown in Figure 65 over the voltage applied across the superlattice [74]. The T_α of the 5-period sample is symmetric for both bias directions, whereas clear asymmetric behavior is observed for all other samples, and the asymmetry grows with the period number. The observed asymmetric behavior is assigned to the onset of diffusive transport. If a positive bias is applied to the collector, the scattered electrons contribute additionally with the coherent electrons to the collector current, resulting in an increase of the transfer ratio. For the negative bias only coherent electrons traverse the superlattice, and scattered electrons flow back to the base according to the applied electric field. Therefore the presence of scattering destroys the symmetry of the transmission with respect to the field direction. It is concluded that the transition from coherent transmission to diffusive transport occurs visibly between 5 and 10 periods for the structures used. The main scattering mechanism is interface roughness in these undoped structures at a temperature of 4.2 K. The average velocity in the miniband is on the order of $\Delta d/2\hbar = 1.5 \times 10^7$ cm/s. If the scattering time τ_s is taken to be 1 ps, then the mean free path $l_{coh} = v\tau_s = 150$ nm, which can be seen as the coherent length.

As for the device application, it seems that is a long way to go. Hadjazi et al. [75] reported high-frequency operation up to 60 GHz, limited by the instrumentation and sample processing. The coherent electromagnetic radiation originating from Bloch oscillations in a biased superlattice structure was directly detected in the time domain [76].

Another application of SPSLs is mid-infrared quantum cascade lasers (QCLs). Faist et al. [77, 78] first reported a QCL made of an $Al_{0.48}In_{0.52}As/Ga_{0.47}In_{0.53}As$ superlattice structure designed by band structure engineering. A strong narrowing of the emission spectrum, above threshold, provided direct evidence of laser action at a wavelength of 4.2 μm with peak powers of 8 mW in pulsed operation. A schematic conduction band energy diagram of a portion of the 25-period (active region plus injector) section of the QCL is shown in Figure 66 [78]. The structure is under positive bias condition and an electric field of 76 kV/cm. The active region comprises three quantum wells, which is engineered so that at the threshold field (76 kV/cm), the ground states of the 4.7- and 4.0-nm-thick quantum wells in the active region have anticrossed to achieve an energy separation resonant with the optical phonon energy ($\sim$36 meV). As a result, the $n = 2$ state has a short lifetime ($\tau_2 \sim \tau_{21} = 0.4$ ps), whereas the lifetime of level 3 is $\tau_3 = 1.3$ ps. An additional 0.9-nm-thick GaInAs quantum well coupled to the active region by a 1.5-nm barrier selectively enhances the amplitude of the wave function of level 3 in the 5.0-nm injection barrier (Fig. 66). This maximizes the injection efficiency by increasing the overlap between the $n = 3$ wave function and the ground-state wave function g of the injector. The superlattice in the injector region is a graded-gap superlattice as shown in Figure 62, designed so that under positive bias conditions the miniband acts like a funnel (dark shaded region) for injecting electrons. This has

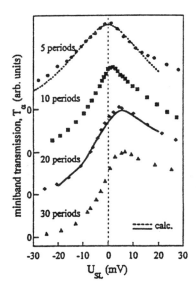

Fig. 65. Miniband transmission versus electric field of the 5-, 10-, 20-, and 30-period samples. Data for the 5-period sample (dots) is compared with a one-dimensional calculation (dashed line) for coherent transmission, whereas data for the 20-period sample (diamonds) are compared with a calculation including interface roughness (solid line). Reproduced with permission from [74], copyright 1998, American Physical Society.

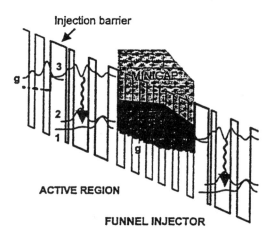

Fig. 66. Schematic conduction band diagram of a portion of a QCL under positive bias at an electric field of 76 kV/cm. The wavy line indicates the transition (3-2) responsible for laser action. The moduli squared of the relevant wave functions are shown. Reproduced with permission from [78], copyright 1996, American Institute of Physics.

the effect of funneling electrons into the ground state g of the injection region and minimizing the electron thermal excitation to higher states of the miniband. Figure 67 is the emission spectrum of a 500-μm-long laser at various drive currents [77]. The spectrum below a drive current of 600 mA is broad, indicative of spontaneous emission. Above a drive current of 850 mA, corresponding to a threshold current density $J_{th} = 15$ kA/cm^2, the signal increases abruptly by orders of magnitude, accompanied by a dramatic line-narrowing. This is a direct manifestation of laser action. In a high-resolution spectrum, well-defined, nearly equally spaced longitudinal modes are observed. The mode spacing $\Delta\nu = 2.175$ cm^{-1} is in good agreement with the calculated one ($1/2nL = 2.13$ cm^{-1}, where L is the length of the laser). A peak pulsed power of 200 mW and an average power of 6 mW are obtained at 300 K and at a wavelength $\lambda = 5.2$ μm. The devices also operate in continuous wave up to 140 K.

We should go back to our subject: GaAs-AlAs SPSLs. The quantum cascade light-emitting diode (QCLED) was realized with the GaAs/AlGaAs material system [79, 80]. Electroluminescence powers up to a few nanowatts at 6.9 μm have been measured. The structure design is completely similar to that of the QCL (Fig. 66). The emission from 100×100 μm^2 mesas was coupled out of the sample through a polished 45o facet. Strong injection occurs for bias voltages above $V_b = 7.5$ V, corresponding to a current on the order of $I = 100$ mA (current density $j = 1$ kA/cm^2). Typical emission spectra are shown in Figure 68 for two samples of ref. 80 at 10 K and room temperature, respectively [80]. The main emission peak (sample 1: 1450 cm^{-1} or 6.9 μm; sample 2: 1380 cm^{-1} or 7.25 μm) is due to the 3-2 transition. The second weak peak for sample 1 is due to the 3-1 transition resulting from a finite overlap of the respective wave functions. The width of the luminescence line is found to be as narrow as 14 (17) meV for sample 1 (sample 2) at 10 K and broadens to 20 meV at room temperature. The total optical power is estimated to be 15 nW. This is a clear demonstration of quantum cascade transitions in the GaAs/AlGaAs system.

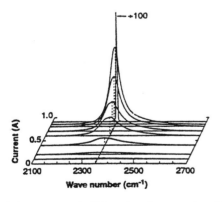

Fig. 67. Emission spectrum of a 500-μm-long laser at various drive currents. The emission wavelength, $\lambda = 4.26\mu$m, is in agreement with the calculated wavelength for the $n = 3$ to $n = 2$ transition. Reproduced with permission from [77], copyright 1994.

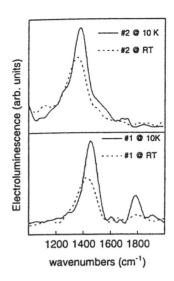

Fig. 68. Electroluminescence spectra at 10 K and room temperature (RT) for samples 1 and 2. Reproduced with permission from [80], copyright 1998, Materials Research Society.

The terahertz (THz) frequency range (1–10 THz, or 30–300 μm) is among the most underdeveloped electromagnetic spectra, mainly because of the lack of compact, coherent solid-state sources. The Bloch oscillation in superlattices mentioned before is a possible way. The QCLs operate generally at mid-infrared frequencies (3–5 μm and 8–12 μm). Xu et al. [81] reported an electrically pumped periodic three-level system designed and fabricated with an AlGaAs/GaAs multiple quantum well structure. Under appropriate biases, the structure emits radiation at a frequency about 7 THz, as a result of diagonal (or interwell) intersubband transition. Because of the diagonal intersubband transition the emission peak width is broad, and the emission power is estimated to be 0.2 μW. Rochat et al. [82] investigated the intersubband electroluminescence in a quantum cascade structure based on a vertical transition designed for far-infrared ($\lambda = 88$ μm) emission. The structure consists of 35 periods, and each period consists of four GaAs quantum wells separated by thin $Al_{0.5}Ga_{0.85}As$ tunnel barriers. As in a mid-infrared quantum structure, each period consists of an undoped active region, in which spontaneous emission occurs, and a graded-gap injector. The active region consists of a 28-nm GaAs quantum well coupled through a 2.5-nm $Al_{0.5}Ga_{0.85}As$ barrier to an 18-nm GaAs well. The emission occurs in the 28-nm well through a vertical transition, i.e., an optical recombination between two states with strong spatial overlap. Compared with diagonal transitions, the vertical transition has the advantage of large oscillator strength and narrower line width because it is less influenced by the interface roughness. Optical spectra of the emitted radiation for various injected currents are shown in Figure 69 [82]. The luminescence spectrum mainly consists of one narrow peak centered at a wavelength of $\lambda = 88$ μm. This peak is identified to correspond to the $n = 2$ to $n = 1$ transition in the 28-nm well. The full width at half-maximum of the peak is 0.7 meV. When the drive current exceeds 80 mA at

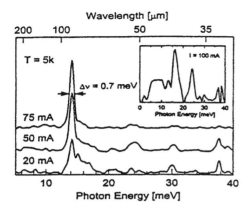

Fig. 69. Optical spectra of the emitted radiation at 5 K for various injected currents. The inset shows the spectrum of the device driven by a current above the region of negative differential resistance. Reproduced with permission from [82], copyright 1998, American Institute of Physics.

$T = 5$ K, the optical spectrum becomes extremely broad (> 30 me V), as shown in the inset of Figure 69. This is due to the fact that the ground state g of the injector is resonant with the $n = 2$ state and a negative differential resistance occurs. Such a good correlation between the electric and spectral characteristics in the same device is proof that the electroluminescence arises from a resonant tunneling injection into the $n = 2$ state and not from heating of the electron gas.

Ulrich et al. [83] also reported spontaneous terahertz emission from a quantum cascade structure. The structure is made up of 50 periods of a chirped $Al_{0.15}Ga_{0.85}As/GaAs$ superlattice, as shown in Figure 70 [83]. The radiative transition occurs in the widest (26 nm) quantum well. At an appropriate electric field (22 mV/period), electrons are resonantly injected from the first miniband of the injector (four wells) into the second state (E_2) of the transition well. The terahertz-emission spectrum at 5 K is like that in Figure 69 with a peak at 17.3 meV ($\lambda = 72$ μm), consistent with the calculated energy difference $E_2 - E_1$ of 17.1 meV. The line width is 2.1 meV. They have not observed emission from transitions other than $E_2 - E_1$. There is a monotonous increase in signal with the inject current up to the breakup of the first miniband. The increase in light output with current is sublinear. Ulrich et al. also found that the edge

outcoupling is about twice as efficient as the grating-coupled surface emission.

11. SUMMARY

The problem of CBM in SPSLs has attracted wide experimental and theoretical attention for about 15 years; now it is at a full stop. To solve this problem, nearly all experimental and theoretical methods (as a Chinese proverb says, 18 sorts of arms) were used. In the process the conclusion becomes more and more definite, at the same time the material quality and the experimental and theoretical methods have been improved dramatically. Many physical properties of SPSLs can be used to design and fabricate devices, for example, super-high-frequency devices, mid-infrared and terahertz laser and light-emitting devices, etc. But there is still a long way to go. In this process an elaborate energy band engineering design based on the accumulated physical knowledge is needed, and some new phenomena and problems will be explored.

Acknowledgment

This work is supported by Hong Kong RGC (Research Grant Council) grant HKUST 6135/97P.

REFERENCES

1. A. C. Gossard, P. M. Petroff, W. Wiegman, R. Dingle, and A. Savage, *Appl. Phys. Lett.* 29, 323 (1976).
2. N. Sano, H. Kato, M. Nakayama, C. Chika, and N. Tereauchi, *Jpn. J. Appl. Phys.* 23, 1640 (1984).
3. T. Isu, D. S. Jiang, and K. Ploog, *Appl. Phys. A* 43, 75 (1987).
4. A. Ishibashi, Y. Mori, M. Itabashi, and N. Watanabe, *J. Appl. Phys.* 58, 2691 (1985).
5. R. Cingolani and K. Ploog, *Adv. Phys.* 40, 535 (1990).
6. Y. T. Lu and L. J. Sham, *Phys. Rev. B: Solid State* 40, 5567 (1989).
7. L. J. Sham and Y. T. Lu, *J. Lumin.* 44, 207 (1989).
8. W. K. Ge, W. D. Schmidt, M. D. Sturge, L. N. Pfeiffer, and K. W. West, *J. Lumin.* 59, 163 (1994).
9. J. Zak, A. Casher, M. Glück, and Y. Gur, "The Irreducible Representations of Space Groups." Benjamin, New York, 1969.
10. D. Z.-Y. Ting and Y. C. Chang, *Phys. Rev. B: Solid State* 36, 4359 (1987).
11. M. D. Sturge, E. Finkman, and M. C. Tamargo, *J. Lumin.* 40/41, 425 (1988).
12. B. A. Wilson, C. E. Bonner, R. C. Spitzer, P. Dawson, K. J. Moore, and C. T. Foxon, *J. Vac. Sci. Technol., B* 6, 1156 (1988).
13. E. Finkman, M. D. Sturge, and M. C. Tamago, *Appl. Phys. Lett.* 49, 1299 (1986).
14. J. Nagle, M. Garriga, W. Stolz, T. Isu, and K. Ploog, *J. Phys. (Paris) Colloq.* 48, C5-495 (1987).
15. K. J. Moore, P. Dawson, and C. T. Foxon, *Phys. Rev. B: Solid State* 38, 3368 (1988).
16. J. Schulman and T. C. McGill, *Phys. Rev. B: Solid State* 19, 6341 (1979).
17. J. B. Xia, *Phys. Rev. B: Solid State* 38, 8358 (1988).
18. M. A. Gell, D. Ninno, M. Jaros, and D. C. Herbert, *Phys. Rev. B: Solid State* 34, 2416 (1986).
19. K. J. Moore, G. Duggan, P. Dawson, and C. T. Foxon, *Phys. Rev. B: Solid State* 38, 5535 (1988).

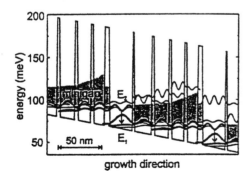

Fig. 70. Schematic conduction band diagram of a quantum cascade structure designed for the emission of terahertz radiation. Reproduced with permission from [83], copyright 1999, Elsevier Science.

20. D. S. Jiang, K. Kelting, T. Isu, H. J. Queisser, and K. Ploog, *J. Appl. Phys.* 63, 845 (1988).

21. M. C. Muñoz, V. R. Velasco, and F. G. Moliner, *Phys. Rev. B: Solid State* 39, 1786 (1989).

22. R. Cingolani, L. Baldassarre, M. Ferrara, and K. Ploog, *Phys. Rev. B: Solid State* 40, 6101 (1989).

23. G. H. Li, D. S. Jiang, H. X. Han, and Z. P. Wang, *Phys. Rev. B: Solid State* 40, 10430 (1989).

24. H. Kato, Y. Okada, M. Nakayama, and Y. Watanabe, *Solid State Commun.* 70, 535 (1989).

25. S. Gapalan, N. E. Christensen, and M. Cardona, *Phys. Rev. B: Solid State* 39, 5165 (1989).

26. H. Fujimoto, C. Hamaguchi, T. Nakazawa, K. Taniguchi, and K. Imanishi, *Phys. Rev. B: Solid State* 41, 7593 (1990).

27. M. Holtz, R. Cingolani, K. Reimann, R. Muralidharan, K. Syassen, and K. Ploog, *Phys. Rev. B* 41, 3641 (1990).

28. J. B. Xia and Y. C. Chang, *Phys. Rev. B: Solid State* 42, 1781 (1990).

29. T. Matsuoka, T. Nakazawa, T. Ohya, K. Taniguchi, and C. Hamaguchi, *Phys. Rev. B: Solid State* 43, 11,798 (1991).

30. T. Nakayama, F. Minami, and K. Inoue, *J. Lumin.* 53, 380 (1992).

31. M. Nakayama, K. Imazawa, I. Tanaka, and H. Nishimura, *Solid State Commun.* 88, 43 (1993).

32. D. E. Aspnes and J. E. Rowe, *Phys. Rev. B: Solid State* 5, 4022 (1972).

33. D. E. Aspnes, *Surf. Sci.* 37, 418 (1973).

34. M. V. Klein, M. D. Sturge, and E. Cohen, *Phys. Rev. B: Solid State* 25, 4331 (1982).

35. J. Ihm, *Appl. Phys. Lett.* 50, 1068 (1987).

36. M. D. Sturge, E. Cohen, and R. A. Logan, *Phys. Rev. B: Solid State* 27, 2362 (1983).

37. W. K. Ge, M. D. Sturge, W. D. Schmidt, L. N. Pfeiffer, and K. W. West, *Appl. Phys. Lett.* 57, 55 (1990).

38. M. Nakayama, K. Imazawa, K. Suyama, I. Tamaka, and H. Nishimura, *Phys. Rev. B: Solid State* 49, 13,564 (1994).

39. M. H. Meynadier, R. E. Nahory, J. M. Worlock, M. C. Tamarga, J. L. de Miguel, and M. D. Sturge, *Phys. Rev. Lett.* 60, 1338 (1988).

40. D. J. Wolford and J. A. Bradley, *Solid State Commun.* 53, 1069 (1985).

41. D. J. Wolford, T. F. Kuech, J. A. Bradley, M. A. Gell, D. Ninno, and M. Jaros, *J. Vac. Sci. Technol., B* 4, 1043 (1986).

42. M. A. Gell, D. Ninno, M. Jaros, D. J. Wolford, T. F. Keuch, and J. A. Bradley, *Phys. Rev. B: Solid State* 35, 1196 (1987).

43. G. Danan, B. Etienne, F. Mollot, R. Planel, A. M. Jean-Louis, F. Alexandre, B. Jusserand, G. Le Roux, J. Y. Marzin, H. Savary, and B. Sermage, *Phys. Rev. B: Solid State* 35, 6207 (1987).

44. M. S. Skolnick, G. W. Smith, I. L. Spain, C. R. Whitehouse, D. C. Herbert, D. M. Whittaker, and L. J. Reed, *Phys. Rev. B: Solid State* 39, 11,191 (1989).

45. J. Bleuse, G. Bastard, and P. Voisin, *Phys. Rev. Lett.* 60, 220 (1988).

46. M. Morifuji, M. Yamaguchi, K. Taniguchi, and C. Hamaguchi, *Phys. Rev. B: Solid State* 50, 8722 (1994).

47. M. Yamaguchi, H. Nakasawa, M. Morifuji, K. Taniguchi, H. Hamaguchi, C. Gmachi, and E. Gornik, *Semicond. Sci. Technol.* 9, 1810 (1994).

48. H. W. van Kesteren, E. C. Cosman, F. J. A. M. Greidanus, P. Dawson, K. J. Moore, and C. T. Foxon, *Phys. Rev. Lett.* 61, 129 (1988).

49. H. W. van Kesteren, E. C. Cosman, P. Dawson, K. J. Moore, and C. T. Foxon, *Phys. Rev. B* 39, 13426 (1989).

50. N. Miura, Y. Shimamato, Y. Imanaka, H. Arimoto, H. Nojiri, H. Kunimatsu, K. Uchida, T. Fukuda, K. Yamanaka, H. Momose, N. Mori, and C. Hamaguchi, *Semicond. Sci. Technol.* 11, 1586 (1996).

51. T. Fukuda, K. Yamanaka, H. Momose, N. Mori, C. Hamaguchi, Y. Imanaka, Y. Shimamoto, and N. Miura, *Surf. Sci.* 361/362, 406 (1996).

52. P. Lefebvre, B. Gil, H. Matheiu, and R. Planel, *Phys. Rev. B: Solid State* 39, 5550 (1989).

53. P. Lefebvre, B. Gil, H. Matheiu, and R. Planel, *Phys. Rev. B: Solid State* 40, 7802 (1989).

54. W. K. Ge, J. L. Mackay, L. N. Pfeiffer, and K. W. West, *J. Lumin.* 50, 133 (1991).

55. R. Schwabe, F. Pietag, V. Gottschalch, G. Wagner, M. Di Ventra, A. Bitz, and J. L. Staehli, *Phys. Rev. B: Solid State* 56, R4329 (1997).

56. D. B. Laks and A. Zunger, *Phys. Rev. B: Solid State* 45, 11,411 (1992).

57. F. Calle, D. J. Mowbray, D. W. Niles, M. Cardona, J. M. Calleja, and K. Ploog, *Phys. Rev. B: Solid State* 43, 9152 (1991).

58. O. Günther, C. Janowitz, G. Jungk, B. Jenichen, R. Hey, L. Däweritz, and K. Ploog, *Phys. Rev. B: Solid State* 52, 2599 (1995).

59. R. M. A. Azzam and N. M. Bashara, "Ellipsometry and Polarized Light" p. 274. North-Holland, Amsterdam, 1987.

60. E. Caruthers and P. J. Lin-Chung, *Phys. Rev. B: Solid State* 17, 2705 (1978).

61. J. B. Xia and A. Baldereschi, *Chin. J. Semicond.* 8, 574 (1987).

62. S. H. Wei and A. Zunger, *J. Appl. Phys.* 63, 5794 (1988).

63. I. Morrison, L. D. L. Brown, and M. Jaros, *Phys. Rev. B: Solid State* 42, 11,818 (1990).

64. K. A. Mäder and A. Zunger, *Phys. Rev. B: Solid State* 50, 17,393 (1994).

65. J. C. Slater and G. F. Koster, *Phys. Rev.* 94, 1498 (1954).

66. J. M. Jancu, R. Scholz, F. Beltram, and F. Bassani, *Phys. Rev. B: Solid State* 57, 6493 (1998).

67. R. Scholz, J. M. Jancu, F. Baltram, and F. Bassani, *Phys. Status Solidi B* 217, 449 (2000).

68. L. Esaki and R. Tsu, *IBM J. Res. Dev.* 14, 61 (1970).

69. A. Sibille, J. F. Palmier, C. Minot, and F. Mollot, *Appl. Phys. Lett.* 54, 165 (1989).

70. A. Sibille, J. F. Palmier, C. Minot, and F. Mollot, *Phys. Rev. Lett.* 64, 265 (1989).

71. R. Tsu and G. Döhler, *Phys. Rev. B: Solid State* 12, 680 (1975).

72. B. Deveaud, J. Shah, T. C. Damen, B. Lambert, and A. Regreny, *Phys. Rev. Lett.* 58, 2582 (1987).

73. C. Rauch, G. Strasser, K. Unterrainer, E. Gornik, and B. Brill, *Appl. Phys. Lett.* 70, 649 (1997).

74. C. Rauch, G. Strasser, K. Unterrainer, W. Boxleitner, E. Gornik, and A. Wacker, *Phys. Rev. Lett.* 81, 3495 (1998).

75. M. Hadjazi, J. F. Palmier, A. Sibille, H. Wang, E. Paris, and F. Mollot, *Electron. Lett.* 29, 648 (1993).

76. C. Waschke, H. G. Roskos, R. Schwedler, K. Leo, H. Kurz, and K. Köhler, *Phys. Rev. Lett.* 70, 3319 (1993).

77. J. Faist, F. Capasso, D. L. Sivco, C. Sirtori, A. L. Hutchinson, and A. Y. Cho, *Science* 264, 553 (1994).

78. J. Faist, F. Capasso, C. Sirtori, D. L. Sivco, J. N. Baillargeon, A. L. Hutchinson, S. G. Chu, and A. Y. Cho, *Appl. Phys. Lett.* 68, 3680 (1996).

79. G. Strasser, P. Kruck, M. Helm, J. N. Heyman, L. Hvozdara, and E. Gornik, *Appl. Phys. Lett.* 71, 2892 (1997).

80. G. Strasser, S. Gianordoli, L. Hvozdara, H. Bichl, K. Unterrainer, E. Gornik, P. Kruck, M. Helm, and J. N. Heyman, in "Infrared Applications of Semiconductors II" (D. L. McDaniel, Jr., M. O. Manasreh, R. H. Miles, and S. Sivanathan, Eds.), *Mater. Res. Soc. Symp. Proc.* 484, 165 (1998).

81. B. Xu, Q. Hu, and M. R. Melloch, *Appl. Phys. Lett.* 71, 440 (1997).

82. M. Rochat, J. Faist, M. Beck, U. Oesterle, and M. Ilegems, *Appl. Phys. Lett.* 73, 3724 (1998).

83. J. Ulrich, R. Zobl, N. Finger, K. Unterrainer, G. Strasser, and E. Gornik, *Physica B* 272, 216 (1999).

Chapter 4

SPIN WAVES IN THIN FILMS, SUPERLATTICES AND MULTILAYERS

Zhang Zhi-Dong

Shenyang National Laboratory for Materials Science and International Centre for Material Physics, Institute of Metal Research, Academia Sinica, Shenyang 110015, People's Republic of China

Contents

1. GENERAL INTRODUCTION

Spin waves (i.e., *magnons*) are made up of elemental excitons in magnetic materials, from which one could derive thermodynamic properties, such as magnetization, specific heat, etc., and their dynamical behaviors. The concept of spin waves as low-lying excitations in ordered magnetic materials was due to Bloch [1]. When the spins deviate slightly from the orientations of their ground states, the disturbances will travel with a wavelike behavior through a crystal. This is analogous to the vibrational modes in a solid (i.e., phonons), which represent small-amplitude oscillations of atoms about their equilibrium positions.

In the past five decades, there has been a rapidly growing interest in the properties of spin waves in ordered magnetic materials, either bulk or thin films. Specifically, dramatic advances in experimental techniques for preparing high-quality thin films and superlattices have provided an apportunity to understand the behaviors of spin waves in these limited magnetic materials. It is well known that any excitation, such as a spin wave, in a crystalline solid that is effectively infinite, must satisfy the symmetry requirement in accordance with Bloch's theorem [2]. The excitation amplitudes of the spin waves have a plane wave variation with a periodic function satisfying the periodicity properties of the crystal lattice. The spin waves in finite systems, such as thin films and multilayers, must have their own behaviors with respect to the breakdown of symmetry [3]. There is no translational symmetry operator in one of the three dimensions (i.e., the direction perpendicular to the plane of the films), and the translational symmetry remains in the plane of the atomic layers parallel to the surface. In the case of thin films or multilayers, therefore, the contribution of the plane wave variation to the amplitudes of spin waves originates only from the propagation parallel to the surface. To describe the properties of linear spin waves propagating along the direction normal to magnetic films of finite thickness, one has to consider the effect of symmetry breaking and the boundary conditions on the film surfaces [4]. The finite thickness causes the spin wave spectrum to be discrete; the spectrum consists of separate dispersion branches. The effect of the boundaries results in the dependence of the spin wave eigenfrequencies on the magnitude of the wave vector. Another effect caused by the

Handbook of Thin Film Materials, edited by H.S. Nalwa
Volume 5: Nanomaterials and Magnetic Thin Films

ISBN 0-12-512913-0/$35.00

presence of the boundaries is the existence of the surface spin waves. The surface spin waves are localized near the surface, rather than spread through the whole crystal, as for the bulk spin waves of an infinite medium. However, in this case, bulk spin waves with a wavelike behavior in three dimensions may also exist in the vicinity of a surface, of course, which must satisfy boundary conditions at the surfaces. This means that one can find localized spin waves as well as bulk spin waves in either a magnetic film or multilayer. The number of localized spin waves (i.e., surface or interface spin waves) depends on the number of surfaces and/or interfaces in the limited system. As a special case of multilayer, a superlattice gives a structure with an overall quasi-periodicity of the composition and thickness along one dimension. The spin wave spectrum of superlattices shows a subband structure, with some bulk spin wave behavior. This is due to the fact that the periodicity length of the superlattices is equal to the sum of the thicknesses of component unit layers, and the translations along this direction still constitute a symmetry operation of the structure. In some limited cases, mixed interface and bulk modes may exist in superlattices.

The spin wave theories are divided into two groups, macroscopic phenomenological and quantum microscopic. The macroscopic phenomenological theories usually start from Maxwell's equations of electromagnetism, considering the boundary conditions of the systems. The macroscopic approach treats the magnetization as a field vector $\mathbf{M}(\mathbf{r})$ depending on the position $\mathbf{r}$. The total energy, including the exchange and dipole–dipole interactions, is expressed in terms of the magnetization. The magnetization is written as the sum of a static part M_0 and a fluctuating part $\mathbf{m}(\mathbf{r})$, and the magnetic field due to the dipole–dipole interactions can be expressed as the sum of a demagnetizing field and a fluctuating term $\mathbf{h}_d(\mathbf{r})$. To solve for spin wave frequencies, one needs to use the torque equation of motion for the magnetization,

$$\frac{\partial M}{\partial t} = \gamma M \times \mathbf{H}_{\text{eff}} \tag{1.1}$$

where the total effective field $\mathbf{H}_{\text{eff}}$ is composed of terms for the applied field, the exchange field, and the dipole field. The fluctuating part of the dipole field $\mathbf{h}_d(\mathbf{r})$ and the magnetization $\mathbf{m}(\mathbf{r})$ must satisfy the magnetostatic form of Maxwell's equations,

$$\begin{aligned} \operatorname{div}\big[h_d(r) + 4\pi m(r)\big] &= 0 \\ \operatorname{curl} h_d(r) &= 0 \end{aligned} \tag{1.2}$$

This continuum approach is more powerful in studying the spin waves at relatively long wavelengths compared with the microscopic lattice parameter. It is also more convenient for including dipole–dipole interactions via Maxwell's equations. The quantum microscopic theories usually start from the Heisenberg Hamiltonian, followed by a standard method based on the second quantization [5]. The Hamiltonian is usually written in terms of the spins at individual sites. The Heisenberg Hamiltonian of the system may be written as

$$H = -\sum_{i,j} J_{ij} S_i \bullet S_j \tag{1.3}$$

In many cases, it is sufficient to take into account the nearest-neighbor exchange interactions only, because the exchange interactions are short range. There are several alternative ways of proceeding with the calculation. Among them, the one most commonly used is the method based the Holstein–Primakoff transformation [6–8], the Fourier transformation, and the Bogoliubov transformation [9]. Spin operators in the Hamiltonian are transformed by the Holstein–Primakoff transformation to be in terms of boson operators. The Fourier transformation transfers the boson operators from a position to a wave vector representation. The Bogoliubov transformation eliminates the nondiagonal terms in the Hamiltonian to obtain the frequencies of the elementary excitations, i.e., the spin waves. Another method is based on the Green's function for solving the equations of motion of spins [10–12]. After defining the suitable Green's functions, within the Bogoliubov and Tyablikov decoupling approximation [10], one obtains the Fourier components of the Green's functions as well as the spin wave spectra. Using the spectral theorem and Callen's technique [12], one finally obtains the magnetization of each sublattice or layer. For simplicity, one may also apply the technique of Green's functions after the linear approximation.

It is well known that there are exchange interactions and dipole–dipole interactions, which may contribute to properties of spin waves in materials. The exchange interaction is short range, electrostatic in origin, because of the quantum mechanical exchange. Usually, one considers the exchange interaction between neighboring electronic spins or their corresponding magnetic moments. The exchange interactions predominantly determine the magnetic ordering and the static properties of magnetic materials. General accounts of exchange interactions can be found in most textbooks on magnetism [13–15]. Although an individual dipole–dipole interaction is typically very much smaller than the exchange interaction between neighboring sites, the influence of the dipole–dipole interactions becomes important for the dynamic properties (i.e., spin waves) at long enough wavelengths. The relative importance of the contributions from exchange interactions and dipole–dipole interactions is governed largely by the magnitude of the wave vector $\mathbf{q}$ of the spin wave [4]. Before the theories for spin waves in limited systems are introduced in detail, it is helpful to briefly discuss spin waves in bulk magnetic materials, for different regions of spin wave excitations in terms of the magnitude $|\mathbf{q}|$ their wave vector (as shown in Table I). Typically the exchange interactions are dominant for spin waves in the exchange region with $|\mathbf{q}| > 10^6$ cm^{-1}. This exchange region includes most of the Brillouin zone with a zone boundary wave vector of magnitude about 10^8 cm^{-1}. Actually, one may also relate to the angle α formed by adjacent spins. The exchange contribution to the spin-wave energy is dominant and the dipolar coupling between the spins can be neglected, if the angles α between the adjacent spins differ considerably from those of completely parallel or antiparallel configurations (such as $\alpha = 0$ for the ferromagnetic equilibrium state). The dipole–dipole interactions become significantly dominant for spin waves with smaller values of $|\mathbf{q}|$ (typically $< 10^5$ cm^{-1}) because they extend over

Table I. The Different Regions of Spin Wave Excitations in Terms of the Magnitude $|\mathbf{q}|$ of Their Wave Vector

Region	Wavevector range	Theory
Exchange region	$|\mathbf{q}| > 10^6 \text{ cm}^{-1}$	Microscopic discrete models
Dipole exchange region	$10^6 \text{ cm}^{-1} > |\mathbf{q}| > 10^5 \text{ cm}^{-1}$	Microscopic discrete models; macroscopic continuum models including the macroscopic dipolar field and the macroscopic effective exchange field
Magnetostatic region	$10^5 \text{ cm}^{-1} > |\mathbf{q}| > 30 \text{ cm}^{-1}$	Microscopic discrete models; macroscopic models with Maxwell's equations of electromagnetism, neglecting the effects of retardation
Electromagnetic region	$|\mathbf{q}| < 30 \text{ cm}^{-1}$	Microscopic discrete models; macroscopic models with full form of Maxwell's equations of electromagnetism, including terms of retardation

After [4], with appropriate theories. The numbers are approximate for ferromagnetic materials. For comparison, a Brillouin zone boundary wave vector is approximately of magnitude 10^8 cm^{-1}.

a longer range. In this long-wavelength limit ($\mathbf{q}a \ll 1$), the exchange contribution to the spin-wave energy is negligible because the angles α are very small ($\alpha \ll 1$). The dipole–dipole region, which can be treated either by a microscopic approach or by a macroscopic approach in terms of Maxwell's equations of electromagnetism, could be divided into two subregions, namely, the magnetostatic region and the electromagnetic region. The magnetostatic region typically corresponds to the wave vector range $30 \text{ cm}^{-1} < |\mathbf{q}| < 10^5 \text{ cm}^{-1}$. In this region, the inequality of $|\mathbf{q}| \gg \omega/c$ (ω is the frequency and c is the velocity of light) holds, so that the effects of retardation can be neglected. At smaller $|\mathbf{q}|$ the full form of Maxwell's equations, including retardation, must be used for analyzing spin waves in the electromagnetic region ($|\mathbf{q}| < 30 \text{ cm}^{-1}$). In this case, one obtains so-called magnetic polaritons, which are mixed states of magnons and photons. Although the dipolar–dipolar region can be treated microscopically, it is usually more convenient to follow the macroscopic approach. The dipole exchange region, typically corresponding to $10^5 \text{ cm}^{-1} < |\mathbf{q}| < 10^6 \text{ cm}^{-1}$, is an intermediate regime, in which the exchange terms and dipolar terms may be comparable. This crossover region is of particular interest from a theoretical point of view because the microscopic model of the exchange-dominated modes and the continuum model of magnetostatic waves can be unified to give consistent results. The region with comparable exchange and dipolar contributions can be treated within the macroscopic continuum approximation, because the corresponding wavelengths are long enough. This allows us to pass from the discrete to the continuum description and to include the dipolar–dipolar coupling by the macroscopic dipolar field and the exchange interaction by a macroscopic effective exchange field. The numbers for the typical wave vectors are approximate for ferromagnetic materials, which may vary for different magnetic materials [4]. The same regions can usually be identified for the limited magnetic systems, although the interplay between the exchange and the dipolar effects may sometimes be subtler as the surface geometries are investigated.

The purpose of this chapter is to give a comprehensive review introducing a theoretic investigation of the spin waves of thin films, superlattices, and multilayers. This article does not

focus on a review of the experimental data. There are several excellent reviews on the topic of spin waves in thin films, superlattices, and multilayers [4, 16–30]. Readers are referred to these review papers. For the spin waves in bulk materials, readers may refer to the corresponding reviews in the field [13–15, 20–23, 31]. Our interest is restricted to linear spin waves, not nonlinear ones. In the nonlinear region, the spin waves can exhibit a range of additional phenomena, including instabilities, auto-oscillations, chaotic behavior, and the formation of envelope solitons. Readers should refer to the corresponding review articles [32–34]. Theoretical studies of spin waves in thin films, superlattices, and multilayers are reviewed in Sections 2 and 3. Both the macroscopic phenomenological spin wave theory and the quantum microscopic spin wave theory are introduced. In Section 4, different methods used for spin waves in magnetostatic and exchange limits, the different spin wave spectra with respect to periodicity and different boundary conditions, and the character of spin waves in limited systems are discussed. Section 5 is for concluding remarks.

2. SPIN WAVES IN THIN FILMS

2.1. Introduction

The first experimental observation of ferromagnetic resonance in thin films was carried out in the later 1950s [35]. Since then, a great amount of interest has been devoted to spin waves in thin ferromagnetic films [36–47]. Different experiments such as magnetization measurement [42], ferromagnetic resonance [36], the Mössbauer effect [46], neutron scattering [45], and low energy electron diffraction (LEED) in antiferromagnetic samples [43, 47] have been performed to detect spin waves and their effects. A number of theoretical studies have dealt with the spin wave spectrum of thin films. Among them, there are macroscopic phenomenological theories and quantum microscopic theories, which are represented in Sections 2.2 and 2.3.

2.2. Macroscopic Phenomenological Theories

2.2.1. Dipolar Modes (Magnetostatic Modes)

Early in 1955, it was pointed out by Anderson and Suhl [48] that, for samples of finite size, the dispersion relation is shape dependent because of the effects of the demagnetizing field. Imposition of the usual magnetostatic boundary conditions at the surface of a spheroid was considered by Walker [49–51]; this leads to a characteristic equation for the mode frequencies and Walker's equation for the magnetostatic potential ψ inside the magnetic film, which is magnetized along the z direction:

$$(1 + \chi_1)\left(\frac{\partial^2 \psi^i}{\partial x^2} + \frac{\partial^2 \psi^i}{\partial y^2}\right) + \frac{\partial^2 \psi^i}{\partial z^2} = 0 \qquad (2.1)$$

This equation could be derived directly from Maxwell's equations (Eq. (1.2)) and the definition of the magnetostatic potential ψ (i.e., $h_d = \nabla\psi$) [4, 15, 52–54]. Outside the magnetic film, the magnetostatic potential satisfies the Laplace equation [16],

$$\nabla^2 \psi^e = 0 \qquad (2.2)$$

In Walker's theory, the characteristic distances involved are assumed to be so great that exchange effects may be neglected and to be short enough that propagation effects may also be ignored. The theory is thus truly magnetostatic in scope. It includes the shape of the sample, but not its size. Within this framework it was found that a large number of processional modes are possible in the motion of magnetization. The modes with symmetric and antisymmetric magnetostatic potential across the film thickness have been discussed in detail [4, 15, 54] for the case of static magnetization either perpendicular or parallel to film surfaces. The standing modes are often referred to as "Walker modes," with patterns much like those of normal vibration modes for elastic waves. Dillon [40] explained his observation of magnetostatic modes in thin disks of yttrium iron garnet (YIG) and manganese ferrite in terms of Walker's theory [49–51] for the magnetostatic modes in spheroids.

Actually, the existence and properties of elastic waves on the surface of a solid had been the subject of investigation since the time of Lord Rayleigh [55], who discussed the case of waves at the surface of a semi-infinite isotropic medium. These waves were characterized by an exponential decrease in displacement amplitude with increasing distance from the surface; they are customarily referred to as Rayleigh surface waves. Love [56, 57] treated another type of surface wave that can exist when a macroscopic layer of one material is supported by a substrate of another material. Seismologists such as Stoneley [58] and Press and Ewing [59, 60] were interested in the possible interpretation of seismic waves in terms of various types of surface waves in either a semi-infinite solid or a material made up of many layers. The existence of surface elastic waves in anisotropic media was studied by Stoneley [58], who showed that in cubic crystals Rayleigh-type surface waves exist only for certain values of the three elastic constants c_{11}, c_{12}, and c_{44}, but not for others. In parallel research areas, a detailed mathematical investigation of the effect of localized irregularities on

lattice vibrations was carried out by Montroll and Potts [61–63], who derived a correct term for the normal modes of the Debye frequency spectrum [64], and a theoretical work on semiconductors via Wannier wave functions was published by Koster and Slater [65, 66]. By considering the material as a lattice of interacting discrete particles rather a continuum, Lifshitz and Rosenzweig [67] employed a technique comparable to that used by Montroll and Potts [61–63] and found that two types of surface modes may exist in diatomic crystals, one analogous to Rayleigh waves and another derived from the optical branch and having no analogy in continuum theory, which was also found by Wallis [68]. Using Stoneley's theory [58], Gazis et al. carried out a theoretical investigation of surface elastic waves in cubic crystals [69, 70] and determined the range of elastic constants for which Rayleigh-type surface waves exist on a (100) free surface. Meanwhile, the problem of scattering of conduction electrons or phonons in metals by localized point imperfections as well as line and surface defects was treated by Seeger [71], Callaway [72], and Brown [73], using the Wannier-function–Green's-function technique of Koster and Slater [65, 66]. It was found that the localized perturbation must exceed a certain minimum strength to give rise to either bounds or scattering resonance in the electron or phonon spectrum [71–73]. There are other studies of surface waves in anisotropic media [74–84].

Turning again to magnetic thin films, Damon and Eshbach [52, 53] studied surface spin waves for ferromagnetic slabs and thin films in the limit of very long wavelengths, where the exchange interactions can be ignored and the dipolar interaction predominates. Following Walker's approach [49–51], they found that in this region, the surface wave frequency lies above that of the bulk modes, namely between $\omega = \gamma (B_i H_i)^{1/2}$ and $\omega = \gamma (H_i + 2\pi M)$, which corresponds to an imaginary part of the wave vector component. It was found that the complete spin wave spectrum consists of a set of surface spin waves in addition to the spin wave band usually considered. In the thin slab magnetized in its planes, the mode configuration clearly changes from a volume distribution to a surface wave as the frequency is increased above the extrapolated spin wave region. The backward volume wave is reciprocal in-plane propagation at any angle relative to the field and magnetization direction, whereas the surface wave is nonreciprocal in-plane propagation at angles greater than some critical angle relative to the field and magnetization. The mode density variation shows that the volume modes connect smoothly to the spin wave spectrum of a medium of infinite extent. Surface modes also exist, even at very short wavelengths, but become statistically less important as the wavelength decreases.

Wigen et al. [85] studied dynamic pinning in thin-film spin wave resonance. The essential feature of their model is that the effective internal field must differ by a small amount in the two regions. The difference in the magnetization of the surface layer, giving a slightly different demagnetization field, is only one way to produce the dynamic pinning effect. A different state of strain in the surface layer through the magnetoelastic effect or a change in an anisotropy field that is perpendicular to the film will produce the same results. Portis [86] proposed a generalization of the Wigen model in which the volume magnetization is nonuniform within the films and posited the low-lying spin wave mode around a parabolic maximum in magnetization, which are identical to the energy eigenfunctions of the harmonic oscillator in a certain approximation.

A modified formulation of the Damon–Eshbach theory of magnetostatic waves for in-plane magnetized isotropic or anisotropic films was presented by Hurben and Patton [87, 88]. Some new results, relative to the usual backward volume wave and nonreciprocal surface wave, were obtained. These included (1) the volume mode critical angle effect; (2) non-reciprocal volume mode properties; (3) volume mode profiles with quasi-pinning at the film surface, even in the case without explicit boundary conditions on the surface spins; (4) surface mode suppression; and (5) mode conversion and a reentrant mode character for the lowest order dispersion branch. They explained physically the out-of-band surface mode frequencies in terms of the dipole field for circularly polarized modes and demonstrated the continuous nature of the change from volume mode to surface mode character for near-uniform mode at low wave number due to competing dipole field effects [87]. They also discussed several critical propagation angles for the inverts or reverts between different mode dispersion branches [88].

2.2.2. Exchange Modes

Landau and Lifshitz [89] first showed how the exchange effect might be included in the equation of motion of the magnetization, and the solution of this equation for rather general boundary conditions was the subject of a detailed investigation by Macdonald [90–92]. The motion of the magnetization vector, M, in a ferromagnetic material may often be described by the well-known Landau–Lifshitz dynamical equation [89, 93, 94],

$$\frac{\partial M}{\partial t} = \gamma M \times H + \frac{2A\gamma}{M_s^2} M \times \nabla^2 M - \frac{\lambda}{M_s^2} M \times (M \times H) \quad (2.3)$$

where $\gamma = ge/2mc = g\mu_B/h$ is the magnetomechanical ratio, g is the Landé splitting factor, and μ_B is the Bohr magneton. $M_s = |M|$ is the saturation magnetization. The magnetic field, H, includes the applied field as well as the demagnetizing field. The term containing the exchange stiffness constant, A, measures the exchange torque density due to any nonuniformity in the orientation of M (the form of this form depends on the crystal structure of the material). The third term in Eq. (2.3) provides a slightly simplified phenomenological description of damping due to the relaxation effects. A general version of Eq. (2.3) would contain torque density terms due to magneto-crystalline anisotropy and stress, in accordance with various systems [90–92]. This equation is the origin of much work studying the spin waves in thin films as well as multilayers.

Macdonald [90–92] and Kittel and Herring [37–39] independently suggested that the exchange effects arising from the second term of Eq. (2.3) can play a role in ferromagnetic

resonance. The spin exchange effects in ferromagnetic resonance [93, 94] were treated by Macdonald [90–92] in the case of a plane-parallel metallic sheet of arbitrary thickness with material of arbitrary impedance abutting its rear surface. In his work, a classical treatment of the domain energy terms of a homogeneous ferromagnetic solid leads to a formula for the internal field contributions from these terms. Modifications in the resonance condition of the ferromagnetic resonance arising from self energy, exchange energy, magnetocrystalline anisotropy, and applied or intrinsic stress were obtained and then applied to various crystalline anisotropy and stress conditions of interest in ferromagnetic resonance experiments. Kittel dealt with the spin wave interpretation in a ferromagnetic thin film [37, 38] and demonstrated that it is possible to excite the exchange and magnetostatic spin waves in a ferromagnet with a uniform rf field. This can be valid provided that spins on the surface of the specimen experience anisotropy interactions are different from those acting on spins in the interior. Actually, the local symmetry of a spin at the surface is always lower than the symmetry of a spin in the interior, so that the terms in the energy, which vanish because of the symmetry at interior points, will not vanish at the surface. In this sense, the spin waves were studied by examining in detail the motion of the interior spins and the end spins $m = 1$ and N (in a line of N spins along the z axis). In [37, 38], the classical Hamiltonian density as well as the classical motion equation were used, but as Kittel suggested, a parallel atomic treatment may be given, starting from the spin wave theory with the Heisenberg exchange interaction for magnons.

In [4], the exchange-dominated spin waves were studied based on the equation for circularly polarized amplitudes of variable magnetization $m^\pm (= m^x \pm im^y)$ in the form

$$\left[\frac{d^2}{dz^2} + \lambda_\pm^2\right] m^\pm(r) = 0 \quad (2.4)$$

where

$$\lambda_\pm^2 = -\frac{\omega_H \pm \omega}{\alpha \omega_M} \quad (2.5)$$

with the boundary condition for the two in-plane components, m^x and m^y, of the variable magnetization.

2.2.3. Dipole Exchange Modes

At the present time there is experimental evidence that there are wide ranges of film thicknesses and wave vectors for which both dipole–dipole and exchange interactions must be included in the theoretical treatment [16, 95–101]. If both dipole–dipole and exchange interactions are taken into account, the solution of the boundary problem for spin wave modes in a ferromagnetic film is much more complicated than the preceding magneto-static or exchange limit. The dipole exchange spin wave spectra in films were calculated by many authors in the framework of the magnetostatic potential method [102–107] and the tensorial Green function method [108–111]. The exact results obtained by these methods are rather cumbersome and usually have to be analyzed numerically. However, it is possible to reach an explicit approximate dispersion equation by the latter method.

The authors of [24, 102, 104–107, 110, 112–129] have used a macroscopic continuum technique, based on the simultaneous solution of Maxwell's equations (without retardation) and the linearized Landau–Lifshitz equation of motion for magnetization, together with the usual electrodynamic boundary conditions and with exchange boundary conditions (the surface spin pinning conditions) [16]. As mentioned above, there are two main approaches adopted for solving eigenvalue boundary problems in the phenomenological dipole exchange theory [16]. One, named the partial wave (PW) approach, is based on a representation of the problem variables (for instance, the variable magnetization, magnetic field, or magnetic potential) inside the ferromagnetic film as a sum (or linear combination) of partial wave contributions, i.e., plane waves [24, 102, 104–107, 110, 112–123]. These plane waves are characterized by total wave vectors $\mathbf{q}$ consisting of two parts: the in-plane propagation vector $\mathbf{Q}$ and the quantized wave vectors $\mathbf{q}_n$ across the film. Taking the example that the ferromagnetic film is magnetized to saturation normally to its surface along the z axis and spin waves can propagate in the film plane xy, we introduce the 2D in-plane wave vector $\mathbf{Q} = (q_x, q_y)$ and the 2D in-plane position vector $\mathbf{R} = (x, y)$. Then we have $\mathbf{q}^2 = \mathbf{Q}^2 + \mathbf{q}_{zn}^2$. Introducing a scalar magnetic potential ψ such that $h_d = \nabla\psi$, we rewrite the initial system of Eqs. (1.1) and (1.2) after linearization as a sixth-order differential equation for ψ [16],

$$\left[F^2 + F - \frac{\omega^2}{\omega_M^2}\right]\nabla^2\psi - F\nabla_z^2\psi = 0 \qquad (2.6)$$

where the operator F is defined by

$$F = \frac{\omega}{\omega_M} - \alpha\nabla^2 \qquad (2.7)$$

and other parameters are $\omega_H = \gamma H_i$, $\omega_M = 4\pi\gamma M_0$. A time dependence like $\exp(i\omega t)$ is assumed. Owing to the property of translational invariance in the xy plane, we seek a solution of Eq. (2.6) inside the film in the form

$$\psi(r) = \psi_Q(z)\exp(-iQR) \qquad (2.8)$$

Solutions for the distribution of the magnetic potential across the film $\psi_Q(z)$ must be found as a combination of three pairs of waves,

$$\psi_Q(z) = \sum_{n=1}^{3}\left[A_n\exp(iq_{zn}z) + B_n\exp(-iq_{zn}z)\right] \qquad (2.9)$$

where the wave numbers q_{zn} are the roots of the characteristic bicubic equation

$$q^2\left[\left(\alpha q^2 + \frac{\omega_H}{\omega_M}\right)^2 - \frac{\omega^2}{\omega_M^2}\right] + Q^2\left[\alpha q^2 + \frac{\omega_H}{\omega_M}\right] = 0 \quad (2.10)$$

The investigation of q_{zn} as the roots of the characteristic equation and their dependence on the film parameters and on the direction and value of constant bias magnetic field is one of the major achievements of the PW approaches [102, 104, 106,

107, 114]. Outside the film the scalar magnetic potential obeys Laplace's equation. With the imposition of electrodynamic and exchange boundary conditions on the problem variables, a system of homogeneous linear algebraic equations for the unknown coefficients is found. The condition for the vanishing of the determinant of the system leads to the dispersion relation for the dipole exchange spin waves in the film. A similar study could be carried out for the case of parallel magnetization. Readers should refer to the original papers and/or the review articles [16, 102, 104, 106, 107, 114].

The other approach is called the spin wave mode (SWM) approach [16]. It is based on a solution of an integrodifferential equation for variable magnetization that follows from the initial system of equations consisting of the linearized equation of motion for the magnetization, Maxwell's equations of magnetostatics, and the electrodynamic boundary conditions [124–129]. Let us consider a ferromagnetic film of thickness L in the ξ direction with the surfaces corresponding to the planes $\xi = L/2$ and $\xi = -L/2$ and with infinite dimensions in the $\rho\xi$ plane [16]. The film is magnetized to saturation by a static external magnetic field in an arbitrary direction. The axis ζ is always assumed to be oriented along the direction of propagation of the spin waves:

$$m(\xi, \zeta, t) = m_Q(\xi)\exp\left[i(\omega t - Q\zeta)\right] \qquad (2.11)$$

The first step of the SWM approach is to solve Maxwell's equations of magnetostatics for a ferromagnetic film. A connection between the variable magnetization of Eq. (2.11) and the dipole field [16],

$$h(\xi, \zeta, t) = h_Q(\xi)\exp\left[i(\omega t - Q\zeta)\right] \qquad (2.12)$$

is found in the integral form,

$$h_Q(\xi) = 4\pi\int_{-\infty}^{\infty}\widehat{G}_{\xi\rho\zeta}(\xi, \xi')m_Q(\xi')\,d\xi' \qquad (2.13)$$

The tensorial Green's function in Eq. (2.13), defined in the coordinate system $\xi\rho\zeta$, is

$$\widehat{G}_{\xi\rho\zeta}(\xi, \xi') = \begin{bmatrix} G^P - \delta(\xi - \xi') & 0 & iG^Q \\ 0 & 0 & 0 \\ iG^Q & 0 & -G^P \end{bmatrix} \qquad (2.14)$$

where the components inside the film are

$$\begin{aligned} G^P &= \frac{Q}{2}\exp\left(-Q|\xi - \xi'|\right) \\ G^Q &= G^P\,\mathrm{sgn}\left(\xi' - \xi\right) \end{aligned} \qquad (2.15)$$

and the components outside the film are

$$\begin{aligned} G^P &= \frac{Q}{2}\exp\left(-Q|\xi' - \xi|\right) \\ G^Q &= -G^P\,\mathrm{sgn}(\xi) \end{aligned} \qquad (2.16)$$

Equations (2.13)–(2.16) give a solution of Maxwell's equations of magnetostatics satisfying electromagnetic boundary conditions. As the second step of the SWM approach, the integral form for the dipole field, Eq. (2.13), is substituted in

Table II.   Summary of Theoretical Study on Spin Waves in Thin Films, Based on Macroscopic Phenomenological Theories

Modes	Models and/or the shape of samples	Spin wave spectra and/or the topic or technique of the study	References
Dipolar modes	Finite size of samples	Shape dependence	[48]
Dipolar modes	A spheroid	Walker's modes	[49–51]
Dipolar modes	Thin disks of YIG and manganese ferrite	Walker's modes	[40]
Dipolar modes	Surfaces	Surface spin waves	[52, 53]
Dipolar modes	Thin films	Dynamic pinning	[85]
Dipolar modes	Wigen model with nonuniform magnetization	Low-lying mode with harmonic oscillator	[86]
Dipolar modes	In-plane magnetized isotropic films	Volume model critical angle effect	[87]
Dipolar modes	In-plane magnetized anisotropic films	Mode conversion and reentrant	[88]
Dipolar modes	Epitaxial Co films	Anisotropy effects	[222]
Exchange modes	Exchange effects in thin films	Landau–Lifshitz dynamic equations	[89]
Exchange modes	A plane-parallel metallic sheet	A classical treatment of the domain energy	[90–92]
Exchange modes	A ferromagnetic thin film	Anisotropy interaction of spins on the surface	[37–39]
Exchange modes	Thin films	A review	[4]
Dipole exchange modes	Partial wave approach/magnetostatic potential method	A sum (or linear combination) of partial wave contributions	[24, 102–107, 110, 113–123]
Dipole exchange modes	Spin wave mode approach/tensorial Green's function method	Maxwell's equations and electrodynamic boundary	[108–111, 124–129]
Dipole exchange modes	Thin films	A review	[16]

the linearized Landau–Lifshitz equation of motion, Eq. (1.1) for the magnetization. The vector of variable magnetization in Eq. (2.11) could be considered two-dimensional in the coordinate system xyz:

$$m_Q(\xi) = m^x(\xi)\hat{x} + m^y(\xi)\hat{y} = \begin{bmatrix} m^x \\ m^y \end{bmatrix} \qquad (2.17)$$

Then we write the integrodifferential equation for the Fourier amplitude Eq. (2.17) of the spin wave Eq. (2.11) as

$$\widehat{F} \cdot m_Q(\xi) = \int_{-L/2}^{L/2} \widehat{G}_{xy}(\xi, \xi') m_Q(\xi') \, d\xi' \qquad (2.18)$$

where the linear matrix–differential operator has four components,

$$\widehat{F} = \begin{bmatrix} F^{xx}, F^{xy} \\ F^{yx}, F^{yy} \end{bmatrix} \qquad (2.19)$$

where

$$\begin{aligned} F^{xx} = F^{yy} &= \frac{\omega_H}{\omega_M} + \alpha Q^2 - \alpha \nabla_\xi^2 \\ F^{yx} = -F^{xy} &= i\frac{\omega}{\omega_M} \end{aligned} \qquad (2.20)$$

G_{xy} in Eq. (2.18) is a two-dimensional part of matrix G_{xyz}, obtained from the matrix $G\xi\rho\zeta$ in Eq. (2.14) with the use of orthogonal transformations for rotations through angles ϕ and $(\pi/2 - \theta)$ [16, 128].

The integrodifferential equation (2.18) is the principal relation of the SWM approach. To solve the integrodifferential equation, an expansion of the variable magnetization in the infinite series of complete vector functions satisfying the exchange

boundary conditions (i.e., spin wave modes) is used. The natural form of the exact dispersion relation for dipole exchange spin waves is expressible in terms of a vanishing infinite determinant and an infinite convergent series. One of the useful features of the SWM approach is that it provides the possibility of deriving exact and approximate dispersion relations in various forms. One of the exact forms coincides completely with the dipole exchange dispersion relation of the PW approach, constructing a bridge between different theories. For detail discussion of the dipole exchange modes, readers should refer to the review article [16]. For convenience, Table II gives a summary of theoretical study on spin waves in thin films, based on macroscopic phenomenological theories.

2.3. Quantum Microscopic Theories

After the previous calculations for finite or semi-infinite isotropic elastic continua by Breger and Zhukhovitskii [130–132] and for a semi-infinite elastically isotropic plate of a solid by Stratton [133, 134] and Dupuis et al. [135], Maradudin and Wallis [136] presented a lattice-dynamical calculation of the surface contribution to the low-temperature specific heat of a crystal. The free boundary surfaces of a crystal were treated as a perturbation of an unperturbed crystal in which the atomic displacements satisfy the cyclic boundary condition [137]. Wallis et al. [138] examined the effect of a surface on some properties of the Heisenberg ferromagnet by employing an equation of motion technique and a certain Green's function. It was pointed out that in the low-temperature limit there are solutions to the equations of motions in which the spin deviation decays exponentially from the surface. These surface magnons are normal modes of the crystal in the presence of a surface, which will be thermally excited at a finite temperature and thus contribute to the specific heat. The presence of the surface magnons will cause the spin deviation to increase as the boundary is approached from within the crystal. Fillipov also showed that a surface spin wave branch would exist for the (100) surface of a simple cubic ferromagnet in the nearest-neighbor exchange model if the exchange parameter coupling spins on the surface were different from the bulk exchange [139]. Mills and Maradudin found that, in the low-temperature limit, half of the contribution of the surface spin waves to the specific heat is canceled by a decrease in the bulk spin wave part [140]. Mills and Saslow [141] investigated the effect of a free (100) surface on the low-temperature properties of a Heisenberg antiferromagnet of the CsCl structure. It was found that a surface magnon branch occurs in the excitation spectrum, with an excitation energy less than the frequency of a bulk magnon of the same wavelength. The surface modes may be observed in the infrared absorption spectrum of the material and will affect the low-temperature thermodynamic properties of the system. The influence of the surface on the infrared absorption spectrum, the specific heat, and the low-temperature form of the parallel susceptibility and the mean sublattice deviation were examined by means of a Green's function method.

Evidently, when the wavelength of the surface magnon is sufficiently short that the dominant contribution to the excitation energy comes from exchange interactions, the surface wave frequency lies below the bulk band. In the dipolar-dominated regime, the surface mode lies above the bulk band. Aware of this contrast, Benson and Mills [142–144] formulated a variational principle suitable for the discussion of long-wavelength spin waves in ferromagnets, which fully reproduces the bulk spin wave dispersion relation and the magnetostatic surface-mode dispersion relation for a semi-infinite geometry with magnetization parallel to the surface. They applied the variational principle to a discussion of the effect of exchange on the Damon–Eshbach branch [52, 53] and found a new surface branch lying between the bulk manifold and the Damon–Eshbach branch.

In a series of papers [120–122] on ferromagnetic resonance in thin films and disks, Sparks developed a theory for the frequencies, the linewidths, and the mode intensities of the ferromagnetic normal modes of these systems. The effect of an inhomogeneous interfield and saturation magnetization on the frequencies of high-order modes was considered. Effects of surface-spin pinning on the frequencies, the linewidths, and the mode intensities were discussed. Sparks also used a physical interpretation of the eigenvalue equation to obtain the normal mode frequencies (eigenvalue) and the spin configurations (eigenvectors) of ferromagnetic and antiferromagnetic surface waves [123].

De Wames and Wolfram investigated the properties of surface spin wave for a Heisenberg ferromagnet, using a Green's function method and keeping only nearest-neighbor interactions [145]. The exchange constant coupling spins parallel and normal to the crystal surface were allowed to differ from their bulk value. The problem was reduced to determining the roots of a cubic equation, from which the surface spin wave energies and eigenvectors were easily obtained. It was found that there is a mode with a maximum amplitude on the second layer of spins, in contrast to the mode discussed earlier, in which the amplitude has a maximum on the first layer. The dispersion curves may be truncated and may lie below or above the bulk spectrum, depending on the values of the exchange bonds at the surface.

Lévy and Motchane tried to find the exact eigenvalues and eigenvectors corresponding to the spin waves in a finite Heisenberg ferromagnet [146]. The presence of the surface was taken into account in the following two ways. (1) The exchange integral between the nearest neighbors is not the same on the surface and in the bulk. (2) There is surface anisotropy energy. For simplicity, the dipolar interactions and the bulk anisotropy were neglected and the spin Hamiltonian was treated in a quasi-harmonic approximation equivalent to the usual random phase approximation (R.P.A.). They found that the uniform mode is not generally an eigenmode, but is so only for this ideal Heisenberg ferromagnet, and the spin wave energy spectrum presents a gap that modifies the magnon-specific heat and magnetization at low temperatures. The former conclusion was in good agreement with the argument of Wigen et al. [85], Sokolov and Tavger [147, 148], and Puszarski [149]. Puszarski [149] proposed a quantum theory of spin wave resonance for arbitrary types of lattice, arbitrary orientation of the film surface, and arbitrary configuration of the static field. Lévy et al. [150] developed a Green's function decoupling method for spin 1/2 in the thin-film case, which is similar to that used by Bogoliubov and Tyablikov [10] for infinite samples. The spin wave spectrum and the localized magnetization were studied over a wide range of temperatures by the inclusion of a new term in the Heisenberg Hamiltonian the surface anisotropy coupling. The pinning effect and the appearance of surface spin waves were connected to the "surface condition" by means of a complex phase shift characteristic of the waves.

Hoffmann et al. [151] estimated the effect of dynamic pinning [152] due to two ferromagnetic layers upon the spin wave resonance in a sandwiched film with a higher magnetization, assuming the interface interaction energy to be of an energy type. Kooi et al. [152] discussed the intensities and frequencies of the spin wave resonance lines, the spin wave–phonon interaction, the critical angle and several other experiments, in relation to the surface-spin-pinning model and the volume inhomogeneity model for magnetic films. Davies studied the effect of nonuniform magnetization on the spin wave spectrum in thin ferromagnetic films [153].

Dobrzynski studied the phonon and magnon properties of the crystal surface [154]. The magnetic properties of a semi-infinite ferromagnetic crystal were considered in the presence of a surface pinning field, with the use of simple models and mathematical tools [154–156]. For a simple cubic ferromagnet in the shape of a film of thickness t with (100) surfaces [155, 156], constructed of n atomic layers, the Hamiltonian of a system in which the spins are coupled via nearest-neighbor exchange interactions J may be written as

$$H = -J \sum_{i,j} S_i \bullet S_j \qquad (2.21)$$

Equation (2.21) is reduced from Eq. (1.3) as $J_{ij} \equiv J$. Choosing and assuming that the full translational symmetry in the x and y directions remains in the presence of two surfaces, one seeks the solutions of the Bloch form in the x and y directions. Then the eigenfunction for a spin wave in the thin film can be written as

$$|\mathbf{k}_{\parallel\alpha}\rangle = \frac{1}{\sqrt{N_s}} \sum_{l_a} e^{i\mathbf{k}_{\parallel} \bullet l_a} f_\alpha(l_z) a^+(l)|0\rangle \qquad (2.22)$$

where $\mathbf{k}_{\parallel} = k_x \mathbf{x} + k_y \mathbf{y}$. The function $f_\alpha(l_z)$ was assumed to be taken as a linear combination of the two functions $e^{ik_z l_z a}$ and $e^{-ik_z l_z a}$, namely,

$$f_\alpha(l_z) = \alpha e^{ik_z l_z a} + \beta e^{-ik_z l_z a} \qquad (2.23)$$

where $k_z \geq 0$. The ratio of the constants α and β, as well as the possible values of k_z, is determined by requiring the solution in Eq. (2.22) to satisfy the equations of motion for the spins in the surface layers at $l_z a = 0$ and $l_z a = t$. In this way, the dependences of spin deviation with distance from the surface of the film for the first few low-lying modes were illustrated. A form for the surface-specific heat valid in the spin-wave regime was derived by employing a Green's function technique. The effect of a surface pinning field on the thermodynamic properties of the semi-infinite Heisenberg ferromagnet was considered [154]. It was pointed out that the differences between the thermodynamic quantities of finite and infinite systems occur not only because one may have normal modes of the system in which the excitation is localized near the surface, but also because the perturbation produced by the presence of the surface alters the distribution in frequency of the bulk modes [155, 156].

Corciovei et al. [19] reviewed various magnetic properties of ferromagnetic thin films, such as critical phenomena and phase transitions and elementary excitations, with an interest in saturation magnetization, Curie temperature, spin wave spectra, dispersion laws, and resonant excitations. Kittel succeeded in accounting for some basic types of magnetic domain configurations in thin films [157]. To study the spin waves, a simple model of thin films with pinned surfaces was discussed. The pinning mechanism was discussed by Kittel [37, 38] and Pincus [158] for linear chains of spins and used by others [86, 159–165] to account for thin film properties.

As usual, the spin wave analysis of the saturation magnetization at low temperatures can focus on the part $\mathbf{H}_{SW}$, which has a quadratic form in the bosonic variables, whereas all of the terms involving more than two bosonic operators $\mathbf{b}_{fp}$ and $\mathbf{b}_{fp}^+$ can be neglected. Here $p = 1, 2, \ldots, q$ labels the planar sublattices (monolayers), and f is a lattice vector in the plane of the sublattice. Because of the perfect periodicity in the film plane, one assumes that the bosonic operators $\mathbf{b}_{fp}$ and $\mathbf{b}_{fp}^+$ have a plane wavelike character and uses their Fourier expansions only in the plane of sublattices. Then the effective spin wave

Hamiltonian $\mathbf{H}_{SW}$ becomes a quadratic combination of bosonic amplitudes $\mathbf{b}_{pv}$ and $\mathbf{b}_{pv}^+$ of the form [19]

$$H_{SW} = \sum_v \sum_{p,p'} \left[\Gamma_{pp'}^v b_{pv}^+ b_{pv} + \tfrac{1}{2} \left(\Lambda_{pp'}^{v*} b_{pv}^+ b_{p',-v}^+ \right. \right.$$
$$\left. \left. + \Lambda_{pp'}^v b_{p,-v} b_{p'v} \right) \right] \qquad (2.24)$$

The coefficients $\Gamma_{pp'}^v$, $\Lambda_{pp'}^{v*}$, and $\Lambda_{pp'}^v$ follow from the composition of the model Hamiltonian, which includes plane Fourier transforms of the coupling parameter, like the exchange constant, the uniaxial anisotropy constant, and the dipolar interaction constant.

The diagonalization of quadratic forms likes Eq. (2.24) was widely studied by Bogoliubov and Tyablikov for bosons [166]. For ferromagnetic thin films without the quadratic anisotropic terms for axes perpendicular to the magnetization, the coefficients $\Lambda_{pp'}^{v*}$ and $\Lambda_{pp'}^v$ vanish, so that a simple form of a canonical transformation [19, 166]

$$b_{pv} = \sum_\tau u_{p\tau}^v \xi_{\tau v} \qquad (2.25)$$

can be written, where $\xi_{\tau v}$ and $\xi_{\tau v}^+$ are also bosonic variables. Their dynamical equation results in the eigenequations involving the spin-wave amplitudes $u_{\tau p}^v$,

$$E_{\tau v} u_{\tau p}^v = \sum_{p'=1}^q \Gamma_{pp'} u_{p'\tau}^v \qquad (2.26)$$

which obey the orthogonality rules

$$\sum_{p=1}^q u_{p\tau}^{v*} u_{p\tau'}^v = \delta_{\tau\tau'} \qquad (2.27)$$

The effective spin wave Hamiltonian can be diagonalized to the form

$$H_{SW} = \sum_\tau \sum_v E_{\tau v} \xi_{\tau v}^+ \xi_{\tau v}. \qquad (2.28)$$

The canonical transformation (2.25) indicates that the spin wave modes in the thin films depend on more than the plane wave vector v. A new index, τ, which is used to label the perpendicular standing component of the spin wave, is needed to diagonalize the spin wave Hamiltonian. The corresponding spectrum of the energy cannot be determined by the cyclic condition, but by the actual surface boundary conditions in the perpendicular direction, following from the model [19, 167–170].

Furthermore, Bailey and Vittoria deduced the existence of uniaxial magnetic surface anisotropy in permalloy films by comparing the theoretical and experimental spin wave resonance spectra of several films with different thicknesses [171]. Vittoria et al. formulated a unified theory for calculations of magnetoacoustic efficiencies and ferromagnetic transmission resonance for ferromagnetic conducting films [172]. Bhagat and Lubitz reported experimental results for ferromagnetic resonance in single-crystal specimens of the 3D transition metals [173]. The linewidth data were analyzed to obtain

Table III. Summary of Theoretical Study on Spin Waves in Thin Films, Based on Quantum Microscopic Theories

Modes	Models and/or the shape of samples	Spin wave spectra and/or the topic or technique of the study	References
Exchange modes/surface magnons	Heisenberg ferromagnet	Effect of a surface/Green's function	[138]
Exchange modes	A simple-cubic ferromagnet	(100) surface	[139]
Exchange modes	A Heisenberg antiferromagnet of CsCl–type structure	A free (100) surface	[141]
Magnetostatic surface mode	Ferromagnets	Effect of exchange	[142–144]
Ferromagnetic and antiferromagnetic surface waves	Thin films and disks	Effects of an inhomogeneous interfield and surface-spin pinning	[120–123]
Exchange modes/surface magnons	A Heisenberg ferromagnet	Green's function method	[145]
Exchange modes	A finite Heisenberg ferromagnet with anisotropic interaction or surface anisotropy energy	Presence of a gap in the spin wave spectrum/ a quasi-harmonic approximation	[146]
Exchange modes	Arbitrary type of lattice, orientation of surface and configuration of the state field		[149]
Exchange modes	Spin 1/2 in thin films	Pinning effect/Greens function decoupling method	[150]
Exchange modes	A sandwiched film with two ferromagnetic layers	Effect of dynamic pinning	[151, 152]
Exchange modes/surface magnons	Surface-spin-pinning model and volume inhomogeneity model for magnetic films	Spin-wave-phonon interaction and critical angle	[152]
Exchange modes	Ferromagnetic thin films	Effect of nonuniform magnetization	[153]
Exchange modes	A semi-infinite ferromagnetic crystal	Presence of surface pinning field	[154]
Exchange modes	A simple-cubic ferromagnetic film	(100) surface/Green's function technique	[155, 156]
Exchange modes	Thin films with pinned surfaces	Pinning mechanism	[86, 159–165]
Exchange modes	Permalloy films with uniaxial magnetic surface anisotropy		[171]
Exchange modes	Ferromagnetic thin films	Surface exchange integrals/Green's functions	[174–176]
Exchange modes	Ferromagnetic films	A review	[19]

the temperature dependence of the ferromagnetic relaxation frequency. With the use of a Green's function, Diep-The-Hung and Levy [174–176] calculated the critical temperature and temperature-dependent surface magnetization of ferromagnetic thin films as a function of surface exchange integrals.

The theoretical study of spin waves in thin films, based on quantum microscopic theories, is summarized in Table III.

In the late 1970s, Spronken et al. presented a method for calculating the power absorbed in the ferromagnetic resonance of a sample consisting of metallic multilayers with arbitrary directions of the applied static field [177, 178]. Their method may be used to predict the behavior of metallic multilayers with ferromagnetic or with antiferromagnetic interfacial coupling and to investigate resonance in metallic films with ferromagnetic surface layers or with inhomogeneous ferromagnetic surface layers. Theirs was pioneering work in the theoretical study of spin waves in superlattices and multilayers; papers on this subject have been appearing in great numbers. We deal with these closely related areas in the next section.

3. SPIN WAVES IN SUPERLATTICES AND MULTILAYERS

3.1. Introduction

Recently, layered composite materials such as multilayers and superlattices have become of great interest because magnetic properties of these composite materials may be distinctly different from those of their bulk counterparts. An exciting development in materials science in the early 1980s was the appearance of artificial, periodic layered material structure in which the separation between magnetic layers could be controlled to a precision on the order of intra-atomic distances [179–208]. The so-called magnetic superlattices are defined as periodic layered structures with alternating layers that have different magnetic and/or electrical properties. The magnetic properties of the layered structures have been found, by magnetization [179–181], neutron scattering [182, 192], ferromagnetic resonance [183, 185], Brillouin scattering [184], light scattering [186], and Mössbauer [192] and magnetoresistivity [201–203] measurements, to be different from those of a single magnetic layer. Spin waves in superlattices and multilayers have their own behaviors, which are different with those in bulk materials or even in thin films. As an elementary excitation, differences in spin wave spectra evidently affect magnetic properties, resulting in novel magnetic phenomena in superlattices and multilayers. We present the macroscopic phenomenological theories and the quantum microscopic theories, respectively, in Sections 3.2 and 3.3.

3.2. Macroscopic Phenomenological Theories

The spin waves of the superlattices or multilayers in both the dipolar limit and the exchange limit have been studied with the use of macroscopic phenomenological theories [187, 189, 211–235, 245–260].

3.2.1. Dipolar Modes

The nature of the spin wave spectrum of superlattice and multilayer systems has been studied theoretically [187, 188, 209, 211–215] within the framework of a description valid for modes whose eigenfunctions vary slowly on the scale of the lattice constant. In this approach the dominant contribution to the spin wave energy comes from dipolar and Zeeman energies and the exchange effects are ignored.

Camley et al. [209] presented an analysis of the spin wave spectrum of a semi-infinite stack of ferromagnetic layered films, within a formalism that includes the Zeeman and dipolar energies. Two features unique to the stack were found to appear in the spectrum. Each film, in isolation, possesses surface spin waves on its boundaries (Damon–Eshbach waves [52, 53]). In the layered geometry these interact to form a band of excitations of the array, which has a nonvanishing component of the wave vector normal to the stack. The position of the peak in the spectrum associated with scattering from this band of modes is controlled by dispersion introduced by interfilm interactions. Under certain conditions, the semi-infinite stack possesses a surface spin wave, the eigenfunction of which is a linear superposition of individual film states, with an amplitude that decays to zero as one moves from the surface into the interior. The giant magnetoresistence (GMR) effect was interpreted by Camley, Barnaś, and their co-authors by solving the Boltzmann transport equation with spin-dependent scattering at the interfaces [210, 211].

The spin wave spectrum of a ferromagnetic multilayer was derived by Grünberg and Mika [212–215] in the dipolar-magnetostatic limit, either for the in-plane magnetization or for the alternating directions of magnetization. The theory for the spin wave modes of a ferromagnetic single layer by Damon and Eshbach [52, 53] was extended to double-layer as well as multilayer systems. The existence of an additional singular solution depending on the ratio of the thicknesses of magnetic and nonmagnetic layers was found in the limit of an infinitely large number of magnetic layers. The qualitative differences for an even and an odd number of single layers that constitute the multilayer were also found. The solution was classified as bulklike or surfacelike with respect to the whole multilayers.

Magnetostatic calculations were performed for infinite and semi-infinite superlattices with alternating ferromagnetic/nonmagnetic or antiferromagnetic/nonmagnetic layers in the case of the static magnetization perpendicular to the surface [216]. The spectrum of bulk and surface modes turns out to be drastically different from the case of magnetization parallel to the surface. For this orientation of the static magnetization (or sublattice magnetization), there can be many surface modes of the semi-infinite superlattice. Good agreement between theory and experiment was obtained for magnetostatic surface wave propagation in an yttrium-iron-garnet double layer [217].

The magnetostatic modes of a lateral antiferromagnetic/nonmagnetic superlattice were studied for two different geometries [218]. Comparing with common antiferromagnetic/nonmagnetic superlattices, the surface-mode features appear, and especially for $f_1 < 0.5$ (f_1 is the magnetic fraction in the superlattice), the surface magnetostatic modes can still be seen.

Wang and Tilley also studied the magnetostatic modes of a lateral ferromagnetic/nonmagnetic superlattice in a transverse field [219]. Li et al. discussed magnetostatic modes and the consistency condition for the existence of surface waves in semi-infinite magnetic/nonmagnetic superlattices in an arbitrary-angle magnetization geometry [220]. The surface and bulk spin waves in a thick bcc cobalt film were found to have comparable frequencies and are strongly mixed [221]. Exchange and anisotropy effects on spin waves in epitaxial Co films were investigated by Grimsditch et al. [222]. Spin wave resonance studies of polycrystalline and epitaxial permalloy-chromium layered structures were carried out by Schmool et al. [223].

Aubry [224] studied the devil's staircase [225] and order without periodicity in classical condensed matter. The existence of incommensurate structures proves that a crystalline ordering is not always the most stable one for nonquantum matter. Some properties of the structure, obtained by minimizing a free energy, were investigated by the Frenkel–Kontorova [226] and related [227] models. It was shown that an incommensurate structure could be either quasi-sinusoidal with a phason mode or built out of a sequence of quasi-distant defects (discommensurations) that are locked to the lattice by the Peierls force. In that situation, the variation of the commensurability ratio with physical parameters forms a complete devil's staircase, with interesting physical consequences.

Feng et al. studied the magnetostatic modes in a Fibonacci multilayer consisting of alternating magnetic and nonmagnetic layers [228]. The constant of motion, depending on both the in-plane wave vector and the frequency of the mode, was explicitly obtained and used to describe general features of the frequency spectra. Spin wave spectra and procession amplitudes of magnetization for a finite Fibonacci multilayer were numerically calculated by the transfer matrix method. For a given in-plane wave vector, the distribution of frequency exhibits a triadic Cantor structure with large gaps in the low-frequency region, small gaps in the high-frequency region, and many "isolated" modes in the gaps. The gaps strongly depend on the in-plane wave vector and the thicknesses of the magnetic and nonmagnetic layers. Three types of states exist in the quasi-periodic direction: extended states in the high-frequency region near the upper band edge, critical states in the triadic subbands, and surface modes in gaps. Besides the conventional Damon–Eshbach surface mode localized at two opposite surfaces of the multilayer for positive and negative wave vectors, respectively, another kind of surface mode was found, which are still localized at the same sides of the multilayer, even when the wave vector is reversed [228].

The localization and scaling properties of magnetostatic modes in quasi-periodic magnetic superlattices of Fibonacci, Thue–Morse, and period-doubling sequences were studied by Anselmo et al. for a geometry where the magnetization is perpendicular to the interfaces of the superlattices [229]. Ignatchenko et al. studied the spectrum and damping of the spin waves for a periodic multilayer structure with a randomly modulated period [230]. The energy gap in the spectrum, which is characteristic for periodical structures, transforms into an inflexion of the dispersion curve with increasing random modulation. Han investigated spin wave modes in inhomogeneous ferromagnetic films at an arbitrary magnetic field direction [231].

A theory of periodic multilayers consisting of two alternating ferromagnetic materials with different transition temperatures was presented by Schwenk et al. [251], based on an inhomogeneous Ginzburg–Landau functional where the Ginzburg–Landau coefficients were chosen to model the alternating layers and interface interactions. Meanwhile, the spin wave spectrum of layers of ferromagnetic thin films separated by nonmagnetic layers was derived on the basis of linearized micromagnetic equations by Vayhinger and Kronmüller [252]. For a typical anisotropic magnetic superlattice, magnetic polaritons and magnetostatic spin waves in the Voigt geometry were analyzed by Zhou and Gong [253, 254], in the framework of the vector potential representation. Special emphasis was placed to the frequency window of the polaritons characteristic of this superlattice.

Thibaudeau and Caillé [232] studied the interface and guided magnetic polaritons for a three-layer system composed of a semi-infinite ferromagnet and a semi-infinite antiferromagnet separated by a dielectric layer of finite thickness. The single interface between a ferromagnet and an antiferromagnet was studied as a limit case. Inversion of the magnetostatic limit of the interface polaritons with the bulk magnetic polaritons was predicted. The evanescent and guided modes in the case of the two interfaces were predicted, and their characteristics were investigated. Emtage and Daniel [233] studied the character of magnetostatic waves and spin waves in layered ferrite structures and found an excitation similar to the surface wave on a continuous medium and a dense continuum of other excitations that is related to a system of modes localized on the gaps.

Camley and Stamps [234] reviewed the study of long-wavelength dipolar spin waves in the magnetostatic limit in magnetic multilayered structures using the Green's function method. The Green's function method was used to examine the dynamic excitations of magnetic systems whose geometry includes one or more surfaces or interfaces. They illustrated some basic techniques for calculating the response functions for both a semi-infinite homogeneous material and a semi-infinite superlattice. The Green's function equation was defined, according to the torque equations governing the spin motion (i.e., the homogeneous equations of motion for the magnetic scalar potential). The equations for the Green's function were solved in the presence of boundaries by the following processes: (1) finding a particular solution to the inhomogeneous equation in the material; (2) finding solutions to the homogeneous equation both inside and outside the material; and (3) matching the complete solutions (inside it is the sum of the particular and homogeneous solutions) at the boundaries. They also illustrated the application of Green's functions with two different problems: the nonlinear mixing of bulk and surface magnetostatic spin waves and calculations of Brillouin scattering intensities for light scattering from spin wave excitations in semi-infinite ferromagnets and superlattices.

Tilley [235] gave an overview of spin wave polaritons in films and multilayers, for the region of small wave vector $q \sim \omega/c$, in which retardation effects are important and Maxwell's equations in their full form must be used. The surface polaritons, i.e., electromagnetic modes localized at the plane interface between two media, were discussed, and the discussion was extended to the case of film polaritons and superlattice polaritons. Green's functions for magnetic polaritons in superlattices were studied [236]. When retardation effects become important, an effective medium approach to magnetic superlattices is a useful simplification [237]. This method depends on showing that the superlattice behaves, in some sense, like a single medium but with an effective (or averaged) permeability tensor. For a detailed description of magnetic polaritons, readers should refer to the review article [235].

To have a better understanding of spin waves in superlattices, one may also compare the results obtained for their analogous waves, such as elastic waves in polytype superlattices [238–244].

3.2.2. Exchange Modes

Spin wave spectra of exchange-coupled epitaxial double layers of bcc ferromagnetic materials with (100) and (110) surfaces were analyzed theoretically by Barnaś and Grünberg [245] within the classical continuum model. The spin wave spectrum is reciprocal in the case of parallel alignment of the single-film magnetization and strongly nonreciprocal in the antiparallel case. The special cases of a magnetic field applied along the easy and hard directions were discussed in detail. Results of the numerical calculations were compared with available experimental data obtained by Brillouin light scattering in Fe/Au/Fe and Fe/Cr/Fe structures. Vohl et al. [246] investigated the spin wave spectra of dipolar and exchange-coupled double-layer systems both theoretically and experimentally. Their theoretical formalism is based on a macroscopic description using the Landau–Lifshitz [89] equation of motion and the Rado–Weertman [93, 247] and Hoffmann [248] boundary conditions.

It has been found that the spin wave spectrum of a double-layer structure can be strongly sensitive to the interlayer exchange coupling.

Vittoria [249] suggested two mathematical approaches by which ferromagnetic resonance (FMR) fields and line shapes may be calculated. The first approach [177, 178] extends previous magnetic-metal-film calculations [250] to the magnetic layered problem. This approach matches the number of boundary surface conditions to the number of internal-field amplitudes. The surface impedance may then be expressed in terms of internal-field amplitudes. An alternative approach is to introduce a transfer-function matrix that relates the microwave fields at the two surfaces of a given magnetic layer [249]. The surface impedance of a layered structure is then expressed in terms of the four transfer-function matrix elements corresponding to the surface fields of the layered structure instead of internal microwave fields corresponding to all of the layers within the layered structure.

The magnetic properties of bilayers of bcc Fe (001) separated by bcc Cu(001) films of variable thickness were investigated by Cochran et al. [255], by means of Brillouin light scattering. The experimental data were compared with the results of a model of the exchange coupling or the dipole–dipole coupling between the two thin films. Furthermore, a theory was given for energy dissipation in a sliding crystal surface, by Sokoloff [256], using the equation of the vibrational motion.

3.2.3. Dipole Exchange Modes

The ground-state spin configuration of antiferromagnetically coupled ferromagnetic thin films was determined by competition between anisotropies, interlayer exchange, and the applied field. A theory for spin waves was presented by Stamps [257], in which both exchange and dipolar interactions were accounted for. It was shown how dramatic changes in the spin wave frequencies could be observed by carefully choosing the experimental geometry. Many details of the ground-state spin configuration were obtained by studying the behavior of the spin wave frequencies as functions of propagation direction and applied field strength in these geometries. The results were summarized from the point of view of possible light-scattering experiments.

Kostylev et al. [258, 259] developed a rigorous theory of the parallel pump instability threshold of spin waves in ferromagnetic films, taking into account dipole and exchange interactions, the dipole–dipole hybridization of spin wave branches, and magnetocrystalline and induced anisotropies. The form of the threshold curve for ferromagnetic films depends strongly on the film thickness, film saturation magnetization, and the direction and the value of the internal static magnetic field of the film, which is not universal, as it is for bulk materials. This behavior is due to the discreteness of the spin wave spectrum and the peculiarities of the spin wave polarization in ferromagnetic films. Expressions for the parallel pump instability threshold obtained in a general form can be used to calculate the threshold of a single ferromagnetic layer or a multilayered

structure containing an arbitrary number of ferromagnetic films. The minimum threshold field at a given frequency corresponds to the generation of spin wave pairs with maximum ellipticity of polarization. In the case of perpendicularly (or tangentially) magnetized films, the maximum ellipticity is inherent for short (or long)-wavelength ($kL \gg$ (or $\ll$) 1) spin waves. They also discussed the influence of the dipole–dipole hybridization of parametrically generated spin waves and of the nonreciprocal propagation properties. The dispersions of dipolar and exchange-dominated spin waves were calculated for in-plane-magnetized films with a (111) crystal orientation [260].

For convenience, a summary of the theoretical study on spin waves in superlattices and multilayers, based on macroscopic phenomenological theories, is tabulated in Table IV.

3.3. Quantum Microscopic Theories

A number of investigations have focused on the spin waves of the superlattices or multilayers in the exchange limit, with the use of various quantum microscopic theories [18, 261–318].

3.3.1. Green's Function Method

Various methods developed for layered structures are based on Green's function calculations, known as surface Green's function matching [261] and interface response theory [18]. Dobrzynski [18] presented a general theory of "interface responses" in discrete composite d-dimensional systems for operators with two-body interactions. The interface responses of all of the internal and external interfaces of any composite system are linear superposition of the responses to a coupling operator of all individual interfaces and of the responses to a cleavage operator of the corresponding ideal free surfaces of the same but noninteracting subsystems.

Herman et al. [262] investigated the electronic and magnetic structures of ultrathin cobalt–chromium superlattices. Lattice-matched Co/Cr superlattice models were constructed to study the exchange coupling and spin distributions at atomically abrupt ferromagnetic/antiferromagnetic interfaces. The superlattice spin distributions were determined by carrying out first-principle self-consistent spin-polarized linearized muffin-tin-orbital electronic-structure calculations.

Dobrzynski et al. [263] presented a theoretical Green's function study of the bulk and surface magnons of a semi-infinite stack of two different ferromagnetic films, within an atomic Heisenberg model, including exchange effects between the first nearest neighbors and neglecting dipolar and Zeeman energies. The superlattice has a larger periodicity in the direction perpendicular to the slabs and therefore many magnon branches in the folded Brillouin zone. The surface-localized modes may appear within the extra gaps between these folded bulk bands. Hinchey and Mills [264] studied theoretically the basic magnetic response characteristics of superlattice structures formed from alternating layers of ferromagnetic and antiferromagnetic materials, each described through the use of a localized spin

Table IV. Summary of Theoretical Study of Spin Waves in Superlattices and Multilayers, Based on Macroscopic Phenomenological Theories

Modes	Models and/or the shape of samples	Spin wave spectra and/or the topic or technique of the study	References
Dipolar modes	A semi-infinite stack of ferromagnetic layered films	A band of excitations of array/a surface spin wave	[209]
Dipolar magnetostatic limit	A ferromagnetic multilayer for alternating directions of magnetizations	Qualitative differences for an even and an odd number of single layers	[212–215]
Magnetostatic modes	Infinite and semi-infinite superlattices with alternating magnetic/nonmagnetic layers	Magnetization perpendicular to the surface/many surface modes	[216]
Magnetostatic modes	A lateral antiferromagnetic/nonmagnetic superlattice	Two different geometries	[218]
Magnetostatic modes	A lateral ferromagnetic/nonmagnetic superlattice	In a transverse field	[219]
Magnetostatic modes	Semi-infinite magnetic/nonmagnetic superlattices	In an arbitrary-angle magnetization geometry	[220]
	A thick bcc cobalt film	Strongly mixed surface and bulk modes	[221]
Extended states, critical states, and surface modes	Fibonacci multilayers	Transfer matrix method/Cantor structure with gaps and isolated modes	[228]
Magnetostatic modes	Quasi-periodic superlattice of Fibonacci, Thue–Morse, and period-doubling sequences	Magnetization perpendicular to the interfaces	[229]
Magnetostatic modes	Periodic multilayer with randomly modulated period	Energy gap in the spectrum/damping of spin waves	[230]
Magnetostatic modes	Inhomogeneous ferromagnetic film	At arbitrary magnetic field direction	[231]
Magnetostatic modes	Periodic multilayers consisting of two alternating ferromagnetic materials	Inhomogeneous Ginzburg–Landau functions	[251]
Magnetostatic modes	Ferromagnetic layers separated by nonmagnetic layers	Linearized micromagnetic equations	[252]
Magnetostatic modes and magnetic polaritons	Anisotropic magnetic superlattices in the Voigt geometry	Vector potential representation/frequency window of the polaritons	[253, 254]
Interface and guided magnetic polaritons	A three-layer system with semi-infinite ferromagnet and antiferromagnet separated by a finite dielectric layer	Evanescent and guided modes/inversion of the magnetostatic limit of interface polaritons with bulk polaritons	[232]
Magnetostatic modes	Layered ferrite structures	An excitation similar to the surface wave and a dense continuum of other excitations localized on the gap	[233]
Magnetostatic limit	Magnetic multilayered structures	A review of the Green's function method	[234]
Spin wave polaritons	Films, superlattices, and multilayers, in full form of Maxwell's equations	A review for retardation effects, surface polaritons	[235]
Spin wave polaritons	Superlattices	Green's functions for magnetic polaritons	[236]
Spin wave polaritons	Magnetic superlattices	An effective-medium approach	[237]
Exchange modes	Double layers of bcc ferromagnetic materials with (100) and (110) surfaces/ compared with Fe/Au/Fe and Fe/Cr/Fe	Classical continuum model/reciprocal and nonreciprocal modes	[245]
Dipolar and exchange-coupled modes	Double layers with Rado–Weertman and Hoffmann boundary conditions	Landau–Lifshitz equation of motion/ sensitive to the interlayer exchange	[246]
Exchange modes	Magnetic layers	Approach to match the boundary surface conditions with internal-field amplitudes	[177, 178, 249]
Exchange modes	Magnetic layers	Transfer-function matrix approach	[249]
Exchange and dipolar modes	Bilayers of bcc Fe (011) separated by bcc Cu (001) films		[255]
Dipole exchange modes	Antiferromagnetically coupled ferromagnetic thin films	Ground-sate spin configurations/competition between anisotropies, interlayer exchange and applied field	[257]
Dipole exchange modes	Ferromagnetic films	Parallel pump instability threshold of spin waves/discreteness of the spectrum	[258, 259]
Dipole exchange modes	In-plane-magnetized films with (111) crystal orientation		[260]

model. The geometry explored is one in which the antiferromagnet consists of sheets parallel to each interface within which the spins are aligned ferromagnetically. The study of the classical (mean-field) ground state as a function of magnetic field showed that a sequence of spin-reorientation transitions occurs, particularly for superlattices within which the antiferromagnetic constituent consists of an even number of layers. They presented calculations of the spin wave spectrum and the infrared absorption spectrum for the various phases.

A system of localized spins interacting by periodically modulated exchange interactions was studied by Morkowski and Szajek [265]. The spin wave spectrum was determined by solving numerically equations of motion for the magnon Green's function. It was found that the Bloch law for the low-temperature dependence of magnetization is obeyed by the model with coefficients weakly depending on the modulation wave vector. Chen et al. used the tridiagonal matrix method and the random-phase approximation to obtain a closed form of the Green's functions for two different semi-infinite ferromagnets coupled across an interface [266]. These Green's functions were then used to calculate the interface spin wave modes and the localized modes associated with an isolated impurity at or near the interface. Various effects are exhibited on the modes because of the interplay between the bulk and interface modes and the strength of the impurity-related interactions [266].

Wei et al. also studied the spin waves of layered Heisenberg ferrimagnets [267] and the disordered ground-state properties of a double-layer Heisenberg antiferromagnet [268]. In [268], they considered a layered Heisenberg ferrimagnetic model on a simple-cubic lattice with intralayer lattice parameter a and interlayer lattice parameters a and c. The Hamiltonian is as in Eq. (1.3), but with $J_{ij} = -J$ if sites i and j are in the same layer and $J_{ij} = -J_\perp$ if sites i and j are in two nearest-neighbor layers. To analyze the layered Heisenberg ferrimagnets, according to Callen [12], the Green's functions are introduced as [268]

$$G_A(E, i_1, i_2) = \langle\langle S_{i_1}^+ | \exp(p S_{i_2}^z) S_{i_2}^- \rangle\rangle_E \quad (3.1)$$

$$F_A(E, j_2, i_2) = \langle\langle S_{j_2}^- | \exp(p S_{i_2}^z) S_{i_2}^- \rangle\rangle_E \quad (3.2)$$

$$G_B(E, j_1, j_2) = \langle\langle S_{j_1}^+ | \exp(q S_{j_2}^z) S_{j_2}^- \rangle\rangle_E \quad (3.3)$$

$$F_B(E, i_2, j_2) = \langle\langle S_{i_2}^- | \exp(q S_{j_2}^z) S_{j_2}^- \rangle\rangle_E \quad (3.4)$$

where p and q are parameters. For convenience, it is necessary to introduce the transformation $S_j^z \to -S_j^z$, $S_j^+ \to S_j^-$ and $S_j^- \to S_j^+$, as done by Cheng and Pu [269]. Using the technique of equation of motion for the Green's functions, within the Bogoliubov and Tyablikov decoupling approximation [10], one obtains the Fourier component of the Green's functions [268],

$$G_A(E, k) = \frac{F_A(p)}{E_A^+ - E_A^-} \left(\frac{E_A^+ + 4J(2+\delta)\langle S^z \rangle_A}{E - E_A^+} - \frac{E_A^- + 4J(2+\delta)\langle S^z \rangle_A}{E - E_A^-} \right) \quad (3.5)$$

$$G_B(E, k) = \frac{F_B(q)}{E_B^+ - E_B^-} \left(\frac{E_B^+ + 4J(2+\delta)\langle S^z \rangle_B}{E - E_B^+} - \frac{E_B^- + 4J(2+\delta)\langle S^z \rangle_B}{E - E_B^-} \right) \quad (3.6)$$

where

$$F_A(p) = \langle[S_i^+, \exp(p S_i^z) S_i^-]\rangle \quad (3.7)$$

$$F_B(q) = \langle[S_j^+, \exp(q S_j^z) S_j^-]\rangle \quad (3.8)$$

$$E_A^\pm = 2J(2+\delta)\langle S^z \rangle_A \{(\alpha - 1) \pm [(1-\alpha)^2 + 4\alpha(1-\eta_k^2)]^{1/2}\} \quad (3.9)$$

$$E_B^\pm = 2J(2+\delta)\langle S^z \rangle_A \{(1-\alpha) \pm [(1-\alpha)^2 + 4\alpha(1-\eta_k^2)]^{1/2}\} \quad (3.10)$$

$$\eta_k = \frac{\cos(k_x a) + \cos(k_y a) + \delta \cos(k_z c)}{2 + \delta} \quad (3.11)$$

$$\delta = \frac{J_\perp}{J} \quad (3.12)$$

$$\alpha = \frac{\langle S^z \rangle_B}{\langle S^z \rangle_A} \quad (3.13)$$

Here $\langle S^z \rangle_A$ and $\langle S^z \rangle_B$ are sublattices magnetizations per site in sublattices A and B, respectively. From Eqs. (3.9) and (3.10), one has that $E_A^+ = -E_B^-$ and $E_B^+ = -E_A^-$; the spin wave spectrum has two branches for ferrimagnets. When $k \to 0$, $E_A^+ = -E_B^- \to 0$, which represents the acoustic branch, whereas $E_B^+ = -E_A^-$ remains finite and represents the optical branch. Using the spectrum theorem and Callen's technique [12], one finally obtains the magnetization of each sublattice. Wei et al. also applied the Green's function technique after performing the liner spin wave approximation to calculate the spin wave spectra of layered Heisenberg ferrimagnets, and similar results were obtained [267, 270]. Qiu and Zhang dealt with the spin waves in the multisublattice system and the corresponding superlattices with an elementary unit in which the number of spins is larger than two [271]. The effects of the competition between the exchange constants with different signs on the spin wave spectra and the physical properties of the multisublattices and/or superlattices were discussed.

Magnetic properties, such as spin wave excitations and the phase transition, of antiferromagnetic superlattices at finite temperatures were investigated theoretically by Azaria and Diep [272–274], with a multisublattice Green's function technique that takes into account the quantum nature of Heisenberg spins. The results of various physical quantities and magnetic transition temperature were shown for different sets of intraplane and interplane interactions. Zero-point spin contractions and the crossover of layer magnetization at low temperatures due to quantum fluctuations were found, which have no counterpart in ferromagnetic superlattices. Monte Carlo simulations of the corresponding classical Heisenberg model were also performed to compare with the quantum case.

A general recursion method was developed by Mathon [275] for calculating the exact local spin wave Green's function in arbitrary ferromagnetic interfaces, superlattices, and disordered layers. The complete response function of an arbitrary layer magnetic insulator structure can be generated from a single matrix element of the Green's function in the surface plane of a magnetic overlayer. Surface spin waves on the (001) surface layer of a semi-infinite CsCl-type (bcc) Heisenberg ferrimagnet with a single-ion anisotropy (uniaxial and nonuniaxial) were studied by Shen and Li [276], with the Green's function technique.

3.3.2. Transfer Matrix Method

A superlattice consisting of alternating layers of two simple-cubic Heisenberg ferromagnets was considered by Albuquerque et al. [277], who showed that the transfer matrix method leads to a compact expression for the spin wave dispersion relation of the magnetic superlattice. It was argued that a spin that is not in an interface layer has the same nearest-neighbor environment and therefore the same equation of motion as a spin in the corresponding bulk medium. The spin wave amplitudes were given within each component by a linear combination of the positive- and negative-going solutions for the bulk medium. For a traveling wave in the superlattice, the wave vectors in the x and y directions must be real, whereas those in the z direction are only real when the frequency lies within a pass band for the corresponding bulk mode and can be either imaginary or complex outside the pass band.

General dispersion equations were derived by Barnaś [278–281] for exchange, magnetostatic, and retarded waves in infinite and semi-infinite ferromagnetic superlattices, with the transfer matrix method. Magnetostatic modes and magnetic polaritons in the Voigt configuration were discussed. In these papers, the elementary unit of the superlattices was taken to be of N different ferromagnetic layers. The equations for bulk waves were expressed through the diagonal elements of the $\mathbf{T}$ matrix, whereas those for surface waves were expressed through the $\mathbf{T}$ matrix and a column matrix $\mathbf{D}$ that describes surface boundary conditions. Following the papers by Barnaś [29, 278–281] and Lévy et al. [282, 283], we describe the transfer matrix method here in detail.

We consider a model simple–cubic layered structure with ferromagnetic exchange coupling between nearest neighbors and elementary units composed of N_A and N_B atomic planes of materials A and B. The elementary unit is repeated periodically along the x axis of the coordinate system. The Hamiltonian of the system in the exchange–dominated region can be written as [29]

$$H = -\frac{1}{2} \sum_l \sum_\rho \sum_{\delta_\parallel} J_l S_{l,\rho} \bullet S_{l,\rho+\delta_\parallel}$$
$$- \sum_l \sum_\rho J_{l,l+1} S_{l,\rho} \bullet S_{l+1,\rho} + \mu_B H_0 \sum_l \sum_\rho g_l S_{l,\rho}^z \quad (3.14)$$

where l denotes the spins in the elementary unit in the x direction. δ indicates that only the exchanges between the nearest neighbors are taken into account, and $\delta_\parallel$ is the position vector to the nearest neighbors within one plane parallel to the y–z

plane of the coordinate system. J_l is the exchange constant of the spins within the lth atomic plane, which is equal to J_A (J_B) within A (B) material. $J_{l,l+1}$ is the exchange integral for the spins lying in the lth and $(l + 1)$st atomic planes, which is assumed to equal J_I only between the two materials A and B. Otherwise it equals J_A (or J_B) within material A or (B). A time and two-dimensional space Fourier transformation,

$$S_{l,\rho}^+(t) = \frac{1}{(2\Pi)^3} \int dE \int S_l^+(q, E) \exp\left[i\left(q \bullet \rho - \frac{Et}{\hbar}\right)\right] dq \quad (3.15)$$

could be introduced for the spin raising operators $S_{l,\rho}^+$ [29], where q is a two-dimensional wavevector in the plane yz. The quantum mechanical equation of motion for the spin raising operators,

$$i\hbar \frac{\partial S_{l,\rho}^+(t)}{\partial t} = \left[S_{l,\rho}^+, H\right] \quad (3.16)$$

determining the dynamical properties of the system, leads to the equation for the spin wave amplitudes $u_l = S_l^+(q, E)|0\rangle$,

$$(E - E_l)u_l + S_l J_{l,l+1} u_{l+1} + S_l J_{l,l-1} u_{l-1} = 0 \quad (3.17)$$

where $S_l = S_A(S_B)$ is the spin number for magnetic atoms in the lth atomic plane of the material A (B). $E_l = E_A (E_B)$ is for all internal atomic planes of the material A (B), whereas $E_l = E_A^i(E_B^i)$ for the interfacial atomic planes of the material A(B), with

$$E_{A(B)} = g\mu_B H_0 + 4(1 - \gamma_\parallel)S_{A(B)}J_{A(B)} + 2S_{A(B)}J_{A(B)} \quad (3.18)$$
$$E_{A(B)}^i = g\mu_B H_0 + 4(1 - \gamma_\parallel)S_{A(B)}J_{A(B)} + S_{A(B)}J_{A(B)} + S_{B(A)}J_I \quad (3.19)$$

where the parameter $\gamma_\parallel$ is defined as

$$\gamma_\parallel = \frac{1}{2}\left[\cos(q_y a) + \cos(q_z a)\right] \quad (3.20)$$

and $\gamma_\parallel \in \langle -1, 1 \rangle$ for q ranging throughout the whole two-dimensional Brillouin zone. Equation (3.17) is a recurrence equation for the spin wave amplitudes u_l, from which one could determine the amplitude for a given atomic plane, provided the amplitudes at the two preceding atomic planes are available. This equation can be rewritten in a convenient matrix form as [29]

$$\begin{pmatrix} u_{l+1} \\ u_l \end{pmatrix} = M_l \begin{pmatrix} u_l \\ u_{l-1} \end{pmatrix} \quad (3.21)$$

where M_l has different values $M_{A(B)}$, $M_{A(B)}^{il}$, $M_{A(B)}^{ir}$, respectively, corresponding to the positions of the spins in the system, with

$$M_A = \begin{bmatrix} \dfrac{E_A - E}{S_A J_A} & -1 \\ 1 & 0 \end{bmatrix} \quad (3.22)$$

$$M_A^{ir} = \begin{bmatrix} \dfrac{E_A^i - E}{S_A J_I} & -\dfrac{J_A}{J_I} \\ 1 & 0 \end{bmatrix} \quad (3.23)$$

$$M_A^{il} = \begin{bmatrix} \dfrac{E_A^i - E}{S_A J_A} & -\dfrac{J_I}{J_A} \\ 1 & 0 \end{bmatrix} \qquad (3.24)$$

The matrices M_B, M_B^{ir}, and M_B^{il} are given by similar formulae, but with the index B instead of A. Applying Eq. (3.21) to all N atomic planes in the elementary unit, one obtains the relation

$$\begin{pmatrix} u_{mN+1} \\ u_{mN} \end{pmatrix} = M \begin{pmatrix} u_{(m-1)N+1} \\ u_{(m-1)N} \end{pmatrix} \qquad (3.25)$$

where the transfer matrix M for the whole elementary bilayer is equal to

$$M = M_B^{ir}(M_B)^{N_B-2} M_B^{il} M_A^{ir}(M_A)^{N_A-2} M_A^{il} \qquad (3.26)$$

The translational symmetry along the superlattice direction results in a matrix M that is independent of the bilayer index m and obeys the condition

$$\det M = 1 \qquad (3.27)$$

Generally, the eigenmodes of the system can be described by the respective eigenvalues and eigenvectors of the matrix M. One could distinguish several possibilities for the eigenvalues and the corresponding modes [29]. For a detailed description, readers may refer to [29, 278–283].

An equivalent form of the transfer matrix formalism can be obtained by considering a vector with the components u_l and $u_l - u_{l-1}$, instead of u_l and u_{l-1} in Eq. (3.21) [282, 283]. The eigenvalue problem for layered structures reduces effectively to a one-dimensional problem, because of the translational symmetry in the plane yz. The transfer matrix formalism in site representation, as described above, could be transferred to the transfer matrix formalism in plane-wave-like (complex wavevectors) representation [29, 278–283]. Another method, the so-called interface rescaling method, is to some extent similar to the transfer matrix approach. This method includes a series of rescalings of the respective boundary conditions and reduces the eigenvalue problem of the composite system to that of one of its individual constituents [284–286]. The rescaled boundary conditions contain all of the information about the dynamical properties of the removed part of the system. The dynamical properties of the elementary unit as a whole are contained in the matrix M (or T in the plane-wave-like representation), reducing the whole composite system to some effective homogeneous system with new dynamical parameters. This can be concluded directly from Eq. (3.25), which gives

$$u_{(m+1)N+1} + (M_{11} + M_{22})u_{mN+1} + u_{(m-1)N+1} = 0 \quad (3.28)$$

This equation has a form similar to that for a homogeneous system, but with an effective parameter $M_{11} + M_{22}$ instead of $(E - E_0)/JS$.

A calculation of the spin wave propagation in magnetic sandwiches was given by Mercier and Lévy [287–289] and their co-workers [282], as a function of its thickness and of a parameter that describes either the roughness of the interface or the decay of the exchange through this layer. Unpinning at the interlayer from the very beginning of the coupling and the spin

wave spectra for magnetic sandwiches were derived. Surface roughness in magnetic thin films tends to unpin spin waves at the external surface [290]. This unpinning effect increases with the roughness height as long as this height remains small. The spin wave intensity spectrum is quite sensitive to roughness, and new surface spin waves appear because of this roughness.

The spin wave spectrum was calculated by a transfer matrix approach of the propagation, layer by layer, which had already been introduced for spin waves in layered media [291]. Long-range couplings in multilayers were accounted for as complex couplings for the spin wave resonance in saturating applied fields [292]. Large linewidths were found for some modes, and the transition from coupling to decoupling of the magnetic films appears to depend upon the spin wave wavelength [292]. The spin wave spectrum of infinite, semi-infinite, and finite ferromagnetic superlattices with arbitrary elementary units was analyzed theoretically by Barnaś and Hillebrands in the exchange-dominated region [281, 293]. The general dispersion equations for spin wave modes were also derived in the framework of the transfer matrix formalism. Some numerical results were given for structures with $N = 1, 2$, and 6 atomic planes in the elementary unit.

The mathematical eigenproblem formalism specific to bilayer films was treated by Puszkarski [292, 294] by the invented interface-rescaling approach. An exact solution of the eigenproblem, valid for quite arbitrary interface and surface conditions, was obtained. The majority of the roots of the characteristic equations are real, which corresponds to bulk solutions, but complex roots can also occur, corresponding to localized solutions. The interface effects were discussed with regard to the concept of effective pinning parameters and dynamical coupling parameters for the spin waves at the interface [292]. It was found that for certain values of the ratio of the interface exchange integral components, the constituent magnetic sublayers may be considered as effectively decoupled with respect to the spin wave properties.

The junction of two semi-infinite magnetic superlattices, each formed by alternating layers of two simple-cubic Heisenberg ferromagnets, was studied by Seidov et al. [295]. The dispersion equation for localized spin waves propagating over the contact layer and damping in the layers was derived by the transfer matrix method. Static properties of a magnetic superlattice consisting of spin 1/2 Ising layered media intercalated by periodic spin 1/2 Heisenberg interfaces were studied by Valadares and Plascak [296]. Phase diagrams and different profiles of magnetization determined by the parameters of the model were discussed within a variation approach based on Bogoliubov's inequality for the free energy. A formulation of ferromagnetic spin waves in quasi-periodic superlattices at low temperature was given by Pang and Pu [297]. The spectra, wave functions, and some related physical quantities of these systems were discussed, and specified extended states were found.

Akjouj et al. [298] studied composite materials with two interfaces formed of one ferromagnetic slab sandwiched between two different semi-infinite ferromagnets based on the Heisenberg model. The response function was obtained. Later, the eigenproblem of a composite system of two different layer subsystems was considered by Puszkarski and Dobrzynski [284, 285], with the assumption that they were coupled at their interface. They addressed mainly finite-layered composite materials (multilayers) that do not show a periodicity. For such materials, when studying their physical properties within a matrix representation approach, one is generally faced with the necessity of dealing with large composite matrices. A direct numerical analysis of such a matrix would lead to a huge numerical computation. Alternative analytical approaches were proposed. One approach to the analysis of a composite matrix consists of the calculation of its inverse, a response function or Green's function. This can be achieved from the knowledge of the inverse bulk or slab matrices of individual constituent subsystems [18]. The other calculates the eigenvalues and eigenvectors of the finite composite system by reducing of its eigenproblem to that of one of its individual constituent subsystems [286, 299–304]. In their papers, Puszkarski and Dobrzynski [284, 285] used the interface response theory, starting from the inverse matrices (or surface response functions) for each single slab. General formulas for the eigenvectors and eigenvalues of the total system, as well as its response function, were derived and expressed in terms of matrix elements of the individual subsystem response functions and the interface coupling parameter.

3.3.3. Operator Transformation Method

In this subsection, explicit attention will be paid to a general description of the spin wave spectrum of Heisenberg superlattices and multilayers, in terms of creation and annihilation operators. In these cases, the effective spin wave Hamiltonian $\mathbf{H}_{SW}$ may be written in a quadratic combination of bosonic amplitudes $\mathbf{b}_{pv}$ and $\mathbf{b}_{pv}^+$ of the form [19]

$$\mathbf{H}_{SW} = \sum_{v} \sum_{p,p'} \left[\Gamma_{pp'}^v \mathbf{b}_{pv}^+ \mathbf{b}_{p'v} + \frac{1}{2} \left(\Lambda_{pp'}^{v*} \mathbf{b}_{pv}^+ \mathbf{b}_{p',-v}^+ \right. \right.$$
$$\left. \left. + \Lambda_{pp'}^v \mathbf{b}_{p,-v} \mathbf{b}_{p'v} \right) \right] \qquad (3.29)$$

The Hamiltonian (3.29) has the same form as (2.24). However, in the case of multilayers and superlattices, the simple form of a canonical transformation (2.25) for ferromagnetic thin films cannot be found. According to Bogoliubov and Tyablikov [166], a more complex form for the canonical transformation could be written as

$$\mathbf{b}_{pv} = \sum_{\tau} \left(\mathbf{u}_{p\tau}^v \xi_{\tau v} + \mathbf{v}_{p\tau}^v \xi_{\tau, -v}^+ \right) \qquad (3.30)$$

where $\xi_{\tau v}$ and $\xi_{\tau v}^+$ are also bosonic variables. Their dynamical equation results in the eigenequations involving the spin wave amplitude $u_{\tau p}^v$ and $v_{\tau p}^v$,

$$E_{\tau v} u_{\tau p}^v = \sum_{p'} \Gamma_{pp'}^v u_{p'\tau}^v + \sum_{p'} \Lambda_{pp'}^{v*} v_{p'\tau}^v \qquad (3.31a)$$

$$-E_{\tau v} v_{\tau p}^v = \sum_{p'} \Lambda_{pp'}^v u_{p'\tau}^v + \sum_{p'} \Gamma_{pp'}^{v*} v_{p'\tau}^v \qquad (3.31b)$$

They obey the orthogonality rules

$$\sum_{p=1}^q \left(u_{p\tau}^{v*} u_{p\tau'}^v - v_{p\tau}^{v*} v_{p\tau'}^v \right) = \delta_{\tau\tau'} \qquad (3.32a)$$

$$\sum_{p=1}^q \left(u_{p\tau}^v u_{p\tau'}^{-v} - v_{p\tau}^v v_{p\tau'}^{-v} \right) = 0 \qquad (3.32b)$$

Then the effective spin wave Hamiltonian can be diagonalized to the form

$$\mathbf{H}_{SW} = E_0 + \sum_{\tau} \sum_{v} E_{\tau v} \xi_{\tau v}^+ \xi_{\tau v} \qquad (3.33)$$

It is clear that Eqs. (3.30)–(3.33) are the general forms of Eqs. (2.25)–(2.28). Similar to the spin wave modes in thin films, those in multilayers and superlattices also depend on both the plane wave vector v and the new index τ labeling the perpendicular standing component of the spin wave. The corresponding spectrum of the energy can be determined only by the actual surface boundary conditions in the perpendicular direction, following from the model [19, 167–170]. However, usually it is hard to solve analytically the equation group above, and a numerical procedure and a computer is needed.

A superlattice is a special case of multilayer in which there is an overall quasi-periodicity of the composition and thickness along one dimension. The periodicity length of the superlattices is equal to the sum of the thicknesses of component unit layers, and the translations along this direction still constitute a symmetry operation of the structure. These characters result in some special structures for a spin wave spectrum, such as the subbands and the mixed modes. An analytical procedure is necessary for a better understanding of the dynamic properties, such as spin waves, in superlattices. In the author's previous work [305, 306], spin wave spectra of a four-sublattice Heisenberg ferromagnet or ferrimagnet with different exchange constants ($J_{ab} = J_{cd} \neq J_{bc} = J_{da}$) and a three-sublattice Heisenberg system with different exchange constants ($J_{ab} = J_{bc} \neq J_{ca}$) were determined explicitly by performing the standard Holstein–Primakoff transformation and an extended Bogoliubov transformation. The method, used for the multisublattice systems [305, 306], was extended to make it appropriate for deriving analytical solutions of the spin wave spectra of the corresponding magnetic superlattices [306].

The infinite Heisenberg superlattices studied are formed from two ferromagnetic materials with the simple-cubic lattice, which couple antiferromagnetically at the interfaces. In the layered structure an elementary unit consisting of q (100) atomic planes is repeated periodically along the stacking direction parallel to the x axis of the coordinate system. The system can be described by the Hamiltonian [306],

$$\mathbf{H} = -\frac{1}{2} \sum_{l=1}^q \sum_{\rho,\delta} J_{l,\rho;l,\rho+\delta} S_{l,\rho} \bullet S_{l,\rho+\delta}$$
$$= -\frac{1}{2} \sum_{l=1}^q \sum_{\rho} \sum_{\delta_{\parallel}} J_l S_{l,\rho} \bullet S_{l,\rho+\delta_{\parallel}}$$
$$- \sum_{l=1}^q \sum_{\rho} J_{l,l+1} S_{l,\rho} \bullet S_{l+1,\rho} \qquad (3.34)$$

where l denotes the spins in the elementary unit in the x direction. δ indicates that only the exchanges between the nearest neighbors are taken into account, and $\delta_\parallel$ denotes the nearest neighbors within one plane parallel to the y–z plane of the coordinate system. When the number q of the planes in the magnetic unit cell, for instance, is chosen to be 4, the linearized Holstein–Primakoff transformation [6–8] and the Fourier transforms of the boson operators in the reduced Brillouin zone make it possible to rewrite the Hamiltonian in the forms

$$
\mathbf{H} = H_0 + S \sum_{l=1}^{4} \sum_k \left[Z_x|J'| + ZJ(1 - \gamma_{k_\parallel}) \right] b_{l,k}^+ b_{l,k}
$$
$$
- ZS \sum_k \gamma_{k_x} \left[J \left(b_{1,k}^+ b_{2,k} + b_{1,k} b_{2,k}^+ \right) \right.
$$
$$
+ J' \left(b_{2,k}^+ b_{3,k} + b_{2,k} b_{3,k} \right)
$$
$$
\left. + J \left(b_{3,k}^+ b_{4,k} + b_{3,k} b_{4,k}^+ \right) + J' \left(b_{4,k}^+ b_{1,k}^+ + b_{4,k} b_{1,k} \right) \right]
$$
$$
(3.35)
$$

where $k_\parallel$ is a two-dimensional wave vector parallel to the interfaces and k_x is that along the x direction. There should be a difference between the imaginaries of γ_{kx} and γ_{kx} because the inversion symmetry is broken down, although the translation symmetry is retained along the direction normal to the plane. This difference was neglected for simplicity when we derived Eq. (3.35) because in most cases we are interested in a long wavelength range. Then one needs to perform an extended Bogoliubov transformation [9] for eliminating the nondiagonal terms in the Hamiltonian for the four-layer superlattices as follows [306]:

$$
\begin{bmatrix} \xi_{1,k}^+ \\ \xi_{2,k}^+ \\ \xi_{3,k} \\ \xi_{4,k} \end{bmatrix} = \begin{bmatrix} a_{1k} & a_{2k} & a_{3k} & a_{4k} \\ a_{2k} & a_{1k} & a_{4k} & a_{3k} \\ a_{3k} & a_{4k} & a_{1k} & a_{2k} \\ a_{4k} & a_{3k} & a_{2k} & a_{1k} \end{bmatrix} \begin{bmatrix} b_{1,k}^+ \\ b_{2,k}^+ \\ b_{3,k} \\ b_{4,k} \end{bmatrix}
\quad (3.36)
$$

with its reversed matrix having parameters A_{jk} ($j = 1, 2, 3$, and 4). The commutation relation of the new operators leads to one equation [306],

$$
a_{1k}^2 + a_{2k}^2 - a_{3k}^2 - a_{4k}^2 = 1 \quad (3.37)
$$

It is obvious that an index τ is needed to label the perpendicular standing component of the spin waves and to diagonalize the spin wave Hamiltonian (3.35). To eliminate the nondiagonal terms of the new operators, one needs to establish the three equations as Eqs. (3.12a)–(3.14a) in [306]. The relations between the parameters a_{ik} and A_{jk} are the same as those in Eqs. (27)–(31) in [305]. The problem becomes to solve the equation group of the parameters a_{ik} and A_{jk}, which consists of eight equations and eight unknowns. This equation group (denoted as E_1 in [306]) is solved by the procedure represented in Appendix C in [306], so that the transformation can be carried out. After performing the transformation, one obtains the final form of the Hamiltonian for the present four-layer superlattice [306],

$$
\mathbf{H} = H_0 + H_0' + H_1 \quad (3.38)
$$

where one has

$$
H_0 = -2NS^2 \left(ZJ + Z_x|J'| \right) \quad (3.39)
$$
$$
H_0' = 4S \sum_k \left[\left(ZJ(1 - \gamma_{k_\parallel}) + Z_x|J'| \right) \left(A_{3k}^2 + A_{4k}^2 \right) \right.
$$
$$
- Z \left(J(A_{1k}A_{2k} + A_{3k}A_{4k}) \right.
$$
$$
\left. \left. + J'(A_{1k}A_{4k} + A_{2k}A_{3k}) \right) \gamma_{k_x} \right] \quad (3.40)
$$

and

$$
H_1 = S \sum_k \left[\left(ZJ(1 - \gamma_{k_\parallel}) + Z_x|J'| \right) \right.
$$
$$
\times \left(A_{1k}^2 + A_{2k}^2 + A_{3k}^2 + A_{4k}^2 \right)
$$
$$
- 2Z \left(J(A_{1k}A_{2k} + A_{3k}A_{4k}) \right.
$$
$$
\left. + J'(A_{2k}A_{3k} + A_{4k}A_{1k}) \right) \gamma_{k_x} \right]
$$
$$
\times \left[\xi_{1,k}^+ \xi_{1,k} + \xi_{2,k}^+ \xi_{2,k} + \xi_{3,k}^+ \xi_{3,k} + \xi_{4,k}^+ \xi_{4,k} \right] \quad (3.41)
$$

They are the energies for the initial state, the zero-point vibrating, and the spin waves, respectively.

The spin wave spectra obtained above depend on the strength of the exchange constants J and J'. For the four-layer superlattice, as shown in Eq. (3.41), the degeneracy of the spin wave spectra is still retained, and the number of the degeneracy of the spin wave spectra is 4. The reader may refer to [306] for the diagonalizing procedure and the spin wave spectrum of a three-layer superlattice.

More recently, Parkov et al. [307] reported that the diagonalizing procedure of Eqs. (3.29)–(3.33) could be used to deal with the multisublattice model proposed by Zhang [305, 306]. Considering the periodicity condition along the stacking direction parallel to the x axis of the coordinate system, the energy spectrum for spin waves of the multisublattices may be derived directly from the vanishing of determination of the parametric determinant of Eq. (3.31). A clearer form for the spin wave spectrum was obtained, so that a clearer physical significance was achieved. Obviously, this procedure could be extended to solve the spin wave spectrum of the superlattices.

3.3.4. Other Methods

Gokhale and Mills investigated the spin excitations of an ultrathin ferromagnetic film, in which the magnetic moment-bearing electrons are itinerant in character, within the frame of the one-band Hubbard model treated in mean-field theory [308]. By means of the random-phase approximation, they found the spin wave modes of standing wave character and a spectrum of Stoner excitation influenced strongly by size effects. Although the dispersion relations of the standing spin wave modes appear as expected from a localized-spin model, there are substantive differences between the itinerant- and localized-spin cases.

Spin waves in metallic multilayered structures consisting of two different ferromagnets were investigated by Świrkowicz within the framework of the multiband model [309]. Band structure and ground-state properties were calculated within the framework of the Hartree–Fock approximation. Spin wave

Table V.   Summary of Theoretical Study on Spin Waves in Superlattices and Multilayers, Based on Quantum Microscopic Theories

Method	Models and/or the shape of samples	Spin wave spectra and/or the topic or technique of the study	References
Green's functions	Layered structures	Green's function matching	[261]
Interface response theory	Discrete composite d-dimensional systems	Linear superposition of the responses	[18]
First-principle self-consistent calculations	Lattice matched Co/Cr model	Electronic and magnetic structures	[262]
Green's functions	A semi-infinite stack of two different ferromagnetic films with a Heisenberg model	Folded bulk bands, extra gaps, and surface-localized modes	[263]
	Superlattice of alternating ferromagnetic and antoferromagnetic layers	Basic magnetic response characteristics/ classical ground states and spin reorientations	[264]
Green's functions	Periodically modulated exchange interactions	Bloch law	[265]
Green's functions	Two different semi-infinite ferromagnets coupled across an interface/with an isolated impurity at or near the interface	Tridiagonal matrix method and the random-phase approximation	[266]
Green's functions	Layered Heisenberg ferromagnets	Linear spin wave approximation	[267, 270]
Green's functions	A double-layer Heisenberg antiferromagnet	Bogoliubov–Tyabilkov decoupling approximation	[268]
Green's functions	Multilayers and superlattices	Linear spin wave approximation/ competition between exchange constants	[271]
Green's functions and Monte Carlo simulations	Antiferromagnetic superlattices	Spin wave excitations, phase transitions, and quantum fluctuations	[272–274]
Green's functions	Arbitrary ferromagnetic interfaces, superlattices, and disordered layers	General recursion method/response function	[275]
Green's functions	Semi-infinite CsCl-type (bcc) Heisenberg ferromagnet	Surface modes on the (001) layer/a single ion anisotropy	[276]
Transfer matrix	A superlattice consisting of alternating two simple-cubic Heisenberg ferromagnetic layers	Either imaginary or complex wave vectors outside the pass band, but only real ones within a pass band	[277]
Transfer matrix	Infinite and semi-infinite ferromagnetic superlattices	Magnetostatic modes and magnetic polaritons in the Voigt configuration	[278–283]
Transfer matrix	Ferromagnetic superlattices	A review	[29]
Interface rescaling	A composite system of two different layer subsystems/rescaled boundary conditions		[284–286]
Transfer matrix	Magnetic sandwiches with the roughness of interfaces and the decay of the exchange	Unpinning at the interlayers	[282, 287–289]
Transfer matrix	Magnetic thin films with the roughness	Unpinning effect at the external surfaces	[290]
Transfer matrix	Layered media		[291]
Transfer matrix	Multilayers with long-ranged couplings/spin wave resonance in saturating applied field	Large linewidths for some modes and transition from coupling to decoupling of the magnetic films	[292]
Transfer matrix	Infinite, semi-infinite, and finite ferromagnetic superlattices with arbitrary elementary units	Numerical results for $N = 1, 2$, and 6 atomic planes in the elementary unit	[281, 293]
Invented interface rescaling	Bilayer films for arbitrary interface and surface conditions	Real bulk solutions and complex roots for localized solutions	[292, 294]
Transfer matrix	Junction of two semi-infinite superlattices, each formed by alternating two simple-cubic Heisenberg ferromagnets	Localized spin waves, propagating order of the contact layer, and damping in the layers	[295]
Transfer matrix	A superlattice of spin 1/2 Ising layered media intercalated by periodic spin 1/2 Heisenberg interfaces	A variation approach based on Bogoliubov's inequality for the free energy	[296]
Transfer matrix	Quasi-periodic superlattices	Ferromagnetic spin waves and specified extended states	[297]

Table V.   (Continued)

Method	Models and/or the shape of samples	Spin wave spectra and/or the topic or technique of the study	References
Transfer matrix	Composite materials with two interfaces formed of one ferromagnetic slab sandwiched between two different semi-infinite Heisenberg ferromagnets	Response function	[298]
Operator transformation	Thin films, superlattices, and multilayers	Bogoliubov–Tyablikov transformation and the dynamic equations of spin wave amplitudes	[19, 167–170]
Operator transformation	Heisenberg three and four sublattices and corresponding superlattices	Extended Bogoliubov transformation	[305, 306]
Operator transformation	Heisenberg three and four sublattices	Bogoliubov–Tyablikov transformation	[307]
Mean-field theory	Ferromagnetic film with itinerant electrons	One-band Hubbard model with the random–phase approximation	[308]
Multiband model	Metallic multilayered structures consisting of two different ferromagnets	Hartree–Fock approximation/band, subband structures, and ground state properties	[309]
Multiband model	Thin films composed of ferromagnetic and nonferromagnetic materials	Band structure and subband structure	[310]
Multiband model	Ultrathin film with uniform or nonuniform magnetization across the film	Band structure and subband structure	[311]
A microscopic approach	Superlattices formed from two ferromagnets coupled antiferromagnetically at the interfaces	Phase transition/a macroscopic Landau–Ginzburg and a microscopic approach	[312]
A Ruderman-Kittel approach	Heisenberg ferromagnetic films with RKKY coupling;	Effect of the s-d indirect exchange interaction	[313, 314]
	2D antiferromagnetic Heisenberg model with easy plane symmetry;	Spin wave and vortex excitations/dynamic correlation functions	[316–318]
	yttrium iron garnet (YIG) and Fe/Cr/Fe films	Spin wave instability and solitons	[319–321]

dispersion relations and the amplitudes of the modes were calculated, and the subband structure of the magnon spectrum was found. Świrkowicz further studied spin waves in thin films composed of ferromagnetic and nonmagnetic materials [310] and ultrathin film with uniform or nonuniform magnetization across the film [311].

Magnetic-field- and temperature-dependent equilibrium structures of magnetic superlattices formed from two ferromagnetic materials that couple antiferromagnetically at the interfaces were studied by Camley and Tilley [312]. Phase transitions in the superlattices were investigated by both a macroscopic Landau–Ginzburg and a microscopic approach.

The effect of the s-d indirect exchange interaction via a nonferromagnetic region of a double ferromagnetic film on the hysteresis curve was studied by Schmidt [313]. Calculations were performed for the discrete Heisenberg model of ferromagnetic films, with the Ruderman–Kittel–Kasuya–Yosida (RKKY) coupling between the magnetic films. Bruno and Chappert [314] presented a Ruderman–Kittel approach to the problem of oscillatory exchange coupling between ferromagnetic layers separated by a nonmagnetic metal spacer. Their model provides a very simple explanation for the occurrence of long periods as well as multiperiodic oscillations. The electrical transport properties of magnetic multilayered structures were found by Levy et al. [315] to be dominated by three ingredients: (1) the scattering within layers that changes from one layer to another, (2) the additional scattering resistivity due to the roughness of the interfaces between layers, and (3) the resistivity that depends on the orientation of the magnetization of the magnetic layers. The third point suggests that the spin waves may play a role in the giant magnetoresistence effect.

The dynamics of the two-dimensional antiferromagnetic Heisenberg model with easy plane symmetry were investigated by Völkel et al. [316–318]. A phenomenology of spin wave and vortex excitations was developed, and their contributions to the dynamical correlation functions were calculated. Spin wave instability and solitons in yttrium iron garnet (YIG) and Fe/Cr/Fe films were investigated by Patton and co-workers [319–321], both experimentally and theoretically.

Table V summarizes the theoretical study on spin waves in superlattices and multilayers, based on quantum microscopic theories.

4. DISCUSSION

There are basically two fundamental reasons for why we study spin excitations in magnetic materials, especially in magnetic thin films, multilayers, and superlattices [30]. First, all of the parameters that characterize the magnetic materials—exchange interaction, dipolar interaction, magnetization, anisotropy, surface effects, impurities, and magnetic structure—play a role in determining the frequency of the allowed spin waves. Therefore, measurements of spin waves are very useful in determining the fundamental parameters of magnetic structures. Second, spin waves are the fundamental excitations of magnetic materials, which determine the low-temperature behavior of the magnetization. The low-frequency spin waves are excellent indicators of magnetic phase transitions, because these transitions are often indicated by a soft mode or by substantial changes in frequency.

As introduced above, the spin waves in magnetic thin films, multilayers, and superlattices have been investigated with the use of a macroscopic phenomenological spin wave theory based on the Landau–Lifshitz gyromagnetic equation of motion and with the use of a quantum microscopic spin-wave theory, with the multisublattice Green's function technique, the local transfer matrix method, or the operator transformation method.

In practice, most of this work is approximate and numerical. The macroscopic spin wave theory based on the Landau–Lifshitz gyromagnetic equation of motion is phenomenological, neglecting the quantum microscopic effects of the magnetic systems. The multisublattice Green's function technique is a powerful method for the quantum microscopic spin wave theory. In most cases, it is within the framework of the perturbation theory, treating the effects of the surfaces or interfaces as a perturbation of the energy of the bulk spin waves. Besides the Green's function method, the local transfer matrix method is one of the commonly used methods that is founded upon the local character of the spin wave equation. In the case of a layer-to-nearest-layer coupling, if the wave amplitude (and its gradient) on two neighboring layers is (are) known, this wave amplitude (and its gradient) is (are) known everywhere. With this very local definition, the transfer matrix method was developed in mechanics by Poincaré [322]. It is usually difficult to derive the analytical expression for spin wave dispersion in the case of many spins in unit cells. For numerical calculations, if there is any error, it can be quite strongly amplified far from the initial layer. However, in some cases with few spins in the unit cells, the spin wave spectra obtained by this method are not consistent with those derived by the operator transformation method. In general, the methods, such as the macroscopic Landau–Lifshitz theory and the quantum microscopic perturbation theory, are not appropriate for the case in which the volume factor of the interfaces is dominant.

The operator transformation method shows the possibility of explicitly studying spin–wave spectra of magnetic superlattices, in terms of creation and annihilation operators. The methods as well as the transformations developed for the multisublattice Heisenberg magnets can be used to study the spin wave spectra of the magnetic superlattices. The spin waves in the magnetic superlattices can be dealt with by means of a one-dimensional model, by taking advantage of the periodic condition in the plane and performing the two-dimensional Fourier transformation. The superlattices have a larger periodicity in the direction perpendicular to the slabs, so that the one-dimensional Fourier transformation can also be performed along the direction perpendicular to the plane. The Hamiltonian for multisublattice systems can be treated as a one-dimensional one. The analytical results were derived within the Holstein–Primakoff approximation, namely, with less approximation, followed by the extended Bogoliubov transformation [305, 306] or the Bogoliubov–Tyablikov [19, 307] transformation, which reveal some natural characters, such as quantum fluctuation, degeneracy, etc., of the spin wave spectra of the superlattices.

However, a difficulty of the operator transformation method is that the procedure for calculating the analytical solutions of the spin wave spectra of magnetic superlattices with a large number of different layers in the elementary unit becomes very complicated. In those cases, it is hard to solve an equation group that consists of a lot of equations and unknowns and to perform the extended Bogoliubov transformation [305, 306] or the Bogoliubov–Tyablikov transformation [19, 307]. On the other hand, with the linear spin wave approximation, the Green's function technique can obtain the results for a superlattice [271], like those derived analytically by the Bogoliubov–Tyablikov transformation [307]. Nevertheless, for spin waves in magnetic multilayers, the actual surface boundary conditions in the perpendicular direction must be taken into account in accordance with the models [19, 167–170]. This makes the analytical procedure much more difficult, and normally, a numerical procedure is still needed.

The spin wave spectra depend on the strength of the different exchange constants in the magnetic thin films, superlattices, and multilayers. The dispersions of the spin waves in the magnetic thin films, superlattices, and multilayers are related not only by the two-dimensional wave vector k_l, but also by the one-dimensional wave vector k_x [305, 306]. Thus they depend on the values of γ_{kl} and γ_{kx}. In the four-layer superlattice, as shown in Eq. (3.41), the degeneracy of the spin wave spectra exists and the number of the degeneracy of the spin wave spectra is 4. For the three-layer superlattice [306], the degeneracy of the spin wave spectra is partially removed and the number of the degeneracy of the spin-wave spectra is 2. This is ascribed to the different values of the exchange constants and the asymmetry of the system. This also implies that the number of the spin wave spectra is the same as that of the magnetic layers in the elementary unit of the superlattice. This fact can easily be seen from the Hamiltonian (3.41) and the canonical transformation described by the matrices (3.36).

On the other hand, there is another type of splitting of the energy level for the four-layer superlattice. The spin wave modes in the superlattices depend on more than the plane wave vector k. It is true that the index τ, which is used to label the perpendicular standing component of the spin waves, is needed to diagonalize the spin-wave Hamiltonian. There are four different spin wave dispersions in the superlattice with the elementary unit of the four different layers. The number of total modes for the spin waves of the four-layer superlattices is 16 (i.e., the number of energy levels times that of their degeneracy [305, 306]). This might be ascribed to the square of the number of spins in the elementary unit. The same is true for the superlattice with the elementary unit of the three different layers. The total number for the spin wave modes should be 9. (The number 8 suggested in [306] for the spin wave modes in the three-layer superlattice was due to a misunderstanding.) The fact that the number of total spin wave modes is equal to the square of the number of the spins in the elementary unit could be confirmed by evaluating the situation of the superlattices with larger numbers of spins in an elementary unit. For instance, we can evaluate the transformation matrix for an eight-layer superlattice and immediately find that the number of the total spin wave modes is equal to 64. The results for degeneracy of the spin wave spectra of the superlattices was confirmed by the work of Parkov et al. [307], which showed that for each μ (denotes the layers in the elementary unit), one obtains a group of equations (3.31) with which to solve the energy spectrum. The degeneracy of each energy level should be equal to the value of the sublattices in the multisublattice system.

In the case of the superlattices, the cyclic condition and translation invariance in the normal direction are still retained so that the spin wave spectra in this direction can be described in the form of the plane waves. The superlattice has a larger periodicity in the direction perpendicular to the slabs and therefore many magnon branches in the folded Brillouin zone. It was found by Dobrzynski et al. that the surface-localized modes appear within the extra gaps existing between the folded bulk bands [263]. It has commonly been assumed that a spin that is not in an interface layer has the same nearest-neighbor environment and therefore the same equation of motion as a spin in the corresponding bulk medium [277]. The spin wave amplitudes were given within each component by a linear combination of the positive- and negative-going solutions for the bulk medium. For a traveling wave in the superlattice, the wave vectors in the directions within the plane must be real. Those in the direction normal to the plane are only real when the frequency lies within a pass band for the corresponding bulk mode and can be either imaginary or complex outside the pass band. The surface modes lie outside the bulk modes. When the wavelength of the surface magnon is sufficiently short so that the dominant contribution to the excitation energy comes from exchange interactions, the surface wave frequency lies below the bulk band. In the dipolar dominant regime, the surface mode lies above or between the bulk manifold and the Damon–Eshbach branch. The situation in the thin films is similar to that in the superlattice. The corresponding spectrum of the energy in the thin films cannot be determined by the cyclic condition, but by the actual surface boundary conditions in the perpendicular direction, following from the model [19, 167–170].

However, the surface-localized modes can exist only in the limit of the systems with few surfaces or interfaces. In this limit, the effect of the surfaces or interfaces is so weak that it can be treated as a perturbation of the energy of the bulk spin waves. In the system in which the fraction of the interfaces is comparable to that of the bulk, such a perturbation approach is not appropriate for the spin-wave study, and such surface-localized modes cannot be found. The effects of the interfaces and the bulk on the spin wave spectrum are mixed, so that the

dispersions of the interface and the bulk modes cannot be distinguished. Furthermore, a spin in a superlattice, which is not in an interface layer, does not have the same nearest-neighbor environment and the same equation of motion as a spin in the corresponding bulk medium. This means that the assumption used in [277] is not valid for a large fraction of the interfaces. The quantum correlation between the spins in different layers of a magnetic superlattice may result in the complication of the spin wave spectra of the superlattice.

The question of the boundary conditions, required by the second-order differential equation Eq. (1.1) or Eq. (2.3), was widely discussed in the literature [19, 93]. There are three types of boundary conditions for thin films: (1) *"free"* surfaces, requiring the vanishing of the normal derivatives of the magnetization displacement on the surfaces with the condition that $\partial \mathbf{m}/\partial x|_{x=0} = 0$ and $\partial \mathbf{m}/\partial x|_{x=L} = 0$; (2) completely *"pinned"* surfaces, meaning the vanishing of the displacement itself on the surfaces with the condition $\mathbf{m}(x = 0) = 0$ and $\mathbf{m}(x = L) = 0$; (3) one *"pinned"*/one *"free"* surface—a mixture of the above cases, i.e., $\partial \mathbf{m}/\partial x|_{x=0} = 0$ and $\mathbf{m}(x = L) = 0$. The definition of the "free" or "pinned" surface could be extended to the case of multilayers and superlattices. Of course, the numbers of the surfaces in multilayers and superlattices are larger than 2, and there are various combinations of the "free" and "pinned" surfaces, leading to the complexity of the surface conditions on the spin wave excitations.

The presence of the surfaces (and/or interfaces) can also be taken into account by considering the difference of the exchange integrals between the nearest neighbors on the surfaces (and/or at the interfaces) and in the bulk. The symmetry breaking at the surfaces (or interfaces) leads to the difference in the exchange coupling. One may also introduce the anisotropy interaction for the spins located on the surfaces (and/or the interfaces). An alternative approach is to introduce the different terms of anisotropy energy for the spins on the surfaces and those far from the surfaces. The difference in the anisotropy energy may originate from the Néel surface anisotropy energy, the induced anisotropy, and the indirectional anisotropy [19].

There are several characteristics of the spin wave spectra for limited systems, such as the thin films and multilayers.

One is the discreteness of the spin wave spectra. The spin wave spectra in the bulk materials and superlattices are continuous, because of the infiniteness of the systems. The continuum spin wave spectra disappear along the direction perpendicular to the plane of the thin films and multilayers, because of the finite number of the spins along this direction. The gap between the sequence points of the spin wave spectra can be estimated by π/Na (a is a lattice constant and N is the number of spins along the direction perpendicular to the plane). However, if the number N of spins in the thin films and multilayers is large enough, the spectra may be treated as quasi-continuum ones.

The second characteristic is the direct result of the loss of the cyclic condition and the translation symmetry. In bulk materials, the cyclic condition as well as the translation symmetry in three dimensions lead to the plane wave character of the spin waves. In the case of the superlattices, these conditions are still retained not only in the plane, but also along the direction normal to the plane, so that the spin wave spectra can be described in the form of plane waves. For bulk materials and superlattices, there are infinite numbers of Brillouin zones that can be denoted as the first, second, third Brillouin zones, and so on. In the cases of thin films and multilayers, spin waves in the form plane waves may still exist, which must be consistent with the boundary conditions. But the Brillouin zone would not repeat along the direction perpendicular to the plane because of the loss of the cyclic condition and the translation symmetry. Actually, the break down of the translation symmetry results in a wavevector that is not a "good" quantum number. In the limit case of many magnetic layers long the normal direction (for instance, semi-infinite systems), one could treat the wavevector as a "pseudo-good" quantum number and deal with the problem approximately by using the bulk theory. In the limit case of a few layers, such as ultrathin films, such a concept, as well as the corresponding approximation, is no longer valid.

The third characteristic is the effect of damping effects of the spin waves. It has been well known that the spin waves are elementary excitations, which are quasi-particles with limited lifetimes. For infinite systems, such as bulk materials and superlattices, such effects could be neglected. For finite systems, such as thin films and multilayers, the damping effects of spin waves become more pronounced when the number of the spins along the normal direction decreases. Because of damping effects, according to the quantum uncertainty principle, it is difficult to determine precisely the energies of high-lying (or even low-lying) spin wave modes. Under this limit, the wavevectors of a spin wave may distribute in a wide range, because of the uncertainty principle, so that it can be described only in terms of wave packets. The damping effects are closely related to the discreteness of the spin wave spectra. The lifetime of a quasiparticle, such as a spin wave, along the normal direction of the thin films and multilayers, may be very short and difficult to detect or measure experimentally, if the number of the spins is small enough.

The characteristics listed above (and maybe many others) for the spin waves in the limited systems show the complexity of the problems we face. These characteristics create difficulties for studying and understanding the nature of the spin waves in these magnetic systems. Therefore, greater efforts in this area will be needed to reveal the deeper physical significance and to illustrate the clearer physical picture of spin waves in magnetic thin films, superlattices, and multilayers.

5. CONCLUDING REMARKS

Understanding spin wave properties in magnetic thin films, superlattices, and multilayers has been one of the most active topics in the area of condensed matter physics. This is mainly due to the fact that the spin wave is a basic elementary excitation that governs the thermodynamic properties (including magnetic properties) and dynamic behaviors of these magnetic materials at low temperatures. In this work, we try to give an overview of the theoretical studies on this subject in the last half-century. It is understandable that because of the limits of the author's knowledge, many important works as well as advances in the literature may not be covered in this review. Nevertheless, as introduced above in the last three sections, there has been great progress in this field since the 1950s.

The spin waves in magnetic thin films, multilayers, and superlattices have been investigated with macroscopic phenomenological spin-wave theory based on the Landau–Lifshitz gyromagnetic equation of motion, Maxwell's equations, and electrodynamic boundary conditions. Various techniques, such as Green's functions, magnetostatic potential method, the effective-medium approach, micromagnetism, and the transfer-function approach, have been used to derive spin wave spectra in macroscopic phenomenological fashion. Shape dependence, finite size effects, anisotropy effects, retardation effects, critical angle effects, dynamic pinning, and ground states have been discussed, based on the models for various types of geometries, symmetries, and magnetism of the films. Extensive new phenomena have been discovered, such as surface modes, mixed surface and bulk modes, Cantor structure with gaps and isolated modes, reciprocal and nonreciprocal modes, spin wave instability, the discreteness of the spectrum and the damping of spin waves, etc.

Spin waves in magnetic thin films, multilayers, and superlattices have also been studied with the quantum microscopic spin wave theory, using the multisublattice Green's function technique, the local transfer-matrix method, interface response theory, the operator transformation method, the mean field theory, and the multiband model. The effects of surfaces, dynamic spin pinning and unpinning; inhomogeneous interfield, nonuniform magnetization; and the roughness of the interface (and/or surface) have been studied for arbitrary types of lattice, orientation of surface and configuration of the state field, the interaction, and the anisotropy, respectively. A number of novel magnetic phenomena have been found, such as folded bulk bands, subband structure, extra gaps, surface-localized modes, quantum fluctuations, spin-reorientation transitions, transitions from coupling to decoupling of the films, spin wave instability, and dynamic correlation, etc. It has been found that in addition to the wave vector $\mathbf{k}$, an index τ is needed to labeling the perpendicular standing components of spin waves in thin films, superlattices, and multilayers.

Although there has been great progress in the theoretical study (and experimental) study of spin waves in magnetic thin films, superlattices, and multilayers, it is understood that there are still many open questions in this field, remaining to be explored in further investigations. As indicated at the beginning of this chapter, all of the theories have their own limitations and are appropriate only for limited regions of spin wave excitations in terms of the magnitude $|\mathbf{q}|$ of the spin wave vector (Table I). As showed in Section 4, all of the theories have disadvantages in dealing with the problem of spin wave excitations in the thin films, superlattices, and multilayers, because of the approximation used. This is also due to the difficulty of deriving and/or illustrating the spin wave spectra analytically in a simplified form for such complicated systems. Everyone dealing with spin wave spectra needs to keep this fact in mind when applying one of the previous theories and/or developing a the new theory or technique. One also needs to pay careful attention to the special characters and features of the spin waves, which might occur only in the thin films, superlattices, and multilayers, not in bulk materials.

Following the advances in the techniques for growing artificial structures, more precisely control of growth at the atomic level will result in more prefect interfaces and surfaces of high-quality thin films, superlattices, and multilayers. The advances in the techniques for magnetic measurements will also improve the accuracy of the experiments, detecting the information of the spin wave excitations. All of them will be quite helpful for providing a clearer physical picture of spin wave excitations in thin films, superlattices, and multilayers. It is expected that a great number of new models and theoretical techniques will be proposed and developed in the future, which will reveal more remarkable phenomena of spin waves in these systems. In this new century we shall have a much better understanding of the nature of spin wave excitations in thin films, superlattices, and multilayers.

ACKNOWLEDGMENTS

This work has been supported by the National Natural Science Foundation of China (grants 59421001 and 59725103) and the Science and Technology Commission of Liaoning and Shenyang.

REFERENCES

1. F. Bloch, *Z. Phys.* 61, 206 (1930).
2. C. Kittel, "Introduction to Solid State Physics," 6th ed. Wiley, New York, 1986.
3. M. G. Cottam and D. R. Tilley, "Introduction to Surface and Superlattice Excitations." Cambridge Univ. Press, Cambridge, UK, 1989.
4. M. G. Cottam and A. N. Slavin, in "Linear and Nonlinear Spin Waves in Magnetic Films and Superlattices" (M. G. Cottam, Ed.), Chap. 1. World Scientific, Singapore, 1994.
5. C. Kittel, "Quantum Theory of Solids," 2nd ed. Wiley, New York, 1987.
6. T. Holstein and H. Primakoff, *Phys. Rev.* 58, 1098 (1940).
7. P. W. Anderson, *Phys. Rev.* 86, 694 (1952).
8. R. Kubo, *Phys. Rev.* 87, 568 (1952).
9. N. N. Bogoliubov, *J. Phys. USSR* 11, 23 (1947).
10. N. N. Bogoliubov and S. V. Tyablikov, *Sov. Phys.-Dokl.* 4, 604 (1959).
11. R. Tahir-Kheli and D. ter Haar, *Phys. Rev.* 127, 88 (1962).
12. H. B. Callen, *Phys. Rev.* 130, 890 (1963).
13. D. C. Mattis, "The Theory of Magnetism." Harper & Row, New York, 1965.
14. D. Wagner, "Introduction to the Theory of Magnetism." Pergamon, Oxford, 1972.
15. A. I. Akhiezer, V. G. Bar'yakhtar, and S. V. Peletminskii, "Spin Waves." North-Holland, Amsterdam, 1968.
16. B. A. Kalinikos, in "Linear and Nonlinear Spin Waves in Magnetic Films and Superlattices" (M. G. Cottam, Ed.), Chapt. 2. World Scientific, Singapore, 1994.
17. H. Puszkarski, *Surf. Sci. Rep.* 20, 45 (1994).
18. L. Dobrzynski, *Surf. Sci. Rep.* 6, 119 (1986).
19. A. Corciovei, G. Costache, and D. Vamanu, *Solid State Phys.* 27, 237 (1972).
20. D. L. Mills, in "Light Scattering in Solids V, Superlattices and Other Microstructures" (M. Cardona and G. Gütherodt, Eds.), p. 13. Springer-Veralg, Berlin, 1989.
21. M. H. Grimsditch, in "Light Scattering in Solids V, Superlattices and Other Microstructures" (M. Cardona and G. Gütherodt, Eds.), p. 285. Springer-Veralg, Berlin, 1989.
22. P. Grünberg, in "Light Scattering in Solids V, Superlattices and Other Microstructures" (M. Cardona and G. Gütherodt, Eds.), p. 303. Springer-Veralg, Berlin, 1989.
23. M. G. Cottam and D. R. Tilley, "Introduction to Surface and Superlattice Excitations." Cambridge Univ. Press, Cambridge, UK, 1989.
24. T. Wolfram and R. E. DeWames, *Prog. Surf. Sci.* 2, 233 (1972).
25. H. Puszkarski, *Prog. Surf. Sci.* 9, 191 (1979).
26. C. E. Patton, *Phys. Rep.* 103, 252 (1984).
27. P. Grünberg, *Prog. Surf. Sci.* 18, 1 (1985).
28. R. E. Camley, *Surf. Sci. Rep.* 7, 103 (1987).
29. J. Barnaś, in "Linear and Nonlinear Spin Waves in Magnetic Films and Superlattices" (M. G. Cottam, Ed.) Chapt. 3. World Scientific, Singapore, 1994.
30. R. E. Camley, *J. Magn. Magn. Mater.* 200, 583 (1999).
31. C. Kittel, *Rev. Mod. Phys.* 21, 541 (1949).
32. S. M. Rezende, A. Azevedo, and F. M. de Aguiar, in "Linear and Nonlinear Spin Waves in Magnetic Films and Superlattices" (M. G. Cottam, Ed.), Chapt. 7. World Scientific, Singapore, 1994.
33. P. E. Wigen, in "Linear and Nonlinear Spin Waves in Magnetic Films and Superlattices" (M. G. Cottam, Ed.), Chapt. 8. World Scientific, Singapore, 1994.
34. A. N. Slavin, B. A. Kalinikos, and N. G. Kovshikov, in "Linear and Nonlinear Spin Waves in Magnetic Films and Superlattices" (M. G. Cottam, Ed.), Chapt. 9. World Scientific, Singapore, 1994.
35. H. S. Jarrett and R. K. Waring, *Phys. Rev.* 111, 1223 (1958).
36. M. H. Seavey, Jr., and P. E. Tannenwald, *Phys. Rev. Lett.* 1, 168 (1958).
37. C. Kittel, *Phys. Rev.* 110, 1295 (1958).
38. C. Kittel, *Phys. Rev.* 110, 836 (1958).
39. C. Kittel and C. Herring, *Phys. Rev.* 77, 725 (1950).
40. J. F. Dillon, Jr., *J. Appl. Phys.* 31, 1605 (1960).
41. V. R. J. Jelitto, *Z. Naturforsch. A: Phys. Sci.* 19, 1567 (1964).
42. C. A. Neugebauer, "Physics of Thin Films," Vol. 3, p. 30. Academic Press, New York, 1964.
43. P. W. Palmberg, R. E. De Wames, and L. A. Vredevoe, *Phys. Rev. Lett.* 21, 682 (1968).
44. B. R. Tittmann and R. E. De Wames, *Phys. Lett. A* 30A, 499 (1969).
45. R. J. Birgeneau, H. J. Guggenheim, and G. Shirane, *Phys. Rev. Lett.* 22, 720 (1969).
46. W. Zinn, *Czech. J. Phys. B* 21, 391 (1971).
47. K. Hayakawa, K. Namikawa, and S. Miyake, *J. Phys. Soc. Jpn.* 31, 5 (1971).
48. P. W. Anderson and H. Suhl, *Phys. Rev.* 100, 1788 (1955).
49. L. R. Walker, *Phys. Rev.* 105, 390 (1957).
50. L. R. Walker, *J. Appl. Phys.* 29, 318S (1958).
51. S. Geschwind and L. R. Walker, *J. Appl. Phys.* 30, 163S (1959).
52. R. W. Damon and J. R. Eshbach, *J. Phys. Chem. Solids* 19, 308 (1961).
53. J. R. Eshbach and R. Damon, *Phys. Rev.* 118, 1208 (1960).
54. R. W. Damon and H. van de Vaart, *J. Appl. Phys.* 36, 3453 (1965).
55. Lord Rayleigh, *Proc. London Math. Soc.* 17, 4 (1887).
56. A. E. H. Love, "Some Problems of Geodynamics." Cambridge Univ. Press, London, 1911.
57. A. E. H. Love, "Theory of Elasticity." Dover, New York, 1944.
58. R. Stoneley, *Proc. R. Soc. London, Ser. A* 232, 447 (1955).
59. F. Press and M. Ewing, *Trans. Am. Geophys. Union* 32, 677 (1951).
60. W. M. Ewing, W. S. Jardetzky, and F. Press, "Elastic Waves in Layered Media." McGraw-Hill, New York, 1957.
61. E. W. Montroll and R. B. Potts, *Phys. Rev.* 100, 525 (1955).
62. E. W. Montroll and R. B. Potts, *Phys. Rev.* 102, 72 (1956).
63. E. W. Montroll, *J. Chem. Phys.* 18, 183 (1950).

64. P. Debye, *Ann. Phys. (Leipzig)* 39, 789 (1912).
65. G. F. Koster and J. C. Slater, *Phys. Rev.* 95, 1167 (1954).
66. G. F. Koster, *Phys. Rev.* 95, 1436 (1954).
67. I. M. Lifshitz and L. N. Rosenzweig, *J. Exp. Theor. Phys.* 18, 1012 (1948).
68. R. F. Wallis, *Phys. Rev.* 105, 540 (1957).
69. D. C. Gazis, R. Herman, and R. F. Wallis, *Phys. Rev.* 119, 533 (1960).
70. D. C. Gazis and R. F. Wallis, *Surf. Sci.* 5, 482 (1966).
71. A. Seeger, *J. Phys. Radium* 23, 616 (1962).
72. J. Callaway, *J. Math. Phys.* 5, 783 (1964).
73. R. A. Brown, *Phys. Rev.* 156, 889 (1967).
74. J. L. Synge, *J. Math. Phys.* 35, 323 (1957).
75. T. C. Lim and G. W. Farnell, *J. Appl. Phys.* 39, 4319 (1968).
76. T. C. Lim and G. W. Farnell, *J. Acoust. Soc. Am.* 45, 845 (1968).
77. K. A. Ingebrigtsen and A. Tonning, *Phys. Rev.* 184, 942 (1969).
78. S. A. Thau and J. W. Dally, *Int. J. Eng. Sci.* 7, 37 (1969).
79. R. E. Allen, G. P. Alldredge, and F. W. de Wette, *Phys. Rev. B: Solid State* 4, 1648 (1971).
80. R. E. Allen, G. P. Alldredge, and F. W. de Wette, *Phys. Rev. B: Solid State* 4, 1661 (1971).
81. G. P. Alldredge, R. E. Allen, and F. W. de Wette, *Phys. Rev. B: Solid State* 4, 1682 (1971).
82. R. E. Allen, F. W. de Wette, and A. Rahman, *Phys. Rev.* 179, 887 (1969).
83. R. E. Allen and F. W. de Wette, *Phys. Rev.* 188, 1320 (1969).
84. R. E. Allen, G. P. Alldredge, and F. W. de Wette, *Phys. Rev. Lett.* 23, 1285 (1969).
85. R. E. Wigen, C. F. Kooi, M. R. Shanaberger, and T. D. Rossing, *Phys. Rev. Lett.* 9, 206 (1962).
86. A. M. Portis, *Appl. Phys. Lett.* 2, 69 (1963).
87. M. J. Hurben and C. E. Patton, *J. Magn. Magn. Mater.* 139, 263 (1995).
88. M. J. Hurben and C. E. Patton, *J. Magn. Magn. Mater.* 163, 39 (1995).
89. L. Landau and E. Lifshitz, *Physik. Z. Sowjetunion* 8, 153 (1935).
90. J. R. Macdonald, Ph.D. Thesis, Oxford, 1950. Spin exchange effects in ferromagnetic resonance.
91. J. R. Macdonald, *Proc. Phys. Soc., London A* 64, 968 (1951).
92. J. R. Macdonald, *Phys. Rev.* 103, 280 (1956).
93. G. T. Rado and J. R. Weertman, *Phys. Rev.* 94, 1386 (1954).
94. G. T. Rado and J. R. Weertman, *J. Phys. Chem. Solids* 11, 315 (1959).
95. J. D. Adam, T. W. O'Keefe, and R. W. Patterson, *J. Appl. Phys.* 50, 2446 (1979).
96. Yn. V. Gulyaev, A. S. Bugaev, P. E. Zul'berman, I. A. Ignat'ev, A. G. Konovalov, A. V. Logovskoi, A. M. Mednikov, B. P. Nam, and E. I. Niko-laev, *Sov. Phys. JETP Lett.* 30, 565 (1979) [*Pis'ma Zh. Eksp. Teor. Fiz.* 30, 600 (1979)].
97. Yu. F. Ogrin and A. V. Lugovskoi, *Sov. Tech. Phys. Lett.* 9, 421 (1983) [*Pis'ma Zh. Tekh. Fiz.* 9, 421 (1983)].
98. A. S. Andreev, Yu. V. Gulyaev, P. E. Zil'berman, V. B. Kravchenko, A. V. Lugovskoi, Yu. F. Ogrin, A. G. Temiryazev, and L. M. Filimonova, *Sov. Phys. JETP* 59, 586 (1984) [*Zh. Eksp. Teor. Fiz.* 59, 1005 (1984)].
99. B. A. Kalinikos, N. G. Kovshikov, and A. N. Slavin, *Sov. Phys. JETP Lett.* 38, 413 (1983) [*Pis'ma Zh. Eksp. Teor. Fiz.* 38, 343 (1983)].
100. V. I. Dmitriev, B. A. Kalinikos, and N. G. Kovshikov, *Sov. Phys. Tech. Phys.* 30 (1985). [*Zh. Tekh. Fiz.* 55, 2051 (1985)].
101. B. A. Kalinikos, N. G. Kovshikov, and A. N. Slavin, *Sov. Phys. JETP* 94, 303 (1988). [*Zh. Eksp. Teor. Fiz.* 94, 157 (1988)].
102. V. V. Gann, *Sov. Phys. Solid State* 8, 2537 (1967) [*Fiz. Tverd. Tela* 9, 3167 (1966)].
103. L. V. Mikhailovskaya and R. G. Khlebopros, *Sov. Phys. Solid State* 11, 1381 (1969) [*Fiz. Tverd. Tela* 11, 2854 (1969)].
104. R. E. De Wames and T. Wolfram, *J. Appl. Phys.* 41, 987 (1970).
105. T. Wolfram and R. E. De Wames, *Phys. Rev. B: Solid State* 4, 3125 (1971).
106. M. Spark, *Phys. Rev. Lett.* 24, 1178 (1970).
107. B. N. Filippov, *Phys. Met. Metall.* 32, 911 (1971) [*Fiz. Met. Metalloved.* 29, 1131 (1970)].
108. B. A. Kalinikos and A. N. Slavin, *J. Phys. C: Solid State Phys.* 19, 7013 (1986).
109. O. G. Vendik and D. N. Chartorizhskii, *Sov. Phys. Solid State* 12, 2537 (1970) [*Fiz. Tverd. Tela* 12, 1538 (1970)].
110. O. G. Vendik, B. A. Kalinikos, and D. N. Chartorizhskii, *Sov. Phys. Solid State* 19, 222 (1977) [*Fiz. Tverd. Tela* 19, 387 (1977)].
111. B. A. Kalinikos, *Sov. Phys. J.* 24, 719 (1981) [*Izvestiya VUZov Fizika* 24, 42 (1981)].
112. T. Wolfram and R. E. De Wames, *Solid State Commun.* 8, 191 (1970).
113. B. N. Filippov and I. G. Titjakov, *Phys. Met. Metall.* 35 (1973) [*Fiz. Met. Metalloved.* 35, 28 (1973)].
114. R. G. Khlebopros and L. V. Mikhailovskaya, *Bull. Acad. Sci. USSR Phys. Ser.* 36 (1972) [*Izv. AN SSSR Ser. Fiz.* 36, 1522 (1972)].
115. L. V. Mikhailovskaya and R. G. Khlebopros, *Sov. Phys. Solid State* 16, 46 (1974) [*Fiz. Tverd. Tela* 16, 77 (1974)].
116. Yu. V. Gulyaev, P. E. Zil'berman, and A. V. Lugovskoi, *Sov. Phys. Solid State* 23, 660 (1981) [*Fiz. Tverd. Tela* 23, 1136 (1981)].
117. A. V. Lugovskoi and P. E. Zil'berman, *Sov. Phys. Solid State* 24, 259 (1982) [*Fiz. Tverd. Tela* 24, 458 (1982)].
118. P. E. Zil'berman and A. V. Lugovskoi, *Sov. Phys.Tech. Phys.* 57 (1987). [*Zh. Tekh. Fiz.* 57, 3 (1987)].
119. G. T. Rado and R. J. Hicken, *J. Appl. Phys.* 63, 3885 (1988).
120. M. Sparks, *Phys. Rev. B: Solid State* 1, 3831 (1967).
121. M. Sparks, *Phys. Rev. B: Solid State* 1, 3856, (1967).
122. M. Sparks, *Phys. Rev. B: Solid State* 1, 3869 (1967).
123. M. Sparks, *Phys. Rev. B: Solid State* 1, 4439 (1967).
124. O. G. Vendik and D. N. Chartorizhskii, *Sov. Phys. Solid State* 11, 1957 (1970) [*Fiz. Tverd. Tela* 11, 2420 (1969)].
125. O. G. Vendik and D. N. Chartorizhskii, *Sov. Phys. Solid State* 12, 1209 (1970) [*Fiz. Tverd. Tela* 12, 1538 (1970)].
126. B. A. Kalinikos, *IEE Proc. Pt. H* 127, 4 (1980).
127. B. A. Kalinikos and S. I. Miteva, *Sov. Phys. Tech. Phys.* 51, (1981) [*Zh. Tekh. Fiz.* 51, 2213 (1981)].
128. B. A. Kalinikos and A. N. Slavin, *J. Phys. C: Solid State Phys.* 19, 7013 (1986).
129. B. A. Kalinikos, M. P. Kostylev, N. V. Kozhus', and A. N. Slavin, *J. Phys.: Condens. Matter.* 2, 9861 (1990).
130. A. Brager and A. Schuchowitzky, *J. Phys. Chem. U.S.S.R.* 20, 1459 (1946).
131. A. Brager and A. Schuchowitzky, *Acta Physiochim. U.S.S.R.* 21, 1001 (1946).
132. A. Brager and A. Schuchowitzky, *J. Chem. Phys.* 14, 569 (1946).
133. R. Stratton, *Philos. Mag.* 44, 519 (1953).
134. R. Stratton, *J. Chem. Phys.* 37, 2972 (1962).
135. M. Dupuis, R. Mazo, and L. Onsager, *J. Chem. Phys.* 33, 1452 (1960).
136. A. A. Maradudin and R. F. Wallis, *Phys. Rev.* 148, 945 (1966).
137. M. Born, *Proc. Phys. Soc., London* 54, 362 (1942).
138. R. F. Wallis, A. A. Maradudin, I. P. Ipatova, and A. A. Klochikhin, *Solid State Commun.* 5, 89 (1967).
139. B. N. Fillipov, *Sov. Phys. Solid State* 9, 1048 (1967).
140. D. L. Mills and A. A. Maradudin, *J. Phys. Chem. Solids* 28, 1855 (1967).
141. D. L. Mills and W. M. Saslow, *Phys. Rev.* 171, 488 (1968).
142. H. Benson and D. L. Mills, *Phys. Rev.* 178, 839 (1969).
143. H. Benson and D. L. Mills, *Phys. Rev.* 188, 849 (1969).
144. D. L. Mills, *J. Phys. (Paris)* 31, C1-33 (1970).
145. R. E. De Wames and T. Wolfram, *Phys. Rev.* 185, 720 (1969).
146. J. C. Lévy and J. L. Motchane, *J. Vac. Sci. Technol.* 9, 721 (1972).
147. V. M. Sokolov and B. A. Tavger, *Sov. Phys. Solid State* 10, 1412 (1968).
148. V. M. Sokolov and B. A. Tavger, *Sov. Phys. Solid State* 10, 1793 (1968).
149. H. Puszkarski, *Phys. Lett.* 30A, 227 (1969).
150. J. C. S. Lévy, J. L. Motchane, and E. Gallais, *J. Phys. C: Solid State Phys.* 7, 761 (1974).
151. F. Hoffmann, A. Stankoff, and H. Pascard, *J. Appl. Phys.* 41, 1022 (1970).
152. C. F. Kooi, E. P. Wigen, M. R. Shanabarger, and J. V. Kerrigan, *J. Appl. Phys.* 35, 791 (1964).
153. J. D. Davies, *J. Appl. Phys.* 35, 804 (1964).
154. L. Dobrzynski, *Ann. Phys. (Leipzig)* 4, 637 (1969).
155. L. Dobrzynski and D. L. Mills, *Phys. Rev.* 186, 538 (1969).
156. L. Dobrzynski and D. L. Mills, *J. Phys. Chem. Solids* 30, 1043 (1969).
157. C. Kittel, *Phys. Rev.* 70, 965 (1946).
158. P. Pincus, *Phys. Rev.* 118, 658 (1960).

159. R. F. Soohoo, *Phys. Rev.* 131, 594 (1963).

160. T. Ayukawa, *J. Phys. Soc. Jpn.* 18, 970 (1963).

161. I. Davis, *J. Appl. Phys.* 36, 3520 (1965).

162. L. Wojtczak, *Rev. Roum. Phys.* 12, 577 (1967).

163. H. Puszkarski, *Phys. Status Solidi* 22, 355 (1967).

164. A. Corciovei and D. Vamanu, *IEEE Trans. Magn.* MAG-5, 180 (1969).

165. A. Corciovei and D. Vamanu, *Rev. Roum. Phys.* 15, 473 (1970).

166. N. N. Bogoliubov and S. V. Tyablikov, *JETP* 19, 256 (1949).

167. A. Corciovei, *Phys. Rev.* 130, 2223 (1963).

168. A. Corciovei, *Rev. Roum. Phys.* 10, 3 (1965).

169. L. Wojtczak, *Acta Phys. Pol.* 28, 25 (1965).

170. L. Wojtczak, *Phys. Status Solidi* 13, 245 (1966).

171. G. C. Bailey and C. Vittoria, *Phys. Rev. B: Solid State* 8, 3247 (1973).

172. C. Vittoria, M. Rubinstein, and P. Lubitz, *Phys. Rev. B: Solid State* 12, 5150 (1975).

173. S. M. Bhagat and P. Lubitz, *Phys. Rev. B: Condens. Matter.* 10, 179 (1974).

174. Diep-The-Hung and J. C. S. Lévy, *Surf. Sci.* 80, 512 (1979).

175. J. C. S. Lévy and Diep-The-Hung, *Surf. Sci.* 83, 267 (1979).

176. Diep-The-Hung, *Phys. Status Solidi B* 103, 809 (1981).

177. G. Spronken, A. Friedmann, and A. Yelon, *Phys. Rev. B: Solid State* 15, 5141 (1977).

178. G. Spronken, A. Friedmann, and A. Yelon, *Phys. Rev. B: Solid State* 15, 5151 (1977).

179. E. M. Gyorgy, J. F. Dillon, Jr., D. B. McWhan, L. W. Rupp, Jr., L. R. Testardi, and P. J. Flanders, *Phys. Rev. Lett.* 45, 57 (1980).

180. Z. Q. Zheng, C. M. Falco, J. B. Ketterson, and I. K. Schuller, *Appl. Phys. Lett.* 38, 424 (1981).

181. Z. Q. Zheng, J. B. Ketterson, C. M. Falco, and I. K. Schuller, *J. Appl. Phys.* 53, 3150 (1982).

182. G. P. Felcher, J. W. Cable, J. Q. Zheng, J. B. Ketterson, and J. E. Hilliard, *J. Magn. Magn. Mater.* 21, L198 (1980).

183. B. J. Thaler, J. B. Ketterson, and J. E. Hilliard, *Phys. Rev. Lett.* 41, 336 (1978).

184. M. Grimsditch, M. R. Khan, A. Kueny, and I. K. Schuller, *Phys. Rev. Lett.* 51, 498 (1983).

185. R. Krishnan, *J. Magn. Magn. Mater.* 50, 189 (1985).

186. A. Kueny, M. R. Khan, I. K. Schuller, and M. Grimsditch, *Phys. Rev. B: Condens. Matter.* 29, 2879 (1984).

187. T. Jarlborg and A. J. Freeman, *J. Appl. Phys.* 53, 8041 (1982).

188. N. K. Flevaris, J. B. Ketterson, and J. E. Hilliard, *J. Appl. Phys.* 53, 8046 (1982).

189. T. Shinjo, K. Kawaguchi, R. Yamamoto, N. Hosaito, and T. Takada, *J. Phys. (Paris) Colloq.* 45, C5-367 (1984).

190. A. J. Freeman, Jian-Hua Xu, S. Ohnishi, and T. Jarlborg, *J. Phys. (Paris) Colloq.* 45, C5-369 (1984).

191. R. Krishnan and W. Jantz, *Solid State Commun.* 50, 533 (1984).

192. N. Hosoito, K. Kawaguchi, T. Shinjo, T. Takada, and Y. Endoh, *J. Phys. Soc. Jpn.* 53, 2659 (1984).

193. P. F. Carcia and A. Suna, *J. Appl. Phys.* 54, 2000 (1983).

194. H. Sakakima, R. Krishnan, and M. Tessier, *J. Appl. Phys.* 57, 3651 (1985).

195. F. Herman, P. Lambin, and O. Jepsen, *J. Appl. Phys.* 57, 3654 (1985).

196. C. F. Majkrzak, J. D. Axe, and P. Böni, *J. Appl. Phys.* 57, 3657 (1985).

197. H. K. Wong, H. Q. Yang, J. E. Hilliard, and J. B. Ketterson, *J. Appl. Phys.* 57, 3660 (1985).

198. L. R. Sill, M. B. Brodsky, S. Bowen, and H. C. Hamaker, *J. Appl. Phys.* 57, 3663 (1985).

199. P. Grünberg, *J. Appl. Phys.* 57, 3673 (1985).

200. M. J. Pechan, M. B. Salamon, and I. K. Schuller, *J. Appl. Phys.* 57, 3678 (1985).

201. M. N. Baibich, J. M. Broto, A. Fert, F. Nguyen Van Dau, F. Petroff, P. Etienne, G. Creuzet, A. Friederich, and J. Chazelas, *Phys. Rev. Lett.* 61, 2472 (1988).

202. G. Binasch, P. Grünberg, F. Saurenbach and W. Zinn, *Phys. Rev. B: Condens. Matter.* 39, 4828 (1989).

203. F. Saurenbach, J. Barnaś, G. Binasch, M. Vohl, P. Grünberg, and W. Zinn, *Thin Solid Films* 175, 317 (1989).

204. T. Morishita, Y. Togami, and K. Tsushima, *J. Magn. Magn. Mater.* 54–57, 789 (1986).

205. M. B. Stearns, C. H. Lee, and S. P. Vernon, *J. Magn. Magn. Mater.* 54–57, 791 (1986).

206. Y. Yafet, *J. Appl. Phys.* 61, 4058 (1987).

207. J. J. Krebs, B. T. Jonker, and G. A. Prinz, *J. Appl. Phys.* 63, 3467 (1988).

208. B. Heinrich, K. B. Urquhart, J. R. Dutcher, S. T. Purcell, J. F. Cochran, A. S. Arrott, D. A. Steigerwald, and W. F. Egelhoff, Jr., *J. Appl. Phys.* 63, 3863 (1988).

209. R. E. Camley, T. S. Rahman, and D. L. Mills, *Phys. Rev. B: Condens. Matter.* 27, 261 (1983).

210. R. E. Camley and J. Barnaś, *Phys. Rev. Lett.* 63, 664 (1989).

211. J. Barnaś, A. Fuss, R. E. Camley, P. Grünberg, and W. Zinn, *Phys. Rev. B: Condens. Matter.* 42, 8110 (1990).

212. P. Grünberg and K. Mika, *Phys. Rev. B: Condens. Matter.* 27, 2955 (1983).

213. K. Mika and P. Grünberg, *Phys. Rev. B: Condens. Matter.* 31, 4465 (1985).

214. P. Grünberg, *J. Appl. Phys.* 51, 4338 (1980).

215. P. Grünberg, *J. Appl. Phys.* 52, 6824 (1981).

216. R. E. Camley and M. G. Cottam, *Phys. Rev. B: Condens. Matter.* 35, 189 (1987).

217. A. I. Voronko, P. M. Vetoshko, V. B. Volkovoy, A. D. Boardman, J. W. Boyle, and R. F. Wallis, *Appl. Phys. Lett.* 69, 266 (1996).

218. H. Gao and X. Z. Wang, *Phys. Rev. B: Condens. Matter.* 55, 12424 (1997).

219. X. Z. Wang and D. R. Tilley, *J. Phys.: Condens. Matter.* 9, 5777 (1997).

220. B. Li, J. Yang, J. L. Shen, and G. Z. Yang, *Phys. Rev. B: Condens. Matter.* 50, 9906 (1994).

221. X. Liu, R. L. Stamps, R. Sooryakumar, and G. A. Prinz, *Phys. Rev. B: Condens. Matter.* 54, 11903 (1996).

222. M. Grimsditch, E. E. Fullerton, and R. L. Stamps, *Phys. Rev. B: Condens. Matter.* 56, 2617 (1997).

223. D. S. Schmool, J. S. S. Whiting, A. Chambers, and E. A. Wilinska, *J. Magn. Magn. Mater.* 131, 385 (1994).

224. S. Aubry, *J. Phys. (Paris)* 44, 147 (1983).

225. B. Mandelbrot, "Fractal." Freeman, San Francisco, 1977.

226. T. Kontorova and Ya. I. Frenkel, *Zh. Eksp. Teor. Fiz.* 89, 1340 (1938).

227. F. C. Frank and J. Van Der Merwe, *Proc. R. Soc. London, Ser. A* 198, 205 (1949).

228. J. W. Feng, G. J. Jin, A. Hu, S. S. Kang, S. S. Jiang and D. Feng, *Phys. Rev. B: Condens. Matter.* 52, 15312 (1995).

229. D. H. A. L. Anselmo, M. G. Cottam, and E. L. Albuquerque, *J. Phys.: Condens. Matter.* 12, 1041 (2000).

230. V. A. Ignatchenko, R. S. Iskhakov, and Yu. I. Mankov, *J. Magn. Magn. Mater.* 140–144, 1947 (1995).

231. Z. Q. Han, *J. Magn. Magn. Mater.* 140–144, 1995 (1995).

232. C. Thibaudeau and A. Caillé, *Phys. Rev. B: Condens. Matter.* 32, 5907 (1985).

233. P. R. Emtage and M. R. Daniel, *Phys. Rev. B: Condens. Matter.* 29, 212 (1984).

234. R. E. Camley and R. L. Stamps, in "Linear and Nonlinear Spin Waves in Magnetic Films and Superlattices" (M. G. Cottam, Ed.), Chapt. 5. World Scientific, Singapore, 1994.

235. D. R. Tilley, in "Linear and Nonlinear Spin Waves in Magnetic Films and Superlattices" (M. G. Cottam, Ed.), Chapt. 4. World Scientific, Singapore, 1994.

236. N. Raj and D. R. Tilley, *Phys. Status Solidi B* 152, 135 (1989).

237. N. Raj and D. R. Tilley, *Phys. Rev. B: Condens. Matter.* 36, 7003 (1987).

238. R. E. Camley, B. Djafari-Rouhani, L. Dobrzynski, and A. A. Maradudin, *Phys. Rev. B: Condens. Matter.* 27, 7318 (1983).

239. J. Sapriel and B. Djafari-Rouhani, *Surf. Sci. Rep.* 10, 189 (1989).

240. R. A. Brito-Orta, V. R. Velasco, and F. García-Moliner, *Surf. Sci.* 187, 223 (1987).

241. L. Fernández, V. R. Velasco, and F. García-Moliner, *Europhys. Lett.* 3, 723 (1987).

242. L. Fernández, V. R. Velasco, and F. García-Moliner, *Surf. Sci.* 188, 140 (1987).

243. E. H. El Boudouti, B. Djafari-Rouhani, and A. Nougaoui, *Phys. Rev. B: Condens. Matter.* 51, 13801 (1995).
244. L. Fernández-Alvarez and V. R. Velasco, *J. Phys. Condens. Matter.* 8, 6531 (1996).
245. J. Barnaś and P. Grünberg, *J. Magn. Magn. Mater.* 82, 186 (1989).
246. M. Vohl, J. Barnaś, and P. Grünberg, *Phys. Rev. B: Condens. Matter.* 39, 12003 (1989).
247. G. T. Rado, *Phys. Rev. B: Condens. Matter.* 26, 295 (1982).
248. F. Hoffmann, *Phys. Status Solidi* 41, 807 (1970).
249. C. Vittoria, *Phys. Rev. B: Condens. Matter.* 32, 1679 (1985).
250. G. C. Bailey and C. Vittoria, *Phys. Rev. B: Condens. Matter.* 8, 3247 (1973).
251. D. Schwenk, F. Fishman, and F. Schwabl, *Phys. Rev. B: Condens. Matter.* 38, 11618 (1988).
252. K. Vayhinger and H. Kronmüller, *J. Magn. Magn. Mater.* 72, 307 (1988).
253. C. Zhou and C. D. Gong, *Phys. Rev. B: Condens. Matter.* 39, 2603 (1989).
254. C. Zhou and C. D. Gong, *Phys. Rev. B: Condens. Matter.* 38, 4936 (1988).
255. J. F. Cochran, J. Rudd, W. B. Muir, B. Heinrich, and Z. Celinski, *Phys. Rev. B: Condens. Matter.* 42, 508 (1990).
256. J. B. Sokoloff, *Phys. Rev. B: Condens. Matter.* 42, 760 (1990).
257. R. L. Stamps, *Phys. Rev. B: Condens. Matter.* 49, 339 (1994).
258. B. A. Kalinikos and N. V. Kozhus', *Sov. Phys. Tech. Phys.* 34, 1105 (1989).
259. M. P. Kostylev, B. A. Kalinikos, and H. Dötsch, *J. Magn. Magn. Mater.* 145, 93 (1995).
260. G. Gubbiotti, G. Carlotti, and B. Hillebrands, *J. Phys.: Condens. Matter.* 10, 2171 (1998).
261. F. Garcia-Moliner and V. Velasco, *Phys. Scr.* 34, 257 (1986).
262. F. Herman, P. Lambin, and O. Jepsen, *Phys. Rev. B: Condens. Matter.* 31, 4394 (1985).
263. L. Dobrzynski, B. Djafari-Rouhani, and H. Puszkarski, *Phys. Rev. B: Condens. Matter.* 33, 3251 (1986).
264. L. L. Hinchey and D. L. Mills, *Phys. Rev. B: Condens. Matter.* 33, 3329 (1986).
265. J. A. Morkowski and A. Szajek, *J. Magn. Magn. Mater.* 71, 299 (1988).
266. N. N. Chen, M. G. Cottam, and A. F. Khater, *Phys. Rev. B: Condens. Matter.* 51, 1003 (1995).
267. Guo-zhu Wei, Rong-ke Qiu, and An Du, *Phys. Lett. A* 205, 335 (1995).
268. Guo-zhu Wei and An Du, *J. Phys.: Condens. Matter.* 7, 8813 (1995).
269. C.C. Cheng and F.C. Pu, *Acta Phys. Sin.* 20, 624 (1964).
270. R. K. Qiu, G. Z. Wei, and A. Du, *Phys. Status Solidi B* 205, 645 (1998).
271. R. K. Qiu and Z. D. Zhang, *J. Phys.: Condens. Matter.* 13, 4165 (2001).
272. P. Azaria, *J. Phys. C: Solid State Phys.* 19, 2773 (1986).
273. P. Azaria and H. T. Diep, *J. Appl. Phys.* 61, 4422 (1987).
274. H. T. Diep, *Phys. Rev. B: Condens. Matter.* 40, 4818 (1989).
275. J. Mathon, *J. Phys.: Condens. Matter.* 1, 2505 (1989).
276. Wen-Zhong Shen and Zhen-Ya Li, *Phys. Rev. B: Condens. Matter.* 47, 2636 (1993).
277. E. L. Albuquerque, P. Fulco, E. F. Sarmento, and D. R. Tilley, *Solid State Commun.* 58, 41 (1986).
278. J. Barnaś, *J. Phys. C: Solid State Phys.* 21, 1021 (1988).
279. J. Barnaś, *J. Phys. C: Solid State Phys.* 21, 4097 (1988).
280. J. Barnaś, *J. Phys.: Condens. Matter.* 2, 7173 (1990).
281. J. Barnaś, *Phys. Rev. B: Condens. Matter.* 45, 10427 (1992).
282. D. Mercier, J. C. S. Lévy, M. L. Watson, J. S. S. Whiting, and A. Chambers, *Phys. Rev. B: Condens. Matter.* 43, 3311 (1991).
283. J. C. S. Lévy and H. Puszkarski, *J. Phys.: Condens. Matter.* 3, 5247 (1991).
284. H. Puszkarski and L. Dobrzynski, *Phys. Rev. B: Condens. Matter.* 39, 1819 (1989).
285. H. Puszkarski and L. Dobrzynski, *Phys. Rev. B: Condens. Matter.* 39, 1825 (1989).
286. H. Puszkarski, *Acta Phys. Pol. A* 74, 701 (1988).
287. D. Mercier and J. C. S. Lévy, *J. Magn. Magn. Mater.* 93, 557 (1991).
288. D. Mercier and J. C. S. Lévy, *J. Magn. Magn. Mater.* 139, 240 (1995).
289. D. Mercier and J. C. S. Lévy, *J. Magn. Magn. Mater.* 145, 133 (1995).
290. D. Mercier and J. C. S. Lévy, *J. Magn. Magn. Mater.* 163, 207 (1996).
291. R. P. van Stapele, F. J. A. M. Greidanus, and J. W. Smits, *J. Appl. Phys.* 57, 1282 (1985).
292. H. Puszkarski, *J. Magn. Magn. Mater.* 93, 290 (1991).
293. J. Barnaś and B. Hillebrands, *Phys. Status Solidi B* 176, 465 (1993).
294. H. Puszkarski, *Solid State Commun.* 72, 887 (1989).
295. Y. M. Seidov, N. F. Hashimzade, and R. I. Tagiyeva, *J. Magn. Magn. Mater.* 136, 88 (1994).
296. E. C. Valadares and J. A. Plascak, *Physica A* 153, 252 (1988).
297. Gen-di Pang and Fu-cho Pu, *Phys. Rev. B: Condens. Matter.* 38, 12649 (1988).
298. A. Akjouj, B. Sylla, P. Zielinski, and L. Dobrzynski, *Phys. Rev. B: Condens. Matter.* 37, 5670 (1988).
299. H. I. Zhang, K. H. Lee, and M. H. Lee, *J. Phys. C: Solid State Phys.* 19, 699 (1986).
300. H. I. Zhang, K. H. Lee and M. H. Lee, *J. Phys. C: Solid State Phys.* 19, 709 (1986).
301. M. H. Lee, H. I. Zhang, and K. H. Lee, *Phys. Lett. A* 90, 435 (1982).
302. B. Djafari-Rouhani, L. Dobrzynski, and P. Masri, *Ann. Phys. (Paris)* 6, 259 (1981).
303. B. Djafari-Rouhani, P. Masri, and L. Dobrzynski, *Phys. Rev. B: Solid State* 15, 5690 (1977).
304. S. C. Ying, *Phys. Rev. B: Solid State* 3, 4160 (1971).
305. Zhi-dong Zhang, *Phys. Rev. B: Condens. Matter.* 53, 2569 (1996).
306. Zhi-dong Zhang, *Phys. Rev. B: Condens. Matter.* 55, 12408 (1997).
307. M. Parkov, M. Škrinjar, D. Kapor, and S. Stojanović, *Phys. Rev. B: Condens. Matter. and Mater. Phys.* 62, 6385 (2000).
308. M. P. Gokhale and D. L. Mills, *Phys. Rev. B: Condens. Matter.* 49, 3880 (1994).
309. R. Świrkowicz, *J. Magn. Magn. Mater.* 121, 134 (1993).
310. R. Swirkowicz, *J. Magn. Magn. Mater.* 140–144, 1945 (1995).
311. R. Swirkowicz, *J. Magn. Magn. Mater.* 163, 70 (1996).
312. R. E. Camley and D. R. Tilley, *Phys. Rev. B: Condens. Matter.* 37, 3413 (1988).
313. W. Schmidt, *J. Magn. Magn. Mater.* 84, 119 (1990).
314. P. Bruno and C. Chappert, *Phys. Rev. B: Condens. Matter.* 46, 261 (1992).
315. P. M. Levy, Shufeng Zhang, and A. Fert, *Phys. Rev. Lett.* 65, 1643 (1990).
316. A. R. Völkel, G. M. Wysin, A. R. Bishop, and F. G. Mertens, *Phys. Rev. B: Condens. Matter.* 44, 10066 (1991).
317. A. R. Völkel, F. G. Mertens, A. R. Bishop, and G. M. Wysin, *Phys. Rev. B: Condens. Matter.* 43, 5992 (1991).
318. F. G. Mertens, A. R. Bishop, G. M. Wysin and C. Kawabata, *Phys. Rev. B: Condens. Matter.* 39, 591 (1989).
319. P. Kabos, G. Wiese, and C.E. Patton, *Phys. Rev. Lett.* 72, 2093 (1994).
320. Ming Chen, M. A. Tsankov, J. M. Nash, and C. E. Patton, *Phys. Rev. B: Condens. Matter.* 49, 12773 (1994).
321. P. Kabos, C. E. Patton, M. O. Dima, D. B. Church, R. L. Stamps, and R. E. Camley, *J. Appl. Phys.* 75, 3553 (1994).
322. H. Poincaré, *Acta Math.* 13, 1 (1890).

Chapter 5

QUANTUM WELL INTERFERENCE IN DOUBLE QUANTUM WELLS

Zhang Zhi-Dong

Shenyang National Laboratory and International Centre for Materials Physics, Institute of Metal Research, Academia Sinica, Shenyang 110015, Peoples' Republic of China

Contents

Handbook of Thin Film Materials, edited by H.S. Nalwa
Volume 5: Nanomaterials and Magnetic Thin Films

ISBN 0-12-512913-0/$35.00

1. INTRODUCTION

Since the discovery of interlayer exchange interaction between ferromagnetic layers separated by a nonmagnetic spacer [1, 2] and of the oscillations of interlayer exchange coupling [3], the mechanism of interlayer exchange coupling has been a subject of intense investigation in the last decade. Among the mechanisms of the interlayer exchange coupling, the quantum well states (QWS) have attracted considerable attention [4, 5]. It is believed that the spin-polarized QWS form in layered magnetic nanostructures because of the spin-dependent scattering of the electrons at the magnetic/nonmagnetic interfaces [6–10]. Work on Co/Ni/Co/Cu/Co/Ni/Co [11], Fe/Cr [12], and Fe/Au [12] systems revealed that the coupling strength actually depends on the magnetic layer thickness. Moreover, the coupling was found to oscillate with the thickness of a Cu cap layer on top of a Co/Cu/Co sandwich [13]. All of these results imply that each layer in an entire multilayer stack is relevant to the magnetic coupling.

To have a better understanding of the mechanism of the exchange coupling, one needs to investigate in detail quantum well interference (QWI) in double quantum wells (DQWs). Several photoemission experiments have been conducted that reveal the interference behaviors of electrons at the Fermi surface [14–16]. However, it is necessary to carry out theoretical investigations to interpret the experimental data and to gain a deeper insight into this phenomenon.

It is well known that the energy of electrons moving in a quantum well is quantized according to quantum mechanics. If the electrons move in two or more quantum wells separated by spacers, the quantum well interference between the wells may either be neglected when the thickness of the spacers is thick enough or be very pronounced when the wells are close enough. It is easy to find the solution of the quantization of the quantum well states in a single quantum well in any textbook on quantum mechanics. However, to the author's knowledge, there has been no detailed introduction or systematically theoretical study of quantum well interference in double quantum well systems.

The first principle calculation should be the first task for this purpose. The total energy calculations have been performed either within semiempirical tight-binding models [17] or *ab initio* schemes [18–22]. Unfortunately, this kind of approach meets many difficulties, due to the tiny energy difference between the parallel and antiparallel configurations of the magnetization and to the limitations of computation time. The coupling strengths obtained are typically one order of magnitude larger than the experimental coupling strengths [23]. To circumvent these difficulties, various models have been proposed to interpret the interlayer exchange coupling. They include the

Ruderman–Kittel–Kasuya–Yosida (RKKY) model [24–28], the free electron model [29–33], the hole confinement model (i.e., the single-band tight-binding model) [34, 35], and the Anderson (or sd-mixing) model [36–38].

The effective mass model [39–41], based on the assumption of one electron, is the simplest model dealing with this problem. Although the effective mass model is a rather crude theory with approximations neglecting the interaction and the correlation between the electrons, it still gives results that are in good qualitative agreement with the experimental results. One of advantages of the effective mass model is that it allows us to derive the analytical solutions of the probability of electrons moving in the wells and thus gives a clearer physical picture than other models could give. In this work, we restrict ourselves to the topic of the transition-metal systems. Although the model uses an approximation that may not be well for transition metals with iterate character, it can either fit or explain the experimental results qualitatively. This chapter simply tries to illustrate the extensive features of quantum well interference in double quantum well systems and to show the existence of a possible correlation between quantum well states and exchange coupling in these systems.

Before studying the DQW in detail, first we briefly introduce single quantum wells (SQWs) in Section 2. The bound states and the resonant scattering states are introduced, and the asymmetry of the SQW due to the effect of vacuum is discussed. Second, we describe in Section 3 various models for exchange coupling in magnetic nanostructures. The first two sections construct the theoretical background and the starting point of the present investigation into DQWs. In Section 4 we concentrate on the bound states for symmetric DQWs. The quantization condition for the bound states of DQWs is derived analytically. In Section 5 we are interested in the resonant scattering states of symmetric DQWs. In addition to the behavior of QWI across the nonmagnetic layers, we also study QWI across DQWs consisting of magnetic layers. The results show that there might be an oscillation of the exchange coupling between the magnetic layers if the quantum well mechanism for the exchange coupling is valid. We pay attention mainly to the resonant scattering states of several asymmetric DQWs in Sections 6 and 7. The numerical results are in good agreement with the photoemission data. The asymmetry of DQWs is discussed, in comparison with symmetric DQWs. In Section 8 the advantages and limitations of the model and the relation between quantum well effects and exchange coupling are discussed. In this chapter we focus our attention only on quantum interference in quantum well systems formed by the transition metals Co, Cu, Ni, and Fe. We do not study quantum interference in other quantum well systems formed by other transition metals, semiconducting materials, and/or oxides. We do not pay attention to transport properties and biquadratic coupling. Instead, we list in the discussion section some relevant literature for these related topics and/or areas for the convenience of the reader. Section 9 contains concluding remarks. The parameters of the wave functions derived are represented in Appendixes A–H.

2. SINGLE QUANTUM WELLS

2.1. Introduction

The well-known elementary example of quantization in one-dimensional SQWs has been described in quantum mechanics textbooks [42–46]. The simplest quantum well to be dealt with consists of a single layer of material A embedded between two thick layers of material B. The thickness of the layers B is much greater than the penetration length of the confined wave function. The material B has a bandgap larger than the material A, and the band discontinuities are such that the carriers are confined in the A material. In the approximation of the envelope wave function [47–49], the energy levels in the conduction band can be calculated, and the electron wave function takes approximately the form

$$\Psi = \sum_{A,B} e^{i\mathbf{k}_\perp \cdot r} \mathbf{u}_{ck}^{A,B}(r) \psi_n(x) \tag{1}$$

where x is the growth direction, $\mathbf{k}_\perp$ is the transverse electron wave vector, and $\mathbf{u}_{ck}(r)$ is the Bloch wave function in the A or B material. $\psi_n(x)$ is the envelope wavefunction, which can be determined by the Schrödinger-like equation

$$\left(-\frac{\hbar^2}{2m^*(x)} \frac{\partial^2}{\partial x^2} + V_c(x) \right) \psi_n(x) = \varepsilon_n \psi_n(x) \tag{2}$$

where $m^*(x)$ is the electron effective mass of the A or B material. $V_c(x)$ represents the energy level of the bottom of the conduction bands, and ε_n is the so-called confinement energy of the carriers. This is the so-called effective mass model.

There are two types of states, bound states at negative energies, and scattering states at positive energies, as the energy origin is chosen to be the band minimum in the asymptotic region. At negative energies, the quantum interference at discrete energies is constructive, resulting in the bound states. The bound states consist of the waves bouncing back and forth in the well with tails exponentially decaying into the asymptotic region. There are scattering states at all positive energies. The scattering states for a SQW consist of a plane wave incident on the well from either side, a reflected wave with reduced amplitude on the same side of the well, the scattering waves in both directions, and a transmitted wave on the other side of the well. The oscillatory component to the reflection probability is caused by transmission resonances. If the energies are close to those at which an integer number of wavelengths fits inside the well, the state undergoes increased multiple scattering in the well and transmits with unit probability. The energies of the resonances follow the progression of the bound state energies [4]. There

is no great difference between the probability densities for the resonant and the nonresonant states. The differences are mainly due to the changing amount of interference between left- and right-going waves, depending on the amount of reflections.

We shall describe infinitely deep wells and finite wells in Sections 2.1 and 2.3. We shall discuss in Section 2.4 a variety of the SQW, in which one of the two thick layers of the material B is replaced by a vacuum.

2.2. Infinitely Deep Wells

For an infinitely deep well (Fig. 1A), the solution of Eq. (2) is very simple. Because the wave function must be zero in the confining layer B and at the interface because of the continuity condition, the solution can evidently only be in the form of $\sin(n\pi x/d)$, where n is an odd or even integer. The confining energy ε_n is then simply $n^2(\pi^2 \hbar^2 / 2m^* d^2)$. All of the states are bound states, and there is an infinity of bound states.

2.3. Finitely Deep Eells

2.3.1. Bound States

The Schrödinger-like equation (2) for the finitely deep well (Fig. 1B) with the continuity boundary condition can be solved exactly to yield the wavefunctions and energies. The wavefunction for bound states can be expressed as

$$\psi_n(x) = \begin{cases} A e^{\kappa_B x}, & x \leq -\frac{d}{2} \\ B e^{ik_A x} + C e^{-ik_A x}, & -\frac{d}{2} \leq x \leq \frac{d}{2} \\ D e^{-\kappa_B x}, & x \geq \frac{d}{2} \end{cases} \tag{3}$$

where

$$\varepsilon_n = \frac{\hbar^2 k_A^2}{2m_A^*} - V_0, \qquad -V_0 < \varepsilon < 0 \tag{4a}$$

$$\varepsilon_n = -\frac{\hbar^2 \kappa_B^2}{2m_B^*} \tag{4b}$$

The center of the well is taken as the center of coordinates. The relations among the parameters A, B, C, and D are represented in Appendix A. The continuity condition at the boundaries of the well yields

$$\frac{k_A}{m_A^*} \tan\left(\frac{k_A d}{2} \right) = \frac{\kappa_B}{m_B^*} \tag{5a}$$

and

$$\frac{k_A}{m_A^*} \cotan\left(\frac{k_A d}{2} \right) = -\frac{\kappa_B}{m_B^*} \tag{5b}$$

for even and odd solutions of the wavefunctions of Eq. (2), respectively. Equations (5a) and (5b) can be solved numerically or graphically. A very simple graphic type of the solutions can be developed if $m_A^* \approx m_B^*$. Equations (5a) and (5b) can be transformed into implicit equations in k_A alone, by using Eqs. (4a)

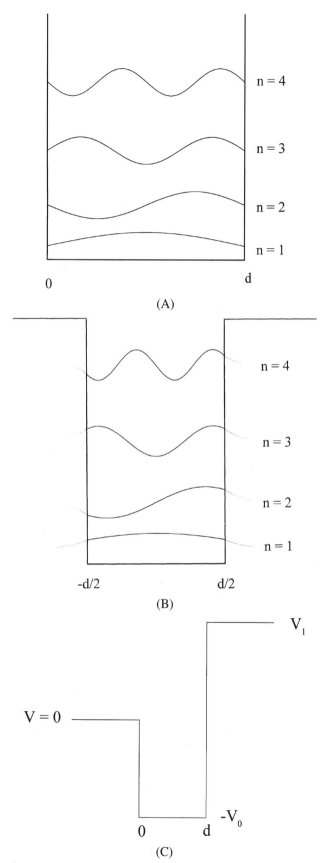

Fig. 1. (A) Infinite deep quantum-well energy levels and wave functions. (B) First four bound energy levels and wave functions in a finite quantum well. (C) A single quantum well with metal/vacuum interface.

and (4b), in the form

$$\cos\left(\frac{k_A d}{2}\right) = \frac{k_A}{k_0} \tag{6a}$$

for $\tan k_A d/2 > 0$ and

$$\sin\left(\frac{k_A d}{2}\right) = \frac{k_A}{k_0} \tag{6b}$$

for $\tan k_A d/2 < 0$. $k_0^2 = 2m^*V_0/\hbar^2$. Equations (6a) and (6b) can be visualized graphically. It is evident that there is always one bound state. The number of bound states is determined by

$$1 + \text{INT}\left[\left(\frac{2m_A^*V_0 d^2}{\pi^2\hbar^2}\right)^{1/2}\right]$$

where INT[z] indicates the integer part of z. The solutions in Eqs. (6a) and (6b) reduce to the limiting case of the infinitely deep quantum well if one takes $k_0 = \infty$.

2.3.2. Resonant Scattering States

The wavefunctions for free states with a positive energy $E > 0$ are

$$\psi_n(x) = \begin{cases} Ae^{ik_B x} + Be^{-ik_B x}, & x \leq -\frac{d}{2} \\ Ce^{ik_A x} + De^{-ik_A x}, & -\frac{d}{2} \leq x \leq \frac{d}{2} \\ Ee^{ik_B x}, & x \geq \frac{d}{2} \end{cases} \tag{7}$$

where k_A and k_B are the wavevectors of the electrons in the A and B materials, respectively. The continuity conditions yield the relations among the parameters A, B, C, D, and E as represented in Appendix B. For Eq. (B.2), it is seen that the amplitude B of the reflection wave is normally not equal to zero, with the exception of $k_A^2 = k_B^2$ or $kd = n\pi$, $n = 1, 2, 3, \ldots$. The former condition is equivalent to saying that there is no potential well, i.e., $V_0 = 0$, which is not meaningful. The latter condition can be satisfied only for particles with some special values of energies. In this case, the wave that reflects from the well at $x = d/2$ and then penetrates through the well at $x = -d/2$ will be canceled by the wave that reflects from the well at $x = -d/2$, because of the quantum interference. All of the incident waves penetrate the quantum well.

It should be noted that there is always the solution of the free states for any positive energy. This is the characteristic of the free states. One obtains the probability of the electrons being found inside the well, i.e., in the material A, as.

$$\int_{-d/2}^{d/2} |\psi(x)|^2\, dx = \frac{|E|^2}{4k_A^3}\left[2(k_A^2 + k_B^2)k_A d \right.$$
$$\left. + (k_A^2 - k_B^2)\sin 2k_A d\right] \tag{8}$$

when $|E|^2 = 1$, i.e., $2k_A d = 2n\pi$, the probability is maximized.

2.4. Effect of Vacuum

2.4.1. Bound States

If one of the two thick layers of material B is replaced by a vacuum (Fig. 1C), the wavefunction for the bound states may be written as

$$\psi_n(x) = \begin{cases} Ae^{\kappa_B x}, & x \le 0 \\ Be^{ik_A x} + Ce^{-ik_A x}, & 0 \le x \le d \\ De^{-\kappa x}, & x \ge d \end{cases} \quad (9)$$

where

$$\varepsilon_n = -\frac{\hbar^2 \kappa_B^2}{2m_B^*} \quad (10a)$$

$$\varepsilon_n = \frac{\hbar^2 k_A^2}{2m_A^*} - V_0, \qquad -V_0 < \varepsilon_n < 0 \quad (10b)$$

$$\varepsilon_n = -\frac{\hbar^2 \kappa^2}{2m^*} + V_1 - V_0 \quad (10c)$$

where V_1 is the potential for a vacuum. For the continuity condition, one has

$$e^{i\phi_1} = \frac{B}{C} = \frac{k_A - i\kappa_B}{k_A + i\kappa_B} \quad (11a)$$

$$e^{i\phi_2} = \frac{C}{B} e^{-2ik_A d} \quad (11b)$$

where ϕ_1 and ϕ_2 are the phase gains upon reflections at $x = 0$ and $x = d$, respectively.

$$\tan\left(\frac{\phi_1}{2}\right) = -\frac{\kappa_B}{k_A} \quad (12)$$

and ϕ_2 is calculated with the imaging potential,

$$\phi_2 = \pi\sqrt{\frac{3.4}{W}} - \pi \quad (13)$$

Here W is the working potential of the electron to a vacuum. Then one immediately obtains the quantization condition,

$$2k_A d + \phi_1 + \phi_2 = 2n\pi \quad (14)$$

The relation among the parameters A, B, C, and D in Eq. (9) is given in Appendix C. The probability of the electron being found in the well is

$$\int_0^d |\psi(x)|^2 dx = \frac{|A|^2}{4k_A \cos^2(\phi_1/2)} \big[2k_A d + \sin(2k_A d + \phi_1) - \sin\phi_1 \big] \quad (15)$$

where $|A|^2$ is determined by normalization:

$$|A|^2 = \big(4k_A \kappa_B \kappa \cos^2(\phi_1/2) \big) \big(2k_A \kappa \cos^2(\phi_1/2) + 2k_A \kappa_B \kappa d + \kappa_B \kappa (\sin(2k_A d + \phi_1) - \sin\phi_1) + 2k_A \kappa_B \cos^2(k_A d + \phi_1) \big)^{-1} \quad (16)$$

2.4.2. Resonance States

The wavefunction for resonance states is described by

$$\psi_n(x) = \begin{cases} e^{ik_B x} + re^{-ik_B x}, & x \le 0 \\ Be^{ik_A x} + Ce^{-ik_A x}, & 0 \le x \le d \\ De^{-\kappa x}, & x \ge d \end{cases} \quad (17)$$

where

$$\varepsilon_n = \frac{\hbar^2 k_B^2}{2m_B^*} \quad (18a)$$

$$\varepsilon_n = \frac{\hbar^2 k_A^2}{2m_A^*} - V_0 \quad (18b)$$

$$\varepsilon_n = -\frac{\hbar^2 \kappa^2}{2m^*} + V_1 - V_0 \quad (18c)$$

The continuity condition leads to

$$e^{i\phi_2} = \frac{C}{B} e^{-2ik_A d} = \frac{k_A - i\kappa}{k_A + i\kappa} \quad (19)$$

where ϕ_2 is the phase gain upon reflections at $x = d$. In this case, the phase gain ϕ_1 upon reflections at $x = d$ equals zero. The parameters r, B, C, and D in Eq. (17) are dealt with in Appendix D.

The probability that the electrons are found in the well is

$$\int_0^d |\psi(x)|^2 dx = \frac{2k_A d + \big[\sin(2k_A d + \phi_2) - \sin\phi_2 \big]}{k_A \left[\left(1 + \frac{k_A^2}{k_B^2} \right) - \left(\frac{k_A^2}{k_B^2} - 1 \right) \cos(2k_A d + \phi_2) \right]} \quad (20)$$

The probability is maximized when

$$2k_A d + \phi_2 = 2n\pi \quad (21)$$

It is therefore concluded that in all cases, the probability of the electron being found inside the well is maximized when the quantization condition

$$2k_A d + \phi_1 + \phi_2 = 2n\pi \quad (22)$$

is satisfied. It is evident that the quantization conditions in single wells are linear. In the remainder of this chapter, we shall be interested in the quantum behaviors of electrons in DQWs. It is obvious that quantum interference in DQWs should be much more complex than that in SQWs. It would be interesting to determine whether the quantization conditions are still linear in DQWs. Before a detailed discussion on QWI in DQWs, we shall introduce briefly in the next section the mechanisms for interlayer exchange coupling. Among these are the RKKY model [24–28], the free electron model [29–33], the hole confinement model [34, 35], the Anderson sd-mixing model [36–38], and the quantum well mechanism.

3. MECHANISMS FOR EXCHANGE COUPLING

3.1. Introduction

The interlayer exchange interaction between ferromagnetic layers separated by a nonmagnetic spacer was first observed by Grünberg et al. in the Fe/Cr/Fe system [1]. Exchange coupling was subsequently discovered, by Parkin et al. [3], to oscillate between ferromagnetic and antiferromagnetic layers. Parkin showed that the oscillations of the interlayer exchange coupling as a function of the spacer thickness occurs with almost all nonmagnetic transition metals as spacer materials [50]. There has been intense effort applied to understanding this phenomenon since these discoveries. A great number of theoretical studies have been performed, which are essentially two classes of approaches: total energy calculations [17–22] and model calculation. Bruno represented a detailed and comprehensive discussion of the various aspects of the theory of interlayer magnetic coupling [5]. Interlayer exchange coupling was described in terms of quantum interference due to confinement in ultrathin layers. Stiles showed that the Fermi surface of the spacer–layer material is responsible for the oscillatory coupling, based on a simple model of quantum interference [4]. We shall introduce briefly total energy calculations in Section 3.2. Several models have been developed to gain a better understanding of interlayer exchange coupling. We shall introduce the RKKY model [24–28], the free electron model [29–31, 31–33], the tight-binding model [34, 35], and the Anderson sd-mixing model [36–38], respectively in Sections 3.3–3.6. Finally, we discuss in detail the quantum well mechanism for interlayer exchange coupling in Section 3.7. According to Bruno, the results obtained by the quantum well theory could be applied to various models above.

3.2. Total Energy Calculations

The idea of total energy calculations is to compute the total energy of a system for configurations of parallel and antiparallel alignment of the magnetization in neighboring magnetic layers and to identify the energy difference with the interlayer exchange coupling. The calculations could be carried out either within semiempirical tight-binding models [17] or first-principle schemes [18–22, 51–57]. Several different first-principle calculations have been reported that have revealed oscillatory behavior [18, 21, 51–57].

This kind of approach is actually very difficult, because the energy difference between the parallel and antiparallel configurations is tiny (on the order of 1 meV or less), in comparison with the large total energy. The numerical convergence of the calculations is a serious problem of this approach, although it is very simple and straightforward in principle. Furthermore, the computation time increases very rapidly with the size of the unit cell, restricting the applicability of the total energy calculations to small layer thicknesses. This makes the investigation of long-period oscillations problematic. That interpretation of the calculation results relies on various models is another difficulty of the total energy calculations. This is mainly due to the fact that

a Fourier analysis of the exchange coupling versus the spacer thickness is needed to identify the oscillation periods. Becasue of these problems, the coupling strengths obtained by the total energy calculations are typically one order of magnitude larger (or even more) than the experimental coupling strengths [23], although this kind of approach has a certain success in simulating the oscillation periods. Therefore, it is still a serious challenge to elucidate interlayer exchange coupling by total energy calculations.

3.3. RKKY Model

The RKKY model has been used to describe magnetic layers as arrays of localized spins interacting with conduction electrons by a contact exchange potential. The RKKY model was originally proposed by Ruderman and Kittel to explain the indirect coupling between nuclear spins via conduction electrons, and then was extended by Kasuyu and Yosida to the case of electronic magnetic moments. The RKKY energy can be obtained from second-order perturbation theory, with the use of the s-d Hamiltonian as the perturbation, and the unperturbed one-electron states are described by Bloch functions periodic in the lattice. The indirect exchange interaction is long range and oscillatory. This model has been successfully applied to interpretation of the exchange interaction between the local magnetic moments of rare earth ions either in rare earth elements or in rare earth–transition metal intermetallics. The effective interaction between localized spins is written, with the use of second-order perturbation theory as

$$V_{ij} = J(R_{ij})S_i \bullet S_j \tag{23}$$

Within the free electron approximation, the exchange integral $J(R)$ is given by

$$J(R) = \frac{4A^2 m k_F^4}{(2\pi)^3 \hbar^2} K(2k_F R) \tag{24}$$

with

$$K(x) = \frac{x \cos x - \sin x}{x^4} \tag{25}$$

Yafet [24] studied the interlayer interaction of two-dimensional layers with a uniform distribution of spins, of areal density N_S, in which the spins within a layer are assumed to be aligned. He found, by using the second-order perturbation, for the free electron approximation,

$$J_1 = \frac{m A^2 S^2 N_S^2 k_F^2}{4\pi^2 \hbar^2} Y(2k_F D) \tag{26}$$

with

$$Y(x) = \frac{x \cos x - \sin x}{2x^2} - \frac{1}{2} \int_x^{+\infty} \frac{\sin y}{y} dy \tag{27}$$

Chappert and Renard [25] and Coehoorn [28] discussed the effect of discrete lattice spacing, and Bruno and Chappert [26, 27] treated the general case of an arbitrary Fermi surface. Bruno discussed Yafet's RKKY model from the point of view of a general quantum well theory. Skomski investigated the RKKY interaction between well-separated magnetic particles in a non-magnetic metallic matrix [58]. It is understood that the actual electron states in multilayers are not eigenfunctions of the lattice transition operators. Electrons propagating along the growth direction of the multilayers feel potential discontinuities when moving from the ferromagnetic or the nonmagnetic slabs. Thus the electron wave functions, although periodic in two dimensions, have to satisfy matching conditions at the interfaces, as shown in Section 2.

3.4. Free Electron Model

Many variants of the free electron model have been proposed [29–33]. The advantage of this simple model is that the calculation can be performed analytically, providing a physically transparent illustration of the various aspects of the problem. For the free electron model, the zero of energy is taken at the bottom of the majority band of the ferromagnetic layers. The potential of the minority band is given by the exchange splitting Δ, and the spacer, of thickness D, has a potential equal to U. The magnetizations of the ferromagnetic layers of thickness L are at an angle θ with respect to each other. In accordance with the position of the Fermi level, this model describes the case of a metallic spacer for $\varepsilon_F > U$ or of an insulating spacer for $\varepsilon_F < U$. The energy difference between the ferromagnetic and antiferromagnetic configurations has been found to relate to the (complex) amplitudes of the reflections of electrons at the interfaces between the ferromagnetic and nonmagnetic layers. The reflection coefficients for the free electron model can be calculated as the process in standard textbooks of quantum mechanics [42–46] as well as that shown in the last section.

3.5. Hole Confinement Model

The hole confinement model is essentially a tight-binding model with spin dependent potential steps. The single-band tight-binding model has been proposed by Edwards et al. [34, 35] and Li et al. [59] for the study of interlayer exchange coupling.

The Hamiltonian for a simple–cubic lattice is written as

$$H = H_0 + V_A + V_B \tag{28}$$

where the Hamiltonian of the pure spacer material is

$$H_0 = \sum_{R_\parallel R_\perp \sigma} \left[\varepsilon_0 c^*_{R_\parallel R_\perp \sigma} c_{R_\parallel R_\perp \sigma} + t_\parallel \sum_{d_\parallel} c^*_{R_\parallel R_\perp \sigma} c_{(R_\parallel + d_\parallel) R_\perp \sigma} \right.$$
$$\left. + t_\perp \sum_{d_\perp} c^*_{R_\parallel R_\perp \sigma} c_{(R_\parallel + d_\perp) R_\perp \sigma} \right] \tag{29}$$

and the perturbation due to the magnetic layers F_A (F_B) is given by

$$V_{A(B)} = \sum_\sigma \sum_{(R_\parallel R_\perp) \in F_{A(B)}} \left(\varepsilon^\sigma_{A(B)} - \varepsilon_0 \right) c^*_{R_\parallel R_\perp \sigma} c_{R_\parallel R_\perp \sigma} \tag{30}$$

where $d_\parallel$ or $d_\perp$ is a vector joining nearest neighbors, and c^* and c are creation and annihilation operators, respectively. This model is the tight-binding analog of the free electron model with spin-dependent potential steps. Bruno extended this model to a more general situation, which may be applicable for a tetragonal distortion of the lattice [5]. The Hamiltonian H_0 was rewritten in terms of the Bloch functions,

$$H_0 = \sum_{k_\parallel k_\perp \sigma} \varepsilon_{k_\parallel k_\perp} c^*_{k_\parallel k_\perp \sigma} c_{k_\parallel k_\perp \sigma} \tag{31}$$

with

$$\varepsilon_{k_\parallel k_\perp \sigma} = \varepsilon_0 - 2t_\parallel \left[\cos(k_z a) + \cos(k_y a) \right] - 2t_\perp \cos(k_\perp a) \tag{32}$$

Bruno gave the general expression of the interlayer exchange coupling for semi-infinite magnetic layers.

The exchange coupling between magnetic layers across non-magnetic superlattices was investigated by Ferreira [60] with a multiorbital tight-binding model. It was shown that the alteration of the oscillation periods reflects the deformation of the corresponding Fermi surface only when the coupling is investigated as a function of the number of superlattice cells separating the magnetic layers. A change in the composition of the superlattice enables one to deform the corresponding Fermi surface and consequently tune the oscillation periods of the coupling.

A rigorous quantum calculation of the current perpendicular-to-plane giant magnetoresistance (CPP GMR) of a Co/Cu/Co (001) trilayer without impurity scattering was performed by Mathon et al. [61, 62]. The torque method was formulated in terms of Green's functions in a tight-binding representation [63]. Symmetry dependence of QWS in thin metallic overlayers was studied by means of the interface Green's function technique [64–66]. This method was also used to investigate the formation of the magnetic induced moments and their relation to the QWS in bcc (001) Fe/Cu$_n$/Fe trilayers [67]. The conductances per spin in the ferromagnetic and antiferromagnetic configurations of the magnetic layers were computed from the Kubo formula [68, 69]. The electronic structure of the Cu and Co layers was described by fully realistic s, p, d tight-binding bands fitted to the *ab initio* band structure of Cu and ferromagnetic fcc Co. Luce et al. calculated the nonlinear magneto-optical response due to the QWS, by using an electronic tight-binding theory [70].

3.6. The Anderson Model

The Anderson (or sd-mixing) model was originally proposed to describe the magnetic behavior of isolated impurities in a nonmagnetic host material. Caroli, using perturbation theory, studied the exchange coupling between two impurities. Interlayer coupling was investigated by Wang et al. [36], Shi et al. [37], and Bruno [38] with the Anderson model within the frame of perturbation theory. Bruno treated the Anderson model within the nonperturbation theory [5, 38]. The Hamiltonian of the spacer material, described by a three-dimensional array of "s states," is

$$H_s = \sum_{k_\parallel k_\perp \sigma} \varepsilon^s_{k_\parallel k_\perp} c^*_{k_\parallel k_\perp \sigma} c_{k_\parallel k_\perp \sigma} \tag{33}$$

The magnetic layers F_A and F_B consist of two-dimensional arrays of localized "d states" embedded in the host material. The Hamiltonian of the magnetic layers consists of three terms, corresponding to the two-dimensional band energy due to inplane hopping, the on-site repulsive Coulomb interaction, and the sd mixing. For a detailed description, readers may refer to Bruno's paper [5]. Shen and Pu tried to understand the mechanism of interlayer magnetic coupling, based on rigorous results for both the antiferromagnetic Heisenberg model and the Hubbard model [71]. They found intrinsic periodicity associated with QWS in a magnetic sandwich, using the standard s-d interaction Hamiltonian, considering the Heisenberg interaction between the nearest-neighbor magnetic atoms [72].

3.7. Quantum Well Model

A one-dimensional SQW is the simplest model that illustrates oscillatory exchange coupling. Oscillatory exchange coupling arises because the energy that it takes to fill the electron states up to the Fermi level oscillates as a function of the well thickness [4]. The period of the oscillation is governed by the wave vector at the Fermi energy in the quantum well. Thus for the magnetic sandwich structures the oscillatory exchange coupling arises for the same reasons as the oscillatory energy in the quantum well. The energy oscillates for different configurations of ferromagnetic or antiferromagnetic alignment of the magnetization. The difference in the oscillatory energies is the oscillatory exchange coupling. There are three different oscillatory energies, one each for spin-up and spin-down electrons in a ferromagnetically alignment, and one for either spin in an antiferromagnetic alignment. For the SQW, all of the oscillatory energies have the same period because the period is determined only by the Fermi surface of the bulk spacer–layer material. However, the reflection probabilities may be different, because of the difference in the potential barriers, which lead to the difference in the amplitudes of the oscillatory energies. The exchange coupling was found by Stiles [4] to be equal to the difference between the sum of the energies for the spin-up and spin-down electrons in a ferromagnetically aligned sandwich and twice the energy for either spin in an antiferromagnetically aligned sandwich,

$$J_\infty(d) = \frac{\hbar v_F}{2\pi d}\left[\left|r_\uparrow^\uparrow\right|^2 + \left|r_\downarrow^\uparrow\right|^2 - 2\left|r_\uparrow^\uparrow r_\downarrow^\uparrow\right|\right]\sin(2k_F d + \phi_0) \quad (34)$$

where $r_\uparrow^\uparrow$ ($r_\downarrow^\uparrow$) is the reflection amplitude for a spin-up (-down) electron in the well material reflecting from an up magnetization barrier, and $r_\uparrow^\uparrow = r_\downarrow^\downarrow$ and $r_\downarrow^\uparrow = r_\uparrow^\downarrow$. It is concluded that the exchange coupling in the magnetic sandwich has the same origin as all other oscillation behavior in metals, the response of the electrons at the Fermi surface. The strength of the exchange coupling depends both on the geometry of the Fermi surface and on the reflection amplitudes for electrons scattering from the interfaces between the spacer layers and the magnetic layers. The effects of the existence of an underlying lattice, in turn, a reciprocal lattice with Brillouin zones, were discussed by Stiles [4]. To test this and related models, Stiles calculated the

extremal spanning vectors and the associated Fermi-surface geometrical factors for a large set of spacer–layer materials and interface orientations. The periods calculated by these models are consistent with the measured oscillation periods, but the numbers for the calculated periods are greater than the experimental data.

Bruno found the exchange coupling energy per unit area [5]:

$$E_F - E_{AF} = \frac{1}{4\pi^3}\text{Im}\int d^2k_\parallel$$
$$\times \int_{-\infty}^{\varepsilon_F}\ln\left[\frac{(1 - r_A^\uparrow r_B^\uparrow e^{2ik_\perp D})(1 - r_A^\downarrow r_B^\downarrow e^{2ik_\perp D})}{(1 - r_A^\uparrow r_B^\downarrow e^{2ik_\perp D})(1 - r_A^\downarrow r_B^\uparrow e^{2ik_\perp D})}\right]d\varepsilon \quad (35)$$

According to this formula, the variation in the exchange coupling versus spacer thickness depends only on the space material (via the wave vectors $k_\perp$). The strength and phase of the exchange coupling are determined by the spin asymmetry of the reflection coefficients at the paramagnet-ferromagnet interfaces, depending on the degree of matching of the band structure on the two sides of the interface.

The exchange coupling between two magnetic planes was analyzed by Barnaś [73] within an exact approach, and the results were compared with those obtained perturbatively [40, 74, 75]. Particular attention was paid to the quantum well effects due to confinement by the external boundaries as well as by internal potential barriers. Barnaś also studied the influence of out boundary conditions on the interlayer exchange coupling in magnetic sandwich structures [76] within the framework of the theory developed by Bruno [5, 77, 78]. Two parallel calculations of the exchange coupling in a Co/Cu/Co(001) trilayer were carried out either in the framework of the quantum well formalism or with an alternative expression referred to as the cleavage formula [79]. Electronic structure, photoemission, and magnetic dichroism of ultrathin films of fcc (001) Co and Cu were calculated by a relativistic Green's function theory within the framework of one-electron effective quasi-particle potential [80, 81].

The exchange coupling in magnetic multilayers was explained by an indirect RKKY exchange interaction, taking the quantum size effect into account [40, 41, 82]. A perturbative theory was derived for the RKKY-like exchange coupling between two ferromagnetic layers separated by a nonmagnetic slab. For simplicity, the potential was taken to be either the one-electron potential of the bulk ferromagnet or that of the nonmagnetic metal. As usual, the multilayers are represented by quantum wells or quantum barriers, depending on the relative alignment of the constituents' one-electron potentials [83, 84]. Each eigenvalue for the quantum well wavefunction is twofold degenerate, corresponding to two waves traveling in opposite directions [40]. For the SQWs, the calculation of the wavefunction is straightforward, following the procedure in the last section (also in quantum mechanics textbooks [42–46]). Lee and Chang [41] found that the RKKY coupling integrated over the magnetic layer is a strong oscillatory function of the magnetic layer thickness with a period related to the nesting vector of the Fermi surface for the magnetic material. They also discussed the coupling across a quantum barrier and the effect of the different effective masses. It was found that the RKKY coupling through the quantum barrier is mainly due to the tunneling effect and that an effective mass ratio different from 1 gives rise to quantum size effects and causes a phase shift. In this chapter we focus on the electrons with the same effective mass, rather than the effect of different effective masses on the exchange coupling.

Most models above for the oscillatory exchange coupling predict the same periods if they use the same band structure for the spacer–layer material. However, the calculations of the strength of the exchange coupling are still approximate, because of the drastic approximation of the reflection probabilities. Although the self-consistent electronic structure calculations do not use the drastic approximation, they are very difficult to converge, even for extremely thin layers.

In the following sections, we study in detail QWI in various DQWs with the effective mass model. The wavefunction of the electrons in the DQWs will be derived analytically. The results will also be plotted for a clearer illustration of the probability of electrons in quantum wells and for a better understanding of QWI in DQWs. Special attention is paid to the effect of the symmetry of the quantum wells and of each layer in the quantum well systems. The phase accumulation method is applied to determine whether the linear quantization condition is still valid for each layer or for the whole quantum well in the DQW systems.

4. SYMMETRIC DOUBLE QUANTUM WELLS: BOUND STATES

4.1. Introduction

In this section we study the behaviors of the bound state of a symmetric DQW, which is constructed by inserting a single layer of material A between two thick layers of material B. The thickness of the layers B is smaller than or comparable to the penetration length of the confined wave function. The surfaces of the layers B may be exposed to a vacuum or placed in contact with another material with a high potential. In the following, we assume that the effective masses of the electrons in all of the materials are the same. Section 4.2 describes the model and gives analytical results.

4.2. Model and Analytical Results

The wavefunction $\psi(x)$ of an electron in the bound states of a one-dimensional DQW (as shown in Fig. 2) is

$$\psi(x) = \begin{cases} Ae^{\kappa x}, & x \leq 0 \\ Be^{ik_1 x} + Ce^{-ik_1 x}, & 0 \leq x \leq d_0 \\ De^{ik_2 x} + Ee^{-ik_2 x}, & d_0 \leq x \leq d_0 + d_1 \\ Fe^{ik_1 x} + Ge^{-ik_1 x}, & d_0 + d_1 \leq x \leq d_0 + d_1 + d_2 \\ He^{-\kappa x}, & x \geq d_0 + d_1 + d_2 \end{cases}$$

(36)

where k_1 and k_2 are the wave vectors of the electrons in the materials A and B, respectively. The continuity conditions at

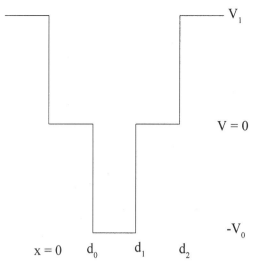

Fig. 2. A symmetrical double quantum well.

the interfaces of different materials yield the parameters A–H as represented in Appendix E. One could define the phase gains upon the reflections at the interfaces also as shown in Appendix E. From the definitions of the phase gains, one immediately obtains [85]

$$e^{i(2k_1 d_0 + \phi_1 + \phi_2)} = 1 \tag{37a}$$

$$e^{i(2k_2 d_1 + \phi_3 + \phi_4)} = 1 \tag{37b}$$

$$e^{i(2k_1 d_2 + \phi_5 + \phi_6)} = 1 \tag{37c}$$

These three relations are equal to

$$2k_1 d_0 + \phi_1 + \phi_2 = 2n\pi \tag{38a}$$

$$2k_2 d_1 + \phi_3 + \phi_4 = 2m\pi \tag{38b}$$

$$2k_1 d_2 + \phi_5 + \phi_6 = 2l\pi \tag{38c}$$

where n, m, and l are integers. It is proved that the three relations for the phase accumulation correspond to the same quantization condition as follows:

$$k_1 k_2 \cos k_2 d_1 \sin\left[k_1(d_0 + d_2) + \phi_2\right]$$
$$+ k_2^2 \sin k_2 d_1 \cos\left(k_1 d_0 + \frac{\phi_2}{2}\right) \cos\left(k_1 d_2 + \frac{\phi_2}{2}\right)$$
$$- k_1^2 \sin k_2 d_1 \sin\left(k_1 d_0 + \frac{\phi_2}{2}\right) \sin\left(k_1 d_2 + \frac{\phi_2}{2}\right) = 0 \quad (39)$$

When $d_0 = d_2$, the double quantum well is completely symmetric. One may obtain a relation by the symmetry of the DQW because the probability of the electrons being in $[0, d_0]$ and $[d_0 + d_1, d_0 + d_1 + d_2]$ should be same. One has

$$\frac{1}{k_2^2}\left[k_2 \cos k_2 d_1 \cos\left(k_1 d_2 + \frac{\phi_2}{2}\right)\right.$$
$$\left. - k_1 \sin k_2 d_1 \sin\left(k_1 d_2 + \frac{\phi_2}{2}\right)\right]^2$$

$$+ \frac{1}{k_1^2} \left[k_2 \sin k_2 d_1 \cos \left(k_1 d_2 + \frac{\phi_2}{2} \right) \right.$$

$$\left. + k_1 \cos k_2 d_1 \sin \left(k_1 d_2 + \frac{\phi_2}{2} \right) \right]^2 = 1 \qquad (40)$$

Equations (39) and (40) can be reduced to be the same condition [85],

$$k_1 k_2 \sin(2k_1 d_2 + \phi_2) \cos k_2 d_1$$

$$+ k_2^2 \sin k_2 d_1 \cos^2 \left(k_1 d_2 + \frac{\phi_2}{2} \right)$$

$$- k_1^2 \sin k_2 d_1 \sin^2 \left(k_1 d_2 + \frac{\phi_2}{2} \right) = 0 \qquad (41)$$

It is concluded that the phase accumulation relation for each well could be expressed in a linear form for the symmetric DWQ. However, the quantization condition of the DQW is actually nonlinear, as shown in Eqs. (39), (40), and (41). This is the character of the bound states of the DQW, indicating that the quantization of the bound states in the DQW is much more complex than that in the SQW. In the remainder of this chapter, we shall be interested only in the resonant scattering states in either symmetric or asymmetric DQW, because it is believed that the probability of the resonant scattering states corresponds to the results of the photoemission experiments. The nonlinear quantization condition is the characteristic of the DQW, a direct result of QWI in the DQW. We discuss such nonlinear behavior in the next several sections.

5. SYMMETRIC DOUBLE QUANTUM WELLS: RESONANT SCATTERING STATES

5.1. Introduction

The DQW structure is normally treated with the usual tight-binding perturbation model [44] or the quasi-classical approximation [45, 46]. However, in the double-wedged samples with film thicknesses on the monolayer (ML) scale, these approximations would be not valid in the ML limit. To gain a systematic understanding of QWI in magnetic nanostructures, in this section, we focus our attention on the resonant scattering states, rather than the bound states, because it is believed that the former (i.e., the QWS above the Fermi level) are observed by photoemission [6–10, 14]. We represent the analytical solution of the resonant scattering states for a DQW system. At first, a DQW constructed of magnetic layers separated by nonmagnetic layers is studied. Numerical results are given for Co/Cu/Co/Cu/Co(100), Co/Cu/Ni/Cu/Co(100), and Co/Cu/Fe/Cu/Co(100). Moreover, there are some interesting questions to be answered: Does the interlayer exchange interaction exist between ferromagnetic layers separated by a ferromagnetic spacer? If so, does it oscillate with the thickness of the ferromagnetic spacer layer? We shall give some numerical results for the DQW system, consisting of ferromagnetic layers separated by ferromagnetic spacers.

5.2. Model and Analytical Results

The wavefunction $\psi(x)$ of an electron in the resonant scattering states of a one-dimensional DQW (Fig. 3) is

$$\psi(x) = \begin{cases} e^{ik_0 x} + r e^{-ik_0 x}, & x \leq 0 \\ A e^{ik_1 x} + B e^{-ik_1 x}, & 0 \leq x \leq \frac{d_1}{2} \\ C e^{ik_2 x} + D e^{-ik_2 x}, & \frac{d_1}{2} \leq x \leq \frac{d_1}{2} + d_2 \\ E e^{ik_1 x} + F e^{-ik_1 x}, & \frac{d_1}{2} + d_2 \leq x \leq d_1 + d_2 \\ t e^{ik_0 x}, & x \geq d_1 + d_2 \end{cases} \qquad (42)$$

Here k_0, k_1, and k_2 are the wave vectors of the electron in the corresponding ferromagnetic layers. In this work, we choose Fe, Co, Cu, and Ni as the materials that make up the DQW, because their lattice constants are nearly the same and high-quality single-crystal multilayers can be grown with them. d_1 and d_2 are of the thickness of the ferromagnetic spacer layers. The parameters r, t, and $A–F$ depend on the QWIs at different interfaces, which are shown in Appendix F. It is worth noting that the present DQW is symmetric. Following the normal quantum theoretical procedure for a SQW [44–46], one finds the relations among these parameters by using the continuum boundary conditions. It should be noted that there are two degeneracy resonant scattering states in the quantum well [40]. Each eigenvalue for the quantum well wavefunction is twofold degenerate, corresponding to two waves traveling in opposite directions [40]. One is transmitted from the left to the right of the DQW, and the other moves in the opposite direction. In the remainder of this chapter, we pay attention only to the former and calculate the probability of the presence of the electron in the different layers. The probability of the presence of the electron in the latter has the opposite sequence for different layers. However, the probability in the whole DQW range of $[0, d_1 + d_2]$ is the same for the two degeneracy resonant scattering states. We list here the formulas for the probability in the

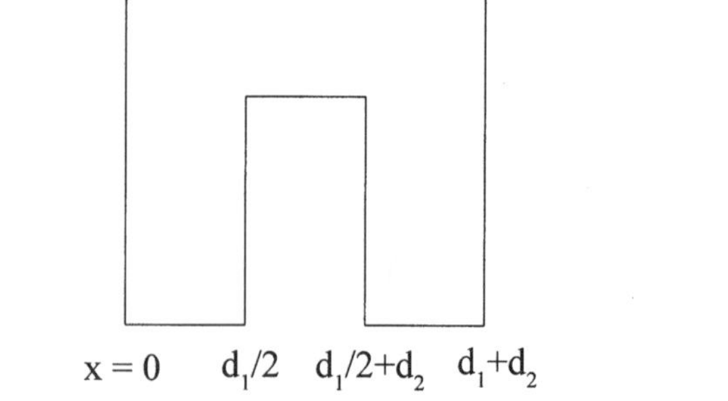

Fig. 3. A symmetrical double quantum well. The dashed line represents the Fermi level.

range $[0, d_1 + d_2]$:

$$\int_0^{d_1+d_2} |\psi(x)|^2\, dx = \int_0^{d_1/2} |\psi(x)|^2\, dx$$
$$+ \int_{d_1/2}^{d_1/2+d_2} |\psi(x)|^2\, dx + \int_{d_1/2+d_2}^{d_1+d_2} |\psi(x)|^2\, dx \quad (43)$$

The last two terms are equal to $s|t|^2$. Here $|t|^2$ is

$$|t|^2 = 64 k_0^2 k_1^4 k_2^2 \big\{ 4k_0^2 k_1^2 \big[(k_1+k_2)^2 \cos(k_1 d_1 + k_2 d_2) $$
$$- (k_1-k_2)^2 \cos(k_1 d_1 - k_2 d_2) \big]^2 $$
$$+ \big[(k_0^2 + k_1^2) \big[(k_1+k_2)^2 \sin(k_1 d_1 + k_2 d_2) $$
$$- (k_1-k_2)^2 \sin(k_1 d_1 - k_2 d_2) \big] $$
$$+ 2(k_1^2-k_0^2)(k_2^2-k_1^2) \sin k_2 d_2 \big]^2 \big\}^{-1} \quad (44)$$

and s is described by

$$s = \frac{1}{2}\bigg\{ \frac{1}{2}\bigg[\Big(1 + \frac{k_0^2}{k_1^2}\Big) d_1 - \Big(1 - \frac{k_0^2}{k_1^2}\Big) \frac{\sin k_1 d_1}{k_1} \bigg]$$
$$+ \bigg[\Big(\frac{k_0^2}{k_1^2} + \frac{k_1^2}{k_2^2}\Big) \sin^2 \frac{k_1 d_1}{2} + \Big(1 + \frac{k_0^2}{k_2^2}\Big) \cos^2 \frac{k_1 d_1}{2} \bigg] d_2$$
$$- \Big(\frac{k_1}{k_2} - \frac{k_0^2}{k_1 k_2}\Big)(1 - \cos 2k_2 d_2) \frac{\sin k_1 d_1}{2k_2}$$
$$+ \bigg[\Big(\frac{k_0^2}{k_1^2} - \frac{k_1^2}{k_2^2}\Big) \sin^2 \frac{k_1 d_1}{2} + \Big(1 - \frac{k_0^2}{k_2^2}\Big) \cos^2 \frac{k_1 d_1}{2} \bigg]$$
$$\times \frac{\sin 2k_2 d_2}{2k_2} \bigg\} \quad (45)$$

The first four terms in Eq. (45) correspond to the probability of the presence of the electron in $[d_1/2+d_2, d_1+d_2]$, and others in Eq. (45) are for the probability of the presence of the electron in $[d_1/2, d_1/2 + d_2]$. The first term in Eq. (43) for the probability of the presence of the electron in $[0, d_1/2]$ is determined by

$$\int_0^{d_1/2} |\psi(x)|^2\, dx = \frac{d_1}{4k_1^2}\bigg[(k_0^2 + k_1^2)\Big(1 + \frac{u^2+w^2}{v^2}\Big)$$
$$+ 2(k_1^2 - k_0^2)\frac{u}{v}\bigg] + \frac{\sin k_1 d_1}{4k_1^3}\bigg[\Big[(k_1^2 - k_0^2)\Big(1 + \frac{u^2+w^2}{v^2}\Big)$$
$$+ 2(k_0^2 + k_1^2)\frac{u}{v}\bigg]\bigg] + \frac{k_0}{k_1^2}\,[1 - \cos k_1 d_1] \quad (46)$$

where u, v, and w are represented in Appendix F. As indicated above, the probability of the first degeneracy resonant scattering state in the range of $[0, d_1/2]$, $[d_1/2, d_1/2+d_2]$, and $[d_1/2+d_2, d_1+d_2]$ is just that of the second degeneracy resonant scattering state in the range of $[d_1/2 + d_2, d_1 + d_2]$, $[d_1/2, d_1/2 + d_2]$, and $[0, d_1/2]$, respectively. We do not mention such difference between the two degeneracy resonant scattering states in the DQW systems in the following sections, but we study one of them.

5.3. Numerical Results

5.3.1. Exchange Coupling across the Nonmagnetic Layers

Oscillatory exchange coupling across nonmagnetic layers has been well known since the discovery of this phenomenon. In this section, the purpose is to illustrate in detail the oscillatory periods of the calculated probability of the presence of the electrons in the symmetrical DQWs. It is expected that the study of DQWs would give a better understanding of what happens in SQWs. We first study quantum well interference in a DQW system of Co/Cu/Ni/Cu/Co(100). The probability of the presence of the electron in the range $[0, d_1 + d_2]$ is represented in Figure 4. All probabilities given in this chapter are only modified by the thickness of the layers in the corresponding range and are not normalized by the total densities of all of the layers in the system. They are plotted in arbitrary units. The peaks in blue correspond to the maxima of the probabilities, and those in red are for the minima of the probabilities. The oscillation in the intensity corresponds to the formation of the QWS in the layers.

The oscillation periodicity of the magnetic coupling [28, 86] and the density of the QWS [4, 10, 14] are given by $\pi/(k_{BZ}-k_F)$, i.e., π/k^{eff}, instead of π/k_F. k_{Fe}^{eff} and k_{Ni}^{eff} are taken to be 0.47 Å^{-1} and 0.57 Å^{-1}, respectively. k_{F1}^{Co} and k_{F2}^{Co} are 0.44 Å^{-1} and 0.74 Å^{-1}, which are close to the Fermi wavevector of k_{F1}^{eff} (0.42 Å^{-1}) for Co [14, 87] and for distorted fcc Co [11, 14]. k_{BZ}^{Fe}, k_{BZ}^{Co}, and k_{BZ}^{Ni} are in good agreement with those for the calculated band structure along the [100] direction [88–92]. The color-filled contours of the probability are obtained by summing directly the probabilities calculated with k_{F1}^{Co} and k_{F2}^{Co}, respectively. As shown below, the probabilities calculated with k_{F1}^{Co} and k_{F2}^{Co} have to be summed to have a good agreement with the experimental data.

The oscillation of the probabilities of the electron is clearly illustrated in Figure 4. The oscillation periodicity modulated by the Cu layers in the absence of a Ni layer is 10.5 Å, i.e., 5.83 ML, in good agreement with experimental results [14]. The oscillation periodicity modulated by the Cu layers is modified by the introduction of Ni layers. The oscillation periodicity of the maxima of the probabilities varies with increasing thickness of the Ni layers. The oscillation periodicity of the minima of the probabilities becomes about 20.5 Å (11.4 ML). The oscillation periodicity of the maxima of the probabilities, modulated by the Ni layers, is 5.87 Å, i.e., 3.26 ML.

For a detail view of the resonant scattering states, the probabilities of the presence of the electron in $[0, d_1/2]$, $[d_1/2, d_1/2+d_2]$, and $[d_1/2 + d_2, d_1 + d_2]$ are shown, respectively, in Figures 5–7. As shown in Figure 5, the oscillation periodicity modulated by the Cu layers in the absence of a Ni layer is the same as that in Figure 4 (10.5 Å, i.e., 5.83 ML). The oscillation periodicity of the minima of the probabilities, modulated by the Ni layers, is slightly lower than in Figure 4 (about 5.73 Å, i.e., 3.19 ML). A different feature is that in Figure 5; the regular oscillation of the minima of the probabilities modulated by the Cu layers is disturbed, and it is hard to account for the oscillation periodicity. The islands of the maxima of the probabilities,

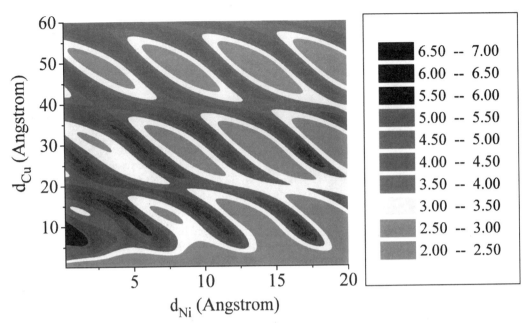

Fig. 4.   Probabilities of the electron in the range $[0, d_1 + d_2]$ of Co/Cu/Ni/Cu/Co(100) (color-filled contours) as a function of thicknesses of the spacer layers. The parameters used during the calculation were $k_{Cu}^{eff} = 0.32$ Å^{-1}, $k_{Ni}^{eff} = 0.57$ Å^{-1}, $k_{F1}^{Co} = 0.44$ Å^{-1}, and $k_{F2}^{Co} = 0.74$ Å^{-1}.

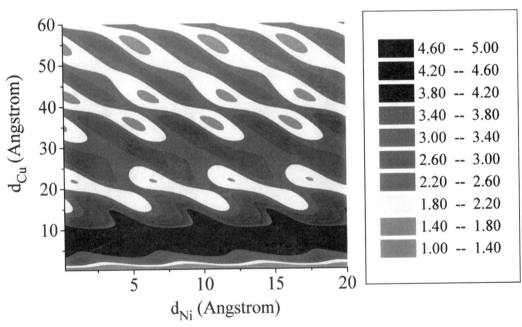

Fig. 5.   Probabilities of the electron in the range $[0, d_1/2]$ of Co/Cu/Ni/Cu/Co(100) (color-filled contours) as a function of thicknesses of the spacer layers. The parameters used during the calculation are the same as those used in Figure 4.

represented in Figure 6, have constant oscillation periodicities of 20.0 Å and 5.59 Å (11.1 ML and 3.11 ML), respectively, modulated by the Cu and Ni layers. The oscillation periodicity modulated by the Cu layers without a Ni layer varies in the range of 9.5–10.0 Å (i.e., 5.28–5.56 ML). The valley for the minima of the probabilities oscillates in a periodicity of 5.59 Å (3.11 ML) upon the modulation of the Ni layers.

The probabilities of the presence of the electron in the wells of Co/Cu/Co/Cu/Co(100) and Co/Cu/Fe/Cu/Co(100) are rep-

resented in Figures 8 and 9, respectively. The probabilities in these two DQWs have a feature similar to those in the wells of Co/Cu/Ni/Cu/Co(100), as shown in Figure 4. The oscillation periodicity modulated by the Cu layers in the absence of a Co or Fe layer is 10.5 Å (i.e., 5.83 ML). The oscillation periodicity modulated by the Cu layers is also modified by the introduction of Co or Fe layers. The oscillation periodicity of the maxima of the probabilities varies with increasing thickness of the Co or Fe layers. The oscillation periodicity of the maxima of

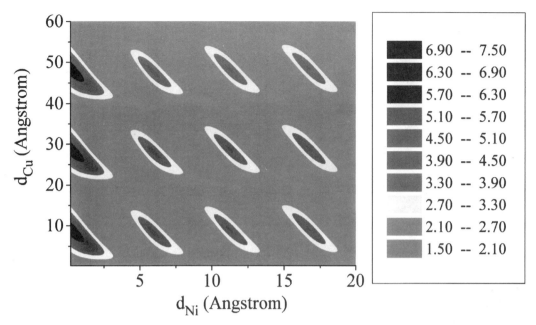

Fig. 6. Probabilities of the electron in the range $[d_1/2, d_1/2 + d_2]$ of Co/Cu/Ni/Cu/Co(100) (color-filled contours) as a function of thicknesses of the spacer layers. The parameters used during the calculation are the same as those used in Figure 4.

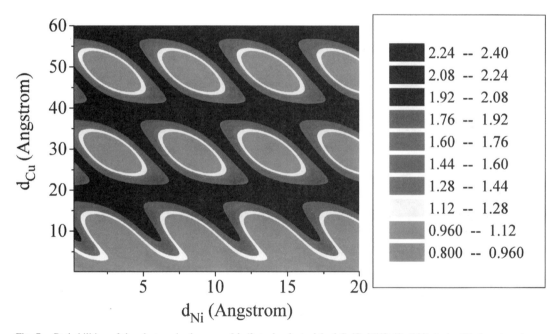

Fig. 7. Probabilities of the electron in the range $[d_1/2 + d_2, d_1 + d_2]$ of Co/Cu/Ni/Cu/Co(100) (color-filled contours) as a function of thicknesses of the spacer layers. The parameters used during the calculation are the same as those used in Figure 4.

the probabilities, modulated by the Co layers, is 4.31 Å (i.e., 2.39 ML) (see Fig. 8). The oscillation periodicity modulated by the Co layers is dominated by the Fermi wavevector k_{F2}^{Co} of 0.74 Å^{-1}. The oscillation periodicity of the maxima of the probabilities, modulated by the Fe layers, is 6.67 Å (i.e., 3.70 ML) (see Fig. 9).

From the results above, it is concluded that the probabilities of the presence of the electron in the wells of Co/Cu/Ni/Cu/Co(100), Co/Cu/Co/Cu/Co(100), and Co/Cu/Fe/ Cu/Co(100) oscillate with a similar tendency. This is due to the fact that these DQWs have the same structure, except for their different center layer. The oscillation periodicity modulated by the Cu layers is same, whereas those modulated by the center (i.e., Ni, Co, Fe) layers vary, depending on the Fermi wavevector of the electrons in the layers. The shorter the Fermi wavevector of the electrons in the center layers is, the longer is the oscillation periodicity modulated by the center layers.

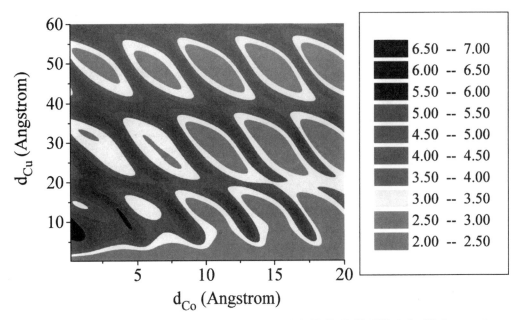

Fig. 8.   Probabilities of the electron in the range $[0, d_1 + d_2]$ of Co/Cu/Co/Cu/Co(100) (color-filled contours) as a function of thicknesses of the spacer layers. The parameters used during the calculation were $k_{Cu}^{eff} = 0.32$ Å^{-1}, $k_{F1}^{Co} = 0.44$ Å^{-1}, and $k_{F2}^{Co} = 0.74$ Å^{-1}.

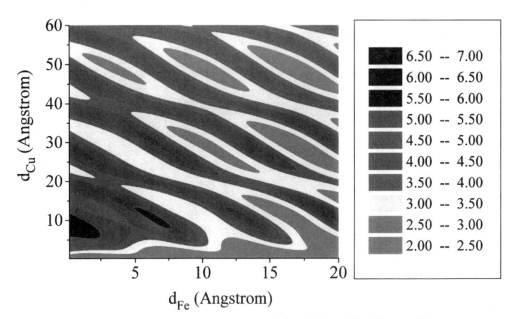

Fig. 9.   Probabilities of the electron in the range $[0, d_1 + d_2]$ of Co/Cu/Fe/Cu/Co(100) (color-filled contours) as a function of thicknesses of the spacer layers. The parameters used during the calculation were $k_{Cu}^{eff} = 0.32$ Å^{-1}, $k_{Fe}^{eff} = 0.47$ Å^{-1}, $k_{F1}^{Co} = 0.44$ Å^{-1}, and $k_{F2}^{Co} = 0.74$ Å^{-1}.

5.3.2. The Exchange Coupling across Magnetic Layers

In this section, we are interested in exchange coupling between ferromagnetic layers separated by a ferromagnetic layer [96]. The probabilities of six DQWs (i.e., Fe/Co/Ni/Co/Fe(100), Fe/Ni/Co/Ni/Fe(100), Co/Fe/Ni/Fe/Co(100), Co/Ni/Fe/Ni/Co (100), Ni/Fe/Co/Fe/Ni(100), and Ni/Co/Fe/Co/Ni(100)) are calculated. In Figures 10–15, the color-filled contours of the probabilities in the range of $[0, d_1 + d_2]$ are represented as a function

of the thickness of the spacer layers, according to Eqs. (43)–(46), respectively, for these six multilayer systems.

The probability of Fe/Co/Ni/Co/Fe(100) is plotted in Figure 10. The long oscillation periodicity of the probabilities modulated by the Co layers when there is no Ni layer is 17.5 Å (i.e., 9.72 ML). There are two other oscillation periodicities of 4.5 Å (2.5 ML) and 9.0 Å (5.0 ML). Ni layers modify the oscillation periodicity of the probabilities modulated by the Co layers. The oscillation periodicity modulated by the Ni layers

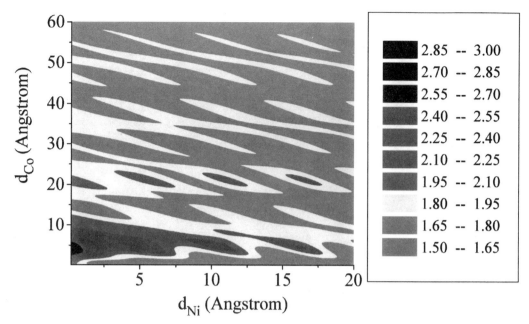

Fig. 10. Probabilities of the electron in Fe/Co/Ni/Co/Fe(100) (color-filled contours) as a function of thicknesses of the spacer layers. The parameters used during the calculation were $k_{Fe}^{eff} = 0.47$ Å^{-1}, $k_{Ni}^{eff} = 0.57$ Å^{-1}, $k_{F1}^{Co} = 0.44$ Å^{-1}, and $k_{F2}^{Co} = 0.74$ Å^{-1}.

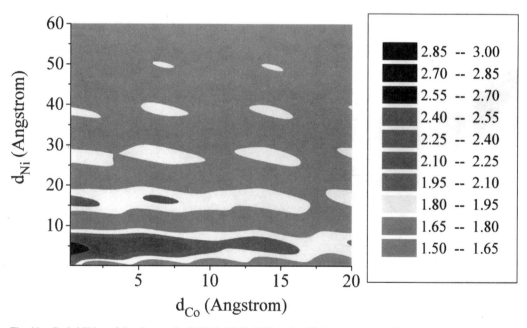

Fig. 11. Probabilities of the electron in Fe/Ni/Co/Ni/Fe(100) (color-filled contours) as a function of thicknesses of the spacer layers. The parameters used during the calculation are the same as those in Figure 10.

varies in the range of 5.30–5.99 Å (i.e., 2.94–3.33 ML), which is also modified by Co layers.

QWI results in a different feature of the probabilities of Fe/Ni/Co/Ni/Fe(100) compared with Fe/Co/Ni/Co/Fe(100). As illustrated in Figure 11 for Fe/Ni/Co/Ni/Fe(100), the oscillation periodicity of the probabilities modulated by the Ni layers when there is no Co layer is 11.5–12.0 Å (i.e., 6.39–6.67 ML). Increasing the thickness of the Ni layer weakens the amplitude of the probabilities. The oscillation periodicity of the maxima of

the probabilities modulated by the Co layers varies in the range of 6.43–8.25 Å (i.e., 3.57–4.58 ML), as the thickness of the Ni layer increases.

The probability of Co/Fe/Ni/Fe/Co(100) is presented in Figure 12. The long oscillation periodicity of the probabilities modulated by the Fe layers when there is no Ni layer is 14.0 Å (i.e., 7.78 ML). When the thickness of the Fe layers increases to 40 Å, a short oscillation periodicity of 7.0 Å (3.89 ML) appears. The oscillation periodicity of the probabilities modulated

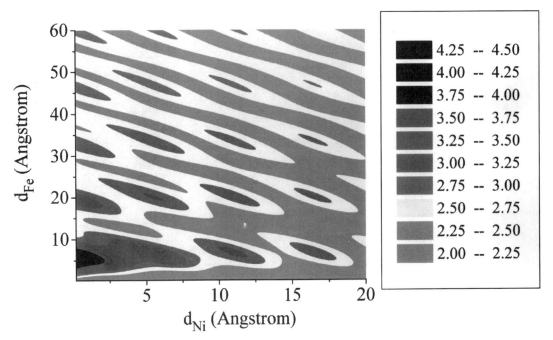

Fig. 12. Probabilities of the electron in Co/Fe/Ni/Fe/Co(100) (color-filled contours) as a function of thicknesses of the spacer layers. The parameters used during the calculation are the same as those in Figure 10.

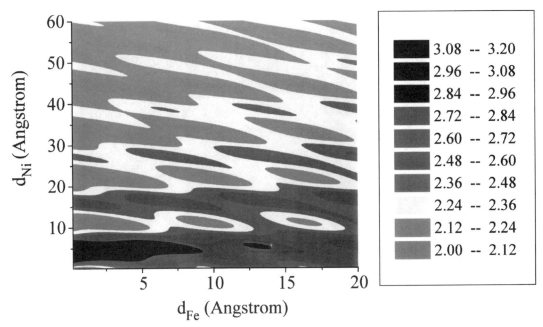

Fig. 13. Probabilities of the electron in Co/Ni/Fe/Ni/Co(100) (color-filled contours) as a function of thicknesses of the spacer layers. The parameters used during the calculation are the same as those in Figure 10.

by the Fe layers changes slightly as the thickness of the Ni layer increases. The short oscillation periodicity appears at a thinner Fe thickness when the thickness of the Ni layers increases. However, the oscillation periodicity modulated by the Ni layers is about 5.83 Å (i.e., 3.24 ML).

As shown in Figure 13 for Co/Ni/Fe/Ni/Co(100), the oscillation periodicity of the probabilities modulated by the Fe (or Ni) layers is 7.08 Å (i.e., 3.94 ML) (or 11.5 Å, i.e., 6.39 ML) when there is no Ni (or Fe) layer. The probability is strong in the case

of a Ni layer with small thickness, which is weakened by introducing Ni layers. The regular oscillation of the probabilities is disturbed by the introduction of both Fe and Ni layers.

The probability of Ni/Fe/Co/Fe/Ni(100) has a structure different from those above. It can be seen from Figure 14 that the oscillation periodicity of the probabilities modulated by the Fe or Co layers is 14.0 Å (7.78 ML) and 4.58 Å (2.55 ML). The oscillation of probabilities becomes less evident when the thickness of either the Fe or the Co layer increases.

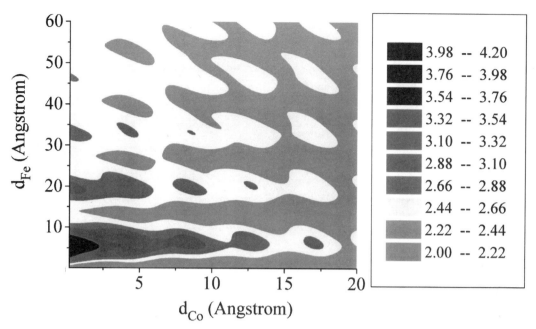

Fig. 14. Probabilities of the electron in Ni/Fe/Co/Fe/Ni(100) (color-filled contours) as a function of thicknesses of the spacer layers. The parameters used during the calculation are the same as those in Figure 10.

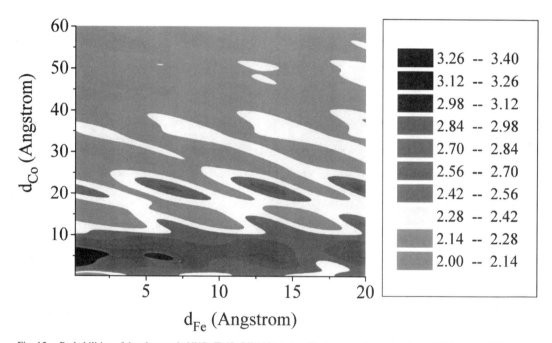

Fig. 15. Probabilities of the electron in Ni/Co/Fe/Co/Ni(100) (color-filled contours) as a function of thicknesses of the spacer layers. The parameters used during the calculation are the same as those in Figure 10.

Figure 15 presents the probabilities of Ni/Co/Fe/Co/Ni(100). The oscillation of the probabilities modulated by the Co layers is in the range of 15.5–17.0 Å (i.e., 8.61–9.44 ML), and that modulated by the Fe layers increases gradually from 5.83 Å (3.24 ML) to 6.94 Å (3.86 ML) as the thickness of the Fe layers increases. The appearance of the Co layer not only weakens the probabilities, but also alters the position of the probabilities, so that it is hard to evaluate the oscillation periodicity of the probabilities in the range of the thicker Co.

From the results above, all of the values derived for the oscillation periodicity of the probabilities modulated by various ferromagnetic layers are self-consistent, indicating the correctness of our analytical formula (i.e., Eqs. (43)–(46)). The present work shows that QWI of DQWs depends sensitively on the relative strengths of the wavevectors of ferromagnetic layers and on the structure of the DQWs. If the quantum well picture is correct for the mechanism of interlayer exchange coupling, interlayer exchange coupling between ferromagnetic layers separated by a ferromagnetic spacer would not only exist, but would also oscillate, as the thickness of the spacer layer increases. This might correspond to the oscillation only in the strength of the ferromagnetic couplings, and the oscillation might be much weaker. It is not the same as the oscillation from antiferromagnetic to ferromagnetic coupling in the case of nonmagnetic spacer layers. This prediction could be confirmed experimentally by means of magnetic coupling measurements using the surface magneto-optical Kerr effect (SMOKE) and photoemission experiments.

In this chapter, for simplicity, we do not consider spin-dependent effects. Neglecting the difference between the bands and/or the potential barriers for spin-up and spin-down electrons, we have illustrated successfully the oscillation of the probability of the electrons in DQW systems. The oscillation periodicity of the probability due to QWI is the same as that of exchange coupling and GMR observed experimentally for Co/Cu/Co(100) films. This shows that the spin-dependent scattering might not be the only mechanism for the oscillation of exchange coupling and GMR. Pure quantum interference is another possible mechanism for the oscillation of exchange coupling and GMR. However, as indicated in Section 3.7, the exchange coupling in the magnetic sandwich (or multilayers or superlattices) has the same origin as all other oscillation behavior in metal, the response of the electrons at the Fermi surface. We emphasize here that spin-dependent scattering is not a necessary condition for the existence of the oscillation of exchange coupling and GMR.

A recent work supports our point of view [94]. Manyala et al. argued that magnetoresistance can arise by a different mechanism in certain ferromagnets—quantum interference effects rather than simple scattering [94]. Magnetoresistance in disordered, low-carrier-density magnets $Fe_{1-y}Co_ySi$ and $Fe_{1-x}Mn_xSi$, where the same electrons are responsible for both the magnetic properties and electrical conduction, differs from the mechanism for the magnetoresistance in low-carrier-density ferromagnets, such as $Tl_2Mn_2O_7$, where the electrons involved in electrical conduction are different from those responsible for magnetism [95, 96].

6. ASYMMETRIC DOUBLE QUANTUM WELL Cu/Co/Ni/Co(100)

6.1. Introduction

One recent advance is the observation of QWI in a magnetic nanostructure Cu/Co/Ni/Co(100) system by photoemission [14], providing more information on the interaction between the QWS. In [14], a simple model was used to explain QWI, treating Cu as quantum well and Ni as a reflecting layer. For a better understanding of QWI, a more refined model, namely, a DQW model, is required. It would also be interesting to determine whether the phase accumulation method used in [9, 14] for the quantization condition of a SQW is valid in DQW systems. In this section, we represent the analytical solution of an asymmetric DQW system, which can be used to explain

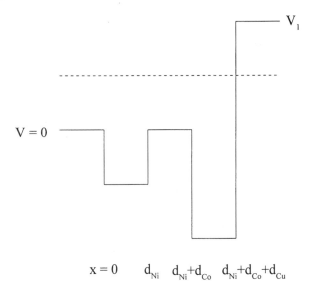

Fig. 16. An asymmetrical double quantum well Cu/Co/Ni/Co(100). The dashed line represents the Fermi level.

QWI in magnetic nanostructures, such as Cu/Co/Ni/Co(100) [14]. We focus our attention on the resonant scattering states rather than the bound states, because it is believed that the former (i.e., the QWS above the Fermi level) are what one observes by photoemission [6–10, 14]. Nevertheless, a similar procedure could be carried out for the latter. In Section 6.2 the model and the analytical results are described. The analytical solutions for the probability of the presence of electrons in the atomic layers are plotted numerically in Section 6.3. In Sections 6.4 and 6.5 we try to discuss the quantization condition and the special feature of the density of states, respectively. Some parameters for the wave function are represented in Appendix G.

6.2. Model and Analytical Results

The DQW structure is normally treated with the usual tight-binding perturbation model [44] or with the quasi-classical approximation [45, 46]. However, in the double-wedged samples with film thicknesses on the ML scale [14], these approximations would be not valid in the ML limit. For a systematic understanding of QWI in magnetic nanostructures, we treat the Cu/Co/Ni/Co(100) system with a DQW model described as follows. The wavefunction $\psi(x)$ of an electron in the resonant scattering states of one-dimensional DQW (Fig. 16) is

$$\psi(x) = \begin{cases} e^{ik_{Co}x} + re^{-ik_{Co}x}, & x \leq 0 \\ Ae^{ik_{Ni}x} + Be^{-ik_{Ni}x}, & 0 \leq x \leq d_{Ni} \\ Ce^{ik_{Co}x} + De^{-ik_{Co}x}, & d_{Ni} \leq x \leq d_{Ni} + d_{Co} \\ Ee^{ik_{Cu}x} + Fe^{-ik_{Cu}x}, & d_{Ni} + d_{Co} \leq x \leq d_{Ni} \\ & \qquad + d_{Co} + d_{Cu} \\ Ge^{-\kappa x}, & x \geq d_{Ni} + d_{Co} + d_{Cu} \end{cases}$$

$$(47)$$

where k_{Co}, k_{Ni}, and k_{Cu} are the wave vectors of the electron in Co, Ni, Cu layers, respectively, and d_{Co}, d_{Ni}, and d_{Cu} have thicknesses that are the same as those of the Co, Ni, and Cu layers. Parameters r and A–G depend on QWI at different interfaces. The origin of x is defined to be at the Co(36Å)/Ni(15Å) interface as shown in Figure 3a of [14] so that $x = d_{Ni} + d_{Co} + d_{Cu}$ corresponds to the Cu/vacuum interface. It is worth noting that the present DQW is asymmetric because it consists of the Co, Ni, Cu layers with different thicknesses, and the potential energies of Cu/vacuum and Co/Ni are not same. Following the normal procedure of the quantum theory for a SQW [44–46], one finds the relations among these parameters (see Appendix G) by using the continuum boundary conditions.

Then one can calculate the probability of the presence of an electron in the different layers. We first list here the formulas that are useful for the probability in the Cu wedged layer, for comparison with the experimental photoemission intensity [14]. The probability in the Cu wedge is determined by

$$
\int_{d_{Ni}+d_{Co}}^{d_{Ni}+d_{Co}+d_{Cu}} |\psi(x)|^2 dx
$$
$$
= 2|E|^2 \left[d_{Cu} + \frac{1}{2k_{Cu}} \left[\sin(2k_{Cu}d_{Cu} + \phi_B) - \sin \phi_B \right] \right] \quad (48)
$$

where ϕ_B is the phase gain of the electron wavefunction upon reflection at the Cu/vacuum interface, which can be determined from the Cu work function [14, 97]. $|E|^2$ is related to $|A|^2$ by

$$
|E|^2 = |A|^2 k_{Co}^2 k_{Ni}^2 \left\{ k_{Ni}^2 \left[k_{Co} \cos k_{Co} d_{Co} \cos \left(k_{Cu} d_{Cu} + \frac{\phi_B}{2} \right) \right. \right.
$$
$$
\left. - k_{Cu} \sin k_{Co} d_{Co} \sin \left(k_{Cu} d_{Cu} + \frac{\phi_B}{2} \right) \right]^2
$$
$$
+ k_{Co}^2 \left[k_{Co} \sin k_{Co} d_{Co} \cos \left(k_{Cu} d_{Cu} + \frac{\phi_B}{2} \right) \right.
$$
$$
\left. \left. + k_{Cu} \cos k_{Co} d_{Co} \sin \left(k_{Cu} d_{Cu} + \frac{\phi_B}{2} \right) \right]^2 \right\}^{-1} \quad (49)
$$

and $|A|^2$ is described by

$$
|A|^2 = \frac{1 + T^2}{1 + (k_{Ni}T/k_{Co})^2} \quad (50)
$$

Here T is defined by (G.9) in Appendix G.

Similarly, the probabilities in the Co wedge and the Ni layer are

$$
\int_{d_{Ni}}^{d_{Ni}+d_{Co}} |\psi(x)|^2 dx = |E|^2 \left\{ \left[\left(1 + \frac{k_{Cu}^2}{k_{Co}^2} \right) \right. \right.
$$
$$
\left. + \left(1 - \frac{k_{Cu}^2}{k_{Co}^2} \right) \cos(2k_{Cu}d_{Cu} + \phi_B) \right] d_{Co}
$$
$$
+ \left[\left(1 - \frac{k_{Cu}^2}{k_{Co}^2} \right) + \left(1 + \frac{k_{Cu}^2}{k_{Co}^2} \right) \cos(2k_{Cu}d_{Cu} + \phi_B) \right]
$$
$$
\times \frac{\sin 2k_{Co}d_{Co}}{2k_{Co}} - \frac{k_{Cu}}{k_{Co}^2} (1 - \cos 2k_{Co}d_{Co})
$$
$$
\times \sin(2k_{Cu}d_{Cu} + \phi_B) \right\} \quad (51)
$$

and

$$
\int_0^{d_{Ni}} |\psi(x)|^2 dx = \frac{2}{1 + (k_{Ni}T/k_{Co})^2} \left[(1 + T^2)d_{Ni} \right.
$$
$$
+ (1 - T^2) \frac{\sin 2k_{Ni}d_{Ni}}{2k_{Ni}}
$$
$$
\left. + \frac{T}{k_{Ni}} (1 - \cos 2k_{Ni}d_{Ni}) \right] \quad (52)
$$

respectively.

6.3. Numerical Results

In Figure 17, the color-filled contour of the probability per Cu layer according to Eqs. (48)–(50) is represented as a function of the thicknesses of the Co and Cu layers. The oscillation in the intensity corresponds to the formation of the QWS in the Cu layer. The oscillation periodicity of the magnetic coupling [28, 86] and the density of the QWS [4, 10, 14] are given by $\pi/(k_{BZ} - k_F)$, i.e., π/k^{eff}, instead of π/k_F. In Figure 17, k_{Cu}^{eff}, k_{Ni}^{eff}, and k_{Co}^{eff} are taken to be 0.32 Å^{-1}, 0.57 Å^{-1}, and 0.74 Å^{-1}, respectively. From $k_F^{Cu} = 1.44$ Å^{-1} and $k_F^{Ni} = 1.21$ Å^{-1}, one obtains the values $k_{BZ}^{Cu} = 1.76$ Å^{-1} and $k_{BZ}^{Ni} = 1.78$ Å^{-1}. k_{BZ}^{Ni} is the same as the value taken from the band structure of Ni along the [100] direction, and k_{BZ}^{Cu} is slightly larger than that (1.74 Å^{-1}) for the calculated band structure of Cu along the [100] direction [88–92]. k_{Co}^{eff} is in good agreement with k_{F2}^{eff} for Co in [14], which corresponds to the Fermi vector of distorted fcc Co [11]. However, the present value and that of Kawakami et al. are larger than that in [11]. This might be due to the fact that the Ni layers with a smaller lattice, compared with Cu and Co layers, decrease the lattices of the covering Cu and Co layers slightly. QWI is clearly shown in this figure, illustrating the periodic modulation of the Cu QWS intensities as a function of the Co thickness. Six peaks show up as the Cu thickness increases from 0 Å to 60 Å, corresponding to the six QWS with $\nu = 1$–6, as noted in Eq. (2) and Figure 3b of [14]. The oscillation periodicity along the Cu layers is 9.84 Å (i.e., 5.47 ML), in good agreement with the experimental results of Qiu et al. [14, 98]. It is governed by the "belly" of the Cu Fermi surface [14, 99]. With a certain value of the Cu thickness for the maxima, five peaks emerge as the Co thickness is increased from 0 Å to 20 Å. The oscillation periodicity modulated by the Co layers is 4.30 Å (i.e., 2.39 ML). So there is a total of 30 peaks in Figure 17. Each peak rotates its direction to deviate from the horizontal direction by about 29° (note the different scales of the axes for the Co and Cu thicknesses in the figures), because of the QWI.

The color-filled contour of the probability per Cu layer as a function of thickness of the Co and Cu layers (Fig. 18) shows a stronger QWI. All parameters used for Figure 18 are the same as those for Figure 17, except for $k_{F1}^{Co} = 0.44$ Å^{-1} instead of k_{Co}^{eff}, which is close to the Fermi wavevector of k_{F1}^{eff} (0.42 Å^{-1}) for Co [14, 100]. The oscillation periodicity modulated by the Co layers is 7.23 Å (i.e., 4.02 ML), and the oscillation periodicity modulated by the Cu layers remains unchanged, compared with Figure 17. Each peak in Figure 18 rotates its direction

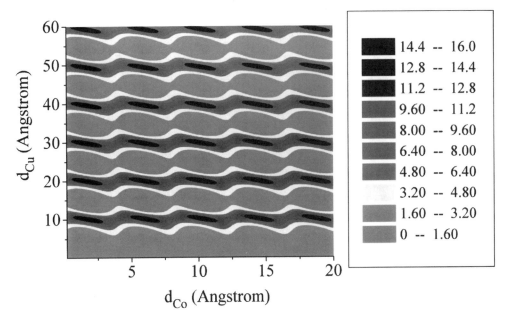

Fig. 17. Probabilities of the electron in the Cu wedge of Cu/Co/Ni/Co(100) (color-filled contours) as a function of the Cu and Co thicknesses. The parameters used during the calculation were $\phi_B = 2.86$, $k_{Cu}^{eff} = 0.32$ Å^{-1}, $k_{Ni}^{eff} = 0.57$ Å^{-1}, and $k_{Co}^{eff} = 0.74$ Å^{-1}.

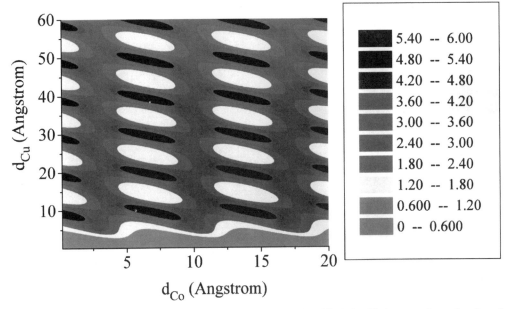

Fig. 18. Probabilities of the electron in the Cu wedge of Cu/Co/Ni/Co(100) (color-filled contours) as a function of the Cu and Co thicknesses. The parameters used during the calculation were $\phi_B = 2.86$, $k_{Cu}^{eff} = 0.324$ Å^{-1}, $k_{Ni}^{eff} = 0.57$ Å^{-1}, and $k_{F1}^{Co} = 0.44$ Å^{-1}.

to deviate from the horizontal direction by about 44°, so that the peaks with different quantization numbers ν align nearly obliquely, since of the strong QWI. This indicates that the QWI of DQWs depends sensitively on the relative strengths of the wavevectors of metals. In other words, QWI also depends sensitively on the structure of the DQW. The results above suggest that the linear quantization condition might not be appropriate for asymmetric DQWs because QWI may result in different features of the probability.

In [14], the observed QWS photoemission intensity in the Cu layers of the DQW was fitted by the modulation of the two wavevectors ($k_{F1} = 0.42$ Å^{-1} and $k_{F2}^{eff} = 0.72$ Å^{-1}) for Co layers. In Figure 19, we show the color-filled contour of the probability per Cu layer as a function of the thicknesses of the Co and Cu layers, obtained by summing directly the probabilities in Figures 17 and 18. Because the amplitudes of the probabilities in Figure 17 (with $k_{Co}^{eff} = 0.74$ Å^{-1}) are much higher than those in Figure 18 (with $k_{F1}^{Co} = 0.44$ Å^{-1}), Figure 19 keeps most of

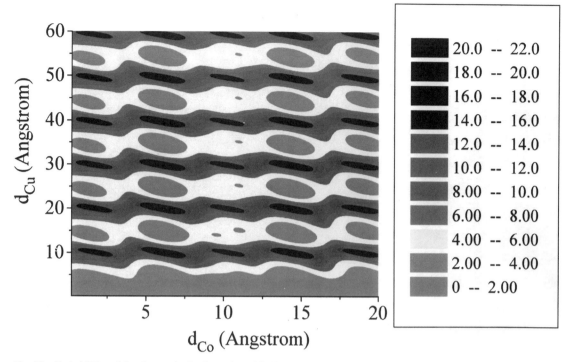

Fig. 19. Probabilities of the electron in the Cu wedge of Cu/Co/Ni/Co(100) (color-filled contours) as a function of the Cu and Co thicknesses. The parameters used during the calculation were $\phi_B = 2.86$, $k_{Cu}^{eff} = 0.32$ Å^{-1}, $k_{Ni}^{eff} = 0.57$ Å^{-1}, $k_{Co}^{eff} = 0.74$ Å^{-1}, and $k_{F1}^{Co} = 0.44$ Å^{-1}.

the main features of Figure 17, with slight modifications of the probability with $k_{F1}^{Co} = 0.44$ Å^{-1}. The deviation angle of the peaks from the horizontal direction is about 34°. The oscillation periodicity modulated by the Co layers is modified to be 4.39 Å, 4.18 Å, 4.39 Å, and 4.59 Å, respectively, corresponding to 2.44 ML, 2.32 ML, 2.44 ML, and 2.55 ML. Although the oscillation periodicity modulated by the Cu layers is unchanged, parts of the minima in the probability are altered. The results in Figure 19 are in good agreement with the experimental data (fig. 3b in [14]) and give more detailed information on the probability of the DQW system.

It should be noted that the phase gain ϕ_B we used to plot all of the figures above is 2.86. That derived from the Cu work function by Eq. (13) (or Eq. (5) in [9]) should be -0.38. There is a difference in π between the two values. If one were to choose a value of -0.38 for the phase gain ϕ_B and kept other parameters in Figure 19 unchanged, the position of all of the peaks of the probability would shift by about 3 Å to the thinner Cu thickness. The change in the phase gain ϕ_B does not alter the periodicity of the oscillations but affects the intensities of the peaks. The intensities of the peaks with a higher quantization number ν would be weakened. Meanwhile, the amplitudes of the peaks in the middle column would be decreased also. The color-filled contour of the probability with $\phi_B = -0.38$ would be in even better agreement with the experimental observation [14], except for the positions of the peaks.

For comparison, we would like to give one more figure, to show how the probability depends sensitively on the relative

strengths of the wavevectors of metals making up the DQW. If we were to choose $k_F^{Cu} = 1.443$ Å^{-1} and $k_F^{Ni} = 1.21$ Å^{-1} instead of k_{Cu}^{eff} and k_{Ni}^{eff}, the probability in Figure 20 would become a structure quite different from what we plotted in Figure 19. The difference between the probabilities in Figures 19 and 20 is due mainly to the difference in the wavevectors used for the Cu and Ni layers. The oscillation periodicity modulated by the Co layers is modified, because of the two different wavevectors for the Co layer. The oscillation periodicity modulated by the Cu layers is 2.19 Å (i.e., 1.21 ML), much shorter than that observed experimentally [14, 98]. This confirms that the oscillations of the interlayer exchange coupling correspond to the effective Fermi vectors k^{eff} of the nonmagnetic layers.

To gain a better understanding of the quantum interference, we plotted in Figures 21 and 22 the probabilities of the presence of an electron in the Co wedge and the Ni layers, in accordance with Eqs. (51) and (52). The parameters used during the calculations are the same as those for Figure 19. It is seen that the probabilities of the presence of an electron in the Co wedge and the Ni layers have structures quite different from those of the probability of the presence of an electron in the Cu wedge as shown in Figure 19, as well as in the experimental data in figure 3b in [14]. However, the oscillation periodicity modulated by the Cu layers of the probability in either the Co wedge or the Ni layers remains keeps unchanged, in comparison with that in the Cu wedge. In contrast, the oscillation periodicity modulated by the Co layers of the probability in either the Co wedge or the Ni layer varies remarkably.

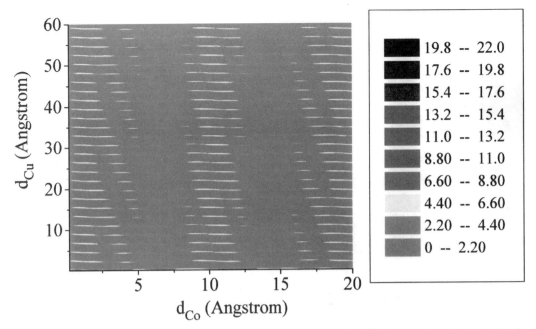

Fig. 20.   Probabilities of the electron in the Cu wedge of Cu/Co/Ni/Co(100) (color-filled contours) as a function of the Cu and Co thicknesses. The parameters used during the calculation were $\phi_B = -0.38$, $k_F^{Cu} = 1.443$ Å^{-1}, $k_F^{Ni} = 1.21$ Å^{-1}, $k_{Co}^{eff} = 0.72$ Å^{-1}, and $k_{F1}^{Co} = 0.42$ Å^{-1}.

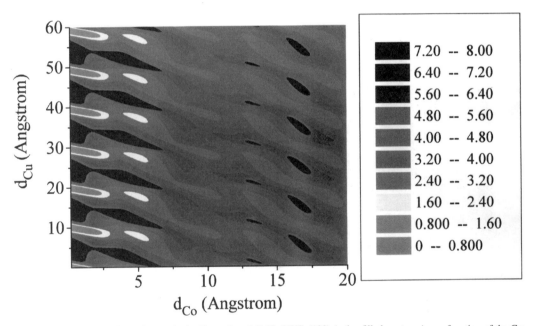

Fig. 21.   Probabilities of the electron in the Co wedge of Cu/Co/Ni/Co(100) (color-filled contours) as a function of the Cu and Co thicknesses. The parameters used during the calculation are the same as those in Figure 19.

6.4. The Quantization Condition

Quantitative understanding of the SQW spectra has been developed by the phase accumulation method (PAM) [9], which gives the linear quantization condition for the SQW. As shown in the last section, QWI results in different features of the probability, so that the linear quantization condition might be not appropriate for the asymmetric DQW. It would be interesting to determine whether the phase accumulation method used in [9, 14] for the quantization condition of SQW is valid in DQW systems. From the parameters listed in Appendix G and the probability in Section 6.2, one expects that the phase gains at the Co/Ni and the Co/Cu interfaces might not be constant in the quantization condition for the asymmetric DQW systems. This leads to the possibility that the validity of the linear quantization condition is broken under certain conditions of the asymmetric DQW systems. In this section we try to derive and discuss the quantization condition in the present asymmetric DQW.

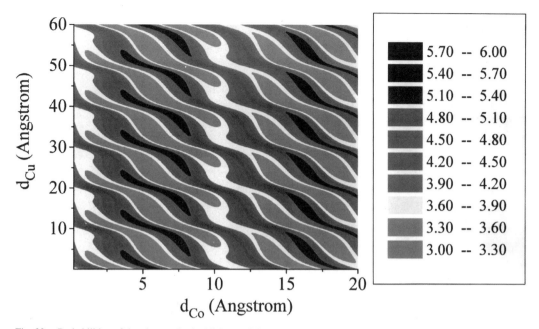

Fig. 22. Probabilities of the electron in the Ni layer of Cu/Co/Ni/Co(100) (color-filled contours) as a function of the Cu and Co thicknesses. The parameters used during the calculation are the same as those in Figure 19.

From the parameters in Appendix G, one easily proves that the relation

$$e^{i(2k_{Ni}d_{Ni}+\phi_1+\phi_2)} \equiv 1 \tag{53}$$

automatically exists, where ϕ_1 and ϕ_2 are the phase gains of the electron wavefunction upon reflection at the Co (substrate)/Ni interface and the Ni/Co(wedge) interface, respectively, which are given by

$$\phi_1 = \arctan\left[\frac{2T}{T^2-1}\right] \tag{54}$$

and

$$\phi_2 = \arctan\left[\frac{2T}{1-T^2}\right] - 2k_{Ni}d_{Ni} \tag{55}$$

These two phase gains at the Co (substrate)/Ni interface and the Ni/Co(wedge) interface depend on the function of T and on the values of $k_{Co}d_{Co}$, $k_{Ni}d_{Ni}$, and $k_{Cu}d_{Cu}$. They are not constant indeed. However, Eq. (53) automatically exists, indicating that these two phase gains actually do not contribute any restriction to the quantization condition for the Ni layers and that the linear quantization condition is valid for the Ni layers. Because the Ni layer was kept constant in Kawakami's photoemission work [14], in the present work we do not show the modulation of the probability by the Ni layers.

As the condition

$$2k_{Cu}d_{Cu} + \phi_B = n\pi \tag{56}$$

is satisfied, one can prove the relation

$$e^{i(2k_{Cu}d_{Cu}+\phi_3+\phi_B)} = 1 \tag{57}$$

where ϕ_3 is the phase gain of the electron wavefunction upon reflection at the Co(wedge)/Cu(wedge) interface, which is rep-

resented by

$$\phi_3 = \arctan\Big[\big(k_{Co}k_{Cu}\sin(2k_{Cu}d_{Cu}+\phi_B)\big)$$
$$\times \big(k_{Co}^2\cos^2(k_{Cu}d_{Cu}+\phi_B/2)$$
$$- k_{Cu}^2\sin^2(k_{Cu}d_{Cu}+\phi_B/2)\big)^{-1}\Big] \tag{58}$$

Equations (56) and (57) immediately give $\phi_3 = n\pi$. This indicates that the phase gain ϕ_3 at the Co(wedge)/Cu(wedge) interface keeps constants of $n\pi$ under the quantization condition for the Cu layers. The linear quantization condition is still valid for the Cu layers, in good agreement with the oscillation periodicity modulated by the Cu layers of the probabilities of the presence of an electron in the Cu wedge, the Co wedge and the Ni layers.

The relation

$$e^{i(2k_{Co}d_{Co}+\phi_2+\phi_3)} = 1 \tag{59}$$

exists only under the conditions

$$2\tan 2k_{Ni}d_{Ni}(k_{Co}\cos^2 k_{Co}d_{Co} + k_{Ni}\sin^2 k_{Co}d_{Co})$$
$$- (k_{Co} - k_{Ni})\sin 2k_{Co}d_{Co} = 0 \tag{60a}$$

when $\phi_3 = (2n+1)\pi$ and

$$2\cot an 2k_{Ni}d_{Ni}(k_{Co}\cos^2 k_{Co}d_{Co} + k_{Ni}\sin^2 k_{Co}d_{Co})$$
$$+ (k_{Co} - k_{Ni})\sin 2k_{Co}d_{Co} = 0 \tag{60b}$$

when $\phi_3 = 2n\pi$. It is clear from Eq. (59) and/or Eq. (60) that because ϕ_2 depends on the function of T (and the values of $k_{Co}d_{Co}$, $k_{Ni}d_{Ni}$, and $k_{Cu}d_{Cu}$), the linear condition is invalid for the quantization of the Co wedge. According to Eq. (60), the quantization in the Co wedge is affected by the value of $k_{Ni}d_{Ni}$ of the Ni layers and by the value of ϕ_3, which is determined by the quantization condition of Eqs. (56) and (57) in the Cu layers. This is a direct result of QWI in the asymmetric DQW. The nonlinear quantization condition of the Co wedge is consistent with the rapid change in the oscillation periodicity of the probabilities modulated by the Co layers. Nevertheless, from the discussion above, the phase accumulation method used in [9, 14] for SQWs can be applied to discuss the quantization condition in DQW systems.

6.5. The Special Feature of the Probabilities

It can be seen from Eqs. (G.1)–(G.9) that all of the parameters in the wavefunction contain the parameter T. This means that T is a common feature of the wavefunction of the electrons in the different layers, because of the quantum interference as well as the resonance. This is also true for the probabilities, as shown in Eqs. (48)–(52). It is therefore interesting to evaluate the function of T. We could prove the following relations for T:

$$\frac{\partial T}{\partial d_{Cu}} = -\frac{k_{Ni}k_{Co}^2 k_{Cu}^2}{t} \neq 0 \qquad (61)$$

and

$$\frac{\partial T}{\partial d_{Co}} = \left[-k_{Ni}k_{Co}^2 \left[k_{Co}^2 \cos^2\left(k_{Cu}d_{Cu} + \frac{\phi_B}{2} \right) \right. \right.$$
$$\left. \left. + k_{Cu}^2 \sin^2\left(k_{Cu}d_{Cu} + \frac{\phi_B}{2} \right) \right] \right] t^{-1} \neq 0 \quad (62a)$$

Here t is the square of the denominator of T, as shown in Eq. (G.9). Equation (62a) would be reduced to

$$\frac{\partial T}{\partial d_{Co}} = -\frac{k_{Ni}k_{Co}^2 k_{Cu}^2}{t} \neq 0 \qquad (62b)$$

if Eq. (56), i.e., the quantization condition for the Cu layers, were satisfied. One needs to study the equilibrium condition for evaluating the minima (or maxima) of the probability in Eqs. (48), (51), and (52). However, Eqs. (61) and (62) result in equilibrium conditions that are reduced to three very complicated equations that are difficult to solve analytically.

As shown in Sections 6.2 and 6.3 and Appendix G, the wave functions as well as the probability have their own behaviors in the different layers, which are mainly due to the quantum interference in the asymmetrical nanostructures. The difference between the wave vectors of the different metals contributes strongly to the quantum interference. Comparing Eq. (48) with Eq. (51), one finds that the difference in the probabilities between the Cu and the Co wedges is due to the different functions in the brackets in the equations. It is thought here that the functions in the brackets determine the main features of the probability, and $|E|^2$ gives their fine structures. Thus one could study the characters of these functions by assuming a constant $|E|^2$.

The function in brackets in Eq. (48) corresponds to the agreement between the main features of the probability in the Cu wedge and the experimental photoemission intensity. The derivative of this function with respect to d_{Cu} leads to

$$2k_{Cu}d_{Cu} + (\pi + \phi_B) = 2n\pi \qquad (63)$$

for the equilibrium condition in the case where $|E|^2$ is treated as being independent of k_{Cu}. This corresponds to the oscillation periodicity modulated by the Cu layers, which just fits with our numerical results in Section 6.3, where the phase gain ϕ_B used for plotting Figure 19 differs from that derived from the Cu work function by Eq. (5) in [9]. There is indeed a difference of π between the two values, which originates from the phase gain at other interfaces. It should be noted that Eq. (63) is not derived directly from the quantization condition, but the equilibrium condition of the function in brackets in Eq. (48) for the probability in the Cu wedge. The oscillation periodicity of the probability has a similar relation to the quantization condition (see Eqs. (57) and (63)). The derivative of this function with respect to d_{Co} is zero, suggesting that the oscillation periodicity modulated by the Co layers is controlled only by the factor $|E|^2$.

The derivative of the function in brackets in Eq. (51) with respect to d_{Cu} leads to more solutions. We would like to list only two special solutions as examples:

$$2k_{Cu}d_{Cu} + \phi_B = n\pi$$
$$2k_{Co}d_{Co} = 2n\pi \qquad (64)$$

and

$$2k_{Cu}d_{Cu} + \phi_B = \left(n + \frac{1}{2}\right)\pi$$
$$2(k_{Co}^2 - k_{Cu}^2)k_{Co}d_{Co} + (k_{Co}^2 + k_{Cu}^2)\sin 2k_{Co}d_{Co} = 0 \qquad (65a)$$

Its derivative with respect to d_{Co} results in Eq. (64) and

$$2k_{Cu}d_{Cu} + \phi_B = \left(n + \frac{1}{2}\right)\pi$$
$$k_{Co}^2 \cos^2 k_{Co}d_{Co} + k_{Cu}^2 \sin^2 k_{Cu}d_{Cu} \qquad (65b)$$
$$- 2k_{Co}k_{Cu}\sin 2k_{Co}d_{Co} = 0$$

This is in good agreement with the probability in the Co wedge in Figure 21, which shows more minima and maxima. However, it is clear that the oscillation periodicity of the probability in the Co wedge, modulated by the Cu layers, still satisfies the linear quantization condition. This is in good agreement with the DOS plotted in Figure 21. On the other hand, the oscillation periodicity modulated by the Co layers in Figure 21 is controlled not only by the factor $|E|^2$, but also by the function in brackets in Eq. (51).

According to Eq. (53), the linear quantization condition is valid for the Ni layers. Because the Ni layer was kept constant in Kawakami's photoemission work [14], in this work we do not show how the probability is modulated by the Ni layers, i.e., the Ni layer dependence of the probability. As shown in Figure 22, the oscillation periodicity of the probability in the Ni layers, modulated by the Cu layers, also satisfies the linear quantization condition. The function for the probability in Eq. (52), as well as a nonlinear quantization condition for the modulation of the

Co layers, results in a complex structure of the probability in the Ni layers in Figure 22. We do not discuss it in detail here.

7. ASYMMETRIC DOUBLE QUANTUM WELLS Cu/Ni/Cu/Co(100) AND Cu/Co/Cu/Co(100)

7.1. Introduction

In Section 5, we derived the probability for an electron in a symmetrical DQW. The probability in the spacer layers between two magnetic layers is expected to be related to the exchange coupling. The symmetry is a common character of the nanostructures grown for studying the exchange coupling by means of SMOKE. For a photoemission study on the multilayers, usually the top magnetic layer is not grown, so that the data for the photoemission come from the nonmagnetic layer. Actually, this leads to the asymmetrical behavior of the nanostructure, which should differ from the behavior of the symmetrical nanostructures. It is interesting to study the probability in an asymmetrical DQW, which is comparable to probability in the symmetrical DQW studied in Section 5.

In this section, we present the analytical solution of a DQW system with asymmetry and compare it with those obtained for the symmetrical DQW. The results are helpful for understanding the relation between the probability and the exchange coupling obtained by photoemission and SMOKE experiments. We focus our attention on the resonant scattering states, rather than the bound states, because it is believed that the former (i.e., the QWS above the Fermi level) are observed by photoemission [6–10, 14]. In Section 7.2, the model and analytical results are represented. In Section 7.3, the numerical results are plotted for a better understanding of the probability in the DQW. The parameters derived for the wavefunction are shown in Appendix H.

7.2. Model and Analytical Results

The wavefunction $\psi(x)$ of an electron in the resonant scattering states of a one-dimensional DQW is (Fig. 23):

$$\psi(x) = \begin{cases} e^{ik_0x} + re^{-ik_0x}, & x \leq 0 \\ Ae^{ik_1x} + Be^{-ik_1x}, & 0 \leq x \leq \frac{d_1}{2} \\ Ce^{ik_2x} + De^{-ik_2x}, & \frac{d_1}{2} \leq x \leq \frac{d_1}{2} + d_2 \\ Ee^{ik_1x} + Fe^{-ik_1x}, & \frac{d_1}{2} + d_2 \leq x \leq d_1 + d_2 \\ Ge^{-\kappa x}, & x \geq d_1 + d_2 \end{cases} \quad (66)$$

Here k_0, k_1, and k_2 are the wave vectors of the electron in the corresponding ferromagnetic layers. In this work, we choose Co and Ni as the magnetic materials and, Cu as the nonmagnetic material, making up the DQW, because their lattice constants are nearly the same, and high-quality single-crystal multilayers can be grown with them. The nanostructures of vacuum/Cu/Ni/Cu/Co(100) and vacuum/Cu/Co/Cu/Co(100) will be studied in the detail in the next section. The interface between the vacuum and the magnetic layer is defined to be of $x = d_1 + d_2$. d_1 and d_2 are the same thickness as the nonmagnetic Cu and the ferromagnetic spacer layers. The parameters

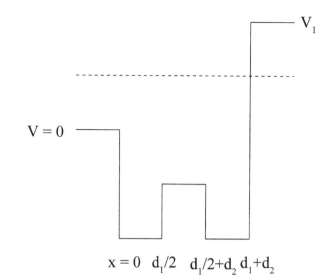

$x = 0 \quad d_1/2 \quad d_1/2{+}d_2 \quad d_1{+}d_2$

Fig. 23. An asymmetrical double quantum well, which is derived from the symmetrical double quantum well shown in Figure 3, by the introduction of a metal/vacuum interface. The dashed line represents the Fermi level.

r and A–G depend on the QWIs at different interfaces. Following the normal quantum theoretical procedure for a SQW [44–46], one finds the relations among these parameters by using the continuum boundary conditions (see Appendix H). One can then calculate the probability of the electron in the different layers. We list here the formulas for the probability in each layer. For the layer in the range $[d_1/2 + d_2, d_1 + d_2]$, one has

$$\int_{d_1/2+d_2}^{d_1+d_2} |\psi(x)|^2 \, dx = |E|^2 \big[d_1 + (\sin(k_1d_1 + \phi_B) - \sin\phi_B)/k_1\big] \quad (67a)$$

For the layer in the range $[d_1/2, d_1/2 + d_2]$, one has

$$\int_{d_1/2}^{d_1/2+d_2} |\psi(x)|^2 \, dx$$

$$= \left[\left(\cos\left(\frac{k_1d_1}{2}\right) + t\sin\left(\frac{k_1d_1}{2}\right)\right)^2 + \frac{k_1^2}{k_2^2}\left(t\cos\left(\frac{k_1d_1}{2}\right) - \sin\left(\frac{k_1d_1}{2}\right)\right)^2\right]$$

$$\times \big[2d_2 + (\sin(k_2d_1 - \phi_C) - \sin(k_2d_1 - 2k_2d_2 - \phi_C))/k_2\big] \bigg/ \left(1 + \left(\frac{k_1T}{k_0}\right)^2\right) \quad (67b)$$

For the layer in the range $[0, d_1/2]$, one has

$$\int_{0_2}^{d_1/2} |\psi(x)|^2 \, dx = |A|^2 \big[d_1 + \big(\sin(2k_1d_1 + \phi_D) - \sin(k_1d_1 + \phi_D)\big)/k_1\big] \quad (67c)$$

Here $|E|^2$ is related to $|A|^2$ by

$$|E|^2 = 4|A|^2 \left\{\left(2\cos(k_2d_2) + \left(\frac{k_2}{k_1} - \frac{k_1}{k_2}\right)\right.\right.$$

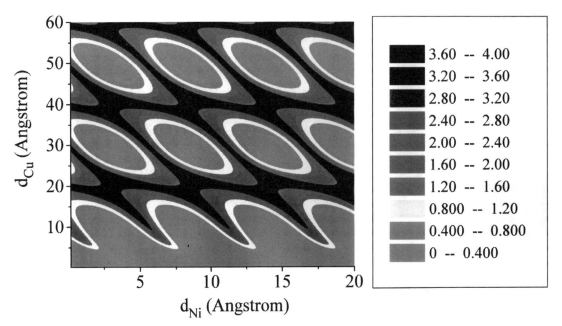

Fig. 24. Probability in the range of $[d_1/2 + d_2, d_1 + d_2]$ for a vacuum/Cu/Ni/Cu/Co(100) system (color-filled contours). The parameters used during the calculation are the same as those in Figure 19.

$$\times \sin(k_2 d_2) \sin(k_1 d_1 + \phi_B)\Bigg)^2$$
$$+ \Bigg(\sin(k_2 d_2)\left(\left(\frac{k_1}{k_2} + \frac{k_2}{k_1}\right) + \left(\frac{k_2}{k_1} - \frac{k_1}{k_2}\right)\right.$$
$$\left.\times \cos(k_1 d_1 + \phi_B)\right)\Bigg)^2\Bigg\}^{-1} \tag{68}$$

$|A|^2$ is described by

$$|A|^2 = \frac{1 + t^2}{1 + (k_1 t/k_0)^2} \tag{69}$$

t is defined in Eq. (H.8) in Appendix H. ϕ_B is the phase gain of the electron wavefunction upon reflection at the metal/vacuum interface, which can be determined from the work function of the electron at the metal/vacuum interface. ϕ_C and ϕ_D are

$$\phi_C = \arctan\Big[\big(2k_1 k_2(\cos(k_1 d_1/2)$$
$$+ t\sin(k_1 d_1/2))(t\cos(k_1 d_1/2) - \sin(k_1 d_1/2))\big)$$
$$\times \big(k_2^2(\cos(k_1 d_1/2) + t\sin(k_1 d_1/2))^2$$
$$- k_1^2(t\cos(k_1 d_1/2) - \sin(k_1 d_1/2))^2\big)^{-1}\Big]$$
$$- 2k_2 d_2 + k_2 d_1 \tag{70}$$

and

$$\phi_D = \arctan\left[\frac{2t}{1 - t^2}\right] - 2k_1 d_1 \tag{71}$$

7.3. Numerical Results

The probabilities of the electrons in two nanostructures, i.e., vacuum/Cu/Ni/Cu/Co(100) and vacuum/Cu/Co/Cu/Co(100), are calculated. In Figures 24–27, the color-filled contours of the

probability in the range of $[d_1/2 + d_2, d_1 + d_2]$, $[d_1/2, d_1/2 + d_2]$, $[0, d_1/2]$, and $[0, d_1 + d_2]$ are presented as functions of the thicknesses of the spacer layers, according to Eqs. (67a)–(67c), for a vacuum/Cu/Ni/Cu/Co(100) system. The probability in the range of $[d_1/2 + d_2, d_1 + d_2]$ shown in Figure 24 has the evident periodicities modulated by both Cu and Ni layers. The distances between the neighboring maxima of the probability in the absence of the Ni layers are 9.0 Å (5.0 ML) and 10.5 Å (5.83 ML), respectively. The Ni layers change the distances between the neighboring maxima of the probability, modifying the oscillation periodicity of the probability modulated by Cu layers. The oscillation periodicity of the minima of the probability modulated by the Cu layers is 20.0 Å (11.1 ML) after the introduction of the Ni layers. On the other hand, the oscillation periodicity of the minima of the probability modulated by the Ni layers is 5.55 Å (3.09 ML).

As shown in Figure 25, the probability in the range of $[d_1/2, d_1/2 + d_2]$ looks like water waves. The distances between the neighboring maxima of the probability in the absence of the Ni layers are same as those in Figure 24. The Ni layers adjust the distances between the neighboring maxima of the probability as well as the oscillation periodicity of the probability modulated by Cu layers. However, it is hard to see any minimum valley of the probability in Figure 25 and to determine the corresponding oscillation periodicity. It is difficult to illustrate the oscillation of the probability in the range of $[0, d_1/2]$, because the probability of electrons depends sensitively on the thicknesses of the Cu and Ni layers and varies in a wide range of numbers (see Fig. 26). The probability of electrons in the range of $[0, d_1 + d_2]$ (Fig. 27) has a feature similar to that in the range of $[d_1/2 + d_2, d_1 + d_2]$. This means that the probability of electrons in the range of $[d_1/2 + d_2, d_1 + d_2]$ is the main contribution to the probability in the whole DQW $[0, d_1 + d_2]$. The electrons are

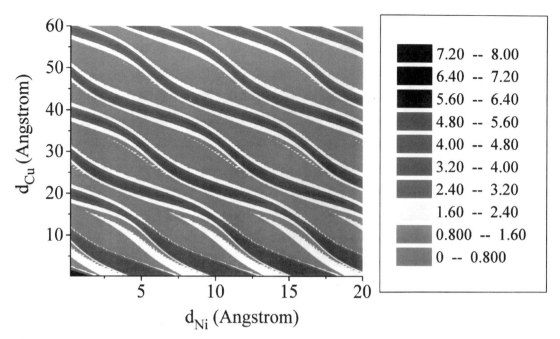

Fig. 25. Probability in the range of $[d_1/2, d_1/2 + d_2]$ for a vacuum/Cu/Ni/Cu/Co(100) system (color-filled contours). The parameters used during the calculation are the same as those in Figure 19.

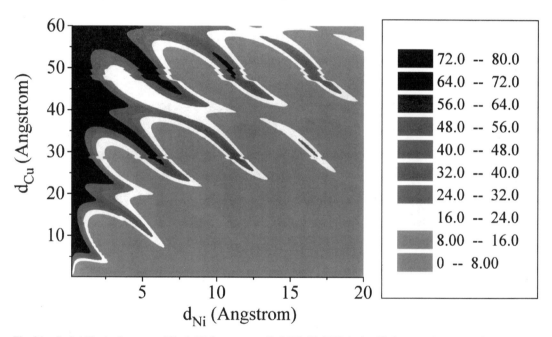

Fig. 26. Probability in the range of $[0, d_1/2]$ for a vacuum/Cu/Ni/Cu/Co(100) (color-filled contours) system. The parameters used during the calculation are the same as those in Figure 19.

mainly confined to the well of $[d_1/2 + d_2, d_1 + d_2]$, which is close to the surface/vacuum interface.

Figure 28 is the color-filled contour of the probability in the range of $[d_1/2 + d_2, d_1 + d_2]$ for a vacuum/Cu/Co/Cu/Co(100) system, which is similar to the contour shown in Figure 24. The main differences are that the oscillation periodicity of the minima of the probability modulated by Co layers is 4.31 Å (i.e., 2.39 ML), and that there are some asymmetries for some valleys of the probabilities.

It is interesting to compare the results obtained in this section for vacuum/Cu/Ni/Cu/Co(100) and vacuum/Cu/Co/Cu/Co(100) with those in Section 5 for Co/Cu/Ni/Cu/Co(100) and Co/Cu/Co/Cu/Co(100). It is clear that the probability in Figure 27 has a feature similar to that in Figure 4. The oscillation periodicities of the probabilities of the electrons in these two DQWs are exactly the same. This suggests that the behaviors of the electrons in the resonant scattering states in these two DQWs are very close, regardless of the appearance of the vacuum. Although

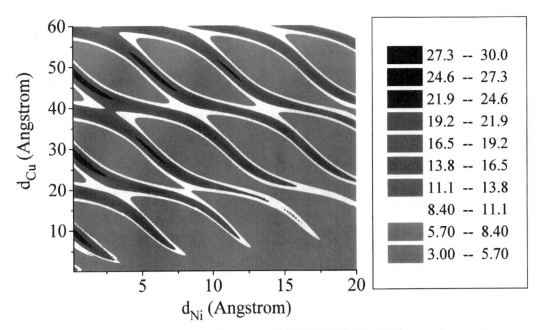

Fig. 27. Probability in the range of $[0, d_1 + d_2]$ for a vacuum/Cu/Ni/Cu/Co(100) (color-filled contours) system. The parameters used during the calculation are the same as those in Figure 19.

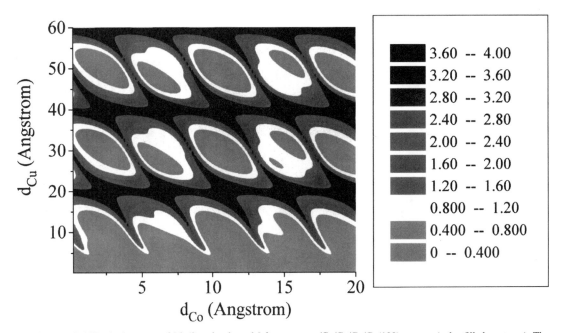

Fig. 28. Probability in the range of $[d_1/2 + d_2, d_1 + d_2]$ for a vacuum/Cu/Co/Cu/Co(100) system (color-filled contours). The parameters used during the calculation were $k_{Cu}^{eff} = 0.32 \text{ Å}^{-1}$, $k_{Co}^{eff} = 0.74 \text{ Å}^{-1}$, and $k_{F1}^{Co} = 0.44 \text{ Å}^{-1}$.

in vacuum/Cu/Ni/Cu/Co(100) the existence of the vacuum introduces asymmetry into the DQW system, the probability of the electrons in the resonance scattering states of the symmetric Co/Cu/Ni/Cu/Co(100) is actually asymmetric (see Section 5). Therefore, the asymmetry originating from the vacuum does not seriously affect the probability of the DQW in the Cu layers. Meanwhile, Figures 7 and 24 have nearly the same features, and the oscillation periodicities of the probabilities illustrated in these two figures are also the same. This supports the observation that the photoemission intensity of the surface Cu layers in

vacuum/Cu/Ni/Cu/Co(100) also corresponds to the probability of the electrons in the Cu interlayer in Co/Cu/Ni/Cu/Co(100). It also suggests that that the observed photoemission intensity of the surface Cu layers in vacuum/Cu/Ni/Cu/Co(100) oscillates as the probability of the electrons in the Cu interlayer of Co/Cu/Ni/Cu/Co(100) does and thus as the exchange coupling does. However, it should be noted that the features in Figures 25 and 26 differ from those in Figures 5 and 6, which are ascribed to the effects of either the vacuum or the asymmetry of the DQW.

The same conclusion can be derived for vacuum/Cu/Co/Cu/Co(100) and Co/Cu/Co/Cu/Co(100), by comparing Figure 28 with Figure 8.

The above results are in good agreement with the recent experimental results in [16]. The observed photoemission intensity on Cu layers of vacuum/Cu/$Cu_{70}Ni_{30}$/Cu/Co(100) was found to have the same oscillation periodicity as the strength of the saturated field measured by SMOKE for Co/Cu/$Cu_{70}Ni_{30}$/Cu/Co(100) [16]. The observed photoemission intensity on Cu layers of vacuum/Cu/$Cu_{70}Ni_{30}$/Cu/Co(100) corresponds to the exchange coupling in the Cu spacer layers in Co/Cu/$Cu_{70}Ni_{30}$/Cu/Co(100) [16]. All of these results strongly support the quantum well mechanism of exchange coupling in ultrathin films, multilayers, and superlattices.

8. DISCUSSION

In this section we discuss the advantages and limitations of the model and make some remarks on the quantum well effects and exchange coupling in Sections 8.1 and 8.2, respectively. In Section 8.3, we introduce some advances in related topics and areas for the convenience of the reader.

8.1. Advantages and Limitations of the Model

The one-dimensional model used in this work treats each layer in the nanostructure as infinitely large in two directions vertical to the growth direction. Thus this treatment is only for normal incidence, dealing with the states at the center of the 2D Brillouin zone. Therefore, the short oscillation periodicity that corresponds to the "neck" of the Cu Fermi surface [14, 25] does not appear in the calculation because it only occurs as nonincident. Under the experimental conditions in [14], the photoemission energy spectrum essentially measures the local density of state (DOS) of the sample. The normal emission geometry selects electron states with crystal momentum normal to the film surface. It is believed that the quantum well states at the Fermi level correspond to antiferromagnetic coupling, and the normal emission in the photoemission selects only long-period oscillations [8].

The parameters chosen for the analytical formulas are those taken from the band calculation for bulk materials, which might not be suitable in the case of ultrathin films or multilayers. It is expected that with decreasing film thickness to the ML scale, the wavevector of the electron would be changed by the size effect and the surface. However, the simple model used in this work gives a result in fair agreement with that of the experimental work [14], suggesting that the simplicity is valid even in ultrathin films. Thin means that for the present system, the size effects on the wavevectors could be neglected in a certain sense.

This chapter has not paid attention to spin-dependent scattering of the electrons at the interfaces. If one takes spin-dependent scattering into account, one would use different wavevectors for the electrons with spin-up and spin-down spins during the calculation and give a meticulous calculation. Nevertheless, the present work shows that the oscillation of the probability of the electrons in DQWs can be reproduced well, even when the effect of the spin-dependent scattering is not considered. This is due to the fact that as indicated in Section 3.7, the exchange coupling in the magnetic sandwich or multilayers has the same origin as all other oscillation behavior in metals, the response of the electrons at the Fermi surface. As suggested already in Section 5.3.2, spin-dependent scattering is not the only mechanism for the oscillation of exchange coupling and GMR. Quantum interference is a possible mechanism for these phenomena, in addition to the spin-dependent scattering of the electrons.

The first-principles calculations provide an opportunity to understand the behavior of the electrons in the ML limit. In the first-principles spin-density functional calculations, the self-consistency in each atomic plane was achieved separately in the ferromagnetic and antiferromagnetic configurations of the magnetic layers [18, 20, 51]. Another approach based on the first principles is to use atomic potentials that are independent of the magnetic configuration and to calculate the total energy difference by comparing sums of one-electron energies, i.e., the stationary-phase approximation (SPA) method [21, 102, 103]. Nordström et al. studied QWS in a double well, by using the first principles [21, 101]. They presented an *ab initio* calculation of spin-polarized QWS within Cu/Co [101] and of the magnetic interaction energies for Co/Cu/Co [21], by using a Korringe–Kohn–Rostoker (KRR)–Green's function method. They chose the width of the vacuum layer to be large enough so that its two boundaries would not interact, and the method was based on the local spin-density approximation. They gave the change in partial density of states at the Γ point, both for spin-down and spin-up states, and dealt with not only the resonant states but also the bound states. As indicated in [101], a quantitative comparison with the experimental oscillation was difficult because the tetragonal distortion of the cobalt layer was ignored and, moreover, the experiments depend on the roughness of the cobalt layer. Butler et al. [104] performed the first-principles calculations of electrical conductivity and GMR of Co/Cu/Co spin valves, by using a method that is the same as that used by Nordström et al. The self-consistent electronic structure calculations carried out by Mirbt et al. are also within the local spin-density approximation [105]. Excellent agreement was achieved [79] in two parallel calculations of the exchange coupling in a Co/Cu/Co(001) trilayer, based on the analytical quantum well theory [34, 106] and the full numerical calculation [79].

In a review for physics of simple metal clusters [107], Brack commented on the advantage and limitation of the first-principles, semiclassical, and classical approaches, which is also appropriate for the physics of ML and multilayers. It is impossible to solve exactly the Schrödinger equation for the exact Hamiltonian for a real system. The Born–Oppenheimer hypothesis neglecting the dynamics of the nuclei, the pseudopotential, and Hartree–Fock approximation are the basis of the quantum-chemical approaches, such as the first-principles calculation. The first-principles calculation has a clear and simple concept for treating the many-body problem, but in practice it is computationally very complex and requires sophisticated approximations even for small molecules. The pure *ab initio* treatment of all electrons is limited to very small clusters ($N < 20$). For larger systems, such as multilayers (even in the ML limit), further simplifications using density-functional methods and/or pseudopotential must be made [107].

The present simple model is actually a crude effective mass theory, in which one no longer varies many electronic single-particle wave functions, but one single wave function. The difference in the DOS of spin-down and spin-up states is also neglected. One of the advances of the present model is an enormous gain in simplicity so that the analytical result could be derived. A clear picture of quantum interference could be achieved, and it can still give significant results for average properties of the considered system, although it is within the crude effective mass theory. Another advantage is a greater flexibility, allowing closer contact with the measured data. It is rather remarkable that one is able to fit quite well the experimental data of the local DOS with a model that uses only a very oversimplified single-particle Hamiltonian. However, the cost of such simplicity is a less fundamental description because the contact with the many-body interaction is lost, and the parameters of the model have to be determined by fits to experimental observations or by using the results of the band calculations. The limitations of the present model are hard to exceed, and its physical applicability is difficult to judge. The present model cannot compete quantitatively with the *ab initio* quantum-chemical methods in experimental details of physical properties with a few atoms (maybe also in the ML limit) [107].

In general, each approach has its merits and limitations [107]. The question remains, when and where to use them for the systems in which both methods can be applied. Ultimately, this dispute can only be resolved empirically [107]. Nevertheless, the present work suggests that the effective mass theory could be used to fit and to interpret the experimental data of the multilayers, even in the ML limit. This model is an easy-to-use and rather effective tool for such systems.

8.2. Quantum Well Effects and Exchange Coupling

The excellent agreement between the probability in Figure 19 and the photoemission result of Kawakami et al. result [14] confirms that what the photoemission experiment detects is just the information on the local density of states (or the probability of the electrons) from the confined surface layers. Furthermore, a comparison of the different structures of the probabilities in the Co wedge and the Ni layers (shown in Figs. 21 and 22) with those of the probability in the Cu layers, obtained both theoretically and experimentally, indicate that photoemission cannot measure the information of the probabilities in the interlayers. All of the results in the present work suggest that each layer in an entire multilayer stack is relevant to magnetic coupling, thus favoring quantum well coupling. This is in good agreement with the photoemission experiment by Kawakami et al., which

indeed identifies the local density of states at the Fermi level oscillations as a function of the ferromagnetic layer thickness due to the quantum interference effect [14]. The probability of the electrons depends sensitively on the details of the nanostructures, consisting of various metals. It is worth noting that the better agreement between the calculation and the experimental results is achieved after the probabilities in Figures 17 and 18 are summed together. This illustrates that the contribution to the probabilities could be summed directly for the electrons with different wavevectors at the Fermi surfaces. The obvious disagreement between the probability in Figure 20 and the photoemission result of Kawakami et al. [14] reveals the fact that the oscillations of the interlayer exchange coupling correspond to the effective Fermi vectors k^{eff} of the nonmagnetic layers.

Comparing the calculation results for vacuum/Cu/Ni/Cu/Co (100) and Co/Cu/Ni/Cu/Co(100) (or for vacuum/Cu/Co/Cu/Co (100) and Co/Cu/Co/Cu/Co(100)), we derived the conclusion in the last section that the observed photoemission intensity of the surface Cu layers in vacuum/Cu/Ni/Cu/Co(100) oscillates as the probability of the electrons in the Cu interlayer of Co/Cu/Ni/Cu/Co(100) does and thus as the exchange coupling does. This confirms the experimental results that the observed photoemission intensity on Cu layers of vacuum/Cu/Cu$_{70}$Ni$_{30}$/Cu/Co(100) has the same oscillation periodicity as the strength of the saturated field measured by SMOKE for Co/Cu/Cu$_{70}$Ni$_{30}$/Cu/Co(100) [16]. All of these results strongly support the quantum well mechanism of exchange coupling in ultrathin films, multilayers, and superlattices.

As indicated already in Section 5.3.2, spin-dependent scattering is not the only mechanism for the oscillation of exchange coupling and GMR. Quantum interference is a possible mechanism for these phenomena, in addition to the spin-dependent scattering of the electrons. One of the interesting results from this work is the oscillation of the probability of electrons in a DQW consisting of ferromagnetic layers, without nonmagnetic spacer layers. This suggests that the interlayer exchange coupling between ferromagnetic layers separated by a ferromagnetic spacer would exist and even oscillate with increasing thickness of the ferromagnetic spacer layers. It is worth confirming this phenomenon experimentally by SMOKE and photoemission. Our theoretical prediction has been supported by recent experimental work [94]. Manyala et al. reported magnetoresistance in disordered, low-carrier-density magnets, Fe$_{1-y}$Co$_y$Si and Fe$_{1-x}$Mn$_x$Si, and argued that it can be ascribed to a different mechanism in certain ferromagnets—quantum interference effects rather than simple scattering [94].

In our present work, we have restricted ourselves to the (001) orientation of the layers. For the (111) orientation, the theoretical study showed that the oscillation period obtained by the quantum well theory [34, 35, 108] is different from that obtained by the RKKY theory [26]. The reason for this is as follows. In the RKKY theory [26], the period is determined by two extremal portions of the Fermi surface connected by a perpendicular spanning vector. In contrast, in the quantum well theory [34, 35, 108], only one extremal portion of the Fermi surface with a tangent plane parallel to the layers is involved (for a single-sheet Fermi surface). On the one hand, the energy denominator in a second-order perturbation theory, such as RKKY, contains two one-electron energies with wave vectors differing by a spanning vector. The oscillation period is then determined by the minimum of the energy denominator with respect to the spanning vector perpendicular to the layers. On the other hand, the exchange coupling in the nonperturbative quantum well theory is determined by a single one-electron energy [34, 35, 108]. However, the quality of the RKKY and quantum well periods was proved by Edwards et al. [109] for any single-band tight-binding model with hopping normal to the layers restricted to the nearest-neighbor planes. One may refer to the experimental results for magneto-optical and photoemission studies of magnetic quantum well states [14, 98, 110–122].

8.3. Related Topics

In the present work, we have studied quantum interference only for QWS formed by the transition metals Co, Cu, Ni, and Fe. In this section, for the convenience of the reader, we list some relevant literature for some related topics, such as quantum interference in other quantum well systems formed by other transition metals, semiconducting materials, and/or oxides; transport properties; biquadratic coupling; etc.

Antiferromagnetic interlayer exchange coupling was discovered in Fe/Cr/Fe structures by Grünberg et al. [1], and the Fe/Cr system is particularly important for our understanding of the mechanism of interlayer exchange coupling [123–132]. However, we have not dealt with quantum interference and exchange coupling in Cr-based systems, because the situation in Cr-based systems is more complicated. First, bulk chromium is known to exhibit incommensurate spin-density-wave antiferromagnetism [133]. Second, there is a so-called biquadratic interlayer coupling that is proportional to the cosine of the angle squared and can favor a 90° alignment of the magnetizations of the ferromagnetic layers. The biquadratic interlayer coupling was discovered in Fe/Cr/Fe (001) magnetic trilayers by Rührig et al. [134] and was found to occur in a number of other systems [135–137]. Several theories have been proposed to account for biquadratic coupling, some of which are intrinsic to the electronic structure of the multilayer system [31, 78, 138–144], and some of which reply on effects other than the electronic structure [143, 145].

We also do not pay attention to QWI in other metal sandwiches or superlattices, such as Fe/Mo/Fe [146, 147], Fe/Mn/Fe [148, 149], Fe/Au (or Ag)/Fe or Co/Au (or Ag)/Co [150–159], Ag/Cu [160, 161], Ag/Au [162, 163], and Pd, Pt overlayers [164], Na overlayers [165], Co/Ru superlattices [166], etc. We have not dealt with the problem in spin-valve systems [167], because in this case the quantum interference between more quantum wells has to be taken into account.

In addition to the interest in the mechanism of exchange coupling, the transport properties of layered materials have been the subject of several theoretical investigations. Most of the studies are based on the model of free electrons with random point scatterers (FERPS). Using this model, Fuchs [168] and, later, Sondheimer [169] obtained a solution to the semiclassical Boltzmann equation with boundary conditions appropriate to free electrons in a thin film. Barnaś et al. extended this approach to films with layers that have different scattering rates [170]. Hood and Falicov treated the effects of potential steps at interfaces [171]. The semiclassical model has been widely applied in modified form to several multilayers and spin-valve systems [172–179]. Levy and others applied the more rigorous quantum-mechanical Kubo–Greenwood formula to this model [180–191]. The quantum-mechanical model of Levy, Zhang, and Fert was used to explain the GMR data of the Fe/Cr [192, 193] and Co/Cu [194] systems. Vedyayev et al. developed an analytical quantum-statistical theory to study the quantum effects in GMR due to interfaces in magnetic sandwich [195].

Itoh et al. used a simple-cubic, single-band, tight-binding model together with the coherent potential approximation to model transport in multilayers and calculate the GMR [196]. Furthermore, there have been a few applications of first-principles techniques for calculating GMR in multilayers [197–201]. The structure of the quantum wells is different from that of electrons of opposite spin. The oscillation of exchange coupling is connected to the change in the number of quantum states for ferromagnetic and antiferromagnetic ordering, depending on the thicknesses of the magnetic and nonmagnetic layers. Changing the density of states at the Fermi level leads to GMR [202].

The recent observations of large magnetoresistance effects [203–210] at room temperature in tunnel junctions of the form MOM′ (M and M′ = magnetic metals, O = oxide tunnel barrier) has stimulated the study of the conductance and QWS in multilayered metal/oxide structures [211–215].

On the other hand, the discovery of the oscillatory exchange interaction between ferromagnets through a nonmagnetic metallic spacer has excited investigation of the exchange coupling between ferromagnets through an amorphous semiconducting spacer. This was verified first by Toscano et al. [216] in a sandwich structure of Fe/Si/Fe, and later by Füllerton et al. [217, 218] and Foiles et al. [219] in Fe/Si superlattices. The exchange coupling is always ferromagnetic for Fe/SiO/Fe [220] and Fe/Ge/Fe [221] trilayers and changes from ferromagnetic to antiferromagnetic for Fe/Si superlattices. The oscillation of the exchange was found to depend on the thickness of the spacer for Fe/Si/Fe [220–222]. These phenomena have been explained by the models proposed by Slonczewski [212], Bruno [5, 77, 223], and Xiao and Li [224], respectively.

The study of magnetic metal thin films, superlattices, and multilayers follows advances in modern growth techniques such as molecular beam epitaxy, chemical beam epitaxy, and metal organic chemical vapor deposition, which make possible the realization of high-quality semiconducting heterostructures consisting of layers of different semiconductors with sharp interfaces and controlled layer thickness. A great deal of work has been devoted the understanding of the nature of the electronic, excitonic, and impurity states as well as QWS in semiconducting heterostructures. One may refer to the relevant references for a comparison [225–231].

9. CONCLUDING REMARKS

An introduction to QWI in DQWs is presented, based on the one-electron effective mass model. The quantization condition in several SQWs is described before the DQWs are discussed. The quantum well mechanism for the exchange coupling between ferromagnetic layers separated by a nonmagnetic spacer, in either superlattices or multilayers, is introduced briefly. The probability that electrons are found in several DQWs is derived analytically, providing more detailed information on QWI. The theoretical results in this work are in good agreement with the observed photoemission intensity at the Fermi edge.

QWI in DQWs depends sensitively on the relative magnitudes of the wavevectors of metals as well as the amplitudes of the density of states induced by different wavevectors. The phase accumulation method is still appropriate for interpreting QWI in DQW systems. The phase accumulation relation could be linear for each well but nonlinear for the whole system in symmetric DWQs. In the asymmetric DQW system, the linear quantization condition is valid only for some layers, whereas the quantization condition for others may be nonlinear because of QWI. The nonlinear quantization condition is a characteristic of DQWs, which distinguishes it from SQWs.

Neglecting the spin-dependent effects, we have illustrated successfully the oscillation of the probability of electrons in the DQW system. The oscillation periodicity of the probability due to QWI is in good agreement with the experimental data for that of exchange coupling and GMR. We argue that pure quantum interference is a possible mechanism for the oscillation of exchange coupling and the GMR, in addition to spin-dependent scattering. Because this mechanism is only the response of the electrons at the Fermi surface, the probability of electrons in DQWs constructed from ferromagnetic layers oscillates as the thickness of the spacer layer increases. This indicates that the interlayer exchange coupling between ferromagnetic layers separated by a ferromagnetic spacer oscillates as the thickness of the spacer layer increases.

We have discussed the advantages and the limitations of the present model, in comparison with those of the first-principles approach. The present work suggests that the effective mass theory could be used to fit and to interpret the experimental data on multilayers, even in the ML limit. This model is an easy-to-use and rather effective tool for such systems. In some cases, this simple model could be used to predict new phenomena in ultrathin films.

All of the results suggest that each layer in an entire multilayer stack is relevant to magnetic coupling, thus favoring quantum well coupling, in good agreement with photoemission experiments. This work strongly supports the quantum well picture of magnetic coupling, confirming that oscillations of the interlayer exchange coupling correspond to the effective Fermi vectors k^{eff} of the layers.

Acknowledgments

This work has been supported by the National Natural Sciences Foundation of China under grant no. 59725103 and by the Sciences and Technology Commission of Shenyang and Liaoning. The author thanks Prof. Dr. Z. Q. Qiu of the Department of Physics, University of California at Berkeley, for helpful discussions.

APPENDIX A

There are two possibilities for the relations among the parameters in the wave function in Eq. (3) for the bound states of the single well. For an even solution, one has $C = B$ and

$$A = D = 2Be^{\kappa_B d/2} \cos\left(\frac{k_A d}{2}\right) \tag{A.1}$$

For an odd solution, one has $C = -B$ and

$$A = -D = -2iBe^{\kappa_B d/2} \sin\left(\frac{k_A d}{2}\right) \tag{A.2}$$

APPENDIX B

The parameters in the wavefunction in Eq. (7) for the free states of the single well are

$$A = \left[\frac{(k_A + k_B)^2}{4k_A k_B}e^{i(k_B - k_A)d} - \frac{(k_A - k_B)^2}{4k_A k_B}e^{i(k_A + k_B)d/2}\right]E \tag{B.1}$$

$$B = \left[i\frac{k_A^2 - k_B^2}{2k_A k_B}\sin k_A d\right]E \tag{B.2}$$

$$C = \frac{k_A + k_B}{2k_A}e^{i(k_B - k_A)d/2}E \tag{B.3}$$

$$D = \frac{k_A - k_B}{2k_A}e^{i(k_A + k_B)d/2}E \tag{B.4}$$

APPENDIX C

The relations among the parameters A, B, C, and D in Eq. (9) are

$$B = \frac{1}{2}\left(1 - \frac{\kappa_B}{k_A}i\right)A = \frac{e^{i\phi_1/2}A}{2\cos(\phi_1/2)} \tag{C.1}$$

$$C = \frac{1}{2}\left(1 + \frac{\kappa_B}{k_A}i\right)A = \frac{e^{-i\phi_1/2}A}{2\cos(\phi_1/2)} \tag{C.2}$$

$$D = \frac{e^{\kappa d}\cos\left(k_A d + \frac{\phi_1}{2}\right)A}{\cos(\phi_1/2)} \tag{C.3}$$

where

$$\tan\frac{\phi_1}{2} = -\frac{\kappa_B}{k_A} \qquad (C.4)$$

APPENDIX D

The relations among the parameters r, B, C, and D in Eq. (17) are

$$B = \frac{1}{2k_A}\big[(k_A + k_B) + (k_A - k_B)r\big] \qquad (D.1)$$

$$C = \frac{1}{2k_A}\big[(k_A - k_B) + (k_A + k_B)r\big] \qquad (D.2)$$

$$D = e^{\kappa d}\left[(1 + r)\cos k_A d + i\frac{k_B}{k_A}(1 - r)\sin k_A d\right] \qquad (D.3)$$

APPENDIX E

The parameters in Eq. (36) are

$$A = B(1 + e^{-i\phi_2}) \qquad (E.1)$$

$$B = \frac{F}{4k_1k_2}e^{ik_1d_1}\big\{4k_1k_2\cos k_2d_1 \\
+ i\sin k_2d_1\big[-2(k_1^2 + k_2^2) \\
+ 2(k_1^2 - k_2^2)e^{i(2k_1d_2+\phi_2)}\big]\big\} \qquad (E.2)$$

$$C = \frac{F}{4k_1k_2}e^{ik_1(2d_0+d_1)}\big\{4k_1k_2\cos k_2d_1 e^{i(2k_1d_2+\phi_2)} \\
+ i\sin k_2d_1\big[2(k_2^2 - k_1^2) \\
+ 2(k_1^2 + k_2^2)e^{i(2k_1d_2+\phi_2)}\big]\big\} \qquad (E.3)$$

$$D = \frac{F}{2k_2}e^{i(k_1-k_2)(d_0+d_1)}\big[(k_1 + k_2) + (k_2 - k_1)e^{i[2k_1d_2+\phi_2]}\big] \qquad (E.4)$$

$$E = \frac{F}{2k_2}e^{i(k_1+k_2)(d_0+d_1)}\big[(k_2 - k_1) + (k_1 + k_2)e^{i[2k_1d_2+\phi_2]}\big] \qquad (E.5)$$

$$G = Fe^{i[2k_1(d_0+d_1+d_2)+\phi_2]} \qquad (E.6)$$

$$H = Fe^{(\kappa+ik_1)(d_0+d_1+d_2)}(1 + e^{i\phi_2}) \qquad (E.7)$$

The phase gain ϕ_2 at $x = 0$ is

$$e^{i\phi_2} = \frac{B}{C} = \frac{k_1 - i\kappa}{k_1 + i\kappa} \qquad (E.8)$$

The phase gain ϕ_1 at $x = d_0$ is

$$e^{i\phi_1} = \frac{Ce^{-ik_1d_0}}{Be^{ik_1d_0}} = \frac{1 + iR}{1 - iR} \qquad (E.9)$$

where

$$R = \big(k_2^2 \sin k_2d_1 \cos(k_1d_2 + \phi_2/2) \\
+ k_1k_2 \cos k_2d_1 \sin(k_1d_2 + \phi_2/2)\big) \\
\times \big(k_1k_2 \cos k_2d_1 \cos(k_1d_2 + \phi_2/2) \\
- k_1^2 \sin k_2d_1 \sin(k_1d_2 + \phi_2/2)\big)^{-1} \qquad (E.10)$$

The phase gain ϕ_3 at $x = d_0 + d_1$ is

$$e^{i\phi_3} = \frac{Ee^{-ik_2(d_0+d_1)}}{De^{ik_2(d_0+d_1)}} \\
= \frac{k_2 \cos(k_1d_2 + \phi_2/2) + ik_1 \sin(k_1d_2 + \phi_2/2)}{k_2 \cos(k_1d_2 + \phi_2/2) - ik_1 \sin(k_1d_2 + \phi_2/2)} \qquad (E.11)$$

One also obtains the phase gain ϕ_4 at $x = d_0$,

$$e^{i\phi_4} = \frac{De^{ik_2d_0}}{Ee^{-ik_2d_0}} \\
= \frac{k_2 \cos(k_1d_0 + \phi_2/2) + ik_1 \sin(k_1d_0 + \phi_2/2)}{k_2 \cos(k_1d_0 + \phi_2/2) - ik_1 \sin(k_1d_0 + \phi_2/2)} \qquad (E.12)$$

The phase gain ϕ_5 at $x = d_0 + d_1 + d_2$ is

$$e^{i\phi_5} = \frac{Ge^{-ik_1(d_0+d_1+d_2)}}{Fe^{ik_1(d_0+d_1+d_2)}} = \frac{k_1 - i\kappa}{k_1 + i\kappa} = e^{i\phi_2} \qquad (E.13)$$

The phase gain ϕ_6 at $x = d_0 + d_1$ is

$$e^{i\phi_6} = \frac{Fe^{ik_1(d_0+d_1)}}{Ge^{-ik_1(d_0+d_1)}} = \frac{1 + iR'}{1 - iR'} \qquad (E.14)$$

where

$$R' = \big(k_2^2 \sin k_2d_1 \cos(k_1d_0 + \phi_2/2) \\
+ k_1k_2 \cos k_2d_1 \sin(k_1d_0 + \phi_2/2)\big) \\
\times \big(k_1k_2 \cos k_2d_1 \cos(k_1d_0 + \phi_2/2) \\
- k_1^2 \sin k_2d_1 \sin(k_1d_0 + \phi_2/2)\big)^{-1} \qquad (E.15)$$

APPENDIX F

The parameters in Eq. (42) are

$$r = \frac{1}{k_1 - k_0}\big[2k_1 A - (k_0 + k_1)\big] \qquad (F.1)$$

$$A = \frac{te^{i[k_0(d_1+d_2)]}}{8k_1^2k_2}\big[(k_1 + k_2)^2(k_0 + k_1)e^{i[-k_1d_1-k_2d_2]} \\
- (k_1 - k_2)^2(k_0 + k_1)e^{i[-k_1d_1+k_2d_2]} \\
- 2i(k_1^2 - k_2^2)(k_0 - k_1)\sin k_2d_2\big] \qquad (F.2)$$

$$B = \frac{te^{i[k_0(d_1+d_2)]}}{8k_1^2k_2}\big[(k_1 - k_2)^2(k_0 - k_1)e^{i[k_1d_1-k_2d_2]} \\
- (k_1 + k_2)^2(k_0 - k_1)e^{i[k_1d_1+k_2d_2]} \\
- 2i(k_1^2 - k_2^2)(k_0 + k_1)\sin k_2d_2\big] \qquad (F.3)$$

$$C = \frac{te^{-ik_2[d_1/2+d_2]+ik_0(d_1+d_2)}}{4k_1k_2}\Big[(k_0+k_1)(k_1+k_2)e^{-i(k_1d_1/2)}$$
$$+ (k_0-k_1)(k_1-k_2)e^{i(k_1d_1/2)}\Big] \qquad (F.4)$$

$$D = -\frac{te^{ik_2[d_1/2+d_2]+ik_0(d_1+d_2)}}{4k_1k_2}\Big[(k_0+k_1)(k_1-k_2)e^{-i(k_1d_1/2)}$$
$$+ (k_0-k_1)(k_1+k_2)e^{i(k_1d_1/2)}\Big] \qquad (F.5)$$

$$E = \frac{t}{2k_1}(k_0+k_1)e^{i[(k_0-k_1)(d_1+d_2)]} \qquad (F.6)$$

$$F = \frac{t}{2k_1}(k_1-k_0)e^{i[(k_0+k_1)(d_1+d_2)]} \qquad (F.7)$$

$$u = (k_0^4-k_1^4)\big[(k_1+k_2)^2\sin(k_1d_1+k_2d_2)$$
$$- (k_1-k_2)^2\sin(k_1d_1-k_2d_2)\big]^2$$
$$+ 4(k_0^4+k_1^4)(k_1^2-k_2^2)$$
$$\times \sin k_2d_2\big[(k_1+k_2)^2\sin(k_1d_1+k_2d_2)$$
$$- (k_1-k_2)^2\sin(k_1d_1-k_2d_2)\big]$$
$$+ 4(k_0^4-k_1^4)(k_1^2-k_2^2)^2\sin^2 k_2d_2 \qquad (F.8)$$

$$v = 4k_0^2k_1^2\big[(k_1+k_2)^2\cos(k_1d_1+k_2d_2)$$
$$- (k_1-k_2)^2\cos(k_1d_1-k_2d_2)\big]^2$$
$$+ \big[(k_0^2+k_1^2)\big[(k_1+k_2)^2\sin(k_1d_1+k_2d_2)$$
$$- (k_1-k_2)^2\sin(k_1d_1-k_2d_2)\big]$$
$$+ 2(k_0^2-k_1^2)(k_1^2-k_2^2)\sin k_2d_2\big]^2 \qquad (F.9)$$

$$w = 2k_0k_1\big[(k_1+k_2)^2\cos(k_1d_1+k_2d_2)$$
$$- (k_1-k_2)^2\cos(k_1d_1-k_2d_2)\big]$$
$$\times \big[(k_0^2-k_1^2)\big[-(k_1+k_2)^2\sin(k_1d_1+k_2d_2)$$
$$+ (k_1-k_2)^2\sin(k_1d_1-k_2d_2)\big]$$
$$- 2(k_0^2+k_1^2)(k_1^2-k_2^2)\sin k_2d_2\big] \qquad (F.10)$$

APPENDIX G

The parameters in the wavefunction in Eq. (47) are

$$A = \frac{(1+k_{Ni}T^2/k_{Co})+i(k_{Ni}/k_{Co}-1)T}{1+(k_{Ni}T/k_{Co})^2} \qquad (G.1)$$

$$B = \frac{(1-k_{Ni}T^2/k_{Co})+i(k_{Ni}/k_{Co}+1)T}{1+(k_{Ni}T/k_{Co})^2} \qquad (G.2)$$

$$C = \frac{E}{2}\exp\big[i(k_{Cu}-k_{Co})(d_{Ni}+d_{Co})\big]$$

$$\times \Big[\Big(1+\frac{k_{Cu}}{k_{Co}}\Big)+\Big(1-\frac{k_{Cu}}{k_{Co}}\Big)$$
$$\times \exp\big[i(2k_{Cu}d_{Cu}+\phi_B)\big]\Big] \qquad (G.3)$$

$$D = \frac{E}{2}\exp\big[i(k_{Cu}+k_{Co})(d_{Ni}+d_{Co})\big]$$
$$\times \Big[\Big(1-\frac{k_{Cu}}{k_{Co}}\Big)+\Big(1+\frac{k_{Cu}}{k_{Co}}\Big)$$
$$\times \exp\big[i(2k_{Cu}d_{Cu}+\phi_B)\big]\Big] \qquad (G.4)$$

$$E = Ak_{Co}k_{Ni}\exp\big[i(k_{Ni}d_{Ni}-k_{Cu}(d_{Ni}+d_{Co}))\big]$$
$$\times \exp\Big[-i\Big(k_{Cu}d_{Cu}+\frac{\phi_B}{2}\Big)\Big]$$
$$\times \Big\{k_{Ni}\Big[k_{Co}\cos k_{Co}d_{Co}\cos\Big(k_{Cu}d_{Cu}+\frac{\phi_B}{2}\Big)$$
$$- k_{Cu}\sin k_{Co}d_{Co}\sin\Big(k_{Cu}d_{Cu}+\frac{\phi_B}{2}\Big)\Big]$$
$$- ik_{Co}\Big[k_{Co}\sin k_{Co}d_{Co}\cos\Big(k_{Cu}d_{Cu}+\frac{\phi_B}{2}\Big)$$
$$+ k_{Cu}\cos k_{Co}d_{Co}\sin\Big(k_{Cu}d_{Cu}+\frac{\phi_B}{2}\Big)\Big]\Big\}^{-1} \qquad (G.5)$$

$$F = E\exp\big[i(2k_{Cu}(d_{Ni}+d_{Co}+d_{Cu})+\phi_B)\big] \qquad (G.6)$$

$$G = \frac{2(1-i\kappa/k_{Cu})E}{1+(\kappa/k_{Cu})^2}\exp\big[(ik_{Cu}+\kappa)(d_{Ni}+d_{Co}+d_{Cu})\big] \qquad (G.7)$$

$$r = \frac{1-(k_{Ni}T/k_{Co})^2+i2k_{Ni}T/k_{Co}}{1+(k_{Ni}T/k_{Co})^2} \qquad (G.8)$$

Here T in the above formula is defined as

$$T = -\big[k_{Co}(k_{Co}\sin k_{Co}d_{Co}\cos k_{Ni}d_{Ni}$$
$$+ k_{Ni}\cos k_{Co}d_{Co}\sin k_{Ni}d_{Ni})\cos(k_{Cu}d_{Cu}+\phi_B/2)$$
$$+ k_{Cu}(k_{Co}\cos k_{Co}d_{Co}\cos k_{Ni}d_{Ni}$$
$$- k_{Ni}\sin k_{Co}d_{Co}\sin k_{Ni}d_{Ni})\sin(k_{Cu}d_{Cu}+\phi_B/2)\big]$$
$$\times \big[k_{Co}(k_{Co}\sin k_{Co}d_{Co}\sin k_{Ni}d_{Ni}$$
$$- k_{Ni}\cos k_{Co}d_{Co}\cos k_{Ni}d_{Ni})\cos(k_{Cu}d_{Cu}+\phi_B/2)$$
$$+ k_{Cu}(k_{Co}\cos k_{Co}d_{Co}\sin k_{Ni}d_{Ni}$$
$$+ k_{Ni}\sin k_{Co}d_{Co}\cos k_{Ni}d_{Ni})\sin(k_{Cu}d_{Cu}+\phi_B/2)\big]^{-1}$$
$$\qquad (G.9)$$

APPENDIX H

The parameters in Eq. (66) are

$$r = \frac{1}{k_1-k_0}[2k_1A-(k_0+k_1)] \qquad (H.1)$$

$$A = \frac{E}{4k_1k_2} e^{ik_1d_2} \{ e^{-ik_2d_2} [(k_1+k_2)^2 + (k_2^2-k_1^2)e^{i(k_1d_1+\phi_2)}]$$
$$+ e^{ik_2d_2} [-(k_1-k_2)^2 + (k_1^2-k_2^2)e^{i(k_1d_1+\phi_2)}] \} \quad \text{(H.2)}$$

$$B = \frac{E}{4k_1k_2} e^{ik_1(d_1+d_2)} \{ 4k_1k_2 \cos k_2d_2 \, e^{i(k_1d_1+\phi_2)}$$
$$- 2i \sin k_2d_2 [(k_1^2-k_2^2)$$
$$- (k_1^2+k_2^2)e^{i(k_1d_1+\phi_2)}] \} \quad \text{(H.3)}$$

$$C = \frac{E}{2k_2} e^{i[k_1-k_2][d_1/2+d_2]} [(k_1+k_2) - (k_1-k_2)e^{i[k_1d_1+\phi_2]}]$$
$$\text{(H.4)}$$

$$D = \frac{E}{2k_2} e^{i[k_1+k_2][d_1/2+d_2]} [(k_2-k_1) + (k_1+k_2)e^{i[k_1d_1+\phi_2]}]$$
$$\text{(H.5)}$$

$$F = E e^{i[2k_1(d_1+d_2)+\phi_2]} \quad \text{(H.6)}$$

$$G = \frac{2k_1(k_1-i\kappa)E}{k_1^2+\kappa^2} e^{(ik_1+\kappa)(d_1+d_2)} \quad \text{(H.7)}$$

t in Eq. (69) is defined as

$$t = (k_1k_2 \cos(k_2d_2) \sin(k_1d_1+\phi_B/2) + \sin(k_2d_2)$$
$$\times [k_2^2 \cos((k_1d_1+\phi_B)/2) \cos(k_1d_1/2)$$
$$- k_1^2 \sin((k_1d_1+\phi_B)/2) \sin(k_1d_1/2)])$$
$$\times (k_1k_2 \cos(k_2d_2) \cos(k_1d_1+\phi_B/2)$$
$$- \sin(k_2d_2)[k_2^2 \cos((k_1d_1+\phi_B)/2) \sin(k_1d_1/2)$$
$$+ k_1^2 \sin(k_1d_1+\phi_B)/2) \cos(k_1d_1/2)])^{-1} \quad \text{(H.8)}$$

REFERENCES

1. P. Grünberg, R. Schreiber, Y. Peng, M. B. Brodsky, and H. Sower, *Phys. Rev. Lett.* 57, 2442 (1986).
2. M. N. Baibich, J. M. Broto, A. Fert, F. Nguyen Van Dau, F. Petroff, P. Etienne, G. Creuzet, A. Friederich, and J. Chazelas, *Phys. Rev. Lett.* 61, 2472 (1988).
3. S. S. P. Parkin, N. More, and K. P. Roche, *Phys. Rev. Lett.* 64, 2304 (1990).
4. M. D. Stiles, *Phys. Rev. B: Condens. Matter* 48, 7238 (1993).
5. P. Bruno, *Phys. Rev. B: Condens. Matter* 52, 411 (1995).
6. A. C. Ehrlich, *Phys. Rev. Lett.* 71, 2300 (1993).
7. A. L. Wachs, A. P. Shapiro, T. C. Hsieh, and T.-C. Chiang, *Phys. Rev. B: Condens. Matter* 33, 1460 (1986).
8. J. E. Ortega and F. J. Himpsel, *Phys. Rev. Lett.* 69, 844 (1992).
9. N. V. Smith, N. B. Brookes, Y. Chang, and P. D. Johnson, *Phys. Rev. B: Condens. Matter* 49, 332 (1994).
10. P. Segovia, E. G. Michel, and J. E. Ortega, *Phys. Rev. Lett.* 77, 3455 (1996).
11. P. J. H. Bloemen, M. T. Johnson, M. T. H. van de Vorst, R. Coehoorn, J. J. de Vries, R. Jungblut, J. aan de Stegge, A. Reinders, and W. J. M. de Jonge, *Phys. Rev. Lett.* 72, 764 (1994).
12. S. N. Okuno and K. Inomata, *Phys. Rev. Lett.* 72, 1553 (1994).
13. S. N. Okuno and K. Inomata, *Phys. Rev. B: Condens. Matter* 51, 6139 (1995).
14. R. K. Kawakami, E. Rotenberg, E. J. Escorcia-Aparicio, H. J. Choi, T. R. Cummins, J. G. Tobin, N. V. Smith, and Z. Q. Qiu, *Phys. Rev. Lett.* 80, 1754 (1998).

15. R. K. Kawakami, E. Rotenberg, Hyuk J. Choi, Ernesto J. Escorcia-Aparicio, M. O. Bowen, J. H. Wolfe, E. Aronholz, Z. D. Zhang, N. V. Smith, and Z. Q. Qiu, *Nature (London)* 398, 132 (1999).
16. Z. D. Zhang, Hyuk J. Choi, R. K. Kawakami, Enresto Escorcia-Apericio, Martin Bowen, Jason Wolfe, E. Rotenberg, N. V. Smith, and Z. Q. Qiu, *Phys. Rev. B: Condens. Matter* 61, 76 (2000).
17. D. Stoeffler and F. Gautier, *Phys. Rev. B: Condens. Matter* 44, 10389 (1991).
18. M. van Schilfgaarde and F. Herman, *Phys. Rev. Lett.* 71, 1923 (1993).
19. S. Krompiewski, U. Krey, and J. Pirnay, *J. Magn. Magn. Mater.* 121, 238 (1993).
20. S. Krompiewski, F. Süss, and U. Krey, *Europhys. Lett.* 26, 303 (1994).
21. P. Lang, L. Nordström, R. Zeller, and P. H. Dederichs, *Phys. Rev. Lett.* 71 1297 (1993).
22. J. Kudrnovský, V. Drchal, I. Turek, and P. Weinberger, *Phys. Rev. B: Condens. Matter* 50, 16105 (1994).
23. A. Fert and P. Bruno, in "Ultrathin Magnetic Structures" (B. Heinrich and J. A. C. Bland, Eds.), Vol. 2, Chap. 2.2, p. 82. Springer-Verlag, Berlin, 1994.
24. Y. Yafet, *Phys. Rev. B: Condens. Matter* 36, 3948 (1987).
25. C. Chappert and J. P. Renard, *Europhys. Lett.* 15, 553 (1991).
26. P. Bruno and C. Chappert, *Phys. Rev. Lett.* 67, 1602 (1991).
27. P. Bruno and C. Chappert, *Phys. Rev. B: Condens. Matter* 46, 261 (1992).
28. R. Coehoorn, *Phys. Rev. B: Condens. Matter* 44, 9331 (1991).
29. J. Barnaś, *J. Magn. Magn. Mater.* 111, L215 (1992).
30. R. P. Erickson, K. B. Hathaway, and J. R. Cullen, *Phys. Rev. B: Condens. Matter* 47, 2626 (1993).
31. J. C. Slonczewski, *J. Magn. Magn. Mater.* 126, 374 (1993).
32. E. Bruno and B. L. Gyorffy, *Phys. Rev. Lett.* 71, 181 (1993).
33. E. Bruno and B. L. Gyorffy, *J. Phys.: Condens. Matter* 5, 2109 (1993).
34. D. M. Edwards, J. Mathon, R. B. Muniz, and M. S. Phan, *Phys. Rev. Lett.* 67, 493 (1991).
35. J. Mathon, M. Villeret, and D. M. Edwards, *J. Phys.: Condens. Matter* 4, 9873 (1992).
36. Y. Wang, P. M. Levy, and J. L. Fry, *Phys. Rev. Lett.* 65, 2732 (1990).
37. Z. P. Shi, P. M. Levy, and J. L. Fry, *Phys. Rev. Lett.* 69, 3678 (1992).
38. P. Bruno, *J. Magn. Magn. Mater.* 116, L13 (1992).
39. B. A. Jones and C. B. Hanna, *Phys. Rev. Lett.* 71, 4253 (1993).
40. M. C. Muñoz and J. L. Pérez-Díaz, *Phys. Rev. Lett.* 72, 2482 (1994).
41. B. C. Lee and Y. C. Chang, *Phys. Rev. B: Condens. Matter* 51, 316 (1995).
42. D. Bohm, "Quantum Mechanics." Prentice–Hall, Englewood Cliffs, NJ, 1951.
43. E. Merzbacher, "Quantum Mechanics." Wiley, New York, 1961.
44. C. Weisbuch and B. Vinter, "Quantum Semiconductor Structures, Fundamentals and Applications." Academic Press, San Diego, 1991.
45. L. D. Landau and E. M. Lifshitz, "Quantum Mechanics." Pergamon, Oxford, 1958.
46. I. I. Gol'dman and V. D. Krivchenkov, "Problems in Quantum Mechanics." Pergamon, Oxford, 1961.
47. M. Altarelli, in "Heterojunctions and Semiconductor Superlattices" (G. Allan et al., Eds.), p. 12. Springer-Verlag, Berlin/New York, 1985.
48. G. Bastard, *Phys. Rev. B: Condens. Matter* 24, 5693 (1981).
49. G. Bastard, *Phys. Rev. B: Condens. Matter* 25, 7594 (1982).
50. S. S. P. Parkin, *Phys. Rev. Lett.* 67, 3598 (1991).
51. F. Herman, J. Sticht, and M. van Schilfgaarde, *J. Appl. Phys.* 69, 4783 (1991).
52. S. Mirbt, H. L. Skriver, M. Aldén, and B. Johansson, *Solid State Commun.* 88, 331 (1993).
53. A. M. N. Niklasson, S. Mirbt, H. L. Skriver, and B. Johansson, *Phys. Rev. B: Condens. Matter* 53, 8509 (1996).
54. A. M. N. Niklasson, S. Mirbt, H. L. Skriver, and B. Johansson, *Phys. Rev. B: Condens. Matter* 56, 3276 (1997).
55. P. Lang, L. Nordström, K. Wildberger, R. Zeller, P. H. Dederichs, and T. Hoshino, *Phys. Rev. B: Condens. Matter* 53, 9092 (1996).
56. P. van Gelderen, S. Crampin, and J. E. Inglesfield, *Phys. Rev. B: Condens. Matter* 53, 9115 (1996).

57. S. Krompiewski, F. Süss, and U. Krey, *J. Magn. Magn. Mater.* 164, L263 (1996).

58. R. Skomski, *Europhys. Lett.* 48, 455 (1999).

59. L. M. Li, B. Z. Li, and F. C. Pu, *J. Phys.: Condens. Matter* 6, 1941 (1994).

60. M. S. Ferreira, *J. Phys.: Condens. Matter* 9, 6665 (1997).

61. J. Mathon, A. Umerski, and M. Villeret, *Phys. Rev. B: Condens. Matter* 55, 14378 (1997).

62. J. Mathon, *Phys. Rev. B: Condens. Matter* 54, 55 (1996).

63. D. M. Edwards, A. M. Robinson, and J. Mathon, *J. Magn. Magn. Mater.* 140–144, 517 (1995).

64. J. L. Pérez-Díaz and M. C. Muñoz, *J. Magn. Magn. Mater.* 156, 165 (1996).

65. J. L. Pérez-Díaz and M. C. Muñoz, *Phys. Rev. B: Condens. Matter* 50, 8824 (1994).

66. H. L. Skriver and N. M. Rosengaard, *Phys. Rev. B: Condens. Matter* 43, 9538 (1991).

67. A. M. N. Niklasson, S. Mirbt, H. L. Skriver, and B. Johansson, *J. Magn. Magn. Mater.* 156, 128 (1996).

68. P. A. Lee and D. S. Fisher, *Phys. Rev. Lett.* 47, 882 (1981).

69. G. D. Mahan, "Many-Particle Physics." Plenum, New York, 1981.

70. T. A. Luce, W. Hübner, and K. H. Bennemann, *Phys. Rev. Lett.* 77, 2810 (1996).

71. S. Q. Shen and F. C. Pu, *Phys. Lett. A* 210, 135 (1996).

72. Z. C. Wang, S. F. Wang, S. H. Shen, S. C. Zou, and Z. M. Zhang, *J. Phys.: Condens. Matter.* 8, 6381 (1996).

73. J. Barnaś, *J. Magn. Magn. Mater.* 154, 321 (1996).

74. G. M. Genkin and M. V. Sapozhnikov, *J. Magn. Magn. Mater.* 139, 179 (1995).

75. B. A. Jones and C. B. Hanna, *Phys. Rev. Lett.* 71, 4253 (1993).

76. J. Barnaś, *J. Magn. Magn. Mater.* 167, 209 (1997).

77. P. Bruno, *J. Magn. Magn. Mater.* 121, 248 (1993).

78. P. Bruno, *J. Magn. Magn. Mater.* 164, 27 (1996).

79. J. Mathon, M. Villeret, A. Umerski, R. B. Muniz, J. d'Albuquerque e Castro, and D. M. Edwards, *Phys. Rev. B: Condens. Matter* 56, 11797 (1997).

80. J. Henk, T. Scheunemann, S. V. Halilov, and R. Feder, *Phys. Status Solidi B* 192, 325 (1995).

81. S. V. Halilov, E. Tamura, H. Gollisch, D. Meinert, and R. Feder, *J. Phys.: Condens. Matter* 5, 3859 (1993).

82. B. A. Jones and C. B. Hanna, *Phys. Rev. Lett.* 71, 4253 (1993).

83. A. Shik, "Quantum Wells, Physics and Electronics of Two Dimensional Systems." World Scientific, Singapore, 1997.

84. R. Turton, "The Quantum Dots, A Journey into the Future of Microelectronics." W. H. Freeman, Oxford, 1995.

85. Yu Ming-hui and Zhang Zhi-dong, to be published.

86. D. M. Deaven, D. S. Rokhsar, and M. Johnson, *Phys. Rev. B: Condens. Matter* 44, 5977 (1991).

87. P. Van Gelderen, S. Crampin, and J. E. Inglesfield, *Phys. Rev. B: Condens. Matter* 53, 9115 (1996).

88. V. L. Moruzzi, J. F. Janak, and A. R. Williams, "Calculated Electronic Properties of Metals." Pergamon, Oxford, 1978.

89. G. Hörmandinger and J. B. Pendry, *Surf. Sci.* 295, 34 (1993).

90. G. J. Mankey, R. F. Willis, and F. J. Himpsel, *Phys. Rev. B: Condens. Matter* 48, 10284 (1993).

91. P. T. Coleridge and I. M. Templeton, *Phys. Rev. B: Condens. Matter* 25, 7818 (1982).

92. M. R. Halse, *Philos. Trans. R. Soc. London* 265, 1507 (1969).

93. Zhi-dong Zhang, submitted for publication.

94. N. Manyala, Y. Sidls, J. F. DiTusa, G. Aeppll, D. P. Young, and Z. Fisk, *Nature (London)* 404, 581 (2000).

95. Y. Shimakawa, Y. Kubo, and T. Manako, *Nature (London)* 379, 53 (1996).

96. P. Majumdar and P. B. Littlewood, *Nature (London)* 395, 479 (1998).

97. V. S. Fomenko, "Handbook of Thermionic Properties" (G. V. Samsanov, Ed.). Plenum Press Data Division, New York, 1966.

98. Z. Q. Qiu, J. Pearson, and S. D. Bader, *Phys. Rev. B: Condens. Matter* 46, 8659 (1992).

99. M. Johnson, S. T. Purcell, N. W. E. McGee, R. Coehoorn, J. aan de Stegge, and W. Hoving, *Phys. Rev. Lett.* 68, 2688 (1992).

100. P. van Gelderen, S. Crampin, and J. E. Inglesfield, *Phys. Rev. B: Condens. Matter* 53, 9115 (1996).

101. L. Nordström, P. Lang, R. Zeller, and P. H. Dederichs, *Europhys. Lett.* 29 395 (1995).

102. L. Nordström, P. Lang, R. Zeller, and P. H. Dederichs, *Phys. Rev. B: Condens. Matter* 50, 13058 (1994).

103. V. Drchal, J. Kudrnovsky, I. Turek, and P. Weinberger, *Phys. Rev. B: Condens. Matter* 53, 15036 (1996).

104. W. H. Butler, X. G. Zhang, D. M. C. Nicholson, and J. M. MacLaren, *Phys. Rev. B: Condens. Matter* 52, 13399 (1995).

105. S. Mirbt, A. M. N. Niklasson, B. Johansson, and H. L. Skriver, *Phys. Rev. B: Condens. Matter* 54, 6382 (1996).

106. D. M. Edwards and J. Mathon, *J. Magn. Magn. Mater.* 93, 85 (1991).

107. M. Brack, *Rev. Mod. Phys.* 65, 677 (1993).

108. D. M. Edwards, J. Mathon, R. B. Muniz, and M. S. Phan, *J. Phys.: Condens. Matter* 3, 4941 (1991).

109. D. M. Edwards, J. Mathon, and R. B. Muniz, *Phys. Rev. B: Condens. Matter* 21, 16066 (1994).

110. S. D. Bader and D. Q. Li, *J. Magn. Magn. Mater.* 156, 153 (1996).

111. P. D. Johnson, *J. Magn. Magn. Mater.* 148, 167 (1995).

112. C. Hwang and F. J. Himpsel, *Phys. Rev. B: Condens. Matter* 52, 15368 (1995).

113. E. J. Escorcia-Aparicio, R. K. Kawakami, and Z. Q. Qiu, *Phys. Rev. B: Condens. Matter* 54, 4155 (1996).

114. K. Garrison, Y. Chang, and P. D. Johnson, *Phys. Rev. Lett.* 71, 2801 (1993).

115. C. Carbone, E. Vescovo, O. Rader, W. Gudat, and W. Eberhardt, *Phys. Rev. Lett.* 71, 2805 (1993).

116. P. Segovia, E. G. Michel, and J. E. Ortega, *Phys. Rev. Lett.* 77, 3455 (1996).

117. R. Nakatani, K. Hoshino, H. Hoshiya, and Y. Sugita, *J. Magn. Magn. Mater.* 166, 261 (1997).

118. A. Ishii, T. Aisaka, and K. Yamada, *Surf. Sci.* 357–358, 331 (1996).

119. Z. Q. Qiu, J. Pearson, and S. D. Bader, *Phys. Rev. B: Condens. Matter* 46, 8195 (1992).

120. D. Q. Li, M. Freitag, J. Pearson, Z. Q. Qiu, and S. D. Bader, *Phys. Rev. Lett.* 72, 3112 (1994).

121. A. M. Zeltser and N. Smith, *J. Appl. Phys.* 79, 9224 (1996).

122. Y. Saito, S. Hashimoto, and K. Inomata, *IEEE Trans.* MAG-28, 2751 (1992).

123. A. Fert, P. Grünberg, A. Barthélémy, F. Petroff, and W. Zinn, *J. Magn. Magn. Mater.* 140–144, 1 (1995).

124. Y. Wang, P. M. Levy, and J. L. Fry, *Phys. Rev. Lett.* 65, 2732 (1990).

125. J. Unguris, R. J. Celotta, and D. T. Pierce, *Phys. Rev. Lett.* 67, 140 (1991).

126. S. T. Purcell, W. Folkerts, M. T. Johnson, N. W. E. McGee, K. Jager, J. van de Stegge, W. B. Zeper, and W. Hoving, *Phys. Rev. Lett.* 67, 903 (1991).

127. S. T. Purcell, M. T. Johnson, N. W. E. McGee, R. Coehoorn, and W. Hoving, *Phys. Rev. B: Condens. Matter* 45, 13064 (1992).

128. J. Unguris, R. J. Celotta, and D. T. Pierce, *Phys. Rev. Lett.* 69, 1125 (1992).

129. E. E. Fullerton, M. J. Conover, J. E. Mattson, C. H. Sowers, and S. D. Bader, *Phys. Rev. B: Condens. Matter* 48, 15755 (1993).

130. A. Davies, J. A. Stroscio, D. T. Pierce, and R. J. Celotta, *Phys. Rev. Lett.* 76, 4175 (1996).

131. B. Heinrich, J. F. Cochran, D. Venus, K. Totland, C. Schneider, and K. Myrtle, *J. Magn. Magn. Mater.* 156, 215 (1996).

132. D. Stoeffler and F. Gautier, *Phys. Rev. B: Condens. Matter* 44, 10389 (1991).

133. E. Fawcett, *Rev. Mod. Phys.* 60, 209 (1988).

134. M. Rührig, R. Schäfer, A. Hubert, R. Mosler, J. A. Wolf, S. Demokritove, and P. Grünberg, *Phys. Status Solidi A* 125, 635 (1991).

135. A. Fuss, S. Demokritove, P. Grünberg, and W. Zinn, *J. Magn. Magn. Mater.* 103, L221 (1992).

136. Z. Celinski, B. Heinrich, and J. F. Cochran, *J. Magn. Magn. Mater.* 145, L1 (1995).

137. U. Köbler, K. Wagner, R. Wiechers, A. Fuss, and W. Zinn, *J. Magn. Magn. Mater.* 103, 236 (1992).

138. V. V. Kostyuchenko and A. K. Zvezdin, *J. Magn. Magn. Mater.* 176, 155 (1997).

139. D. M. Edwards, J. M. Ward, and J. Mathon, *J. Magn. Magn. Mater.* 126, 380 (1993).

140. J. Barnaś, *J. Magn. Magn. Mater.* 123, L21 (1993).

141. H. Fujiwara, W. D. Doyle, A. Matsuzono, and M. R. Parker, *J. Magn. Magn. Mater.* 140–144, 519 (1995).

142. R. P. Erickson, K. B. Hathaway, and J. R. Cullen, *Phys. Rev. B: Condens. Matter* 47, 2626 (1993).

143. J. C. Slonczewski, *Phys. Rev. Lett.* 67, 3172 (1991).

144. J. C. Slonczewski, *J. Appl. Phys.* 73, 5957 (1993).

145. S. Demokritov, E. Tsymbal, P. Grünberg, W. Zinn, and I. K. Schuller, *Phys. Rev. B: Condens. Matter* 49, 720 (1994).

146. Z. Q. Qiu, J. Pearson, A. Berger, and S. D. Bader, *Phys. Rev. Lett.* 68, 1398 (1992).

147. Z. Q. Qiu, J. Pearson, and S. D. Bader, *J. Appl. Phys.* 73, 5765 (1993).

148. D. Stoeffler and F. Gautier, *J. Magn. Magn. Mater.* 121, 259 (1993).

149. S. T. Purcell, M. T. Johnson, N. W. E. McGee, R. Coehoorn, and W. Hoving, *Phys. Rev. B: Condens. Matter* 45, 13064 (1992).

150. T. Katayama, Y. Suzuki, M. Hayashi, and A. Thiaville, *J. Magn. Magn. Mater.* 126, 527 (1993).

151. T. Katayama, Y. Suzuki, and W. Geerts, *J. Magn. Magn. Mater.* 156, 158 (1996).

152. S. Crampin, S. De Rossi, and F. Ciccacci, *Phys. Rev. B: Condens. Matter* 53, 13817 (1996).

153. R. Mégy, A. Bounouh, Y. Suzuki, P. Beauvillain, P. Bruno, C. Chappert, B. Lecuyer, and P. Veillet, *Phys. Rev. B: Condens. Matter* 51, 5586 (1995).

154. J. E. Ortega, F. J. Himpsel, G. J. Mankey, and R. F. Willis, *Phys. Rev. B: Condens. Matter* 47, 1540 (1993).

155. P. Bruno, Y. Suzuki, and C. Chappert, *Phys. Rev. B: Condens. Matter* 53, 9214 (1996).

156. A. Marty, B. Gilles, J. Eymery, A. Chamberod, and J. C. Joud, *J. Magn. Magn. Mater.* 121, 57 (1993).

157. Z. Q. Qiu, J. Pearson, and S. D. Bader, *Phys. Rev. B: Condens. Matter* 49, 8797 (1994).

158. A. Fuss, S. Demokritov, P. Grünberg, and W. Zinn, *J. Magn. Magn. Mater.* 103, L221 (1992).

159. S. S. P. Parkin, R. F. C. Farrow, T. A. Rabedeau, R. F. Marks, G. R. Harp, Q. Lam, C. Chappert, M. F. Toney, R. Savoy, and R. Geiss, *Europhys. Lett.* 22, 455 (1993).

160. M. A. Mueller, T. Miller, and T. C. Chiang, *Phys. Rev. B: Condens. Matter* 41, 5214 (1990).

161. E. D. Hansen, T. Miller, and T. C. Chiang, *Phys. Rev. B: Condens. Matter* 55, 1871 (1997).

162. A. Beckmann, M. Klaua, and K. Meinel, *Phys. Rev. B: Condens. Matter* 48, 1844 (1993).

163. W. E. McMahon, T. Miller, and T. C. Chiang, *Phys. Rev. B: Condens. Matter* 54, 10800 (1996).

164. D. Hartmann, W. Weber, A. Rampe, S. Popovic, and G. Güntherodt, *Phys. Rev. B: Condens. Matter* 48, 16837 (1993).

165. A. Carlsson, D. Claesson, S. Å. Lindgren, and L. Walldén, *Phys. Rev. B: Condens. Matter* 52, 11144 (1995).

166. D. Stoeffler and F. Gautier, *J. Magn. Magn. Mater.* 140–144, 529 (1995).

167. B. Dieny, *J. Magn. Magn. Mater.* 136, 335 (1994).

168. K. Fuchs, *Proc. Cambridge Philos. Soc.* 34, 100 (1938).

169. E. H. Sondheimer, *Adv. Phys.* 1, 1 (1952).

170. J. Barnaś, A. Fuss, R. E. Cameley, P. Günberg, and W. Zinn, *Phys. Rev. B: Condens. Matter* 42, 8110 (1990).

171. R. Q. Hood and L. M. Falicov, *Phys. Rev. B: Condens. Matter* 46, 8287 (1992).

172. J. L. Duvail, A. Fert, L. G. Pereira, and D. K. Lottis, *J. Appl. Phys.* 75, 7070 (1994).

173. A. Barthélémy and A. Fert, *Phys. Rev. B: Condens. Matter* 43, 13124 (1991).

174. B. L. Johnson and R. E. Camley, *Phys. Rev. B: Condens. Matter* 44, 9997 (1991).

175. R. E. Camley and J. Barnaś, *Phys. Rev. Lett.* 63, 664 (1989).

176. B. Dieny, *Europhys. Lett.* 17, 261 (1992).

177. B. Dieny, *J. Phys.: Condens. Matter* 4, 8009 (1992).

178. L. M. Falicov and R. Q. Hood, *J. Magn. Magn. Mater.* 121, 362 (1993).

179. R. Q. Hood, L. M. Falicov, and D. R. Pann, *Phys. Rev. B: Condens. Matter* 49, 368 (1994).

180. P. M. Levy, S. Zhang, and A. Fert, *Phys. Rev. Lett.* 65, 1643 (1990).

181. S. Zhang, P. M. Levy, and A. Fert, *Phys. Rev. B: Condens. Matter* 45, 8689 (1992).

182. H. E. Camblong, S. Zhang, and P. M. Levy, *Phys. Rev. B: Condens. Matter* 47, 4735 (1993).

183. H. E. Camblong and P. M. Levy, *Phys. Rev. Lett.* 69, 2835 (1992).

184. H. E. Camblong and P. M. Levy, *J. Appl. Phys.* 73, 5533 (1993).

185. D. A. Greenwood, *Proc. Phys. Soc. London* 71, 585 (1958).

186. A. Vedyayev, C. Cowache, N. Ryzhanova, and B. Dieny, *J. Phys.: Condens. Matter* 5, 8289 (1993).

187. A. Vedyayev, B. Dieny, and N. Ryzhanova, *Europhys. Lett.* 19, 329 (1992).

188. X. G. Zhang and W. H. Butler, *Phys. Rev. B: Condens. Matter* 51, 10085 (1995).

189. J. Barnaś, *J. Magn. Magn. Mater.* 131, L14 (1994).

190. B. Bulka and J. Barnaś, *Phys. Rev. B: Condens. Matter* 51, 6348 (1995).

191. R. Kubo, *J. Phys. Soc. Jpn.* 12, 570 (1957).

192. M. A. M. Gijs and M. Okada, *Phys. Rev. B: Condens. Matter* 46, 2908 (1992).

193. M. A. M. Gijs and M. Okada, *J. Magn. Magn. Mater.* 113, 105 (1992).

194. S. K. J. Lenczowski, M. A. M. Gijs, J. B. Giesbers, R. J. M. van de Veerdonk, and W. J. M. de Jonge, *Phys. Rev. B: Condens. Matter* 50, 9982 (1994).

195. A. Vedyayev, M. Chshiev, and B. Dieny, *J. Magn. Magn. Mater.* 184, 145 (1998).

196. H. Itoh, J. Inoue, and S. Maekawa, *Phys. Rev. B: Condens. Matter* 51, 342 (1995).

197. W. H. Butler, X. G. Zhang, D. M. C. Nicholson, and J. M. MacLaren, *Phys. Rev. B: Condens. Matter* 52, 13399 (1995).

198. T. Oguchi, *J. Magn. Magn. Mater.* 126, 519 (1993).

199. R. K. Nesbet, *J. Phys.: Condens. Matter* 6, L449 (1994).

200. K. M. Schep, P. J. Kelly, and G. E. Bauer, *Phys. Rev. Lett.* 74, 586 (1995).

201. W. H. Butler, *Phys. Rev. B: Condens. Matter* 31, 3260 (1985).

202. V. M. Uzdin and N. S. Yartseva, *J. Magn. Magn. Mater.* 156, 193 (1996).

203. M. Julliere, *Phys. Lett. A* 54, 225 (1975).

204. S. Maekawa and U. Gäfvert, *IEEE Trans. Magn.* MAG-18, 707 (1982).

205. T. Miyazaki and N. Tezuka, *J. Magn. Magn. Mater.* 139, L231 (1995).

206. J. S. Moodera, L. R. Kinder, T. M. Wang, and R. Meservey, *Phys. Rev. Lett.* 74, 3273 (1995).

207. N. Tezuka and T. Miyazaki, *J. Appl. Phys.* 79, 6262 (1996).

208. T. Miyazaki, S. Kumagai, and T. Yaoi, *J. Appl. Phys.* 81, 3753 (1997).

209. W. J. Gallagher, S. S. P. Parkin, Y. Lu, X. P. Bian, A. Marley, K. P. Roche, R. A. Altman, S. A. Rishton, C. Jahnes, T. M. Shaw, and G. Xiao, *J. Appl. Phys.* 81, 3741 (1997).

210. K. Inomata, S. N. Okuno, Y. Saito, and K. Yusu, *J. Magn. Magn. Mater.* 156, 219 (1996).

211. J. Inoue and S. Maekawa, *Phys. Rev. B: Condens. Matter* 53, R11927 (1996).

212. J. C. Slonczewski, *Phys. Rev. B: Condens. Matter* 39, 6995 (1989).

213. A. Vedyayev, N. Ryzhanova, C. Lacroix, L. Giacomoni, and B. Dieny, *Europhys. Lett.* 39, 219 (1997).

214. X. Zhang, B. Z. Li, G. Sun, and F. C. Fu, *Phys. Rev. B: Condens. Matter* 56, 5484 (1997).

215. A. Vedyayev, N. Ryzhanova, R. Vlutters, and B. Dieny, *J. Phys.: Condens. Matter* 10, 5799 (1998).

216. S. Toscano, B. Briner, H. Hipster, and M. Landolt, *J. Magn. Magn. Mater.* 114, L6 (1992).

217. E. E. Füllerton, J. E. Mattson, S. R. Lee, C. H. Sowers, Y. Y. Huang, G. Felcher, and S. D. Bader, *J. Magn. Magn. Mater.* 117, L301 (1992).

218. E. E. Füllerton, J. E. Mattson, S. R. Lee, C. H. Sowers, Y. Y. Huang, G. Felcher, and S. D. Bader, *J. Appl. Phys.* 73, 6335 (1993).

219. C. L. Foiles, M. R. Franklin, and R. Loloee, *Phys. Rev. B: Condens. Matter* 50, 16070 (1994).

220. B. Briner and M. Landolt, *Z. Phys. B: Condens. Matter* 92, 137 (1993).

221. B. Briner, U. Ramsperger, and M. Landolt, *Phys. Rev. B: Condens. Matter* 51, 7303 (1995).

222. B. Briner and M. Landolt, *Phys. Rev. Lett.* 73, 340 (1994).

223. P. Bruno, *Phys. Rev. B: Condens. Matter* 49, 13231 (1994).

224. Ming-wen Xiao and Zheng-zhong Li, *Phys. Rev. B: Condens. Matter* 54, 3322 (1996).

225. M. de Dios-Leyva, A. Bruno-Alfonso, A. Matos-Abiague, and L. E. Oliveira, *J. Phys.: Condens. Matter* 9, 8477 (1997).

226. G. Bastard, E. E. Mendez, L. L. Chang, and L. Esaki, *Phys. Rev. B: Condens. Matter* 26, 1974 (1982).

227. R. L. Greene, K. K. Bajaj, and D. E. Phelps, *Phys. Rev. B: Condens. Matter* 29, 1807 (1984).

228. M. F. Pereira, Jr., I. Galbraith, S. W. Koch, and G. Duggan, *Phys. Rev. B: Condens. Matter* 42, 7084 (1990).

229. L. C. Andreani and A. Pasquarello, *Phys. Rev. B: Condens. Matter* 42, 8928 (1990).

230. G. Bastard, *Phys. Rev. B: Condens. Matter* 24, 4714 (1981).

231. W. T. Masselink, Y. C. Chang, and H. Morkoc, *Phys. Rev. B: Condens. Matter* 28, 7373 (1983).

Chapter 6

ELECTRO-OPTICAL AND TRANSPORT PROPERTIES OF QUASI-TWO-DIMENSIONAL NANOSTRUCTURED MATERIALS

Rodrigo A. Rosas

Departamento de Física, Universidad de Sonora, Hermosillo, Sonora, México

Raúl Riera, José L. Marín, Germán Campoy

Departamento de Investigación en Física, Universidad de Sonora, Hermosillo, Sonora, México

Contents

Handbook of Thin Film Materials, edited by H.S. Nalwa
Volume 5: Nanomaterials and Magnetic Thin Films
Copyright © 2002 by Academic Press
All rights of reproduction in any form reserved.

1. INTRODUCTION

In the last 30 years, the study of the quasi-two-dimensional systems, also called semiconductor heterostructures by layers, has acquired a special significance. This is due to the development of growing techniques which allow us to set systems with a totally quantum character. These techniques are characterized by allowing the growth of extremely thin layers, controlling the process with great detail, specifically the thickness of the layers, the content of impurities, and the abruptness of the interfaces. In the technique of growth called molecular beam epitaxy (MBE) for example, the ultrahigh vacuum atmosphere allows the implementation of powerful analytic techniques *in situ* and specification of the characterizations of the sequence of growth in the camera, assuring that each step in the process is correctly done.

In fact, on the basis of such systems by layers, luminous high power emitters, ultrarapid photodetectors, high-speed transistors, amplifiers with low noise levels, and other electronic devices have been designed and built. The most interesting fact of all this is the possibility of designing a heterostructure in such a way that it will exhibit certain properties which could not be obtained if we had started from homostructures. An example of this situation is the semiconductor laser based on a double heterostructure, for which confinement of the carriers can be made to increase as much as that of the luminous wave (whose name in the literature is bandgap engineering).

1.1. Definition of the Quasi-Two-Dimensional Nanostructured Materials

(a) Confined systems: In the broadest sense of the word, a confined system, also called a low-dimensional system, is any quantum system in which the carriers are free to move in only two, one, or even zero dimensions. They could be real and ideal systems. In these systems the spatial dimensions are of the order of the De Broglie wavelength of the carriers, whose

movement becomes quantized; those dimensions are now said to be of "quantum scale." Several methods for obtaining the confined systems exist, namely, the reduction of some spatial directions to quantum scale by the application of any kind of fields or by limiting the borders in the synthesis of the materials.

(b) Quasi-two-dimensional nanostructured materials: They are real confined systems manufactured with materials such as semiconductors, metals, dielectrics, magnetics, etc., in which any kind of carrier, for example electrons, atoms, or excitons, is confined in one direction, keeping free the other two directions for the movement (see Fig. 1). They are generally represented by the acronym Q2D.

1.2. Classification of the Quasi-Two-Dimensional Nanostructured Materials

There exist several forms of classifying the quasi-two-dimensional systems, the most universal being the one that considers the number of heterostructures or layers of which they are formed. Four types of quasi-two-dimensional systems were investigated: (i) single heterostructure (SH) (see Fig. 2), (ii) double heterostructure or quantum well (QW) (see Fig. 1), (iii) multiple heterostructures or multiple quantum wells (see Fig. 3), and (iv) superlattices (SL) (see Fig. 4). Of all these systems, most theoretical and experimental investigations that appear in the literature are about QW and SL.

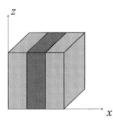

Fig. 1.   Quasi-two-dimensional system (single square quantum well).

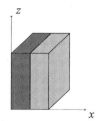

Fig. 2. Single heterostructure.

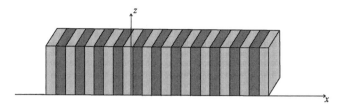

Fig. 4. Superlattice.

The quasi-two-dimensional systems were the first confined systems studied. They are real systems, built by using heterostructures between different materials (polar semiconductors initially). The main characteristic of these systems is that the energy spectrum in the spatial direction of confinement is always discrete if the particle remains totally confined and discrete and continuous if the particle remains partially confined.

In general the Q2D are fabricated with two different materials A and B, which are semiconductors with bandgaps, ε_g^A and ε_g^B, respectively. In the Q2D growth process, one of the materials is deposited upon the other one and a potential barrier is formed in the interface, due to the difference of bandgaps. According to the gap positions and values, three types of Q2D can be obtained, which will be identified as types I, II, and III (see Fig. 5).

In the type I Q2D, the B material is a barrier for both the valence and conduction electrons, which are localized within the same A material. Examples of type I Q2Ds are the QWs GaAs–Ga(Al)As, $Ga_{0.47}Al_{0.53}As$–InP, GaSb–AlSb, etc.

In the type II Q2Ds (also called staggered nanostructures) one material acts as a well for conduction electrons but as a barrier for valence electrons. Examples of the type II Q2Ds are the QW InP–$Al_{0.53}In_{0.48}As$, where the electrons are mostly in the InP material and the holes in $Al_{0.53}In_{0.47}As$, and InAs–GaSb, where the electrons are mostly in the InAs semiconductor and the holes are mostly in the GaSb semiconductor. In type II Q2Ds, we deal with interface excitons where the interacting particles are spatially separated, a situation which is reminiscent of the bound impurity states created by impurities placed in the barriers of the quantum structures.

Of the Q2D, only types I and II are of interest in the literature; this is the reason why only this system will be considered in the future.

2. METHODS OF SYNTHESIS AND FABRICATION OF QUASI-TWO-DIMENSIONAL NANOSTRUCTURED MATERIALS

There is an increasing interest in producing structures of a nanometer size for use in device fabrication and testing the laws of physics in nanometer sized structures. This is a growing field and nowadays it is very important to investigate new techniques for nanometer device fabrication.

The most used materials for production of Q2D nanostructured materials are, in the first place, semiconductors and metals, followed by dielectric, magnetic, and superconductor compounds. They are manufactured using organic materials, polymers, biomaterials, etc.

Nowadays there are many methods of synthesis and production of Q2D nanostructured materials, at a laboratory scale, but the most important are those where high quality quantum nanostructures are achieved. These methods produce materials that have the precision of one atomic layer, with abruptly grown interfaces with one monolayer roughness, and a great control for purity and doped impurities.

In addition to this technology, in order to guarantee the high quality of the Q2D nanostructured materials, the investigation laboratories need to count with several characterization techniques *in situ*. Independently of the grown Q2D systems, the most used and known techniques for the production or fabrication of the different kinds of quasi-two-dimensional nanostructured materials are MBE, metal–organic chemical vapor deposition (MOCVD), and lithography, also known as nanolithography.

2.1. Molecular Beam Epitaxy

MBE is a method by means of which a semiconductor, such as GaAs, is grown in an ultrahigh vacuum environment by

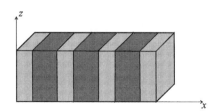

Fig. 3. Multiple quantum well.

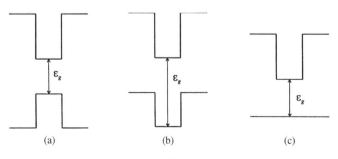

Fig. 5. Types of Q2D systems. (a) Type I, (b) type II, (c) type III.

supplying molecular and/or atomic beams of Ga, As, and other constituents onto a cleaned crystalline substrate [1–9]. Because one can prepare semiconductor films of high quality by MBE with a thickness precision of one atomic layer or better, MBE is one of the best methods for the preparation of a very flat surface of one semiconductor upon which another semiconductor of an arbitrary thickness can be deposited. This feature allows one to control the composition profile and thereby to implement an artificial potential $V(z)$ as a function of position z along the direction of stacking. A number of studies have shown, however, that interfaces of GaAs–AlAs (or AlGaAs) grown by MBE are generally abrupt but have a roughness Δ of one monolayer (ML), $a = 2.83$ Å. Hence it is very important to clarify the origin and nature of such roughness and to establish a method for controlling the atomic structure of interfaces. In addition, one should note that the atomic structure of an interface is usually a frozen image of the freshly grown surface and therefore provides valuable information on the microscopic process of epitaxial growth. On the other hand, the high purity of the Q2D systems obtained by MBE indicates that the mobility is dominated mainly through scattering by phonons. The advantage of epitaxial growth by MBE, with respect to other methods or technologies, is that they possess multiple techniques *in situ*. That technique permits an absolute control of the ground parameters of the grown structure, such as thickness, high purity, abruptness and flatness of interface, and impurities in doped heterostructures, in addition to the ultrahigh vacuum environment that it utilizes.

2.2. Metal–Organic Chemical Vapor Deposition

The MOCVD method is the most useful and conventional process for fabricating thin layer semiconductor heterostructures. In general the MOCVD has a reaction chamber that is basically dependent upon the heating system which is fitted for long growth runs. It is possible to use a radio frequency (RF) induction coil as an alternative heater. There is a possibility of dissociation of gases which produces radicals due to a plasma discharge at reduced pressures, and then it is necessary to employ a large-scale electric power supply in connection with a precise electric circuit. However, in the case of synthesis of the II–VI compound semiconductors, it should be pointed out that there are several problems concerning the difficulty in precisely controlling the II/VI mole ratio. This is due to the high dissociation pressure of VI elements, and also to the premature reaction taking place in the vapor phase prior to epitaxial deposition on the substrate. In order to avoid such problems, Yamada [10] has used new metal–organic sources diluted in a He gas, namely, dimethylzinc (DMZn) with a concentration of 1.06% in a He gas, for the growth of cubic-structured $Cd_xZn_{1-x}S$–ZnS strained-layer multiple quantum wells [11–13]. A mixture of 10% of H_2S in a H_2 gas was used as the sulphur source. Its MOCVD apparatus consists of a main reaction chamber and a sample exchange supplemental chamber with a transfer tube. A special feature is that they employ a halogen lamp capable of heating substrates to up to 600°C by

infrared radiation. Two chambers are constructed for all VI materials with each inner wall coated with TiN film to avoid accidental impurity contamination. In order to control the gas flow ratio of II/VI, they used a personal computer connected to a relay box linked to an interface which was capable of switching both the piezo valves of the MFC (mass flow controller) and the air valves on or off. Then, all of gas sources flowed directly into the MFCs, so that the bubbling system by a carrier gas was unnecessary for the metal–organic (DMZn and DMCd) as well as the H_2S gas flows [14–15]. Therefore, their process using the low-pressure MOCVD system with all gaseous sources has proven to be advantageous for precise control of II/VI mole ratio as well as lower temperature growth (350°C) of the II–VI compound hetrostructures.

2.3. Lithography

It is instructive to consider the conventional but advanced forms of lithography as practiced today. As it has developed for the semiconductor industry, lithography is the creation of a pattern in a resist layer, usually an organic polymer film, upon a substrate material. A latent image, consisting of a chemical change in the resist, is created by exposure of the desired area to some kind of radiation. The pattern is developed by selectively removing either the exposed areas of resist (for a positive resist) or the unexposed areas (for a negative resist). These processes are shown schematically in Figure 6. The development of a polymeric resist is usually done with a solvent that brings out the pattern based on the solubility difference created by exposure.

2.3.1. Overview of Lithography

Electrons, ions, and photons can all be used for exposure in the lithography process. Because of intrinsic scattering mechanisms, in no case can the energy provided by the exposing radiation be confined to arbitrarily small volumes. First of all, the radiation or particle beam cannot be confined arbitrarily because of the diffraction, Coulomb repulsion among particles, optical limitations, or other effects. Second, the dissociation energy of a resist layer spreads in the film, because of intrinsic scattering mechanisms in the solid.

The figures of merit for a resist/developer system include the following aspects: the sensitivity to the exposure dose,

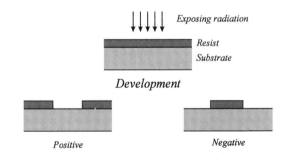

Fig. 6. Schematic diagram of the basic lithographic steps of resist exposure and development.

dependence of the resist response; the resolution, dictating the minimum feature size; and the suitability of the resist to the pattern transfer. While the sensitivity and contrast of a system are vital in determining the speed and process latitude, it is the resolution and pattern transfer utility that are most important to Q2D fabrication at a research level.

Inorganic resist materials exist and have been applied in the high-resolution field. Related processes include vapor or plasma developments that operate based on differences in gas phase or plasma reactivity rather than liquid development based on solubility differences. Self-developing resists are of some importance in the high-resolution area. For these materials the exposure to radiation volatilizes the resist, creating on it a pattern without a development step. By this method, some years ago structures with dimensions of the order of one unit cell were made [16].

2.3.2. Photon Lithography

The use of photons for high-resolution lithographic exposure is a broad field that includes most of the lithography in use today. This is the technique most widely used by the semiconductor industry and illustrates the general technique of lithography. It can be taken to include X-ray lithography and deep and near-UV photolithography. Only the shortest wavelengths are directly relevant to high-resolution fabrication. The projection and proximity printing processes to be described are analogous to processes that could be used at nanometer dimensions for parallel printing with electrons or ions.

In projection lithography a mask containing the desired pattern is diminished by an optical system and projected on the resist layer. This is desirable since there is no damage by the contact between the mask and wafer, and the fabrication of the mask is not so critical since it is diminished. The resolution is limited, however, by the diffraction and the quality of the optics. The diffraction limit for the minimum resolvable grating period is usually taken as the Rayleigh criterion for the overlap of the diffraction peaks, which is approximately

$$\text{minimum line resolvable spacing} \approx \frac{\lambda}{\text{NA}} \qquad (1)$$

where λ is the wavelength and NA is the numerical aperture of the optical system, which is of the order of one. The shortest wavelength to be used with conventional optic is about 193 nm from an excimer laser source. The use of wavelengths much shorter than this is limited by absorption in optical materials. While the technology is constantly improving, with better optics, improved contrast enhanced resists, and sources of shorter wavelength, the limit of far-field diffraction is a fundamental one and prevents the use of characteristic sizes smaller than the light wavelength. Photolithography is essentially ruled out for fabrication at dimensions below 100 nm.

Projection systems employing X-rays would reduce diffraction effects, but the problem of fabricating optics for use in the X-ray region remains. Work in this area is advancing, with the development of multilayer mirrors and X-ray mask technology.

Proximity printing allows the 1:1 replication of a mask pattern. The effects of near-field diffraction can be reduced arbitrarily by decreasing the distance between mask and substrate. Qualitatively, the minimum linewidth is

$$d \approx \sqrt{\lambda s} \qquad (2)$$

where s is the separation between mask and substrate.

Proximity printing can be done effectively by X-ray exposure where the shorter wavelengths, of the order of 1 nm, allow much greater mask–substrate separation. With short wavelength X-rays, diffraction can be effectively ignored for attainable gaps of the order of a few micrometers. However, the exposure of polymeric resist by X-rays takes place primarily through the generation of photoelectrons which then expose the resist by the same mechanisms as an electron beam. The range of the photoelectrons in the resist will limit the resolution. Since the photoelectrons can have energies up to that of the X-rays, their range can be greater than 100 nm for the shorter wavelength X-rays. It can be qualitatively seen that there is a trade-off between diffraction and the range of photoelectrons in limiting the resolution. Perhaps the most significant problem with X-ray printing is the creation of a durable and stable mask of high contrast. All materials absorb X-rays, so the transparent area of the mask must be very thin or nonexistent. One type of mask is an absorbing metal such as gold supported on a thin film, as for example SiN_4. Some of the smallest features replicated by X-rays were 17.5 nm in extent, fabricated by the use of edge evaporated metal as a mask. Some other high-resolution process is necessary to generate the mask features; in most cases this means electron beam lithography.

2.3.3. Electron Beam Lithography

It has now been developed substantially and has been extensively used for the generation of photolithographic masks. Schemes for electron beam 1:1 printing similar to those using X-rays or ions have been explored but generally not for the creation of Q2D systems. Because of the easy availability of small electron sources and the high quality of electron optics, an electron beam can be focused to dimensions less than 1 nm and scanned with great accuracy.

In scanned electron beam lithography, a small electron source is imaged on the substrate through a series of electromagnetic lenses in an electron-optic column. The beam can be rapidly scanned over the sample to trace out any desired pattern. In the normal exposure of organic polymer resist, the electron beam causes the breaking of bonds or induces additional chemical bonds to be formed. This results in different molecular sizes in the exposed versus the unexposed areas, and this can be translated into differences in solubility in an appropriate developer.

The problem with electron beams is to confine the deposition of energy to a small area in the resist. There are several scattering processes in the material that lead to the deposition of energy at distances remote from the initial impact point of the electron. This leads to the well-known proximity and resolution limiting effects.

The scattering effect of longest range is the backscattering of electrons, resulting in large-angle scattering with energies near that of the incident electron. The range of the backscattered electrons can be of many micrometers for typical electron beam energies. This range has been determined to have a power law dependence, proportional to (Energy)$^{1.7}$, over a wide electron energy range. This presents the greatest problems for energies where the electron range is comparable to the feature spacing [17].

Because of small-angle scattering events, forward scattering tends to spread the incident beam passing through the resist. The scattering angles decrease with electron energy, and the spread of the beam is obviously more important for thicker resist layers.

With electron beams it is possible to expose areas so small that the molecular size of the resist can become significant. This can be of the order of 10 nm for large organic molecules. The grain size of a polycrystalline layer to be patterned can also become larger than the size of the desired features. Clearly there are many new material considerations to be addressed at nanometer dimensions. Another important consideration for the ultrasmall size range is the time taken to write a pattern by a scanning serial process. This time is inversely related to the size of the pixel. The smaller the electron beam size, the lower the current available for exposure and the longer the needed exposure time. The resists of highest resolution also require higher exposures. Fortunately the areas involved in exposures for devices are small. The time for writing is insignificant for experimental devices where only a few objects are being written, and the benefits and successes of scanned electron beam lithography have been very important.

A recently studied alternative approach to scanned electron bean lithography is the use of a technology related to the scanning tunneling microscope to scan a low energy electron beam in close proximity to the surface. The use of low energy beams allows only thin conducting layers to be modified by the electron beam. There are exciting possibilities on the manipulation of individual atoms, as demonstrated by IBM researchers. For dimensions larger than this, the flexibility and capabilities of scanned beams as opposed to scanned source are, as yet, more significant in the fabrication of devices for research.

2.3.4. Ion Beam Lithography

The use of ion beams is diverse, with a great range of ion species and energies. Both collimated beams of large area and scanned focused beams have been applied. The interaction of the ion beam with a solid can consist in an exposition of the resist, which is most appropriate for high energy in low-mass ions. It is also possible to etch away a material, to change the chemistry by implantation, to selectively deposit films, or to induce structural changes within a selected area [18].

For ion lithographic printing, the effects of diffraction are negligible, because of the extremely small de Broglie wavelengths. There is a challenge in fabricating masks of high contrast, since the ions are absorbed in all solids. The use of ion channeling masks consisting of absorbers on oriented single crystals and of shadow masks supported by grids has been demonstrated. These masks are usually defined by electron beam lithography. Considerations of ion scattering in solids are similar to those for electrons but are less severe.

Scanned focused beams can be used to draw patterns of any desired shape. The limits on the size of the ion source and the quality of focusing optics have been the factors limiting a focused beam resolution. The quality of ion optics is poor compared to light optics and aberrations have limited the size of the spot. Advances with liquid metal field emitters have been significant in producing bright ion sources. Scanned focused systems with liquid metal sources have been demonstrated with beam diameters as small as about 20 nm.

In the exposure of the resist by scanned ion beams there is no proximity effect of the type seen for electron beams, and the energy is transferred very efficiently to the resist. By appropriate choices of resist processes, it is possible to expose thin imaging layers very efficiently without ion damage to the substrate. Resist features as small as 20 nm have been obtained by exposure using focused systems with liquid metal–ion sources. Direct ion milling with a scanned focused beam or chemically assisted ion milling processes can be used to etch materials directly with impressive resolution approaching the size of the beam. However, the rates achieved limit this approach to removal of small volumes of material.

By implantation-enhanced disordering within selected areas, lateral quantum structures have been created in GaAs–AlGaAs heterostructures. These techniques allow the lateral modulation of the composition and the bandgap, without exposing surfaces.

2.4. Other Techniques

There exist other methods of production of Q2D nanostructured materials that do not require such advanced technologies as MBE, MOCVD, and lithography, differing in the environment in which Q2Ds appear, growth conditions, size range, and size distribution, as well as physical and chemical stability and reliability.

2.4.1. Composite Semiconductor-Glass Films

Composite semiconductor-glass films are used to develop Si and Ge Q2Ds embedded in a SiO_2 matrix. The method is based on a planar magnetron radio frequency sputtering of Si or Ge in a hydrogen or argon atmosphere on a silicon substrate with a thin film of native silicon oxide. The latter is held on an electrode while another electrode buried by permanent magnets holds in the silicon or germanium target. The size of the layer embedded in SiO_2 films can be controlled by the substrate temperature, RF power, and the pressure of the environmental gas; the samples thus obtained are suitable for both transmission and emission optical studies as well as for X-ray and transmission electron microscopy studies. Semiconductor-glass composite films have a concentration of monocrystals, on the order of 10 to 30%, the film thickness being about few micrometers. Nanocrystals of Si [19–21] and Ge [22, 23], have been obtained and investigated by this technique. A substrate can be heated during the sputtering deposition to provide size control. The average size obeying the $t^{1/3}$ law inherent to the limited diffusion growth in a glass matrix has been reported for Ge crystallites [24].

2.4.2. Nanocrystals on Crystal Substrates

The self-organized growth of nanocrystals on a crystal substrate resulting from a strain-induced 2D → 3D transition has been discovered. It provides wide opportunities in developing high quality nanocrystals of the most important industrial semiconductors: III-C compounds, silicon, and germanium. This growth mode takes place in the case of the submonolayer heteroepitaxy with a noticeable lattice misfit of the monocrystal substrate and the growing layer. In this case the growing monolayer exhibits a coherent growth, that is, the structure of the layer reproduces the structure of the substrate. This means that the monolayer experiences a strong pressure because of the lattice misfit. If this strain is compressive with respect to the monolayer, the latter becomes unstable and tends to a strain-relaxed arrangement. Remarkably, the strain relaxation occurs via the 2D → 3D transition, after which the strained monolayer divides into a number of hutlike crystallites. These crystallites have a well-defined pyramidal shape with a hexagonal basement. This effect was found both for metal–organic chemical vapor deposition and molecular beam epitaxy.

A number of nanocrystals have been developed to date in this way: InAs and InGaAs on GaAs substrate [25-28], InP on GaInP surface [29], Si$_{0.5}$Ge$_{0.5}$ on Si [30], (ZnCd)Se on ZnSe [31], and GaSb in a GaAs matrix [32]. In the latter case, originally developed crystallites on the GaAs surface were buried by the GaAs deposition and a number of successive cycles have been performed to produce layers of the buried GaSb nanocrystals sandwiched between layers of GaAs. The lattice misfit values for various structures are the following: 7% for InAs/GaAs, 4% for Ge/Si, InP/GaInP and InP/GaAs. The characteristic size of nanocrystals was found to correlate with the misfit value: the larger the misfit, the smaller the crystallites. InAs pyramids on GaAs have typical sizes of 12–20 nm base diameter and heights of 3–6 nm, whereas InP nanocrystals on GaInP have base sizes of about 50 nm and heights on the order of 10–20 nm.

Although the mechanisms of self-organized growth are poorly understood at present, there is no doubt that this approach opens wide prospects for application of nanocrystals in novel optical and optoelectronic devices. Not only have high quality crystallites with an efficient intrinsic emission been developed, but also strong evidence of the high size/shape uniformity of these structures has been reported. For example, by means of high-resolution electron microscopy, InP pyramids on a InGaAs surface were found to have well-defined

{111}, {110}, and {001} facets [29]. Furthermore, a quantum dot ensemble fabricated by successive layer-on-layer growth was found to possess a spatially regular arrangement of the dots [33, 34]. Taking into account all these features, it can be concluded that self-assembled nanocrystals may lead to revolutionary changes in the field, providing an efficient way to the further progress in micro-, nano-, and optoelectronics.

3. ELECTRONIC STATES OF THE IDEALIZED QUASI-TWO-DIMENSIONAL NANOSTRUCTURED MATERIALS

We shall begin with the single heterostructure because it is the simplest quasi-two-dimensional system formed by only one interface.

3.1. Single Heterostructure

A single heterostructure is the region obtained as a product of an interface between two different materials, for example two semiconductors (see Fig. 2). Let us consider an ideal one-dimensional movement in the $\hat{e}_z$ direction of a carrier of mass m near the interface, with a confinement potential of triangular form, including an infinite barrier at the interface.

For the confinement potential, it is given that

$$V_b(z) = \begin{cases} F_0 z & z > 0 \\ \infty & z = 0 \end{cases} \tag{3}$$

where $F_0 = $ const. and $\psi_l(z)$ are wavefunctions for the carrier movement in the $\hat{e}_z$-direction and satisfy the Schrödinger equation

$$\psi_l^{''}(z) + \frac{2m}{\hbar^2}[\varepsilon_l - V_b]\psi_l(z) = 0 \tag{4}$$

If $b = (2mF_0/\hbar^2)$ is introduced and the following change of variable is made,

$$\xi = b\left(z - \frac{\varepsilon_l}{F_0}\right) \tag{5}$$

then the Schrödinger equation is transformed to

$$\left\{\frac{d^2}{d\xi^2} - \xi\right\}\psi_l(\xi) = 0 \tag{6}$$

which is the well known Airy differential equation. The solution of this equation satisfying $\psi \to 0$ when $\xi \to \infty$ is

$$\psi_l(\xi) = B_l A_i(\xi) \tag{7}$$

where $A_i(\xi)$ are the Airy functions and B_l are constants determined by the normalization condition.

For $0 < z < \varepsilon_l/F_0$, $\xi < 0$ and $A_i(\xi)$ is an oscillating function with zeroes on the ξ axis. Let us denote by ξ_{0l}, $l = 1, 2, 3, \ldots$, the zeros of $A_i(\xi)$. Then the energy levels are given by

$$\varepsilon_l = \frac{F_0}{b}|\xi_{0l}| \tag{8}$$

$\psi_l(\xi)$ is normalized, that is

$$\int_0^\infty \psi_l^*(z)\psi_l(z)\,dz = 1 \qquad (9)$$

leading to

$$B_l^2 \int_{-|\xi_{0l}|}^\infty A_i^2(\xi)\,d\xi = b \qquad (10)$$

From the well known formula

$$\int A_i^2(\xi)\,d\xi = \xi A_i^2 - A_i'^2 \qquad (11)$$

the normalization constant B_l is given by

$$B_l^2 = \frac{b}{A_l'^2(-|\xi_{0l}|)} \qquad (12)$$

3.2. Double Heterostructure or Single Quantum Well

Let us consider the one-dimensional motion of a particle of mass m is subjected to a potential energy $V_b(z)$ (see Fig. 7) given by the expression

$$V_b(z) = \begin{cases} 0 & |z| > \dfrac{L}{2} \\[2mm] -V_0 & |z| < \dfrac{L}{2} \end{cases} \qquad (13)$$

where L is the thickness of the potential well (hereafter called the quantum well).

3.2.1. Bound Motions ($-V_0 \le \varepsilon < 0$)

(i) Classical analysis: For energies ε such that $-V_0 \le \varepsilon < 0$, the classical solutions consist of bound particle motions which take place between $\pm L/2$. The particle never penetrates the barrier since $\varepsilon \le 0$ and $|z| > L/2$ would correspond to an imaginary velocity, which is meaningless in classical mechanics. Inside the well the velocity of the carrier is constant and equal to

$$\nu_z = \pm\sqrt{\frac{2}{m}(\varepsilon + V_0)} \qquad (14)$$

At the well boundaries the carrier is perfectly reflected and its velocity changes instantaneously

from $\pm|\nu_z|$ to $\mp|\nu_z|$. The classical motion is thus periodic in time with a period $T = 2L/|\nu_z|$.

Any energy $\varepsilon \ge -V_0$ is allowed; i.e., it can be associated with a possible motion.

(ii) Quantum-mechanical analysis: The quantum-mechanical motion is described by the wavefunction $\psi(z, t)$, which is a solution of the time-dependent Schrödinger equation

$$i\hbar\frac{\partial\psi}{\partial t} = \widehat{H}\left(z, \hat{p}_z = \frac{\hbar}{i}\frac{\partial}{\partial z}\right)\psi \qquad (15)$$

where $\widehat{H}$ is the Hamiltonian of the classical problem

$$\widehat{H} = \frac{\hat{p}_z^2}{2m} + V_b(z) \qquad (16)$$

Since $\widehat{H}$ does not depend explicitly on the time t, $\psi(z, t)$ factorizes into

$$\psi(z, t) = \chi(z)\exp\left(-i\frac{\varepsilon t}{\hbar}\right) \qquad (17)$$

provided that $\chi(z)$ satisfies the eingenvalue problem

$$\widehat{H}(z, \hat{p}_z)\chi(z) = \varepsilon\chi(z) \qquad (18)$$

or more explicitly,

$$\left[-\frac{\hbar^2}{2m}\frac{d^2}{dz^2} + V_b(z)\right]\chi(z) = \varepsilon\chi(z) \qquad (19)$$

$\chi(z)$ must fulfill the following boundary conditions:

(a) $\chi(z)$ is continuous everywhere;
(b) by integrating the last equation around any z_0, and for the specific potential of the square quantum well, $d\chi/dz$ is continuous everywhere;
(c) $\lim_{z\to\infty}|\chi(z)|$ is finite.

Since $V_b(z)$ is piecewise constant, exact solutions of Eq. (19) can be obtained. For both $|z| < L2$ and $|z| > L2$, $\chi(z)$ is the sum of two plane waves with opposite wavevectors. Inside the quantum well these waves propagate and the characteristic wavevectors are $\pm k_w$ where

$$k_w = \sqrt{\frac{2m}{\hbar^2}(\varepsilon + V_0)} \qquad (20)$$

The waves are evanescent outside the quantum well and thus the modules of the imaginary wavevectors, associated with the evanescent waves, are

$$k_b = \sqrt{-\frac{2m\varepsilon}{\hbar^2}} \qquad (21)$$

It should be noted that $V_b(z)$ is an even function in z. Thus the wavefunction $\chi(z)$ can be chosen either as even or odd in

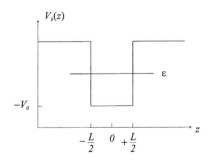

Fig. 7.   Single quantum well potential.

z. To account for this symmetry property, $\chi(z)$ can be written as

$$\chi(z) = A \cos k_w z \quad \varepsilon = -V_0 + \frac{\hbar^2 k_w^2}{2m} \quad \text{for even states}$$

$$\chi(z) = A \sin k_w z \quad \varepsilon = -V_0 + \frac{\hbar^2 k_w^2}{2m} \quad \text{for odd states,}$$

(22)

inside the well. Outside the well the form of the wavefunction is

$$\chi(z) = B \exp\left[-k_b\left(z - \frac{L}{2}\right)\right]$$
$$+ C \exp\left[k_b\left(z - \frac{L}{2}\right)\right] \qquad z \geq \frac{L}{2}$$

$$\chi(z) = E \exp\left[k_b\left(z + \frac{L}{2}\right)\right]$$
$$+ D \exp\left[-k_b\left(z + \frac{L}{2}\right)\right] \qquad z \leq \frac{-L}{2}$$

(23)

However, since $|\chi(z)|$ should not diverge when $z \rightarrow \pm\infty$, $C = D = 0$.

For even states, $B = E$ and for odd states $B = -E$. If the boundary conditions at $z = L2$ are applied, it is easily found that the energy ε satisfies the transcendental equations

$$k_w \tan\left(k_w \frac{L}{2}\right) = k_b \qquad \text{for even states}$$

$$k_w \cot \tan\left(k_w \frac{L}{2}\right) = -k_b \qquad \text{for odd states}$$

(24)

These equations are satisfied only for discrete values of the energy ε. Thus, in marked contrast with the continuous classical spectrum, the quantum-mechanical spectrum ($\varepsilon < 0$) is discrete.

Let us define P_b as the integrated probability of finding the carrier in the barrier ($|z| > L/2$) while it is in the ground state ($\varepsilon = \varepsilon_1$):

$$P_b = \int_{|z| >} \frac{L}{2} \chi_1^2(z)\, dz$$

(25)

In Figure 8 P_b is plotted versus L for a quantum well problem corresponding to $V_0 = 224$ meV, $m = 0.067m_0$ and $V_0 = 150$ meV, $m = 0.4m_0$, respectively (these values are for GaAs). It can be observed that the barrier penetration by the carrier is very small when $L > 100$ Å. For these L's one may consider to a good approximation that the carrier is completely confined in the well.

Equation (24) can only be solved numerically. However, one may easily show algebraically that the solution ε_{n+1} of these equations ($n = 0, 1, \ldots$) is such that $d\varepsilon_{n+1}/dL < 0$. It can also be noticed that the solution $\varepsilon_{n+1} = 0$, i.e., $k_b(\varepsilon_{n+1}) = 0$, which happens just as the $(n+1)$th bound state enters the well, corresponds to the condition

$$k_w(\varepsilon_{n+1} = 0) L = n\pi \qquad n = 0, 1, \ldots$$

(26)

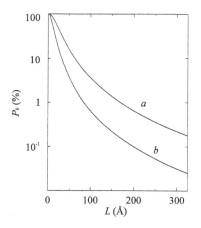

Fig. 8. Integrated probability for single quantum well.

These solutions are thus evenly spaced in L with a period $\pi/k_w(\varepsilon = 0)$ In other words a quantum well of thickness L admits $n(L)$ bound states where

$$n(L) = 1 + \left[\sqrt{\frac{2mV_0L^2}{\pi^2\hbar^2}}\right]$$

(27)

where $[x]$ denote the integer part of x (see Fig. 9). The one-dimensional quantum well supports an infinite number of bound levels if V_0 is infinite. Under this condition the solutions of Eq. (24) are

$$k_w L = p\pi \qquad p = 1, 2, \ldots$$

(28)

and thus, with the zero of energy coinciding with the bottom of the well,

$$\varepsilon_p = \frac{\hbar^2 \pi^2}{2mL^2} p^2 \qquad p \geq 1$$

(29)

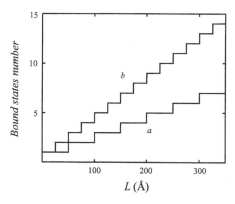

Fig. 9. Number of quantum well bound states of a square well. (a) $V_0 = 224$ meV, $m_e^* = 0.067m_0$. (b) $V_0 = 150$ meV, $m_e^* = 0.4m_0$.

The normalized even and odd wavefunctions of the infinitely deep well are

$$\chi_{2p+1}(z) = \sqrt{\frac{2}{L}} \cos\left[(2p+1)\frac{\pi z}{L}\right] \qquad |z| \le \frac{L}{2} \quad p \ge 0$$

$$\chi_{2p+2}(z) = \sqrt{\frac{2}{L}} \sin\left[(2p+2)\frac{\pi z}{L}\right] \qquad |z| \le \frac{L}{2} \quad p \ge 0 \tag{30}$$

3.2.2. Unbound Motions ($\varepsilon \ge 0$)

For positive energies, the classical motion is unbound. A carrier moving from the left to the right of Figure 7 has a constant velocity $\hbar \tilde{k}_b / m$ in the far left side of the barrier. At $z = -L/2$ it accelerates instantaneously, moves across the well with the constant velocity $\hbar k_w / m$ until the interface $z = L/2$ is reached, then instantaneously decelerates, and finally escapes to $z = +\infty$ at the constant velocity $\hbar \tilde{k}_b / m$. A second classical motion occurs at the same energy: the carrier now comes from the far right side barrier, impinges on the quantum well, and finally escapes to $z = -\infty$. In this motion the velocities are opposed to those found in the previous description.

The quantum–mechanical motion is characterized by a continuous spectrum of allowed energy states. Each eigenvalue ε is twofold degenerate because the carrier motion occurs either from left to right or from right to left in the quantum well. The carrier wavevectors in the well and in the barrier are real, corresponding in both cases to propagating states with

$$\tilde{k}_b = \sqrt{\frac{2m}{\hbar^2}\varepsilon} \qquad k_w = \sqrt{\frac{2m}{\hbar^2}(\varepsilon + V_0)} \tag{31}$$

An electron, which impinges from $z = -\infty$ on the quantum well, is partially reflected and partially transmitted at the interface $z = -L/2$. Inside the well, the eigenstate is a combination of plane waves characterized by wavevectors $\pm k_w$. The occurrence of a wave with wavevector $-k_w$ accounts for the partial reflection at the second interface $z = +L/2$. For $z \ge L/2$ the carriers are also partly transmitted and escape toward $z = \infty$ with wavevector $+k_b$. Thus

$$\chi(z) = \begin{cases} \exp\left[i\tilde{k}_b\left(z+\frac{L}{2}\right)\right] \\ \quad + r\exp\left[-i\tilde{k}_b\left(z+\frac{L}{2}\right)\right] & z \le -\frac{L}{2} \\ \alpha\exp(ik_w z) + \beta\exp(-ik_w z) & |z| \le \frac{L}{2} \\ t\exp\left[i\tilde{k}_b\left(z-\frac{L}{2}\right)\right] & z \le \frac{L}{2} \end{cases} \tag{32}$$

Writing the continuities of $\chi(z)$ and $d\chi(z)/dz$ at both interfaces it is obtained, after some manipulation, that

$$t(\varepsilon) = \left\{\cos k_w L - \frac{i}{2}\left(\xi + \frac{1}{\xi}\right)\sin k_w L\right\}^{-1}$$

$$r(\varepsilon) = \frac{i}{2}\left(\xi - \frac{1}{\xi}\right)\sin k_w L \tag{33}$$

$$\times \left\{\cos k_w L - \frac{i}{2}\left(\xi + \frac{1}{\xi}\right)\sin k_w L\right\}^{-1}$$

where $\xi = k_w / \tilde{k}_b$.

Let $T(\varepsilon)$ and $R(\varepsilon)$ denote the transmission and reflection coefficients of the well:

$$T(\varepsilon) = |t(\varepsilon)|^2 \qquad R(\varepsilon) = |r(\varepsilon)|^2 \tag{34}$$

Then

$$R(\varepsilon) + T(\varepsilon) = 1 \tag{35}$$

$$T(\varepsilon) = \left[1 + \frac{1}{4}\left(\xi - \frac{1}{\xi}\right)^2 \sin^2 k_w L\right]^{-1} \tag{36}$$

The last two equations look familiar. Actually they are nothing but the reflection and transmission coefficients of a Fabry–Pérot dielectric slab. This is not accidental but arises from the close analogy between the one-dimensional Schrödinger equation and the equation governing the propagation of an electromagnetic wave in a medium characterized by a position-dependent refractive index $n(z)$.

The transmission coefficient has two remarkable features:

(i) If $Lk_w(\varepsilon = 0) \ne p\pi$, then the transmission coefficient vanishes at the onset of the continuum. It could be expected that a carrier with a vanishingly small velocity in the barrier would easily fall into the potential well. However, just the reverse is true because the electron does not penetrate the well ($\alpha = \beta = 0$). This behavior reflects the wave-like nature of the electron, which is most strikingly revealed when the potential has sharp discontinuities.

(ii) The transmission coefficient is not a smooth function of the energy. Instead $T(\varepsilon)$ reaches unity whenever

$$k_w L = p\pi \tag{37}$$

This corresponds to constructive interference inside the quantum well slab, whose effective thickness $2L$ should fit an integer number of carrier wavelengths. The discrete energies, which fulfill Eq. (37), are called transmission resonance.

3.2.3. Density of States

It is important to know how many quantum states $|\nu\rangle$ per unit energy are available around a given energy ε. This quantity, the density of states, is equal to

$$\rho(\varepsilon) = \sum_\nu \delta(\varepsilon - \varepsilon_\nu) \tag{38}$$

where ε_ν is the energy associated with the state $|\nu\rangle$. It often proves convenient to rewrite Eq. (38) in slightly different terms,

$$\rho(\varepsilon) = \sum_\nu \delta(\varepsilon - \varepsilon_\nu)$$
$$= \sum_\nu \langle \nu | \delta(\varepsilon - \widehat{H}) | \nu \rangle = \text{Trace} \left[\delta(\varepsilon - \widehat{H}) \right] \tag{39}$$

where $\widehat{H}$ is the Hamiltonian operator whose eigenvalues are ε_ν. The Trace has the property that its value is independent of the basis $|l\rangle$ in which it is calculated, even if the states $|l\rangle$ are not eigenstates of $\widehat{H}$.

In some cases one can also rewrite $\delta(\varepsilon - \widehat{H})$ as

$$\delta(\varepsilon - \widehat{H}) = -\frac{1}{\pi} \lim_{\eta \to \infty} \text{Im}\, (\varepsilon - \widehat{H} + i\eta)^{-1} \tag{40}$$

where $\text{Im}\, z$ stands for "imaginary part of z." Using Eqs. (39) and (40), $\rho(\varepsilon)$ is found to be equal to

$$\rho(\varepsilon) = -\frac{1}{\pi} \lim_{\eta \to \infty} \text{Im} \sum_l \langle l | (\varepsilon - \widehat{H} + i\eta)^{-1} | l \rangle \tag{41}$$

For the problem of a square quantum well, the discrete spectrum is labeled by the index n of the bound state and, since the electrons have spin, by a spin quantum number $\sigma_z = \pm 1/2$. Thus $|\nu\rangle = |n, \sigma_z\rangle$ and

$$\rho(\varepsilon) = 2 \sum \delta(\varepsilon - \varepsilon_n) \qquad \varepsilon < 0 \tag{42}$$

where the factor of 2 gives account for the spin degeneracy of each level n.

In fact it is known that a carrier motion is three-dimensional. Thus a more realistic description of one-dimensional quantum well structures is provided by solutions of

$$-\frac{\hbar^2}{2m} \left[\frac{\partial^2}{\partial x^2} + \frac{\partial^2}{\partial y^2} + \frac{\partial^2}{\partial z^2} \right] \psi(\vec{r}) + V_b(z)\psi(\vec{r}) = \varepsilon\psi(\vec{r}) \tag{43}$$

As the Hamiltonian is the sum of $x, y,$ and z dependent contributions, it is known that one can look for eigenfunctions which are separable in x, y, z. Moreover, the carrier motion is free along x and y directions. Thus

$$\psi(\vec{r}) = \frac{1}{\sqrt{S}} \exp \left[ik_x x + ik_y y \right] \chi(z) \tag{44}$$

where $S = L_x L_y$ is the sample area. With Eq. (44) is found that

$$\varepsilon = \varepsilon_n + \frac{\hbar^2 k_\perp^2}{2m} \tag{45}$$

where ε_n is one of the eigenvalues of the one-dimensional Schrödinger equation for the quantum well problem (Eq. (19)) and $k_\perp^2 = k_x^2 + k_y^2$. Thus, one may associate a two-dimensional subband, which represents the kinetic energy arising from the in plane motion of the carrier with each of the quantum well

bound states ε_n. The density of states associated with the motion described by Eq. (43) is

$$\rho(\varepsilon) = 2 \sum_{k_x k_y n} \delta \left[\varepsilon - \varepsilon_n - \frac{\hbar^2 k_\perp^2}{2m} \right] \qquad \varepsilon_n < 0 \tag{46}$$

The sample area S is assumed to be of macroscopic size. Applying cyclic boundary conditions to the x and y motions, it is found that

$$k_x = m \frac{2\pi}{L_x} \qquad k_y = n \frac{2\pi}{L_y} \tag{47}$$

Since L_x, L_y are very large, any summation $\sum_{k_x k_y} a(k_x, k_y)$ can be converted into an integral:

$$\sum_{k_x k_y} a(k_x, k_y) \xrightarrow[L_x L_y \to \infty]{} \frac{L_x L_y}{(2\pi)^2} \int dk_x \, dk_y \, a(k_x, k_y) \tag{48}$$

Finally

$$\rho(\varepsilon) = \frac{mS}{\pi\hbar^2} \sum_n Y[\varepsilon - \varepsilon_n] \qquad \varepsilon_n < 0 \tag{49}$$

where $Y(x)$ is the step function

$$Y(x) = \begin{cases} 1 & \text{if} \qquad x > 0 \\ 0 & \text{if} \qquad x < 0 \end{cases} \tag{50}$$

The density of states is thus staircase-shaped (see Fig. 10). Two facts are worth noting:

(i) $\rho(\varepsilon) = 0$ for $\varepsilon < \varepsilon_1$. A classical description would have led to allowed motions for any ε. Here the effect of the confinement energy associated with the localization of the particle along the z direction is found again.

(ii) $\rho(\varepsilon)$ exhibits jumps of finite amplitude whenever the energy passes through the edge of a two-dimensional subband. Such behavior contrasts with the smoothly varying $\rho(\varepsilon)$ which is found for the free particles $(\rho(\varepsilon) = A\sqrt{\varepsilon})$. The reduction of the dimensionality (i.e., two versus three) is witnessed by the increasingly singular behavior of the density states.

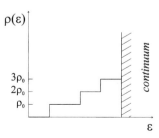

Fig. 10. Density of states: $\rho_0 = m_e^* S / \pi \eta^2$. The energy zero is taken at the bottom of the well.

In practice each step is smoothed out by defects. An empirical description of the broadening effects is obtained by replacing the step function $Y(\varepsilon - \varepsilon_n)$ by

$$Y(\varepsilon - \varepsilon_n) \underset{\text{broadening}}{\rightarrow} \frac{1}{\pi}\left\{\frac{\pi}{2} + \text{Arctan}\left[\frac{(\varepsilon - \varepsilon_n)}{\Gamma_n}\right]\right\} \quad (51)$$

where Γ_n is the broadening parameter.

3.3. Symmetric Square Double Wells: Tunneling Coupling between Wells

Let us consider two equivalent one-dimensional quantum wells of height V_0, width L, and separated by a distance h (Fig. 11). Each of the wells possesses $n_{\max}$ bound states when they are isolated ($n_{\max} \geq 1$). A localized wavefunction, which may be associated with each of these bound states, exponentially decays far away from the well. In the limit of infinite h, the bound states of the discrete spectrum ($0 \geq \varepsilon \geq -V_0$) are twofold degenerate: the particle can be found in either well with the same probability. At finite h, however, the referred states are no longer eigenstates of the coupled wells Hamiltonian, which is given by

$$\hat{H} = \frac{\hat{p}_z^2}{2m} + V_b(z - z_1) + V_b(z - z_2) \quad (52)$$

with

$$V_b(z - z_i) = \begin{cases} 0 & |z - z_i| \geq \frac{L}{2} \\ & \qquad\qquad i = 1, 2 \quad (53) \\ -V_0 & |z - z_i| \leq \frac{L}{2} \end{cases}$$

Let $\chi_1(z - z_2)$ be the ground state eigenfunction of the isolated well centered at $z = z_2$. Despite the exponential decay displayed by $\chi_1(z - z_2)$, when $|z - z_2| \geq L/2$, $\chi(z_1 - z_2)$ is different from zero. Thus $\hat{H}\chi_1(z - z_2)$ is not proportional to $\chi_1(z - z_2)$. However, if h is large enough, one can expect the coupling between the wells (due to the tunnel effect in the middle barrier) to be small enough to admit only those states which are twofold degenerate at $h = \infty$ and not the ground states with excited states. From the exact solution of Eq. (52),

$$\psi(z) = \sum_\nu a_\nu \chi_\nu(z - z_1) + b_\nu \chi_\nu(z - z_2) \quad (54)$$

where ν runs over the discrete and continuous spectra, only a linear combination of the ground states of the isolated wells is retained for the lowest lying states of the double quantum well,

$$\psi(z) = \alpha\chi_1(z - z_1) + \beta\chi_1(z - z_2) \quad (55)$$

obtaining

$$\begin{bmatrix} \varepsilon_1 + s - \varepsilon & (\varepsilon_1 - \varepsilon)r + t \\ (\varepsilon_1 - \varepsilon)r + t & \varepsilon_1 + s - \varepsilon \end{bmatrix}\begin{bmatrix} \alpha \\ \beta \end{bmatrix} = 0 \quad (56)$$

with solutions

$$\varepsilon = \varepsilon_1 \mp \frac{t}{1 \mp r} + \frac{s}{1 \mp r} \quad (57)$$

where ε_1 is the energy of the ground bound state of the wells when they are isolated and

$$\begin{aligned} r &= \langle\chi_1(z - z_1)|\chi_1(z - z_2)\rangle = \langle\chi_1(z - z_2)|\chi_1(z - z_1)\rangle \\ s &= \langle\chi_1(z - z_1)|V_b(z - z_2)|\chi_1(z - z_1)\rangle \\ &= \langle\chi_1(z - z_2)|V_b(z - z_1)|\chi_1(z - z_2)\rangle \quad (58) \\ t &= \langle\chi_1(z - z_1)|V_b(z - z_1)|\chi_1(z - z_2)\rangle \\ &= \langle\chi_1(z - z_2)|V_b(z - z_2)|\chi_1(z - z_1)\rangle \end{aligned}$$

The quantities r, s, t are called overlap (r), shift (s), and transfer (t) integrals, respectively. The overlap integral is positive whereas both the transfer and shift integral are negative. It is usual to neglect r, although it is not quantitatively a very good approximation. In this case

$$\varepsilon = \varepsilon_1 \mp t + s \quad (59)$$

s is thus interpreted as the shift from the ground state of each well due to the presence of the other. This shift does not lift the twofold degeneracy prevailing at infinite h (see Fig. 12). This degeneracy is lifted by the transfer term t. The ground state corresponds to the symmetric combination $\alpha = \beta$ of isolated quantum well wavefunction, whereas the excited state corresponds to $\alpha = -\beta$.

Finally, the exact solution of the double, symmetric quantum well problem will be stated. These eigenstates are the solutions of the transcendental equation

$$2\cos k_w L + \left(\xi - \frac{1}{\xi}\right)\sin k_w L$$

$$\pm\left(\xi + \frac{1}{\xi}\right)\exp(-k_b h)\sin k_w L = 0 \quad (60)$$

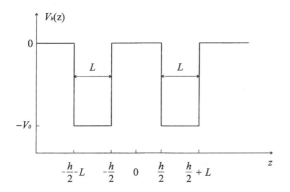

Fig. 11.   Potential barrier for symmetric double square well.

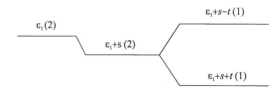

Fig. 12.   Shift and lifting of the degeneracy of the twofold degenerated ground state of isolated quantum well.

where

$$\xi = \frac{k_b}{k_w} \qquad k_b = \sqrt{\frac{-2m\varepsilon}{\hbar^2}} \qquad k_w = \sqrt{\frac{2m}{\hbar^2}(\varepsilon + V_0)} \quad (61)$$

The minus (plus) sign in Eq. (60) refers to the symmetric (antisymmetric) states with respect to the center of the structure.

The previous equation admits at least one solution for any h (like any one-dimensional problem with an attractive and symmetric potential). When h decreases from infinity to zero this bound state, which is a symmetric one, continuously evolves from the ground state of a well with thickness L to the ground state of a well with thickness $2L$. An antisymmetric state remains bound for any h if the quantum well of thickness $2L$ binds at least two states (i.e., if $Lk_w(\varepsilon = 0) \geq \pi$). If the quantum well of thickness $2L$ binds only one state, then the ground antisymmetric state of the double well, bound at very large h, becomes unbound when $h < h_c$, where

$$2\cot[Lk_w(\varepsilon = 0)] = h_c k_w(\varepsilon = 0) \quad (62)$$

As in the problem of a single well, the bound states of the double well are transformed in transmission resonance when their confinement energy exceeds the top of the confining barrier. The resonance can be narrow when the ratio of the wavevectors k_b/k_w is $\ll 1$.

3.4. Superlattices

3.4.1. Superlattice Dispersion Relations

Let us consider an infinite sequence of quantum wells of thickness L and separation h (Fig. 13). The potential energy $V_b(z)$ is a periodic function of z with a period d ($d = L + h$):

$$V_b(z) = \sum_{n=-\infty}^{\infty} V_b(z - nd) \quad (63)$$

$$V_b(z - nd) = \begin{cases} -V_0 & \text{if} \quad |z - nd| \leq \dfrac{L}{2} \\ 0 & \text{if} \quad |z - nd| > \dfrac{L}{2} \end{cases} \quad (64)$$

The solutions of the one-dimensional Schrödinger equation are known exactly for each layer: well-acting (A) layer or barrier-acting (B) layer. For positive energies it is obtained that

$$\chi(z) = \alpha \exp[ik_w(z - nd)] + \beta \exp[-ik_w(z - nd)]$$
$$|z - nd| \leq \frac{L}{2}$$

$$\chi(z) = \gamma \exp\left[i\tilde{k}_b\left(z - nd - \frac{d}{2}\right)\right]$$
$$+ \delta \exp\left[-i\tilde{k}_b\left(z - nd - \frac{d}{2}\right)\right] \quad (65)$$
$$\left|z - nd - \frac{d}{2}\right| \leq \frac{h}{2}$$

and

$$\varepsilon = \frac{\hbar^2 \tilde{k}_b^2}{2m} = -V_0 + \frac{\hbar^2 k_w^2}{2m} \quad (66)$$

Now the periodicity of $V_b(z)$ will be exploited. Consider the translation operator τ_d which is such that for any function $f(z)$, $\tau_d f(z) = f(z + d)$. The operator τ_d commutes with the Hamiltonian $\hat{H}$ of the particle. Thus, the eigenfunctions of $\hat{H}$, which are also eigenfunctions of τ_d, can be found. The τ_d eigenvalues can be written $\exp(iqd)$ where q is an arbitrary complex number. Moreover, $\tau_{nd} = [\tau_d]^n$ is a worth relation.

Thus the functions $\chi_q(z)$, which are eigenfunctions of both $\hat{H}$ and τ_d, must fulfill not only the Schrödinger equation but also [35]

$$\tau_{nd}\chi_q(z) = \chi_q(z + nd) = \exp[iqnd]\chi_q(z) \quad (67)$$

In a way equivalent to Eq. (67), the χ_q's can be written in the form

$$\chi_q(z) = u_q(z)\exp[iqz] \qquad u_q(z + d) = u_q(z) \quad (68)$$

i.e., the χ_q's are the product of a plane wave by a periodic function in z. Equations (67) and (68) constitute the Bloch–Floquet theorem; χ_q is the Bloch function and u_q is its associated periodic part [35].

Let us impose cyclic boundary conditions on the eigenstates of $\hat{H}$. Supposing that the length Nd of the crystal is very large and assuming that the crystal maps onto itself end to end, $\chi_q(z)$ must be such that

$$\chi_q(z + Nd) = \chi_q(z) \quad (69)$$

which means that:

$$qNd = 2p\pi \qquad p \text{ integer} \quad (70)$$

Consequently q must be real. N independent values q are allowed, the spacing between any two of these consecutive values being $2\pi/Nd$. Without loss of generality q can be restricted to the segment $[-\pi/d, +\pi/d[$. The q space is called the reciprocal space and the segment $[-\pi/d, +\pi/d[$, which is a particular unit cell of the reciprocal space, is the first Brillouin zone. In the limit of large N, the discreteness

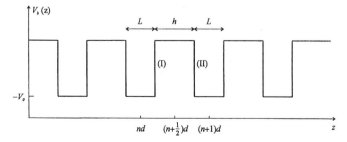

Fig. 13. Density of states of the superlattice. $\rho_0 = m_e^* NS/\pi \eta^2$.

of the allowed q values becomes unimportant and any summation of the form $\sum_{q \in \text{first B.Z.}} g(q)$ can be converted into an integral

$$\sum_{q \in \text{first B.Z.}} g(q) \xrightarrow[N \to \infty]{} \frac{Nd}{2\pi} \int_{-\pi/d}^{+\pi/d} g(q)\, dq \qquad (71)$$

where B.Z. stands for Brillouin zone.

Let us express the continuity of χ_q and $d\chi_q/dz$ at the interface I of Figure 13:

$$\begin{aligned}
&\alpha \exp\left(ik_w \frac{L}{2}\right) + \beta \exp\left(-ik_w \frac{L}{2}\right) \\
&\quad = \gamma \exp\left(-i\tilde{k}_b \frac{h}{2}\right) + \delta \exp\left(i\tilde{k}_b \frac{h}{2}\right) \\
&ik_w\left[\alpha \exp\left(ik_w \frac{L}{2}\right) - \beta \exp\left(-ik_w \frac{L}{2}\right)\right] \\
&\quad = i\tilde{k}_b\left[\gamma \exp\left(-i\tilde{k}_b \frac{h}{2}\right) - \delta \exp\left(i\tilde{k}_b \frac{h}{2}\right)\right]
\end{aligned} \qquad (72)$$

At interface II of Figure 13 it is fulfilled that

$$\begin{aligned}
&\chi_q\left[z = (n+1)d - \frac{L}{2} + 0\right] \\
&\quad = \gamma \exp\left(i\tilde{k}_b \frac{h}{2}\right) + \delta \exp\left(-i\tilde{k}_b \frac{h}{2}\right)
\end{aligned} \qquad (73)$$

To express $\chi_q[z = (n+1)d - \frac{L}{2} + 0]$ one can take advantage of the Bloch theorem, writing

$$\begin{aligned}
&\chi_q\left[z = (n+1)d - \frac{L}{2} + 0\right] \\
&\quad = \exp(iqd)\chi_q\left[z = nd - \frac{L}{2} + 0\right]
\end{aligned} \qquad (74)$$

Using the same method for the derivative, a 4×4 homogeneous system in $\alpha, \beta, \gamma, \delta$ is finally obtained. No trivial solutions exist only if the equation is satisfied (see [35])

$$\begin{aligned}
\cos(qd) &= \cos(k_w L)\cos(\tilde{k}_b h) \\
&\quad - \frac{1}{2}\left(\xi + \frac{1}{\xi}\right)\sin(k_w L)\sin(\tilde{k}_b h)
\end{aligned} \qquad (75)$$

where

$$\xi = \tilde{k}_b/k_w \qquad (76)$$

Equations (75) and (76) are implicit equations linking the allowed energy ε and the wave vector q; i.e., they are the dispersion relations $\varepsilon(q)$ of the Bloch waves. For a given value of q, these equations admit an infinite number of solutions. Thus, in order to label the various solutions, a subband index n is affixed to ε and χ. A one-dimensional Bloch state depends on two orbital quantum numbers n, q and will thus be written as $|nq\rangle$. The allowed subbands ε_{nq} are separated by forbidden gaps. For energies corresponding to these energy gaps, there

are no allowed Bloch states; i.e., there are no allowed eigenstates of $\widehat{H}$ which fulfill the cyclic boundary conditions of Eq. (69). Evanescent states, corresponding to imaginary q's, do, however, exist and they are only allowed if the superlattice has a finite length along the z axis.

The magnitude of the forbidden gaps that separate two consecutive superlattice subbands $|n, q\rangle$, $|n+1, q\rangle$ decreases when n increases. In fact, at large $\varepsilon(\varepsilon >> V_0)$, $\tilde{k}_b \approx k_w$ and Eq. (75) reduces to $k_w = q + 2j\pi/d$. The superlattice potential is hardly felt by highly energetic electrons, or equivalently the electron wavelength $(2\pi/k_w)$ becomes much smaller than the characteristic length of the obstacles (d) and thus the effects of diffraction become negligible.

It can be noticed that these energies ε, which are such that $k_w L = p\pi$, always correspond to an allowed superlattice state (see Eq. (75)). The superlattice states, which correspond to the continuum states of the isolated well, can thus be viewed as the result of the hybridization of the virtual bound states of isolated wells.

Up to now only the case of Bloch states associated with states propagating in both kinds of layers (wells and barriers), $\varepsilon_{nq} \geq 0$, has been considered. Now the case corresponding to $-V_0 \leq \varepsilon_{nq} \leq 0$ will be examined. The wave is evanescent in the barrier and thus

$$\tilde{k}_b = ik_b \qquad k_b = \sqrt{\frac{-2m\varepsilon}{\hbar^2}} \qquad (77)$$

$$\xi \to i\zeta \qquad \zeta = \frac{k_b}{k_w} \qquad (78)$$

Allowed Bloch states must then fulfill

$$\begin{aligned}
\cos(qd) &= \cos(k_w L)\cosh(k_b h) \\
&\quad - \frac{1}{2}\left[-\zeta + \frac{1}{\zeta}\right]\sin(k_w L)\sinh(k_b h)
\end{aligned} \qquad (79)$$

In the limit of infinitely thick barriers the right side of Eq. (79) will diverge like $\exp(k_b h)$, unless the multiplicative coefficient in front of it vanishes. This occurs if

$$\cos(k_w L) - \frac{1}{2}\left[-\zeta + \frac{1}{\zeta}\right]\sin(k_w L) = 0 \qquad (80)$$

which coincides with the transcendental equation whose solutions give the bound states of isolated quantum wells of thickness L. Actually Eq. (80) is the product of both Eqs. (22); i.e., all the levels, odd or even, are obtained at the same time. This is why the symmetric property (evenness in z) of $V_b(z)$ in the single well analysis was directly exploited.

Thus, the superlattice subbands of negative energies appear as the hybridization of isolated quantum well bound states due to the tunnel coupling between the wells across the finite barriers. If these barriers are thick (and/or high) but finite, the product $k_b h$ is large and the subbands are narrow (the subband width, which is proportional to the transfer integral, decreases roughly exponentially with h at fixed k_b). The function $F(\varepsilon)$ which appears on the right side of Eq. (79) shows large variations with ε and the energy segments where $|F(\varepsilon) < 1|$ is narrow (see Fig. 14). In order to obtain approximate, but convenient, subband dispersion relations, let us denote by $\varepsilon_1, \varepsilon_2, \ldots, \varepsilon_n$ the isolated quantum well bound state solutions of Eq. (80). In addition, let us expand $F(\varepsilon)$ in the vicinity of the ε's, say ε_j. To first order in $\varepsilon - \varepsilon_j$ the dispersion relations of the jth subband can be obtained in the form

$$\varepsilon_j(q) = \varepsilon_j + s_j + 2t_j \cos(qd) \tag{81}$$

where

$$s_j = \frac{-F(\varepsilon_j)}{[F'(\varepsilon)]_{\varepsilon=\varepsilon_j}} \qquad 2t_j = \frac{1}{[F'(\varepsilon)]_{\varepsilon=\varepsilon_j}} \tag{82}$$

and where the prime denotes the derivative with respect to ε. Equation (81) is simply a tight-binding result. Explicit expressions for t_j and s_j can be obtained:

(i) by expanding $\chi_{jq}(z)$ in terms of the "atomic" functions $\varphi_{loc}^{(j)}(z-nd)$, where $\varphi_{loc}^{j}(z-nd)$ is the jth bound state wavefunction of the quantum well centered at $z_n = nd$ when it is considered as isolated,

$$\chi_{jq}(z) = \frac{1}{\sqrt{N}} \sum_n \exp(iqnd)\varphi_{loc}^{(j)}(z-nd) \tag{83}$$

and

(ii) by retaining only nearest-neighbor interactions:

$$t_j = \int \varphi_{loc}^{(j)}(z) V_b(z) \varphi_{loc}^{(j)}(z-d)\, dz \tag{84}$$

$$s_j = \sum_{n \neq 0} \int \varphi_{loc}^{(j)}(z) V_b(z-nd) \varphi_{loc}^{(j)}(z)\, dz \tag{85}$$

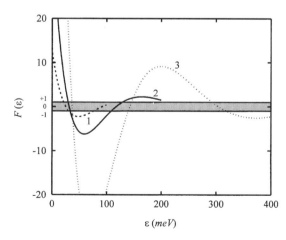

Fig. 14. $F(\varepsilon)$ versus ε. Three barrier heights are considered: curve (1) $V_0 = 0.1$ eV; curve (2) $V_0 = 0.2$ eV; curve (3) $V_0 = 0.4$ eV. $m_e^* = 0.067m_0$, $L = 100$ Å, $h = 50$ Å. Energy zero is taken at the bottom of the well. Each curve is interrupted at the top of the well. The shaded area corresponds to the allowed superlattice state: $|F(\varepsilon)| < 1$.

3.4.2. Symmetry Properties of the Eigenfunctions

For $q = 0$ or $q = \pi/d$, the Bloch wave is stationary $[\chi_{n,q=0}(z+d) = \chi_{n,q=0}(z); \quad \chi_{n,q=\pi/d}(z+d) = -\chi_{n,q=\pi/d}(z)]$. The wavefunction repeats itself identically ($q = 0$) or changes sign ($q = \pi/d$) when going from one elementary cell to the next. Let us consider the ratio $r = \alpha/\beta$ (see Eq. (65)). It is found that [35]

$$r = -\frac{(1-\xi)}{(1+\xi)} \times \frac{[\exp(i\tilde{k}_b h) - \exp i(k_w L + qd)]}{[\exp i(k_w L + \tilde{k}_b h) - \exp(iqd)]} \tag{86}$$

One can show that for $q = 0$, $q = \pi/d$ the ratio r is real and such that $|r|^2 = 1$; thus $r = \pm 1$. This shows that for $q = 0$ and $q = \pi/d$ the Bloch function is either odd or even with respect to the centers of the well-acting layers. The same conclusion holds for the parity with respect to the center of the barrier-acting layers. These analytical results arise from symmetry properties of the superlattice Hamiltonian: let Δ_A, Δ_B be two parallel axes separated by $d/2$. The product of two symmetries with respect to these two axes is equal to a translation τ_d, the vector $\vec{d}(|\vec{d}| = d)$ being perpendicular to Δ_A, Δ_B. Applying this geometrical property to the axes Δ_A, Δ_B, which bisect two consecutive segments A and B of a superlattice (Fig. 13), it is found that

$$R_B R_A = \tau_{\vec{d}} \qquad R_A R_B = -\tau_{\vec{d}} \tag{87}$$

where $\tau_{\vec{d}}$ is the translation characterized by $\vec{d} = d\hat{e}_z$ with $d = L + h$. It is known that the Bloch states are eigenstates of $\tau_{\vec{d}}$ with an eigenvalue $\exp(iqd)$. In addition, however, the commutators

$$[V_b(z), R_A] = [V_b(z), R_B] = 0 \tag{88}$$

and also

$$\left[\frac{\hat{p}_z^2}{2m}, R_A\right] = \left[\frac{\hat{p}_z^2}{2m}, R_B\right] = 0 \tag{89}$$

Thus R_A, R_B commute with the superlattice Hamiltonian $\widehat{H}$. However $R_A R_B$ do not commute with each other since

$$[R_A, R_B] = -\tau_{\vec{d}} + \tau_{-\vec{d}} \tag{90}$$

It is thus impossible to diagonalize simultaneously $\widehat{H}$, $\tau_{\vec{d}}$, R_A, and R_B in the general case. The points $q = 0$ and $q = \pi/d$ of the Brillouin zone are noticeable exceptions since the commutator $[R_A, R_B]$, evaluated over the Bloch states, vanishes for these two values of q. For $q = 0$, $R_B R_A \chi_{n,q=0}(z) = +\chi_{n,q=0}(z)$. If $\chi_{n,q=0}(z)$ is even (odd) with respect to the center of the A layers, then it should also be even (odd) with respect to the centers of the B layers. For $q = \pi/d$, $R_B R_A[\chi_{n,q=\pi/d}(z)] = -\chi_{n,q=\pi/d}(z)$. Thus if $\chi_n, q = \pi/d(z)$ is even (odd) with respect to the center of the A layers, it should be odd (even) with respect to the centers of the B layers.

3.4.3. Superlattice Density of States

Let us reintroduce the free motion in the layer plane $(x, y$ directions) and label a superlattice with $n, q, \vec{k}_\perp, \sigma_z$ where $\sigma_z = \pm 1/2$. The eigenenergies

$$\varepsilon(n, q, \vec{k}_\perp, \sigma_z) = \frac{\hbar^2 k_\perp^2}{2m} + \varepsilon_n(q) \tag{91}$$

The density of states is then equal to

$$\rho(\varepsilon) = \sum_{n,q,k_\perp,\sigma_z} \delta\left[\varepsilon - \varepsilon_{nq} - \frac{\hbar^2 k_\perp^2}{2m}\right] \tag{92}$$

or

$$\rho(\varepsilon) = \frac{SNd}{\pi^2}\frac{m}{\hbar^2}\sum_n \int_0^{\frac{\pi}{d}} dq\, Y[\varepsilon - \varepsilon_{nq}] = \sum_n \rho_n(\varepsilon) \tag{93}$$

where Nd denotes the length of the superlattice, $\rho_n(\varepsilon)$ is the density of states associated with the nth subband, and $Y(x)$ is the step function.

In general the subband ε_{nq} has a finite width when q describes the first Brillouin zone: $\varepsilon_{min} \leq \varepsilon_{nq} \leq \varepsilon_{max}$. For $\varepsilon < \varepsilon_{min}$, $\rho_n(\varepsilon) = 0$ whereas for $\varepsilon > \varepsilon_{max}$, $\rho_n(\varepsilon)$ is a constant:

$$\rho_n(\varepsilon) = N\frac{mS}{\pi\hbar^2} \qquad \varepsilon > \varepsilon_{max} \tag{94}$$

It can be observed that the plateau value $NmS/\pi\hbar^2$ is equal to N times the density of states associated with a single bound state of a given quantum well. From Eq. (94) it can be deduced that when ε falls into the bandgaps of the superlattice dispersion relations, the density of states is quantized in units of $NmS/\pi\hbar^2$. To study what happens when ε corresponds to the allowed superlattice states, let us take the simple case,

$$\varepsilon_{nq} = \varepsilon_n + s_n - 2|t_n|\cos(qd) \tag{95}$$

which, as shown previously, corresponds to a tight-binding description of the superlattice subbands. In the vicinity of $q = 0$ it is found that

$$\varepsilon_{nq} \approx \varepsilon_n + s_n - 2|t_n| + |t_n|q^2 d^2 \tag{96}$$

whereas in the vicinity of $q = \pi/d$

$$\varepsilon_{nq} = \varepsilon_n + s_n + 2|t_n| - |t_n|\left[q - \frac{\pi}{d}\right]^2 d^2 \tag{97}$$

Notice that in the vicinity of both $q = 0$ and $q = \pi/d$ the carrier dispersion relation is quadratic in q. In both cases the effective mass along the $\hat{e}_z$ axis is proportional to the modulus of the transfer integral $|t_n|$. The smaller $|t_n|$ (thick and/or high barrier), the heavier the effective mass. In the limit of vanishing $|t_n|$ (isolated quantum wells) the subband dispersion is flat (discrete level). For uncoupled discrete levels the effective mass for the propagation along $\hat{e}_z$ is finite, although the carrier is characterized in each layer by a finite mass m. The reason for this is that the carrier oscillates back and forth in the well and on average does not move along the $\hat{e}_z$ axis. Thus its

effective mass is infinite. Using Eq. (95) the density of states is readily evaluated,

$$\rho_n(\varepsilon) = \begin{cases} 0 & \text{if } \varepsilon < \varepsilon_n + s_n - 2|t_n| \\ \frac{\rho_0}{\pi}\text{Arc}\cos\left\{\frac{-\varepsilon + \varepsilon_n + s_n}{2|t_n|}\right\} & \text{if } |\varepsilon - \varepsilon_n - s_n| < 2|t_n| \\ \rho_0 & \text{if } \varepsilon > \varepsilon_n + s_n + 2|t_n| \end{cases} \tag{98}$$

where

$$\rho_0 = \frac{mNS}{\pi\hbar^2} \tag{99}$$

In the vicinity of $\varepsilon_n + s_n \pm 2|t_n|$, the derivatives of $\rho_n(\varepsilon)$ are infinite. These two van Hove singularities correspond respectively to a three-dimensional minimum ($q = 0$) where

$$\frac{\partial^2 \varepsilon}{\partial q^2} > 0 \qquad \frac{\partial^2 \varepsilon}{\partial k_x^2} > 0 \qquad \frac{\partial^2 \varepsilon}{\partial k_y^2} > 0$$

and to a saddle point ($q = \pi/d$) where

$$\frac{\partial^2 \varepsilon}{\partial q^2} < 0 \qquad \frac{\partial^2 \varepsilon}{\partial k_x^2} > 0 \qquad \frac{\partial^2 \varepsilon}{\partial k_y^2} > 0$$

The energy dependence of $\rho_n(\varepsilon)$ is sketched in Figure 15.

3.5. Idealized Q2D Systems when the Schrödinger Equation that Characterizes the Problem is Nonseparable

When the Schrödinger equation for a given quantum system is nonseparable in nature, then some approximation techniques must be used to study its solutions. In a similar situation, for a confined quantum system the variational method might constitute an economical and physical appealing approach. It has been shown that the variational method is useful for studying this class of systems when their symmetries are compatible with those of the confining boundaries.

Only the application of the variational method to the case of a symmetric quantum system confined by impenetrable potentials will be considered as a simple example of a more general situation.

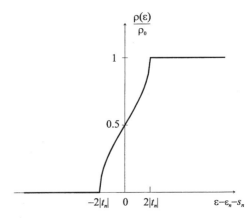

Fig. 15.   Density of states of the superlattice. $\rho_0 = m_e^* NS\ \pi\hbar^2$.

3.5.1. General Description of the Method

For the moment, let us assume that, for a given system, the time-independent Schrödinger equation can be written as

$$\widehat{H}\psi_n = E_n\psi_n \tag{100}$$

where $\widehat{H} = -(1/2)\nabla^2 + V(q)$ is the associated Hamiltonian for this system, $\{q\}$ is the set of coordinates on which V depends, and $\hbar = m = 1$.

An approximate solution of Eq. (100) may be obtained by replacing ψ_n by a trial wavefunction χ_n, which possesses a similar behavior at the "origin" as well as asymptotically at infinity. An estimate of the system energy is then obtained by minimizing the functional

$$\int \chi_n^* \widehat{H} \chi_n \, d\tau \tag{101}$$

with the additional restriction

$$\varepsilon_n = \min \int \chi_n^* \widehat{H} \chi_n \, d\tau = \delta_{n,n'} \tag{102}$$

As is shown in any quantum mechanics textbook, one find that

$$\varepsilon_n = \min \int \chi_n^* \widehat{H} \chi_n \, d\tau \geq E_n \tag{103}$$

i.e., a poor guess of χ_n leads to a poor estimate of ε_n, which means that the qualitative knowledge of the system under study has not been properly included.

In practical calculations, the functions χ_n depend upon the coordinates, say $\{q_1, q_2, \ldots, q_s\}$ and on a set of unknown parameters $\{\alpha_1, \alpha_2, \ldots, \alpha_k\}$ such that $\varepsilon_n = \varepsilon_n\{\alpha_1, \alpha_2, \ldots, \alpha_k\}$. In order to determine the set $\{\alpha_k\}$ one must solve the system of equations

$$\frac{\partial \varepsilon_n}{\partial \alpha_i} = 0 \qquad i = 1, 2, \ldots, k \tag{104}$$

from which ε_n can be readily obtained. In doing this ε_n and χ_n are obtained as approximate energy and wavefunction for the state n, respectively.

Let us assume a quantum system confined within a domain D, for which the Hamiltonian can be written as

$$\widehat{H} = \widehat{H}_0 + V(q) \tag{105}$$

where $\widehat{H}_0$ is the Hamiltonian for the free system, and

$$V(q) = \begin{cases} +\infty & q \notin D \\ 0 & q \in D \end{cases} \tag{106}$$

is the confining potential.

The solution of Eq. (105) must satisfy

$$\psi(q) = 0 \qquad q \in \partial D \tag{107}$$

where ∂D is the boundary of D.

In order to use the variational method, a trial wavefunction that satisfies the boundary condition imposed by Eq. (107) must be constructed. This function can be defined as

$$\chi(q) = g(q)\psi_0(q, \alpha) \tag{108}$$

where $\psi_0(q, \alpha)$ has a structure similar to the solution of the Schrödinger equation for $\widehat{H}_0$, but the good quantum number(s) that define the energy of the free system is (are) replaced by a parameter (or a set of parameters) α, because the confining conditions impose new quantization rules.

The function $g(q)$ must satisfy the boundary condition given by Eq. (107). This technique can be used when the symmetry of the confining potential is compatible with the symmetry of the system under study.

When the latter is not the case, the shape of the trial wavefunction given by Eq. (108) must be slightly changed, since the symmetry is broken by the confinement potential. However, in spite of the lack of compatibility between both symmetries, the system and confining domain often are related through a coordinate transformation of the form

$$q = q(q') \tag{109}$$

where q' are the coordinates of the center of symmetry of D and q are the coordinates of the center of symmetry of the system.

Even in the aforementioned case, it is still possible to construct the trial wavefunction as in Eq. (108), except that the coordinates q of the system under study must be written in terms of the coordinates q' of D using the rule given in Eq. (109). Under such conditions the trial wavefunction for this asymmetric situation can be written as

$$\chi(q(q')) = g(q(q))\psi_0(q(q'), \alpha) \tag{110}$$

Once this has been done, the energy functional may be constructed and minimized with respect to α, as usual.

4. QUASI-PARTICLE STATES IN THE QUASI-TWO-DIMENSIONAL STRUCTURES

In nanostructured semiconductors the concept of a quasi-particle is related to electrons, holes, and excitons, and their characteristic lengths that define the confinement degree are the de Broglie wavelength for electrons and holes, λ_e, λ_h, and the effective Bohr radius of excitons, a_B^*. These magnitudes may be considerably larger than the lattice constant a_L. Therefore, it is possible to create a mesoscopic structure, which in one dimension is comparable to or even less than $\lambda_e, \lambda_h, a_B^*$ but still larger than a_L. In that structure the elementary excitations will experience quantum confinement, resulting in a finite motion along of a confinement axis and an infinite motion in other directions. This way, one deals with the so-called quasi-two-dimensional nanostructured systems that in a broadly accepted classification are single heterostructures, single quantum wells, multiple quantum wells, and superlaticces.

The confinement degree of quasi-particles is related to the magnitude a_B^* and the quantum confinement size d; two regimes can be readily distinguished: a weak confinement regime and a strong confinement regime. The weak and strong confinement regimes in the quasi-particles are more important for the exciton than for the independent electron and hole. In the latter both regimes are important when, besides the confinement potential, another external potential acts in the Q2D system, so a comparison between the two different regimes and the value of the external potential can be established.

The weak confinement regime corresponds to the case when the Q2D confinement size (d) is small but still a few times larger than a_B^* for the electron–hole pair, the mathematical condition being $a_B^* \ll d$. When this happens, the electron and hole are correlated; the exciton can be envisioned as a quasi-particle moving around inside the Q2D confinement with only a little energy increment due to it. In this case, the infinite potential well model (within the single band in the effective mass approximation) gives a reasonable description of the experimentally observed shift in the exciton ground state energy.

The strong confinement regime corresponds to the condition $a_B^* \gg d$. In this case the confining effect dominates over the Coulomb potential, and the electron and hole should be viewed predominantly as individual particles in their respective single particle ground states with only little spatial correlation between them. In this regime the exciton in the Q2D system "feels" the boundary effects strongly and the inclusion of a finite height for the confining potential barrier has become an important requirement in order to account for recent experiments on the optical properties of small layers.

4.1. Envelope Function Approximation

In this section we shall deal with the determination of eingenstates in the Q2D systems. The emphasis will be focused on a simple description of these eingenstates.

First, the assumptions used to derive the envelope function scheme will be presented. The reader will perceive that when the $\vec{k}$ wavevector of the carrier is zero, the algebra is relatively easy. Let us see, in a general way, the method fundamentals. The description is based on the dispersion relations of the host materials.

Remember that the electronic state in the periodic field is described by a Bloch function

$$\phi_{\vec{k}n}(\vec{r}) = \exp(i\vec{k} \cdot \vec{r}) u_{\vec{k}n}(\vec{r}) \tag{111}$$

which is periodic in "k" space,

$$\phi_{\vec{k}n}(\vec{r}) = \phi_{\vec{k}+\vec{g},n}(\vec{r}) \tag{112}$$

where $\vec{g}$ is a reciprocal lattice vector and n is the band number. Therefore, this function can be expressed as a Fourier series in reciprocal space:

$$\phi_{\vec{k}n}(\vec{r}) = \frac{1}{\sqrt{N}} \sum_{\vec{l}} a_n(\vec{r}, \vec{l}) \exp(i\vec{k} \cdot \vec{l}) \tag{113}$$

where $\vec{l}$ is a direct lattice vector. The coefficients of this expansion,

$$a_n(\vec{r}, \vec{l}) = \frac{1}{\sqrt{N}} \sum_{\vec{k}} \phi_{\vec{k}n}(\vec{r}) \exp(-i\vec{k} \cdot \vec{l}) \tag{114}$$

are denominated the Wannier functions and can also be written as

$$a_n(\vec{r}, \vec{l}) = \frac{1}{\sqrt{N}} \sum_{\vec{k}} u_{\vec{k}n}(\vec{r}) \exp[i\vec{k} \cdot (\vec{r} - \vec{l})] \tag{115}$$

Let us remember that N is the number of possible values of $\vec{k}$ for the nth band in the first Brillouin zone.

One can appreciate that the Wannier functions do not depend on the $\vec{k}$ wavevector and possess the following properties:

1. There are as many Wannier functions as points in the direct lattice and therefore as many values of $\vec{k}$ in the first Brillouin zone.

   Indeed, in Eq. (113) there are as many terms (and consequently as many values of $a_n(\vec{r}, \vec{l})$) as different values of $\vec{l}$ (actually N).

2. The Wannier functions are located at the proximity of each $\vec{l}$ node of the direct lattice.

3. The Wannier functions obey also the relation

$$a_n(\vec{r}, \vec{l}) = a_n(\vec{r} - \vec{l}) \tag{116}$$

which is almost evident if the periodicity of $u_{\vec{k}n}(\vec{r})$ in Eq. (115) is taken account.

4. The Wannier functions satisfy the relations

$$a_n(\vec{r} - \vec{s}, \vec{l}) = a_n(\vec{r}, \vec{l} + \vec{s}) \tag{117}$$

which is derived immediately from Eq. (116).

5. The Wannier functions are orthonormalized:

$$\int a_{n'}^*(\vec{r}, \vec{l}) a_n(\vec{r}, \vec{l}) d\vec{r} = \delta_{nn'} \delta_{\vec{l}\vec{l}'} \tag{118}$$

That is,

$$\int a_{n'}^*(\vec{r}, \vec{l}) a_n(\vec{r}, \vec{l}) d\vec{r}$$
$$= \frac{1}{N} \sum_{\vec{k}'} \sum_{\vec{k}} \exp[-i(\vec{k} \cdot \vec{l} - \vec{k}' \cdot \vec{l})]$$
$$\times \int \phi_{\vec{k}'n'}^*(\vec{r}) \phi_{\vec{k}n}(\vec{r}) d\vec{r} \tag{119}$$
$$= \frac{1}{N} \delta_{nn'} \sum_{\vec{k}} \exp[-i\vec{k} \cdot (\vec{l} - \vec{l})] = \delta_{nn'} \delta_{\vec{l}\vec{l}'}$$

4.1.1. Effective Hamiltonian

Now, the solution of the nonstationary Schrödinger equation for one electron in the crystal, under the influence of an external potential, will be looked at; i.e.,

$$\{\widehat{H} + \widehat{U}\}\psi(\vec{r}, t) = i\hbar \frac{\partial \psi(\vec{r}, t)}{\partial t} \tag{120}$$

where $\widehat{H}$ is the Hamiltonian of the electron in the periodic field of the crystal and $\widehat{U}$ is the perturbation potential associated to the external field.

Let us look for a solution of the form

$$\psi(\vec{r}, t) = \sum_n \sum_{\vec{l}} f_n(\vec{l}, t) a_n(\vec{r}, \vec{l}) \tag{121}$$

where $f_n(\vec{l}, t)$ is named the envelope function of the nth band. Notice that the solution involves all the bands.

Substituting Eq. (121) in (120), after multiplying by $a_{n'}^*(\vec{r} - \vec{l})$ and integrating, it is obtained that

$$\sum_n \sum_{\vec{l}} \int a_{n'}^*(\vec{r} - \vec{l})[\widehat{H} + \widehat{U}]a_n(\vec{r} - \vec{l})f_n(\vec{l}, t)d\vec{r}$$
$$= i\hbar \sum_n \sum_{\vec{l}} \frac{\partial f_n(\vec{l}, t)}{\partial t} \int a_{n'}^*(\vec{r} - \vec{l})a_n(\vec{r} - \vec{l})d\vec{r} \tag{122}$$

Considering the normalization condition of the Wannier functions and making the sum in the right side of the last equation, it is obtained that

$$\sum_n \sum_{\vec{l}} f_n(\vec{l}, t) \int a_{n'}^*(\vec{r} - \vec{l})[\widehat{H} + \widehat{U}]a_n(\vec{r} - \vec{l})d\vec{r}$$
$$= i\hbar \frac{\partial f_{n'}(\vec{l}', t)}{\partial t} \tag{123}$$

Noting that for the nonperturbed electron

$$\widehat{H}\phi_{\vec{k}n}(\vec{r}) = \varepsilon_n(\vec{k})\phi_{\vec{k}n}(\vec{r}) \tag{124}$$

applying the $\widehat{H}$ operator to the Wannier functions (Eq. (114)), and taking into account Eq. (113), it is obtained that

$$\widehat{H}a_n(\vec{r} - \vec{l}) = \frac{1}{\sqrt{N}} \sum_{\vec{k}} \varepsilon_n(\vec{k})\phi_{\vec{k}n}(\vec{r})\exp(-i\vec{k}\cdot\vec{l})$$
$$= \frac{1}{N} \sum_{\vec{k}} \varepsilon_n(\vec{k})\exp(-i\vec{k}\cdot\vec{l}) \sum_{\vec{l}} a_n(\vec{r} - \vec{l})$$
$$\times \exp(i\vec{k}\cdot\vec{l}) \tag{125}$$
$$= \frac{1}{N} \sum_{\vec{l}} a_n(\vec{r} - \vec{l}) \sum_{\vec{k}} \varepsilon_n(\vec{k})\exp[i\vec{k}\cdot(\vec{l} - \vec{l})]$$
$$= \sum_{\vec{l}} \varepsilon_{n, \vec{l} - \vec{l}} a_n(\vec{r} - \vec{l})$$

where

$$\varepsilon_{n, \vec{l} - \vec{l}} = \frac{1}{N} \sum_{\vec{k}} \varepsilon_n(\vec{k})\exp[-\vec{k}\cdot(\vec{l} - \vec{l})] \tag{126}$$

has been introduced. This is the Fourier transform of the energy in $\vec{k}$ space, which is showing the overlap integral of $\widehat{H}$ between two Wannier functions in $\vec{l}$ and $\vec{l}$ for the same band; that is,

$$\int a_{n'}^*(\vec{r} - \vec{l})\widehat{H}a_n(\vec{r} - \vec{l})d\vec{r}$$
$$= \sum_{\vec{l}} \varepsilon_{n, \vec{l} - \vec{l}} \int a_{n'}^*(\vec{r} - \vec{l})a_n(\vec{r} - \vec{l})d\vec{r} \tag{127}$$
$$= \sum_{\vec{l}} \varepsilon_{n, \vec{l} - \vec{l}} \delta_{nn'}\delta_{\vec{l}\vec{l}} = \varepsilon_{n, \vec{l} - \vec{l}}\delta_{nn'}$$

Substituting (127) in (123), it is obtained that

$$\sum_n \sum_{\vec{l}} \{\delta_{nn'}\varepsilon_{n, \vec{l} - \vec{l}} + U_{nn'}(\vec{l}, \vec{l})\}f_n(\vec{l}, t) = i\hbar \frac{\partial f_{n'}(\vec{l}, t)}{\partial t} \tag{128}$$

where

$$U_{nn'}(\vec{l}, \vec{l}) = \int a_{n'}^*(\vec{r} - \vec{l})\widehat{U}a_n(\vec{r} - \vec{l})d\vec{r} \tag{129}$$

is the matrix element of the perturbation potential.

Considering that the Fourier series for energy in the $\vec{k}$ space is

$$\varepsilon_n(\vec{k}) = \sum_{\vec{l}} \varepsilon_{n\vec{l}}\exp(i\vec{k}\cdot\vec{l}) \tag{130}$$

the $\varepsilon_{n\vec{l}}$ coefficients are given by Eq. (126), and considering that

$$\vec{k} \to -i\nabla \tag{131}$$

the following product it can be obtained:

$$\varepsilon_n(\vec{k})f(\vec{r}) = \sum_{\vec{l}} \varepsilon_{n\vec{l}}\exp[i(-i\nabla)\cdot\vec{l}]f(\vec{r})$$
$$= \sum_{\vec{l}} \varepsilon_{n\vec{l}}\exp(\vec{l}\cdot\nabla)f(\vec{r}) \tag{132}$$

But, as

$$\exp(\vec{l}\cdot\nabla) \approx 1 + \vec{l}\cdot\nabla + \frac{1}{2}(\vec{l}\cdot\nabla)^2 \tag{133}$$

then

$$\varepsilon_n(-i\nabla)f(\vec{r})$$
$$= \sum_{\vec{l}} \varepsilon_{n\vec{l}}\left\{f(\vec{r}) + \vec{l}.\nabla f(\vec{r}) + \frac{1}{2}\vec{l}^2\nabla^2 f(\vec{r}) + \cdots\right\} \tag{134}$$
$$= \sum_{\vec{l}} \varepsilon_{n\vec{l}}f(\vec{r} + \vec{l})$$

where it has been taken into account that the term in curl brackets is the Taylor series expansion for the function $f(\vec{r} + \vec{l})$.

Using Eq. (134) it can be obtained that

$$\sum_{\vec{l}} \varepsilon_{n', \vec{l} - \vec{l}} f_{n'}(\vec{l}, t) = \sum_{\vec{l}} \varepsilon_{n', \vec{l} - \vec{l}} f_{n'}(\vec{l} + \vec{l} - \vec{l}, t)$$
$$= \varepsilon_{n'}(-i\nabla)f_{n'}(\vec{r}, t)|_{\vec{r} = \vec{l}} \tag{135}$$

Then, Eq. (128) can be rewritten as

$$
\begin{aligned}
&\left[\varepsilon_{n'}(-i\nabla)f_{n'}(\vec{r},t)-i\hbar\frac{\partial f_{n'}(\vec{r},t)}{\partial t}\right]_{\vec{r}=\vec{l}} \\
&+\sum_{n'}\sum_{\vec{l}}U_{nn'}(\vec{l},\vec{l})f_n(\vec{l},t)=0
\end{aligned}
\tag{136}
$$

which is the expression for the envelope function, also called the Schrödinger equation in the Wannier representation.

Suppose that the electron moves only in one band. This means that the external potential is weak enough to not induce interband transitions; therefore

$$
U_{nn'}(\vec{l},\vec{l})=\delta_{nn'}U_{nn}(\vec{l},\vec{l})
\tag{137}
$$

Let consider that the potential varies slowly with the position; that is,

$$
a_L\frac{\Delta U}{U}\ll 1
\tag{138}
$$

a_L being the lattice constant. Then

$$
U_{nn}(\vec{l},\vec{l})=U(\vec{r},t)\delta_{\vec{l}\vec{l}'}
\tag{139}
$$

where Eqs. (118) and (129) have been taken into account.

Introducing

$$
U_{nn}(\vec{l},\vec{l})=U(\vec{r},t)|_{\vec{r}=\vec{l}}
\tag{140}
$$

and substituting Eqs. (137) and (139) in Eq. (136), it is obtained that

$$
\left[\varepsilon_n(-i\nabla)+U(\vec{r},t)-i\hbar\frac{\partial}{\partial t}\right]f_n(\vec{r},t)=0
\tag{141}
$$

where it is clear that $\vec{r}=\vec{l}$. $f_n(\vec{r},t)$ is formally treated as a continuous function that varies slowly with the position around $\vec{l}$ (according to Eq. (138)).

Indeed, let us compare Eq. (121) with Eq. (113) (considering Eq. (116)).

If it is applied a weak external field which varies slowly in distances of the order of the lattice constant, the electron wavefunction, in a determined band, is expressed as (according to Eq. (121))

$$
\phi_{\vec{k}n}(\vec{r},t)=\sum_{\vec{l}}a_n(\vec{r}-\vec{l})f_n(\vec{l},t)
\tag{142}
$$

where $f_n(\vec{l},t)$ is defined from (141). It has been considered that the relation must be satisfied

$$
f_n(\vec{r},t)|_{\vec{r}=\vec{l}}=f(\vec{l},t)
\tag{143}
$$

where it has been formally passed from the discrete variable l, to the continuous $\vec{r}$, which is acceptable if $f_n(\vec{l},t)$ varies slowly in distances of the order of the lattice constant, around the $\vec{l}$ node.

In absence of external field ($U(\vec{r},t)=0$), it is simply obtained that

$$
f_n^0(\vec{l},t)=\exp\left[i\left(\vec{k}\cdot\vec{l}-\frac{\varepsilon(\vec{k})}{\hbar}t\right)\right]
\tag{144}
$$

which, if substituted in Eq. (142), leads to the equation of the stationary states.

Notice that instead of the original Eq. (120), now a Schrödinger equation for the envelope function has been obtained, where the $\widehat{H}+\widehat{U}$ Hamiltonian has been substituted by the effective Hamiltonian,

$$
\widehat{H}_{\text{ef}}=\varepsilon_n(-i\nabla)+\widehat{U}
\tag{145}
$$

which, as one can appreciate, depends on the dispersion relation $\varepsilon_n(\vec{k})$.

Thus, the envelope function can be understood like the effective wavefunction for the electron in the crystal.

Fortunately, in most physical phenomena occurring in semiconductors and metals, the participating electrons are those situated in the bottom or the top of the bands (that is, in their extremes). For these electrons is convenient to introduce the effective mass and the effective Hamiltonian, which can be written as follows:

$$
\widehat{H}_{\text{ef}}=-\left[\frac{\hbar^2}{m_{nx}^*}\frac{\partial^2}{\partial x^2}+\frac{\hbar^2}{m_{ny}^*}\frac{\partial^2}{\partial y^2}+\frac{\hbar^2}{m_{nz}^*}\frac{\partial^2}{\partial z^2}\right]+\widehat{U}
\tag{146}
$$

In the isotropic crystal case, the effective masses are equal in any direction and Eq. (146) is rewritten as

$$
\widehat{H}_{\text{ef}}=-\frac{\hbar^2\nabla^2}{2m_n^*}+\widehat{U}
\tag{147}
$$

where m_n^* is the effective mass of the n band.

In this case it can be said that the band is parabolic since the dispersion relation is given by

$$
\varepsilon_n(\vec{k})=\frac{\hbar^2k^2}{2m_n^*}
\tag{148}
$$

which is, besides, isotropic.

So for a constant potential (in the time) one can write the effective Schrödinger equation for stationary states as

$$
\begin{aligned}
&\left\{-\frac{\hbar^2}{2}\left[\frac{1}{m_{nx}^*}\frac{\partial^2}{\partial x^2}+\frac{1}{m_{ny}^*}\frac{\partial^2}{\partial y^2}+\frac{1}{m_{nz}^*}\frac{\partial^2}{\partial z^2}\right]+\widehat{U}(\vec{r})\right\}f_n(\vec{r}) \\
&=\varepsilon f_n(\vec{r})
\end{aligned}
\tag{149}
$$

an expression which is widely used.

This method is known as the effective mass approximation.

4.2. Envelope Function Description of Quasi-Particle States for Q2D Systems

The envelope function model is particularly adequate to study quasi-particle properties of low-dimensional semiconductor structures. This is a dynamic method for the determination of the energy bands of particles in solid state physics; and the

Q2D systems are considered dynamic systems since they are always under the action of either a barrier potential or an interface potential. The envelope function scheme and the effective mass approximation were clearly discussed for the first time by Bastard [36, 37].

In order to obtain the Q2D energy states, in the envelope function scheme, the goal will be to find the boundary conditions which the slowly varying part of the Q2D wavefunctions must fulfill at the heterointerfaces.

Other approaches to the Q2D eigenenergies have been proposed which are in essence more microscopic than the envelope function description. The tight-binding model is successfully used for Q2Ds of any size, although it may have difficulties in handling self-consistent calculations arising when charges are present in the nanostructures. Another microscopic approach is the pseudo-potential formalism, which is very successful for bulk materials. The advantage of these microscopic approaches is their capacity to handle any Q2D energy levels, i.e., those close to or far from the Γ edge. This occurs because these models reproduce the whole bulk dispersion relations. The envelope function approximation has no such generality. Basically, it is restricted to the vicinity of the high-symmetry points in the host's Brillouin zone (Γ, X, L). However, it is invaluable due to its simplicity and versatility. It often leads to analytical results and leaves the user with the feeling that he or she can trace back, in a relatively transparent way, the physical origin of the numerical results. Besides, most of the Q2D energy levels relevant to actual devices are relatively close to a high-symmetry point in the host's Brillouin zone.

In the following it will be assumed that A and B materials constituting the Q2D are perfectly lattice matched and that they crystallize with the same crystallographic structure. In order to apply the envelope function model two key assumptions will be made:

(1) Inside each material the wavefunction is expanded in terms of the Wannier functions (with $\vec{l} = 0$) of the edges under consideration

$$\psi(\vec{r}) = \sum_n f_n^A(\vec{r}) u_{n,\vec{k}_0}^A(\vec{r}) \tag{150}$$

if $\vec{r}$ corresponds to an A material and

$$\psi(\vec{r}) = \sum_n f_n^B(\vec{r}) u_{n,\vec{k}_0}^B(\vec{r}) \tag{151}$$

if $\vec{r}$ corresponds to a B material, where $\vec{k}_0$ is the point in the Brillouin zone around which are built the nanostructure states. The summation over n is over as many edges as are included in the analysis.

(2) The periodic parts of the Bloch functions are assumed to be the same in each kind of material that constitutes the nanostructure:

$$u_{n,\vec{k}_0}^A(\vec{r}) \equiv u_{n,\vec{k}_0}^B(\vec{r}) \tag{152}$$

Thus the nanostructure wavefunction will be written as

$$\psi(\vec{r}) = \sum_n f_n^{A,B}(\vec{r}) u_{n,\vec{k}_0}(r) \tag{153}$$

and the objective will be to determine $f_n^{A,B}(\vec{r})$. The envelope function $f_n^{A,B}(\vec{r})$ depends of the kind of low-dimensional system.

These quasi-two-dimensional systems are formed by multiple heterostructures as the single quantum well, multiple quantum wells, and superlattices. The envelope function is in this case

$$f_n^{A,B}(\vec{r}_\perp, z) = \frac{1}{\sqrt{S}} \exp(i\vec{k}_\perp \cdot \vec{r}_\perp) \chi_n^{A,B}(z) \tag{154}$$

where S is the sample area and $\vec{k}_\perp = (k_x, k_y)$ is a two-dimensional wavevector which is the same in the A and B layers in order to preserve translational invariance in the plane (x, y).

Although $\vec{k}_\perp$ could span theoretically the whole plane section of the host's Brillouin zone, it is seldom in practice larger than $\approx 1/10$ of its size.

It will also be assumed that for all n, $\chi_n^{A,B}(z)$ varies slowly at the scale of the host's unit cell. Thus the heterostructure wavefunction $\psi(\vec{r})$ is a sum of the products of rapidly varying functions.

The heterostructure Hamiltonian can be written in the form

$$\widehat{H} = \frac{\hat{p}_x^2}{2m_x^*} + \frac{\hat{p}_y^2}{2m_y^*} + \frac{\hat{p}_z^2}{2m_z^*} + V_A(\vec{r})Y_A + V_B(\vec{r})Y_B \tag{155}$$

where $Y_A(Y_B)$ are step functions which are unity if $\vec{r}$ corresponds to an A layer (to a B layer) and $V_A(V_B)$ is the potential in the A (B) layer.

The envelope function $\chi_n(z)$ in the z direction of an A/B layer is the solution of the Schrödinger equation

$$\left\{ -\frac{\hbar^2}{2m_{nz}^*} \frac{d^2}{dz^2} + V_n(z) \right\} \chi_n(z) = \varepsilon_n \chi_n(z) \tag{156}$$

where $V_n(z)$ is the heterostructure potential determined by the band discontinuity: $m_{nz}^* \equiv m_n^*(z) = m_{n,A}^*(m_{n,B}^*)$ and $V_n(z)=0$ in the A(B) layer.

Let us see the general case of a superlattice. In a given A/B superlattice there is a new periodicity along the growth direction z. The SL period d (sum of the layer thickness of the A and B layers, $d_A + d_B$) is longer than the lattice constants of the constituents, so that a new mini-Brillouin zone is formed in the wavevector range of $-\pi/d \leq k_z \leq \pi/d$. The mini-Brillouin zone formation results in the generation of miniband structures in the conduction and valence bands.

The miniband structures of the Γ_6 conduction band and the Γ_8 valence band (heavy hole, $|J, M_J\rangle = |3/2, \pm 3/2\rangle$, and light hole, $|3/2, \pm 1/2\rangle$), are considered. The total wavefunction in each layer can be written in the form

$$\psi_{\vec{k}}^{A,B}(r) = \sum_n f_{n,\vec{k}}^{A,B}(\vec{r}) u_{n,\vec{k}=0}(\vec{r}) \tag{157}$$

where the following form of the envelope function can be used:

$$F_{n,\vec{k}}^{A,B}(\vec{r}) = \exp(ik_{n,x}^{A,B}x)\exp(ik_{n,y}^{A,B}y)f_{n,\vec{q}}^{A,B}(z) \qquad (158)$$

Then $f_{n,\vec{q}}^{A,B}(z)$ is the envelope function determined by the SL potential and the effective masses, and $\vec{q}$ is the SL wavevector parallel to z. The periodicity of the SL results in the Bloch theorem for the envelope function,

$$f_{n,\vec{q}}^{A,B}(z+md) = \exp(imqd)f_{n,\vec{q}}^{A,B}(z) \qquad (159)$$

where m is an integer.

The envelope function of an A/B SL is the solution of the Schrödinger equation

$$\left\{-\frac{\hbar^2}{2m_n^*(z)}\frac{d^2}{dz^2} + V_n(z)\right\}f_{n,\vec{q}}(z) = \varepsilon_n f_{n,\vec{q}}(z) \qquad (160)$$

where $V_n(z) = 0(\Delta\varepsilon_n)$ in the A(B) layer. For convenience, the band index n will be omitted in the following treatment. The envelope function can be written as

$$f_{\vec{q}}(z) = \begin{cases} f_{\vec{q}}^A(z) = \alpha_{\vec{q}}^A\exp(ik_A z) + \beta_{\vec{q}}^A\exp(-ik_A z) \\ \qquad\text{in the A layer} \\ f_{\vec{q}}^B(z) = \alpha_{\vec{q}}^B\exp(ik_B z) + \beta_{\vec{q}}^B\exp(-ik_B z) \\ \qquad\text{in the B layer} \end{cases} \qquad (161)$$

$$k_A = \sqrt{2m_A^*\varepsilon}/\hbar \qquad k_B = \sqrt{2m_B^*(\varepsilon - \Delta\varepsilon)}/\hbar \qquad (162)$$

Considering the continuity of the probability density $|f_{\vec{q}}(z)|^2$, for a stationary state, the following boundary condition on the envelope function at the A/B interface for $(z_A - z_B) \to 0$ is obtained:

$$f_{\vec{q}}^A(z_A) = f_{\vec{q}}^B(z_B) \quad\text{and}\quad \frac{1}{m_A^*}\frac{df_{\vec{q}}^A(z_A)}{dz} = \frac{1}{m_B^*}\frac{df_{\vec{q}}^B(z_B)}{dz} \qquad (163)$$

Finally, the miniband dispersion relation, which is similar to the well known Kronig–Penney relation, is given by

$$\cos(qd) = \cos(k_A d_A)\cos(k_B d_B)$$
$$- \frac{1}{2}\left(\frac{k_A m_B^*}{k_B m_A^*} + \frac{k_B m_A^*}{k_A m_B^*}\right)\sin(k_A d_A)\sin(k_B d_B) \qquad (164)$$

This equation has been widely used to calculate the miniband energies.

Figure 16 shows (a) the potential structure and the envelope function profiles of the electron and heavy hole in a GaAs(3.2 nm)/AlAs(0.9 nm)/AlAs(0.9 nm) SL and (b) the dispersion relations of the electron and heavy hole minibands calculated through Eq. (164). In Figure 16a the solid and dashed lines indicate the envelope functions at the mini-Brillouin-zone center ($q = 0$, Γ point) and the mini-Brillouin-zone edge ($q = \pi/d$, π point), respectively. In this calculation, the following effective mass parameters are used [38]: $m_e^* = 0.0665m_0$, $m_{hh}^* = 0.34m_0$, and $m_{lh}^* = 0.117m_0$ for the electron, heavy hole, and light hole of GaAs, and $m_e^* =$

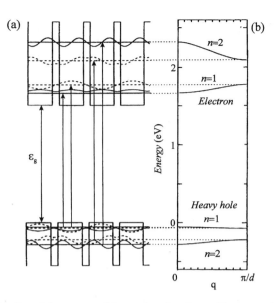

Fig. 16. Potential structure and the envelope function profiles in superlattice.

$0.15m_0$, $m_{hh}^* = 0.4m_0$, and $m_{lh}^* = 0.18m_0$ for AlAs, where m_0 is the free-electron mass. The miniband dispersions are tailored by changing the layer thickness, potential height, and effective masses that are structural parameters in crystal growth. Figure 17 shows the $n = 1$ and $n = 2$ electron miniband energies at Γ and π points in a GaAs(3.2 nm)/AlAs(d_B nm) SL as a function of the AlAs layer thickness. This figure exhibits a typical example of the miniband tailoring. The miniband width increases with the decrease of the AlAs (barrier) thickness, which results from the increase of the coupling strength of wavefunctions due to the increase of the tunneling probability. In the case of disappearance of the miniband width, the electronic state corresponds to a quasi-two-dimensional state in an isolated QW.

The z dependence of the effective mass means, in principle, that it is different for different layers but in general may vary even in a given medium if no abrupt interfaces are considered. Then the kinetic energy term in Eq. (160) should be substituted by the more general Ben Daniel–Duke [39] hermitian

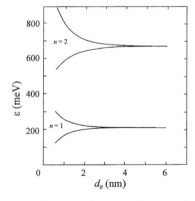

Fig. 17. Electron miniband energies at Γ and π points in a superlattice.

expression

$$-\frac{\hbar^2}{2}\frac{\partial}{\partial z}\frac{1}{m^*(z)}\frac{\partial}{\partial z} \tag{165}$$

Finally, Eq. (160) for the envelope function can be generalized as

$$\left\{-\frac{\hbar^2}{2}\frac{\partial}{\partial z}\frac{1}{m^*(z)}\frac{\partial}{\partial z}+V(z)\right\}\chi(z)=\varepsilon\chi(z) \tag{166}$$

4.3. Excitonic States in Q2D Semiconductors

A description based on noninteracting electrons and holes as the only elementary excitations correspond to the so-called single particle presentation. Actually, electrons and holes as charged particles do interact via the Coulomb potential and form an extra quasi-particle that corresponds to the hydrogenlike bound state of an electron–hole pair which is given the name of exciton. There are two limiting types of exciton: the Wannier exciton and the Frenkel exciton. Here the attention is paid only to the former, which is a weakly bound state of a system formed by an electron and a hole. Because the spatial extent of the Wannier exciton is much larger than a lattice constant, its wavefunction is more sensitively affected by a spatial geometry than the one of the Frenkel excitons. The Wannier exciton has three degrees of freedom: the center of mass motion, the electron–hole (e–h) relative motion, and the spin configuration. The latter two are internal degrees of freedom. In three-dimensional perfectly rigid crystals, the center of mass motion is well described by a plane wave with a three-dimensional wavevector $\vec{k}=(k_x,k_y,k_z)$. Wavefunctions of an electron–hole relative motion are sensitive to the type of Q2D semiconductor and determine its optical properties; in this fundamental aspect resides the importance of excitons as quasi-particles. To simplify the problem, the effective mass approximation and also the envelope function approximation will be employed. Exchange and spin-orbit interactions are neglected here.

It is known, from the basic concepts of bulk semiconductor physics, that a single spherical conduction band can approximate the band structure, with the dispersion relation

$$\varepsilon_c(\vec{k})=\varepsilon_g+\frac{\hbar^2k^2}{2m_c^*} \tag{167}$$

which is separated by the bandgap ε_g from a single spherical valence band with dispersion relation

$$\varepsilon_v(\vec{k})=-\frac{\hbar^2k^2}{2m_v^*} \tag{168}$$

where the N electrons occupying the N available places in the valence band have been considered as the initial states (ground state), whereas all conduction band levels are empty. An excited state of a bulk semiconductor is built when one photon with energy near the bandgap is absorbed, creating an e–h pair (exciton); the wavevector of the hole is $\vec{k}_h=-\vec{k}_e$. This pair travels freely over the whole crystal. Because

the e–h pair is weakly bound (exciton binding energy $\ll\varepsilon_g$) and spreads over many lattice sites, the e–h interaction must be screened by the static constant κ of the semiconductor. The electron–hole interaction can be described in the effective mass approximation and envelope function scheme by the equation

$$\left\{-\frac{\hbar^2}{2m_e^*}\nabla_e^2-\frac{\hbar^2}{2m_h^*}\nabla_h^2-\frac{e^2}{\kappa|\vec{r}_e-\vec{r}_h|}\right\}f(\vec{r}_e,\vec{r}_h)$$
$$=(\varepsilon-\varepsilon_g)f(\vec{r}_e,\vec{r}_h) \tag{169}$$

Since the coulombic term only affects the relative e–h coordinates, it is convenient to set

$$\vec{r}=\vec{r}_e-\vec{r}_h;\ \vec{R}=\frac{m_e^*\vec{r}_e+m_h^*\vec{r}_h}{m_e^*+m_h^*} \tag{170}$$

where $\vec{R}$ is the center of mass vector of the e–h pair. Equation (169) is written as

$$\left\{-\frac{\hbar^2}{2(m_e^*+m_h^*)}\nabla_R^2-\frac{\hbar^2}{2\mu^*}\nabla_r^2-\frac{e^2}{\kappa r}\right\}f(\vec{r},\vec{R})$$
$$=(\varepsilon-\varepsilon_g)f(\vec{r},\vec{R}) \tag{171}$$

where

$$\mu^*=\frac{m_e^*m_h^*}{m_e^*+m_h^*} \tag{172}$$

is the reduced effective mass of the e–h pair. The envelope function $f(\vec{r},\vec{R})$ can be factorized as

$$f(\vec{r},\vec{R})=\frac{1}{\sqrt{\Omega}}\exp(i\vec{K}\cdot\vec{R})\phi(\vec{r}) \tag{173}$$

where Ω is the volume of the crystal and $\vec{K}$ is the center of mass wavevector, the total energy being

$$\varepsilon_n(\vec{K})=\varepsilon_g+\frac{\hbar^2K^2}{2(m_e^*+m_h^*)}+\varepsilon_n \tag{174}$$

Then, the equation for the new envelope function $\phi(\vec{r})$ is finally obtained,

$$\left\{-\frac{\hbar^2}{2\mu^*}\nabla_r^2-\frac{e^2}{\kappa r}\right\}\phi(\vec{r})=\varepsilon_n\phi(\vec{r}) \tag{175}$$

The bound states correspond to $\varepsilon_n<0$ whereas positive ε_n's correspond to unbound e–h pairs. The ground state wavefunction is the $1s$ hydrogenic wavefunction

$$\phi(\vec{r})=\frac{1}{[\pi a_B^{*3}]^{1/2}}\exp\left(-r/a_B^*\right) \tag{176}$$

where a_B^* is the exciton effective Bohr radius,

$$a_B^*=\frac{\hbar^2\kappa}{\mu^*e^2}=\kappa\frac{m_0}{\mu^*}\times0.53\ \text{Å} \tag{177}$$

m_0 being the free-electron mass.

The binding energy of the ground state ($K = 0$ and $n = 1$) is the exciton effective Rydberg

$$R^*_{y,\,\text{exc}} = \frac{e^2}{2\kappa a^*_B} = \frac{\mu^* e^4}{2\kappa^2 \hbar^2} = \frac{\mu^*}{m_0} \frac{1}{\kappa^2} \times 13.6 \text{ eV} \qquad (178)$$

The forming process of an exciton band is shown in the diagram of Figure 18.

The reduced e–h mass is smaller than the free-electron mass, and the dielectric constant κ is several times larger than that of the vacuum. That is why the exciton Bohr radius is significantly larger and the exciton Rydberg energy is significantly smaller than the corresponding values for the hydrogen atom. Absolute values of a^*_B for the common semiconductors range in the interval 10–100 Å, and the exciton Rydberg energy takes values of 1–100 meV, approximately.

The dispersion relation can be expressed as

$$\varepsilon_n(\vec{K}) = \varepsilon_g - \frac{R^*_{y,\,\text{exc}}}{n^2} + \frac{\hbar^2 K^2}{2(m^*_e + m^*_h)} \qquad (179)$$

The exciton energy spectrum consists of subbands (see Fig. 19).

Exciton gas can be described as a gas of bosons with the energy distribution function obeying the Bose–Einstein statistics

$$F(\varepsilon) = \left[\exp\!\left(\frac{\varepsilon - \eta}{k_B T}\right) - 1 \right]^{-1} \qquad (180)$$

where η is the chemical potential. For a given temperature T, the concentration of excitons n_{exc} and of free electrons and holes $n = n_e = n_h$ are related via the ionization equilibrium equation known as the Saha equation:

$$n_{\text{exc}} = n^2 \left(\frac{2\pi \hbar^2}{k_B T} \frac{1}{\mu^*}\right)^{3/2} \exp\!\left(\frac{R^*_{y,\,\text{exc}}}{k_B T}\right) \qquad (181)$$

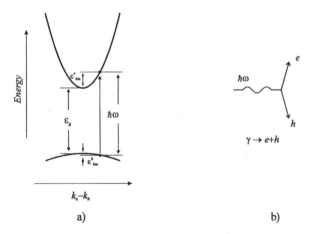

a)   b)

Fig. 18. A process of a photon absorption resulting in a creation of one electron–hole pair in different presentations. In a diagram including dispersion curves for conduction and valence band this event can be shown as a vertical transition exhibiting simultaneous energy and momentum conservation (a). This event also may be treated as a conversion of a photon into electron and hole (b).

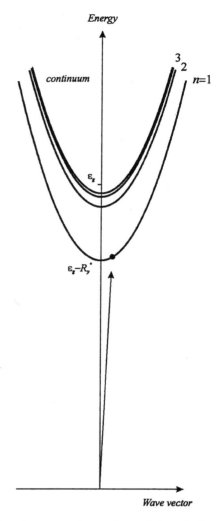

Fig. 19. Dispersion curves of an exciton and the optical transition corresponding to a photon absorption and exciton creation. Dispersion curves correspond to the hydrogenic set of energies $\varepsilon_n = \varepsilon_g - R^*_{y,\,\text{exc}}/n^2$ at $K = 0$ and a parabolic $\varepsilon(K)$ dependence for every ε_n, describing the translational center of mass motion.

For $k_B T \gg R^*_{y,\,\text{exc}}$ most excitons are ionized and the properties of the electron subsystem of the crystal are determined by the free electrons and holes. At $k_B T \leq R^*_{y,\,\text{exc}}$ a significant part of electron–hole pairs exists in the bound state.

Only the excitonic states in quantum well structures, which consist of a slab of A material embedded between two semi-infinite layers of B material, will be studied. Let us assume that both B layers have the same conduction m^*_e and valence m^*_h effective masses, as well as the same dielectric constant κ.

Two cases can be considered: type I and type II quantum wells.

4.3.1. Type I Quantum Wells

Both electrons and holes are located within the same layers (in this case A material). The A bulk bandgap and the barrier heights for electron and hole are denoted by ε_g, V_e, V_h, respectively. Assuming that the growth axis is in the z direction (confining direction) and the conduction and valence bands dispersion relations in both materials are spherical and quadratic in $\vec{k}$, the equation for the envelope function and excitonic states can be generalized into

$$\left\{-\frac{\hbar^2}{2(m_e^* + m_h^*)}\nabla_{R,\perp}^2 - \frac{\hbar^2}{2\mu}\nabla_{r,\perp}^2 - \frac{e^2}{\kappa\sqrt{r^2 + (z_e - z_h)^2}}\right.$$

$$\left. -\frac{\hbar^2}{2m_e^*}\nabla_{ze}^2 - \frac{\hbar^2}{2m_h^*}\nabla_{zh}^2 + V_e Y\left(z_e^2 - \frac{L^2}{4}\right) \right. \qquad (182)$$

$$\left. + V_h Y\left(z_h^2 - \frac{L^2}{4}\right)\right\} f(\vec{r}_e, \vec{r}_h) = (\varepsilon - \varepsilon_g)f(\vec{r}_e, \vec{r}_h)$$

The envelope function $f(\vec{r}_e, \vec{r}_h)$ can be partially factorized as

$$f(\vec{r}_e, \vec{r}_h) = \frac{1}{\sqrt{S}}\exp\left(i\vec{K}_\perp \cdot \vec{R}_\perp\right)\phi(z_e, z_h, \vec{r}) \qquad (183)$$

where $\hbar\vec{K}_\perp$ is the center of mass momentum, which corresponds to the free motions in the layer planes and $\phi(z_e, z_h, \vec{r})$ is the new envelope function. Two kinds of variational procedures have been attempted to solve Eq. (182), either by using a gaussian basis set [40] or nonlinear variational parameters [40–43]. In the latter case, advantage is taken of the prevailing effects of the confining barriers on the z_e, z_h motions. Thus it can be written that [44]

$$\phi(z_e, z_h, \vec{r}) = N_1 \chi_{1,e}(z_e)\chi_{1,h}(z_h)\exp\left(-\frac{r}{\lambda}\right) \qquad (184)$$

or

$$\phi(z_e, z_h, \vec{r})$$
$$= N_2 \chi_{1,e}(z_e)\chi_{1,h}(z_h)\exp\left(-\frac{1}{\lambda}\sqrt{r^2 + (z_e - z_h)^2}\right) \qquad (185)$$

where N_1 and N_2 are two normalization constants and λ is the variational parameter. The ground state of the exciton is assumed to consist of an electron in the lowest conduction subband e_1 and a hole in the lowest hole subband h_1.

At this point one should point out that there exist two kinds of excitons, which in this simplified approach are completely decoupled. The first kind corresponds to the pairing between an electron and a heavy hole (hh), which will be termed the heavy hole excitons, while the second corresponds to the pairing between an electron and a light hole (lh) which will be termed the light hole excitons.

In the simplified model used in this subsection the light hole exciton is always less bound than the heavy hole exciton because:

(i) The light holes in the lh_1 subband are spatially less confined than the heavy holes in hh_1 subband. Thus the wavefunction χ_{lh_1} is less localized than χ_{hh_1}, implying a weaker effective electron–hole interaction in the layer plane for the light hole exciton and consequently a weaker binding.

(ii) The bulk light hole Rydberg exciton is smaller than the bulk heavy Rydberg hole since $m_{lh}^* < m_{hh}^*$.

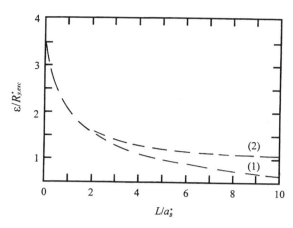

Fig. 20. Exciton binding energy in units of the effective Rydberg versus the ratio of the type I quantum well thickness to the effective Bohr radius. Curves labeled (1) and (2) correspond respectively to an exciton wavefunction which is separable or nonseparable in r and z_e, z_h.

The exciton binding energy versus the quantum well thickness L increases, reaches a maximum, and then drops to the effective Rydberg energy of the barrier material. If the confining barriers V_e, V_h are infinitely high the binding energy $\varepsilon(L)$ increases monotonically to $4R_{y,\text{exc}}^\infty$ where $R_{y,\text{exc}}^\infty$ is $R_{y,\text{exc}}^{hh}$ or $R_{y,\text{exc}}^{lh}$, depending on whether one is dealing with the light hole exciton or the heavy hole exciton.

This is illustrated in Figure 20 where the results obtained by using a separable (Eq. (184)) or a nonseparable (Eq. (185)) exciton wavefunction in r and z_e, z_h correspond to the curves labeled (1) and (2), respectively. In Figure 20 the center of mass energy is disregarded.

4.3.2. Exciton in Type II Quantum Wells

In these structures the electrons and holes are spatially separated. Also, due to the fact that a double heterostructure and not a superlattice is being considered, the quantum states of one kind of particle will not be confined along the growth axis (in this case B material). To be specific, let us consider that the A material mostly confines the electrons and repels the holes whereas the cladding B layers repel the electrons and attract the holes (see Fig. 21). If the same simplifying assumptions which were made in (a) concerning the band structure are taken, the exciton equation can be written as

$$\left\{-\frac{\hbar^2}{2m_e^*}\nabla_e^2 + V_e Y\left(z_e^2 - \frac{L^2}{4}\right) - \frac{\hbar^2}{2m_h^*}\nabla_h^2 \right.$$

$$\left. - V_h Y\left(-\frac{L^2}{4} + z_h^2\right) - \frac{e^2}{\kappa|\vec{r}_e - \vec{r}_h|}\right\} \qquad (186)$$

$$\times f(\vec{r}_e, \vec{r}_h) = (\varepsilon - \varepsilon_A)f(\vec{r}_e, \vec{r}_h)$$

In the absence of electron–hole interaction, the envelope function can be written as

$$f_0(\vec{r}_e, \vec{r}_h) = \frac{1}{S}\exp i\left[\vec{k}_{e,\perp}\cdot\vec{r}_e + \vec{k}_{h,\perp}\cdot\vec{r}_h\right]\chi_n(z_e)\xi(z_h) \qquad (187)$$

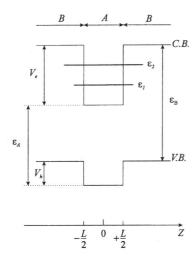

Fig. 21. Conduction and valence band edge profiles in a type II quantum well.

where $\chi_n(z_e)$ is one of the electron quantum well bound states and $\xi(z_h)$ is an extended hole state labeled by a wavevector k_{zh} corresponding to a hole which stems from $z_h = -\infty$ and impinges on the A slab. This hole then tunnels the A slab and escapes to $z_h = +\infty$ in the right-hand B layer. Another hole state corresponding to $-k_{zh}$ has the same energy as $\xi(z_h)$. To build the exciton the interest is put on the hole levels which are near the top of the valence band of the B layer. The characteristic penetration length of a heavy hole state ($m_{hh}^* = 0.4m_0$) in the A layer is $l_{hh} \approx 10$ Å (if $V_h = 0.1$ eV). Thus, if the A layer is thick enough (say, $L \geq 30$ Å) the hole tunneling can be neglected and it can be assumed that the A slab is impenetrable for the holes. In this case

$$\xi_{\pm}(z_h) = \sqrt{\frac{2}{\ell}} \left\{ \sin k_{zh}\left(z_h - \frac{L}{2}\right) Y\left(z_h - \frac{L}{2}\right) \right.$$
$$\left. \pm \sin k_{zh}\left(z_h + \frac{L}{2}\right) Y\left(-z_h - \frac{L}{2}\right) \right\} \quad (188)$$

and $\xi_{\pm}(z_h) = 0$ if $|z_h| < L/2$, where L is the A slab thickness and $\ell(\gg L)$ is the macroscopic length of the structure. In Eq. (188) $k_{zh} > 0$ and $\pm$ refers to the parity of the heavy hole state with respect to the center of the A layer ($-$ corresponds to even states and $+$ to odd states).

Equations (170) and (183) can be used to eliminate the in plane center of mass coordinates and search for a variational solution of the relative motion of the e–h pair. It will be seen later that only parity conserving transitions are optically allowed in symmetric heterostructures. A trial wavefunction for the lowest lying exciton state (photon energy $\hbar\omega_l = \varepsilon_A - V_h + \varepsilon_1$ in Fig. 21) is thus:

$$\phi(r, z_e, z_h)$$
$$= N\chi_1(z_e) \exp\left(-\frac{r}{\lambda}\right) \int_0^{\infty} dk_{zh}\, \alpha(k_{zh})\xi_-(z_h) \quad (189)$$

where the summation over k_{zh} is imposed by the continuous nature of the valence spectrum, N is a normalization constant, λ is a variational parameter, and $\alpha(k_{zh})$ is a function to be determined. To write Eq. (189) it has been assumed that the electron size quantization is large enough to neglect the admixture between ε_1 and $\varepsilon_2, \varepsilon_3, \ldots$ due to the coulombic interaction, thus forcing the z_e dependence of the exciton wavefunction to be identical to $\chi_1(z_e)$. On the other hand, given that the hole spectrum is bulklike, a hole wave packet concentrated near the interfaces ($\pm L/2$) is formed to maximize the electrostatic interaction with the electron charge distribution. It should also be noted that the excitons formed between the valence band and higher electron subbands $\varepsilon_2, \varepsilon_3, \ldots$ are likely to be more tightly bound than the lowest lying exciton level, since χ_2, χ_3 leak more heavily in the B layers than χ_1.

The exciton binding energy in GaSb–InAs–GaSb double heterostructures [42] has been calculated. These heterostructures are an extreme case of type II quantum wells inasmuch as the top of the GaSb valence lies above (by ≈ 0.15 eV) the bottom of the InAs conduction band (see Fig. 22). The electron confinement in the InAs slab was assumed to be complete ($\chi_1(z_e) = \sqrt{2/L}\cos(\pi z_e/L)$) and the z_h dependence of the exciton wavefunction was taken as a symmetric combination of Fang–Howard wavefunction [44]

$$\phi(\vec{r}, z_e, z_h)$$
$$= N\sqrt{\frac{2}{L}} \cos\left[\frac{\pi z_e}{L}\right] \frac{b^{3/2}}{2}$$
$$\times \left\{ \left(z_h - \frac{L}{2}\right) \exp\left[-\frac{b}{2}\left(z_h\frac{L}{2}\right)\right] Y\left[z_h - \frac{L}{2}\right] \right.$$
$$\left. - \left(z_h + \frac{L}{2}\right) \exp\left[\frac{b}{2}\left(z_h + \frac{L}{2}\right)\right] Y\left(-z_h - \frac{L}{2}\right) \right\}$$
$$\times \exp\left(-\frac{r}{\lambda}\right) \quad (190)$$

where b and λ are variational parameters. The mean distance from the hole to the A–B interface is $3/b$. Figure 23 shows the results of the variational calculation of the exciton binding energy in the GaSb–InAs–GaSb double heterostructures [42]; there $R_{y,\,exc}^{\infty} = 1.3$ meV and $a_B^{\infty} = 370$ Å are the bulk InAs exciton Rydberg energy and Bohr radius, respectively.

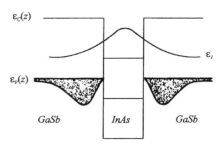

Fig. 22. Conduction and valence band edge profiles in GaSb–InAs–GaSb double heterostructure. The solid line enclosing the shaded area represents the deformed hole envelope function due to the electron–hole interaction.

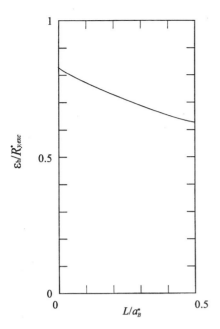

Fig. 23. Exciton binding energy versus the InAs thickness for GaSb–InAs–GaSb double heterostructure.

In Figure 23 the strong influence of the spatial separation between the charges in the excitonic binding is clearly seen: the exciton is less bound in InAs-based type II quantum wells than in bulk InAs, despite the pronounced size quantization of the electrons in the InAs slab.

From Figure 23 one can conclude that excitonic effects are unimportant in type II quantum wells.

4.4. Band Structure on Realistic Q2D Nanostructures

4.4.1. Effective Mass Hamiltonian and Boundary Conditions for the Valence Bands of Q2D Semiconductor

The envelope function and effective mass approximations have been employed extensively in the determination of electronic states in semiconductor heterostructures such as quantum wells and superlattices following the early work by Nedorezov [45] and Bastard [36]. In this approach, the periodic interatomic potential is eliminated from the description and is replaced by a smoothly varying macroscopic potential determined by the band offset of the materials. The wavefunction of the system is assumed to be a linear combination of the product of a slowly varying envelope function together with a Bloch function appropriate to a bulk carrier. The latter changes rapidly on an interatomic scale, with a period of unit cell length, and can be eliminated from the description [35]. Indeed, the approach offers a simple but successful way to interpret experimental data from "large systems" such as wide quantum wells or long period superlattices [46–49]. It is to be noted, however, that these successes rely, to some extent, on the judicious choice of important parameters such as the effective masses of the carriers.

In contrast with the large system, the envelope function approach was shown to be less effective in describing the electronic properties of "small systems" such as narrow quantum wells, or short period superlattices, if one employs the same input parameters as with the large system [50–52]. Indeed, when the characteristic dimensions of the systems decrease to values comparable with the interatomic length scale, e.g., less than 20 Å, there exist dramatic differences between the results based on the envelope function approximation and the experimental data [53]. It is evident that some of the basic approximations used in the envelope function approach are not valid for these small systems. For example, the envelope function and the macroscopic potential change on a scale comparable with those of the Bloch function and the periodic interatomic potential, hence they can no longer be regarded as slowly varying in this case. Furthermore, the effective mass parameters (or equivalently, the Luttinger parameters), which incorporate the effect of microscopic potential change in the parameterization scheme, are normally treated as constants derived from band structure properties associated with the corresponding bulk materials. One can expect such a treatment to become increasingly inappropriate as the characteristic dimension of the system decreases. Although some progress has been made with regard to the general formalism of the envelope function and effective mass theory for heterostructures [54–57], the central problems mentioned above related to the values of the effective masses to be employed in a given calculation remain largely unresolved.

Although computationally costly, solving the Schrödinger equation with microscopic atomic potentials is widely employed in band structure calculations such as the semi-empirical pseudo-potential or the tight-binding method. The developments that have made use of these methods extend from band structure calculations of bulk materials to those of microstructures [58–60]. In particular, Jaros et al. [59, 61] and Zunger et al. [60, 62–65] have developed two different schemes for pseudo-potential calculations of semiconductor heterostructures, which can handle efficiently both small and large systems. These pseudo-potential calculations are, of course, free from the approximations and restraints which the envelope function approach suffers from. Consequently, a comparison of the predictions of both methods permits a direct evaluation of the limitation of the envelope function approach and the modifications to it that are needed in order to have an agreement with the exact results of the pseudo-potential theory.

In conclusion, while the use of the effective mass theory in homogeneous bulk semiconductors is very well established, its application to heterostructures has been the subject of considerable debate, especially with regard to the boundary conditions connecting the envelope functions across an abrupt heterojunction [66–68]. The controversy has now been conclusively settled with the development by Burt of an exact envelope function theory for semiconductor microstructures [55]. The exact theory gives a general solution for the effective mass Hamiltonian in terms of momentum matrix elements and energy bandgaps which applies even at an abrupt interface. This work is concerned with writing the general multiband effective mass Hamiltonian given by Burt in an explicit form for use in valence band problems. This provides an unambiguous prescription for the boundary conditions at an interface; it will be shown that the correct boundary conditions differ substantially from those currently found in the literature.

The specific problem of concern here is the valence band structure of quantum wells or superlattices composed of semiconductors of the zinc-blende structure (which have a valence band maximum at the Γ point, neglecting the small linear k splitting terms) [69]. The bulk band structure [70] of these semiconductors is described in terms of the Luttinger mass parameters γ_1, γ_2 and γ_3, which are determined by the $\vec{k} \cdot \vec{p}$ interaction of the Γ_{15} valence bands with all other states of symmetry $\Gamma_1, \Gamma_{15}, \Gamma_{12}$, or Γ_{25}. These states are compatible (to lowest order) with the $s, p, d,$ and f orbitals, respectively, of the constituent atoms. Since the contribution of the f orbitals to the valence electronic structure of semiconductors is insignificant [71], the Γ_{25} states will be neglected here.

Each of the remaining ($\Gamma_1, \Gamma_{15},$ and Γ_{12}) interactions will generate terms in the valence band Hamiltonian with a different form (i.e., a different ordering of the differential operators with respect to the band parameters). It is therefore convenient to write the Luttinger parameters in a manner that explicitly reveals the contribution of each symmetry type [70],

$$\gamma_1 = -1 + 2\sigma + 4\pi + 4\delta$$
$$\gamma_2 = \sigma - \pi + 2\delta \qquad (191)$$
$$\gamma_3 = \sigma + \pi - \delta$$

where the dimensionless quantities $\sigma, \pi,$ and δ are defined by

$$\sigma = (1/3m_0) \sum_j^{\Gamma_1} \left| \langle X | \hat{p}_x | u_j \rangle \right|^2 / (\varepsilon_j - \varepsilon_v)$$

$$\pi = (1/3m_0) \sum_j^{\Gamma_{15}} \left| \langle X | \hat{p}_y | u_j \rangle \right|^2 / (\varepsilon_j - \varepsilon_v) \qquad (192)$$

$$\delta = (1/6m_0) \sum_j^{\Gamma_{12}} \left| \langle X | \hat{p}_x | u_j \rangle \right|^2 / (\varepsilon_j - \varepsilon_v)$$

Here $\hat{p}$ is the momentum operator, and ε_v is the valence band energy in the absence of spin-orbit splitting; the sum is over the basis states u_j of all remote bands of a given symmetry. These states are functions that have the periodicity of the zinc-blende lattice and the angular dependence shown in Table I. To calculate $\sigma, \pi,$ and δ in terms of the experimentally determined Luttinger parameters, the following identities are used:

$$\bar{\gamma} = \frac{1}{2}(\gamma_3 + \gamma_2) \quad \mu = \frac{1}{2}(\gamma_3 - \gamma_2) \quad \sigma = \bar{\gamma} - \frac{1}{2}\delta$$
$$\pi = \mu + \frac{3}{2}\delta \quad \delta = \frac{1}{9}(1 + \gamma_1 + \gamma_2 - 3\gamma_3) \qquad (193)$$

Table I.   Basis Functions

Representation	Basis functions
Γ_1	$x^2 + y^2 + z^2$ or xyz
Γ_{15}	x, y, z
Γ_{12}	$2z^2 - x^2 - y^2, 3(x^2 - y^2)$

Taking GaAs as a typical case, $\sigma \approx 4\pi$ and $\pi \approx 4\delta$. (The only qualitatively different semiconductor is Si, in which the conduction band p states actually lie below the s state [71]; hence $\sigma \approx \pi$.)

Using these parameters and the basis functions in Table I, it is straightforward to evaluate the effective Hamiltonian given by the exact envelope function theory [55]. However, the spin-orbital interaction splits the Γ_{15} states, so it is more convenient to work with eigenfunctions of the total angular momentum $|J, m_j\rangle$ which diagonalize this interaction [72]:

$$\left| \frac{3}{2}, \frac{3}{2} \right\rangle = \frac{1}{\sqrt{2}} \left| (X + iY) \uparrow \right\rangle$$

$$\left| \frac{3}{2}, -\frac{3}{2} \right\rangle = \frac{1}{\sqrt{2}} \left| (X - iY) \downarrow \right\rangle$$

$$\left| \frac{3}{2}, \frac{1}{2} \right\rangle = \frac{1}{\sqrt{6}} \left| (X + iY) \downarrow \right\rangle - \sqrt{\frac{2}{3}} \left| Z \uparrow \right\rangle$$

$$\left| \frac{3}{2}, -\frac{1}{2} \right\rangle = -\frac{1}{\sqrt{6}} \left| (X - iY) \uparrow \right\rangle - \sqrt{\frac{2}{3}} \left| Z \downarrow \right\rangle \qquad (194)$$

$$\left| \frac{1}{2}, \frac{1}{2} \right\rangle = \frac{1}{\sqrt{3}} \left| (X + iY) \downarrow \right\rangle + \sqrt{\frac{1}{3}} \left| Z \uparrow \right\rangle$$

$$\left| \frac{1}{2}, -\frac{1}{2} v \right\rangle = -\frac{1}{\sqrt{3}} \left| (X - iY) \uparrow \right\rangle + \sqrt{\frac{1}{3}} \left| Z \downarrow \right\rangle$$

In the exact envelope function theory, these functions are required to be the same throughout the structure, independently of the material composition. In this basis, the effective mass Hamiltonian takes the form

$$\widehat{H} = \begin{vmatrix} P+Q & 0 & -S_- & R & (1/\sqrt{2})S_- & \sqrt{2}R \\ 0 & P+Q & -R^+ & -S_+ & -\sqrt{2}R^+ & (1/\sqrt{2})S_+ \\ -S^+ & -R & P-Q & C & \sqrt{2}Q & \sqrt{3/2}\Sigma_- \\ R^+ & -S_+^+ & C^+ & P-Q & -\sqrt{3/2}\Sigma_+ & \sqrt{2}Q \\ (1/\sqrt{2})S^+ & -\sqrt{2}R & \sqrt{2}Q & -\sqrt{3/2}\Sigma_+^+ & P+\Delta & -C \\ \sqrt{2}R^+ & (1/\sqrt{2})S_+^+ & \sqrt{2/3}\Sigma_-^+ & \sqrt{2}Q & -C^* & P+\Delta \end{vmatrix}$$
$$(195)$$

where, in atomic units ($\hbar = m_0 = 1$) and with the hole energy positive,

$$P = \varepsilon_v(z) + \frac{1}{2}(\gamma_1 k_{//}^2 + k_z \gamma_1 k_z)$$

$$Q = \xi(z) + \frac{1}{2}(\gamma_2 k_{//}^2 - 2k_z \gamma_2 k_z)$$

$$R = -(\sqrt{3}/2)\bar{\gamma}k_-^2 + (\sqrt{3}/2)\mu k_+^2$$

$$S_\pm = \sqrt{3}k_\pm [(\sigma - \delta)k_z + k_z \pi]$$

$$\Sigma_{\pm} = \sqrt{3}k_{\pm}\{[(1/3)(\sigma-\delta)+(2/3)\pi]k_z \qquad (196)$$
$$+ k_z[(2/3)(\sigma-\delta)+(1/3)\pi]\}$$
$$C = k_-[k_z(\sigma-\delta-\pi)-(\sigma-\delta-\pi)k_z]$$
$$k_{//}^2 = k_x^2 + k_k^2 \qquad k_{\pm} = k_x \pm ik_y \qquad k_z = -i\partial/\partial z$$

Here the quantum well or superlattice is assumed to be grown on a (001) substrate, and the possibility of lattice mismatch has been taken into account in the matrix elements P and Q [73]. In these equations, $\varepsilon_\nu(z)$ is the bulk Γ_8 valence band profile (including hydrostatic but not shear strain), Δ is the spin-orbit splitting, and ξ is the shear strain splitting energy. Note the asymmetry with respect to k_z of $S_{\pm}$, which will be of fundamental importance in the boundary conditions. Also the matrix element C, which is zero in bulk material, introduces a coupling of the states $|J,\pm m_j\rangle$ in the light hole (lh) and split-off (SO) bands. It arises because any change in material composition breaks the local inversion symmetry of the lattice (since the inversion asymmetry of the zinc-blende lattice has already been neglected). A similar term [36] will couple the heavy hole (hh) states if the spin-orbit interaction is included in the Luttinger parameters. Finally, note that $S_{\pm} = \Sigma_{\pm} = \sqrt{3}\gamma_3 k_{\pm} k_z$ in bulk material.

For theoretical work on a quantum well or superlattice, the Hamiltonian given by Eq. (195) is not the most convenient, since a change of basis will reduce it to block diagonal form [74, 75]. One possible choice for the transformation is

$$|u_1\rangle = \alpha\left|\frac{3}{2},-\frac{3}{2}\right\rangle - \alpha^*\left|\frac{3}{2},\frac{3}{2}\right\rangle$$

$$|u_2\rangle = \beta\left|\frac{3}{2},\frac{1}{2}\right\rangle + \beta^*\left|\frac{3}{2},-\frac{1}{2}\right\rangle$$

$$|u_3\rangle = \beta\left|\frac{1}{2},\frac{1}{2}\right\rangle + \beta^*\left|\frac{1}{2},-\frac{1}{2}\right\rangle$$

$$|u_4\rangle = \alpha\left|\frac{3}{2},-\frac{3}{2}\right\rangle + \alpha^*\left|\frac{3}{2},\frac{3}{2}\right\rangle \qquad (197)$$

$$|u_5\rangle = \beta\left|\frac{3}{2},\frac{1}{2}\right\rangle - \beta^*\left|\frac{3}{2},-\frac{1}{2}\right\rangle$$

$$|u_6\rangle = \beta\left|\frac{1}{2},\frac{1}{2}\right\rangle - \beta^*\left|\frac{1}{2},-\frac{1}{2}\right\rangle$$

where

$$\alpha = (1/\sqrt{2})\exp(i\pi/4)\exp[i(\eta+\phi/2)]$$
$$\eta = (1/2)\arctan[(\gamma_3/\gamma_2)\tan(2\phi)] \qquad (198)$$
$$\beta = (1/\sqrt{2})\exp(i3\pi/4)\exp[i(-\eta+\phi/2)]$$
$$\phi = \arctan(k_y/k_x)$$

Note that $\eta = \phi$ for $\vec{k}_{//}$ along the (100) and (110) directions. In other directions, as the basis functions must remain independent of the material composition, an average value is used for η. (This choice is consistent with the approximations made in deriving the effective mass Hamiltonian, from

the exact envelope function equation [55].) The Hamiltonian in the new basis is transformed in a pair of 3×3 blocks,

$$\widehat{H}_\pm = \begin{vmatrix} P+Q & R\mp iS & \sqrt{2}R\pm\frac{i}{\sqrt{2}}S \\ R\pm iS^+ & P-Q\mp iC & \sqrt{2}Q\mp i\sqrt{\frac{3}{2}}\Sigma \\ \sqrt{2}R\mp\frac{i}{\sqrt{2}}S^+ & \sqrt{2}Q\pm i\sqrt{\frac{3}{2}}\Sigma^+ & P+\Delta\pm iC \end{vmatrix} \quad (199)$$

where the upper and lower blocks correspond to the upper and lower signs. Here P and Q are the same as before, but R, S, Σ, and C are altered:

$$R = -\left(\sqrt{3}/2\right)\gamma_\phi k_{//}^2 \qquad \gamma_\phi = \sqrt{\bar{\gamma}^2+\mu^2-2\bar{\gamma}\mu\cos^4\phi},$$

$$S = \sqrt{3}k_{//}[(\sigma-\delta)k_z+k_z\pi]$$

$$C = k_{//}[k_z(\sigma-\delta-\pi)-(\sigma-\delta-\pi)k_z] \qquad (200)$$

$$\Sigma = \sqrt{3}k_{//}\left\{\left[\frac{1}{3}(\sigma-\delta)+\frac{2}{3}\pi\right]k_z+k_z\left[\frac{2}{3}(\sigma-\delta)+\frac{1}{3}\pi\right]\right\}$$

In order to eliminate off-block-diagonal terms involving the matrix elements Σ and C, it was necessary to assume that $\eta = \phi$. Therefore, the Hamiltonian given by Eq. (199) is strictly applicable only along the (100) and (110) directions; nonetheless, it provides an excellent approximation to the in plane angular dependence of Eq. (195), which is entirely neglected in the conventional axial approximation [74]. Also, the transformation given by Eq. (197) was chosen so that the upper and lower block basis functions transform into one another upon reflection in the $z = 0$ plane. Therefore, if the structure under consideration has reflection symmetry, the eigenfunctions of the upper and lower block Hamiltonians will be reflections on one another, and their corresponding eigenenergies will be degenerate.

To determine the boundary conditions [55] on the envelope functions, one integrates the effective mass equation,

$$\widehat{H}\vec{F}(\vec{r}) = \varepsilon\vec{F}(\vec{r}) \qquad (201)$$

across an interface, where $\vec{F}(\vec{r})$ is a three-component envelope function vector and $\widehat{H}$ is either of the blocks in (199). The boundary conditions require the continuity of $\vec{F}$ and $\widehat{B}\vec{F}$, where

$$\widehat{B} = \begin{vmatrix} (\gamma_1-2\gamma_2)\partial/\partial z & \pm2\sqrt{3}\pi k_{//} & \mp\sqrt{6}\pi k_{//} \\ \mp2\sqrt{3}(\sigma-\delta)k_{//} & (\gamma_1+2\gamma_2)\partial/\partial z & -2\sqrt{2}\gamma_2\partial/\partial_z \\ & \pm2(\sigma-\delta-\pi)k_{//} & \pm\sqrt{2}(2\sigma-2\delta+\pi)k_{//} \\ & -2\sqrt{2}\gamma_2\frac{\partial}{\partial_z} & \\ \pm\sqrt{6}(\sigma-\delta)k_{//} & \mp\sqrt{2}(\sigma-\delta+2\pi)k_{//} & \gamma_1\frac{\partial}{\partial_z}\mp2(\sigma-\delta-\pi)k_{//} \end{vmatrix}$$
$$(202)$$

Note that since $\gamma_1-2\gamma_2 = -1+6\pi$, only the small π terms appear in the hh equation, while in the lh and SO equations, the coefficients are dominated by the large value of σ. Thus in all three cases the interband coupling due to the derivative boundary condition is roughly of order $k_{//}/k_z$.

The boundary conditions found in the current literature are significantly different. Prior to the development of the exact envelope function theory, the only guidelines available for the construction of the effective mass Hamiltonian were that it must be hermitian and that it must reduce to the correct bulk form away from an interface. This ambiguity allows for an infinite set of possible boundary conditions, but the most common choice in the literature is obtained from a "symmetrized" Hamiltonian [72, 73] in which all the linear operators in k_z are written in the form $(1/2)[k_z f(z) + f(z)k_z]$. Applying the symmetrization procedure to the Hamiltonian given by Eq. (199) is equivalent to substituting $\gamma_3/2$ for both $\sigma - \delta$ and π in the matrix elements S, Σ and C, thus changing the matrix $\widehat{B}$ to

$$\widehat{B} = \begin{vmatrix} (\gamma_1 - 2\gamma_2)\frac{\partial}{\partial_z} & \pm\sqrt{3}\gamma_3 k_{//} & \mp\sqrt{\frac{3}{2}}\gamma_3 k_{//} \\ \mp\sqrt{3}\gamma_3 k_{//} & (\gamma_1 + 2\gamma_2)\frac{\partial}{\partial_z} & -2\sqrt{2}\gamma_2\frac{\partial}{\partial_z} \pm \frac{3}{\sqrt{2}}\gamma_3 k_{//} \\ \pm\sqrt{\frac{3}{2}}\gamma_3 k_{//} & -2\sqrt{2}\gamma_2\frac{\partial}{\partial_z} \mp \frac{3}{\sqrt{2}}\gamma_3 k_{//} & \gamma_1\frac{\partial}{\partial_z} \end{vmatrix}$$

(203)

The most important change caused by the symmetrization occurs in the boundary condition for the hh band, where the large quantity σ is incorporated into the coupling terms. The use of symmetrized boundary conditions will therefore overestimate the magnitude of this interband coupling, which can lead to qualitative errors in the band structure, as is shown below. The changes in the lh and SO boundary conditions are not very significant since the resulting underestimate of the coupling causes only small numerical modifications.

An example of the consequences of the symmetrized boundary conditions is presented in Figure 24, where the zone center effective masses of the heavy hole subbands for an $In_{0.2}Ga_{0.8}As$–GaAs quantum well are plotted versus quantum well width. Results are given for three sets of boundary conditions: correct, symmetrized, and uncoupled. (In the approximation in which the Luttinger parameters are independent of position, the boundary conditions are not coupled.) On a physical basis, one would expect the coupled solution for the effective mass to differ from the uncoupled solution by only a small amount (i.e., something on the order of the relative change in Luttinger parameters at the interface), and indeed this is true for the correct boundary conditions. However, the exaggeration of the interband coupling due to the symmetrized boundary conditions causes the mass of the hh_2 subband to

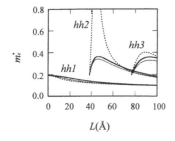

Fig. 24. Zone center heavy hole subband effective masses versus quantum well width for an $In_{0.2}Ga_{0.8}As$–GaAs quantum well. Three different boundary conditions are shown: correct (solid line), symmetrized (dashed line), and uncoupled (dotted line). The symmetrized hh_2 mass reaches a peak value of 1.42, more than triple the mass obtained from the correct boundary conditions.

become a very volatile function of both the physical dimensions of the quantum well and the precise numerical values chosen for the Luttinger parameters. This nonphysical behavior clearly demonstrates that the choice of Hamiltonian is not arbitrary, and that the correct boundary conditions are required for physically reasonable solutions.

In conclusion, the effective mass Hamiltonian given by the exact envelope function theory has been evaluated for the valence band. In the boundary conditions obtained from this Hamiltonian, the heavy hole states are coupled to the light hole and split-off states only through the interaction with remote bands of p symmetry. The symmetrized boundary conditions implicitly include the much larger interaction with s states. This exaggerates the interband coupling; consequently, nonphysical solutions are obtained for the heavy hole subbands in which the quantum well effective mass is very sensitive to the change in Luttinger parameters at the well–barrier interface.

4.4.2. Effective Mass and Luttinger Parameters in Semiconductor Quantum Wells

Calculations based on the envelope function approximation are described. Two classes of model are employed within the envelope function approach [76]:

(1) The single band effective mass model: although the couplings between electrons in different bands are not included explicitly in this kind of model, it still gives a good description of band edge states for a large quantum well system provided an appropriate choice of the effective masses is made.

(2) The multiband $\vec{k} \cdot \vec{p}$ model: perturbation theory in which a small set of coupled center zone states of the different bands are used to give a description of the band structures and other related electronic properties of bulk materials and corresponding heterostructures.

In the class (1) model, the Schrödinger equation for a quantum well is reduced to a one-dimensional problem which for definiteness was taken to have the form

$$\left\{ -\frac{1}{2}\frac{\partial}{\partial z}\frac{1}{m^*(z)}\frac{\partial}{\partial z} + V_{ext}(z) \right\} f_i(z) = \varepsilon_i f_i(z)$$

(204)

where $V_{ext}(z)$ is the quantum potential which represents the band edge changes of the bulk carriers between the two materials of the quantum well structure. Under zero bias conditions, $V_{ext}(z)$ is constant in each material and its change at the interface is defined in terms of the band offset. The envelope function of the ith state of a carrier in the quantum well is $f_i(z)$ and $m^*(z)$ is the effective mass (taken as a constant within each material comprising the quantum well). It is worth noting that utilization of Eq. (204) automatically implies a boundary condition of the form

$$\frac{1}{m^*}\frac{df_i(z)}{dz}\bigg|_{\text{interface 1}} = \frac{1}{m^*}\frac{df_i(z)}{dz}\bigg|_{\text{interface 2}}$$

(205)

i.e., $f_i(z)$ and $(1/m^*)df_i(z)/dz$ are continuous at the interface of the quantum well.

In the class (2) model, the states of the quantum well are expanded in terms of linear combinations of the periodic parts of the zone center Bloch function appropriate to bulk materials: i.e.,

$$f(\vec{r}) = \exp(i\vec{k}_\perp \cdot \vec{r}) \sum_{l=1}^{N} u_{l\Gamma}(\vec{r}) f_l(z), \qquad (206)$$

where $\vec{k}_\perp$ is the in plane wavevector, $u_{l\Gamma}(\vec{r})$ is the part of the zone center Bloch function with the period of the bulk unit cell, and $f_l(z)$ is the envelope function. At this stage, two assumptions are made: (1) $u_{l\Gamma}(\vec{r})$ takes the same form in both well and barrier materials. (2) $f_l(z)$ varies slowly on the scale of the unit cell appropriate to bulk material.

Substituting Eq. (206) into the Schrödinger equation for the quantum well and utilizing the property of $u_{l\Gamma}(\vec{r})$, a set of N coupled partial differential equations can be established. In the present work, the $\vec{k} \cdot \vec{p}$ model developed by Ekenberg et al. [77] for the quantum well system is used and emphasis is set on the hole energy levels. In particular, the Luttinger parameters are directly related to the band edge effective masses through the relation [78]

$$\frac{m_0}{m_{hh}^*[100]} = \gamma_1 - 2\gamma_2 \qquad \frac{m_0}{m_{lh}^*[100]} = \gamma_1 + 2\gamma_2$$

$$\frac{m_0}{m_{hh}^*[111]} = \gamma_1 - 2\gamma_3 \qquad \frac{m_0}{m_{lh}^*[111]} = \gamma_1 + 2\gamma_2$$

$$\frac{m_0}{m_e^*} = \alpha + \frac{\varepsilon_p(\varepsilon_g + \frac{2}{3}\Delta)}{\varepsilon_g(\varepsilon_g + \Delta)} \qquad (207)$$

$$\frac{m_0}{m_{3-0}^*} = \gamma_1 - \frac{\varepsilon_p \Delta}{3\varepsilon_g(\varepsilon_g + \Delta)}$$

where γ_1, γ_2, γ_3, α, and ε_p are the Luttinger parameters [35]. ε_g and Δ are the bandgap and the spin-orbit gap, respectively. The m^*'s are the effective masses along the various different directions. It is well known that the Luttinger parameters reflect the effects of a microscopic potential in the form of the $V(\vec{k})$ and that, in general, they are energy dependent. Consequently, even if the Luttinger formalism could be applied to quantum well structures (a feature which has been recently shown is questionable), the energy dependence of the parametrization scheme would need to be incorporated into the formalism, particularly for narrow wells.

5. EFFECT OF STATIC EXTERNAL ELECTRIC AND MAGNETIC FIELDS ON THE QUASI-PARTICLE ENERGY LEVELS IN THE Q2D SYSTEMS

In this section we deal with the effects of a static external electric field on the energy levels (Stark effects) of the different Q2Ds, which have importance due to the possibility of making fast electro-optical devices.

5.1. Stark Effect in Quasi-Two-Dimensional Structures

In the following discussion band structure effects are ignored. The carriers are characterized by an isotropic effective mass, which is assumed to be position independent.

Two kinds of Stark effects can be found in quasi-two-dimensional systems, depending on whether the electric field $\vec{E}$ is applied parallel to the growth (z) axis (longitudinal Stark effect) or perpendicular to it (transverse Stark effect). Usually, the Q2D nanostructures designed for Stark effect experiments are insulated. The field is applied either via a Schottky barrier [79, 80] or by inserting the structure into the intrinsic part of a positive-intrinsic-negative junction [81] (longitudinal Stark effect). In order to study the transverse Stark effect, contacts have to be diffused throughout the structure. As a result of these techniques, the magnitude of the electric field is difficult to assess precisely and may, moreover, show inhomogeneities. In some experiments, the electric field was estimated with $\approx$ 30% accuracy [80]. In others, the uncertainty was ± 10 kV/cm over the whole field range investigated [81, 82]. Longitudinal field in excess of 10^5 V/cm could be applied on multiple quantum well structures either by biasing a $p.i.n$ diode [82] or by using a Schottky contact [80]. These fields bend the band by 0.1 eV.

5.1.1. Transverse Stark Effect in a Quantum Well ($\vec{E}//\hat{e}_x$)

Classically, due to the presence of the confining barriers (the unbound quantum well states are not considered), the z motion is bound ($|z| < L/2$) and periodic upon the time. Along the electric field direction (x axis), the carrier is uniformly accelerated whereas along the y direction the particle is free. The total energy of the carrier is a constant of the motion and can be split into contributions from the x, y, z motions, respectively:

$$\varepsilon = \varepsilon_x + \varepsilon_y + \varepsilon_z = \frac{\hat{p}_x^2}{2m_e^*} + eEx + \frac{\hat{p}_y^2}{2m_e^*} + \frac{\hat{p}_z^2}{2m_e^*}$$
$$+ V_b\left(z^2 - \frac{L^2}{4}\right) \qquad (208)$$

The allowed classical motions should be such that $p_x^2 \geq 0$; thus the x motion is limited to

$$x < \frac{1}{eE}\varepsilon_x \qquad (209)$$

The envelope function of the stationary states for the quantum motion can be factorized into

$$f(\vec{r}) = \varphi(x)\chi_n(z)\exp(ik_y y) \qquad (210)$$

where $\chi_n(z)$ is the nth quantum well bound state wavefunction (energy ε_n) which has been previously determined and the plane wave takes care of the free motion along the y axis. Consequently

$$\varepsilon = \varepsilon_x + \frac{\hbar^2 k_y^2}{2m_e^*} + \varepsilon_n \qquad (211)$$

and $\varphi(x)$ is the solution of the one-dimensional Schrödinger equation

$$\left\{ -\frac{\hbar^2}{2m_e^*}\frac{d^2}{dx^2} + eEx \right\}\varphi(x) = \varepsilon_x\varphi(x) \qquad (212)$$

with the boundary condition that $|\varphi(x = \pm\infty)|$ be finite. The change of variable

$$x = \frac{1}{eE}\varepsilon_x - \left(\frac{\hbar^2}{2m_e^*eE}\right)^1/3\eta \qquad (213)$$

results in the transformation of Eq. (212) into the Airy equation

$$\frac{d^2\varphi}{d\eta^2} + \eta\varphi = 0 \qquad |\varphi(\pm\infty)| \text{ bound} \qquad (214)$$

whose solutions are

$$\varphi(\eta) = \begin{cases} \sqrt{\dfrac{|\eta|}{3\pi}}K_{1/3}\left(\dfrac{2}{3}|\eta|^{3/2}\right) & \eta \le 0 \\[2mm] \dfrac{1}{3}\sqrt{\pi\eta}\left[J_{-1/3}\left(\dfrac{2}{3}\eta^{3/2}\right) + J_{1/3}\left(\dfrac{2}{3}\eta^{3/2}\right)\right] & \eta \ge 0 \end{cases} \qquad (215)$$

where $J_\nu(\eta)$ and $K_\nu(\eta)$ are the Bessel functions of order ν with real and imaginary arguments, respectively. The wavefunction decays exponentially in the classically forbidden region ($\eta \le 0$):

$$\varphi(\eta) \approx |\eta|^{-1/4}\exp\left[-\frac{2}{3}|\eta|^{3/2}\right] \qquad \eta \to -\infty \qquad (216)$$

In the classically accessible region the wavefunction oscillates as

$$\varphi(\eta) \approx |\eta|^{-1/4}\sin\left[\frac{2}{3}|\eta|^{3/2} - \frac{\pi}{4}\right] \qquad \eta \to \infty \qquad (217)$$

Since the potential energy eEx is negative and arbitrarily large for negative x, the whole energy spectrum of Eq. (214) is continuous and an allowed motion may be associated with any value of the energy ε_x. This means that below the $E = 0$ edge (i.e., $\varepsilon_x = 0$), allowed energy states can be found at finite E. Symmetrically, above the $E = 0$ edge of valence electrons one may find allowed valence levels at finite E. In other words, the electric field suppresses the bandgap (see Fig. 25).

5.1.1.1. Excitons in Electric Field Applied in the Layer Plane

In a purely two-dimensional system, the exciton Hamiltonian in the presence of an electric field ($\vec{E}//\hat{e}_x$) reads

$$\begin{aligned}\widehat{H}_{\text{exc}} = {} & \frac{\hat{p}_{x_e}^2}{2m_e^*} + \frac{\hat{p}_{y_e}^2}{2m_e^*} + \frac{\hat{p}_{x_h}^2}{2m_h^*} + \frac{\hat{p}_{y_h}^2}{2m_h^*} \\ & - \frac{e^2}{\kappa\sqrt{(x_e - x_h)^2 + (y_e - y_h)^2}} + eE(x_e - x_h) \\ & + V_{be} + V_{bh} \end{aligned} \qquad (218)$$

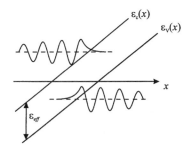

Fig. 25. Conduction and valence band profiles along the $\hat{e}_x$ direction for the transverse Stark effect geometry. ε_{eff} is an effective bandgap separating two conduction and valence subbands, which arise from the size quantization along the $\hat{e}_z$ direction. Two Airy functions are drawn for conduction and valence states, respectively.

As usual, $\widehat{H}_{\text{exc}}$ can be separated into contributions from the center of mass and relative motions:

$$\widehat{H}_{\text{exc}} = \frac{\hat{p}_x^2 + \hat{p}_y^2}{2(m_e^* + m_h^*)} + \frac{\hat{p}_x^2 + \hat{p}_y^2}{2\mu^*} - \frac{e^2}{\kappa r} + eEx + V_{\text{bexc}} \qquad (219)$$

If center of mass degrees of freedom are discarded, the ground state of the relative motion at $E = 0$ is bound by $4R_{y,\text{exc}}^*$, where $R_{y,\text{exc}}^*$ is the three-dimensional effective Rydberg energy, the two-dimensional Bohr radius being equal to one-half of the three-dimensional one a_B^*. Figure 26 shows the potential energy profile along the x axis for the relative motion at both $E = 0$ and $E \ne 0$. It can be seen that if the potential energy difference induced by the field over one Bohr radius ($eEa_B^*/2$) is comparable to the zero field binding energy ($4R_{y,\text{exc}}^*$), the relative motion becomes unbound on the negative x side. This result is qualitatively the same as that found in bulk materials, where it is known that a critical field $E_c \approx R_{y,\text{exc}}^*/ea_B^*$ ionizes the exciton.

Since quasi-two-dimensional excitons are more tightly bound than the corresponding bulk ones, they will be field-ionized at larger fields. Typically, a factor of 2–4 on E_c is

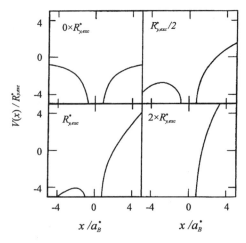

Fig. 26. The potential energy profiles along the $\hat{e}_x$ direction for the relative motion of a two-dimensional exciton subjected to a static electric field $E//\hat{e}_x$ are shown for four different magnitudes of the applied field $E : eEa_B^* = 0$, $0.5R_{y,\text{exc}}^*$, $R_{y,\text{exc}}^*$, and $2R_{y,\text{exc}}^*$ respectively.

gained by decreasing the quantum well thickness: the excitonic binding is increased by a factor of 2–3 and the effective exciton Bohr radius shrinks from a_B^* to $\approx 0.7\, a_B^*$. Using appropriated numbers for GaAs, it is found that $E_c \approx 1.9 \times 10^4$ V/cm in GaAs quantum wells. Chemla et al. [82] have performed a detailed study of electroabsorption effects in multiple GaAs–Ga$_{1-x}$Al$_x$As quantum well structures at room temperature. Their results for the transverse configuration are shown in Figure 27. Clearly at $E = 1.6 \times 10^4$ V/cm the light hole excitonic resonance has been washed out and the heavy hole excitonic structure is already significantly blurred. At $E = 4.8 \times 10^4$ V/cm the absorption spectrum is featureless.

5.1.1.2. Longitudinal Stark Effect in a Quantum Well ($\vec{E} // \hat{e}_z$)

The potential energy profile for such a configuration is shown in Figure 28. The genuine feature of the longitudinal configuration is the appearance of a finite dipole D between the electron and the hole of a photocreated e–h pair. Classically, the lowest lying state of the conduction electron is to remain immobile at the left-hand corner of the tilted quantum well. Since valence electrons have a negative mass, the topmost valence state corresponds to a carrier which remains immobile on the right-hand corner. Compared with the classical results, the wave mechanical description brings to light several new features.

(i) The confinement over a distance L leads to size quantization. At zero field the lowest lying conduction and valence states are ε_{e1} and ε_{h1} (the electron and hole ground states confinement energies), respectively.

(ii) The expectation value of the dipole operator $D = |e|(z_h - z_e)$ can be smaller or larger than L. This is due to the fact that the wavefunction leaks into the barrier and spreads over the whole well.

(iii) In principle this problem has no stationary bound states since the potential energy eEz becomes arbitrarily large and negative (positive) for the conduction (valence) electrons when $z \to -\infty\ (z \to +\infty)$.

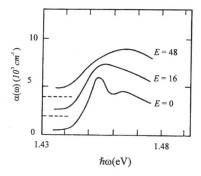

Fig. 27. Room temperature electroabsorption in GaAs–Ga$_{0.68}$Al$_{0.3}$As multiple quantum wells. $E // \hat{e}_x$ (kV/cm).

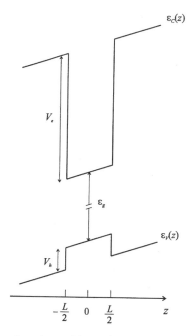

Fig. 28. Position dependence of the conduction and valence bands edges of a type I quantum well subjected to a longitudinal electric field $E(E // \hat{e}_x)$.

Over a large field range, however, the field induced ionization of the quantum well remains of secondary importance. Let us respectively denote the confinement energy, conduction barrier height, and characteristic penetration length of the ground conduction state in the barrier at zero field by ε_{e1}^0, V_{eb}, ν_e^{-1}, respectively. The conduction electron escape toward $z = -\infty$ will remain unimportant if the barrier drop over the penetration length $eE\nu_e^{-1}$ is much smaller than the effective barrier height for the ε_{e1}^0 state $V_{eb} - \varepsilon_{e1}^0$ [83],

$$eE\nu_e^{-1} \ll V_{eb} - \varepsilon_{e1}^0 \tag{220}$$

In this way, the effective barrier height does not change in the barrier region where the conduction electron is mostly localized (the integrated probability of finding the carrier between $-\nu_e^{-1} - (L/2)$ and $-L/2$ is 86% of the integrated probability of finding it over the whole left-hand side barrier).

Figure 29 sketches the qualitative evolution of the ground state wavefunction of the conduction electrons in a quantum well, which are subjected to a longitudinal field. The zero of the energy is taken at the center of the well, which is also the origin of coordinates. At zero field (Fig. 29a), the average electron position in the ground states $\langle z_e \rangle$ is zero. By increasing E, $\langle z_e \rangle$ becomes negative. At low field (Fig. 29b), $\langle z_e \rangle$ is proportional to E. The ground state energy ε_{e1} experiences a downward shift $\Delta\varepsilon_{e1}$ that is equal to half of the product of the induced dipole $-e\langle z_e \rangle$ by the field. This is the domain of the quadratic Stark shift. At larger fields, the carriers tend to accumulate near the left-hand side barrier (Fig. 29c). The induced dipole tends to saturate, as does $\Delta\varepsilon_{e1}$. If the barrier is infinitely high, this situation prevails for arbitrarily large E.

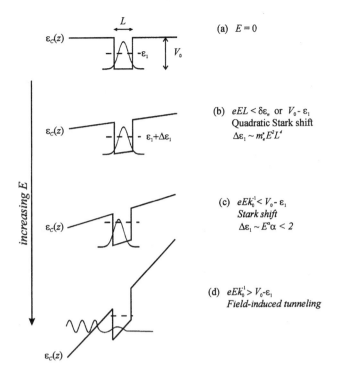

Fig. 29. Qualitative evolution with increasing field strength ($E//\hat{e}_z$) of the ground bound state of a quantum well subjected to a longitudinal electric field. (a) $E = 0$, (b) quadratic Stark shift, (c) carrier accumulation regime, and (d) field ionization of the quantum well.

Instead of spreading over the whole well the carrier becomes essentially localized by a triangular well. Its energy varies as

$$\varepsilon_{e1}(E) = -eE\frac{L}{2} + eEL\left(\frac{\hbar^2}{2m_e^*eEL^3}\right)^{1/3}\eta_0 \qquad (221)$$

where η_0 is the first positive zero of the Airy function ($\eta_0 \approx 2.33$) [84]. It is worth pointing out that ε_{e1} becomes negative when E is sufficiently large and the carrier is swept out from the quantum well (Fig. 29d).

The conduction electron Hamiltonian which describes the longitudinal Stark effect in a quantum well reads

$$\widehat{H} = \widehat{H}_\perp + \widehat{H}_{//} = \frac{\hat{p}_x^2 + \hat{p}_y^2}{2m_e^*} + \frac{\hat{p}_z^2}{2m_e^*} + eEz_e + V_{eb} \qquad (222)$$

The in plane motion ($\widehat{H}_\perp$) is free and decouples from the z motion ($\widehat{H}_{//}$). Let $\chi_j^0(z)$ and ε_j^0 denote the zero field eigenfunctions and eigenenergies of $\widehat{H}_{//}$. At weak field the quadratic Stark shift of the ground state may be sought by a perturbation calculus

$$\Delta\varepsilon_{e1} = -e^2E^2\sum_{n\neq 1}\frac{|z_{1n}|^2}{\varepsilon_n^0 - \varepsilon_1^0} \qquad (223)$$

where

$$z_{1n} = \int_{-\infty}^{+\infty}\chi_1^0 z\chi_n^0\,dz \qquad (224)$$

The summation in Eq. (223) runs over all the quantum well states (bound or unbound). Thus $\Delta\varepsilon_{e1}$ is quite tedious to

evaluate. However, it appears that the dominant contribution is due to the $n = 1 \rightarrow n = 2$ virtual transitions when the $n = 2$ state is bound [83]. Since $\varepsilon_{e2}^0 - \varepsilon_{e1}^0 \approx \hbar^2/2m_e^*L^2$, $\Delta\varepsilon_{e1} \approx E^2m_e^*L^4$; thus it increases with L rapidly. However, the field range where Eq. (223) is justified is such that $\Delta\varepsilon_{e1} \ll \varepsilon_{e2}^0 - \varepsilon_{e1}^0$. Thus it narrows quite rapidly with L and is such that $E^2m_e^{*2}L^6$ is constant.

Finally, it should be noticed that the exact eigenstates of $\widehat{H}_1$ might be obtained. Within each layer, $\chi_1(z)$ is a linear combination of the two independent solutions of the Airy equation. Thus, the eigenstates are obtained by matching conditions at the $z = \pm L/2$ interfaces. However, the Airy functions are too complicated for convenient use.

5.1.2.1 Longitudinal Stark Effect for an Exciton

Only the lowest lying heavy hole ($\varepsilon_1 - hh_1$) and light hole ($\varepsilon_1 - lh_1$) excitons will be considered. The electric field polarizes ε_1 and hh_1 (lh_1) along opposite directions and thus weakens the excitonic binding. However, and this is the major advantage of the longitudinal configuration, the exciton dissociation is considerably hindered by the conduction and valence potential barriers. It is expected that the exciton binding energy $\varepsilon_b(E)$ shows little variation with E and reflects the carrier accumulation at the interfaces. However, if the field becomes so large that the carriers are swept out of the quantum well (Fig. 29d), the exciton will finally be field ionized. In Figure 27 it was roughly seen that 20 kV/cm is enough to wash out the excitonic resonance in the transverse configuration. Chemla et al. [82] have demonstrated that fields as large as 100 kV/cm can be applied in the longitudinal configuration without destroying the excitonic binding.

The quadratic excitonic Stark shift is followed by smoother field dependence, which is reminiscent of the carrier accumulation regime. The electron and the hole tend to accumulate on each side of the quantum well. Thus, D tends to saturate. If the carrier accumulation were complete (impenetrable barriers), the excitonic binding energy would extrapolate to that of an exciton whose electron and hole move on two different planes separated by L.

However, it is known that the carrier wavefunctions leak more and more heavily into the finite barriers when E increases. Ultimately, the carriers are swept out of the quantum well and the exciton finally is field-ionized.

The exciton Hamiltonian in the presence of a longitudinal electric field is written as

$$\widehat{H}_{\text{exc}} = \varepsilon_g + \frac{\hat{p}_\perp^2}{2\mu^*} + \frac{\hat{p}_{ze}^2}{2m_e^*} + \frac{\hat{p}_{zh}^2}{2m_h^*} + V_{eb} + V_{hb}$$
$$+ eE(z_e - z_h) - \frac{e^2}{\kappa[r^2 + (z_e - z_h)^2]^{1/2}} \qquad (225)$$

where the center of mass kinetic energy term has been dropped. The approximated solutions of Eq. (225) when e^2/κ is set equal to zero are already known: $\varepsilon_g + \varepsilon_1(E) + hh_1(E)$ for the lowest lying state of an e–h pair. Although the z and (x, y)

motions are not exactly separable, it has already seen previously that the exciton z motion is forced by the quantum well barriers, while the coulombic term binds the e–h pair in the layer plane. Thus, a reasonable trial wavefunction for the heavy hole exciton ground state will be [85]

$$f_{\text{exc}}(\vec{r}_e, \vec{r}_h) = A\chi_{\varepsilon_1}(z_e)\chi_{hh_1}(z_h)\exp\left(-\frac{r}{\lambda}\right) \qquad (226)$$

where λ is the variational parameter, r is the in plane e–h distance, A is a normalization constant, and $\chi_{\varepsilon_1}(z_e)\chi_{hh_1}(z_h)$ are the field-dependent wavefunctions for the electron and hole which move in the quantum well tilted by the electric field. Since the exciton wavefunction is taken as separable in r and z_e, z_h, the required minimization amounts to finding the minimum over λ of

$$\hbar\omega_{hh}(\lambda) = \varepsilon_g + \varepsilon_1(E) + hh_1(E) + \frac{\hbar^2}{2\mu\lambda^2}$$
$$- \frac{2e^2}{\kappa\lambda}\int_0^\infty x\,dx\,\exp(-x) \qquad (227)$$
$$\times \int\int_{-\infty}^{+\infty}\frac{dz_e\,dz_h\,\chi_{\varepsilon_1}^2(z_e)\chi_{hh_1}^2(z_h)}{\sqrt{x^2 + \frac{4}{\lambda^2}(z_e - z_h)^2}}$$

Note, however, that as $\varepsilon_1(E), hh_1(E)$ are not known exactly (except at $E = 0$), there is no guarantee that $[\varepsilon_g + \varepsilon_1(E) + hh_1(E) - \text{Min}_\lambda(\hbar\omega_{hh}(\lambda))]$ will be a lower bound to $\varepsilon_{hh}(E)$. Figures 30 and 31 show the E dependence at fixed L or the L dependence at fixed E of the heavy hole exciton binding energy ε_b^{hh} in GaAs–Ga$_{0.68}$Al$_{0.32}$As quantum wells. In the $\varepsilon_b^{hh}(E)$ curves, one clearly sees that the quadratic Stark shift decreases at low field, becoming steeper when L increases and becomes observable on a narrower field range. This regime is followed by a weaker E dependence, reminiscent of the carrier accumulation regime.

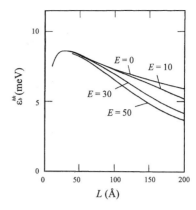

Fig. 31. Calculated variations of the heavy hole exciton binding energy ε_b^{hh} versus the quantum well thickness L for different field strengths in GaAs–Ga$_{0.68}$Al$_{0.32}$As ($E//\hat{e}_z$).

5.2. Effect of Static External Magnetic Field on the Quasi-Particle Energy Levels in the Q2D Structures

5.2.1. Magnetic Field Effect in the Quasi-Two-Dimensional System

The study of the magnetic field effects (Landau quantization) gives access to fine characterization methods of conducting in the quasi-two-dimensional structures through cyclotron resonance or Shubnikov–de Haas experiments [86]. At the same time, they prove to be difficult to understand. As will be seen below, a magnetic field $\vec{B}$ completely reorganizes the energy spectrum of quasi-two-dimensional electrons by splitting the continuum of the low-dimensional structures density of states ($B = 0$) into highly degenerate levels which are separated by finite bandgaps ($B \neq 0$). Thus, what was a metallic system at $B = 0$ is now either metallic or insulating at finite B depending on whether the Fermi level sits in the gaps or in the levels. The existence of finite bandgaps is responsible for the quantization of the Hall conductivity into multiples of the fundamental constant $e^2/\hbar$. The quantized Hall effect has become one of the most spectacular experimental effects discovered in solid state physics in the last three decades.

In the following discussion band structure effects are ignored. The carriers are characterized by an isotropic effective mass, which is assumed to be position independent.

In this topic, the motion of a quasi-two-dimensional electron under the action of a static and uniform magnetic field $\vec{B}$, which forms an angle θ with respect to the growth z axis of the structure (see Fig. 32), is considered. $\vec{B}$ is thus given by

$$\vec{B} = (0, B\sin\theta, B\cos\theta) \qquad (228)$$

It can be shown that the Lorentz force $(q/c)\vec{v} \times \vec{B}$ acting on a particle of charge q can be derived by using the Hamilton equations

$$\frac{d\vec{r}}{dt} = \frac{\partial H}{\partial \vec{p}} \qquad \frac{d\vec{p}}{dt} = -\frac{\partial H}{\partial \vec{r}} \qquad (229)$$

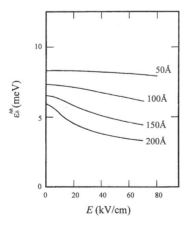

Fig. 30. Calculated variations of the heavy hole exciton binding energy ε_b^{hh} versus the electric field $E(E//\hat{e}_z)$ in GaAs–Ga$_{0.68}$Al$_{0.32}$As quantum wells of different thicknesses.

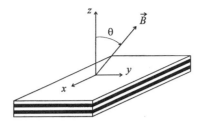

Figure 32

provided the particle momentum $\vec{p}$ is replaced by Π in the classical Hamiltonian. The vector Π is equal to

$$\Pi = \vec{p} - \frac{q}{c}\vec{A} \tag{230}$$

where $\vec{A}$ is the vector potential associated to the magnetic field $\vec{B}$ by the expression

$$\vec{B} = \nabla \times \vec{A} \tag{231}$$

and q is the algebraic charge of the carrier ($q = -e$ for electrons). Unfortunately $\vec{A}$ is only a mathematical device which in fact is not very convenient since, if $\vec{A}$ is replaced by $\vec{A}' = \vec{A} + \nabla f$ with f an arbitrary differentiable scalar function, the magnetic field $\vec{B}' = \nabla \times \vec{A}'$ coincides with $\vec{B}$. Thus, there is a large ambiguity in the choice of the vector potential that corresponds to a given field $\vec{B}$. Since only the $\vec{B}$ field retains a physical significance, the results should only depend on $\vec{B}$ and not on the precise choice of $\vec{A}$, i.e. they should be gauge invariant. In the following

$$\vec{A} = (Bz \sin\theta, Bx \cos\theta, 0) \tag{232}$$

will be used, as well as the notation

$$B \sin\theta = B_\perp \qquad B \cos\theta = B_{//} \tag{233}$$

$$\omega_{c\perp} = \frac{eB_\perp}{m_e^* c} \qquad \omega_{c//} = \frac{eB_{//}}{m_e^* c} \tag{234}$$

The electron spin σ interacts with the field adding to the particle energy the quantity

$$g^*\mu_B \sigma \cdot \vec{B} \tag{235}$$

where $\mu_B = e\hbar/2m_0 c$ is the Bohr magneton and g^* is the effective Landé g factor of the quasi-particle ($g^* = 2$ for electrons in the vacuum, $g^* > 0$ or $g^* < 0$ in the semiconductors). The three cartesian components of the vector operator σ have the eigenvalues $\pm 1/2$.

The electron Hamiltonian is expressed by

$$\widehat{H} = \frac{1}{2m_e^*}\left(-i\hbar\frac{\partial}{\partial x} + \frac{e}{c}B_\perp z\right)^2 + \frac{1}{2m_e^*}\left(-i\hbar\frac{\partial}{\partial y} + \frac{e}{c}B_{//}x\right)^2$$
$$+ \frac{\hat{p}_z^2}{2m_e^*} + V_b(z) + g^*\mu_B\sigma \cdot \vec{B} \tag{236}$$

Thus, the envelope wavefunction for $\widehat{H}$ can be written as a product of functions depending on separated orbital and spin variables

$$f(\vec{r}, \sigma) = f(\vec{r})\alpha(\sigma) \tag{237}$$

Suppose σ is quantized along $\overline{B}$. Let us denote by $|\uparrow\rangle$ and $|\downarrow\rangle$ the two eigenvectors of σ_x and by $|\nu\rangle$ the eigenstates of the orbital part of $\widehat{H}$, associated with the eigenvalue ε_V. The eigenstates and eigenenergies of Eq. (236) are obtained in the following form:

$$f(\vec{r}, \sigma) = \langle\vec{r}\,|\,\nu\rangle \otimes |\uparrow\rangle \qquad \varepsilon_\nu + \frac{1}{2}g^*\mu_B B \tag{238}$$

$$f(\vec{r}, \sigma) = \langle\vec{r}\,|\,\nu\rangle \otimes |\downarrow\rangle \qquad \varepsilon_\nu - \frac{1}{2}g^*\mu_B B \tag{239}$$

The spin term is rotationally invariant. Therefore, if σ is quantized along a fixed direction (say z) which is not necessarily colinear to $\vec{B}$, it will obtain (after diagonalizing $g^*\mu_B\sigma \cdot \vec{B}$) the same spin eigenvalues $\pm(1/2)g^*\mu_B B$. Thus the spin splitting $g^*\mu_B B$ depends on the modulus of $\vec{B}$ but not on its direction.

Since the spin effects are now well known, let us put the attention on the search of the eigenvalues ε_ν. The spin effects, however, will have to be taken into account when evaluating the carrier density of states.

The enveloped wavefunction for the orbital part of $\widehat{H}$, considering that the motion in y is independent, can be written as

$$f_\nu(\vec{r}) = \frac{1}{\sqrt{L_y}}\exp(ik_y y)\varphi_\mu(x, z) \tag{240}$$

Generally ($\theta \neq 0$, $V_b(z)$ arbitrary), the dependence on the variables x and z cannot be separated. However, except when θ is close to $\pi/2$, the separability between x and z motions appears to be a good zero order approximation. In fact, let us rewrite $\widehat{H}$ in the form

$$\widehat{H} = \widehat{H}_{//} + \widehat{H}_\perp + \delta\widehat{H} \tag{241}$$

where

$$\widehat{H}_{//} = \frac{\hat{p}_x^2}{2m_e^*} + V_b(z) \tag{242}$$

$$\widehat{H}_\perp = \frac{\hat{p}_x^2}{2m_e^*} + \frac{1}{2m_e^*}\left(\hbar k_y + \frac{e}{c}B_{//}x\right)^2 \tag{243}$$

$$\delta\widehat{H} = \delta\widehat{H}_1 + \delta\widehat{H}_2 = \frac{1}{2m_e^* c^2}e^2 B_\perp^2 z^2 + \hat{p}_x z\frac{eB_\perp}{m_e^* c} \tag{244}$$

and where Eq. (240) was used. If the splitting of the original Hamiltonian proves to be correct, the zero order energies (i.e., those of $\widehat{H}_{//} + \widehat{H}_\perp$) will only depend on $\widehat{B}_{//}$ and not on $\widehat{B}_\perp$. This means that, in contrast to the spin term, the eigenvalues ε_ν will depend on the component of $\vec{B}$ parallel to the growth axis.

An immediate experimental check of this property for a given heterostructure is to rotate the sample with respect to B and to see whether or not the physical property (of orbital origin) which is under study depends on $B_{//}$, i.e., on the angle θ. For a three-dimensional isotropic system with vanishing $V_b(z)$, it is very easy to show that the eigenvalues of $\widehat{H}$ only depend on the modulus of $\vec{B}$ and not on its direction. Thus, provided the decoupling of $\widehat{H}$ into a main separable term and a small nonseparable one is valid, a clear proof of the existence of quasi-two-dimensional electron gas in a given heterostructure is obtained by studying the angular dependence of magnetic field dependent properties.

Let us then examine the validity of the decoupling procedure. First, the eigenstates of $\widehat{H}_{//} + \widehat{H}_\perp$ are investigated. The wavefunctions $\varphi_\mu(x, z)$ factorize into,

$$\varphi_\mu(x, z) = \chi_m(z)\varphi_n(x) \tag{245}$$

where

$$\widehat{H}_{//}\chi_n(z) = \xi_m\chi_m(z) \tag{246}$$

and

$$\begin{aligned}\widehat{H}_\perp\varphi_n(x) &= \left\{\frac{\hat{p}_x^2}{2m_e^*} + \frac{1}{2m_e^*}\left(\hbar k_y + \frac{eB_{//}}{c}x\right)^2\right\}\varphi_n(x) \\ &= \varepsilon_n\varphi_n(x)\end{aligned} \tag{247}$$

$\chi_m(z)$ are the wavefunctions for the z motion which correspond to the mth state (bound or unbound) of energy ξ_m for the heterostructure Hamiltonian at zero magnetic field. The functions $\varphi_n(x)$ are the eigensolutions of an harmonic oscillator problem (frequency ω_q) centered at

$$x_0 = -\frac{\lambda^2 k_y}{\cos\theta} \tag{248}$$

where λ is the magnetic length

$$\lambda = \sqrt{\frac{\hbar c}{eB}} \tag{249}$$

The eigenvalues of Eq. (247) are k_y independent and evenly spaced in $\hbar\omega_q$. They are the celebrated Landau levels

$$\varepsilon_n(k_y) = \left(n + \frac{1}{2}\right)\hbar\omega_q \qquad n = 0, 1, 2, \ldots \tag{250}$$

The eigenfunctions of Eq. (242) which correspond to the eigenvalue $(n + 1/2)\hbar\omega_q$ can be expressed in terms of the Hermite polynomials

$$\begin{aligned}\varphi_n(x) &= \frac{(\cos\theta)^{1/4}}{\sqrt{\lambda}}\frac{1}{2^n n!\sqrt{\pi}}\exp\left[-\frac{\cos\theta}{2\lambda^2}(x - x_0)^2\right] \\ &\times H_n\left[\left(\frac{x - x_0}{\lambda}\right)\sqrt{\cos\theta}\right]\end{aligned} \tag{251}$$

where

$$H_n(x) = (-1)^n \exp(x^2)\frac{d^n}{dx^n}\exp(-x^2) \tag{252}$$

It can be verified that

$$\langle n - 1|x|n\rangle = \int_{-\infty}^\infty \varphi_{n-1}(x)x\varphi_n(x)dx = \frac{\lambda}{\sqrt{\cos\theta}}\sqrt{\frac{n}{2}} \tag{253}$$

$$\langle n|x|n\rangle = x_0 \tag{254}$$

$$\left\langle n - 1\left|\frac{\partial}{\partial x}\right|n\right\rangle = \frac{\sqrt{\cos\theta}}{\lambda}\sqrt{\frac{n}{2}} \tag{255}$$

Thus, the classification of the eigenstates of $\widehat{H}_{//} + \widehat{H}_\perp$ was successful. These eigenstates consist of Landau level ladders which are attached to each of the zero field heterostructure bound (or unbound) states. Each of the levels is degenerate with respect to k_y or x_0, the center of the nth harmonic oscillator function. Thus, apart from the spin quantum number, it can be written that

$$|\nu\rangle = |m, n, k_y\rangle \tag{256}$$

which corresponds to the energy

$$\varepsilon_\nu = \xi_m + \left(n + \frac{1}{2}\right)\hbar\omega_{c//} \tag{257}$$

The remarkable point is that the carrier motion is entirely quantized, $V_b(z)$ has quantized the z component of the carrier motion, while the $B_{//}$ component, which is parallel to the growth axis, has quantized the in plane motion.

Each of the eigenvalues ε_ν is, however, enormously degenerate. This orbital degeneracy g_{orb} can be calculated as follows: the center of the harmonic oscillator function has to be in the crystal. Thus

$$-\frac{L_x}{2} < x_0 < \frac{L_x}{2} \tag{258}$$

where the dimension L_x is so large that the crystal boundaries were justifiably ignored in Eqs. (241)–(244).

The allowed k_y values are uniformly distributed and separated by the interval $2\pi/L_y$ (periodic boundary condition applied to the y axis). Thus, the degeneracy of the eigenvalue ε_ν is equal to

$$g_{orb} = \frac{L_x L_y}{2\pi\lambda^2}\cos\theta \tag{259}$$

Notice that g_{orb} is exact at $\theta = 0$ ($\delta\widehat{H}_1 = \delta\widehat{H}_2 = 0$ in this case) but only approximate at $\theta \neq 0$ since the effects of $\delta\widehat{H}_1$, $\delta\widehat{H}_2$ have not yet been evaluated.

The effects of $\delta\widehat{H}$ on the zeroth order eigenstates $|m, n, k\rangle$ have now to be investigated. The first term $\delta\widehat{H}_1$ does not affect the separability between x and z (as a matter of fact $\delta\widehat{H}_1$ could have been included in $\widehat{H}_{//}$). The only effect is that the different quantum states of the z motion are admixed. For a state $|m\rangle$, which is sufficiently separated in energy from the others (e.g., the ground state), $\delta\widehat{H}_1$ can be treated by first order perturbation theory. Thus ξ_m is changed by $\delta\xi_m$, where

$$\delta\xi_m = \frac{e^2}{2m_e^* C^2}B_\perp^2\langle\chi_m|z^2|\chi_m\rangle \tag{260}$$

To ensure the convergence of this perturbative scheme $\langle\chi_n|\delta\widehat{H}_1|\chi_m\rangle$ should remain much smaller than the level spacings $\xi_n - \xi_m$. For the states of a particle confined by a potential well of range $\approx L$, $\xi_n \approx \hbar^2\pi^2 n^2/2m_e^* L^2$ whereas $\langle\chi_n|z^2|\chi_m\rangle \approx L^2$. Thus, $\langle\chi_1|\delta\widehat{H}_1|\chi_2\rangle/(\xi_2 - \xi_1) \approx L^4\sin^2\theta/3\pi^2\lambda^4$ and the perturbative approach will be justified for the ground state if the Landau quantization along the z direction remains much smaller than the size quantization arising from $V_b(z)$.

Suppose, for instance, that the heterostructure consists of a quantum well of thickness L, clad between impenetrable barriers. For the ground state it is obtained that

$$\delta\xi_1 = \frac{e^2 B^2}{4m_e^* c^2}\sin^2\theta L^2\left(\frac{1}{6} - \frac{1}{\pi^2}\right) \tag{261}$$

If $B = 10T$, $\theta = \pi/4$, $L = 100$ Å, $m_e^* = 0.0067 m_0$, it is obtained that $\delta\xi_1 \approx 0.2$ meV, which is much smaller than $\xi_2 - \xi_1 \approx 168$ meV. Note, however, that $\delta\xi_1 \approx \xi_2 - \xi_1$ if $L \geq 500$ Å. In the latter situation, the first order perturbation becomes insufficient. A better approach would include $\delta\widehat{H}_1$ into $\widehat{H}_{//}$ from the very beginning. However, the failure of the perturbative approach indicates that the whole decoupling procedure underlying Eqs. (241) and (242) becomes dubious. This is what happens when the system is weakly two-dimensional, because the bound states supported by $V_b(z)$ are either too dense or (when $V_b(z)$ is finite) too close to the continuum. A better zero order Hamiltonian is the whole magnetic field dependent term (bulklike). The perturbation becomes $V_b(z)$ and, if B is large enough, the effects associated with $V_b(z)$ can be treated within the subspace spanned by all the states with a given n. For more complete treatments of tilted magnetic effects in superlattices, see [87, 88].

The second kind of perturbing terms, $\delta\widehat{H}_2$, can be rewritten as $z\hat{p}_x\omega_{c\perp} \cdot \delta\widehat{H}_2$ has no diagonal elements nonvanishing in the $|m, n, k_y\rangle$ basis since $\langle n|\hat{p}_z|n\rangle = 0$. In second order of perturbation theory, it is obtained that

$$\delta(\xi_m + \varepsilon_n) = \sum_{m', n' \neq m, n} \frac{\omega_{c\perp}^2|\langle\chi_m|z|\chi_{m'}\rangle|^2|\langle\varphi_n|\hat{p}_x|\varphi_{n'}\rangle|^2}{\xi_m - \xi_{m'} + (n-n')\hbar\omega_q} \tag{262}$$

Thus, $\delta\widehat{H}_2$ couples Landau levels of different subbands whose indexes n', n differ by one. In addition, if $V_b(z)$ is even in z, only subbands of opposite parities are coupled by $\delta\widehat{H}_2$. Note that $\delta[\xi_m + \varepsilon_n]$ can become very large (but cannot be described by means of perturbative treatments) when the energy denominators in Eq. (262) vanish. Practically, this means that two consecutive Landau levels, belonging to two adjacent subbands, intersect each other. For the sake of definiteness, let us assume that the involved levels and subbands are $n = 0$, $n = 1$ and $m = 1$, $m = 2$, respectively. Notice that at $\theta = 0$ the two levels $|1, 1, k_y\rangle$ and $|2, 0, k_y\rangle$ can cross since $\delta\widehat{H}_2$ vanishes. Let us denote by B_c the crossing fields at $\theta = 0$: $B_c = (\xi_2 - \xi_1)m_e^* c/\hbar e$. At finite θ, the crossing field would take place at $B_c/\cos\theta$. However, the $\delta\widehat{H}_2$ matrix element that connects the two levels is now nonvanishing and increases with θ. Thus, the crossing is suppressed and replaced by an anti-crossing

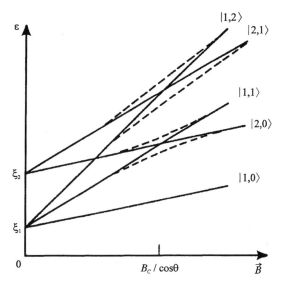

Fig. 33. Schematic fan charts of Landau levels attached to two subbands 1 and 2 with confinement energies ξ_1 and ξ_2, respectively. The dashed lines represent the effect of the anti-crossing induced by $\delta\widehat{H}_2$.

(see Fig. 33). Retaining the dominant $\delta\widehat{H}_2$ contributions and neglecting $\delta\widehat{H}_1$, it can be written that

$$|f\rangle \approx \alpha|1, 1, k_y\rangle + \beta|2, 0, k\rangle \qquad B \approx B_c/\cos\theta \tag{263}$$

The eigenvalues are the solutions of

$$\begin{vmatrix} \xi_1 + \frac{3}{2}\hbar\omega_q - \varepsilon & \langle 2, 0, k_y|\delta\widehat{H}_2|1, 1, k_y\rangle \\ \langle 1, 1, k_y|\delta\widehat{H}_2|2, 0, k_y\rangle & \xi_2 + \frac{1}{2}\hbar\omega_q - \varepsilon \end{vmatrix} = 0 \tag{264}$$

That is,

$$\varepsilon_\pm = \frac{\xi_1 + \xi_2}{2}\hbar\omega_q$$
$$\pm\sqrt{\left(\frac{\xi_2 - \xi_1 - \hbar\omega_q}{2}\right)^2 + |\langle 2, 0, k_y|\delta\widehat{H}_2|1, 1, k_y\rangle|^2} \tag{265}$$

The magnitude of the anti-crossing (which is the energy difference $\delta\varepsilon = \varepsilon_+ - \varepsilon_-$ evaluated at $B = B_c/\cos\theta$) is

$$|\delta_\varepsilon| = \frac{1}{\lambda(B_c)}\langle\chi_1|z|\chi_2\rangle\frac{\hbar e B_c}{m_e^* c}\tan\theta\sqrt{2} \tag{266}$$

which varies as $B_c^{3/2}$. At the anti-crossing the mixing between the two interacting levels is complete: $|\alpha| = |\beta| = 1/\sqrt{2}$. Note finally the close formal analogy between this type of anti-crossing and those arising when the cyclotron energy $\hbar\omega_q$ becomes equal to the energy of the longitudinal optical phonon energy $\hbar\omega_{LO}$ (magnetopolaron [89]).

5.2.1.1. Magnetic Field Density of States

Since the effects of $\delta\widehat{H}_1$ are only approximately known, the attention in this paragraph will be restricted to the case of $\theta = 0$, taking into account the spin splitting effect. As the z

and (x, y) motions are separable, the total density of states can be decomposed in contributions which arise from different subbands m,

$$\rho(\varepsilon) = \sum_m \rho_m(\varepsilon) \qquad (267)$$

where

$$\rho_m(\varepsilon) = \sum_{\sigma_z, k_y, n} \delta \left\{ \varepsilon - \xi_m - \left(n + \frac{1}{2} \right) \hbar \omega_q - g^* \mu_B \sigma_z \right\}$$

$$\sigma_z = \pm \frac{1}{2} \qquad (268)$$

Since the eigenenergies are k_y independent, the summation over k_y can immediately be performed. These results in the apparition of the orbital degeneracy factor g_{orb}. Thus

$$\rho_m(\varepsilon) = \frac{S}{2\pi \lambda^2} \sum_{n, \sigma_z} \delta \left\{ \varepsilon - \xi_m - \left(n + \frac{1}{2} \hbar \omega_q - g^* \mu_B B \sigma_z \right) \right\} \qquad (269)$$

As the carrier motion is entirely quantized, the density of states is zero except at the discrete energies

$$\varepsilon_{mn\sigma_z} = \xi_m \left(n + \frac{1}{2} \right) \hbar \omega_q + g^* \mu_B B \sigma_z \qquad (270)$$

where it is infinite. The results of Eqs. (268)–(270) are quite remarkable when compared with the zero field situation:

$$\rho_{m\sigma_z}^{B=0}(\varepsilon) = \frac{m_e^* S}{2\pi \hbar^2} Y(\varepsilon - \xi_m) \qquad (271)$$

At zero field the quasi-two-dimensional electron gas is metallic, having a density of states with no gaps for $\varepsilon > \xi_1$. A finite magnetic field dramatically alters this situation, replacing the continuum Eq. (271) by pointlike singularities (Eq. (269)) separated by gaps. Thus, depending on the respective locations of the characteristic electron energy η (the Fermi energy for a degenerate electron gas at $T = 0$ K) and $\varepsilon_{mn\sigma_z}$, the electron gas will behave either like an insulator ($\eta \neq \varepsilon_{mn\sigma_z}$) or a metal ($\eta = \varepsilon_{mn\sigma_z}$). Notice that similar effects cannot be found in bulk materials: the carrier free motion along the field prevents the density of states from vanishing at any energy larger than $\hbar \omega_c |g_c^*| \mu_B B / 2$.

Of course, it is known that imperfections (impurities, interface defects, etc.) will alter the deltalike singularities of $\rho_m(\varepsilon)$, rounding off the peaks and adding states into the gaps. However, it remains generally accepted that if B is large enough (i.e., $B \geq$ few Teslas in good GaAs–Ga$_{1-x}$Al$_x$As heterostructures at low temperature), the previous conclusions retain their validity. At least the following meaning can be given them: for energies belonging to a certain bandwidth of finite (yet undefined) extension, the density of states is large and corresponds to conducting states, whereas for other energy segments the states are localized. It means that they are unable to contribute to the electrical conduction at $T = 0$ K, for vanishingly small electric fields.

Some words have to be added concerning the broadening effects. The present status of the art is far from being satisfactory. First, there is, to the authors knowledge, no general consensus on the microscopic origin of these effects. Most likely it is sample dependent: in some samples, charged impurities play a dominant part while in others, interface roughness or alloy scattering (if it exists) may dominate. Second, even if it is assumed that the disorder potential has a simple algebraic form (e.g., uncorrelated delta like scatterers), it proves to be difficult, if not impossible, to obtain precise information on $\rho_m(\varepsilon)$ by any methods other than numerical simulations. The arising complications are twofold:

(a) The unperturbed density of states is highly singular. This precludes the use of non-self-consistent perturbative treatments of the disorder potential.
(b) When the required self-consistency is included, calculations are performed up to a finite order in the perturbation approach. By doing so, the nature of the states that occur in the tails of the broadened delta function are poorly treated: a localized level requires the inclusion of the electron disorder potential to an infinite order.

6. DYNAMICS OF THE LATTICE IN Q2D SYSTEMS: PHONONS AND ELECTRON–PHONON INTERACTIONS

In order to construct the crystal Hamiltonian, we can divide the crystal in two subsystems related to the electrons of the outer shells and to the remaining ionic cores, consisting of the nuclei and the electrons of the inner shells. The adiabatic approximation has provided an opportunity to consider various events in the electron subsystem in terms of electrons, holes, and excitons. However, elementary excitations in the electron subsystem do interact with the lattice, and we are going to consider optical manifestations of these interactions in this section. Note that the lattice properties of Q2D do not differ too much from those of the bulk crystals as electron properties do, because in the majority of cases the Q2D size is considerably larger than the lattice period.

6.1. Lattice Oscillations

Ionic cores in a crystal experience continuous oscillations near the equilibrium positions, which coincide with the lattice sites. The amplitudes of these oscillations increase with increasing the temperature but remain much smaller than the lattice period. Therefore, the snapshots of a real solid state lattice are different for every instant and never coincide with the ideal geometric lattice.

In a crystal consisting of N elementary cells, each with n atoms, $(3nN - 6)$ elementary oscillations exist, which are called normal modes. This number is nothing but the sum of all the degrees of freedom of all particles, $3nN$ minus six degrees of freedom inherent in a solid body, i.e., three translational plus three rotational for the whole crystal. Elementary oscillations are independent of each other and may coexist. Every crystal state can be described in terms of a superposition of normal modes. As the value of N is rather large in a bulk crystal, the six in the above sum can be omitted. This is justified for Q2D as well, since we agreed that a Q2D is still large as compared with the lattice period.

All $3nN$ normal modes can be classified in terms of $3n$ different groups (branches) containing N oscillations each. Three of them are called acoustic branches whereas the remaining $(3n-3)$ modes are referred to as optical branches.

To show the difference between acoustic and optical modes, consider a model ionic lattice in the form of a linear chain of solid particles connected by elastic springs. If all the particles are identical (monoatomic chain, Fig. 34a), then the elementary cell contains a single particle that possesses only one degree of freedom. Accordingly, a unique branch of normal oscillations exists, with the dispersion law,

$$\omega(q) = 2\sqrt{\frac{\gamma}{m}}\left|\sin\frac{qa}{2}\right| \tag{272}$$

where γ is a force constant representing the harmonic character of the oscillations in the model under consideration and m is the atomic mass. Note that $\omega = 0$ for $q = 0$.

In the diatomic chain (Fig. 34b) the elementary cell contains two atoms $(n = 2)$ with different masses, m and M. It has two branches of oscillation, with the more complicated dispersion laws

$$\omega_{op}^2 = \gamma\left(\frac{1}{m} + \frac{1}{M}\right) + \gamma\left[\left(\frac{1}{m} + \frac{1}{M}\right)^2 - \frac{4}{mM}\sin^2 qa\right]^{1/2} \tag{273}$$

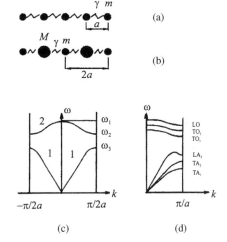

(a)

(b)

(c)   (d)

Fig. 34. Model linear (a) monoatomic and (b) diatomic chains; (c) dispersion laws for acoustic (curve 1) and optical (curve 2) branches of oscillation modes in a diatomic chain within the first Brillouin zone; (d) sketch of the dispersion relation of lattice vibrations for a three-dimensional anisotropic lattice with partly ionic binding. LO, TO, LA, and TA labels corresponds to longitudinal optical, transverse optical, longitudinal acoustic, and transverse acoustic modes, respectively.

and

$$\omega_{oc}^2 = \gamma\left(\frac{1}{m} + \frac{1}{M}\right) - \gamma\left[\left(\frac{1}{m} + \frac{1}{M}\right)^2 - \frac{4}{mM}\sin^2 qa\right]^{1/2} \tag{274}$$

which are plotted in Figure 34c for the first Brillouin zone,

$$\left(-\frac{\pi}{2a} < q < \frac{\pi}{2a}\right) \tag{275}$$

Equation (273) describes the optical branch, whereas Eq. (274) corresponds to the acoustic one. The main difference of these two branches is connected with their features when $q \to 0$. The optical branch at $q = 0$ is characterized by the finite frequency

$$\omega_1 = \sqrt{2\gamma\left(\frac{1}{m} + \frac{1}{M}\right)} \tag{276}$$

whereas the acoustic branch has $\omega = 0$. Nonzero energy with $q = 0$ becomes possible because in the case of an optical mode, different atoms are displaced in antiphase. At the boundary of the first Brillouin zone $(q = \pm\pi/2a)$ one has

$$\omega_2 = \sqrt{\frac{2\gamma}{m}} \tag{277}$$

That is, only the light atoms oscillate. In the case of acoustic modes at the boundary of the first Brillouin zone only the heavy atoms oscillate, resulting in

$$\omega_3 = \sqrt{\frac{2\gamma}{M}} \tag{278}$$

If two particles in the elementary cell are charged (i.e., the binding is at least partly ionic), then their oscillations can be described as oscillating electric dipoles. As there are eigenmodes for negligible , these modes can be generated by photons via resonant absorption. This fact provides a justification for the term "optical" branch.

In a three-dimensional lattice with more than one atom in the primitive cell there are always three types of acoustic modes and three types of optical modes, namely one longitudinal and two transversal. The latter may be degenerate in isotropic structures, whereas in an anisotropic lattice the degeneracy is lifted (Fig. 34d).

6.2. Concept of Phonons

In terms of quantum mechanics within the framework of harmonic oscillations, the Hamiltonian of an atomic lattice can be written in the form

$$\widehat{H} = \sum_i \left(\frac{P_i^2}{2M} + \frac{M\omega_i^2}{2}x_i^2\right) \tag{279}$$

Therefore, the general Schrödinger equation

$$\widehat{H}\psi = \varepsilon\psi \tag{280}$$

reduces to the set of equations

$$\left(\frac{P_i^2}{2M} + \frac{M\omega_i^2}{2}x_i^2\right)\varphi(x_i) = \varepsilon_i\varphi(x_i) \qquad (281)$$

where

$$\psi = \prod_i \varphi(x_i) \qquad \varepsilon = \sum_i \varepsilon_i \qquad (282)$$

Equation (281) is the standard harmonic oscillator equation, which results in the energy spectrum

$$\varepsilon_i = \hbar\omega_i\left(n_i + \frac{1}{2}\right) \qquad n = 0, 1, 2, \ldots \qquad (283)$$

The state of a given oscillator corresponding to a normal mode of a given branch is determined by the quantum number n_i. A set of n_i numbers for all modes of all branches provides an identification of the lattice states.

The total energy of the lattice then reads

$$\varepsilon = \frac{1}{2}\sum_i \hbar\omega_i + \sum_i n_i\hbar\omega_i \qquad (284)$$

The first term in Eq. (284) is the zero energy, which has a finite constant value, whereas the second term is a variable describing the excitation of the system. Various excited states are characterized by different sets of n_i numbers. In other words, every vibrational state of a lattice can be described in terms of the state of an ideal gas of noninteracting quasiparticles called phonons. The phonon number in every state can take any value; that is, a phonon gas can be described by the Bose–Einstein statistic. The total number of phonons in a crystal is not fixed because it is simply the sum of all the n_i numbers. Thus, an ensemble of phonons has a zero chemical potential, and the phonon distribution function reduces to Plank's formula

$$n_i = \left(\exp\frac{\hbar\omega_i}{k_BT} - 1\right)^{-1} \qquad (285)$$

already known for photons. In a similar way to the case of the photons, the probability of processes with phonon absorption has the coefficient n_i, whereas the process with phonon emission is characterized by the coefficient $(n_i + 1)$. Equations (284) and (285) combined with the dispersion laws provide calculations of all the thermodynamic properties of the lattice in terms of the ideal gas of bosons. In the case of nonharmonic oscillations, a gas of phonons should be considered as a nonideal gas.

The exciton–phonon interaction is of considerable importance in polar semiconductors. In semiconductor nanocrystals, the exciton–phonon interaction determines the shape of the absorption and emission spectra. At low temperatures ($k_BT \ll R_{y,\,exc}^*$) in perfect crystals, exciton–photon coupling dominates over exciton–phonon scattering. In this case coupled exciton–photon excitation (polariton) propagates throughout the crystal and the absorption coefficient concept fails to describe the specific features of light–crystal interaction [90, 91]. Both the

integral exciton absorption and the lineshape are controlled by the exciton–phonon dynamics, giving rise to the temperature-dependent oscillator strength (i.e., a failure of the absorption coefficient concept) and to the non-Lorentzian lineshape (i.e., non-Markovian phase relaxation). At high temperature ($k_BT > R_{y,\,exc}^*$) the exciton–phonon interaction results in a noticeable absorption at photon energies lower than the bandgap energy and the 1S-exciton resonance. The exponential absorption tail (Urbach rule) observed for a large number of perfect single crystals is a direct consequence of the exciton–phonon interaction.

The role of the exciton–phonon interaction in Q2Ds is of great importance as well. Contrary to the situation for bulk nanocrystals, it does not determine the integral exciton absorption because of negligible exciton–phonon coupling when $a \ll \lambda$. However, phonons still control the intrinsic dephasing and energy relaxation processes and therefore determine the absorption and emission linewidths and the luminescence Stokes shift. Several aspects of exciton–phonon interactions are still the subject of extensive research.

The II–VI and III–V semiconductors compounds have two ions by primitive cell; therefore there are three optical branches [one longitudinal optical (LO), and two transversal optical (TO)] and three acoustic branches [one longitudinal acoustic (LA) and two transversal acoustic (TA)]. In the center of the Brillouin zone (Γ), the TO and TA branches are twofold degenerate, but in some determined directions they are unfolded. When $\bar{q} \approx 0$ (long wavelength approach), the acoustic branches correspond to the two ions of the cell moving in the same direction, while in optical branches they are moving in opposite directions. In the latter case if the bonding is polar, an electric field appears coupled to the movement of the ions. The presence of this field makes the optical oscillations (especially those of long wavelength) get involved in numerous optical processes.

Frequently, in the Q2D systems, the investigations that include the participation of acoustic phonons are carried out considering that the acoustic phonons are not confined; thus the Fröhlich Hamiltonian (bulk crystals) is also used to describe the electron–acoustic phonon interaction Hamiltonian. However, many proposed applications of mesoscopic electronic structures involve carrier transport at low temperatures and low carrier energies. Frequently, the regime of interest is such that the dimensional confinement modifies substantially the phase space. In this low-temperature, low energy regime [92–97] acoustic phonons play an enhanced role in carrier scattering and may dominate over the scattering of carriers by optical phonons. Furthermore, in nanometer scale structures it is possible that phase space restrictions may weaken or forbid optical phonon scattering processes that would normally dominate in bulk structures. In recent years, there has appeared extensive literature about the role of dimensional confinement in modifying optical phonon modes and their interactions with charge carriers (see, for example, [98–102] and the numerous references therein). However, there are relatively few treatments dealing with the role of dimensional confinement in modifying acoustic phonon modes and their interactions with charge carriers [93–95]. While there is extensive literature on the theory of acoustic modes in conventional waveguide resonators and related structures [96], no efforts have been reported toward a formulation of a theory of acoustic phonons in nanometer scale structures, where both the phonon confinement and the quantum mechanical treatment of the phonon normalization are essential. However, Constantinou has discussed the unnormalized acoustic phonon modes in cylindrical polar semiconductor quantum wires [103].

Therefore, it is not a good choice to use the three-dimensional Fröhlich Hamiltonian to describe the electron–phonon interaction in the Q2D nanostructured materials or to use an unconfined phonon. However, a general theory for the acoustic phonons has not been carried out for all kind of Q2D systems, and there exist only a few works in quantum well wires [104, 105]. For all these reasons only the optical phonons and the electron–optical phonon interaction Hamiltonian will be studied, taking into account other theories and models that are more appropriate and include the confinement effect.

The investigation of polar optical vibrations (phonons) in semiconductor Q2D nanostructures has been a subject of great interest in the last years. The theoretical treatment of several physical properties, such as electron scattering rates, polaron effects, Raman scattering efficiencies, hot-electron phenomena, etc., requires a reliable description of phonon modes and electron–phonon interaction potentials in such structures. When we refer to a confined optical phonon we are thinking only in the Fröhlich optical phonon and not in those obtained by a deformation potential, which in general are not considered as confined. We should remember that the Fröhlich electron–phonon interaction has an intraband character, while the deformation potential has an interband character.

6.3. Phenomenological Models for Long Wavelength Polar Optical Modes in Q2D Systems

The description of the long wavelength polar optical oscillations in a Q2D system is carried out through three formalisms: The phenomenological models of long wavelength, the microscopic calculations, and the treatments based on a combination of the previous two. These microscopic calculations lead to results in close agreement with the Raman scattering experiments [106–109]. In these experiments the vibration amplitude cannot be determined and for this reason the microscopic models are used as for verification of the phenomenological models, which have been developed in two fundamental ways: the dielectric continuous models [110, 111] and the hydrodynamic models [110, 112, 113]. In this respect an exhaustive analysis of the different phenomenological models were reported in [110]. However, the application of this type of approach to mesoscopic structures leads to certain differences when compared to Raman scattering experiments and calculations based on the microscopic models [107–109, 114, 115].

The phenomenological treatment, free of the previous difficulties, has considered the introduction of the called "reformulated slab vibrations" [116], where the results of microscopic calculations are introduced as corrections to a modified phenomenological treatment. Along this line some electron–phonon interaction Hamiltonians have been proposed, as in [117]. The above-mentioned difficulties are also present in the phenomenological treatment applied to a quantum well wire and a freestanding wire [99, 118–124] as well as in the quantum dots [125, 126].

However, it has been demonstrated that the rigorous application of the phenomenological theory, satisfying all the macroscopic physical requirements for a continuous medium, can lead to results that nearly correspond to experimental data and microscopic calculations [127–135]. So, in the same way, the interfaces and confined modes are obtained as solutions of the complete system of four coupled second order differential equations for the vibrational amplitudes and the electrostatic field. These equations are solved with compatible mechanical and electrostatic matching boundary conditions. The crystal translational symmetry is partially or totally broken in the Q2D nanostructures. So the symmetry in the free directions is preserved and the Fröhlich Hamiltonian is applied to them in the unconfined model frame, but not in the confinement directions.

The Raman scattering experiments in layered systems reveal the existence of vibration normal modes concentrated in some of the layers (vicinity of interfaces), which are denominated interface confined modes.

In order to describe the oscillations, the vector field $\vec{u}(\vec{r}, t)$, which represents the relative displacement of two ions of the cell, and a scalar potential $\phi(\vec{r}, t)$ associated with the electric field $\vec{E} = -\nabla\phi$, coupled to the mechanical oscillations of the medium, are used. The electric field $\vec{E}$ shall be treated in the unrecorded limit ($c \to \infty$) and must obey the quasi-stationary Maxwell equations. Other physical parameters of the medium are $\rho = \bar{M}/V_c$ where $\bar{M}$ is the reduced mass of the ions and V_c is the unitary cell volume. $\omega_T(\omega_L)$ is the frequency of the TO (LO) branch in the Γ point, β_L and β_T are two parameters that describe the oscillations dispersion in the vicinity of the Γ point, and κ_0 and κ_∞ are static and high frequency dielectric constants.

Let us consider the medium properties to be piecewise continuous and embedded in a second material. The lagrangian density $L(\vec{u}, \phi)$ is introduced in the form

$$L = \frac{1}{2}\rho\left(\frac{\partial\vec{u}}{\partial t}\right)^2 + \frac{1}{2}\vec{u}\cdot\vec{\gamma}\cdot\vec{u} - \vec{u}\cdot\vec{\alpha}\cdot\nabla\phi$$
$$+ \frac{1}{2}\nabla\phi\cdot\vec{\beta}\cdot\nabla\phi + \frac{1}{8\pi}(\nabla\phi)^2 \qquad (286)$$
$$+ \frac{1}{8}\sum_{i,j,k,l}(\nabla_i u_j + \nabla_j u_i)\lambda_{ijkl}(\nabla_k u_l + \nabla_l u_k)$$

In Eq. (286) the first term represents the kinetic energy density. $\vec{\gamma}$, $\vec{\alpha}$, and $\vec{\beta}$ are tensors which describe the coupling between vibrational amplitudes with themselves, with electric and mechanical fields, and the electric field with itself in the second, third, and fourth terms respectively. The fifth term represents the energy density of the electric field (in the vacuum) and the last term corresponds to the energy density due to internal stress described by the elastic modulus tensor $\vec{\lambda}$.

The movement equations in the limits of high and low frequency allow us to know the above-mentioned tensors [129]

$$\vec{\gamma} = -\rho\omega_T^2 \vec{I}$$
$$\vec{\beta} = \frac{1}{4\pi}(\vec{\kappa}_\infty - \vec{I}) \qquad (287)$$
$$\vec{\alpha}\vec{\gamma}^{-1}\vec{\alpha} = \frac{1}{4\pi}(\vec{\kappa}_\infty - \vec{\kappa}_0)$$

For an isotropic medium all the tensors are diagonal and in particular

$$\vec{\alpha} = \alpha\vec{I} = \sqrt{\frac{(\kappa_0 - \kappa_\infty)\rho\omega_T^2}{4\pi}} \qquad (288)$$

Also, the nonvanishing components are given through two independent parameters. In this case,

$$\lambda_{1111} = \lambda_{2222} = \lambda_{3333} = \rho\beta_L^2$$
$$\lambda_{1122} = \lambda_{1133} = \cdots = \rho(\beta_L^2 - 2\beta_T^2) \qquad (289)$$
$$\lambda_{1212} = \lambda_{2112} = \cdots = \rho\beta_T^2$$

Taking into account Eq. (289), the σ_{ij} stress tensor in isotropic media acquires the following form:

$$\sigma_{ij} = -\rho(\beta_L^2 - 2\beta_T^2)(\nabla \cdot \vec{u})\delta_{ij} - 2\rho\beta_T^2 u_{ij} \qquad (290)$$

From the lagrangian density of Eq. (286), the movement equations are obtained which, assuming a harmonic dependence for $\vec{u}$ and ϕ and writing the divergence $\vec{\sigma}$ on an explicit form, can be expressed as

$$\rho(\omega^2 - \omega_T^2)\vec{u} = \alpha\nabla\phi + \rho\beta_T^2\nabla\times(\nabla\times\vec{u}) - \rho\beta_L^2\nabla\nabla\cdot\vec{u} \quad (291)$$

$$\nabla^2\phi = \frac{4\pi\alpha}{\kappa_\infty}\nabla\cdot\vec{u} \qquad (292)$$

A more general analysis considering the anisotropy of the constituent media can be found in [129].

The dispersion law in the bulk is obtained by finding the Fourier transform in Eq. (291) and through a direct calculation the laws

$$\omega_T^2(\vec{k}) = \omega_T^2 - \beta_T^2\vec{k}^2 \qquad (293)$$

$$\omega_L^2(\vec{k}) = \omega_L^2 - \beta_L^2\vec{k}^2 \qquad (294)$$

are obtained, where $\vec{k}$ is the three dimensional wavevector, These equations allow us to obtain fittings of β_T and β_L starting from the experimental dispersion law.

The normal modes are determined by solving Eqs. (291) and (292) for the constituent parts of the Q2D nanostructure,

imposing matching conditions at the interfaces, consistent with the differential equations. Physical reasons allow us to assert that $\vec{u}$ and ϕ are continuous functions at the S interface:

$$\phi|_{\vec{r}\in S^-} = \phi|_{\vec{r}\in S^+}, \quad \vec{u}|_{\vec{r}\in S^-} = \vec{u}|_{\vec{r}\in S^+} \qquad (295)$$

The indexes $+$ and $-$ indicate that the corresponding functions are evaluated in one or the other side of the surface S. The functions $\vec{u}$ and ϕ must be bounded in each region, given that they describe physical magnitudes. The partial derivatives of $\vec{u}$ and ϕ cannot be continuous in S because the parameters included in the fundamental equations change abruptly in the interface. The discontinuities of the normal derivatives of $\vec{u}$ and ϕ at S can be determined directly from the equations. Integrating Eqs. (291) and (292) in an infinitesimal volume limited by the surface S and applying the Gauss theorem (as in electrodynamics) it is found that

$$\vec{\sigma}\cdot\vec{N}|_{r\in S^-} = \vec{\sigma}\cdot\vec{N}|_{r\in S^+} \qquad (296)$$

$$\vec{D}\cdot\vec{N}|_{r\in S^-} = \vec{D}\cdot\vec{N}|_{r\in S^+} \qquad (297)$$

where $\vec{N}$ is the normal unit vector to the surface. Equation (296) expresses the force flow continuity through the interface, while the Eq. (297) establishes the continuity of the normal component of the electric displacement. In some studies the continuity condition of $\vec{u}$ at the interface is substituted by $\vec{u} = 0$ for $\vec{r} \in S$ [117, 128]. It is our purpose to get a correct (although approximated) description of the structure of GaAs/AlAs. In this system the oscillations in GaAs do not penetrate to the AlAs side because the bandwidth of the polar optical phonons of both materials is much smaller than the gap between them. In such cases, matching conditions such as

$$\vec{u} = 0 \quad \text{in } S \text{ (complete confinement)}$$
$$\text{continuity of } \phi \text{ and } \vec{D}\cdot\vec{N} \quad \text{in } S \qquad (298)$$

are valid.

The eigenfunctions $\vec{u}$ and ϕ satisfy the orthonormality condition [129, 133]

$$\int \rho(\vec{r})\vec{u}_m^*(\vec{r})\cdot\vec{u}_n(\vec{r})d\vec{r} = \Omega_0\delta_{nm} \qquad (299)$$

where Ω_0 is a constant that can be chosen arbitrarily. The completeness of the basis $\{\vec{u}_m\}$ is expressed by the following relation:

$$\sum_m \rho\vec{u}_m(\vec{r})\cdot\vec{u}_m(\vec{r}') = \Omega_0\delta(\vec{r}-\vec{r}'). \qquad (300)$$

6.4. Analysis of the Phenomenological Models for Long Wavelength Polar Optical Modes in a Semiconductor Layered System

The standard type of phenomenological model for this problem consists of a mechanical field equation for the vibration amplitude and (in the quasi-static limit) Poisson's equation. The critical issue concerns the matching boundary conditions.

The apparent incompatibility between mechanical and electrostatic boundary conditions often appearing in the literature is analyzed and clarified. It is shown how the complete solution can be obtained so that the incompatibility is ruled out and the key features of the results, notably the symmetry pattern, agree with the microscopic calculations and with the Raman scattering data.

Considerable attention is being devoted to the study of long wavelength polar optical modes in semiconductor quantum wells and superlattices [136]. Various theoretical models have been proposed [107, 113, 137–142] that represent substantially different and often diverging viewpoints and have met with varying degrees of success in the interpretation of observed experimental facts. Raman scattering data are available for a wide range of superlattices, especially GaAs related superlattices, and show the existence of different types of modes. Some of these modes have amplitudes mainly concentrated in one of the constituent slabs, usually termed confined modes [143–145], and some with the amplitudes spreading to both constituents but tending to concentrate close to the interfaces; hence they are called interface modes [146–148]. The Raman experiments do not give the spatial dependence of the amplitudes, but this is surmised from the fact that the observed frequencies are forbidden in one of the constituent materials (guided or confined modes) or in both (interface modes, in this case they are induced by interface disorder). Moreover the geometry of the Raman scattering experiment can be chosen [149] to detect modes with different symmetries. In particular, one can then experimentally establish whether the amplitude (or the potential) of the observed modes is even or odd with respect to the center of the slabs [144]. These aspects of the experimental data and the key features of the results obtained in microscopic calculations [107, 139, 140] should be taken as the criterion to decide whether a phenomenological model is reasonable or not, while some lack of quantitative accuracy can, to some extent, be expected and tolerated.

The central issue can be introduced by considering just one interface. The essence of a long wavelength phenomenological model is that one uses differential calculus, and then a way must be found to match the solutions at the interface. Of course it can be argued [140, 142] that in principle the differential equations are not valid in the immediate vicinity of the interface, when finite changes take place over microscopic distances. Any phenomenological model of the matching problem must necessarily be an approximation, just as it is in all the existing matching calculations for long wavelength acoustic or piezoelectric modes. Differential calculus is used to match at abrupt interfaces and the model works quite well. The same can be said, for instance, of envelope function matching calculations for electronic states [150, 151].

The question raised here is not whether the standard type of phenomenological model can be rigorously justified. It is clear that formally it cannot. The question is whether one can start from a simple phenomenological model, obtain a solution which has all the general properties one may require, and

reproduce, to a good approximation, experimental results. In this context they show that this can be done.

The key issue concerns the matching boundary conditions. One approach to the problem is with a dielectric model [137] and imposing electrostatic continuity. The pattern of vibrational amplitudes and the electrostatic potential obtained are in open disagreement with experimental evidence [143–148] and with microscopic calculations [107, 139–141].

Moreover, one achieves electrostatic continuity at the expense of mechanical discontinuity. Alternatively, one can start from a mechanical equation of motion for the vibration amplitude [138] and then impose the condition of mechanical continuity.

This yields the correct symmetry pattern for the vibration amplitudes, but it appears to produce a discontinuous electrostatic potential. The incompatibility between the separate use of these two approaches has been correctly stressed [145, 146]. The purpose is to show that both requirements, mechanical and electrostatic continuity, can be satisfied by combining the two approaches in a proper way, which accounts for the coupling between the mechanical and electrostatic fields.

One possibility is to start from a phenomenological model for the vibrational wave, and then introduce *ad hoc* matching boundary conditions, forcing the electrostatic continuity into the analysis [140, 152]. The goal is to save the simplicity of the phenomenological model, with parameters determined from a fit to bulk phonon–dispersion relations, to use it in problems of practical interest like the derivation of an electron–phonon interaction Hamiltonian and the calculation of Raman scattering efficiencies.

A different approach has been taken [113] in a study of this problem for a model in which the vibrations are totally confined by infinitely rigid barriers. The mechanical amplitudes then vanish at the boundaries but the electrostatic potential, which is continuous, does not necessarily. This model describes well the confined modes that are totally confined due to the rigid barrier condition and correctly reproduces the observed symmetries, but it is limited by its inherent approximation and cannot describe interface modes. It is the simultaneous account of the mechanical and electrostatic field and the corresponding compatible boundary conditions that makes this approach significantly different from the others. Now this approach will be used, but with the restriction reduced to rigid barriers.

A phenomenological model based on an equation of motion for the vibration amplitude $\vec{u}$, which contains (i) spatial dispersion through terms representing mechanical forces, (ii) a natural frequency for the harmonic oscillatorlike vibrations, and (iii) the coupling to the electrostatic field $\vec{E}$ [153], was accepted from the start. This is the well known equation

$$\rho(\omega^2 - \omega_T^2)\vec{u} - \nabla(\rho\beta_a^2 \nabla \cdot \vec{u}) - \nabla \cdot (\rho\beta_b^2 \nabla \vec{u})$$
$$+ \left[\frac{(\kappa_0 - \kappa_\infty)}{4\pi}\right]^{1/2} \omega_T \vec{E} = 0 \qquad (301)$$

The ansatz for the mechanical forces follows from general considerations [138, 154] and has the form of the divergence of a tensor $\vec{\tau}$. In the case of an isotropic solid, assumed implicitly in Eq. (301),

$$\vec{\tau} = \rho\beta_a^2 \nabla \cdot \vec{u}\vec{I} + \rho\beta_b^2 \nabla\vec{u} \tag{302}$$

In the case of acoustic waves ($\omega_T = 0$), $\vec{\tau}$ is literally the mechanical stress tensor, with β_a^2 and β_b^2 related to the Lamé coefficients λ, μ. In the present case β_a^2 and β_b^2 are phenomenological parameters which take into account the spatial dispersion, i.e., the k dependence of the bulk frequencies up to terms of order k^2. The essence of the phenomenological model is that Eq. (301) is accepted as it stands, even for inhomogeneous systems where the material parameters depend on the position coordinate z normal to the interface, and this includes abrupt interfaces. Having assumed this model we want to see how the mechanical and electrostatic field can be obtained while their coupling is completely taken into account.

To do this, Eq. (301) must be solved simultaneously with Maxwell's equations which, in the quasi-static limit here considered ($c \to \infty$) or from an experimental point of view, for wavelengths small compared with the bulk reststrahlen wavelengths, are reduced to the equations of electrostatics. In the case

$$\vec{E} = -\nabla\phi \tag{303}$$

and, since there is no external free charge

$$\nabla \cdot \vec{D} = 0 \tag{304}$$

However, written in this way, the coupling between the two fields is not apparent. From the general relation

$$\vec{D} = \vec{E} + 4\pi\vec{P} \tag{305}$$

and the particular constitutive relationship of this model [153]

$$\vec{P} = \left[\frac{\rho(\kappa_0 - \kappa_\infty)}{4\pi}\right]^{1/2} \omega_T \vec{u} + \frac{(\kappa_\infty - 1)}{4\pi}\vec{E} \tag{306}$$

the coupling of the mechanical and electrical fields is made to appear explicitly in the electrostatic field equation by noting that the source of ϕ is the polarization charge, so that

$$\nabla^2\phi = 4\pi\nabla \cdot \vec{P} \tag{307}$$

which, using the above expressions and the Lyddane–Sachs–Teller relations, yields

$$\nabla^2\phi = \left[4\pi\rho(\kappa_\infty^{-1} - \kappa_0^{-1})\right]^{1/2} \omega_L \nabla \cdot \vec{u} \tag{308}$$

Furthermore, substituting Eq. (303) in Eq. (301)

$$\rho(\omega^2 - \omega_T^2)\vec{u} - \nabla(\rho\beta_a^2 \nabla \cdot \vec{u}) - \nabla \cdot (\rho\beta_b^2 \nabla\vec{u})$$
$$- \left[\frac{(\kappa_0 - \kappa_\infty)}{4\pi}\right]^{1/2} \omega_T \nabla\phi = 0 \tag{309}$$

The coupling between $\vec{u}$ and ϕ is now explicitly written in the field equations. It is stressed that $\vec{u}$ consists in general of a longitudinal and a transverse field

$$\vec{u} = \vec{u}_L + \vec{u}_T \tag{310}$$

with $\nabla \times \vec{u}_L = 0$ and $\nabla \cdot \vec{u}_T = 0$. Only the $\vec{u}_L$ part of $\vec{u}$ appears explicitly in Eq. (308), but this $\vec{u}_L$ is in general affected by the coupling to $\vec{u}_T$. This is another aspect which is often obscure in the literature. In a bulk homogeneous medium described by this model the longitudinal and transverse modes decouple and one easily obtains from Eqs. (308) and (309) independent dispersion relations for the bulk LO and TO modes. It is the boundary conditions that, in general, couple the longitudinal and transverse parts, as is a common experience with problems involving matching between isotropic media. It suffices to study ordinary Rayleigh surface waves, Eq. (291), to see this. In some cases one can decouple them, notably with some geometry specifically designed for a Raman scattering experiment that detects vibrating modes in some symmetry direction and with $\vec{q} = 0$, $\vec{q}$ being the two-dimensional wavevector parallel to the surface. In other situations one can make an approximation, for example in the study of electron–phonon scattering, which involves only the $\vec{u}_L$ part of $\vec{u}$ and is such that only very low wave vectors have an appreciable weight [107].

However, the issues raised here can be clarified without the need to decouple LO and TO modes. What is found in Eq. (309) is a field equation for the total $\vec{u}$, given that the $\vec{u}_L$ part of it is coupled to ϕ, while Eq. (308) says that ϕ is the potential created by the polarization charge

$$\rho_{\text{pol}} = -\left[\frac{\rho}{4\pi}(\kappa_\infty^{-1} - \kappa_0^{-1})\right]^{1/2} \omega_L \nabla \cdot \vec{u} \tag{311}$$

where $\vec{u}$ is the solution of the mechanical field equation (309). When seen from the point of view of electron–phonon coupling, this is the potential that the external test charge of an electron sees in a system where the vibration $\vec{u}$ produces a polarization charge ρ_{pol} given by Eq. (311). Hence, it is very important to obtain a complete solution that simultaneously describes both $\vec{u}$ and ϕ correctly and, according to the laws of electrostatics, requires continuity of ϕ everywhere. The phenomenological model embodied in Eqs. (308) and (309) guarantees this, if it is accepted that the differential equations describing the amplitudes $\vec{u}$ and ϕ must be continuous at the matching interface:

$$\vec{u}(z = +0) = \vec{u}(z = -0); \quad \phi(z = +0) = \phi(z = -0) \tag{312}$$

The Fourier transform in two dimensions introduces a dependence on the transverse wavevector $\vec{q}$ and left differential equations in the position variable z. Integrating Eq. (309) from $-\eta$ to $+\eta$ and letting $\eta \to 0$, given that ϕ is continuous, yields matching boundary conditions involving the first derivatives of the components of $\vec{u}$. If the force per unit volume is written in the form of the divergence of the tensor, written down in Eq. (302), then these conditions take the form

$$\tau_{zj}(z = +0) = \tau_{zj}(z = -0), \qquad (j = x, y, z) \quad (313)$$

In the case of acoustic waves this expresses the continuity of the forces per unit area transmitted across the surface. In the present problem one can interpret this in the same way, always within the frame of the approximation embodied in the model. The z component of Eq. (313) reads

$$\rho(\beta_a^2 + \beta_b^2)\frac{\partial u_z}{\partial z}\bigg|_{z=+0} = \rho(\beta_a^2 + \beta_b^2)\frac{\partial u_z}{\partial z}\bigg|_{z=-0} \quad (314)$$

which one can, in the same spirit, interpret as the continuity of the hydrostatic pressure [138], although the present analysis shows that this does not require the decoupling of $\vec{u}_L$ and $\vec{u}_T$ in the approximation one may choose to call hydrodynamic. All that matters is that given Eq. (309), the three continuity conditions of Eq. (313) are mathematically derived. The explicit form of these is readily obtained from Eq. (309). Note that Eq. (314) can be mapped, onto the Bastard boundary condition [150, 151], by replacing $\rho(\beta_a^2 + \beta_b^2)$ with $1/m$, m being the effective mass.

It is also easily seen, by using Eq. (306), that integration of Eq. (308), from $-\eta$ to $+\eta$ with $\eta \to 0$, yields

$$D_z(z = +0]) = D_z(z = -0) \quad (315)$$

as expected from the general principles of electrostatics. What is stressed here is that the matching boundary conditions, Eqs. (313) and (315), follow mathematically and unambiguously once the phenomenological model has been accepted.

The issue initially raised has been formally settled. Equations (308) and (309) represent a system of four second order differential equations, so eight matching boundary conditions are needed, which are contained in Eqs. (312), (313), and (315). This defines an eigenvalue problem whose solution is uniquely determined by the four differential equations and eight independent matching boundary conditions. The physical interpretation of these conditions and the approximations embodied in the phenomenological model are a separate issue. Mathematically the problem is uniquely and correctly specified while the coupling between $\vec{u}$ and $\vec{E}$ fields is evident and explicit in the analysis.

The confusion arises when one decouples $\vec{u}_L$ and $\vec{u}_T$, and the two field equations. If $\vec{u}$ is only longitudinal, then so is $\vec{P}$ (Eq. (306)), so $\vec{D}$ is only $\vec{D}_L$ and its curl vanishes. Since its divergence also vanishes, Eq. (304), then $\vec{D}$ vanishes, yielding

$$\vec{E} = -\left[4\pi\rho(\kappa_\infty^{-1} - \kappa_0^{-1})\right]^{1/2}\omega_L\vec{u} \quad (316)$$

Substituting this in Eq. (309) one obtains a field equation only for $\vec{u}$. The other field equation is often taken to be Eq. (304), which involves the dispersive dielectric functions. However, the two field equations are apparently decoupled and this has led to their independent use as optional alternatives.

The contradiction arises because $\vec{E}$ and $\vec{u}$ are really coupled fields: they must always fulfill Eq. (316). This is what causes the apparent incompatibility of boundary conditions: if one chooses to work starting from the mechanical equation, it can be seen that it imposes mechanical continuity and uses Eqs. (316) and (303) to obtain ϕ by inspection, so that ϕ is discontinuous. The point is that this is not the correct way to solve the complete problem. Instead one should determine ϕ by solving Eq. (308) with correct electrostatic matching boundary conditions, even when one studies a purely longitudinal field. The physical implications of this for the electron–phonon interaction have just been stressed.

The main objection one can make to the procedure just followed is the assumption of isotropy in Eq. (301). Cubic crystals, for instance, are known to be anisotropic to second order in the components of $\vec{k}$. It is possible to generalize Eq. (302) so as to include this anisotropy and this can be done easily by replacing Eq. (301) with

$$\rho(\omega^2 - \omega_T^2)\vec{u} - \nabla \cdot \vec{\tau} - \left[\frac{(\kappa_0 - \kappa_\infty)}{4\pi}\right]^{1/2}\omega_T\nabla\phi = 0 \quad (317)$$

where for the tensor $\vec{\tau}$,

$$\vec{\tau} = \rho\begin{pmatrix} \beta_a^2\nabla\cdot\vec{u} & (\beta_c^2 - \beta_a^2)\frac{\partial u_y}{\partial y} & (\beta_c^2 - \beta_a^2)\frac{\partial u_z}{\partial z} \\ (\beta_c^2 - \beta_a^2)\frac{\partial u_x}{\partial x} & \beta_a^2\nabla\cdot\vec{u} & (\beta_c^2 - \beta_a^2)\frac{\partial u_z}{\partial Z} \\ (\beta_c^2 - \beta_a^2)\frac{\partial u_x}{\partial X} & (\beta_c^2 - \beta_a^2)\frac{\partial u_y}{\partial y} & \beta_a^2\nabla\cdot\vec{u} \end{pmatrix}$$
$$+ \rho\beta_b^2\nabla\cdot\vec{u} \quad (318)$$

is used instead of (302) (in matrix notation).

In this case the parameters $\beta_a^2, \beta_b^2, \beta_c^2$ are obtained by fitting the corresponding bulk phonon dispersion relation [$\beta_c = \beta_a$ leads back to the isotropic case of Eq. (302)].

The mechanical vibration amplitude has been treated [155] in a way that is similar in spirit to the analysis proposed here, although it differs in detail. A dynamical matrix is written down inspired by the form of the effective Hamiltonian for envelope function calculations of electronic structures. Due to crystal symmetry the mathematical form is the same as that obtained from mechanical considerations. The parameters of this dynamical matrix are found by fitting to the results of a microscopic calculation based on a rigid-ion model that gives a correct description of the optical vibrations. The discussion is then centered on the equation of motion for the mechanical vibration amplitudes and the corresponding matching boundary conditions, for which two extreme cases are studied. These correspond to $\Delta\omega_0$, the difference in resonant frequency at zero wavevector, being either much larger or much smaller than the bulk bandwidths. In the first case the amplitude is assumed to vanish at some chosen plane. In the second case, after some plausible approximations, a kind of continuity equation is derived which expresses energy conservation and this is used to establish matching boundary conditions.

Although this is close in spirit to the present analysis, it differs in two important aspects, namely: (i) the model does not depend on whether $\Delta\omega_0$ is large or small and thus can be directly used for GaAs/Al$_x$Ga$_{1-x}$As systems with arbitrary x; (ii) the differential equation of ϕ, its coupling to the mechanical differential equation, and the compatibility between the matching boundary conditions for both fields are discussed explicitly. This also allows for an explicit study of the electron–phonon interaction Hamiltonian in which one can easily see how to introduce screening due to the electron gas if one has a model for its dielectric function. This is a very important matter for transport calculations in populated quantum wells with external modulation doping.

Finally it is interesting to consider the particular case of a GaAs well with AlAs barriers using the method just presented. In this case the frequency difference is very large and it is a good approximation to assume infinitely high rigid barriers. That is, $\vec{u}(z = \pm d/2 = 0)$ and $\vec{u} \equiv 0$ for $|z| > d/2$, d being the well width.

The case of (001) interfaces and vibrations in the (001) direction with $\vec{q} = 0$, where the transverse part $\vec{u}_T \equiv 0$ for all $z \in (-d/2, d/2)$, was considered; it was only necessary to study a longitudinal field. From the vanishing of $\nabla \times \vec{u}$ one can reduce the mechanical field equation to one differential equation for u_z with the amplitude vanishing for $|z| \geq d/2$. Equation (308) now reads

$$\frac{d^2\phi}{dz^2} = \left[4\pi\rho(\kappa_\infty^{-1} - \kappa_0^{-1})\right]^{1/2} \omega_L \frac{du_z}{dz} \qquad (319)$$

with ϕ continuous. The condition (315) is in this case automatically satisfied. The complete solutions, Eqs. (309) and (319), are [113]

$$u_z(z) = A_n \sin\left(\frac{n\pi}{d}(z + d/2)\right) \qquad n = 1, 2, \ldots \qquad (320)$$

and

$$\phi_n(z) = C_n \begin{cases} [1 - (-1)^n] & z < -d/2 \\ 2\cos\left[\left(z + \frac{n\pi}{d}\right)\right] - [1 + (-1)^n] & |z| < d/2 \\ [(-1)^n - 1] & z > d/2 \end{cases} \qquad (321)$$

where

$$C_n = A_n \frac{\left[4\pi\rho(\kappa_\infty^{-1} - \kappa_0^{-1})\right]^{1/2} \omega_L}{(n\pi/d)} \qquad (322)$$

and A_n is a normalization constant. These results agree rather well with those of microscopic calculations [107, 139–141] and there is no incompatibility. The solution yields a continuous electrostatic potential and the symmetry pattern agrees with experimental evidence. Raman scattering data [144, 146–148] give information about LO modes, with ϕ odd corresponding to n odd in Eq. (321), when the geometry of the experiment is $\bar{Z}(Y, X)Z$ and ϕ even–n even–when it is $\bar{Z}(X, X)Z$ and this pattern is the one found from Eq. (321) and from microscopic calculations.

Studies based on the so-called dielectric model use the boundary condition Eq. (315) and impose the vanishing of ϕ for $|z| \geq d/2$. This yields

$$\phi(z) \approx \begin{cases} \cos\left(\frac{n\pi}{d}z\right) & n = 1, 3, \ldots \\ \sin\left(\frac{n\pi}{d}z\right) & n = 2, 4, \ldots \end{cases} \qquad (323)$$

which gives the opposite symmetry pattern. This happens at the expense of mechanical discontinuity. On the contrary, calculations based on a decoupled field equation for $\vec{u}$ and the vanishing of E_z for $|z| \geq d/2$ yield the correct vibration amplitudes so ϕ is then discontinuous. An ad hoc way out of this dilemma can be found [152] in which ϕ is forced to be continuous by adding in each layer a term independent of z but depending on n. These constants are exactly the terms involving $(-1)^n$ contained in Eq. (321), which is the solution naturally obtained without the need to resort to any artificial stuff, when the mechanical and electrostatic field equations are correctly coupled and solved.

This clarifies the issue raised at the beginning. The mechanical field equation is only a model and hence an approximation, but it works quite well for the vibration amplitudes, while ϕ must be obtained from (308). This provides the basis for getting, in a rather simple way, an electron–phonon interaction Hamiltonian based on a reasonable model, which works rather well in the cases where it has been tested and also leads to solutions which meet all general requirements, with no incompatibility between the mechanical and electrostatic matching boundary conditions.

6.5. Polaron Properties in a Semiconductor Q2D Nanostructure

This subsection is concerned with the properties of an electron moving in a conduction band of a nanostructure. Such an electron is continually in interaction with its surroundings, such that a good approximation can be described in terms of a continuous (macroscopic) polarization field. The electron moves in the potential produced by the polarization charges, while the polarization is influenced in turn by the electron Coulomb field. The polarization field can undergo harmonic oscillations at a certain frequency ω, so that its energy can only be changed in multiples of $\hbar\omega$. For all values of ω a full quantum treatment of the problem is essential. Here there is a quantum field theory in which a low energy electron should be pictured surrounded by its self-induced polarization field. This assembly of electron and field in equilibrium with each other is usually called a polaron.

This is the simplest known physical example of a consistent nonrelativistic theory of an interaction between a particle and a quantized field, and it is interesting from a methodological point of view, quite apart from its physical implications. Here we shall be concerned mainly with the mathematical problem posed by the polaron, confining the attention to the wavefunction, energy, and effective mass.

6.5.1. Weak Coupling Polaron

The weak coupling polaron problem in the Q2D systems has been the subject of various theoretical discussions often based upon simplified models for the phonon spectrum and the electron–phonon interaction used in the calculations. According to the standard perturbation treatment, the electron self-energy due to the electron–LO phonon interaction in the weak coupling approximation is given by

$$\varepsilon_\nu - \varepsilon_\nu^{(0)} = \langle \nu | \hat{H}_{eph}^\pm | \nu' \rangle + \sum_{\nu' \neq \nu} \frac{\left| \langle \nu' | \hat{H}_{eph}^\pm | \nu \rangle \right|^2}{\varepsilon_\nu - \varepsilon_{\nu'}^{(0)} \pm \hbar\omega_{\vec{q}}} \quad (324)$$

where $\varepsilon_\nu^{(0)}$ are the unperturbed electron energies of the electron–phonon interaction and $\hbar\omega_{\vec{q}}$ is the phonon energy; the sign "+" indicates absorption and the sign "−" indicates emission of phonons. The first term on the right side of Eq. (324) is neglected so that the self-energy effect is of second order in the perturbation theory.

In this case $|\nu\rangle \equiv |n\vec{k}\rangle |N_{\vec{q}}\rangle$ and $\hat{H}_{eph}$ describes the electron–phonon interaction Hamiltonian in which different models and approaches are used.

The expression (324) contains all matrix elements between initial and intermediate states involving electron transitions induced by electron–LO phonon interaction. For any model used in the description of the electron–phonon interaction, and any kind of Q2D nanostructure considered, in obtaining the matrix element of Eq. (324), it has to be calculated terms of the form

$$\langle N_{q'} | \hat{b}_q | N_q \rangle \qquad \langle N_{q'} | \hat{b}_q^+ | N_q \rangle \quad (325)$$

where $\hat{b}_q$ and $\hat{b}_q^+$ are the annihilation and creation phonon operators respectively and $|N_q\rangle$ are the phonon wavefunctions. Therefore

$$\langle N_{q'} | \hat{b}_q | N_q \rangle = \sqrt{N_q}\delta(N_{q'} - N_q + 1)$$
$$\langle N_{q'} | \hat{b}_q^+ | N_q \rangle = \sqrt{N_q + 1}\delta(N_{q'} - N_q - 1) \quad (326)$$

N_q being related with the phonon number average in such a form that it can be written $N_q = [\exp(\hbar\omega_q/k_B T) - 1]^{-1}$, which equals zero when the temperature is zero. This means that at $T = 0$ K the LO-phonon absorption is neglected (in fact, it is sufficient to require $k_B T \ll \hbar\omega_q$).

6.5.2. LO-Phonon Confinement and Polaron Effect in a Quantum Well

Now, polaron binding energy and effective mass are calculated for the case of a GaAs/AlAs quantum well with the use of a phenomenological model for the long wavelength polar optical oscillations that account for the coupling between the longitudinal and transverse components of the vibrations as well as for the coupling between mechanical and electrical fields. It is shown that the contribution of GaAs-like modes

to the electron–phonon interaction in that system gives good polaronic corrections.

In this section, polaron binding energy and effective mass in a GaAs/AlAs double heterostructure (DHS) are calculated with the use of the long wavelength phonon theory of [128, 129, 156]. The consequent electron–phonon interaction Hamiltonian is directly written in the spirit of the general ideas stated in [157].

6.5.2.1. Review of the Model

In the theory presented in [128], a DHS grown along the z-axis is considered, with interfaces at $z = \pm L/2$. In each layer one deals with an isotropic continuum characterized by a reduced mass density ρ, a static (high frequency) dielectric constant $\kappa_0(\kappa_\infty)$, and a transverse frequency of the bulk material for $\vec{q} \to 0$, ω_T. β_L and β_T are two constants describing quadratic dispersion laws for small q of the longitudinal and transversal oscillation modes in the bulk, respectively. For $|z| > L/2$, semi-infinite media of the same nature are assumed.

The field of mechanical oscillations $\vec{u}(\vec{r})$ and the electrostatic potential $\varphi(\vec{r})$ are solutions of the coupled differential equation

$$\rho(\omega^2 - \omega_T^2)\vec{u}(\vec{r}) = \sigma\nabla\varphi(\vec{r}) + \rho\beta_L^2\nabla(\nabla \cdot \vec{u}(\vec{r}))$$
$$- \rho\beta_T^2\nabla \times (\nabla \times \vec{u}(\vec{r})) \quad (327)$$

$$\nabla^2\varphi(\vec{r}) = \frac{4\pi\sigma}{\kappa_\infty}\nabla \cdot \vec{u}(\vec{r}) \quad (328)$$

where $\sigma^2 = \omega_T^2\rho(\kappa_0 - \kappa_\infty)/4\pi$.

The matching boundary conditions at the interfaces are derived in a coherent way from the differential equations. They are:

- Continuity of $\vec{u}$ and φ at the interfaces.
- Continuity at the interfaces of the vector.

$$\tau_N = \tau \cdot \vec{N}_i, \quad (329)$$

where τ is a certain "stress tensor" whose components are given by

$$\tau_{ij} = -\rho(\beta_L^2 - 2\beta_T^2)\delta_{ij}\nabla \cdot \vec{u} - \rho\beta_T^2(\nabla_i\vec{u}_j + \nabla_j\vec{u}_i) \quad (330)$$

and $\vec{N}$ is the unit vector normal to the interface.

- Continuity at the interfaces of the normal component of the induction vector

$$\vec{D} = 4\pi\alpha\vec{u} - \kappa_\infty\nabla\varphi \quad (331)$$

These boundary conditions are both mechanical and electrical in character and were not artificially imposed, but derived from the field equations. It must be remarked that the field, Eqs. (327) and (328), and the corresponding boundary conditions are not restricted to be fulfilled in a particular geometry for the heterostructure, but they in fact apply to anyone.

For the AlAs/GaAs/AlAs double heterostructure, the condition $\vec{u} = 0$ at the interfaces seems to give a very good description of the oscillation modes. This can be justified by the fact that the AlAs side of the heterostructure does not mechanically vibrate in the range of the GaAs-like oscillations, since the bandwidth of GaAs and AlAs bulk optical phonons is much smaller than the gap between them. For that reason, AlAs behaves like a completely rigid medium GaAs for the mechanical oscillations, which do not significantly penetrate into AlAs. The other boundary conditions to be fulfilled in this case will be:

- Continuity of φ at the interfaces.
- Continuity of $\kappa_\infty \delta\varphi/\delta N$ at the interfaces.

In order to look for the solutions of Eqs (327) and (328) for the GaAs-like oscillations, $\vec{u}$ and $\varphi(\vec{r})$ are written, respectively, as

$$\vec{u}(\vec{r}) = \vec{U}(z)\exp(i\vec{q}\cdot\vec{r}_\perp) \tag{332}$$

$$\varphi(\vec{r}) = f(z)\exp(i\vec{q}\cdot\vec{r}_\perp) \tag{333}$$

$\vec{r}_\perp = (x, y)$, and $\vec{q}$ is the in-plane wavevector for the oscillations. $\vec{U}(z) = (X(z), Y(z), Z(z))$, such that $\vec{U}(z = \pm L/2 = 0)$.

According to the calculation procedure of [128], the component $X(z)$ of $\vec{U}$ is completely uncoupled to the other two components and to $f(z)$. It has a purely transverse character and corresponds to the so-called "shear horizontal modes," which are of a mechanical type. The coupled oscillation modes are divided in two groups depending on the symmetry of the electrostatic potential with respect to $z = 0$. By introducing the dimensionless variables $\xi = qL/2$ and $\eta = 2z/L$, the dispersion relations $\omega(q)$ of the odd potential states (OPS) are the nonspurious solutions of the transcendental equation

$$\left(\omega_{\text{LO}}^2 - \omega^2\right)[k_L\cos k_L + k_L][k_T\sin k_T\cosh\xi + \xi\cos k_T\sinh\xi]$$
$$- \left(\omega_{TO}^2 - \omega^2\right)[\xi\cos k_T - k_T\sin k_T]$$
$$\times [k_L\cos k_L\sinh\xi - \xi\sin k_L\cosh\xi] = 0 \tag{334}$$

where ω_{LO} and ω_{TO} are the frequencies of the longitudinal optical modes and the transversal optical modes in bulk GaAs, respectively. On the other hand

$$k_T = \sqrt{\frac{\omega_{\text{TO}}^2 - \omega^2}{4\beta_T^2}L^2 - \xi^2} \tag{335}$$

$$k_L = \sqrt{\frac{\omega_{\text{LO}}^2 - \omega^2}{4\beta_L^2}L^2 - \xi^2} \tag{336}$$

Here, the components Y, Z and the function $f(z)$ are

$$Y^{\text{OPS}}(\xi, \eta) = \begin{cases} \xi\sin k_L\eta + k_T\dfrac{S_L}{C_T}\left(\dfrac{\omega_{\text{LO}}^2 - \omega^2}{\omega_{\text{TO}}^2 - \omega^2}\right) \\ \quad \times \sin k_T\eta + S_L\left(\dfrac{\omega_{\text{LO}}^2 - \omega_{\text{TO}}^2}{\omega_{\text{TO}}^2 - \omega^2}\right) \\ \quad \times \exp(-\xi)\sinh\xi\eta \qquad |\eta| < 1 \\ 0 \qquad\qquad\qquad\qquad\qquad |n| > 1 \end{cases} \tag{337}$$

$$Z^{\text{OPS}}(\xi, \eta) = \begin{cases} k_L\sin k_L\eta - \xi\dfrac{S_L}{C_T}\left(\dfrac{\omega_{\text{LO}}^2 - \omega^2}{\omega_{\text{TO}}^2 - \omega^2}\right)\sin k_T\eta \\ + S_L\left(\dfrac{\omega_{\text{LO}}^2 - \omega_{\text{TO}}^2}{\omega_{\text{TO}}^2 - \omega^2}\right)\exp(-\xi)\cosh\xi\eta \quad |\eta| < 1 \\ 0 \qquad\qquad\qquad\qquad\qquad\qquad |\eta| > 1 \end{cases} \tag{338}$$

$$f^{\text{OPS}}(\xi, \eta) = \frac{2\pi\sigma L}{\kappa_\infty}A^{\text{OPS}}(\xi)\begin{cases} \sin k_L\eta - \dfrac{S_L}{\xi}\exp(-\xi) \\ \quad \times\sinh\xi\eta \qquad |\eta| < 1 \\ S_0\exp(-|\eta|\xi)\dfrac{\eta}{|\eta|} \quad |\eta| > 1 \end{cases} \tag{339}$$

Now, $\sigma, \kappa_0, \kappa_\infty$, and ρ correspond to the bulk GaAs. Furthermore,

$$\left(\frac{1}{A^{\text{OPS}}(\xi)}\right)^2 = \int_{-1}^1\left[\left|Y^{\text{OPS}}(\xi, \eta)\right|^2 + \left|Z^{\text{OPS}}(\xi, \eta)\right|^2\right]d\eta \tag{340}$$

$$S_L = \sqrt{\xi^2 + k_L^2}\sin(k_L + \theta_L) \tag{341}$$

$$C_T = \sqrt{\xi^2 + k_T^2}\cos(k_T + \theta_T) \tag{342}$$

with $\tan\theta_L = (k_L/\xi)$ and $\tan\theta_T = (k_T/\xi)$. The quantity S_0 is

$$S_0 = \sin k_L\cosh\xi - \frac{k_L}{\xi}\cos k_L\sinh\xi \tag{343}$$

The even potential states (EPS) have oscillation frequencies given by the non-spurious solutions of the equation

$$\left(\omega_{\text{LO}}^2 - \omega^2\right)[\xi\cos k_L - k_L\sin k_L][k_T\cos k_T\sinh\xi - \xi\sin k_T\cosh\xi]$$
$$- \left(\omega_{\text{TO}}^2 - \omega^2\right)[\xi\sin k_T + k_T\cos k_T]$$
$$\times [k_L\sin k_L\cosh\xi + \xi\cos k_L\sinh\xi] = 0 \tag{344}$$

and

$$Y^{\text{EPS}}(\xi, \eta) = \begin{cases} \xi\cos k_L\eta - k_T\dfrac{C_L}{S_T}\left(\dfrac{\omega_{\text{LO}}^2 - \omega^2}{\omega_{\text{TO}}^2 - \omega^2}\right) \\ \quad \times\cos k_T\eta + C_L\left(\dfrac{\omega_{\text{LO}}^2 - \omega_{\text{TO}}^2}{\omega_{\text{TO}}^2 - \omega^2}\right) \\ \quad \times\exp(-\xi)\cosh\xi\eta \qquad |\eta| < 1 \\ 0 \qquad\qquad\qquad\qquad\qquad |\eta| > 1 \end{cases} \tag{345}$$

$$Z^{\text{EPS}}(\xi, \eta) = \begin{cases} k_L \sin k_L \eta + \xi \dfrac{C_L}{S_T}\left(\dfrac{\omega_{\text{LO}}^2 - \omega^2}{\omega_{\text{TO}}^2 - \omega^2}\right) \\ \qquad \times \sin k_T \eta + C_L\left(\dfrac{\omega_{LO}^2 - \omega_{TO}^2}{\omega_{\text{TO}}^2 - \omega^2}\right) \\ \qquad \times \exp(-\xi)\sin h\xi\eta \qquad |\eta| < 1 \\ 0 \qquad\qquad\qquad\qquad\qquad |\eta| > 1 \end{cases}$$

(346)

$$f^{\text{EPS}}(\xi, \eta) = \frac{2\pi\sigma L}{\kappa_\infty} A^{\text{EPS}}$$

$$\times (\xi)\begin{cases} \cos k_L \eta - \dfrac{C_L}{\xi}\exp(-\xi)\cosh\xi\eta & |\eta| < 1 \\ C_0 \exp(-|\eta|\xi) & |\eta| > 1 \end{cases}$$

(347)

$$\left(\frac{1}{A^{\text{EPS}}(\xi)}\right)^2 = \int_{-1}^1 [|Y^{\text{EPS}}(\xi, \eta)|^2 + |Z^{\text{EPS}}(\xi, \eta)|^2]d\eta \quad (348)$$

$$C_L = \sqrt{\xi^2 + k_L^2}\,\cos(k_L + \theta_L) \tag{349}$$

$$S_L = \sqrt{\xi^2 + k_T^2}\,\sin(k_T + \theta_T) \tag{350}$$

$$C_0 = \cos k_L \sinh \xi + \frac{k_L}{\xi}\sin k_L \cosh \xi \tag{351}$$

The coupled modes are neither purely longitudinal (L) nor purely transverse (T). What can be said is that some modes are predominantly "L" (quasi-L) or predominantly "T" (quasi-T), in such a way that all modes have to be considered by their contribution to the electron–phonon interaction. The influence of the electric potential φ leads to a certain "interface" character in some modes, but pure interface modes do not exist at all. Interface modes would be obtained in the limit of a very wide DHS ($L \to \infty$) and they correspond to those modes described in the dielectric continuum model (DCM). For $q = 0$ the modes decouple, giving rise to a set of pure "L" modes with eigenfrequencies

$$\omega_m^2 = \omega_{\text{LO}}^2 - \beta_L^2\left(\frac{m\pi}{L}\right)^2 \qquad m = 1, 2, \ldots \tag{352}$$

and a set of pure "T" modes, with eigenfrequencies

$$\omega_m^2 = \omega_{\text{TO}}^2 - \beta_T^2\left(\frac{m\pi}{L}\right)^2 \qquad m = 1, 2, \ldots \tag{353}$$

As was previously mentioned, the solutions for the GaAs-like modes obtained within this theory exhibit very close agreement with the experimental data and the microscopic calculations. In [128] dispersion relations for these modes are shown for values of L equal to 20 and 56 Å and, also, the dispersion relation for the AlAs modes of a 20 Å GaAs /AlAs/GaAs DHS. At the same time, the amplitudes $Z(z)$ obtained in that work for the GaAs-like modes are depicted together with the ones coming from the microscopic calculation of [107] for $q = 0$ and 0.15 Å^{-1}, the agreement obtained

being, in fact, remarkable. On the other hand, the general features of the dispersion laws obtained in [128] are later reproduced in the modified DCM version introduced in [158].

6.5.2.2. Polaronic Corrections

The electron–phonon interaction Hamiltonian arising from the theory of the previous subsection is of the form [157]

$$\widehat{H}_{\text{eph}} = \sum_{\xi, m} C_m(\xi)\Phi_m(\xi, \eta)[\exp(i\xi \cdot \vec{R})\hat{b}_{\xi m} + \exp(-i\xi \cdot \vec{R})\hat{b}_{\xi m}^+]$$

(354)

where $\xi = L\vec{q}/2$ and $\vec{R} = 2\vec{r}_\perp/L$. Here, $\Phi_m(\xi, \eta)$ are the functions located to the right of the braces in Eqs. (339) and (347), while

$$C_m(\xi) = \left(\frac{\pi\hbar e^2 L \omega_{\text{LO}}^2}{\kappa^* S \omega_m(\xi)}\right)^{1/2} A_m(\xi) \tag{355}$$

In Eq. (355), $1/\kappa^* = 1/\kappa_\infty - 1/\kappa_0$ and S is the system area in the (x, y) plane.

In [159], a Green function many-body formalism for the electron–polar optical phonon interaction in semiconductor heterostructures was developed. The problem is reduced to an effective electron–electron interaction through a summation over phonon variables. For the case of a planar layered double heterostructure of width L in the limits of $T = 0$ and of a single particle, the expressions for the polaronic corrections to the effective mass and energy of an electron in the conduction band are obtained, in second order Rayleigh–Schrödinger perturbation theory, from the real part of the first order retarded self-energy [160]

$$\Sigma^{(1)}(\vec{k}, l) = -\frac{4m_w^* S}{\pi\hbar^2}\sum_{l_1 m}\int_0^\infty d\xi \int_0^\pi d\vartheta \left[(\xi M_{ll_1}^m(\xi)) \Big/ \right.$$
$$\left(4\xi^2 + \frac{2m_w^* L^2}{\hbar}\omega_m(\xi) - \frac{2m_w^* L^2}{\hbar^2}(\varepsilon_l - \varepsilon_{l_1})\right.$$
$$\left.\left. - 4L\xi k \cos \vartheta\right)\right]$$

(356)

Here ε_l is the energy corresponding to a given subband l in the quantum well and $M_{ll_1}^m$ are the electron–phonon matrix elements.

As usual, the integrand can be expanded in a series of the electron in plane wavevectors, neglecting the terms of order greater than two, and the expression for the electron total energy in this approximation is reduced to

$$E_{\vec{k}, l} = \varepsilon_l - \varepsilon_{l_1} + \frac{\hbar^2 k^2}{2m^*}(1 - \mu_l) \tag{357}$$

where

$$\varepsilon_l = 2\alpha\beta\hbar\omega_{LO}\sum_{l_1}\sum_m \int_0^\infty \frac{d\xi}{w_m(\xi)}$$
$$\times \frac{\xi^3 I_{ll_1}^m(\xi)}{\xi^2 + \gamma^2(\varepsilon_{l_1} - \varepsilon_l) + \beta^2 w_m(\xi)} \tag{358}$$

and

$$\mu_l = 4\alpha\beta^3 \sum_{l_1} \sum_m \int_0^\infty \frac{d\xi}{w_m(\xi)}$$

$$\times \frac{\xi^3 I_{ll_1}^m(\xi)}{(\xi^2 + \gamma^2(\varepsilon_{l_1} - \varepsilon_l) + \beta^2 w_m(\xi))^3} \qquad (359)$$

are the polaron corrections for the binding energy and the effective mass respectively, corresponding to a given subband l in the quantum well. The effective mass is

$$m_{\text{pol}}^{(l)} = \frac{m^* - w}{1 - \mu_l} \qquad (360)$$

The function $w_m(\xi) = \omega_m(\xi)/\omega_{\text{LO}}$ has been introduced together with the quantities $\gamma^2 = m_w^* L^2/2\hbar^2$ and $\beta^2 = (m_w^* \omega_{\text{LO}} L^2/2\hbar)$. m_w^* is the electron effective mass in the conduction band of the GaAs and α is the corresponding Fröhlich constant. Summation over the index m represents the inclusion of the contributions from all phonon modes in the system. The functions

$$I_{ll_1}^m(\xi) = A_m^2(\xi) \left| \int_{-\infty}^\infty \psi_{l_1}^*(\eta) \Phi_m(\xi, \eta) \psi_l(\eta) d\eta \right|^2 \qquad (361)$$

are evaluated by choosing a particular scheme for the electronic states in the QW. If a finite potential barrier model is adopted, the description would be more realistic, but a disadvantage will result from the difficulty in the incorporation of all upper levels (specifically, the continuum sates) because they are not analytically known. For that reason, the use of a finite QW model usually implies the application of the so-called leading term approximation (LTA), in which only the polaronic corrections corresponding to the values $l = l_1 = 1$ are calculated [110, 113]. Then, one only needs to know the —dimensionless—wavefunction $\psi_1(\eta)$,

$$\psi_1(\eta) = B_1 \begin{cases} \cos \kappa_w \eta & |\eta| < 1 \\ \cos \kappa_w \exp(\kappa_b(1 - |\eta|)) & |\eta| > 1 \end{cases} \qquad (362)$$

where

$$\kappa_w = \sqrt{\frac{m_w^* L^2}{2\hbar^2} \varepsilon_1} \qquad \kappa_b = \sqrt{\frac{m_b^* L^2}{2\hbar^2}(V_b - \varepsilon_1)} \qquad (363)$$

ε_1 is the solution of the equation

$$\tan\left[\frac{(2m_w^* \varepsilon_1)^{1/2}}{2\hbar} L\right] = \left[\frac{m_w^*(V_b - \varepsilon_1)}{m_b^* \varepsilon_1}\right]^{1/2} \qquad (364)$$

and the normalization constant of the state is

$$B_1 = \left(\frac{2\kappa_w \kappa_b}{\kappa_w + 2\kappa_w \kappa_b + \kappa_w \cos 2\kappa_w + \kappa_b \sin 2\kappa_w}\right)^{1/2} \qquad (365)$$

The quantity m_b^* is the conduction band effective mass of the AlAs. Here, like in [113], it is assumed that the electron is in the Γ band. Even when it is known that in the GaAs/AlAs system the lowest conduction band is really the X band,

this extra complication is going to be avoided. The value of the confinement potential will be taken as $V_b = 915$ meV.

Then, in the LTA, the expressions for the polaron corrections in the finite barrier QW model are

$$\varepsilon_1 = 2\alpha\beta\hbar\omega_{\text{LO}} \sum_m \int_0^\infty \frac{d\xi}{w_m(\xi)} \frac{\xi I_{ll}^m(\xi)}{\xi^2 + \beta^2 w_m(\xi)} \qquad (366)$$

and

$$\mu_1 = 4\alpha\beta^3 \sum_m \int_0^\infty \frac{d\xi}{w_m(\xi)} \frac{\xi^3 I_{ll}^m(\xi)}{[\xi^2 + \beta^2 w_m(\xi)]^3} \qquad (367)$$

with

$$I_{ll}^m(\xi) = B_1^4 (A_m^{\text{EPS}}(\xi))^2$$

$$\times \left\{ \frac{2\kappa_w \sin 2\kappa_w}{4\kappa_w^2 - k_L^2} \cos k_L + k_L \left(\frac{1}{k_L^2} - \frac{\cos 2\kappa_w}{4\kappa_w^2 - k_L^2}\right) \sin k_L \right.$$

$$- \frac{C_L}{\xi} \exp(-\xi) \left[\frac{2\kappa_w \sin 2\kappa_w}{4\kappa_w^2 + \xi^2} \cosh \xi\right.$$

$$\left. + \left(\frac{\xi \cos 2\kappa_w}{4\kappa_w^2 + \xi^2} + \frac{1}{\xi}\right) \sinh \xi\right] + 2C_0 \exp(-\xi) \frac{\cos^2 \kappa_w}{2\kappa_b + \xi}\right\}^2 \qquad (368)$$

Now, if the electronic states in the GaAs well are described within an infinite potential barrier model, it is possible to sum over all higher states when polaron corrections for $l = 1$ are calculated. The wavefunctions corresponding to the even states are

$$\psi_l^{(p)}(\eta) = \begin{cases} \cos \frac{l\pi}{2} \eta & |\eta| < 1 \\ 0 & |\eta| > 1 \end{cases} \qquad (369)$$

with energies

$$\varepsilon_l^{(p)} = \frac{\pi^2 \hbar^2 l^2}{2m_w^* L^2} \qquad l = 1, 3, 5, \ldots \qquad (370)$$

and, for the states of odd symmetry with respect to $\eta = 0$,

$$\psi_l^{(i)}(\eta) = \begin{cases} \sin \frac{l\pi}{2} \eta & |\eta| < 1 \\ 0 & |\eta| > 1 \end{cases} \qquad (371)$$

with

$$\varepsilon_l^{(p)} = \frac{\pi^2 \hbar^2 l^2}{2m_w^* L^2} \qquad l = 2, 4, 6, \ldots \qquad (372)$$

In this case, the expressions for the polaron corrections in the lowest subband will be

$$\varepsilon_1^{(l)} = 2\alpha\beta\hbar\omega_{\text{LO}} \sum_l \sum_m \int_0^\infty \frac{d\xi}{w_m(\xi)}$$

$$\times \frac{\xi I_{lm}(\xi)}{\xi^2 + \pi^2(l^2 - 1) + \beta^2 w_m(\xi)} \qquad (373)$$

$$\mu_1^{(l)} = 4\alpha\beta^3 \sum_l \sum_m \int_0^\infty \frac{d\xi}{w_m(\xi)}$$

$$\times \frac{\xi^3 I_{lm}(\xi)}{[\xi^2 + \pi^2(l^2-1) + \beta^2 w_m(\xi)]^3} \qquad (374)$$

$$I_{lm} = \begin{cases} F_{lm}(\xi) & l \text{ odd} \\ G_{lm}(\xi) & l \text{ even} \end{cases} \qquad (375)$$

where

$$F_{lm}(\xi) = 4(A_m^{\mathrm{EPS}}(\xi))^2$$

$$\times \left\{ 2k_L \left[\frac{1}{4k_L^2 - (l-1)^2\pi^2} - \frac{1}{4k_L^2 - (l+1)^2\pi^2} \right] \right.$$

$$\times \sin k_L + C_L(\exp(2\xi) - 1)$$

$$\left. \times \left[\frac{1}{4\xi^2 + (l+1)^2\pi^2} - \frac{1}{4\xi^2 + (l-1)^2\pi^2} \right] \right\}^2 \quad (376)$$

and

$$G_{lm}(\xi) = 4(A_m^{\mathrm{OPS}}(\xi))^2$$

$$\times \left\{ 2k_L \left[\frac{1}{4k_L^2 - (l-1)^2\pi^2} - \frac{1}{4k_L^2 - (l+1)^2\pi^2} \right] \right.$$

$$\times \cos k_L + S_L(\exp(2\xi) + 1)$$

$$\left. \times \left[\frac{1}{4\xi^2 + (l-1)^2\pi^2} - \frac{1}{4\xi^2 + (l+1)^2\pi^2} \right] \right\}^2 \quad (377)$$

In the calculation the values of the different parameters were taken to be $\omega_{\mathrm{LO}}(\mathrm{GaAs}) = 291.9$ cm^{-1}, $\omega_{\mathrm{TO}}(\mathrm{GaAs}) = 273.8$ cm^{-1}, $\kappa_0(\mathrm{GaAs}) = 13.18$, $\kappa_\infty(\mathrm{GaAs}) = 10.89$, $\beta_L(\mathrm{GaAs}) = 3.2 \times 10^5$ cm/s, $\beta_T(\mathrm{GaAs}) = 3.3 \times 10^5$ cm/s, $m_w^* = m_e^*(\mathrm{GaAs}) = 0.0665m_0$, and $m_b^* = m_e^*(\mathrm{AlAs}) = 0.15m_0$ (m_0 is the bare electron mass and $\alpha(\mathrm{GaAs}) = 0.068$).

Figure 35 shows the polaron binding energy (in units of $\hbar\omega_{\mathrm{LO}}$) as a function of the GaAs QW width; i.e.,

$$E_r = \frac{\varepsilon_1}{\hbar\omega_{\mathrm{LO}}} \qquad (378)$$

Curves (a) to (c) are respectively, the relative binding energy E_r in the LTA for a finite barrier QW, the relative binding energy in the LTA for an infinite barrier QW, and the relative polaron binding energy for an infinite barrier QW obtained after summing over all states $l \geq 1$ in the well. It can be seen that in the three cases, this magnitude exhibits an increasing behavior as L increases in the range under consideration. It is possible to infer that the growing tendency would reverse for $L \approx 200$ Å, but it will still remain well above the bulk value $E_r^{(\mathrm{bulk})} = \alpha$, which would be recovered for a sufficiently wide well. The range of values chosen for L obeys two main reasons. The first is related to the fact that small values of the well width have been a subject of much experimental and theoretical interest. The second reason is entirely of a practical character: for higher values of L, the number of solutions of Eqs. (334) and (344) increases very rapidly and the calculation process then becomes really tedious.

It is interesting to discuss the results reported for the lower values of L. Relative polaron binding energy is not reported for L below 20 Å for in that region a macroscopic continuum model for the GaAs long wavelength oscillations would certainly not work well and its validity is doubtful. When $L \to 0$, only the bulk AlAs is present, and in this subsection the contribution to the electron–phonon interaction coming from the electric potential associated to the barrier modes has not been included. On the other hand, it is known that the LTA fails for small values of L, because continuum states are then much more relevant. From Figure 35 it seems that for 20 Å $\leq L < 40$ Å LTA and the full calculation in the case of the infinite barrier QW give almost the same result. However, it must be remembered that the perfect confinement approximation is also much less realistic for smaller L [161]. Indeed, the comparison of curves (a) and (b), both corresponding to the LTA, shows that the infinite QW model could be suitable to study qualitative properties, but in order to have an adequate quantitative information, a finite barrier model should be employed. It must be observed that the biggest difference between (a) and (b) occurs precisely in a region of L where LTA is supposed to work better [162].

Figure 36 shows the relative polaron effective mass

$$M_r = \frac{m_{\mathrm{pol}}^1 - m_w^*}{m_w^*} \qquad (379)$$

as a function of L. Again, curve (a) represents the result obtained for a finite barrier QW in the LTA, curve (b) accounts for the result in the LTA for an infinite barrier QW, and (c) shows the same quantity for an infinite barrier QW with the contributions from all levels in the well. As usual, mass correction reverses its increasing behavior at lower values of L. As for the case of E_r, it is also possible to make a similar analysis for lower L with the distinction that the coincidence between LTA and the full calculation in the infinite QW case extends to higher values of the well width.

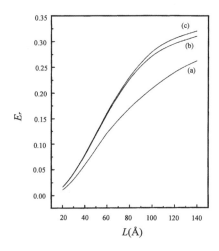

Fig. 35. Polaron relative binding energy for the lowest subband as a function of quantum well width.

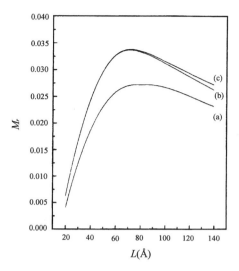

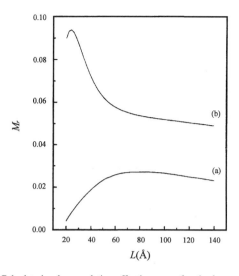

Fig. 36.   Polaron relative effective mass for the lowest subbands as a function of quantum well width.

Fig. 38.   Calculated polaron relative effective mass for the lowest subbands as a function of quantum well width.

The phonon model considered in this subsection predicts a maximum for the increment of the ground level parabolic conduction effective mass in the GaAs layer, for values of the well width around 80 Å (the bulk value of M_r is 0.011) due to the interaction between the electrons and the GaAs-like phonons. This may constitute an important element when studying the influence of that interaction on optical and transport properties, for instance.

Another important feature arising from the calculation can be observed in Figures 37 and 38. In both cases, curves (a) correspond to the polaron corrections for a finite barrier QW in the LTA, according to the model discussed in the previous subsection. Curves (b) correspond to the same quantities but calculated within the DCM [161–165] and include the contributions from the symmetric interface phonon modes

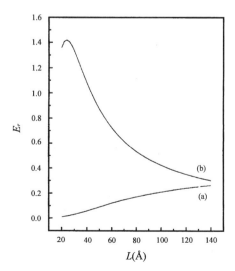

Fig. 37.   Calculated polaron relative binding energy for the lowest subbands as a function of quantum well width.

and from the confined LO-phonon modes appearing in that approach. The interesting thing here is that the values associated to the curves (b) are in fact multiplied by a factor of 0.25, in order to have an appropriate scale to show both results together. That means that the theory developed in [128, 130] to describe the long wavelength polar optical vibrations in layered GaAs/AlAs heterostructures gives values for the polaron effective mass and binding energy which are quite lower than those reported in previous works when only the GaAs-like phonon modes are considered. In other words, the calculation of the present work accounts for a considerably less intense electron–phonon interaction in comparison with DCM. From the results obtained in [161, 162], it follows that the main contribution to the polaron corrections in the lowest subband comes from the "interface modes." As was discussed in the previous subsection, the approach used along this subsection does not allow pure interface modes, but only a certain "interface" character that is associated with the more dispersing modes. Thus, the only way to conciliate the results of this model with those obtained in the DCM or the modified dielectric continuum model (MDCM) is to include the contribution from the AlAs-like modes with frequencies higher than ω_{LO}, which fit closely to DCM interface modes according to what is reported in [158]. This means that a different mathematical manipulation of the field Eqs. (327) and (328) will be needed, with a complete matching at the interfaces, as has been already done for GaAs/Al$_x$Ga$_{1-x}$ As single [166, 167] and double heterostructures [168]. Inclusion of AlAs modes will certainly raise the values of the polaron corrections, but what remains to be seen is if the contribution from those barrier modes would be significant for wide enough GaAs wells and, therefore, if the conclusion about the difference between the results here reported and the results from the DCM will be substantially modified; or one could always expect lower strength electron–phonon interaction when dealing with the phonon model developed in [128, 130].

7. THEORY OF QUANTUM TRANSPORT IN Q2D SYSTEMS

7.1. Introduction

Over the last 25 years, remarkable progress has been achieved in the study of semiconductor nanostructures. In particular, the two-dimensional electron gas confined in quantum wells and interfacial inversion layers with the typical thickness of 10 nm has been extensively studied due to its importance in the physics of low-dimensional electron systems and also in the development of high performance devices such as microwave transistors, Stark modulators, and low threshold semiconductor lasers. The transport of less tightly confined electrons normal to the double barrier and multiple barrier structures has been also widely studied to disclose such unique phenomena as the tunneling, the Bloch oscillation, and the resonant infrared photoconductive response caused by the intersubband excitation.

In the transport study, the artificial control of the electron wavefunction $\psi(x, y, z)$ in engineered Q2D structures plays very important roles. This is because the probabilities of quantum mechanical processes, such as scattering and optical transitions from the initial state i to the final state f induced by the interaction potential V_{int}, are usually determined by the square of the matrix element $\langle \psi_f | V_{int} | \psi_i \rangle$ and are therefore strongly dependent on the modification of the wavefunctions. On this respect, the envelope function approach will be used to discuss, in a unified way, several novel attempts to manipulate the scattering and transition processes in Q2D semiconductors.

7.1.1. Inhibition of Impurity Scattering in Q2D Nanostructured Materials

The spatial separation of the impurity potential V_{imp} from the wavefunctions $\psi(z)$ of the two-dimensional electron gas in a GaAs channel is the essence of the modulating doping scheme [169, 170]. As this approach results in the reduction of impurity scattering rates and, therefore, enhances the low field mobility of the electron, it is commonly used in preparing high-speed field-effect transistors [169, 170].

In polar semiconductors, whether they are bulk or quantum well systems, in general the interaction of electrons with phonons dominates the scattering mechanisms; for example at high temperatures and high electric fields, the electron–optical phonon interaction determines the main scattering mechanism. It also governs the intersubband relaxation process. Another of the fundamental scattering mechanisms is the electron–electron interaction, which is known to play a key role in establishing the thermal distribution of electrons and also in breaking the phase memory (or coherence) of electron waves.

The transport theory in Q2D systems depends on the directions of the electron movement, if it is free and with confined directions, then the transport properties are completely different in both directions. There exist free and confined directions for the electron movement in quantum wells. Therefore two fundamental types of quantum transport will be considered:

for free directions the mobility or conductivity is determined from the quasi-classical Boltzmann theory or the Kadanoff–Baym method, and for the confined direction it is necessary to apply a quantum transport theory via resonant tunneling. The quantum electron-wave transport caused by electric and magnetic fields will also be considered.

7.2. Electrical Conductivity in the Free Directions of Quasi-Two-Dimensional Electron Gas

In the previous sections the energy levels of quasi-two-dimensional electron gases at thermal and electrical equilibrium have been studied. It was found that the carrier motion along the growth axis is bound, whereas the in plane motion is free.

Under equilibrium conditions there is no permanent electrical current but transient local electrical currents do exist (since the electrons have a nonzero in plane momentum). However, the unavoidable electron scattering by the various defects and imperfections randomize the momentum directions on a time scale $\approx 10^{-12}$ s. As a result, no macroscopic permanent current flows in the layer plane. For electrical transport along the z-axis this is impossible, since the matrix element $\langle \chi_1 | \hat{p}_z | \chi_1 \rangle$ vanishes for a bound state.

Suppose that a weak, constant, and homogeneous electric field $\vec{E} = E \hat{e}_x$ has been applied along the x-axis. In addition to the scattering events, now the carrier experiences an acceleration along $\vec{E}$, resulting in a nonzero drift velocity. In the steady state situation and for weak electric fields, the areal density of electrical current $\vec{J} = J \hat{e}_x$ will be proportional to $\vec{E}$: the system has then reacted to a weak disturbance by a flux proportional to the disturbance (ohmic regime). Since only systems that are isotropic in the layer plane will be considered, the proportionality between the electrical current and the external field will not depend on the field orientation and thus will be a c-constant,

$$\vec{J} = \sigma \vec{E} \tag{380}$$

where σ is the two-dimensional ohmic conductivity of the electron gas.

In this context the means to calculate σ at low temperatures shall be studied as much as how to relate σ to the material parameters.

In multiple quantum well structures and superlattices, the levels of the isolated wells hybridize. There is thus a possibility of electrical conduction along the growth axis. In fact, in the pioneering work of Esaki and Tsu [171], they advocated the possibility of a negative differential resistance occurring in superlattices. In comparison with the in plane transport, very little effort has been devoted to the transport along the growth axis (also called vertical transport). Some of the features of this longitudinal transport, which appears to be so crucially dependent on the quality of the structures, will be discussed.

7.2.1. Static Conductivity of a Quasi-Two-Dimensional Electron Gas

The same assumptions as in the previous subsections are retained. Thus, it will be hereafter assumed that the carriers can be considered as independent and that (x, y) and z components of their motion can be decoupled. Apart from the spin, a quantum state is labelled by n, the subband indexes $(\varepsilon_1, \varepsilon_2, \ldots)$, and $\vec{k}_\perp = (k_x, k_y)$, the two-dimensional wavevector, which characterizes the in-plane motion. The energy and envelope wavefunction associated with $|n\vec{k}_\perp\rangle$ are thus written as

$$\varepsilon_n(\vec{k}_\perp) = \varepsilon_n + \frac{\hbar^2 k_\perp^2}{2m_{ne}^*} \tag{381}$$

$$\langle \vec{r} \,|\, n\vec{k}_\perp \rangle = \frac{1}{\sqrt{S}} \exp(i\vec{k}_\perp \cdot \vec{r}_\perp) \chi_n(z) \tag{382}$$

where $\vec{r}_\perp = (x, y)$, S is the sample area, and m_{ne}^* is the electron effective mass for the in plane motion of the nth subband.

At thermal and electrical equilibrium, the distribution functions of the states $|n\vec{k}_\perp\rangle$ (which are the diagonal elements of the one-electron density matrix) are the Fermi–Dirac functions

$$f_n^{(0)}(\vec{k}_\perp) = \frac{1}{1 + \exp\left[\beta(\varepsilon_n(\vec{k}_\perp) - \mu)\right]} \tag{383}$$

where

$$\beta = \frac{1}{k_B T} \tag{384}$$

and k_B is the Boltzmann constant, T is the electron temperature, and μ is the chemical potential.

It is of interest to find the linear electrical response of the quasi-two-dimensional electron gas to a weak, constant, and homogeneous electric field $\vec{E}$ applied parallel to the x-axis

$$\vec{E} = E\hat{e}_x \tag{385}$$

The one-electron Hamiltonian, including defects, is

$$\widehat{H} = \widehat{H}_0 + eEx + \widehat{H}_{\text{def}} \tag{386}$$

where $\widehat{H}_0$ is the unperturbed Hamiltonian whose eigenstates are $|n\vec{k}_\perp\rangle$ and $\widehat{H}_{\text{def}}$ includes the local fluctuations of the electrostatic potentials due to the defects (notice that the spatial averages of these potentials are included in $\widehat{H}_0$ and give rise to band bending).

$\widehat{H}_{\text{def}}$ is not invariant under translations in the layer plane. Thus, it broadens the $|n\vec{k}_\perp\rangle$ state and degrades the momentum of the carriers. $\widehat{H}_{\text{def}}$ has the property of being the sum of the contributions arising from randomly distributed scatterers located at $\vec{R}_i$

$$\widehat{H}_{\text{def}} = \sum_{\vec{R}_i} \widehat{H}_{\text{def}}(\vec{r} - \vec{R}_i) \tag{387}$$

where $\vec{R}_i$ are random variables.

A long time ago Kohn and Luttinger [172, 173] showed how to establish a rigorous connection between a fully quantum treatment of electrical transport (i.e., based on the solution of the Liouville equation for the density matrix) and the more familiar approach based on the solution of the Boltzmann equation [174]. The latter approach will be followed here.

Due to the presence of eEx and $\widehat{H}_{\text{def}}$ terms in Eq. (386), the distribution functions of the $|n\vec{k}_\perp\rangle$ states are no longer the $f_n^{(0)}$'s. Instead, each of the f_n's is permanently decreased (increased) by transitions from (to) the state $|n\vec{k}_\perp\rangle$ which arise from the other states $|n'\vec{k}_\perp'\rangle$. As a result, in the stationary regime $\left[(\partial f_n/\partial t) = 0\right]$, the $f_n(\vec{k}_\perp)$ are the solutions of the coupled Boltzmann equations,

$$-\frac{e}{m_{ne}^*} \hbar \vec{k}_\perp \cdot \vec{E} \frac{\partial f_n}{\partial \varepsilon_n(\vec{k}_\perp)}$$
$$= \sum_{n', k_\perp'} W_{n'k_\perp' \to nk_\perp} f_{n'}(\vec{k}_\perp')\left[1 - f_n(\vec{k}_\perp)\right]$$
$$- W_{nk_\perp \to n'k_\perp'} f_n(\vec{k}_\perp)\left[1 - f_{n'}(\vec{k}_\perp')\right] \tag{388}$$

where $W_{\alpha \to \beta}$ is the transition probability per unit time that a transition induced by $\widehat{H}_{\text{def}}$ takes place from the state $|\alpha\rangle$ to the state $|\beta\rangle$. From microreversibility it is known that [175]

$$W_{\alpha \to \beta} = W_{\beta \to \alpha} \tag{389}$$

The rates $W_{\alpha \to \beta}$ will be evaluated in the Born approximation. Thus

$$W_{\alpha \to \beta} = \frac{2\pi}{\hbar} \left|\langle \alpha |\widehat{H}_{\text{def}}| \beta \rangle\right|^2 \delta(\varepsilon_\alpha - \varepsilon_\beta) \tag{390}$$

When the scatterers are quite efficient, the Born approximation may not be good enough and a more refined treatment, the self-consistent Born approximation [176], may be necessary. In the latter scheme the collision broadening is also included in the energy levels ε_α, resulting in the replacement of the delta function by a Lorentzian function, whose width and level shift have to be self-consistently determined.

Retaining Eq. (390) for simplicity, the Boltzmann equation is rewritten in the form

$$-\frac{e}{m_{ne}^*} \hbar \vec{k}_\perp \cdot \vec{E} \frac{\partial f_n}{\partial \varepsilon_n(\vec{k}_\perp)}$$
$$= \frac{2\pi}{\hbar} \sum_{n', k_\perp'} \left\langle \left|\langle n\vec{k}_\perp |\widehat{H}_{\text{def}}| n'\vec{k}_\perp' \rangle\right|^2 \right\rangle_{\text{average}}$$
$$\times \left[f_{n'}(\vec{k}_\perp') - f_n(\vec{k}_\perp)\right] \delta\left[\varepsilon_{n'}(\vec{k}_\perp') - \varepsilon_n(\vec{k}_\perp)\right] \tag{391}$$

The bracket $\langle \cdots \rangle_{\text{average}}$ means that an average over the $\vec{R}_i$'s has been performed,

$$\langle A(\vec{R}_1, \vec{R}_2, \ldots, \vec{R}_N)\rangle_{\text{average}}$$
$$= \int A(\vec{R}_1, \vec{R}_2, \ldots, \vec{R}_N) P(\vec{R}_1, \vec{R}_2, \ldots, \vec{R}_N)$$
$$\times d^3 R_1 d^3 R_2 \cdots d^3 R_N \tag{392}$$

where $P(\vec{R}_1, \vec{R}_2, \ldots, \vec{R}_N)$ is the probability density of finding a scatterer at $\vec{R}_1$, another at $\vec{R}_2$, etc.

In the limit of diluted impurities the scattering sites become noncorrelated. Thus $P(\vec{R}_1, \vec{R}_2, \ldots, \vec{R}_N)$ factorizes in the form

$$P(\vec{R}_1, \vec{R}_2, \ldots, \vec{R}_N) = \prod_i P(\vec{R}_i) \quad (393)$$

and $P(\vec{R}_i)$ reduces to Ω_μ^{-1} where Ω_μ is the macroscopic volume of the heterostructure which is occupied by the μth kind of scatterers, e.g., $\Omega_\mu = Sl_d$ in the case of the ionized donors in a single modulation-doped heterojunction. Assuming noncorrelated scattering sites, the average over the impurity sites of a double sum like

$$S_{\mu\mu} = \sum_{\vec{R}_i^\mu} \sum_{\vec{R}_j^\mu} \langle n\vec{k}_\perp | \widehat{H}_{\text{def}}(\vec{R}_i^\mu) | n'\vec{k}_\perp' \rangle \langle n'\vec{k}_\perp' | \widehat{H}_{\text{def}}(\vec{R}_j^\mu) | n\vec{k}_\perp \rangle \quad (394)$$

reduces to a sum of the diagonal elements in i and j:

$$S_{\mu\mu} = \frac{N_\mu}{\Omega_\mu} \int d^3 R_\mu |\langle n\vec{k}_\perp | \widehat{H}_{\text{def}}(\vec{r} - \vec{R}_\mu) | n'\vec{k}_\perp' \rangle|^2 \quad (395)$$

The averaged sums $S_{\mu\nu}$ of crossed terms involving scatterers of different species vanish. Equation (395) represents a considerable simplification since it can be already anticipated that each of the defect species will contribute separately to the right-hand side of Eq. (391).

The assumption of diluted impurities is almost always used in the analysis of conductivity phenomena. For the specific case of modulation-doped heterojunctions it is probably an excellent approximation for the residual impurities, whose volume concentration is usually very low ($\approx 10^{14} \, \text{cm}^{-3}$). Its validity, as far as the deliberately introduced donors are concerned, is open to question since the donor concentration is usually quite large ($\approx 10^{18} \, \text{cm}^{-3}$). The validity of the assumption of diluted impurities has not been critically examined.

A solution to Eq. (391) in the form

$$f_j(\vec{k}_\perp) = f_j^{(0)}(\vec{k}_\perp) + \frac{e}{m_{je}^*} \hbar \vec{k}_\perp \cdot \vec{E} \frac{\partial f_j^{(0)}}{\partial \varepsilon_j(\vec{k}_\perp)} \tau_j[\varepsilon_j(\vec{k}_\perp)] \quad (396)$$

is looked at and Eq. (391) is made linear with respect to the field. This means that the τ_j's as well as the conductivity σ will be field independent. Siggia and Kwok [177], and Ando and Mori [178], have shown that the ansatz of (Eq. (396)) is a solution of Eq. (391), provided that the τ_j's fulfill the linear relations

$$-\sum_{\vec{k}_\perp \sigma} \frac{k_\perp^2}{m_{ie}^*} \frac{\partial f_i^{(0)}}{\partial \varepsilon_{\vec{k}_\perp}} = \sum_j K_{ij} \tau_j \quad (397)$$

where

$$K_{ij} = \frac{2\pi}{\hbar} \sum_{\vec{k}_\perp \vec{k}_\perp' \sigma\sigma'} \delta_{ij} \Biggl\{ \frac{\vec{k}_\perp}{m_{ie}^*} \cdot (\vec{k}_\perp' - \vec{k}_\perp) \delta[\varepsilon_i(\vec{k}_\perp) - \varepsilon_i(\vec{k}_\perp')]$$

$$\times \langle |\langle i\vec{k}_\perp \sigma | \widehat{H}_{\text{def}} | i\vec{k}_\perp' \sigma' \rangle|^2 \rangle_{\text{average}} \frac{\partial f_i^{(0)}}{\partial \varepsilon_{\vec{k}_\perp}}$$

$$- \frac{k_\perp^2}{m_{ie}^*} \frac{\partial f_i^{(0)}}{\partial \varepsilon_{\vec{k}_\perp}} \sum_l (1 - \delta_{il})$$

$$\times \delta[\varepsilon_i(\vec{k}_\perp) - \varepsilon_l(\vec{k}_\perp')] \langle |\langle i\vec{k}_\perp \sigma | \widehat{H}_{\text{def}} | l\vec{k}_\perp' \sigma' \rangle|^2 \rangle_{\text{average}} \Biggr\}$$

$$+ (1 - \delta_{ij}) \frac{\vec{k}_\perp \cdot \vec{k}_\perp'}{m_{je}^*} \delta[\varepsilon_i(\vec{k}_\perp) - \varepsilon_j(\vec{k}_\perp')]$$

$$\times \langle |\langle i\vec{k}_\perp \sigma | \widehat{H}_{\text{def}} | j\vec{k}_\perp' \sigma' \rangle|^2 \rangle_{\text{average}} \frac{\partial f_i^{(0)}}{\partial \varepsilon_{\vec{k}_\perp}} \quad (398)$$

and where the spin variables σ, σ' have been explicitly introduced. When $T = 0$ K Eqs. (397) and (398) can be further simplified. In this limit, the chemical potential μ coincides with the Fermi energy ε_F and

$$\frac{\partial f_i^{(0)}}{\partial \varepsilon_{\vec{k}_\perp}} = -\delta[\varepsilon_F - \varepsilon_i(\vec{k}_\perp)] \quad (399)$$

$$\sum_{\vec{k}_\perp \sigma} \hbar^2 \frac{k_\perp^2}{m_{ie}^*} \frac{\partial f_i^{(0)}}{\partial \varepsilon_{\vec{k}_\perp}} = -\frac{2m_{ie}^* S}{\pi \hbar^2} (\varepsilon_F - \varepsilon_i) \quad (400)$$

Thus, Eqs. (397) and (398) can be written as

$$\varepsilon_F - \varepsilon_i = \sum_j K_{ij}(\varepsilon_F) \tau_j(\varepsilon_F) \quad (401)$$

with

$$K_{ij}(T = 0) = \frac{\pi^2 \hbar^3}{m_{ie}^* S} \sum_{\vec{k}_\perp \vec{k}_\perp' \sigma\sigma'} \Biggl[\delta_{ij} \Bigl\{ \frac{\vec{k}_\perp}{m_{ie}^*} \cdot (\vec{k}_\perp - \vec{k}_\perp') \delta[\varepsilon_F - \varepsilon_i(\vec{k}_\perp')]$$

$$\times \delta[\varepsilon_F - \varepsilon_i(\vec{k}_\perp)] \langle |\langle i\vec{k}_\perp \sigma | \widehat{H}_{\text{def}} | i\vec{k}_\perp' \sigma' \rangle|^2 \rangle_{\text{average}}$$

$$+ \frac{k_\perp^2}{m_{ie}^*} \delta[\varepsilon_F - \varepsilon_i(\vec{k}_\perp)]$$

$$\times \sum_l (1 - \delta_{il}) \delta[\varepsilon_F - \varepsilon_l(\vec{k}_\perp')]$$

$$\times \langle |\langle i\vec{k}_\perp \sigma | \widehat{H}_{\text{def}} | l\vec{k}_\perp' \sigma' \rangle|^2 \rangle_{\text{average}} \Bigr\}$$

$$- (1 - \delta_{ij}) \frac{\vec{k}_\perp \cdot \vec{k}_\perp'}{m_{je}^*} \delta[\varepsilon_F - \varepsilon_i(\vec{k}_\perp)] \delta[\varepsilon_F - \varepsilon_j(\vec{k}_\perp')]$$

$$\times \langle |\langle i\vec{k}_\perp \sigma | \widehat{H}_{\text{def}} | j\vec{k}_\perp' \sigma' \rangle|^2 \rangle_{\text{average}} \Biggr] \quad (402)$$

The areal density of electrical current, averaged over the distribution function $f_{n'}$, is equal to

$$\langle \vec{J} \rangle = -\frac{e}{S} \sum_{i, \vec{k}_\perp, \sigma} \frac{\hbar \vec{k}}{m_{ie}^*} f_i[\varepsilon_i(\vec{k}_\perp)] \quad (403)$$

As expected, the zeroth order terms $f_l^{(0)}$ do not contribute to $\langle \vec{J} \rangle$, therefore

$$\langle \vec{J} \rangle = \frac{e^2 \hbar^2}{S} \sum_{i, \vec{k}_\perp, \sigma} \vec{k}_\perp \frac{\vec{k}_\perp \cdot \vec{E}}{\left(m_{ie}^*\right)^2} \tau_i(\varepsilon_F) \delta[\varepsilon_F - \varepsilon_i(\vec{k}_\perp)] \quad (404)$$

at $T = 0$ K. Exploiting the in plane isotropy of the dispersion relations $\varepsilon_i(\vec{k}_\perp)$, the formula (404) can be re-expressed in terms of the areal density of carriers $n_e^{(i)}$ in the ith subband:

$$\eta_e^{(i)} = \frac{m_{ie}^*}{\pi \hbar^2}(\varepsilon_F - \varepsilon_i) \quad (405)$$

The Ohm law is found

$$\langle \vec{J} \rangle = \sigma \vec{E} \quad (406)$$

where

$$\sigma = \sum_i \sigma_i \quad (407)$$

and

$$\sigma_i = \frac{n_e^{(i)} e^2 \tau_i(\varepsilon_F)}{m_{ie}^*} \quad (408)$$

The carrier mobility of the ith subband at zero temperature is given by

$$\mu_i = \frac{e}{m_{ie}^*} \tau_i(\varepsilon_F) \quad (409)$$

In terms of the total areal density of carriers n_e

$$n_e = \sum_i n_e^{(i)} \quad (410)$$

An effective mobility $\vec{\mu}$ can be defined, related to σ by

$$\sigma = n_e e \vec{\mu} \quad (411)$$

Thus,

$$\vec{\mu} = \sum_i \mu_i \frac{n_e^{(i)}}{n_e} \quad (412)$$

The very fact that $\tau_i(\varepsilon_F)$ is not simply proportional to $\varepsilon_F - \varepsilon_i$ but should be derived by inverting the matrix K_{ij} clearly shows that elastic intersubband transitions contribute to $\langle \vec{J} \rangle$. It is only in the electric quantum limit that Eq. (401) can be simplified to display a more transparent structure.

7.2.2. Electrical Conductivity in the Electric Quantum Limit

When only the ε_1 subband is populated, the only nonvanishing K_{ij} term is K_{11}. Let us define $\vec{q}_\perp$ by

$$\vec{q}_\perp = \vec{k}_\perp' - \vec{k}_\perp \quad (413)$$

$\hbar \vec{q}_\perp$ being the momentum gained by the carrier during a collision against the defects (see Fig. 39). Assume for simplicity

Figure 39

that $\widehat{H}_{\text{def}}$ is spin-conserving (i.e., does not involve spin-orbit couplings). K_{11} can then be rewritten as

$$K_{11} = -\frac{2\pi^2 \hbar^3}{\left(m_{1e}^*\right)^2 S} \sum_{\vec{k}_\perp, \vec{q}_\perp} \vec{k}_\perp \cdot \vec{q}_\perp \delta[\varepsilon_F - \varepsilon_1(\vec{k}_\perp + \vec{q}_\perp)]$$
$$\times \delta[\varepsilon_F - \varepsilon_1(\vec{k}_\perp)] \langle |\langle 1\vec{k}_\perp \sigma | \widehat{H}_{\text{def}} | 1\vec{k}_\perp' \sigma \rangle|^2 \rangle_{\text{average}} \quad (414)$$

However, from Figure 39, $\vec{k}_\perp \cdot \vec{q}_\perp = -2k_\perp^2 \sin^2 \theta / 2$.

Thus, by performing the summation over $\vec{k}_\perp$ and $\vec{k}_\perp'$, it is obtained that

$$K_{11}(T=0) = \frac{m_{1e}^* S}{2\pi \hbar^3}[\varepsilon_F - \varepsilon_1] \int_0^{2\pi} d\theta (1 - \cos \theta)$$
$$\times \langle |\langle 1k_\perp \sigma | \widehat{H}_{\text{def}} | 1(\vec{k}_\perp + \vec{q}_\perp)\sigma \rangle|^2 \rangle_{\text{average}}^{q_\perp = 2k_F |\sin(\theta/2)|} \quad (415)$$

and

$$\frac{1}{\tau_1(\varepsilon_F)} = \frac{m_{1e}^* S}{2\pi \hbar^3} \int_0^{2\pi} d\theta (1 - \cos \theta)$$
$$\times \langle |\langle 1k_\perp \sigma | \widehat{H}_{\text{def}} | 1(\vec{k}_\perp + \vec{q}_\perp)\sigma \rangle|^2 \rangle_{\text{average}}^{q_\perp = 2k_F |\sin(\theta/2)|} \quad (416)$$

where $\vec{k}_F$ is the Fermi wavevector

$$\varepsilon_F - \varepsilon_1 = \frac{\hbar^2 k_F^2}{2m_{1e}^*} \quad (417)$$

Expression (416) shows that the velocity relaxation frequency $1/\tau_1$ is the angular average over all possible elastic collisions which occur on the Fermi circle, of the product of the transition probability for an elastic scattering to change the carrier velocity orientation by the angle θ, by the relative velocity decrement $k_F(1 - \cos \theta)/k_F$.

7.2.3. Intersubband Scattering

Before evaluating $\widehat{H}_{\text{def}}$ for specific defect potentials, let us qualitatively describe the mobility anomalies taking place when the Fermi energy coincides with the onset of the first excited subband ε_2. Experimentally, two techniques have been used to modify ε_F in GaAlAs–GaAs heterojunctions. η_e has been changed by using the persistent photoconductivity effect or by biasing the heterostructure via, for instance, a Cu electrode placed on the GaAs side of the heterojunctions [179].

It was assumed that the ε_2 density of states is a perfect staircase and that $T = 0$ K. When $\varepsilon_F = \varepsilon_2 - \eta, \eta \to 0$,

$$\mu(\varepsilon_F) = \frac{e\tau_1}{m_{1e}^*} \qquad \tau_1 = \frac{(\varepsilon_F - \varepsilon_1)}{K_{11}} \quad (418)$$

On the other hand, when $\varepsilon_F = \varepsilon_2 + \eta, \eta \to 0$,

$$\mu(\varepsilon_F) = \frac{e\tau_1}{m_{1e}^*} \qquad \tau_1 \cong \frac{(\varepsilon_F - \varepsilon_1)}{K_{11} - K_{21}K_{12}/K_{22}} \qquad (419)$$

The second subband does not yet contribute to the conductivity (because $n_2 \to 0$) but the τ_1 expression has changed compared to Eq. (418) due to the nonvanishing K_{12} and the change in the expression for K_{11}. Let us examine K_{12} and K_{11} in some detail. From Eq. (402),

$$K_{12} = -\frac{2\pi^2\hbar^3}{Sm_{1e}^*m_{2e}^*}\sum_{\vec{k}_\perp,\vec{k}_\perp'}\vec{k}_\perp \cdot \vec{k}_\perp'\delta\left[\varepsilon_F - \varepsilon_1 - \frac{\hbar^2 k_\perp^2}{2m_{1e}^*}\right]$$

$$\times\delta\left[\varepsilon_F - \varepsilon_2 - \frac{\hbar^2 k_\perp'^2}{2m_{2e}^*}\right]\left\langle\left|\langle 1\vec{k}_\perp\sigma|\widehat{H}_{\text{def}}|2\vec{k}_\perp'\sigma\rangle\right|^2\right\rangle_{\text{average}} \quad (420)$$

Since $\varepsilon_F \approx \varepsilon_2, \vec{k}_\perp' \to 0$ and the $\widehat{H}_{\text{def}}$ matrix element is finite. Thus $K_{12} \approx 0$ because of the factor $\vec{k}_\perp'$ sitting in front of the delta functions.

As for K_{11}, the Eq. (402) can be rewritten as

$$K_{11}(\varepsilon_F = \varepsilon_2 + \eta)$$

$$= K_{11}(\varepsilon_F = \varepsilon_2 - \eta) + \frac{2\pi^2\hbar^3}{Sm_{1e}^*}$$

$$\times\sum_{\vec{k}_\perp}\frac{\vec{k}_\perp}{m_{1e}^*}\delta\left[\varepsilon_F - \varepsilon_1 - \frac{\hbar^2 k_\perp^2}{2m_{1e}^*}\right] \quad (421)$$

$$\times\sum_{\vec{k}_\perp'}\delta\left(\varepsilon_F - \varepsilon_2 - \frac{\hbar^2 k_\perp'^2}{2m_{2e}^*}\right)\left\langle\left|\langle 1\vec{k}_\perp\sigma|\widehat{H}_{\text{def}}|2\vec{k}_\perp'\sigma\rangle\right|^2\right\rangle_{\text{average}}$$

The important difference between Eqs. (420) and (421) is the absence in Eq. (421) of a $\vec{k}_\perp'$ term multiplying the delta function, so the sum over $\vec{k}_\perp'$ is nonvanishing. It amounts to the density of states of the ε_2 subband, which is constant if the $\widehat{H}_{\text{def}}$ matrix elements are $\vec{q}$ independent. Therefore, at the onset of the population of the second subband, $\tau_1(\varepsilon_F)$ and thus $\mu(\varepsilon_F)$ have a discontinuity (a drop since $\delta K_{11} > 0$) (see Fig. 40) which is entirely due to the peculiar shape of the two-dimensional density of states [178, 180].

Notice that for bulk materials (whose density of states vanishes at the band extreme), the same kind of reasoning would lead to a continuous $\mu(\varepsilon_F)$; only the derivative $d\mu/d\varepsilon_F$ would be singular. For more singular densities of states (e.g., those of quasi-one-dimensional materials), K_{11} would diverge and thus $\mu(\varepsilon_F)$ would vanish at the onset of the occupation of a one-dimensional subband (Fig. 41). However, near such singularities the simple Born approximation used is no longer justified (the scattering is no longer weak) and self-consistent treatments are required to provide sensible results.

In practice, the mobility drop can never be achieved in quasi-two-dimensional materials due to thermal and collisional broadenings, which blur the ε_2 onset. Nevertheless, pronounced decreases in mobility have been observed in GaAs–GaAlAs heterojunctions and rather detailed calculations of this mobility drop have been performed at $T = 0$ K.

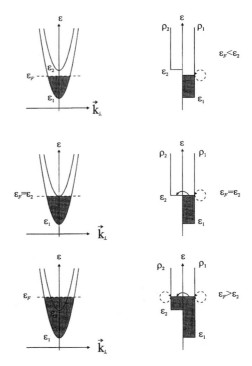

Fig. 40. Interband scattering processes in a quasi-two-dimensional electron gas at $T = 0$ K. Three situations have been sketched: $\varepsilon_F < \varepsilon_2$, $\varepsilon_F = \varepsilon_2$, $\varepsilon_F > \varepsilon_2$. The mobility exhibits a drop when $\varepsilon_F = \varepsilon_2$ due to the finite value of the density of states $\rho_2(\varepsilon_F)$ at $\varepsilon_F = \varepsilon_2$.

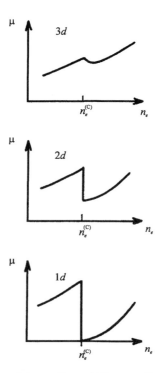

Fig. 41. Comparison between the mobility anomalies which take place at the onset of occupancy of an excited subband (or band) in three-dimensional (upper panel), quasi-two-dimensional (middle panel), and quasi-one-dimensional (lower panel) materials, respectively.

7.2.4. Screening in a Two-Dimensional Electron Gas

In modulation-doped heterostructures, the two-dimensional electron gas is heavily degenerate $((\varepsilon_F - \varepsilon_1)/k_B T \gg 1)$ at low temperatures. The interaction between a given electron and an impurity, or an optical phonon, or an alloy, etc. (i.e., with an external perturbation that is essentially of electromagnetic origin), is strongly modified by the presence of the other electrons. In fact, the impurity, or optical phonon, or alloy induces a spatially dependent charge density in the electron gas and the actual electron–electromagnetic external perturbation interaction should account for this spatial rearrangement of the electron gas. This modification of the electron–electromagnetic external perturbation interaction with respect to that of the bare one is called the screening. In the following, a linear response formalism to calculate the screening effects shall be adopted. This means that the electron–electromagnetic external potential will be assumed weak enough to modify the electron wavefunctions only to first order in additional $\widehat{V}$.

Suppose then that an electromagnetic external perturbation gives rise to a bare electrostatic potential $\phi_0(\vec{r})$. The latter induces a charge distribution $\eta_1(\vec{r})$, which itself gives rise to the electrostatic potential $\phi_1(\vec{r})$. The net electrostatic potential is $\phi_0(\vec{r}) + \phi_1(\vec{r})$, and an electron of charge $-e\,(e > 0)$, instead of experiencing only the self-consistent potential due to the heterojunction band bending (traslationally invariant in the layer plane), has the perturbed Hamiltonian

$$\widehat{H}(\vec{r}, z; t) = \widehat{H}_0(\vec{r}, z) + \widehat{V}(\vec{r}, z; t) \tag{422}$$

where the perturbation is

$$\widehat{V}(\vec{r}, z; t) = \widehat{V}(\vec{r}, z) \exp\left[(i\omega + \delta)t\right] \tag{423}$$

with $\delta \to +0$ for adiabatic connection (this limit must be taken after making $t \to -\infty$). Here $\vec{r} = (x, y)$ and ω is the scattering field frequency.

The unperturbed Hamiltonian in the effective mass approximation is given by

$$\widehat{H}_0 = -\frac{\hbar^2}{2m_{ne}^*}\left(\nabla_{\vec{r}}^2 + \nabla_z^2\right) + V_b(z) \tag{424}$$

with $V_b(z)$ the confining potential according to the chosen model, i.e., infinite triangular potential for a single heterostructure, infinite or finite barrier potential for a quantum well, self-consistent potential for a single heterostructure and quantum well, etc.

The statistical operator is defined as

$$\hat{\rho}(t) = \sum_{\vec{k}_\perp, n} \left|n\vec{k}_\perp\right\rangle f_{\vec{k}_\perp, n}(t)\left\langle n\vec{k}_\perp\right| \tag{425}$$

where $f_{\vec{k}_\perp, n}$ is the probability of occupation of the state $\left|n\vec{k}_\perp\right\rangle$. The time dependence of operators is written explicitly, but the dependence on the coordinates is considered implicitly.

In thermodynamic equilibrium,

$$\hat{\rho}_0\left|n\vec{k}_\perp\right\rangle = f_{\vec{k}_\perp, n}^{(0)}\left|n\vec{k}_\perp\right\rangle \tag{426}$$

where $f_{\vec{k}_\perp, n}^{(0)} \equiv f_n^{(0)}(\vec{k}_\perp) = f_n^{(0)}(\varepsilon_{\vec{k}_\perp})$ is the Fermi–Dirac distribution function.

In thermodynamic nonequilibrium,

$$\hat{\rho}(t) = \hat{\rho}_0 + \hat{\rho}_1(t) \tag{427}$$

Remember that in the canonical ensemble (since the number of electrons does not change),

$$\hat{\rho}(t) = \exp\left[\frac{1}{k_B T}(\Omega - \widehat{H}(t))\right] \tag{428}$$

where Ω is the thermodynamic potential

One can write

$$\hat{\rho}(t) = \exp\left[\frac{1}{k_B T}(\Omega - \widehat{H}_0 - \widehat{V}(t))\right]$$
$$= \hat{\rho}_0 \exp\left[-\frac{1}{k_B T}\widehat{V}(t)\right] \tag{429}$$

with

$$\hat{\rho}_0 = \exp\left[-\frac{1}{k_B T}(\Omega - \widehat{H}_0)\right] \tag{430}$$

If the perturbation is small enough $(V(t) \ll k_B T)$ then

$$\hat{\rho}(t) = \hat{\rho}_0\left\{1 - \frac{1}{k_B T}\widehat{V}(t)\right\} = \hat{\rho}_0 - \frac{1}{k_B T}\hat{\rho}_0\widehat{V}(t) \tag{431}$$

Comparing with the expression above,

$$\hat{\rho}_1(t) = -\frac{1}{k_B T}\hat{\rho}_0\widehat{V}(t) \tag{432}$$

i.e., $\hat{\rho}_1(t)$ is linear in $\widehat{V}(t)$.

The movement equation for the statistical operator states is as follows:

$$i\hbar\frac{\partial\hat{\rho}(t)}{\partial t} = \left[\widehat{H}(t), \hat{\rho}(t)\right] \tag{433}$$

Substituting Eqs. (422) and (427) in Eq. (433) and bearing in mind that $\hat{\rho}_1(t) \approx \widehat{V}(t)$, keeping only terms of first order in the perturbation expansion, it can be written

$$i\hbar\frac{\partial\hat{\rho}_1(t)}{\partial t} = \left[\widehat{H}_0, \hat{\rho}_1(t)\right] + \left[\widehat{V}(t), \hat{\rho}_0\right] \tag{434}$$

This is the linearized movement equation for the nonequilibrium part of the statistical operator.

Notice that

$$\hat{\rho}_1(t) = -\frac{1}{k_B T}\hat{\rho}_0\widehat{V}\exp\left[(i\omega + \delta)t\right]$$
$$= \hat{\rho}_1 \exp[(i\omega + \delta)t] \tag{435}$$

where obviously $\hat{\rho}_1 = -\hat{\rho}_0\widehat{V}/k_B T$. Thus

$$i\hbar\frac{\partial\hat{\rho}_1(t)}{\partial t} = i\hbar(i\omega + \delta)\hat{\rho}_1 \exp\left[(i\omega + \delta)t\right]. \tag{436}$$

Substituting Eqs. (423), (435), and (436) in Eq. (434) a time-independent equation

$$(-\hbar\omega + i\hbar\delta)\hat{\rho}_1 = [\widehat{H}_0, \hat{\rho}_1] + [\widehat{V}, \hat{\rho}_0] \qquad (437)$$

is obtained, which can be written as

$$(\hbar\omega - i\hbar\delta)\hat{\rho}_1 = \hat{\rho}_1\widehat{H}_0 - \widehat{H}_0\hat{\rho}_1 + \hat{\rho}_0\widehat{V} - \widehat{V}\hat{\rho}_0 \qquad (438)$$

Taking the matrix elements between the electron states $|n\vec{k}_\perp\rangle$ and $|n\vec{k}'_\perp\rangle$ it is obtained that

$$(\hbar\omega - i\hbar\delta)\langle n'\vec{k}'_\perp|\hat{\rho}_1|n\vec{k}_\perp\rangle$$
$$= \langle n'\vec{k}'_\perp|\hat{\rho}_1\widehat{H}_0|n\vec{k}_\perp\rangle - \langle n'\vec{k}'_\perp|\widehat{H}_0\hat{\rho}_1|n\vec{k}_\perp\rangle \qquad (439)$$
$$+ \langle n'\vec{k}'_\perp|\hat{\rho}_0\widehat{V}|n\vec{k}_\perp\rangle - \langle n'\vec{k}'_\perp|\widehat{V}\hat{\rho}_0|n\vec{k}_\perp\rangle$$

Since these are orthonormalized states, the following relation can be written:

$$\langle n'\vec{k}'_\perp n'\vec{k}'_\perp \mid n\vec{k}_\perp\rangle = \delta_{\vec{k}'_\perp, \vec{k}_\perp}\delta_{n'n} \qquad (440)$$

The unit operator can be defined as

$$\sum_{\vec{k}_\perp, n} |n\vec{k}_\perp\rangle\langle n\vec{k}_\perp| = \hat{1}. \qquad (441)$$

Equation (439) for the matrix elements can be rewritten as

$$(\hbar\omega - i\hbar\delta)\langle n'\vec{k}'_\perp|\hat{\rho}_1|n\vec{k}_\perp\rangle$$
$$= \langle n'\vec{k}'_\perp|\hat{\rho}_1|n\vec{k}_\perp\rangle\langle n\vec{k}_\perp|\widehat{H}_0|n\vec{k}_\perp\rangle$$
$$- \langle n'\vec{k}'_\perp|\widehat{H}_0|n'\vec{k}'_\perp\rangle\langle n'\vec{k}'_\perp|\hat{\rho}_1|n\vec{k}_\perp\rangle \qquad (442)$$
$$+ \langle n'\vec{k}'_\perp|\hat{\rho}_0|n'\vec{k}'_\perp\rangle\langle n'\vec{k}'_\perp|\widehat{V}|n\vec{k}_\perp\rangle$$
$$- \langle n'\vec{k}'_\perp|\widehat{V}|n\vec{k}_\perp\rangle\langle n\vec{k}_\perp|\hat{\rho}_0|n\vec{k}_\perp\rangle$$

but it is known that

$$\langle n\vec{k}_\perp|\widehat{H}_0|n\vec{k}_\perp\rangle = w_{n,\vec{k}_\perp} = \varepsilon_{\vec{k}_\perp} + \varepsilon_n$$
$$\langle n\vec{k}_\perp|\hat{\rho}_0|n\vec{k}_\perp\rangle = f^{(0)}_{\vec{k}_\perp} = f^{(0)}_n(\varepsilon_{\vec{k}_\perp}) \qquad (443)$$

Thus,

$$(\hbar\omega - i\hbar\delta)\langle n'\vec{k}'_\perp|\hat{\rho}_1|n\vec{k}_\perp\rangle$$
$$= \{\varepsilon_{\vec{k}_\perp} + \varepsilon_n - \varepsilon_{\vec{k}'_\perp} - \varepsilon_{n'}\}\langle n'\vec{k}'_\perp|\hat{\rho}_1|n\vec{k}_\perp\rangle$$
$$+ \{f^{(0)}_{n'}(\varepsilon_{\vec{k}'_\perp}) - f^{(0)}_n(\varepsilon_{\vec{k}_\perp})\}\langle n'\vec{k}'_\perp|\widehat{V}|n\vec{k}_\perp\rangle \quad (444)$$

and hence

$$\langle n'\vec{k}'_\perp|\hat{\rho}_1|n\vec{k}_\perp\rangle$$
$$= \frac{f^{(0)}_{n'}(\varepsilon_{\vec{k}'_\perp}) - f^{(0)}_n(\varepsilon_{\vec{k}_\perp})}{\varepsilon_{\vec{k}'_\perp} + \varepsilon_{n'} - \varepsilon_{\vec{k}_\perp} - \varepsilon_l + \hbar\omega - i\hbar\delta}\langle n'\vec{k}'_\perp|\widehat{V}|n\vec{k}_\perp\rangle \qquad (445)$$

a basic relation between matrix elements of the nonequilibrium part of the statistical operator and the perturbation Hamiltonian. It is remarked that this relation was obtained under

the assumption that the perturbation potential is small enough compared with the thermal energy, i.e., $V << k_BT$. Furthermore, since the thermal energy is nothing else but the statistical average value of the kinetic energy of the electrons, this very assumption allows the use of the Born approximation (first-order perturbation theory) to calculate the matrix elements of the perturbation potential, namely Fermi's Golden Rule.

It must be understood that the perturbation is

$$\widehat{V} = \widehat{V}_0 + \widehat{V}_1 \qquad (446)$$

where $\widehat{V}_0$ is the bare scattering potential (external perturbation) and $\widehat{V}_1$ is the screening potential (induced perturbation). The bare interaction corresponds to the scattering mechanism considered. Elastic scattering of electrons occurs when they interact with acoustic phonons, ion impurities, interface roughness, or some other defects. Each of these mechanisms is usually modeled in the same way and one deals with static screening ($\omega = 0$). On the other hand, the inelastic scattering of electrons by optical phonons depends on the starting model for this branch of lattice vibrations, leading to different interaction Hamiltonians, and one must deal with dynamical screening ($\omega \neq 0$).

Actually, in these cases the external perturbation comes from an essentially electromagnetic interaction. Therefore, it produces local changes in the "up to now" uniform density of electrons, giving rise to induced charges which act upon each electron via corresponding induced electric fields.

The induced perturbation, as a rule, reduces the effect of the external one. Because of that, it is usually named screening perturbation (there is a certain "antiscreening" effect reported by some authors) [181].

Now it is well understood that the perturbation potential is

$$\phi = \phi_0 + \phi_1 \qquad (447)$$

where ϕ_0 is the external electrostatic potential and ϕ_1 is the screening electrostatic potential.

Since the screening potential is associated with the induced charge, it satisfies the Poisson equation

$$\nabla^2\phi_1 = -\frac{4\pi e^2}{\kappa}\eta_1 \qquad (448)$$

with κ being the dielectric constant of the medium, not including the electron contribution.

Here, the density of electrons is

$$\eta = \eta_0 + \eta_1 \qquad (449)$$

where η_0 is the uniform density of electrons in the thermodynamic equilibrium state and η_1 is the "induced" density of electrons in the thermodynamic nonequilibrium state.

Multiplying by $\exp(-i\vec{q}\cdot\vec{r})$ and integrating on $\vec{r}$ both sides of Eq. (448) it is obtained that

$$\int [\nabla^2\phi_1(\vec{r}, z)]\exp(-i\vec{q}\cdot\vec{r})d\vec{r}$$
$$= -\frac{4\pi e^2}{\kappa}\int \eta_1(\vec{r}, z)\exp(-i\vec{q}\cdot\vec{r})d\vec{r} \qquad (450)$$

where

$$\nabla^2 = \frac{\partial^2}{\partial \vec{r}^2} + \frac{\partial^2}{\partial z^2} = \frac{\partial^2}{\partial x^2} + \frac{\partial^2}{\partial y^2} + \frac{\partial^2}{\partial z^2}$$

$$\int \cdots d\vec{r} = \int_{-\infty}^{\infty} \int_{-\infty}^{\infty} \cdots dx\, dy \qquad (451)$$

On the left side of Eq. (450) we have

$$\int [\nabla^2 \phi_1(\vec{r}, z)] \exp(-i\vec{q} \cdot \vec{r})\, d\vec{r}$$

$$= \int [\nabla_{\vec{r}}^2 \phi_1(\vec{r}, z)] \exp(-i\vec{q} \cdot \vec{r})\, d\vec{r}$$

$$+ \int [\nabla_z^2 \phi_1(\vec{r}, z)] \exp(-i\vec{q} \cdot \vec{r})\, d\vec{r} \qquad (452)$$

Integrating twice by parts the first term on the right side of Eq (452), it is obtained that

$$\int [\nabla_{\vec{r}}^2 \phi_1(\vec{r}, z)] \exp(-i\vec{q} \cdot \vec{r}) d\vec{r}$$

$$= (i\vec{q})^2 \int \phi_1 \exp(-i\vec{q} \cdot \vec{r}) d\vec{r} \qquad (453)$$

since the Coulomb potential and its derivative vanish at infinity.

The second integral on the right side of Eq. (452) is given by

$$\int [\nabla_z^2 \phi_1(\vec{r}, z)] \exp(-i\vec{q} \cdot \vec{r}) d\vec{r}$$

$$= \nabla_z^2 \int \phi_1(\vec{r}, z) \exp(-i\vec{q} \cdot \vec{r}) d\vec{r} \qquad (454)$$

This way, Eq. (452) is transformed in

$$\int [\nabla^2 \phi_1(\vec{r}, z)] \exp(-i\vec{q} \cdot \vec{r}) d\vec{r}$$

$$= [-q^2 + \nabla_z^2] \int \phi_1(\vec{r}, z) \exp(-i\vec{q} \cdot \vec{r}) d\vec{r} \quad (455)$$

The two-dimensional Fourier transforms are given by

$$\tilde{\phi}_q(z) = \frac{1}{S} \int \phi_1(\vec{r}, z) \exp(-i\vec{q} \cdot \vec{r}) d\vec{r}$$
$$\tilde{\eta}_q(z) = \frac{1}{S} \int \eta_1(\vec{r}, z) \exp(-i\vec{q} \cdot \vec{r}) d\vec{r} \qquad (456)$$

This way, the Poisson equation for the two-dimensional Fourier transform is obtained in the following form:

$$[-q^2 + \nabla_z^2] \tilde{V}_q(z) = -\frac{4\pi e^2}{\kappa} \tilde{\eta}_q(z) \qquad (457)$$

The average of the density of particles operator is

$$\eta = Sp\{\hat{\rho}\hat{\eta}\} = Sp\{[\hat{\rho}_0 + \hat{\rho}_1]\hat{\eta}\}$$
$$= Sp\{\hat{\rho}_0\hat{\eta}\} + Sp\{\hat{\rho}_1\hat{\eta}\} = \eta_0 + \eta_1 \qquad (458)$$

Consider the induced density of electrons in the thermodynamic nonequilibrium states,

$$\eta_1 = Sp\{\hat{\rho}_1\hat{\eta}\} = \sum_{\vec{k}_\perp, n} \langle n\vec{k}_\perp | \hat{\rho}_1 \hat{\eta} | n\vec{k}_\perp \rangle \qquad (459)$$

Taking into account Eq. (441) and the orthonormality of eigenstates it can be written that

$$\sum_{\vec{k}_\perp, n} |n\vec{k}_\perp\rangle \langle n\vec{k}_\perp | . n'\vec{k}_\perp' \rangle = \sum_{\vec{k}_\perp, n} |n\vec{k}_\perp\rangle \delta_{\vec{k}_\perp \vec{k}_\perp'} \delta_{ll'} = |n'\vec{k}_\perp'\rangle \quad (460)$$

Now, Eq. (459) can be rewritten as

$$\eta_1 = \sum_{\vec{k}_\perp', n'} \sum_{\vec{k}_\perp, n} \langle n'\vec{k}_\perp' | \hat{\rho}_1 | n\vec{k}_\perp \rangle \langle n\vec{k}_\perp | \hat{\eta} | n'\vec{k}_\perp' \rangle \qquad (461)$$

The density of particles operator at the point $(\vec{r}, z)$ is defined as

$$\hat{\eta}(\vec{r}, z) = \delta(\vec{r} - \vec{r}')\delta(z - z') \qquad (462)$$

Thus, in the coordinate representation,

$$\langle n\vec{k}_\perp | \hat{\eta} | n'\vec{k}_\perp' \rangle = \int \int \psi_{\vec{k}_\perp n}^*(\vec{r}', z')\delta(\vec{r} - \vec{r}')$$

$$\times \delta(z - z')\psi_{\vec{k}_\perp' n'}(\vec{r}', z') d\vec{r}' dz' \qquad (463)$$

That is,

$$\langle n\vec{k}_\perp | \hat{\eta} | n'\vec{k}_\perp' \rangle = \psi_{\vec{k}_\perp n}^*(\vec{r}, z)\psi_{\vec{k}_\perp' n'}(\vec{r}, z) \qquad (464)$$

Then Eq. (461) is transformed in

$$\eta_1(\vec{r}, z) = \sum_{\vec{k}_\perp n} \sum_{\vec{k}_\perp' n'} \langle n'\vec{k}_\perp' | \hat{\rho}_1 | n\vec{k}_\perp \rangle \psi_{\vec{k}_\perp n}^*(\vec{r}, z)\psi_{\vec{k}_\perp' n'}(\vec{r}, z) \quad (465)$$

Their two-dimensional Fourier transform is considered as

$$\tilde{\eta}_{\vec{q}}(z) = \frac{1}{S} \int \eta_1(\vec{r}, z) \exp(-i\vec{q} \cdot \vec{r}) d\vec{r}$$

$$= \sum_{\vec{k}_\perp n} \sum_{\vec{k}_\perp' n'} \langle n'\vec{k}_\perp' | \hat{\rho}_1 | n\vec{k}_\perp \rangle$$

$$\times \int \psi_{\vec{k}_\perp n}^*(\vec{r}, z)\psi_{\vec{k}_\perp' n'}(\vec{r}, z) \exp(-i\vec{q} \cdot \vec{r}) d\vec{r} \qquad (466)$$

but

$$\int \psi_{\vec{k}_\perp n}^*(\vec{r}, z)\psi_{\vec{k}_\perp' n'}(\vec{r}, z) \exp(-i\vec{q} \cdot \vec{r}) d\vec{r}$$

$$= \frac{1}{S} \int \exp(-i\vec{k}_\perp \cdot \vec{r})\chi_n^*(z)\exp(i\vec{k}_\perp' \cdot \vec{r})\chi_{n'}(z)$$

$$\times \exp(-i\vec{q} \cdot \vec{r}) d\vec{r} = \frac{1}{S} \chi_n^*(z)\chi_{n'}(z)\delta_{\vec{k}_\perp', \vec{k}_\perp + \vec{q}} \qquad (467)$$

This way the Poisson equation results in

$$\left[-q^2 + \frac{d^2}{dz^2} \right] \tilde{\phi}_{\vec{q}}(z) = -4\pi J(z) \qquad (468)$$

where

$$J(z) = \frac{e^2}{\kappa S} \sum_{\vec{k}_\perp, n} \sum_{\vec{k}_\perp', n'} \langle n'\vec{k}_\perp' | \hat{\rho}_1 | n\vec{k}_\perp \rangle \chi_n^*(z)\chi_{n'}(z)\delta_{\vec{k}_\perp', \vec{k}_\perp + \vec{q}} \quad (469)$$

and must be solved together with the matching boundary conditions corresponding to the considered problem.

Single heterostructures and quantum wells must be treated. Usually the two materials forming the semiconductor heterostructures have approximately the same dielectric constant, e.g., $Al_xGa_{1-x}As$ and $GaAs$.

Now, the difference between the corresponding dielectric constants is simply neglected and thus the Eqs. (468) and (469) must be solved together with the simplest boundary conditions:

$$\tilde{\phi}_{\vec{q}}(-\infty) = 0; \qquad \tilde{\phi}_{\vec{q}}(+\infty) = 0$$
$$\frac{d}{dz}\tilde{\phi}_{\vec{q}}(-\infty) = 0; \qquad \frac{d}{dz}\tilde{\phi}_{\vec{q}}(+\infty) = 0 \qquad (470)$$

For this kind of problem Eq. (468) can be written in the form

$$\frac{d^2}{dz^2}G(z, z') + \alpha^2 G(z, z') = -4\pi\delta(z - z') \qquad (471)$$

with the solution

$$G(z, z') = \frac{2\pi i}{\alpha}\exp(i\alpha|z - z'|) \qquad (472)$$

named the Green function of the problem.

In this case $\alpha = iq$. Then

$$G(z, z') = \frac{2\pi i}{q}\exp(iq|z - z'|) \qquad (473)$$

Thus, the solution of the Poisson equation (Eq. (468)) is

$$\tilde{\phi}_{\vec{q}}(z) = \int G(z - z')J(z')dz'$$
$$= \frac{2\pi}{q}\int J(z')\exp(-q|z - z'|)dz' \qquad (474)$$

which fulfills the matching boundary conditions, as can easily be shown.

Consider

$$|z - z'| = \begin{cases} z - z' & \text{when } z - z' > 0 \quad \text{i.e., } z > z' \\ z' - z & \text{when } z - z' < 0 \quad \text{i.e., } z < z' \end{cases} \qquad (475)$$

Then

$$\exp(-q|z - z'|) = \begin{cases} \exp[-q(z - z')] & z > z' \\ \exp[-q(z' - z)] & z < z' \end{cases} \qquad (476)$$

obviously vanishes when $z \to +\infty$ or $z \to -\infty$ if z' is fixed (and finite)

This way,

$$\frac{d}{dz}\exp(-q|z - z'|) = \begin{cases} -q\exp[-q(z - z')] & z > z' \\ q\exp[-q(z' - z)] & z < z' \end{cases} \qquad (477)$$

also vanishes when $z \to \pm\infty$ at fixed z'.

Finally the solution of Poisson equation can be written as follows:

$$\tilde{\phi}_{\vec{q}}(z) = \frac{2\pi}{q}\frac{e^2}{\kappa S}\sum_{\vec{k}_\perp, n}\sum_{\vec{k}'_\perp n'}\langle n'\vec{k}'_\perp|\hat{\rho}_1|n\vec{k}_\perp\rangle\delta_{\vec{k}'_\perp, \vec{k}_\perp + \vec{q}}$$
$$\times \int \exp(-q|z - z'|)\chi_n^*(z')\chi_{n'}(z')dz' \qquad (478)$$

Taking into account the relation given by Eq. (445), Eq. (478) can be written as

$$\tilde{\phi}_{\vec{q}}(z) = \frac{2\pi}{q}\frac{e^2}{\kappa S}\sum_{\vec{k}_\perp, n}\sum_{\vec{k}'_\perp n'}\frac{f_{n'}^{(0)}(\varepsilon_{\vec{k}'_\perp}) - f_n^{(0)}(\varepsilon_{\vec{k}_\perp})}{\varepsilon_{\vec{k}'_\perp} + \varepsilon_{n'} - \varepsilon_{\vec{k}_\perp} - \varepsilon_n + \hbar\omega - i\hbar\delta}$$
$$\times \langle n'\vec{k}'_\perp|\hat{\phi}|n\vec{k}_\perp\rangle\delta_{\vec{k}'_\perp, \vec{k}_\perp + \vec{q}}$$
$$\times \int \exp(-q|z - z'|)\chi_n^*(z')\chi_{n'}(z')dz' \qquad (479)$$

Now it is convenient to note that the action of a perturbation operator reduces to multiply the eigenstate by the scalar potential, i.e.,

$$\langle n'\vec{k}'_\perp|\hat{\phi}|n\vec{k}_\perp\rangle \equiv \langle n'\vec{k}'_\perp|\phi|n\vec{k}_\perp\rangle \qquad (480)$$

The action of the delta-Kronecker in the sum on $\vec{k}'_\perp$ results in the following:

$$\tilde{\phi}_{\vec{q}}(z) = 2\frac{2\pi}{q}\frac{e^2}{\kappa S}\sum_{\vec{k}_\perp, n}\sum_{n'}\frac{f_{n'}^{(0)}(\varepsilon_{\vec{k}_\perp + \vec{q}}) - f_n^{(0)}(\varepsilon_{\vec{k}_\perp})}{\varepsilon_{\vec{k}_\perp + \vec{q}} + \varepsilon_{n'} - \varepsilon_{\vec{k}_\perp} - \varepsilon_n + \hbar\omega - i\hbar\delta}$$
$$\times \langle n'\vec{k}_\perp + \vec{q}|\hat{\phi}|n\vec{k}_\perp\rangle$$
$$\times \int \exp(-q|z - z'|)\chi_n^*(z')\chi_{n'}(z')dz' \qquad (481)$$

The factor 2 is due to the spin. It will also be taken into account when $\sum_{\vec{k}_\perp}$ is transformed to an integral on the energy. Notice that

$$\langle n'\vec{k}_\perp + \vec{q}|\hat{\phi}|n\vec{k}_\perp\rangle$$
$$= \frac{1}{S}\int \exp[-i(\vec{k}_\perp + \vec{q})\cdot\vec{r}]\chi_{n'}^*(z')\phi(\vec{r}, z')$$
$$\times \exp(i\vec{k}_\perp \cdot \vec{r})\chi_n(z')d\vec{r}\,dz'$$
$$= \int \chi_{n'}^*(z')\left\{\frac{1}{S}\int \phi(\vec{r}, z')\exp(-i\vec{q}\cdot\vec{r})d\vec{r}\right\}\chi_n(z')dz'$$
$$= \int \chi_{n'}^*(z')\phi_{\vec{q}}(z')\chi_n(z')dz' \qquad (482)$$

That is,

$$\langle n'\vec{k}_\perp + \vec{q}|\hat{\phi}(\vec{r}, z')|n\vec{k}_\perp\rangle = \langle n'|\phi_{\vec{q}}(z')|n\rangle \qquad (483)$$

Thus,

$$\tilde{\phi}_{\vec{q}}(z) = 2\frac{2\pi}{q}\frac{e^2}{\kappa S}\sum_{n,n'}\sum_{\vec{k}_{\perp}}\frac{f_{n'}^{(0)}(\varepsilon_{\vec{k}_{\perp}+\vec{q}}) - f_n^{(0)}(\varepsilon_{\vec{k}_{\perp}})}{\varepsilon_{\vec{k}_{\perp}+\vec{q}} + \varepsilon_{n'} - \varepsilon_{\vec{k}_{\perp}} - \varepsilon_n + \hbar\omega - i\hbar\delta}$$

$$\times \langle n'|\phi_{\vec{q}}(z'')|n\rangle \int \exp(-q|z - z'|)$$

$$\times \chi_n^*(z')\chi_{n'}(z')dz' \tag{484}$$

Notice that z'' and z' are just integration variables in two different integrals. The left side corresponds to the screening potential, while in the right one they are matrix elements of the perturbation potential. So it is convenient to calculate the matrix elements in the form of

$$\langle m'|\tilde{\phi}_{\vec{q}}(z)|m\rangle$$

$$= 2\frac{2\pi}{q}\frac{e^2}{\kappa S}\sum_{n,n'}\sum_{\vec{k}_{\perp}}\frac{f_{n'}^{(0)}(\varepsilon_{\vec{k}_{\perp}+\vec{q}}) - f_n^{(0)}(\varepsilon_{\vec{k}_{\perp}})}{\varepsilon_{\vec{k}_{\perp}+\vec{q}} + \varepsilon_{n'} - \varepsilon_{\vec{k}_{\perp}} - \varepsilon_n + \hbar\omega - i\hbar\delta}$$

$$\times \langle n'|\phi_{\vec{q}}(z'')|n\rangle \int \chi_{m'}^*(z)$$

$$\times \{\exp(-q|z - z'|)\chi_n^*(z')\chi_{n'}(z')dz'\}\chi_m(z)dz \tag{485}$$

Introducing the polarizability

$$\prod_{nn'}(q, \omega) = \sum_{\vec{k}_{\perp}}\frac{f_{n'}^{(0)}(\varepsilon_{\vec{k}_{\perp}+\vec{q}}) - f_n^{(0)}(\varepsilon_{\vec{k}_{\perp}})}{\varepsilon_{\vec{k}_{\perp}+\vec{q}} + \varepsilon_{n'} - \varepsilon_{\vec{k}_{\perp}} - \varepsilon_n + \hbar\omega - i\hbar\delta} \tag{486}$$

and the form factor

$$F_{m'mnn'}(q) = \int dz \int dz' \exp(-q|z - z'|)$$

$$\times \chi_{m'}^*(z)\chi_m(z)\chi_n^*(z')\chi_{n'}(z') \tag{487}$$

then

$$\langle m'|\tilde{\phi}_{\vec{q}}(z)|m\rangle = 2\frac{2\pi e^2}{q\kappa S}\sum_n\sum_{n'}\prod_{nn'}(q, \omega)$$

$$\times F_{m'mnn'}(q)\langle n'|\phi_{\vec{q}}(z)|n\rangle \tag{488}$$

The susceptibility can also be introduced;

$$X_{m'mnn'}(q, \omega) = \frac{e^2}{q\kappa S}\prod_{nn'}(q, \omega)F_{m'mnn'}(q) \tag{489}$$

Then write

$$\langle m'|\tilde{\phi}_{\vec{q}}(z)|m\rangle = 4\pi\sum_n\sum_{n'}X_{m'mnn'}(q, \omega)\langle n'|\phi_{\vec{q}}(z)|n\rangle \tag{490}$$

Turning to the expression for the perturbation potential

$$\phi(\vec{r}, z) = \phi_0(\vec{r}, z) + \phi_1(\vec{r}, z) \tag{491}$$

where $\phi_0(\vec{r}, z)$ and $\phi_1(\vec{r}, z)$ are the external and the screening potential, respectively.

Multiplying (491) by $\exp(-i\vec{q}\cdot\vec{r})/.S$ and integrating in $\vec{r}$ it is obtained that

$$\frac{1}{S}\int\phi(\vec{r}, z)\exp(-i\vec{q}\cdot\vec{r})d\vec{r}$$

$$= \frac{1}{S}\int\phi_0(\vec{r}, z)\exp(-i\vec{q}\cdot\vec{r})d\vec{r}$$

$$+ \frac{1}{S}\int\phi_1(\vec{r}, z)\exp(-i\vec{q}\cdot\vec{r})d\vec{r} \tag{492}$$

That is,

$$\tilde{\phi}_{\vec{q}}(z) = \tilde{\phi}_{\vec{q}}^0(z) + \tilde{\phi}_{\vec{q}}^1(z) \tag{493}$$

which is the relation expected for the two-dimensional Fourier transform.

Now multiplying Eq. (493) from the left by $\chi_{l'}^*(z)$, to the right by $\chi_l(z)$, and integrating in z, it is obtained that

$$\int\chi_{m'}^*(z)\tilde{\phi}_{\vec{q}}(z)\chi_m(z)dz = \int\chi_{m'}^*(z)\tilde{\phi}_{\vec{q}}^0(z)\chi_m(z)dz$$

$$+ \int\chi_{m'}^*(z)\tilde{\phi}_{\vec{q}}^1(z)\chi_m(z)dz \tag{494}$$

For the matrix elements can also be written

$$\langle m'|\tilde{\phi}_{\vec{q}}(z)|m\rangle = \langle m'|\tilde{\phi}_{\vec{q}}^0(z)|m\rangle + \langle m'|\tilde{\phi}_{\vec{q}}^1(z)|m\rangle \tag{495}$$

Notice that

$$\langle m'\vec{k}_{\perp}'|\tilde{\phi}_{\vec{q}}(z)|\vec{k}_{\perp}m\rangle = \langle m'|\langle\vec{k}_{\perp}'|\tilde{\phi}_{\vec{q}}(z)|\vec{k}_{\perp}\rangle|m\rangle \tag{496}$$

and

$$\langle\vec{k}_{\perp}'|\phi(\vec{r}, z)|\vec{k}_{\perp}\rangle = \frac{1}{S}\int\exp(-i\vec{k}_{\perp}'\cdot\vec{r})\phi(\vec{r}, z)\exp(i\vec{k}_{\perp}\cdot\vec{r})d\vec{r}$$

$$= \frac{1}{S}\int\phi(\vec{r}, z)\exp[-i(\vec{k}_{\perp}' - \vec{k}_{\perp})\cdot\vec{r}]d\vec{r}$$

$$= \tilde{\phi}_{\vec{k}_{\perp}'-\vec{k}_{\perp}}(z) \tag{497}$$

Actually in this problem $\vec{k}_{\perp}' - \vec{k}_{\perp} = \vec{q}$ because of the momentum conservation in a plane parallel to the interfaces. Anyway it can be accepted as a variables change, so that

$$\langle\vec{k}_{\perp}'|\phi(\vec{r}, z)|\vec{k}_{\perp}\rangle = \tilde{\phi}_{\vec{q}}(z) \tag{498}$$

is the two-dimensional Fourier transform.

The matrix element for the perturbation potential can be written as follows:

$$\langle m'|\tilde{\phi}_{\vec{q}}(z)|m\rangle = \sum_n\sum_{n'}\langle n'|\tilde{\phi}_{\vec{q}}(z)|n\rangle\delta_{m'n'}\delta_{mn} \tag{499}$$

Now the relation between matrix elements is expressed in the form

$$\langle m'|\tilde{\phi}_{\vec{q}}^0(z)|m\rangle = \langle m'|\tilde{\phi}_{\vec{q}}(z)|m\rangle - \langle m'|\tilde{\phi}_{\vec{q}}^1(z)|m\rangle \tag{500}$$

and taking into account Eqs. (488) and (490),

$$\langle m'|\tilde{\phi}_{\vec{q}}^0(z)|m\rangle = \sum_n \sum_{n'} (\delta_{mn}\delta_{m'n'} - 4\pi X_{m'mnn'}(q,\omega))$$
$$\times \langle n'|\tilde{\phi}_{\vec{q}}(z)|n\rangle \qquad (501)$$

or

$$\langle m'|\tilde{\phi}_{\vec{q}}^0(z)|m\rangle = \sum_n \sum_{n'} \in_{m'mnn'}(q,\omega)\langle n'|\tilde{\phi}_{\vec{q}}(z)|n\rangle \qquad (502)$$

where

$$\in_{m'mnn'}(q,\omega) = \delta_{m'n'}\delta_{mn} - 4\pi X_{m'mnn'}(q,\omega) \qquad (503)$$

is known as the dielectric function for a quasi-two-dimensional system.

Bearing in mind the expression for the susceptibility it is found that

$$\in_{m'mnn'}(q,\omega) = \delta_{m'n'}\delta_{mn}$$
$$- \frac{4\pi e^2}{q\kappa S}\prod_{nn'}(q,\omega)F_{m'mnn'}(q,\omega) \qquad (504)$$

where the polarizability and the form factor were previously defined.

7.2.5. Dielectric Function in Single Heterostructures

Consider the case of a single heterostructure and take into account the different dielectric constants at both sides of the interface.

Now the Poisson equation (448) is transformed in

$$\nabla \cdot [\kappa(z)\nabla\phi_1(\vec{r},z)] = -4\pi e^2 \eta_1(\vec{r},z) \qquad (505)$$

where

$$\kappa(z) = \begin{cases} \kappa_b & \text{for } z < 0 \quad \text{i.e., in the barrier} \\ & \text{(e.g., Al}_x\text{Ga}_{1-x}\text{As)} \\ \kappa_w & \text{for } z > 0 \quad \text{i.e., in the well} \\ & \text{(e.g., GaAs)} \end{cases} \qquad (506)$$

Therefore

$$\nabla^2\phi_1(\vec{r},z) = -\frac{4\pi e^2}{\kappa_b}\eta_1(\vec{r},z) \qquad z < 0$$
$$\nabla^2\phi_1(\vec{r},z) = -\frac{4\pi e^2}{\kappa_w}\eta_1(\vec{r},z) \qquad z > 0 \qquad (507)$$

Following the same procedure developed above, the corresponding equations for the two-dimensional Fourier transforms can be found.

This way

$$\frac{d^2}{dz^2}\tilde{\phi}_{\vec{q}}(z) - q^2\tilde{\phi}_{\vec{q}}(z) = -4\pi J_b(z) \qquad \text{for } z < 0$$
$$\frac{d^2}{dz^2}\tilde{\phi}_{\vec{q}}(z) - q^2\tilde{\phi}_{\vec{q}}(z) = -4\pi J_w(z) \qquad \text{for } z > 0 \qquad (508)$$

where

$$J_b(z) = \frac{e^2}{\kappa_b S}\sum_{\vec{k}_\perp, n\vec{k}'_\perp, n'}\langle n'\vec{k}'_\perp|\hat{\rho}_1|n\vec{k}_\perp\rangle\chi_n^*(z)\chi_{n'}(z)\delta_{\vec{k}'_\perp, \vec{k}_\perp + \vec{q}}$$

$$J_w(z) = \frac{e^2}{\kappa_w S}\sum_{\vec{k}_\perp, n\vec{k}'_\perp, n'}\langle n'\vec{k}'_\perp|\hat{\rho}_1|n\vec{k}_\perp\rangle\chi_n^*(z)\chi_{n'}(z)\delta_{\vec{k}'_\perp, \vec{k}_\perp + \vec{q}} \qquad (509)$$

with the boundary conditions

$$\begin{array}{ll} \tilde{\phi}_{\vec{q}}(-\infty) = 0 & \tilde{\phi}_{\vec{q}}(+\infty) = 0 \\ \frac{d}{dz}\tilde{\phi}_{\vec{q}}(-\infty) = 0 & \frac{d}{dz}\tilde{\phi}_{\vec{q}}(+\infty) = 0 \end{array} \qquad (510)$$

and the matching conditions

$$\tilde{\phi}_{\vec{q}}(-0) = \tilde{\phi}_{\vec{q}}(+0) \qquad \kappa_b\frac{d}{dz}\tilde{\phi}_{\vec{q}}(-0) = \kappa_w\frac{d}{dz}\tilde{\phi}_{\vec{q}}(+0) \qquad (511)$$

Dealing with the original equation,

$$\int (\nabla\kappa(z)\nabla\phi_1(\vec{r},z))\exp(-i\vec{q}\cdot\vec{r})$$
$$d\vec{r} = -4\pi e^2 \int \eta_1(\vec{r},z)\exp(-i\vec{q}\cdot\vec{r})d\vec{r} \qquad (512)$$

Now

$$\nabla\kappa(z)\nabla\phi_1(\vec{r},z) = \kappa(z)\frac{\partial^2}{\partial r^2}\phi_1(\vec{r},z)$$
$$+ \frac{\partial}{\partial z}\kappa(z)\frac{\partial}{\partial z}\phi_1(\vec{r},z) \qquad (513)$$

Thus developing the derivatives in the left side and introducing the Fourier transforms it is obtained that

$$\frac{d}{dz}\kappa(z)\frac{d}{dz}\tilde{\phi}_{\vec{q}}(z) - q^2\tilde{\phi}_{\vec{q}}(z) = -4\pi I(z) \qquad (514)$$

where

$$I(z) = \frac{e^2}{S}\sum_{\vec{k}_\perp, n}\sum_{\vec{k}'_\perp}\langle n'\vec{k}'_\perp|\hat{\rho}_1|n\vec{k}_\perp\rangle\chi_n^*(z)\chi_{n'}(z)\delta_{\vec{k}'_\perp, \vec{k}_\perp + \vec{q}} \qquad (515)$$

with the same boundary and matching conditions.

Let us write the following two equations:

$$\kappa_b\frac{d^2}{dz^2}\tilde{\phi}_{\vec{q}}(z) - q^2\tilde{\phi}_{\vec{q}}(z) = -4\pi I(z) \qquad z < 0$$
$$\kappa_w\frac{d^2}{dz^2}\tilde{\phi}_{\vec{q}}(z) - q^2\tilde{\phi}_{\vec{q}}(z) = -4\pi I(z) \qquad z > 0 \qquad (516)$$

7.2.6. Static Dielectric Function in the Size Quantum Limit

It will be shown first that the dielectric function calculated in Eqs. (5) and (6) for a quasi-two-dimensional system reduces in the appropriate limits to the dielectric functions which have been obtained for purely two-dimensional or three-dimensional systems [182, 183].

In the size quantum limit, all the electrons are confined to the ground state of the system where $n = 1$, $n' = 1$, $l =$

$l' = 1$, since the active layer is so thin that the energy differences between the different subbands are very large. In this limit, intersubband transitions cannot take place at low temperatures. Since the condition for size quantum limit is that the de Broglie wavelength of the electron be larger than the thickness of the active layer, i.e., $kL < 1$ [184], it is found that in this limit Eqs. (503) and (504) reduce to

$$\lim_{\delta \to +0} \in_{1111} (q, \omega)$$
$$= \left[1 - \lim_{\delta \to +0} \sum_{\vec{k}_{\perp}} \frac{4\pi e^2}{\kappa_w LS} \right.$$
$$\left. \times \frac{f_1^0(\varepsilon_{\vec{k}_{\perp}+\vec{q}}) - f_1^0(\varepsilon_{\vec{k}_{\perp}})}{\varepsilon_{\vec{k}_{\perp}+\vec{q}} - \varepsilon_{\vec{k}_{\perp}} + \hbar\omega - i\hbar\delta F_{1111}(q)} \right] \quad (517)$$

Since $\omega = 0$ in the static case, then the form factor is given by

$$F_{1111}(q) = \left[\frac{1}{2} \left(\frac{2}{q^2} + \frac{1}{q^2 + (\frac{2\pi}{L})^2} \right) - \frac{2q}{L} \left(\frac{2}{q^2} + \frac{1}{q^2 + (\frac{2\pi}{L})^2} \right)^2 \right.$$
$$\left. \times \left(\frac{\kappa_w}{\kappa_b} + \coth \frac{qL}{2} \right)^{-1} \right] \quad (518)$$

$$\varepsilon_{\vec{k}_{\perp}} = \frac{\hbar^2 k_{\perp}^2}{2m_{1e}^*} + \varepsilon_0 \quad (519)$$

For a pure two-dimensional system, the thickness of the active layer goes to zero and all the electrons are restricted to move in a plane. Therefore, for such a system, F_{1111} in Eq. (518) reduces to $\kappa_w L / 2\kappa_b q$, where the relation $\coth x = x^{-1} + x/3$ has been used, since $x = qL/2$ goes to zero as L goes to zero. From Eq. (517), the dielectric function for a strictly two-dimensional system is

$$\lim_{\delta \to +0} \in_{2D} (q, \omega)$$
$$= \left[1 - \lim_{\delta \to +0} \sum_{\vec{k}_{\perp}} \frac{2\pi e^2}{\kappa_b qS} \frac{f_1^0(\varepsilon_{\vec{k}_{\perp}+\vec{q}}) - f_1^0(\varepsilon_{\vec{k}_{\perp}})}{\varepsilon_{\vec{k}_{\perp}+\vec{q}} - \varepsilon_{\vec{k}_{\perp}} + \hbar\omega - i\hbar\delta} \right] \quad (520)$$

The same result can also be obtained if one starts with the wavefunction for a two-dimensional gas, given by

$$\lim_{d \to 0} \chi_{\vec{k}_{\perp} n}(\vec{r}, z) = \frac{1}{\sqrt{S}} \exp(i\vec{k}_{\perp} \cdot \vec{r}) \delta\left(z - \frac{L}{2} \right) \quad (521)$$

in the self-consistent field calculation for $\in_{2D}$. Note that in Eq. (520), $\varepsilon_{\vec{k}_{\perp}} = \hbar^2 k_{\perp}^2 / 2m_{1e}^*$. From this it can be seen that the results for a quasi-two-dimensional system with the carriers confined to the lowest quantized subband differs from those for a strictly two-dimensional system by a factor of $\gamma(Q) = 2\kappa_w q F_{1111}(q)/L\kappa_b$. This factor is shown in Figure 42 as a function of $Q = qL$ for the case where the surrounding media is vacuum ($\kappa_b = 1$) and for the case where the surrounding media is GaAlAs ($\kappa_b = 11.6$). When Q approaches zero, γ approaches unity and the dielectric function for a quasi-two-dimensional system reduces to that of a two-dimensional

system, as expected. However, for finite L, the results for a quasi-two-dimensional system will differ from those for a two-dimensional system. In Figure 42, the value of L is taken to be 100 Å and it is seen that $\gamma(Q)$ decreases sharply as Q increases. Therefore, there can be a significant difference between the results for a quasi-two-dimensional system and those for a two-dimensional system. To examine this deviation more carefully, an analytic form for the static dielectric function of a quasi-two-dimensional system when the electron gas is completely degenerate must be used, obtaining

$$\in_{Q2D} (q, 0) = \begin{cases} 1 + \left(\frac{2}{qa_B^*} \right) \gamma(Q) & q \leq 2k_F \\ 1 + \left(\frac{2}{qa_B^*} \right) \gamma(Q) \\ \cdot \left\{ 1 - \left[1 - \left(\frac{2k_F}{q} \right)^2 \right]^{1/2} \right\} & q \leq 2k_F \end{cases} \quad (522)$$

where $a_B^* = \kappa_b \hbar^2 / m_e^* e^2$ is the effective Bohr radius, $k_F = (2\pi\eta)^{1/2}$ is the two-dimensional Fermi wavevector for a degenerate electron gas, and η is the two-dimensional electron surface density. When L approaches zero, Eq. (522) reduces to the expression for $\in_{2D} (q)/\kappa_b$ obtained by Stern [182], as expected.

The other extreme limit is the one when L approaches infinity, where it is expected to obtain the dielectric function for a three-dimensional electron gas. In this limit, it is obtained that

$$F_{\infty} = \frac{1}{2} \left\{ \frac{\delta_{l'-l, \pm(n'-n)} - \delta_{l'+l, \pm(n'-n)}}{q^2 + (n'-n)^2(\frac{\pi}{L})^2} \right.$$
$$\left. - \frac{\delta_{l'-l, \pm(n'+n)} - \delta_{l'+l, \pm(n'+n)}}{q^2 + (n'+n)^2(\frac{\pi}{L})^2} \right\} \quad (523)$$

Note that the first and last Kronecker delta functions in Eq. (523) can be obtained if the equation

$$\lim_{d \to \infty} \chi_{\vec{k}_{\perp}}(\vec{r}, z) = \frac{1}{\sqrt{SL}} \exp(i\vec{k}_{\perp} \cdot \vec{r}) \exp(ik_z z)$$
$$-\infty < k_z < +\infty \quad (524)$$

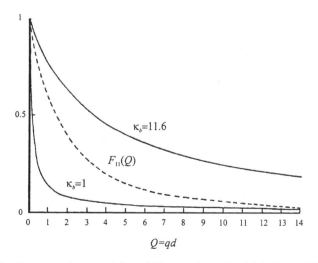

Fig. 42. γ as a function of $Q = qd$. The form factor $F_{1111}(q)$ is shown with the dotted line ($\kappa_w = 13.1$ (GaAs)).

is used in the self-consistent field calculation of the dielectric function in the three-dimensional limit, and the other two Kronecker delta functions in Eq. (523) have the opposite sign because in the z direction the incident and reflected waves have to be out of phase by π, which is a direct result of the boundary conditions at $z = 0$ and $z = L$. Thus, in the three-dimensional limit all the negative signs in the Kronecker delta functions have to be replaced by positive signs because, in this limit, the two waves propagating in the opposite directions are in phase (see Eq. (524)). After summing over all possible values of n and n' in Eqs. (503) and (504), a $\in_{3D}(q, \omega)$ function is obtained, which is consistent with Lindhard's result [183,185,186], for the dielectric function in the bulk limit.

In Figure 43, the static dielectric function $\in (q, 0)$ is shown as a function of q for quasi-two-dimensional and three-dimensional systems and also for the case where the Thomas–Fermi approximation is used, respectively. It is assumed that GaAs is the active layer, with $\kappa_w = 13.1$ and $L = 100$ Å, while the surrounding layers are GaAlAs with $\kappa_b = 11.6$. It is well known that the static dielectric function for a three-dimensional system has a logarithmic singularity at $q = 2k_F$, where $k_F = (3\pi^3 \bar{\eta})^{1/3}$ and $\vec{\eta} = \eta/L$ is the carrier density per unit volume. Although this singularity in $\in_{3D}(q)$ is very weak, it is important in some of the properties of metals and is called the Kohn effect [187, 188]. This Kohn effect becomes stronger in quasi-two-dimensional and two-dimensional systems, as can be seen from Figure 43. It is seen that $\in_{Q2D}(q)$ is always smaller than $\in_{2D}(q)$ because $\gamma(Q) < 1$ as was mentioned previously in connection with Figure 42. From Figure

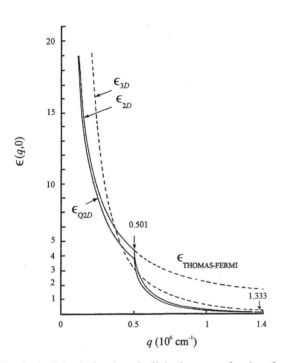

Fig. 43. Static dielectric function $\epsilon(q, 0)$ is shown as a function of q for quasi-two-dimensional and three-dimensional systems and also for the case where we use the Thomas–Fermi approximation, respectively. Arrows indicate the location of the Kohn effect for various systems. $\eta = 10^{10}$ cm^{-2}; $T = 0$ K, and $\omega = 0$.

43, it is seen that the result for the two-dimensional dielectric function using the Thomas–Fermi approximation is only valid for $q < 2k_F$. For small q, the dielectric functions for quasi-two-dimensional and two-dimensional electron gases do not differ too much, but for large q they differ by a larger amount and therefore using the dielectric function of a two-dimensional system for a quasi-two-dimensional system will overestimate the amount of screening. In addition, from Figure 42 and Eq. (522) it is seen that when κ_b decreases, $\gamma(Q)$ does also and that leads to a corresponding decrease in the amount of screening for large q.

To see how the dielectric function affects the calculation of the scattering rate, the momentum relaxation time due to the scattering of electrons by ionized impurities in the active layer, i.e., by background impurities, is considered. In the following, only the ground state ($n = 1$) is assumed to be populated by electrons so that we are in size quantum limit. The momentum relaxation time as given by Mori and Ando [178] is

$$\tau^{-1} = \frac{2m_e^* e^4}{\hbar^3 \kappa_w^2} \left(\frac{N_s}{\eta} \right) \int_0^{\pi/2} dx \left(\frac{F_{11}(q)}{\in^2(q)} \right) \text{(cgs)} \quad (525)$$

where

$$F_{11}(q) = \text{form factor}$$
$$= \frac{4}{L^3} \int_0^L dz' \left(\int_0^L dz \exp(-q|z - z'|) \sin^2 \frac{\pi z}{L} \right)^2 \quad (526)$$

$$q = 2k_F \sin x = 2\sqrt{2\pi\eta} \sin x. \quad (527)$$

Here, N_s is the surface density of background impurities, which has been assumed to be uniform, z' is the location of the impurity in the active layer, and the subscript in the form factor $F(q)$ denotes that it is being evaluated using the ground state wavefunctions. The form factor $F_{11}(Q)$ is shown in Figure 42 by a dotted line for the case where the thickness of the active layer is 100 Å. As can be seen from this figure, $F_{11}(Q)$ decreases as Q increases. For a strictly two-dimensional electron gas with an unscreened Coulomb potential due to the ionized impurities, the factor $F_{11}(q)/\in^2(q)$ in Eq. (525) is equal to unity and this gives the maximum scattering rate

$$(\tau^{-1})^D_{\text{bare potential}} = \frac{\pi m_e^* e^4}{\hbar^3 \kappa_w^2} \left(\frac{N_s}{\eta} \right) \quad (528)$$

This result is consistent with the results obtained in [189–191]. Since for a quasi-two-dimensional system $F_{11}(q) < 1$ and $\in (q) > 1$, it is expected that the scattering rate be decreased over that for an unscreened potential. However, $\in_{Q2D}(q) < \in_{2D}$ for a quasi-two-dimensional system; therefore, in order to see whether the scattering rate is increased or decreased for a Q2D electron gas with a screened Coulomb potential, over a 2D system with a screened potential, the factor $F_{11}(q)/\in^2(q)$ is plotted as a function of x, where $q = 2k_F \sin x$, in Figure 44 for various values of the electron surface density η. In these numerical calculations, the Eqs. (518), (519), and (522) have

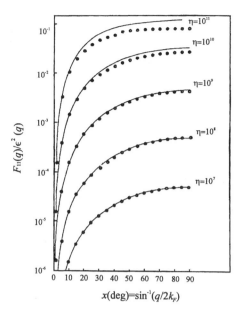

Fig. 44. Factor $F_{11}(q)/\epsilon^2(q)$ is plotted as a function of x for various values of the electron surface density η.

been used for $\in (q)$ and Eq. (526) for $F_{11}(q)$. For comparison, the results using the Thomas–Fermi approximation are plotted for the same values of η, using the dotted lines. From Figure 44, it is seen that both models predict almost the same scattering rate for small η, i.e., for $\eta < 10^8$ cm^{-2} corresponding to $k_F L < 0.025$. However, for larger values of η, the two models begin to predict different scattering rates. When $\eta = 10^{11}$ cm^{-2}, the scattering rates predicted by the different models can differ by as much as 25%. In the limit of low electron densities, both $\gamma(Q)$ and $F_{11}(Q)$ can be expanded to lowest order in Q, thus obtaining in the limit $k_F L < 1$,

$$\lim_{k_F d \to 0} (\tau^{-1})^{\text{Q2D}}_{\substack{\text{screened} \\ \text{potential}}} = N_s \hbar \kappa_b^2 \pi^2 / m_e^* \kappa_w^2 \tag{529}$$

This result is equivalent to that obtained by Mori and Ando [178] if $\kappa_b/\kappa_w = 1$ is setted, as in their calculations. However, for high surface electron densities, their use of the Thomas–Fermi approximation for the dielectric function would tend to underestimate the scattering rate.

7.2.7. Screening of Intersubband Scattering in Semiconductor Heterostructures

A brief summary of the self-consistent field method leading to a four-index matrix dielectric function [192], as the one of the random phase approximation, for a quasi-two-dimensional electron gas is reminded. Calculation of matrix elements of screened interaction through those corresponding to bare interaction requires the inversion of this matrix. A method to invert such a four-index matrix is explained and easy algorithms to carry out this inversion are presented. General expressions for the form factor concerning electron–electron interaction in single heterostructures and quantum wells are derived when different dielectric constants at each side of an interface are

considered and, besides, the envelope functions in the confinement direction can penetrate into the barrier regions. These problems are tightly connected with numerical calculations for multisubband transport in inversion layers and semiconductor heterostructures as well as for other phenomena in this kind of systems where screening of involved interactions in intersubband scattering must be taken into account.

Among the electronic properties of a quasi-two-dimensional electron gas formed in semiconductor heterostructures and inversion layers, transport phenomena have been widely studied both theoretically and experimentally.

Most of the works developing concrete calculations has employed the so-called size quantum limit approximation, where only the lowest subband and intrasubband transitions are considered. Multisubband transport, including intersubband transitions, has been treated following the quasi-classical formalism of the Boltzmann transport equation [176–178, 180, 193–199] and the quantum formalism of the Kadanoff–Baym ansatz [200–202]. Theoretical approaches have been mainly concerned with elastic scattering mechanisms [176–178, 180, 193, 194, 200–202], while inelastic scattering mechanisms have been included in a few works [195–199]. In numerical calculations carried out for momentum relaxation rates or mobilities, screening of the involved interactions has been disregarded [197, 198, 200] or considered in different ways [178, 180, 193–196, 199, 201].

Some works suggest using the random phase approximation result, which gives a four-index matrix dielectric function, to take into account the screening effect [176–178, 193]. But the needed matrix elements of the screened interaction expressed through the known matrix elements of bare interaction require the inversion of such a four-index matrix, a problem not explicitly considered in [176–178, 193]. Other works employ a strictly two-dimensional or a size quantum limit result of the quasi-two-dimensional case, originally obtained in a self-consistent field approach, for the dielectric function in order to consider the screening effect even in intersubband scattering [180, 194–196, 201]. This seems to be a coarse approximation, but it avoids the inversion of the four-index matrix dielectric function. A mathematical formalism has been developed to find directly the elements of the inverted four-index matrix dielectric function [203], avoiding in this way the inversion of the generally known random phase approximation result.

The best dielectric function for a quasi-two-dimensional electron gas to be employed in numerical calculation related to transport phenomena is the random phase approximation result, which could be further enhanced in two ways: introducing a factor multiplying the polarizability, on the line developed by Hubbard or Singwi and Sjolander for three-dimensional electron gas [160], or improving the polarizability itself, as Maldague did for a two-dimensional electron gas [204]. None of those would change the four-index matrix form of dielectric function in the quasi-two-dimensional case; therefore, it is also important to know how to invert this matrix. This question has not been explicitly considered before and deserves a rigorous approach, which could further provide a concrete procedure to include screening of involved interactions in intersubband scattering. The present subsection aims to solve this practical question connected not only with multisubband transport, where it would be necessary to explain experimental results [179, 199, 205, 206] accurately, but also with hot electron transport, polaronic effects, Raman scattering, and so on in such kinds of systems.

The random phase approximation result includes the so-called form factor concerning electron–electron interaction, which takes into account the inhomogeneous character of the quasi-two-dimensional electron gas in the confinement direction through the envelope functions corresponding to the confining potential model [207]. It also depends on whether the Coulomb interaction is assumed to occur in a homogeneous or an inhomogeneous medium. In the former case the difference between dielectric constants at both sides of each interface is neglected, and a general expression for this form factor has been widely employed in calculations [198]. In the latter case, considering the different dielectric constants at both sides of an interface a quite general expression has been reported for inversion layers and single heterostructures [176], but it has not been derived for quantum wells. A particular expression, assuming an infinite barrier confining potential model, has been obtained for a quantum well [184]. These two expressions have in common that they have been derived for envelope functions in the confinement direction, which do not penetrate into the barriers at the interfaces. But consideration of the finite barrier at each interface or self-consistent calculations lead to penetration of these enveloped functions into the barrier region, a fact that cannot be disregarded in calculations [207]. Since a further enhancement of the random phase approximation result, as suggested just above, would not change the structure of this form factor, it is also important for the derivation of general expressions for single heterostructures and quantum wells. The present subsection also aims to solve this question of both theoretical and practical interest.

The form factor concerning electron–electron interaction given by Eq. (487) is obtained by neglecting the difference between dielectric constants of material forming the inversion layer, single heterostructure, or quantum well. Considering the different dielectric constants at both sides of each interface such a general expression has been reported for inversion layers and single heterostructures (see Eq. (337) in [176]), but it has not been derived for quantum wells. A particular expression, assuming infinite barrier confining potential, was first obtained in [184].

7.2.8. Matrix Elements of the Screened Interaction

If the perturbation is sufficiently small, linearization of the equation of motion for the statistical operator and solution of the Poisson equation, relating induced perturbation with induced density of electrons, yield the following relation between two-dimensional Fourier components of the bare interaction and the screened interaction previously obtained:

$$\tilde{\phi}^0_{nn'}(q, \omega) = \sum_{l,l'} \epsilon_{nn'll'}(q, \omega) \tilde{\phi}_{ll'}(q, \omega) \qquad (530)$$

Expression (530) gives each matrix element of the bare interaction $\tilde{\phi}^0_{nn'}$ through the matrix elements of the screened interaction $\tilde{\phi}_{ll'}$. However, what is required in concrete calculations is each $\tilde{\phi}_{mm'}$ through $\tilde{\phi}^0_{nn'}$, since the latter matrix elements correspond to a generally known interaction Hamiltonian. Hereafter the arguments (q, ω) are omitted for the sake of briefness in writing equations.

The inverted four-index matrix dielectric function ϵ^{-1} must be such that

$$\sum_{n,n'} \epsilon^{-1}_{mm'nn'} \epsilon_{nn'll'} = \sum_{n,n'} \epsilon_{mm'nn'} \epsilon^{-1}_{nn'll'} = e_{mm'll'} \qquad (531)$$

where the four-index unit matrix e must obviously satisfy the following relation:

$$\phi_{mm'} = \sum_{l,l'} e_{mm'll'} \phi_{ll'} \qquad (532)$$

From Eq. (530), taking into account Eqs. (531) and (532), one obtains

$$\sum_{n,n'} \epsilon^{-1}_{mm'nn'} \phi^0_{nn'} = \sum_{l,l'} \sum_{n,n'} \epsilon^{-1}_{mm'nn'} \epsilon_{nn'll'} \phi_{ll'}$$
$$= \sum_{l,l'} e_{mm'll'} \phi_{ll'} = \phi_{mm'} \qquad (533)$$

i.e., the needed expression,

$$\phi_{mm'} = \sum_{nn'} \epsilon^{-1}_{mm'nn'} \phi^0_{nn'} \qquad (534)$$

Nevertheless the problem of calculating $\phi_{mm'}$ is not yet solved, since the problem of inverting the four-index matrix dielectric function ϵ stays an open question.

7.2.9. Inversion of a Four-Index Matrix

Let us introduce the four-index matrix a (small letter) as

$$B_{nn'} = \sum_{ll'} a_{nn'll'} A_{ll'} \qquad (535)$$

i.e., relating the elements of two-index matrices A and B (capital letters). Notice that this is exactly the situation one has in Eq. (530): $A_{ll'}$ and $B_{nn'}$ can represent the matrix elements of the screened and bare interactions, respectively, while $a_{nn'll'}$ can be identified with the elements of the four-index matrix dielectric function.

The unit (or identity) matrix e must be such that

$$B_{nn'} = \sum_{ll'} e_{nn'll'} B_{ll'} \qquad (536)$$

On the other hand, one can write

$$B_{nn'} = \sum_{ll'} \delta_{nl} \delta_{n'l'} B_{ll'} \qquad (537)$$

with $\delta_{nl} = 1$ if $n = l$ and $\delta_{nl} = 0$ if $n \neq l$. Then

$$\sum_{ll'} (e_{nn'll'} - \delta_{nl}\delta_{n'l'})B_{ll'} = 0 \tag{538}$$

For fixed n and n', and for arbitrary $B_{ll'}$, i.e., assuming at least one of $B_{ll'} \neq 0$, one obtains

$$e_{nn'll'} = \delta_{nl}\delta_{n'l'} \tag{539}$$

the expression of unit matrix elements, which is quite reasonable comparing with Eq. (503) and bearing in mind the random phase approximation result in the three-dimensional case.

The corresponding inverted matrix b must be introduced as

$$A_{ll'} = \sum_{mm'} b_{ll'mm'}B_{mm'} \tag{540}$$

where $b_{ll'mm'}$ can be identified with the elements of the inverted four-index matrix dielectric function appearing in Eq. (534). Substitution of Eq. (540) in Eq. (535) gives

$$B_{nn'} = \sum_{mm'}\sum_{ll'} a_{nn'll'}b_{ll'mm'}B_{mm'} \tag{541}$$

and, according to Eq. (536),

$$\sum_{ll'} a_{nn'll'}b_{ll'mm'} = e_{nn'mm'} \tag{542}$$

Let assume each index varies from 1 until N; i.e., $n, n', l, l' = 1, 2, \ldots, N$. This is a quite reasonable assumption, since they can represent subband indices, and then it corresponds to the N-subband approximation in Eqs. (530) and (534), which is the usually accepted situation allowing numerical procedures to carry out actual calculations. Then Eq. (542) represents N^2 systems of linear equations in N^2 unknowns each of them. Indeed, for fixed m and m', say $m = m_i$ and $m' = m'_j$, one has a system given by

$$\sum_{ll'} a_{nn'll'}b_{ll'm_im'_j} = e_{nn'm_im'_j} \tag{543}$$

with $a_{nn'll'}$ the N^4 coefficients, $b_{ll'm_im'_j}$ the N^2 unknowns, and $e_{nn'm_im'_j}$ the N^2 constants.

Now one has the right of labeling the four-index matrix elements $a_{nn'll'}$ as two-index matrix elements α_{ij} of the coefficient matrix of the system, the elements $b_{ll'm_im'_j}$ as the β_{jk_m} elements of the unknown column vector, and the elements $e_{nn'm_im'_j}$ as the ϵ_{ik_m} elements of the constant column vector. Then the system (543) can be written as

$$\sum_j \alpha_{ij}\beta_{jk_m} = \epsilon_{ik_m} \tag{544}$$

Note that $k_{nn'll'}$ is fixed. Let us display the above explained procedure to obtain the coefficient matrix α of the system (543):

$$\begin{pmatrix}
\alpha_{11} = a_{1111} & \cdot & \alpha_{1N} = a_{111N} & \alpha_{1,N+1} = a_{1121} & \cdot & \alpha_{1N^2} = a_{11NN} \\
\alpha_{21} = a_{1211} & \cdot & \alpha_{2N} = a_{121N} & \alpha_{2,N+1} = a_{1221} & \cdot & \alpha_{2N^2} = a_{12NN} \\
\cdot & \cdot & \cdot & \cdot & \cdot & \cdot \\
\cdot & \cdot & \cdot & \cdot & \cdot & \cdot \\
\alpha_{N1} = a_{1N11} & \cdot & \alpha_{2N} = a_{121N} & \alpha_{N,N+1} = a_{1N21} & \cdot & \alpha_{NN^2} = a_{1NNN} \\
\cdot & \cdot & \cdot & \cdot & \cdot & \cdot \\
\alpha_{N^21} = a_{NN11} & \cdot & \alpha_{N^2N} = a_{NN1N} & \alpha_{N^2,N+1} = a_{NN21} & \cdot & \alpha_{N^2N^2} = a_{NNNN}
\end{pmatrix} \tag{545}$$

Notice that this order of labeling is not arbitrary, but it appears in a logical way exactly as it should be. The use of greek letters to denote this and other two-index matrices, obtained by changing the labeling of four-index matrix elements, intends to distinguish them formally from the properly two-index matrices. Furthermore, the i, j, k indexes are employed to remark these cannot be subband indexes like l, m, n.

Similarly the unknown and constant columns are, respectively,

$$\beta_{\tilde{k}_m} = \begin{pmatrix}
\beta_{1k_m} = b_{11m_im'_j} \\
\beta_{21k_m} = b_{12m_im'_j} \\
\cdot \\
\cdot \\
\beta_{Nk_m} = b_{1Nm_im'_j} \\
\cdot \\
\cdot \\
\beta_{N^2k_m} = b_{NNm_im'_j}
\end{pmatrix} \quad \epsilon_{\tilde{k}_m} = \begin{pmatrix}
\epsilon_{1k_m} = e_{11m_im'_j} \\
\epsilon_{21k_m} = e_{12m_im'_j} \\
\cdot \\
\cdot \\
\epsilon_{Nk_m} = e_{1Nm_im'_j} \\
\cdot \\
\cdot \\
\epsilon_{N^2k_m} = e_{NNm_im'_j}
\end{pmatrix} \tag{546}$$

This way the system (543) can be written in a compact form:

$$\alpha\beta_{\tilde{k}_m} = \epsilon_{\tilde{k}_m} \tag{547}$$

Following the same procedure for every pair of remaining m and m' one finally obtains all the N^2 systems

$$\alpha\beta_1 = \epsilon_1$$
$$\cdot$$
$$\alpha\beta_N = \epsilon_N$$
$$\alpha\beta_{N+1} = \epsilon_{N+1} \tag{548}$$
$$\cdot$$
$$\alpha\beta_{N^2} = \epsilon_{N^2}$$

Let us introduce now the two-index unknown matrix β as

$$\beta = (\beta_1 \cdots \beta_N \, \beta_{N+1} \cdots \beta_{N^2}) \tag{549}$$

or better displayed as

$$
\begin{pmatrix}
\beta_{11} = b_{1111} & \cdot & \beta_{1N} = b_{111N} & \beta_{1,N+1} = b_{1121} & \cdot & \beta_{1N^2} = b_{11NN} \\
\beta_{21} = b_{1211} & \cdot & \beta_{2N} = b_{121N} & \beta_{2,N+1} = b_{1221} & \cdot & \beta_{2N^2} = b_{12NN} \\
\cdot & \cdot & \cdot & \cdot & & \cdot \\
\cdot & \cdot & \cdot & \cdot & & \cdot \\
\cdot & \cdot & \cdot & \cdot & & \cdot \\
\beta_{N1} = b_{1N11} & \cdot & \beta_{2N} = b_{121N} & \beta_{N,N+1} = b_{1N21} & \cdot & \beta_{NN^2} = b_{1NNN} \\
\cdot & \cdot & \cdot & \cdot & & \cdot \\
\cdot & \cdot & \cdot & \cdot & & \cdot \\
\cdot & \cdot & \cdot & \cdot & & \cdot \\
\beta_{N^21} = b_{NN11} & \cdot & \beta_{N^2N} = b_{NN1N} & \beta_{N^2,N+1} = b_{NN21} & \cdot & \beta_{N^2N^2} = b_{NNNN}
\end{pmatrix} \quad (550)
$$

i.e., formed by the unknown column vectors, and the two index constant matrix ϵ as

$$
\epsilon = (\epsilon_1 \cdots \epsilon_N \ \epsilon_{N+1} \cdots \epsilon_{N^2}) \quad (551)
$$

i.e., formed by the constant column vectors. It can be better displayed as follows:

$$
\epsilon = \begin{pmatrix}
\epsilon_{11} = e_{1111} & \cdot & \epsilon_{1N} = e_{111N} & \epsilon_{1,N+1} = e & \cdot & \epsilon_{1N^2} = e_{11NN} \\
\epsilon_{21} = e_{1211} & \cdot & \epsilon_{2N} = e_{121N} & \epsilon_{2,N+1} = e_{1221} & \cdot & \epsilon_{2N^2} = e_{12NN} \\
\cdot & \cdot & \cdot & \cdot & & \cdot \\
\cdot & \cdot & \cdot & \cdot & & \cdot \\
\cdot & \cdot & \cdot & \cdot & & \cdot \\
\epsilon_{N1} = e_{1N11} & \cdot & \epsilon_{2N} = e_{121N} & \epsilon_{N,N+1} = e_{1N21} & \cdot & \epsilon_{NN^2} = e_{1NNN} \\
\cdot & \cdot & \cdot & \cdot & & \cdot \\
\cdot & \cdot & \cdot & \cdot & & \cdot \\
\cdot & \cdot & \cdot & \cdot & & \cdot \\
\epsilon_{N^21} = e_{NN11} & \cdot & \epsilon_{N^2N} = e_{NN1N} & \epsilon_{N^2,N+1} = e_{NN21} & \cdot & \epsilon_{N^2N^2} = e_{NNNN}
\end{pmatrix}
$$
$$(552)$$

Notice that according to Eq. (539) all elements are zero except in the diagonal, where they equal unity; i.e., $\epsilon_{ij} = \delta_{ij}$. This way the expected two-index unity matrix has been obtained.

Now all the systems (548) can be written in a compact form as a matrix equation

$$
\alpha\beta = \epsilon \quad (553)
$$

whose solution is

$$
\beta = \alpha^{-1} \quad (554)
$$

i.e., the inverse of α. This statement is guaranteed by the well known theorem of linear algebra for two-index square matrices: $\alpha\beta = \epsilon$ if and only if $\alpha\beta = \epsilon$, so $\beta = \alpha^{-1}$.

The algorithm to invert a two-index matrix can be established following well known methods, so the problem of inverting a four-index matrix, such as the random phase approximation result Eq. (8), is already solved. It is beyond the purpose of this subsection to demonstrate that the set of four-index "square" matrices with elements of an arbitrary field, say the complex field C, constitutes a vector space. But once this is done, the above justification of procedures given α and β matrices to change labeling of four-index matrix elements into two-index matrix elements can be understood as a demonstration of the following theorem: the N^2-dimension vector space of four-index "square" matrices is isomorphic with the N^2-dimension vector space of two-index square matrices. This

theorem is completely analogous to the one of linear algebra: the N^2-dimension vector space of two-index square matrices is isomorphic with the N^2-dimension vector space of one-index matrices (column vectors or row vectors), which is actually expressed in Eq. (546).

7.2.10. Procedure to Take into Account the Screening Effect

The method just explained in the previous subsection can be resumed as follows:

(1) label the four-index matrix elements of a, say the dielectric function ϵ, as two-index matrix elements of α exactly in the order appearing in Eq. (545);
(2) invert the obtained two-index matrix α and denote the inverted matrix as β;
(3) label the two-index matrix elements of β as four-index matrix elements of b, say the inverted dielectric function ϵ^{-1}, exactly in order appearing in β matrix.

The algorithm to change the labeling of the four-index matrix elements $a_{nn'll'}$ into two-index matrix elements α_{ij} can be written in FORTRAN as follows:

```
I = 0
DO 1 N = 1, NMAX
DO 1 N1 = 1, NMAX
I = I + 1
J = 0
DO 1 L = 1, NMAX
DO 1 L1 = 1, NMAX
J = J + 1
ALPHA(I,J) = A(N,N1,L,L1)

1 CONTINUE
```

Here $n \to N$, $n' \to N1$, $l \to L$, $l' \to L1$, $N \to NMAX$, etc.

A concrete algorithm to invert α (i.e., to find β) should be found in available FORTRAN libraries or suitable books on numerical recipes; otherwise, it can also be established following well known rules. Finally, the algorithm to renumber β_{ij} as $b_{nn'll'}$ is similar to the one just presented, but now

$$
B(N, N1, L, L1) = BETA(I, J) \text{ instead of } ALPHA(I, J)
$$
$$
= A(N, N1, L, L1)
$$

This way the screening of involved interactions in intersubband scattering can be taken into account as follows:

(a) calculating the four-index matrix elements of the dielectric function according to Eq. (530);
(b) finding the four-index matrix elements of the inverted dielectric function following the method resumed just above in this subsection;
(c) calculating the screened matrix elements according to Eq. (534).

This procedure is considerably easier than that of [202], because, first, it makes use of the well known random phase approximation result for the dielectric function and, second, the algorithms to find the four-index matrix elements of the inverted dielectric function are very simple.

7.2.11. Form Factor for Electron–Electron Interaction

The derivation of a general expression for the form factor concerning electron–electron interaction, similar to Eq. (487), but considering different dielectric constants at both sides of each interface in a quantum well, requires the solution of an electrostatic problem, which arises when one follows the self-consistent field approach to find the four-index matrix dielectric function of a quasi-two-dimensional electron gas.

The Poisson equation relating induced perturbation with induced density of electrons results in the equation for the two-dimensional Fourier component of the screening potential,

$$\widehat{L}\phi^1(z) = -4\pi J(z) \qquad (555)$$

where the lineal differential operator is

$$\widehat{L} = \frac{d^2}{dz^2} - q^2 \qquad (556)$$

and the induced charge density is given by

$$J(z) = \frac{e^2}{\kappa(z)S}\sum_{\vec{k}_\perp,l}\sum_{\vec{k}'_\perp,l'}\langle l'\vec{k}'_\perp|\hat{\rho}_1|l\vec{k}_\perp\rangle\chi_l^*(z)\chi_{l'}(z)\delta_{\vec{k}'_\perp,\vec{k}_\perp+\vec{q}} \qquad (557)$$

where

$$\kappa(z) = \begin{cases} \kappa_w & 0 < z < L \\ \kappa_b & z < 0 \text{ and } z > L \end{cases} \qquad (558)$$

Equation (555) must be solved together with the boundary conditions at $z \to \pm\infty$ and the matching conditions at $z = 0$ and $z = L$:

$$\phi^1(\pm\infty) = 0 \qquad (559)$$

$$\phi^1(0^-) = \phi^1(0^+) \qquad \phi^1(L^-) = \phi^1(L^+) \qquad (560)$$

$$\kappa_b\frac{d}{dz}\phi^1(0^-) = \kappa_w\frac{d}{dz}\phi^1(0^+)$$

$$\kappa_b\frac{d}{dz}\phi^1(L^-) = \kappa_w\frac{d}{dz}\phi^1(L^+) \qquad (561)$$

The solution of the mathematical problem, Eq. (555), with Eqs. (559)–(561) is sought employing the Green function method. Thus one has

$$\phi^1(z) = \int dz' J(z')G(z, z') \qquad (562)$$

where the Green function must satisfy the equation

$$\widehat{L}G(z, z') = -4\pi\delta(z - z') \qquad (563)$$

together with the same kind of boundary and matching conditions:

$$G(\pm\infty, z') = 0 \qquad (564)$$

$$G(0^-, z') = G(0^+, z') \qquad G(L^-, z') = G(L^+, z') \qquad (565)$$

$$\kappa_b\frac{d}{dz}G(0^-, z') = \kappa_w\frac{d}{dz}G(0^+, z')$$

$$\kappa_b\frac{d}{dz}G(L^-, z') = \kappa_w\frac{d}{dz}G(L^+, z') \qquad (566)$$

The solution of Eqs. (563)–(566) is looked for in the form

$$G(z, z') = \frac{2\pi}{q}\Gamma(z, z') \qquad (567)$$

where

$$\Gamma(z, z') = \begin{cases} \Gamma_L(z, z') & z' < 0 \\ \Gamma_w(z, z') & 0 < z' < L \\ \Gamma_R(z, z') & z' > L \end{cases} \qquad (568)$$

The Green function can be interpreted as reflecting the interaction with a source charge placed at z'. The presence of the interfaces is taken into account by means of the two image charges corresponding to the source charge. The test charge interacts with an image charge only if they are located at different media and, besides, at different sides of the interface. Notice that here one is dealing with dielectrics and there are only two images of the source, which are due to the polarization of the media, but not a series of images as in the case of conductors.

This way, when the source charge is on the left side of the quantum well ($z' < 0$),

$$\Gamma_L(z, z') = \begin{cases} \exp(-q|z - z'|) + A_L\exp[q(z + z')] & z < 0 \\ B_L\exp[-q(z - z')] \\ \quad + C_L\exp[q(z - z' - 2L)] & 0 < z < L \\ D_L\exp[-q(z - z')] & z > L \end{cases} \qquad (569)$$

where the constants are found from the matching conditions, Eqs. (565) and (566),

$$A_L = \frac{1}{E}(\kappa_b^2 - \kappa_w^2)[1 - \exp(-2qL)]$$

$$B_L = \frac{2}{E}\kappa_b(\kappa_w + \kappa_b) \qquad (570)$$

$$C_L = \frac{2}{E}\kappa_b(\kappa_w - \kappa_b)\exp(2qz')$$

$$D_L = \frac{4}{E}\kappa_b\kappa_w \qquad (571)$$

with

$$E = (\kappa_b + \kappa_w)^2 - (\kappa_b - \kappa_w)^2\exp(-2qL) \qquad (572)$$

When the source charge is inside of the quantum well ($0 < z' < L$),

$$\Gamma_w(z, z') = \begin{cases} A_w\exp[q(z - z')] & z < 0 \\ \exp(-q|z - z'|) + B_w\exp[-q(z + z')] \\ \quad + C_w\exp[q(z + z' - 2L)] & 0 < z < L \\ D_w\exp[-q(z - z')] & z > L \end{cases} \qquad (573)$$

where, from the matching conditions,

$$A_w = \frac{2}{E}\kappa_w\{(\kappa_w + \kappa_b) + (\kappa_w - \kappa_b)\exp[2q(z' - L)]\} \quad (574)$$

$$B_w = \frac{1}{E}(\kappa_w - \kappa_b)$$
$$\times \{(\kappa_w + \kappa_b) + (\kappa_w - \kappa_b)\exp[2q(z' - L)]\} \quad (575)$$

$$C_w = \frac{1}{E}(\kappa_w - \kappa_b)\{(\kappa_w + \kappa_b) + (\kappa_w - \kappa_b)\exp[-2qz']\} \quad (576)$$

$$D_w = \frac{2}{E}\kappa_w\{(\kappa_w + \kappa_b) + (\kappa_w - \kappa_b)\exp[-2qz']\} \quad (577)$$

Finally, when the source charge is on the right side of the quantum well ($z' > L$),

$$\Gamma_R(z, z') = \begin{cases} A_R \exp[-q(z - z')] & z < 0 \\ B_R \exp[-q(z - z' + 2L)] & \\ +C_R \exp[-q(z' - z)] & 0 < z < L \\ \exp(-q|z - z'|) + D_R & \\ \times \exp[-q(z + z' - 2L)] & z > L \end{cases} \quad (578)$$

where, from the matching conditions,

$$A_R = \frac{4}{E}\kappa_w\kappa_b$$

$$B_R = \frac{2}{E}\kappa_b(\kappa_w - \kappa_b)\exp[-2q(z' - L)] \quad (579)$$

$$C_R = \frac{2}{E}\kappa_b(\kappa_w + \kappa_b)$$

$$D_R = \frac{1}{E}(\kappa_b^2 - \kappa_w^2)[1 - \exp(-2qL)] \quad (580)$$

Notice that if one takes the limit case $L \to \infty$ and $\kappa_w \to \kappa_b$ the following values are obtained: $A_L = 0$, $B_L = 1$, $C_L = 0$, $D_L = 1$; $A_w = 1$, $B_w = 0$, $C_w = 0$, $D_1 = 1$; $A_R = 1$, $B_R = 0$, $C_R = 1$, $D_R = 0$. This means that from Eqs. (569), (573), or (578) one recovers the homogeneous case.

Following the Green function method, as it has just been used for a quantum well, one can find for a single heterostructure that the Green function corresponding to the electrostatic problem similar to Eqs. (555) and (559)–(561) (now one must omit matching conditions at $z = L$ since this interface does not exist) has exactly the form given by Eq. (567), but now

$$\Gamma(z, z') = \begin{cases} \Gamma_-(z, z') & z' < 0 \\ \Gamma_+(z, z') & z' > 0 \end{cases} \quad (581)$$

When the source charge is to the left of the interface ($z' < 0$),

$$\Gamma_-(z, z') = \begin{cases} \exp[-q|z - z'|] + A_- \exp[q(z + z')] & z' < 0 \\ B_- \exp[q(z' - z)] & z' > 0 \end{cases} \quad (582)$$

where

$$A_- = \frac{\kappa_b - \kappa_w}{\kappa_b + \kappa_w} \qquad B_- = \frac{2\kappa_b}{\kappa_b + \kappa_w} \quad (583)$$

Finally, when the source charge is to the right of the interface ($z' > 0$),

$$\Gamma_+(z, z') = \begin{cases} C_+ \exp[q(z - z')] & z' < 0 \\ \exp(-q|z - z'|) + D_+ \exp[q(z + z')] & z' > 0 \end{cases} \quad (584)$$

where

$$C_+ = \frac{2\kappa_w}{\kappa_b + \kappa_w} \qquad D_+ = \frac{\kappa_w - \kappa_b}{\kappa_b + \kappa_w} \quad (585)$$

Notice that if one takes the limit case $\kappa_w \to \kappa_b$ the following values are obtained: $A_- = 0$, $B_- = 1$; $C_- = 1$, $D_- = 0$. This means that from Eqs. (29) or (31) one recovers the homogeneous case.

Finally, note that taking the limit $L \to \infty$ from Eq. (569) one obtains Eq. (582), while from Eqs. (573) or (578) one obtains Eq. (584).

Following the self-consistent field approach, straightforward calculations lead to

$$F_{nn'll'}(q) = \int_{-\infty}^{+\infty} dz \int_{-\infty}^{+\infty} dz' \, \Gamma(z, z'; q)\varphi_{n'}^*(z)\varphi_n(z)\varphi_{l'}^*(z')\varphi_{l'}(z) \quad (586)$$

which is the more general expression for the form factor one can write in order to use any envelope functions in the confinement direction. Of course, the $\Gamma(z, z'; q)$ function has to be specified in accordance with the results presented just above in this subsection.

In this subsection it has been demonstrated that a specific order to change labeling of four-index matrix elements into two-index matrix elements, which appeared in a strictly logical way and is not arbitrary at all, reduces the inversion of a four-index matrix to the inversion of a two-index matrix. Moreover, the algorithms to obtain the inverted four-index matrix can be easily established. Thus the inversion of the four-index matrix dielectric function, required for calculation of matrix elements of screened interaction through those corresponding to bare interaction, is already a solved problem. Furthermore, a concrete procedure to take into account the screening effect has been provided. General expressions for the form factor concerning electron–electron interaction in single heterostructures and quantum wells when different dielectric constants at each side of an interface are considered and, besides, the envelope functions in the confinement direction that can penetrate into the barrier regions have been derived. For this purpose the Green function method and the image method to solve the electrostatic problem arising in the confinement direction have been employed; these methods seems to be the more suitable ones, as will be discussed elsewhere.

This problem is connected not only with multisubband transport in semiconductor heterostructures or inversion layers, but also with any calculation concerning phenomena in quasi-two-dimensional systems where screening of involved interactions in intersubband scattering must be take into account.

7.3. Quantum Transport in the Confinement Direction in a Quasi-Two-Dimensional System. Vertical Transport

The idea of vertical transport, i.e., transport occurring along the growth axis of a microstructure, was the reason Esaki and Tsu put forward the concept of semiconductor superlattices [171]. Let us consider a one-dimensional superlattice and assume that the carrier dispersion relations can be written as

$$\varepsilon_n(q, \vec{k}_\perp) = \frac{\hbar^2 k_\perp^2}{2m_{1e}^*} + \varepsilon_n(q) \qquad (587)$$

with

$$\varepsilon_n\left(q + \frac{2\pi}{d}\right) = \varepsilon_n(q) \qquad (588)$$

A static field $\vec{E}$, of small intensity, is applied parallel to the growth axis. If it is assumed that the field does not induce intersubband transitions and thus restrict the attention to the lowest miniband, the semi-classical equations of motion become exact. They are

$$\hbar \frac{d\vec{k}}{dt} = -e\vec{E} \qquad (589)$$

$$v = \frac{1}{\hbar} \frac{\partial \varepsilon_1(\vec{k})}{\partial \vec{k}} \qquad (590)$$

The x, y motions are readily calculated:

$$\vec{k}_\perp = \vec{k}_0 \qquad \vec{v}_\perp = \frac{\hbar \vec{k}_0}{m_{1e}^*} \qquad (591)$$

The z motion is more interesting, for it leads to an oscillatory behavior of the velocity component ν_z upon time (Bloch oscillator), despite the fact that the electric field is static [208]:

$$q = q_0 - \frac{eEt}{\hbar} \qquad (592)$$

$$\nu_z(t) = -\frac{1}{eE} \frac{\partial \varepsilon_1}{\partial t}\left[q_0 - \frac{eEt}{\hbar}, \vec{k}_0\right] \qquad (593)$$

The velocity is periodic upon time with a frequency

$$\nu = \frac{eEd}{2\pi\hbar} \qquad (594)$$

For a field strength of 10^4 V/cm and a superlattice period of 200 Å, one obtains $\nu \approx 5 \times 10^{12}$ Hz. It is quite remarkable that the ν_z periodicity upon time arises from Eq. (588), i.e.,

from the spatial periodicity of the superlattice. The realization of a Bloch oscillator would considerably broaden the available spectrum of millimeter devices.

The point is that to exhibit oscillations the carrier must not experience any scattering event during the time elapse $(2\pi\nu)^{-1}$. Thus, the scattering time τ_s should be such that $2\pi\nu\tau_s > 1$. The mobility relaxation time τ_μ is an upper bound to τ_s. Bloch oscillations would thus become unobservable if $2\pi\nu\tau_\mu \leq 1$. For GaAs–GaAlAs superlattices this means a mobility smaller than or equal to $\approx 8 \times 10^2$ cm^2/V/s. This mobility, although relatively low, is still 80 times larger than the one recently reported by Palmier et al. [209].

Another physical reason, which may lead to a failure of the Bloch oscillator, is the field-induced intersubband scattering which was explicitly neglected in Eqs. (589) and (590).

The vertical transport in GaAs–GaAlAs superlattices has been interpreted [209] in terms of phonon-assisted hopping between the localized levels in one well to those in the neighboring one. The occurrence of such a hopping type of transport (instead of the standard conduction in minibands) has been proved by two observations.

(i) The order of magnitude of the experimental mobility was found to be much lower than predicted by assuming conduction through the delocalized Bloch states.

(ii) The experimental mobility increases with temperature, which dismisses any interpretation of the data in terms of phonon-limited relaxation time but strongly favors an interpretation in terms of phonon-assisted tunneling between wells. If the conduction were to occur via extended Bloch states, the electron–phonon scattering would increase the relaxation frequency with increasing T as more phonons are available at higher temperatures. In turn, the mobility μ, which is proportional to τ_μ, would decrease with increasing T, as is invariably observed in the in plane transport. The hopping type of transport describes the electrical conductivity as arising from phonon-assisted hopping between wells. Then, by the Fermi Golden Rule, the conductivity is proportional to the scattering frequency and increases with increasing temperature.

In other GaAs–GaAlAs superlattices Chomette et al. have reported optical evidence of vertical transport at low temperatures [210], They purposely introduced GaAs enlarged wells into short period GaAs–GaAlAs superlattices and studied at low temperatures the excitonic photoluminescence which arises from both the superlattice and the enlarged wells. If the coupling between wells were negligible, the intensities of both types of luminescence would roughly be in proportion to the number of enlarged wells to superlattices periods. These results are far from what was measured. The observed data were explained by means of miniband conduction. Rate equations involving the trapping time from the superlattice to the enlarged wells and the carrier lifetimes in the superlattices and the enlarged wells have been used to calculate the photo-luminescence intensities. A reasonable agreement was found between the calculated trapping time and the one deduced from the experiments.

7.3.1. Resonant Tunneling

Another vertical transport mechanism, which is physically different from both defect-limited conduction in the miniband of a superlattice and the phonon-assisted hopping between wells, is the transport via resonant tunneling. As in the case of the Bloch oscillator, there is no contribution from perturbative agents (impurities, defects,...) at the lowest order in the resonant tunneling. The latter is a quantum effect of intrinsic origin, which should be bettter observed in perfect heterostructures. The physics of resonant tunneling is still a matter of active research, especially its time-dependent aspect [211].

Let us summarize some basic considerations.

First it will be assumed that the carrier motions parallel and perpendicular to the growth axis are decoupled. Thus the total carrier energy can be split into

$$\varepsilon = \varepsilon_L + \frac{\hbar^2 k_\perp^2}{2m_{1e}^*} \tag{595}$$

where ε_L is the carrier energy associated with the longitudinal motion. Under this assumption, the tunneling problem becomes quasi-one-dimensional. Let us consider the simplest device structure aimed at exploiting the resonant tunneling effect. It consists of a quantum well, in principle of arbitrary shape (but in the following assumed to be rectangular), flanked by two finite barriers (see Fig. 45). Electrical contacts are placed on each side of the structure.

For $\varepsilon_L \leq V_0$, the transmission coefficient of a single barrier is an exponentially decaying function of the barrier thickness b and of the barrier height V_0,

$$T_{(1)}(\varepsilon_L) = \left| t_{(1)}(\varepsilon_L) \right|^2$$
$$= \left[1 + \frac{1}{4}\left(\xi + \frac{1}{\xi}\right)^2 \sinh^2(k_b b) \right]^{-1} \quad \varepsilon_L \leq V_0 \tag{596}$$

where

$$\xi = \frac{k_w}{m_w^*} \times \frac{m_b^*}{k_b} \tag{597}$$

and

$$k_w = \sqrt{\frac{2m_w^*}{\hbar^2}\varepsilon_L} \qquad k_b = \sqrt{\frac{2m_b^*}{\hbar^2}(V_0 - \varepsilon_L)} \tag{598}$$

In the case of a double symmetrical barrier, the transmission $T_{(2)}(\varepsilon_L)$ will be even lower for most ε_L. However, $T_{(2)}(\varepsilon_L)$ will reach unity for certain discrete energies ε_{iL} (see Fig. 46). If L denotes the quantum well thickness, these energies are the roots of the equation

$$\cos k_w L - \frac{1}{2}\left[\xi - \frac{1}{\xi}\right]\sin k_w L = 0 \tag{599}$$

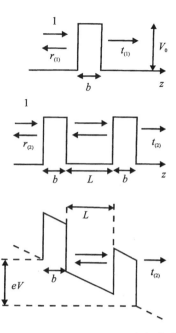

Fig. 45. Conduction band profile of an unbiased single (upper panel) and double (middle panel) barrier structure. When a voltage v is applied between the left and right hand side electrodes, the double barrier structure is no longer symmetrical (lower panel).

i.e., the ε_{iL}'s coincide with the bound states of a quantum well of thickness L and barrier height V_0 (see Section 3).

For these particular energies, the transmitted and reflected electron beams interfere constructively inside the quantum well and build up a virtual bound state. The state is only virtually bound due to the finite thickness of the cladding barriers. As in the previous discussion (Section 3) of the virtual bound states which occur in the continuum of a single quantum well, the virtual bound states of the double barrier structure can equally be viewed as transmission resonances (see again what happens in a laser). The widths $\delta\varepsilon_i$ of the ith resonance ε_{iL} are decreasing functions of the barrier thickness b (see Fig. 46). Thus, the contrast between the maximum and minimum values of $T_{(2)}(\varepsilon)$ increases with increasing b.

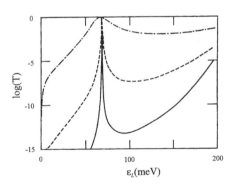

Fig. 46. Natural logarithm of the transmission versus the energy ε_L of the longitudinal motion. $V_0 = 0.2$ eV, $m_w^* = 0.07m_0$, $m_b^* = 0.088m_0$, $d = 50$ Å. Three barrier thicknesses have been considered: $b = 20$ Å (dot-dashed line), 50 Å (dashed line), 80 Å (solid line). The transmission peaks coincide with the only bound state of the isolated well ($\varepsilon_1 \cong 68$ meV).

These transmission resonances can be exploited to produce sharp peaks in the current–voltage characteristic of a biased double barrier (see Fig. 45). Whenever the Fermi level ε_F of the metallic electrode coincides with one of the ε_{iL}, one may expect a current peak. For other biases the current is very small. This implies that, by tuning the bias voltage V between the two electrodes, one finds negative differential resistance regions once a virtual bound state has passed through ε_F: a further increase in V leads to a sharp drop in the current.

The exact bias value at which a current peak develops is difficult to assess simply from the calculation of transmission resonances in unbiased double barrier structures.

If the bias is vanishingly small (which seldom happens in practice) the current peaks take place when

$$eV = 2\varepsilon_{iL} \tag{600}$$

However, the biased structure in general

(i) is no longer symmetrical
(ii) can no be longer described in terms of field-independent virtual bound states.

As pointed out by Ricco and Azbel [211] the piling up of charges inside the well under the resonance condition produces a band bending. In turn, this shifts the resonant state energy, making the achievement of resonances a complicated self-consistent, time-dependent problem.

The first clear-cut resonant tunneling effects in GaAs–GaAlAs double barriers were demonstrated by Chang et al. (see Fig. 47) in 1974 [212]. Since then, improved MBE growth has made possible the observation of narrower peaks in the current–voltage characteristics. Recently, the resonant tunneling of holes in high quality AlAs–GaAs–AlAs heterostructures was reported by Mendez et al. [213] (see Fig. 48). Although one should expect to see peaks associated with GaAs heavy

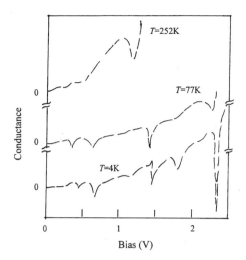

Fig. 48. Conductance versus voltage bias, for representative temperatures, of an undoped AlAs–GaAs–AlAs double barrier structure sandwiched between p^*–GaAs regions. $d = 50$ Å; $b = 80$ Å.

and light hole virtual bound states, the actual peaks in the current–voltage characteristics did not fit these assumptions. This calls for detailed calculations of the resonant tunneling of carriers with complicated subband structures.

The resonant tunneling effects in multiple barriers were calculated by Tsu and Esaki [214]. If N is the number of identical barriers and $(N-1)$ is thus the number of identical wells, the double barrier virtual bound states which fulfill Eq. (599) split into $(N-1)$ states where the transmission is unity. At the limit $N \rightarrow \infty$, one recovers the superlattice minibands, where the transmission is unity for energies corresponding to the allowed superlattice states and zero for energies corresponding to the superlattice bandgaps.

7.4. Magnetoconductivity of a Quasi-Two-Dimensional Electron Gas

7.4.1. Macroscopic Derivation

Let us consider a Q2D electron gas subjected to a strong magnetic field. The interest is in its linear response to a weak static electric field $\vec{E} \perp \hat{e}_z$, i.e., applied in the layer plane of the heterostructure. The electron gas responds to this external disturbance with an electrical current, which, since E is assumed to be small, will be proportional to $\vec{E}$. Thus, it can be written that the x, y components of the areal current density $\vec{J}$ are related to $\vec{E}$ by

$$J_x = \sigma_{xx}E_x + \sigma_{xy}E_y \qquad J_x = \sigma_{yx}E_x + \sigma_{yy}E_y \tag{601}$$

Since the host materials are cubic, $\sigma_{xx} = \sigma_{yy}$ and the Onsager relationships [215, 216] show that

$$\sigma_{xy} = -\sigma_{yx} \tag{602}$$

The conductivity tensor $\vec{\sigma}$ is the quantity, which is most naturally calculated. Practically, however, this is the resistivity tensor

$$\vec{p}\,\vec{\sigma}^{-1} \tag{603}$$

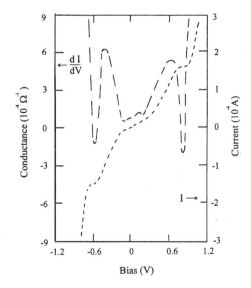

Fig. 47. Current–voltage and conductance–voltage characteristics of a double barrier structure of GaAs between two $Ga_{0.7}Al_{0.3}As$ barriers sandwiched between n^*–GaAs regions. $d = 50$ Å; $b = 80$ Å; $T = 77$ K.

which is the most accessible to the experiments. If Eq. (601) is inverted,

$$\rho_{xx} = \rho_{yy} = \frac{\sigma_{xx}}{\sigma_{xx}^2 + \sigma_{xy}^2} \qquad \rho_{yx} = -\rho_{xy} = \frac{\sigma_{xy}}{\sigma_{xy}^2 + \sigma_{xx}^2} \qquad (604)$$

Let us consider a sample with a "spider" geometry. It consists of a long bar ($L_x \gg L_y$) with side contacts (see Fig. 49). The voltage V_x is imposed via a battery and the voltage V_H is measured across the small dimension of the bar. Since J_y vanishes, then

$$E_x = \rho_{xx} J_x \qquad E_y = \rho_{yx} J_x \qquad (605)$$

If one assumes that the current densities J_x, J_y are uniform in the sample (the spatial variation of $\vec{J}$ is still a matter of controversy) it can be written that

$$J_x = \frac{I_x}{L_y} \qquad V_H = E_y L_y \qquad V_x = E_x L_x \qquad (606)$$

Thus

$$V_x = \frac{\rho_{xx}}{L_y} L_x I_x = R_{xx} I_x \qquad V_H = \rho_{yx} I_x = R_{yx} I_x \qquad (607)$$

where V_x, V_H are expressed in volts, I_x in amperes, and R_{xx}, R_{yx} in ohms. Finally

$$R_{xx} = \frac{L_x}{L_y} \frac{\sigma_{xx}}{\sigma_{xx}^2 + \sigma_{xy}^2} \qquad R_{yx} = \frac{\sigma_{xy}}{\sigma_{xx}^2 + \sigma_{xy}^2} \qquad (608)$$

The resistance R_{xx} involves the geometrical factor L_x/L_y and thus is sample dependent. In literature, people often get rid of this inconvenience by quoting the resistance which would be measured in samples with a square shape but which are otherwise identical to their investigated samples. Thus, to recall this feature, R_{xx} measurements are given in units of ohm square (Ω_0).

On the other hand, the Hall resistance R_{yx} is independent of the sample shape. This explains why several Bureaus of Standards, in the hope of defining a new resistance standard, are interested in accurate R_{yx} data [217]. The former will be independent of sample size and thus free of dilatation effects which often plague the usual ohm standards which

are, presently, based on resistance of platinum rods. Finally let us stress once more that both R_{xx} R_{yx} as well as ρ_{xx}, ρ_{xy} are expressed in ohms. Nothing is wrong with the latter property. Only the resistivities ρ_{xx}, ρ_{xy} are two-dimensional resistivities. It is known that, in the dimension equation,

$$\rho = \frac{m_e^*}{\eta_e e^2 \tau} \qquad (609)$$

as in bulk materials (τ has the dimension of a relaxation time). However, in Eq. (609) η_e is an areal electron concentration instead of a volumetric concentration as usually found in bulk materials.

7.4.2. Microscopic Discussion

The linear response theory, originally derived by Kubo, has been used to evaluate the components σ_{xx} and σ_{yx} of the static magnetoconductivity tensor in the independent electron approximation [218]. It was found that at $T = 0$ K

$$\sigma_{xx} = \frac{\pi \hbar e^2}{(m_e^*)^2 S} \text{ Trace } \{\delta(\varepsilon_F - \widehat{H}) \prod_x \delta(\varepsilon_F - \widehat{H}) \prod_x\} \qquad (610)$$

$$\sigma_{yx} = \frac{-ie^2 \hbar}{(m_e^*)^2 S} \int_{-\infty}^{+\infty} f(\varepsilon) d\varepsilon$$
$$\times \text{Trace} \left\{ \delta(\varepsilon - \widehat{H}) \left[\prod_y \frac{1}{(\varepsilon - \widehat{H})^2} \prod_x - \prod_x \frac{1}{(\varepsilon - \widehat{H})^2} \prod_y \right] \right\} \qquad (611)$$

where $f(\varepsilon)$ is the Fermi–Dirac distribution function which, at zero temperature, is a step function of the argument $\varepsilon_F - \varepsilon$ where ε_F is the Fermi level. In Eqs. (610) and (611) $\prod$ is $\hat{p} + (e/c)\vec{A}$ and $\widehat{H}$ is the quasi-two-dimensional Hamiltonian which includes disorder,

$$\widehat{H} = \frac{\hat{p}_x^2}{2m_e^*} + \frac{1}{2m_e^*}\left(\hat{p}_y + \frac{e}{c}Bx\right)^2 + \varepsilon_1 + g^*\mu_B B \sigma_z + V_{\text{disorder}}(\vec{\rho})$$
$$\widehat{H} = \widehat{H}_0 + V_{\text{disorder}}(\vec{\rho}) \qquad (612)$$

where

$$V_{\text{disorder}}(\vec{\rho}) = \int dz \chi_1^2(z) V_{\text{disorder}}(\vec{r}) \qquad (613)$$

and ε_1 is the confinement energy of the ground subband (wavefunction χ_1) which arises from the size quantization along the growth axis of the heterostructures. In Eqs. (612) and (613) $\vec{\rho}(x, y)$, $\vec{r} = (x, y, z)$. To make sense, Eq. (612) requires a pronounced two-dimensionality of the electron gas, i.e., that the excited subbands $\varepsilon_2 \cdots$ must be far enough away from ε_1 to allow a factorization of the carrier wavefunction $\psi(\vec{r})$ in $\chi_1(z)\varphi(\vec{\rho})$. The assumed strong two-dimensionality also allows us to treat the case of the tilted magnetic fields with respect to the growth axis. Only the magnetic field B

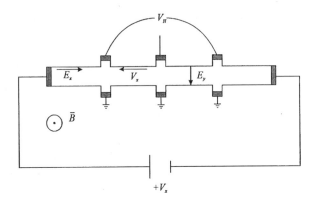

Fig. 49. Sample geometry used to measure the resistivity tensor $\tau(B)$.

which appears in Eq. (612) has to be understood as the projection of B along the z-axis. The matrix elements of $\prod_x$, $\prod_y$ on the Landau levels basis (eigenstates of $\widehat{H}_0$) are

$$
\left\langle n, k_y, \sigma_z \left| \prod_x \right| n', k'_y, \sigma'_z \right\rangle
$$
$$
= \frac{i\hbar}{\lambda\sqrt{2}} \left[\sqrt{n}\,\delta_{n', n-1} - \sqrt{n+1}\,\delta_{n', n+1} \right] \delta_{k'_y, k_y} \delta_{\sigma'_z, \sigma_z} \quad (614)
$$

$$
\left\langle n, k_y, \sigma_z \left| \prod_{\ddot{y}} \right| n', k'_y, \sigma'_z \right\rangle
$$
$$
= \frac{\hbar}{\lambda\sqrt{2}} \left[\sqrt{n}\,\delta_{n', n-1} + \sqrt{n+1}\,\delta_{n', n+1} \right] \delta_{k'_y, k_y} \delta_{\sigma'_z, \sigma_z} \quad (615)
$$

In a perfect heterostructure $V_{\text{disorder}}(\vec{\rho})$ vanishes. Thus the states $|n, k_y, \sigma_z\rangle$ are eigenstates of $\widehat{H}$. Consequently the matrix elements

$$
\langle n, k_y, \sigma_z | \delta(\varepsilon - \widehat{H}) | n', k'_y, \sigma'_z \rangle \quad (616)
$$

are equal to

$$
\langle n, k_y, \sigma_z | \delta(\varepsilon - \widehat{H}) | n', k'_y, \sigma'_z \rangle = \delta(\varepsilon - \varepsilon_{n\sigma_z}) \delta_{nn'} \delta_{k'_y k_y} \delta_{\sigma'_z \sigma_z} \quad (617)
$$

Equations (615)–(617) immediately lead to

$$
\sigma_{xx} = 0; \qquad \sigma_{yx} = \frac{\eta_e e^2}{m^*_e \omega_c} \quad (618)
$$

Thus, for a perfect system the microscopic evaluation of the conductivity tensor leads to the same result as the classical Drude type of calculations.

Equations (618) are not surprising but merely serve as a reminder that, in the absence of scattering, a carrier subjected to crossed electric and magnetic fields actually drifts along a direction perpendicular to these two fields. The electrical transport along the electric field is entirely due to collisions, i.e., to a nonvanishing $V_{\text{disorder}}(\vec{\rho})$.

It has been already stated that very little is known about the exact shape of $V_{\text{disorder}}(\vec{\rho})$. Its spatial average over a macroscopic area should, however, vanish. Thus, in the (x, y) plane, $V_{\text{disorder}}(\vec{\rho})$ shows uncorrelated fluctuations, which are either repulsive or attractive for electrons. (Notice that for very large magnetic fields both kinds of fluctuations, when considered as isolated, are capable of supporting bound states.) Taking the randomness of the disorder potential into account implies that Eqs. (610) and (611) should be completed by a spatial average over $V_{\text{disorder}}(\vec{\rho})$ and, by taking the limit $S \to \infty$. If $V_{\text{disorder}}(\vec{\rho})$ can be written as the sum of one-defect potentials centered at $\vec{\rho}_i$,

$$
V_{\text{disorder}}(\vec{\rho}) = \sum_{\rho_i} W(\vec{\rho} - \vec{\rho}_i) \quad (619)
$$

The spatial average signifies an average over the positions $\vec{\rho}_i$ of the defects, which are assumed to be uncorrelated.

7.4.2.1. Magnetic Field Dependence of σ_{xx} (Shubnikov–de Hass Effect)

Let us evaluate σ_{xx} by taking the eigenstates $|\nu\rangle$ of $\widehat{H}$ as the basis vectors

$$
\sigma_{xx} = \frac{\pi e^2 \hbar}{(m^*_e)^2 S} \sum_{\mu, \nu} \delta(\varepsilon_F - \varepsilon_\mu) \delta(\varepsilon_F - \varepsilon_\nu) \left| \left\langle \nu \left| \prod_x \right| \mu \right\rangle \right|^2 \quad (620)
$$

Equation (620) shows that the zero temperature conductivity σ_{xx} is proportional to the number of pairs of states which have exactly the Fermi energy ($\varepsilon_F = \varepsilon_\mu = \varepsilon_\nu$), weighted by the square of the $\prod_x$ matrix element between them.

Let us now suppose that the scheme of broadened Landau levels whose tail states are localized is adopted. Then, if the Fermi energy stays within the tail states, the matrix element $\langle \nu | \prod_x | \mu \rangle$ will decay (say exponentially) with $|\vec{\rho}_\mu - \vec{\rho}_\nu|$ which is the distance separating the two localized states $|\mu\rangle$, $|\nu\rangle$. In addition, due to the randomness of $V_{\text{disorder}}(\vec{\rho})$ the number of states with exactly the same energy $\varepsilon_\mu = \varepsilon_\nu$ is extremely small. All these facts lead us to the conclusion that σ_{xx} is a very small at $T = 0$ K. More sophisticated analysis [219] shows that σ_{xx} does indeed vanish when the limit of infinite S is taken. The previous argument does not hold for extended states of the broadened Landau levels at least because $\langle \nu | \prod_x | \mu \rangle$ is $\approx 1/\lambda$ for those states. Thus, when the magnetic field is swept and the Fermi energy passes trough the consecutive Landau levels, σ_{xx} undergoes oscillations and is zero when ε_F is in the localized levels (or in a region where the density of states vanishes) and positive when ε_F coincides with the extended states of the spectrum.

Such behavior is analytically derived within the framework of the self-consistent Born approximation. In such a model, one obtains [220]

$$
\sigma_{xx} = 0 \quad \text{if} \quad \left| \varepsilon_F - \left(n + \frac{1}{2} \right) \hbar\omega_c - \varepsilon_1 - g^* \mu_B B \sigma_z \right| > \Gamma_n \quad (621)
$$

$$
\sigma_{xx} = \frac{e^2}{\pi^2 \hbar} \left(n + \frac{1}{2} \right)
$$
$$
\times \left\{ 1 - \left[\frac{\varepsilon_F - \left(n + \frac{1}{2} \right) \hbar\omega_c - \varepsilon_1 - g^* \mu_B B \sigma_z}{\Gamma_n} \right]^2 \right\} \quad (622)
$$

elsewhere. Here $2\Gamma_n$ is the width of the nth broadened Landau level whose density of states is given by Eq. (42). Despite the incorrect treatment of the tail states, the formula (Eq. (622)) is very helpful in explicitly showing the oscillatory behavior of σ_{xx} upon B.

These oscillations are well known (Shubnikov–de Haas effect) [86, 218] and have been observed in a variety of bulk materials and heterostructures. They are of quantum origin (being absent in a Drude-type analysis of the magnetoconductivity tensor) and evidence the magnetic field induced singularities of the density of states.

The σ_{xx} maxima (or equivalently the ρ_{xx} maxima since $\sigma_{xx} \ll \sigma_{xy}$ in the first of Eqs. (604)) are evenly spaced in $1/B$. The fields $B_{n\sigma_z}$ where σ_{xx} is maximum are such that

$$\varepsilon_F - \varepsilon_1 = \left[n + \frac{1}{2} + \frac{g^* \mu_B \sigma_z m_e^*}{\hbar e} \right] \frac{\hbar e}{m_e^*} B_{n\sigma_z} \qquad (623)$$

This means that a plot of $B_{n\sigma_z}^{-1}$ versus integer n should be a straight line (Fig. 50) whose slope S_B yields direct information on $\varepsilon_F - \varepsilon_1$ or, equivalently, on the areal carrier concentration η_c, which is present in the heterostructure:

$$\eta_c = \frac{m_e^*}{\pi \hbar^2} (\varepsilon_F - \varepsilon_1) \qquad (624)$$

$$S_B = \frac{\hbar e}{m_e^*} \frac{1}{(\varepsilon_F - \varepsilon_1)} = \frac{e}{\pi \hbar \eta_e} \qquad (625)$$

The determinations of η_e by means of the Shubnikov–de Hass effect are very accurate. In addition they do not require any information concerning the sample geometrical details (shape, distance between contacts, etc.).

7.5. Magnetic Field Dependence of σ_{xy}: Quantized Hall Effect

In the self-consistent Born approximation, there should exist corrections to the perfect Hall effect ($\sigma_{xy} = -\eta_e e^2 / m_e^* \omega_c$). However, it is not very interesting to discuss these corrections insofar as the most spectacular effects are completely missed by any theory, which neglects the localized states in the tails of the Landau levels.

What was experimentally observed is the appearance of plateaus in the ρ_{xy} curves versus B (Fig. 51) at low temperatures ($T \leq 4.2$ K). The ρ_{xy} plateaus are exactly (within at least one part per million) quantized, being subharmonics of $\hbar/e^2$. The so-called quantized Hall effect, discovered by von Klitzing et al. [221] in Si-metal-oxide semiconductor field effect transistors (Si-MOSFETs), can be termed a universal effect. It has also been observed in other two-dimensional systems: $Ga_{1-x}Al_xAs$–GaAs [222], InP–$Ga_{0.47}In_{0.53}$ As [223], AlSb–InAs [224], and GaSb–InAs [225]. Its occurrence is independent of the band structure of the host materials: the quantized Hall effect can be observed when the carriers are electrons belonging to a single conduction band ($Ga_{1-x}Al_xAs$–GaAs) [222] or to a multi-valley conduction band (Si-MOSFET) [221]. It is also observed when the carriers are holes belonging to a degenerate valence band ($Ga_{1-x}Al_x$ As–GaAs) [226] and even when there is a mixed conduction (GaSb–InAs) [225].

The width ΔB of a given ρ_{xy} plateau increases as the temperature decreases, reaching an amplitude which is close to the maximum one at very low T (milli-Kelvin range) [227]. It appears established that ρ_{xy} develops plateaus when the Fermi

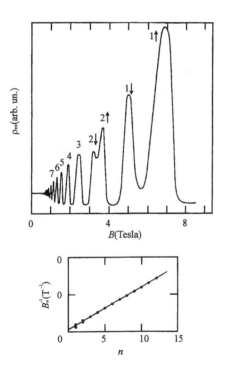

Fig. 50. Oscillatory behavior displayed by $\rho_{xx}(B)$ upon B (Shubnikov–de Haas effect) in a GaAs–$Ga_{0.7}Al_{0.3}$As heterojunction at $T = 2$ K (upper panel). The inverses of the magnetic fields $1/B_n$ corresponding to maxima of $\rho_{xx}(B)$ are plotted versus consecutive integers n (lower panel). The slope of the straight line B_n^{-1} versus n directly yields the areal concentration of quasi-two-dimensional electrons n_c. $B//\hat{e}_z$, $\eta = 3.93 \times 10^{11}$ cm^{-2}.

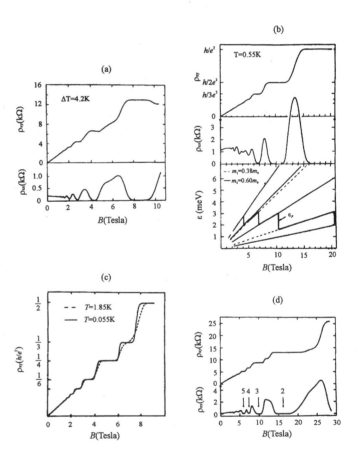

Fig. 51. The quantized Hall effect in III–V heterostructures. The ρ_{xy} curves plotted versus are shown for: (a) n–$Ga_{1-x}Al_xAs$–GaAs, (b) p–$Ga_{1-x}Al_xAs$–GaAs, (c) p–$Ga_{1-x}Al_xAs$–GaAs, and n–GaSb–InAs–GaSb.

level coincides with the localized states of the broadened Landau levels. In fact, in the plateau regime for ρ_{xy}, ρ_{xx} (and σ_{xx}) they are almost zero. The deviation from the zero resistance state exponentially decreases as temperature decreases. Such a feature evidences a hopping type of transport along the electric field. The details of the hopping processes remain uncertain: the hopping laws

$$\sigma_{xx}(T) \approx \exp\left[-\left(\frac{T_0}{T}\right)^\alpha\right] \qquad (626)$$

fit the data reasonably well but the exponent α is sample-dependent [225, 228, 229]. Perhaps the disorder potential responsible for the existence of localized states is markedly different in different types of heterostructures.

That σ_{xy} could be exactly quantized in units of a fundamental constant e^2/h irrespective of the band structure, defects, etc. clearly called for a theoretical explanation which would go beyond the usual perturbative approaches of the magneto-conductivity tensor and which would exploit the genuine characteristic of two-dimensional magnetotransport, which is the existence of nonconducting states separating conducting ones.

It was Aoki and Ando [230] and Prange [231] who first pointed out that localized levels do not contribute to σ_{xy} and that, moreover, delocalized states, in the presence of localized ones, conduct more efficiently, thus ensuring the pinning of σ_{xy} to a multiple of e^2/h. The theoretical breakthrough was made by Laughlin [232] who showed that the gauge invariance of the electromagnetism implies the quantization of σ_{xy}, at $T = 0$ K, provided that the Fermi level sits in the localized levels ($\sigma_{xx} = 0$). For a detailed discussion of the theoretical aspects of the quantized Hall effect, see [233].

There is no available theory of the quantized Hall effect at finite temperature. Such a theory is, however, very desirable in order to ascertain the magnitude and temperature dependence of the corrections to the $T = 0$ K value, and thus the accuracy of new resistance standards based on the quantized Hall effect.

7.5.1. Fractionally Quantized Hall Effect

When the Fermi level has passed the 0^- Landau level, nothing is expected to occur. A wealth of extra features was in fact detected for fractional occupancies of the remaining 0^+ sublevel (Fig. 52) [234, 235]. These features are only seen in samples with very high zero-field mobilities. Like the "normal" quantized Hall effect, the fractionally quantized Hall effect does not depend on the band structure of the host materials. It has been observed in both electrons [234–236] and holes [237] in quasi-two-dimensional systems. Of the fractional p/q defined by

$$\rho_{xy} = \frac{q}{p}\frac{h}{e^2} \qquad (627)$$

only those with odd denominators have been unambiguously observed. The fractionally quantized Hall effect has been explained by Laughlin [238, 239] in terms of a condensation

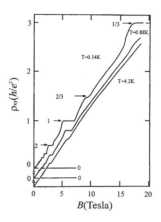

Fig. 52. Fractionally quantized Hall effect in a n–Ga$_{1-x}$Al$_x$As–GaAs single heterojunction. The plateaus labeled 2/3 and 1/3 take place when the fractional occupation of the last (0^+) Landau level is equal to 2/3 and 1/3, respectively. $\eta = 1.48 \times 10^{11}$ cm^{-2}.

of the electron (or hole) system into a microscopic collective ground state due to electron–electron (or hole–hole) interaction. This ground state is separated from the excited one by $\approx 0.03e^2/\lambda$, where λ is the magnetic length defined in Eq. (22). That a repulsive interaction between carriers may give rise to a condensation is intimately linked to the quasi-two-dimensionality of the electron motion. In a bulk material two such electrons can never form a bound state, whereas the two-dimensional motion prevents the escape of the two electrons at infinite distance from each other. Laughlin's wavefunction of the condensed ground state has been generalized by Halperin [240]. The exact microscopic nature of the ground state is still disputed [241].

7.5.2. Cyclotron Resonance

As can be shown, the frequency-dependent magnetoconductivity tensor $\sigma_{\alpha\beta}(\omega)$ (which describes the linear response of the electron gas to a periodically varying electric field) exhibits singularities at $\omega = \omega_c$. Let us work in the Drude model (classical description) and assume that

$$\vec{E}(\vec{r}, t) = E\hat{e}_x \cos\omega t \qquad (628)$$

The equation of motion of the complex velocity

$$\xi = v_x + iv_y \qquad (629)$$

is

$$\frac{d\xi}{dt} = -\frac{\xi}{\tau} + i\omega_c\xi - \frac{eE}{m_e^*}\cos\omega t \qquad (630)$$

where a viscous friction force $-m_e^*\nu/\tau$ accounts for the scattering mechanisms. In the permanent regime, ξ oscillates at the frequency ω and the $\hat{e}_x$ component of the areal current density $\vec{j} = -\eta_e e\vec{\nu}$ is found to behave like

$$j_x = \frac{\eta_e e^2\tau}{2m_e^*}E\left\{\frac{\cos\omega t + \tau(\omega - \omega_c)\sin\omega t}{1 + (\omega - \omega_c)^2\tau^2}\right.$$
$$\left. + \frac{\cos\omega t + \tau(\omega + \omega_c)\sin\omega t}{1 + (\omega + \omega_c)^2\tau^2}\right\} \qquad (631)$$

Over a period $2\pi/\omega$, the average power taken to the electric field is

$$\langle P \rangle = \frac{\omega}{2\pi} \int_0^{2\pi/\omega} j_x E \cos \omega t \, dt \qquad (632)$$

or

$$\langle P \rangle = \frac{\eta_e e^2 \tau}{4m_e^*} E^2 \left\{ \frac{1}{1+(\omega-\omega_c)^2\tau^2} + \frac{1}{1+(\omega+\omega_c)^2\tau^2} \right\} \qquad (633)$$

There are two contributions to $\langle P \rangle$. The first one presents a resonance at $\omega = \omega_c$, while the second one is a decreasing function of ω_c or ω. The resonant term arises from the $\exp(+i\omega t)/2$ part of the electric field. Indeed, the linear vibration (Eq. (628)) can be rewritten as the sum of two clockwise and counterclockwise circular vibrations:

$$E(t) = \begin{cases} E\cos\omega t \\ 0 \end{cases} = \begin{cases} \frac{1}{2}E\cos\omega t \\ \frac{1}{2}E\sin\omega t \end{cases} + \begin{cases} \frac{1}{2}E\cos(-\omega)t \\ \frac{1}{2}E\sin(-\omega)t \end{cases} \qquad (634)$$

The first kind of circular vibration is, for $\omega = \omega_c$, constantly in phase with the electron motion. As a result, it transfers a finite amount of energy to the electron during a cyclotron period. On the other hand, the second type of circular vibration partly accelerates and partly decelerates the electron and, on average, does not transfer energy to the electron. Note that, ultimately, the carrier does not store the electromagnetic energy. Instead, this energy is given up to the lattice via the viscous friction force. In the limit $\tau \to \infty$, $\langle P \rangle$ would be zero unless $\omega = \omega_c$. Figure 53 shows a plot of $\langle P \rangle$ versus ω_c/ω for three values of $\omega\tau$. Two features are remarkable:

(i) $\langle P \rangle$ displays an asymmetric lineshape at low $\omega\tau$ and a rather symmetric one at large $\omega\tau$.

(ii) $\langle P \rangle$ gets narrower for larger $\omega\tau$.

Suppose a heterostructure with a dc mobility of 10^4 cm^2/V/s is considered. The condition $\omega_c\tau \geq 1$, which ensures

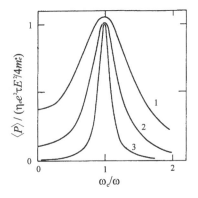

Fig. 53. The power absorbed by a classical electron gas from an electromagnetic wave (angular frequency ω), averaged over a period of the cyclotron motion, is plotted versus the dimensionless ratio ω_c/ω for $\omega\tau = 2.1(1)$, $4.1(2)$, $10.1(3)$, respectively.

that the electron has completed a revolution around $\vec{B}$ without being scattered, is fulfilled as soon as $B \geq 1$ Tesla. The cyclotron resonance absorption $\omega = \omega_c$ is observed in the far infrared or millimeter parts of the electromagnetic spectrum for the actual masses of III–V heterostructures. Two experimental techniques are used: either to sweep the magnetic field while keeping ω fixed, or to work at fixed B by sweeping the frequency ω. The first technique requires the use of far-infrared lasers or carcinotrons as monochromatic sources and a magnet to produce the field. It is more practical than the second one, which requires the use of Fourier interferometers combined with a magnet. By reporting the photon energies $\hbar\omega_n$ versus the observed resonance fields B_n on a $(B, \hbar\omega)$ diagram one gets a straight line (for parabolic band structures) whose slope $\hbar e/m_e^*$ yields a very precise determination of the carrier effective mass m_e^*.

It is clear that the Drude model cannot describe very accurately all the experimental observations in quasi-two-dimensional electron gases. In particular, the quantity τ, which accounts for the scattering effects, has, in practice, no reason to be ω- and B-independent as postulated in the Drude model.

A better model of the cyclotron resonance absorption has been derived by Ando and is based on the microscopic calculation of $\sigma_{\alpha\beta}(\omega)$ within the framework of the self-consistent Born approximation [242]. This model predicts resonance absorptions of the electromagnetic wave, not only at $\omega = \omega_c$ but also at the harmonics $\omega = n\omega_c$. In addition, when the width ΔB at half maximum of the main cyclotron resonance line ($\langle P \rangle$ versus B at fixed ω) is large enough (i.e., when the dc mobility is not too high) to allow the Fermi level to be crossed by several Landau levels (i.e., when η_e is large enough), the cyclotron resonance lineshape is calculated to exhibit extra oscillations which are ω-independent. These Ando oscillations are of the Shubnikov–de Hass type. They have been observed in Si-MOS structures [243], as well as in modulation-doped Ga$_{1-x}$Al$_x$As–GaAs heterostructures [244].

Apart from the Si-MOSFETs the most complete study of the cyclotron absorption has been performed in Ga$_{1-x}$Al$_x$As–GaAs heterostructures [244–247]. The first output of these studies is the effective mass of the carrier in the GaAs channel. It has been found that this mass is slightly enhanced over the band edge conduction mass of bulk GaAs($0.067\,m_0$). The mass enhancement is, however, modest (few percents) and is in general attributed to the weak nonparabolicity of the GaAs conduction band. It has not yet been proved that part of the enhancement does not originate from the matching conditions for the wavefunction and its derivative at the interface between GaAs and Ga$_{1-x}$Al$_x$As.

The most striking results of cyclotron resonance experiments pertain to the cyclotron resonance linewidth and lineshape. If interference effects are carefully excluded from the experiments, the lineshape of the cyclotron absorption is rather symmetric in Ga$_{1-x}$Al$_x$As–GaAs heterostructures. For heterostructures with modest mobilities ($\leq 10^5$ cm^2/V/s) the width at half maximum of the absorption curves follows the predictions of Ando's model [242]. In high mobility heterostructures it has been shown [247] that the cyclotron resonance linewidth ΔB plotted versus the resonance field displays oscillations, whereas Ando's model (at least with pointlike scatterers) would rather predict a smooth increase of ΔB. Such oscillations have been tentatively correlated with the oscillations of the screening of the scattering potentials by the carriers.

In fact, if the resonance field corresponds to a situation where the Fermi level stays in the localized levels of a broadened Landau level or in the gap separating two such levels, the screening is inefficient, since the heterostructure is, for that particular field, an insulator. If, on the other hand, the resonance field is such that the Fermi level coincides with the extended states of a Landau level, the screening effects are very pronounced due to the large polarizability of the electron gas for that particular configuration. When the screening effects are weak, both the damping effects and the cyclotron resonance linewidth are large and vice versa.

When the host materials of the heterostructure are markedly nonparabolic (e.g., GaSb–InAs [248], InSb MOS structures [249]) many subbands are often populated (both effects arise from the smallness of the effective mass). These subbands have different cyclotron resonance effective masses and the ground subband displays the heavier mass. Thus, the cyclotron resonance absorption shows structures which can be associated with the cyclotron resonance absorption of each individual subband.

8. OPTICAL PROPERTIES OF Q2D NANOSTRUCTURED MATERIALS

In this section the optical processes in Q2D systems that can be interpreted in terms of creation and annihilation of a single electron–hole pair or exciton within a Q2D system are considered. Size-dependent absorption and emission spectra and their fine structures as well as size-dependent radiative lifetime will be discussed for the different Q2D nanostructured materials. Nontrivial aspects of exciton–phonon interactions that manifest themselves in homogeneous linewidths and/or intraband relaxation processes will be outlined.

Optical spectroscopy of semiconductors has been a lively subject for many decades. It has helped to shape the concepts of energy gaps, impurity states, and resonant states, among others. Of course, spectroscopic techniques had been in use for some considerable time before semiconductors were developed, in the study of atomic spectra for example.

In spectroscopic terms, the size of the energy gap is the key difference between metals, insulators, and semiconductors. In both intrinsic luminescence and near-gap absorption, the main features of the spectrum can be associated with the magnitude of the energy gap, since the photons reflect the energy spacing between the states taking part in the optical process. This will be discussed in more detail in the following.

The states involved in optical transitions must be allowed states, and hence an understanding of the density of allowed states and its correlation with optical transitions is helpful. In the case of three-dimensional or bulk materials, the density of allowed states follows a $\varepsilon^{1/2}$ dependence on the energy of the carrier and the occupation of the energy levels is governed by Fermi–Dirac statistics. Lowering the dimensionality to two leads to a steplike density of allowed states. Not only must the states be allowed for a given optical process to occur, but they also need to be connected by selection rules. For example, the quantum numbers of the wavefunction of the electron and hole may have to be same, and for luminescence in two dimensions this would be translated to the requirement $\Delta l = 0$ where l is the orbital angular momentum quantum number of the states.

The strength of optical transitions is usually described by the oscillator strength, which in three-dimensional space is dimensionless and of order unity. The concept of oscillator strength conveys the probability that the transition can occur and is proportional to the number of $\vec{k}$ states coupled to a given energy range. As discussed by Weisbuch and Vinter [250], the oscillator strength per atom in two dimensions does not increase, but the new form of the density of allowed states arising from the two-dimensional confinement results in a shared k_z for electrons and holes in the same subband. To have the same wavevector k_z helps to concentrate the oscillator strength, when compared to the three-dimensional case. This suggests that the emission becomes weaker as the dimensions are reduced, contrary to expectations based on the enhancement of the oscillator strength, predicted from a treatment of "bottleneck" effects in energy and momentum relaxation.

The optical characterization of semiconductors has been reviewed several times. A recent publication edited by Stradling and Klipstein [251] covers most techniques in a clear introductory manner. It includes luminescence, scanning electron microscopy, and localized vibrational mode spectroscopy. A more advanced survey of techniques suitable for semiconductor quantum wells and superlattices is given by Fasol et al. [252] which includes time-resolved spectroscopy, Raman scattering, and the effects of a magnetic or an electric field.

The application of external perturbations such as electric fields and their impact on the absorption and emission spectra has been discussed in Bastard's book [35]. In general, a parameter, which can be changed in a controllable manner and affects at least one of the optical properties, can be used to study the optical behavior; with such a technique the background signal is subtracted from the processes under study. Pollak and Glembocki [253] have given an excellent review of photoreflectance, electroreflectance, piezoreflectance, analysis of lineshape, and interference effects in semiconductors and quantum wells. The information that can be obtained is usually directly related to the energy of the optically active electronic levels and the scattering mechanism involved.

Time-resolved spectroscopy is used among other applications to measure the energy relaxation of quasi-particles such as excitons in quantum wells, to probe hot phonon distributions and the lifetime of electron–hole liquids and plasmas, and to measure trapping times of carriers by impurities or other crystalline defects. There are several techniques in use such as pump-and-probe, streak cameras, etc. A useful series of publications covering applications of time-resolved spectroscopy to the study of three- and two-dimensional semiconductors has been edited by Alfano [254]. Techniques, experimental arrangements, tools for data analysis, and results are covered.

In general, the application of large magnetic fields results in confinement of the electron and hole wavefunctions. Magnetic fields were used to obtain fingerprints of the two-dimensional electron gas in the early days of low-dimensional semiconductor structures, confirming its existence by monitoring the positions of Landau levels as a function of the tilt angle between the magnetic field and the normal to the two-dimensional layer. Similar oscillatory behavior has been seen in magnetophotoluminescence; in this case a Landau ladder appears associated with each occupied electronic subband. In undoped structures, magnetophotoluminescence excitation has allowed the measurement of the binding energies of excitons in two-dimensional semiconductors, as discussed by Heitmann [255]. Another parameter that can be obtained from the separation between Landau levels at a given field is the effective mass of the electrons. A comprehensive coverage of the use of high magnetic fields can be found in the series on the proceedings of the Conferences on Applications of High Magnetic Fields in Semiconductor Physics. Magneto-Raman spectroscopy is also a very powerful technique; in doped structures it reveals the spacing of energy levels and effective masses and also allows probing the existence of more exotic quasi-particles, such as the rotons, recently demonstrated by Pinczuk et al. [256].

One of the crucial problems in semiconductor physics and particularly in these new low-dimensional semiconductor structures is the presence of ionized impurities, which play a fundamental role in transport mechanisms at low temperature and optical applications. This represents the basis for technological fabrications of the new electronic and optical devices of nanometric dimensions. The investigation of hydrogen atom and hydrogenic impurities confined in different Q2D nanostructured materials has attracted the attention of several research groups in the last few years, mainly because of the diverse technological applications possessed by devices based on such structures.

From the physical point of view the energy levels of hydrogenic impurities and excitonic states pose certain similarities, however, from the mathematical point of view the excitonic states are more complicated. In the hydrogenic impurity states the ion or core is bound to the electron through the coulombic potential but it remains in a fixed site, while in the formation of exciton the electron–hole pair is correlated through a coulombic interaction, but it moves through the whole Q2D system.

8.1. Absorption (One Electron Approximation) in Q2D Systems

An optical absorption experiment consists of measuring the attenuation of a light beam, which passes through a sample of thickness d. If the incident electromagnetic wave is monochromatic and propagates perpendicularly to the sample surface (area S), which is supposed to be flat, it can be shown [257, 258] that the transmitted intensity is given by

$$I_d = \frac{I_0(1-R)^2 \exp(-\alpha d)}{1 - R^2 \exp(-2\alpha d)} \tag{635}$$

where I_0 is the intensity of the incident beam, R is the sample reflectance, and α is the absorption coefficient at the angular frequency ω of the electromagnetic wave. For a weakly absorbing media, R is given by

$$R = R_0 = \left(\frac{n-1}{n+1}\right)^2 \tag{636}$$

where n is the refractive index of the sample. For absorption measurements, which probe the vicinity of the bandgap of the heterostructures, the ω-dependence of and R_0 in Eqs. (635) and (636) can often be neglected.

To calculate the absorption coefficient $\alpha(\omega)$, the transition probability per unit time W for a photon to disappear is first of all evaluated and, as a result, $\alpha(\omega)$ can then be obtained from the knowledge of W.

8.1.1. Electron–Photon Interaction

In order to determine the electron–photon interaction Hamiltonian, we will start from the Hamiltonian that represents the electrons motion in the crystal under the action of an electromagnetic field,

$$\widehat{H} = \frac{1}{2m_0}\left(\hat{p} - \frac{|e|}{c}\vec{A}\right)^2 + e\varphi + V(\vec{r}) \tag{637}$$

where $\vec{A}$ and φ are the vector and scalar potentials of the electromagnetic field, respectively. Considering the Landau gauge, the Hamiltonian can be divided in two parts: the first corresponds to the electrons motion in the crystal (the first two terms) and the second (the last two terms) represents the electron–radiation interaction:

$$\widehat{H} = \frac{1}{2m_0}\hat{p}^2 + V(\vec{r}) - \frac{|e|}{m_0 c}\hat{p}\cdot\vec{A} + \frac{e^2}{2m_0 c}A^2 \tag{638}$$

Working at first order of perturbation theory, the term that contains A^2 can be neglected, therefore

$$\widehat{H}_{er} = -\frac{|e|}{m_0 c}\hat{p}\cdot\vec{A} \tag{639}$$

The vector potential of the radiation has the form of a plane wave with wavevector $\vec{\kappa}$ and frequency ω,

$$\begin{aligned}
\vec{A} &= A_0\vec{e}_r \cos[(\vec{\kappa}\cdot\vec{r}) - \omega t] \\
&= \frac{1}{2}A_0\vec{e}_r\{\exp[-i(\vec{\kappa}\cdot\vec{r})]\exp(i\omega t) \\
&\quad + \exp[i(\vec{\kappa}\cdot\vec{r})]\exp(-i\omega t)\}
\end{aligned} \tag{640}$$

$$\vec{E} = -\frac{1}{c}\frac{\partial \vec{A}}{\partial t} = A_0 \vec{e}_r \frac{\omega}{c} \text{sen}[(\vec{\kappa}\cdot\vec{r}) - \omega t] \qquad (641)$$

where $\vec{e}_r$ is the polarization vector of the radiation. Considering N photons and a volume V,

$$\frac{1}{V}N\hbar\omega = \frac{\overline{E^2}}{4\pi} = \frac{A_0^2\omega^2}{8c^2\pi} \qquad (642)$$

Then

$$A_0 = 2c\sqrt{\frac{2N\hbar\pi}{\omega V}} \qquad (643)$$

Substituting Eqs. (640) and (643) in Eq. (639) it is obtained that

$$\widehat{H}_{er} = -\frac{e}{m_0}\sqrt{\frac{2N\hbar\pi}{\omega V}}\exp[-i(\vec{\kappa}\cdot\vec{r})](\vec{e}_r \cdot \hat{p}) \qquad (644)$$

Under typical experimental conditions, the photon wavevector $\vec{\kappa}$ can safely be neglected: the wave amplitude varies over considerably larger distances (the wavelength $2\pi/\kappa \approx 1\mu m$) than any characteristic dimension of electronic origin. Thus, under this electric dipole approximation $\widehat{H}_{er}$, considering $N = 1$ reduces to

$$\widehat{H}_{er} = -\frac{e}{m_0}\sqrt{\frac{2\hbar\pi}{\omega V}}(\vec{e}_r \cdot \hat{p}) \qquad (645)$$

The mass of the free electron that appears in Eq. (645) will be substituted by the effective mass in the case of interaction with the secondary radiation field.

Thus, to first order in $\vec{A}$ (linear absorption) and neglecting any contributions arising from the spin-orbit terms, the one-electron Hamiltonian of a Q2D system can be written as

$$\widehat{H} = \widehat{H}_0 + \widehat{H}_{er} \qquad (646)$$

where $\widehat{H}_0$ is the Q2D system Hamiltonian in the absence of an electromagnetic wave. In Eq. (646), the effects associated with the time varying magnetic field of the wave have been neglected.

The electromagnetic perturbation is time-dependent. It will thus induce transitions between the initial states $|i\rangle$ and the final states $|f\rangle$, where $|\nu\rangle$ denotes an eigenstate of $\widehat{H}_0$ with energy ε_ν. In the following, $\varepsilon_f > \varepsilon_i$ will be taken.

The Fermi Golden Rule shows how to calculate the transition probability per unit time $\widetilde{W}_{if}$ that an electron, under the action of $\widehat{H}_{er}$, makes a transition from the state $|i\rangle$ to the state $|f\rangle$. Since $\varepsilon_f > \varepsilon_i$, only the $+\omega$ component of $\vec{A}$ induces transitions, so

$$\widetilde{W}_{if} = \frac{2\pi}{\hbar}|\langle f|\widehat{H}_{er}|i\rangle|^2\delta(\varepsilon_f - \varepsilon_i - \hbar\omega) \qquad (647)$$

In the case where levels $|i\rangle$ and $|f\rangle$ are partially or completely occupied, one has to account for the impossibility of allowing transitions from either an empty level or toward a filled one. The perturbation of the statistical distributions of the levels $|\nu\rangle$ of the Q2D nanostructure, by the electromagnetic wave, is assumed to be negligible. The probability per unit time that an electron makes a $|i\rangle \rightarrow |f\rangle$ transition or, equivalently, that a photon disappears is thus equal to

$$W_{if} = \widetilde{W}_{if}f(\varepsilon_i)[1 - f(\varepsilon_f)] \qquad (648)$$

In Eq. (648) $f(\varepsilon_\nu)$ is the mean occupancy of the level $|\nu\rangle$,

$$f(\varepsilon_\nu) = \{1 + \exp[\beta(\varepsilon_\nu - \mu)]\}^{-1} \qquad (649)$$

where $\beta = (k_BT)^{-1}$, T is the temperature, and μ is the chemical potential of the electrons.

As a result of the electronic transitions $|i\rangle \rightarrow |f\rangle$, the electromagnetic wave loses the energy $\hbar\omega\widetilde{W}_{if}\,dt$ during the time interval dt.

The transitions $|f\rangle \rightarrow |i\rangle$ can also be induced by the wave (stimulated emission); they occur at the rate of

$$\widetilde{W}_{fi} = \frac{2\pi}{\hbar}|\langle i|\widehat{H}_{er}|f\rangle|^2\delta(\varepsilon_i - \varepsilon_f - \hbar\omega) \qquad (650)$$

transitions per unit time and contribute to the creation of photons in a quantity

$$W_{fi} = \widetilde{W}_{fi}f(\varepsilon_f)[1 - f(\varepsilon_i)] \qquad (651)$$

per unit time. The net energy loss of the electromagnetic wave per unit time associated with $|i\rangle \leftrightarrow |f\rangle$ transitions is thus

$$\hbar\omega[W_{if} - W_{fi}] \qquad (652)$$

Finally, by summing over all states $|i\rangle, |f\rangle$, the decrease in the wave energy per unit time is obtained:

$$W(\omega) = \frac{2\pi}{\hbar}\frac{2\hbar\pi e^2}{m_0^2\omega V}\sum_{i,f}\delta[\varepsilon f - \varepsilon_i - \hbar\omega]|\langle f|\vec{e}_r\cdot\hat{p}|i\rangle|^2$$
$$\times[f(\varepsilon_i) - f(\varepsilon_f)] \qquad (653)$$

The absorption coefficient $\alpha(\omega)$ is given by

$$\alpha(\omega) = \frac{W(\omega)}{nc}$$
$$= \gamma\sum_{i,f}\frac{1}{m_0}|\vec{e}_r\cdot\vec{p}_{if}|^2\delta(\varepsilon_f - \varepsilon_i - \hbar\omega)$$
$$\times[f(\varepsilon_i) - f(\varepsilon_f)], \qquad (654)$$

with

$$\gamma = \frac{4\pi^2 e^2}{ncm_0\omega V} \qquad \text{and} \qquad \vec{p}_{if} = \langle i|\hat{p}|f\rangle \qquad (655)$$

where c is the vacuum speed of light and n is the refraction index of the quantum well. In Eq. (654) the spontaneous emission due to the classical treatment of the electromagnetic field has been neglected.

γ has the dimension of $[\text{length}]^{-1}$ and $\vec{P}_{if}$, which is proportional to the dipolar matrix element $e\vec{r}_{if}$, will govern the selection rules for the allowed optical transitions. To exploit the general expression of $\alpha(\omega)$, it is neccessary to have a model for the energy levels. The simplest one is chosen and the wavefunctions of the initial state $|i\rangle$ are written in the form

$$\Psi_i(\vec{r}) = u_{\nu_i}(\vec{r}) f_i(\vec{r}) \qquad (656)$$

where $u_{\nu_i}(\vec{r})$ is the periodic part of the Bloch functions at the zone center (assumed to be the same in both types of layers) for the band $\nu_{i^*} f_i(\vec{r})$ implicitly contains the effective envelope function that describes the confined motion of the electron in the subband i corresponding to the extreme ν_i.

An expression similar to Eq. (656) holds for the final state. This expression of the wavefunction is given for each kind of Q2D system.

As the interest is focused in optical transitions from one subband derived from a given host band (say a valence subband) to a subband derived from another host band (say a conduction subband), Eq. (656) is sufficient. On the other hand, if we are dealing with transitions between the subbands derived from the same host band (say both initial and final states are conduction subbands), Eq. (656) becomes insufficient even at the lowest order. In the search of consistency, the eigenfunctions that include $\vec{k} \cdot \hat{p}_{\nu\nu}/(\varepsilon_\nu - \varepsilon_{\nu'})$ corrections to the next order should be considered, but keeping a coupling Hamiltonian between the carriers and light given by Eq. (645), or keeping Eq. (656) for the wavefunction, but using an effective coupling Hamiltonian between carriers and the electromagnetic wave. Here, in order to evaluate $\alpha(\omega)$ for the case of intraband transitions, only the parabolic dispersion relations of the host bands will be considered. The effective coupling procedure therefore amounts to replacing $1/m_0$ in Eq. (645) by $1/m_e^*$, taking into account the considered confinement directions. This means that, under most circumstances, the selection rules for the intraband transitions will be the same, irrespective of the use of a bare or effective coupling between carrier and light. Thus, in the derivation of the intraband transition selection rules, Eq. (656) for the wavefunction will be kept, but m_e^* will replace m_0 where appropriate. This is relevant for the evaluation of the magnitude of the absorption coefficient.

Let us denote by V_0 the volume of the elementary cell of the materials involved in the considered structure. Thus

$$\vec{e}_r \cdot \vec{p}_{if} = \vec{e}_r \cdot \int_\nu \Psi_i^*(\vec{r}) \hat{p} \Psi_f(\vec{r}) d^3 r \qquad (657)$$

$$\vec{e}_r \cdot \vec{p}_{if} = \vec{e}_r \cdot \langle u_{\nu_i} | \hat{p} | u_{\nu_f} \rangle \int_V f_i^*(\vec{r}) d^3 r$$
$$+ \delta_{if} \vec{e}_r \cdot \int_V f_i^*(\vec{r}) \hat{p} f_f(\vec{r}) d^3 r \qquad (658)$$

where

$$\langle u_{\nu_1} | \hat{p} | u_{\nu_f} \langle = \int_{V_0} u_{\nu_i}^*(\vec{r}) \hat{p} u_{\nu_f}(\vec{r}) d^3 r \qquad (659)$$

In Eqs. (657)–(659), we have been taken advantage of the rapid variations of u_{ν_i}, u_{ν_f} over $\vec{k}^{-1}$ or over the characteristic lengths of variation of f_i, f_f.

The allowed optical transitions split in two categories: on the one hand, the intraband transitions ($\nu_i = \nu_f$) which involve the dipole matrix elements between envelope functions and on the other hand the interband transitions which occur between subbands originating from different extrema. The selection rules for the latter have two origins:

(i) The overlap integral between envelope functions selects the quantum numbers of the initial and final subbands.
(ii) The atomiclike dipole matrix element $\langle u_{\nu_i} | \hat{p} | u_{\nu_f} \rangle$ gives rise to the selection rules on the polarization of the light wave.

Let us examine the intraband and interband transitions for Q2D nanostructures in closer detail.

8.1.2. Intraband Transitions in Quasi-Two-Dimensional Systems

The dipole matrix element between envelope functions is

$$\langle f_i | \vec{e}_r \cdot \hat{p} | f_f \rangle = \frac{1}{S} \int d^3 r \, \chi_{n_i}^*(z) \exp(-i\vec{k}_\perp \cdot \vec{r}_\perp)$$
$$\times [e_x \hat{p}_x + e_y \hat{p}_y + e_z \hat{p}_z]$$
$$\times \chi_{n_f}(z) \exp(i\vec{k}_\perp \cdot \vec{r}_\perp) \qquad (660)$$

or

$$\langle f_i | \vec{e}_r \cdot \hat{p} | f_f \rangle = (e_x \hbar k_x + e_y \hbar k_y) \delta_{n_i, n_f} \delta_{\vec{k}_\perp, \vec{k}_\perp}$$
$$+ e_z \delta_{\vec{k}_\perp, \vec{k}_\perp} \int dz \, \chi_{n_i}^*(z) \hat{p}_z \chi_{n_f}(z) \qquad (661)$$

The polarizations e_x, e_y give rise to allowed transitions only if both the initial and final states coincide (i.e., if $\omega = 0$). The intrasubband absorption ($n_i = n_f$), which in the static limit cannot be reasonably treated without including scattering mechanisms, is the two-dimensional analogue of the free carrier absorption. For perfect heterostructures, the free carrier absorption is forbidden in quasi-two-dimensional electron gases for the same reason as in bulk materials, i.e., the impossibility of conserving the energy and momentum simultaneously during the photon absorption by an electron. Free carrier absorption may be induced by defects (impurities, phonons) capable of providing the momentum necessary for the electron transition. The effect of reduced dimensionality on the free carrier absorption has been calculated [259].

The polarization e_z, which corresponds to an electromagnetic wave propagating in the layer plane with an electric field vector parallel to the growth axis of the structure, leads to optical transitions which are allowed between subbands, provided that the in plane wavevector of the carrier is conserved (vertical transitions in the $\vec{k}_\perp$ space). In addition, if the heterostructure Hamiltonian has a definite parity, the initial and final subbands should be of opposite parities. The occurrence of intraband transitions without the participation of defects is specific to the quasi-two-dimensional materials and originates from the existence of a z-dependent potential in $\hat{H}_0$. This potential sets up the momentum necessary for the intraband

Table II. Selection Rules for Intraband (Intrasubband) Transition

Polarization	e_x	e_y	e_z
Propagation parallel to $\hat{e}_z$	Forbidden if $\omega \neq 0$	Forbidden if $\omega \neq 0$	Impossible
Propagation parallel to $\hat{e}_x$	Impossible	Forbidden if $\omega \neq 0$	Allowed
Propagation parallel to $\hat{e}_y$	Forbidden if $\omega \neq 0$	Impossible	Allowed

absorption when the electric field vector of the light is parallel to z.

In Table II are shown the selection rules for intraband (intersubband) transitions.

Now we evaluate the order of magnitude of such allowed intraband transitions in a rectangular quantum well whose lowest subband (confinement energy ε_1) is occupied with N_e electrons.

We will first consider transitions sending the ε_1 electrons deep into the quantum well continuum, a possible loss mechanism in quantum well lasers. For these highly excited continuum states the reflection–transmission phenomena which take place at $z = \pm L/2$ may be neglected, where L is the quantum well thickness. Thus

$$\chi_f(z) = \frac{1}{\sqrt{\ell}} \exp(ikz) \qquad (662)$$

The definition of ℓ should be done in an empirical way; for a perfect heterostructure ℓ is identified as the length over which the heterostructure energy levels are well defined, i.e., a length which, if changed slightly, does not appreciably alter the energy levels and their associated wavefunctions. For a single quantum well structure with thickness L one should take $\ell \gg L$. For a superlattice comprising N periods (thickness h) one may take $\ell \geq Nh$ if $N \gg 1$. On the other hand, ℓ should not be too large to ensure the assumption of a plane electromagnetic wave.

Besides,

$$\chi_i(z) = \sqrt{\frac{2}{L}} \cos\left(\frac{\pi z}{L}\right) \qquad |z| \leq \frac{L}{2} \qquad (663)$$

where, for simplicity, the exponential tails of the bound state wavefunction in the barrier have been neglected. $\langle \chi_i | \hat{p}_z | \chi_f \rangle$ is evaluated to obtain

$$\langle \chi_i | \hat{p} \chi_f | \rangle = \sqrt{\frac{2}{L\ell}} \cos\left(\frac{kL}{2}\right) \left[\frac{1}{k + \pi/L} - \frac{1}{k - \pi/L} \right] \hbar k \qquad (664)$$

The quantity

$$B = \frac{1}{m_e^*} \sum_{i,f} |\bar{e}_r \cdot \bar{p}_{if}|^2 \delta(\varepsilon_f - \varepsilon_i - \hbar\omega)[f(\varepsilon_i) - f(\varepsilon_f)] \qquad (665)$$

is also easily evaluated if the continuum levels have a negligible population. It is found that

$$B = \frac{2k_0 L N_e}{\pi} \cos^2\left(\frac{k_0 L}{2}\right) \left[\frac{1}{\pi + k_0 L} - \frac{1}{k_0 L - \pi} \right] \qquad (666)$$

where

$$k_0^2 = \frac{2m_e^*}{\hbar^2}(\hbar\omega - V_0 + \varepsilon_1) \qquad (667)$$

and V_0 is the barrier height. The absorption coefficient is therefore

$$\alpha(\omega) = \frac{8\pi e^2 \eta_e}{ncm_e^* \ell \omega} x \cos^2\left(\frac{x}{2}\right) \left[\frac{1}{x + \pi} \frac{1}{x - \pi} \right] \qquad (668)$$

with $x = k_0 L$, $\eta_e = N_e/S$. At large ω, $\alpha(\omega)$ decreases like $\omega^{-5/2}$. It can be noticed that the absorption coefficient is inversely proportional to ℓ, but it does not explicitly depend on S. These results are not surprising given that both initial and final states are delocalized in the layer plane. The carrier interacts with the electromagnetic wave all over the area S. On the other hand, since the initial state is confined within the quantum well, only a fraction L/ℓ of the electromagnetic energy is available for absorption.

As a second example, let us now consider the optical absorption that promotes the carrier from the ground (ε_1) to the first excited (ε_2) subbands of a quantum well.

To lighten the algebra, the wavefunctions are again approximated with those of an infinitely deep well; i.e.,

$$\chi_i(z) = \chi_1(z) = \sqrt{\frac{2}{L}} \cos\left(\frac{\pi z}{L}\right) \qquad (669)$$

$$\chi_f(z) = \chi_2(z) = \sqrt{\frac{2}{L}} \sin\left(\frac{2\pi z}{L}\right) \qquad (670)$$

Thus

$$\langle \chi_i | \hat{p}_z | \chi_f \rangle = \frac{8i\hbar}{3L} \qquad (671)$$

and

$$B = \frac{64\hbar^2}{9m_e^* L^2}(N_1 - N_2)\delta(\varepsilon_2 - \varepsilon_1 - \hbar\omega) \qquad (672)$$

where N_1 and N_2 are the number of electrons in the initial and final subbands whose confinement energies are ε_1 and ε_2, respectively. The absorption coefficient associated with $|1\rangle \rightarrow |2\rangle$ transitions is then equal to

$$\alpha(\omega) = \frac{256\pi^2 e^2 \hbar^2}{9nc(m_e^*)^2 L^2 \omega \ell}(\eta_1 - \eta_2)\delta(\varepsilon_2 - \varepsilon_1 - \hbar\omega) \qquad (673)$$

where η_1 and η_2 are the areal density of carriers found in the ε_1 and ε_2 subbands, respectively.

The delta singularity originates from the exact parallelism between the in plane dispersion relations of the initial and final subbands. Introducing nonparabolicity or any other effect which alters this parallelism would change the delta peak into a line of finite energy extension. It may also be noticed that the same ℓ^{-1} dependence for α as in Eq. (668) has been obtained. In both examples, the physical absorption mechanism takes place within the quantum well thickness, whereas the electromagnetic energy is delocalized over the whole heterostructure.

To get an order of magnitude of the absorption coefficient associated with $|1\rangle \to |2\rangle$ transitions, let us take $m_e^* = 0.07 m_0$, $n = 3.6$, $\eta_1 - \eta_2 = 10^{11}$ cm^{-2}, $\ell = 1$ μm, $\hbar\omega = \varepsilon_2 - \varepsilon_1 = 3\hbar^2\pi^2/2m_e^* L^2$ and replace the delta function by a Lorentzian function with a full width at half maximum $2\Gamma = 4$ meV. $\alpha \approx 67$ cm^{-1} is obtained. The absorption coefficient associated with $|1\rangle \to$ continuum transitions is very small ($\alpha \approx 10^{-3}$ cm^{-1}) since x in Eq. (668) is usually ≈ 10 for material parameters adapted to GaAs wells. Note finally that collective excitations have not been taken into account.

The selection rules for intraband transitions in perfect heterostructures were summarized in Table II.

8.1.3. Interband Transitions

On the one hand, the selection rules associated with the wave polarizations and, on the other hand, the selection rules on the subband index which arise from the envelope functions overlap have to be discussed here. In addition, interband transitions raise the question of the respective location of the initial and final states. It is clear that the optical absorption in a type II system (e.g., InAs–GaSb), where electrons and holes are spatially separated, should be weaker than in type I structures (e.g., GaAs–GaAlAs) where electrons and holes are essentially localized in the same layer.

A major complication arises due to the intricate valence subband dispersions in the layer plane. The decoupling of $m_j^* = \pm 3/2$ and $m_j^* = \pm 1/2$ states at $\vec{k}_\perp = 0$ is well known. Strictly speaking, it is necessary to treat the problem of the in plane dispersion relations at $\vec{k}_\perp \neq 0$ rigorously, in order to be able to evaluate $\alpha(\omega)$. In fact, recent publications [260, 261] have dealt with this very complicated problem. Here, this difficulty will be bypassed by assigning an arbitrary in plane effective mass to the valence subbands while retaining the $m_j^* = \pm 3/2$, $m_j^* = \pm 1/2$ decoupling. This will give us a reasonable description of the onsets of the absorptions. It can be noticed parenthetically that if this model predicts that a given interband transition is allowed in the vicinity of $\vec{k}_\perp = 0$, it cannot in practice become forbidden at $\vec{k}_\perp = 0$. Symmetrically, any forbidden transition found in this simplified approach might become allowed due to the valence subbands mixing between the $m_j^* = \pm 3/2$, $m_j^* = \pm 1/2$, states at $\vec{k}_\perp \neq 0$. But it will present a smoother absorption edge and will be weaker than an allowed transition. The major weakness of this simplified model lies in its inability to account for peaks in the joint density of states, which arise away from the zone center when the initial and final subbands are nearly parallel. We believe, however, this situation to be rare.

Below, the subband dispersion relations are thus adopted,

$$\varepsilon_{hh_n}(\vec{k}_\perp) = -\varepsilon_g - hh_n - \frac{\hbar^2 k_\perp^2}{2M_n^*} \qquad (674)$$

$$\varepsilon_{lh_n}(\vec{k}_\perp) = -\varepsilon_g - lh_n - \frac{\hbar^2 k_\perp^2}{2m_n^*} \qquad (675)$$

$$\varepsilon_{e,n}(\vec{k}_\perp) = \varepsilon_n + \frac{\hbar^2 k_\perp^2}{2m_{en}^*} \qquad (676)$$

where M_n^* and m_n^* can eventually become negative if the hh_n (or lh_n) subband has a positive curvature in the vicinity of $\vec{k}_\perp = 0$. The matrix element is introduced,

$$\prod = \frac{-i}{m_0}\langle S|\hat{p}_x|X\rangle = \frac{-i}{m_0}\langle S|\hat{p}_y|Y\rangle = \frac{-i}{m_0}\langle S|\hat{p}_z|Z\rangle \qquad (677)$$

which is related to the Kane matrix element $\varepsilon_P (\approx 23$ eV$)$ by

$$\varepsilon_P = 2m_0 \prod^2 \qquad (678)$$

8.1.3.1. Polarization Selection Rules

Examination of the periodic parts of the Bloch functions for the Γ_6, Γ_7, Γ_8 bands leads to the selection rules for the electromagnetic wave polarization. These are summarized in Table III for $hh_n \to \varepsilon_m$, $lh_n \to \varepsilon_m$, and $(\Gamma_7)_n \to \varepsilon_m$ transitions, respectively.

Table III shows that for interband transitions:

(a) The heavy hole $\to$ electron transitions are three times more intense than the light hole $\to$ electron transitions. The $\Gamma_7 \to$ electron transitions have an intermediate strength (2/3 of those involving heavy holes).

(b) For propagation parallel to z, corresponding to the electric field of the wave in the layer planes, the three possible types of transitions are allowed. On the other hand, when the propagation occurs in the layer planes, the polarization e_z is forbidden for the $hh_n \to \varepsilon_m$ transitions. This may have some influence on the lasing action in quantum well systems since, in this case, the light propagates perpendicularly to the growth axis. Anyhow, absorption (or spontaneous emission) studies in this geometry should show a strong dependence of the spectra on the exciting light polarization for the $hh_n \to \varepsilon_m$ transitions. An example of such dependence is presented in Figure 54.

8.1.3.2. Selection Rules on the Envelope Function Quantum Numbers: Evaluation of $\langle f_i|f_f\rangle$

It is neccessary to evaluate $\langle f_i|f_f\rangle$, which is equal to

$$\langle f_i|f_f\rangle = \langle \chi_n^{(h)}|\chi_m^{(e)}\rangle\langle \vec{k}_\perp|\vec{k}'_\perp\rangle$$

$$= \int dz\, \chi_n^{*(h)}(z)\chi_m^{(e)}(z) \qquad (679)$$

$$\times \frac{1}{S}\int d^2r_\perp \exp[i(\vec{k}'_\perp - \vec{k}_\perp)\cdot\vec{r}_\perp]$$

Table III. Selection Rules for Interband (Intersubband) Transitions Obtained from the Absolute Value of the Matrix Elements $S \uparrow \vec{e}_r \cdot \hat{p} u_\alpha$, $\alpha = \Gamma_8, \Gamma_7$; $\vec{k}$

Polarization	e_x	e_y	e_z	Type of transitions
Propagation parallel to $\hat{e}_z$	$\frac{\Pi}{2}$	$\frac{\Pi}{2}$	Impossible	$hh_n \to \varepsilon_m$
Propagation parallel to $\hat{e}_x$	Impossible	$\frac{\Pi}{2}$	Forbidden	$hh_n \to \varepsilon_m$
Propagation parallel to $\hat{e}_y$	$\frac{\Pi}{2}$	Impossible	Forbidden	$hh_n \to \varepsilon_m$
Propagation parallel to $\hat{e}_z$	$\frac{\Pi}{6}$	$\frac{\Pi}{6}$	Impossible	$lh_n \to \varepsilon_m$
Propagation parallel to $\hat{e}_x$	Impossible	$\frac{\Pi}{6}$	$\frac{2\Pi}{6}$	$lh_n \to \varepsilon_m$
Propagation parallel to $\hat{e}_y$	$\frac{\Pi}{6}$	Impossible	$\frac{2\Pi}{6}$	$lh_n \to \varepsilon_m$
Propagation parallel to $\hat{e}_z$	$\frac{\Pi}{3}$	$\frac{\Pi}{3}$	Impossible	$(\Gamma_7)_n \to \varepsilon_m$
Propagation parallel to $\hat{e}_x$	Impossible	$\frac{\Pi}{3}$	$\frac{\Pi}{3}$	$(\Gamma_7)_n \to \varepsilon_m$
Propagation parallel to $\hat{e}_y$	$\frac{\Pi}{3}$	Impossible	$\frac{\Pi}{3}$	$(\Gamma_7)_n \to \varepsilon_m$

Thus, the first selection rule is the conservation of the in plane wavevector in the optical transitions: .

$$\vec{k'}_\perp - \vec{k}_\perp = 0 \qquad (680)$$

It arises from the translational invariance of the heterostructure Hamiltonian in the layer plane.

Evaluation of the selection rules on the subband indexes is obtained by calculating $\langle \chi_n^{(h)} | \chi_m^{(e)} \rangle$. At this point, a distinction between the systems of types I and II should be made.

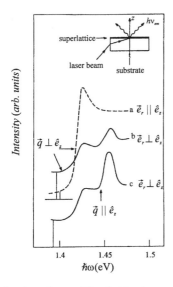

Fig. 54. Polarization dependence of the photoluminescence excitation spectrum in a $In_{1-x}Ga_xAs$–GaAs superlattice at $T = 77$ K. In this experiment the excitation spectrum mimics the absorption coefficient. For an electromagnetic wave propagating in the layer plane ($\vec{q} \perp \hat{e}_z$) two absorption peaks are seen in the $\vec{e}_r \perp \hat{e}_z$ polarization while only one shows up in the $\vec{e}_r // \hat{e}_z$ polarization. This is in agreement with the predictions of previous table. Moreover, when $\vec{q} // \hat{e}_z$ and thus $\vec{e}_r \perp \hat{e}_z$ two peaks are observed which coincide in energy with those seen in the $\vec{e}_r \perp \hat{e}_z$, $\vec{q} \perp \hat{e}_z$ polarization.

Type I Systems (GaAs–GaAlAs; GaInAs–InP; GaInAs–AlInAs,. . .). These heterostructures are such that conduction and valence electrons are essentially confined in the same layers. For a type I symmetrical quantum well, the envelope functions have a definite parity with respect to the center of the well. The overlap integral $\langle \chi_n | \chi_m \rangle$ is thus nonzero only if $n + m$ is even.

In addition, if the valence and conduction quantum wells are rectangular and infinitely deep, only the transitions that fulfill $n = m$ are allowed. In the GaAs–GaAlAs rectangular wells the transitions $\Delta n = n - m = 0$ are much stronger than those corresponding to Δn even ($\neq 0$). The transition $hh_3 \to \varepsilon_1$ has been identified in absorption [262] and photoluminescence excitation spectroscopy [263]. Other GaAs–GaAlAs nonrectangular wells, e.g., the pseudo-parabolic wells [264] or separate confinement heterostructures [265], have been optically studied. The former structures are obtained by imposing a quadratic variation of the conduction and valence edges by beam chopping during the epitaxial growth, whereas the latter structures consist of embedding a narrow GaAs–$Ga_{1-x_1}AlAs$ quantum well into $Ga_{1-x_2}Al_{x_2}As$ barriers with $x_2 > x_1$. Both kinds of heterostructures have permitted the observation of more optical transitions than those found in the plain GaAs–GaAlAs rectangular quantum wells. This has led to a significant reappraisal of the band discontinuities in the GaAs–GaAlAs system (for reviews see [266, 267]).

The selection rule $n + m$ even is in fact very strong and only very asymmetrical wells can lead to transitions with $n + m$ odd. Investigation of intentionally designed asymmetrical wells might yield a better determination of the band offsets, since more optical transitions are allowed.

Type II Systems (InAs–GaSb, InP–As_{0.48}In_{0.52}As). In these structures, the electrons and holes are spatially separated. Let us consider the case of a type II quantum well, whose band profiles are shown in Figure 55. The well-acting (barrier-acting) materials, for electrons, are denoted by A (B), and their respective bandgaps are denoted by $\varepsilon_A, \varepsilon_B$. The A material is a barrier for the holes and since a quantum well structure is being considered, the valence spectrum is continuous and twice degenerate. To simplify the discussion only photon energies such that $\hbar\omega < \varepsilon_A$ will be considered and $\varepsilon_1 < \Delta_n u$ will be assumed. For a given bound state ε_n of the conduction band, a transition should involve only one of the two degenerate valence states: if $\chi_m^{(e)}$ is even in z, the initial state should also be even in z in order to participate in the optical transitions. It is also clear that the overlap of the conduction and valence envelope functions is only due to the exponential tails of the conduction and valence states in the layers B and A, respectively: in the limit of large valence and conduction band discontinuities, the overlap integrals $\langle \chi_n^{(h)} | \chi_m^{(e)} \rangle$ tend to zero. In addition, $\langle \chi_n^{(h)} | \chi_m^{(e)} \rangle$ should increase with increasing n and m: the tails of the conduction and valence envelopes in their respective barriers become more and more important. Notice that the opposite is true in type I systems.

8.1.3.3. Order of Magnitude of the Absorption Coefficient; Comparison between Type I and Type II Systems

As an example, let us consider the optical transitions between the ground heavy hole and electron subbands of a rectangular type I quantum well (e_x polarization). In Eq. (654) the quantity $|\vec{e}_r \cdot \vec{p}_{if}|^2/m + 0$ is equal to

$$\frac{1}{m_0}|\vec{e}_r \cdot \vec{p}_{if}|^2 = \frac{\varepsilon_P}{4}|\langle \chi_1^{(h)}|\chi_1^{(e)}\rangle|^2 \tag{681}$$

and the absorption coefficient $\alpha_{hh_1} \to \varepsilon_1(\omega)$ is equal to

$$\begin{aligned}
&\alpha_{hh_1 \to \varepsilon_1}(\omega) \\
&= \gamma \frac{\varepsilon_P}{4}|\langle \chi_1^{(h)}|\chi_1^{(e)}\rangle|^2 2 \sum_{\vec{k}_\perp} \delta \\
&\times \left[\varepsilon_A + \varepsilon_1 + hh_1 + \frac{\hbar^2 k_\perp^2}{2}\left(\frac{1}{m_e^*} + \frac{1}{M_e^*}\right) - \hbar\omega \right]
\end{aligned} \tag{682}$$

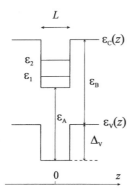

Fig. 55. Conduction and valence band edge profiles in a type II quantum well.

By performing the summation upon it is obtained that

$$\begin{aligned}
\alpha_{hh_1 \to \varepsilon_1}(\omega) &= \frac{\pi e^2 \varepsilon_P}{ncm_0 \omega \ell \hbar^2} \frac{m_e^* M_1^*}{m_e^* + M_1^*} \\
&\times |\langle \chi_1^{(h)}|\chi_1^{(e)}\rangle|^2 Y(\hbar\omega - \varepsilon_A - \varepsilon_1 - hh_1)
\end{aligned} \tag{683}$$

In Eqs. (682) and (683), population effects for both initial and final states have been neglected (undoped quantum wells), ε_A is the bandgap of the bulk well-acting material, and $Y(x)$ is the step function. For a $lh_1 \to \varepsilon_1$ transition a similar expression is obtained with M_1^*, hh_1 replaced by $\varepsilon + P$ and by $\varepsilon_P/3$ (see Table III).

The absorption coefficient has a staircaselike shape at the onset of the absorption (see Fig. 56), as expected from the joint density of states of the two two-dimensional subbands. For $hh_n \to \varepsilon_m$ transitions, $n, m \neq 1$, the magnitude of the step will involve the square of $\langle \chi_n^{(h)}|\chi_m^{(e)}\rangle$ which is smaller than one and also the in plane effective mass of the nth heavy hole subband in the vicinity of the zone center, i.e., M_n^*, will replace M_1^* in Eq. (683). It has previously seen that some of the M_n^*'s may be negative (i.e., may have electronlike behavior near $\vec{k}_\perp = 0$). This, in turn, affects the magnitude of $\alpha_{hh_n \to \varepsilon_m}$. Experimentally, it is often found that the edges of higher lying heavy hole–electron transitions are less marked than the $hh_1 \to \varepsilon_1$ one, which is consistent with an increased broadening of the excited subband states.

However, the most significant effect in Eq. (683) is the blue shift of the quantum well fundamental absorption edge, with respect to that of the bulk material A. Photons start to be absorbed at the energy $\varepsilon_1 + hh_1$, above ε_A. This blue shift can be tuned by varying the quantum well thickness. For instance, bulk GaAs starts absorbing in the infrared part of the spectrum ($\varepsilon_a = 1.5192$ eV at low temperature), while narrow GaAs–Ga$_{1-x}$Al$_x$As quantum wells ($L \approx 30$ Å) can be designed to start absorbing light only in the red part of the spectrum ($\varepsilon_A + \varepsilon_1 + hh_1 \approx 1.7275$ eV if $x = 0.5$). The blue shift of the fundamental absorption edge has been observed in a number of heterostructures, for example, in GaAs–GaAlAs [262], GaInAs–InP [268], GaInAs–AlInAs [269], GaSb–AlSb [270], and GaInAs–GaAs [271].

Finally, let us calculate the magnitude of $\alpha_{hh_1 \to \varepsilon_1}$ in the case of a single quantum well embedded in a structure of total thickness $\ell = 1$ μm. For $\hbar\omega = 11.6$ eV, $m_e^* = 0.067m_0$, $M_1^* \gg m_e^*$, $n = 3.6$, $\varepsilon_P = 23$ eV and taking $\langle \chi_1^{(h)}|\chi_1^{(e)}\rangle \approx 1$, then

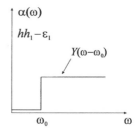

Fig. 56. Staircase lineshape of the absorption coefficient of a type I quantum well.

$\alpha_{hh_1 \to \varepsilon_1} \cong 60$ cm^{-1} is obtained. Such a faint absorption can hardly be measured in practice, for it leads to an attenuation of the light beam intensity of only $1 - \exp(-\alpha_{hh_1 \to \varepsilon_1} \ell) \approx \alpha_{hh_1 \to \varepsilon_1} \ell = 0.6\%$. Therefore, one has to use multiple quantum wells to enhance the absorption. These structures are obtained by separating N identical wells with $N - 1$ thick barriers, resulting in an absorption enhanced by a factor of N over that of a single well. The barriers have to be thick enough to prevent any tunnel coupling between the wells, a condition more difficult to achieve when higher lying transitions ($n, m > 1$) are involved. In practice N should be ≥ 10 in order to obtain a sizeable absorption.

We stress again the fact that the low absorption coefficient is due to the different spatial locations of the electromagnetic energy (delocalized over ℓ) and the electrons that actually absorb light (essentially localized within the quantum well).

For a type II quantum well, the only difficulty consists in evaluating the overlap integral between valence states (which are extended in and labelled by a wavevector $k + \nu$) and conduction states (which are localized in the well). For that purpose, a rather crude approach is used: the valence barrier Δ_ν is assumed high enough to be considered as impenetrable by the valence electrons. In this case, two linearly independent valence envelope functions correspond to each k_ν. They are

$$\chi_r^{(h)} = \frac{2}{\sqrt{\ell}} \sin\left[k_\nu \left(z - \frac{L}{2}\right)\right] Y\left[z - \frac{L}{2}\right] \quad (684)$$

$$\chi_l^{(h)} = \frac{-2}{\sqrt{\ell}} \sin\left[k_\nu \left(z + \frac{L}{2}\right)\right] Y\left[-z - \frac{L}{2}\right] \quad (685)$$

These functions are normalized to 1 over $\ell/2 (L \ll \ell)$ and for heavy holes

$$\varepsilon_\nu(k_\nu, \vec{k}_\perp) = -\varepsilon_A + \Delta_\nu - \frac{\hbar^2 k_\nu^2}{2m_\nu^*} - \frac{\hbar^2 k_\perp^2}{2M_1^*} \quad (686)$$

where m_ν^* is the heavy hole mass along the z direction. Note that in principle m_ν^* should be set equal to M_1^* in Eq. (686), since each of the B layers is a bulk material. The notation m_ν^* and M_1^* is kept in order to be able to make a comparison with the absorption coefficient in type I structures later on.

Now functions $\chi_e(z)$ and $\chi_0(z)$ are introduced, which are even and odd with respect to z:

$$\chi_e^{(h)}(z) = \frac{1}{\sqrt{2}}[\chi_r^{(h)}(z) + \chi_l^{(h)}(z)] \quad (687)$$

$$\chi_0^{(h)}(z) = \frac{1}{\sqrt{2}}[\chi_r^{(h)}(z) - \chi_l^{(h)}(z)] \quad (688)$$

The interest is focused in calculating the absorption coefficient at the onset of the absorption, i.e., that associated with valence $\to \varepsilon_1$ transitions. The envelope function $\chi_1^{(h)}(z)$ of the electrons in the ε_1 subband can thus be written

$$\chi_1^{(e)}(z) = A_c \cos(K_c z) \quad |z| \leq \frac{L}{2} \quad (689)$$

$$\chi_1^{(e)}(z) = B_c \exp\left[-\kappa_c\left(z - \frac{L}{2}\right)\right] \quad z > \frac{L}{2} \quad (690)$$

$$\chi_1^{(e)}(-z) = \chi_1^{(e)}(z) \quad (691)$$

The overlap integral $\langle \chi_0^{(h)} | \chi_1^{(e)} \rangle$ vanishes while $\langle \chi_0^{(h)} | \chi_1^{(e)} \rangle$ is equal to

$$\langle \chi_0^{(h)} | \chi_1^{(e)} \rangle = 2\sqrt{\frac{2}{\ell}} B_c \frac{k_\nu}{k_\nu^2 + \kappa_c^2} \quad (692)$$

The coefficient B_c can easily be related to the probability $P_b(\varepsilon_1)$ of finding the electron in the B layers, while in the ε_1 states

$$P_b(\varepsilon_1) = \frac{B_c^2}{\kappa_c} \quad (693)$$

The quantity B defined in Eq. (665), with m_e^* replaced by m_0 (since interband transitions are being considered), can now be calculated. It is equal to

$$B = \frac{P_b(\varepsilon_1) \kappa_c \varepsilon_P}{\pi^2 \hbar^2} \frac{m_e^* M_1^* S}{m_e^* + M_1^*} \int_0^{k_{\max}} \frac{dk_\nu k_\nu^2}{(\kappa_c^2 + k_\nu^2)^2} \quad (694)$$

where

$$k_{\max}^2 = \frac{2m_\nu^*}{\hbar^2}(\hbar\omega - \varepsilon_A + \Delta_\nu - \varepsilon_1) \quad (695)$$

and finally the absorption coefficient is found to be equal to

$$\alpha(\omega)_{hh \to \varepsilon_1} = \frac{2e^2 \varepsilon_P P_b(\varepsilon_1)}{ncm_0 \omega \hbar^2 \ell} \frac{m_e^* M_1^*}{m_e^* + M_1^*}\left[\frac{-x}{1+x^2} + \arctan x\right] \quad (696)$$

$$x = \frac{k_{\max}}{\kappa_c}$$

The onset of the absorption is located at

$$\hbar\omega_0 = \varepsilon_A - \Delta_\nu + \varepsilon_1 \quad (697)$$

In the vicinity of this onset, $\alpha_{hh \to \varepsilon_1}$ behaves like $(\omega - \omega_0)^{3/2}$ (Fig. 57) instead of displaying the staircaselike behavior characteristic of type I quantum wells. Thus, the indirect optical transitions in real space have the same smoothing effect on the absorption edge as the indirect transitions in reciprocal space in the case of bulk materials. The physical origin of the absorption (exponential tail of the ε_1 wavefunction outside its confining layer) is confirmed by the presence of the proportionality factor P_b in Eq. (696). At large ω, i.e., $k_{\max} \gg \kappa_c$, the $\alpha(\omega)_{hh \to \varepsilon_1}$ levels are set off at the value

$$\alpha_{hh \to \varepsilon_1}^\infty = \frac{\pi e^2 \varepsilon_P}{ncm_0 \omega \ell \hbar^2} P_b(\varepsilon_1) \frac{m_e^* M_1^*}{m_e^* + M_1^*} \quad (698)$$

which is just $P_b(\varepsilon_1)$ times the magnitude of the plateau value for absorption in type I quantum wells (see Eq. (683)). This means that the use of multiple wells to enhance the single well absorption is even more imperative in type II than in type I structures. Since $P_b(\varepsilon_1)$ is usually only a few percent, it appears that hundreds of wells may be necessary to measure the absorption in type II quantum wells.

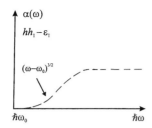

Fig. 57. Absorption lineshape of a type II quantum well.

Finally, it may be added that light hole $\to \varepsilon_1$ ($lh \to \varepsilon_1$) transitions have the same threshold as heavy hole $\to \varepsilon_1$ ($hh \to \varepsilon_1$) transitions (Eq. (697)). Thus only the sum of the two contributions will actually be measured. In addition, transitions from the valence levels to the excited conduction subbands ε_i, $i > 1$, will be more intense than $hh \to \varepsilon_1$ or $lh \to \varepsilon_1$ transitions. This arises from the smaller κ_c and larger $P_b(\varepsilon_i)$ for the excited subbands.

8.2. Absorption: A Simplified Description of Excitonic Effects

A convenient picture of an exciton is an electron and a hole orbiting around each other. This picture is quite difficult to derive from first principles, because an exciton is in fact a many-body effect.

The excitonic effects are seldom negligible in practice. They most clearly manifest themselves with the appearance of sharp peaks in the interband absorption spectra, whereas, as discussed in the previous paragraph, noninteracting electron models at most only predict staircaselike structures starting at $\hbar\omega_{\mathrm{onset}}$. In the exciton (or electron–hole pairs) picture, the pair energies larger than $\hbar\omega_{\mathrm{onset}}$ correspond to dissociated pairs. It is often forgotten that the electron–hole attraction also alters the optical response of the pair in the continuum [272]. This alteration is so deep that, in bulk materials, for instance, the calculated one-electron absorption is never correct. Even deep in the continuum, where the large electron–hole relative velocity makes the noninteracting particle model its best, it falls short of accounting for the correct absorption coefficient, i.e., the coefficient calculated by including the electron–hole interaction (although approaching it asymptotically from below [272]). It will be seen later how this enhancement can be traced back algebraically. Physically, it arises from the long range nature of the coulombic attraction between the electron and the hole, even when they do not form a bound state.

The exciton picture emerges after lengthy many-body calculations [273]. All of these start by attempting to calculate an energy difference between the ground state of a crystal, which consists of N electrons occupying a filled valence band, and an excited state of the crystal obtained by optically promoting an electron from the aforesaid filled band to an empty conduction band. If there were no electron–electron interactions, this energy difference $\Delta\varepsilon$ would simply be the sum of the bandgap ε_g (eventually including valence and conduction confinement energies in the case of heterostructures) and the

difference between the kinetic energies in the final and excited states; i.e., for parabolic bands

$$\Delta\varepsilon = \varepsilon_g + \frac{\hbar^2 k_e^2}{2m_e^*} - \left(\frac{-\hbar^2 k_\nu^2}{2m_\nu^*} \right) \qquad (699)$$

The onset of the interband absorption would occur at $\hbar\omega_{\mathrm{onset}} = \varepsilon_g$. The existence of lower excited states of the crystal is the origin of peaked structures below ε_g. They are obtained by rearranging the conduction and valence electronic densities (in a coupled way). To the excited state whose energy with respect to the ground state is $\Delta\varepsilon$ (Eq. (699)), one associates valence and conduction charge distributions which are uniform in space (at the scale of the envelope functions) and which are spatially uncorrelated. By making wave packets of conduction states with various $\vec{k}_e$ and wave packets of valence states with various $\vec{k}_\nu$, where $\vec{k}_e$ and $\vec{k}_\nu$ are correlated ($\vec{k}_\nu = \vec{k}_e - \vec{K}$, $\vec{K}$ fixed) in a unique way, the potential energy gained is greater (by a decrease of the repulsive electron–electron interaction) than the lost kinetic energy (by spatially correlating the electrons).

The following discussion is a survey of excitonic absorptions in semiconductor heterostructures and of their associated selection rules. This discussion does not pretend to be complete: although relatively well documented on the experimental side, the optical properties of excitons in semiconductor heterostructures have not received as much attention on the theoretical side.

The same decoupling procedure for the excitons as for the free particle states (Eqs. (656) and (674)–(676)) will be adopted. Thus, in this crude model there exist two kinds of decoupled excitons: those formed between heavy holes and electrons (heavy hole excitons) and those formed between light holes and electrons (light hole excitons). This decoupled exciton theory, first used by Miller et al. [43] and Greene and Bajaj [40] in the calculation of exciton binding energies, neglects off-diagonal coupling terms in the valence band Hamiltonian. For this purpose, it proves to be very convenient, for it lightens the already heavy algebra involved in the derivation of the excitonic absorption.

Let us consider $hh_1 \to \varepsilon_1$ excitonic transitions in a type I single quantum well. Other situations may be treated with minor modifications. The ground state of this quantum well consists of N valence electrons occupying the hh_1 subband and an empty ε_1 conduction subband. The dispersion relations of hh_1 and ε_1 are taken as parabolic upon the in-plane wavevector $\vec{k}_\perp$ (see Eqs. (674), (676)), and the one-electron wavefunctions are respectively

$$\psi_{\vec{k}_{\perp\nu}}^{(h)}(\vec{r}) = u_{hh}(\vec{r})\chi_1^{(h)}(z)\frac{1}{\sqrt{S}}\exp(i\vec{k}_{\perp\nu} \cdot \vec{r}_\perp) \qquad (700)$$

$$\psi_{\vec{k}_{\perp e}}^{(e)}(\vec{r}) = u_e(\vec{r})\chi_1^{(e)}(z)\frac{1}{\sqrt{S}}\exp(i\vec{k}_{\perp e} \cdot \vec{r}_\perp) \qquad (701)$$

where u_{hh} and u_e are the heavy hole (i.e., $u_{hh} \equiv u_{3/2,\pm 3/2}$) and electron (i.e., $u_e \equiv iS\uparrow$ or $iS\downarrow$) periodic parts of the Bloch function at the zone center of the host layers. It is clear that the quadratic dispersion law for hh_1 is unable to give rise to a filled band. It is understood that at large enough $\vec{k}_{\perp\nu}$ this dispersion flattens so that ultimately the hh_1 energy will be a periodic function upon $\vec{k}_{\perp\nu}$, with a period given by the in plane basis vectors of a suitable Brillouin zone. Actually, the knowledge of the whole dispersion of the hh_1 subband is not needed, for the only interest is in the Wannier excitons that are weakly bound (with respect to the bandgaps). Their wavefunctions are built from hh_1 and ε_1 states which involve $\vec{k}_{\perp e}, \vec{k}_{\perp\nu}$ values much smaller than the in plane size of the Brillouin zone. It is assumed that, for these states, the quadratic laws are valid.

8.2.1. Absorption in the Absence of Coulombic Interactions; Equivalence with the One-Electron Model

In the absence of e^2/r_{ij} terms in $\widehat{H}_0$ (see Eq. (638)), the exact initial and final states of $\widehat{H}_0$ are Slater determinants [175, 272, 273]. The initial state is unique and corresponds to the N electrons which occupy the hh_1 subband:

$$\psi_i(\vec{r}_1, \vec{r}_2, \ldots, r_N)$$

$$= \frac{1}{\sqrt{N!}} \begin{vmatrix} \psi_{\vec{k}_{\perp 1}}^{(h)}(\vec{r}_1) & \psi_{\vec{k}_{\perp 1}}^{(h)}(\vec{r}_2) & \cdots & \psi_{\vec{k}_{\perp 1}}^{(h)}(\vec{r}_N) \\ \psi_{\vec{k}_{\perp\nu}}^{(h)}(\vec{r}_1) & \psi_{\vec{k}_{\perp\nu}}^{(h)}(\vec{r}_2) & \cdots & \psi_{\vec{k}_{\perp\nu}}^{(h)}(\vec{r}_N) \\ \vdots & \vdots & & \vdots \\ \psi_{\vec{k}_{\perp N}}^{(h)}(\vec{r}_1) & \psi_{\vec{k}_{\perp N}}^{(h)}(\vec{r}_2) & \cdots & \psi_{\vec{k}_{\perp N}}^{(h)}(\vec{r}_N) \end{vmatrix} \quad (702)$$

A final state ψ_f is obtained by removing a hh_1 electron with wavevector $\vec{k}_{\perp\nu}$ and sending it into the ε_1 subband, where it acquires an in plane wavevector $\vec{k}_{\perp e}$. Its wavefunction is obtained from Eq. (702) by changing the line $\psi_{\vec{k}_{\perp\nu}}^{(h)}(\vec{r}_i)$ into $\psi_{\vec{k}_{\perp e}}^{(e)}(\vec{r}_i)$, $i = 1, 2, \ldots, N$.

The Slater determinants can be rewritten in the more compact form

$$\langle \vec{r}_1, \vec{r}_2, \ldots, \vec{r}_N | \Psi_i \rangle = \frac{1}{\sqrt{N!}} \sum_P (-1)^P P \prod_{\vec{k}_{\perp n}} \psi_{\vec{k}_{\perp n}}^{(h)}(\vec{r}_n) \quad (703)$$

$$\langle \vec{r}_1, \vec{r}_2, \ldots, \vec{r}_N | \vec{r}_f \rangle = \frac{1}{\sqrt{N!}} \sum_P (-1)^P \psi_{\vec{k}_{\perp e}}^{(e)}(\vec{r}_e) \prod_{\vec{k}_{\perp n} \neq \vec{k}_{\perp\nu}} \psi_{\vec{k}_{\perp n}}^{(h)}(\vec{r}_n) \quad (704)$$

where P is the permutation operator exchanging the particles with themselves. The case $N = 2$ leads to

$$\langle \vec{r}_1, \vec{r}_2 | \psi_i \rangle = \frac{1}{\sqrt{2}} \left[\psi_{\vec{k}_{\perp 1}}^{(h)}(\vec{r}_1)\psi_{\vec{k}_{\perp 2}}^{(h)}(\vec{r}_2) - \psi_{\vec{k}_{\perp 2}}^{(h)}(\vec{r}_1)\psi_{\vec{k}_{\perp 1}}^{(h)}(\vec{r}_2) \right] \quad (705)$$

$$\langle \vec{r}_1, \vec{r}_2 | \psi_f \rangle = \frac{1}{\sqrt{2}} \left[\psi_{\vec{k}_{\perp 1}}^{(h)}(\vec{r}_1)\psi_{\vec{k}_{\perp e}}^{(e)}(\vec{r}_2) - \psi_{\vec{k}_{\perp 1}}^{(h)}(\vec{r}_2)\psi_{\vec{k}_{\perp e}}^{(e)}(\vec{r}_1) \right] \quad (706)$$

The dipolar matrix element of the $\psi_i \to \psi_f$ transition can therefore be written as

$$\langle \psi_i | \sum_n \widehat{H}_{er}^n | \psi_f \rangle$$

$$= \frac{1}{N!} \sum_n \sum_{PP'} (-1)^{P+P'} \int d^3 r_1 \cdots d^3 r_N \left[P \prod_{\vec{k}_{\perp n}} \psi_{\vec{k}_{\perp n}}^{(h)*}(\vec{r}_n) \right]$$

$$\times \left(-\frac{e}{m_0}\sqrt{\frac{2\hbar\pi}{\omega V}} \right) \vec{e}_r \cdot \hat{p}_n \left[P\psi_{\vec{k}_{\perp e}}^{(e)}(\vec{r}_e) \right] \prod_{\vec{k}_{\perp n} \neq \vec{k}_{\perp\nu}} \psi_{\vec{k}_{\perp n}}^{(h)}(\vec{r}_n) \quad (707)$$

Fortunately this complicated expression simplifies into [175, 272, 273]

$$\left\langle \psi_i | \sum_n \widehat{H}_{er}^n | \psi_f \right\rangle = -\frac{e}{m_0}\sqrt{\frac{2\hbar\pi}{\omega V}} \vec{e}_r \cdot \int d^3 r \ \psi_{\vec{k}_{\perp\nu}}^{(h)*}(\vec{r})\hat{p}\psi_{\vec{k}_{\perp e}}^{(e)}(\vec{r}) \quad (708)$$

due to the identity

$$\left\langle \psi_{\vec{k}_{\perp n}}^{(h)} \Big| \psi_{\vec{k}_{\perp n}}^{(h)} \right\rangle = \left\langle \psi_{\vec{k}_{\perp n}}^{(h)} | \hat{p} | \psi_{\vec{k}_{\perp n}}^{(h)} \right\rangle = 0 \quad (709)$$

In the case $N = 2$, Eq. (707) can be checked by direct computation. Using Eqs. (705) and (706), one obtains eight terms; of these, six vanish owing to Eq. (709) and the last two, which are identical, cancel the $1/2!$. Finally W_{if} is simply

$$W_{if} = \frac{2\pi}{\hbar} \left(-\frac{e}{m_0}\sqrt{\frac{2\hbar\pi}{\omega V}} \right)^2 \left| \left\langle \psi_{\vec{k}_{\perp\nu}}^{(h)} | \vec{e}_r \cdot \hat{p} | \psi_{\vec{k}_{\perp e}}^{(e)} \right\rangle \right|^2$$

$$\times \delta \left[\varepsilon_A + \varepsilon_1 + hh_1 + \frac{\hbar^2 k_{\perp e}^2}{2m_e^*} + \frac{\hbar^2 k_{\perp\nu}^2}{2M_1^*} - \hbar\omega \right] \quad (710)$$

This expression is exactly the same as that obtained in the one electron picture, which after all is not very surprising. If the dipolar matrix elements in Eq. (710) were written in terms of both the dipole matrix element between u_{hh} and u_e and the overlap integral of the envelope functions, the polarization selection rules, on the one hand, and the selection rules on the envelope functions quantum numbers on the other hand could be rederived.

In particular it would be found that only "vertical" transitions are optically allowed; i.e., $\vec{k}_{\perp\nu} = \vec{k}_{\perp e}$.

There is, however, another way to derive this selection rule. Let us consider the translation operator $\hat{\tau}_\perp$ which is such that

$$\hat{\tau}_\perp f(\vec{r}_1, \vec{r}_2, \ldots, \vec{r}_N) = f(\vec{r}_1 + \vec{a}_\perp, \vec{r}_2 + \vec{a}_\perp, \ldots, \vec{r}_N + \vec{a}_\perp) \quad (711)$$

with $\vec{a}_\perp$ an in plane lattice vector. The operators $\hat{\tau}_\perp$ and $\widehat{H}_0$ commute, as well as $\hat{\tau}_\perp$ and $\sum_n \widehat{H}_{er}^n$. The initial state $|\psi_i\rangle$ is an eigenstate of $\hat{\tau}_\perp$ with an eigenvalue that can always be written as $\exp(i\vec{K}_\perp^{in} \cdot \vec{a}_\perp)$.

On the other hand, from Eq. (702) it can be directly obtained that

$$\hat{\tau}_\perp \psi_i(\vec{r}_1, \vec{r}_2, \ldots, \vec{r}_N) = \exp\left[i\left(\sum_n \vec{k}_{\perp n}\right) \cdot \vec{a}_\perp\right] \psi_i(\vec{r}_1, \vec{r}_2, \ldots, \vec{r}_N) \tag{712}$$

Thus

$$\vec{K}_\perp^{in} = \sum_n \vec{k}_{\perp n} \tag{713}$$

However, for a filled band

$$\sum_n \vec{k}_{\perp n} = 0 \qquad \text{i.e., } \vec{K}_\perp^{in} = 0 \tag{714}$$

The final state is also an eigenstate of $\hat{\tau}_\perp$, with eigenvalue $\exp(i\vec{K}_\perp \cdot \vec{a}_\perp)$. By directly applying $\hat{\tau}_\perp$ on $|\psi_f\rangle$ it is obtained that

$$\hat{\tau}_\perp \psi_i(\vec{r}_1, \vec{r}_2, \ldots, \vec{r}_N) = \exp[i(\vec{k}_{\perp e} - \vec{k}_{\perp v}) \cdot \vec{a}_\perp] \psi_i(\vec{r}_1, \vec{r}_2, \ldots, \vec{r}_N) \tag{715}$$

Thus

$$\vec{K}_\perp = \vec{k}_{\perp e} - \vec{k}_{\perp v} \tag{716}$$

From now on, the final state will be denoted by $\psi_{\vec{k}_{\perp e}, \vec{k}_{\perp e} - \vec{K}_\perp}$. The selection rule on the envelope function quantum number means $\vec{K}_\perp = 0$. This is not surprising in view of the commutativity between $\sum_n \widehat{H}_{er}^n$ and $\hat{\tau}_\perp$. As the initial state was characterized by $\vec{K}_\perp^{in} = 0$, it could only be coupled by $\sum_n \widehat{H}_{er}^n$ to an excited state with the same $\vec{K}_\perp$ values, resulting in the selection rule $\delta_{\vec{K}_\perp, 0}$. In short, it has been obtained that

$$\begin{aligned} W_{if} = \frac{2\pi}{\hbar} &\left(-\frac{e}{m_0}\sqrt{\frac{2\hbar\pi}{\omega V}}\right)^2 |\langle u_{hh}|\vec{e}_r \cdot \hat{p}|u_e\rangle|^2 \\ &\times \left|\left\langle \chi_1^{(h)} \middle| \chi_1^{(h)} \chi_1^{(e)} \chi_1^{(e)}\right\rangle\right|^2 \delta_{\vec{K}_\perp, 0} \\ &\times \delta\left[\varepsilon_A + \varepsilon_1 + hh_1 + \frac{\hbar^2 k_{\perp e}^2}{2}\left(\frac{1}{m_e^*} + \frac{1}{M_1^*}\right) - \hbar\omega\right] \end{aligned} \tag{717}$$

From such an expression, one could easily derive the staircaselike absorption spectrum, Eq. (683) which is characteristic of subband → subband interband transitions in type I quantum wells.

8.2.2. Absorption in the Presence of Electron–Electron Interactions

In the presence of coulombic terms, single Slater determinants are no longer exact eigenstates of $\widehat{H}_0$. However, they remain the best single determinant approximation of the ground state if the one-electron functions $\psi_{\vec{k}_{\perp v}}^{(h)}(\vec{r})$ are solutions of the coupled integro-differential Hartree–Fock equations. This approximation for $|\psi_i\rangle$ is kept. For the final state a single determinant is not an exact solution also. If a single determinant of

the form given by Eq. (704) is kept, a Hartree–Fock solution will be obtained, which describes a situation where a valence electron, uniformly spread over the area S, is replaced by a conduction electron whose charge density is also uniformly spread over S. The removal of a valence electron hardly affects the self-consistent Hartree–Fock potential experienced by the remaining $N - 1$ valence electrons (the charge density is of the order of $1/S$). Similarly the single conduction electron only slightly modifies the self-consistent potential. This gives rise to an excited energy that differs from the ground state energy by almost [i.e., to the order of $(1/S)$] the same energy $\Delta\varepsilon$ as calculated in Eq. (699). A lower lying excited state can be constructed by expanding $|\psi_f\rangle$ on the set of the single determinants $|\psi_{\vec{k}_{\perp e}, \vec{k}_{\perp e} - \vec{K}_\perp}\rangle$. As $\widehat{H}_0$ commutes with $\hat{\tau}_\perp$, it is known that such an expansion can be restricted to a summation over $\vec{k}_{\perp e}$, $\vec{K}_\perp$ being a good quantum number. Thus, even with the inclusion of electron–electron interactions (within the Hartree–Fock approximation), it can be written that

$$\langle \vec{r}_1, \vec{r}_2, \ldots, \vec{r}_N | \psi_f\rangle = \sum_{\vec{k}_{\perp e}} A(\vec{k}_{\perp e}) \psi_{\vec{k}_{\perp e}, \vec{k}_{\perp e} - \vec{K}_\perp}(\vec{r}_1, \vec{r}_2, \ldots, \vec{r}_N) \tag{718}$$

If now it is reasoned in terms of electron–hole states instead of $(N - 1) + 1$ electrons states, the wavefunction $\psi_{\vec{k}_{\perp e}, \vec{k}_{\perp e} - \vec{K}_\perp}$ corresponds to an electron–hole pair with a total wavevector $\vec{k}_{\perp e} - (\vec{k}_{\perp e} - \vec{K}_\perp) = \vec{K}_\perp$ (since the hole wavevector is minus that of the missing electron). Thus, the physical meaning of Eq. (718) is that we are looking for an excited state of the heterostructure, which consists of a wave packet of electron–hole pairs, which have the same total wavevector $\vec{K}_\perp$. This wave packet is the exciton. The exciton can be seen either as a particular excitation of $(N - 1) + 1$ electrons or as a particular combination of electron–hole pairs. The exciton is a truly delocalized excitation. This can be checked by calculating the average in plane electronic density $(1/N)\sum_m \delta(\vec{r}_\perp - \vec{r}_{\perp m})$ in the excited state Eq. (718) and by comparing it with that found in the ground state, Eq. (702). Both are found equal to $1/S$ (if only the slow spatial variations, which occur at the scale of the envelope function, are retained). What may give rise to localized states in the exciton are the internal degrees of freedom of the pairs. This is due to the attraction between the electron and the hole (or equivalently to the diminished electron–electron repulsion) in the excited state, Eq. (718). The electron–hole attraction, together with the restricted class of electron–hole pairs that was retained in Eq. (718) (i.e., which all have the same total wavevector $\vec{K}_\perp$), can be described in a better way if the exciton amplitude is used:

$$\beta(\vec{r}_\perp) = \frac{1}{\sqrt{S}} \sum_{\vec{k}_{\perp e}} A(\vec{k}_{\perp e}) \exp(i\vec{k}_{\perp e} \cdot \vec{r}_\perp) \tag{719}$$

The physical interpretation of $\beta(\vec{r}_\perp)$ is that $|\beta(\vec{r}_\perp)|^2$ is the areal probability density of finding the electron and the hole separated from each other by $\vec{r}_\perp$, while having a total wavevector $\vec{K}_\perp$. Equations (718) and (719) are the starting points of the exciton formalism, fully developed in, for instance, Knox's textbook [273]. For these purposes, it is sufficient to know that $\beta(\vec{r}_\perp)$ (after considerable manipulations and numerous assumptions) is the solution of

$$\left[\frac{p_\perp^2}{2\mu^*} - \frac{e^2}{\kappa} \iint \frac{dz_e dz_h |\chi_1^{(e)}(z_e)|^2 |\chi_1^{(h)}(z_h)|^2}{\sqrt{r_\perp^2 + (z_e - z_h)^2}} \right. \tag{720}$$
$$\left. + \varepsilon_A + \varepsilon_1 + hh_1 \right]\beta(\vec{r}_\perp) = \varepsilon\beta(\vec{r}_\perp)$$

where μ^* is the exciton reduced effective mass and κ is the relative dielectric constant of the heterostructure. The energy difference between the excited state (Eq. (718)) and the ground state is equal to

$$\Delta\varepsilon = \varepsilon_\nu + \frac{\hbar^2 K_\perp^2}{2(m_e^* + M_1^*)} \tag{721}$$

where ε_ν is one of the eigenvalues of Eq. (720). The ground state of Eq. (720) has been extensively discussed previously. It is known that a simple trial wavefunction for this ground state is

$$\beta_{1s}(\vec{r}_\perp) = \sqrt{\frac{2}{\pi a_{B2D}^{*2}}} \exp\left(-\frac{r_\perp}{a_{B2D}^*}\right) \tag{722}$$

where a_{B2D}^* is the effective Bohr radius of the quasi-two-dimensional exciton. The $1s$ exciton state lies at an energy lower than $\varepsilon_g + \varepsilon_1 + hh_1$ by a quantity ε_b, which is the exciton binding energy. Thus, the minimum energy separation between the ground state and the excited state (characterized by the wavevector $\vec{K}_\perp$) is

$$\Delta\varepsilon_{min} = \varepsilon_A + \varepsilon_1 + hh_1 - \varepsilon_b + \frac{\hbar^2 K_\perp^2}{2(m_e^* + M_1^*)} \tag{723}$$

which can be lower than $\varepsilon_A + \varepsilon_1 + hh_1$ if $\vec{K}_\perp$ is small enough.

There exist an infinite number of bound states for Eq. (720). They correspond to $\beta_\nu(\vec{r}_\perp)$ functions, which decay to zero at large $\vec{r}_\perp$. Unbound solutions also exist. They form a continuum and are characterized by $\beta_\nu(\vec{r}_\perp)$ exciton amplitudes that extend over the whole area S, and by energies $\varepsilon_\nu > \varepsilon_A + \varepsilon_1 + hh_1$. In the interacting electron–hole problem these unbound solutions are the counterparts of the plane wave solutions in the noninteracting electron–hole problem.

Now W_{if} is evaluated to obtain the selection rules for the excitonic absorption from the ground state to the final state, characterized on the one hand by $\vec{K}_\perp$ and on the other by a quantum number ν, that labels the solutions of Eq. (720). This is indicated by affixing a subscript ν to the $A(\vec{k}_{\perp e})$ of Eq. (718):

$$\left\langle \psi_i \left| \sum_n \widehat{H}_{er}^n \right| \psi_f \right\rangle = \sum_{\vec{k}_{\perp e}} A(\vec{k}_{\perp e})\left\langle \psi_i \left| \sum_n \widehat{H}_{er}^n \right| \psi_{\vec{k}_{\perp e} - \vec{K}_\perp} \right\rangle \tag{724}$$

The matrix element appearing on the right hand side of Eq. (724) has already been evaluated in the previous subsection. This leaves us with

$$\left\langle \psi_i \left| \sum_n \widehat{H}_{er}^n \right| \psi_f \right\rangle = -\frac{e}{m_0}\sqrt{\frac{2\hbar\pi}{\omega V}} \langle u_{hh}|\vec{e}_r \cdot \hat{p}|u_e\rangle \langle \chi_1^{(h)}|\chi_1^{(e)}\rangle$$
$$\times \delta_{\vec{K}_\perp, 0} \sum_{\vec{k}_{\perp e}} A_\nu(\vec{k}_{\perp e}) \tag{725}$$

which, together with Eq. (719), leads to

$$W_{if} = \frac{2\pi}{\hbar}\left(-\frac{e}{m_0}\sqrt{\frac{2\hbar\pi}{\omega V}}\right)^2 |\langle u_{hh}|\vec{e}_r \cdot \hat{p}|u_e\rangle|^2 |\langle \chi_1^{(h)}|\chi_1^{(e)}\rangle|^2$$
$$\times \delta_{\vec{K}_\perp, 0} S |\beta_\nu(0)|^2 \delta(\varepsilon_\nu - \hbar\omega) \tag{726}$$

where ε_ν is the energy of the eigenstates $|\nu\rangle$ of Eq. (720). Again, if the electron–hole interaction is neglected, the result of the noninteracting electron model is recovered: in this case $\beta_\nu(\vec{r}_\perp)$ and ε_ν are given by

$$\beta_\nu(\vec{r}_\perp) = \frac{1}{\sqrt{S}}\exp(i\vec{k}_\perp \cdot \vec{r}_\perp)\varepsilon_\nu$$
$$= \frac{\hbar^2 k_\perp^2}{2\mu^*} + \varepsilon_1 + hh_1 + \varepsilon_A \tag{727}$$

and the staircaselike shape of the absorption is readily derived.

In the case of excitonic absorption, the selection rule $\delta_{\vec{K}_\perp, 0}$ is very important since it considerably restricts the possibilities of creating (or annihilating) excitons. It expresses the delocalized nature of the exciton, an entity which corresponds to a delocalized Bloch state of the excited crystal, despite the fact that the exciton amplitude $\beta_\nu(\vec{r}_\perp)$ can be a spatially localized function of $\vec{r}_\perp$. The $\delta_{\vec{K}_\perp, 0}$ selection rule can also be associated with the conservation of the in plane wavevector $\vec{K}_\perp$ which, being zero in the ground state, must also be zero in the excited state if the dipolar approximation is used to describe the interaction between the electromagnetic wave and the carriers. Going beyond the dipolar approximation would give rise to the selection rule $\delta_{\vec{K}_\perp, \vec{\kappa}_\perp}$, where $\vec{\kappa}_\perp$ is the in plane projection of the photon wavevector.

As seen from Eq. (726) the excitonic absorption fulfills the same polarization selection rules as the band-to-band absorption model that neglects the electron–electron interaction. This is because the electron–electron interaction has more symmetry (spherical symmetry) than the heterostructure potential. Looking at Table III, it can be noticed that the heavy hole excitonic transitions are forbidden (within this model) when the electromagnetic wave propagates in the layer plane and when its electric field vector is polarized along the growth z-axis.

An important new fact brought by the electron–hole interaction is the appearance of the term $|\beta_\nu(0)|^2$ in Eq. (726). It means that only excitons with nonzero amplitude at $\vec{r}_\perp = 0$ can absorb light. Since the effective two-dimensional potential that appears in Eq. (720) is radial, the $\beta_\nu(\vec{r})$ functions can be classified according to the eigenvalues of $\widehat{L}$ and $\widehat{L}_z$, where $\widehat{L}$ is the angular momentum. In addition, a radial quantum number n (discrete for the bound states, continuous for the extended states) labels the eigenstates. Thus, only the ns excitons (corresponding to $\widehat{L} = 0$) can be optically created.

The only situation where Eq. (720) admits exact solutions corresponds to a purely two-dimensional case (i.e., $|\chi_1^{(e)}(z)|^2 = |\chi_1^{(h)}(z)|^2 = \delta(z)$). In this case ε_b is equal to $4R_{y,exc}^*$, where $R_{y,exc}^*$ is the three-dimensional exciton Rydberg. The magnitude of $|\beta_{ns}(0)|^2$, and thus of the ns excitonic absorption, is proportional to $(8/\pi) a_B^{*2} (2n-1)^3$ where $n = 1, 2, \ldots$ and a_B^* is the bulk exciton Bohr radius [274]. This shows that the magnitude of the ns exciton absorption decreases very rapidly with n. However, the levels are denser. Both effects combine leading to a finite absorption when $\hbar\omega$ approaches $\varepsilon_A + \varepsilon_1 + hh_1$ from below. The absorption associated with the dissociated electron–hole hole pairs is also enhanced with respect to the noninteracting electron model (see Fig. 58). It approaches the one-electron result when $\hbar\omega$ exceeds $\varepsilon_A + \varepsilon_1 + hh_1$ by many ε_b.

Let us rewrite the $hh_1 \rightarrow \varepsilon_1$ excitonic absorption coefficient in the quasi-two-dimensional case,

$$\underset{hh_1 \rightarrow \varepsilon_1}{\alpha(\omega)} = \frac{2\pi^2 e^2 \varepsilon_P}{ncm_0 \omega \ell} |\langle \chi_1^{(h)} | \chi_1^{(e)} \rangle|^2 \sum_\nu |\beta_\nu(0)|^2 \delta(\hbar\omega - \varepsilon_\nu) \quad (728)$$

where $\varepsilon_P, \ell, n, \ldots$ have the same meaning as in Section 8.1. Focusing attention on the quantum well thickness (L) dependence of the magnitude of the $1s$ exciton peak, to account for the broadening mechanisms, the delta functions in Eq. (728) are replaced by a Gaussian function (which sometimes fits the observed lineshapes better than a Lorentzian one) and Eq. (722) is used to evaluate $\beta_{1s}(0)$. For a single quantum well, the peak value is obtained,

$$\ell\alpha_{hh_1 \rightarrow \varepsilon_1}^{peak}(\omega) = \frac{4\pi^2 e^2 \varepsilon_P}{ncm_0 \omega \lambda^2} |\langle \chi_1^{(h)} | \chi_1^{(e)} \rangle|^2 \frac{1}{\Gamma\sqrt{2\pi}} \quad (729)$$

where $2\Gamma\sqrt{2}$ is the full width at e^{-1} amplitude of the Gaussian peak. For a GaAs–GaAlAs quantum well a representative figure for $\ell\alpha_{hh_1 \rightarrow \varepsilon_1}^{peak}$ is obtained with $n = 3.6$, $\hbar\omega = 1.6$ eV, $\varepsilon_P = 23$ eV, $a_{B2D}^* = 70$ Å, $\Gamma\sqrt{2\pi} = 3$ meV, and $\langle \chi_1^{(h)} | \chi_1^{(e)} \rangle \approx 1$.

$\ell\alpha_{hh_1 \rightarrow \varepsilon_1}^{peak} \approx 0.19$ is obtained. This is ≈ 32 times larger than the band-to-band plateau value obtained in the noninteracting electron model of the $hh_1 \rightarrow \varepsilon_1$ absorption. Precise absorption measurements of the continuum edge are difficult, however, since it is often (energy) located in the low energy tail of the $1s\,\varepsilon_1 - lh_1$ exciton.

When the quantum well thickness varies, two factors affect the peak value of the $1s$ excitonic absorption.

(i) a_{B2D}^* decreases with decreasing L. This strengthens the excitonic absorption and witnesses an increase in the two-dimensionality of the exciton. In a correlative way, the energy distance between the $1s$ exciton peak and the continuum increases.

(ii) The broadening parameter Γ also increases. As shown by Weisbuch et al. [275] the layer thickness fluctuations make Γ vary like L^{-3}.

Thus, depending on which mechanism prevails, the peak value of the excitonic absorption can increase or decrease. Masumoto et al. [276] have recently reported a complete study on the thickness dependence of the excitonic absorption in GaAs–GaAlAs multiple quantum wells.

The exciton continuum absorption ($\hbar\omega > \varepsilon_A + \varepsilon_1 + hh_1$) is also enhanced by the electron–hole attraction compared with that calculated with a noninteracting electron model. Note, however, that this enhancement cannot be explicitly calculated because the solutions of the Schrödinger equation for the quasi-two-dimensional exciton (Eq. (720)) are not known analytically. Experimentally, a small hump is often visible in high quality GaAs quantum wells. It is located in energy between $1s\,hh_1 - \varepsilon_1$ and $1s\,lh_1 - \varepsilon_1$ exciton peaks (Fig. 59) and can be attributed either to the $2s\,hh_1 - \varepsilon_1$ broadened exciton absorption [43] or to the onset of the $hh_1 - \varepsilon_1$ exciton continuum [40]. These two energies are so close (1–2 meV) that the small hump probably arises from both of them.

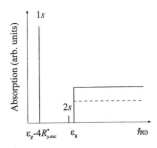

Fig. 58. Theoretical (unbroadened) absorption spectrum of a purely two-dimensional semiconductor (solid line). $R_{y,exc}^*$ is the three-dimensional exciton binding energy. Only the $1s$ and $2s$ exciton bound states contributing to the absorption have been represented. The dashed line plateau corresponds to the absorption coefficient of dissociated and uncorrelated electron–hole pairs.

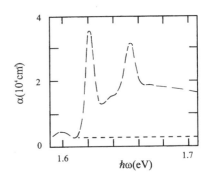

Fig. 59. Measured absorption coefficient of a 76–33 Å GaAs multiple quantum well sample at $T = 2$ K. Between the two main absorption peaks corresponding to excitonic $\varepsilon_1 - hh_1$, $\varepsilon_1 - lh_1$, transitions, respectively, one sees a weaker absorption due to either the $\varepsilon_1 - hh_1$ $2s$ excitons or to the $\varepsilon_1 - hh_1$ exciton continuum.

8.3. Photoluminescence

8.3.1. Introduction

The optical spontaneous emission, for instance, photoluminescence, cannot be predicted when the electromagnetic field is described classically. However, it is possible to relate its intensity exactly to that of absorption by means of the Einstein coefficients [257, 258, 277]. Here, we are interested in presenting the theoretical considerations about the lineshape of some recombination processes in semiconductor heterostructures rather than calculating the photoluminescence intensities. When possible, these considerations will be illustrated by examples. In contrast with a widespread belief, the information that can be extracted from the photoluminescence concerning the energy levels of a given heterostructure is often very scarce and depends to a large extent on an a priori knowledge of that heterostructure. It is true that photoluminescence experiments are much easier to perform than absorption experiments. It is also true that they are much more complicated to interpret.

Absorption and photoluminescence seem to be, at first sight, symmetrical processes. On the one hand a photon is absorbed promoting an electron from the level $|i\rangle$ to the level $|f\rangle$ and photons that have not been absorbed are detected. On the other hand, the emission involves the transition $|i\rangle \to |f\rangle$ with $\varepsilon_i > \varepsilon_f$ and the photons, which are emitted at the energy $\varepsilon_i - \varepsilon_f$, can be observed. Such a process is feasible only if the system has been initially excited in the state $|i\rangle$. Thus, it is not at equilibrium and one of the ways in which this can be reached is by optical emission. The radiative channel is in competition with the nonradiative relaxation processes (phonon emission, capture by deep centers, Auger effect, etc.) which send the excited carriers to lower states from which they can emit photons or relax nonradiatively, etc. It appears therefore that luminescence is different from absorption because, instead of having 100% efficiency (as in the case of absorption, where one absorbed photon creates one electron–hole pair), spontaneous emission is only one of the mechanisms that might occur. This is a matter of the lifetimes of the level $|i\rangle$ with respect to the radiative and nonradiative relaxations. To illustrate these remarks let us consider the four-level system shown in Figure 60. The system is pumped by $|1\rangle \to |4\rangle$ transitions and only the luminescence signal of levels $|3\rangle$ and $|2\rangle$ at energies $\varepsilon_3 - \varepsilon_1$ and $\varepsilon_2 - \varepsilon_1$ respectively will be considered. The parameters τ_{ij} and T_{i1} are time constants which characterize interexcited levels, nonradiative channels, and radiative channels from level $|i\rangle$ to level $|1\rangle$, respectively. The effect of the final state population will be neglected. As level $|3\rangle$ is assumed to be populated nonradiatively from $|4\rangle$ it suffices, since levels $|1\rangle$ and $|4\rangle$ are of no interest, to write the rate equations

$$\frac{dn_3}{dt} = \frac{n_0}{\tau_p} - \frac{n_3}{\tau_{32}} - \frac{n_3}{T_{31}} \tag{730}$$

$$\frac{dn_2}{dt} = +\frac{n_3}{\tau_{32}} - \frac{n_2}{\tau_{21}} - \frac{n_2}{T_{21}} \tag{731}$$

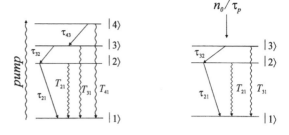

Figure 60

In the steady state, $dn/dt = 0$ and therefore

$$n_3 = \frac{n_0}{\tau_p} \frac{T_{31}}{1 + T_{31}/\tau_{32}} \tag{732}$$

$$n_2 = \frac{n_0}{\tau_p} \frac{T_{31}}{1 + T_{31}/\tau_{32}} \frac{T_{21}}{\tau_{32}} \frac{1}{1 + T_{21}/T_{21}\tau_{21}} \tag{733}$$

The luminescence signals at energies $\hbar\omega_{31} = \varepsilon_3 - \varepsilon_1$ and $\hbar\omega_{21} = \varepsilon_2 - \varepsilon_1$ are respectively proportional to n_3 and n_2. It is easy to see from Eqs. (732) and (733) that, if $\tau_{32} \ll T_{31}$, almost all the population of level $|3\rangle$ is used to populate $|2\rangle$ and does not participate in the emission which occurs at $\hbar\omega_{31}$. Even if the density of states of level $|3\rangle$ is more important (or, in this "atomic" scheme, if its degeneracy is larger) than that of $|2\rangle$, the spontaneous emission from $|3\rangle$ can be considerably smaller than that from $|2\rangle$. This is to be compared with an absorption experiment, which would give a larger absorption at energy $\hbar\omega_{31}$ than at energy $\hbar\omega_{21}$ if the oscillator strengths of the two transitions are the same and the degeneracy of $|3\rangle$ is larger than that of $|2\rangle$. In fact, in bulk semiconductors at low temperature, the luminescence very often involves impurity levels (impurity-to-band, donor–acceptor transitions,...) while the density of states of the impurity levels is much smaller (by several orders of magnitude) than the density of states of the free states in the bands. These impurity levels, which are so prevalent in the photoluminescence, are almost invisible in absorption spectra. While the absorption is sensitive to the density of states, the photoluminescence gives information on this quantity, which is completely distorted by the relaxation effects. This occurs to such an extent that it favors the lowest lying excited states of the materials, even if their density of states is very small.

Let us now consider the transient phenomena (i.e., time-resolved photoluminescence) and assume that the pump rate of level $|3\rangle$ is described by a function $g(t)$ that is arbitrary, except that $g(t)$ vanishes if $t < 0$. If the quantity $T_{21}^{-1} + \tau_{21}^{-1} - T_{31}^{-1} - \tau_{32}^{-1}$ is denoted by Ω_{23} the time dependences are obtained,

$$n_2(t) = \frac{1}{\tau_{32}\Omega_{23}} \left[\exp\left(-\frac{t}{T_3}\right) \int_{-\infty}^{t} g(t') \exp\left(\frac{t'}{T_3}\right) dt' - \exp\left(-\frac{t}{T_2}\right) \int_{-\infty}^{t} g(t') \exp\left(\frac{t'}{T_2}\right) dt' \right] \tag{734}$$

$$n_3(t) = \exp\left(-\frac{t}{T_3}\right) \int_{-\infty}^{t} g(t') \exp\left(\frac{t'}{T_3}\right) dt' \tag{735}$$

where

$$T_3^{-1} = T_{31}^{-1} + \tau_{32}^{-1} \qquad T_2^{-1} = T_{21}^{-1} + \tau_{21}^{-1} \qquad (736)$$

In the case where the pump state is the delta spike $n_0\delta(t)$ it is obtained that

$$n_2(t) = \frac{n_0 Y(t)}{\tau_{32}\Omega_{23}}\left[\exp\left(-\frac{t}{T_3}\right) - \exp\left(-\frac{t}{T_2}\right)\right] \quad (737)$$

$$n_3(t) = n_0 Y(t)\exp\left(-\frac{t}{T_3}\right) \qquad (738)$$

The population $n_2(t)$ first increases with time (as long as the filling of $|2\rangle$ from state $|3\rangle$ is larger than the radiative and nonradiative depopulations) and then decreases (see Fig. 61). It is very important to notice that the luminescence decay time at energy $\varepsilon_2 - \varepsilon_1$ only yields the total lifetime T_2 of the excited level. This involves both the radiative and nonradiative lifetimes. The time-resolved luminescence experiments thus need to be cautiously interpreted if conclusions are to be drawn on the radiative lifetimes of the excited levels. In addition, level $|3\rangle$ is never pumped at a rate $n_0\delta(t)$. The actual function $g(t)$ displays an apparatus-limited time width, as does the detection system, which counts the photons emitted at the energy $\varepsilon_2 - \varepsilon_1$ or $\varepsilon_3 - \varepsilon_1$. This implies that deconvolutions of the experimental signal $n_2(t)$ are required in order to distinguish between the "true" $n_2(t)$ and the time dependence caused by instrument limitations.

8.3.1.1. Excitation Spectroscopy

A technique closely associated with photoluminescence is the photoluminescence excitation spectroscopy. The detection spectrometer is set at some energy inside the emitted photoluminescence band ($\varepsilon_2 - \varepsilon_1$ in the "atomic" model of Fig. 62) and the energy of the exciting light is scanned. As a result, the various excited levels of the solids $|2\rangle$, $|3\rangle$, ... are populated at rates $g_2, g_3, \ldots$, respectively. These rates are proportional to the absorption coefficients $\alpha(\varepsilon_2), \alpha(\varepsilon_3), \ldots$. Once populated, the excited levels relax either radiatively or nonradiatively toward lower energies. In particular, a fraction of their populations ends up in the lowest lying excited state $|2\rangle$. In

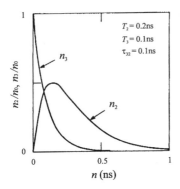

Fig. 61. Calculated time evolution of the n_2 and n_3 populations in a three levels system. The level 3 has been assumed to be populated by a $\delta(t)$ pulse.

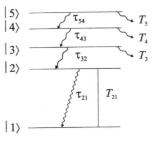

Figure 62

the steady state and for the situation depicted in Figure 62 one obtains

$$n_2 = \frac{g_2}{T_{21}^{-1} + \tau_{21}^{-1}} \qquad (739)$$

for an excitation energy equal to $\varepsilon_2 - \varepsilon_1$,

$$n_2 = \frac{g_3}{\left(T_{21}^{-1} + \tau_{21}^{-1}\right)\left(1 + \tau_{32}T_3^{-1}\right)} \qquad (740)$$

for an excitation energy equal to $\varepsilon_3 - \varepsilon_1$, and

$$n_2 = \frac{g_4}{\left(T_{21}^{-1} + \tau_{21}^{-1}\right)\left(1 + \tau_{32}T_3^{-1}\right)\left(1 + \tau_{43}T_4^{-1}\right)} \qquad (741)$$

$$n_2 = \frac{g_5}{\left(T_{21}^{-1} + \tau_{21}^{-1}\right)\left(1 + \tau_{32}T_3^{-1}\right)\left(1 + \tau_{43}T_4^{-1}\right)\left(1 + \tau_{54}T_5^{-1}\right)} \quad (742)$$

for excitation energies equal to $\varepsilon_4 - \varepsilon_1$ and $\varepsilon_5 - \varepsilon_1$, respectively. In Eqs. (739)–(742) τ_{ij} is now the relaxation time associated with the nonradiative path $i \to j$ and T_i is the decay time due to all the mechanisms which empty $|i\rangle$, apart from those which contribute to the population of $|i-1\rangle$.

The previous model is too crude to be applied as such to semiconductor heterostructures, but it helps us to understand the different kinds of information given by the photoluminescence excitation spectra. It should be noticed that the photoluminescence signal at energy $\varepsilon_2 - \varepsilon_1$, which is proportional to n_2, has an amplitude governed by two competing factors:

   (i) the absorption coefficients of the exciting light of energy $\varepsilon_2 - \varepsilon_1$ via the rates
   (ii) the relative orders of magnitude of the τ_{ij}'s and the T_i's.

In the limiting case where all the τ_{ij}'s are much shorter than the T_i's, the excitation spectrum gives information equivalent to that provided by absorption (see Fig. 63). The advantage of the former technique over the latter is its sensitivity. Like photoluminescence, excitation spectroscopy does not require of thick samples; e.g., it can easily be performed on a single GaAs quantum well, whereas the absorption experiments, as shown previously, require multiple quantum wells.

The excitation spectroscopy is, moreover, selective, allowing the physical origin of the photoluminescence signal in the heterostructure to be traced back. Let us suppose that the photoluminescence band contains several peaks. By setting the detection wavelength at each of these peaks, one obtains

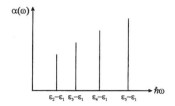

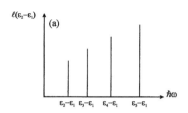

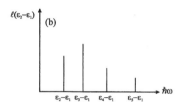

Fig. 63. Comparison between absorption and luminescence excitation spectra in a five levels system. Upper figure: absorption coefficient; middle and lower figures: luminescence excitation spectra at $\varepsilon_2 - \varepsilon_1$. Case (a) $\tau_{ij} \ll T_i$: the excitation spectrum mimics the absorption spectrum; case (b) resonant relaxation $\tau_{32} \ll T_3$ and $\tau_{ij} = T_i$, $i \neq 3$.

excitation spectra whose shape is characteristic of the various excited levels that, after relaxation, ultimately give rise to these different photoluminescence peaks. If the excitation spectra are independent of the detection wavelength, one may safely assume that the whole luminescence band has a single physical origin. If, on the other hand, the excitation spectra are markedly different, one can conclude that the various photoluminescence lines have different physical origins. As is often the case in heterostructures, these differences arise from the different locations of the recombining levels.

In another limiting case, the excitation spectrum is very different from the absorption spectrum. This happens when, for instance, the τ_{ij}'s are all of the same order of magnitude or larger than the T_i's but one, say τ_{32}, which is much shorter than T_3 (resonant relaxation from $|3\rangle$ to $|2\rangle$). In this case, the excitation spectrum shows a peak at $\varepsilon_3 - \varepsilon_1$, which is not found in the absorption spectrum (see Fig. 63). This peak arises because the levels $|4\rangle$ and $|5\rangle$, once populated, contribute little to the population of $|3\rangle$ and $|2\rangle$, leading to a reduced photoluminescence signal (at energy $\varepsilon_2 - \varepsilon_1$) over what could have been inferred uniquely from the rates g_4, g_5.

Thus excitation spectra may or may not be equivalent to absorption spectra. There is no significant luminescence band involving the two-dimensional electron gas confined near the GaAs–GaAlAs heterointerface. This is due to the very quick relaxation of holes far away in the acceptor depletion length. Strong luminescence peaks are associated with the radiative recombination of the bulk GaAs buffer layer. The resonant

photon energies, which produce structures in the excitation spectrum, are

$$\hbar\omega = \varepsilon_g + n\hbar\omega_{LO}(1 + m_e^*/m_{lh}^*) \tag{743}$$

$$\hbar\omega = \varepsilon_g + n\hbar\omega_{LO}(1 + m_e^*/m_{hh}^*) \tag{744}$$

The other limiting regime, where the excitation spectrum roughly mimics the absorption spectrum, is exemplified in the GaAs–GaAlAs quantum wells and superlattices (see [275]). In the vast majority of these heterostructures any trace of peaks or structures in the excitation spectra can be fairly associated with an absorption process. The lack of structures associated with optical phonons, which are so prevalent in bulk GaAs (see above), has not been clearly explained. It calls either for a strongly diminished electron–phonon coupling in quantum wells or for a very fast carrier–carrier interaction, which would overshadow the resonant emission.

8.3.2. Quantum Well Luminescence (Steady State)

Through the Einstein relationships the spontaneous emission can be related to the absorption. However, since the interest is only in the lineshape of the luminescence spectrum and not in its absolute magnitude, all the proportionality constants will be discarded and the spontaneous emission and absorption will be treated on the same footing. The stationary distribution functions of the initial ($|i\rangle$) and final ($|f\rangle$) states will be denoted by $f(\varepsilon_i)$ and $f(\varepsilon_f)$. Thus the luminescence signal $\ell(\omega)$, associated with that particular $|i\rangle \to |f\rangle$ transition, will be proportional to

$$\ell_{i \to f}(\omega) \propto W_{if} f(\varepsilon_i)[1 - f(\varepsilon_f)] \tag{745}$$

where W_{if} is the transition probability per unit time that the carrier undergoes a transition $|i\rangle \to |f\rangle$ due to the effect of the (dipolar) coupling Hamiltonian between the light and the carrier:

$$W_{if} = \frac{2\pi}{\hbar} \left(-\frac{e}{m_0}\sqrt{\frac{2\hbar\pi}{\omega V}}\right)^2 |\langle i|\vec{e}_r \cdot \hat{p}|f\rangle|^2$$
$$\times \delta(\varepsilon_i - \varepsilon_f - \hbar\omega) \tag{746}$$

The optical transition $|i\rangle \to |f\rangle$ is symmetrical to the absorption one ($|i\rangle \to |f\rangle$) and therefore follows the same selection rule. As for the absorption, the dipole matrix element can be split into two parts (see Eqs. (657)–(659)),

$$\langle i|\vec{e}_r \cdot \hat{p}|f\rangle = \langle u_i|\vec{e}_r \cdot \hat{p}|u_f\rangle \int d^3 r f_i^*(\vec{r}) f_f(\vec{r}) \tag{747}$$

The first matrix element on the right hand side of Eq. (747) gives the selection rule on the polarization of the emitted light (in the layer plane of the layers or along the growth axis; see Table III) and the overlap integral selects the subband indexes that govern interband recombination.

In a quantum well with band edge profiles whose band edges are symmetric in z it is found that, as with absorption, the only optical allowed transitions between subbands preserve the parity of the z-dependent envelope functions. Thus, using the same labeling as in Sections 8.1 and 8.2, $n + m$ should be even. In addition, if the broadening is neglected, the radiative recombination occurs with conservation of the in plane wavevector of the carrier ($\vec{k}_\perp^{(i)} = \vec{k}_\perp^{(f)}$). Similarly, in superlattices the optical transitions take place vertically in the superlattice Brillouin zone, ($\vec{k}_\perp^{(i)} = \vec{k}_\perp^{(f)}$), $q_z^{(i)} = q_z^{(f)}$.

The explicit form of the stationary distribution functions $f(\varepsilon_i)$ and $f(\varepsilon_f)$ is often not known. Thus, they are (usually) taken as being Fermi–Dirac distribution functions characterized by T_e, μ_e for conduction electrons and T_h, μ_h for the valence electrons:

$$f(\varepsilon_i) = \{1 + \exp[\beta_e(\varepsilon_i - \mu_e)]\}^{-1} \quad \beta_e = (k_B T_e)^{-1} \quad (748)$$

$$1 - f(\varepsilon_f) = \{1 + \exp[-\beta_h(\varepsilon_f - \mu_h)]\}^{-1}$$
$$\beta_h = (k_B T_h)^{-1} \quad (749)$$

This formulation assumes that the photocreated conduction (valence) electrons thermalize among themselves much quicker than they recombine and that they acquire a temperature $T_e(T_h)$. This temperature is often different from the lattice temperature T and depends on the density of injected carriers and external perturbations (e.g., heating by an in plane electric field [278]). The reader should be aware that a great part of the photoluminescence spectra interpretations are based on the assumptions of thermalized carriers. This assumption is very convenient but is seldom justified by detailed calculations. With this restriction in mind, let us discuss some emission mechanisms in more detail.

8.3.2.1. Band-to-Band Emission

For an emitted light that propagates along the z-axis, it is known (see Table III) that $hh_n \leftrightarrow \varepsilon_m$ or $lh_n \leftrightarrow \varepsilon_m$ transitions are allowed, provided that $n + m$ is even and that the heterostructure potential is even in z. For light emitted in the layer plane, the e_z polarization is forbidden for the $\varepsilon_m \leftrightarrow hh_n$ transitions and allowed for the $lh_n \leftrightarrow \varepsilon_m$ transitions. The quadratic dispersions (Eqs. (674)–(676)) for the various subbands are retained and the luminescence lineshape of the subband-to-subband transitions are computed. For the $\varepsilon_m \leftrightarrow hh_n$ recombination it is obtained that

$$\ell_{m \to n}(\omega) \propto \langle u_{hh} | \vec{e}_r \cdot \hat{p} | u_e \rangle^2 |\langle \chi_n^{(h)} | \chi_m^{(e)} \rangle|^2$$
$$\times \int d^2 k_\perp \delta\left(\varepsilon_g + \varepsilon_m + hh_n + \frac{\hbar^2 k_\perp^2}{2\mu_{nm}^*} - \hbar\omega\right)$$
$$\times \frac{1}{1 + \exp\left[\beta_e\left(\varepsilon_m + \frac{\hbar^2 k_\perp^2}{2m_e^*} - \mu_e\right)\right]}$$
$$\times \frac{1}{1 + \exp\left[-\beta_h\left(-\varepsilon_g - hh_n - \frac{\hbar^2 k_\perp^2}{2M_n^*} - \mu_h\right)\right]} \quad (750)$$

where μ_{nm}^* is the reduced electron–hole mass ($\mu_{nm}^{*-1} = m_e^{*-1} + M_n^{*-1}$).

Let us introduce the chemical potentials η_e, η_h of the electrons and holes measured from ε_m and hh_n, respectively (see Fig. 64):

$$\eta_m^{(e)} = \mu_e - \varepsilon_m \qquad \eta_n^{(h)} = -\mu_h - hh_n - \varepsilon_g \quad (751)$$

They are related to the steady state areal concentrations of electrons (n_m) and holes (p_n) in the ε_m and hh_n subbands by

$$\eta_m^{(e)} = k_B T_e \ln\left[-1 + \exp\left(\frac{\pi \hbar^2 n_m}{k_B T_e m_e^*}\right)\right] \quad (752)$$

$$\eta_n^{(h)} = k_B T_h \ln\left[-1 + \exp\left(\frac{\pi \hbar^2 p_n}{k_B T_h M_n^*}\right)\right] \quad (753)$$

The total steady state concentrations n, p of the electrons and holes are, respectively,

$$n = \sum_m n_m \qquad p = \sum_n p_n \quad (754)$$

Let us denote by y the excess emitted photon energy over $\varepsilon_g + \varepsilon_m + hh_n$:

$$\hbar\omega = \varepsilon_g + \varepsilon_m + hh_n + y \quad (755)$$

Then,

$$\ell_{m \to n}(y) \propto \frac{1}{1 + \exp\left[\beta_e\left(y\frac{\mu_{nm}}{m_e^*} - \eta_n^{(e)}\right)\right]}$$
$$\times \frac{1}{1 + \exp\left[\beta_h\left(y\frac{\mu_{nm}}{M_n^*} - \eta_n^{(h)}\right)\right]} Y(y) \quad (756)$$

It is interesting to notice that this lineshape is obtained analytically whatever the degeneracies of the electron and hole gases (namely, whatever $\beta_e, \beta_h, \mu_e, \mu_h$). This is a particular feature of two-dimensional systems, which display a constant density of states if the dispersion relations are quadratic in $\vec{k}_\perp$.

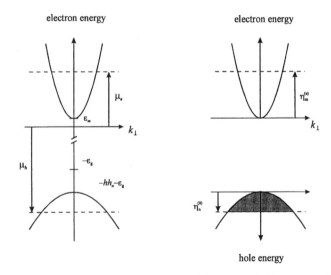

Fig. 64. Definition of the reduced electron ($\eta_m^{(e)}$) or hole ($\eta_n^{(h)}$) Fermi levels in terms of the conduction (μ_e) and valence (μ_h) Fermi levels.

Several limiting cases can be considered and two are discussed below.

(i) Nondegenerate electrons and holes $\left(\eta_n^{(e)} < 0, \eta_n^{(h)} < 0\right)$: Such distributions are obtained at low concentrations of injected carriers and/or high temperatures,

$$\ell(y) \propto Y(y) \exp\left(-\frac{y}{k_B T^*}\right) \qquad (757)$$

with

$$\frac{1}{k_B T^*} = \frac{1}{k_B T_e} \frac{M_n^*}{M_n^* + m_e^*} + \frac{1}{k_B T_h} \frac{m_e^*}{M_n^* + m_e^*} \qquad (758)$$

The steplike onset reflects the quasi-two-dimensional nature of the carrier motions. In practice it is rounded off by band tailing, damping, etc. However, the characteristic feature is the exponential decay at large y (i.e., ω), which may allow T^* to be deduced from the luminescence spectrum. In addition, electrons and holes are often assumed to be in thermal equilibrium with each other ($T_e = T_h = T^*$), which is likely for delocalized carriers. When $k_B T^*$ becomes comparable to the energy separation between two consecutive subbands of a given band, the photoluminescence spectrum displays several lines. For instance in GaAs–GaAlAs quantum wells with a GaAs layer thickness $\geq 100\,\text{Å}$, one observes at low carrier injection and room temperature two luminescence lines associated with the $\varepsilon_1 \rightarrow hh_1$ and $\varepsilon_1 \rightarrow lh_1$ recombinations, respectively. By decreasing T^*, the $\varepsilon_1 \rightarrow lh_1$ feature disappears.

(ii) Degenerate electrons and holes $\left(\beta_e \eta_m^{(e)} > 3, \left(\beta_h \eta_n^{(h)}\right) > 3\right)$: This situation corresponds to high injected carrier concentrations and/or low temperatures. The lineshape is again very simple:

$$\ell(y) \propto Y(y) Y\left(\eta_m^{(e)} - \frac{\mu_{nm}}{m_e^*} y\right) Y\left(\eta_n^{(h)} - \frac{\mu_{nm}}{M_n^*} y\right) \qquad (759)$$

The photoluminescence signal versus ω is rectangular but the edges are actually smoothed by the damping (low energy side) and finite temperature effects (high energy side). The high energy cutoff corresponds to $y_{\max}$ where

$$y_{\max} = \text{Inf}\left[\frac{m_e^*}{\mu_{nm}} \eta_m^{(e)}, \frac{M_n^*}{\mu_{nm}} \eta_n^{(h)}\right] \qquad (760)$$

The previous limiting cases are illustrated in Figures 65 and 66, where plots of the photoluminescence lineshape versus energy are calculated for band-to-band emission in a GaAs–Ga$_{0.67}$Al$_{0.33}$As quantum well (thickness 55 Å, at different temperatures and for different concentrations of injected carriers. Only $\varepsilon_1 \rightarrow hh_1$ and $\varepsilon_1 \rightarrow lh_1$ recombinations have been considered and equal carrier concentrations $(n = p)$ and temperatures $\left(T_e = T_h\right)$ have been assumed in the calculations.

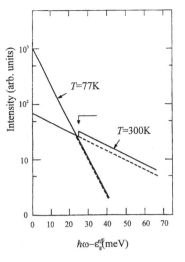

Fig. 65. Calculated band-to-band recombination lineshape in a GaAs–Ga$_{0.67}$Al$_{0.33}$As quantum well for nondegenerate populations of electrons and holes. At $T = 77$ K only the ε_1–hh_1 recombination is significant while at $T = 300$ K the thermal population of the hh_1 subband is sufficient enough to allow the existence of a sizeable ε_1–lh_1 recombination. $\varepsilon_g^{\text{eff}}$ denotes the effective bandgap of the quantum well, viz. $\varepsilon_g(\text{GaAs}) + \varepsilon_1 + hh_1$. For $\hbar\omega < \varepsilon_g^{\text{eff}}$ there is no band-to-band emission. $lh_1 - hh_1 = 25$ meV, $n = p = 2 \times 10^{10}$ cm^{-2}.

Band-to-band recombination takes place when excitonic effects can be discarded. This may be the case in undoped GaAs quantum wells at room temperature (see the discussion in Section 8.3.2.2) but the situation is controversial [279, 280]. In Figure 67 an example in which the radiative recombination does not involve excitons, because their binding energies are insignificant, is shown. In InAs–GaSb type II superlattices at room temperature [281] the spatial separation between the electrons and holes has weakened the excitonic binding so much that the excitons, if any, are ionized into free electron–hole pairs. The theoretical lineshape in these "true" superlat-

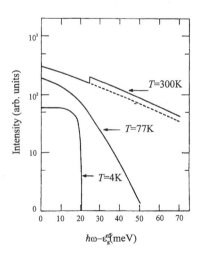

Fig. 66. Same as in Figure 65 except that the recombining populations have much larger densities. At high temperature the electrons and holes approximately follow a Boltzmann distribution while at low temperature the electron and hole gases are degenerate, which affects the luminescence lineshape. For $\hbar\omega < \varepsilon_g^{\text{eff}}$ there is no band-to-band emission. $n = p = 5 \times 10^{11}$ cm^{-2}.

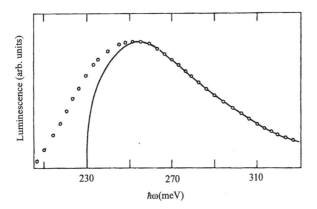

Fig. 67. Calculated and measured band-to-band recombination in an InAs–GaSb superlattice. Solid line: theory. Open circles: data in [35]; 30–50 Å, $T_{\mathrm{eff}} = 370$ K, and $\varepsilon_g = 230$ meV.

tices is modified with respect to the formula given in Eq. (756) due to the subband dispersions along the growth axis and the peculiar selection rules prevailing in type II superlattices (see [282]): one of the two van Hove singularities of the joint density of states is suppressed. From Figure 67 it can be seen that the high energy part of the spectrum is well reproduced by the calculations whereas the low energy side cannot be interpreted by a model which assumes an ideal superlattice. Recombination that involves impurities or defects has to be invoked to account for the low energy side of the recombination spectrum.

Notice that the effective carrier temperature is significantly higher than the lattice temperature. This hot carrier luminescence may have arisen due to the large kinetic energies supplied to the injected carriers by the exciting laser light which is well in excess of the effective bandgap of the superlattice.

A second example about the radiative recombination of n-type modulation-doped GaAs quantum wells is presented in [283]. The large electron concentration effectively decreases the excitonic binding in these structures and consequently band-to-band emission predominates. The luminescence band contains three lines. The two high energy lines have been interpreted in terms of $\varepsilon_1 \to h_1$ and $\varepsilon_2 \to h_1$ recombinations. To account for the spectral position of the lines, a "renormalized" GaAs bandgap had to be invoked, the shrinkage of this gap being equal to 20 meV.

It is worth noticing that it is possible to observe the luminescence spectra calculated in this subsection (and later on) only if the absorption of the emitted photons by the sample can be neglected. If $\ell_0(\omega)$ is the luminescence spectrum originating from a point M at a distance z from the sample surface, the spectrum effectively observed corresponds to photons that traveled through the sample to reach the surface and is given by

$$\ell(\omega) = \ell_0(\omega)[1 - R]\exp[-z\alpha(\omega)] \tag{761}$$

where $\alpha(\omega)$ and R are the absorption coefficient and the reflectivity of the considered sample, respectively. It is sensible to assume that the spontaneous emission is homogeneous,

which means that $\ell_0(\omega)$ is independent of z, so that the apparent luminescence spectrum is given by

$$\ell(\omega) = (1 - R)\ell_0(\omega)\frac{[1 - \exp(-h\alpha(\omega))]}{h\alpha(\omega)} \tag{762}$$

where h is the distance that a photon should travel in the sample. The luminescence spectrum is thus modified by photon reabsorption, the low energy side being favored with respect to the high energy one (see Fig. 68). For photons which propagate along the growth axis of the heterostructure this effect is weak and for single quantum wells negligible, since $h \leq 1\mu$m and $\alpha(\omega) \approx 60$ cm^{-1}. However, the situation may be different for photons that propagate in the layer planes because, in this case, h is of the order of several millimeters.

8.3.2.2. Excitonic Recombination

The excitonic recombination fulfills the same selection rules as the excitonic absorption; i.e., $\vec{k}_\perp = 0$ and $n + m$ even if the heterostructure potential is symmetric in z (see Section 8.2.2). In addition, only the ns excitons can radiatively recombine. This means that the excitonic luminescence emitted below the $\varepsilon_m - hh_n$ band edge should consist of monochromatic lines. Most likely, a single line caused by the $1s$ exciton attached to the $\varepsilon_1 - hh_1$ bandgap will be observed, owing to the fast relaxation of exciton states toward the ground ones. There have already had difficulties to interpret the low energy side of the band-to-band recombination line of noninteracting electrons, all luminescence signals below the $\varepsilon_1 - hh_1$ edge arising from defects.

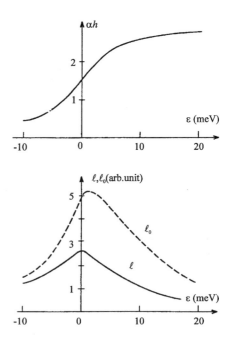

Fig. 68. Calculated reabsorption effects on a photoluminescence line. Upper figure: absorption coefficient α times sample thickness versus energy (measured from the effective bandgap). Lower figure: band-to-band luminescence intensities without reabsorption effect (ℓ_0) and including reabsorption effects (ℓ). In both figures a phenomenological damping coefficient of $\Gamma = 5.1$ meV and $k_B T = 10$ meV have been assumed.

However, the high energy side at least could be successfully described by models which neglect defects. In the case of excitonic recombination, a proper theory should take into account the broadening effects from the very beginning. The broadening effects are twofold. First they provide bound exciton states below $\varepsilon_g + \varepsilon_1 + hh_1 - R^*$ and second they relax the $\vec{k}_\perp = 0$ selection rules, allowing a finite luminescence above $\varepsilon_g + \varepsilon_1 + hh_1 - R^*$ to be predicted. At low temperatures the broadening mechanisms are either the exciton–defect interactions or the exciton–acoustical phonon interactions. Careful studies should be able to discriminate between both kinds of effects.

The second difficulty with excitons is the hybrid statistics that they obey. Diluted excitons can be treated as bosons to a first approximation, essentially because they arise from the pairing of two fermions, the electron and the hole. Excitons retain an increasing fermion nature when their density increases to the point where the concept of bound electron–hole pairs fades away to be replaced by that of an electron hole fluid, i.e., interacting fermions systems.

These two difficulties (relaxation of the $\vec{k}_\perp$ selection rule and the nature of the exciton statistics) have been the subject of a considerable body of literature in bulk materials (see, e.g., [277, 284]). They have not been studied much in heterostructures, from both the experimental and the theoretical points of view. Therefore, a presentation of some experimental results will be considered.

One of the dominant features of the optical properties of GaAs–GaAlAs quantum wells is the strong intensity of the "free" exciton luminescence at low temperatures with respect to the intensity of the lines involving impurities. This is the opposite of what is usually observed in bulk GaAs. Possibly, this may be related to the very small effective volume in which the carriers and the electromagnetic waves interact in the case of quantum wells. Once created, the photon has fewer chances of being reabsorbed to create another exciton than in bulk materials, where the effective volume of interaction is the whole crystal.

The main arguments used to assign the photoluminescence line to excitonic recombination in GaAs quantum wells is its energy position and the nature of the recombining species, as supported by the spin orientation measurements [285]. In "good" samples, the maximum of the photoluminescence line is very close to (and sometimes coincides with) the maximum of the $hh_1 \rightarrow \varepsilon_1$ absorption line (or photoluminescence excitation line). In such a case, the Stokes shift s, i.e., the energy separation between these two lines (luminescence and absorption), is equal to zero, but in other samples s can be as large as several meV's. The existence of such a shift seems to be independent of the residual doping level of the GaAs (usually p- and n-type in MBE and MOCVD grown materials, respectively). Thus it seems reasonable to conclude that the observed excitons are not bound to extrinsic defects (acceptors or donors). A model of exciton trapping on intrinsic interface defects [286] has been proposed, which correlates s to the sizes of the defects (extension along the z-axis and in

the layer plane). This model predicts that beyond $L \approx 100$ Å in GaAs–Ga$_{0.47}$Al$_{0.53}$As quantum wells, the trapped exciton binding energy becomes smaller than 1 meV. Thus, in this model the Stokes shift could only be observed in thin quantum wells. The luminescence line would be associated with trapped excitons with a low density of states ($\leq 10^{10}$ cm^{-2}), as compared to that of the delocalized excitons ($\approx$ a few times 10^{11} cm^{-2}), while the absorption, which is essentially sensitive to the large density of states regions of the energy spectrum, would exhibit features due to the delocalized excitons. The thermally activated detrapping of excitons has in fact been observed [287]. In should be noted that the previous model was designed for defects whose in plane dimensions are not too large (≤ 500 Å in GaAs–GaAlAs quantum well). It is clear that the excitonic photoluminescence in quantum wells with defects, which extend over 1000 Å or more in the layer plane, is better described by models which consider free excitons moving in "microquantum wells" whose areas are equal to the defect areas and whose thickness is equal to the local quantum well thickness in the defect. In other words, the criterion for deciding which model is more appropriate is to count the number $\aleph$ of bound states (for the exciton center of mass) that the defect supports. For $\aleph \leq 5$ the motion of the exciton center of mass is size-quantized and one recovers a photoluminescence due to excitons bound to these defects. For $\aleph \geq 5$, microquantum well models are preferable and if the diffusion of excitons from one large defect to another is difficult, structures in the excitonic photoluminescence should be observed. These structures appear at the local exciton energies in these various islands. Typically, with one monolayer fluctuations of the quantum thickness L, the photoluminescence shows three peaks due to the excitons which move in microquantum wells of thickness $L, L \pm a$, where $a = 2.83$Å is the thickness of a GaAs monolayer.

Such structures have been observed [288] in high quality MBE grown GaAs–GaAlAs multiple quantum wells and superlattices. It is now admitted that the depth of interface fluctuations in GaAs–GaAlAs quantum wells grown by MBE can be reduced to one monolayer. On the other hand, the in plane extensions of the interface defects depend significantly on the growth conditions.

Thus, a variety of experimental results concerning the Stokes shift and the shape of the excitonic photoluminescence line may and do occur. A series of sophisticated optical measurements (transient gratings, Rayleigh scattering, hole burning, etc.) have been undertaken by Sturge et al. [289, 290] to ascertain the nature of exciton states in quantum wells. The aim is to discover whether quasi-two-dimensional excitons can become localized by a weak disorder, the localization being the result of constructive interferences of the exciton wavefunctions by randomly located defects. These experiments tend to show that, at low temperature, excitons are localized rather high in energy and even that a large part of the absorption spectrum arises from these localized excitons.

One currently characterizes the quality of quantum well structures by the width δ at half-maximum of this line. When

δ is smaller than some meV's the sample is claimed to be good. However, it is worth noticing that quantum wells that exhibit broad lines display a very intense luminescence, showing the arbitrary nature of the criterion. Empirically the width δ and the Stokes shift are correlated: the wider the luminescence, the larger the shift.

A model of the width γ of the luminescence excitation spectrum has been proposed by Weisbuch et al. [275] If the local quantum well thickness is $L + \delta L$ inside an interface defect, the exciton absorption occurs at an energy given by

$$\hbar\omega(L + \delta L) = \varepsilon_1(L + \delta L) + hh_1(L + \delta L) - \varepsilon_b(L + \delta L)$$
$$+ \varepsilon_g \qquad (763)$$

Thus the linewidth γ is given by

$$\gamma \approx \delta L \left[\frac{dE_1}{dL} + \frac{dhh_1}{dL} \right] \qquad (764)$$

since the variation of the exciton binding is negligible if $\delta L \ll L$. As the confinement energies vary like L^{-2}, γ should vary like L^{-3}. This behavior has been observed in GaAs–GaAlAs multiple quantum wells (see Fig. 69). Recently, Singh et al. [291] have proposed a model of the photoluminescence linewidth δ. These authors have correlated δ to the distribution of fluctuations in the well thickness, neglecting, however, the effect of carrier relaxation toward lower energy states.

The nature of the radiative recombination at room temperature in GaAs quantum wells is rather controversial. Taking into account the dissociation of the excitons, it seems unlikely that this recombination is entirely due to excitons. However Dawson et al. [280] and Bimberg et al. [279] have interpreted their steady state photoluminescence and time-resolved cathodoluminescence lines emitted by the GaAs well and the GaAs substrate in terms of excitonic recombination.

Investigation of the time dependence of the photoluminescence of GaAs–GaAlAs quantum well gives results differing

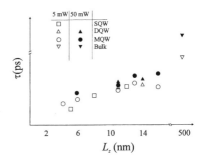

Fig. 70. The photoluminescence decay times τ (at $1/e$ of the maximum) of several GaAs–Ga$_{1-x}$Al$_x$As heterostructures are plotted versus the GaAs layer thickness L_z at $T = 10$ K. SQW, DQW, and MQW stand for single, double, and multiple quantum wells, respectively.

significantly from one group to the other. The simple model described in Eqs. (734)–(738) would lead us to interpret these differences in terms of sample-dependent nonradiative lifetimes. Nonetheless, it seems agreed that the characteristic time of the exciton luminescence decay at low temperature shortens when the quantum well thickness L is decreased (see, e.g., [292, 293]). If one identifies the decay time with the radiative lifetime τ_r and assumes the recombination to be excitonic, the lifetime reduction can be understood from the squeezing of the in plane exciton Bohr radius a^*_{B2D} with decreasing L. Indeed, the inverse of the radiative lifetime τ_r^{-1}, which is proportional to the transition probability per unit time that a photon is emitted, contains (as in the case of excitonic absorption) an enhancement factor $|\beta_{1s}(0)|^2$ that varies (see Eq. (722)) like $a^{*-2}_{B2d}(L)$. In the ideal situation where valence and conduction barriers are infinitely high and where all valence band mixings are neglected, the in plane extension a^*_{B2D} decreases from a^*_B to $a^*_B/2$ when L decreases from infinity to zero. Thus, the excitonic luminescence lifetime should decrease by a factor of four between these two limits. It so happens that the photoluminescence decay time decreases approximately by a factor of four from bulk GaAs to GaAs quantum well with $L = 40$ Å (see Fig. 70). This excellent agreement between theory and experiment is probably fortuitous.

9. ELECTRON RAMAN SCATTERING IN Q2D SYSTEMS

Raman scattering is a powerful tool that permits the investigation of several physical properties of the Q2D systems. In particular, the electronic structure of the semiconductors and nanostructures can be studied considering several polarizations for the incident and emitted radiation. In connection with this type of experiment, the calculations of the differential cross-sections for Raman scattering remain a fundamental and interesting source in order to obtain a better understanding of the Q2D characterized by their mesoscopic dimensions. Between the different processes of Raman scattering involved in this type of investigation, electron Raman scattering is a

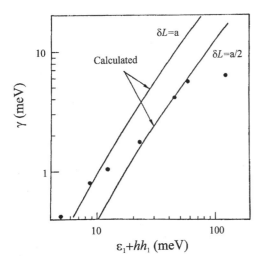

Fig. 69. Calculated and measured excitation spectrum linewidth as a function of the confinement energy (viz. $\varepsilon_1 + hh_1$) in GaAs–Ga$_{1-x}$Al$_x$As multiple quantum wells.

useful technique that provides direct information on the structure of energy bands and optical properties of the investigated systems.

Interband electron Raman scattering processes can be qualitatively described in the following way [294]: an external photon is absorbed from the incident radiation field and after that a virtual electron–hole pair is created in an intermediate crystal state by means of an electron interband transition involving the crystal valence and conduction bands. The electron (hole) in the conduction (valence) band is subject to a second intraband transition with the emission of a secondary radiation photon. Therefore, in the final state there is a real electron–hole pair in the crystal and a long wavelength photon of the secondary radiation field. The effects of externally applied fields on this kind of processes were investigated in [295, 296]. Particularly the case of an external magnetic field with the electron (or hole) transitions between Landau subbands was considered in [297].

The fundamental aim of the present section is to analyze the main features of electron Raman scattering emission and excitation spectra in semiconductor Q2Ds, where the electron (hole) system essentially bears a low-dimensional character of the Q2D. Due to electron confinement the conduction (valence) band is split in a subband system and transitions between them determine the electron Raman scattering processes. The general relations needed for calculating the electron Raman scattering differential cross-section in the "two-step" model for the light scattering processes are presented. Only two bands are considered (a valence and a conduction band), which are split in an infinite subband system under the assumption of a Q2D nanostructure with barriers of infinite height. The effective mass and envelope function approximations are also assumed and the emission and excitation spectra are reported.

9.1. Model and Applied Theory

The electron Raman scattering cross-section can be written in the form [298]

$$\frac{d^2\sigma}{d\omega_s\,d\Omega} = \frac{V^2\omega_s^2\eta(\omega_s)}{8\pi^3C^4\eta(\omega_l)}W(\omega_s,\vec{e}_s) \tag{765}$$

where c is the light velocity in vacuum, $\eta(\omega)$ is the material refraction index, ω_s is the secondary radiation frequency, $\vec{e}_s$ is the (unit) polarization vector for the secondary radiation field, and $W(\omega_s,\vec{e}_s)$ is the transition rate for the emission of secondary radiation (with frequency ω_s and polarization $\vec{e}_s$) in the solid angle $d\Omega$, when there is in the volume V a photon (with frequency ω_l) from the incident radiation field. $W(\omega_s,\vec{e}_s)$ is calculated assuming a refraction index equal to unity (because the material refraction index is explicitly included in the prefactor (in Eq. (765)) by means of Fermi's Golden Rule,

$$W(\omega_s,\vec{e}_s) = \frac{2\pi}{\hbar}\sum_f |M_1 + M_2|^2 \delta(\varepsilon_f - \varepsilon_l) \tag{766}$$

where

$$M_j = \sum_a \frac{\langle f|\widehat{H}_{js}|a\rangle\langle a|\widehat{H}_{jl}|i\rangle}{\varepsilon_l - \varepsilon_a + i\Gamma_a} + \sum_b \frac{\langle f|\widehat{H}_{jl}|b\rangle b|\widehat{H}_{js}|i\rangle}{\varepsilon_l - \varepsilon_b + i\Gamma_b} \tag{767}$$

and $j = 1, 2$ denotes electron or hole contributions, respectively. ε_i and ε_f are the energies of the initial $|i\rangle$ and final $|f\rangle$ states of the system, while ε_a and ε_b correspond to the system intermediate state energies. Γ_a and Γ_b are the corresponding lifetime broadenings. $\widehat{H}_{l(s)}$ is the interaction of the incident (emitted) radiation field with the crystal, which was chosen previously in the form

$$\widehat{H}_{jl(s)} = \frac{|e|}{m}\sqrt{\frac{2\pi\hbar}{V\omega}}\,\vec{e}_{l(s)}\cdot\hat{p} \tag{768}$$

where e and $m = m_0(m^*)$ are the charge and free (effective) mass for the incident (emitted) radiation.

In Figure 71 Feynmann diagrams describing the considered processes are shown. Diagrams (a) and (c) describe the emission of a photon (with frequency ω_s) by a conduction electron and can be directly related to the two terms on the right side of Eq. (767). Diagrams (b) and (d) can be similarly interpreted but correspond to a hole in the valence band. In Figure 72 band diagrams are shown; Figure 72a and b is equivalent to Figure 71a and b, respectively.

In Eq. (767) the intermediates states represent the electron–hole pair in a virtual state, while the intermediates states $|b\rangle$ are related to the interference state, which can be neglected for semiconductors with a large enough energy gap (for example, the GaAs, CdTe, etc.).

The wavefunctions and energy levels of the electrons and holes system in the Q2D are described by using the Schrödinger equation in the envelope function approximation. This equation is limited to the case in which the minimum of the conduction band and the maximum of the valence band are located in the point Γ (case of interest). In these systems the width or the radius of confinement is lesser than the Bohrs exciton radius, the confinement energy of the electron–hole

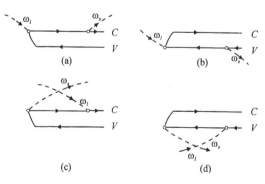

Fig. 71. Diagrams contributing to electron Raman scattering for interband intersubband transitions. The final state is characterized by an electron in the subband N_1 with momentum $\hbar\vec{k}_e$, a hole in the subband N_2 with momentum $\hbar\vec{k}_h$, and a secondary radiation photon of frequency ω_s. Diagrams (a) and (b) show contributions from the electron or the hole, respectively. Diagrams (c) and (d) show interference processes.

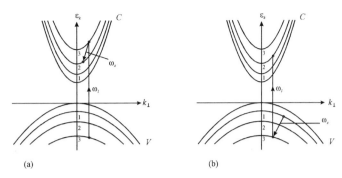

Fig. 72. Band diagrams contributing to electron Raman scattering. They are equivalent to the diagrams of Figure 71a and 71b, respectively.

pair being dominant over the coulombic interaction energy. Therefore, in a first approach it is considered that the electron and hole form an uncorrelated pair. In this model it will be assumed that in semiconductor Q2D systems in the initial state, the conduction band is totally empty and the valence band is completely full, so that the presence of impurities does not exist. However, in semiconductors with the structure of the of zinc blende, the heavy and light hole bands are degenerated in the point Γ. Nevertheless, in Q2Ds the mixture of states of light and heavy holes diminishes when the confinement radius or width decreases, coinciding with the calculations made by using the multiband formalism; this permits one, in first approach, to consider that the light and heavy hole can be considered as decoupled.

It is also assumed that these are parabolic conduction and valence bands, of isotropic nature. The parabolicity of bands is fulfilled around the Γ point. Due to degeneration, the valence band does not fulfill the isotropic feature. Nevertheless, good results can be obtained if the effective mass on several directions (cubic symmetry crystals) is averaged.

In theoretical investigation about electron Raman scattering, it must be taken into account that:

1. The electron–photon Hamiltonian is given by Eq. (768) and the wavefunctions corresponding to the $|a\rangle$ and $|i\rangle$ states are given by (in the envelope function approximation) $\psi(\vec{r}) = \chi(\vec{r})u(\vec{r})$.
2. The crystal has a volume $V = N_1\Omega_0$, N_1 and Ω_0 being the number of cells and the volume of the unitary cell, respectively.
3. The functions $u(\vec{r})(\chi(\vec{r}))$ vary quickly (slowly) in the volume.
4. The Bloch functions between different bands are orthogonal.
5. The matrix element that connects the conduction and valence bands in the center of the Brillouin zone, $\vec{k} = 0$, is given by $\vec{P}_{cv}(0) = (1/\Omega_0)\int u_1^*(\vec{r})\hat{p}u_2(\vec{r})dV$.
6. The variation of the functions in the whole crystal volume allows one to write the matrix element characterizing the electron–incident photon interaction

in the following form:

$$\langle a|\widehat{H}_{1l}|i\rangle = \frac{|e|}{m_0}\sqrt{\frac{2\pi\hbar}{V\omega_l}}\,\vec{e}_l \cdot \vec{P}_{cv}(0)\int_V \vec{\chi}_1^{'*}(\vec{r})\chi_2(\vec{r})dV \tag{769}$$

Physically, a quantum transition from the valence band to the conduction band will occur. The possible selection rules will depend on the explicit form of the wavefunctions that are involved in the integral. For the holes, an expression similar to Eq. (768) is obtained, but the indexes in the terms under the integral sign must be exchanged.

Considering the previous 1, 2, and 3 conditions and knowing that the integral of an odd operator between states of the same parity is equal to zero, and that the Bloch functions are normalized in the same band, the matrix element that characterizes the electron-emitted photon interaction can be written as

$$\langle f|\widehat{H}_{js}|a\rangle = (-1)^{j+1}\frac{|e|}{m_j^*}\sqrt{\frac{2\pi\hbar}{V\omega_s}}\,\vec{e}_s \cdot \int_V \vec{\chi}_j^*(\vec{r})\hat{p}\chi_j(\vec{r})dV \tag{770}$$

Physically, a quantum transition from the intermediate state $|a\rangle$ to the final state $|f\rangle$ is obtained. The selection rules will depend again on the form of the wavefunctions.

The particular type of Q2D nanostructure requires a specific knowledge of wavefunctions and energy in the different states.

9.2. Electron Raman Scattering in a Quantum Well

The electron wavefunction in the effective mass approximation and infinite potential barrier can be written as

$$\psi_{\vec{k}_{1\perp}, N_1}(\vec{r}) = \sqrt{\frac{2}{V}}\exp(i\vec{k}_{1\perp}\cdot\vec{r}_\perp)\sin\left(\frac{N_1\pi}{L}z\right)u_c(\vec{r}) \tag{771}$$

$$\psi_{\vec{k}_{2\perp}, N_2}(\vec{r}) = \sqrt{\frac{2}{V}}\exp(i\vec{k}_{2\perp}\cdot\vec{r}_\perp)\sin\left(\frac{N_2\pi}{L}z\right)u_v(\vec{r}) \tag{772}$$

The electron energies are given by

$$\varepsilon_c(\vec{k}_{1\perp}, N_1) = \frac{\hbar^2 k_{1\perp}^2}{2m_1^*} + \frac{\pi^2\hbar^2}{2m_1^2 L^2}N_1^2 \quad N_1 = 1, 2, \ldots \tag{773}$$

$$\varepsilon_v(\vec{k}_{2\perp}, N_2) = -\frac{\hbar^2 k_{2\perp}^2}{2m_2^*} - \frac{\pi^2\hbar^2}{2m_2^2 L^2}N_2^2 - \varepsilon_g \quad N_2 = 1, 2, \ldots \tag{774}$$

where "c" ("v") denotes conduction (valence) band and ε_g is the gap energy for the bulk semiconductors.

The initial state energy is

$$\varepsilon_i = \hbar\omega_l \tag{775}$$

The final state is characterized by an electron with wavevector $\vec{k}_{1\perp}$ occupying the N_1 subband of the original conduction band, a hole with momentum $\vec{k}_{2\perp}$ in the N_2 subband of the original valence band, and a secondary radiation photon with frequency ω_s.

The final energy of the system is thus given by

$$\varepsilon_f = \hbar\omega_s + \varepsilon_g + \frac{\hbar^2 k_{1\perp}^2}{2m_1^*} + \frac{\hbar^2 k_{2\perp}^2}{2m_2^*} + \frac{\pi^2\hbar^2}{2m_2^*L^2}N_2^2 + \frac{\pi^2\hbar^2}{2m_1^*L^2}N_1^2 \tag{776}$$

According to diagrams of Figure 71, the intermediate state electron energies ε_a and ε_b are

$$\varepsilon_a = \frac{\hbar^2 k_\perp^2}{2\mu^*} + \varepsilon_g + \frac{\hbar^2\pi^2}{2L^2}\left(\frac{N_1^2}{m_1^*} + \frac{N_2^2}{m_2^*}\right) \tag{777}$$

$$\varepsilon_b = \hbar\omega_l + \hbar\omega_s + \frac{\hbar^2 k_\perp^2}{2\mu^*} + \varepsilon_g + \frac{\hbar^2\pi^2}{2L^2}\left(\frac{N_1^2}{m_1^*} + \frac{N_2^2}{m_2^*}\right) \tag{778}$$

In Eqs. (777) and (778) the momentum conservation law and $\vec{k}_{1\perp} = -\vec{k}_{2\perp} = \vec{k}_\perp$ were explicitly assumed.

The electron–hole pair does not change its total momentum during the absorption or emission of a photon (photon momentum is neglected).

From Eq. (775) to (778) and using energy and momentum conservation laws, the denominators in Eq. (767) can be evaluated:

$$\varepsilon_i - \varepsilon_a = \bar{\omega}_s + \frac{\pi^2\hbar^2}{2m_1^*L^2}(N_1^2 - N_2'^2) \tag{779}$$

$$\varepsilon_i - \varepsilon_b = -\hbar\omega_l + \frac{\pi^2\hbar^2}{2m_1^*L^2}(N_1^2 - N_2'^2) \tag{780}$$

From the latter results it can be deduced that for semiconductors with a large enough energy gap ε_g (e.g., for GaAs) the contribution from the interference diagram (c) of Figure 71 can be neglected if it is compared with the contribution of diagram (a) [299].

Similar expressions can be written for the hole intermediate state energies, and analogous conclusions can be made: diagram (d) in Figure 71 is negligible compared with diagram (b).

Thus, for the determination of the cross-section, just the contribution of the first term in the right side of Eq. (767) must be considered during the calculation of M_j.

9.2.1. Differential Cross-Section

From Eqs. (768) and (771), (772) it can be obtained that

$$\langle f|\widehat{H}_{js}|a\rangle = \frac{|e|\hbar}{m_j^*}\sqrt{\frac{2\pi\hbar}{V\omega_s}}\left\{\vec{e}_s\cdot\vec{k}_{\perp e}\delta_{N_j'N_j} + (-1)^{j+1}\frac{2ie_{sz}}{L}\right.$$
$$\left.\times\frac{N_j'N_j}{N_j^2 - N_j'^2}(1-\delta_{N_j'N_j})[(-1)^{N_j'+N_j}-1]\right\} \quad j=1,2 \tag{781}$$

$$\langle f|\widehat{H}_{jl}|i\rangle = \frac{|e|}{m_0}\sqrt{\frac{2\pi\hbar}{V\omega_l}}\vec{e}_l\vec{P}_{cv}(0)\begin{cases}\delta_{N_1'N_2} & \text{for } j=1 \\ \delta_{N_1N_2'} & \text{for } j=1\end{cases} \tag{782}$$

where m_0 is the bare electron mass and $\vec{P}_{cv}(0)$ the interband momentum matrix element (evaluated at $\vec{k}=0$).

The matrix elements from Eq. (766) are given by

$$M_1 + M_2 = \frac{2\pi e^2\hbar^2}{m_0 V}\sqrt{\frac{1}{\omega_l\omega_s}}(\vec{e}_l\cdot\vec{P}_{cv}(0))\left\{\frac{\vec{e}_s\cdot\vec{k}_\perp}{\hbar\omega_s}\frac{\delta_{N_1N_2}}{\mu^*}\right.$$
$$+\frac{2ie_{sz}}{L}\frac{N_2N_1}{N_1^2-N_2^2}(1-\delta_{N_1N_2})[(-1)^{N_1+N_2}-1]$$
$$\times\left[\frac{m_1^{*-1}}{\hbar\omega_s + \frac{\pi^2\hbar^2}{2m_1^*L^2}(N_1^2-N_2^2)}\right.$$
$$\left.\left.+\frac{m_2^{*-1}}{\hbar\omega_s - \frac{\pi^2\hbar^2}{2m_2^*L^2}(N_1^2-N_2^2)}\right]\right\} \tag{783}$$

After substitution of Eq. (783) in (765) and (766), it is obtained that

$$\frac{d^2\sigma}{d\Omega\,d\omega_s} = \frac{9}{8}\sigma_0\left\{\sum_{N_1}\left[\frac{\hbar\omega_l-\varepsilon_g}{\varepsilon_0} - \frac{\hbar\omega_s}{\varepsilon_0} - N_1^2\right](\cos^2\theta + 1)\right.$$
$$+\frac{8}{\pi^2}\sin^2\theta\left(\frac{\hbar\omega_s}{\varepsilon_0}\right)^2\sum_{N_1N_2}\frac{N_2^2N_1^2}{(N_1^2-N_2^2)^2}(1-\delta_{N_1N_2})$$
$$\times[(-1)^{N_1+N_2}-1]^2$$
$$\times\left[\frac{1}{(1+\beta^{-1})\frac{\hbar\omega_s}{\varepsilon_0}+(N_1^2-N_2^2)}\right.$$
$$\left.\left.+\frac{1}{(1+\beta)\frac{\hbar\omega_s}{\varepsilon_0}-(N_1^2-N_2^2)}\right]^2\right\} \tag{784}$$

where

$$\varepsilon_0 = \frac{\pi^2\hbar^2}{2\mu^*L^2} \qquad \beta = \frac{m_2^*}{m_1^*} \tag{785}$$

and

$$\sigma_0 = \frac{4\sqrt{2}Ve^4|\vec{e}_l\cdot\vec{P}_{cv}(0)|^2\eta(\omega_s)\mu^{*1/2}\varepsilon_0^{3/2}}{9\pi^2 m_0^2\hbar^4 c^4\eta(\omega_l)\omega_l\omega_s} \tag{786}$$

In Eq. (784), summation is over all polarizations of the secondary radiation field, while θ is the angle between the secondary radiation photon wavevector and the growth direction of the quantum well (the z-axis). Summations over the subband labels (N_1 and N_2 in Eq. (784)) must be done under the following conditions:

$$\hbar\omega_l - \hbar\omega_s - \varepsilon_g - \frac{\pi^2\hbar^2}{2\mu^*L^2}N_1^2 \geq 0 \tag{787}$$

$$\hbar\omega_l - \hbar\omega_s - \varepsilon_g - \frac{\pi^2\hbar^2}{2L^2}\left(\frac{N_1^2}{m_1^*} + \frac{N_2^2}{m_2^*}\right) \geq 0 \tag{788}$$

The electron Raman scattering differential cross-section for a semiconductor quantum well (Eq. (784)), presents singular peaks for the secondary radiation frequency ω_s such that

$$\omega_s = \omega_s^e(N_1; N_2) \quad \text{or} \quad \omega_s = \omega_s^h(N_1; N_2) \tag{789}$$

where

$$\omega_s^e(N_1; N_2) = \frac{\pi^2 \hbar}{2m_1^* L^2}(N_2^2 - N_1^2) \qquad (790)$$

$$\omega_s^h(N_1; N_2) = \frac{\pi^2 \hbar}{2m_2^* L^2}(N_1^2 - N_2^2) \qquad (791)$$

These frequency values correspond to conduction electron intersubband transitions ($\omega_s^e(N_1; N_2)$) or hole intersubband transitions in the valence band ($\omega_s^h(N_1; N_2)$). Intersubband transitions obey a selection rule requiring that $N_1 + N_2$ be odd, otherwise the transition is forbidden. The latter result is in close analogy with the case of electron Raman scattering in bulk semiconductors subject to a magnetic field [297], where singularities in the emission spectra are also observed due to transitions between Landau subbands satisfying the selection rule $N' = N \pm 1$ (with $N, N' = 0, 1, 2, \ldots$).

In the case of a semiconductor quantum well, the frequencies ω_s must lie in the interval

$$0 < \hbar\omega_s \leq \hbar\omega_l - \varepsilon_g - \varepsilon_0 \qquad (792)$$

From Eq. (792), it can be deduced that secondary radiation is emitted only for $\hbar\omega_l > \varepsilon_g + \varepsilon_0$; otherwise the electron Raman scattering cross-section is equal to zero.

Let remark that Eq. (784) approaches the corresponding bulk semiconductor expression [300]

$$\frac{d^2\sigma}{d\Omega\, d\omega_s} = \frac{3}{2}\sigma_0\left(\frac{\hbar\omega_l - \varepsilon_g - \hbar\omega_s}{\varepsilon_0}\right)^{3/2} \qquad (793)$$

when $L \to \infty$. It can also be noticed that the first term in the differential cross-section (Eq. (784)), is free from singularities and related to transitions between valence and conduction subbands with $N_1 = N_2$.

In Figure 73 the electron Raman scattering differential cross-section in relative units as a function of $\hbar\omega_s/\varepsilon_0$, the so-called emission spectrum, has been shown. The masses where chosen for the case of bulk GaAs ($m_1^* = 0.0665m_0$ and

$m_2^* = 0.45m_0$ (heavy hole)). In Figure 73, $\theta = \pi/2$ has been chosen. It can be observed that for $\hbar\omega_l - \varepsilon_g = 10\varepsilon_0$ the differential cross-section displays four singularities at $\hbar\omega_s^h(2, 1)$, $\hbar\omega_s^h(2, 3)$, $\hbar\omega_s^e(1, 2)$, and $\hbar\omega_s^e(3, 2)$, while for $\hbar\omega_l - \varepsilon_g = 6\varepsilon_0$ only two singularities are observed at $\hbar\omega_s^h(2, 1)$ and $\hbar\omega_s^e(1, 2)$. It should be noted that the ω_s values for which singularities are found do not depend on the incident radiation frequency ω_l and only depend on the energy differences between the valence and conduction subbands. In general, for the model of a quantum well with infinite potential barriers the position peaks $\omega_s(N_1, N_2)$ will change with the variations of the quantum well width according to the law $\approx L^{-2}$. For higher energies $\hbar\omega_l$ of the incident radiation photon a larger number of singular peaks shall be observed in the emission spectra as can be deduced from Eqs. (790) and (791). For a fixed value of ω_l, a certain steplike behavior at given values of ω_s can be observed in the emission spectra. The maximum number of steps is determined by Eqs. (787) and (788) when $\omega_s = 0$ is setted and the selection rule $N_1 + N_2 = 2N + 1$ ($N = 1, 2, \ldots$) is used. The position of the steps in the spectra depends on the value chosen for ω_l.

In Figure 74 is shown the differential cross-section as a function of $(\hbar\omega_l - \varepsilon_g)\varepsilon_0^{-1}$ for three values of $\hbar\omega_s/\varepsilon_0$ (the so-called excitation spectrum). As can be observed, for increasing values of $(\hbar\omega_l - \varepsilon_g)\varepsilon_0^{-1}$, new steps will be appearing because new subbands will be accessible to electrons and holes in the

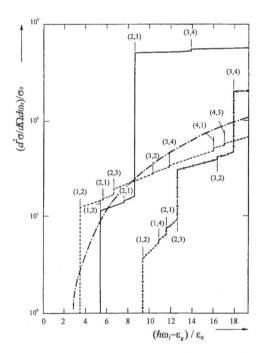

Fig. 74. Electron Raman scattering cross-section (in units of σ_0) as a function of $(\hbar\omega_l - \varepsilon_g)\varepsilon_0^{-1}$ for three values of $\hbar\omega_s/\varepsilon_0$: $\hbar\omega_s = 4\varepsilon_0$ (solid line), $\hbar\omega = 2\varepsilon_0$ (dashed line), and $\hbar\omega_s = 8\varepsilon_0$ (cross-dashed line). Steplike behavior due to transitions between valence and conduction subbands is indicated by (N_1, N_2) (as in Fig. 73). The dash-dotted curve shows the electron Raman scattering cross-section for a bulk semiconductor with $\hbar\omega_s = 2\varepsilon_0$.

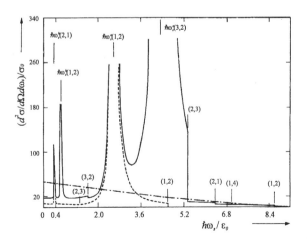

Fig. 73. Electron Raman scattering cross-section (in units of σ_0) as a function of $\hbar\omega_s/\varepsilon_0$ for $\hbar\omega_l - \varepsilon_g = 10\varepsilon_0$ (solid line), $\hbar\omega_l - \varepsilon_g = 6\varepsilon_0$ (dashed line), and $\hbar\omega_l - \varepsilon_g = 10\varepsilon_0$ (dash-dotted line) for a bulk semiconductor.

quantum well. The step positions are given by

$$\hbar\omega_l = \hbar\omega_s + \varepsilon_g + \frac{\pi^2\hbar^2}{2L^2}\left(\frac{N_1^2}{m_1^*} + \frac{N_2^2}{m_2^*}\right) \quad (794)$$

For increasing values of ω_l there will be more terms contributing to the double sum in Eq. (784), which will produce steplike increments in the differential cross-section. A similar result was obtained for the light absorption coefficient in a semiconductor quantum well [262].

Now let consider the first term in Eq. (784). This term weakly depends on ω_l and its increase can be connected with the participation of a higher number of subbands of the valence or conduction type. The latter dependence can be observed as the slope changes in the curves of Figure 75, for $\hbar\omega_l = \hbar\omega_s + \varepsilon_g + (\pi^2\hbar^2/2\mu^*L^2)N_1^2$ and different values of $\hbar\omega_s/\varepsilon_0$.

In order to give numerical estimations of the secondary radiation intensities predicted by these results, the quantum efficiency $(1/V)d\sigma/d\Omega$ for some fixed values of ω_l has been computed. Therefore, $\varepsilon_g = 1.43\,\text{eV}$, $\eta(\omega_l) = 0.899$, $\eta(\omega_s) = 3.46$ [301], and $\hbar\omega_l = \hbar\omega_g = 1.182\varepsilon_0$ have been chosen. By integrating Eq. (784) from $\hbar\omega_s = 0.15\varepsilon_g$ to $\hbar\omega_s = 0.091\varepsilon_g$, with $L = 10$ nm, the result is $(1/V)d\sigma/d\Omega = 2.84 \times 10^{-6}f_{cv}^x\text{sr}^{-1}\text{cm}^{-1}$, where $f_{cv}^x = 2/(m_0\varepsilon_g)|(P_{cv})_x|^2$ is the oscillator strength in the x^a-direction. For GaAs $f_{cv}^x = 6.7599$ [302] and $(1/V)d\sigma/d\omega = 1.92 \times 10^{-5}\text{sr}^{-1}\text{cm}^{-1}$.

Finally, some comments should be done. Electron Raman scattering in semiconductor quantum wells can be used for the determination of the subband structure in real heterostructures of this kind. The fundamental features of the differential cross-section, as described in this subsection, should not change very much in the real case. It can be easily proved that the singular

peaks in the cross-section will be present irrespective of the model used for the subband structure and shall be determined for values of $\hbar\omega_s$ equal to the energy difference between two subbands: $\hbar\omega_s^{e(h)} = \varepsilon_\alpha^{e(h)} - \varepsilon_\beta^{e(h)}$, where $\varepsilon_\alpha^{e(h)} > \varepsilon_\beta^{e(h)}$ are the respective electron (hole) energies in the subbands.

Similarly, it will have a steplike dependence in the differential cross-section for $\hbar\omega_l = \hbar\omega_s + \varepsilon_g + \varepsilon_\alpha + \varepsilon_\beta$.

A complete study of electron Raman scattering in semiconductor quantum well heterostructures should also be concerned with the case when phonon assisted transitions are involved. In this case the intermediate states in Eq. (767) are real and the electron–phonon interaction must be explicitly included in the calculations. This latter point requires independent work. All the results obtained can be extended to the case of a superlattice, where the calculations would be concerned with the miniband structure created by superlattice superperiodicity.

9.3. Resonant Raman Scattering in Quantum Wells in High Magnetic Fields: Fröhlich and Deformation Potential Interaction

A theoretical study of one-phonon resonant Raman scattering in a quantum well in high magnetic fields was performed in [303, 304]. The Raman scattering efficiency is calculated as a function of magnetic field, quantum well thickness, and laser frequency. The basic theory is first developed assuming parabolic masses in the plane perpendicular to the growth direction of the QW. Selection rules for Fröhlich and deformation potential allowed scattering are given, and a compact analytical expression for the Raman scattering efficiency is obtained for infinite barriers. The double-resonance conditions are derived as functions of the magnetic field or well thickness. The heavy hole–light hole valence band admixture has been taken into account through a 4×4 Luttinger Hamiltonian. It was shown that phonons couple the heavy component of a band with the light component of the other via a deformation potential, giving rise to the selection rule $\Delta n = \pm 2$ ($\Delta N = 0$ for the Landau quantum number). The Raman polarizability has been calculated for a 100 Å GaAs/AlAs multiple QW for different magnetic fields as a function of laser energy and magnetic field. For the case of Fröhlich, the phonon confinement is studied in thin QW and a comparison among the different phonon modes is presented: only even phonon modes couple via Fröhlich interaction for $q_\perp = 0$. Comparison with experimental results allows us to classify the leading double resonance as due to transitions between the excitons formed with the first and second heavy hole subbands and the first electron subband with Landau quantum number $N = 1$.

Raman scattering observations have allowed the study of the electronic structure and phonon modes in SL and QW semiconductors [146, 149]. Furthermore, resonant Raman scattering under high magnetic fields permits exploration of details of their conduction and valence bands. A simplified theoretical model, neglecting to a first approximation the valence band degeneracy of III–V compounds, allows understanding of the physics involved in the Raman process [305]

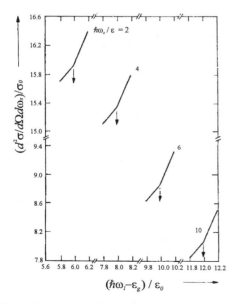

Fig. 75. Electron Raman scattering cross-section (in units of σ_0) as a function of $(\hbar\omega_l - \varepsilon_g)\varepsilon_0^{-1}$ for four values of the parameter $\hbar\omega_s/\varepsilon_0$ and $\theta = \pi 2$. The slope change for the transition $\hbar\omega_l - \varepsilon_g = \hbar\omega_s + 4\omega_0$ is indicated by an arrow.

and a simple expression was obtained for the double-resonance condition. A more realistic approach was introduced by using the theory of Trebin et al. [306] for calculating the correct energy of the Landau levels [307]. In a QW the complication generated by the magnetic field and by the valence band degeneracy increases considerably the complexity of the one-phonon Raman scattering, as can be seen from the few experiments on resonant Raman scattering in magnetic fields in QW reported in the literature [308–311].

In quantum well systems the structure of Raman profiles as a function of the magnetic field is even more complicated than that in the bulk [307], due to the contribution of different subbands to the transitions. The structure obtained by analyzing the Raman intensities as a function of the magnetic field for a fixed wavelength or vice versa is difficult to interpret. Recent experimental results [309, 310] on a 100 Å GaAs/AlAs multiple quantum well (MQW) show that the behavior of one of the Raman peaks as a function of the magnetic field presents a structure which can be assigned to double resonant exciton-band transitions.

9.3.1. Scattering Efficiency

The scattering efficiency per unit crystal length and solid angle for the creation of a phonon of frequency ω_0 (Stokes process) is [312]

$$\frac{dS}{d\Omega} = \frac{\omega_l \omega_s^3 n_s}{c^4 n_l} \frac{\hbar}{2MV_c \omega_0} |\vec{e}_s^* \cdot \overleftrightarrow{R} \cdot \vec{e}_l|^2 [n(\omega_0) + 1] \quad (795)$$

where n_j, ω_j, and $\vec{e}_j (j = s, l)$ are the refractive index, frequency, and polarization vector of the laser (l) and scattered (s) light fields and $\overleftrightarrow{R}$ is the Raman tensor for the optical phonons near the center of the Brillouin zone. $n(\omega_0)$ is the Bose–Einstein phonon occupation factor for phonons of frequency ω_0 in thermal equilibrium at a temperature T, c is the speed of light, V_c is the volume of the unit cell, and M is the reduced mass of the atoms in the unit cell. Using Fermi's Golden Rule, the scattering efficiency can be written in terms of the amplitude probability W_{FI} of the process [313]

$$\frac{dS}{d\Omega} = \frac{\omega_l \omega_s^3 n_l n_s}{(2\pi)^2 c^4} \frac{V}{(\hbar\omega_l)^2} |W_{FI}(\omega_s, \vec{e}_s; \omega_l, \vec{e}_l)|^2 [n(\omega_0) + 1] \quad (796)$$

where $|I\rangle$ and $|F\rangle$ are the initial and final states corresponding to the scattering of a photon of frequency ω_l by a phonon of frequency ω_0, giving a scattered photon of frequency ω_s, and $V = L^2 d$ is the volume of the QW. For a one-phonon process, the Raman efficiency can be obtained in third order of perturbation theory. Considering only the resonant terms, W_{FI} can be written as

$$W_{FI} = \sum_{\mu, \beta} \frac{\langle F|H_{ER}|\mu\rangle \langle \mu|H_{EP}|\beta\rangle \langle \beta|H_{ER}|l\rangle}{(\hbar\omega_s - \varepsilon_\mu + i\Gamma_\mu)(\hbar\omega_l - \varepsilon_\beta + i\Gamma_\beta)} \quad (797)$$

where $|\mu\rangle (|\beta\rangle)$ refers to the intermediate uncorrelated electron–hole pair state, $\varepsilon_\mu (\varepsilon_\beta)$ and $\Gamma_\mu (\Gamma_\beta)$ being the corresponding energy and lifetime broadening. In the framework of second quantization, the electron-radiation (H_{ER}) and electron–phonon (H_{EP}) interaction Hamiltonians are given by

$$H_{ER} = \sum_{\kappa, e} (\hat{a}_{\kappa, e} + \hat{a}_{-\kappa, e}^+) \frac{|e|}{m_0} \sqrt{\frac{2\pi\hbar}{V\omega n^2}} \vec{e}_r \cdot \vec{p} \quad (798)$$

$$H_{EP} = \sum_{\mu, \beta} [S_\mu^\beta(\vec{q}) \hat{d}_\mu^+ \hat{d}_\beta \hat{b}_{\vec{q}}^+] + C.C. \quad (799)$$

$\hat{a}_{\kappa, e}^+ (\hat{a}_{-\kappa, e})$, $\hat{d}_\mu^+ (\hat{d}_\beta)$, and $\hat{b}_{\vec{q}}^+ (\hat{b}_{-\vec{q}})$ are the creation (annihilation) operators for photons, electron–hole pairs, and phonons, respectively, $\vec{e}_r$ represents the polarization of the photon field with wavevector κ, $\vec{q}$ is the phonon wavevector, and $\vec{p}$ is the momentum operator. $S_\mu^\beta(\vec{q})$ is the electron–hole-pair phonon coupling, which for the deformation-potential interaction can be written as [313]

$$S_\mu^\beta(\vec{q}) = \frac{u_0\sqrt{3}}{2a_0} \langle\mu|\vec{D}_e(\vec{r}_e) \exp(-i\vec{q} \cdot \vec{r}_e) - \vec{D}_h(\vec{r}_h)$$
$$\times \exp(-i\vec{q} \cdot \vec{r}_h)|\beta\rangle \quad (800)$$

where u_0 is the zero-point amplitude of the relative sublattice displacement, $u_0 = \sqrt{\hbar V_c / 2VM\omega_0}$, a_0 is the lattice constant, and $\vec{D}_\alpha (\alpha = e, h)$ is the deformation potential as defined by Bir and Pikus [314]. In Eq. (800) the confined phonons are not considered, but extended ones are. In order to introduce the different confined-phonon modes in the quantum well, the deformation potential must be multiplied by a modulation function $\phi_{q_z}(z)$ ($q_z = p\pi/d$, with p an integer) that takes into account the phonon confinement [113]. In this case the scattering efficiency will be different for the different confined-phonon modes, decreasing rapidly with increasing p. This fact must be taken into account, for thin quantum wells ($d < 50$ Å) or short period superlattices. The Raman efficiency of semiconductor structures can be calculated through Eq. (800), provided that d is assumed so large that a number of phonons up to a large value of p cannot be resolved.

For the Fröhlich electron–hole pair phonon interaction, the coupling constant $S_\mu^\beta(\vec{q})$ has the expression [113]

$$S_\mu^\beta(\vec{q}) = \frac{1}{\sqrt{V}} \sqrt{\frac{\omega_{LO}}{\omega_q}} \frac{C_F^*}{|\vec{q}|} \langle\mu|\Phi_{\vec{q}}^*(z_e) \exp(-i\vec{q}_\perp \cdot \rho_e) - \Phi_{\vec{q}}^*(z_h)$$
$$\times \exp(-i\vec{q}_\perp \cdot \rho_h)|\beta\rangle \quad (801)$$

where C_F is the Fröhlich constant,

$$C_F = -i\sqrt{2\pi e^2 \hbar\omega_{LO}\left(\frac{1}{\kappa_\infty} - \frac{1}{\kappa_0}\right)} \quad (802)$$

κ_0 and κ_∞ are the static and optical dielectric constants, and $\omega_q^2 = \omega_{LO}^2 - \mu^2 q_z^2$ represents the dispersion of the LO-phonon frequency, μ being a constant to be determined by fitting the experimental phonon-dispersion relations, ρ and $\vec{q}_\perp$ are the components of space coordinate and momentum in the plane of the layers, and $V = L^2 d$, the volume of the QW. In Eq. (801) the modulation function $\Phi_q(z)$ takes into account the confinement of the phonon in the quantum well [113].

9.3.2. Raman Scattering in a Semiconductor QW

9.3.2.1. Basic Hamiltonian and Wavefunctions

The system studied here consists of a type I QW in a homogeneous magnetic field applied along the growth direction. This direction will be taken as the z-axis, and the Landau gauge $(\nabla \cdot \vec{A} = 0)$ will be chosen, with $\vec{A} = B(0, x, 0)$ for the vector potential.

In the envelope function approximation, the one-electron Hamiltonian describes the band electrons

$$H_0 = \frac{\hbar^2}{2}\left[\frac{\hat{k}_\perp^2}{m_\perp} + \hat{k}_z\left(\frac{1}{m(z)}\hat{k}_z\right)\right] + \frac{1}{2}\mu_B\sigma_z g + V_0\Theta\left(\frac{d}{2} - |z|\right) \tag{803}$$

where $m_\perp$ and $m(z)$ are the effective masses in the plane and the QW direction, respectively, g is the Landé factor under consideration, σ_z is the Pauli matrix, and μ_B is the Bohr magneton. $\hat{k} = (1/\hbar)(-i\hbar\nabla + e\vec{A})$ is the wave vector operator, which includes the magnetic interaction, V_0 represents the band offset, and $\Theta(z)$ is the Heavyside step function. In the case of a type I QW, V_0 is positive for the conduction band and negative for the valence band.

Following the usual procedure for the motion of the electrons in a homogeneous magnetic field, the creation and annihilation operators are defined as

$$\hat{a}^+ = \frac{R}{\sqrt{2}}(\hat{k}_x + i\hat{k}_y) \quad \hat{a} = \frac{R}{\sqrt{2}}(\hat{k}_x - i\hat{k}_y) \quad \hat{N} = \hat{a}^+\hat{a} \tag{804}$$

where $R = \sqrt{\hbar/eB}$ is the magnetic length (radius of the cyclotron orbit).

With these definitions, the Hamiltonian H_0 can be written as

$$H_0 = \hbar\omega_c\left(\hat{N} + \frac{1}{2}\right) + \frac{\hbar^2}{2}\hat{k}_z\left(\frac{1}{m(z)}\hat{k}_z\right) + V_0\Theta\left(\frac{d}{2} - |z|\right)$$
$$+ \frac{1}{2}g\sigma_z\mu_B B \tag{805}$$

where $\omega_c = \hbar/R^2 m_\perp$ is the cyclotron frequency.

The motion of the electron in the band is quantized in the xy plane by a harmonic-oscillator potential and in the z direction by the QW potential. The corresponding energy will be given by

$$\varepsilon = \hbar\omega_c\left(N + \frac{1}{2}\right) + \varepsilon_l + gm_s\mu_B B \tag{806}$$

where N is the Landau quantum number, $m_s = \pm 1/2$ for the two spin states, and ε_l is obtained from the equation

$$\left[-\frac{\hbar^2}{2}\frac{d}{dz}\frac{1}{m(z)}\frac{d}{dz} + V_0\Theta\left(\frac{d}{2} - |z|\right)\right]\varphi_l(z) = \varepsilon_l\varphi_l(z) \tag{807}$$

taking as boundary conditions the continuity of the function φ_l and the current $[1/m(z)](\partial\varphi_l/\partial z)$ at the QW interfaces. A more appropriate boundary condition for the light hole band involves the continuity of $\Psi_{LH} - (1/\sqrt{2})\Psi_{SH}$ ($\Psi_{LH, SH}$ are the light hole and split-off-hole wave functions) instead of that of Ψ_{LH}. The effect of these boundary conditions on the confinement energies is $\approx 10\%$ and will be neglected here.

For even and odd states, the bound state conditions are

$$\frac{\chi^B}{m^B} = \frac{\chi^A}{m^A(\varepsilon_l)} \times \begin{cases} \tan(\chi^A d/2) & \text{even states} \\ -\cot(\chi^A d/2) & \text{odd states} \end{cases} \tag{808}$$

with

$$\chi^A = \sqrt{2\varepsilon_l m^A(\varepsilon_l)/\hbar^2} \tag{809}$$

and

$$\chi^B = \sqrt{2(V_0 - \varepsilon_l)m^B(\varepsilon_l)/\hbar^2} \tag{810}$$

$m^A(m^B)$ being the effective mass in the well (barrier). In the calculation of the energy sublevels, the dependence of the mass of the well m_A on the energy has been taken into account (see Section 9.3.2.3). In this approximation, the complete wave function for the Hamiltonian given by Eq. (803) will be

$$\Psi_{nl} = \frac{e^{ik_y y}}{\sqrt{L}}\varphi_l(z)u_n(x - x_0)v_0(\vec{r}) \tag{811}$$

where $u_n(x - x_0)$ is the wavefunction of the one-dimensional harmonic oscillator centered at $x_0 = \hbar k_y/eB$ and v_0 is the Bloch function at $\vec{k} = 0$. Using the above wavefunctions, the matrix elements that appear in the expression of the scattering amplitude can be calculated as

$$\langle\beta|H_{ER}|I\rangle = \frac{|e|}{m_0}\sqrt{\frac{2\pi\hbar}{V\omega n^2}}\vec{e}_r \cdot \vec{P}_{cv}\int\varphi_{l_e}^*(z)\varphi_{l_h}(z)dz$$
$$\times \delta_{n_e, n_h}\delta_{k_{ye}, k_{yh}\pm k_y} \tag{812}$$

where the $+(-)$ sign stands for the laser (scattered) light and $\vec{P}_{cv}$ is the momentum matrix element between the valence and conduction bands. For the electron–lattice interaction matrix elements

$$\langle\mu|H_{EP}|\beta\rangle = -\langle v|\vec{D}_h|v'\rangle\frac{u_0\sqrt{3}}{2a_0}e^{-iq_z x_{0,h}}\delta_{n_{e'}, n_e}\delta_{k'_{kye}, k_{ye}}\delta_{k'_{yh}, k_{yh}}$$
$$- q_y\delta_{l'_e, l_e}\int\varphi_{l'_h}^*(z)\varphi_{l_h}(z)e^{-iq_z z}\,dz\int u_{n'h}^*(x)u_{nh}$$
$$\times (x + R^2 q_y)e^{-iq_z x}dx \tag{813}$$

In this model, phonons cannot couple the electron band via the deformation potential because of the bulk symmetry and the matrix element $\langle c|\vec{D}_e|c\rangle$ equals zero (i.e., intraband transitions are forbidden).

The matrix elements corresponding to the Fröhlich electron and hole–lattice interaction are

$$\langle \mu | H_{EP} | \beta \rangle = \frac{1}{\sqrt{V}} \sqrt{\frac{\omega_{LO}}{\omega_q}} \frac{C_F^*}{q} \Big\{ F_{l_e', l_e} f_{N_e', N_e}^e e^{-iq_x x_{0,e}} \delta_{N_h', N_h}$$

$$\times \delta_{k_{yh}', k_{yh} + q_y} \delta_{l_h', l_h} - F_{l_h', l_h} f_{N_h', N_h}^h e^{-iq_x x_{0,h}} \delta_{N_e', N_e}$$

$$\times \delta_{k_{ye}', k_{ye} + q_y} \delta_{l_e', l_e} \Big\} \delta_{v', v} \delta_{c', c} \tag{814}$$

where the functions F_{l_1, l_2} and f_{N_1, N_2}^α represent

$$F_{l_1, l_2}(q) = \int_{-\infty}^{+\infty} \varphi_{l_1}^*(z) \varphi_{l_2}(z) \Phi_q^*(z) dz \tag{815}$$

$$f_{N_1, N_2}^h(q_x, q_y) = f_{N_1, N_2}^e(q_x, -q_y)$$

$$= \int_{-\infty}^{+\infty} u_{N_1}^*(x) u_{N_2}(x + q_y R^2) e^{-iq_x x} dx \tag{816}$$

The fact that the Fröhlich Hamiltonian does not act on the Bloch part of the wave functions is implicit in Eq. (814). The processes will be purely intraband as in the bulk case. In the following $\delta_{c', c}$ and $\delta_{v', v}$ will be omitted for simplicity.

Taking the wavevectors of the laser and scattered light $\kappa_l \approx \kappa_s \approx 0$, and calculating the sum over $k_{yh(e)}$, the following is obtained:

$$\sum_{k_{yh(e)}} e^{-iR^2 q_x k_{yh(e)}} = \frac{L^2}{2\pi R^2} \delta_{q_x, 0} \tag{817}$$

Taking this result into account and making the approximation $q \approx 0$, the amplitude probability W_{FI} for the case deformation potential takes the form

$$W_{FI} = -K_{ls} \sum_{v, v'} e_s^* \cdot P_{cv'}^* \langle v | \vec{D}_h | v' \rangle \vec{e}_l \cdot \vec{P}_{cv}$$

$$\times \sum_{N, l_e} \sum_{l_h, l_h'} \frac{G_{l_e', l_h'}^* G_{l_h', l_h} G_{l_e, l_h}}{(\hbar \omega_s - \varepsilon_\mu + i\Gamma_\mu)(\hbar \omega_l - \varepsilon_\beta + i\Gamma_\beta)} \tag{818}$$

with

$$K_{ls} = \frac{e^3 u_0 \sqrt{3} B}{m_0^2 2 a_0 n_l n_s \sqrt{\omega_l \omega_s} d} \tag{819}$$

and the overlap integral of the well functions

$$G_{l_1, l_2} = \int \varphi_{l_1}^*(z) \varphi_{l_2}(z) dz \tag{820}$$

The energy of the intermediate states is

$$\varepsilon_{\beta(\mu)} = \varepsilon_g(c, v(v')) + \varepsilon_{l_h(l_h')} + \varepsilon_{l_e} + \hbar \bar{\omega}_c(\bar{\omega}_{c'}) \left(N + \frac{1}{2} \right)$$

$$\pm \frac{\mu_B}{2} B(g_e + g_{h(h')}) \tag{821}$$

$\bar{\omega}_c$ being the cyclotron frequency with the reduced mass $\bar{m}$, $\varepsilon_g(c, v)$ the energy gap between the valence and conduction states involved in the Raman process, and $n_e = n_h =$

N following from the orthogonality of the Landau eigenfunctions. If the electronic wavefunctions $\varphi_l(z)$ are chosen as $\sqrt{(2/d)} \sin[(\pi l/d)(z + d/2)]$, with $\varepsilon_l = \pi^2 \hbar^2 l^2 / 2md^2$ (infinitely deep wells), $G_{l_1, l_2} = \delta_{l_1, l_2}$ and from Eqs. (795), (796) and (818), a simple expression for the Raman tensor is obtained,

$$\vec{e}_s^* \cdot \vec{R} \cdot \vec{e}_l = -\frac{a_0^2}{16\pi} \frac{e^3 \sqrt{3}}{m_0^3 \hbar \omega_l \sqrt{\hbar \omega_l \hbar \omega_s}} \frac{B}{d} \frac{1}{\varepsilon_0 \varepsilon_0'}$$

$$\times \sum_{v, v'} e_s^* \cdot P_{cv'}^* \langle v | \vec{D}_h | v' \rangle \vec{P}_{cv} \cdot \vec{e}_l$$

$$\times \sum_N \frac{1}{A_l^2 - A_s^2} \Bigg[\frac{1}{2A_s} (\pi \coth(\pi A_s) - \frac{1}{A_s})$$

$$- \frac{1}{2A_l} (\pi \coth(\pi A_l) - \frac{1}{A_l}) \Bigg] \tag{822}$$

where

$$A_{l(s)} = \left[\hbar \omega_{l(s)} - \varepsilon_g(c, v(v')) + \hbar \bar{\omega}_c(\bar{\omega}_{c'}) \left(N + \frac{1}{2} \right) \right.$$

$$\left. \pm \frac{\mu_B}{2} B(g_e + g_{h(h')}) + i\Gamma(\Gamma') \right] \Big/ [\varepsilon_0(\varepsilon_0')] \tag{823}$$

and $\varepsilon_0(\varepsilon_0') = \hbar^2 \pi^2 / 2\bar{m}(\bar{m}')d^2$. Some remarks concerning these equations are in order. Equation (819) gives the same dependence of the Raman scattering efficiency on B as in the case of bulk semiconductors [305] ($dS/d\Omega \propto B^2$). For quantum wells, however, the scattering efficiency also depends on the well width ($dS/d\Omega \propto d^2$). This dependence can be attributed to the spatial confinement of electrons and holes in the QW direction and the reduction of the motion in the xy plane produced by the magnetic field. The probability amplitude of the first order Raman intensity is inversely proportional to the volume occupied by electrons and holes, i.e., $L^2 d$, while the density of states is proportional to $L^2 B^2$ [see Eq. (817)], which gives $W_{FI} \propto B/d$. From Eqs. (818) and (821), a set of incoming and outgoing resonances follows at the frequencies:

$$\hbar \omega_{l(s)}(l_e, l_h, N) = \varepsilon_g + \varepsilon_{l_e} + \varepsilon_{l_h} + \hbar \bar{\omega}_c \left(N + \frac{1}{2} \right)$$

$$\pm \frac{\mu_B}{2} (g_e + g_h) B \tag{824}$$

In view of the number of parameters in Eq. (824), it is possible to have different resonances occurring at the same energy (accidental degeneracy). An accidental degeneracy can be eliminated by changing the QW width or the magnetic field (see Fig. 76). The fluctuation in the strength of the resonances between electron and hole subbands versus B contrasts with the monotonic behavior observed in bulk semiconductors [305, 307].

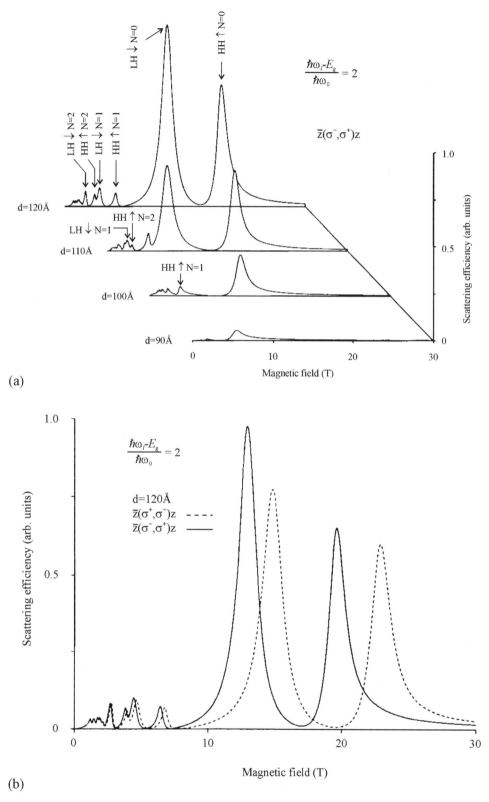

(a)

(b)

Fig. 76. Raman scattering efficiency in arbitrary units calculated with Eqs. (796) and (818) as a function of B for four different well widths. All peaks represent incoming resonances between the first electron subband ($l_e = 1$) and the first two valence subbands ($l_h = 1$). Transitions corresponding to heavy holes are labeled by HH and those of light holes by LH.

The probability amplitude in the Fröhlich case can be written as

$$W_{FI} = \frac{K'_{ls}}{q\sqrt{\omega_q}} \sum_{N,m_s} \sum_{l_e,l'_h} \sum_{l_h,l'_h} G'^*_{l'_e,l'_h} G_{l_e,l_h}$$
$$\times \frac{\delta_{l'_h,l_h} F_{l'_e,l_e}(q) - \delta_{l'_e,l_e} F_{l'_h,l_h}(q)}{(\hbar\omega_s - \varepsilon'_\mu + i\Gamma_\mu)(\hbar\omega_l - \varepsilon'_\beta + i\Gamma_\beta)} \quad (825)$$

with

$$K'_{ls} = \frac{e^3 B C^*_F \sqrt{\omega_{LO}}}{\sqrt{V} m_0^2 n_l n_s \sqrt{\omega_l \omega_s} d} \qquad G'_{l_1,l_2} = e \cdot P_{cv} G_{l_1,l_2} \quad (826)$$

The energy of the intermediate states is

$$\varepsilon'_{\bar\beta(\mu)} = \varepsilon_g(c(c'), v(v')) + \varepsilon_{l_h(l'_h)} + \varepsilon_{l_e(l'_e)} + \hbar\bar\omega_c(\bar\omega_{c'})\left(N + \frac{1}{2}\right)$$
$$+ \mu_B m_s B(g_{e(e')} + g_{h(h')}) \quad (827)$$

According to Eqs. (825) and (827), the incoming and outgoing resonances will occur at frequencies

$$\hbar\omega_{l(s)}(l_e, l_h, N) = \varepsilon_g + \varepsilon_{l_e} + \varepsilon_{l_h} + \hbar\bar\omega_c\left(N + \frac{1}{2}\right)$$
$$+ \mu_B m_s(g_e + g_h)B \quad (828)$$

9.3.2.2. Selection rules

The selection rules for the deformation potential model are obtained through the condition that the matrix element

$$\vec{e}^*_s \cdot \vec{P}^*_{cv'} \langle v|\vec{D}_h|v'\rangle \vec{e}_l \cdot \vec{P}_{cv} G^*_{l_e,l'_h} G'_{l'_h,l_h} G_{l_e,l_h} \quad (829)$$

be different from zero. This depends not only on which bands can be coupled by the deformation potential interaction but also on the polarizations of the scattering configuration. The selection rules for the case of backscattering are derived from a QW grown in the (001) direction of a cubic semiconductor crystal in the Faraday configuration ($B\|\kappa_l\|z$). The deformation potential matrix for LO phonons (TO phonons do not couple in this configuration) is given by (carried over from the bulk) [313]

$$\vec{D}_h = d_0 \begin{pmatrix} 0 & 1 & 0 \\ 1 & 0 & 0 \\ 0 & 0 & 0 \end{pmatrix} \quad (830)$$

where d_0 is the deformation potential constant of the Γ_{15} valence band. The Bloch functions of the Γ_6 conduction band

and Γ_7, Γ_8 valence bands $|J, J_z\rangle$, where J is the total angular momentum and J_z its z component, can be written as

$$c\uparrow = \left|\frac{1}{2}, \frac{1}{2}\right\rangle = |S\uparrow\rangle$$

$$v^+_{LH} = \left|\frac{3}{2}, \frac{1}{2}\right\rangle = \frac{1}{\sqrt{6}}(x+iy)\downarrow - \left(\frac{2}{3}\right)^{1/2} z\uparrow$$

$$c\downarrow = \left|\frac{1}{2}, -\frac{1}{2}\right\rangle = |S\downarrow\rangle$$

$$v^-_{LH} = \left|\frac{3}{2}, -\frac{1}{2}\right\rangle = \frac{1}{\sqrt{6}}(x-iy)\uparrow + \left(\frac{2}{3}\right)^{1/2} z\downarrow$$

$$v^+_{HH} = \left|\frac{3}{2}, \frac{3}{2}\right\rangle = \frac{1}{\sqrt{2}}(x+iy)\uparrow \qquad (831)$$

$$v^+_{SO} = \left|\frac{1}{2}, \frac{1}{2}\right\rangle = \frac{1}{\sqrt{3}}(x+iy)\downarrow + \left(\frac{1}{\sqrt{3}}\right) z\uparrow$$

$$v^-_{HH} = \left|\frac{3}{2}, -\frac{3}{2}\right\rangle = \frac{1}{\sqrt{2}}(x-iy)\downarrow$$

$$v^-_{SO} = \left|\frac{1}{2}, -\frac{1}{2}\right\rangle = \frac{1}{\sqrt{3}}(x-iy)\uparrow - \left(\frac{1}{\sqrt{3}}\right) z\downarrow$$

The polarization of the light is referred to as the fixed z-axis, with right ($+$) and left ($-$) polarized light corresponding, respectively, to the $\vec{e} = \sigma_+ = (\hat{e}_x + i\hat{e}_y)/\sqrt{2}$ and $\vec{e} = \sigma_- = (\hat{e}_x - i\hat{e}_y)/\sqrt{2}$ polarization vectors.

The nonzero matrix elements of Eq. (829) are those that couple the bands showed in Table IV.

The couplings mediated by the deformation potential conserve the Landau number, $\Delta N = 0$, the z component of the angular momentum changes by two units, $\Delta J_z = \pm 2$, while the spin remains unchanged ($\Delta m_s = 0$). For the functions G_{l_1,l_2} to be nonzero, the interband couplings must conserve the parity of the quantum well functions.

The couplings mediated by the Fröhlich interaction conserve the Landau number, $\Delta N = 0$, the third component of the angular momentum, $\Delta J_z = 0$, and the spin ($\Delta m_s = 0$).

Owing to wavevector conservation, the optical phonons excited in the backscattering process propagate parallel to the z direction. Furthermore, the function $\Phi_{\bar{q}}(z)$ [20, 21] has a definite parity with respect to the bisector plane of the well, being even for even modes ($p = 2, 4, 6, \ldots$) and odd for odd

Table IV. Selection Rules for Deformation Potential Interaction

$z(\sigma^-, \sigma^+)zc$	$z(\sigma^+, \sigma^-)zc$
$v^+_{HH} \to v^-_{LH}(\uparrow)$	$v^-_{HH} \to v^+_{LH}(\downarrow)$
$v^+_{LH} \to v^-_{HH}(\downarrow)$	$v^-_{LH} \to v^+_{HH}(\uparrow)$
$v^+_{SO} \to v^-_{HH}(\downarrow)$	$v^-_{SO} \to v^+_{HH}(\uparrow)$
$v^+_{HH} \to v^-_{SO}(\uparrow)$	$v^-_{HH} \to v^+_{SO}(\downarrow)$

modes ($p = 1, 3, 5, \dots$): For $p = 2, 4, 6, \dots$

$$\Phi_{\vec{q}} \equiv \Phi_p(z)$$

$$= \begin{cases} 0 & z < -d/2 \\ 2\left[(-1)^{p/2}\cos\dfrac{p\pi}{d}z - 1\right] & -d/2 < z < d/2 \\ 0 & z > d/2 \end{cases} \quad (832)$$

If $p = 1, 3, 5, \dots$

$$\Phi_{\vec{q}} \equiv \Phi_p(z)$$

$$= \begin{cases} 2 & z < -d/2 \\ 2(-1)^{(p+1)/2}\cos\dfrac{p\pi}{d}z & -d/2 < z < d/2 \\ -2 & z > d/2 \end{cases} \quad (833)$$

Equation (814) shows that when the electron is scattered, the hole remains in the same subband. That means that both electron subbands must have the same parity, in order for $G'_{l'_e, l_h}$ and G_{l_e, l_h} not to vanish. If the electron subbands l'_e and l_e have the same parity, the function $\Phi_p(z)$ must be even. In that case $F_{l'_e, l_e}$ is different from zero. Odd modes are thus forbidden for Fröhlich interaction. It can be concluded then that only even confined phonon modes are excited in the Raman process under consideration.

Once the condition for $F_{l', l}$ to be nonzero is known, the selection rules for the optical transitions are derived from the momentum matrix elements to be the same as in bulk materials, with the additional possibility of intersubband transitions. The only nonzero matrix elements are [305]

$$\langle c\uparrow|\sigma_- \cdot \vec{P}|v_{HH}^+\rangle = \langle c\downarrow|\sigma_+ \cdot \vec{P}|v_{HH}^-\rangle = ip$$

$$\langle c\uparrow|\sigma_+ \cdot \vec{P}|v_{LH}^-\rangle = \langle c\downarrow|\sigma_- \cdot \vec{P}|v_{LH}^+\rangle = \frac{i}{\sqrt{3}}p \quad (834)$$

$$\langle c\uparrow|\sigma_+ \cdot \vec{P}|v_{SO}^-\rangle = \langle c\downarrow|\sigma_- \cdot \vec{P}|v_{SO}^+\rangle = -i\sqrt{\frac{2}{3}}p$$

where $p = \langle x|p_x|c\rangle$, and the selection rules for backscattering and circularly polarized light are given in Table V.

In the crossed configuration $[\bar{z}(\sigma^\pm, \sigma^\mp)z]$ the Raman efficiency is zero.

9.3.2.3. Calculation of the Raman Efficiency: Deformation Potential Interaction

In order to obtain quantitative information on resonant profiles, the Raman-scattering efficiency has been calculated for the model proposed in Section 9.3.2.1. The parameters used for this purpose are given in Table VI. As a first step it is

Table V. Selection Rules for Fröhlich Interaction

$z(\sigma^-, \sigma^-)z$	$z(\sigma^+, \sigma^+)z$
$v_{HH}^+ \to v_{HH}^+(\uparrow)$	$v_{HH}^+ \to v_{HH}^+(\downarrow)$
$v_{LH}^+ \to v_{LH}^+(\downarrow)$	$v_{LH}^- \to v_{LH}^-(\uparrow)$
$v_{SO}^+ \to v_{SO}^+(\downarrow)$	$v_{SO}^- \to v_{SO}^-(\uparrow)$

Table VI. Parameters Used to Calculate the Landau Levels in a GaAs/AlAs QW

Parameters	Values	Reference
ε_g (GaAs)	1520 meV	315
ε_g (AlAs)	2766 meV	315
$P = 2\pi/a$	0.65 a.u.	
m_{hh} (GaAs)	0.34 m_0	315
m_{hh} (AlAs)	0.55 m_0	315
ω_{LO}	36 meV	316
g_e	−0.44	323
$g_{3/2}$	7.2	324
$g_{1/2}$	−2.4	324

necessary to calculate the well subbands. This is done using Eq. (808) with the band offsets given in Table VI. If $k_x = k_y = 0$ in the Kane 8×8 Hamiltonian [317] the eigenvalue determinant can be decoupled into two 3×3 (with spin up and down) for electrons, light holes, and split-off holes, and two 1×1 (with spin up and down) for heavy holes. Taking $k_z \to 0$ the usual $\vec{k} \cdot \vec{p}$ expressions for the masses are obtained. If $k_z [k_z = \sqrt{2m(\varepsilon_l)\varepsilon_l}]$ is kept and the free-electron mass in the $\vec{k} \cdot \vec{p}$ expressions is neglected, the masses for electrons and light holes can be obtained by iteration from the expressions

$$m_{LH}(k_z) = \left[\frac{2}{3}p^2 \frac{1}{\varepsilon_g + \frac{k_z^2}{2m_{LH}}} + \left(2 - \frac{\frac{k_z^2}{2m_{LH}}}{\Delta - \frac{k_z^2}{2m_{LH}}}\right)\right]^{-1} \quad (835)$$

$$m_e(k_z) = \left[\frac{2}{3}p^2 \left(\frac{2}{\varepsilon_g + \frac{k_z^2}{2m_e}} + \frac{1}{\varepsilon_g + \Delta + \frac{k_z^2}{2m_e}}\right)\right]^{-1} \quad (836)$$

Introducing these expressions in Eq. (808), the energy levels are obtained in a self-consistent way [318]. Once the energy ε_1 and the envelope wavefunctions φ_l of the subbands are known, the wavefunctions of this problem [Eq. (811)] can be written down and the energy levels of the intermediate states ε_μ and the G_{l_1, l_2} factors can be calculated through Eqs. (821) and (820), respectively. The masses in the barrier have been obtained through Eqs. (821) and (820) with $k_z = 0$. If the masses (or Luttinger parameters) of the well and barrier are very different, it is not a good approximation to use simply the mass of the well for the calculation of the wavefunction [319, 320]. Applying the selection rules given by Table IV, the scattering efficiency can be calculated from Eqs. (818) and (796) in the two backscattering configurations $\bar{z}(\sigma^+, \sigma^-)z$ and $\bar{z}(\sigma^-, \sigma^+)z$. The dependence of the lifetime on the magnetic field and the well parameters will govern the relative intensity of the Raman peaks related to different transitions. In the present work the lifetime broadenings of the different Landau levels have been taken to be 1 meV, for simplicity. The spin quantization is taken into account through the g factors that appear in Eq. (821). If the experimental g factors are known, they can be used directly for the calculation of the energy levels. The difference between the two spin components for heavy and light hole subbands is used as g factors in the limit $B \to 0$ for a fixed n, $g_{3/2} = 6\xi$, and $g_{1/2} = -2\xi$ with $\xi = 1, 2$ (taken from GaAs [321]). These values are too high when used for the $n = 0$ level because of excitonic contributions [322] but they should be reasonable for higher Landau numbers [323].

Figure 76a shows the Raman scattering efficiency in arbitrary units for the $\bar{z}(\sigma^-, \sigma^+)z$ scattering configuration as a function of the magnetic field for four different well widths. The calculations have been carried out for a given laser energy equal to $\varepsilon_g + 2\hbar\omega_0$. According to the selection rules of Table IV, an incoming resonance is produced with the states v_{HH}^+ and v_{LH}^+ in the $\bar{z}(\sigma^-, \sigma^+)z$ scattering configuration, while in the other scattering configuration, $\bar{z}(\sigma^+, \sigma^-)z$, the resonance takes place with v_{HH}^- and v_{LH}^-. In the following only to the first heavy and light subbands will be referred because the other nearby subbands have odd parity and the overlap integrals are zero. For the first two well widths (90 and 100 Å) the incoming resonances corresponding to the $N = 0, 1, 2, \ldots, N$ levels of the heavy hole states can be observed. It can be seen how the $N = 0$ resonance happens at a large magnetic field due to the fact that the level becomes less deep with increasing d. The intensity of the peaks must decrease when d increases (inversely proportional to d^2) but the peaks appear at a large magnetic field and the change in B is large, which produces an increase of intensity. In the plot for 100 Å, the peak corresponding to the incoming resonance of the light hole state appears close to zero and is not seen in the drawing. In the plots corresponding to larger widths (110–120 Å, the incoming resonance with the light hole state $N = 0$ shifts strongly with B. Two facts are noticeable. The first is the strong shift with B as compared to the heavy hole transitions. Because of this the level corresponding to the light hole state has a lower slope as a function of the magnetic field than the one corresponding to the heavy hole state. A small shift in the energy scale (due to a change in the well width) produces a large shift in the magnetic field. The second noticeable fact is the intensity of the scattering efficiency at the light hole resonances, which is larger than that of the heavy hole ones. This is a general fact related to the selection rules in a QW. The incoming heavy hole resonances are outgoing for the light hole states, which are lower in energy. The incoming light hole resonances, however, have their outgoing counterpart in the heavy hole states, thus producing a small outgoing denominator in Eq. (818). Also, the overlap integrals between electrons and holes are longer for light hole states [Eq. (820)]. The increase of the intensity of the heavy hole state $N = 0$ with B is monotonic and mainly due to the shift to high magnetic fields. In Figure 76b the scattering efficiency for two scattering configurations $\bar{z}(\sigma^-, \sigma^+)z$ and $\bar{z}(\sigma^+, \sigma^-)z$ in a 120 Å QW is compared. The slight shift observed in the spectrum between both configurations is due to the spin quantization.

Figure 77 shows the Raman scattering efficiency in both scattering configurations as a function of the laser frequency for $d = 80$, 100, and 120 Å and a fixed magnetic field (10 T). It can be seen that the intensity of the peaks increases when d decreases. Different incoming and outgoing transitions are

obtained when increasing the laser frequency; the respective peak positions move to lower energy when d increases. As in Figure 76a, the incoming resonances (in) involving the light holes display stronger intensities compared with the heavy hole ones. From Figures 76 and 77, it is evident that there is a strong selection rule in both scattering configurations, $\bar{z}(\sigma^+, \sigma^-)z$, due to the spin quantization and the coupling of different valence bands. The difference between both scattering configurations is reduced when the magnetic field tends to zero, similar to the case of the bulk [305].

9.3.3. Raman Efficiency within the Simplified Model: Fröhlich Interaction

In order to obtain the Raman scattering efficiency $dS/d\Omega$, it is neccessary to calculate the Raman polarizability corresponding to one phonon mode and add over all phonon modes with the restriction $\omega_q^2 > 0$. The resulting final expression can be written as

$$\frac{dS}{d\Omega} = S_0 \sum_{p \, \text{even}} \frac{1}{p^2 \omega_q}$$
$$\times \left| \sum_{N, m_s} \sum_{l'_e, l_e} \sum_{l'_h, l_h} \frac{G^{'*}_{l'_e, l'_h} G'_{l_e, l_h} [\delta_{l'_h, l_h} F_{l'_e, l_e} - \delta_{l'_e, l_e} F_{l'_h, l_h}]}{(\hbar\omega_s - \varepsilon'_\mu + i\Gamma_\mu)(\hbar\omega_l - \varepsilon'_\beta + i\Gamma_\beta)} \right|^2 \quad (837)$$

where the constant S_0 is

$$S_0 = \frac{\omega_s^2 n_s e^6 B^2 |C_F|^2 \omega_{\text{LO}}}{\omega_l^2 n_l \pi^4 \hbar^2 c^4 m_0^4} \quad (838)$$

In thin QWs ($d \leq 50$ Å) the different phonon modes are going to have well-defined, different frequencies because of the bulk dispersion. In that case, the $dS/d\Omega$ must be calculated for a fixed p. For wide enough quantum wells ($d \geq 50$ Å) such that the separation between individual confined modes is less than their width, the nondispersive approximation can be made, setting $\omega_q = \omega_{\text{LO}}$ and adding all the efficiencies corresponding to the different phonon modes, in order to obtain the total Raman efficiency. The number of phonon modes necessary to add is not high because of the factor p^2 that appears in Eq. (837).

It the scattering cross section $dS/d\Omega$ has been calculated within this model for QWs of different thicknesses as a function of laser frequency for the scattering configuration $\bar{z}(\sigma^+, \sigma^+)z$, whose selection rules are given in Table V. For this purpose the energy subbands (including the QW confined electronic sublevels and their masses in a self-consistent way) and the corresponding well wavefunctions $\varphi_l(z)(l = l_e, l_{lh}, l_{hh})$ were first calculated, as explained in [324]. The matrix elements were then evaluated with the wavefunctions of Eq. (811), and the scattering efficiency was finally obtained by means of Eq. (837). The parameters used in the calculation are those for GaAs/AlAs QWs, given in Table VI. The lifetimes used in calculating $dS/d\Omega$ were taken from a least-squares fit to a photoluminescence excitation profile

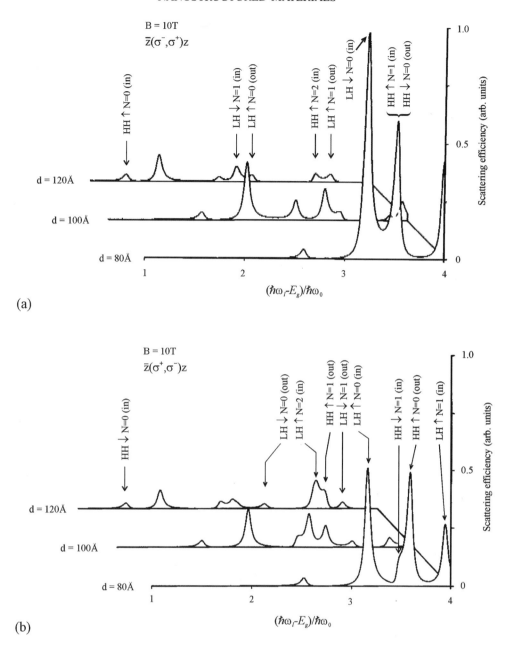

Fig. 77.   Raman scattering efficiency, according to Eqs. (796) and (818), as a function of $(\hbar\omega_l - \varepsilon_g)/\hbar\omega_0$ for three different well widths in arbitrary units. For clarity, the notation of the different transitions contains the electron spin and the type of resonance (incoming or outgoing).

[309]. They are assumed to increase quadratically with the Landau number N. The numerical expression used is

$$\Gamma(N) = 1.2 - 0.47(N+1) + 0.26(N+1)^2 \qquad (839)$$

independently of the light or heavy character of the band.

Figure 78 shows the scattering efficiency as a function of laser energy for 135, 145, and 150 Å QWs in the $\bar{z}(\sigma^+, \sigma^+)z$ configuration with the magnetic field kept fixed at 8 T. The selection rules given in Table V show that states only of the same valence band and the same parity are coupled by the Fröhlich interaction. The peaks in the spectra correspond to intrasubband transitions. For $d = 145$ Å the energy difference between the first and third heavy hole well states equals that of the phonon. Because the Fröhlich interaction couples these

two QW levels, a double resonance for the $HH3 \to HH1$ transition is obtained at the laser energy $\varepsilon_g + 1.94\hbar\omega_{LO}$. In the parabolic QW model, energy levels of the same subband depend linearly on the magnetic field with the same slope for levels with the same Landau number, so that the double resonance condition is independent of the magnetic field. This condition is fulfilled simultaneously for the different Landau levels of the bands involved in the transition. In Figure 78 can be seen the double resonant peaks corresponding to Landau levels $N = 0$, 1, and 2 [labeled as (in) (out)]. The difference in intensity is due to the dependence of the linewidth on the Landau quantum number. When the well thickness is changed, the splitting between $HH1$ and $HH3$ changes and the double resonance is lost as the incoming and outgoing resonances

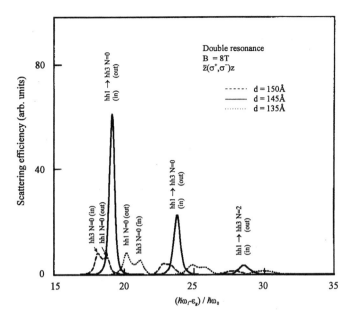

Fig. 78. Calculated Raman scattering efficiency, according to Eq. (837), in the configuration $z(\sigma^+, \sigma^+)z$ close to a double resonance. The magnetic field has been kept at 8 T.

separate. The energy at which the resonances appear depends linearly on the magnetic field:

$$
\hbar\omega_l = \varepsilon_g + \varepsilon_{l_e} + \varepsilon_{l_h} + \hbar e B\left(\frac{1}{m_{e\perp}} + \frac{1}{m_{h\perp}}\right)(N + \tfrac{1}{2})
$$
$$
+ \mu_B(g_e + g_h)m_{se}B \tag{840}
$$

9.3.3.1. Double-Resonance Conditions: Deformation Potential Interaction

When the incoming and outgoing resonances coincide, the Raman scattering efficiency is strongly enhanced. This situation is achieved when the energy difference between two allowed electronic transitions equals the energy of the emitted phonon ($\hbar\omega_{LO}$). Changing either the magnetic field or the width of the QW, one can tune the splitting of the Landau and QW levels to match the required energy. If these levels can be coupled through the deformation potential interaction, a double resonance will occur.

The double-resonance condition (DRR) follows from Eq. (824). The magnetic field needed to produce DRRs is given by

$$
B_N(d) = \frac{\hbar\omega_0 + \varepsilon_{l_h} - \varepsilon'_{l_h}}{2\mu_B m_0(N + \tfrac{1}{2})\left(\frac{1}{m_{h\perp}} - \frac{1}{m'_{h\perp}}\right) - \mu_B(g_h - g'_h)m_{s_e}} \tag{841}
$$

The laser energy at which the double resonance occurs is

$$
\hbar\omega_l = \varepsilon_g + \varepsilon_{l_e} + \varepsilon_{l_h} + \hbar e B_N\left(\frac{1}{m_{h\perp}} + \frac{1}{m'_{h\perp}}\right)(N + \tfrac{1}{2})
$$
$$
+ \mu_B(g_e - g_h)m_{s_e}B_N \tag{842}
$$

where B_N is the magnetic field defined in Eq. (841). The above equation establishes the width d_c of the QW needed for double

resonance. For an infinite well this width is

$$
d_c(B) = \left(\left[\hbar^2\pi^2/2(1/m_h - 1/m'_h)l^2\right]\right/
$$
$$
\left[\hbar\omega_0 + \mu_B B(g_h - g'_h)m_{s_e} - \hbar(N + 1/2)e\right.
$$
$$
\left.\times B(1/m_{h\perp} - 1/m'_{h\perp})\right]^{1/2}\right) \tag{843}
$$

where l labels the different infinite well subbands. Equation (843) can be a good approximation for d_c on the first levels of finite but deep quantum wells.

Figure 79 shows how the Raman scattering efficiency in $\bar{z}(\sigma^+, \sigma^-)z$ changes close to a double resonance as a function of the magnetic field. In the low magnetic field region, the incoming resonance with the light hole (ν^-_{LH}) is reached before the outgoing one with the heavy hole (ν^+_{HH}), while for magnetic fields above the double resonance, the difference in energy between the two levels is larger than the energy of a phonon. This makes the outgoing resonance with the heavy hole appear before the incoming resonance with the light hole. The reason for this sequence of resonances is that, at low magnetic fields, the difference between the heavy hole level (ν^+_{HH}) and the light hole level (ν^-_{LH}) is smaller than the frequency of the phonon. When the magnetic field increases, the level $\nu^+_{HH}(N = 0)$ increases in energy and the $\nu^-_{LH}(N = 0)$ decreases, producing a double resonance. In the other scattering configuration, however, $\nu^-_{HH}(N = 0)$ decreases with the magnetic field and $\nu^+_{LH}(N = 0)$ increases, and the double resonance should take place at very high magnetic fields.

9.3.4. Luttinger Hamiltonian: Mixing Effects

The valence band states of the zinc-blende-type semiconductor quantum wells in the presence of a magnetic field cannot be taken as a product of pure Bloch and harmonic-oscillator wavefunctions, due to the degeneracy of the Γ_8 valence band. A more general theory, which treats the coupling of the bands through a $\vec{k} \cdot \vec{P}$ Hamiltonian, yields states in which the different degenerate Bloch bands of the bulk are present to a considerable degree. The Landau ladders are not equidistant any more, and the effective mass is a function of the magnetic field and the Landau level.

The theory developed in the preceding section can be easily extended using a more realistic Hamiltonian and applying the selection rules found in Section 9.3.2.3 multiplied by the corresponding admixture coefficients. Using the harmonic-oscillator operators defined in Eq. (804) and taking a QW grown in the (001) crystal direction, the components of a 4×4 Luttinger Hamiltonian [325] can be written in atomic units as

$$
H_{11} = \frac{1}{2R^2}(\gamma_1 + \gamma_2)(1 + 2\widehat{N}) - \frac{1}{2}\frac{\partial}{\partial z}(\gamma_1 - 2\gamma_2)\frac{\partial}{\partial z}
$$
$$
+ \frac{3}{2}\frac{\eta}{R^2} + V_0\Theta\left(\frac{d}{2} - |z|\right)
$$

$$
H_{12} = \frac{i\sqrt{6}}{R}\gamma_3\hat{a}\frac{\partial}{\partial z}
$$

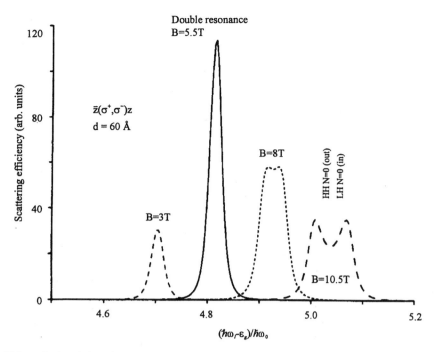

Fig. 79. Raman scattering efficiency in the $z(\sigma^+, \sigma^-)z$ configuration near a double resonance as a function of the magnetic field. The calculations have been done for a 60 Å well.

$$H_{22} = \frac{1}{2R^2}(\gamma_1 - \gamma_2)(1 + 2\widehat{N}) - \frac{1}{2}\frac{\partial}{\partial z}(\gamma_1 + 2\gamma_2)\frac{\partial}{\partial z}$$
$$+ \frac{1}{2}\frac{\eta}{R^2} + V_0\Theta(\frac{d}{2} - |z|)$$

$$H_{24} = H_{13}$$

$$H_{33} = \frac{1}{2R^2}(\gamma_1 - \gamma_2)(1 + 2\widehat{N}) - \frac{1}{2}\frac{\partial}{\partial z}(\gamma_1 + 2\gamma_2)\frac{\partial}{\partial z}$$
$$- \frac{1}{2}\frac{\eta}{R^2} + V_0\Theta\left(\frac{d}{2} - |z|\right)$$

$$H_{34} = -H_{12}$$

$$H_{44} = \frac{1}{2R^2}(\gamma_1 + \gamma_2)(1 + 2\widehat{N}) - \frac{1}{2}\frac{\partial}{\partial z}(\gamma_1 - 2\gamma_2)\frac{\partial}{\partial z}$$
$$- \frac{3}{2}\frac{\eta}{R^2} + V_0\Theta\left(\frac{d}{2} - |z|\right)$$

$$H_{14} = H_{23} = 0 \tag{844}$$

where γ_1, γ_2, γ_3, and η are the Luttinger parameters. The axial approximation has been used, taking $\gamma_2 \approx \gamma_3 \approx \bar{\gamma} = (\gamma_2 + \gamma_3)/2$. Owing to the cylindrical symmetry, for a magnetic field along z, the exact solution of the Hamiltonian operator (50) for energies $\varepsilon < V_0$ is

$$\Psi_{\alpha, n} = \frac{1}{\sqrt{L}} \begin{pmatrix} e^{ik_{y1}y} \phi_1^{\alpha n}(z) u_{n-3}(x - x_0)\nu_1 \\ e^{ik_{y2}y} \phi_2^{\alpha n}(z) u_{n-2}(x - x_0)\nu_2 \\ e^{ik_{y3}y} \phi_3^{\alpha n}(z) u_{n-1}(x - x_0)\nu_3 \\ e^{ik_{y4}y} \phi_4^{\alpha n}(z) u_n(x - x_0)\nu_4 \end{pmatrix} \tag{845}$$

Here ν_1, ν_2, ν_3, and ν_4 refer to $|3/2, +3/2\rangle$, $|3/2, +1/2\rangle$, $|3/2, -1/2\rangle$, and $|3/2, -3/2\rangle$ respectively, $u_p \equiv 0$ if $p < 0$,

and the function $\phi_i^{\alpha n}(z)$ is a linear combination of the QW functions $\varphi_{l_i}(z)$: i.e.,

$$\phi_i^{\alpha n}(z) = \sum_{l_i} c_i^{\alpha n} \varphi_{l_i}(z) \tag{846}$$

In this case the energy levels cannot be classified according to their heavy or light hole character, because of the mixing of the different bands. The label α, however, corresponds to their heavy or light character in the $B \to 0$ limit, and the quantum number n refers to the highest Landau level of its different harmonic-oscillator components (Landau orbital quantum number) in Eq. (845). Using Eq. (797) and the wavefunctions (811) and (845), for electrons and holes, respectively, the scattering amplitude corresponding to the one-phonon Raman process in a semiconductor QW when electron–phonon interaction corresponds to the deformation potential is now given by

$$W_{FI} = K_{l_s} \sum_{N, l_e} \sum_{\alpha, \alpha'} \frac{G^*_{l_e, \alpha'}(e_s)F_{\alpha', \alpha}G_{l_e, \alpha}(e_l)}{(\hbar\omega_s - \varepsilon_\alpha + i\Gamma_\alpha)(\hbar\omega_l - \varepsilon_{\alpha'} + i\Gamma_{\alpha'})} \tag{847}$$

where

$$G_{l_e, \alpha} = \sum_n \sum_{i=1}^{4} \langle c|\vec{e} \cdot \vec{P}_{cv}|v_i\rangle \delta_{n-4+i, N} \int \varphi_{l_e}(z)^* \phi_i^{\alpha n}(z)dz \tag{848}$$

$$F_{\alpha', \alpha} = \sum_{n, n'} \sum_{i=1}^{4}\sum_{j=1}^{4} \langle v_j|\vec{D}_h|v_i\rangle \delta_{n'+j, n+i}$$
$$\times \int \phi_j^{\alpha' n'*}(z) \phi_i^{\alpha n}(z)dz \tag{849}$$

As in the previous model, the transitions mediated by the deformation potential are possible in the crossed configuration, with polarizations $\bar{z}(\sigma^-, \sigma^+)z$ and $\bar{z}(\sigma^+, \sigma^-)z$

[the $\bar{z}(\sigma^{\pm}, \sigma^{\pm})z$ configurations are forbidden]. The first configuration couples levels with $\Delta n = 2$ and the second one couples those with $\Delta n = -2$, i.e., with $\Delta N = 0$. The orbital Landau number involved in these transitions can differ by 0, 1, 2, or 3 units from the conduction band Landau level N because of N mixing in the valence bands. In the model without mixing presented in the last section an additional selection rule was added which involved the conservation of the parity of the quantum well functions. Now this selection rule is relaxed, and its importance will depend on the mixing of the different subbands at a given magnetic field. Nevertheless, according to the selection rule given by Table IV, for light polarizations $e = \sigma^-$ or $e = \sigma^+$, so Eq. (848) contains only one element.

9.3.5. GaAs/AlAs QW: Comparison with Experimental Results

As was mentioned before, there are only a few experiments on resonant Raman scattering in QW under high magnetic fields. In [309, 310] the authors studied a 100 Å GaAs/AlAs MQW. In this section the theoretical results are applied to this particular structure, although the conclusions obtained can be, in a straightforward manner, carried over to other similar heterostructures, in particular GaAs/Ga$_x$Al$_{1-x}$As QWs.

9.3.5.1. Scattering Efficiency and Selection Rules: Fröhlich Interaction

The scattering amplitude is derived from the wavefunctions (51). The probability amplitude is found to be

$$W_{Fi} = \frac{K_{ls}}{q\sqrt{\omega_q}} \sum_{N, m_s l_e, l'_e \alpha, \alpha'} G^*_{l'_e, \alpha'} G_{l_e, \alpha}$$

$$\times \frac{\delta_{\alpha', \alpha} F_{l'_e, l_e}(q_z) - \delta_{l'_e, l_e} F_{\alpha', \alpha}(q_z)}{(\hbar\omega_s - \varepsilon_\mu + i\Gamma_\mu)(\hbar\omega_l - \varepsilon_\beta + i\Gamma_\beta)} \quad (850)$$

with the following definitions:

$$F_{\alpha', \alpha} = \sum_n \sum_{i=1}^4 \delta_{n-4+i, N} \int \phi_i^{\alpha' n*}(z) \phi_i^{\alpha n}(z) \Phi_q^*(z) dz \quad (851)$$

The energies of the intermediate states are now

$$\varepsilon_{\beta(\mu)} = \varepsilon_g(c(c'), v(v')) + \varepsilon_{\alpha(\alpha')} + \varepsilon_{l_e(l'_e)} + \hbar\omega_c(\omega_{c'})(N + \tfrac{1}{2})$$
$$+ \mu_B m_s B g_{e(e')} \quad (852)$$

In Eqs. (848) and (851), for a particular value of the spin m_s and polarization e, i is fixed (see Table V), n is then fixed by the δ function, and the expressions reduce to

$$G_{l_e, \alpha} = \sum_{l_e} C_{l_i}^{\alpha, N+4-i} G_{l_e, l_i} \quad (853)$$

$$F_{\alpha', \alpha} = \sum_{l_i} \sum_{l_j} C_{l_i}^{\alpha', N+4+i*} C_{l_j}^{\alpha, N+4-i} F_{l_i, l_j} \quad (854)$$

The transitions mediated by the Fröhlich interaction are only possible with polarizations $\bar{z}(\sigma^+, \sigma^+)z$ and $\bar{z}(\sigma^-, \sigma^-)z$. In both configurations, the coupled levels are such that $\Delta n = \Delta N = 0$. As before, only intraband coupling is allowed.

9.3.5.2. Raman Scattering in GaAs/AlAs QWs

The simplest, unambiguous way to label the hole Landau levels is by means of (n, α), the Landau oscillator quantum number, and a new quantum number α arising from the diagonalization of the Hamiltonian. Unfortunately, this labeling does not explicitly display the physical information that it actually contains. It is known that the deformation potential interaction is interband and couples light and heavy hole states, while Fröhlich interaction is intraband and couples either light–light or heavy–heavy states; this restricts the Landau sublevels involved in the process.

For $B = 0 (\bar{k}_\perp = 0)$, the Luttinger Hamiltonian is diagonal and the different well subbands are purely light or purely heavy. For small B (also small n and thin QWs) the eigenvectors that appear in Eq. (845) must have only one component different from zero. In this case the state can be labeled by the well subband index l_i instead of the new quantum number α. For instance, the first heavy hole subband can be labeled with $n = 0$ as $l_4(0)$. It is more common to use HH^- instead of 4. Thus, the label would be $HH1(0^-)$. This notation has a clear physical meaning only for small magnetic fields. As an example of the loss of physical meaning, it can be observed (see Figure 4 of [324]) that the $LH1$ and $HH2$ QW subbands are very close to each other in a 100 Å GaAs/AlAs QW and that there is an anticrossing around $B = 10$ T, with the paradox that a pure light level at $B = 15$ T would be labeled $HH2(3^+)$.

Figure 80 shows the Raman efficiency as a function of laser energy for $B = 0$ and $d = 100$ Å for the $z(\sigma^+, \sigma^+)z$ scattering configuration. The other configuration yields a similar profile, although for the other spin components. Figure 80 clearly depicts the difference in the intensity between the heavy and light contributions to the Raman profile resulting mainly from

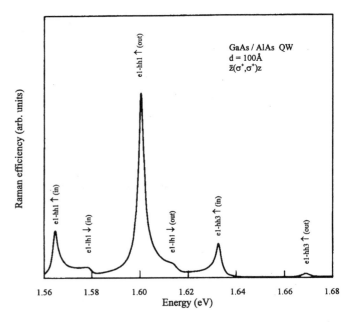

Fig. 80. Calculated Raman scattering efficiency, according to the scattering amplitude given by Eq. (850), for a GaAs/AlAs QW with $d = 100$ Å and $\hbar\omega_{LO} = 36$ meV at $B = 0$ in the $z(\sigma^+, \sigma^+)z$ scattering configuration.

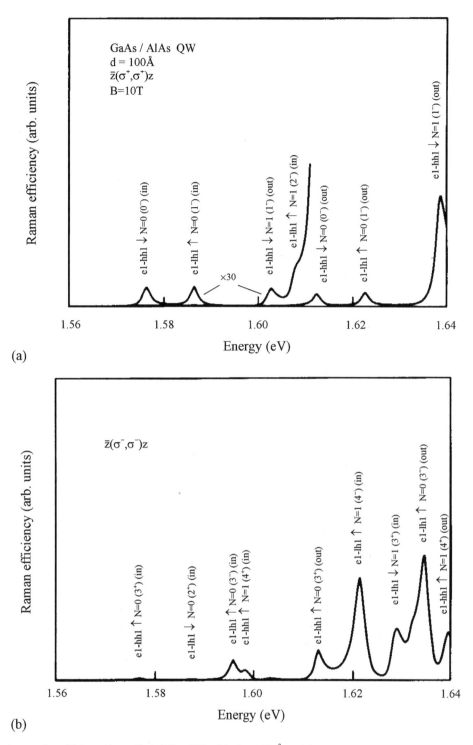

Fig. 81. Magneto-Raman scattering efficiency for a GaAs/AlAs QW with $d = 100$ Å and $\hbar\omega_{\mathrm{LO}} = 36$ meV at $B = 10$ T for both (a) $z(\sigma^+, \sigma^+)z$ and (b) $z(\sigma^-, \sigma^-)z$ scattering configurations.

the closeness of the electron and light hole masses. The overlap integrals $F_{l',l}$ subtracted in Eq. (837) are nearly equal, giving a small contribution to the Raman scattering efficiency. The $E1 - HH1$ outgoing resonance is larger than the incoming one because when the scattered light resonates with the $E1 - HH1$ electronic transition, the laser light is close to the $HH3$ level. The same fact makes the $E1 - HH3$ incoming

transition larger than the outgoing one. The $E1 - HH2$ resonance is forbidden by parity.

In Figure 81, the calculated Raman efficiency is represented as a function of the laser energy for $B = 10$ T in a QW of thickness 100 Å for both scattering configurations (a) $\bar{z}(\sigma^+, \sigma^+)z$ and (b) $\bar{z}(\sigma^-, \sigma^-)z$. According to Table V for $\bar{z}(\sigma^+, \sigma^+)z$ the resonances are either incoming or outgoing

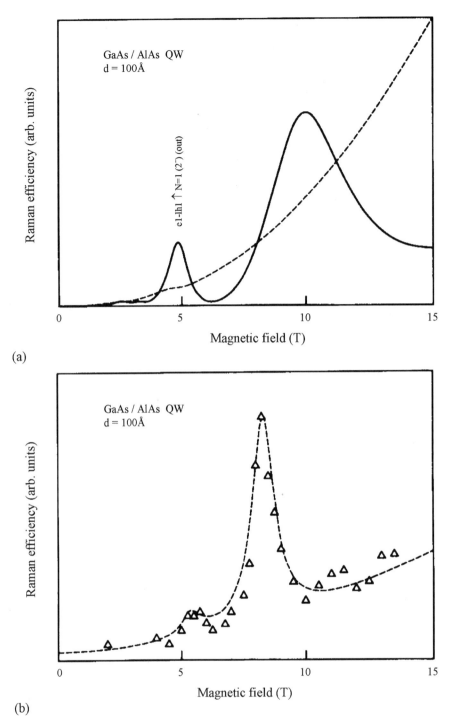

Fig. 82. (a) Raman intensity as a function of the magnetic field for a calculated for $\bar{z}(\sigma^-, \sigma^-)z$ (solid line) and $\bar{z}(\sigma^+, \sigma^+)z$ (dashed line) configurations with Eq. (850). The curve has been obtained for ω_l following the outgoing transition $E1 - LH1\,(1^-)$. (b) Experimental data points [310]; the dashed line is a guide to the eye.

and involve $v_{LH}^- \rightarrow v_{LH}^-$ or $v_{HH}^+ \rightarrow v_{HH}^+$ matrix elements of the electron–phonon interaction. The number of possible transitions is notably reduced by these selection rules. Outgoing transitions are larger than incoming ones due, as in the case $B = 0$, to the energy denominators. The light and heavy components of the Landau levels make new resonances possible as compared with the parabolic approximation (Fig. 81b). This

is the case for $LH1 \uparrow N = 0(3^-)$ (in) and (out) resonances, in which the transitions take place via the $+3/2$(i.e., spin up) light hole band (with weights 0.32 and 0.54, respectively).

Figure 82 shows the Raman efficiency as a function of magnetic field for $\bar{z}(\sigma^-, \sigma^-)z$ (solid line) and $\bar{z}(\sigma^+, \sigma^+)z$ (dashed line) configurations. The laser energy is so that the transition $E1 - LH1(1^-)$ remains outgoing resonant. The first peak around 5 T(solid line) appears when the $E1 - LH1(2^-)$ transition crosses the $E1 - LH1(1^-)$ transition. The second peak is due to the virtual transition $E1 - HH3(0^-) \rightarrow E1 - HH1(0^-)$ and it appears because the different slope with respect to the transition that the laser is following makes the product of energy denominators a minimum. The dashed line shows that no resonances show up in the other scattering configuration. Figure 82b displays the experimental data [310] supposedly obtained in the $\bar{z}(\sigma^+, \sigma^+)z$ scattering configuration. If the configuration of these measurements were the opposite of the nominal one (as a result of an experimental error), the first observed peak would be qualitatively explained if it were corrected for excitonic effects taking a smaller binding energy for upper levels. After this correction, the theoretical peak moves to higher values of the magnetic field. The same arguments shift the second theoretical peak to the left.

Figure 83 shows the calculated scattering efficiency as a function of laser energy for a well width of 25 Å. Since this well is very narrow, the different confined phonon modes can be resolved. The difference in intensity between $p = 2$ and $p = 4$ modes is rather large, but it saturates quickly for larger p's.

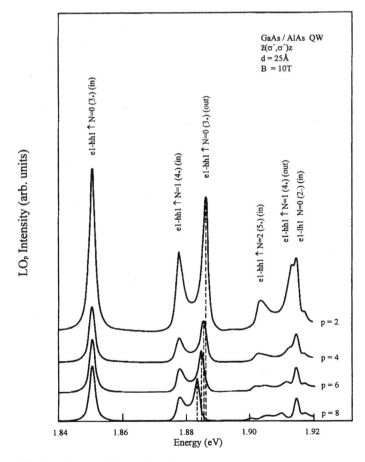

Fig. 83. Raman efficiency for different phonon modes, $p = 2, 4, 6$, and 8. The broken lines represent the shift in outgoing resonances due to the phonon dispersion.

As can be seen, only even phonon modes are excited. This is due to the fact that in the present analysis the same effective mass Hamiltonian for the diamond and zinc-blende lattices has been assumed, which means neglecting the very small linear terms which are odd under inversion in the zinc-blende structure. Then, the eigenstates of the Hamiltonian have a definite parity since only the envelope functions are being used; i.e., the asymmetry of the periodic part of the Bloch functions is neglected.

10. PHYSICAL EFFECTS OF IMPURITY STATES AND ATOMIC SYSTEMS CONFINED IN Q2D NANOSTRUCTURED MATERIALS

In contrast with bulk materials, the Q2D systems are characterized by a lack of translational invariance along the confinement direction. Thus, the impurity binding energy explicitly depends on the precise location of the impurity. It is important to know where the impurity atom is placed in the Q2D, and the binding energy of the donor state associated with the impurity atom will depend on whether or not the impurity atom sits at the center of the Q2D.

The second important feature of coulombic problems in Q2Ds is the variation of the impurity binding energy with the characteristic dimension of the well. When the Q2D confinement dimension (confinement size) decreases, the impurity binding energy increases, as long as the penetration of the unperturbed Q2D wavefunction in the barrier remains small. This may seem surprising at first sight since the extra kinetic energy is intuitively associated with the localization of a particle in a finite region of space. Moreover, it is actually true that the energy of the ground bound state of the impurity increases as the confinement size decreases, when it is measured from some fixed reference. However, the onset of the impurity continuum that (for donors) coincides with the energy of the Q2D bound state also moves up with decreasing the confinement size. The binding energy finally increases since, in the bound impurity states, the carrier is kept near to the attractive center by Q2D walls and thus experiences a larger potential energy than in the absence of a Q2D. On the other hand, the onset of the impurity continuum does not benefit from any extra potential energy gain, as the continuum states are hardly affected by the coulombic potential.

When the Schrödinger equation for a given quantum system is nonseparable in nature, some approximation techniques must be used to study its solution. In a similar situation, for a confined quantum system, the variational method might constitute an economical and physically appealing approach [326]. It has been shown [327] that the variational method is useful for studying this class of systems when their symmetries are compatible with those of the confining boundaries. In general, in the problems related with impurities and atoms confined in the different Q2Ds, these considerations are fulfilled.

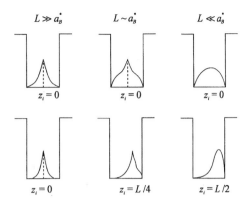

Fig. 84. Evolution of the shape of the impurity wavefunction in a quantum well of decreasing well thickness for an on-center impurity (upper part) and with the impurity position in a thick well (lower part).

10.1. Hydrogenic Impurity in Quasi-Two-Dimensional Systems

In Figure 84 the shape of the impurity wavefunction is sketched, keeping the in plane distance between the carrier and the attractive center $\rho(\rho = \sqrt{x^2 + y^2})$ equal to zero. Several quantum well thicknesses (upper part of Fig. 84) and several impurity positions in a thick well (lower part of Fig. 84) have been considered. For thick wells ($L \gg a_B^*$) the on-center impurity wavefunction resembles that of the bulk $1s$ states. On the other hand, the on-edge impurity wavefunction approaches the shape of a $2p_z$ wavefunction if $(L/a_B^*) \gg 1$. This is because the barrier potential forces the impurity wavefunction to almost vanish at $z = \pm L/2$. If $L/a_B^* \ll 1$ and if the barrier height is very large, the electron z motion becomes forced by the quantum well potential. Thus, along the z-axis the impurity wavefunction looks like the ground state wavefunction $\chi_1(z)$ of the well.

10.1.1. Approximate Solutions of the Hydrogenic Impurity Problem

In all that follows donorlike impurities will be considered, unless otherwise specified. The conduction bands of both host materials of the quantum well are assumed to be isotropic and parabolic in $\vec{k}$. Effective mass jumps at the interfaces are neglected, as well as the differences in the relative dielectric constants of the two host materials. The impurity envelope functions are the solutions of the effective Hamiltonian

$$\widehat{H} = \widehat{H}_0 + V_I = \frac{\widehat{p}_z}{2m_e^*} + \frac{\widehat{p}_x + \widehat{p}_y^2}{2m_e^*} \\ + V_0 Y\left(z^2 - \frac{L^2}{4}\right) - \frac{e^2}{\kappa\sqrt{\rho^2 + (z - z_i)^2}} \quad (855)$$

where V_0 is the barrier height and $Y(x)$ is the step function ($Y(x) = 1$ if $x > 0$; $Y(x) = 0$ if $x < 0$). The impurity position along the growth axis is z_i. The (x, y) origin is at the impurity site because all impurity positions are equivalent in the layer plane; $\vec{\rho}$ is the projection of the electron position vector in the layer plane ($\rho(x, y)$).

In the absence of impurity, the eigenstates of $\widehat{H}_0$ are separable in (x, y) and z,

$$\widehat{H}_0|\nu, \vec{k}_\perp\rangle = \left(\varepsilon_\nu + \frac{\hbar k_\perp^2}{2m_e^*}\right)|\nu, \vec{k}_\perp\rangle \quad (856)$$

where ν labels the quantum well eigenstates (energy ε_ν), i.e., the quantum well bound ($\varepsilon_\nu < V_0$) and unbound states ($\varepsilon_\nu > V_0$) and $\vec{k}_\perp = (k_x, k_y)$. Since the $|\nu, \vec{k}_\perp\rangle$ basis is complete, the impurity wavefunction ψ_{loc} can be always expanded in the form

$$|\psi_{\text{loc}}\rangle = \sum_{\nu, \vec{k}_\perp} c(\nu, \vec{k}_\perp)|\nu, \vec{k}_\perp\rangle \quad (857)$$

The coulombic potential couples a given subband ν, as well as a given vector $\vec{k}_\perp$, with all others. The intersubband coupling (especially the one with the subbands of the quantum well continuum) is difficult to handle. In a quasi-two-dimensional situation we would like to set $c(\nu, \vec{k}_\perp) \equiv c_{\nu_0}(\vec{k}_\perp)\delta_{\nu, \nu_0}$, i.e., to neglect intersubband coupling [328, 329]. This procedure is convenient as the impurity wavefunction displays a separable form:

$$\langle\vec{r}|\psi_{\text{loc}}\rangle = \chi_{\nu_0}(z)\varphi(\vec{\rho}) \quad (858)$$

The wavefunction $\varphi(\vec{\rho})$ is the solution of the two-dimensional Schrödinger equation

$$\left[\frac{\widehat{P}_x + \widehat{P}_y^2}{2m_e^*} + V_{\text{eff}}(\rho)\right]\varphi(\vec{\rho}) = (\varepsilon - \varepsilon_{\nu_0})\varphi(\vec{\rho}) \quad (859)$$

where V_{eff} is the effective in plane coulombic potential,

$$V_{\text{eff}}(\rho) = \frac{-e^2}{\kappa}\int_{-\infty}^{+\infty} dz\, \chi_{\nu_0}^2(z)\frac{1}{\sqrt{\rho^2 + (z - z_i)^2}} \quad (860)$$

which is ν_0 and z_i dependent. A solution of Eq. (859) can be sought variationally, the simplest choice for the ground state being the nodeless one-parameter trial wavefunction,

$$\varphi_0(\rho) = \frac{1}{\lambda}\sqrt{\frac{2}{\pi}}\exp(-\rho/\lambda) \quad (861)$$

where λ is the variational parameter. The bound state energy is obtained through minimization of the function

$$\varepsilon_{\nu_0}(z_i, \lambda) = \frac{\hbar}{2m_e^*\lambda^2} - \frac{2e^2}{\kappa\lambda}\int_0^\infty xe^{-x} \\ \times \int_{-\infty}^{+\infty}\frac{dz\,\chi_{\nu_0}^2(z)}{\sqrt{x^2 + \frac{4}{\lambda^2}(z - z_i)^2}}\,dx + \varepsilon_{\nu_0} \quad (862)$$

and the binding energy is

$$\varepsilon_{b\nu_0}(z_i) = \varepsilon_{\nu_0} - \min_\lambda \varepsilon_{\nu_0}(z_i, \lambda) \quad (863)$$

One should be aware that the decoupling procedure (Eqs. (858)–(863)) runs into difficulties in the limits $L \to \infty$ and $L \to 0$. In the former case the energy difference $\varepsilon_{\nu_0+1} - \varepsilon_{\nu_0}$

becomes very small and many subbands become admixed by the coulombic potential. Clearly, for infinite L one cannot describe the bulk $1s$ hydrogenic bound state by a separable wavefunction. In the other case ($L \to 0$) one finds similar problems due to the energy proximity between the ground quantum well subband ε_1 and the top V_0 of the well. Consequently $\chi_1(z)$ leaks more and more heavily in the barrier. At $L = 0$ the only sensible result is to find a $1s$ bulk hydrogenic wavefunction corresponding to the barrier-acting material (in this model the latter has a binding energy equal to R_y^*). Once again this state cannot have a wavefunction like that of Eq. (858). If V_0 is infinite the result at $L = 0$ is qualitatively different. The quantum well only has bound levels whose energy separation increases like L^{-2} when L decreases. The smaller the well thickness, the better the separable wavefunction becomes. At $L = 0$, one obtains a true two-dimensional hydrogenic problem whose binding energy is $4R_y^*$ whereas $\lambda = a_B^*/2$.

To circumvent the previous difficulties and obtain the exact limits at $L = 0$ and $L = \infty$ for any V_0 one may use [330–332]

$$\psi_{\text{loc}}(\vec{r}) = N\chi_1(z)\exp\left[-\frac{1}{\lambda}\sqrt{\rho^2 + (z - z_i)^2}\right] \quad (864)$$

where N is a normalization constant, λ is the variational parameter, and here attention is focused on the ground bound state attached to the ground quantum well subband ε_1. Calculations are less simple than in the case of the separable wavefunction (Eq. (858)). Comparing the binding energy deduced from Eqs. (858)–(864) one finds, for infinite V_0, that the separable wavefunction gives almost the same results as the nonseparable one, if $L/a_B^* \leq 3$. This is the range where the quantum size effects are important for most materials.

Other variational calculations have been proposed [333, 334]. For example, instead of using a nonlinear variational parameter one uses a finite basis set of fixed wavefunctions (often Gaussian ones) in which $\widehat{H}$ is numerically diagonalized. The numerical results obtained by using a single, nonlinear variational parameter compare favorably with these very accurate treatments.

10.1.2. Results for the Ground Impurity State Attached to the Ground Subband

Figures 85 to 87 give a sample of some calculated results for the impurity ground state attached to the ground subband. Two parameters control the binding energy:

(i) Thickness dependence of the impurity binding energy: the dimensionless ratio L/a_B^* indicates the amount of two-dimensionality of the impurity state. If $L/a_B^* \geq 3$ or $L/a_B^* \leq 0.2$ and $V_0 \approx 3$ eV in GaAs–Ga$_{1-x}$Al$_x$As, the problem is almost three-dimensional. This is either because the subbands are too close ($L/a_B^* \geq 3$) or because the quantum well continuum is too close ($L/a_B^* \leq 0.2$). The on-center donor binding energy increases from $R_y^*(L \to \infty)$ to reach a maximum ($L/a_B^* \leq 1$) whose exact L-location and amplitude

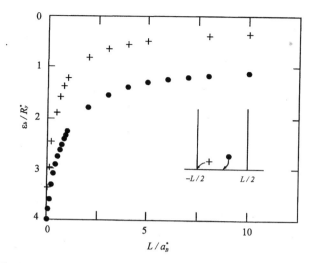

Fig. 85. Calculated dependence of the on-edge (crosses) and on-center (circles) hydrogenic donor binding energy versus the well thickness L in a quantum well with an infinite barrier height.

depend on V_b. Finally, it decreases to the value R_y^* at $L = 0$ [333, 334]. If V_0 is infinite, the maximum is only reached at $L = 0$ and has a value of $4R_y^*$ [330].

(ii) Position dependence of the impurity binding energy: the impurity binding energy monotonically decreases when the impurity location z_i moves from the center to the edge of the well and finally deep into the barrier. In Figure 88 it is shown that this decrease is rather slow for $z_i > L/2$; for instance, a donor placed 150 Å away from a GaAs–Ga$_{0.7}$Al$_{0.3}$As quantum well ($L = 94.8$ Å) still binds a state by $\approx 0.5R_y^*$ which is ≈ 2.5 meV [331].

10.1.3. Excited Subbands: Continuum

The procedure followed for the bound state attached to the ε_1 subband can be generalized for excited subbands $\varepsilon_2, \varepsilon_3, \ldots$ as well as for the quantum well continuum.

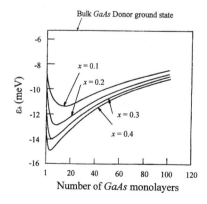

Fig. 86. Dependence of on-center hydrogenic donor binding energy in GaAs–Ga$_{1-x}$Al$_x$As quantum wells versus the GaAs slab thickness L. $V_b(x) = 0.85 [\varepsilon_g (\text{Ga}_{1-x}\text{Al}_x\text{As}) - \varepsilon_g(\text{GaAs})]$. One monolayer is 2.83 Å thick.

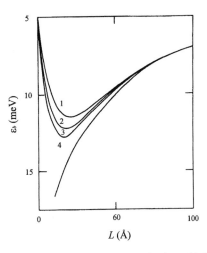

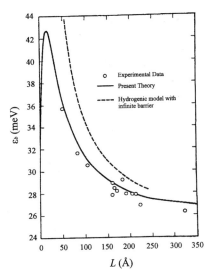

Fig. 87. Dependence of the on-edge hydrogenic donor binding energy in GaAs–Ga$_{1-x}$Al$_x$As quantum wells versus the GaAs slab thickness L. $m_e^* = 0.067 m_0$. $\kappa = 13.1$. $V_0 = 212$ meV (curve 1), 318 meV (curve 2), 424 meV (curve 3), and infinite (curve 4).

Fig. 89. Dependence of the on-center carbon binding energy versus well width in GaAs–Ga$_{1-x}$Al$_x$As quantum wells. $V_b(x) = 0.15$ [ε_g(Ga$_{1-x}$Al$_x$As)$-\varepsilon_g$(GaAs)] is the assumed hole confining barrier height. The open circles are the experimental values obtained by Miller et al. [154].

However, it becomes rapidly cumbersome, since a correct variational procedure for excited states requires the trial wavefunctions to be orthogonal to all the states of lower energies. For separable wavefunctions (Eq. (858)), this requirement is automatically fulfilled and one may safely minimize $\varepsilon_{\nu_0}(z_i, \lambda)$ to obtain a lower bound of $\varepsilon_{b\nu_0}(z_i)$. The new feature associated with the bound states attached to excited subbands is their finite lifetime. This is due to their degeneracy with the two-dimensional continua of the lower lying subbands. This effect, however, is not very large [328, 329], since it arises from the intersubband coupling induced by the coulombic potential: if the decoupling procedure is valid, the lifetime of the quasi-bound state calculated with Eq. (858) should be long.

For impurities located in the barriers (an important practical topic with regard to the modulation-doping technique), it has been seen that they weakly bind a state below the ε_1 edge. There exists (at least) a second quasi-discrete level attached to the barrier edge, i.e., with energy $\approx V_0 - R_y^*$. This state is reminiscent of the hydrogenic ground state of the bulk barrier. Due to the presence of the quantum well slab, it becomes a resonant state since it interferes with the two-dimensional continua of the quantum well subbands $\nu(\varepsilon_\nu < V_0)$. Its lifetime τ can be calculated using the expression

$$\frac{\hbar}{2\tau} = \frac{2\pi}{\hbar} \sum_{\nu, \vec{k}_\perp} \left| \left\langle \tilde{\varphi}(\vec{r}, z_i) \left| \frac{-e^2}{\kappa \sqrt{\rho^2 + (z - z_i)^2}} \right| \nu \vec{k}_\perp \right\rangle \right|^2 \times \delta\left(V_0 - R_y^* - \varepsilon_\nu - \frac{\hbar^2 k_\perp^2}{2 m_e^*} \right) \quad (865)$$

where $\tilde{\varphi}$ is the $1s$ bulk hydrogenic wavefunction of the barrier. If the distance d separating the impurity from the quantum well edge is much larger than a_B^*, Eq. (865) simplifies to

$$\frac{\hbar}{2\tau} \approx 16 R_y^* \sum_\nu P_\nu \left(\frac{R_y^*}{V_0 - \varepsilon_\nu} \right)^{3/2} \exp\left(-\frac{2d}{a_B^*} \sqrt{\frac{V_0 - \varepsilon_\nu}{R_y^*}} \right) \quad (866)$$

where P_ν is the total integrated probability of finding the carrier in the νth state (energy ε_ν) in any of the two barriers of the quantum well. One sees from Eq. (866) that the lifetime τ is strongly dominated by the escape processes to the excited subband ε_ν, whose energy is nearest to $V_0 - R_y^*$. As an example let us take a GaAs–GaAlAs quantum well, with thickness $L = 50$ Å, $V_0 = 0.2$ eV and assume that $d = 3a_B^*$ (i.e., $d \approx 300$ Å). Then $\tau \approx 3 \times 10^{-6}$ s is reached. The quasi-bound state can thus be considered, to a reasonable approximation, as stationary ($\hbar/\tau \approx 4 \times 10^{-8} R_y^*$).

10.1.4. Excited Impurity Levels Attached to the Subband

The Schrödinger equation (Eq. (855)) has several bound states below ε_1. Their binding energies have been calculated by

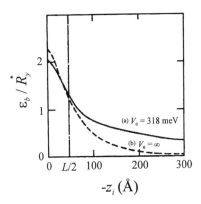

Fig. 88. Dependence of the hydrogenic donor binding energy in a quantum well versus the impurity position z_i (a) in the case of a finite barrier well ($V_0 = 318$ meV) and (b) in the case of an infinite barrier well [147]. There is an interface at $-z_i = L/2$. $L = a_B^* = 94.8$ Å.

several groups [333, 334]. The calculated energy difference between the on-center donor ground state (quasi-1s) and the excited states (quasi-2p_x, 2p_y) agrees with the far-infrared absorption and magneto-absorption data [335]. On-edge donor levels have also been investigated.

10.1.5. Acceptor Levels in a Quantum Well

The problem of acceptor levels in semiconductor quantum wells is much more intricate than the equivalent donor problem. This is due to the degenerate nature of the valence bands in cubic semiconductors. In quantum wells this degeneracy is lifted (light and heavy holes have different confinement energies).

However, the energy separation between the heavy and light hole subbands is seldom comparable to the bulk acceptor binding energies. Thus, many subbands are coupled by the combined actions of the coulombic potential and the quantum well confining barrier potential. No simple decoupling procedures appear manageable.

Masselink et al. [336] used variational calculations to estimate the binding energy of the acceptor level due to carbon (a well known residual impurity in MBE grown GaAs layers). Their results, shown in Figure 89, agree remarkably with Miller et al.'s experiments [337]. One notes in Figure 89 the same trends versus quantum well thickness as displayed by coulombic donors (Figs. 86 and 87). The binding energy first increases when L decreases (increasing tendency to two-dimensional behavior), then saturates, and finally drops to the value of the acceptor binding energy of the bulk barrier. Finally, it should be noted that the relative increase in binding is smaller for acceptors than for donors. This is because the bulk acceptor Bohr radius is much smaller than that of the bulk donor.

REFERENCES

1. A. Y. Cho and J. R. Arthur, *Progr. Solid State Chem.* 10, 157 (1975).
2. R. Ueda (Ed.), "Collected Papers of the 2nd Int. Symp. On Molecular Beam Epitaxy and Clean Surface Techniques," Japanese Society of Applied Physics, Tokyo, 1982.
3. K. Ploog and K. Graf, "Molecular Beam Epitaxy of III–V Compounds— A Comprehensive Bibliography," Springer-Verlag, Berlin, 1984.
4. J. Arthur (Ed.), "Proc. 3rd Int. Conf. on MBE," San Francisco, 1985, *J. Vac. Sci. Technol. B* 3, 509 (1985).
5. E. H. C. Parker (Ed.), "Technology and Physics of MBE," Plenum, New York, 1985.
6. C. T. Foxon and J. J. Harris (Eds.), "Proc. 4th Int. Conf. on MBE, York, UK," *J. Cryst. Growth* 81, 1 (1987).
7. R. J. Marik (Ed.), "Molecular Beam Epitaxy in III–V Semiconductor Materials and Devices." Elsevier, Amsterdam/New York, 1989.
8. Y. Shiraki and H. Sakaki (Eds.), "Proc. 5th Int. Conf. on MBE," Sapporo, 1988, *J. Cryst. Growth* 95, 1 (1989).
9. C. W. Tu and J. S. Harris (Eds.), "Proc. 6th Conf. on MBE," La Jolla, USA, 1990, *J. Cryst. Growth* 111, (1991).
10. Y. Yamada, in "Optical Properties of Low-Dimensional Materials," (T. Ogawa and Y. Kanemitsu, Eds.), p. 202. World Scientific, Singapore, 1995.
11. Y. Endoh and T. Taguchi, *Mater. Res. Soc. Symp. Proc.* 161, 211 (1990).
12. T. Taguchi and Y. Endoh, *Japanese J. Appl. Phys.* 30, L952 (1991).
13. T. Taguchi, Y. Endoh, and Y. Nozue, *Appl. Phys. Lett.* 59, 3434 (1991).
14. Y. Yamada, T. Taguchi, and A. Hiraki, *Tech. Rep. Osaka Univ.* 39, 211 (1989).
15. T. Taguchi, Y. Kawakami, and Y. Yamada, *Phys. B* 191, 23 (1993).
16. A. Muray, M. Isaacson, and I. Adesida, *Appl. Phys. Lett.* 45, 1289 (1984).
17. L. D. Jackel, R. E. Howard, P. M. Mankiewich, H. G. Craighead, and R. W. Epworth, *Appl. Phys. Lett.* 45, 698 (1984).
18. R. L. Kubena, F. P. Stratton, J. W. Ward, G. M. Atkinson, and R. J. Joyce, *J. Vac. Sci. Technol. B* 7, 1798 (1989).
19. S. Furukawa and T. Miyasato, *Phys. Rev. B* 38, 5726 (1988).
20. S. Hayashi, Y. Kanzawa, M. Kataoka, T. Nagareda, and K. Yamamoto, *Z. Phys. D* 26, 144 (1993).
21. M. Yamamoto, R. Hayashi, K. Tsunetomo, K. Kohno, and Y. Osaka, *Japanese J. Appl. Phys.* 30, 136 (1991).
22. S. Hayashi, M. Fujii, and K. Yamamoto, *Japanese J. Appl. Phys.* 28, L1464 (1989).
23. Y. Maeda, *Phys. Rev. B* 51, 1658 (1995).
24. R. Hayashi, M. Yamamoto, K. Tsunemoto, K. Kohno, Y. Osaka, and H. Nasu, *Japanese J. Appl. Phys.* 29, 756 (1990).
25. R. Nötzel, N. N. Ledentsov, L. Däweritz, M. Hohenstein, and K. Ploog, *Phys. Rev. Lett.* 67, 3812 (1991).
26. S. Fafard, D. Leonard, J. L. Merz, and P. M. Petroff, *Appl. Phys. Lett.* 65, 1388 (1994).
27. J. Y. Marzin, J. M. Gerard, A. Izrael, D. Barrier, and G. Bastard, *Phys. Rev. Lett.* 73, 716 (1994).
28. P. D. Wang, N. N. Ledentsov, C. M. Sotomayor-Torres, P. S. Kop'ev, and V. M. Ustinov, *Appl. Phys. Lett.* 64, 1526 (1994).
29. K. Georgsson, N. Carlsson, L. Samuelson, W. Seifert, and L. R. Wallenberg, *Appl. Phys. Lett.* 67, 2981 (1995).
30. D. E. Jesson, K. M. Chen, and S. J. Pennycook, *Mater. Res. Soc. Bull.* 21/4, 31 (1995).
31. M. Lowisch, M. Rabe, B. Stegeman, F. Henneberger, M. Grundmann, V. Tuerck, and D. Bimberg, *Phys. Rev. B* 54, R11074 (1996).
32. F. Hatami, N. N. Ledentsov, M. Grundmann, J. Bohrer, F. Heinrichsdorff, M. Beer, D. Bimberg, S. S. Ruvimov, P. Werner, U. Gosele, J. Heydenreich, U. Richter, S. V. Ivanov, B. Ya. Meltser, P. S. Kop'ev, and Zh. I. Alferov, *Appl. Phys. Lett.* 67, 656 (1995).
33. C. Priester and M. Lannoo, *Phys. Rev. Lett.* 75, 93 (1995).
34. J. Tersoff, C. Teichert, and M. G. Lagally, *Phys. Rev. Lett.* 76, 1675 (1996).
35. G. Bastard, "Wave Mechanics Applied to Semiconductor Heterostructures," Monographs of Physics Series. Wiley, New York, 1991.
36. G. Bastard, *Phys. Rev. B* 24, 5693 (1981).
37. G. Bastard, *Phys. Rev. B* 25, 7584 (1982).
38. D. F. Nelson, R. C. Miller, C. W. Tu, and S. K. Sputz, *Phys. Rev. B* 36, 8063 (1987).
39. D. J. Ben Daniel and C. B. Duke, *Phys. Rev.* 152, 683 (1996).
40. R. L. Greene and K. K. Bajaj, *Solid State Commun.* 45, 831 (1983).
41. R. L. Greene, K. K. Bajaj, and D. E. Phelps, *Phys. Rev. B* 29, 1807 (1984).
42. G. Bastard, E. E. Mendez, L. L. Chang, and L. Esaki, *Phys. Rev. B* 26, 1974 (1982).
43. R. C. Miller, D. A. Kleinman, W. T. Tsang, and A. C. Gossard, *Phys. Rev. B* 24, 1134 (1981).
44. F. F. Fang and W. E. Howard, *Phys. Rev. Lett.* 16, 797 (1966).
45. S. S. Nedorezov, *Sov. Phys. Solid State* 12, 1814 (1971).
46. D. L. Smith and C. Mailhiot, *Rev. Modern Phys.* 62, 173 (1990).
47. D. L. Smith and C. Mailhiot, *Phys. Rev. B* 33, 8345 (1986).
48. S. Schmitt-Rink, D. S. Chemla, and D. A. B. Miller, *Adv. Phys.* 38, 89 (1989).
49. T. Stirner, P. Harrison, W. E. Hagston, and J. P. Goodwin, *Phys. Rev. B* 50, 5713 (1994).
50. G. Bastard, J. A. Brum, and R. Ferreria, in "Solid State Physics, Advances in Research and Applications" (H. Ehrenreich and D. Turnbul, Eds.). Academic Press, New York, 1991.

51. S. R. Jackson, J. E. Nicholls, W. E. Hagston, P. Harrison, T. Stirner, J. H. C. Hogg, B. Lunn, and D. E. Ashenford, *Phys. Rev. B* 50, 5392 (1994).
52. P. Harrison, F. Long, and W. E. Hagston, *Superlattices Microstruct.* 19, 123 (1996).
53. M. V. Rama Krishna and R. A. Friensner, *Phys. Rev. Lett.* 67, 629 (1991).
54. M. G. Burt, *Semicond. Sci. Technol.* 3, 739 (1988).
55. M. G. Burt, *J. Phys. Condens. Matter* 4, 6651 (1992).
56. M. G. Burt, *Phys. Rev. B* 50, 7518 (1994).
57. B. A. Foreman, *Phys. Rev. B* 52, 12241 (1995).
58. P. Vogl, H. P. Hjalmarson, and J. Dow, *J. Phys. Chem. Solids* 44, 365 (1983).
59. M. Jaros, K. B. Wong, and M. A. Gell, *Phys. Rev. B* 31, 1205 (1985).
60. L.-W. Wang and A. Zunger, *J. Chem. Phys.* 100, 2394 (1994).
61. M. A. Gell, D. Ninno, M. Jaros, and D. C. Herbert, *Phys. Rev. B* 34, 2416 (1986).
62. K. Mäder, L.-W. Wang, and Z. Zunger, *Phys. Rev. Lett.* 74, 2555 (1995).
63. K. Mäder, L.-W. Wang, and Z. Zunger, *J. Appl. Phys.* 78, 6639 (1995).
64. L.-W. Wang, A. Zunger, and K. Mäder, *Phys. Rev. B* 53, 2010 (1996).
65. D. M. Wood and A. Zunger, *Phys. Rev. B* 53, 7948 (1996).
66. R. A. Morrow and K. R. Brwnstein, *Phys. Rev. B* 30, 678 (1984).
67. R. A. Morrow, *Phys. Rev. B* 35, 8074 (1987).
68. S. Nojima, *Japanese J. Appl. Phys.* 31, L1401 (1992).
69. B. A. Foreman, *Phys. Rev. B* 48, 4964 (1993).
70. E. O. Kane, in "Handbook on Semiconductors," (W. Paul, Ed.). North-Holland, Amsterdam, 1982.
71. J. R. Chelikowsky and M. L. Cohen, *Phys. Rev. B* 14, 556 (1976).
72. R. Eppenga, M. F. H. Schuurmans, and S. Colak, *Phys. Rev. B* 36, 1554 (1987).
73. C. Y.-P. Chao and S. L. Chuang, *Phys. Rev. B* 46, 4110 (1992).
74. D. A. Broido and L. J. Sham, *Phys. Rev. B* 31, 888 (1985).
75. U. Ekenberg, W. Batty, and E. P. O'Reilly, *J. Phys. Colloq.* 48, C5-553 (1987).
76. F. Long and W. E. Hagston, *J. Appl. Phys.* 82, 3414 (1997).
77. U. Ekenberg and M. Altarelli, *Phys. Rev. B* 35, 7585 (1987).
78. D. Gershoni, C. H. Henry, and D. A. Baraff, *IEEE J. Quantum Electron.* 29, 2433 (1993).
79. E. E. Mendez, G. Bastard, L. L. Chang, L. Esaki, H. Morkoc, and R. Fisher, *Phys. B* 117/118, 711 (1983).
80. C. Alibert, S. Gaillard, J. A. Brum, G. Bastard, P. Frijlink, and R. Erman, *Solid State Commun.* 53, 457 (1985).
81. T. H. Wood, C. A. Burrus, D. A. B. Miller, D. S. Chemla, T. C. Damen, A. C. Gossard, and W. Wiegmann, *Appl. Phys. Lett.* 44, 16 (1984).
82. D. S. Chemla and D. A. B. Miller, in "Heterojunction B and Discontinuities: Physics and Device Applications" (F. Capasso and G. Margaritondo, Eds.). North-Holland, Amsterdam, 1987.
83. G. Bastard, E. E. Mendez, L. L. Chang, and L. Esaki, *Phys. Rev. B* 28, 3241 (1983).
84. M. Abramowitz and I. A. Stegun, "Handbook of Mathematical Functions," Dover, New York, 1970.
85. J. A. Brum and G. Bastard, *Phys. Rev. B* 31, 3893 (1985).
86. P. N. Argyres and L. M. Roth, in "Physics of III–V Compounds," Vol. 1 (R. K. Willardson and A. C. Beer, Eds.). Academic Press, New York, 1966.
87. J. C. Maan, *Springer Ser. Solid-States Sci.* 53, 183 (1984).
88. T. Ando, *J. Phys. Soc. Japan* 50, 2978 (1981).
89. S. Das Sarma, *Surf. Sci.* 142, 341 (1984).
90. H. Haug and S. W. Koch, "Quantum Theory of the Optical and Electronic Properties of Semiconductors." World Scientific, Singapore, 1990.
91. C. Klingshirn, "Semiconductor Optics." Springer-Verlag, Berlin, 1995.
92. T. Kawamura and S. Das Sarma, *Phys. Rev. B* 45, 3612 (1992).
93. B. Hillebrands, S. Lee, G. I. Stegeman, H. Cheng, J. E. Potts, and F. Nizzoli, *Phys. Rev. Lett.* 60, 832 (1988).
94. J. Seyler and M. N. Wybourne, *Phys. Rev. Lett.* 69, 1427 (1992).
95. H. Benistry, C. M. Sotomayor-Torres, and C. Weisbuch, *Phys. Rev. B* 44, 10945 (1991).
96. A. Auld, "Acoustic Fields and Waves." Wiley, New York, 1973.
97. P. J. Price, *Ann. Phys.* 133, 217 (1981).
98. N. Mori and T. Ando, *Phys. Rev. B* 40, 6175 (1989).
99. M. A. Stroscio, *Phys. Rev. B* 40, 6428 (1989).
100. K. W. Kim and M. A. Stroscio, *J. Appl. Phys.* 68, 6289 (1990).
101. K. W. Kim, M. A. Stroscio, A. Bhatt, V. V. Mitin, and R. Mickevicius, *J. Appl. Phys.* 70, 319 (1991).
102. M. A. Stroscio, K. W. Kim, G. J. Iafrate, M. Dutta, and H. L. Grubin, *Philos. Mag. Lett.* 65, 173 (1992).
103. N. C. Constantinou, in "Phonons in Semiconductor Nanostructures," NATO ASI Ser. E, Vol. 236 (J.-P. Leburton, J. Pascual, and C. M. Sotomayor-Torres, Eds.). Kluwer, Boston, 1993.
104. M. A. Stroscio and K. W. Kim, *Phys. Rev. B* 48, 1936 (1993).
105. S. M. Komirenko, K. W. Kim, M. A. Stroscio, and V. A. Kochelap, *Phys. Rev. B* 58, 16360 (1998).
106. H. Akera and T. Ando, *Phys. Rev. B* 40, 2914 (1989).
107. H. Rücker, E. Molinari, and P. Lugli, *Phys. Rev. B* 44, 3463 (1991).
108. H. Rücker, E. Molinari, and P. Lugli, *Phys. Rev. B* 45, 6747 (1992).
109. E. Molinari, S. Baroni, P. Giannozzi, and S. Gironcoli, *Phys. Rev. B* 45, 4280 (1992).
110. K. J. Nash, *Phys. Rev. B* 46, 7723 (1992).
111. G. Weber, *Phys. Rev. B* 46, 16171 (1992).
112. R. Chen, D. L. Lin, and T. F. George, *Phys. Rev. B* 41, 1435 (1990).
113. C. Trallero-Giner and F. Comas, *Phys. Rev. B* 37, 4583 (1988).
114. S.-F. Ren, H. Chu, and Y.-C. Chang, *Phys. Rev. B* 37, 8899 (1988).
115. F. Bechsted and H. Gerecke, *Phys. Status Solidi B* 156, 151 (1989).
116. S. Rudin and T. L. Reinecke, *Phys. Rev. B* 41, 7713 (1990); S. Rudin and T. L. Reinecke, *Phys. Rev. B* 43, 9228 (1991) (errata).
117. T. Tsuchiya and T. Ando, *Phys. Rev. B* 47, 7240 (1993).
118. N. C. Constantinou and B. K. Ridley, *Phys. Rev. B* 41, 10622 (1990).
119. M. A. Stroscio, K. W. Kim, M. A. Littlejohn, and H. Chuang, *Phys. Rev. B* 42, 1488 (1990).
120. P. A. Knipp and T. L. Reinecke, *Phys. Rev. B* 45, 9091 (1992).
121. P. A. Knipp and T. L. Reinecke, *Phys. Rev. B* 48, 57000 (1993).
122. R. Enderlein, *Phys. Rev. B* 47, 2162 (1993).
123. M. Watt, C. M. Sotomayor-Torres, H. E. G. Arnot, and S. P. Beaumont, *Semicond. Sci. Technol.* 5, 285 (1990).
124. X. F. Wang and X. L. Lei, *Phys. Rev. B* 49, 4780 (1994).
125. M. C. Klein, F. Hache, D. Ricard, and C. Flytzanis, *Phys. Rev. B* 42, 11123 (1990).
126. S. Nomura and T. Kobayashi, *Phys. Rev. B* 45, 1305 (1992).
127. M. P. Chamberlain, M. Cardona, and B. K. Ridley, *Phys. Rev. B* 48, 14356 (1993).
128. F. Comas, R. Pérez-Alvarez, C. Trallero-Giner, and M. Cardona, *Superlattices Microstruct.* 14, 95 (1993).
129. F. Comas and C. Trallero-Giner, *Phys. B* 192, 394 (1993).
130. R. Pérez-Alvarez, F. García-Moliner, V. R. Velasco, and C. Trallero-Giner, *J. Phys. Condens. Matter* 5, 5389 (1993).
131. F. Comas, C. Trallero-Giner, and A. Cantarero, *Phys. Rev. B* 47, 7602 (1993).
132. C. Trallero-Giner, *Phys. Scripta T* 55, 50 (1994).
133. C. Trallero-Giner and F. Comas, *Philos. Mag. B* 70, 583 (1994).
134. F. Comas, A. Cantarero, C. Trallero-Giner, and M. Moshinsky, *J. Phys. Condens. Matter* 7, 1789 (1995).
135. E. Roca, C. Trallero-Giner, and M. Cardona, *Phys. Rev. B* 49, 13704 (1994).
136. C. Trallero-Giner, F. García-Moliner, V. R. Velasco, and M. Cardona, *Phys. Rev. B* 45, 11944 (1992).
137. V. M. Fomin and E. P. Pokatilov, *Phys. Status Solidi B* 132, 69 (1985).
138. M. Babiker, *J. Phys. C* 19, 683 (1986).
139. E. Molinari, A. Fasolino, and K. Kunc, *Superlattices Microstruct.* 2, 398 (1986).
140. K. Huang and B. Zhu, *Phys. Rev. B* 38, 13377 (1988).
141. Z. V. Popovic, M. Cardona, E. Richter, D. Strauch, L. Tapfer, and K. Ploog, *Phys. Rev. B* 41, 5904 (1990).
142. R. Enderlein, *Phys. Rev. B* 43, 14513 (1991).

143. J. E. Zucker, A. Pinczuk, D. S. Chemla, A. C. Gossard, and W. Wiegmann, *Phys. Rev. B* 29, 7065 (1984).

144. A. K. Sood, J. Menéndez, M. Cardona, and K. Ploog, *Phys. Rev. Lett.* 54, 2111 (1985).

145. M. V. Klein, *IEEE J. Quantum Electron.* QE-22, 1760 (1986).

146. M. Cardona, *Superlattices Microstruct.* 7, 183 (1990).

147. D. J. Mowbray, M. Cardona, and K. Ploog, *Phys. Rev. B* 43, 1598 (1991).

148. D. J. Mowbray, M. Cardona, and K. Ploog, *Phys. Rev. B* 43, 11815 (1991).

149. "Light Scattering in Solids V" (M. Cardona and G. Güntherodt, Eds.), Topics of Applied Physics Vol. 66. Springer-Verlag, Heidelberg, 1989.

150. "Heterojunctions and Semiconductor Superlattices," (G. Allan, O. Bastard, N. Bocarra, M. Lannoo, and M. Voos, Eds.). Springer-Verlag, Heidelberg, 1985.

151. G. Allan, O. Bastard, N. Bocarra, M. Lannoo, and M. Voos, *Phys. Status Solidi B* 167, 115 (1991).

152. B. Ridley, *Phys. Rev. B* 39, 5282 (1989).

153. M. Born and K. Huang, "Dynamical Theory of Crystal Lattices." Clarendon, Oxford, 1988.

154. L. Landau and E. M. Lifshitz, "Course of Theoretical Physics," Vol. 7, Theory of Elasticity. Pergamon, Oxford, 1970.

155. T. Tsuchiya, H. Akera, and T. Ando, *Phys. Rev. B* 39, 6025 (1989).

156. M. E. Mora-Ramos and D. A. Contreras-Solorio, *Phys. B* 253, 325 (1998).

157. C. Trallero-Giner, F. Comas, and F. García-Moliner, *Phys. Rev. B* 50, 1755 (1994).

158. S. N. Klimin, E. P. Pokatilov, and V. M. Fomin, *Phys. Status Solidi B* 190, 441 (1995).

159. M. E. Mora-Ramos, Ph.D. Thesis, University of Havana, 1995.

160. G. D. Mahan, "Many-Particle Physics." Plenum, New York, 1990.

161. R. Zheng, Sh. Ban, and X. X. Liang, *Phys. Rev. B* 49, 1796 (1994).

162. G. Q. Hai, F. M. Peeters, and J. T. Devreese, *Phys. Rev. B* 42, 11063 (1990).

163. D. L. Lin, R. Chen, and T. F. George, *J. Phys. Condensed Matter* 3, 4645 (1991).

164. T. Lu and Y. Zheng, *Phys. Rev. B* 53, 1438 (1996).

165. J. J. Shi, X. Q. Zhu, Z. X. Liu, Sh. H. Pan, and X. Y. Li, *Phys. Rev. B* 55, 4670 (1997).

166. A. Chubykalo, V. R. Velasco, and F. García-Moliner, *Surf. Sci.* 319, 184 (1994).

167. M. E. Mora-Ramos and C. A. Duque, *Phys. Status Solidi B* 200, 159 (2000).

168. R. Pérez-Alvarez, V. R. Velasco, and F. García-Moliner, *Phys. Scripta* 51, 526 (1995).

169. R. Dingle, H. L. Stormer, A. C. Gossard, and W. Wiegmann, *Appl. Phys. Lett.* 33, 665 (1978).

170. T. Miura, S. Hiyamizu, T. Fujii, and K. Nanbu, *Japanese. J. Appl. Phys.* 19, L225 (1980).

171. L. Esaki and R. Tsu, *IBM J. Res. Dev.* 14, 61 (1970).

172. W. Kohn and J. M. Luttinger, *Phys. Rev.* 108, 590 (1957).

173. J. M. Luttinger and W. Kohn, *Phys. Rev.* 109, 1892 (1958).

174. D. Calecki and J. Tavernier, "Introduction aux phénoménes de transport linéaires dans les semiconducteurs." Mason. Paris, 1969.

175. A. Messiah, "Mécanique Quantique." Dunod, Paris, 1959.

176. T. Ando, A. B. Fowler, and F. Stern, *Rev. Modern. Phys.* 54, 437 (1982).

177. E. D. Siggia and P. C. Kwok, *Phys. Rev. B* 2, 1024 (1970).

178. S. Mori and T. Ando, *J. Phys. Soc. Japan* 48, 865 (1980).

179. H. L. Stormer, A. C. Gossard, and W. Wiegmann, *Solid State Commun.* 41, 707 (1982).

180. G. Fishman and D. Calecki, *Phys. Rev. B* 29, 5778 (1984).

181. F. J. Fernández-Velicia, F. García-Moliner, and V. R. Belasco, *Phys. Rev. B* 53, 2034 (1996).

182. F. Stern and W. Howard, *Phys. Rev. Lett.* 18, 546 (1967).

183. C. Kittel, "Quantum Theory of Solids." Wiley, New York, 1963.

184. J. Lee and H. N. Spector, *J. Appl. Phys.* 54, 6989 (1983).

185. H. Ehrenreich and M. H. Cohen, *Phys. Rev.* 115, 786 (1959).

186. J. Lindhard, *Kgl. Danske Videnskab. Selskab. Mat. Fys. Medd.* 28, 8 (1954).

187. J. M. Ziman, "Principles of the Theory of Solids," Cambridge Univ. Press, London, 1972.

188. E. J. Woll, Jr. and W. Kohn, *Phys. Rev.* 126, 1693 (1962).

189. V. Arora and F. Awad, *Phys. Rev. B* 23, 5570 (1981).

190. B. Fell, J. T. Chen, J. Hardy, M. Prasad, and S. Fujita, *J. Phys. Chem. Solids* 34, 221 (1978).

191. J. Lee, H. Spector, and V. Arora, *Appl. Phys. Lett.* 42, 363 (1983).

192. H. León, R. Riera, J. L. Marín, and E. Roca, *Phys. Low-Dim. Struct.* 7/8, 59 (1996).

193. S. Mori and T. Ando, *Phys. Rev. B* 19, 6433 (1979).

194. G. Fishman and D. Calecki, *Phys. B* 117/118, 744 (1983).

195. W. Walukiewicz, H. E. Ruda, J. Lagowski, and H. C. Gatos, *Phys. Rev. B* 29, 4818 (1984).

196. W. Walukiewicz, H. E. Ruda, J. Lagowski, and H. C. Gatos, *Phys. Rev. B* 30, 4571 (1984).

197. R. B. Darling, *IEEE J. Quantum Electron.* 24, 1628 (1988).

198. H. León and F. Comas, *Phys. Status Solidi B* 160, 105 (1990).

199. K. Inoue and T. Matsuno, *Phys. Rev. B* 47, 3771 (1993).

200. O. Ziep, M. Suhrke, and R. Keiper, *Phys. Status Solidi B* 134, 789 (1986).

201. C. Pratsch and M. Suhrke, *Phys. Status Solidi B* 154, 315 (1989).

202. H. León, F. Comas, and M. Suhrke, *Phys. Status Solidi B* 159, 731 (1990).

203. F. J. Fernández-Velicia, F. García-Moliner, and V. R. Velasco, *J. Phys. (Math. Gen.) A* 28, 391 (1995).

204. P. F. Maldague, *Surf. Sci.* 73, 296 (1978).

205. H. L. Stormer, A. L. Gossard, W. Weigmann, and K. Baldwin, *Appl. Phys.* 39, 912 (1981).

206. C. Nguyen, K. Ensslin, and H. Kroemer, *Surf. Sci.* 267, 549 (1992).

207. H. León, F. García-Moliner, and V. R. Velasco, *Thin Solid Films* 226, 38 (1995).

208. N. W. Ashcroft and N. D. Mermin, "Solid State Physics." Holt, Rinehart and Winston, New York, 1976.

209. J. F. Palmier, H. Le Person, C. Minot, A. Chomette, A. Regreny, and D. Calecki, *Superlattices Microstruct.* 1, 67 (1985).

210. A. Chomette, B. Deveaud, J. Y. Emery, A. Regreny, and B. Lambert, in "Proceedings of the 17th International Conference on the Physics of Semiconductors," San Francisco 1984 (J. D. Chadi and W. A. Harrison, Eds.). Springer-Verlag, New York, 1985.

211. B. Ricco and M. Ya. Azbel, *Phys. Rev. B* 29, 1970, (1984).

212. L. L. Chang, L. Esaki, and R. Tsu, *Appl. Phys. Lett.* 24, 593, (1974).

213. E. E. Mendez, W. I. Wang, B. Ricco, and L. Esaki, *Appl. Phys. Lett.* 47, 415 (1985).

214. R. Tsu and L. Esaki, *Appl. Phys. Lett.* 22, 562 (1973).

215. L. Onsager, *Phys. Rev.* 37, 405 (1931).

216. L. Onsager, *Phys. Rev.* 38, 2265 (1931).

217. K. Von Klitzing, in "Festkörperprobleme Advances in Solid State Physics" (J. Treusch, Ed.). Vieweg, Branschweig, 1981.

218. R. Kubo, S. B. Miyake, and N. Hashitsume, *Solid State Phys.* 17, 269 (1965).

219. H. Aoki and T. Ando, *Phys. Rev. Lett.* 54, 831 (1985).

220. T. Ando and Y. Uemura, *J. Phys. Soc. Japan* 36, 959 (1974).

221. K. Von Klitzing, G. Dorda, and M. Pepper, *Phys. Rev. Lett.* 45, 494 (1980).

222. D. C. Tsui and A. C. Gossard, *Appl. Phys. Lett.* 38, 550 (1981).

223. Y. Guldner, J. P. Hirtz, J. P. Vieren, P. Voisin, M. Voos, and M. Razeghi, *J. Phys. Lett. France* 43, L-613 (1982).

224. C. A. Chang, E. E. Mendez, L. L. Chang, and L. Esaki, *Surf. Sci.* 142, 598 (1984).

225. E. E. Mendez, L. L. Chang, C. A. Chang, L. F. Alexander, and L. Esaki, *Surf. Sci.* 142, 215 (1984).

226. H. L. Stormer, A. M. Chang, Z. Schlesinger, D. C. Tsui, A. C. Gossard, and W. Wiegmann, *Phys. Rev. Lett.* 51, 126 (1983).

227. M. A. Paalanen, D. C. Tsui, A. C. Gossard, and J. C. M. Hwang, *Solid State Commun.* 50, 841 (1984).

228. G. Ebert, K. Von Klitzing, C. Probst, E. Schuberth, K. Ploog, and G. Weimann, *Solid State Commun.* 45, 625 (1983).

229. M. A. Paalanen, D. C. Tsui, and A. C. Gossard, *Phys. Rev. B* 25, 5566 (1982).

230. H. Aoki and T. Ando, *Solid State Commun.* 38, 1079 (1981).

231. R. E. Prange, *Phys. Rev. B* 23, 4802 (1981).

232. R. Laughlin, *Phys. Rev. B* 23, 5632 (1981).

233. R. B. Laughlin, *Springer Ser. Solid-State Sci.* 53, 272 (1984).

234. D. C. Tsui, H. L. Stormer, and A. C. Gossard, *Phys. Rev. Lett.* 48, 1559 (1982).

235. D. C. Tsui, H. L. Stormer, J. C. M. Hwang, J. S. Brooks, and M. J. Maughton, *Phys. Rev. B* 28, 2274 (1983).

236. E. E. Mendez, M. Heiblum, L. L. Chang, and L. Esaki, *Phys. Rev. B* 28, 4886 (1983).

237. H. L. Stormer, A. M. Chang, D. C. Tsui, J. C. M. Hwang, A. C. Gossard, and W. Wiegmann, *Phys. Rev. Lett.* 50, 1953 (1983).

238. R. B. Laughlin, *Phys. Rev. B* 27, 3383 (1983).

239. R. B. Laughlin, *Phys. Rev. Lett.* 50, 1395 (1983).

240. B. I. Halperin, *Helv. Phys. Acta* 56, 75 (1983).

241. R. Tao and D. J. Thouless, *Phys. Rev. B* 28, 1142 (1983).

242. T. Ando, *J. Phys. Soc. Japan* 38, 989 (1975).

243. G. Abstreiter, J. P. Kothaus, J. F. Koch, and G. Dorda, *Phys. Rev. B* 14, 2480 (1976).

244. P. Voisin, Y. Guldner, J. P. Vieren, M. Voos, J. C. Maan, P. Delescluse, and N. T. Linh, *Phys. B* 117/118, 634 (1983).

245. H. L. Stormer, R. Dingle, A. C. Gossard, W. Wiegmann, and M. D. Sturge, *Solid State Commun.* 29, 705 (1979).

246. E. Gornik, W. Seidenbush, R. Lassnig, H. L. Stormer, A. C. Gossard, and W. Wiegmann, *Springer Ser. Solid-State Sci.* 53, 183 (1984).

247. Th. Englert, J. C. Maan, Ch. Uihlein, D. C. Tsui, and A. C. Gossard, *Phys. B* 117/118, 631 (1983).

248. Y. Guldner, J. P. Vieren, P. Voisin, M. Voos, J. C. Maan, L. L. Chang, and L. Esaki, *Solid State Commun.* 41, 755 (1982).

249. A. Darr, A. Huber, and J. P. Kotthaus, "Physics of Narrow Gap Semiconductors" (J. Rauluszkiewicz, M. Gorska, and E. Kaczmarek, Eds.). P. W. N.—Polish Scientific, Warsaw, 1978.

250. C. Weisbuch and B. Vinter, "Quantum Semiconductor Structures: Fundamentals and Applications." Academic Press, Boston, 1991.

251. R. A. Stradling and P. C. Klipstein, "Growth and Characterisation of Semiconductors." Hilger, Bristol, 1990.

252. G. Fasol, A. Fasolino, and P. Lugli (Eds.), "Spectroscopy of Semiconductor Microstructures." Plenum, New York, 1989.

253. F. H. Pollak and O. J. Glembocki, *Proc. SPIE* 946, 2 (1988).

254. R. R. Alfano, *Proc. SPIE* 793 (1986), 942 (1988), 1282 (1990).

255. D. Heitmann, in "Physics of Nanostructures, Proceedings of the Thirty-Eighth Scottish Universities Summer School in Physics, St. Andrews" (J. H. Davies and A. R. Long, Eds.). Institute of Physics, Bristol, 1992.

256. A. Pinczuk, J. P. Valladares, D. Heiman, L. N. Pfeiffer, and K. W. West, *Surf. Sci.* 229, 384 (1990).

257. M. Born and E. Wolf, "Principles of Optics." Pergamon, Oxford, 1964.

258. J. Pankove, "Optical Processes in Semiconductors." Dover, New York, 1975.

259. H. N. Spector, *Phys. Rev. B* 28, 971 (1983).

260. G. D. Sanders and Y.-C. Chang, *J. Vac. Sci. Technol. B* 3, 1285 (1985).

261. J. N. Schulman and Y.-C. Chang, *Phys. Rev. B* 31, 2056 (1985).

262. R. Dingle, in "Festkörperprobleme" (H. J. Queisser, Ed.), Adv. Solid State Phys., Vol. 15. Pergamon/Vieweg, Braunschweig, 1975.

263. R. C. Miller, D. A. Kleinman and A. C. Gossard, *Phys. Rev. B* 29, 7085 (1984).

264. R. C. Miller, A. C. Gossard, D. A. Kleinman, and O. Munteanu, *Phys. Rev. B* 29, 3740 (1984).

265. M. H. Meynadier, C. Delalande, G. Bastard, M. Voos, F. Alexandre, and J. L. Lievin, *Phys. Rev. B* 31, 5539 (1985).

266. G. Duggan, *J. Vac. Sci. Technol. B* 3, 1224 (1985).

267. H. Kroemer, *Surf. Sci.* 174, 299 (1986).

268. P. Voisin, M. Voos, and M. Razeghi, unpublished.

269. A. F. S. Penna, J. Shah, A. Pinczuk, D. Sivco, and A. Y. Cho, *Appl. Phys. Lett.* 46, 184 (1985).

270. P. Voisin, C. Delalande, M. Voos, L. L. Chang, A. Segmuller, C. A. Chang, and L. Esaki, *Phys. Rev. B* 30, 2276 (1984).

271. J. Y. Marzin, M. Quillec, E. V. K. Rao, G. Leroux, and L. Goldstein, *Surf. Sci.* 142, 509 (1984).

272. J. D. Dow, in "Optical Properties of Solids, New Developments" (B. O. Seraphin, Ed.). North-Holland, Amsterdam, 1976.

273. R. S. Knox, "Theory of Excitons," Solid State Phys. Suppl. 5. Academic Press, New York, 1963.

274. F. L. Lederman and J. D. Dow, *Phys. Rev. B* 13, 1633 (1976).

275. C. Weisbuch, R. Dingle, A. C. Gossard, and W. Wiegmann, *Solid State Commun.* 38, 709 (1981).

276. Y. Masumoto, M. Matsuura, S. Tarucha, and H. Okamoto, *Phys. Rev. B* 32, 4275 (1985).

277. H. B. Bebb and E. W. Williams, in "Semiconductors and Semimetals," Vol. 8, (R. K. Willardson and A. C. Beer, Eds.). Academic Press, New York, 1972.

278. J. Shah, A. Pinczuk, H. L. Stormer, A. C. Gossard, and W. Wiegmann, *Appl. Phys. Lett.* 42, 55 (1983).

279. D. Bimberg, J. Christen, A. Steckenborn, G. Weimann, and W. Schlapp, *J. Lumin.* 30, 562 (1985).

280. P. Dawson, G. Duggan, H. I. Ralph, and K. Woodbridge, *Phys. Rev. B* 28, 7381 (1983).

281. P. Voisin, Thése de Doctorat d'Etat, Paris, 1983.

282. P. Voisin, G. Bastard, and M. Voos, *Phys. Rev. B* 29, 935 (1984).

283. A. Pinczuk, J. Shah, H. L. Stormer, R. C. Miller, A. C. Gossard, and W. Wiegmann, *Surf. Sci.* 142, 492 (1984).

284. T. M. Rice, "Solid State Physics" (H. Ehrenreich, F. Seitz, and D. Turnbull, Eds.), Vol. 32. Academic Press, New York, 1977.

285. C. Weisbuch, R. C. Miller, R. Dingle, A. C. Gossard, and W. Wiegmann, *Solid State Commun.* 37, 219 (1981).

286. G. Bastard, C. Delalande, M. H. Meynadier, P. M. Frijlink, and M. Voos, *Phys. Rev. B* 29, 7042 (1984).

287. C. Delalande, M. H. Meynadier, and M. Voos, *Phys. Rev. B* 31, 2497 (1985).

288. B. Deveaud, J. Y. Emery, A. Chomette, B. Lambert, and M. Baudet, *Superlattices Microstruct.* 1, 205 (1985).

289. M. D. Sturge, J. Hegarty, and L. Goldner, in "Proceedings of the 17th International Conference on the Physics of Semiconductors," San Francisco, 1984 (J. D. Chadi and W. A. Harrison, Eds.). Springer-Verlag, New York, 1985.

290. J. Hegarty, M. D. Sturge, C. Weisbuch, A. C. Gossard, and W. Wiegmann, *Phys. Rev. Lett.* 49, 930 (1982).

291. J. Singh, K. K. Bajaj, and S. Chaudhuri, *Appl. Phys. Lett.* 44, 805 (1984).

292. E. O. Gobel, J. Kuhl, and R. Hoger, *J. Lumin.* 30, 541 (1985).

293. D. Bimberg, J. Christen, A. Steckenborn, G. Weimann, and W. Schlapp, *Appl. Phys. Lett.* 44, 84 (1984).

294. R. Riera, F. Comas, C. Trallero-Giner, and S. T. Pavlov, *Phys. Status Solidi B* 148, 533 (1988).

295. R. F. Wallis and D. L. Mills, *Phys. Rev. B* 2, 3312 (1970).

296. F. Comas, C. Trallero-Giner, I. G. Lang, and S. T. Pavlov, *Fiz. Tverd. Tela* 25, 57 (1985).

297. A. V. Goltsev, I. G. Lang, and S. T. Pavlov, *Phys. Status Solidi B* 94, 37 (1979).

298. E. L. Ivchenko, I. G. Lang, and S. T. Pavlov, *Fiz. Tverd. Tela* 19, 2751 (1977).

299. F. Comas, C. Trallero-Giner, and R. Perez-Alvarez, *J. Phys. C* 19, 6479 (1986).

300. A. V. Goltsev, I. G. Lang, and S. T. Pavlov, *Fiz. Tverd. Tela* 20, 2542 (1978).

301. B. O. Seraphin and H. E. Bennet, "Semiconductor and Semimetal," Vol. 3 (R. K. Willardson and A. C. Beer, Eds.). Academic Press, New York, 1967.

302. H. C. Weisbugh, *Phys. Rev. B* 15, 823 (1977).

303. A. Cros, A. Cantarero, C. Trallero-Giner, and M. Cardona, *Phys. Rev. B* 45, 6106 (1992).

304. A. Cros, A. Cantarero, C. Trallero-Giner, and M. Cardona, *Phys. Rev. B* 46, 12627 (1992).

305. C. Trallero-Giner, T. Ruf, and M. Cardona, *Phys. Rev. B* 41, 3028 (1990).

306. H. R. Trebin, U. Rössler, and R. Ranvaud, *Phys. Rev. B* 20, 686 (1979).

307. T. Ruf, C. Trallero-Giner, R. T. Phillips, and M. Cardona, *Phys. Rev. B* 41, 3039 (1990).

308. M. Dahl, J. Kraus, B. Muller, G. Schaak, and G. Weimann, in "Proceedings of the 3rd International Conference on Phonon Physics" (S. Hunklinger, W. Ludwig, and G. Weiss, Eds.), p. 758. World Scientific, Singapore, 1990.

309. F. Meseguer, F. Calle, C. López, J. M. Calleja, C. Tejedor, K. Ploog, and F. Briones, *Superlatt. Microstruct.* 10, 217 (1991).

310. F. Calle, J. M. Calleja, F. Meseguer, C. Tejedor, L. Viña, C. López, and K. Ploog, *Phys. Rev. B* 44, 1113 (1991).

311. R. Borroff, R. Merlin, J. Pamulapati, P. K. Bhattacharya, and C. Tejedor, *Phys. Rev. B* 43, 2081 (1991).

312. M. Cardona, in "Light Scattering in Solids II" (M. Cardona and G. Güntherodt, Eds.), Topics in Applied Physics, Vol. 50, p. 19. Springer-Verlag, Heidelberg, 1982.

313. A. Cantarero, C. Trallero-Giner, and M. Cardona, *Phys. Rev. B* 39, 8388 (1989).

314. G. E. Pikus and G. L. Bir, *Fiz. Tverd. Tela* 1, 154 (1959).

315. "Heterojunctions and Semiconductor Superlattices" (G. Allan, G. Bastard, N. Boccara, M. Lannoo, and M. Voos, Eds.). Springer-Verlag, Heidelberg, 1986.

316. A. K. Sood, W. Kauschke, J. Menéndez, and M. Cardona, *Phys. Rev. B* 35, 2886 (1987).

317. E. O. Kane, in "Handbook on Semiconductors" (W. Paul, Ed.), Vol. 1, p. 193. North-Holland, Amsterdam, 1983.

318. G. Bastard and J. A. Brum, *IEEE J. Quantum Electron.* 22, 1625 (1986).

319. M. Altarelli, in "Heterojunctions and Semiconductor Superlattices" (G. Allan, O. Bastard, N. Boccara, M. Lannoo, and M. Voos, Eds.), p.12. Springer-Verlag, Heidelberg, 1985.

320. M. Altarelli, in "Interfaces, Quantum Wells and Superlattices," NATO Advanced Study Institute Ser. B (C. R. Leavens and R. Taylor, Eds.), p. 43, Plenum, New York, 1987.

321. R. L. Aggarwal, *Phys. Rev. B* 2, 446 (1970).

322. V. F. Sapega, M. Cardona, K. Ploog, E. L. Ivchenko, and D. N. Mirlin, *Phys. Rev. B* 45, 4320 (1991).

323. G. E. W. Bauer and T. Ando, *Phys. Rev. B* 38, 6015 (1988).

324. A. Cros, A. Cantarero, C. Trallero-Giner, and M. Cardona, *Phys. Rev. B* 45, 11395 (1992).

325. J. M. Luttinger, *Phys. Rev.* 102, 1030 (1956).

326. J. L. Marín, R. Rosas, and A. Uribe, *Amer. J. Phys.* 63, 460 (1995).

327. J. L. Marín and S. A. Cruz, *Amer. J. Phys.* 59, 931 (1991).

328. C. Priester, G. Allan, and M. Lannoo, *Phys. Rev. B* 28, 7194 (1983).

329. C. Priestier, G. Allan, and M. Lannoo, *Phys. Rev. B* 29, 3408 (1984).

330. G. Bastard, *Phys. Rev. B* 24, 4714 (1981).

331. K. Tanaka, M. Nagaoka, and T. Yamabe, *Phys. Rev. B* 28, 7068 (1983).

332. S. Chaudhuri, *Phys. Rev. B* 28, 4480 (1983).

333. C. Mailhiot, Y.-C. Chang, and T. C. McGill, *Phys. Rev. B* 26, 4449 (1982).

334. R. L. Greene and K. K. Bajaj, *Solid State Commun.* 45, 825 (1983).

335. R. J. Wagner, B. V. Shanabrook, J. E. Furneaux, J. Comas, N. C. Jarosik, and B. D. McCombe, in "GaAs and Related Compounds 1984" (B. De Crémoux, Ed.), Institute of Physics Conference Series 74. Hilger, Bristol, 1985.

336. W. T. Masselink, Y.-C. Chang, and H. Morkoc, *Phys. Rev. B* 28, 7373 (1983).

337. R. C. Miller, A. C. Gossard, W. T. Tsang, and O. Munteanu, *Phys. Rev. B* 25, 3871 (1982).

Chapter 7

MAGNETISM OF NANOPHASE COMPOSITE FILMS

D. J. Sellmyer, C. P. Luo, Y. Qiang
Behlen Laboratory of Physics and Center for Materials Research and Analysis, University of Nebraska, Lincoln, Nebraska, USA

J. P. Liu
Department of Physics and Institute for Micromanufacturing, Louisiana Tech University, Ruston, Louisiana, USA

Contents

1. INTRODUCTION AND SCOPE

The subject of magnetism and applications of nanocomposite thin films is one of significant interest from both basic and technological view points. Such materials contain clusters, grains, or dots with characteristic dimensions in the 3- to 100-nm range. It has become clear that such structures are essential for new practical materials and devices in data storage, spin-electronics, field sensors, memory devices, and high-performance magnetic materials.

Particularly important issues in nanoscale structures and novel phenomena include: (1) overcoming problems in miniaturization as feature sizes decrease below 100 nm, (2) developing experimental tools to characterize nanostructures and phenomena, and (3) developing new techniques for synthesis and design. Since new phenomena and properties appear at the nanoscale, the concept of "materials and structures by design" is particularly important to nanoscale sciences and engineering.

There are a number of outstanding physics and materials questions that focus specifically on nanoscale magnetism and

Handbook of Thin Film Materials, edited by H.S. Nalwa
Volume 5: Nanomaterials and Magnetic Thin Films
Copyright © 2002 by Academic Press

ISBN 0-12-512913-0/$35.00

structures. Several of these are as follows. In general, we do not understand the magnetic properties of nanostructured magnetic elements and arrays of such elements. For example, the nanofabrication of specific elements or building blocks with specified properties (magnetization anisotropy, etc.) is a major challenge. Related to this is the need for the building blocks to consist of complex entities—molecules or compounds—rather than simple elements so as to control the properties. The nature of the domain structure and its influence on switching behavior is not clear in most situations. There is much to be learned about the stability of both magnetization and structure of nanomagnets at high temperatures. This includes the understanding and the control of exchange and magnetostatic interactions between grains or clusters. Quantum mechanical studies of complex two-phase nanostructures including electronic structure, magnetic moments, and anisotropy, Curie temperature, and high-temperature properties are still in their infancy. Spin-polarized electron tunneling and transport in nanomagnets and their relevance to electronic devices and memory are just beginning to be explored. Finally, the understanding of extrinsic properties such as coercivity of complex magnetic nanostructures and materials is still mainly qualitative.

Much of the interest in nanostructured magnetic materials is driven by the computing and information-storage industry. Annual revenues for all forms of magnetic recording are of the order of $100 billion. The areal density in gigabits per square inch (Gb/in^2) for hard disk storage has been increasing at the rate of about 100% per year. Presently, the most advanced systems record at about 20 Gb/in^2 and several companies have demonstrated recording above 50 Gb/in^2. This remarkable progress has raised much interest and has stimulated research on how far the areal density can be increased before thermal stability sets in.

A second arena of nanostructured magnetic materials is that of high-performance soft and hard magnetic materials. Annual revenues here are about $5 billion, mostly for electric motor and generator technologies. Again, understanding and controlling the nanostructure and properties of two-phase materials is crucial to creating new magnets with extended ranges of applications. Research discussed herein is focused on materials directly related to both high-density data storage and permanent magnet applications, and has the potential to significantly impact both types of technology.

This contribution is organized as follows. The next section includes a description of nanoparticulate composites consisting of high-anisotropy grains embedded in nonmagnetic amorphous matrices. These films have high coercivities and the nanograins can be controlled in orientation and in their exchange interactions. The following section gives a brief description of the techniques of cluster assembly of nanocomposite materials. Examples of magnetic nanoclusters such as Co embedded in nonmagnetic hosts such as Cu and SiO$_2$ are discussed. This method, only recently applied to magnetic nanocomposites, holds great potential for designing both the properties of the particles and their interactions. The third class of material discussed is that of exchanged

coupled, hard–soft, two-phase materials. Prototype thin-film permanent-magnet structures have been fabricated with near world-record energy products, i.e., about 53 MGOe. The control of the grain sizes of both the hard and soft phases and their interactions is crucial to the development of future high-energy product magnets, particularly in bulk form. Finally, brief concluding notes are given along with possible future directions. It should be emphasized that this chapter is focused specifically on the three topics in nanocomposite films discussed previously, and mainly reviews work originating from our own laboratory.

2. NANOCOMPOSITE THIN FILMS

2.1. Introduction

Magnetic granular solids consist of nanometer-size magnetic metal particles embedded in an immiscible medium that may be either insulating or metallic. These nanocomposites have intricate structure on the nanometer scale and extra degrees of freedom with which the magnetic properties can be tailored for applications and for exploration of new physical phenomena. The relevant extra degrees of freedom of nanocomposite thin films are the size of the magnetic particles, which can be controlled by processing conditions, and the volume fraction of the magnetic granules, which can be varied experimentally between 0 and 1. Work on this type of nanostructured films was pioneered by Abeles et al. [1]. In the following years, these films have attracted considerable attention because they exhibit a wide variety of interesting properties in magnetism and have opened up many possibilities for them to be used in various technological applications. For instance, nanoparticles of magnetic transition metals (Co, Fe, Ni) and their compounds embedded in either an insulating or metallic matrix show peculiar magnetic and magnetotransport properties like enhanced coercivity [2], superparamagnetism [3], giant magnetoresistance [4, 5], or tunnel magnetoresistance [6]. The potential applications of the nanocomposite thin films as data-storage media [7–9] stand out as a promising area for development. In this case, superior magnetic properties can be tailored through manipulation of the size and volume fraction of the magnetic particles. In addition to this, advantages are also derived from the structure of the film in which the matrix provides chemical and mechanical protection to the magnetic particles.

In this section, we primarily discuss the magnetic properties of a variety of granular magnetic composite films. The results of single element (Fe or Co) granules embedded in a nonmagnetic matrix, and the CoPt- and FePt-based high-anisotropy nanocomposite thin films are presented.

2.2. Fabrication Methods of Nanocomposite Thin Films

Nanocomposite thin films can be synthesized by a variety of deposition methods, such as coevaporation, sol–gel processes [10], ion implantation [11], and sputtering. Of these, sputtering has been the most versatile and the most applied method.

Sputtering is often carried out by using a single composite target, or by cosputtering. Since the nanostructure of the nanocomposite is strongly influenced by the processing conditions, deposition parameters such as deposition rate, sputtering pressure, substrate temperature must be optimized. The granule size is also dependent on the volume fraction of the granule material; usually it increases with the granule volume fraction [1]. The tandem deposition method, which consists of many sequential depositions on a substrate of small amounts of a metal and an insulator from separate deposition sources, sometimes followed by thermal annealing, is also applicable in some cases [9, 12]. By this method, the metal particle size depends on the amount of materials of each deposition. Therefore, the granule size can be controlled independent of the granule volume fraction.

Granular materials consisting of small magnetic particles embedded in a metallic matrix can be fabricated by phase precipitation because some metallic elements such as Cu and Ag are essentially immiscible with Fe and Co. The precipitation of Co particles in dilute Co-Cu alloy was reported by Becker [13] in the 1950s. The amount of magnetic material available for the formation of small particles had been limited by the small equilibrium solubility (usually $< 5\%$) in the dilute alloys. However, many homogeneous metastable alloys can be formed at almost all compositions by using nonequilibrium deposition methods such as sputtering. By deposition on a heated substrate or post deposition annealing of the metastable alloy, granular films with a high concentration of magnetic particles can be obtained [14–16].

A developed method of making nanocomposites is that of annealing a composite film containing a thermally labile component at the appropriate temperature. The labile component decomposes into the magnetic particles, which are dispersed in the thermally stable component. Attractive candidate components for thermally labile nanocomposite films are nickel and cobalt nitrides, which can be made by reactive sputtering [17, 18].

2.3. Theoretical Background

2.3.1. Critical Particle Size

The particle sizes in nanocomposite films are usually in the range from a few nanometers to tens of nanometers. Frenkel and Dorfman were the first to predict that a ferromagnetic particle, below a critical particle size, would consist of a single magnetic domain [19]. On the grounds of energy considerations, the exchange energy favors uniform magnetization, while the magnetostatic energy tends to form an inhomogeneous magnetization configuration that lowers the magnetostatic energy. The exchange forces are short-range forces. Hence, in sufficiently small particles they predominate over the long-range magnetostatic forces. Therefore, a ferromagnetic particle, below a critical particle size, remains uniformly magnetized in arbitrary external fields. The critical size (D_c) is

dependent on the shape of the particle. For a high anisotropy, spherical particle, the critical diameter D_c is given by [20],

$$D_c = \frac{9\gamma}{2\pi M_s^2} \cong 1.44 \frac{\gamma}{M_s^2} \qquad (1)$$

where $\gamma \cong 4\sqrt{AK_1}$ is the specific domain wall energy, A is the exchange constant, and K_1 is the first-order anisotropy constant. In nanocomposite thin films, the size of the magnetic particles is typically in the order of 10 nm, usually below the critical size D_c. Therefore, they are usually single-domain particles.

As pointed out by Neel [21], if single-domain particles become small enough, thermal fluctuations cause the magnetic moments to undergo a sort of Brownian rotation. The particles behave like paramagnetic material, except the spontaneous magnetization is much larger than the ordinary paramagnetic material. Therefore, they are called superparamagnetic particles. For an assembly of aligned uniaxial particles with volume V and anisotropy constant K, the magnetization decays exponentially with time [22],

$$M(t) = M_s \exp(-t/\tau) \qquad (2)$$

where τ is the relaxation time given by

$$1/\tau = f_0 \exp(-KV/kT) \qquad (3)$$

where f_0 is a frequency factor of the order of 10^9 s^{-1}. When $kT \gg KV$, the magnetization reaches the thermal-equilibrium value very quickly (10^{-9} s). When $kT \ll KV$, the relaxation time τ can be much longer than the measuring time, so the equilibrium state cannot be observed. The critical superparamagnetic transition size is roughly taken at $KV = 25kT$, corresponding to $\tau = 100$ s.

For magnetic recording media, the recording density has been increasing at a compound annual rate of 60% for many years. In the last couple of years, the recording density has increased even more than 100% annually in laboratory demonstrations. As the density increases, the bit size decreases. To keep a reasonably large signal-to-noise ratio, which is proportional to the number of grains in a bit, the grain size must decrease. As the grain size approaches the superparamagnetic limit, the grain becomes magnetically unstable and it cannot store any information. This is a fundamental limit for the achievable recording density. Thermal stability has attracted considerable attention as the recording density approaches 100Gb/in^2 [23–25], and soon we will reach the density limit for the current Co-alloy commercial media. To further increase the recording density, high-anisotropy media [26] and/or the perpendicular-recording paradigm [27] might have to be used.

2.3.2. Magnetization Processes

Magnetization processes are sensitive to the structures of magnetic materials. In nanocomposite films, when the magnetic particles are isolated and small enough, the magnetization of the particle rotates coherently. When the particles

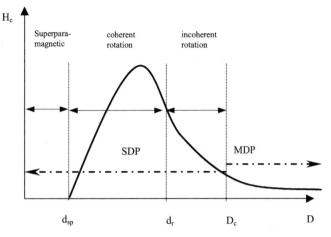

Fig. 1. A schematic qualitative relation between the coercivity and the magnetic grain size.

are larger and/or interparticle interaction occurs, incoherent magnetization processes occur. Therefore, the magnetic properties of nanocomposite thin films are sensitive to the grain size. Figure 1 shows the qualitative relationship between the coercivity H_c and the particle diameter D. The shape of the curve is strongly dependent on the chemical homogeneity of the particle and its structure [28]. In this figure, d_{sp} is the grain diameter below which the single-domain particle (SDP) becomes superparamagnetic. The boundary between coherent and incoherent rotation is given by d_r. The boundary between single-domain particle and multiple-domain particle (MDP) is given as D_c. In our nanocomposite films, the grain diameter is mainly in the single-domain particle range. The magnetization reversal mechanism can be either coherent rotation [29] or incoherent rotation (curling), depending on the size and the shape of the grain.

2.3.2.1. Coherent Rotation—The Stoner–Wohlfarth Theory

Stoner and Wohlfarth introduced a simple model for magnetization reversal in single-domain particles in 1948 [29]. They assume the magnetization in these particles is always homogeneous; i.e., the magnetization rotates coherently. For a single-domain particle with uniaxial anisotropy constant K,

magnetization M_s, and volume V in an external field H, as shown in Figure 2, the total energy is given by

$$E = KV \sin^2(\theta - \varphi) - M_s VH \cos(\theta) \quad (4)$$

where θ and φ are the angles between the magnetization vector and the applied field, the easy axis and the applied field, respectively. The total energy is a function of the angle θ. A necessary condition for an energy minimum is $dE/d\theta = 0$, which can be written as

$$\frac{2K}{M_s} \sin(\theta - \varphi) \cos(\theta - \varphi) + H \sin(\theta) = 0 \quad (5)$$

The stability of the magnetic state require that $d^2E/d\theta^2 > 0$:

$$\frac{2K}{M_s} \cos 2(\theta - \varphi) H \cos \theta > 0 \quad (6)$$

It is easy to solve Eq. (5) for the special cases $\varphi = 0°$ (applied field parallel to the easy axis) and $\varphi = 90°$ (applied field perpendicular to the easy axis). Figure 3 shows the hysteresis loops calculated with Eq. (5). For $\varphi = 0°$, starting from positive saturation, the magnetization remains along the easy axis until the applied field reaches $-H_A = -(2K/M_s)$, then the magnetization reverses to the opposite direction. In this case, the coercivity is equal to the anisotropy field H_A. When the applied field is perpendicular to the easy axis ($\varphi = 90°$), the particle only rotates reversibly with the applied field. For the intermediate cases ($0° < \varphi < 90°$), the magnetization reversal process consists of both reversible and irreversible processes. First, the magnetization undergoes reversible rotation away from the field direction as the field is reduced from the initial saturating state, reaches the easy axis at $H = 0$, and then rotates farther away from the initial-field direction toward the reverse direction. At a reversal field equal to the nucleation field (or switching field) where $d^2E/d\theta^2 = 0$, the M–H loop becomes steep corresponding to large irreversible angular changes in the magnetization direction.

For a collection of noninteracting S–W particles, the hysteresis loop can be calculated by summing up the contribution of each particle, and the distribution of the easy axis can be

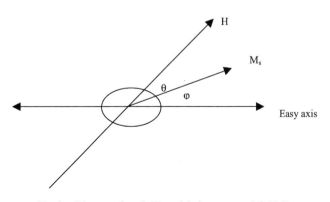

Fig. 2. Diagram of an S–W particle in an external field H.

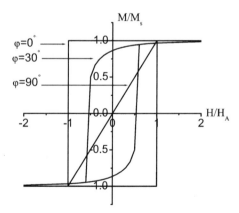

Fig. 3. Hysteresis loops for different orientations of the easy axis of an S–W particle.

used to account for the texture of the material [30]. For a random distribution of easy axes in space, the remanence ratio is $M_r/M_s = 0.5$, the coercivity is $H_c = 0.48H_A$, and the remanent coercivity $H_{cr} = 0.52H_A$. In the case of very thin films with the easy axes predominantly oriented in the film plane, i.e., the so-called two-dimensional random distribution as in longitudinal recording media, the remanence ratio is 0.637, the coercivity is $0.51H_A$, and the remanence coercivity is $0.55H_A$.

2.3.2.2. Incoherent Magnetization Reversal

The Stoner–Wohlfarth model assumes that the magnetization remains uniform during the reversal processes. However, small particles, in general, do not have spatially uniform magnetizations, especially if the particles are not ellipsoidal in shape. Then, nonuniform magnetization reversal will occur. To understand more complicated reversal mechanisms, a more sophisticated theory is needed. Micromagnetism has been very successful in theoretical descriptions of the magnetization reversal process. Micromagnetism is based on the variational principle that the magnetization directions are chosen so that the total energy reaches an absolute or relative minimum. There are four kinds of energy considered in micromagnetic theory:

(1) external-field energy (Zeeman energy): $-\vec{M} \cdot \vec{H}_{ext}$;
(2) magnetocrystalline energy: $K \sin^2\theta$;
(3) magnetostatic energy: $-2\pi\vec{M} \cdot \vec{H}_d$;
(4) exchange energy: $A[(\nabla m_x)^2 + (\nabla m_y)^2 + (\nabla m_z)^2]$
    where $\vec{m} \equiv \vec{M}/M_s$;

By calculating all the energy terms of a specific sample and by finding the total energy minimum, the configurations of the magnetization are therefore determined. The magnetostatic energy and the exchange energy compete with each other since the magnetostatic energy favors nonuniform magnetization, while the exchange energy favors uniform magnetization. As the exchange energy is short ranged and strong, while the magnetostatic energy is long ranged and weak, the magnetization reversal processes are dependent on the size of the particles. For small particles, uniform magnetization reversal is more favorable, while for larger particles incoherent magnetization reversal (such as curling) takes over. The critical particle radius and the nucleation field for curling in small spheres are [31, 32],

$$R_0 = \sqrt{\frac{A}{2M_s^2}} \tag{7}$$

$$H_n = -H_A - \frac{k_c M_s}{2S^2} + N_\| M_s \tag{8}$$

where $\mu = R/R_0$ is the reduced radius, $N_\|$ is the demagnetization factor in the applied field direction, and k_c is a geometrical factor and ranges from 1.08 (long cylinder) to 1.42 (thin plate). For $S > 1$, the magnetization reverses by curling and for $S < 1$ the magnetization reverses by coherent rotation.

2.3.3. Interparticle Interactions

In nanocomposite thin films, the magnetic particles are interacting with each other both magnetostatically and by exchange coupling. These interactions couple the magnetization of adjacent grains and cause them to reverse simultaneously. For magnetic recording media, this leads to the increase of medium noise. Therefore, characterizing the interaction between grains is a very important issue. It is known that the hysteresis loop is affected by the interaction between grains. It is known that the coercivity of nanocomposite thin films is dependent on the packing density. Hughes reported that the remanence ratio S ($S = M_r/M_s$) and the coercivity squarness S^* defined as

$$S^* = \frac{H_c}{M_r} \cdot \left(\frac{dM}{dH}\right)_{H=H_c} \tag{9}$$

both increase with exchange interaction [33]. However, the hysteresis loops do not tell one much about the degree of interaction involved. In 1958, Wohlfarth pointed out that there exists a simple relationship (Eq. (10)) between the direct current (dc) demagnetization remanent magnetization and the isothermal remanent magnetization for a noninteracting single-domain particle system [34],

$$2m_r(H) = 1 - m_d(H) \tag{10}$$

where m_r and m_d are the reduced isothermal and dc demagnetization remanent magnetization. A plot of m_r vs m_d should thus give a straight line for noninteracting system. This is called the Henkel plot because Henkel was the first to apply such a plot to permanent-magnet materials. He observed deviations from the straight line expected from Eq. (10) and attributed the deviations to magnetic interactions [34]. The Henkel plot lacks field information. Today, most people use the δm plot to investigate the particle interaction [35, 36]. δm is defined as

$$\delta m(H) = m_d(H) - (1 - 2m_r(H)) \tag{11}$$

Positive value of δm usually indicates exchange-coupling interaction, while negative value of δm usually indicates magnetic dipole interactions.

Although the δm plot has been extensively used in characterizing particle interaction, it does not give a direct indication of the magnitude of the interaction. Che and Bertram have developed a simple phenomenological model which express the interaction field in terms of the remanent magnetization as [37]:

$$f(m_r) = \alpha m_r + \beta(1 - m_r^2) \tag{12}$$

The first term is a mean field contribution that depends on the global magnetization and local exchange interaction. The second term is introduced to account for interaction field fluctuations from the mean field. This model has been used to quantify the interaction field in recording media. The parameter α and β can be determined by numerical simulation of the δm curve [38].

In micromagnetic modeling, the magnetostatic interaction field is taken as the ratio of the saturation magnetization to the magnetocrystalline anisotropy field [39]:

$$h_m = \frac{M_s}{H_A} \qquad (13)$$

For high-anisotropy material, h_m is small, which indicates the magnetocrystalline energy dominates. The exchange coupling interaction field is defined as

$$h_e = \frac{A^*}{K \cdot \Delta^2} \qquad (14)$$

where Δ is the center-to-center distance between adjacent particles. The exchange strength between particles has to be distinguished from that within the particle and is denoted by A^*. In magnetic recording media, strong exchange interaction will cause adjacent grains to switch together, resulting in decrease of coercivity and increase of medium noise. To reduce the exchange-coupling interaction, an increase of the distance between grains is essential. Nanocomposite thin films have advantages in this respect because the concentration of the magnetic grains can be controlled.

2.3.4. Magnetic Viscosity

Magnetization reversal may be realized by changing applied field or by thermal activation. At constant field, thermal activation alone may lead to significant variation of magnetization in some cases. Time dependence of the magnetization under constant field is referred to as magnetic viscosity or magnetic aftereffect. The magnetic viscosity due to thermal activation is a general property of all ferromagnetic materials. In bulk material, the timescale for magnetic viscosity is so large that magnetic viscosity often cannot be observed. However, in thin-film and fine particle system, magnetic viscosity becomes important. For instance, in magnetic recording media, as the grain size approaches the superparamagnetic limit, the magnetization is strongly time dependent. Thermally activated magnetization reversal has been investigated intensively, because it might be the limiting factor for the achievable density of conventional magnetic recording media. In this section, we review the theory of magnetic viscosity.

2.3.4.1. Time Dependence of Coercivity

Street and Woolley first pointed out that magnetization reversal might be activated by the thermal energy $k_B T$ over an energy barrier ΔE [40]. For a single energy barrier, an exponential relaxation of the magnetization is expected, with a relaxation time τ given by the Arrhenius–Néel law [41],

$$\tau^{-1} = f_0 \exp(-\Delta E/k_B T) \qquad (15)$$

where f_0 is an attempt frequency of the order of 10^9 Hz.

For a Stoner–Wohlfarth particle, the energy of the particle in an external field is given by Eq. (4). As shown in Figure 4, there are two local energy minima in the general case: the

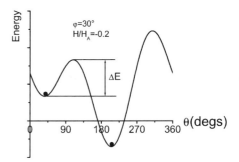

Fig. 4.   Energy barrier of a Stoner–Wohlfarth particle.

minimum having lower energy corresponding to the magnetization aligned with the applied field; the other energy minimum with higher energy corresponding to the magnetization vector antiparallel to the applied field direction. In the case the applied field is in the easy axis direction ($\varphi = 0$), the energy barrier is usually given by

$$\Delta E = KV(1 - H/H_A)^2 \qquad (16)$$

In a more general case,

$$\Delta E = KV(1 - H/H_0)^\alpha \qquad (16a)$$

where α is close to $3/2$ and H_0 is the switching field [42]. For an assembly of noninteracting S–W particles aligned at the field direction at time $t = 0$, at time t the number of particles whose magnetization is not yet reversed is given by

$$n(t) = N_0 \exp(-t/\tau) \qquad (17)$$

where N_0 is the total number of particles. By definition, half the particles have reversed magnetization at the remanent coercivity. The time dependence of the remanent coercivity is given by [43, 44]:

$$H_{cr} = H_A \left\{ 1 - \left[\frac{k_B T}{KV} \ln\left(\frac{f_0 t}{\ln 2} \right) \right]^{1/2} \right\} \qquad (18)$$

Equation (18) requires that all the particles are aligned at the field direction initially. In magnetic thin films such as recording media, the grains are randomly oriented in the film plane. In this case, the remanent coercivity can be approximated by [42, 45]:

$$H_{cr} = 0.556 H_A \left\{ 1 - 0.977 \left[\frac{k_B T}{KV} \ln\left(\frac{f_0 t}{\ln 2} \right) \right]^{2/3} \right\} \qquad (19)$$

Equation (19) assumes there is no magnetic interaction between grains. In the case exchange-coupling interaction occurs, the prefactor (0.566) will be reduced [45].

Experimental measurement of the time dependence of coercivity of recording media has been carried out by several methods. Stinnett et al. used a microstrip to apply field pulses and measured the time dependence of the magnetization [46]. Moser, Weller, and Best recorded di-bits with pulsed fields from a recording head [47]. The coercivity is identified with the current needed to reach 50% of the maximum di-bit signal. The experimental data follow the theory generally well, except at very short time (10^{-9} s) the coercivity increases much faster than the theory predicts. This presumably indicates the onset of the gyromagnetic effects.

2.3.4.2. Time Dependence of Magnetization

In the case of a single energy barrier ΔE, the time dependence of magnetization is given by [21, 48],

$$M(t) = M(\infty) + [M(0) - M(\infty)] \exp(-t/\tau) \quad (20)$$

where $M(\infty)$ is the equilibrium magnetization, $M(0)$ is the magnetization at $t = 0$. This kind of magnetization decay has been seldom observed because in real cases there always exists a distribution of energy barriers [49]. In most materials, it is found experimentally that [40],

$$M(t) = M_0 - Sv \ln(t/t_0) \quad (21)$$

Here, M_0 is a constant, Sv is called the magnetic viscosity coefficient. Street, Woolley, and Smith proposed the empirical relationship [50],

$$Sv = \chi_{\text{irr}} H_f \quad (22)$$

where χ_{irr} is the irreversible susceptibility, and H_f is the fluctuation field defined as

$$H_f = -\frac{k_B T}{(\partial \Delta E/\partial H)_T} \quad (23)$$

Further, Wohlfarth proposed that H_f might be related to the volume of the material which was involved in activation processes, i.e., the so-called activation volume V^* [51],

$$H_f = -\frac{k_B T}{M_s V^*} \quad (24)$$

and

$$V^* = -\frac{1}{M_s} \left(\frac{\partial \Delta E}{\partial H} \right)_T \quad (25)$$

By measuring the magnetic viscosity coefficient Sv and the irreversible susceptibility χ_{irr}, the fluctuation field and activation volume can be determined.

2.3.4.3. Sweep-Rate Dependence of Coercivity

Another way of determining the activation volume is by measuring the sweep rate dependence of the coercivity. We have already seen the time dependence of H_{cr} in Eqs. (18) and (19). Therefore, the measured value of H_c is dependent on the timescale of the measurements: a shorter timescale gives a larger H_c than a measurement done over a longer timescale. However, there is a difference between the time dependence of H_{cr} and the sweep-rate dependence of H_c which is a continuous process. The sweep-rate dependence of H_c was first calculated by Chantrell, Coverdale, and O'Grady [52] and is

given by

$$H_c(R) = H_A \left[1 - \left(\frac{k_B T}{KV} \ln \frac{k_B T f_0 H_A}{2KVR(1 - H_c(R)/H_A)} \right)^{1/2} \right] \quad (26)$$

where $R = dH/dt$ is the sweep rate. This is a complex equation as $H_c(R)$ also appears in the argument. An improved description of $H_c(R)$ was given by El-Hilo et al. as [53]:

$$H_c(R) = H_A \left[1 - \left(\frac{k_B T}{KV} \ln \frac{k_B T f_0 H_A}{2KVR} \right)^{1/2} \right] \quad (27)$$

By applying a binomial expansion, H_c can be written as

$$H_c(R) = A_0 + A_1 \ln(R/R_0) + \cdots \quad (28)$$

where R_0 is the initial sweep rate. Experimentally, H_c has been found linearly dependent on $\ln(R)$ [54, 55]. The slope of the plot of H_c vs $\ln(R)$ is found to be related to the activation volume V^* [54, 56]. Therefore, the activation volume can be determined from the slope of the H_c vs $\ln(R)$ plot.

The fluctuation field and activation volume have been extensively studied in permanent-magnet materials and recording media. The Barbier plot indicates that H_f is related to H_c by the empirical equation [57]

$$\log(H_f) = \log(H_c) + C \quad (29)$$

where C is a constant. From the slope of the Barbier plot and Eq. (24) it appears that

$$H_c \sim (V^*)^{-x} \quad \text{where } x = 0.73 \quad (30)$$

The activation volume can be obtained by measuring the viscosity coefficient or the sweep-rate dependence of H_c. However, the interpretation of the measured activation volume is not straightforward due to the lack of information about the distribution of energy barriers. It is further complicated by the complex magnetization reversal processes that are closely related to the structure of the material. For single-domain particles with the energy barrier given by [49, 56],

$$\Delta E = E_0 - M_s V H \quad (31)$$

where E_0 is a constant, the activation volume is the actual particle volume. For S–W particles with energy barrier given by Eq. (16), the activation volume is given by

$$V^* = V(1 - H/H_A) \quad (32)$$

The activation volume is field dependent. The magnetic viscosity coefficient Sv is also dependent on the applied field and usually has maximum value near the coercive field. In high-anisotropy particulate materials with volumes less than about 10^{-18} cm^3, when $H \sim H_c \ll H_A$, V^* often is close to the actual particle volume. The field dependence of the activation volume is important in understanding and assessing the thermal stability of high-density recording media [58]. The activation volume and the fluctuation field are also affected by the interaction between particles. It has been found that the fluctuation field decreases with the increase of interactions, which indicates that particle interaction has the tendency to stabilize the magnetization [59]. The interaction between particles might have a positive effect on the thermal stability of recording media; unfortunately it is also the major cause of medium noise. Therefore, the reduction of particle interaction is essential to the achievement of extremely high areal density in recording media.

2.4. Structure and Properties of Magnetic Nanocomposites

Magnetic nanocomposite films consist of fine magnetic grains dispersed in a nonmagnetic matrix. The properties of nanocomposite materials are dependent on the volume fraction (x_v) of the granules. For small values of x_v, the granules are isolated from each other by the matrix, while for large values of x_v, the granules form an infinite network. The percolation threshold (x_p) in the nanocomposite system is the granule volume fraction at which an infinite granule network first forms (typically about 0.5). The properties of the nanocomposite changes drastically around the percolation threshold. For instance, in typical metal-insulator nanocomposite films, the resistivity of the nanocomposite decreases sharply at the percolation threshold [1]. In magnetic nanocomposite films, as pointed by Chien [2], the magnetic granules are isolated single domain particles exhibiting hard magnetic properties when x_v is below the percolation threshold x_p, while the magnetic granules form a network and behave like a multidomain structure, exhibiting soft magnetic property when x_v is above x_p. In this section, the structure and the magnetic properties of a variety of nanocomposite thin films are given.

2.4.1. Transition-metal Based Nanocomposite Films

2.4.1.1. Fe:SiO₂ Nanocomposite Films

Nanocomposite thin films consisting of fine transition metal particles (Fe, Co, Ni) dispersed in a nonmagnetic matrix have been studied extensively [60–67]. One of the unique features of these nanocomposite films is the enhanced properties around the percolation threshold. For the Fe:SiO₂ granular system consisting of fine Fe particles embedded in a SiO₂ matrix, enhanced properties such as coercivity and anisotropy have been found around the percolation threshold [60, 61]. Large coercivity enhancement, reaching a peak value of 2500 Oe, was found at the percolation threshold x_p by Xiao and Chien [60]. When $x_v > x_p$, the coercivity drops quickly. The bulk magnetocrystalline anisotropy of Fe particles cannot account for the coercivity enhancement [7]. Holtz, Lubitz, and Edelstein found that the magnetic anisotropy was also enhanced at x_p [61]. A lot of factors contribute to the enhancement of coercivity and magnetic anisotropy, including particle shape effect [61], surface effects due to the bonding between the Fe particles and the silica matrix [2].

Experimentally, the percolation threshold x_p has been found to lie in the range of $0.5 - 0.6$ for a variety of granular solids

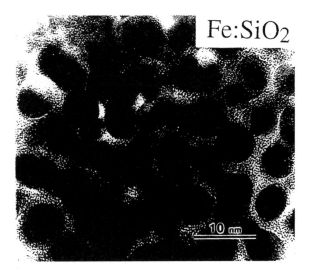

Fig. 5. HRTEM image for the $Fe_{80}(SiO_2)_{20}$ film with a thickness of 700 Å (after Malhotra et al. [62]).

[1]. The x_p was also found dependent on the film thickness [62, 63, 66]. The x_p increases as the film thickness decreases, due to the reduced dimensionality [63].

The thickness dependence of the electrical and magnetic properties of the Fe:SiO₂-nanocomposite films was studied by Malhotra et al. [62]. The films were prepared by radio frequency (rf) magnetron sputtering from a composite target of $Fe_{80}(SiO_2)_{20}$ (55 vol% of Fe). The structure of the films was studied by transmission electron microscopy (TEM). For the 700-Å thick film, the high-resolution TEM image shows that the film consists of fine Fe particles, in the size range from 46 to 66 Å, dispersed in the amorphous SiO₂ matrix, as shown in Figure 5. The Fe crystallites are mostly isolated by the SiO₂ matrix.

Figure 6 shows the thickness dependence of the resistivity of the Fe:SiO₂-granular films. A drastic change of resistivity has been observed around the film thickness of 700 Å. The

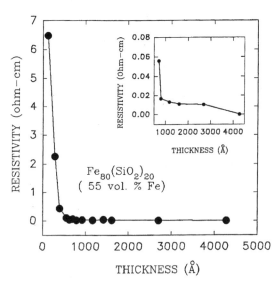

Fig. 6. Resistivity vs thickness for the $Fe_{80}(SiO_2)_{20}$ films (after Malhotra et al. [62]).

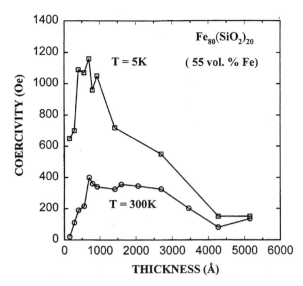

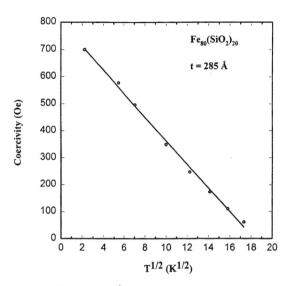

Fig. 7. Magnetic coercivities at 300 and 5 K vs thickness for the $Fe_{80}(SiO_2)_{20}$ films (after Malhotra et al. [62]).

Fig. 9. H_c vs $T^{1/2}$ for the 285-Å thick $Fe_{80}(SiO_2)_{20}$ film (after Malhotra et al. [62]).

rapid decrease of resistivity indicates that the Fe particles form a connecting network when the film thickness is over 700 Å.

The coercivities of the films have also been found to be dependent on the film thickness, as shown in Figure 7. The coercivity first increases with the increase of film thickness, reaches the maximum values of 400 Oe at 300 K and 1200 Oe at 5 K at the film thickness of 700 Å, then decreases with the further increase of film thickness. Figure 8 shows the dependence of coercivity on temperature for films with different thickness. For thick films (above 1000 Å), the coercivities almost decrease linearly with temperature, while for the film with thickness of 285 Å, a $T^{1/2}$ dependence of coercivities has been observed (Fig. 9). The $T^{1/2}$ dependence of coercivity is characteristic of a system of superparamagnetic particles below the blocking temperature [43]. When the film thickness

is small, the Fe particles are isolated and the size of the particles are close to the superparamagnetic limit, therefore the coercivity is small. As the film thickness increases, the number of connected particles increases, resulting in the increase of interaction between particles and the change of magnetic properties.

2.4.1.2. Co:C Nanocomposite Films

Co and C form an ideal magnetic nanocomposite system since Co and C are immiscible, and the metastable Co carbides (Co_2C and Co_3C) decompose easily into Co and C [68]. The properties of Co:C have been investigated [65, 67], and its potential application as recording media has been suggested by Hayashi et al. [65].

Co:C films have been deposited on to water-cooled 7059 glass substrate by cosputtering from a pure Co target and a graphite target. The Co concentration varied from 50 to 80 at.% and the thickness of the films varied from 5 to 1000 nm. The as-deposited films were annealed in a vacuum at various temperatures for 2 h to form the Co:C nanocomposite.

The as-deposited films are amorphous Co-C alloys, as indicated by the X-ray diffraction pattern in Figure 10. Magnetic measurements indicate that the as-deposited films are nonmagnetic, which is in agreement with the amorphous structure. The Co_2C phase formed as the majority phase when annealed at 250 °C. As further increase of the annealing temperature (T_A), Co_2C starts to decompose and isotropically oriented hexagonal close-packed (hcp) Co crystallites are formed. Co crystallites of about 10 nm in size have been observed in the film annealed at 350 °C, with Co concentration of 60 at.% and film thickness of 100 nm, as shown by the high-resolution TEM image (Fig. 11). The grain grows bigger as the annealing temperature increases. As shown in Figure 12, the grain volume steadily increases with the increase of T_A.

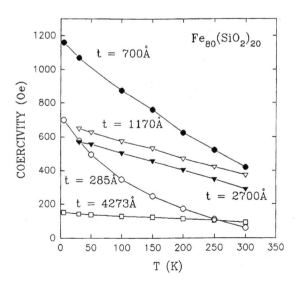

Fig. 8. Coercivities vs temperatures of the $Fe_{80}(SiO_2)_{20}$ films (after Malhotra et al. [62]).

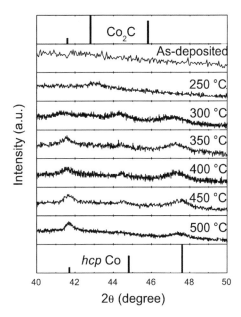

Fig. 10. XRD pattern of the 100-nm thick Co:C films with 60 at.% Co annealed at various temperatures (after Yu, Liu, and Sellmyer [67]).

The saturation magnetization (M_s) of the films increases with the increase of T_A, reaches a steady value as T_A reaches 400 °C, as shown in Figure 12. Since the Co_2C is nonmagnetic, M_s is directly related to the amount of Co crystallites formed in the film. The small value of M_s indicates there is small amount of Co crystallites in the film when $T_A = 250$ °C. As T_A increases, more Co crystallites are formed until $T_A = 400$ °C, when the Co_2C has decomposed completely.

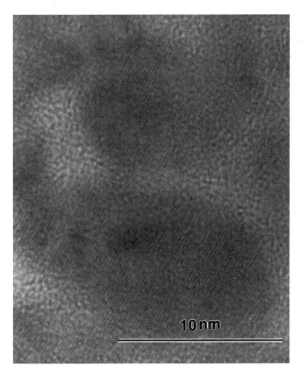

Fig. 11. High-resolution TEM image of a 100-nm thick Co:C film with 60 at.% of Co annealed at 350°C (after Yu, Liu, and Sellmyer [67]).

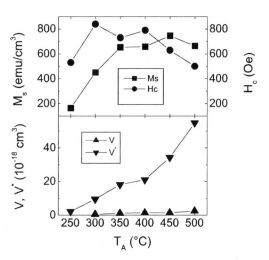

Fig. 12. M_s, H_c, V, and V^* of the 100-nm thick Co:C films with 60 at.% Co annealed at various temperatures (after Yu, Liu, and Sellmyer [67]).

The coercivities of the films first increase with T_A, has a maximum value of about 800 Oe when $T_A = 300$ °C (Fig. 12). Further increase of T_A results in the decrease of H_c. As the Co crystallites grow bigger with the increase of T_A, neighboring Co crystallites start to contact with each other when T_A reaches 300 °C. Hence, the percolation threshold is reached and further increase of T_A results in the formation of a effective multidomain structure and the decrease of H_c.

The coercivity, the saturation magnetization, and the grain volume are also dependent on the Co concentration, as shown in Figure 13. As expected, M_s almost increases linearly with the increase of Co concentration. The coercivity has a maximum value at the percolation threshold when the Co concentration is about 60 at.%.

The activation volumes (V^*) of the films were obtained by measuring the magnetic viscosity coefficient and the irreversible susceptibility. As shown in Figure 13, V^* increases rapidly when the Co concentration exceeds the percolation threshold, suggesting strong exchange coupling between grains.

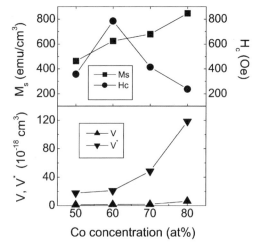

Fig. 13. M_s, H_c, V, and V^* of 100 nm Co:C films annealed at 400 °C with various Co concentration (after Yu, Liu, and Sellmyer [67]).

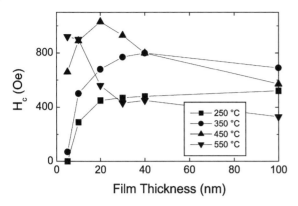

Fig. 14. Thickness dependence of H_c of Co:C films with 60 at.% Co annealed at various temperatures (after Yu, Liu, and Sellmyer [67]).

The thickness dependence of coercivities for films annealed at different temperatures is shown in Figure 14. All the films have a Co concentration of 60 at.%. As T_A increases, the percolation threshold (defined at the maximum coercivity) shifted to lower film thickness. Similar results have been reported in the Co:Ag granular system [63]. When the film thickness is comparable to the particle size, the particles form a two-dimensional network in the film plane when the percolation threshold is reached. Therefore, the decrease of percolation threshold is due to the reduced dimensionality in very thin films [63].

2.4.2. High-Anisotropy Nanocomposite Systems

High-anisotropy nanocomposite thin films have gained considerable attention because of their potential application as extremely high-density recording media. The high-anisotropy constant allows thermally stable magnetic grains with size below 10 nm, which is very attractive as the recording density approaches 100 Gbits/in^2 and beyond. A lot of work has been reported on nanocrystalline rare-earth transition metal films [69–71]. However, the rare-earth system has the problem of chemical stability. CoPt and FePt films with an L1$_0$ structure have shown high anisotropy and huge coercivity [72–74]. Recording experiments have been carried out in ZrO$_x$-doped CoPt-L1$_0$ media [75] and FePt-B media [76]. Some of the common problems with these high-anisotropy media are the unfavorable high processing temperatures that may adversely affect grain growth. In nanocomposite films such as CoPt:C [8], FePt:SiO$_2$ [9], FePt:B$_2$O$_3$ [77], and CoPt:Ag [78], the grain growth can be suppressed by the nonmagnetic matrix, and their properties can be tailored for extremely high-density recording applications.

2.4.2.1. The Ordered L1$_0$ Phase of FePt and CoPt

According to the phase diagrams of FePt and CoPt alloys, the ordered face-centered tetragonal (fct) structure (L1$_0$ phase) exists in the composition range from 40 to 60 at.% Fe or Co. The crystal structure of this fct phase is shown in Figure 15. It has a superlattice structure consisting of Fe and Pt monatomic layers stacked alternately along the c-axis, which is the mag-

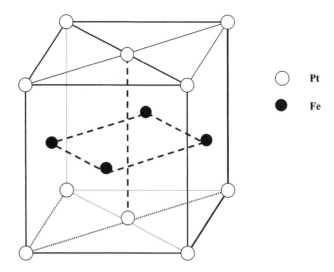

Fig. 15. Crystal structure and unit cell of FePt L1$_0$ phase.

netic easy axis. The axial ratio (c/a) is slight less than 1 $(c/a \sim 0.96\text{-}0.98)$. For the disordered face-centered cubic (fcc) phase, the Fe and Pt atoms randomly occupy the lattice sites. The long-range order parameter is defined as [79],

$$S_0 = (r_{Fe} - x_{Fe})/y_{Pt} = (r_{Pt} - x_{Pt})/y_{Fe} \qquad (33)$$

where $x_{Fe(Pt)}$ is the atomic fraction of Fe(Pt) in the sample, $y_{Fe(Pt)}$ is the fraction of Fe(Pt) sites, and $r_{Fe(Pt)}$ is the fraction of Fe(Pt) sites occupied by the correct atom. The long-range order parameter can be determined from the X-ray diffraction pattern [80–81].

The ordered fct phase is the low-temperature phase. Therefore, it can be obtained by thermally annealing the disordered fcc phase. Direct formation of the ordered phase has also been obtained by deposition of the films on heated substrates [82]. Figure 16 shows the x-ray diffraction patterns of the as-deposited and the annealed FePt film. The as-deposited film has the disordered fcc structure. After annealing, the fct

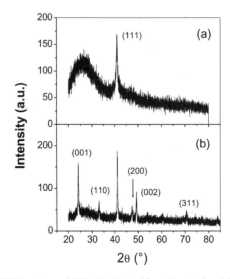

Fig. 16. XRD patterns of FePt-thin films: (a) as-deposited and (b) annealed at 550 °C for 30 min.

structure has formed, as indicated by the (001) and (002) superlattice peaks. By comparing the intensity ratio of the superlattice peaks to the fundamental peaks of the annealed films with that of a perfectly ordered sample, the long-range order parameter can be calculated [83].

The ordered fct phase has very high-anisotropy energy. In bulk materials, the anisotropy constants have been found equal to 5×10^7 and 7×10^7 erg cc^{-1} for CoPt and FePt, respectively. K values greater than 10^8 erg cc^{-1} have been found in fully ordered FePt films grown by molecular-beam epitaxy (MBE) [84]. Very large coercivities have been obtained in CoPt and FePt thin films prepared by magnetron sputtering and postdeposition annealing [72, 74].

FePt and CoPt thin films deposited by magnetron sputtering tend to grow with a (111) texture [75], placing the c-axes of the grains at an angle of about 36° above the film plane. By applying the epitaxial growth technique, FePt can grow on MgO single-crystal substrates with the c-axes of the grains either in the film plane [85] or perpendicular to the film plane [86], depending on the requirements of the applications.

2.4.2.2. CoPt:C-Nanocomposite Thin Films

Yu and co-workers [8] prepared CoPt:C composite films on water-cooled Si(100) substrates by cosputtering Co, Pt, and C. The films have equiatomic Co-Pt ratio with various C concentrations. The thickness of the films varied from 3 to 100 nm. The as-deposited films were annealed in vacuum for 1 h at various temperatures.

The as-deposited films are magnetically soft with H_c less than 100 Oe. They most likely consist of disordered fcc CoPt and/or amorphous CoPt in addition to a C matrix. Since the fct CoPt is the low-temperature phase, appropriate annealing leads to the transition from the fcc structure to the fct structure in

the CoPt:C films. As shown in Figure 17, the X-ray diffraction (XRD) patterns indicate that the fct structure has not formed until the annealing temperature reaches 600 °C. Higher T_A results in the more complete transition from fcc to fct, at the same time the grains grow larger.

A bright-field TEM image of a 100-nm thick CoPt:C film with 30 vol% C annealed at 650 °C is shown in Figure 18a. It clearly shows that the film consists of CoPt particles surrounded by a C matrix. Electron diffraction analysis confirmed that these CoPt particles had the fct structure. Some of the CoPt particles are well isolated by C, as shown by the high-resolution TEM image (Fig. 18b), and some are aggregated.

The grain size and the coercivity of the film are dependent on the annealing temperature, as shown in Figure 19. As expected, the grains get larger as T_A increases. The sharp increase of H_c near $T_A = 650$ °C can only be explained by the more complete transition from fcc to fct structure. The slow increase of H_c as T_A further increases can be attributed to the grain growth. Since the fct CoPt phase has slightly less magnetization than the fcc phase, the gradual decrease of M_s as

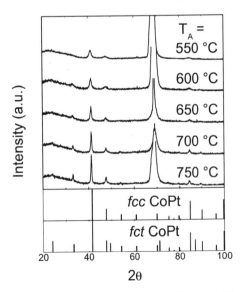

Fig. 17. XRD patterns of 100-nm thick CoPt:C films with 30 vol% of C annealed at various temperatures T_A. The standard XRD patterns for the fct and fcc CoPt phases are also plotted as a reference (after Yu et al. [18]).

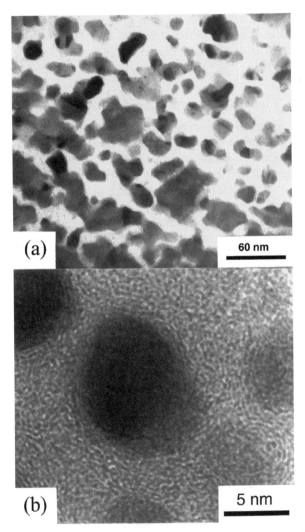

Fig. 18. (a) Bright-field and (b) high-resolution TEM images of a 100-nm thick CoPt:C film with 30 vol% of C annealed at 650 °C (after Yu et al. [8]).

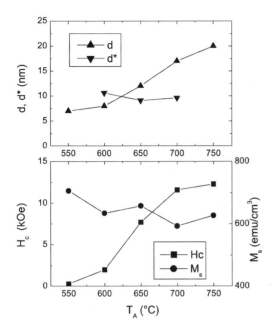

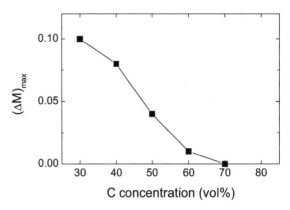

Fig. 21. Normalized maximum ΔM values of the 100-nm thick CoPt:C films with various C concentrations (after Yu et al. [8]).

Fig. 19. Annealing temperature T_A dependence of grain size d, activation unit size d^*, H_c, and M_c of the 100-nm thick CoPt:C films with 30 vol% of C (after Yu et al. [8]).

T_A increases also suggests the increasing conversion from fcc to fct phase.

The effect of C concentration has been investigated for films annealed at 650 °C. As shown in Figure 20, M_s decreases almost linearly with increasing C concentration due to the dilution effect. The grain size d also decreases as C concentration increases because C suppresses the growth of CoPt grains. The primary reason for the decrease of H_c with the increase of C concentration may be the decrease of anisotropy constant

K_u, resulting from the incomplete transition from fcc to fct phase. The decrease of grain size may also contribute to the decrease of H_c.

To investigate the magnetization reversal processes and the thermal stability, the activation volume (V^*) has been derived by measuring the viscosity coefficient and the irreversible susceptibility. As shown in Figure 20, the activation volume is dependent on the C concentration. For films with C concentration over 60 vol%, V^* is close to the actual grain volume and decreases with the increase of C concentration. This suggests that each particle may reverse independently because of the reduced intergrain exchange coupling. For films with C concentration less than 60 vol%, V^* is almost a constant and much less than actual grain volume. This suggests nonuniform magnetization reversal processes occur in these films.

The interparticle interaction has been investigated by the δm plot. The maximum δm value has been plotted as a function of the C concentration, as shown in Figure 21. The maximum δm value decreases as C concentration increases. This is understandable as the C concentration increases, the distance between adjacent CoPt particles increases, hence the exchange interaction decreases. When the C concentration is over 60 vol%, the CoPt particles are fairly isolated and the exchange coupling between them is reduced. Therefore, the δm value approaches zero.

Recording experiments have been carried out on a 10-nm thick sample with 50 vol% of C. The hysteresis loop and magnetic parameters of the sample are shown in Figure 22. The read-back signals of individual tracks at various linear densities are shown in Figure 23. Clear read-back signals at linear densities of at least 5000 fc mm^{-1} can be obtained. This test sample has a 20-nm thick C overcoat. Higher linear densities are expected if the C overcoat thickness is reduced to less than 10 nm.

2.4.2.3. FePt:SiO$_2$- and FePt:B$_2$O$_3$-Nanocomposite Thin Films

FePt:SiO$_2$- and FePt:B$_2$O$_3$-nanocomposite films have been obtained by annealing the as-deposited FePt/SiO$_2$ and

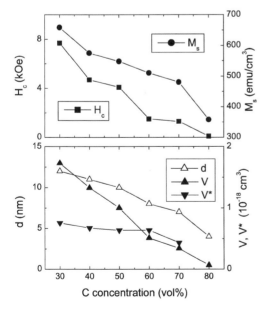

Fig. 20. H_c, M_s, d, $V(= d^3)$, and V^* of the 100-nm thick CoPt:C films annealed at 650 °C with various C concentrations (after Yu et al. [8]).

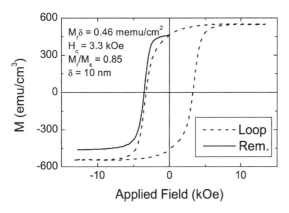

Fig. 22. Hysteresis loop and dc remanence curve of a 10-nm CoPt:C film with 50 vol% of C annealed at 600 °C (after Yu et al. [8]).

FePt/B_2O_3 multilayers. FePt/SiO_2 and FePt/B_2O_3 multilayers were deposited on glass substrates by dc- and rf-magnetron sputtering. A composite FePt target was made by putting some Fe chips on the Pt target. The composition of the FePt layer was adjusted to near equiatomic composition by adjusting the number of Fe chips. The concentration of FePt in the films can be controlled by the ratio of the FePt-layer thickness over the SiO_2- or B_2O_3-layer thickness. The as-deposited films were annealed in vacuum at temperatures from 450 to 650 °C. After annealing, nanocomposite films consisting of $L1_0$ FePt particles embedded in a SiO_2 or B_2O_3 matrix have been obtained.

As in the case of CoPt:C films, the as-deposited multilayers have a disordered fcc structure, as indicated by the X-ray diffraction pattern in Figure 24a. These multilayers are magnetically soft since the fcc structure has very low anisotropy, as shown by the hysteresis loop in Figure 24b.

The structures of the annealed films are sensitive to the annealing temperatures and FePt concentration. The X-ray diffraction patterns of the 650 °C annealed FePt/SiO_2 multilayers are shown in Figure 25. All the films have the ordered fct structure, as indicated by the (001), (110), and (002) superlattice peaks. The textures of the films are dependent on the

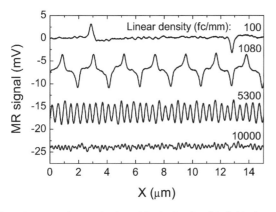

Fig. 23. Magnetoresistance head read-back signals of individual tracks at various linear densities of a 10-nm thick CoPt:C film with 50 vol% of C annealed at 600 °C (X denotes the distance along the track direction) (after Yu et al. [8]).

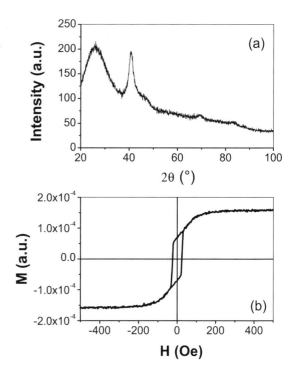

Fig. 24. (a) XRD pattern and (b) hysteresis loop of an as-deposited (FePt 50 Å/SiO_2 15 Å)$_{10}$ multilayer.

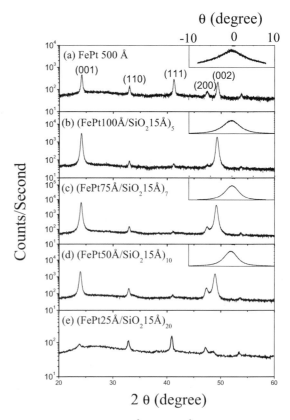

Fig. 25. XRD scans of (FePt x Å/SiO_2 15 Å)$_n$ multilayers annealed at 650 °C for 2 h. The insets are the (001) peak rocking curves (after Luo and Sellmyer [9]).

FePt-layer thickness, or the FePt concentration. As compared with the FePt-single-layer film, the (001) texture is dominant in the films with FePt-layer thickness in the range from 50 to 100 Å. This (001) texture indicates that the c-axes of the FePt grains are mostly aligned along the film-normal direction. When the FePt-layer thickness decreases to 25 Å, the (001) texture becomes weaker, and the FePt grains are randomly oriented.

The structure of the films is also sensitive to the annealing temperature. When the annealing temperature decreases to 550 °C, the (001) texture also diminishes and the FePt grains are randomly oriented [9].

The hysteresis loops of an annealed FePt/SiO_2 multilayer are shown in Figure 26. The hysteresis loops were measured with the applied field both in the film plane and perpendicular to the film plane. When annealed at 650 °C, the film has (001) texture and therefore perpendicular anisotropy. The hysteresis shows that the film is much more easily saturated in the perpendicular direction than in the in-plane direction. When annealed at 550 °C, similar hysteresis loops have been obtained in both directions, which confirms that the FePt grains are randomly oriented.

The saturation magnetization M_s, coercivity H_c and grain size d of the FePt:SiO_2 films all decrease as SiO_2 concentration increases, similar to the case of the CoPt:C films.

Figure 27 shows the dependence of H_c and d of the FePt:SiO_2 films on the annealing temperatures (T_A). For the FePt single-layer film, large H_c has been obtained at relatively low T_A ($\sim$450 °C); at the same time the grain size also quite large (>25 nm). When SiO_2 was added to the films, H_c and d both decrease. However, H_c does not decrease as fast as the

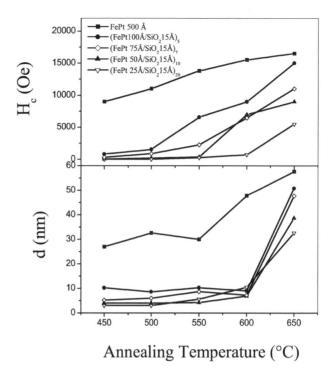

Fig. 27. The dependence of grain size d and H_c on the annealing temperature (after Luo et al. [9]).

grain size d because the structural transition (from disordered fcc to ordered fct) is the primary cause for the increase of H_c. When T_A is equal to or below 600 °C, the grains have limited growth potential. Therefore, films with fairly large H_c and fine grain size (<10 nm) can be obtained. These films have the potential as high-density magnetic recording media.

Comparing with the FePt/SiO_2 films, the FePt/B_2O_3 multilayers require lower annealing temperature to form the fully ordered fct structure because of the low glassy transition temperature of B_2O_3. The X-ray diffraction patterns of the 550 °C annealed FePt/B_2O_3 multilayers are shown in Figure 28. In these samples, the FePt-layer thickness was fixed at 32 Å, while the B_2O_3-layer thickness varies. As we can see, when the B_2O_3-layer thickness increases, the intensity of the (111) peak decreases. As the B_2O_3-layer thickness increases to 12 Å and above, the (111) peak disappears. Only the dominant (001) and (002) peaks appear, indicating the alignment of the c-axes of the FePt grains along the film-normal direction. The FePt grains in the FePt:B_2O_3 films are aligned along the film-normal direction better than in the FePt:SiO_2 films, as indicated by the narrower (001) peak rocking curves and the negligible (110) peaks in the XRD scans of the FePt:B_2O_3 films. When the B_2O_3-layer thickness increases to 16 Å and above, the films retain a layered structure, as indicated by the low-angle diffraction patterns.

The texture of the annealed films also depends on the initial multilayer structure. Figure 29 shows the XRD patterns of the 550 °C annealed films with an initial multilayer structure of $[(Fe\ 10\ \text{Å}/Pt\ 11\ \text{Å})_2/B_2O_3\ x\ \text{Å}]_5$, where x varies from 12 to 30 Å. When $x = 12$ Å, the film has an ordered fct structure

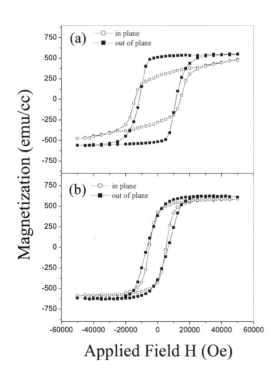

Fig. 26. Hysteresis loops of (FePt 100 Å/SiO_2 15 Å)$_5$ annealed at: (a) 650 °C for 2 h and (b) 550 °C for 30 min (after Luo et al. [9]).

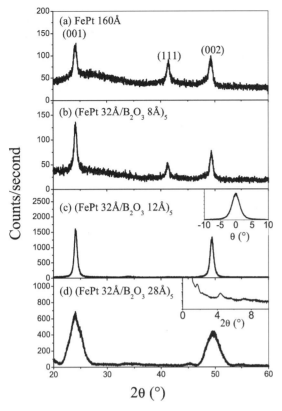

Fig. 28. XRD θ–2θ scans. The insets in (c) and (d) are the (001) peak rocking curve and the low-angle θ–2θ scan, respectively, (after Luo et al. [77]).

with the FePt grains randomly oriented. As x increases, the (001) and (002) peak intensities decrease and the (110) and the (220) peaks become predominant. When x increases to 30 Å, a predominant (110) texture has been obtained, and the c-axes of the FePt grains are aligned in the film plane.

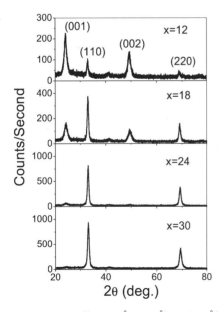

Fig. 29. XRD scans of the $[(Fe\ 10\ \text{Å}/Pt\ 9\ \text{Å})_2/B_2O_3\ x\text{Å}]_5$ multilayers annealed at 550°C for 30 min.

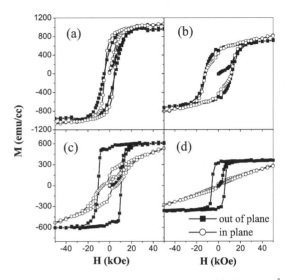

Fig. 30. Hysteresis loops of (a) FePt-single layer; (b) (FePt 32 Å/B_2O_3 8 Å)$_5$; (c) (FePt 32 A/B_2O_3 12 Å)$_5$ and (d) (FePt 32 Å/B_2O_3 48 Å)$_5$. These films were annealed at 550 °C for 30 min. (after Luo et al. [77]).

For magnetic recording media, the easy axes of the magnetic grains are either in the film plane (longitudinal media) or perpendicular to the film plane (perpendicular media). The easy axes of the FePt:B_2O_3 films can be tailored either to be in the film plane or to be perpendicular to the film plane to suit the application. Therefore, these composite films have an advantage over other random high-anisotropy films such as CoPt:C and CoPt:Ag.

Figure 30 shows the hysteresis loops of the annealed (FePt 32 Å/B_2O_3 x Å)$_5$ multilayers. When x is less than 12 Å, the FePt grains are randomly oriented and similar loops have been measured in both directions. As x increases, the in-plane loops diminish while the perpendicular loops remain square, indicating the development of the perpendicular anisotropy due to the (001) texture.

The saturation magnetization M_s of the films decreases linearly with the increase of B_2O_3 concentration, as shown in Figure 31. After normalization by the FePt volume fraction, the M_s of the FePt phase in FePt:B_2O_3 nanocomposite is about 1000 emu cc^{-1}, slightly less than the M_s ($\sim$1100 emu cc^{-1}) of the FePt single-layer film. This might be due to the isolation of FePt grains by the B_2O_3 matrix, resulting in smaller moments for the atoms at the grain surfaces.

By extrapolating the magnetization curves, the anisotropy field H_A and the anisotropy constant K_u (= $M_sH_A/2$) can be obtained, as shown in Figure 32. The H_A values are in the range from 70 to 80 kOe for all films. With $M_s = 1000$ emu cc^{-1}, the K_u value is about $3.5 - 4.0 \times 10^7$ erg cc^{-1}, which is about half of the K_u value for bulk FePt alloys.

The grain size of the FePt:B_2O_3 films is dependent on the B_2O_3 concentration. Figure 33a is a bright-field TEM image of the (FePt 32 Å/B_2O_3 12 Å)$_5$ multilayer annealed at 550 °C for 30 min. The FePt-grain size was found in a wide range from 10 to 30 nm. As the B_2O_3 concentration increases, the grain size decreases. Figure 33b is a high-resolution TEM image of

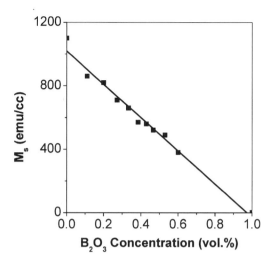

Fig. 31. The dependence of M_s on B_2O_3 concentration in FePt:B_2O_3-nanocomposite films.

the annealed (FePt 32 Å/B_2O_3 20 Å)$_5$ multilayer. It shows that fine FePt grains about 4 nm in size were randomly dispersed in the B_2O_3 matrix.

Figure 34 shows the dependence of H_c and grain size d on the initial B_2O_3-layer thickness. As we can see, a small amount (up to 20 vol%) of B_2O_3 sharply increases the coercivity from 5.2 to 12 kOe. Since the anisotropy constant and the grain size change very little, the sharp increase of H_c may be due to the reduction of intergranular exchange coupling and/or the increase of the number of pinning sites when B_2O_3 is added. However, further increase of B_2O_3 concentration results in the sharp decrease of H_c. This is probably due to the sharp decrease of grain size. As grain size decreases to below 10 nm, the coercivities can be significantly reduced due to the thermal activation effect.

Figure 35 shows the linear logarithmic decay of magnetization M with time. The magnetic viscosity coefficient S is determined from the slope of the M vs $\ln(t)$ curves. The shape of the $S(H)$ curve is coincident with the switching-field distribution curve $\chi_{\text{irr}}(H)$, which suggests that thermal

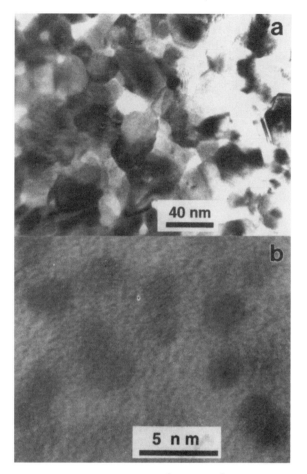

Fig. 33. TEM images of (a) (FePt 32 Å/B_2O_3 12 Å)$_5$ and (b) (FePt 32 Å/B_2O_3 20 Å)$_5$. Both films were annealed at 550 °C for 30 min (after Luo et al. [77]).

activation plays a very important role in magnetization reversal. The activation volume calculated at the coercive field is about 1.1×10^{-18} cm^3, which gives a thermal stability factor ($K_u V / k_B T$) of 900, indicate these films have very good thermal stability.

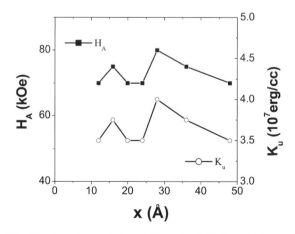

Fig. 32. The dependence of H_A and K_u on the B_2O_3-layer thickness of the annealed (RePt 32 Å/B_2O_3 x Å)$_5$ multilayers (after Luo et al. [77]).

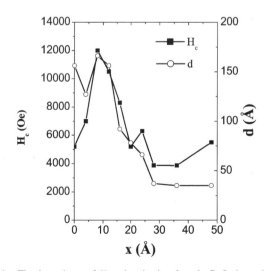

Fig. 34. The dependence of H_c and grain size d on the B_2O_3-layer thickness (after Luo et al. [77]).

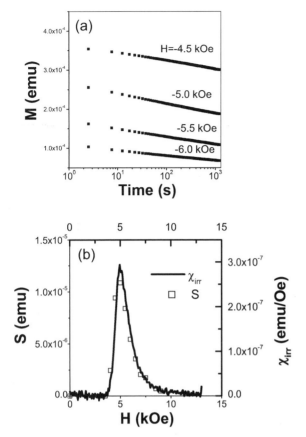

Fig. 35. (a) The decay of magnetization with time; and (b) the field dependence of the viscosity coefficient S and the irreversible susceptibility χ_{irr}.

In summary, nanocomposite thin films exhibit a rich variety of interesting properties that are different from other materials, such as the enhanced magnetic properties at the percolation threshold. These materials have opened up the possibilities for their use in many technological applications. Among them, the potential as high-density recording media has been extensively investigated. However, for potential high-density media, an appropriate choice of matrix material as well as precise control of the particle size and the interparticle distance are both crucial. Some advance has been achieved toward this goal by Sun et al. [87]. In their study, they have been able to synthesize FePt-nanoparticle assemblies with precise control of particle size, composition, and interparticle spacings. This kind of material may contribute to future magnetic recording at areal densities in the terabits per square inch regime.

3. CLUSTER-ASSEMBLED THIN FILMS

3.1. Introduction

One goal of modern condensed-matter science is to use constituents (mainly atoms and molecules) as the elementary bricks to build new materials. In this context, the field of clusters which concerns the aggregation of a few atoms to a few thousands of atoms, appears especially fascinating, since these systems are intermediate between of small molecules and bulk

solids. On a fundamental level, they offer the opportunity to understand more precisely how atomic structure can lead to physical and chemical properties of macroscopic phases. Moreover, it is possible to produce in the form of clusters some atomic arrangements that do not normally exist in nature and may consequently exhibit specific properties that are not due only to the large number of surface atoms.

Nanoclusters can now be produced in sufficient numbers by means of either physical or chemical processes. They can be assembled into materials that can be studied by a variety of conventional experimental methods. These materials have the possibility to take advantage of and to incorporate a number of size-related effects in condensed matter ranging from electronic effects (so-called "quantum size effects") caused by spatial confinement of delocalized valence electrons and altered cooperative ("many body") atomic phenomena, such as lattice vibrations or melting, to the suppression of such lattice-defect mechanisms as dislocation generation and migration in confined grain size. This effort has already stimulated a few commercial dividends, such as ceramics and chemical catalysts with increased efficiency because they have a very high surface-to-volume ratio [88]. Advances with semiconducting nanoclusters are bringing several new electronics and communications applications [89, 90]. Magnetic nanoclusters are very interesting for applications to high-density storage and magnetoresistive heads with the "giant magnetoresistance (GMR)" and "tunneling magnetoresistance (TMR)" effects [91–93]. The unprecedented ability to assemble nanoclusters into new materials with unique or improved properties is thus creating a revolution in our ability to engineer condensed matter for desired utilization. These nanophase materials, assembled using nanoclusters as building blocks, can be synthesized to have a wide variety of controlled optical, electronic, magnetic, mechanical, and chemical properties with attendant useful technological applications.

Cluster physics has developed into a strong and active branch of condensed-matter physics since the last thirty years. Simple free metal clusters have been thoroughly investigated and are well understood, as seen in the reviews of Haberland [88], Ekardt [94], de Heer [95], and Meiwes-Broer [96]. For free clusters, the properties are mainly determined by the delocalized valence electrons, which are considered free electron-like. A jellium model used for study of free clusters shows that effects of the ionic cores on the valence electrons reduce to an effective potential in which the electrons are confined. This results in the electronic shell model, in which the electrons successively fill different energy levels. Predictions based on the shell model are found in good agreement with experiments. Clusters with special "magic" numbers of electrons are predicted to be more stable and to show some special properties compared with others.

The interest in magnetic clusters has grown enormously with increasing attention devoted to the effect of nanosize confinement on the physical properties and with regard to their potential in the nanoscale engineering of materials with very specific properties. The magnetic moments of iron, cobalt, and nickel free clusters in a supersonic beam have been determined in a Stern–Gerlach deflection experiment [97, 98]. The results show that the magnetic moments as a function of size from 20 to 700 atoms decrease from high atomic values to the respective bulk values.

For most technological applications, the properties of embedded clusters are more important than those of free ones. The physics of magnetic phenomena such as giant magnetoresistance, tunneling magnetoresistance, and exchange coupling in cluster-assembled materials is currently under intense investigation. Comparison between the behaviors of the free and embedded clusters would contribute much to the understanding of the specific properties of those materials.

According to the chemical composition, shape and dimensionality, nanostructured materials may be classified into 12 groups as proposed by Gleiter [99] in a comprehensive article including preparation techniques, structures, and properties. Due to dimensionality effects, nanostructured materials have shown unusual properties not only in the case of electronic and magnetic characteristics [91–93], but also in the case of the mechanical properties of nanophase metals [99, 100]. Among various types of nanostructured materials, cluster-assembled materials can conserve the specific properties of the incident free clusters, or change the physical and chemical properties of them via matrix or multilayer interactions.

This chapter is mainly concerned with cluster formation and cluster-assembled materials. The first part of this section is a discussion of cluster formation in cluster sources, size distribution, and cluster-deposition techniques. The mechanism of cluster formation with homogeneous and heterogeneous nucleation in a sputtering-gas-aggregation (SGA) source is discussed in detail. To illustrate the potentialities of the cluster-deposition techniques, the second part is devoted to cluster-assembled magnetic thin films directly related to the incident cluster size and to the well-controlled nanostructures, which are presented with possible technological applications.

3.2. Cluster Formation, Size Distribution, and Deposition Techniques

3.2.1. Cluster Formation

One of the first challenges for researchers in cluster science was to produce clusters of any kind of material in a wide range of sizes. In 1956, Becker, Bier, and Henkens first reported on the formation of clusters by condensation of the respective vapor expanding out of a miniature nozzle into vacuum [101]. Since then, several types of cluster sources have been developed to study the properties of free clusters, such as supersonic-nozzle sources, gas-aggregation sources, laser-vaporization sources, pulsed-arc cluster-ion sources (PACIS), liquid-metal-ion sources and ion-sputtering sources. A few topical review articles summarize activities concerning cluster sources [95, 102–104].

Supersonic nozzle sources are used mostly to produce intense cluster beams of low-boiling-point metals. These sources produce continuous beams with reasonably narrow speed distributions. Clusters with up to several hundreds, (and under suitable conditions), with thousands of atoms per cluster may be produced in adequate abundance. The cluster temperatures are not well known; however, larger clusters are often assumed to be near the evaporation limit.

Gas-aggregation sources are particularly efficient in the production of large clusters ($N > 1000$). The intensities are generally much lower than that of nozzle sources. The overall cluster size distributions can be adjusted within rather broad limits. These sources are used for low- to medium-boiling-point materials (<2000 K). Low cluster temperature (<100 K) can be achieved.

Laser-vaporization sources produce clusters in the size range from the atom to typically several hundreds of atoms per cluster. These sources are pulsed and, although the time-averaged flux is low compared with the supersonic nozzle source, intensities within a pulse are much higher. In principle, these sources can be used for all metals to produce neutral and positively and negatively charged ionized clusters. The cluster temperatures are expected to be near or below the source temperature, depending on the supersonic expansion conditions.

The pulsed-arc cluster-ion source (PACIS) is basically similar to laser-vaporization sources. However, an intense electrical discharge rather than a laser is used to produce the clusters, resulting in intense cluster ion beams.

Ion-sputtering sources are used primarily to produce intense continuous beams of small singly ionized clusters of most metals. The clusters are hot, typically near the evaporation limit.

Liquid-metal-ion sources produce singly and multiply charged clusters of low-melting-point metals. These clusters are hot, since in flight evaporation and fission are observed.

Although the sources mentioned earlier have been used in many laboratories worldwide for some specific requirements, there are some shortcomings, which have to be overcome:

- Most of the sources mentioned before have low beam intensity.
- There is small adjustability of the range of the mean cluster size. Most of the sources can only produce small clusters (laser-vaporization source, PACIS, and ion-sputtering source), and gas-aggregation sources are particularly efficient in the production of large clusters ($N > 1000$); however, it is difficult to get a broad range of sizes using a single technique.
- Some of the cluster sources, such as laser vaporization and PACIS, can only produce a pulsed cluster beam.
- The fraction of ionized clusters in the beam is very low. The intensity of charged clusters is too small for energetic cluster deposition and cluster analysis. Additional ionization methods, such as electron or photon impact, are needed to form more ionized clusters.

To overcome these shortcomings, Haberland and his coworkers have developed a new kind of cluster beam source that combines improved magnetron sputtering with a gas-aggregation tube, the so-called sputtering-gas-aggregation

(SGA) source as discussed in [105–112]. This cluster source can produce a very large range of mean cluster sizes from 200 to 15000 atoms per cluster, and has a high degree of ionization from 20 to 50%, depending on the target materials. Usually the deposition rate is about 3 Å s^{-1} for the total flux of the cluster beam. For one of the charged components, the deposition rate is about 1 Å s^{-1}. Because of its versatility and simplicity, and because the clusters of almost all element–even refractory materials and compounds–can be produced [108], this cluster source has been adopted in many laboratories worldwide.

How can it be understood that this source produces such a high degree of ionization and very narrow cluster-size distribution? What is the mechanism of the cluster formation in this source? To study cluster formation in a gas-aggregation source in a quantitative way, one has to take into account many parameters because the physical details of the gas-aggregation process are quite involved. However, it is a commonly known phenomenon, and has been employed extensively in the cloud chambers used in the early days of nuclear physics [113–115]. Cluster formation by gas aggregation occurs every day in cloud formation in the atmosphere, where the sun leads to evaporation of water molecules into the ambient air. If the water density is high enough, the molecules in the air start to aggregate, produce a small cluster, and then become a large water drop, which is visible to the naked eye, if the drop is larger than the wavelength of visible light.

The cluster formation in the SGA source (Fig. 36) is very similar. The difference in the cluster source is that there are many electrons, ions, and/or energetic particles in the plasma in the near area of surface of target [115], which produce significantly heterogeneous growth processes. The sputter discharge ejects atoms and ions into the rare-gas flux, where they start to aggregate through low-temperature condensation, if their density is high enough. The clusters are charged by many charged or energetic particles in the plasma, so that no additional electron impact is necessary. The mechanism of cluster

growth in a magnetron sputter-gas-aggregation source is discussed later in more detail with theory and experiment.

3.2.2. Theory of Cluster Formation by Gas Aggregation

In this section, cluster formation by gas aggregation with homogeneous and heterogeneous nucleation in sputtered metal-vapor atoms with a rare-gas stream is considered. If cluster formation with an unusually high degree of ionization is to be explained by homogeneous and heterogeneous nucleation theory, the following separate conditions must be shown to be satisfied. (1) An adequate supply of cluster embryos and charge transfer processes must be available. (2) A sufficient number of collisions must occur between sputtered atoms, rare-gas molecules, and growing clusters so that cluster sizes of hundreds to some thousands of atoms can be accumulated.

3.2.2.1. Homogeneous Growth Processes

Classical nucleation theory might be said to have been laid by Gibbs but the theory was not put into any specific form until about the third and fourth decades of the twentieth century, notably by Vollmer and Weber [114] and Becker and Doering [115] for the explanation of experiments in cloud chamber for nuclear physics. The Gibbs free energy is given by $G = U + pV - TS$, where U is the inner energy, p, V, and T are pressure, volume, and temperature of the gas molecules, respectively, and S is the entropy. For the ideal gas system, the free enthalpy G can be expressed by

$$G(p \cdot T) = RT \ln \left[\frac{p \lambda_{th}^3}{k_B T} \right] \qquad (34)$$

where R is the gas constant, k_B the Boltzmann constant, and λ_{th} is the de Broglie wave length. When the pressure p, a supersaturated state, of the sputtered metal vapor at a given temperature T is greater than the saturation pressure p', the change of Gibbs free energy is $\Delta G = RT \ln(p'/p)$. This

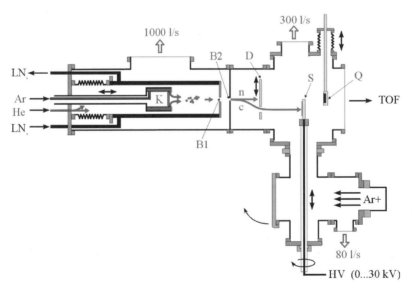

Fig. 36.   The cluster deposition system, with sputtering-gas-aggregation (SGA) source.

thermodynamic system has a spontaneous tendency to form a state of lowest possible Gibbs free energy G, so it forms a spherical embryo of some small radius r, which can be written as $\Delta G = 4\pi r^2 \sigma - \frac{4}{3}\pi r^3 \rho RT \ln(s)$. In this equation, the first term on the right is the intrinsically positive contribution of surface tension, where σ is the specific surface free energy. The second term, in which ρ is the liquid density, and $s = p/p'$ is the vapor supersaturation ratio, represents the contribution to ΔG made by the bulk free energy change. This equation specifies the change ΔG as a function of r and s for a fixed T. In the case $s = 1$, the bulk term vanishes and ΔG rises monotonically with r. For $s > 1$, however, the second term is a negative contribution to ΔG, its magnitude rising as the degree of supersaturation is raised. Its presence assures existence of a maximum ΔG_{max} at a critical embryo radius r^*,

$$\Delta G_{max} = \frac{16\pi\sigma^3}{3(\rho RT \ln s)^2} = \frac{4}{3}\pi r^* \sigma$$

and

$$r^* = 2\sigma/\rho RT(\ln s) \tag{35}$$

This relation reveals the very important rule that the larger the supersaturation ratio s, the smaller the critical radius r^*. To further consider the case where $s \leq 1$, ΔG is a rapidly increasing function of r. The gas-aggregation system does not enter directly into phase-transition nucleation processes. There exists a statistically steady-state population of embryos characterized by the Boltzmann-like distribution function,

$$N_i = N_g \exp(-\Delta G/k_B T) \tag{36}$$

where the number of molecules g per embryo rather than r as the size parameter of the embryo is used; N_g is the number per unit volume of vapor of the embryos of size g.

As noted earlier, the fundamental thermodynamic correlation of this homogeneous nucleation theory is: Raising the degree of supersaturation markedly enhances the probability that fluctuation processes will send some embryos over the top of the activation barrier in a given time because raising s reduces the embryo size r^* and the associated height ΔG_{max} of the top of the barrier.

Now, we consider the rate of cluster nucleation J. According to the generally accepted equation $J = ZWN_i$; the nucleation rate of stable clusters is directly proportional to the equilibrium density of critical embryos N_i, to the rate of collisions per nucleus W, and to a correction Zeldovich factor Z. With the assumption of a thermally stable state, the nucleation rate J for cluster formation can be expressed by $J = K \exp(-\Delta G/k_B T)$, where K is a factor which varies much more slowly with p and T than does the exponential term. Becker and Doering [115] gave K as $K = (p/k_B T)^2(2\sigma/\pi m_{mol}\rho^2)^{1/2}$, where the m_{mol} is the mole mass of the embryo.

The principle of cluster formation in our cluster source is similar: Metal atoms are sputtered into a rare-gas flow, which cools the atoms and, growing clusters, sweeps them out of the condensation region. As almost only single atoms are present at first at the sputter conditions [105], the clustering has to start with the metal dimer formation. Due to the energy and momentum conservation laws, dimer formation is not possible by a two-body collision. The dimers of metal atoms (M_2) are generated by three-body collisions. This process can be written as

$$M + M + Ar \rightarrow M_2 + Ar \tag{37a}$$

$$M_2 + M + Ar \rightarrow M_3 + Ar \tag{37b}$$

$$\cdots\cdots$$

$$M_n + M(M_{n'}) + Ar \rightarrow M_N + Ar \tag{37c}$$

The dimers (M_2) are cooled by collisions with the rare gas and they grow by addition of monomers. When this process has effectively reduced the monomer density, the clusters continue to grow by cluster–cluster coagulation. Clusters of any size can be generated by a variation of the parameters of the gas-aggregation tube, such as temperature T, gas pressure p, sputter rate of target materials and He- to Ar-flux ratio. The most critical variable determining the cluster size is the residence time of the clusters in the aggregation region, i.e., the condensation distance L.

3.2.2.2. Heterogeneous Growth Process

Until now, we have discussed only the homogeneous nucleation of cluster formation. Practically, in the near zone of sputter target there are many electrons, excited Ar* atoms, and Ar$^+$ ions. Among these, a heterogeneous nucleation must happen by the electrostatic interaction and the collisions of the energetic particles. In earlier research of condensation in the Wilson cloud chamber, Das Gupta [113] found the relationship between the supersaturation rate s and the critical radius r^*,

$$\ln s = \frac{1}{\rho RT}\left[\frac{2\sigma}{r^*} - \frac{e^2}{8\pi r^{*4}}\right] \tag{38}$$

where e is the charge of clusters. The two terms on the right-hand side of the equation signify the pressures due to the surface tension and an electric field. The two are in opposite sense, for while the potential energy due to the surface tension (σ) decreases with decreasing radius r^*, that due to electrification decreases on increasing the cluster radius. The former tends to decrease the radius of clusters thereby increasing the supersaturation pressure p, while the latter tends to increase the radius and to decrease its vapor pressure because a charged cluster will therefore always have less vapor pressure than an uncharged one.

Another important point, which must be explained in detail, is that the ionization degree of clusters is commonly found to have more than 30 to 60% in the source without further electron-impact ionization. This degree is nearly 2 orders of magnitude higher than that obtainable from a very efficient electron-impact ionizer.

Usually, sputtered atoms from the surface of metal targets are predominantly neutral, only about 10^{-3} of them are ionized [116]. If one out of every thousand atoms is charged, and these atoms coalesce to form a cluster, then this cluster is charged: $M_{999} + M^+ \rightarrow M_{1000}$. In a plasma discharge, there are many other effective mechanisms to create and to destroy charges. The predominant charged ion near the magnetron sputter cathode will be Ar^+. Because of the large cross section for charge transfer $Ar^+ + M_N \rightarrow Ar + M_N^+$, the argon ion transfers its charge with high efficiency to the metal atoms and clusters. The reverse process is not possible at thermal energies as the first ionization energy of Ar (15.4 eV) is much higher than that of any metal. Another possible mechanism for formation of positive charged clusters is Penning ionization. The electrons in the discharge can excite an argon atom Ar^* to a long-life metastable level. These excited Ar^* atoms have energy of about 11.5 eV, and ionize a neutral cluster on collision: $Ar^* + M_N \rightarrow Ar + M_N^+ + e$. It is well known that large concentrations of metastable rare-gas atoms are present in gas discharges, so that Penning ionization is an effective ionization mechanism for charged-cluster formation.

So far, only the generation of positive charged clusters has been discussed. Experimentally, we have observed roughly an equal flux of both charge polarities. In a plasma, the electron density is very high. An electron colliding with a cluster will have a certain probability to stick: $e + M_N \rightarrow M_N^-$. The internal energy resulting from the electron attachment can either lead to an ejection of atoms from the cluster, or lead to the energy being transferred by collisions to argon atoms.

This cluster source can produce monodispersed nanoclusters with very narrow-size distribution because these strong heterogeneous growth processes happen in the plasma close to the surface of the target. The clusters grow with some aggregation distance L, which is a critical parameter to produce a narrow-size distribution.

According to the charging mechanisms described previously, it is not surprising to have such a high degree of ionization in the cluster source.

3.2.3. Cluster-size Distribution

3.2.3.1. Free-Cluster-Size Distribution Measured by TOF

Figure 37 shows a time-of-flight (TOF) mass spectrometer. It is connected via a small hole from the preaccelerator to the deposition chamber. A gate valve allows the vacuum in the TOF to be isolated from that in the deposition chamber. The three TOF chambers are separately pumped by three turbo-molecular pumps. In the detector chamber, vacuum can be up to 1×10^{-8} mbar. To separate the neutral clusters, the ion lens of the preaccelerator in the deposition chamber accelerates vertically only clusters of one charged polarity. With a potential of 3.3 kV, the charged clusters enter the ion-optic chamber. This charged cluster beam is pulsed electrically to about 2 kV by the WIMAC from Wiley and McLaren type, and bent by the ramp (RAMPE) to enter the drift tube of 3.3-m length. Then, this charged-cluster beam is reflected by the two-step reflector to transfer the other drift tube into the detector chamber. The detector is an ion-to-electron-to-photon converter, colloquially known as an Even-cup. The charged clusters are accelerated up to 30 kV and impinge on the surface of an aluminum target. The ejected secondary electrons are in turn accelerated through a hole in the cup onto the scintillator. The resulting photons are recorded by a photomultiplier (Hammatsu R647-01). The signals are amplified 10 times by a Phillips amplifier. A 100-MHz transient recorder (DSP 2001S, maximum 8192 channels) records the signals and a CAMAC multichannel adder (DSP 4100) averages the arrival time spectra by using up to 6500 events. The digital data are processed by a control program of OS9 system in a computer with a Motorola 68030 CPU, Eltec Company. The spectrum is shown

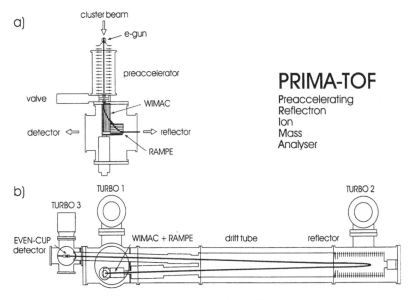

Fig. 37.   Schematic of the time-of-flight (TOF) mass spectroscopy.

in the monitor. At the same time, we can use an oscilloscope to directly observe the change of cluster-size distribution by changing the parameters of the cluster source.

For a spectrum of the neutral background gas, an electron gun, positioned in front of the preaccelerator in the deposition chamber, is pulsed with a potential about 180 V and 0.2 mA current. The generated gas ions are analyzed by the TOF. The standard masses present in an unbaked vacuum system are observed.

This TOF mass spectroscopy has a very high-resolution $Q = m/\Delta m$ of at least 2000. It enables the observation of an individual ion to very large clusters (up to 10^6 amu) and thus helps us with a deeper examination of the mechanism of cluster growth.

Figure 38a shows a typical size distribution of negatively charged Co nanoclusters, as measured *in situ* by TOF. The horizontal axis has been converted from flight time to cluster diameter. The Co clusters have a mean diameter D of 4.4 nm. The lognormal function gives a very good fit to the measured data. Note that no atoms or small clusters are present in the beam if an He gas flux is added. The points give the experimental result, while the dot lines are the lognormal distributions fitted well to the data. Although the lognormal function is often found for heavily clustered materials [101–103], these measurements show that it is also a good description of a nanosized cluster distribution. The mean size of clusters can be easily varied between 200 and more than 15,000 atoms by adjusting the aggregation distance L, the sputter power P, the pressure p in the aggregation tube, and the ratio of He to Ar gas flow rate. Among them, the aggregation distance and the ratio of He to Ar gas flow rate are important parameters for getting a high intensity, very low dispersion of the cluster beam.

The size distribution was checked independently by TEM, AFM, and was found to be in agreement with the TOF data.

3.2.4. Deposited Cluster-Size Distribution

3.2.4.1. TEM Measurements

Since the intricate structure of the thin film deposited by the cluster beam exists only on the nanometer scale, the use of transmission electron microscopy (TEM) is imperative. A TEM result (Fig. 38b) gives an example of cobalt clusters deposited thermally with the same cluster beam but total cluster flux (neutral, negatively, and positively charged clusters) on an amorphous carbon-coated grid of copper. The Co clusters have a mean size of diameter of 4.5 nm, and the size distribution is also fitted very well by a lognormal function. The results measured by TEM and TOF are in good agreement. This means the neutral and charged nanoclusters have nearly the same size distribution.

3.2.4.2. Magnetization Measurements

For ultrafine magnetically ordered particles, there exists a critical size below which the particles can acquire only single

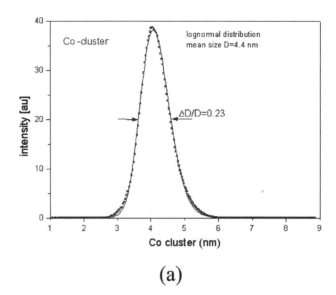

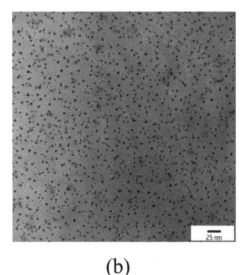

Fig. 38. (a) Experimental size distribution of negatively charged Co clusters measured by TOF. The mean size corresponds to about 3600 Co atoms per cluster or a mean diameter of 4.4 nm, respectively. $\Delta D/D = 0.23$. The thin line is a fit to a lognormal distribution; (b) TEM imaging shows the cobalt clusters deposited thermally with the same cluster beam but the total cluster flux on an amorphous carbon-coated grid of copper. The Co clusters have a mean size of diameter of 4.5 nm.

magnetic domains even in zero magnetic field. The critical size for Co and Fe are of the order of a few hundreds of angstroms, depending on the shape of the particle. The magnetic axis of a single-domain particle is determined by the magnetic anisotropy energy (KV) of the particle, where K is the total magnetic anisotropy energy per unit volume and V is the volume of the particle. At low temperatures, the magnetic axes of the single-domain particles are randomly oriented and frozen. This leads to a zero magnetization of the samples. Under a large external magnetic field, all the magnetic axes will be aligned, giving rise to the saturation magnetization (M_s). When the external field is turned off, one measures the remnant magnetization (M_r) which generally obeys $M_r = M_s/2$, because the magnetic axes are randomly oriented over a hemisphere and $\langle \cos\theta \rangle = 1/2$, if θ is uniformly distributed between 0 and π.

At sufficiently high temperatures, the magnetic anisotropy energy barriers of the single-domain particles are overcome by thermal energy, and superparamagnetism occurs. In the simplest analysis, superparamagnetism relaxation in zero applied fields can be described by the Arrhenius law (Eq. (15)), Superparamagnetic behavior can be observed, using an instrument with a characteristic measuring time (τ), at temperatures above the blocking temperature (T_B), which is defined by

$$T_B = \frac{KV}{k_B[\ln(\tau/\tau_0)]} \quad (39)$$

Below T_B, coercivities of single-domain particles are generally much larger than that of bulk material. The temperature dependence of H_c is dominated by superparamagnetic relaxation that leads to

$$H_c(T) = H_c(0)\left(1 - \sqrt{\frac{T}{T_B}}\right) \quad (40)$$

This $\sqrt{T}$ dependence has been observed in many granular films [107, 108]. Above T_B, in the superparamagnetic state, the magnetization curve $M(H)$ of a noninteracting system with uniform cluster size is described by the Langevin function,

$$M = pM_s L\left(\frac{\mu H}{k_B T}\right) = pM_s\left[\coth\left(\frac{\mu H}{k_B T}\right) - \frac{k_B T}{\mu H}\right] \quad (41)$$

where p is the volume fraction of the magnetic clusters in the film, $\mu = M_s V$ is the magnetic moment of a single cluster with volume V, and H is the external field. Because μ is large ($\approx 10^2 - 10^3 \mu_B$), saturation of the magnetic moments can be accomplished at high fields, but generally not at $T \gg T_B$. In all real systems, there are size distributions. The resultant magnetization should be given by

$$M = pM_s \int_0^\infty L\left(\frac{M_s VH}{k_B T}\right) f(V) \, dV \quad (42)$$

where L is the Langevin function as before and $f(V)$ is the size distribution. Assuming spherical particles of diameter D for simplicity, a lognormal size distribution is most often used as before. Thus, by fitting the experimental data, one can obtain the distribution characterized by the mean size D and the standard deviation σ. Our results show that the size determinations in the samples in which the Co clusters embedded in a metal matrix are in agreement with some of those results by TOF measurements. In those samples, the Co cluster concentration is very low ($<10\%$), where the intercluster interaction is very weak.

3.2.5. Deposition Techniques

A new deposition technique has been developed for depositing nanocluster-assembled materials by combining a cluster beam with two atomic beams from magnetron sputtering. A high intensity, very stable beam of nanoclusters (like Al, Ag, Cu, Co, Mg, Mo, Si, Ti, TiN, TiAlN, and Al_2O_3) is produced by the SGA source. The atomic beams are produced by dc- or rf-magnetron sputter sources for metallic or nonmetallic materials. This technique can be used to deposit simultaneously or alternately mesoscopic thin films or multilayers, and offers the possibility to independently control the incident cluster size and concentration, and thereby the interaction between clusters and cluster-matrix material which is of interest for applications in various fields such as magnetic and semiconductor materials.

Figure 39a shows the principle of the technology. Metal atoms are produced by a high-pressure magnetron discharge. They are clustered by the gas-aggregation technique, so that the mean size is between 10^2 and 10^5 atoms per cluster. The clusters pass through a diaphragm, and are separated by an ion optic into a neutral and charged-cluster beam. The substrate can be floated up to 30 kV. When the substrate is removed, the mass distribution of the charged components can be measured by a time-of-flight mass spectrometer.

To prepare mesoscopic and multilayer systems, two standard magnetron sputter sources located in the deposition chamber allow one to alternatively deposit Co clusters from the cluster beam, metal or nonmetal (Ag, Cu or SiO_2) layers from one of the sputter sources and the magnetic (NiFe) layers from the other one, when the substrate holder S is rotated to the corresponding direction of the beams. An oscillating quartz microbalance, which can be rotated by 360°, is installed in the deposition chamber to measure the deposition rates from all three-beam sources. The nanocluster concentration in the resulting films can be adjusted through the ratio of the deposition rates of the cluster beam and the atomic

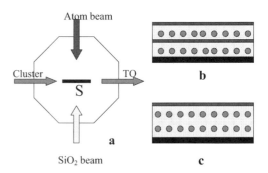

Fig. 39. (a) Principle of the deposition technique for making nanocluster based materials. The cluster beam comes from a sputter-aggregation source, while two atomic beams come from magnetron sputtering sources. This technique can be used to deposit: (b) multilayers of a hybrid structure on the Si substrate. Co clusters are embedded into a silver film, which is then covered by permalloy (NiFe). To increase the GMR sensitivity, this procedure is repeated 15 to 20 times; (c) mesoscopic films deposited by combining a magnetic cluster beam simultaneously with an atomic beam. The clusters are disorderly buried within a nonmagnetic matrix.

beam. Additionally, we can also measure the current on the substrate during the deposition of charged clusters to calculate the deposition rate of Co cluster, giving a control of the rate measured by the microbalance. With this technique, the mean size of the embedded Co clusters controlled by the parameters of the cluster source is thus independent of its concentration in the film. This is in contrast with granular systems prepared by precipitation or codeposition, where the particle size and the concentration depend on each other. Another great advantage of this technique is that we can separately use the neutral, the charged, or the total cluster beam to produce films and to compare the different properties of them. The deposition rate of one charged component of the cluster beam is typically 1–2 Å s^{-1}. The cluster size can be easily controlled by adjusting the parameters of the cluster source, such as the aggregation length, the pressure, the sputter power, and the ratio of helium to argon gas.

Figure 39b shows the structure of a hybrid multilayer deposited by this technique. The film is produced by sputtering a silver film on an Si substrate, depositing cobalt clusters with kinetic energy up to 2 eV per atom into the silver film, and covering the clusters with silver, onto which a permalloy layer is deposited by sputtering. To increase the sensitivity, this procedure is repeated 15 to 20 times. Simulations of energetic cluster impact [111, 112] reveal that a spherical cluster will be smashed to form a flat disk when hitting on the surface with a kinetic energy of about 2 eV atom^{-1}. It is assumed that the demagnetization of such a flat disk of Co clusters will increase the sensitivity of the giant magnetoresistance (GMR) effects.

Figure 39c demonstrates the structure of a mesoscopic thin film deposited at room temperature on a surface of Si substrates by combining the total flux of Co clusters simultaneously with a metal atomic beam, by turning the substrate to 45°. Samples of different thickness from 10 to 5000 nm can be prepared with different mean sizes from 200 to 15000 atoms per cluster (about 2- to 8-nm diameter), and different volume concentrations from 5 to 80% in the matrix (Cu, Ag, Al, Au, SiC, and SiO$_2$). To prevent oxidation, a capping layer is deposited on the surfaces of the films.

The pressure in the deposition chamber is below 10^{-8} mbar, and rises to about 10^{-6} mbar during cluster deposition. This pressure increase is nearly entirely due to argon gas from the gas-aggregation source, which needs a pressure of 0.1 to 1 mbar to work properly.

Thermal loading of the substrate by the deposition process is below 0.01 W cm^{-2}. The beam is highly collimated, and has a very high mass-to-charge ratio (number of atoms per electric charge 10^2 to 10^5). Thus, the charging problem which often arises when depositing ions on a nonconducting surface is much reduced. In fact, it is possible to make a durable metal film on untreated paper or even Teflon [106].

By systematically varying cluster concentrations while keeping their size constant, one can investigate intercluster interactions. Alternatively, a low concentration of varying cluster sizes allows one to investigate the magnetic properties of the individual clusters. For certain combinations of cluster size and concentration, the cluster-assembled materials will change from behaving as a group of individual particles, or as a superparamagnetic system, to a more collective behavior, which might involve spin-glass or random-anisotropy transitions. Previous studies have investigated the similarities and differences of superparamagnetic clusters compared to a spin-glass system [117–121]. Few works have been done in this area systematically since it is difficult to vary the concentration independently of the cluster size for the methods of previous researches.

This new deposition technique permits a fertile exploration of finite-size effects, enhanced and tailored properties of fundamental interest, and technological applications. Some examples of the nanocluster-assembled magnetic materials are given in the following, which cannot be attained by conventional methods of thin film formation, like evaporation, cosputtering, or other nanotechnologies.

3.3. Cluster-Assembled Magnetic Films

3.3.1. Experiments

The magnetic properties of nanocluster-assembled materials have been measured as a function of the size and the concentration of cobalt clusters. Co clusters of a variable mean size of about 200 to 15,000 atoms were generated by the SAG source, and deposited softly at room temperature (RT) on Si(100) substrates simultaneously with an atomic beam (Cu, Ag, Au, or SiO$_2$) from a conventional sputter source. The cluster concentration was adjusted through the ratio of the cluster and atomic beam deposition rates, as measured *in situ* with a quartz microbalance. The cluster-size distribution was monitored by TOF and TEM. Samples were made with volume concentrations of cobalt of 5 and 50%. All the samples were approximately 35-nm thick and were capped with 5 nm (25 nm for the SiO$_2$ case) of matrix material to allow sample transfer in the atmosphere.

For all the samples, the magnetization vs temperature [$M(T)$] and the magnetization vs applied field [$M(H)$] was measured at 5 and 300 K by superconducting quantum interference device (SQUID). The maximum measuring field was $5.5T$. In the $M(T)$ measurements, the samples were cooled in zero fields from 300 to 5 K where the magnetic field was applied (100 Oe). The magnetization was then recorded on heating the samples to 310 K [zero-field-cooled (ZFC)]. After cooling down in the field, M was measured again [field-cooled (FC)].

From the $M(H)$ measurements, all the samples at room temperature are superparamagnetic except 9000 atom clusters at 50%. At 5 K, they are ferromagnetic with hysteresis loops. The FC loops at 5 K did not show any field shift that would indicate an exchange-bias phenomenon from a CoO shell magnetically coupled to the Co cluster core [122–124]. That means the films were not significantly oxidized. The M_s calculations have been made with the given concentrations and thickness.

Magnetic X-ray circular dichroism (MXCD) measurements of the nanomaterials at RT were carried out on the Daresbury synchrotron [109]. Samples were placed in a magnetic field (normal to the surface) and the total yield of electrons measured at wavelengths spanning the L2 and L3 edges. The field was then reversed and the spectrum again recorded. These two spectra can then be subtracted to obtain the dichroism. It is found that the original spectra, i.e., before subtraction, contain no extra peaks resulting from chemical core shifts of the p-states indicating again that there is little or no contamination of the cobalt. General sum rules [125, 126] have been used to obtain the average local spin and orbital components of the magnetism, and the ratio of orbital and spin magnetization. The value for the total average magnetism per atom ($\langle m \rangle = \langle m_S \rangle + \langle m_L \rangle$) has been calculated assuming that the number of d holes is 2 for all the systems under study.

The nanostructure of the clusters was studied by X-ray powder diffraction and TEM. The results show that the Co clusters buried within a Cu matrix have nanocrystalline structure.

3.3.2. Results and Discussion

3.3.2.1. Intercluster Interactions

Magnetic intercluster interactions generally exist in cluster assemblies. They are more or less strong according to the volume concentration. Magnetic dipolar interactions are always present. If the matrix is metallic, Ruderman-Kittel-Kasuya-Yosida (RKKY) interactions occur and depend on $1/L^3$, where L is the distance between centers of clusters, like dipolar interactions. When the matrix is insulating, superexchange interactions could exist at a short-range according to the structure and the nature of the matrix and the bonding at the cluster–matrix interface.

Figure 40 shows the ZFC and FC curves for the 4500 atom samples in the Cu matrix (Co:Cu) at Co concentrations of 10, 20 and 50%. Figure 41 gives the ZFC and FC curves for the 9000 atoms samples in the SiO$_2$ matrix at concentrations of 10, 30 and 50%. On $M(T)$ graphs, T_p, the magnetization peak temperature, corresponds to a peak in the ZFC curve and irreversibility between the ZFC and FC processes occurs at this temperature.

The effect of intercluster interactions is also manifested in the Co cluster size and concentration dependence of T_p in the different matrices (Cu or SiO$_2$) as shown in Figure 42. We observe: (1) the T_p increases with increasing the Co concentration for all the cluster sizes. This behavior is in agreement with the theory which predicts an enhancement of T_p with the concentration due to dipolar interactions which increase the energy barrier [106], but the increase for small clusters (1000 and 4500 atoms) is much faster than that for large one (9000 atoms). (2) At the same concentration in the Cu matrix the large clusters have a large T_p, and the small clusters small T_p, while the increased T_p at less concentration is much larger than that at more concentration. (3) For the same cluster sizes

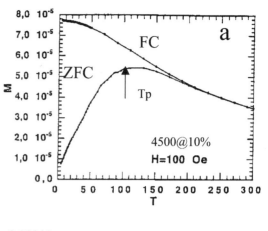

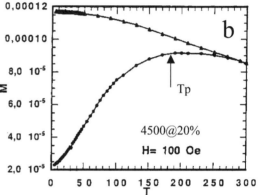

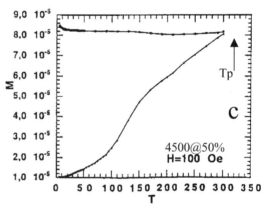

Fig. 40. FC and ZFC curves of the Co 4500 atom clusters in a Cu matrix at the different concentrations (a) 10%, (b) 20%, and (c) 50% (after Qiang et al. [133]).

(9000) in a different matrix (Cu or SiO$_2$), the T_p are significantly different. T_p is higher than 300 K for the Cu matrix, and below 15 K for SiO$_2$ at all concentrations, although the slopes of the concentration dependent curves for both matrices are nearly the same.

For 1000 and 4500 atoms clusters, T_p is increasing with the concentration of Co clusters in the matrix. For the 50% cluster samples, the clusters coalesce to form bigger grains: T_p is much larger for the 1000 atom clusters, and is above 300 K for the 4500 atoms.

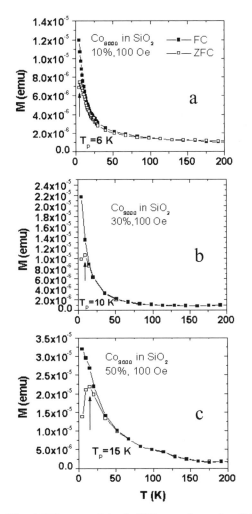

Fig. 41. FC and ZFC curves of the Co 9000 atom clusters in a SiO$_2$ matrix at the different concentrations (a) 10%, (b) 30%, and (c) 50% (after Qiang et al. [133]).

For the 9000-atom clusters in the Cu or Ag matrix, T_p is always above room temperature. We can explain that at 10 and 20%, even if the cluster coalesce giving rise to a larger range of T_p, they remain single-domain grains (the critical size being about 14-nm radius for cobalt), while for 50% they form a continuous granular magnet above the percolation limit, because magnetic loops at RT have shown ferromagnetic behaviors.

For the matrix of SiO$_2$, T_p is below 15 K because there are only dipolar interactions between clusters. A similar result of Co clusters in silica made by ion implantation was also reported [107]. If superexchange coupling is present, the M_s of 9000-atom clusters should be reduced because this coupling is expected to be stronger than dipolar or RKKY interactions, but in fact the M_s for the Co clusters in SiO$_2$ is significant larger than that in Cu or Ag matrices (Fig. 43).

The problem of intercluster interactions is complex because there can be several kinds (dipolar, RKKY exchange, superex-

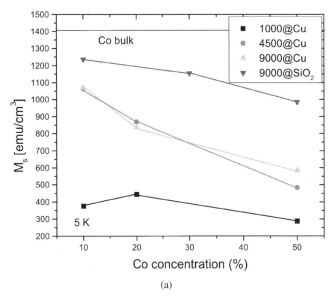

(a)

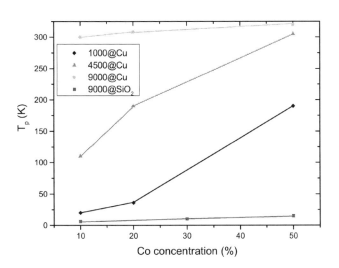

Fig. 42. T_p dependence of the Co concentration with the different cluster sizes and matrices (Cu or SiO$_2$) (after Qiang et al. [133]).

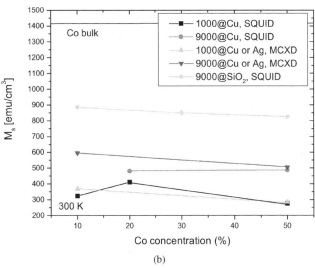

(b)

Fig. 43. M_s dependence of the Co concentration with the different cluster sizes and matrices. (a) 5 K and (b) 300 K (after Qiang et al. [133]).

change, and other more complicated effects [117–121, 127, 128]), and the arrangement of clusters in the assembly is disordered with a volume distribution and easy directions are randomly oriented. In addition, the thermal fluctuations of the cluster magnetic moment also complicate the problem. Many models [119–121], which have been used to discuss the interparticle interactions, are rather simple and cannot completely cover the complex properties determined by the interactions. A more detailed theoretical analysis on this system is needed.

3.3.2.2. Magnetization Reduction

The magnetizations per volume of Co (M_s) obtained from these nanofilms, depending on the concentrations, are significantly below the cobalt bulk value (1440 emu cc^{-1}), as shown in Figure 43a at 5 K and Figure 43b at 300 K. The most notable feature of the data is the dramatic reduction in magnetization particularly for the clusters with average size of 1000 atoms. This could be expected if there are alloying effects at the surfaces of clusters which reduce the magnetization, according to the Slater–Pauling curve [122]. This effect is cluster-size dependent because of melting point reduction [94] due to the ratio of interface to inner atoms. The same results were obtained from our early MXCD measurements on the Co clusters in a Cu matrix [109]. The M_s of Co clusters embedded in Ag particles made by microemulsion technique [129] has shown the reduction of same order as our results. The calculations of small Co clusters coated with Ag or Cu by Guevara, Llois, and Weissmann [130] and Chuanyun et al. [131] have also provided a remarkable qualitative agreement with our experiments.

For 5 K in Figure 43a, it can be seen: (1) M_s is decreasing with an increase of the concentration. As the concentration increases, the distance L_c between the clusters is decreasing, so that the intercluster interactions are becoming stronger, which are reducing the M_s. The system has a spin-glass behavior. (2) M_s of 1000 atom clusters in the Cu matrix is much smaller than that of 4500 and 9000 ones, which both have nearly the same values. This could be explained by the more important ratio of interface atoms (40% for 1000, 24% for 4500, and 20% for 9000). The M_s of large clusters is almost twice that of the M_s of small clusters, while the ratios for large clusters are nearly half of the ratio of the small ones. (3) For the same size of clusters (9000), the M_s for the SiO$_2$ is much larger than the one for the Cu. This could be expected if there are both dipolar and RKKY interactions in the Cu matrix, while only the dipolar interactions exit in the insulating matrix.

For 300 K (Fig. 43b), we observe: (1) M_s is concentration dependent, but decreasing more smoothly than in the case of 5 K. This could suggest that the thermal fluctuations of the cluster magnetic moments at RT are becoming larger. This effect could reach the same order of the effects from dipolar or RKKY interactions, so one cannot see the large difference of M_s between low and high concentrations. (2) M_s of large clusters are bigger than that of small clusters, as in the case of

5 K. (3) For the same size of large clusters (9000), due to the same reason in the case of 5 K the M_s for SiO$_2$ is much larger than that for Cu. (4) M_s for the 9000 or 1000 atom clusters measured by different methods of SQUID and MXCD are in very good agreement.

A very important feature derived from Figure 43a and b is that the M_s of 1000 atom clusters is nearly the same at 5 and 300 K, and decreases less with increasing the concentration. The discrepancy with the bulk value is very large. This could be explained by a large fraction of interface atoms that have an anomalously large ratio of orbital-to-spin moment measured by MXCD. A striking feature is the large values for the ratio of orbital-to-spin magnetization for the 1000 atom samples at 10 and 50%. While the cobalt cluster films gave results similar to those of about 0.1 observed previously [132], values for $\langle m_L \rangle / \langle m_S \rangle$ of 0.84 at 10% and 0.65 at 50% for 1000 atom clusters, and 0.23 at 10% and 0.16 at 50% for 9000 atom clusters were obtained from MXCD measurements on both Co:Cu and Co:Ag systems. Significant antiferromagnetic intercluster interactions would be expected to give enhancements of the orbital-to-spin ratio. This is simply because the exchange coupling directly quenches only the spin contribution to the magnetism.

Detailed calculations of magnetic intercluster interactions are in progress, and these should help to further elucidate the mechanisms underlying this behavior.

3.4. Summary

A cluster-deposition system has been used to make cluster-assembled materials, independently varying the size and the concentration of clusters. The T_p increases with the concentration as expected from theory. T_p is much smaller in the case of the SiO$_2$ matrix, which is consistent with a variety of intercluster spin coupling through RKKY interactions that exist in Cu and Ag matrices, but not in SiO$_2$. M_s of Co clusters is significantly less than that of bulk. For the 1000 atom clusters, the large discrepancy can be explained by a large fraction of interface atoms that have an anomalously large ratio of orbital-to-spin moment measured by MXCD.

4. EXCHANGE-COUPLED NANOCOMPOSITE HARD MAGNETIC FILMS

Since the discovery of experimental evidence of intergrain exchange coupling in a hard–soft magnetic composite [134], which leads to enhanced remanent magnetization, many theoretical studies have been performed on the structure and the performance of this new type of permanent-magnet material. Kneller and Hawig [135] made the first attempt to describe the nanostructure of the composites. Skomski and Coey [136, 137] calculated the dimension of soft-phase grain in the composites for effective exchange interaction. Sabiryanov and Jaswal [138, 139] calculated by first principles the theoretical values of energy products of the FePt/Fe- and the SmCo/Co-layered systems. They found that the energy products for these two systems can reach 90 and 65 MGOe, respectively. Kronmüller et al. [140, 141] have done a systematic study in simulating the exchange-coupled nanocomposite systems.

In the meantime, much experimental work has been devoted to this problem in the past 10 years. The usual way to synthesize the nanocomposites is to make amorphous alloys and then to crystallize them into the desired nanostructured composites by annealing. There are two commonly used processing methods for making the amorphous phases: rapid quenching and mechanical alloying. Melt spinning or ball milling are generally adopted methods.

Many groups have done experiments on melt-spun nanocomposites. Examples include those of Chen, Ni and Hadjipanayis [144] and Ping, Hono, and Hirosawa [142] on the NdFeB system, and Gao et al. [143] on the SmFeCoC system [144]. McCormick et al. [145] and Zhang et al. [146] performed a systematic investigation of mechanical alloying of various nanocomposite rare-earth alloys.

However, energy products of the nanocomposites prepared by rapid quenching and mechanical alloying are not as high as the theoretical predictions. The highest energy products obtained from these works were only around 20 MGOe. This is because these methods have the technical difficulty in controlling the nanostructures, especially the grain size and the texture.

Since 1995, several groups have prepared and have studied the exchange-coupled nanocomposite thin films. Quite encouraging results have been obtained. A review of these works is given in the following.

4.1. CoSm/FeCo Bilayers

Al-Omari and Sellmyer [147] prepared SmCo/FeCo bilayers and multilayers by sputtering. The composition of the bilayer films is $Co_{80}Sm_{20}/Fe_{65}Co_{35}$. Cr underlayers and overlayers were used. The CoSm "hard" phase has coercivity H_c of 2–4 kOe and magnetization M_s of 650 emu cc^{-1} [69, 70]. FeCo was chosen to be the "soft" phase with $M_s = 1934$ emu cc^{-1} and negligible coercivity. All the films studied were deposited on glass substrates by dc (CoSm and Cr) and rf (FeCo) sputtering. X-ray diffraction for the CoSm layer showed no diffraction peaks. This is consistent with other observations by Liu et al. [148, 149] with high-resolution TEM, since the CoSm crystallites are too small (2–5 nm) to show diffraction peaks. The FeCo layer also has been measured by X-ray diffraction and the diffraction patterns for this layer on glass substrate or on Cr underlayer (on glass substrate) showed the body-centered cubic (bcc) structure.

Magnetization measurements of the films show that all the films have in-plane anisotropy. As expected, remanent-magnetization enhancement of the bilayer films was observed upon the deposition of the soft layer onto the hard layer, while the coercivity decreases dramatically with an increase of the thickness of the soft layer.

The maximum energy product $(BH)_{max}$ was calculated for these samples with different soft-layer thickness. The results

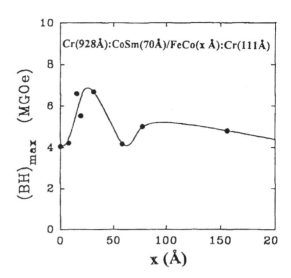

Fig. 44. Dependence of the maximum energy product on the FeCo-layer thickness. The solid line is drawn as a guide to the eye (after Al-Omari and Sellmyer [147]).

are shown in Figure 44. We see that there is an enhancement in $(BH)_{max}$ when $x < 5$ nm; then it decreases as the thickness increases. The enhancement in $(BH)_{max}$ is due to the enhancement in the remanent magnetization. The decrease in $(BH)_{max}$ for $x > 5$ nm is largely due to the decrease in the coercivity. Because the coercivity achieved in the bilayer films is relatively low, the energy products obtained are quite limited.

4.2. Epitaxial CoSm/Fe (or Co) Multilayers

Fullerton et al. [150] did similar work on a similar system. They used carefully chosen substrates and heated substrates during the sputtering so that the controlled epitaxial growth of the films became possible and better magnetic hardening was achieved.

They grew epitaxial Sm-Co films by magnetron sputtering onto Cr-coated MgO substrates [151]. A 20-nm thick Cr buffer layer was deposited onto single-crystal MgO (l00) and (110) substrates at a substrate temperature T_s of 600 °C, resulting in Cr(100) and (211) epitaxial growth, respectively. The Sm-Co films were subsequently deposited with a nominal Sm_2Co_7 composition by cosputtering from separate elemental Sm and Co sources.

Shown in Figure 45 are the X-ray diffraction patterns of 150-nm thick Sm-Co films deposited simultaneously onto Cr(100) and (211) buffer layers. The different crystal orientations lead to distinct magnetic behaviors. The twinning in a-axis Sm-Co films causes the c-axis of different crystallites to lie in two orthogonal in-plane directions [152]. The twinned crystallites are strongly exchange coupled, giving rise to an effective fourfold in-plane anisotropy. On the other hand, the b-axis Sm-Co is uniaxial, showing a square easy-axis loop and a sheared hard-axis loop. The anisotropy fields, estimated from extrapolating the hard-axis loop to saturation, are in the range 20–25T. These values are comparable to those

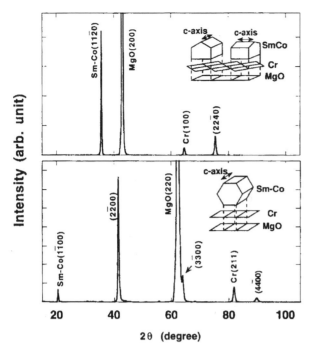

Fig. 45. X-ray diffraction patterns of Sm-Co grown epitaxially via sputtering on Cr-coated single-crystal MgO substrates. The insets illustrate the epitaxial relations (after Fullerton et al. [151]).

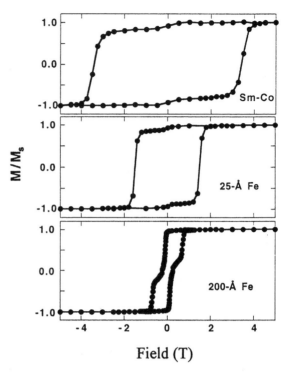

Fig. 46. Room-temperature hysteresis loops of a single Sm-Co film and Sm-Co/Fe bilayers with different Fe-layer thickness (after Fullerton et al. [151, 154]).

reported for bulk $SmCo_5$ (25–44T) [153]. For both orientations, large room-temperature H_c values (>3T) are observed. High-resolution electron microscopy studies show that films of both orientations consist of a mixture of $SmCo_3$, Sm_2Co_7, and $SmCo_5$ polytypoids [152]. The resulting stacking disorder and concomitant variation in local anisotropy constants may give rise to the large H_c values. The difference in the thickness dependence of H_c values for the two crystallographic orientations may arise from additional contributions from the high density of twin boundaries observed in a-axis Sm-Co films. The formation of twin boundaries is controlled by the Sm-Co nucleation and is rather insensitive to film thickness. For the b-axis Sm-Co films, the coercivity varies logarithmically with thickness, increasing to 4.1T for the 7.5-nm films [151]. The anisotropy fields are not strongly thickness dependent, which points toward changes in the microstructure with increasing thickness being responsible for the changes in H_c. The H_c values increase on cooling and values as high as 7T have been measured at 4.2 K.

Coercivity is decreased upon the deposition of the magnetically soft layer onto the hard Sm-Co layer. Figure 46 shows the magnetic hysteresis loops for a single b-axis Sm-Co film and Sm-Co/Fe bilayers with 2.5- and 20-nm Fe. The hysteresis loops are measured with the field H applied along the easy axis direction (MgO[001]). For the Sm-Co single layer, a square loop is observed with a coercive field H_c of 3.4T. The Sm-Co saturation magnetization is in the range 500–600 emu cm^{-3}. For the bilayer with a 2.5-nm Fe layer, the loop shape is similar to that of a single Sm-Co layer. A square easy-axis loop is measured, indicating that the entire Fe layer

is strongly coupled to the underlying Sm-Co film and that the two layers switch as a unit. Compared to the single Sm-Co film, the coercivity of the bilayer is reduced by about 50% to 1.7T. For the bilayer with a 20-nm Fe layer, the loop changes shape quite significantly. The Fe layer nucleates reversal at a field ($H_N = 0.09T$) well below the field required to reverse the Sm-Co layers. The subsequent switching field for the Sm-Co layer (0.6–0.7T) is only 20% of that of a single Sm-Co film. In their experiments, energy products up to 15 MGOe have been obtained.

Fullerton et al. [154] have also carried out numerical simulations based on the equilibrium between the exchange energy, anisotropy energy, and the static magnetic field energy. The simulated results agree well with their experimental results.

4.3. Rapid Thermally Processed Nanocomposite Films

Since 1996, we have studied several nanocomposite systems prepared by plasma sputtering and subsequent heat treatments, including the Fe-Pt, Sm-Co, and Pr-Co alloys [155–162]. At first multilayer films were prepared in a multiple-gun dc- and rf-magnetron sputtering system with base pressure of 2×10^{-7} Torr. The films were deposited onto glass or silicon substrates at room temperature. By choosing suitable multilayer structures of the as-deposited films and consequent heat-treatment processes, the nanostructures of the films have been tailored to achieve desired properties. The thermal processing is found to be the key to control the morphology. Various thermal processes including rapid thermal annealing (RTA) have been investigated. In the RTA process, heating and cooling rates up to 100 K s^{-1} were employed. For the first time, a nearly ideal nanostructure with the soft-phase grains embedded homogeneously in the hard-phase grains has been obtained for the FePt:Fe$_{1-x}$Pt$_x$ ($x \sim 0.3$) system. Effective intergrain exchange coupling has been realized. As a consequence, high-energy products up to 50 MGOe have been achieved. From our experiments, it was found that there are several key issues in preparing the high-energy-product nanocomposites, as we discuss in the following sections.

4.3.1. Magnetic Hardening in the Nanocomposites

Magnetic hardening can be realized upon annealing the nanocomposite systems, because annealing leads to the formation of the hard phase through crystallization from the amorphous phase or from interlayer diffusion in the multilayers. High coercivity can be achieved by carefully choosing the annealing time and the temperature. In our investigations, coercivity up to 20 kOe in the FePt:Fe$_x$Pt$_{1-x}$ ($x \sim 0.7$) and PrCo$_{3.5}$:Co systems was developed. In SmCo$_x$:Co systems, huge coercivity (> 43 kOe) has been obtained. This is quite close to the highest coercivity reported for the single-phase magnets.

As reported by many groups, there is a trade-off between the magnetization and the coercivity in the hard–soft-phase composites. Figure 47 shows H_c and M_s measured as functions of the Co layer thickness in the PrCo$_{3.5}$:Co system.

It is interesting to note that there may be a correlation between the coercivity mechanism and the soft-phase fraction in the composite. Figure 48 shows the initial magnetization curves and the hysteresis loops of the PrCo$_{3.5}$:Co samples with different Co fractions. A commonly accepted interpretation of the initial curves (virgin curves) is that relatively steep curves, as for sample (a), indicate a nucleation- or rotation-type mechanism of coercivity; whereas curves as for sample (d), are typical for wall-pinning controlled reversal. That may be because a softer phase in the composite provides more nucleation sites for the reversed domains. However, in case

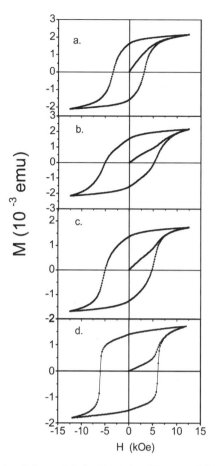

Fig. 48. Hysteresis loops of the heat-treated multilayers with initial structure (PrCo 30 nm/Cox nm)×10. In figure a, $x = 11$ nm, in figure b, $x = 10$ nm, in figure c, $x = 8$ nm, and in figure d, $x = 5$ nm. The magnetization was measured with the field in the film plane (after Liu, Liu, and Sellmyer [157]).

of nanoscale morphology, one may need to consider interaction domains [163, 164], because simple nucleation and pinning models may not be adequate for describing the coercivity mechanism in nanocomposite magnets. As we can see from Figure 49, the domain size (~ 200 nm) is about ten times bigger than the grain size (~ 20 nm). This is a common phenomenon in all the systems we have investigated. A study of the details is needed. Though not yet fully understood, it can be still concluded from Figure 48 that the Co phase fraction in the annealed samples has changed the magnetization-reversal processes.

4.3.2. The Optimized Nanostructure

It is not difficult to understand that there are only three possible two-dimensional morphological configurations for a homogenous two-phase composite, as shown schematically in Figure 50.

To create a texture in the composite which is of vital importance for a permanent magnet, a configuration with small soft-phase grains distributed homogeneously in the hard phase is ideal. Though the reversed picture was first suggested in

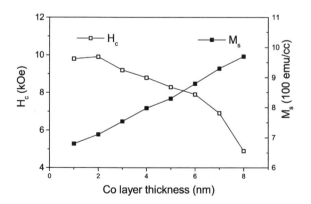

Fig. 47. Dependencies of coercivity and the saturation on Co-layer thickness in the film (PrCo 28 nm/Co x nm)×8 (annealed at 500 °C) (after Liu et al. [156]).

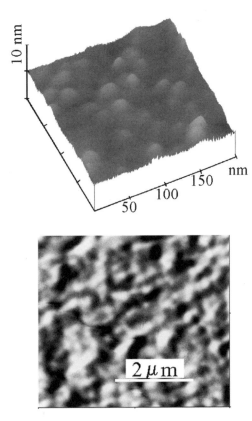

Fig. 49. The AFM (upper) and MFM (lower) images of a (PrCo 30 nm/Co 11 nm)×10 sample with heat treatment at 500 °C (after Liu, Liu, and Sellmyer [157]).

sion of the soft phase in between two neighboring hard grains is under the critical length for effective intergrain exchange coupling. The juxtaposition of each soft-phase grain to hard-phase grains is only guaranteed when the soft-phase grains are embedded in the hard matrix (see Fig. 50). This is also essential for the interphase exchange coupling.

We now address the question regarding why high-energy products have not yet been achieved in mechanically alloyed and melt-spun composites. The reasons may be among the following: (1) The grain size was not small enough (smallest size reported thus far was still bigger than the critical length); (2) The grains of the two phases did not contact each other sufficiently (distribution was not homogenous enough); (3) all the samples were magnetically isotropic (no texture was achieved).

These problems have been partially solved in our experiment on the FePt:Fe$_{1-x}$Pt$_x$ ($x \sim 0.3$) system. The hysteresis loops (Figures 51 and 52) show the nanostructure of a sample with energy product 52.8 MGOe. Very small grain size of the fcc soft phase in the range from 5–8 nm was obtained after the RTA treatment. The energy-dispersive X-ray (EDX) spectroscopy has shown that the composition of this fcc phase was Fe$_{1-x}$Pt$_x$ ($x \sim 0.3$), that is close to the Fe$_3$Pt phase. In the tetragonal FePt phase, the wall thickness is about 4 nm [165], which means that the grain size of the soft phase in the exchange-coupled FePt system should be smaller than 8 nm. The fine soft-phase grains were embedded in the matrix of the

Ref. [135], our experimental results and theoretical considerations have proved that a composite with soft-phase matrix will not work as well as the reversed configuration [151]. The reason is that only in the case that soft-phase grains are distributed in the hard-phase matrix, can significant magnetic anisotropy and high coercivity of the composite be achieved. If we had the reversed situation, i.e., small hard-phase grains are distributed in the matrix of the soft phase, it would be very difficult to align all the small hard-phase grains. An exceptional situation may happen when perfect grain-boundary coherence exists between the hard and soft phases and the crystalline orientation of all hard-phase grains is same. Even in this special case, it would also be very difficult to ensure that the dimen-

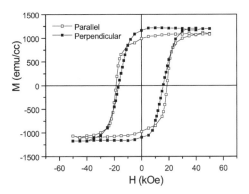

Fig. 51. Hysteresis loops of the annealed (Fe 2.1 nm/Pt 1.5 nm)×16 sample measured in different directions. Plotted on the abscissa is the applied field (after Liu et al. [158]).

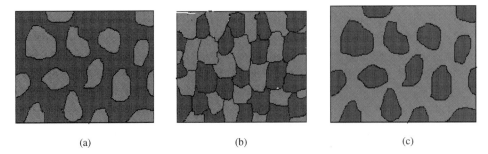

Fig. 50. Three possible two-dimensional configurations of a two-phase composite. (a) The light phase embedded in the dark phase; (b) two phases with comparable grain size; (c) the dark phase embedded in the light phase (after Liu et al. [161]).

fct hard phase with grain size in the range of 50–100 nm. The hard-phase grains were connected to each other all over the sample, which made possible an out-of-plane anisotropy. We noticed that in single-phase FePt films, the as-deposited films had (111) texture (with ⟨111⟩ normal to the film plane). If the annealing temperature was higher than 500 °C, the orientation became (001), i.e., perpendicular anisotropy (easy magnetization direction of the fct phase is ⟨001⟩). In the case of the nanocomposite, the situation was more complicated and determination of the texture was more difficult. However, we observed similar change in anisotropy from in-plane to out-of-plane with increasing annealing temperature. Therefore, it is clear that the out-of-plane anisotropy was not caused by the film growth preference but was developed after sufficient annealing. If the hard phase were not connected all over the sample, the texture hardly could have been achieved, as we discussed earlier.

In the PrCo$_x$:Co and SmCo$_x$:Co systems, nanostructures with the type shown in Figure 50b were obtained. When the soft-phase portion in the composites was small (resulting from the thin Co layers in the multilayers), the grain size of the soft phase was naturally smaller than the hard-phase grains. However, different from the iron–platinum samples, Co grains were found only in the grain boundaries. Because the grain size was well controlled, energy products above 20 MGOe were achieved. These two systems showed in-plane anisotropy, which may be related to the grain orientation formed during film growth. For PrCo$_x$:Co samples, the as-deposited films were amorphous. The crystalline grains were formed during the subsequent heat treatment. Detailed investigation of the anisotropy of the samarium-cobalt and praseodymium-cobalt systems is underway.

4.3.3. Intergrain Exchange Coupling

To investigate the interphase exchange coupling, we have to first check if there are really two phases in the sample, the hard phase and the soft phase. XRD, TEM and EDX have been used to identify the crystal structures of the phases in the nanocomposites studied. In the PrCo$_x$:Co and SmCo$_x$:Co systems, the hcp Co phase was detected in the heat-treated samples. Figure 53 shows the EDX spectra of a PrCo$_x$:Co sample measured in two adjacent grains. One is the matrix grain (a) and another is the soft-phase grain (b). Except for Cr peaks that are from the cover layer and the underlayer, and a trace of Si and C which may be from the substrate and contamination during the specimen preparation, in (a) we detected Co and Pr elements, and in (b) we only detected Co. This clearly shows that there are two phases in the sample, the hard PrCo$_x$ phase and soft Co phase. No interdiffusion was observed and no third phase was found.

Crystal structures of the hard phases in PrCo$_x$:Co and SmCo$_x$:Co systems were found to be different from those in bulk samples [156, 160]. In the SmCo$_{3.5}$:Co system, high-resolution TEM showed the hard phase to be a metastable phase SmCo$_3$ with the hexagonal DO$_{19}$ structure [159]. The reason remains to be explored.

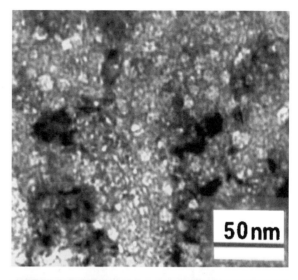

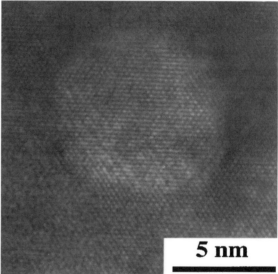

Fig. 52. TEM bright-field image (upper) and high-resolution TEM image (lower) of a heat-treated sample. The "white grains" with diameter less than 10 nm are fcc phase which are embedded in the FePt-matrix grains with lateral grain size from 50 to 100 nm (after Liu et al. [158]).

In the iron–platinum system, unexpectedly, no pure iron phase was found but the fcc Fe$_{1-x}$Pt$_x$ with $x \sim 0.3$, as mentioned before.

The simplest way to see if there is an intergrain exchange coupling in a composite sample with randomly oriented grains is through the hysteresis loops. The ratio of the remanence and the saturation (M_r/M_s) higher than 0.5 is a sign of exchange coupling.

Another way is to measure the "exchange-spring" behavior. We measured the M_R vs H (magnetic field) curves. Here, M_R is different from the remanent magnetization M_r. M_r is only one value at $H = 0$ when the external magnetic field decreases from the first quadrant. M_R can have different values at $H = 0$ when opposite fields release from different points in the second quadrant. Figure 54 shows an example. When the M_R vs H curve is flat, it indicates that M_R does not change, which

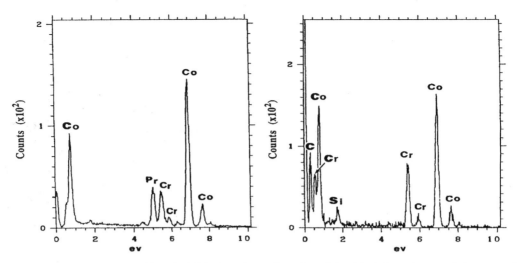

Fig. 53. EDX spectra of a heat-treated PrCo$_x$-Co film. The left is from the matrix grain and the right is from the second phase grain (after Liu, Liu, and Sellmyer [157]).

means that the moment of the soft phase switched back after the external field vanished.

These measurements should of course be combined with the nanostructure observations discussed previously. It has been found in our experiments that when the soft-phase grain size extensively exceeded the critical length, the exchange coupling failed, showing loops with pronounced kinks (shoulders) around $H = 0$, which indicates that the soft phase did not switch together with the hard phase.

A multistep heat treatment has been adopted to improve the situation. Figure 55 shows the loops of the SmCo$_x$:Co sample treated with different procedures: (a) RTA at 500 °C for 20 s (b) anneal at 500 °C for 20 min and (c) a combination of (a) plus (b). It can be seen clearly from Figure 55 that the two-step treatment led to the most effective intergrain exchange coupling (with smallest kinks on the loop). Loop (c) is only a minor loop because our measurement field could only be applied up to 55 kOe. If we could measure this sample to a higher field, the difference between loops (c) and (b) may be more obvious. It turned out in our experiments that no single-step treatment could lead to as good a

result as the two-step one. This is because if the as-deposited sample were treated with standard annealing directly, it could cause excessive grain growth. This would destroy the intergrain exchange coupling. However, sufficient annealing was necessary for high coercivity. To solve this dilemma, a special rapid thermal-annealing process was first applied to the samples. In this process, annealing at above 500 °C was done in 5 s; the heating rate was about 200 °C per second.

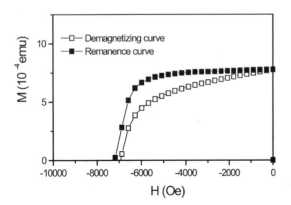

Fig. 54. The demagnetizing curve and the remanence curve of an annealed sample (PrCo 30 nm/Co 8 nm)×10 (after Liu et al. [157]).

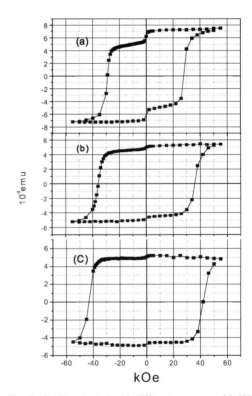

Fig. 55. The SmCo films treated with different processes. (a) 500 °C RTA; (b) 500 °C 20 min anneal; and (c) treatment (a) plus (b) (after Liu et al. [161]).

It is worthwhile to mention the role of grain boundaries in the exchange coupling. It is expected that "cleaner" grain boundaries help the exchange interaction. Some details can be found in [161].

The effective exchange coupling in a nanocomposite at room temperature can be ruined at low temperature. It is interesting to see the magnetization process of the nanocomposite at low temperature, by which we can have a better understanding of the exchange coupling. Figure 56 shows hysteresis loops at different temperatures for a $PrCo_{3.5}$:Co nanocomposite sample with 17% Co in volume. This sample has been given a two-step heat treatment. For the loops measured below room temperature, only the upper half of the loops (demagnetizing curves) are shown. It is clearly seen from the figure that the squareness of the loops becomes increasingly worse with decreasing temperature. The squareness is defined by the ratio of the area closed by the demagnetizing curve and the coordinates in the second quadrant over the area of the rectangle $M_r \times H_c$. In Figure 56, the squareness decreases from 0.89 at 300 K to 0.51 at 5 K. Small squareness is a direct sign of ineffective intergrain exchange coupling in the nanocomposites (excluding other factors which may also have influence on the squareness). A similar phenomenon has been observed by other groups and was interpreted as a decoupling effect caused by enhanced anisotropy at lower temperatures [166, 167]. As we have discussed earlier, to achieve effective intergrain exchange interaction, the dimension of the soft phase in exchange-coupled composites should not be larger than about twice the domain-wall thickness in the hard phase (this length can be regarded as the relevant exchange length). Usually at lower temperatures, the domain-wall thickness becomes smaller because of the enhanced anisotropy in the hard phases [137] which makes the condition for the effective exchange coupling not satisfied anymore, since the grain size does not become smaller with decreasing temperature.

This interpretation can be regarded as a qualitative explanation. A more sophisticated description of the magnetic switching process can be established if we analyze the demagnetization curves more carefully. From the demagnetizing curves at temperatures 50 and 5 K in Figure 56 we can find transition points in the second or third quadrant (as noted by the arrows). Hysteresis loops with these transitions have low squareness. These transition points can be attributed to the switching fields of the hard phase in the composites. Demagnetizing curves with such transition points reflect the fact that switching of magnetic moments of the hard and soft phases has been separated (not cooperative or simultaneously) at low temperatures. The soft phase has been switched at low fields (near 0 Oe). Higher fields (more negative fields) are necessary to reverse the hard phase because of the enhanced anisotropy. This is why the lower the temperature, the higher the transition fields of the hard phase, as we see from the arrows in Figure 56. Similar results were found in other nanocomposites. A detailed study including a theoretical simulation can be found in Ref. [162].

Although the foregoing issues are related to the thermal-processed film materials, they are also general problems for other exchange-coupled nanocomposites. Though some progress has been made, as discussed, the problems on how to obtain optimized nanostructure and therefore high-energy products close to the theoretical predictions remain major challenges for future work.

5. CONCLUDING REMARKS

Magnetism in nanophase composite materials is a phenomenon that depends on several length scales, ranging from less than 1 Å to a length scale of the order of the size of the magnet. We have seen that intrinsic properties such as magnetic anisotropy are highly sensitive to local atomic arrangements so that layered or superlattice structures can be exploited to produce exceedingly large anisotropy constants. These, in turn, are the parameters most important in controlling extrinsic properties such as coercivity that are all important for semihard magnetic recording media and hard permanent-magnet materials.

The concept of enlightened nanostructuring to achieve exchange decoupling of single-domain grains of characteristic length about 10 nm has been emphasized. Because one can control both the properties of the embedded particles and their interactions, there is considerable potential to fabricate magnetic recording media with perpendicular anisotropy, grain sizes of 5–10 nm, and coercivities in the 10- to 15-kOe range. Such films may be able to serve as recording media in the 1

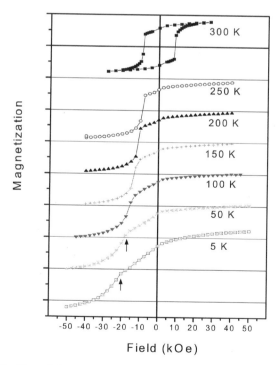

Fig. 56. Magnetic hysteresis loops of a $PrCo_x$-Co nanocomposite at different temperatures (below room temperature only the demagnetizing portions are shown). The arrows indicate the switching points (after Liu et al. [162]).

terabit per square inch regime. Progress has been made here by novel use of multilayering plus carefully controlled thermal processing.

A second major theme of this chapter has been the synthesis and the properties of hard–soft nanocomposites with extremely high-energy products. It was emphasized that control of the nanostructure at a length scale of about 10 nm, and in addition control of the topology of the two-phase material is of utmost importance. The achievement of near record energy products ($\sim$ 53 MGOe) is intriguing and suggests that much further work on synthetic aspects is warranted.

A brief review of the new concept of cluster assembly of magnetic nanocomposites was presented. A major benefit of this method is that in principle one can create two-phase materials of arbitrary constituents, and one can control the size and the properties of each independently. In particular, it is not necessary to employ immiscible materials to fabricate two-phase nanocomposites. Several examples of magnetic cluster glasses were used to show how the character and the strength of the intercluster interactions can be controlled. The exploitation of cluster-assembly techniques to synthesize new magnetic materials is in its earliest stages.

In summary, it is hoped that this chapter has presented a status report on the challenges and the potential of magnetic nanocomposites for future science and applications.

Acknowledgments

The authors are most grateful for the assistance and the many helpful discussions of their colleagues and collaborators including Dr. M. Yu and Dr. M. Yan; Professors H. Haberland, R. Kirby, R. Skomski, R. Sabiryanov, S. Jaswal, S. H. Liou, and Y. Liu. They also are indebted for financial support to the U. S. Department of Energy, National Science Foundation, Defense Advanced Research Projects Agency, U. S. Army Research Office, and the Center for Materials Research and Analysis.

REFERENCES

1. B. Abeles, P. Sheng, M. D. Coutts, and Y. Arie, *Adv. Phys.* 24, 407 (1975).
2. C. L. Chien, *J. Appl. Phys.* 69, 5267 (1991).
3. J. I. Gittleman, B. Abeles, and S. Bozowski, *Phys. Rev. B* 9, 3891 (1974).
4. J. Q. Xiao, J. S. Jiang, and C. L. Chien, *Phys. Rev. Lett.* 68, 3749 (1992).
5. A. E. Berkowitz, J. R. Mitchell, M. J. Carey, A. P. Young, S. Zhang, F. E. Spada, F. T. Parker, A. Hutten, and G. Thomas, *Phys. Rev. Lett.* 68, 3745 (1992).
6. J. Inoue and S. Maekawa, *Phys. Rev. B* 53, R11, 927 (1996).
7. S. H. Liou and C. L. Chien, *Appl. Phys. Lett.* 52, 512 (1988).
8. M. Yu, Y. Liu, A. Moser, D. Weller, and D. J. Sellmyer, *Appl. Phys. Lett.* 75, 3992 (1999); M. Yu, Y. Liu, and D. J. Sellmyer, *J. Appl. Phys.* 87, 6959 (2000).
9. C. P. Luo and D. J. Sellmyer, *Appl. Phys. Lett.* 75, 3162 (1999).
10. S. Roy, D. Das, and D. Chakravorty, *J. Appl. Phys.* 74, 4746 (1993).
11. K. Fukumi, A. Chayahara, K. Kadonó, T. Sakaguchi, Y. Horino, M. Miya, J. Hayakawa, and M. Satou, *Jpn. J. Appl. Phys. B* 30, L742 (1991).
12. E. M. Logothetis, W. J. Kaiser, H. K. Plummer, and S. S. Shinozaki, *J. Appl. Phys.* 60, 2548 (1986).
13. J. J. Becker, *J. Appl. Phys.* 29, 317 (1958).
14. J. R. Childress, and C. L. Chien, *Appl. Phys. Lett.* 56, 95 (1990).
15. Y. X. Zhang, S. H. Liou, R. J. DeAngelis, K. W. Lee, C. P. Reed, and A. Nazareth, *J. Appl. Phys.* 69, 5273 (1991).
16. S. H. Liou, S. Malhotra, Z. S. Shan, D. J. Sellmyer, S. Nafis, J. A. Woollam, C. P. Reed, R. J. DeAngelis, and G. M. Chow, *J. Appl. Phys.* 70, 5882 (1991).
17. L. Maya, T. Thundat, J. R. Thompson, and R. J. Stevenson, *Appl. Phys. Lett.* 67, 3034 (1995).
18. L. Maya, M. Paranthaman, J. R. Thompson, T. Thundat, and R. J. Stevenson, *J. Appl. Phys.* 79, 7905 (1996).
19. J. Frenkel and J. Dorfman, *Nature* 126, 274 (1930).
20. C. Kittel, *Rev. Mod. Phys.* 21, 541 (1949).
21. L. Neel, *Ann. Geophys.* 5, 99 (1949).
22. C. P. Bean and J. D. Livingston, *J. Appl. Phys.* 30, 120S (1959).
23. M. Yu, M. F. Doerner, and D. J. Sellmyer, *IEEE Trans. Magn.* MAG-34, 1534 (1998).
24. D. Weller and A. Moser, *IEEE Trans. Magn.* 35, 4423 (1999).
25. K. O'Grady and H. Laidler, *J. Magn. Magn. Mater.* 200, 616 (1999).
26. D. Weller, A. Moser, L. Folks, M. E. Best, W. Lee, M. F. Toney, M. Schwickert, J.-U. Thiele, and M. F. Doerner, *IEEE Trans. Magn.* 36, 10 (2000).
27. R. Wood, *IEEE Trans. Magn.* 36, 36 (2000).
28. H. Kronmuller, "Science and Technology of Nanostructured Materials," p. 657. Plenum, New York, 1991.
29. E. C. Stoner and E. P. Wohlfarth, *Philos. Trans. R. Soc. London A* 240, 599 (1948).
30. F. Liorzou, B. Phelps, and D. L. Atherton, *IEEE Trans. Magn.* 36, 418 (2000).
31. A. Aharoni, *Phys. Status Solidi* 16, 3 (1966).
32. A. Aharoni, *IEEE Trans. Magn.* 22, 478 (1986).
33. G. F. Hughes, *J. Appl. Phys.* 54, 5306 (1983).
34. O. Henkel, *Phys. Status Solidi* 7, 919 (1964).
35. P. E. Kelly, K. O'Grady, P. I. Mayo, and R. W. Chantrell, *IEEE Trans. Magn.* 25, 3881 (1989).
36. P. I. Mayo, A. Bradbury, R. W. Chantrell, P. E. Kelly, H. E. Jones, and P. R. Bissell, *IEEE Trans. Magn.* 26, 228 (1990).
37. X.-D. Che, and H. N. Bertram, *J. Magn. Magn. Mater.* 116, 121 (1992).
38. P. Huang, J. W. Harrell, M. R. Parker, and K. E. Johnson, *IEEE Trans. Magn.* 30, 4002 (1994).
39. J. G. Zhu and N. H. Bertram, in "Solid State Physics." (H. Ehrenreich and D. Turnbull, Eds.), Vol. 46. Academic Press, San Diego, 1992.
40. R. Street and J. C. Woolley, *Proc. R. Soc. A* 62, 562 (1949).
41. L. Néel, *Adv. Phys.* 4, 191 (1955).
42. R. H. Victora, *Phys. Rev. Lett.* 63, 457 (1989).
43. E. F. Kneller and F. E. Luborsky, *J. Appl. Phys.* 34, 656 (1963).
44. M. P. Sharrock and J. T. McKinney, *IEEE Trans. Magn.* 17, 3020 (1981).
45. H. N. Bertram and H. J. Richter, *J. Appl. Phys.* 85, 4991 (1999).
46. S. M. Stinnett, W. D. Doyle, P. J. Flanders, and C. Dawson, *IEEE Trans. Magn.* 34, 1828 (1998).
47. A. Moser, D. Weller, and M. E. Best, *J. Appl. Phys.* 85, 5018 (1999).
48. M. el-Hilo, S. H. Uren, K. O'Grady, J. Popplewell, and R. W. Chantrell, *IEEE Trans. Magn.* 26, 244 (1990).
49. P. Gaunt, *J. Appl. Phys.* 59, 4129 (1986).
50. R. Street, J. C. Woolley, and P. B. Smith, *Proc. Phys. Soc. B* 65, 679 (1952).
51. E. P. Wohlfarth, *J. Phys. F* 14, L155 (1984).
52. R. W. Chantrell, G. N. Coverdale, and K. O'Grady, *J. Phys. D* 21, 1469 (1988).
53. M. El-Hilo, A. M. de Witte, K. O'Grady, and R. W. Chantrell, *J. Magn. Magn. Mater.* 117, L307 (1992).
54. P. Bruno, G. Bayreuther, P. Beauvillain, C. Chappert, G. Lugert, D. Renard, J. P. Renard, and J. Seiden, *J. Appl. Phys.* 68, 5759 (1990).
55. C. P. Luo, Z. S. Shan, and D. J. Sellmyer, *J. Appl. Phys.* 79, 4899 (1996).
56. R. D. Kirby, M. Yu, and D. J. Sellmyer, *J. Appl. Phys.* 87, 5696 (2000).

57. J. C. Barbier, *Ann Phys. (Paris)* 9, 84 (1954).

58. G. Lauhoff and Takao Suzuki, *J. Appl. Phys.* 87, 5702 (2000).

59. G. Bottoni, D. Candolfo, and A. Cecchetti, *J. Appl. Phys.* 81, 3809 (1997).

60. G. Xiao and C. L. Chien, *Appl. Phys. Lett.* 51, 1280 (1987).

61. R. L. Holtz, P. Lubitz, and A. S. Edelstein, *Appl. Phys. Lett.* 56, 943 (1990).

62. S. S. Malhotra, Y. Liu, J. X. Sheng, S. H. Liou, and D. J. Sellmyer, *J. Appl. Phys.* 76, 6304 (1994).

63. A. Butera, T. J. Klemmer, and J. A. Barnard, *J. Appl. Phys.* 83, 4855 (1998).

64. A. Tsoukatos and G. C. Hadjipanayis, *J. Appl. Phys.* 70, 5891 (1991).

65. T. Hayashi, S. Hirono, M. Tomita, and S. Umemura, *Nature* 38, 772 (1996).

66. J. N. Zhou, A. Butera, H. Jiang, and J. A. Barnard, *J. Appl. Phys.* 84, 5693 (1998).

67. M. Yu, Y. Liu, and D. J. Sellmyer, *J. Appl. Phys.* 85, 4319 (1999).

68. T. J. Konno and R. Sinclair, *Acta Metall. Mater.* 42, 1231 (1994).

69. E. M. T. Velu and D. N. Lambeth, *J. Appl. Phys.* 69, 5175 (1991).

70. E. M. T. Velu and D. N. Lambeth, *IEEE Trans. Magn.* 31, 3249 (1992).

71. S. S. Malhotra, Y. Liu, Z. S. Shan, S. H. Liou, D. C. Stafford, and D. J. Sellmyer, *J. Magn. Magn. Mater.* 161, 316 (1996).

72. S. H. Liou, Y. Liu, S. S. Malhotra, M. Yu, and D. J. Sellmyer, *J. Appl. Phys.* 79, 5060 (1996).

73. Y. Ide, T. Goto, K. Kikuchi, K. Watanable, J. Onagawa, H. Yoshida, and J. M. Cadogan, *J. Magn. Magn. Mater.* 245, 177 (1998).

74. J. P. Liu, C. P. Luo, Y. Liu, and D. J. Sellmyer, *Appl. Phys. Lett.* 72, 483 (1998).

75. K. R. Coffey, M. A. Parker, and J. K. Howard, *IEEE Trans. Magn.* 31, 2737 (1995).

76. N. Li and B. M. Lairson, *IEEE Trans. Magn.* 35, 1077 (1999).

77. C. P. Luo, S. H. Liou, L. Gao, Y. Liu, and D. J. Sellmyer, *Appl. Phys. Lett.* 77, 2225 (2000).

78. S. Stavroyiannis, I. Panagiotopoulos, N. Niarchos, J. A. Christodoulides, Y. Zhang, and G. C. Hadjipanayis, *Appl. Phys. Lett.* 73, 3453 (1998).

79. B. E. Warren, in "X-Ray Diffraction," p. 208 Addison-Wesley, Reading, MA, 1969.

80. P. S. Rudman and B. L. Averbach, *Acta Metall.* 5, 65 (1957).

81. A. Cebollada, D. Weller, J. Sticht, G. R. Harp, R. F. C. Farrow, R. F. Marks, R. Savoy, and J. C. Scott, *Phys. Rev. B* 50, 3419 (1994).

82. M. R. Visokay and Sinclair, *Appl. Phys. Lett.* 66, 1692 (1995).

83. M. Watanabe and M. Homma, *Jpn. J. Appl. Phys.* 35, L1264 (1996).

84. R. F. C. Farrow, D. Weller, R. F. Marks, M. F. Toney, A. Cebollada, and G. R. Harp, *J. Appl. Phys.* 79, 5967 (1996).

85. R. F. C. Farrow, D. Weller, R. F. Marks, M. F. Toney, D. J. Smith, and M. R. McCartney, *J. Appl. Phys.* 84, 934 (1998).

86. J.-U. Thiele, L. Folks, M. F. Toney, and D. K. Weller, *J. Appl. Phys.* 84, 5686 (1998).

87. S. Sun, C. B. Murray, D. Weller, L. Folks, and A. Moser, *Science* 287, 1989 (2000).

88. H. Haberland, Ed. "Clusters of Atoms and Molecules," Springer-Verlag, Berlin/New York, 1994.

89. J. Shi, S. Gider, K. Babcock, and D. D. Awschalom, *Science* 271, 993 (1996).

90. S. Edelstein and R. C. Cammarata, "Nanomaterials: Synthesis, Properties and Applications," IOP, Philadelphia, 1996.

91. M. N. Baibich, J. M. Broto, A. Fert, F. N. Van Dau, F. Petroff, P. Etienne, G. Creuzet, A. Friederich, and J. Chazelas, *Phys. Rev. Lett.* 61, 2472 (1988).

92. E. Berkowitz, J. R. Mitchell, M. J. Carey, A. P. Young, S. Zhang, F. E. Spada, F. T. Parker, A. Hutten, and G. Thomas, *Phys. Rev. Lett.* 68, 3745 (1992).

93. G. Xiao and C. L. Chien, *Appl. Phys. Lett.* 51, 1280 (1987); G. Xiao and C. L. Chien, *J. Appl. Phys.* 63, 4252 (1988).

94. W. Ekardt, Ed., "Metal Clusters," Wiley, New York, 1999.

95. W. A. de Heer, *Rev. Mod. Phys.* 65, 611 (1993).

96. K.-H. Meiwes-Broer (Ed.), "Metal Cluster at Surface," Springer-Verlag, Berlin/New York, 2000.

97. I. M. L. Billas, J. A. Becker, A. Chatelain, and W. A. de Heer, *Phys. Rew. Lett.* 71, 4067 (1993).

98. I. M. L. Billas, A. Chatelain, and W. A. der Heer, *Science* 265, 1682 (1994).

99. H. Gleiter, *Nanostruct. Mater.* 6, 3 (1995).

100. R. W. Siegel, *Nanostruct. Mater.* 6, 205 (1995).

101. E. W. Becker, K. Bier, and W. Henkens, *Z. Phys.* 146, 333 (1956).

102. O. F. Hagena, *Rev. Sci. Instrum.* 64, 2374 (1992).

103. P. Melinon, V. Paillard, V. Dupuis, and A. Perez, *Int. J. Mod. Phys. B* 9, 339 (1995).

104. A. Perez, P. Melinon, V. Dupuist, P. Jensen, B. Prevel, J. Tuaillon, L. Bardotti, C. Martet, M. Treilleux, M. Broyer, M. Pellarin, J. L. Vaille, B. Palpant, and J. Lerme, *J. Phys. D: Appl. Phys.* 30, 709 (1997).

105. H. Haberland, M. Mall, M. Moseler, Y. Qiang, Th. Reiners, and Y. Thurner, *J. Vac. Sci. Technol. A* 12, 2925 (1994).

106. H. Haberland, M. Moseler, Y. Qiang, O. Rattunde, Th. Reiners, and Y. Thurner, *Surf. Rev. Lett.* 3, 887 (1996).

107. H. Haberland, M. Moseler, Y. Qiang, O. Rattunde, Y. Thurner, and Th. Reiners, *Mater. Res. Symp. Proc.* 338, 207 (1996).

108. Y. Qiang, Y. Thurner, Th. Reiners, O. Rattunde, and H. Haberland, *Surf. Coat. Technol.* 100–101, 27 (1998).

109. D. A. Eastham, Y. Qiang, T. H. Maddock, J. Kraft, J.-P. Schille, G. S. Thompson, and H. Haberland, *J. Phys.: Condens. Matter* 9, L497 (1997).

110. Y. Qiang, Ph.D. dissertation, University of Freiburg, Germany, 1997.

111. H. Haberland, Z. Insepov, and M. Moseler, *Z. Phys. D* 26, 229 (1993); *Phys. Rev. B* 51, 11,061 (1995).

112. M. Moseler, O. Rattunde, J. Nordiek, and H. Haberland, *Comput. Mater. Sci.* 10, 452 (1998).

113. N. N. Das Gupta and S.K. Gosh, *Rev. Mod. Phys.* 18, 225 (1946).

114. M. Vollmer and A. Weber, *Z. Phys. Chem.* 119, 277 (1926).

115. R. Becker and W. Doering, *Ann. Phys,* 24, 719 (1935).

116. B. Chapman, "Glow Discharge Processes," J. Wiley, New York, 1982.

117. J. L. Dormann, F. D'Orazio, F. Lucari, E. Tronc, P. Prené, J. P. Jolivet, D. Fiorani, R. Cherkaoui, and M. Noguès, *Phys. Rev. B* 53, 14,291 (1996).

118. J. F. Loeffler, J. P. Meier, B. Doudin, J.-P. Ansermet, and W. Wagner, *Phys. Rev. B* 57, 2915 (1998); L. Deák, G. Bayreuther, L. Bottyán, E. Gerdau, J. Korecki, E. I. Kornilov, H. J. Lauter, O. Leupold, D. L. Nagy, A. V. Petrenko, V. V. Pasyuk-Lauter, H. Reuther, E. Richter, R. Röhloberger, and E. Szilágyi, *J. Appl. Phys.* 85, 1 (1999).

119. J. L. Dormann, *Mater. Sci. Eng. A* 168, 217 (1993).

120. J. L. Dormann and D. Fiorani, *J. Magn. Magn. Mater.* 140–144, 415 (1995).

121. J. L. Dormann, D. Fiorani, and E. Tronc, *Adv. Chem. Phys.* 48, 283 (1997).

122. S. Chikazumi, "Physics of Ferromagnetism," 2nd ed., p. 173. Oxford University Press, Oxford, U.K., 1997.

123. W. H. Mieklejohn and C. P. Bean, *Phys. Rev.* 105, 904 (1957).

124. C. P. Bean and J. D. Livingston, *J. Appl. Phys.* 30, 120S (1959).

125. C. T. Chen, Y. U. Idzerda, H.-J. Lin, N. V. Smith, G. Meigs, E. Chaban, G. H. Ho, E. Pellegrin, and F. Sette, *Phys. Rev. Lett.* 75, 152 (1995).

126. W. L. O'Brien and B. P. Tonner, *Phys. Rev. B* 50, 12,672 (1994).

127. M. El-Hilo, K. O'Grady, and R. W. Chantrell, *J. Magn. Magn. Mater.* 114, 295 (1992).

128. E. Cattaruzza, F. Gonella, G. Mattei, P. Mazzoldi, D. Gatteschi, C. Sangregorio, M. Falconieri, G. Salvetti, and G. Battaglin, *Appl. Phys. Lett.* 73, 1176 (1998).

129. R. D. Sánchez, M. A. López-Quintela, J. Rivas, A. González-Penedo, A. J. Garcia-Bastida, C. A. Ramos, R. D. Zysler, and S. Ribeiro Guevara, *J. Phys: Condens. Matter* 11, 5643 (1999).

130. J. Guevara, A. M. Llois, and M. Weissmann, *Phys. Rev. Lett.* 81, 5306 (1998).

131. C. Y. Xiao, J. L. Yang, K. M. Deng, and K. L. Wang, *Phys. Rev. B* 55, 3677 (1997).

132. M. Tischer, O. Hjortstam, D. Arvanitis, J. Hunter Dunn, F. May, K. Baberschke, J. Trygg, J. M. Wills, B. Johansson, and O. Eriksson, *Phys. Rev. Lett.* 75, 1602 (1995).

133. Y. Qiang, R. Morel, D. Eastham, J. M. Meldrim, J. Kraft, A. Fert, H. Haberland, and D. J. Sellmyer, in "Cluster and Nanostructure Interfaces," (P. Jena, S. U. Khanna, and B. K. Rao, Eds.), p. 217, World Scientific, Singapore, 2000.

134. R. Coehoorn, D. B. de Mooij, and C. de Waard, *J. Magn. Magn. Mater.* 80, 101 (1989).

135. E. F. Kneller and R. Hawig, *IEEE Trans. Magn.* 27, 3588 (1991).

136. R. Skomski and J. M. D. Coey, *Phys. Rev. B* 48, 15,812 (1993).

137. R. Skomski and J. M. D. Coey, "Permanent magnetism," IOP, Bristol, U.K., 1999.

138. R. F. Sabiryanov and S. S. Jaswal, *Phys. Rev. B* 58 12,071 (1998).

139. R. F. Sabiryanov and S. S. Jaswal, *J. Magn. Magn. Mater.* 177–181, 989 (1998).

140. H. Kronmüller, *Phys. Status Solidi B* 144, 385 (1987).

141. H. Kronmüller, R. Fisher, M. Bachmann, and T. Leinewber, *J. Magn. Magn. Mater.* 203, 12 (1999).

142. D. H. Ping, K. Hono, and S. Hirosawa, *J. Appl. Phys.* 83, 7769 (1998).

143. Y. Gao, J. Zhu, Y. Weng, E. B. Park, and C. J. Yang, *J. Magn. Magn. Mater.* 191, 146 (1999).

144. Z. Chen, C. Ni, and G. C. Hadjipanayis, *J. Magn. Magn. Mater.* 186, 41 (1998).

145. P. G. McCormick, W. F. Miao, P. A. I. Smith, J. Ding, and R. Street, *J. Appl. Phys.* 83, 6256 (1998).

146. Z.-D. Zhang, W. Liu, X. K. Sun, X.-G. Zhao, Q.-F. Xiao, Y.-C. Sui, and T. Zhao, *J. Magn. Magn. Mater.* 184, 101 (1998).

147. I. A. Al-Omari and D. J. Sellmyer, *Phys. Rev. B* 52, 3441 (1995).

148. Y. Liu, B. Robertson, Z. S. Shan, S. S. Malhotra, M. J. Yu, S. K. Renukunta, S. H. Liou, and D. J. Sellmyer, *IEEE Trans. Magn.* 30, 4035 (1994).

149. Y. Liu, B. Robertson, Z. S. Shan, S. H. Liou, and D. J. Sellmeyer, *Appl. Phys. Lett.* 77, 3831 (1995).

150. E. E. Fullerton, J. S. Jiang, C. H. Sowers, J. E. Pearson, and S. D. Bader, *Appl. Phys. Lett.* 72, 380 (1998).

151. E. E. Fullerton, J. S. Jiang, C. Rehm, C. H. Sowers, S. D. Bader, J. B. Patel, and X. Z. Wu, *Appl. Phys. Lett.* 71, 1579 (1997). J. S. Jiang, E. E. Fullerton, C. H. Sowers, A. Inomata, S. D. Bader, A. J. Shapiro, R. D. Shull, V. S. Gornakov, and V. I. Nikitenko *IEEE Trans. Magn.* 35, 3229 (1999).

152. M. Benaissa, K. Krishnan, E. E. Fullerton, and J. S. Jiang, *IEEE Trans. Magn.* 34, 1204 (1998).

153. K. J. Strnat and R. M. W. Strnat, *J. Magn. Magn. Mater.* 100, 38 (1991).

154. E. E. Fullerton, J. S. Jiang, M. Grimsditch, C. H. Sowers, and S. D. Bader, *Phys. Rev. B* 58, 12,193 (1998).

155. J. P. Liu, Y. Liu, C. P. Luo, Z. S. Shan, and D. J. Sellmyer, *J. Appl. Phys.* 81, 5644 (1997).

156. J. P. Liu, Y. Liu, Z. S. Shan, and D. J. Sellmyer, *IEEE Trans. Magn.* 33, 3709 (1997).

157. J. P. Liu, Y. Liu, and D. J. Sellmyer, *J. Appl. Phys.* 83, 6608 (1998).

158. J. P. Liu, C. P. Luo, Y. Liu, and D. J. Sellmyer, *Appl. Phys. Lett.* 72, 483 (1998).

159. Y. Liu, R. A. Thomas, S. S. Malhotra, Z. S. Shan, S. H. Liou, and D. J. Sellmyer, *J. Appl. Phys.* 85, 4812 (1998).

160. J. P. Liu, Y. Liu, R. Skomski, and D. J. Sellmyer, *J. Appl. Phys.* 85, 4812 (1999).

161. J. P. Liu, Y. Liu, R. Skomski, and D. J. Sellmyer, *IEEE Trans. Magn.* 35, 3241 (1999).

162. J. P. Liu, Y. Liu, R. Skomski, and D. J. Sellmyer, *J. Appl. Phys.* 87, 6740 (2000).

163. W. Rave, D. Eckert, R. Schäfer, B. Gebel, and K.-H. Müller, *IEEE Trans. Magn.* 32, 4362 (1996).

164. R. K. Mishra and R. W. Lee, *Appl. Phys. Lett.* 48, 733 (1986).

165. T. Klemmer, D. Hoydick, H. Okumura, B. Zhang, and W. A. Soffa, *Sc. Metall. Mater.* 33, 1793 (1995).

166. D. Goll, M. Seeger, and H. Kronmüller, *J. Magn. Magn. Mater.* 185, 49 (1998).

167. G. C. Hadjipanayis, *J. Magn. Magn. Mater.* 200, 373 (1999).

Chapter 8

THIN MAGNETIC FILMS

Hans Hauser, Rupert Chabicovsky, Karl Riedling
*Institute of Industrial Electronics and Material Science, Vienna University of Technology,
A-1040 Vienna, Austria*

Contents

1. MAGNETISM OVERVIEW

Beginning with a phenomenology of magnetism, the physical quantities and their units are described. After a short overview of matter and magnetism, emphasis is placed on magnetically ordered media, including anisotropy, nonlinearity, and hysteresis. Furthermore, both the fundamental properties and the performance characterize selected materials that are important for thin films and applications.

Magnetism is a comparatively new subject of science research. The main general reviews have been published since about the middle of the twentieth century [1–13]. The first part of this chapter is based mainly on these references, which are well suited for further reading.

1.1. Magnetic Quantities and Units

The following section describes a phenomenology of magnetism, tracing the historical development as an introduction.

Handbook of Thin Film Materials, edited by H.S. Nalwa
Volume 5: Nanomaterials and Magnetic Thin Films

ISBN 0-12-512913-0/$35.00

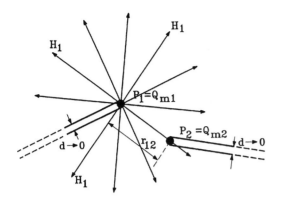

Fig. 1. Long, thin magnets approximate magnetic charges of dipoles; only the field of the pole Q_{m1} is drawn.

The mathematical dualism between "vortex ring and double layer" will lead to two different approaches to the explanation of the magnetic properties of matter. The units of the physical quantities will develop consequently from empirically found electromagnetic laws.

1.1.1. Magnetic Force, Charges, and Electrical Current

At the beginning we will consider magnetic forces based on dipoles. As will be shown later, the magnetic behavior of matter can be explained by moving electrical charges on the atomic level, and we will use the latter explanation for our further considerations.

The basic physical experience when one is dealing with magnetic fields is a fundamental phenomenon: the force $\vec{F}$ between magnetic poles. The Coloumb interaction for two point-like poles with a pole strength $Q_{m,1,2}$ is

$$\vec{F} = \frac{Q_{m1} Q_{m2}}{4 \pi \mu_0 r^2} \vec{e}_r \qquad (1.1)$$

Because single magnetic poles or magnetic charges have not been discovered yet (but they could have been generated at the beginning of the universe), one could imagine an approximation by very long and thin bar magnets in a distance r (unity vector $\vec{e}_r$; see Fig. 1). The proportionality constant, in our case $1/4\pi\mu_0$, depends on the actual system of units and will be explained later.

The first attempt to explain the magnetic force[1] was made by introducing a field $\vec{H}$ of magnetic lines of force, which is thought to originate at a magnetic pole, as shown in Figure 1. Consequently, the force acting on a magnetic pole in a field $\vec{H}$ is

$$\vec{F} = Q_m \vec{H} \qquad (1.2)$$

The magnetic field constant μ_0 has been introduced in the Système International d'Unités (SI) to avoid a coefficient in Eq. (1.2).

Two magnetic poles (approximately a bar magnet with the length l) form a dipole with the Coulomb magnetic moment

$$\vec{j} = Q_m \vec{l} \qquad (1.3)$$

experiencing a torque

$$\vec{T} = \vec{j} \times \vec{H} \qquad (1.4)$$

in a homogeneous field and an additional force in an inhomogeneous field (e.g., $F_x = Q_m l \cdot \partial H / \partial x$ in the x direction). The last two equations describe the fundamentals of the first magnetic sensor. Some sources date the Chinese invention of the compass about 4000 years in the past, and it remained the only technical magnetization application until the nineteenth century.

In 1820, Ørsted found a deviation of a compass needle near a current-carrying conductor. Ampère assumed from these results that a magnetic field H can also originate from moving electrical charges.[2] He formulated the basic law of magnetomotive force,

$$NI = \oint_s \vec{H} \cdot d\vec{s} \qquad (1.5)$$

which equals the ring integral of H over a closed path s to the Ampère windings (N conductors with current I). Therefore, the field H is measured in A/m. With the unit of force (N = kgm/s^2 = VAs/m), Eq. (1.2) yields that Q_m is measured in Vs, in analogy to the electrical charge.

The law of Biot-Savart is used to calculate the field of general conductor arrangements,

$$d\vec{H} = \frac{I}{4\pi r^2} d\vec{s} \times \vec{e}_r \qquad (1.6)$$

The current I flowing through a conductor part $d\vec{s}$ causes a field $d\vec{H}$ at a distance $\vec{r}$. Using Eqs. (1.5) and (1.6), respectively, the circumferential field of a long conductor is $H = I/2\pi r$, the field along the axis of a long cylindrical coil (N turns, length $l \gg$ diameter d) is $H = NI/l$, and along the center x axis of a current loop with diameter d it is

$$H = \frac{Id^2}{\sqrt{(d^2 + 4x^2)^3}} \qquad (1.7)$$

The magnetic field can be visualized by lines of force. The tangent gives the direction, and their density (distance) determines the field strength. This is illustrated by Figure 2 for Ampère's law. Consider that Eqs. (1.5)–(1.7) do not take into account the displacement current and thus are only applicable to slowly varying fields.

---

[1] A comprehensive explanation of the magnetic force is still not available. In modern physics it is treated as the bending of space with up to 13 dimensions.

[2] Jiles [11] has derived the magnetic field as a relativistic correction to the electric field.

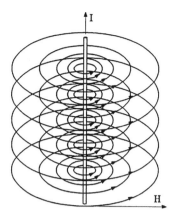

Fig. 2.   Magnetic lines of force of a conductor with current I.

1.1.2. Magnetic Fields and Matter

In general, the reaction of matter in a magnetic field H contributes to the magnetic induction

$$\vec{B} = \underline{\mu} \cdot \vec{H} \qquad (1.8)$$

The permeability tensor $\underline{\mu}$ represents material properties like anisotropy, nonlinearity, inhomogeneity, hysteresis, etc. Only in very special cases is it a simple constant.

1.1.2.1. Flux Density

The basic law related to induction,

$$V = -N\vec{A} \cdot \frac{\partial \vec{B}}{\partial t} \qquad (1.9)$$

has been found in principle by Faraday and Lenz: if the induction changes with time t in an area A enclosed by N turns of a conductor, a voltage V is generated. The magnetic flux,

$$\phi = \int_{A} \vec{B} \cdot d\vec{A} \qquad (1.10)$$

through this area is therefore measured in Vs (Weber, Wb), and the induction B, also called flux density, is measured in Vs/m^2 (Tesla, T). The definition of B is generally given with respect to the Lorentz force,

$$\vec{F} = Q_{e}\vec{v} \times \vec{B} \qquad (1.11)$$

A moving electrical charge Q_{e} with velocity v experiences a deflection force perpendicular to both the $\vec{B}$ and $\vec{v}$ directions. The torque

$$\vec{T} = I\vec{A} \times \vec{B} \qquad (1.12)$$

on a conductor loop is proportional to the circuit area A and the current I; their product is the magnetic moment

$$\vec{m} = I\vec{A} \qquad (1.13)$$

Consequently, the unit of μ is Vs/Am. In a vacuum the effects of the two fields H and B are in principle the same, but they have different units and values:

$$\vec{B} = \mu_{0}\vec{H} \qquad (1.14)$$

The magnetic field constant, $\mu_{0} = 4\pi \cdot 10^{-7}$ Vs/Am (H/m), is only a consequence of the units used (in the former cgs unit system B and H had the same dimension).

In general, the energy per unit volume of a magnetic field is

$$W = \int \vec{H} \cdot d\vec{B} \qquad (1.15)$$

In the case of linear dependence (e.g., Eq. (1.14)), the energy density is

$$W = \frac{\vec{B} \cdot \vec{H}}{2} \qquad (1.16)$$

The flux density is the only magnetic vector quantity that can be measured directly by using Eq. (1.11), and one could propose that all of the other definitions of magnetic fields should be obsolete. But magnetism is a very complex chapter of physics, and we will need other magnetic quantities to find a practical approach to what we are trying to understand as "reality."

1.1.2.2. Polarization, Magnetization, Permeability, and Susceptibility

The flux density B consists of two contributions: the vacuum induction $\mu_{0}\vec{H}$ and the polarization $\vec{J}$ of matter. With

$$\vec{B} = \mu_{0}\vec{H} + \vec{J} \qquad (1.17)$$

we find the same units for J and B. The basic idea is that there are magnetic dipoles with an average moment $\vec{j}$ in a volume V, and

$$\vec{J} = \frac{d\vec{j}}{dV} \qquad (1.18)$$

is the density of these dipoles. As stated before, there are no real dipoles[3] because of the lack of isolated magnetic charges. A completely different approach to the magnetism of matter results from atomic physics: the moving unit charges ($e = -1.60 \cdot 10^{-19}$As) of electrons (mass $m_{e} = 9.11 \cdot 10^{-31}$ kg) are treated as currents—because of orbit and spin—producing a magnetic moment. Using Planck's constant ($h = 6.63 \cdot 10^{-34}$Ws^2), the unit of the quantized orbit moment is the Bohr magneton,

$$\mu_{B} = \frac{eh}{4\pi m_{e}} \approx 9.27 \cdot 10^{-24} \text{Am}^2 \qquad (1.19)$$

---

[3] The difference between a ficitious dipole (producing only a dipole momentum) and a real dipole is that the latter can be devided into two monopoles. But we find only further "dipoles" if a magnetic compass needle is broken into arbitrarily small parts. Therefore, Eqs. (1.1)–(1.4) are fictitious as long as the magnetic charge Q_{m} is not found.

The quantum number of electron spin is 1/2, which leads to twice the gyromagnetic ratio (magnetic moment/mechanical impulse moment) compared with the orbit. When atoms condense to form a solid-state crystal, the orbits are fixed to a large extent with the atomic bond. Therefore, mainly the resulting spin moments—which are partially reduced by next-neighbor interactions—can be rotated by an applied field and contribute to the magnetization:

$$\vec{M} = \frac{d\vec{m}}{dV} \qquad (1.20)$$

It is the density of the average magnetic moments $\vec{m}$ that describes the macroscopic magnetic behavior equivalently to the polarization

$$\vec{J} = \mu_0 \vec{M} \qquad (1.21)$$

and Eq. (1.17) yields

$$\vec{B} = \mu_0(\vec{H} + \vec{M}) \qquad (1.22)$$

The material properties are also often characterized by the relative permeability tensor $\underline{\mu}_r$:

$$\vec{B} = \mu_0 \underline{\mu}_r \vec{H} \qquad (1.23)$$

Therefore, the relation between $\vec{M}$ and $\vec{H}$ can be expressed as

$$\vec{M} = (\underline{\mu}_r - \underline{1}) \cdot \vec{H} = \underline{\kappa} \cdot \vec{H} \qquad (1.24)$$

where $\underline{\kappa}$ is the susceptibility tensor. Using Eqs. (1.12), (1.13), (1.14), (1.20), and the general expression for energy density $W = \int T\, d\varphi/V$ (φ is the angle between $\vec{A}$ and $\vec{B}$), we find the magnetostatic energy density (Zeemann energy) of a magnetized body to be

$$W_H = -\mu_0 \vec{M} \cdot \vec{H} \qquad (1.25)$$

It should be stated that the atomic nucleus also has a magnetic moment (only about 1/2000 of the magnetic electron spin moment). The dependence of its resonance frequency on an applied field is used for materials characterization and diagnostic techniques with nuclear magnetic resonance (NMR).

The sections above were a brief overview of the roots of magnetism. Although the discussion about the aspects of magnetism is still in progress—especially the nature of magnetic fields—the state-of-the-art electromagnetic theory is expressed by Maxwell's [14] equations,

$$\nabla \times \vec{H} = \vec{J}_c + \frac{\partial \vec{D}}{\partial t} \qquad (1.26)$$

$$\nabla \times \vec{E} = -\frac{\partial \vec{B}}{\partial t} \qquad (1.27)$$

$$\nabla \cdot \vec{D} = \varrho_e \qquad (1.28)$$

$$\nabla \cdot \vec{B} = 0 \qquad (1.29)$$

These equations [2] describe the relations between magnetic field $\vec{H}$, current density $\vec{J}_c$, electrical field $\vec{E}$, magnetic flux density $\vec{B}$, dielectric displacement $\vec{D}$, electrical charge density ϱ_e, and time t. Applying the integration law of Stokes ($\int_A \nabla \times \vec{X} = \oint_s \vec{X}\, ds$) for a vector $\vec{X}$ in an area A with a circumference s, Eqs. (1.26) and (1.27) yield Eqs. (1.5) (neglecting the displacement current density $\partial \vec{D}/\partial t$) and (1.9), respectively.

1.1.3. Units and Magnitudes of Magnetic Fields

In addition to SI units, former cgs units are also still found in magnetism literature. The main conversion factors are as follows

Magnetic field intensity H	1 Oe (Oersted) $\overset{\wedge}{=}$ $1000/4\pi$ [A/m]
Magnetic potential	1 Gb (Gilbert, Oe cm) $\overset{\wedge}{=}$ $10/4\pi$ [A]
Flux density B	1 G (Gauss) $\overset{\wedge}{=} 10^{-4}$ [T]
Magnetic flux ϕ	1 M (Maxwell) $\overset{\wedge}{=} 10^{-8}$ [Wb]
Energy density W	1 erg/cm^3 $\overset{\wedge}{=} 10^{-1}$ [J/m^3]
Magnetization M	1 emu/cm^3 $\overset{\wedge}{=} 1000$ [A/m]
Polarization J	$J_{cgs} = J_{SI}/4\pi$
Permeability μ	$\mu_{cgs} = \mu_{SI}/\mu_0$
Susceptibility κ	$\kappa_{cgs} = \kappa_{SI}/4\pi$

There is a wide range of magnitudes of magnetic fields in the universe. To give a feeling of the variety of measurement tasks, some examples illustrate the full scale of the flux density:

Biomagnetic fields (brain, heart)	$B \approx 10$ fT to 1 pT
Galactic magnetic field	$B \approx 0.25$ nT
Vicinity (distance about 1 m) of electrical household appliances	$B \approx 600$ nT
Earth magnetic field (magnetic south is near geographic north)	$B \approx 60\ \mu$T
Vicinity (distance about 10 m) of electrical power machines and cables	$B \approx 0.1$–10 mT
Permanent magnets (surface)	$B \approx 100$ mT to 1 T
Laboratory magnet	$B \approx 2.5$ T
NMR tomography (superconducting magnets with 1 m diameter)	$B \approx 4$ T
Nuclear fusion experiment	$B \approx 10$–20 T
Short pulse laboratory fields	$B \approx 60$–100 T
White dwarf (small star with a density of 1000 kg/cm^3)	$B \approx 1$ kT
Pulsar (tiny star with a density of 10^{10} kg/cm^3)	$B \approx 100$ MT!

The origin of the planetary magnetic fields has been under discussion up to now. Current dynamo theories cannot explain both high efficiency and the reversal of polarity. Recently, a new theory seems to sucessfully describe these open questions by rotating electric dipole domains [15].

1.2. Magnetic States of Matter

The effect of a magnetic field on matter is very diverse. We will only touch diamagnetism and paramagnetism, and we will emphasize the most important phenomena of magnetically ordered media: ferrimagnetism and ferromagnetism.

1.2.1. Weakly Magnetic Materials

If a magnetic field has almost no macroscopic effect in a material, one could believe that the material is "nonmagnetic" ($|\kappa| \ll 1$). In fact, an applied field seems to "shine through" diamagnetic materials, and it is only weakly modified ("falsened") in paramagnetic matter.

1.2.1.1. Diamagnetism and Superconductivity

If a magnetic field is applied to an atom, an electrical field is induced following Eq. (1.27). The forces acting on the electrical charges accelerate the whole electron shell, and the result may be formally described by an additional rotational velocity (Larmor precession). The corresponding magnetic moment is antiparallel to the applied field. With Z electrons per atom and N atoms per cubic meter (effective equilibrium distance $2r$), the susceptibility is

$$\kappa = -\mu_0 N \frac{e^2 Z \langle r^2 \rangle}{6 m_{\mathrm{e}}} \qquad (1.30)$$

The values of κ are usually very small ($\kappa \approx -10^{-5}$). Therefore, the atomic diamagnetic behavior is dominated by other forms of magnetism. A different form of diamagnetism can be observed in metals, where the conduction electrons move in helical trajectories according to Eq. (1.11). Important diamagnetic materials are the noble gases, semiconductors, water, and many metals (Cu, Zn, Ag, Cd, Au, Hg, Pb, Bi, etc.). Diamagnetic materials are used for the design of mechanical parts that must not disturb the magnetic field to be measured.

The magnetic behavior of superconductors can be described as a strong form of diamagnetism. Because of the Meissner–Ochsenfeld effect, the material is completely shielded by nondissipative surface currents ($B \approx 0$) below a critical temperature and a critical applied field and $\kappa \approx -1$ in this case, according to Eqs. (1.22) and (1.24). If the material consists of nonsuperconducting parts (e.g., grain boundaries, inclusions) or the applied field exceeds local critical values, B partially penetrates the material and $|\kappa|$ decreases.

1.2.1.2. Paramagnetism

In the case of a resulting magnetic moment $\vec{m}$ of the atoms that are not compensated within a crystal structure, the magnetic field causes an alignment against the energy $skT/2$ of thermal motion (grade of freedom s ($s = 3$ for the monoatomic case) and Boltzmann constant $k = 1.38 \cdot 10^{-23}\,\mathrm{Ws/K}$). In general, M is a nonlinear Brillouin function of H, but in the case of $mH/kT \ll 1$ (high temperatures ($T \approx 300$ K) or low fields

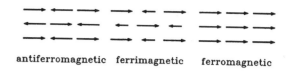

Fig. 3.   Spin alignment due to exchange coupling.

($\mu_0 H \approx 1$ T), as is valid for most standard technical applications) the susceptibility is then

$$\kappa = \mu_0 \frac{Nm^2}{3kT} \qquad (1.31)$$

and the order of magnitude is $\kappa \approx +10^{-3}$. Another kind of paramagnetic behavior can be observed, for example, in metals, where an applied field leads to different energy levels for different orientations of electron spin. Establishing a common Fermi energy level[4] reduces those spin populations that have a negative component with respect to the applied field at the expense of the other (temperature-independent Pauli paramagnetism). Examples of paramagnetic materials are several gases (e.g., oxygen), carbon, and several metals (Mg, Al, Ti, V, Mo, Pd, Pt, etc.). One application of paramagnetic materials is the electron spin resonance (ESR) magnetometer.

1.2.2. Ferromagnetism, Ferrimagnetism, and Antiferromagnetism

Solid-state matter can exhibit a unique feature in special cases of atomic neighborhood order: spontaneous magnetization. Atomic moments (usually spin moments) may be aligned parallel (ferromagnetism) or antiparallel (antiferromagnetism) because of quantum-mechanical exchange coupling.[5] Ferrimagnetism is characterized by two or more sublattices with an antiparallel orientation of their magnetic moments that do not cancel each other. These possibilities of spin alignment are shown in Figure 3. Other examples of magnetic behavior are metamagnetism, helimagnetism, or superparamagnetism.

1.2.2.1. Spontaneous Magnetization

Only a few elements and compounds are ferromagnetic at temperatures above 0°C (e.g., Fe, Ni, Co, Gd, CrO_2). Ferrites consist in most cases of iron oxide with oxides of Mn, Ni, Cu, Mg, Y, etc. (spinels and garnets). The main conditions for establishing a spontaneous magnetization,

$$M_{\mathrm{s}} = NmL\left(\mu_0 m \frac{H + \mu_0 w M_{\mathrm{s}}}{kT}\right) \qquad (1.32)$$

below the Curie temperature $T_{\mathrm{c}} = wN(\mu_0 m)^2/3k$ (the simplified presentation is a result of the classical ferromagnetism

---

[4] The Fermi level is the most probable maximum energy of conduction electrons.
[5] Exchange coupling is based on electrostatic interaction and lowers the system's energy by aligning the uncompensated moments.

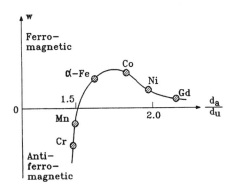

Fig. 4. Exchange energy versus d_a/d_u.

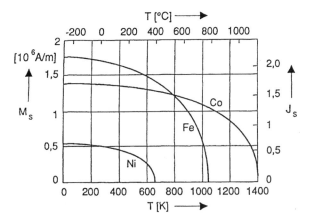

Fig. 5. Dependence of M_s on temperature for the transition elements Co, Fe, and Ni. Reprinted with permissiion from [236], © 2001, Artech.

theory of Pierre Weiss (1907) by means of the Langevin function[6] $L(x) = \coth(x) - 1/x$ to demonstrate the principle) are summarized in the following:

- the resulting atomic magnetic moment m (transition elements: uncompensated spins in 3d or 4f shells);
- a high atomic density N (condensed matter);
- a small range of the ratio of the interatomic distance to the diameter of the shell containing uncompensated spins (crystalline or amorphous solid-state matter), leading to a large hypothetical exchange field H_E proportional (Weiss constant w) to the spontaneous magnetization M_s ($H_E = \mu_0 w M_s$).

Many efforts have been made to find a quantum theory of magnetism, and there are two mutually exclusive models: the localized moment model works satisfactorily for rare earth metals, whereas the itinerant electron model describes quite well the magnetic properties of the 3d transition metals and their alloys. Figure 4 shows the Bethe–Slater curve, depicting the exchange coupling (represented by w) versus atomic distance/uncompensated shell diameter, ratio d_a/d_u. The larger w is, the larger is T_c. The atomic spacing of the antiferromagnetic Mn can be enlarged to become ferromagnetic, e.g., in MnAlCr or MnCuSn (Heusler alloys).

At the Curie temperature the energy kT_c of thermal motion equals the exchange energy and M_s becomes zero. The susceptibility is then given by the Curie–Weiss law, $\kappa = \mu_0 m N / 3k(T - T_c)$ (paramagnetic behavior; see Section 1.2.1.2, but with a temperature bias T_c). The solution of Eq. (1.32) is shown graphically in Figure 5 for the elements Co, Fe, and Ni.

1.2.2.2. Magnetic Anisotropy

Anisotropy is an important feature, utilized in many applications. Only the most important kinds of anisotropy based on atomic or geometric aspects, such as magnetocrystalline, strain, and shape anisotropies, are briefly described. Other forms are surface anisotropy, exchange anisotropy, and diffusion anisotropy.

Magnetocrystalline Anisotropy. The electronic orbits and therefore the magnetic orbital moments are fixed in certain crystallographic directions. The magnetic coupling of spin and orbit moments is why spins are bound to certain directions in the absence of an applied field H. If one tries to rotate the spins out of these easy directions of minimum energy, it is necessary to work against this magnetocrystalline anisotropy energy density W_C. Cubic crystalline materials have a high degree of symmetry. Therefore, it is possible to describe complex anisotropic behavior with a power expansion with only a few constants (K_0 summarizes only direction-independent contibutions), K_1 and K_2. For cubic crystals we may write the energy density,

$$W_C = K_0 + K_1\left(\alpha_1^2 \alpha_2^2 + \alpha_2^2 \alpha_3^2 + \alpha_3^2 \alpha_1^2\right) + K_2\left(\alpha_1^2 \alpha_2^2 \alpha_3^2\right) + \cdots \quad (1.33)$$

where α_i are the cosines of $\vec{M}_s$ with respect to the crystal axes [100], [010], and [001]. For hexagonal crystals W_C only depends on the angle φ between $\vec{M}_s$ and the main axis [0001] of the crystal (in the absence of anisotropy in the (0001) plane):

$$W_C = K_0 + K_1 \sin^2 \varphi + K_2 \sin^4 \varphi + \cdots \quad (1.34)$$

In simplified cases of only one anisotropy constant, it is also usual to define a fictitious anisotropy field,

$$H_k = \frac{2K_1}{\mu_0 M_s} \quad (1.35)$$

Magnetocrystalline anisotropy constants can be determined from the torque $T = \partial W(\varphi)/\partial \varphi$ acting on an oriented single crystal sample (e.g., sphere, circular plate).

Figure 6 illustrates the energy areas (three-dimensional graphs of Eq. (1.33)) for different K_i. Iron has six easy directions (three easy axes) parallel to the edges, and nickel has eight easy directions (four easy axes) parallel to the space diagonals of the cubic crystals. A combination of materials with different signs of K_i can lead to a large number

---

[6] The Langevin function is valid only in the classical limit: if spin quantization is taken into account, integration is replaced by summation, which leads to a more realistic but complicated Brillouin function.

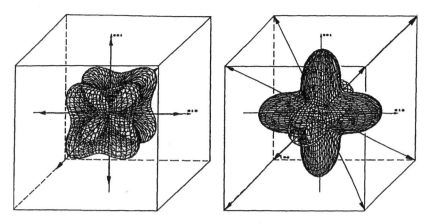

Fig. 6. Energy areas of W_C for iron ($K_1 = +48$ kJ/m³, $K_2 = -9$ kJ/m³, left) and nickel ($K_1 = -4.5$ kJ/m³, $K_2 = +2$ kJ/m³, right).

of easy directions or to an almost complete suppression of anisotropy, which is important for achieving high permeability (e.g., Permalloy).

Figure 7 shows the projection of $\vec{M}_s$ in the direction of the applied field H (energy density W_H of Eq. (1.25)) for different materials and directions of H. These curves can be calculated in principle by tracing the minima of the total energy density $W_T = W_C + W_H$.

Distinct magnetocrystalline anisotropy can be macroscopically found only in single crystalline or grain-oriented materials. Otherwise, the grains are oriented randomly with respect to their easy directions, which leads to macroscopically isotropic behavior, but the magnetization process of polycrystals reflects the values of the anisotropy constants.

Magnetostriction and Strain Anisotropy. An applied field can induce a small anisotropy of the bonding forces and therefore a change in the lattice constants. The macroscopic result of this further manifestation of spin-orbit coupling is a relative length change $\lambda_{\alpha,\beta} = \delta l / l$ of the material at a changing state

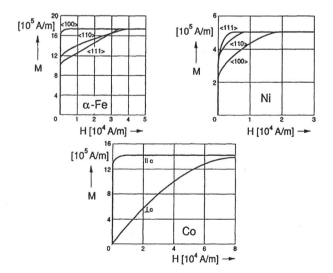

Fig. 7. Projection of $\vec{M}_s$ in the $\vec{H}$ direction for different ferromagnetic crystals. Reprinted with permission from [236], © 2001, Artech.

of magnetization (seldom: volume change) in a measurement direction, defined by the direction cosines β_i with respect to the axes of a cubic crystal,

$$\lambda_{\alpha,\beta} = \tfrac{3}{2}\lambda_{100}\left(\alpha_1^2\beta_1^2 + \alpha_2^2\beta_2^2 + \alpha_3^2\beta_3^2 - \tfrac{1}{3}\right)$$
$$+ 3\lambda_{111}(\alpha_1\alpha_2\beta_1\beta_2 + \alpha_2\alpha_3\beta_2\beta_3 + \alpha_3\alpha_1\beta_3\beta_1) \quad (1.36)$$

where λ_{ijk} are the magnetostriction constants (maximum $\Delta l / l$ in the given crystallographic direction [ijk], occurring after a randomly demagnetized state). In contrast, a mechanical stress $\vec{\sigma}$ can cause a change in the magnetic state, which is established to support the strain by magnetostriction. There is an influence on the easy directions, described by the strain anisotropy energy density,

$$W_S = -\sigma(\lambda_{\alpha,\beta} + \lambda_0) \quad (1.37)$$

where λ_0 is a direction-independent constant and β_i are in this case the direction cosines of the applied stress $\vec{\sigma}$ with respect to the crystal axes. The effect of W_S on the easy directions (the total anisotropy energy density is then $W_A = W_C + W_S$) is shown in Figure 8 for Fe and Ni. If λ_{ijk} has a negative sign, the [ijk] direction becomes harder (higher anisotropy energy) at tensile stress, and vice versa. It is therefore clear that applied mechanical stress can change the magnetic state of a ferromagnetic medium. Inner stresses (e.g., caused by mechanical treatment) will also strongly influence the magnetic behavior of materials.

In the case of polycrystalline materials of isotropic orientation distribution of the crystal axes, the isotropic saturation magnetostriction is

$$\lambda_s = \tfrac{2}{5}\lambda_{100} + \tfrac{3}{5}\lambda_{111} \quad (1.38)$$

In the special case of "isotropic" magnetostriction ($\lambda_{100} = \lambda_{111} = \lambda_s$) the relative length change $\lambda_{\alpha,\beta} = \lambda_\varphi$ depends only on the angle φ between $\vec{\sigma}$ and $\vec{M}_s$:

$$\lambda_\varphi = \tfrac{3}{2}\lambda_s\left(\cos^2\varphi - \tfrac{1}{3}\right) \quad (1.39)$$

Shape Anisotropy. Shape effects play an important role in the design of magnetic devices. A spontaneously magnetized body

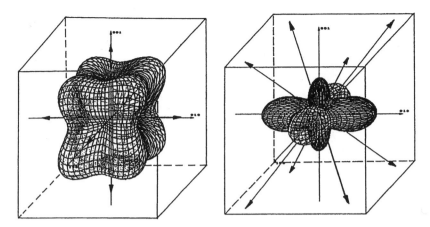

Fig. 8. Magnetocrystalline and strain anisotropy energy areas (compressive stress in the [001] direction, $\sigma < 0$) for iron ($\lambda_{100} = +21 \cdot 10^{-6}$, $\lambda_{111} = -21 \cdot 10^{-6}$, left) and nickel ($\lambda_{100} = -46 \cdot 10^{-6}$, $\lambda_{111} = -24 \cdot 10^{-6}$, right).

exhibits fictitious magnetic charges (north (+) and south (−) poles) at the surface, where the normal component of $\vec{M}$ is discontinuous. According to the beginning of this chapter, these magnetic poles are considered as the origin of a magnetic field[7] $\vec{H}_d$, proportional to $\vec{M}$ (see Fig. 9) and oriented antiparallel to the magnetization within the magnetized body. The energy density of the so-called demagnetizing field of a general ellipsoid is

$$W_{\mathrm{D}} = \frac{\mu_0 M_s^2}{2} \left(N_a \alpha_1^2 + N_b \alpha_2^2 + N_c \alpha_3^2 \right) \qquad (1.40)$$

where N_a, N_b, and N_c are the demagnetizing factors ($N_a + N_b + N_c = 1$) and α_i are the direction cosines of the magnetization with respect to the ellipsoid axes a, b, c.

The energy areas of Eq. (1.40) show that W_{D} is minimum in the axis of a bar or in the plane of a disc (see Fig. 10). The magnetic fields are only homogeneous in an ellipsoid. In this case, the demagnetizing matrix,

$$\underline{N}_{\mathrm{d}} = (N_a, N_b, N_c) \qquad (1.41)$$

defined by

$$\vec{H}_{\mathrm{d}} = -\underline{N}_{\mathrm{d}} \cdot \vec{M} \qquad (1.42)$$

[7] As real magnetic charges do not exist, the field H_d originates from the moving electrical charges producing the magnetic moments of the density M.

can be calculated analytically [16]. Otherwise, it has a very complex structure and depends on the actual susceptibility. If the field H is applied, the inner, effective field,

$$\vec{H}_{\mathrm{i}} = \vec{H} + \vec{H}_{\mathrm{d}} \qquad (1.43)$$

differs from H by H_d (lower if $\kappa > 0$ and greater if $\kappa < 0$).

Visualizing the effect of shape anisotropy in ferromagnetic materials, the total anisotropy energy density, $W_{\mathrm{A}} = W_{\mathrm{C}} + W_{\mathrm{D}}$, of a film of grain-oriented silicon iron (a material widely used for transformer cores and many other soft magnetic applications) is shown in Figure 11. Because of the large value of M_s, W_{D} overcomes W_{C} by about 100 times if the magnetization is perpendicular to the disc plane.

1.2.2.3. Domain Structure

In the preceding section it was assumed that a ferromagnetic body exhibits spontaneous magnetization. But only some hard magnetic materials appear to be macroscopically magnetized in the absence of an applied field. If the sample were magnetized with M_s, the demagnetizing stray field energy density

Stray field energy

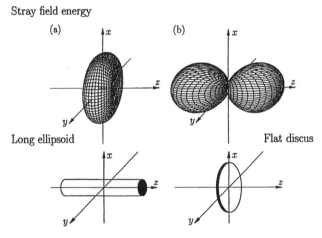

Fig. 10. Demagnetizing stray field energy density of ellipsoid approximations: long bar or fine particle (left) and flat disc or thin film (right).

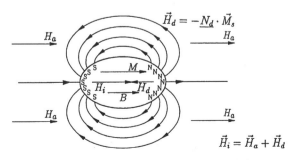

Fig. 9. Demagnetizing field of a magnetized ellipsoid.

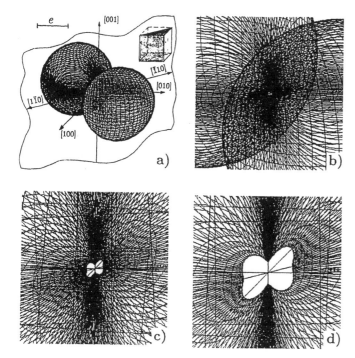

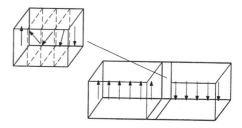

Fig. 13. Spin rotation within a Bloch wall between antiparallel magnetized domains. Reprinted with permission from [236], © 2001, Artech.

applied stresses) and microscopic stray fields [17]. For instance, the domain structure of Figure 12d is only favorable if the magnetostriction is very low and no strain occurs. Otherwise, a compromise structure like that in Figure 12e is established.

Furthermore, the domain walls should provide a steady normal component of M_s, but in small volumes this law can be violated to reduce inner stray fields and stresses. The orientations of the easy directions may also differ for each crystal grain, or they can be randomly distributed in amorphous materials ($K_i \approx 0$). A typical width of a domain is 10–100 μm, and a typical wall thickness is 10–100 nm, but much larger domains may exist in grain-oriented materials, thin films, or wires. Figure 13 shows a simplified 180° Bloch-type domain wall where the magnetic moments are gradually rotated from one easy direction to the antiparallel one.

The sense of magnetization rotation within a domain wall can also alternate to reduce the stray field of the wall. For example, Bloch lines are such tiny zones of alternated rotation, which is defined as the handedness of the domain wall.

1.2.2.4. Magnetization Process

Figure 14 shows schematically the process of increasing the macroscopic magnetization by applying a field to a ferromagnetic material. In the absence of H the spin moments are equally distributed along the four easy directions in this example. At weak fields the domain walls move to increase the volume of the domains with a positive magnetization component with respect to $\vec{H}$ at the expense of the others. This motion is caused by the torque acting on the magnetic moments within the tiny volume of the wall, where exchange energy and anisotropy energy are in a fine balance. These easy domain wall displacements are the reason for the high permeability of soft magnetic materials and determine their technical applications.

Fig. 11. The spatial distribution of the total energy density of ellipsoidally shaped FeSi films. a:b:c = 10:10:1, $K_1 = 29$ kJ/m³, $K_2 = -9$ kJ/m³. (a) Energy scale $e = 500$ kJ/m³. (b–d) Reducing the energy scale to 33% for each picture, the total energy reveals the magnetocrystalline energy in the film plane.

W_D would be very high, as discussed above. Therefore, M_s occupies the available easy directions—given by the magnetocrystalline and strain anisotropy energy minima—to reduce the pole strength and herewith H_d and, consequently, W_D, as shown in Figure 12.

The volumes magnetized uniformly with $\vec{M}_s$ in an easy direction are the magnetic domains, divided by the domain walls. These walls also need energy to be established, which yields a lower domain size limit. The wall energy consists mainly of magnetocrystalline anisotropy and exchange energy (the contributions of stray fields are usually neglected). The wall thickness is a trade-off to minimize the wall energy.

Domain structures and domain walls can have a very complex nature, caused by magnetostriction (both inner and

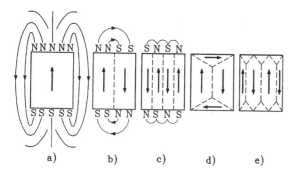

Fig. 12. Reduction of stray fields (a, b, c) by orienation of the spins in domains, magnetized uniformly in easy directions. (d) $\lambda_s \approx 0$. (e) Compromise structure, $|\lambda_s| > 0$. Reprinted with permission from [236], © 2001, Artech.

Fig. 14. The effect of an applied field on a simplified domain structure. Rotation of the domain's magnetization starts before the end of the wall displacements only if a strong stray field can be avoided. Reprinted with permission from [236], © 2001, Artech.

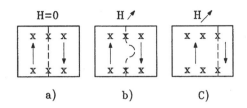

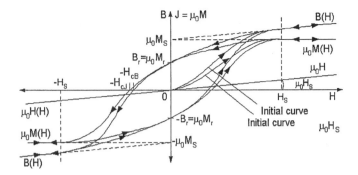

Fig. 15. Reversible (a → b) and irreversible (b → c) domain wall displacements. Reprinted with permission from [236], © 2001, Artech.

Fig. 16. Hysteresis of $\mu_0 M(H)$ and $B(H)$. Reprinted with permission from [236], © 2001, Artech.

The increase in bulk magnetization by domain wall displacements in an applied field (energy density W_H) is balanced by the shape anisotropy energy density W_D and hindered by local stray fields and magnetostriction, for example. If W_H is large enough to compete with the magnetocrystalline anisotropy energy density, the domain's magnetization is rotated toward the field direction, tracing the minima of the total energy density $W_T = W_C + W_S + W_D + W_H$. This simplified magnetization process for single crystals has been described by the phase rule of Néel, Lawton, and Stewart [18] and is the basis for the calculation of magnetization curves of anisotropic materials [19].

In the case of nonideal material structures with grain boundaries, nonmagnetic inclusions, or cracks, for example, the domain walls are pinned, reducing the stray fields of these pinning sites when covering as many imperfections as possible. Figure 15 illustrates the mechanism of irreversible[8] domain wall displacements.

Pinned at two imperfections in this case, the wall is moving reversibly at low H. A further field increase forces the wall to jump until it is pinned again by other impurities. The path of the domain wall differs for each cycle, and these so-called irreversible Barkhausen jumps are energy dissipative because of spin relaxation, magnetomechanical interaction, and microscopic eddy currents. Furthermore, the induction within the moving domain wall's volume due to spin rotation gives rise to the Barkhausen noise. Therefore, we find that the magnetization process depends on the magnetic history experienced by the material: the feature of hysteresis represents the widely known picture of magnetism.

1.2.2.5. Magnetization Curve

The initial magnetization curve is obtained when a field is applied to a previously completely demagnetized sample. Starting with reversible domain wall displacements at weak fields and proceeding with irreversible Barkhausen jumps, the saturation[9] M_s is finally reached by processes of magnetization rotation against the anisotropy energy. After H is reduced

to zero the sample remains magnetized at the remanence M_r. The magnetization becomes zero at the coercive field strength $H = -H_c$. The upper branch of the major hysteresis loop is completed at $M = -M_s$, and the lower branch is obtained by an analogous procedure (see Fig. 16).

Demagnetization can be achieved by several methods, which yield different results. Heating the sample over T_c causes a perfect erasure of the magnetic history. Application of a decaying ac field is the most usual method (see Fig. 17), but it yields no random distribution of the domain magnetization in the easy directions. An improvement is demagnetization with a decaying rotating ac field—this technique is used in geophysics. Demagnetization by mechanical shock waves sometimes happens accidentally to permanent magnets (decaying stress oscillations change the easy directions, which causes domain wall displacements of decreasing amplitude).

The dependence of B on H differs from $\mu_0 M(H)$ by $\mu_0 H$ (see Fig. 16). The total permeability or amplitude permeability,

$$\mu_t = \frac{B_t}{H_t} \qquad (1.44)$$

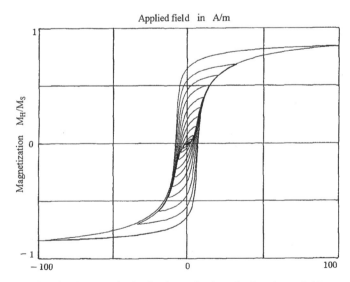

Fig. 17. Demagnetization by the application of a decaying ac field.

---

[8] Rotation processes can also be irreversible in the case of materials with restrained domain structures (permanent magnets, fine particles, thin films, etc.).

[9] The saturation magnetization is almost identical to M_s. It differs from spontaneous magnetization only by the effect of the field on alignment of the magnetic moments ($H \to \infty$) against thermal disorder, as described by Eq. (1.32), which is very small at technical field values.

depends on the operation point B_t, H_t. Important permeabilities are the initial permeability,

$$\mu_i = \lim_{\substack{B \to 0 \\ H \to 0}} \frac{B}{H} \qquad (1.45)$$

which is represented by the tangent line on $B(H)$ at the origin; the maximum permeability,

$$\mu_{max} = \left.\frac{B}{H}\right|_{max} \qquad (1.46)$$

which is the ascent of the tangent line from the origin to $B(H)$; and the reversible permeability,

$$\mu_{rev} = \lim_{\Delta H \to 0} \left.\frac{\Delta B}{\Delta H}\right|_{B_t, H_t} \qquad (1.47)$$

which is the limit of the superposition permeability for small ac fields at the operation point B_t, H_t,

$$\mu_\Delta = \left.\frac{\Delta B}{\Delta H}\right|_{B_t, H_t} \qquad (1.48)$$

The differential permeability,

$$\mu_{diff} = \frac{dB}{dH} \qquad (1.49)$$

is the derivative of the $B(H)$ curve.

Following Eq. (1.15), the energy loss W_L per unit volume and cycle, which is finally converted to heat, corresponds with the area of the hysteresis loop. To yield the power loss one has to multiply W_L by the frequency f. Losses are distinguished between static hysteresis losses and frequency-dependent losses. The latter are distinguished between normal (due to bulk dB/dt) and abnormal (due to additional dB/dt of large-domain wall displacements) eddy current losses.

To reduce the effects of hysteresis, like remanence and coercivity, the magnetization curve can be idealized by the superposition of an ac field of higher frequency (see Fig. 18) and averaging the magnetization by tracing the centers of the minor loops. This is called "magnetic shaking," a technique successfully used to improve the parameters of magnetic shielding or in magnetic recording of analogue signals. The total loss corresponds to the sum of the area of the major loop and the areas of the minor loops in this case. The ideal magnetization curve[10] can be measured by performing a demagnetization at various bias fields (see Fig. 19).

Magnetization curves can be measured inductively at a ring core sample with different windings, by evaluating Eqs. (1.5) and (1.9) for H and B, respectively. If the sample is an open magnetic circuit, the demagnetizing field causes a flattening of the curve by decreasing the measured, apparent permeability,

$$\mu_a \approx \frac{\mu_r}{1 + \mu_r D_d} \qquad (1.50)$$

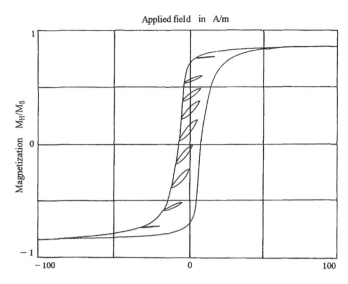

Applied field in A/m

Fig. 18. Minor loops and their permeability μ_Δ, created by the application of a small ac field with a variable bias.

as a function of the relative permeability μ_r (approximation for $\mu_r \gg 1$) and following Eqs. (1.23), (1.24), (1.42), and (1.43). In the simulation of the magnetic behavior of devices, the calculation of hysteresis loops is a very complex task [20], and several models have been developed (e.g., [21–23]).

1.3. Magnetic Materials for Applications

Magnetic materials are classified by their coercivity H_c: soft ($H_c < 1000$ A/m), medium ($H_c = 1$–30 kA/m), and hard ($H_c > 30$ kA/m). To imagine the wide range of properties of modern materials, consider the following example. If the plot of a soft magnetic hysteresis loop ($H_c = 0.1$ A/m) is 1 cm wide, the corresponding width for a typical hard magnetic material ($H_c = 10^6$ A/m) on the same scale would be 100 km! But magnetostrictive properties and thin films are also utilized for

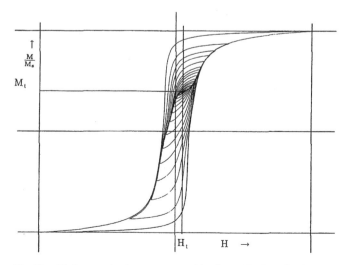

Fig. 19. Ideal magnetization curve created by the application of a decaying ac field with a variable bias.

---

[10] This can be experienced by the "magnetic screwdriver" effect. A long hard magnetic steel rod (easy axis of shape anisotropy) can be magnetized in the bias of the weak earth magnetic field by applying mechanical shock waves (hammer strokes), which affect decaying domain wall displacements, and it can be demagnetized by performing this procedure perpendicular (hard axis) to the earth field.

manifold applications. Typical representatives are covered in the following sections.

1.3.1. Soft Magnetic Materials

Low coercivity and high permeability are achieved in various crystalline, amorphous, and nanocrystalline elements and alloys. General rules for materials design are low magnetocrystalline and strain anisotropy or a large number of easy directions, and negligible inner stresses. This is provided by amorphous materials without crystal structure ($K_i \approx 0$) and nanocrystalline materials, in which the small grain size averages out anisotropy. Anisotropic materials can be used advantageously by designing the flux path parallel to the easy directions (e.g., cores of grain-oriented FeSi steel). Stress-relieving annealing will be necessary after mechanical deformation in most cases to preserve the intrinsic material parameters. Finally, Barkhausen noise must also be considered for the suitability of a material for certain applications. Table I presents the basic properties of typical soft magnetic materials.

1.3.1.1. Crystalline Metals

Metallic crystals without imperfections represent the classical soft magnetic materials. Typical examples are pure iron and nickel. Alloys widely used are FeNi (Permalloy, Supermalloy, Mumetal with very high permeability), FeCo (highest saturation), FeSi (increased resistivity decreases eddy currents), and FeAlSi (Sendust, mechanically hard).

Table I. Maximum Permeability, Coercivity, and Saturation Flux Density (Quasi-static, $T = 300$K) of Selected Soft Magnetic Materials

Material	Composition	μ_{max}	H_c (A/m)	B_s (T)
Cobalt	$Co_{99.8}$	250	800	1.79
Permendur	$Fe_{50}Co_{50}$	5000	160	2.45
Iron	$Fe_{99.8}$	5000	80	2.15
Nickel	$Ni_{99.8}$	600	60	0.61
Silicon-iron[a]	$Fe_{96}Si_4$	7000	40	1.97
Supermendur	$Fe_{49}Co_{49}V_2$	60,000	16	2.40
Ferroxcube $3F3$[b]	Mn-Zn-Ferrite	1800	15	0.50
Manifer 230[c]	Ni-Zn-Ferrite	150	8	0.35
Ferroxplana[d]	$Fe_{12}Ba_2Mg_2O_{22}$	7	6	0.15
Hipernik	$Fe_{50}Ni_{50}$	70,000	4	1.60
78 Permalloy	$Fe_{22}Ni_{78}$	100,000	4	1.08
Sendust	$Fe_{85}Si_{10}Al_5$	120,000	4	1.00
Amorphous[e]	$Fe_{80}Si_{20}$	300,000	3.2	1.52
Mumetal 3	$Fe_{17}Ni_{76}Cu_5Cr_2$	100,000	0.8	0.90
Amorphous[e]	$Fe_{4.7}Co_{70.3}Si_{15}B_{10}$	700,000	0.48	0.71
Amorphous[e]	$Fe_{62}Ni_{16}Si_8B_{14}$	2,000,000	0.48	0.55
Nanocrystalline	$Fe_{73.5}Si_{13.5}B_9Nb_3Cu$	100,000	0.40	1.30
Supermalloy	$Fe_{16}Ni_{79}Mo_5$	1,000,000	0.16	0.79

[a] Nonoriented;
[b] At 100 kHz.
[c] At 100 MHz.
[d] At 1000 MHz.
[e] Annealed.

1.3.1.2. Amorphous Metals

If an alloy is rapidly solidified from melt (typical cooling rate 10^6 K/s), it exhibits a topological disorder without crystalline structure. A typical production process is melt spinning of thin wires and ribbons (5–50-μm thickness). Amorphous materials are alloys mainly based on Fe and Co, with additions of B and Si. The main advantages are high permeability and low losses (maximum magnetization frequency up to 5 MHz). Disadvantages are the lower saturation magnetization ($\approx$1.6 T) compared with crystalline iron alloys and the limited magnetic core design possibilities. Recently, amorphous materials have been successfully produced as tapes thicker than 50 μm and even in a bulk state [24].

1.3.1.3. Nanocrystalline Metals

By annealing an amorphous material at the recrystallization temperature ($\approx 500°$C for FeCoB), nanocrystalline grains with about 10–15-nm diameter can be established. Although these materials exhibit excellent soft magnetic properties ($H_c \to 0$, $\mu_{max} > 10^5$ at relatively high saturation magnetization), the problems of high brittleness and poor mechanical engineering possibilities have prevented wide-scale applications up to now. They are currently used for high-frequency power transformer cores because of their large resistivity.

1.3.1.4. Soft Ferrites

In general, ferrites consist of about 70% iron oxide (Fe_2O_3) and 30% of other metal oxides (e.g., MgO, MnO, NiO, CuO, or FeO). They have become widely available because Fe_2O_3 is a recycling product of steel manufacturing (rust removal). Ferrites for microwave frequencies are almost perfect electrical insulators (resistivity $\rho > 10^7$ Ωm); they consist of oxides of Fe, Ni, Mg, Mn, and Al. Soft magnetic ferrites have a relatively small cubic magnetocrystalline anisotropy.

As the crystal sublattices have antiparallel spin orientations, the resulting M_s can be increased by replacing Mn with the almost nonmagnetic Zn (in the case of inverse spinels; MnZn ferrites). In addition to the spinel structures, the garnets are another group of ferrimagnetic materials (e.g., yttrium iron garnet, YIG). They are used mainly for magneto-optical and resonance applications. Yttrium orthoferrite ($YFeO_3$) has the highest domain wall velocity of magnetically ordered media and both high infrared transparency and Faraday rotation, and it is important for future magneto-optical devices.

Advantages of ferrites are a high electrical resistivity (frequency limit up to 5 GHz), absolutely no oxidation, and the possibility of arbitrarily shaped cores. The main disadvantages are the lower saturation magnetization ($\mu_0 M_s \approx 0.5$ T) compared with ferromagnets and only moderate permeability.

1.3.2. Hard Magnetic Materials

A permanent magnet is used to provide a magnetic field outside its boundary. In the absence of both an additional magnetomotive force and an applied field it is operated in the second

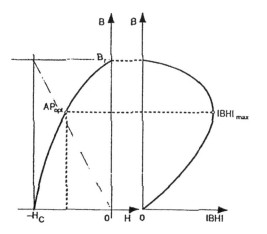

Fig. 20. Demagnetization curve and energy product of permanent magnets. Reprinted with permission from [236], © 2001, Artech.

quadrant of the hysteresis loop (demagnetization curve) as a consequence of Eqs. (1.5), (1.42), and (1.43). A high stability is achieved if the knee of the $B(H)$ curve is far below the operation point—otherwise a changing air gap or strong applied fields could partly demagnetize the magnet, especially at elevated temperatures. In addition to high coercivity and high remanence magnetization, the maximum energy product $|BH|_{max}$ is an important factor in the choice of the optimum operation point (smallest volume), as shown in Figure 20.

Basically, high-remanence and high-coercivity fields can be achieved by the different ways in which the remanent magnetization process is accomplished.

- Domain wall pinning: Coercivity increases as domain wall movements are impeded by intended, artificial imperfections. These pinning-type magnets have in general smaller coercivities and energy products compared with the second group.
- Domain nucleation: The grains are in a monodomain state because of a strong initial applied field. To reverse the magnetization, strong fields are necessary for the nucleation of oppositely magnetized domains, which results in high coercivity values (nucleation-type magnets).
- Irreversible magnetization rotation: Within small, isolated volumes (grains or particles) without domain walls but with high anisotropy, magnetization reversal is only possible through rotation processes. When the spontaneous magnetization rotates beyond the direction of maximum anisotropy energy density (see, for example, the energy area of a bar-like ellipsoid in Fig. 10), its position becomes unstable and the magnetization jumps (switches) at a certain field strength irreversibly to the nearest easy direction. Corresponding to the switching field, the maximum coercivity values are achievable because of high shape and magnetocrystalline anisotropy energies.

1.3.2.1. Alloys

Domain wall pinning is achieved by adding the elements Cr, Co, Ni, Cu, V, or C to steel (structural hardening). These permanent magnets have low H_c and $|BH|_{max}$, but they are very cheap, which permits several applications in low-cost devices.

A first improvement of permanent magnetic properties was achieved with Alnico compounds, consisting of Fe, Co, Al, Ni, Cu, and Ti. A controlled temperature gradient leads to the formation of preferentially oriented needle-like crystals (precipitation hardening). Additionally aligned in an applied field during cooling down from the melting temperature, the thin Fe-Co needles of high anisotropy yield coercivities of up to 180 kA/m and energy density products of about 90 kJ/m³.

1.3.2.2. Sintered Magnets

The state-of-the-art high-performance sintered magnets are SmCo ($H_c \approx 1.2$ MA/m, $B_r \approx 1.1$ T, $|BH|_{max} \approx 250$ kJ/m³, $T_{max} \approx 300°C$) and NdFeB ($H_c \approx 1.0$ MA/m, $B_r \approx 1.3$ T, $|BH|_{max} \approx 350$ kJ/m³, $T_{max} \approx 180°C$). Single-phase $SmCo_5$ can be magnetized to near saturation by small fields (2 kA/m) because of easy domain wall displacements. But the high coercivity (H_c up to 4 MA/m) is achieved only by saturating the grains (against inner stray fields) at strong fields (nucleation type). In contrast, in multi-phase Sm_2Co_{17} alloys the domain walls are strongly pinned at the phase boundaries. The main disadvantages of the rare earth sintered magnets are poor mechanical properties (brittleness) and their high price.

1.3.2.3. Hard Ferrites

Magnetically hard barium and strontium ferrites have a hexagonal crystal structure and possess high magnetocrystalline anisotropy. The coercivity is rather high ($H_c \approx 0.5$ MA/m), but the low remanence of some 100 mT leads to a low energy density product ($|BH|_{max} \approx 35$ kJ/m³). On the other hand, ferrites are cheap, do not oxidize, and can be powdered and embedded in plastics to produce flexible magnets by injection die casting.

1.3.3. Magnetostrictive Materials

Magnetostrictive properties of materials are used in mechanical devices and actuator systems. Applied stresses change the strain anisotropy energy and therefore influence the permeability (see Fig. 8), and applied fields change the magnetic state of the material and the elasticity via magnetostriction. Sensitive materials (large saturation magnetostriction λ_s at low saturation fields H_s) are amorphous wires and ribbons (e.g., $Co_{68}Ni_{10}(Si, B)_{22}$: $\lambda_s = -8 \cdot 10^{-6}$, $H_s = 200$ A/m or $Fe_{50}Co_{50}$: $\lambda_s = +70 \cdot 10^{-6}$, $H_s = 10$ kA/m). The rare earth–iron alloys exhibit the highest magnetostriction at strong fields (e.g., Terfenol $Tb_9Dy_{24}Fe_{67}$: $\lambda_s = +2000 \cdot 10^{-6}$, $H_s = 200$ kA/m).

2. MAGNETISM OF THIN FILMS

This section starts with a classical, phenomenological description of a typical magnetization process. Detailed emphasis is placed on surface/interface anisotropy, exchange anisotropy, and examples of domain processes and magnetization reversal.

2.1. Magnetic Structure

The thickness of a "thin film" is usually on the order of magnitude of a typical domain wall in the bulk material, up to 100 nm. In principle, there are three possibilities for establishing a domain structure (see Fig. 21): If the film is thick enough, the domain structure in Figure 21a is possible.

In the case of only two easy directions perpendicular to the film plane, the structure of Figure 21b could be established. This domain structure is only stable if

$$K_1 > \frac{\mu_0 M_s^2}{2} \tag{2.1}$$

is valid. This is the case in garnets ($K_1 = 10^5$ J/m^3, $M_s = 1.8 \cdot 10^5$ A/m) and orthoferrites ($K_1 = 10^5$ J/m^3, $M_s = 1.0 \cdot 10^3$ A/m), which are important for magneto-optical applications.

In many cases we will have to deal with in-plane magnetization (see Fig. 21c). The spontaneous magnetization of metallic ferromagnets is so large that the demagnetizing stray field energy of the magnetic poles is much larger than the magnetocrystalline anisotropy energy.

If the thickness of the film is below the Bloch-type domain wall width, the rotation of the magnetic moments from one domain to the neighbor domain direction will occur (in a simplified description) in-plane[11] as a Néel-type domain wall. Because of the comparatively high wall energy, thin films often appear uniformly magnetized in the film plane, and magnetization reversal can be achieved by coherent rotation (in real material geometries incoherent rotations also take place), depending on the applied field.

2.2. Coherent Rotation of Magnetization

Consider a spontaneously magnetized (M_s, angle φ with the easy direction) and approximately ellipsoidal ($a > b \gg c$) film with uniaxial anisotropy, where the easy axis is parallel to the long ellipsoid axis a (see Fig. 22). The energy density W_H of the applied field (angle ψ with the easy axis and components $H_x = H\cos\psi$, $H_y = H\sin\psi$)

$$W_H = -\mu_0 M_s H \cos(\psi - \varphi) \tag{2.2}$$

was given by Eq. (1.25), and the direction-dependent terms of the magnetocrystalline anisotropy energy density,

$$W_C = K_1 \sin^2 \varphi \tag{2.3}$$

[11] Depending on film thickness, exchange coupling, and anisotropy, mixed-type walls (e.g., cross-tie walls) are also known.

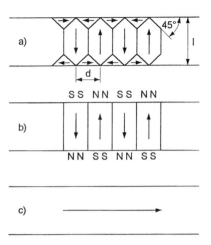

Fig. 21.   Possible domain structures in thin films. Reprinted with permission from [236], © 2001, Artech.

and of the shape anisotropy energy density of demagnetizing stray fields,

$$W_D = \mu_0 M_s \frac{N_b - N_a}{2} \sin^2 \varphi \tag{2.4}$$

were given by Eqs. (1.33) and (1.40), respectively. Using the simplifications $N_b - N_a = N_d$, $H_d = -N_d M_s$ (demagnetizing field), $H_k = 2K_1/\mu_0 M_s$ (anisotropy field) and the characteristic field $H_0 = H_k + H_d$, the zeros of the derivative of the total energy density

$$W_T = W_H + W_C + W_D \tag{2.5}$$

with respect to φ ($dW_T/d\varphi = 0$) give its extrema as a function of the parameter φ:

$$\sin\varphi = \frac{H\sin\psi\cos\varphi}{H_0\cos\varphi + H\cos\psi} \tag{2.6}$$

Depending on the applied field, the magnetization will make an angle φ for which W_T has a minimum—the second derivative of W_T is negative in this case. The magnetization curves can be calculated for the projection of M_s in the H direction:

$$M_H = M_s \cos(\psi - \varphi) \tag{2.7}$$

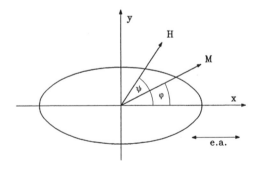

Fig. 22. H and M in a thin-film ellipsoid and the coordinate system.

Stability considerations $(dW_T^2/d^2\varphi > 0)$ lead to the well-known Stoner–Wohlfarth astroid [34] (for $-1 < H_y/H_0 < +1$),

$$|H_x|^{2/3} + |H_y|^{2/3} \leq |H_0|^{2/3} \tag{2.8}$$

for the permitted solutions of φ. If H is within the astroid area, two solutions for M_H are possible. If the field strength is large enough to point out of the astroid area, only one solution is possible and switching occurs. This is shown by the calculated magnetization curves in Figure 23.

In the special case of $\psi = 90°$ the magnetization curve is degenerated to a straight line $M_H = M_s \cdot H/H_0$. In the second special case $\psi = 0°$ the magnetization curve is a rectangle with a width H_0. In reality, the switching often occurs at fields lower than H_0 because of the nucleation of antiparallel magnetized domains.

If there is an angle θ between the easy direction of uniaxial anisotropy and the long ellipsoid axis, the minima of W_T can be calculated by using a new coordinate system x', y' rotated by an angle δ from x, y. The modified characteristic field is

$$H_0 = \sqrt{H_d^2 + H_k^2 + 2H_d H_k \cos 2\theta} \tag{2.9}$$

and the transcendent equation for δ is

$$\tan 2\delta = \frac{H_k \sin 2\theta}{H_d + H_k \cos 2\delta} \tag{2.10}$$

In addition to the special cases of $\theta \approx 0°$ and $\theta \approx 90°$, where the characteristic field is the sum or the difference of H_d and H_k, respectively, the case of $\theta = 45°$ is of interest ($\tan 2\delta = H_k/H_d$ and $H_0 = \sqrt{H_d^2 + H_k^2}$). Much more complex calculations have to be done in the case of cubic anisotropy and three-dimensional field application. Solutions for the total energy minima can be found by numerically tracing the minima of the energy areas, shown in the figures of Section 1.2.2.2.

Finally, the considerations of coherent magnetization rotation can be applied in analogy to thin, ellipsoidal wires in the single-domain state. The easy axis (a) is parallel to the longitudinal direction of the wire, and the hard axes ($b = c$) are perpendicular (in the absence of magnetocrystalline anisotropy). The resulting magnetization curves correspond to those shown in Figure 23.

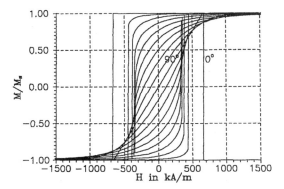

Fig. 23. Magnetization curves $M_H(H)$ of a thin film depending on the angle between $\vec{H}$ and the easy axis.

2.3. Surface Anisotropy and Interface Anisotropy

Magnetic anisotropy in ultrathin films is a subject that has recently attracted a considerable amount of work, from both the experimental and theoretical points of view. The following considerations are mainly based on a review by Givord et al. [35].

2.3.1. Surface Anisotropy

It is well known that a surface/interface anisotropy that is more than an order of magnitude larger than in the bulk may occur in cubic transition metal thin films [36]. In cases where the anisotropy has the appropriate sign and is strong enough, perpendicular magnetization may be established, which is of importance for potential applications.

From an experimental point of view, it is difficult to determine the magnitude of surface/interface anisotropy because other anisotropy terms contribute (e.g., bulk, magnetoelastic, or shape anisotropy). To analyze experimental results an obvious first step is to distinguish between surface and volume contributions. Gradmann and Mueller [37] expressed the measured anisotropy energy density K_t between the film plane and the perpendicular axis as

$$K_t = \frac{2K_s}{d} + K_v \tag{2.11}$$

where K_s is the surface/interface anisotropy (expressed in units of energy per unit area), K_v is the volume anisotropy, and d is the film thickness. By measuring K_t for films of different d, Gradmann and Mueller found that K_t varies linearly as a function of $1/d$, from which they deduced the values of K_v and K_s.

Later, Chappert and Bruno [38] pointed out that the value of K_s thus obtained might not be purely a surface term but might contain a volume magnetoelastic term. Indeed, they showed that the epitaxial strain ϵ of a film is expected to relax proportionally to $1/d$ above a certain critical thickness d_c, and thus the corresponding magnetoelastic term also should vary proportionally to $1/d$. For $d < d_c$, however, ϵ is expected to be constant, and thus the magnetoelastic term should manifest itself as a volume term. In agreement with this, changes in the slope of K_t versus $1/d$ have been observed in several systems at low film thicknesses [39]. In particular, a recent and very detailed study of Ni films deposited on Cu was performed by Jungblut et al. [40], who showed that the change in the slope of $K_t d$ versus d occurs at the thickness at which ϵ becomes constant.

However, all of the above is based on an inherent simplifying assumption that K_v and K_s are thickness independent. This is questionable because the magnetic state of a thin film is thickness dependent. This can be seen by measuring the thickness dependence of the Curie temperature or of the magnetization at a fixed temperature [41].

2.3.2. Interface Anisotropy

2.3.2.1. Phenomenology

In a localized model of magnetism, the thermal variation of anisotropy is due to the thermal disorder of rigid moments. As discussed by Callen and Callen [44], the different constants describing anisotropy (magnetocrystalline and magnetoelastic) are in fact functions of m, the reduced magnetization ($m = M_s(T)/M_s(0)$), at the considered temperature. The strength of the various volume anisotropy terms in a thin film may then be calculated by assuming (i) that at 0 K the various terms have the same values as in the bulk and (ii) that at a finite temperature the value of each term is identical to the bulk value for the same m (but not the same temperature). This leads to

$$K_f(m) = K_b(m) \qquad (2.12)$$

where K_f is a volume anisotropy term for the film (magnetocrystalline, magnetoelastic, or shape) and K_b is the corresponding bulk term. The implication of Eq. (2.12) is obvious: the reduction of m in a thin film due to its finite thickness leads to a corresponding thickness dependence of the anisotropy constants.

The second approach to the evaluation of K_s explicitly accounts for the fact that the thermal variation of anisotropy in itinerant electron (transition metal) systems is determined by both collective excitations (spin waves) and Stoner excitations. A simple description of the thermal variation of anisotropy ($K(T)$) in these systems has been proposed by Ono [45]:

$$K(T) = (K(0) + AT^2)f_{sw}(m) \qquad (2.13)$$

where $K(0)$ is the 0 K anisotropy, AT^2 accounts for the variation in anisotropy due to Stoner excitations, and $f_{sw}(m)$ accounts for the decrease in anisotropy due to thermally excited spin wave excitations. Band structure calculations for thin films [46] suggest that surface perturbations are short ranged, and the film recovers a bulk-like band structure after a monolayer. To a first approximation, one can therefore assume that $K(0)$ and A, which are characteristic of a volume anisotropy term in a thin film, are identical to bulk values. From Eq. (2.13), it is then deduced that at a given T the ratio between a volume anisotropy term in a film ($K_f(T)$) and its counterpart in the bulk ($K_b(T)$) is

$$\frac{K_f(T)}{K_b(T)} = \frac{f_{sw}(m_f)}{f_{sw}(m_b)} \qquad (2.14)$$

where m_f and m_b are the reduced magnetizations of the film and bulk at temperature T, respectively. To a good first approximation for an anisotropy constant of lth order, $f_{sw}(m) = m^{l(l+1)/2}$ [44], thus leading to

$$K_f^l(T) = K_b^l(T)\left(\frac{m_f}{m_b}\right)^{l(l+1)/2} \qquad (2.15)$$

where K_f^l and K_b^l are a film volume and corresponding bulk anisotropy constant of lth order, respectively.

The implication of Eq. (2.15) is similar to that of Eq. (2.12) (i.e., K_f is reduced with respect to K_b), but now K_f is derived from K_b at the same temperature and multiplied by a thickness dependent reduction factor. From [44], it can also be inferred that Eq. (2.15) applies to magnetoelastic constants. For shape anisotropy, a similar relation is obtained where $l(l+1)/2$ is replaced by 2.

2.3.2.2. Electronic Origin

Bruno [42] and Wang et al. [43] have shown that the magnetocrystalline anisotropy of transition metal free monolayers may be understood on the basis of relatively simple considerations derived from examination of their band structure. In this section we extend this discussion to the case of interface anisotropy between a magnetic transition metal M (Fe, Co, or Ni) and a nonmagnetic element X.

Let us first recall the main arguments that apply to the monolayer case. The d-band may be schematically separated into two subbands, formed of the in-plane orbitals (xy, $x^2 - y^2$) and of the out-of-plane orbitals (xz, yz, $3z^2 - r^2$), respectively. Because out-of-plane orbitals cannot form bonds directed toward the surface, the corresponding subband is narrower than in the bulk. In contrast, the in-plane states and, by consequence, the corresponding subband, are essentially unaffected from the bulk to the monolayer. In addition, the out-of-plane states subband is shifted toward higher energy with respect to the in-plane states subband because of the increased electrostatic repulsion for more localized states. The sign of the magnetocrystalline anisotropy as a function of the band filling can then be derived by considering which symmetry states are dominant in the vicinity of the Fermi level. Assuming strong ferromagnetism, only states in the minority band have to be considered. For low band filling the anisotropy is perpendicular as in-plane orbitals dominate. As band filling increases, the contribution of out-of-plane orbitals increases and the anisotropy becomes in-plane. For further filling, the anisotropy decreases and vanishes when the band is full. The results are shown schematically in Figure 24, which reproduces the calculated anisotropy of [47, 48]: Fe (very weakly perpendicular), Co (in-plane), and Ni (weakly in-plane) monolayers.

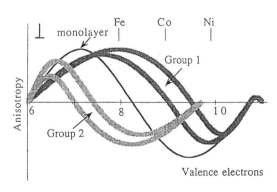

Fig. 24. The sign of free-standing monolayer anisotropy as a function of minority 3d band-filling and the corresponding modifications for interfacing with late transition metals (group 1) and early transition metals (group 2). Reprinted with permission from [35], © 1996, Elsevier Science.

The approach to the anisotropy of M/X multilayers is to consider the modifications that must be made in the above picture when interface anisotropy is discussed. As above, strong ferromagnetism is assumed. In the vicinity of E_F, s↑, s↓, and d↓ are present. It is assumed, as a general rule, that the in-plane orbitals do not depend on the nature of the interface, and only out-of-plane orbitals are important in this discussion [49].

Let us consider the case where X is a late transition element (referred to in the following as group 1: Cu, Ag, Pd, Pt, etc.). At the interface, hybridization occurs between the low-lying s↓X and the s↓M states, which are located in a band crossed by the Fermi level. Through this process, the low-lying energy X states take some M component, and the M states around E_F take some X component. The number of M out-of-plane states around the Fermi level thus tends to be reduced, whatever the exact value of E_F may be. This tends to favor perpendicular anisotropy. In addition, the hybridized M↓ states are fully occupied because they are well below E_F. The total number of electrons to be distributed in the bands that are crossed by E_F (s↑, s↓, and d↓) is thus reduced. This reduction of the number of electrons is equal to the number of states that have been removed by hybridization from the unhybridized Md↓ band. Because the hybridized d↓ band near E_F is not full, the total number of states (s↑, s↓, and d↓) under E_F before hybridization is relatively less reduced than the number of electrons to be distributed. This effect is equivalent to removing some electrons from these bands, and this leads to the modified band filling dependence of anisotropy, as shown in Figure 24. Note that the total number of d↓ electrons is increased in this process because of redistribution between s and d bands. This is in agreement with band structure calculations [49]. The total number of valence electrons is assumed to be constant, however. This is also consistent with band structure calculations which indicate that electron transfer is small [50].

We turn now to the case where X is an early transition element (referred to in the following as group 2: Sc, Ti, Y, rare earth, etc.). The X d-states are now above the transition metal states. The out-of-plane states near E_F take some X component through hybridization, and the high-energy states take some M component. The number of out-of-plane states near E_F is reduced, and there is again a tendency for perpendicular anisotropy to be favored. The M states created by hybridization are now above E_F, and they remain unoccupied. The number of M electrons to be distributed in the bands that are crossed by E_F thus remains the same, whereas the number of available states below E_F is reduced. This is equivalent to filling the minority spin bands around the M atoms. The band filling dependence of anisotropy then obtained is shown in Figure 24. Note that, in this case, the E_F of X is initially well above that of M. As E_F crosses the d-bands of both X and M, the electron transfer is expected to be greater than for group 1 elements. This will increase band filling effects and thus reinforce the conclusions drawn above.

This very simple model is surprisingly successful in describing the anisotropy of X/M interfaces when X is a group 1 element. Perpendicular anisotropy at Fe/X interfaces is pre-dicted. This is in agreement with experimental observations for X = Ag, Au, Cu, and Pd [51]. The same tendency is predicted for Co/X interfaces. Comparison with experiment is difficult because of strain effects; however, perpendicular effective surface anisotropy is observed for X = Au, Cu, Pd, and Pt [51].

For interfaces where X is a group 2 element, in-plane anisotropy is predicted for both Fe and Co. Unfortunately, very few experimental data as well as calculations are available for these systems. It would be of special interest in this context to examine the anisotropy of M/Se, M/Ti, and M/Y interfaces. Note that in-plane anisotropy is effectively found at Fe/W interfaces [51]; however, it is not clear that W may be considered as a group 2 element, because its d-band is approximately half full.

For the case of Ni, in-plane anisotropy is predicted for both X = group 1 and 2. This is in agreement with experimental findings [51]. It appears, in conclusion, that the sign of anisotropy is correctly described by this very simple model for most experimentally known systems. It must be stressed, however, that the above model is, of course, only qualitative. In particular, the splitting between bonding and antibonding states was neglected in the discussion, which restricts the validity of the approach to the cases where the X d-band is almost full or almost empty. In addition, the energy shifts that result from hybridization for states of different symmetries are not taken into account, and the conclusions drawn are certainly not exact in detail.

2.4. Exchange Anisotropy

Exchange anisotropy plays an important role in magnetic multilayers used for modern magneto-electronic and data storage applications. This section covers both experimental and theoretical considerations [64] (see also [25–33]).

2.4.1. Experimental Observations

If different magnetic materials (e.g., ferromagnetic and antiferromagnetic) are in intimate contact with each other or are separated by a layer thin enough to allow spin information to be communicated between the two materials, then exchange coupling between these materials occurs and exhibits the phenomenon of exchange anisotropy (in contrast to magnetostatic coupling which can arise from interfacial stray fields due to roughness of the boundary). This effect was first observed for fine cobalt particles with a 20-nm diameter that were covered by a thin layer of antiferromagnetic CoO [25]. This results in an M/H characteristic displaced by more than 80 kA/m (Fig 25).

This field displaced loop was described to as exhibiting exchange anisotropy as a result of exchange coupling between the moments of Co and CoO. The vertical scale of the M/H loop of the sample reflects only the cobalt magnetization because the antiferromagnetic CoO is essentially unmagnetized at weak fields. The interfacial exchange coupling depends on the submicroscopic details of the interface.

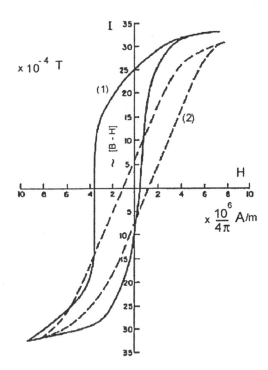

Fig. 25. Hysteresis loops at 77 K of partially oxidized Co particles. Curve (1) shows the resulting loop after cooling in a 80 kA/m field. Curve (2) shows the loop when cooled in a zero field. Reprinted with permission from [64], © 1999, Elsevier Science.

Antiferromagnetic (AFM) materials, such as CoO, magnetically order below their Néel temperatures, T_N, with the spins of the Co^{2+} cations parallel to each other on (111) planes, and with antiparallel spin directions in alternate (111) planes (i.e., zero net moment). There are localized net moments at the interface with a ferromagnetic material (FM), which arise from several sources. The most obvious case is where there are AFM grains with parallel spin planes at the interface, as depicted in Figure 26. Alternatively, even in AFM grains with compensated interfacial spin planes, there can be unequal numbers of parallel and antiparallel spins at the surface of the grain, due to grain size, shape, or roughness. These various origins of localized net AFM moments are depicted in Figure 27. Because the FM is ordered just below its Curie temperature, T_C, which is greater than the T_N of the AFM, a field applied to coupled FM-AFM systems at $T > T_N$ will align the FM magnetization in the field direction, and the AFM spins remain paramagnetic. As the temperature is lowered through T_N, the ordering net localized AFM spins will couple to the aligned FM spins, sharing their general spin direction. For high AFM magnetocrystalline anisotropy, they will not be substantially rotated out of their alignment direction by fields applied at temperatures below T_N if the interfacial AFM spins are strongly coupled to the AFM lattice. This is also a consequence of the fact that the generally compensating antiparallel arrangement of the AFM spins does not result in a strong torque on the spin system when a field is applied (i.e., low susceptibility). Because the localized uncompensated AFM spins are coupled to FM spins at the interface, they exert a strong

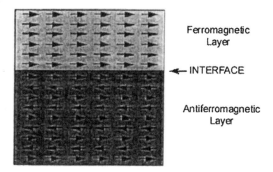

Fig. 26. Schematic of the ideal FM/AFM interface. The FM and AFM layers are single crystal and epitaxial with an atomically smooth interface. The interfacial AFM spin plane is a fully uncompensated spin plane. For this ideal interface, the calculated value of the full interfacial energy density is about two orders of magnitude larger than the experimentally observed values. Reprinted with permission from [64], © 1999, Elsevier Science.

torque on these FM spins, tending to keep them aligned in the direction of the cooling field, i.e., a unidirectional anisotropy. This model explains the shift for the field-cooled particles in curve (1) in Figure 25. There is still a unidirectional bias in each particle when the particles are cooled in a zero field, but there is no net bias because the moments of the particles are randomly arranged, and curve (2) in Figure 25 is symmetric.

2.4.2. Models and Theories

During the past several decades, the pace of activity involving FM-AFM exchange couples has greatly increased. The reason for this is that the effective bias field, H_E, on a thin FM film produced by an interfacial exchange with an AFM

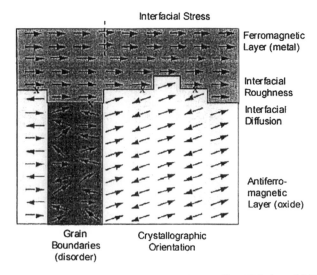

Fig. 27. Interfacial complexities of a polycrystalline FM (metal)/AFM (oxide) interface. In this figure, the interfacial spins prefer to align ferromagnetically. The X marks identify the frustrated exchange bonds, i.e., the interfacial spins that are coupled antiferromagnetically. The interfacial region can undergo a high degree of stress because metals and oxides often have very different lattice parameters. Dislocations (represented by the dashed line) can form during film growth to relieve the stress. Reprinted with permission from [64], © 1999, Elsevier Science.

film has been found to be extremely useful in the information storage industry. Digital data in current high-density magnetic storage disks are sensed by read heads that employ thin FM film devices whose resistances vary with the magnitude and direction of the stray fields above the stored bits 52. This is the phenomenon of magnetoresistance (MR). To linearize the bipolar MR signal, and to minimize the noise produced by discontinuous jumps of domain walls (Barkhausen noise), the FM films must be biased by a magnetic field. Coupling the FM film to an AFM film provides a substantial fraction of the required bias in all current high-density read heads. All of the initial AFM films used for this purpose were metallic Mn alloys, particularly Fe-Mn. Recently, read heads using more robust insulating AFM bias films such as NiO have appeared. By the mid-1970s virtually all of the most significant static and dynamic features of FM-AFM exchange coupling had been recognized, and plausible qualitative models had been developed for a number of issues. What was lacking, and is still missing, is adequate characterization of the chemical states and the exchange interactions at the FM/AFM interface on an atomic scale. There can be no rigorously reliable explanations for the salient features of exchange anisotropy in FM-AFM couples until this extremely complex information becomes available [53–61].

2.4.2.1. Ideal Interfaces

The first simple model of exchange anisotropy examined exchange coupling across an ideal interface as shown in Figure 26. The FM and AFM layers are both single crystalline and epitaxial across an atomically smooth FM/AFM interface. AFM materials have no net moment. The AFM monoxides are composed of atomic planes of ferromagnetically oriented spins with antiparallel alignment between adjacent planes. In Figure 26, the interfacial AFM spin plane is fully uncompensated. During the reversal of the FM magnetization in this ideal model, the spins of the FM layer rotate coherently, and the spins of the AFM layer remain fixed. The energy cost is equal to the interfacial exchange energy. The phenomenological formula of the exchange field is

$$H_E = \frac{\Delta\sigma}{M_{FM}d_{FM}} = \frac{2J_E\mathbf{S}_{FM}\cdot\mathbf{S}_{AFM}}{a^2 M_{FM}d_{FM}}, \qquad (2.16)$$

where $\Delta\sigma$ is the interfacial exchange energy per unit area, J_E is the exchange integral, $\mathbf{S}_{FM}$ and $\mathbf{S}_{AFM}$ are the spins of the interfacial atoms, a is the cubic lattice parameter, and d_{FM} is the thickness of the ferromagnetic layer. In the $Ni_xCo_{(1-x)}O$ system, an atomically smooth (111) oriented single crystal should result in a fully uncompensated interfacial spin plane, as depicted in the ideal model. Therefore, one may expect a FM layer coupled to a (111) oriented $Ni_xCo_{(1-x)}O$ single crystal to exhibit an exchange field magnitude as predicted by Eq. (2.16). Using reasonable parameters, one obtains values of $\Delta\sigma \approx 10\ J/m^2$. As discussed above, the observed exchange fields for single-crystal and polycrystalline AFM films of all types are 2–3 orders smaller. Clearly, this simple ideal

model does not realistically represent the FM/AFM interfacial environment. Phenomena such as interfacial contamination or roughness have been invoked to account for the reduction of interfacial coupling strength. Roughness in the form of interfacial atomic steps could produce neighboring antiparallel spins and thereby reduce the number of interfacial uncompensated spins. Other experimental studies have raised other questions. If only the (111) spin planes are fully uncompensated, one may expect that single crystals of any other crystalline orientation would exhibit a small or no exchange field because of a compensated interfacial spin plane. Polycrystalline films are composed of subunits or grains that possess a distribution of orientations, predominantly not (111). Yet FM films coupled to polycrystalline AFM films often have higher exchange fields than FM films coupled to single crystal (111) films. Perhaps the uncompensated spins originate from the disordered regions within the grain boundaries. These are simply qualitative explanations and do not provide for quantitative analysis and comparison with experimental observations. A model of the exchange bias mechanism must resolve the following discrepancies and questions:

1. What structural and magnetic parameters are responsible for the drastic reduction in the interfacial exchange energy density from the ideal case?
2. What are the origin and role of the interfacial uncompensated AFM spins?
3. How does the magnitude of the exchange field depend upon the AFM grain structure?
4. What determines the temperature dependence of the exchange field?
5. What are the roles of the interfacial exchange integral J_E and the AFM magnetocrystalline anisotropy K_{AF} in unidirectional anisotropy?

In the past decade, a number of models and theories have been proposed to provide qualitative and quantitative descriptions of the exchange biasing mechanism.

2.4.2.2. Interfacial AFM Domain Wall

To explain the discrepancy between the exchange field values predicted by simple theory and experimental observations, Mauri et al. [62] proposed a mechanism that would effectively lower the interfacial energy cost of reversing the FM layer without removing the condition of strong interfacial FM/AFM coupling. They proposed the formation of a planar domain wall at the interface with the reversal of the FM orientation. The domain wall could be either in the AFM or FM, wherever the energy is lower. They examined the case where the domain wall forms at the AFM side of the interface. With the magnetization reversal of the FM layer, the increase in interfacial exchange energy would be equal to the energy per unit area of an AFM domain wall $4\sqrt{A_{AF}K_{AF}}$, where A_{AF} and K_{AF} are the exchange stiffness $\approx J_{AF}/a$ (where J_{AF} is the AFM exchange integral parameter, and a is the AFM lattice parameter) and

AFM magnetocrystalline anisotropy, respectively. Thus, the modified expression for the exchange field would be

$$H_E = 2\frac{\sqrt{A_{AF}K_{AF}}}{M_{FM}d_{FM}} \qquad (2.17)$$

By spreading the exchange energy over a domain wall width $\approx \pi\sqrt{A_{AF}K_{AF}}$ instead of a single atomically wide interface, the interfacial exchange energy is reduced by a factor of $\pi\sqrt{A_{AF}K_{AF}}/a \approx 100$, which would provide the correct reduction to be consistent with the observed values. Smith and Cain [63] fitted the magnetization curves of permalloy/TbCo bilayer films with a similar model with strong interfacial coupling. A numerical calculation over a range of interfacial exchange energies yields the following two limiting cases:

Strong interfacial coupling:

$$H_E = -2\left(\frac{\sqrt{A_{AF}K_{AF}}}{M_S d_{FM}}\right) \qquad (2.18)$$

Weak interfacial coupling:

$$H_E = -\left(\frac{J_E}{M_S d_{FM}}\right) \qquad (2.19)$$

where J_E is the effective interfacial coupling energy (exchange integral), and d_{FM} and M_S are the thickness and saturation magnetization of the FM, respectively. In the case of strong interfacial exchange coupling ($J_E \gg \sqrt{A_{AF}K_{AF}}$), H_E saturates at energies far less than in the case of fully uncompensated interfacial exchange coupling because of the formation of an interfacial AFM domain wall. In the case of weak interfacial exchange coupling ($J_E \ll \sqrt{A_{AF}K_{AF}}$), H_E is limited by the strength of the interfacial exchange coupling J_E. The model offers no insight into the origins of the reduced interfacial coupling, and it does not address the origin of the coupling with compensated spin planes. It highlights the formation of an AFM planar domain wall in the limit of strong interfacial exchange; however, most experimental studies indicate the presence of weak interfacial exchange. Thus, this model does not shed light on the mechanism responsible for the reduced interfacial exchange coupling energy density [64].

2.4.2.3. Random Field Model

Malozemoff [65, 66] rejects the assumption of an atomically perfect uncompensated boundary exchange. He proposed a random-field model in which an interfacial AFM moment imbalance originates from features such as roughness and structural defects. Thus, the interfacial inhomogeneities create localized sites of unidirectional interfacial energy by virtue of the coupling of the net uncompensated moments with FM spins. For interfacial roughness that is random on the atomic scale, the local unidirectional interface energy σ_L is also random, i.e.,

$$\sigma_L = \pm\frac{zJ_E}{a^2} \qquad (2.20)$$

where z is a number of order unity. The random-field theory argues that a net average nonzero interfacial energy will exist, particularly when the average is taken over a small number of sites. Statistically, the average σ in an area of L^2 will decrease as $\sigma \approx (\sigma/\sqrt{N})$, where $N = (l^2/a^2)$ is the number of sites projected onto the interface plane. Given the random field and assuming a single-domain FM film, the AFM film will divide into domain-like regions to minimize the net random unidirectional anisotropy. Unlike in Mauri's model, the AFM domain walls in Malezemoff's model are normal to the interface. Initiation of the AFM domain pattern occurs as the FM/AFM bilayer is cooled through T_N. Although the expansion of the domain size L would lower the random field energy, the in-plane uniaxial anisotropy energy K in the AFM layer will limit the domain size. Anisotropy energy confines the domain wall width to $\pi\sqrt{A/K}(< L)$ and creates an additional surface energy term of the domain wall $4\sqrt{AK}$ (surface tension in a bubble domain). The balance between exchange and anisotropy energy is attained when $L \approx \pi\sqrt{A/K}$. Therefore, the average interfacial exchange energy density $\Delta\sigma$ becomes

$$\Delta\sigma = \frac{4zJ_E}{\pi aL}. \qquad (2.21)$$

Accordingly, the exchange field due to the interfacial random-field energy density is

$$H_E = \frac{\Delta\sigma}{2M_{FM}t_{FM}} = \frac{2z\sqrt{AK}}{\pi^2 M_{FM}t_{FM}} \qquad (2.22)$$

using the equilibrium domain size expression for L. This equation is very similar to the strong interfacial exchange case (Eq. (2.18)) of Mauri's planar interfacial AFM domain wall model. Thus, quantitatively, the random-field model also accounts for the 10^{-2} reduction of the exchange field from the ideal interface model case. In addition to the reduction of the interfacial exchange energy, Malozemoff's model attempts to explain the following magnetic behavior associated with exchange anisotropy: (1) approximately linear fall-off $(1 - T/T_{crit})$ of the exchange field, (2) the magnetic training effect [67] (as the unwinding of the domain state and the annihilation of domains with field cycling), and (3) critical AFM thicknesses for exhibiting an exchange field. However, Malozemoff's model is specifically formulated for single-crystal AFM systems and does not clearly propose how the model can be extended to polycrystalline systems.

2.4.2.4. "Spin-Flop" Perpendicular Interfacial Coupling

The "spin-flop" model introduced by Koon [68] presents a novel solution to a particular problem of exchange biasing. On the basis of his micromagnetic numerical calculations, Koon proposed the existence and stability of unidirectional anisotropy in thin films with a fully compensated AFM interface. His calculations indicate the stability of interfacial exchange coupling with a perpendicular orientation between the directions of the FM and AFM axes, which is a very specific system. The perpendicular interfacial coupling is referred

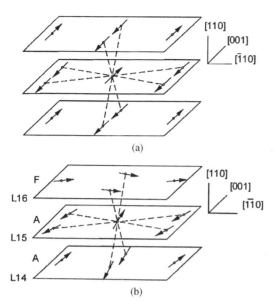

Fig. 28. (a) Magnetic structure of a (110) oriented AFM body-centered tetragonal crystal. The exchange bonds are represented by the dashed lines. (b) Lowest energy spin configuration near the interface plane. The interfacial AFM plane (L15) is fully compensated, and the interfacial FM plane (L16) is oriented perpendicularly (90° coupling). The angles are approximately to scale. Reprinted with permission from [64], © 1999, Elsevier Science.

to as "spin-flop" coupling. To observe perpendicular interfacial coupling, Koon's model specifies the structure and orientation of the AFM layer and the relative orientation between the AFM and FM layer. The model utilizes a single-crystal body-centered tetragonal (BCT) AFM structure as shown in Figure 28a. The BCT structure can be oriented to have a fully uncompensated interfacial spin plane (100) or a fully compensated interfacial spin plane (1 1 0) (shown in Fig. 28a). Koon included uniaxial anisotropy in the AFM crystal along the [001] axis, and the FM layer was modeled with no intrinsic anisotropy.

Koon applied his model to two different cases of the AFM interfacial spin plane: (1) a fully compensated interface and (2) a fully uncompensated interface. For both cases, he calculated the interfacial energy density as a function of the angle between the FM spins and the Néel axis of the AFM spins. The fully uncompensated interface gives the expected results of collinear coupling, a minimum at $\theta = 0°$. The fully compensated interface gives the surprising result of an energy minimum at $\theta = 90°$, indicating perpendicular interfacial coupling between the FM and AFM spins. Figure 28b shows the spin configuration near the interface plane, illustrating the perpendicular orientation of the FM and AFM axes.

The recent work by Schulthess and Butler [69] yielded findings contrary to Koon's calculations. They found an interfacial spin-flopped state similar to Koon's but had contrary conclusions with respect to exchange biasing. Their calculations indicated enhanced uniaxial anisotropy or enhanced coercivity, but not a shifted magnetization curve.

2.4.2.5. Uncompensated Interfacial AFM Spins

The experimental correlation between the interfacial uncompensated CoO spins and the exchange field in polycrystalline CoO/permalloy bilayer films was demonstrated by Takano et al. [70, 71]. They measured the uncompensated spins on the surfaces of antiferromagnetic CoO films as a thermoremanent magnetization (TRM) after field cooling of a series of CoO/MgO multilayers from $T > T_N$. The temperature dependence of the TRM was similar to the temperature dependence of the sublattice magnetization for bulk CoO materials as determined by neutron diffraction [72]. These measurements showed that the spins responsible for this uncompensated AFM moment are interfacial and constitute $\approx 1\%$ of the spins in a monatomic layer of CoO. This low density is consistent with the low exchange fields observed compared with the values predicted by the ideal interface model (see Eq. (25)). The uncompensated AFM TRM also has the same temperature dependence as the exchange field of $Ni_{81}Fe_{19}/CoO$ bilayers after field cooling. Thus, the uncompensated interfacial AFM spins appear to be the spins responsible for unidirectional anisotropy. Takano et al. determined that a linear relationship exists between the strength of the exchange field and the inverse of the CoO crystallite diameter, i.e., $H_E \propto L^{-1}$, where L is the grain diameter. This suggests a structural origin for the density of uncompensated spins. The magnitude and the temperature dependence of the exchange field in FM/CoO bilayers correlate with the density of uncompensated interfacial CoO spins. Takano et al. calculated the density of interfacial uncompensated spins as a function of grain size, orientation, and interfacial roughness of polycrystalline AFM films. Each CoO crystallite was assumed to be a single AFM domain. Figure 29 is a view of the interfacial cross section of an AFM crystal showing that the orientation of the AFM determines the periodicity with which the (111) ferromagnetic spin planes intercept the interface.

Figure 30a is a plan view of the same AFM crystal. The crystalline orientation is reflected in the periodic alternating pattern of four rows. In Figure 30a, an elliptical sampling region simulating a grain with a major axis length L has been

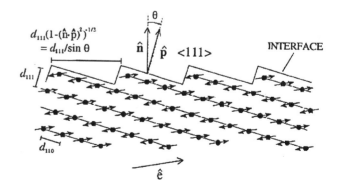

Fig. 29. Schematic of an interface cross section. The film normal is $\hat{n}$, $\hat{p}$ is the normal to the parallel spin plane (111) of the AFM, and $\hat{e}$ is the AFM spin axis (in this case $N_{rows} = 4$). Reprinted with permission from [64], © 1999, Elsevier Science.

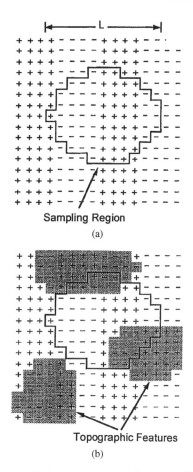

Sampling Region

(a)

Topographic Features

(b)

Fig. 30.   (a) Topographical representation of the interfacial plane: periodic pattern of $N_{rows} = 4$ (as in Fig. 29) with a sample region representing a model crystallite. (b) Topographical representation of the interfacial plane. Eliptical "islands" of monatomic layer thickness were superimposed on the spin map to simulate roughness. Note that adding one atomic layer reverses the direction of the underlying spin. Reprinted with permission from [64], © 1999, Elsevier Science.

superimposed on the spin map. The number of uncompensated spins (ΔN) for a model crystallite was computed by simply adding the total number of spins in each direction contained within the boundaries of the model grain. The fundamental origin of uncompensated AFM moment is scale. Although the large spin map may represent compensated spin regions, one observes small densities of uncompensated spins when sampling small areas within the spin map. Thus, one will observe an exchange bias with polycrystalline AFM films regardless of whether the preferred orientation suggests that the interfacial plane is completely compensated. Interfacial roughness was incorporated into the spin maps by the superimposition of elliptical "islands" of monatomic thickness on the spin map. The effect of adding one atomic layer is to reverse the direction of the spin at each site covered by the island, because successive layers of CoO spins have opposite directions (see Fig. 30b). Takano et al. created a series of spin maps several times larger than the model grain sizes to be sampled. Statistical averages were obtained from sampling 10^6 model crystallites by varying the position, orientation of the major axis, and aspect ratio of the model crystallite.

The following are two primary results of these calculations: (1) a perfectly regular interface without any roughness features results in $(\Delta N) \propto L^{0.5}$; (2) the addition of roughness results in $(\Delta N) \propto L^{0.90-1.04}$. Because the exchange field is proportional to $(\Delta N)/L^2$, the rough case gives $H_E \propto L^{-1}$, in agreement with the experimental results. Using realistic and experimental values, the observed exchange field for a 10-nm CoO biasing film was consistent with an interfacial roughness of only a few "extra" atomic steps across the face of each crystallite. Thus, the model correctly predicts the inverse dependence of the uncompensated spins on grain size and the correct magnitude of H_E. The model indicates that the origins of the uncompensated AFM interfacial moment are (i) the dimensions of the AFM grain boundaries and (ii) the presence of interfacial roughness features. The origin of the exchange biasing mechanism is indeed the uncompensated interfacial AFM spins. The magnitude and temperature dependence of the exchange field can be explained from the measurements of the magnetic properties of the uncompensated AFM spins. The model has demonstrated the structural origins of the uncompensated AFM moment in polycrystalline films [73–80]

2.5. Domain and Domain Wall Configuration

Magnetic domains are regions where the magnitude of the magnetization is constant, but the magnetization direction varies from one domain to the other to reduce the stray field inside and outside of a magnetic structure. This reduction in energy associated with long-range magnetostatic interactions is balanced by the cost in the short-range exchange energy of forming domain walls, which typically have dimensions on the order of a tenth of a micrometer. Domain walls may also have an internal structure (e.g. singularities) that depends on specific properties of the magnetic system, such as film thickness. The domain structure is determined by minimizing the total energy, which consists primarily of contributions from the exchange, anisotropy, self-field, applied field, and magnetostriction. As an extension of the subsections on domain structure and the magnetization process in Section 1, this section will give some details of characteristic domain configurations and domain wall dynamics.

2.5.1. Passage of a 180° Domain Wall

Understanding the nucleation and propagation of a domain wall in the presence of defects, impurities, or inhomogeneities is of crucial importance for applications of magnetic materials. Indeed, these phenomena are known to be responsible for the softness and hardness of magnetic materials [81]. On the other hand, the behavior of domain walls is of interest for fundamental magnetism, being simply related to basic quantities such as the exchange or anisotropy energies. Recently, interest in the domain wall has been extended to the physics of low temperatures, where magnetization reversals can be driven by thermal activation or by macroscopic quantum tunneling (QTM) effects [82]. The fundamental problem is how a domain wall interacts with the potential created by "defects."

Will it be trapped and constitute a "bound state," or will it cross the potential? What is the effect of a magnetic field and of the temperature? Is the depinning process thermally activated, or are there quantum tunneling effects?

To carry out an experimental study of such problems, the first step is to prepare samples in which the potential energy of the domain wall can be defined as well as possible with given profiles and magnitudes. As the potentials due to defects are very difficult to characterize and model in bulk materials, Mangin [83] turned to a one-dimensional problem by inducing uniaxial anisotropy in the plane of a film and looking at the domain wall traveling along the direction perpendicular to the plane of the sample. By creating a trilayer system in which the inner layer is thin and magnetically different from the other two layers, a planar potential barrier for domain walls was constructed. The two materials used to make a sandwich of a thin layer of material B deposited between two layers of material A have their own uniaxial anisotropy constants K_A and K_B, respectively, and a close exchange integral J. As the potential energy of a domain wall can be written as $4(JK)^{1/2}$, the thin layer of material B constitutes a potential barrier for the domain wall if $K_B > K_A$. In that case, the A-B-A trilayer is called a domain wall junction [82].

As has been shown by Gunther and Barbara [82], two extreme situations have to be considered: (i) The width δ of the domain wall is small compared with the thickness e of the B layer. In this case, the barrier width is close to e. (ii) The width of the domain wall is large compared with e. In that case, the spatial extension of the potential barrier is approximately equal to δ. Moreover, the potential energy becomes $U(x) = U_0(x) - \mu_0 M_s H_x$ when a field is applied, where $U_0(x)$ is the potential energy of the barrier when no field is applied. This is a very important point because it leads to a modification of the energy barrier profile when a field H is applied.

Mangin [83] presented experimental results showing the magnetic and thermal dependence of the crossing of a barrier corresponding to the second situation. In his experiments, the barrier is made of a few atomic layers of TbFe, an amorphous alloy deposited between two amorphous GdFe layers of different thicknesses e_1 and e_2. GdFe is a soft material in which the local anisotropy is weak because of the s character of the 4f electron cloud of gadolinium. TbFe is a hard material in which the local anisotropy is large because of the highly anisotropic distribution of 4f electrons in terbium. As shown in [84], the GdFe layer was deposited under appropriate conditions, providing an easy anisotropy axis in the plane of the sample, which leads to a 180° domain wall.

Amorphous GdFe alloy is known to be ferrimagnetic and amorphous TbFe alloy is sperimagnetic.[12] The exchange between iron atoms is positive and large, that between rare earth atoms is weak, and the heavy rare earth–Fe exchange is negative and fairly large. The ordering is dominated by the

Fe-Fe exchange in the film. As the magnetic moment carried by rare earths is large, the magnetization of the rare earth is dominant in $Gd_{60}Fe_{40}$ and $Tb_{45}Fe_{55}$ amorphous alloys. These compositions were chosen to obtain a Curie temperature close to room temperature for an easy demagnetization of the sample if needed.

2.5.1.1. Experimental Procedure

The samples were obtained by coevaporation in a high-vacuum evaporation chamber. The elements were evaporated from separate sources. Substrates were glass plates on which a thin 100-Å amorphous silicon layer was first deposited at 420 K just after outgassing of the chamber. During the deposition of the samples the substrates were kept at the temperature of liquid nitrogen. As reported in [84], an in-plane anisotropy axis could develop along the direction at the intersection of the plane of the sample and the plane perpendicular to the axis of the Gd and Fe sources.

The macroscopic magnetization measurements were first performed down to 1.8 K with a commercial SQUID magnetometer. In the 6–0.1 K range, the magnetic behavior of the sample was determined with six planar micro-bridge dc SQUIDs (of 1 μm diameter, similar to those that had previously been used to determine single-particle magnetization processes [85]) placed on the sample. The micro-SQUID loops collect the flux produced by the sample magnetization. Because of the close proximity of the sample and micro-SQUIDs, Mangin had a very efficient and direct flux coupling. He could detect magnetization reversals corresponding to $10^4 \mu_B$. A field with a maximum amplitude of 0.5 T could be applied in the micro-SQUID plane and be oriented at any angle. Sweeps were performed at rates between 0.01 mT/s and 0.1 T/s.

2.5.1.2. Macroscopic Magnetization Measurements

The following magnetization measurement procedure was performed on the sample GdFe(e_1)/TbFe(e)/GdFe(e_2): (i) A 80 kA/m magnetic field was applied at 300 K along the easy magnetization axis to saturate the magnetization of both GdFe and TbFe. (ii) The sample was cooled under this 80 kA/m field to the measurement temperature T and was kept saturated. (iii) The field was rapidly decreased to zero, reversed, and finally increased step by step from $H = 0$ along the opposite direction (it appears as negative in the figures). At each step a magnetization measurement was performed.

Figure 31 shows the evolution of the magnetization recorded from a GdFe(1000 Å)/TbFe(15 Å)/GdFe(500 Å) sample in the $H = 0$ to $H = -8$ kA/m range. The data were collected for different measurement temperatures between 5 and 150 K. Depending on the temperature range, three different types of behavior were observed:

$T > 120$ K: The magnetization switches in one step from M_s to $-M_s$ at a coercive field $H_{c1}(T)$. This field becomes larger when the temperature decreases. The switch extends over less than 400 A/m.

---

[12] Random anisotropy in amorphous alloys can promote spatial dispersion of the local moments, and the additional presence of more than one magnetic species can lead to a rich variety of magnetic structures with dispersed parallel (speromagnetic) or dispersed antiparallel (sperimagnetic) magnetic sublattices.

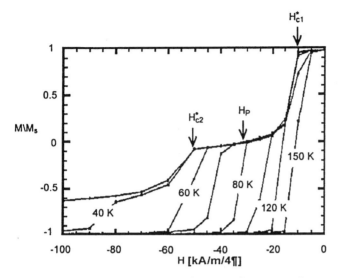

Fig. 31. Magnetization of a GdFe(1000 Å)/TbFe(15Å)/GdFe(500 Å) sample versus field for temperatures ranging from 20 to 150 K. Reprinted with permission from [83], © 1997, Elsevier Science.

$40 \text{ K} < T < 120 \text{ K}$: The switch of the magnetization from $+M_s$ to $-M_s$ that occurred previously at H_{c1} transforms into a decrease of the magnetization, which corresponds to about 45% of $2M_s$. Beyond this field the magnetization decreases slowly up to a second critical field H_P at which the magnetization drops to $-M_s$. It should be noted that H_{c1} is now independent of the temperature and is referred to as H_{c1}^*.

$T < 40 \text{ K}$: As in the previous temperature range, the 45% magnetization drop occurring at a coercive field H_{c1}^* is followed by a decrease of the magnetization. Contrary to the previous case, this slow decrease does not end with a decrease of the magnetization to $-M_s$ at H_P, but a magnetization edge is observed at a field of H_{c2}^* independently of the temperature. Beyond this field there is a damping of the magnetization that extends to -12 kA/m. The fields H_{c1}, H_{c1}^*, H_P, and H_{c2}^* determined on the same sample are shown as a function of temperature in Figure 32. Domains I, II, and III correspond to the temperature ranges presented above.

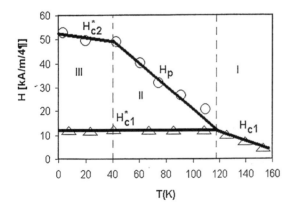

Fig. 32. Thermal evolution of the characteristic fields at which the two drops of magnetization occur in Figure 31. Reprinted with permission from [83], © 1997, Elsevier Science.

The following interpretation of magnetic behavior has been given by Mangin. At high temperatures ($T > 120 \text{ K}$), the reversal of the magnetization first requires the formation of a nucleus in the cross section of the sample. For this, the magnetic moments of the two GdFe layers and of the TbFe layer locally have to reverse simultaneously. The process should be thermally activated. After nucleation, reversal of the magnetization spreads very quickly laterally in the sample, which results in the magnetization switch from M_s to $-M_s$.

When the temperature is lower than 120 K, there is insufficient thermal activation to turn the magnetic moments of the TbFe layer, the anisotropy constant of which is very large compared with that of the GdFe alloy. As a consequence, the nucleus does not cross the whole sample, and the nucleation takes place only in the thicker GdFe amorphous layer. No domain wall appears in the 500-Å GdFe layer because a domain wall whose width should be close to 800 Å in a zero field is larger than the thinner GdFe layer. So, for $H = H_{c1}^*$ a 180° domain wall lies in the 1000-Å GdFe layer, and the direction of the magnetization is kept unchanged in the 500-Å GdFe layer as well as in the TbFe layer.

If the magnetization of the thick (1000 Å) GdFe layer were completely reversed, the magnetization drop would be 66% of $2M_s$. The fact that it is reduced to 45% indicates that there is a wall with a thickness close to 800 Å. In addition, there is a compression effect similar to that previously observed by Dieny et al. [86]. The domain wall crosses the TbFe barrier at H_P and propagates in the 500-Å GdFe layer. This results in the magnetization step to $-M_s$. The phenomenon depends strongly on the temperature and is thermally activated. H_P increases almost linearly when the temperature decreases. This means that the height of the barrier decreases in the same way when the field increases.

When the temperature is lower than 40 K, the domain wall created at H_{c1}^* does not cross the TbFe layer before the nucleation of a new 180° domain wall at H_{c2}^* in the thinner GdFe layer. Indeed, H_{c2}^* is stable with temperature just as H_{c1}^* is, which confirms Mangin's model. The subsequent slow decrease in magnetization characterizes the compression of the two domain walls. From this field, the TbFe is squeezed between two domain walls, and two possibilities have to be considered: in the first, the two walls tend to turn the magnetic moments of the TbFe layer in the same direction. In the second, the walls tend to turn the magnetic moments of TbFe in opposite directions, which stabilizes their orientations (topological defect). Mangin believes that the two geometrical situations coexist in the system, and, because of the second possibility, a high negative field is required for the sample to reach $-M_s$.

To study the height of the energy barrier as a function of the TbFe thickness, a series of GdFe(1000 Å)/TbFe(e)/GdFe(500 Å) samples was prepared whose TbFe layer thicknesses e ranged from 2 to 8 Å. The magnetization evolution is similar to that of Figure 31, even for a 2-Å-thick TbFe layer (with, however, a significant shift toward the lowest temperatures). Therefore, TbFe layers as thin as 2 Å constitute a barrier if the

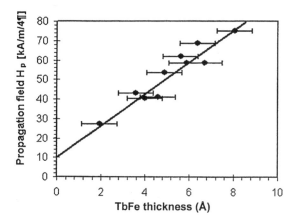

Fig. 33. Propagation field versus the TbFe layer thickness. Reprinted with permission from [83], © 1997, Elsevier Science.

temperature is low enough. Furthermore, a plot of the propagation field H_P as a function of the thickness of the TbFe layer gives a straight line for a given temperature (5 K in Fig. 33), as predicted by Barbara and Uehara [87]. The largest value of e does not appear in this figure because the nucleation in the thinner GdFe layer appears at fields H_{c2}^* smaller than H_P when the TbFe is too thick (for example, for $e = 15$ Å in Fig. 31). According to this model, the propagation field can be expressed as

$$H_P = \frac{\sqrt{JK_B} - \sqrt{JK_A}}{\mu_0 M_s} \frac{e}{\delta} \qquad (2.23)$$

2.5.1.3. Micro-SQUID Measurements

To explore the crossing of a barrier by a domain wall at very low temperatures, a trilayer with very thin TbFe was used, e.g., GdFe(1000 Å)/TbFe(2 Å)/GdFe(500 Å). Macroscopic magnetization measurements were performed above 1.8 K, and the micro-SQUID was used from 7 K down to 0.3 K. Two peaks were observed. The first corresponds to nucleation in the thicker GdFe layer. The second corresponds to H_P, the field at which the barrier is crossed. These peaks are due to the Barkhausen noise that occurs at nucleation and at the crossing of the barrier. It should be noted that there is no noise between these two peaks. Two regimes are observed for this sample: one below 5 K with two peaks, the second above 5 K with a single peak, as in Figure 31 above 120 K. Again, H_P increases when T decreases, which is in favor of a thermally activated process.

2.5.2. Rotation of Ring-Like Domains

Zablotskii and Soika [88] describe observations of the rotation of ring-like domains at various film temperatures. An explanation of rotation of such domains is given on the basis of a model in which the domain walls of the ring are assumed to hold a certain number of Bloch lines (see also [89–95]).

The experiments were carried out in a magneto-optical installation where the domains were observed by the Faraday effect. The sample was an epitaxial garnet film $(TmBi_3)(FeGa)_5O_{12}$ grown on a (111) gallium gadolinium garnet (GGG) substrate. The film thickness was $h = 7.8$ μm, the saturation magnetization was $\mu_0 M_s = 15.6$ mT, the characteristic length was $l = 0.8$ μm, and the Néel point was $T_N = 176°$C.

The domain structures consisting of spiral domains surrounded by an amorphous lattice of cylindrical bubbles were formed by the application of rectangular magnetic field pulses (frequency $\nu = 400$ Hz, pulse length $\tau = 2.6$ μs, amplitude $H_m \leq 12$ kA/m) and a fixed bias field ($H_b = 1.2$ kA/m), both directed perpendicular to the film surface. It should be noted that all of the spiral domains observed later as stationary were thus of a dynamical origin. After the creation of the desired structure, the amplitude of the field pulses was decreased to $H_m = -1.2$ kA/m, and the bias field was gradually increased. With an increase in the bias field strength, the spiral domains decrease their length and the bubbles decrease their diameter. As seen from the pictures, shortening of the spiral domain occurs by shortening of its external tail only. (In particular cases, however, simultaneous motion of both the external and internal ends of the spiral domain can occur.) At a certain value of the bias field, the ring-like domain is formed as a remainder of the spiral domain. The increase in the bias field is stopped at this moment; only the magnetic field pulses with a frequency $\nu = 100$–1000 Hz are applied. The ring-like domain begins to rotate in the same direction in which the shortening tail was moving.

The observed period of a ring-like domain rotation was 5–10 s, and its value was proportional to the ring's inner radius. Ring-like domains are observed to rotate when they are surrounded by a bubble lattice from all sides. They rotate as long as the bubbles are around the ring-like domain from the outside only. The rotation stops at the moment when some bubbles penetrate inside the ring through its cut. Before it stops a ring-like domain usually performs two or three full rotations. Zablotskii and Soika could not produce ring-like domains without a surrounding sea of bubbles. Evidently, an external pressure from the bubble lattice prevents the spiral domain from untwisting into a strip or a wavy labyrinth domain of the same length. The mechanism of the ring-domain rotation can be explained by a qualitative model. It is based on an idealized ring-like domain, assuming the existence of Bloch lines with a linear density n in its domain walls (Fig. 34). A sufficiently large Bloch line density in the walls of a ring-like domain may result from the creation of these domains from spiral domains as described above. Bloch lines are generated by the motion of the butt ends of a spiral domain. The mechanism of the rotation of a ring-like domain can be described by the following model. Both the length and the width of the ring domain are periodically changed by magnetic field pulses. The forces of gyrotropic reaction on the ring domain walls appear because of the moving Bloch lines inside them. However, the gyrotropic reaction forces acting on the ring ends have radial direction and, therefore, they cannot lead to rotation. In contrast, the gyrotropic reactions on both the external

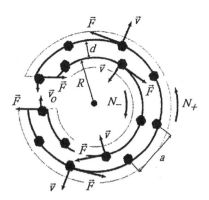

Fig. 34.   Model of an expanding ring-like domain. Bloch lines in its walls are shown by points. Reprinted with permission from [88], © 1998, Elsevier Science.

and internal sides of the ring have tangential directions, and they are caused by the changes in the ring width. It is easy to see from Figure 34 and from the expression for a gyrotropic force,

$$\mathbf{F} = \pm \frac{2\pi M_s h}{\gamma} [\vec{k} \cdot \vec{v}] \qquad (2.24)$$

(here $\vec{v}$ is the velocity of the domain wall, $\vec{k}$ is the normal to the film surface, γ is the gyromagnetic ratio) that the total torque, N_t, is the difference between the torques acting on the external (N_+) and the inner (N_-) sides of the ring-like domain $(N_t = N_+ - N_-)$.

Indeed, when the applied magnetic field increases (decreases), the ring width d decreases (increases), and at each moment the external and internal circumferences of the ring are moving with oppositely directed velocities.

Thus, a qualitative explanation of the domain rotation mechanism is the following. Because we are using the uniform-twist approximation (the handedness of all of the Bloch lines in one ring-like domain is the same) and the assumption $n =$ constant, the number of Bloch lines on the external perimeter of the ring-like domain is larger than on the internal boundary, and the total moment of gyrotropic reactions of all Bloch lines is larger on the external ring side than on the internal ring side. During the application of a magnetic field, the ring domain rotates in one direction. At the end of the field pulse, the reverse rotation occurs, though it never completely cancels the forward rotation.

2.5.3. Domain Wall Mobility in Ultrathin Films

Ultrathin magnetic films have attracted much experimental and theoretical interest in recent years. The magnetocrystalline anisotropies are very well defined across the entire sample under epitaxial growth conditions, and the fact that the film thickness is smaller than the exchange length (3 nm) inhibits to first order any magnetization variations across the thickness of the film. The resulting magnet is therefore greatly simplified, and therefore ultrathin epitaxial magnetic films are an excellent system for studying fundamental magnetic phenomena.

From a technological point of view, thin and ultrathin magnetic films are promising candidates for use in high-density data storage and magnetoelectronic applications.

A question in magnetism that is currently unanswered concerns the microscopic origin of spin precession damping. This has been addressed experimentally on two fronts. The first is through spin resonance experiments, such as ferromagnetic resonance spectroscopy and Brillouin light scattering (BLS) spectroscopy. In these experiments, the damping of the spin precession is measured through the resonance linewidth. The second approach is through the study of the dynamics of domain wall propagation. A typical experiment involves measuring the velocity of propagation of a domain wall as a function of applied field strength. The applied field exerts a driving pressure on the wall. As the domain wall moves the spins within the wall precess, which leads to energy dissipation through spin precession damping. The domain wall thus experiences a velocity-dependent retarding pressure; the final domain wall velocity is determined by the balance of these opposing pressures. Measuring the domain wall velocity is therefore a means of accessing information on the spin precession damping. Overviews of much of this work have been given by Malozemoff and Slonczewski [96] and Bar'yakhtar et al. [97]. Domain wall dynamics are also of wider interest, as they represent an experimental realization of the theoretical problem of a driven interface between two phases in a disordered medium [98–100]. Although domain wall dynamics have been studied extensively in bulk magnetic materials and thick films (see references contained in [96] and [97]), less experimental study has been presented to date on domain wall dynamics in ultrathin films. Kirilyuk et al. [101] have studied the v/H response of out-of-plane magnetized ultrathin Au/Co(8–10 Å)/Au(111) by magneto-optical microscopy. In-plane magnetized ultrathin films are particularly interesting from a dynamical point of view for two reasons. First, the interface atoms of an ultrathin film comprise a significant proportion of the entire system, and so any interface related dissipation mechanisms will become pronounced. Second, the extreme thickness confinement leads to a different type of domain wall, namely a Néel wall instead of the Bloch walls common in bulk materials [102]. R. P. Cowburn et al. [103] have reported the first experimentally measured domain wall v/H response of an in-plane magnetized ultrathin film. They show that the response is linear for velocities of 0–200 m/s, with a mobility comparable to that found in high-quality Fe whiskers.

R. P. Cowburn et al. have grown by molecular beam epitaxy a sample consisting of a GaAs(001) substrate plus a 10-monolayer seed layer of Fe on which were grown $\approx$200 monolayers of Ag(001). The Fe layer of interest to this study was grown next with a thickness of 10 ± 1 monolayers. A further 10 monolayers of Ag were then deposited, followed by a 5-monolayer protective Cr cap. Full details of the growth and structural and magnetic characterization have been published in [104–106].

The domain wall velocity was measured with a novel variation of the magneto-optical Kerr effect–Sixtus-Tonks method [107, 108]. Instead of recording the time of flight between two discrete points, the distance moved through a continuous 20 μm probing length is monitored as a function of time. A distance-time graph for the domain wall is thus obtained directly, from which the velocity can be found. The advantage of this method over the time-of-flight method is that the velocity is found from a straight-line fit to several points, instead of just two, which allows any changes in velocity during the measurement period to be identified. A single 180° Néel domain wall was created at a known position by a highly inhomogeneous pulsed magnetic field. The inhomogeneous field pulse was achieved by charging a 4700-μF capacitor to 30 V and then discharging it into two opposing coils. The peak current was $\approx$ 70A for a duration of $\approx$500 μs, which created a maximum field gradient of $\approx$4 MA/m^2 over a 6-mm distance. The laser beam of a magneto-optical Kerr effect magnetometer operating at room temperature was focused into a 13-μm (full width at half-maximum) spot, the center of which was positioned on the sample surface approximately 30 μm from the domain wall. A single magnetic field pulse of width Δt was then applied to propagate the domain wall toward the focused spot, and the difference between the magnetometer signals before and after the field pulse, Δs, was recorded. The difference signal depends upon how far into the laser spot the domain wall propagated; for the central portion of the focused laser beam, Δs is proportional to the distance. The domain wall was then re-created by the pulsed inhomogeneous field, and the experiment was repeated for a different Δt. A graph of Δs vs. Δt can thus be built up, the gradient of which is proportional to the domain wall velocity for that particular field amplitude. The complete v/H response can then be obtained by repeating with a different pulse amplitude.

The domain wall can be crudely imaged before and after a field pulse by rastering the focused laser beam spot with a linear stepper motor that moves the sample and the coils. This helps to verify that the results are indeed due to a domain wall passing through the laser spot and not to nucleation of new domains and that the domain wall is not significantly and irreversibly deformed because of the collision with a large defect [109].

Figure 35 shows the calibration of Δs (difference in magnetometer signal before and after the field pulse) to true distance as found by using the stepper motor linear drive. The transfer function is highly linear within the portion of the focused laser spot studied, which shows that Δs is a good measure of the domain wall position.

Figure 36 shows a typical Δs-time graph. Δs is proportional to distance, and the constant of proportionality is given by the gradient of the calibration curve of Figure 35. The time values were taken directly from the field pulse width. The gradient of the graph is proportional to the domain wall propagation velocity and gives 257 ± 3 m/s for this particular field amplitude.

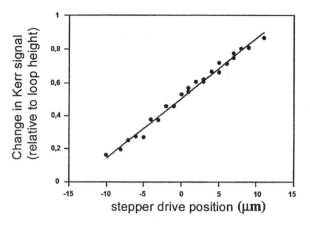

Fig. 35.   The calibration of Δs (difference in magnetometer signal before and after the field pulse) to true distance as found with a stepper motor linear drive. Reprinted with permission from [103], © 1996, American Institute of Physics.

Figure 37 is the measured dependence of domain wall velocity on the applied field amplitude. Each point on the graph corresponds to an entire Δs-time graph like the one shown in Figure 36. The velocity depends linearly on the applied field and can be described by the function $v(H) = \mu(H - H_0)$, where μ is the domain wall mobility and H_0 is closely related to the coercive field. The velocity range studied is too small to show the large departures from linearity ("anomalies") or saturation that occur when the domain wall moves at the speed of sound [110, 111]. The mobility for this data set is found to be 1.35 m^2/sA.

The mobility mean value of 1.46 m^2/sA is comparable to the values for high-quality Fe whiskers [112]; these were 1 m^2/sA for 3.2- and 11.9-μm whiskers and 1.63 m^2/sA for an 8.8-μm whisker. All of these samples are too thin to cause any appreciable eddy current damping [113]. The fact that the mobilities are comparable shows that the Fe/Ag interfaces do not cause substantial dynamic dissipation (i.e., spin precession damping), although the fact that the coercivity is much higher in the ultrathin Fe layer than in the much thicker Fe whisker

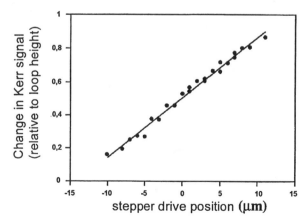

Fig. 36.   A typical Δs-time graph measured at a pulsed field amplitude of approximately 320 A/m. Reprinted with permission from [103], © 1996, American Institute of Physics.

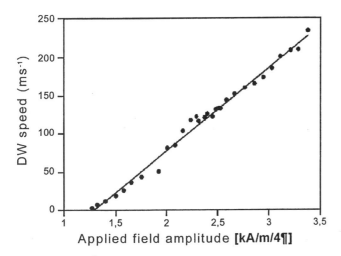

Fig. 37. The experimentally measured dependence of domain wall speed on applied field strength. The gradient of the fitted straight line gives the domain wall mobility as 1.35 m²/sA. Reprinted with permission from [103], © 1996, American Institute of Physics.

($\approx$25 mA/m cf. 0.1 mA/m) is most probably caused by static domain wall pinning at the interfaces.

2.6. Magnetization Reversal

During recent years, magnetic visualization has unambiguously confirmed for some ultrathin films the stability of a multidomain magnetic structure in a zero field and its mobility when an external field is applied [114, 115]. Therefore, understanding the magnetization switch and its dynamics imply the quantification of the nucleation and the wall motion processes depending on the applied magnetic field H and its rate of variation (dH/dt). Raquet [116, 117] presented a general model of the hysteresis loops that makes it possible to consider simultaneously the dynamical effects and the complex competition between domain nucleation and domain wall motion. This model allows the deduction of a general analytical expression for the magnetization $M(H, (dH/dt))$ describing all multidomain reversal processes based on intrinsic and microscopic parameters: the Barkhausen volume, the nucleation rate $R(H)$, the wall velocity $v(H)$, the radius r_c, and the total number of nucleation sites N_0 on the surface of the sample.

2.6.1. Derivation of the Dynamical Magnetization Equation

First, the methodological approach is based on a mathematical calculus of the areas corresponding to the nucleation centers and the propagation of domain walls. This model is developed in light of Fatuzzo's theory [118] concerning the relaxation of the polarization reversal in ferro-electrical crystals. The most fundamental difference between Fatuzzo's theory and the model of Raquet is in the consideration of a variable magnetic field applied to the magnetic layer and not a constant electric or magnetic field as in [118, 119]. This means that the magnetization dependence on the magnetic field H and its rate of

variation (dH/dt) have to be considered. We will demonstrate that the resulting expression $M(H, (dH/dt))$ is simple and allows a fit to experimental magnetization measurements. The detailed calculus of the magnetization is published in [118, 120].

To consider the delicate overlapping of nucleation and the coalescence of domains due to their sideways expansion, the global reduced switched area $A(H, (dH/dt))$ described below, neglecting the possible overlap, is calculated. Then, applying Avrami's theorem [121–123], the real magnetization switched by the field can be deduced from the following expression:

$$M(H, (dH/dt)) = M_s(-2[1 - \exp(-A(H, (dH/dt)))] + 1) \tag{2.25}$$

It is assumed that domains appear at random on the sample surface according to a statistical process [118, 124]. This is defined by the nucleation rate $R(H)$ per unit of time. Let us consider N_0, the total number of nuclei centers, and $N(H, (dH/dt))$, the number of domains appearing at an applied field H. Then, $N(H, (dH/dt))$, in a negative field, is defined by

$$\frac{dN(H, (dH/dt))}{N_0 - N(H, (dH/dt))} = \frac{R(H)dH}{dH/dt} \tag{2.26}$$

Once it appears, the nucleation might expand sideways by radial motion defined by radius $r(H)$ and radial velocity $v(H)$, varying with the applied magnetic field. Let us assume a nucleation appearing in negative field H_1, $0 > H_1 > H$; the switched area by expansion of its center field H, $\sigma(H)_{H_1}$, is assumed to be given by

$$\sigma(H)_{H_1} = \frac{\pi}{T}(r(H)_{H_1} + r_c)^2 - \frac{\pi r_c^2}{T} \tag{2.27}$$

and

$$r(H)_{H_1} = \int_{H_1}^{H} v(H) \frac{dH}{dH/dt} \tag{2.28}$$

where $r(H)_{H_1}$ is the radius in field H for a nucleation site appearing in field H_1. A_T is the global area to be reversed, estimated to be 0.01 cm² for the magneto-optical measurements.

Assigning the above area to the total number of nucleations appearing between a zero field and H, the switched area $A(H, dH/dt)$ can be calculated with the expression

$$A(H, (dH/dt)) = \int_0^H \frac{dN(H, (dH/dt))}{dH}\bigg|_{H_1} \sigma(H)_{H_1} dH_1 \\ + N(H, (dH, dt)) \frac{\pi r_c^2}{A_T} \tag{2.29}$$

Assuming that the elementary mechanisms of the magnetization reversal are thermally activated processes [125, 126], the nucleation rate and the wall motion can be expressed as

$$R(H) = R_0 \exp(-\beta V M_s H) \tag{2.30}$$

and

$$v(H) = v_0 \exp(-\beta V M_s H) \quad (2.31)$$

where R_0 and v_0 are the nucleation rate and the velocity, respectively, of the wall in zero field, $\beta = 1/kT$, and V is the activation volume, called the Barkhausen volume. Introducing the parameters $x = \beta V \mu_0 M_s H$, $dx/dt = \beta V \mu_0 M_s dH/dt$ gives

$$k = \frac{v(H)}{r_c R(H)} = \frac{v_0}{r_c R_0}, \qquad \alpha = \frac{\pi r_c^2 N_0}{A_T} \quad (2.32)$$

where k characterizes the competition between the wall motion and the nucleation (in a first approximation, a predominance of wall motion (nucleation) implies $k \gg 1$ ($k \ll 1$)), and α represents the density of nuclei sites, neglecting possible overlaps. Using Eq. (2.32) and after substituting Eqs. (2.30) and (2.31) into Eq. (2.29), successive integration and term rearrangement allow us to obtain a general analytical expression for the magnetization in a negative magnetic field considering the nucleation, the wall motion, and the dynamics of the applied field:

$$
\begin{aligned}
\frac{M(x,(dx/dt))}{M_s} &= -2(1 - \exp(-2\alpha k^2 \left(\frac{N(x,(dx/dz))}{N_0} \right) \\
&\times \left(1 - \frac{1}{k} + \frac{1}{k^2} \right) \\
&+ \frac{R_0}{2(dx/dt)^2}(\exp(-x) - 1)\left(1 - \frac{1}{k}\right) \\
&+ \frac{R_0^2}{2(dx/dt)^2}(\exp(-x) - 1)^2)) + 1
\end{aligned} \quad (2.33)
$$

Magnetization is defined by five fundamental and intrinsic parameters: $R_0 = R(H=0)$, the nucleation rate in a zero field; $v_0 = v(H=0)$, the wall velocity in zero field; V, the Barkhausen volume; N_0, the total number of nuclei sites; r_c, the radius of the nucleation rate $R(H)$ and the velocity $v(H)$ in a negative field, from Eqs. (2.30) and (2.31). Using R_0, v_0, and r_c, the ratio k is deduced to estimate the main reversal mechanism; finally, the Barkhausen volume is related to the interfacial length between pinning centers, and N_0 corresponds to specific defects responsible for the nucleation (such as atomic steps in the magnetic layer).

2.6.2. Simulation of the Magnetization Reversal

To improve the potential of his model, Raquet [116] first proposed an academic approach that consists of studying the variation of $M(H,(dH/dt))$ versus typical values of V, R_0, v_0, and N_0. Figure 38 shows the drastic dependence of the magnetization (calculated using Eq. (2.18)) on the values of R_0 (Fig. 38a), N_0 (Fig. 38b), v_0 (Fig. 38c), and V (Fig. 38d). For example, an increase in the nucleation rate modifies the magnetic transition $+M_s/-M_s$ (Fig. 38a): the more predominant the nucleation rate becomes compared with the wall motion, the more the coercive field decreases and the more the transition is smooth. One can find here a classical result describing the slope of the magnetic transition versus the main rever-

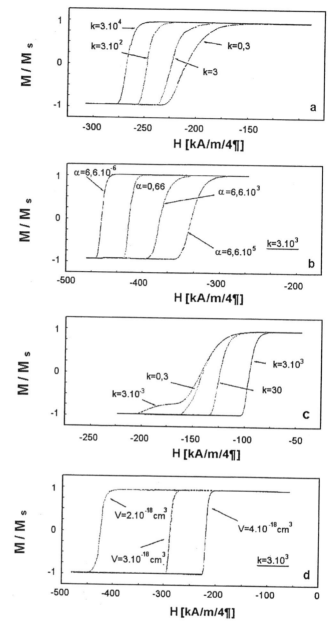

Fig. 38. Variation of $M(H,(dH/dt))$ versus typical values of R_0, N_0, v_0, and V for thin magnetic films. (a) $v_0 = 10^{-13}$ cm/s, $r_c = 3.25 \times 10^{-6}$ cm, $N_0 = 2 \times 10^8$, $V = 3 \times 10^{-18}$ cm^3 for several values of R_0 ($R_0 = 10^{-12,-10,-8,-7,-1}$ s^{-1}) corresponding to $k = 3 \times 10^4$, 3×10^2, 3, and 0.3. (b) $R_0 = 10^{-18}$ s^{-1}, $v_0 = 10^{-20}$ cm/s, $r_c = 3.25 \times 10^{-6}$ cm, $V = 3 \times 10^{-18}$ cm^3 for several values of $N_0(N_0 = 2 \times 10^{3,8,12,14})$ corresponding to $\alpha = 6 \times 10^{-6}$, 0.6, 6×10^3, and 6×10^5. (c) $R_0 = 10^{-4}$ s^{-1}, $N_0 = 6 \times 10^8$, $r_0 = 3.25 \times 10^{-6}$ cm, $V = 3 \times 10^{-18}$ cm^3 for several values of $v_0(v_0 = 10^{-6,-8,-10,-12}$ cm/s, $r_c = 3.25 \times 10^{-6}$ cm, corresponding to $k = 3 \times 10^3$, 30, 0.3, 3×10^{-3}. (d) $R_0 = 10^{-3}$ s^{-1}, $v_0 = 10^{-15}$ cm/s, $r_c = 3.25 \times 10^{-6}$ cm, $N_0 = 2 \times 10^{-8}$ for several values of V ($V = 4 \times 10^{-18}$, 3×10^{-18}, 2×10^{-18} cm^3). Reprinted with permission from [116], © 1997, Elsevier Science.

sal process [127, 128]. Figure 38b shows that the generally unknown value N_0 may induce a large modification in the hysteresis loop: whatever the value of the nucleation rate, few nuclei sites in the magnetic layer tend to favor the wall motion process. Figure 38c shows the variation of the magnetization reversal when the wall motion becomes less efficient compared with the nucleation. The last curve (corresponding to $k \cong 3 \times 10^{-3}$) is an interesting case; it shows that the magnetization can reverse in two steps when the two processes are sufficiently uncoupled energetically: the first step related to the major part of the reversal is only due to the nucleation, but because of the low velocity, the nucleation process cannot reverse the magnetization by itself if the total number of nuclei sites cannot recover all of the area of the sample. That is why one has to reach a higher field to activate the wall motion and complete the switching of the magnetization. Finally, Figure 38d confirms that the Barkhausen volume is a very sensitive and delicate value defining the width of the loop: small volumes need more energy to be reversed. This could be correlated with a higher density of magnetic defects at the interfaces. These simulations confirm that the shape of the hysteresis loop and the coercive field value depend directly on the competition between nucleation and wall motion, the number of nuclei sites, and the Barkhausen volume. The interest in quantifying these parameters is evident.

2.6.3. Application to a Au/Co/Au Sandwich

Raquet applied the model to investigation of the magnetic reversal of a cobalt ultrathin layer, MoS_2/Au(40 nm)/Co(0.8 nm)/Au(3 nm), prepared at room temperature with thermal annealing of the gold buffer at 350°C; the cobalt layer and the gold overlayer were deposited at room temperature. Measurements were made with the use of the polar Kerr effect with a field variation rate of up to 100 MA/sm (for the experimental device, see [128]). By using the analytical expression of the magnetization Eq. (2.33), comparing the theoretical results with the experimental data, and adjusting the five intrinsic parameters V, R_0, v_0, N_0, and r_c, an excellent agreement between experimental and calculated loops has been observed. The weak dispersion of the V, R_0, v_0, N_0, and r_c values has ensured the validity of the investigation. The average Barkhausen volume has been about 0.95×10^{-18} cm^3, which corresponds to a 35-nm characteristic interfacial length L_b in the magnetic layer. From the deduced R_0 and v_0 values, an analytical expression for $v(H)$ and $R(H)$ has been derived; in addition, the ratio k was found to be $k = 0.75$. This means that both nucleation and wall motion take part in the reversal process. Expressed in terms of area, the total number of nuclei is about 2×10^{10} cm^{-2}. This corresponds to an average lateral length between nuclei centers of 70 nm; only a statistical microscopic approach might correlate this density with a specific structural defect. The calculated value α ($\alpha \cong 0.23$) confirms the mixed behavior of the magnetization.

2.6.4. Magnetization Reversal in Thin Cu/Ni/Cu/Si(001) Films

Recently, strong interest has developed in thin films of Cu/Ni/Cu/Si(001) with the prospect of applications such as magneto-optical recording and other magnetic devices, because of the perpendicular magnetic anisotropy (PMA) [129–131] in these films, which extends over a Ni thickness range of about 100 Å. The misfit strain between the crystal lattices of Cu and Ni is understood to cause the PMA. Recently, Bochi et al. reported a Ni thickness dependence of the average domain size in the demagnetized state [132, 133]. P. Rosenbusch et al. [134] presented a detailed characterization of the magnetization reversal mechanism in the Cu/Ni/Cu/Si(001) system exhibiting PMA. They reported measurements of the hysteresis loops, the magnetization relaxation process, and domain images by Kerr microscopy and showed that the magnetic relaxation is consistent with domain propagation. It was demonstrated that a correlation exists between the crystallographic properties and the Barkhausen length.

The sample investigated was of the form 30 Å Cu/50 Å Ni/h Å Cu/Si(001). The Cu buffer layer was wedged in four steps ($h = 600, 1000, 1500, 2000$ Å). The preparation of such a wedged sample has the advantage that any variation in the magnetic properties cannot be due to varying growth conditions, but must be correlated with the crystallographic or structural properties. All films were deposited by electron beam vacuum evaporation at room temperature. The film thickness was estimated with a quartz crystal monitor close to the sample position. Traces of carbon were found by Auger electron spectroscopy (AES). The reflection high-energy electron diffraction (RHEED) images were recorded with a CCD camera, and a line shape analysis of the RHEED streaks was used to evaluate the strain.

For the magnetic characterization a scanning magneto-optic Kerr effect microscope was employed [135, 136]. The light incident normal to the film plane senses the polar effect. Two different objective lenses ($\times 20$ and $\times 50$ magnification) were used to focus the probing laser beam down to about 1 μm. By scanning the sample in 100×100 pixels, images of the domain pattern were obtained. Hysteresis loops and the magnetization relaxation were measured with an unfocused beam of ≈ 1 mm diameter. This ensured averaging over many domains. In addition, 10 measurement cycles were averaged to improve the signal-to-noise ratio and eliminate statistical fluctuations in reversal process.

The hysteresis loops measured at a constant field sweep rate of 22.4 kA/sm are displayed in Figure 39 for each part of the wedge. The loops possess the typical square easy-axis shape indicating PMA, as was expected for a 50-Å Ni film [137]. For all Cu buffer thicknesses the remanence magnetization equals the saturation value. The magneto-optical Kerr effect signal, which is to first order proportional to the magnetization, decreases with the Cu buffer thickness at saturation. The coercive field decreases with increasing Cu buffer thickness.

Enlargement of the field scale in Figure 39 reveals that the field range in which the reversal occurs decreases with increasing Cu buffer thickness. Hysteresis loops measured at a field sweep rate five times slower than the measurements of Figure 39 generally show a coercivity about 7% smaller, indicating that the sample exhibits a strong magnetic aftereffect. For the relaxation measurements the applied field was reversed within 100 ms from positive saturation to a negative field value slightly smaller than H_c. This field was held

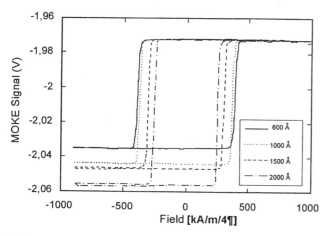

Fig. 39. Polar magneto-optical Kerr effect loops for the stated Cu buffer thickness measured at 22.4 kA/sm. Reprinted with permission from [134], © 1997, Elsevier Science.

constant (within 2%) for 45 s and again $+H_{sat}$ was applied. Saturation before and after the measurement period allows the optical drift within the microscope to be estimated to be below 5%. Figure 40 shows the normalized magneto-optical Kerr effect signal for the 600-Å Cu buffer step as measured under various applied fields. Typically, the magnetization reverses faster in a higher field. For $H = 316$ and 321 kA/m/4π the magnetization terminates after about 25 s at 90% of the saturation value. This incomplete reversal indicates the existence of hard domain wall pinning centers. It was not observed for the thicker Cu buffer layers.

For all Cu buffer thicknesses the demagnetizing time $t_{1/2}$ at which $M = 0$ was found to depend exponentially on the applied field H according to the following relation:

$$t_{1/2} \propto \exp\left(\frac{-V_B M_s \, |H|}{k_B T}\right) \qquad (2.34)$$

Equation 2.34 was derived assuming thermally activated reversal [138, 139], where $k_B T$ is the thermal energy, V_B is the Barkhausen volume, and M_s is the saturation magnetization. From fitting Eq. (2.34) to the experimental data, V_B is obtained. The saturation magnetization was estimated from magnetic circular dichroism measurements. Equivalent to the Barkhausen volume is the more illustrative Barkhausen length, $l_B = \sqrt{V_B/50}$ Å.

Relaxation curves measured under various applied fields are found to lie on a universal curve if M/M_s is plotted versus the reduced time $t/t_{1/2}$. The universal curves for all Cu thicknesses are shown in Figure 41. The curvature becomes stronger as the Cu buffer thickness increases. The data in Figure 41 were fitted by a model developed by Fatuzzo [118] and extended by Labrune et al. [140] for thermally activated magnetization reversal through domain nucleation and domain wall propagation. This model has been used widely to explain relaxation measurements [140, 141]. The model predicts the reduced magnetization $B(t) = [M(t) + M_s]/2M_s$ as a function of time. The solution for $B(\tau) = t/t_{1/2}$ is

$$B(\tau) = \exp(-k^2[2 - 2(c\tau + k^{-1}) + (c\tau + k^{-1})^2 \\ - 2e^{-c\tau}(1 - k^{-1}) - k^{-2}(1 - c\tau)]) \qquad (2.35)$$

with the constant $c = f(\ln 2, k)$, which is a nonanalytic function of k. $k = v/r_c R$ is the ratio between the reversal rates of the two processes, domain nucleation and domain wall propagation, where r_c is the critical radius of a nucleus. v and R are the domain wall velocity and nucleation rate, respectively, with

$$v = f_0 l_B \exp\left(-\frac{E_0^p - M_s V_B \, |H|}{k_B T}\right) \qquad (2.36)$$

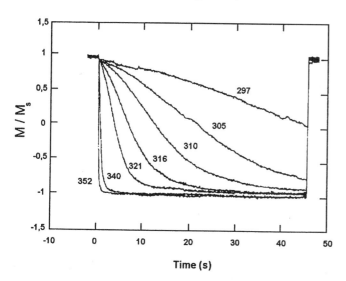

Fig. 40. Measurement of $M(t)$ for 600-Å Cu under a constant reversed field. For $t < 0$ and $t > 45$ s the sample was saturated in the positive direction. The applied field is stated in kA/m/4π. Reprinted with permission from [134], © 1997, Elsevier Science.

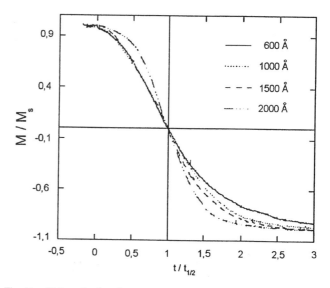

Fig. 41. Universal relaxation curves for all Cu buffer thickness plotted in reduced units. $t_{1/2}$ is the demagnetizing time. Reprinted with permission from [134], © 1997, Elsevier Science.

and

$$R = f_0 \exp\left(-\frac{E_0^n - M_s V_B \, |H|}{k_B T}\right) \qquad (2.37)$$

where $f_0 = 10^9$ Hz is the attempt frequency [142]. $E_0^{n,p}$ are the activation energies for domain nucleation and wall propagation, respectively. Equations (2.36) and (2.37) imply that k is independent of the applied field and specific only to material properties. This is in good agreement, with the experiment, as can be seen from the existence of a universal relaxation curve.

Assuming $r_c = l_B$, $\ln k$ is the difference between E_0^n and E_0^p, relative to the thermal energy:

$$\ln k = \frac{E_0^n - E_0^p}{k_B T} \qquad (2.38)$$

Equation (2.35) fits the data in Figure 41 satisfactorily for all Cu thicknesses. A representative fit is shown for the 2000-Å buffer in Figure 42. For all buffer thicknesses the fitted values for k are bigger than 1 and indicate a domain wall propagation dominated reversal. With increasing Cu buffer thickness k increases. High k for 2000-Å Cu indicates rare nucleation. By far the biggest part of the sample reverses through domain wall propagation. For the thinnest Cu buffer layer, k is on the order of 1, meaning that the areas reversing by nucleation and those reversing by propagation are of similar size. This low k value indicates that the incomplete reversal observed in Figure 40 may be due to pinning. Pinning hinders the reversal of large parts of the sample by domain wall propagation; therefore nucleation events become relatively more important. Alternatively, one can interpret the values of k in terms of the difference between E_0^n and E_0^p as expressed by Eq. (2.38).

With increasing Cu buffer thickness, $(E_0^n - E_0^p)/k_B T$ increases. A high value means rapid domain wall propagation once the activation energy for nucleation is reached and nucleation has occurred.

To confirm this interpretation of the parameter k, the domain structure as it typically evolves in a relaxation measurement has been imaged. After the sample was saturated,

field pulses of about $0.9H_c$ were applied in the reverse direction. The length of the pulses was 70 ms. After each field pulse a domain image was recorded. The pulsing of the field was necessary to freeze the domain pattern for the 4-min scanning period. This technique is well known as quasi-static imaging.

In the demagnetized state, directly after the sample preparation before the application of a magnetic field, no significant difference was observed in the average domain size. To explain the variations in the magnetic properties with the Cu buffer thickness, Rosenbusch analyzed in situ RHEED patterns. The RHEED pattern of the Cu buffer layer step clearly changes as the Cu buffer layer thickness is increased. From 600 to 1000 Å the vertical streaks become narrower. Such narrowing was widely observed for the growth of Cu on Si(001) [143, 144].

After further deposition, the diffraction spots become pronounced. The background intensity decreased and the intensity of the reflection spots increased. For the Cu buffer steps and subsequent Ni layer, the horizontal intensity profile has been fitted by a Gaussian bell curve. The full width at half-maximum clearly decreases for both the Cu buffer layer and the subsequent Ni layer as the Cu buffer becomes thicker. In a simple diffraction model [145, 146], the inverse of the streak width is proportional to the correlation length in the crystal structure. Therefore, the correlation length clearly increases as the Cu buffer layer becomes thicker (Fig. 43). The trend accords with the increasing Barkhausen length, deduced from the relaxation measurements. This fact provides evidence for an interdependence between the crystallographic structure of the buffer layer and the magnetic properties of the overlayer. The exact mechanism causing the domain wall pinning remains unclear. Among the structural properties that have to be considered in the future are misfit dislocations in the Cu buffer layer and in the Ni layer, as well as interface roughness.

In summary, the magnetization reversal in a step wedged Cu/Ni/Cu/Si(001) structure that exhibits PMA occurs through domain nucleation and domain wall propagation. The sample

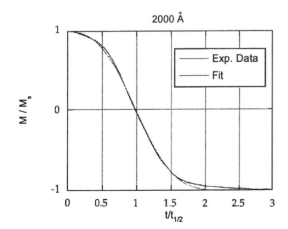

Fig. 42. Universal relaxation curves for the 2000-Å Cu buffer layer fitted by Eq. (2.35). The fitting parameter k equals 580. Reprinted with permission from [134], © 1997, Elsevier Science.

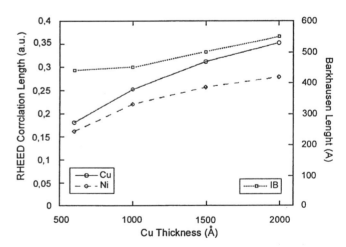

Fig. 43. RHEED correlation length of the Cu buffer layer and the Ni layer, as well as the Barkhausen length plotted versus the Cu buffer thickness. Reprinted with permission from [134], © 1997, Elsevier Science.

shows a strong magnetic aftereffect induced by thermal activation. Pinning has been identified as the origin of the strong variation in the coercive field, and the strength and average distance of pinning centers are related to the thickness of the buffer layer.

3. MAGNETIC FILM CHARACTERIZATION

A great deal of intensive effort and talent has been expended in recent years in developing experimental methods to investigate the domain structure and its dynamics. The increasing complexity of the structure of magnetic recording media has led to more and more powerful methods of observing fabrication processes and magnetic properties in thin and ultrathin magnetic films [132, 147, 148].

3.1. Vibrating Sample Magnetometer

One aspect that is of growing importance for modern media is the analysis of the magnetization vector rather than the magnetization in the direction of the applied field because of the importance of the oblique anisotropy direction. One instrument particularly suitable for this kind of investigation is the vibrating sample magnetometer (VSM). To be able to study the growth process of thin layers, measurement equipment should be sufficiently sensitive for performing measurements on media with a thickness of a few nanometers. This corresponds to a magnetic moment on the order of 10 nAm^2.

3.1.1. Principle of Operation

In a VSM a sample is vibrated in the vicinity of a set of pick-up coils (Fig 44). The flux change caused by the moving magnetic sample causes an induction voltage across the terminals of the pick-up coils that is proportional to the magnetization of the sample,

$$V(t) = C \frac{d\phi}{dt} \tag{3.1}$$

where $\phi(t)$ represents the (changing) flux in the pick-up coils caused by the moving magnetic sample. The proportionality constant C contains the geometry of both sample and pick-up coil.

The signal in the coils is very small (the signal caused by the aforementioned 10 nAm^2 is only a few nanovolts) and therefore extremely sensitive to noise sources. One of the major causes of problems in such a system is vibration of the coils relative to the field applied by the electromagnet. The flux produced by the magnetic sample is approximately 10^{-15} times smaller than the flux produced by the magnet; therefore vibrations must be canceled out by the same factor. Development of this type of instrument is a multidisciplinary job involving electronics, physics, and mechanical aspects.

3.1.2. Vectorial Calibration of a Vibrating Sample Magnetometer

Measurement of the magnetization vector is used for angle-dependent measurements, magnetic anisotropy measurements, and determination of intrinsic magnetic behavior. Measuring the magnetization vector in the xy plane (the rotation plane of the sample) requires a biaxial coil system consisting of two coil sets, as shown in Figure 45, with one coil set for each coordinate. For angle-dependent measurements of the magnetization vector in thin films, an angular calibration of the system is required. The commonly used (conventional) calibration method for a biaxial vector VSM consists of two parts:

- The X-coil (coil set with axis parallel to the applied field) calibration is performed by measuring the signal on the X-coils for every angle of interest while rotating a Ni sample, with known magnetic moment, in a high field.

- The Y-coil (coil set with axis perpendicular to the applied field) calibration is done with the same sample in a zero field or with a perpendicular magnetized remanent sample. For every angle of interest the signal on the Y-coil is measured, and then the remanent sample is rotated back 90° to measure the magnetic moment on the calibrated X-coils.

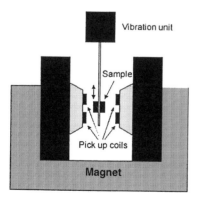

Fig. 44. Schematic of a VSM. The signal in the pick-up coils is caused by the flux change produced by the moving magnetic sample. Reprinted with permission from http://www.el.utwente.nl/tdm/ist/research

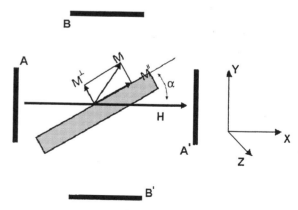

Fig. 45. Top view of a sample within the biaxial coil system, consisting of coil sets A and B. M, $M_\parallel$, $M_\perp$, and α are shown in relation to the sample. Reprinted with permission from [151], © 1999, Elsevier Science.

This calibration method, which is used in some form in many (commercial) vector VSMs, has to be corrected because of a number of invalid assumptions:

- The coil system is equally sensitive for in-plane and perpendicular magnetization components. The results from the calibration of the in-plane component of the magnetization are also used for the measurement of the component perpendicular to the film plane, and the shape of the stray fields resulting from each component is totally different.
- The coil system is orthogonal. One set of coils is considered to have its sensing direction exactly perpendicular to the other set.
- The coils are perfectly aligned with the external magnetic field. One set has its axis exactly aligned with the applied field, and the axis of the other set is perpendicular to the field.

One of the errors resulting from these invalid assumptions is an undesired contribution to the signal of a magnetic moment perpendicular to the measuring direction. This is often considered as an error, sometimes referred to as "cross talk" [149], which is becoming increasingly important when the sample size approaches the sample coil distance, especially when angle-dependent measurements are performed. This cross talk problem leads to errors in the magnetization vector of up to 15% in magnitude and 8° in angle in a configuration where the size of the square sample is half the coil distance.

Several solutions have been proposed to minimize the cross talk problem. Richter [150] introduces a phenomenological correction for the cross talk based on the mean values of magnitude and orientation of the magnetization, which is a good approximation if the deviations are small. Other methods minimize the effect of the invalid assumptions by adjusting the design of the detection system. Bernards [149] uses an extra set of four coils in addition to an eight-coil set-up, Samwel et al. [151] avoid the rotation of the sample with respect to the coils and take the different sensitivities for different components into account.

3.2. Magneto-Optical Methods

The magneto-optical effects were first discovered by Michael Faraday (the Faraday effect is observed in transmission through a material) and J. C. Kerr (the Kerr effect is observed on reflection from a material). It is convention to refer to effects in reflection as Kerr effects and to those in transmission as Faraday effects.

Magneto-optical effects may be observed in nonmagnetic media such as glass when a magnetic field is applied. Indeed, Faraday first discovered his effect in a glass rod with a magnetic field applied along the direction of propagation of an optical beam. The intrinsic effects are usually small in such cases. In magnetic media (ferromagnetic or ferrimagnetic) the effects are much larger, although they are still small and often difficult to detect. In these cases it is usual and convenient to

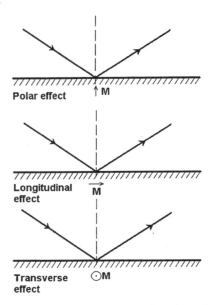

Fig. 46. Diagram illustrating the various configurations of incident light polarization and domain orientation that lead to Kerr contrast. Reprinted with permission © Elsevier Science.

refer to three principal orientations (Fig. 46). These are

- polar orientation,
- longitudinal orientation,
- transverse orientation.

All magneto-optical effects are based on the variation of the medium's refractive index with the local magnetization of the material. This anisotropy is a gyroelectric property of the material and finds expression in the off-diagonal elements of the permittivity tensor (and the electrical susceptibility tensor $\underline{\chi} = \underline{\epsilon}_r - \underline{1}$ of the medium,

$$\underline{\epsilon} = \begin{pmatrix} \epsilon & -jQ\epsilon & 0 \\ jQ\epsilon & \epsilon & 0 \\ 0 & 0 & \epsilon \end{pmatrix} \qquad (3.2)$$

The absolute value of the usually complex magneto-optical parameter Q is proportional to the magnetization M, which can be caused by an external field or originate from the spontaneous magnetization of the material itself. Because of the inertia of the magnetization process at optical frequencies, the gyromagnetic effects described by the permeability tensor $\underline{\mu}(Q)$ are negligible ($\mu_r \approx 1$). The phase of Q is a measure of the change of the ellipticity of either the transmitted or reflected light wave with respect to the incident light wave. Reversing the direction of magnetization results in a 180° phase shift of Q. The optical behavior of the material is completely determined by the real and the imaginary part of ϵ and Q. With respect to the boundary conditions of the particular application, the Maxwell equations combined with Eq. (3.2) describe the reflection and the transmission properties of the considered material. The magneto–optical effects, which describe the influence on the intensity and the polarization status of light waves caused by a magnetic material, are classified either as the Kerr effect (reflected light wave) or as the Faraday effect (transmitted light wave).

The three orientations are defined in terms of the direction of the magnetization vector M with respect to the surface of the material and the plane of incidence of an incident optical beam. In the longitudinal case the magnetization vector is in the plane of the surface and parallel to the plane of incidence. The effect occurs for radiation incident in the P-plane (E-vector parallel to the plane of incidence) or the S-plane (E-vector perpendicular to the plane of incidence). The effect is that radiation incident in either of these linearly polarized states is converted, on reflection, to elliptically polarized light. The major axis of the ellipse is often rotated slightly with respect to the principal plane, and this is referred to as the Kerr rotation. There is an associated ellipticity called the Kerr ellipticity. There is no effect observed at normal incidence. In the polar case the magnetization vector is perpendicular to the plane of the surface. Like the longitudinal case the effect occurs for radiation incident in the P-plane or the S-plane. The complex Kerr rotation angle can be written as follows, since $k \ll r$: $\theta_k + J\varepsilon_k = k/r$.

Similar effects also occur in transmission, although one usually needs to have a thin film to see these, because most magnetic materials are opaque in regions where they are magneto-optically active. The sign and magnitude of these effects are proportional to M and its direction.

The transverse case is quite different from the previous two. First there is only an effect for radiation polarized in the P-plane. Second, in such a case, the reflected radiation remains linearly polarized, and there is only a change in reflected (or transmitted) amplitude such that as M changes sign from $+M$ to $-M$ the reflectivity changes from $R + \Delta R = |r + k|^2$ to $R - \Delta R = |r - k|^2$.

Magneto-optical effects in thin films can be used for domain structure characterization as well as for the utilization of domain wall dynamics for magneto-optical sensors [152, 153].

3.2.1. Magneto-Optical Kerr Spectroscopy

There are two beneficial reasons why a spectroscopic (i.e., wavelength-dependent) tool is preferred to measure the Kerr effect. First the spectroscopic behavior depends on the magneto-optical material and can give insight into the physical processes that cause the magneto-optical effect. Second, to increase storage density, the wavelength of the laser diodes in the applications is reduced. Therefore it is useful to measure the Kerr effect at such wavelengths.

Figure 47 illustrates the measurement setup. For example, to be able to measure from 240 nm ($h\nu = 5.17$ eV; UV radiation) to 920 nm ($h\nu = 1.35$ eV; IR radiation), a xenon arc lamp and a monochromator are applied. After the monochromator, filters are mounted to remove any undesired stray light. The beam is linearly polarized by a polarizer and then modulated between left-handed circularly polarized (LCP) and right-handed circularly polarized (RCP) light by a photoelastic modulator. In this device, a transparent element is alternately compressed and expanded in one direction by piezoactuators. Thus the light speed in one of the optical axes is modulated about the average value (c of the transparent element, which is C_{air}/n). The alternate slowing and speeding of this axis causes the modulation between LCP and RCP light. The light passes through the hollow pole piece of a 3-kW electromagnet (maximum field 1 Tesla) and arrives at the sample, which is positioned between the magnet poles. After reflection, the light travels through an analyzer (another polarizer) and finally is detected by a photomultiplier tube (PMT). Of the more than

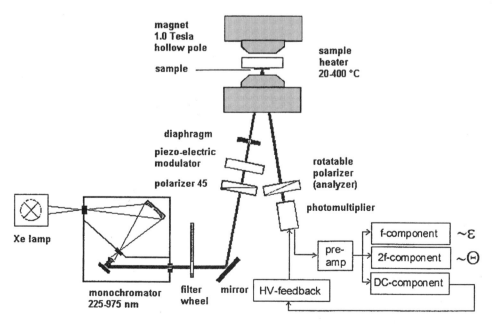

Fig. 47. Schematic of a magneto-optical Kerr measurement setup. Reprinted with permission © IST Group.

10 W of power that the lamp emits as light, only 100 μW remains at the detector. Of this 100 μW, only about 1 μW is the Kerr signal to be measured, and that signal must be resolved in at least a few hundred discrete levels. Therefore, the modulation system and a lock-in amplifier (phase-sensitive detection) is used to lift the Kerr rotation and ellipticity signal from the noise.

Furthermore, it is possible to raise the temperature of the sample to 450°C with a special sample holder. Thus, the Curie temperature can be determined and polar Kerr hysteresis loops can be measured at different wavelengths and different temperatures.

3.2.2. Nonlinear Magneto-optics

The origin of the magneto-optical Kerr effect lies in the spin-orbit coupling that acts like a magnetic field on the current induced by the electromagnetic field of the incident light [154]. This also holds for the nonlinear contributions of the induced current, which are the origin of the second harmonic generation (SHG), a nonlinear optical technique that derives its interface sensitivity from the breaking of symmetry at boundaries between centrosymmetric media [155, 156]. This leads to a magnetization-induced SHG (MSHG) or, indeed, a nonlinear magneto-optical Kerr effect. Based on symmetry arguments, Pan showed that the presence of magnetization would lead to new, nonzero surface contributions to the nonlinear optical response [157, 158], and Hübner and Bennemann independently calculated the nonlinear magneto-optical spectrum of Ni, based on a spin-dependent band structure calculation [159]. They showed that this should lead to observable effects, with magnetic contributions to the nonlinear tensor coefficients of more than 10%. The first experimental evidence for a MSHG effect was given by Reif for a clean Fe(110) surface [160], whereas Spierings et al. showed the first MSHG results from buried Co/Au interfaces [161, 162].

A strong demonstration of the surface and interface sensitivity of the MSHG technique was given by Wierenga et al. [163] by an in situ MSHG and magneto-optical Kerr effect study of the Co/Cu(110) system. These studies also showed that the nonlinear magneto-optical effects are quite sizable: magnetic contrasts between signals with opposite magnetization directions of over 50% were observed. This indicates that the magnetization-induced tensor elements are of the same order of magnitude as the nonmagnetic ones. This is in strong contrast to the linear optical response, where the magnetization induces very small off-diagonal tensor elements.

SHG arises from the nonlinear polarization $P(2\omega)$ induced by an incident laser field $E(\omega)$. This polarization can be written as an expansion in $E(\omega)$:

$$P(2\omega) = \chi^{(2)}E(\omega)E(\omega) + \chi^{(Q)}E(\omega)\nabla E(\omega) + \cdots \quad (3.3)$$

The lowest order term in Eq. (3.3) describes an electric dipole source. Symmetry considerations show that this contribution is zero in a centrosymmetric medium, thus limiting electric

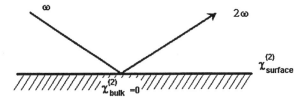

Fig. 48. A schematic view of second harmonic generation. Reprinted with permission from [170], © 1996, Elsevier Science.

dipole radiation to the interface, where inversion symmetry is broken (see Figure 48). The bulk second harmonic can now be described in terms of the much smaller electric quadrupole-like contributions (second term in Eq. 3.3). Because of the large volume difference between interface and bulk this does not necessarily mean that the total bulk second harmonic signal is negligible. Interface sensitivity has to be verified for any given system. In particular, for insulating materials and for semiconductors using below band gap excitation, these bulk effects can be large and sometimes even dominating. The very effective screening in metals usually limits the symmetry breaking to the first atomic layers [164, 165], though sizable bulk contributions have been observed as well [166]. Experimentally, surface sensitivity can often be demonstrated by in situ surface modification, like oxidation or CO adsorption. A complete theoretical treatment of the various contributions to surface SHG is given by Guyot-Sionnest [167]. There are various reviews that discuss the application of SHG in surface science [168] or metals [169].

Magnetization will not affect the inversion symmetry, but will introduce extra nonzero surface tensor elements. Those will change sign when the direction of M is reversed, which leads to a nonlinear polarizability that will depend on M:

$$P(2\omega, \pm M) = (\chi^{(2),+} \pm \chi^{(2),-})E(\omega)E(\omega)$$
$$+ (\chi^{(Q),+} \pm \chi^{(Q),-})E(\omega)\nabla E(\omega) + \cdots \quad (3.4)$$

These new tensor components that are odd in the magnetization will lead to the phenomenon of MSHG. Because M is an axial vector, its orientation is conserved under inversion. This means that magnetization does not induce electric dipole contributions in the bulk and the surface sensitivity is conserved. The time reversal symmetry breaking will induce the new nonzero tensor elements for both $\chi^{(2)}$ and $\chi^{(Q)}$. The derivation of these magnetization-sensitive tensor elements was given in [170]. The results are summarized in Table II [171].

From Table II the effect of the magnetization on the SHG response can be qualitatively understood directly. Take, for example, an s-polarized input beam for the case of a longitudinal configuration. In that situation, there is only one even term (χ_{zyy}) and one odd term (χ_{yyy}) that will contribute to the MSHG response. Because the even term will give rise to a p-polarized output and the odd term will give a s-polarized SHG signal, the total output polarization can be varied by changing the direction of the magnetization. This will lead to a non-

Table II. Nonzero Tensor Elements of the Nonlinear Susceptibility Tensor $\chi^{(2)}(M)$ for the (0 0 1) Cubic Surface, the Magnetization Parallel to the x Axis (Longitudinal Configuration), the y Axis (Transversal Configuration), and the z Axis (Polar Configuration)

	Even in M	Odd in M
Longitudinal $M \parallel x$	$yzy = yyz$	$xyx = xxy$
	$xzx = xxz$	$zyz = zzy$
	zzz	yzz
	zyy	yyy
	zxx	yxx
Transversal $M \parallel y$	$xxz = xzx$	$yxy = yyx$
	$yyz = yzy$	$zxz = zzx$
	zxx	xxx
	zyy	xyy
	zzz	xzz
Polar $M \parallel z$	$xxz = xzx = yyz = yzy$	$xyz = xzy = yxz = yzx$
	$zxx = zyy$	$zxy = zyx$
	zzz	

For simplicity of notation the tensor components are indicated by their indices only.

linear Kerr rotation, similar to the linear case. However, the effects can be much larger than those in the linear case [172, 173]. One of the reasons for this is that the odd component is of the same order of magnitude as the even component.

MSHG offers new possibilities for magnetic domain imaging. The use of an optical response that is governed by a higher rank tensor offers sensitivity to additional combinations of magnetization directions and optical wave vectors and polarization. A symmetry analysis of nonlinear magneto-optical imaging of magnetic domains and domain walls has been presented in [174]. Gradient terms are shown to give rise to MSHG via spatial derivatives of the magnetization. The nonvanishing independent elements of the relevant tensors are derived for cubic media, and different contributions to the MSHG image from domains and domain walls are analyzed for thin magnetic films with symmetry. It is shown that measurements of polarization properties of the MSHG response may yield information about the relative importance of different magnetization-induced contributions and the type of domain walls.

MSHG combines interface specificity with large nonlinear magneto-optical effects, which makes it particularly attractive for the study of magnetic thin films and multilayers. Although the origin of the nonlinear magneto-optical effects is the same that of the linear effect, namely the spin orbit coupling, the effects can be quite different. This is partly related to the different (inhomogeneous) character of the wave equation. This leads to nonlinear Kerr rotations that are about one to two orders of magnitude larger than their linear equivalents. For multilayered systems, MSHG could provide information about domain structures at deeper interfaces. The effects of sample preparation conditions like sputtering pressure, substrate temperature, and deposition speed on the interface magnetic properties can be probed by MSHG.

3.2.3. Nanosecond Resolved Techniques for Dynamical Magnetization Reversal Measurements

Magnetization dynamics of magnetic films in the nanosecond regime is an essential issue for the future of magnetic recording and nonvolatile magnetic memories [175, 176]. Measurements on these time scales involve the ability of generating fast magnetic (or thermal) pulses and detecting the magnetization process by means of high-frequency techniques.

3.2.3.1. Time-Resolved Magneto-Optical Kerr Effect

A simple way to probe dynamically magnetization on the nanosecond scale is by means of visible light. Interactions of light with matter take place in the 10^{-15} s range, so they can be neglected when probing phenomena slower than 10^{-12} s. In addition, the use of light as the probe makes the measurement relatively immune to the strong electromagnetic noise present in high current pulse generation. For example, one can choose magneto-optical Kerr and Faraday effects to probe opaque and transparent materials, respectively.

CNRS–Laboratoire Louis Néel [179] used polarized light from a 5-mW HeNe (or diode) laser shining on a sample placed in or on a micro coil. The polarization of the reflected or transmitted light is then analyzed, and its intensity is detected with a fast Si photodiode (100 MHz). The signal is then measured with a fast digitizing oscilloscope (500 MHz, 1 GS/s) and transferred to a computer. The analyzer is placed at 45° with respect to the polarizer, where one has the highest sensitivity and linearity between the angle and the intensity. Statistical noise is greatly reduced by averaging some measurements taken in the same conditions. To eliminate the synchronous electromagnetic noise, the analyzer is flipped from +45° to −45° between successive measurements, and the difference gives the final value. A resolution of 5 μrad on the polarization plane is achieved in 3 min of acquisition time. In contrast to other dynamic Kerr techniques based on a pump-probe scheme [177, 178], this real-time approach allows single-shot measurements, that are useful for probing nonreproducible magnetic states. Magnetic thin films like garnets, Co, Fe, Ni, TbFe, PtCo multilayers, amorphous GdCo, and TbCo have been dynamically probed with this set-up.

3.2.3.2. Time-Resolved Magneto-Optical Imaging

Three different mechanisms are known to take part in the magnetization reversal dynamics of thin films in the nanosecond scale [180]: domain nucleation, domain wall propagation, and coherent rotation. The first two mechanisms are essentially thermally activated [142, 181] and dominate the reversal in the low-field, low-frequency regime. Coherent rotation is not thermally activated and typically occurs for high-field, high-frequency magnetic fields [180]. To distinguish and study these mechanisms, magneto-optical imaging appears to be a powerful technique for dimensions down to submicron [177].

With the same optical set-up used for dynamical Kerr and Faraday measurements, one can develop a magneto-optical imaging system, with a CCD camera as detector. It is based on a "pump-probe" technique, where the "pump" is the magnetic pulse and the "probe" is a flash of light from a laser diode (850 nm, 5 ns FWHM). This is a new and inexpensive way to replace the picosecond or femtosecond lasers normally used for this kind of experiment, to the detriment of time resolution (subnanosecond laser diodes are already available). With this set-up one is able to visualize magnetic domain motion with a spatial resolution of about 0.7 μm and a time resolution of 5 ns.

Figure 49 shows a series of images from a $(YGdTmBi)_3(FeGa)_5O_{12}$ garnet, taken before and after the application of a pulsed field (0.2 T/50 ns) at room temperature. Once the magnetization of this sample is saturated, reversal occurs randomly, preventing the use of a pump-probe technique. Therefore, the amplitude of the applied field was chosen to avoid complete saturation of the probed region.

One can see that, in this system, the magnetization reversal process on this time scale is dominated by domain wall propagation instead of nucleation or coherent rotation. The domain wall speed after the magnetic pulse duration can easily be calculated from two subsequent images and has an average value of 35 m/s. Domains enlarge to a dimension a bit bigger than the steady state and then relax to a static condition. The coil used in this experiment gives rise to a vertical gradient of field, with its maxima near the borders (top and bottom) and a minimum at the center. This explains why the regions near the borders are completely saturated whereas in the center some domains remain. This also explains the bubble-like domains in the center and ribbon-like domains near the borders.

3.3. Magnetic Force Microscopy

The magnetic force microscope (MFM) (Fig. 50) has been developed from the earlier atomic force microscope, in which

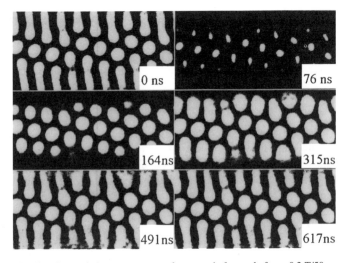

Fig. 49. Dynamic image sequence of a garnet before and after a 0.2 T/50 ns pulse applied at $t = 0$ by a 50 μm × 1000 μm coil. Each frame measures 50 μm × 100 μm. Reprinted with permission from http://lu-w3.polycurs-gre.ft/themes

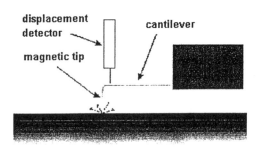

Fig. 50. Principle of MFM. Reprinted with permission © IST Group.

the alternating voltage on the probe tip is replaced with an alternating magnetic field [182]. The tip is used to probe the magnetic stray field above the sample surface. The magnetic tip is mounted on a small cantilever, which translates the force into a deflection that can be measured. The microscope can sense the deflection of the cantilever, which will result in a force image (static mode), or the resonance frequency change of the cantilever, which will result in a force gradient image. The sample is scanned under the tip, which results in a mapping of the magnetic forces or force gradients above the surface. Several improvements have been developed, regarding the tip, also under consideration of applied fields [183–189].

One of the most crucial parts for the image formation process in a MFM is the force sensor formed by a tiny magnetic tip mounted on a flexible beam (cantilever). Artefacts due to the usually unknown magnetic state of the tip, its unknown behavior in the sample's stray field, and its influence on the sample magnetization may lead to image perturbations and misinterpretations. To get quantitative information out of MFM measurements, tips with a well-defined magnetic state are required.

The fact that no sample preparation is necessary and that a lateral resolution below 50 nm can be reached make it a powerful tool for the investigation of submicron magnetization patterns. Because it is possible to apply external magnetic fields during the measurement, the field dependence of domain structures and magnetic reversal processes can be observed. Methods for separating topography and magnetic features allow pure magnetic images to be achieved. Topographic and magnetic details from the same scan can be related to each other. For example, one can use an MFM for the observation of intrinsic magnetic domains and structures of written bits on various recording media.

3.3.1. Intrinsic Domain Studies

Knowledge of the relation between preparation process parameters and magnetic parameters like domain size and shape can be used to tailor better magneto-optic multilayer materials. Figure 51 shows a domain pattern in a Co-Ni/Pt multilayer.

3.3.2. Observations of Perpendicularly Recorded Bits on Co-Cr-Ta Material

Microscopic magnetic observation methods can give a much deeper insight into the magnetic processes during bit writing.

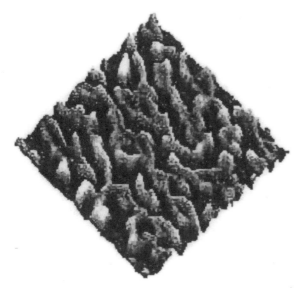

Fig. 51. MFM Image (3.5×3.5 μm) of a domain pattern in a Co-Ni/Pt multilayer. Reprinted with permission © IST Group.

Special attention is paid to a correlation between the writing/reading characteristics of the media, measured in test stages, and the observed magnetic microstructure. An MFM is capable of visualizing magnetic processes that limit the bit density in media. The readback noise for these bits will be determined by the two following effects.

3.3.2.1. Irregularities of the Bit Transitions

In such high writing densities the bits approach the minimum stable domain size of the medium. The bits are broken up by reversed domains, which are formed in the magnetic relaxation process immediately after the bits are written. Because of these irregularities the bits will differ from each other, which is reflected by higher noise.

3.3.2.2. Irregularities of the Track Border

At the track border the domains of the natural domain structure influence the bit formation. Moreover, this will cause irregularities in the track width, which are reflected by higher noise.

3.4. Transmission Electron Microscopy

The modern transmission electron microscope is a powerful tool for the study of a wide range of magnetic materials currently under development. The primary motivation for its use is that many applicable magnetic properties are extrinsic rather than intrinsic to the materials themselves. Hence a detailed knowledge of both physical and magnetic microstructures is essential if the structure–property relation is to be understood and materials with optimized properties produced. Many of the materials of interest are markedly inhomogeneous, with features requiring resolution on a sub-50-nm scale for their detailed investigation. Hence the attraction of transmission

electron microscopy (TEM) is twofold. It offers very high spatial resolution and, because of the large number of interactions that take place when a beam of fast electrons hits a thin solid specimen, detailed insight into composition, electronic, as well as structural and magnetic properties. The resolution that is achievable depends largely on the information sought and may well be limited by the specimen itself. Typical resolutions achievable are 0.2–1.0 nm for structural imaging, 1–3 nm for extraction of composition information, and 2–20 nm for magnetic imaging.

Studies can be made of magnetic structures in the as-grown state, in remanent states, and in the presence of applied fields. From these can be derived basic micromagnetic information; the nature of domain walls, where nucleation occurs; or the importance of domain wall pinning, and many other related phenomena. Furthermore, when the materials of interest are thin films used for information storage in magnetic or magneto-optic recording, the domains themselves are the written "bits" or "marks." In this case imaging studies by TEM can help provide an understanding of bit stability, the factors limiting achievable packing densities, and the sources of media noise. The use of TEM to study magnetic materials is not new; indeed, the first observations of magnetic domain structures were made with the use of the Fresnel or defocus mode 40 years ago [190]. However, since that time a number of different imaging and diffraction modes have been developed, including Foucault, coherent Foucault, differential phase contrast, and a range of holographic techniques as well as low-angle electron diffraction [191–200].

3.4.1. Imaging Magnetic Structures by TEM

The principal difficulty encountered when a TEM is used to study magnetic materials is that the specimen is usually immersed in the high magnetic field (typically 0.5–1.0 T) of the objective lens [201]. This is sufficient to completely eradicate or severely distort most domain structures of interest. A number of strategies have been devised to overcome the problem of the high field in the specimen region [202]. These include (i) simply switching off the standard objective lens; (ii) changing the position of the specimen so that it is no longer immersed in the objective lens field;[13] (iii) retaining the specimen in its standard position but changing the pole pieces [203], once again to provide a nonimmersion environment; or (iv) adding super minilenses in addition to the standard objective lens, which is once again switched off [204, 205]. Magnetic structures are most commonly revealed in the TEM with the use of one of the modes of Lorentz microscopy. This generic name is used to describe all imaging modes in which contrast is generated as a result of the deflection experienced by electrons as they pass through a region of magnetic induction [206]. The Lorentz deflection angle β_L is given by

$$\beta_L = e\lambda t(\vec{B} \wedge \vec{n})/h \qquad (3.5)$$

[13] Hitachi Technical Data EM Sheet No. 47.

where $\vec{B}$ is the induction averaged along an electron trajectory, $\vec{n}$ is a unit vector parallel to the incident beam, t is the specimen thickness, and λ is the electron wavelength. Substituting typical values into Eq. (3.5) suggests that β_L rarely exceeds 100 μrad and, indeed, can be <1 μrad in the case of some magnetic multilayers or films with low values of saturation magnetization. The problem is further exacerbated when the specimen supports perpendicular magnetization. Clearly, under these conditions β_L is zero unless the specimen is tilted with respect to the electron beam to introduce an induction component perpendicular to the direction of electron travel. Given the small magnitude of β_L, there is no danger of confusing magnetic scattering with the more familiar Bragg scattering, where angles are typically in the range of 1–10 mrad.

The description given so far is essentially classical in nature, and much of Lorentz imaging can be understood in these terms. However, for certain imaging modes and, more generally, if a full quantitative description of the spatial variation of induction is required, a quantum mechanical description of the beam–specimen interaction must be sought, which may be considered as a phase modulation of the incident electron wave, with the phase gradient $\triangle\phi$ of the specimen transmittance being given by

$$\triangle\phi = 2\pi et(\vec{B} \wedge \vec{n})/h \qquad (3.6)$$

where e and h are the electronic charge and Planck's constant, respectively. Substituting typical numerical values shows that magnetic films should normally be regarded as strong, albeit slowly varying, phase objects [207]. For example, the phase change involved in crossing a domain wall usually exceeds π rad. So far it has been assumed that the only interaction suffered by the beam is due to the magnetic induction. However, other than in the free space region beyond the edges of a magnetic specimen, allowance must also be made for the effect of the electrostatic potential. Indeed, the electrostatic interaction is generally much the stronger of the two. Hence Eq. (3.6) is incomplete; a fuller form is

$$\nabla\phi = 2\pi et(\vec{B} \wedge \vec{n})/h + \pi\nabla(Vt)/\lambda W_0 \qquad (3.7)$$

where W_0 is the electron energy and V denotes the inner potential. The latter has a rapidly varying component corresponding to the periodicity of the lattice, but there are also variations at boundaries and defects. Furthermore, there is a variation that might be either slow or rapid if there are composition changes throughout the specimen. Whereas for many thin film samples t does not vary appreciably and hence does not contribute to the second term in Eq. (3.7) due to the gradient operator, contributions can also be expected here when there are variations in local specimen topography. A good example of where this is the case is at the edges of small magnetic elements where the thickness changes abruptly, and the large "electrostatic" signal so generated can conceal important magnetic structures in the nearby region.

3.4.2. Fresnel and Foucault Imaging

The most commonly used techniques for revealing magnetic domain structures are the Fresnel (or defocus) and Foucault imaging modes [207]. Both are normally practised in a fixed-beam (or conventional) TEM; schematics of how magnetic contrast is generated are shown in Figure 52. For the purpose of illustration, a simple specimen comprising three domains separated by two 180° domain walls is assumed. In Fresnel microscopy the imaging lens is simply defocused so that the object plane is no longer coincident with the specimen. Narrow dark and bright bands, delineating the positions of the domain walls, can then be seen in an otherwise contrast-free image. The structure shown in the bright band (known as a convergent domain wall image) reflects the fact that a simple ray treatment is incomplete and that detail, in this case interference fringes, can arise because of the wave nature of electrons. For Foucault microscopy, a contrast-forming aperture must be present in the plane of the diffraction pattern, and this is used to obstruct one of the two components into which the central diffraction spot is split because of the deflections suffered as the electrons pass through the specimen. Note that in general, the splitting of the central spot is more complex than for the simple case considered here.

As a result of the partial obstruction of the diffraction spot, domain contrast can be seen in the image. Bright areas cor-

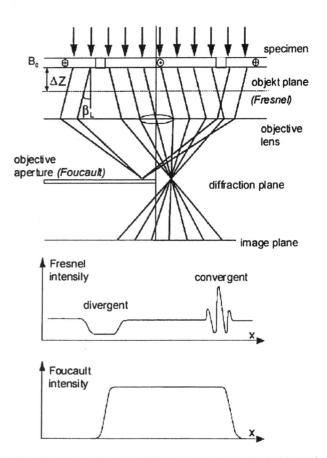

Fig. 52.   Schematic of Fresnel and Foucault imaging. Reprinted with permission © IST Group.

respond to domains where the magnetization orientation is such that electrons are deflected through the aperture, and dark areas correspond to domains where the orientation of magnetization is oppositely directed. The principal advantages of Fresnel and Foucault microscopy together are that they are generally fairly simple to implement, and they provide a clear picture of the overall domain geometry and a useful indication of the directions of magnetization in (at least) the larger domains. Such attributes make them the preferred techniques for in situ experimentation. However, a significant drawback of the Fresnel mode is that no information is directly available on the direction of magnetization within any single domain, and reproducible positioning of the contrast-forming aperture in the Foucault mode is difficult. Moreover, both imaging modes suffer from the disadvantage that the relation between image contrast and the spatial variation of magnetic induction is usually nonlinear [208]. Thus extraction of reliable quantitative data is problematic, especially from regions where the induction varies rapidly. A partial solution to the problem is provided by the novel coherent Foucault imaging mode, which requires a TEM equipped with a field emission gun (FEG) for its successful implementation [209]. Here the opaque aperture is replaced by a thin phase-shifting aperture, an edge of which must be located on the optic axis. An amorphous film containing a small hole works well, and the thickness is chosen so that the phase shift experienced is π rad. The technique can only be applied close to a specimen edge, as a reference beam passing through free space is required. Provided this is the case, the image takes the form of a magnetic interferogram of the kind obtained with electron holography. The fringes map lines of constant flux, and adjacent fringes are separated by a flux of h/e. Hence, as with holography, quantitative data are available, and in the case of coherent Foucault imaging no processing of the recorded image intensity distribution is required. For ultrathin specimens the usefulness of the technique decreases because of the excessive separation of the fringes.

3.4.3. Low-Angle Diffraction

Although it is not an imaging technique, another way of determining quantitative data from the TEM is by directly observing the form that the split central diffraction spot assumes. The main requirement for low-angle diffraction (LAD) is that the magnification of the intermediate and projector lenses of the microscope is sufficient to render visible the small Lorentz deflections. Camera constants (defined as the ratio of the displacement of the beam in the observation plane to the deflection angle itself) in the range 50–1000 m are typically required. Furthermore, high spatial coherence in the illumination system is essential if the detailed form of the low-angle diffraction pattern is not to be obscured. In practice this necessitates that the angle subtended by the illuminating radiation at the specimen be considerably smaller than the Lorentz angle of interest. The latter condition is particularly easy to fulfil in a TEM equipped with a FEG. Thus although LAD does supplement somewhat the deficiencies of Fresnel and Foucault

imaging, the fact that it provides global information from the whole of the illuminated specimen area rather than local information means that alternative imaging techniques must also be utilized.

3.4.4. Differential Phase Contrast Imaging

Differential phase contrast (DPC) microscopy [210, 211] overcomes many of the deficiencies of the imaging modes discussed above. Its normal implementation requires a scanning TEM. It can usefully be thought of as a local-area LAD technique using a focused probe. Figure 53 shows a schematic of how contrast is generated [213]

In DPC imaging the local Lorentz deflection at the position about which the probe is centered is determined with a segmented detector situated in the farfield. Of specific interest are the difference signals from opposite segments of a quadrant detector, as these provide a direct measure of the two components of β_L. By monitoring the difference signals as the probe is scanned in a regular raster across the specimen, directly quantifiable images with a resolution approximately equal to the electron probe diameter are obtained. Providing a FEG microscope is used, probe sizes less than 10 nm can be achieved with the specimen located in field-free space [212]. A full wave-optical analysis of the image formation process confirms the validity of the simple geometric optics argument outlined above for experimentally realizable conditions. The main disadvantages against which this must be set are the undoubted increase in instrumental complexity, the operational difficulty compared with the fixed-beam imaging modes, and the longer image recording times inherent to all scanning techniques. The two difference-signal images are collected simultaneously and are therefore in perfect registration. From them

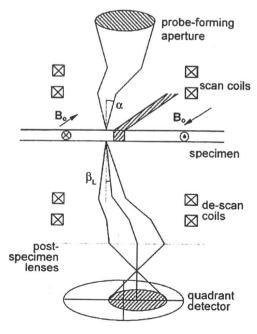

Fig. 53. Schematic of DPC imaging. Reprinted with permission from [213], © 1999, Elsevier Science.

a map of the component of magnetic induction perpendicular to the electron beam can be constructed. In addition, a third image formed by the total signal falling on the detector can be formed. Such an image contains no magnetic information (the latter being dependent only on variations in the position of the bright-field diffraction disc in the detector plane) but is a standard incoherent bright-field image, as would be obtained with an undivided spot detector. Thus a perfectly registered structural image can be built up at the same time as the two magnetic images, a further distinct advantage of DPC imaging. However, the simple analyses given above have assumed that image contrast arises solely as a result of the magnetic induction–electron beam interaction, whereas contrast arising from the physical microstructure is virtually always present as well. The effect of the unwanted contrast of structural origin can be ameliorated to some extent by a suitable choice of operating conditions or modification of the DPC technique itself. It is frequently the case that the physical microstructure is on a significantly smaller scale than its magnetic counterpart. If this is so, the influence of high spatial frequency components in the image can be reduced by replacement of the solid quadrant DPC detector with its annular counterpart [214]. If this is done, not only is the unwanted signal component suppressed considerably, but the signal-to-noise ratio in the magnetic component is significantly enhanced.

4. MAGNETIC THIN FILM PROCESSING

This chapter deals with the deposition and structuring of thin magnetic films. Further details are given in the sections on applications (e.g., describing the technological aspects of anisotropic magnetoresistive sensors).

4.1. Deposition of Magnetic Thin Films

Various techniques are used for the fabrication of thin films, depending on the material and the specific application [215–220]. These techniques can be divided into physical vapor deposition (PVD) and chemical vapor deposition (CVD) methods. Among these various methods, sputtering is the most important deposition technique for magnetic thin films. For example, magnetic films for recording (e.g., FeAlSi, CoNbZr, CoCr, FeNiMo, FeSi, CoNiCr, and CoNiSi) can be fabricated with this technique.

4.1.1. Sputtering

In a common sputtering system positive ions are accelerated from a plasma to a target that is at a negative potential with respect to the plasma (Fig. 54). The ions reach the target surface with an energy given by the potential drop between the target and the plasma. The target surface is considered as the source of material from which films are grown. In addition to the neutral (sputtered) material liberated from the bombarded surface, which eventually condenses as a film, there are numerous other events that can occur at the target surface that

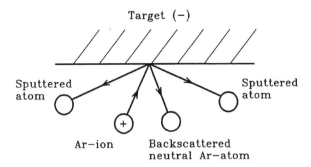

Fig. 54. Principle of sputtering. Reprinted with permission from [217], © 1998, Springer-Verlag.

may influence the growth of films profoundly. These include secondary electron emission, secondary positive and/or negative ion emission, emission of radiation (photons, X-rays), reflection of incident particles, heating, chemical dissociation or reaction, and others.

The principle of sputtering can also be used to clean substrates before film deposition. However, it is assumed that gross surface contaminants have been removed chemically or otherwise before the substrates are put in the vacuum chamber. Furthermore, it should be mentioned that it is very difficult to obtain an atomically clean surface.

4.1.1.1. DC Diode Sputtering

Figure 55 represents a greatly simplified cross section of a dc diode sputtering system. In such a system the cathode electrode is the sputtering target and the substrate holder (anode) faces the target. The holder may be grounded or biased. To establish a dc-diode discharge in argon, the gas pressure must be greater than about 10 mTorr (=1.33 Pa). When the glow discharge is started, positive ions strike the target plate and remove mainly neutral target atoms by momentum transfer, and these condense into thin films. The most critical parameters in every sputtering system are the input power, argon gas pressure, and substrate temperature.

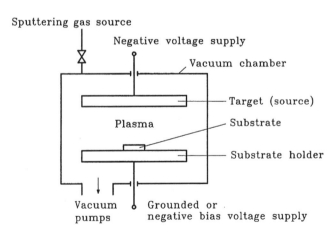

Fig. 55. Principle of a dc diode sputtering system with substrate bias [218].

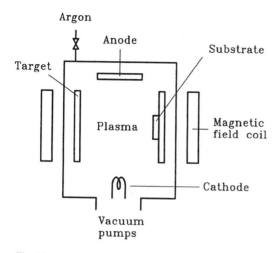

Fig. 56. Principle of a dc triode sputtering system [218].

4.1.1.2. DC Triode Sputtering

In this technique, ions are generated in a low-voltage, high-current arc discharge between a thermionic filament (cathode) and a main anode. A magnetic confinement concentrates the plasma along the cathode–anode axis. The sputtering target is located in this main discharge (Fig. 56). Ions are extracted from the plasma and accelerated toward the negative target. The substrates are facing the target. This technique produces a very high ion density and allows operation at much lower argon pressures than in a diode sputtering system.

4.1.1.3. RF Sputtering

Radio frequencies (rf) used for sputter deposition are in the range of 0.5–30 MHz; 13.56 MHz is a commercial frequency that is often used. Owing to the difference in mobility between electrons and ions, the I–V characteristics of a glow discharge resemble those of a rectifier. The target is coupled to the rf generator through a series capacitor (Fig. 57). Because no charge can be transferred through the capacitor, the voltage

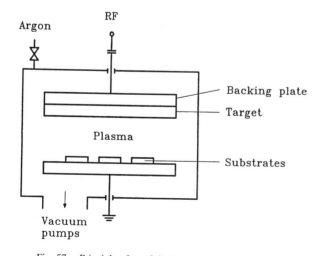

Fig. 57. Principle of an rf diode sputtering system [218].

on the target surface must self-bias negatively until the net current (averaged over each cycle) is zero. This results in a pulsating negative target potential. The corresponding average dc voltage of this potential is nearly equal to the peak voltage applied. rf sputtering can be performed at low gas pressures (<1 mTorr = 0.133 Pa). Because the target is capacitively coupled to the plasma, it makes no difference whether the target surface is electrically conductive or insulating.

4.1.1.4. Magnetron Sputtering

In conventional dc and rf sputtering systems the electrons ejected from the target do not significantly contribute to sustaining of the discharge. Most of these electrons reach the substrate and cause unwanted heating. If a magnetic field is applied parallel to the target surface, the secondary electrons circle around the magnetic field lines and stay near the target surface, thereby increasing the ionization efficiency. The magnetic field is generated by permanent magnets (barium ferrites, alnico alloys, cobalt rare earth alloys) arranged behind the target (Fig. 58). If there is an electric field E perpendicular to the magnetic field B, a drift develops in a direction perpendicular to both E and B. This $E \times B$ drift causes the electrons to move on a closed path parallel to the target surface. This high flux of electrons creates a ring-shaped zone of high-density plasma and a high sputtering rate. The most common magnetron source is the planar magnetron, where the sputter-erosion path is a closed circle or elongated circle.

Both dc and rf magnetron sputtering systems are used for the production of magnetic films. One disadvantage of the magnetron sputtering configurations is that the plasma is confined near the target and is not available to activate reactive gases near the substrate. This disadvantage can be overcome by an unbalanced magnetron configuration where some electrons can escape from the target region [215].

Magnetron sputtering is a high-rate deposition method and can deposit metallic films over large areas at rates comparable to those of electron-beam evaporation without the degree of radiation heating typical for thermal sources. As a further advantage, substrates are not heated up by secondary electrons. These thermal aspects are important if the lift-off technique is used for photolithographic patterning of films.

4.1.1.5. Sputtering with Electrical Substrate Bias

In some instances, a bias potential (usually negative) is applied to the substrate holder, so that the growing film is subject

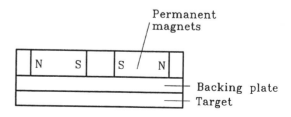

Fig. 58. Principle of an rf magnetron cathode.

to positive ion bombardement (Fig. 55). This negative bias reduces the contamination in sputter-deposited films. Bias sputtering is also often described as a means of improving the surface coverage and planarization of patterned structures. A negative self-bias is also induced on an insulating substrate. Any material body immersed in a glow discharge, unless it is grounded, will acquire a slightly negative potential with respect to ground. The higher the electron energy and flux, the higher the generated negative self-bias.

4.1.1.6. Sputtering with Magnetic Substrate Bias

Sometimes it is desirable to have a magnetic bias on the substrate surface during the deposition of magnetic thin films to influence the film growth to induce a uniaxial anisotropy. For example, NiFe films used for magnetoresistive sensors are deposited with a magnetic substrate bias to achieve the desired easy axis of magnetization. The use of a magnetic field in the vicinity of the target can affect sputtering target performance and may extract electrons from the target [215]. In a dc triode sputtering system the magnetic field coil that is used to concentrate the plasma provides a magnetic bias.

4.1.1.7. Reactive Sputtering

In some cases, gases or gas mixtures other than Ar are used in a sputtering system. In one form of reactive sputtering, the target is a nominally pure metal, alloy, or mixture of species that one desires to synthesize into a compound by sputtering in a pure reactive gas or an inert gas-reactive gas mixture (e.g., O_2, N_2, H_2O, NH_3, H_2S, CH_4, SiH_4, CF_4, in argon). The reaction can occur at the target or in the plasma or at the substrate surface (Fig. 59). Reactive sputtering is also used to replenish constituents of compound targets lost by dissociation. A problem in reactive sputter deposition is to prevent the "poisoning" of the sputtering target by the formation of a compound layer on its surface [215].

4.1.1.8. Sputter Deposition of Composite Magnetic Films

Several types of targets or target arrangements can be used for composite films:

- hot pressed sinter targets
- vacuum cast alloy targets
- chips added to a target
- multiple targets mounted confocal to a stationary substrate
- consecutive coating of a rotating substrate table from several targets

A modified single target can be used for the deposition of alloys and mixtures. For example, a mosaic target may have tiles of several materials. The composition of the film can then be changed by changing the area ratio.

In the case of multiple targets, the flux distribution from each target must be considered. When large area targets are

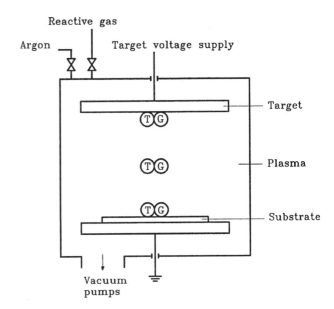

Fig. 59. Principle of reactive sputtering [151, 218]. T, target atom; G, gas molecule.

used, the substrates are rotated in front of the targets (Fig. 60). Layered structures can be achieved by this method.

When magnetic alloys such as CoCrTa, CoNiCrTa, CoCrPt, CoFeTb, or CoCrNiPt are used, material distribution in the target is extremely important [215]. The target should be of uniform composition. Furthermore, the composition of the deposited film may be different from that of the target material (a common example is GdTbFe). The change in composition can be compensated for by changing the target composition empirically.

4.1.1.9. Sputter Deposition of Multilayer Stacks

The multilayer stacks that exhibit the giant magnetoresistance (GMR) effect are typically composed of alternating thin layers of magnetic materials (e.g., NiFe or NiFeCo) and nonmagnetic, conducting materials (e.g., Cu, Ag, Au) [221, 222]. As an example, such a multilayer stack can consist of alternating 10-Å Cu and 20-Å $Ni_{80}Fe_{20}$, for example, with a total film thickness of about 1000 Å. This stack can be fabricated by sputtering. Magnetic and nonmagnetic layers are formed by simultaneous rf and dc discharge, respectively. Layer thicknesses are controlled by regulating the time the shutters are left open. The substrate material can be glass or oxidized silicon, for example.

4.1.1.10. Ion Beam Sputtering

Ion beams are used in thin film deposition in two basic configurations. In one version the ion beam consists of the desired film material and is deposited directly on a substrate. In the second version the ion beam is usually an inert or reactive gas at higher energy and is directed at a target of the desired material, which is sputtered (Fig. 61). Ion beam sputtering

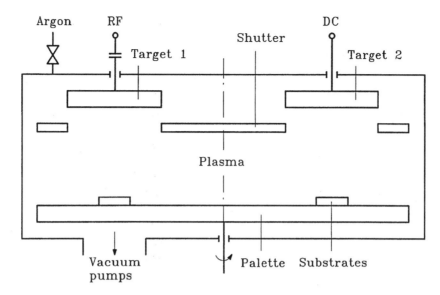

Fig. 60. Consecutive coating of substrates placed on a rotating palette in a diode sputtering system.

has the advantage that the flux and energy of the bombarding ions can be well regulated. It allows better isolation of the substrate from the ion source compared with the conventional diode sputtering configuration. This allows control over the substrate temperature, gas pressure, angle of deposition, etc. This flexibility can lead to unique film properties not obtained with conventional deposition methods [216].

4.1.2. Evaporation

Evaporation is a deposition process that requires a relatively good vacuum (better than 10^{-4} Torr $= 1.33 \times 10^{-2}$ Pa). Atoms or molecules from a thermal vaporization source reach the substrate without collisions with residual gas molecules.

4.1.2.1. Electron Beam-Heated Sources

The energy necessary for the evaporation of high-melting-temperature materials, such as refractory metals or ceramics, can be efficiently provided by a focused electron beam. This method of heating is also very useful for evaporating large quantities of materials. Electrons emitted from a thermionic filament are accelerated by a high voltage (10–20 kV) and focused on the surface of the material to be evaporated (Fig. 62). A magnetic field is used to deflect the beam through 270° to avoid deposition of evaporated material on the filament insulators. Electron gun sources can have multiple pockets, to provide the possibility of vaporizing several

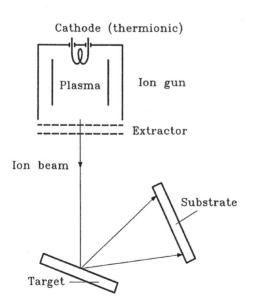

Fig. 61. Principle of an ion beam sputtering system [215, 218] (vacuum chamber not shown).

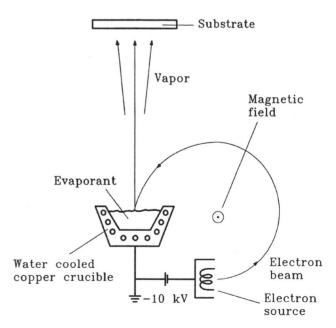

Fig. 62. Principle of an electron beam evaporation system [215] (vacuum chamber not shown).

materials with the same electron source. The magnetic field used in electron gun sources can act as a magnetic substrate bias. Sometimes shielding is necessary to avoid unwanted bias.

4.1.2.2. Resistively Heated Sources

Many materials can be evaporated from high-melting-point sources that are resistively heated by an electrical current. The heated source can be a wire, boat, basket, etc. Typical resistive heater materials are W, Ta, Mo, C, and BN/TiB_2 composite ceramics. The electrically conductive sources are supplied with low voltage and very high current (several hundred amperes). This high current could eventually act as a magnetic substrate bias. Therefore, this evaporation method should not be used for the deposition of magnetic films if such a magnetic bias is not wanted. It should also be mentioned that no proper source material exists for the evaporation of Fe, Co, Ni, Ti, and Zr. In these cases it is better to use electron beam sources or sputtering techniques.

4.1.2.3. Flash Evaporation

Flash evaporation is used to vaporize alloys whose constituents have widely differing vapor pressures. Pellets periodically dropped on a very hot surface are completely vaporized within short periods of time. It is also possible to use a fine-grained powder as the source of material. A new technique is the pulsed laser vaporization of a slowly moving solid surface (see section on laser-induced vaporization).

4.1.3. Molecular Beam Epitaxy

The fabrication of single-crystal films on single-crystal substrates by molecular beam epitaxy (MBE) has increased in importance for several reasons. Ultrathin films only a few atoms thick or multilayers can be grown precisely with this technique. Such structures are often called superlattices. Epitaxial growth of ferromagnetic material on semiconductors (e.g., Fe/GaAs) has been achieved [223].

The availability of MBE led to the investigation of the exchange coupling between thin layers of a ferromagnetic material separated by thin layers of a nonmagnetic material. The film thicknesses used in such multilayer systems were a few tens of angstroms. The specific aspect studied was the change from ferromagnetic (parallel) to antiferromagnetic (antiparallel) coupling of the separated ferromagnetic films.

The issue of ultrathin magnetic films exhibiting perpendicular magnetic anisotropy (PMA) appears to be an exciting challenge in both its applied and fundamental aspects. Such films open a promising path for the race toward higher densities of information storage in magnetic recording media.

In MBE, a vacuum of better than 10^{-9} Torr ($=1.33 \times 10^{-7}$ Pa) is used, and the film material is deposited from a carefully rate-controlled vapor source. The deposition chamber normally contains several analytical instruments for in situ analysis of the growing film.

4.1.4. Laser-Induced Deposition

Laser techniques have grown in importance, for example, for the deposition of magnetic films with precise compositions. Another application of lasers is the fabrication of magnetic nanowires. Understanding the magnetic properties of very small particles is a challenge motivated by the magnetic recording industry.

4.1.4.1. Laser-Induced Vaporization

Laser radiation can be used to vaporize the surface of a material [215]. Laser vaporization or pulsed laser deposition is sometimes called laser ablation deposition. Typically an excimer laser (YAG (yttrium aluminium garnet) or ArF) is used to provide energy in pulses (e.g., 5 ns, 5 Hz in the case of a YAG laser and 20 ns, 50 Hz in the case of an ArF laser). The energy density can be greater than 10 GW/cm². This technique is a special kind of flash evaporation. A constant-composition alloy film can be deposited, whereby a small amount of the alloy material is periodically completely vaporized.

Laser vaporization has been successfully used, for example, for the deposition of amorphous magnetic films [224]. The films of general composition FeCoNiSiB were deposited on glass substrates by laser ablation of amorphous ribbon targets in a high vacuum. A XeCl excimer laser operating at a wavelength of 308 nm with a pulse duration of 30 ns and a repetition rate of 10 Hz served as the radiation source. The radiation was incident at a 45° angle and was focused to produce a 1 mm² spot size. The amorphous ribbons were placed in a holder rotating at 3 Hz (Fig. 63). Substrates were kept at room temperature. The film thickness can be controlled by the number of shots.

4.1.4.2. Laser-Focused Deposition of Magnetic Nanowires

A laser standing wave acts as a linear array of weak cylindrical lenses, which focus Cr atoms emerging from an MBE evaporator into lines. The Cr atoms are deposited on a substrate positioned immediately beneath the standing wave. The spacing between the nodes of the standing wave is half the

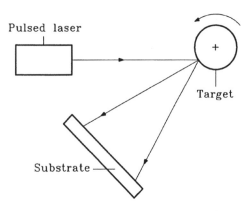

Fig. 63.   Principle of a laser-induced vaporization system (vacuum chamber not shown).

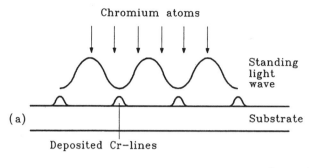

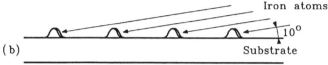

Fig. 64. Principle of the laser-focused deposition of iron magnetic nanowires. (a) Deposition of Cr lines. (b) Oblique deposition of iron. Reprinted with permission from [225], © 1998, American Institute of Physics.

laser wavelength. These Cr lines act as a shadow mask for an oblique evaporation of iron (Fig. 64). The iron lines thus formed are approximately 100 nm wide, 20–40 nm thick, and 0.15 mm long, spaced every 213 nm [225].

4.1.5. Plasma-Enhanced Metalorganic Chemical Vapor Deposition

Soft magnetic ferrite thin films (e.g., NiZnFeO) are indispensable magnetic materials for use in next-generation transformers, inductors, and other devices. Such films can be produced with a high deposition rate by plasma-enhanced metalorganic chemical vapor deposition (MOCVD) [226]. The starting gas sources are the organometallic complexes $Fe(acac)_3$ (acac = $C_5H_7O_2$), $Zn(acac)_2$, and $Ni(acac)_2$; the last is dehydrated at 100°C under reduced pressure (7 Pa). Nitrogen is used as a carrier gas. Furthermore, oxygen is introduced into the reaction chamber (Fig. 65). Film fabrication is performed with an inductively coupled plasma CVD system. Films can be

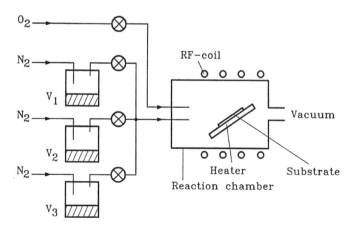

Fig. 65. Principle of plasma-enhanced metalorganic chemical vapor deposition. V1, V2, V3: heated vaporizers containing the different organometallic complexes. Reprinted with permission from [226], © 1994, IEEE.

deposited at a rate of 300 nm/min, with the use of an rf power of 450 W (13.56 MHz) and a substrate temperature of 600°C.

4.1.6. Electroplating

Electroplating can be used for the preparation of magnetic thin films [227–230]. Patterns of NiFe films can be realized through photoresist molds. One example is the deposition of Permalloy ($Ni_{80}Fe_{20}$) [227]. A copper seed layer is first evaporated onto the substrate. After a photoresist pattern is deposited, the sample is placed in a solution containing $NiSO_4$ (0.7 mol/liter), $FeSO_4$ (0.03 mol/liter), $NiCl_2$ (0.02 mol/liter), saccharine (0.016 mol/liter), and boric acid (0.4 mol/liter). Good magnetic properties can be obtained with a typical deposition rate of 120 nm/min. Current densities from 2 to 16 mA/cm^2 are used at room temperature.

An external biasing magnet can be applied with the field lines parallel to the substrate. This bias establishes directions of preferred magnetization (easy axis) within the Permalloy film.

4.2. Dry Etching of Magnetic Thin Films

Significant improvements of lithographic techniques have led to a reduction of the feature size in photo- and electron-sensitive resists. Accordingly, there has been a growing emphasis on the use of dry etching methods [216–218, 220], which have an inherently better resolution capability compared with liquid phase chemical etching. Dry etching mainly includes rf plasma etching, reactive ion etching, and microwave electron cyclotron resonance (ECR) plasma etching, although there are several other dry etching techniques, like sputter etching, ion milling, etc. One example of the use of a dry etching method (inductively coupled plasma etching) is the patterning of thin films of the ferromagnetic Heusler alloy NiMnSb [213].

4.2.1. RF Plasma Etching

Plasma etching employs a glow discharge to generate active species such as atoms or free radicals from a relatively inert molecular gas. High-energy electrons in a low-pressure discharge produce nonequilibrium steady-state conditions that can transform a normally inert molecular gas into a highly reactive medium. Many of the gases used in plasma etching contain one or more halogens (especially F, Cl, and Br) or oxygen. Tetrafluoromethane (CF_4) is a particularly useful gas for this application. The active species diffuse to the substrate, where they react with the surface to produce volatile products. Several different reactor configurations are used. In a parallel-plate plasma etching reactor the substrates are placed flat on the grounded electrode. The upper electrode is connected to the rf power (Fig. 66). This arrangement has the advantage that the substrates can simply be heated or cooled.

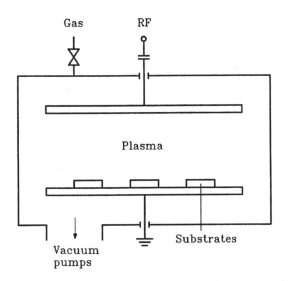

Fig. 66. Principle of a plasma etching system. Reprinted with permission from [217], © 1998, Springer-Verlag.

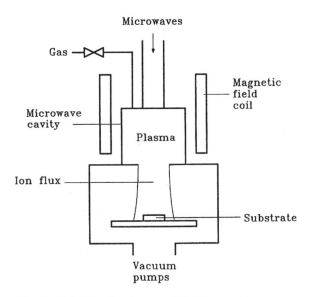

Fig. 68. Principle of a microwave ECR plasma etching system.

4.2.2. Reactive Ion Etching

In a reactive ion etching (RIE) system (Fig. 67) the substrates are placed on the rf powered electrode (13.56 MHz). This allows the grounded electrode to have a significantly larger area because it is in fact the chamber itself. The larger grounded area, combined with the lower operating pressures, leads to significantly higher potentials at the substrate surface, which result in higher energy ion bombardment. The etching gas can be, for example, CF_4, SF_6, CCl_4, BCl_3, or C_2F_6. In a RIE system, not only the chemical etching effect but also the physical sputter effect contributes to the removal of material from the substrate surface. The contribution of sputtering leads to a typical anisotropic etching result. This means that it is possible to produce etched grooves with perpendicular sidewalls. High-resolution patterning of thin magnetic films is necessary to produce ultrasmall magnetic elements [232].

4.2.3. Microwave ECR Plasma Etching

In a microwave plasma etching system the industrial microwave frequency of 2.45 GHz is normally used. If a proper magnetic field is applied (Fig. 68), a resonance condition can be established. The Larmor frequency of the electrons is then equal to the microwave frequency. In this resonance condition (ECR) the ionization becomes extremely high. In the microwave cavity the electron density can be $1–6 \times 10^{11}$ cm^{-3}, and the electron temperature is relatively low (approximately 10 eV) compared with the rf plasma. This results in very effective etching. Typically an ECR discharge is established at a microwave power of 1 kW, a frequency of 2.45 GHz, a magnetic flux density of 80–100 mT, and a gas pressure of 0.1–10 mTorr (0.0133–1.33 Pa) [215]. For example, ECR plasma etching has been used for the patterning of NiFe and NiFeCo thin films [233].

5. APPLICATIONS OF MAGNETIC THIN FILMS

Research activities on magnetic thin films for sensors, actuators, and other micromagnetic devices like inductors, transformers, valves, motors, etc., and, in particular, magnetic data storage devices have been strongly intensified in recent years [217, 234–237]. Thin-film-type miniature sensors have advantages over bulk sensors in their spatial resolution, mass-production capability, and integration capability with semiconductor circuits, etc. Similar advantages apply to other devices based on thin films; the dramatic improvements in the magnetic data storage technology that have occurred in recent years would have been unthinkable without advances in magnetic thin film technology. The following overview gives an introduction to these rapidly growing fields of application. Special emphasis has been placed on anisotropic magnetoresistive sensors because this is the area the authors are most familiar with; a short review of other applications of magnetic thin films has been provided for the sake of completeness.

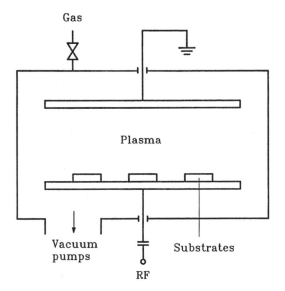

Fig. 67.   Principle of a reactive ion etching system [217].

5.1. Magnetic Sensors

5.1.1. Anisotropic Magnetoresistive Sensors

The resistivity of a thin permalloy film with zero magnetostriction (e.g., 81% Ni, 19% Fe) depends on the angle between the in-plane magnetization and the current [234]. The direction of magnetization can be rotated by an external magnetic field. Therefore, the resistivity is a function of the external field (Fig. 69). The maximum change of resistivity is on the order of a few percent. The easy axis magnetization M_s of the film can be achieved by a magnetic substrate bias during sputter deposition or by subsequent thermal annealing in a magnetic field.

The resistance vs. field characteristic of a single anisotropic magnetoresistive (AMR) element shows a horizontal tangent at zero field. Therefore, in the case of small measuring fields it is necessary to use a bias to improve the sensitivity. Several techniques are known to solve the bias problem [234, 238, 239]. For example, magnetic bias can be achieved by a permanent magnet (hard magnetic film, CoCrPt), a shunt current, or a soft adjacent layer (NiFe, NiFeCr). Further biasing methods are the double AMR element and the exchange bias layer (MnFe, CoO, TbCoFe, CoP, CoNiP, etc.). Another solution of the bias problem is the use of a "herringbone" structure, where M_s is established at 45° with respect to the element length, or of a "barber-pole" scheme. Instead of using magnetic bias to rotate the magnetization to the appropriate angle, the current is made to flow obliquely in the anisotropic magnetoresistive element. The Permalloy film is covered with the so-called barber-pole stripes, which consist of a conductor material (Fig. 70). The geometry of the stripes provides the proper angle between the magnetization and the current.

The AMR effect can be applied to any kind of magnetic sensor. An important application is in magnetoresistive heads, but there are also many other applications, for example, lin-

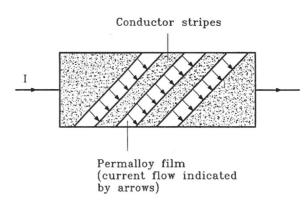

Fig. 70. Magnetoresistive sensor with barber-pole structure. Reprinted with permission from [234], © 1996, Academic Press.

ear position sensors, angular sensors, rotation speed sensors, current sensors, metal detectors, etc.

The following sections cover the effects of electron scattering, depending on the spin direction with respect to spontaneous magnetization in ferromagnetic materials. AMR is utilized in thin films, and GMR is a phenomenon based on exchange coupling in sandwiches of ultrathin magnetic and nonmagnetic layers (superlattices). The recently found colossal magnetoresistance (CMR) also has a potential for future sensor applications.

5.1.1.1. Magnetoresistive Effect

The magnetoresistive effect was discovered in 1857 by Sir W. Thomson [240], but only the last three decades of research and development has permitted its application in industrial sensors and read heads for data storage devices. This recent progress is based on modern microelectronics technology and the demands of miniaturization.

Magnetoresistive sensors are well suited for medium field strengths, for example, earth field navigation or position measuring systems. They can be manufactured at small sizes and low cost (also with on-chip electronics) by the packaging technology of integrated circuits, which is a major prerequisite for mass-market acceptance.

The following sections are based on review books and articles [241, 234, 242], and the references cited there, which are well suited for further reading. A general description of AMR phenomenology, materials characterization, and fabrication aspects is followed by a discussion of linearization and stabilization techniques. The sensor layout aspects are reviewed based on selected examples of industrial products.

5.1.1.2. Anisotropic Magnetoresistance

Mainly three different physical effects contribute to the influence of magnetic fields on the electrical resistivity of solid-state conductors.

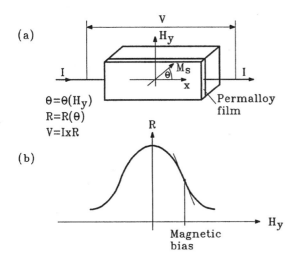

$\theta = \theta(H_y)$
$R = R(\theta)$
$V = I \times R$

Fig. 69. The magnetoresistive effect in a single-domain thin film strip ($Ni_{80}Fe_{20}$). (a) Magnetoresistive element. (b) Resistance as a function of the external field. Reprinted with permission from [234], © 1996, Academic Press.

- The Hall effect is based on the Lorentz force of Eq. (1.11). An increase the resistance is obtained in wide and short conductors[14] because of the deflection of the current paths by the electrical Hall field strength. It is proportional to the electron mobility μ_e and B^2. Because of a low μ_e in metals it is negligible in this case.

- Another contribution is found in paramagnetic and diamagnetic semiconductors and metals (e.g., bismuth). It is caused by band bending at the Fermi surface, and it is also proportional to B^2.

- The third effect appears distinctly in ferri- or ferromagnetic thin films (see Section 1.3.3) with uniform orientation of the spontaneous magnetization, which is parallel to the easy axis of minimum uniaxial anisotropy energy in the absence of an applied field. It is AMR.

Magnetoresistance and Planar Hall Effect. The AMR effect is based on the anisotropic scattering of conduction electrons of the band with uncompensated spins (e.g., 3d orbit for the first transition metals Fe, Co, and Ni) in this exchange split band. The energies of the two states of the magnetic spin moment ($\pm\mu_B$) differ by the quantum-mechanical exchange energy. These electrons are responsible for ferri- and ferromagnetism (see Section 1.2.2).

Theoretical analyses of AMR are given in terms of the electron density in the states diagram and the Fermi level. The difficulty is that the anisotropic part of the resistance depends on the exact three-dimensional shape of the Fermi surface (3D envelope of the Fermi level), which is not known precisely, except for a very few magnetic materials. Theorists, therefore, have not succeeded in calculating the effect to better than one order of magnitude. As a consequence, all materials data have to be found empirically [243]. Not only the magnitude, but the sign of AMR as well cannot be easily predicted. Most of the materials have positive AMR coefficients, which means that the high resistivity state occurs when the spontaneous magnetization $\vec{M}_s$ and the current density $\vec{J}$ are parallel.

The description of the complex behavior of a general magnetoresistor can be simplified by dividing the problem into two parts: first, the relation between resitivity ρ and the direction of M_s, and second, the relation between the applied field H and the magnetization direction.

Resistance and Magnetization. In soft magnetic thin films of single domain state, AMR can be described phenomenologically very simply as a two-dimensional problem. Figure 71 shows the dimensions (length l, width b, and thickness d) of the rectangular thin ferromagnetic film (AMR element) and the coordinate system. An applied field H_y rotates M_s out of the easy axis into the hard axis direction of the uniaxial anisotropy. The resistivity depends on the angle $\theta = \varphi - \psi$ between $\vec{M}_s$ and $\vec{J}$. With

$$\rho(\theta) = \rho_o + (\rho_p - \rho_o)\cos^2\theta = \rho_o + \Delta\rho\cos^2\theta \quad (5.1)$$

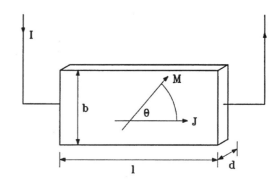

Fig. 71. Current density and spontaneous magnetization in a single-domain thin film strip. Reprinted with permission from [236], © 2001, Artech.

and $\rho = \rho_p$ for $\vec{M}_s$ parallel $\vec{J}$, and $\rho = \rho_o$ for $\vec{M}_s$ and $\vec{J}$ orthogonal, the quotient $\Delta\rho/\rho_o$ is the magnetoresistive coefficient, which may amount to several percent. With the resistance

$$R(\theta) = \rho(\theta)\frac{1}{bd} = R + \Delta R\cos^2\theta \quad (5.2)$$

(see Fig. 72) the voltage in the x direction is

$$U_x = I\frac{1}{bd}(\rho_o + \Delta\rho\cos^2\theta) \quad (5.3)$$

Another effect related to AMR is determined by the tensor property of ρ. Perpendicular to the electrical field E_x, causing the current density J_x, is an electrical field,

$$E_y = J_x\Delta\rho\sin\theta\cos\theta \quad (5.4)$$

Because of its direction, it is known as the planar or extraordinary Hall effect. It must not be confused with the (ordinary) Hall effect because of its different physical origin with respect to both $\vec{B}$ relations and materials. Depending on $\mathrm{sgn}(\theta)$, the planar Hall voltage,

$$U_y = I\frac{\Delta\rho}{d}\sin\theta\cos\theta \quad (5.5)$$

resulting from Eq. (5.4) will be much lower than U_x because of $b \ll l$ in the usual AMR sensor designs. The planar Hall

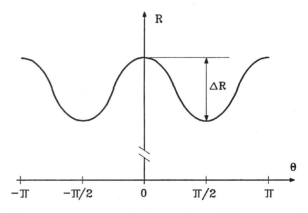

Fig. 72. Dependence of the resistance R on the angle θ between M_s and J. Reprinted with permission from [236], © 2001, Artech.

[14] Field plates are sometimes also called magnetoresistive sensors. They are made of InSb semiconductors and can exhibit a resistance variation of 1:10 in strong fields (≈ 1 T).

effect can be considered for sensors with very small dimensions measuring fields at high spatial resolutions, for example, for magnetic recording applications.

Magnetization and Applied Fields. It should be noted that all magnetic sensors measure the flux density B. The physical effect is the Lorentz force (Eq. (1.11)) acting on moving electrical charges and the torque acting on magnetic moments (e.g., spins in magnetic materials), respectively. As B relates to the applied field H in general outside the sensor ($B = \mu_0 H$), conventionally H is used for the following considerations. Furthermore, magnetic sensors can cause a distortion of the field to be measured. Because of demagnetizing effects, the applied, previously homogeneous field is homogeneous at a certain distance from the sensor only, depending on its geometrical dimensions.

If $\vec{H}$ is acting in the y direction, the angle

$$\theta = \arcsin \frac{H_y}{H_0} \tag{5.6}$$

between $\vec{M}_s$ and $\vec{J}$ (for $-1 < H_y/H_0 < 1$) leads to the field dependence of the resistance,

$$R(H_y) = R_0 + \Delta R \left[1 - \left(\frac{H_y}{H_0} \right)^2 \right] \tag{5.7}$$

for $|H_y| \leq H_0$ and $R(H_y) = R_0$ for $|H_y| > H_0$, shown in Figure 73. The characteristic field H_0 is the sum of the demagnetizing field H_d and the fictitious anisotropy field H_k, which relates to the anisotropy constant K_1 as $H_k = 2K_1/M_s$.

The magnetic field $\vec{H}$, flux density $\vec{B}$, and magnetization $\vec{M}$ are only homogeneous in ellipsoids. Therefore, H_d causes a variation in θ with respect to the y coordinate in general. If H_k is negligible compared with H_d and H_y is not strong enough to rotate M_s completely into the y direction, an analytical solution for θ in rectangular samples is

$$\theta(y) = \arctan \left(\frac{2H_y}{dM_s} \sqrt{\left(\frac{b}{2} \right)^2 - y^2} \right) \tag{5.8}$$

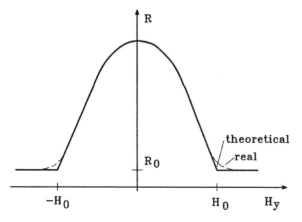

Fig. 73. Dependence of the resistance R on the applied field H_y: theoretical (———) and a 20-nm thin Permalloy film with $H_c \approx 0$ (– – –).

This effect is also shown in Figure 73. Even if $\theta \approx 60°$ in the middle of the sample ($y = 0$), there is still almost no deflection at the edges ($y = -b/2$ and $y = +b/2$). For a first approximation, the demagnetizing matrix is

$$\underline{N_d} \approx \left(\frac{d}{b}, \frac{d}{l}, 1 \right) \tag{5.9}$$

as long as $d \ll b \ll l$ is valid.

Magnetoresistive Films. The most important properties of AMR materials are a large coefficient $\Delta\rho/\rho$ at a large ρ (high signal at a certain resistance within a small area), low temperature dependence of ρ and l (low offset drift), low anisotropy field H_k (high sensitivity), low coercivity H_c in the hard axis direction (high reproducibility), zero magnetostriction ($\lambda_s \approx 0$, magnetic properties independent of mechanical stress), and long-term stability of these properties (no thermally activated effects, e.g., diffusion, recrystallization).

Materials. Besides amorphous ferromagnets, which are characterized by very low H_k and high ρ but only $\Delta\rho/\rho \approx 0.07\%$, the most established materials are crystalline binary and ternary alloys of Ni, Fe, and Co—mainly Permalloy 81Ni/19Fe. This material is characterized by $\mu_0 M_s = 1.1$ T at 300 K, $\rho = 2.2 \cdot 10^{-7}$ Ωm, $\Delta\rho/\rho = 2–4\%$, thermal resistivity coefficient 0.3%/K, $H_k \approx 100–1000$ A/m, $H_c < 10$ A/m, and $\lambda_s \approx 0$. To achieve a material with simultaneously vanishing anisotropy and magnetostriction it is necessary to add about 4% Mo (Supermalloy), but $\Delta\rho/\rho$ is reduced in this case. Therefore, most AMR films are made of Permalloy because of both very low H_k and λ_s at considerable $\Delta\rho/\rho$.

Materials with higher AMR coefficients are 90Ni/10Co (4.9%), 80Ni/20Co (6.5%), 70Ni/30Co (6.6%), and 92Ni/8Fe (5.0%), but they are not suitable for sensor applications, because of either high H_k or high λ_s. Other low-anisotropy materials are 50Ni/50Co (2.2%) and Sendust, which is used for magnetoresistive heads because of mechanical hardness requirements.

Film Processing. The two established methods for the deposition of Permalloy layers are vacuum evaporation and cathode sputtering at very low oxygen partial pressure. The main advantage of the latter procedure is that the composition of the film well corresponds to the composition of the sputter target, which can be vacuum melted or sintered. Both elevated target and substrate temperatures have been proved to yield an AMR coefficient of $\Delta\rho/\rho = 3.93\%$ in a 50-nm thin film [247], which is almost the theoretical bulk value of about 4%.

Thermally activated crystallite growth or the incorporation of residual gas atoms at grain boundaries, which reduces ρ, can also be done by an annealing process in vacuum or hydrogen, but this can increase coercivity. As the resistivity increases with decreasing film thickness, the AMR coefficient decreases at constant $\Delta\rho$ to 3.5% at 20 nm (which is thought to be the optimum thickness for AMR films [248]). This leads to a trade-off between sensor resistance and AMR effect.

To define the easy axis orientation, the deposition and/or annealing has to be done in a homogeneous magnetic bias field of some kA/m, providing a spatial alignment of Ni and Fe atom pairs. This leads to an additional induced anisotropy, which can be used to either decrease or increase the intrinsic H_k, depending on the bias field direction. To avoid unwanted anisotropy and texture, the amorphous substrates (glass or oxidized silicon) have to be very smooth.

The further processing of the film with contact and passivation layers uses well-known microelectronic packaging technologies. As the required dimensional tolerances for AMR sensors are smaller than for other devices (because of bridge unbalance) the masks are made by high-precision electron beam lithography.

Measurements.   The magnetic properties of a film can be determined by measuring the hysteresis loops (component M_H of M_s in the H direction versus H) in different directions of the homogeneous applied field, which can be provided by Helmholtz arrangements or cylindrical coils. Suitable methods are vibrating coil magnetometry (VCM) or reflected light modulation by the magneto-optical Kerr effect. Figure 74 shows such hysteresis loops [244] parallel and perpendicular to the easy axis, which corresponds well to the theory of coherent rotation (see Section 2.2).

Furthermore, ρ and $\Delta\rho(H)$ are determined by a four-point probe, arranged in a square or rectangle, under an applied field of variable direction and strength. Figure 73 shows the resistance dependence on H_y for the sample above.

5.1.1.3. Linearization and Stabilization

The simple magnetoresistive element of Figure 71 has several drawbacks without additional measures: the resistance change at low fields is about zero, the characteristics are nonlinear (even in the vicinity of the inflection point), and the film would exhibit a multidomain state that could cancel out the AMR effect or at least give rise to magnetic Barkhausen noise. Therefore, biasing techniques are required for its linearization and stabilization.

An external magnetic field can be used for both purposes, generated either by permanent magnets or currents in simple coils, but this technique is only advantageous if this field is also required for the given application, for example, for position sensing via field distortion. In the following, only internal, integrated solutions will be discussed.

Perpendicular Bias.   Perpendicular bias is used to establish an angle $\theta > 0$ between $\vec{M}_s$ and $\vec{J}$ in the absence of H. This can be done either by applying a magnetic bias field H_B in the

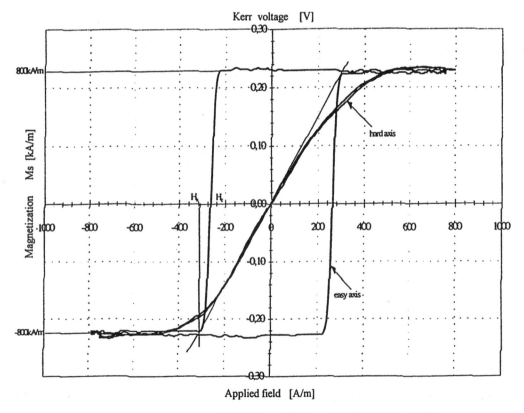

Fig. 74.   Hysteresis loops $M_H(H)$ of a 20-nm thin Permalloy film ($\Delta\rho/\rho = 3.4\%$).

y direction or by rotating the current path out of the easy axis (this "geometrical bias" by herringbone or barber-pole structures is discussed in Section 5.1). Replacing H_y in Eq. (5.7) with $H_y + H_B$ yields

$$R(H) = R_0 + \Delta R \left(\frac{2H_y H_B}{H_0^2} + \frac{H_y^2 + H_B^2}{H_0^2} \right) \quad (5.10)$$

which is linear for $H_y \ll H_0$.

Permanent Magnetic Films. In addition to using an external magnet, a sandwich structure with a hard magnetic film (thickness d_m) can provide a bias field by simple magnetostatic coupling, as shown in Figure 75. These films can be processed by well-established thin-film recording technologies and are made of CoPt, CoNiPt, or CoCrPt. Their remanence should relate to M_s of the AMR film as

$$M_r = M_s \frac{d}{d_m} \sin\theta \quad (5.11)$$

To prevent exchange coupling, electrical shunting, and diffusion processes, a nonmagnetic, electrically insulating intermediate layer should be provided. It is typically 10–20 nm of SiO_2 or Al_2O_3. Disadvantages are the magnetic stray field of the sensor itself and the fact that a postfabrication adjustment of the bias field is impossible.

Exchange Coupled Films. The basic idea is to induce a unidirectional anisotropy by exchange coupling of the AMR film with an antiferromagnetic or a ferromagnetic layer (thickness d_a), in contrast to the uniaxial anisotropy of the AMR film. The direction of anisotropy can be determined by deposition in a field or by subsequent field annealing above the Néel or Curie temperature, respectively. Figure 76 illustrates the principle.

The interfacial exchange coupling depends on the submicroscopic details of the interface. The coupling strength is usually described by an effective exchange field,

$$H_E = \frac{J_E}{\mu_0 M_s d} \quad (5.12)$$

with respect to the interfacial exchange coupling energy J_E per unit area (typically 1 mJ/m²) and the AMR film parameters M_s, K_1, and d. Only if $J_E < K_1 d_a$ is the $M(H)$ curve

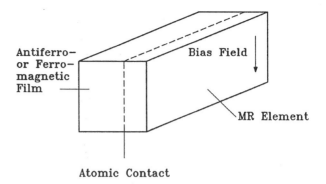

Fig. 76. Exchange coupling of ferro- and antiferromagnetic films causes unidirectional anisotropy represented by an effective exchange (bias) field. Reprinted with permission from [236], © 2001, Artech.

shifted unidirectionally because of the effect of the exchange anisotropy. Otherwise, only the coercivity is increased.

Examples of antiferromagnetic films are MnFe, CoO, and TbCoFe. The materials for the previously described permanent magnet biasing are used for ferromagnetic exchange coupled films. Because of corrosion problems, the latter method is preferred (increasing coercivity), to yield a long term stability. Aside from the problem of complex materials science of exact exchange control on an atomic level, the main advantage is that the antiferromagnetic bias field cannot be reset or changed accidentally.

Shunt Bias. Several methods use shunting of currents or magnetic fields. Figure 77a shows the basic arrangement of an AMR layer, an insulating layer, and a nonmagnetic conductor. The magnetomotive force (see Eq. (1.5)) of the nonmagnetic conductor produces the bias field. A disadvantage of the insulator thickness is the comparatively high shunt current,

$$I_B = M_s d \quad (5.13)$$

of about 30–100 mA needed in this case. A further disadvantage of thick insulators is the reduction of the bias field at the ends of the AMR film (underbiased region). An advantage of this scheme is that the conductor can be used for servo bias or field compensation purposes in a feedback mode. Omitting the insulator layer causes a partial reduction of the resistance variation, and problems can arise because of diffusion of the conductor layer material into the AMR film.

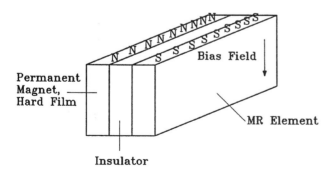

Fig. 75. Perpendicular bias field generation by hard magnetic films. Reprinted with permission from [236], © 2001, Artech.

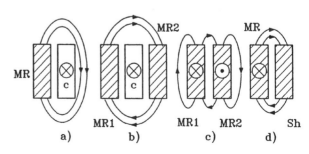

Fig. 77. Shunt bias field generation by (a) AMR film and conductor, (b) double AMR films and conductor, (c) double-element biasing, and (d) soft adjacent layer bias. Reprinted with permission from [236], © 2001, Artech.

Other arrangements are sandwiches of two magnetoresistors of opposite bias, which can be used to form bridge circuits. The current of the scheme in Figure 77b is reduced by the factor of total sandwich thickness/width with respect to Eq. (5.13). The double-element biasing scheme of Figure 77c utilizes the sensing currents of each AMR layer for mutual biasing.

The soft adjacent layer bias arrangement is shown in Figure 77d. Consisting of a high-permeability material (Permalloy or Sendust), the soft magnetic shunt layer is magnetized by the field of the AMR sensor current. According to Figure 9, the magnetic poles at the surface ($\nabla \cdot M \neq 0$) are the origin of a magnetic stray field H_d, which is acting as a bias field for the AMR layer.

A disadvantage of the soft adjacent layer bias scheme is the shunting of the magnetic field H_y (to be measured) because of the two flux paths. This drawback is eliminated by the double-element biasing scheme. Because of the nonlinear mutual magnetostatic interaction between the two soft magnetic layers, the design problems must be solved by micromagnetic numerical techniques.

Longitudinal Bias. The theoretical dependence of resistivity change on the applied field as described in Section 5.1.1.2 is valid under the condition of a single-domain state of the film. Experimental investigations [245] have shown that the magnetization may divide into several domains. This is caused by the large stray field at the ends of a single domain and by favored location of domain walls at imperfections of the film (see Section 1.2.2.3). The main effects are a smaller $\Delta\rho/\rho$ due to antiparallel domain magnetization and hysteresis with Barkhausen noise due to irreversible domain wall displacements.

The deviations of $\vec{M}_s$ from the easy axis (ripple) has been described in terms of internal stray fields, taking into account exchange coupling and different orientation of crystallites. These stray fields are responsible for blocking the free rotation of $\vec{M}_s$, which leads to restrictions for both the direction and amplitude of $\vec{H}$ to yield a characteristic without hysteresis. The theoretical limit for $\vec{H}$ is determined by the Stoner–Wohlfahrth astroid (see Section 1.3.3.2).

Longitudinal bias is necessary to stabilize the single-domain state against perturbation by external fields and thermal and mechanical stresses. This is achieved by pinning the

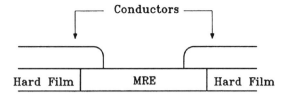

Fig. 79. Hard magnetic films providing a longitudinal bias field. Reprinted with permission from [236], © 2001, Artech.

end-zone domain walls (see Fig. 78) in their initial position or by avoiding these domains by special geometric designs (e.g., triangular endings). In addition to adaptation of the techniques described for perpendicular bias, some other methods are discussed briefly.

Hard Magnetic Films. Thin films with high coercivity (like CoCr, CoPt, SmCo, or ferrites) are sputtered beneath both ends of the AMR film, as shown in Figure 79. Their in-plane magnetization provides a longitudinal bias field and a reduction of the end-zone domains. The performance of this method strongly depends on the microscopic details of the junction between the hard film and the AMR film.

A much simpler possibility is also illustrated in Figure 79. Overcoating the ends of the AMR film with the conductor material causes the current to flow in the single-domain center zone only. A further improvement can be achieved by progressing the AMR film toward the end zones, which increases the coercivity in this region and contributes to the domain wall pinning. This method reduces noise in the current signal, but it cannot prevent blocking of magnetization at strong fields. An alternative method is screen printing of SmCo or hard magnetic ferrite powder onto the rear or the AMR film substrate, which also provides a longitudinal bias field to stabilize the single-domain state.

Exchange Tabs. The end-zone domains are pinned by exchange coupling (as also described for perpendicular bias) to either a ferromagnetic or antiferromagnetic (e.g., MnFe) layer. In the zone of strong exchange coupling (see Fig. 80) the spins are prevented from direction changes; therefore neither domain wall displacement nor rotation of domain magnetization occurs. The direction of exchange anisotropy can be determined by deposition in a magnetic field or by subsequent field annealing above the Néel temperature.

5.1.2. Giant Magnetoresistive Effect

The giant magnetoresistive (GMR) effect is entirely different from the anisotropic magnetoresistive (AMR) effect treated

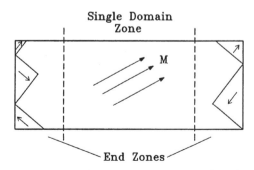

Fig. 78. AMR film with a single-domain center and multidomain end zones. Reprinted with permission from [236], © 2001, Artech.

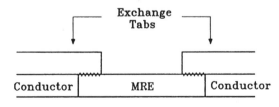

Fig. 80. Exchange coupling in the tabs providing a longitudinal bias field. Reprinted with permission from [236], © 2001, Artech.

before. In the case of GMR, a multilayer system consisting of alternately ferromagnetic (e.g., NiFe) and nonmagnetic (e.g., Cu) material is used (Fig. 81). The magnetizations of the ferromagnetic layers are antiparallel because of exchange coupling at weak magnetic fields but become parallel at higher magnetic fields. This change of orientation (spin-dependent electron scattering at the magnetic–nonmagnetic interfaces) is the reason for the strong change in the electrical resistance [234]. GMR multilayers have been produced by several methods (e.g. MBE, electron-beam evaporation [251], or sputtering [221, 222, 252–255]). Sputtering has been widely used because it gives a higher GMR ratio compared with MBE or vacuum evaporation. Annealing plays an important role because the Cu diffusion along the NiFe grain boundaries is thought to create intralayer magnetic discontinuities in NiFe layers, which elevate interlayer antiferromagnetic coupling, and consequently causes the increase in GMR [251].

GMR sensors can be used not only for magnetic field measurements and data reading applications, but also for other physical quantities, for example, for contactless current detection [256].

5.1.3. Magneto-Impedance Sensors

A very high sensitivity to an external magnetic field is typical for magneto-impedance (MI). Sensors based on thin-film MI essentially consist of two soft ferromagnetic layers that sandwich a nonmagnetic, highly conductive layer (Fig. 82). When an ac current flows through such a multilayer, the impedance is changed by an external dc magnetic field. The impedance of the detecting element can be divided into two components, the resistance (mainly related to the skin effect in the magnetic thin films) and the inductance (related to the permeability of the magnetic thin films). Both components are influenced by the magnetic field [264]. For example, the sensitivity is 340% for a frequency of 10 MHz and a dc magnetic field of 700 A/m in a CoSiB/Cu/CoSiB layer system [265]. The external field is applied parallel to the current, which is flowing in the direction of the length of the film. The MI effect has the potential to be used to develop highly sensitive and quick-response micromagnetic sensors and magnetic heads for high-density magnetic recording. Many other layer systems (mainly amorphous) have already been investigated, such as NiFe/Cu/NiFe, CoFeSiB/Cu/CoSiFeB, CoSiB/Ag/CoSiB, and CoSiB/Ti/CoSiB.

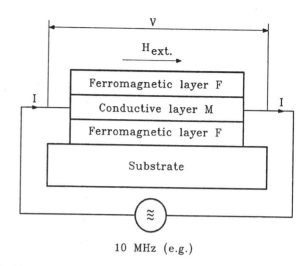

Fig. 82. Schematic drawing of a multilayer magneto-impedance element. (F) CoSiB layers forming a magnetically closed loop structure with magnetic unidirectional anisotropy; (M) Cu. Reprinted with permission from [264], © 2000, Elsevier Science.

The layer system $CoSiB/SiO_2/Cu/SiO_2/CoSiB$ exhibits an extremely high MI ratio [265]. Another version of the MI sensor consists simply of a rectangular stripe of a sputter-deposited amorphous CoNbZr thin film [266].

5.1.4. Micro-Fluxgate Sensor

Miniaturization of the fluxgate sensor utilizing silicon process technology and thin-film technology is driven by advantages such as small size, light weight, low cost, higher resolution, and integration of the supporting electronic circuitry. Fluxgate magnetic sensors consist, for example, of a magnetic ring core (Permalloy), an excitation coil (Cu) wound around it, and at least one sensing coil (Cu) wound around the whole body of the ring core (Fig. 83) [281]. A square-wave pulse current is applied to the excitation coil. The amplitude of the excitation

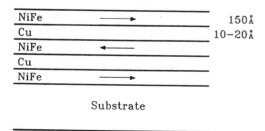

Fig. 81. NiFe-Cu multilayer system with antiferromagnetic exchange coupling (giant magnetoresistive effect).

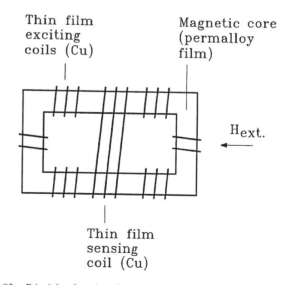

Fig. 83. Principle of a micro-fluxgate magnetic sensor. Reprinted with permission from [281], © 1999, Elsevier Science.

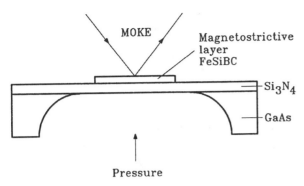

Fig. 84. Schematic illustration of a magnetostrictive pressure sensor. Reprinted with permission from [282], © 2000, Elsevier Science.

current is adjusted so that the excitation field oscillates periodically between the positive and negative saturation regions. An output signal of the sensing coil appears if an applied magnetic field causes a flux shift in the core halves. The direction of the face of the sensing coil determines the direction of the detected magnetic field.

5.1.5. Magnetostrictive Pressure Sensor

A magnetostrictive pressure sensor (Fig. 84) is based on a micromembrane (silicon nitride) coated with a magnetostrictive thin film (e.g., sputtered FeSiBC) [282]. A pressure difference across the diaphragm causes deflection and thus stress in the magnetostrictive layer. This leads to a change in the magnetic properties of the thin film, which can be detected with the use of the magneto-optical Kerr effect (Figure 84). Amorphous thin films, deposited by sputtering or laser ablation, can also be used for magnetoelastic sensing applications (stress, force, torque) [283, 284].

5.2. Magnetic Microactuators

Magnetostrictive films, especially giant magnetostrictive TbFe/FeCo multilayers, present a new and interesting approach to bending transducers as actuators in microsystems (Fig. 85) [277–280]. They offer features like contactless, high-frequency operation, simple actuator designs, and a cost-effective manufacturing technique. Compared with piezoelectric transducers, they make use of their higher energy density, the possibility of biasing with permanent magnets, the

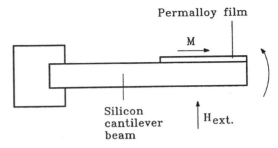

Fig. 86. Schematic illustration of a magnetic actuator using electroplated Permalloy [228].

possibility of remote control operation, and the easier fabrication technique, which requires much lower process temperatures. These properties make them attractive for optical applications like laser scanner devices, for microfluidic devices like pumps or valves, for ultrasonic motors, and for microrelays or microswitches. Furthermore, feedback-controlled variable-frequency resonators with large frequency spans are possible. It should be mentioned that some other materials also show large magnetostriction (e.g., SmFe). Thin films of magnetostrictive materials can be deposited by sputtering techniques.

Another type of magnetic actuator is based on a micromachined cantilever beam coated with a Permalloy film (Fig. 86). This coated beam can be bent by an external magnetic field [228].

5.3. Micro-Inductors with a Closed Ferromagnetic Core

A thin film microinductor consists of an insulating substrate (alumina, glass, silicon), a conductive coil (Cu) with contact pads (NiCr, Cu, Au), a ferromagnetic core (e.g., Permalloy), and isolation layers (Al_2O_3, SiO_2, polymer). The magnetic core can either be placed inside a solenoid coil or can encompass the conductors of a spiral coil (Fig. 87) [275, 276]. Two

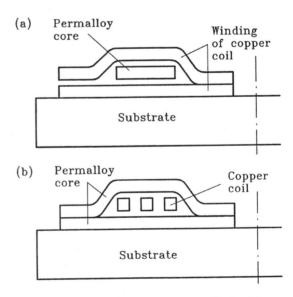

Fig. 87. Schematic illustration of microinductors with closed ferromagnetic core, type a and b (half of a cross section) [275, 276].

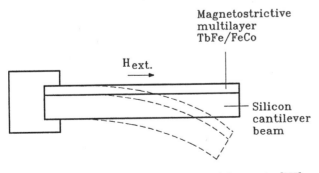

Fig. 85. Schematic illustration of a magnetostrictive actuator [277].

examples of applications are micro-fluxgate magnetic sensors and inductive write heads for magnetic recording. Furthermore, it is possible to encompass each conductor of a spiral or meander-type coil separately with the core material. One feature of this construction is the absence of an insulating layer. When the magnetic core is used in high-frequency devices it is advantageous to use a multilayer structure consisting of alternating magnetic and insulating layers (e.g., CoFeSiB/SiO_2).

Microtransformers and inductors can also be constructed by soft magnetic ferrite thin-film material, such as Ni-Zn ferrite (NiZnFeO) or Mn-Zn ferrite films [226]. Ferrite films can be deposited by plasma-enhanced metal organic chemical vapor deposition (MOCVD) (see Section 4.1.5).

5.4. Magnetic Data Storage

5.4.1. Thin Films for Magnetic Recording

The magnetic thin films used in disks and tape systems can be particulate in nature; the magnetic particles are imbedded in a binder material and then applied to the substrate. However, sputtered or evaporated film media are also widely used in storage systems. Their thinness, low defect level, smoothness, and high induction are required to achieve high recording densities at low flying heights. Two modes can be distinguished: longitudinal and perpendicular recording (Fig. 88). In the case of longitudinal recording the film media should exhibit high magnetization, high coercivity, and a square hysteresis loop. Many film materials can be used for this purpose [270], and in most cases sputtering is the preferred deposition technique (CoRe, CoPt [271], CoNi, CrCo, CoW, FeCoCr, CoSm, Fe_2O_3). Perpendicular magnetic recording is a method used to achieve ultrahigh density recording. The main material for this mode is CoCr (15–20 at.% Cr) [272]. These films must exhibit an easy axis of magnetization perpendicular to the film and a perpendicular magnetic anisotropy. Many deposition techniques have been reported for these films: RF diode sputtering, dc triode sputtering, planar magnetron sputtering, electron-beam evaporation, and others. Some additional layers can improve the quality of recording. CoCr films have been deposited on soft magnetic films such as Permalloy, which provide a yoke function to increase the recording sensitivity.

The magnetic layers can be protected with overcoats of Cr, C, or SiO_2.

5.4.2. Thin Films for Magneto-Optical Recording

A perpendicularly oriented magnetic film (sputtered GdTbFe, TbFeCo, CoPtRe, CoPtNi, etc.) is exposed to a magnetic field in the opposite direction. This field is not strong enough to change the direction of magnetization. However, if the material is heated by a laser, the spontaneous magnetization and consequently the coercivity decrease strongly, and the direction of magnetization changes to the applied field direction (Fig. 89). Readout can be accomplished by the magneto-optical Kerr effect (reflected polarized light is influenced by the magnetization) [218, 273] or by the Faraday effect (transmitted polarized light is influenced by the magnetization). Higher-density recording can be achieved by using a shorter-wavelength light source such as a blue laser, because the minimum spot size of the focused laser beam is directly proportional to the wavelength of the light source [274]. The super-resolution type is a scheme for reading out a recording bit smaller than a focused laser spot with the use of controlled thermo-magnetic properties.

5.4.3. Inductive Write Heads

Miniaturized inductive write heads for high-resolution recording can be fabricated by thin-film technologies [218, 267]. Such write heads are often combined with magnetoresistive read heads, as shown in Figure 90 [234, 268]. The magnetic circuit (NiFe), coil (Cu), magnetic shields (NiFe), magnetoresistive sensor (NiFe), and various insulation layers (Al_2O_3, SiO_2) are produced by sputtering and patterned by photolithography and etching. Electroplating can also be used for some production steps (magnetic layers, copper coils, gold bonding pads). The substrate material is, for example, AlTiC (or Al_2O_3-TiC). One wafer can carry about 300 such write/read heads, which are separated at the end of the production process.

Another type of write head is the single-pole head, which is used for prototype perpendicular recording. The single-pole

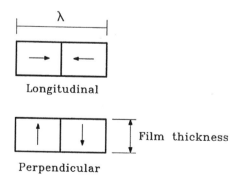

Fig. 88. Schematic illustration of the transition region for longitudinal and perpendicular recording.

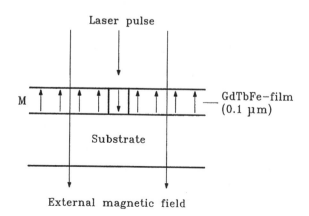

Fig. 89. Schematic illustration of magneto-optical recording [218].

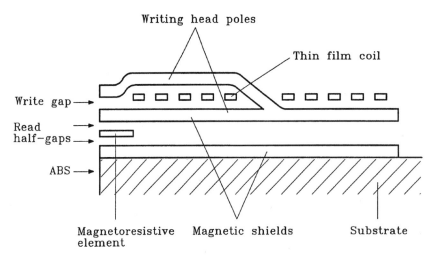

Fig. 90. Schematic cross-sectional view of a thin-film inductive write head combined with a shielded magnetoresistive read head. ABS, air-bearing surface. Reprinted with permission from [234], © 1996, Academic Press.

head consists of a main pole, a thin-film coil with a very small number of turns, and a return path pole (Fig. 91) [269]. It can also be combined with a magnetoresistive read head.

5.4.4. Magnetoresistive Read Heads

The principle of a magnetoresistive head is shown in Figure 92. The magnetic field produced by the recorded magnetization pattern in the tape or disk rotates the magnetization in a properly biased magnetoresistive element. Small changes in the magnetization angle cause almost proportional changes in the resistance [234]. The output signal voltage is proportional to the resistance change times the measuring current.

There are also several versions of dual-element magnetoresistive heads that avoid the influence of thermal fluctuations and external magnetic noise in the magnetoresistive element.

The information storage densities are currently progressing toward 5 Gbit/in^2 with AMR heads and more than 10 Gbit/in^2 with spin-valve heads [249, 250].

5.4.5. Spin-Valve Magnetic Heads

The structure of a typical spin-valve GMR element is shown in Figure 93 [234]. There are just two ferromagnetic layers of Permalloy, which are called the pinned and the free layer. The pinned layer is strongly exchange coupled to an underlying antiferromagnet [255, 257–261]. If the easy axis of the free layer is perpendicular to the magnetization of the pinned layer, the resistance change is proportional to the magnetic field change. Typically, the antiferromagnetic layer is 100 Å thick and is made of MnFe, CoO, NiFeTb, or PdPtMn. A recent advance in spin-valve design used to replace the typical antiferromagnetic materials has been the implementation of an artificial or synthetic antiferromagnetic (AAF or SAF) subsystem [229, 262, 263]. These new spin-valve structures consist of a ferromagnetic layer separated by a Cu spacer layer from the AAF subsystem, which itself consists of a few strongly coupled Co/Cu bilayers or Co/Ru/Co. Their advantage over the antiferromagnetic layers is an improved corrosion resistance and higher processing temperatures.

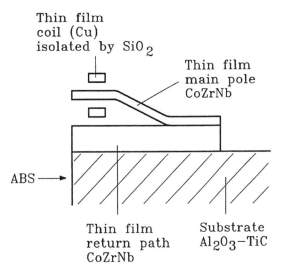

Fig. 91. Schematic cross-sectional view of a single-pole writing head [269]. ABS, air-bearing surface. Reprinted with permission from [269], © 1999, IEEE.

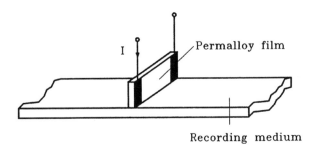

Fig. 92. Schematic illustration of a magnetoresistive head (shields and bias not shown).

(a)

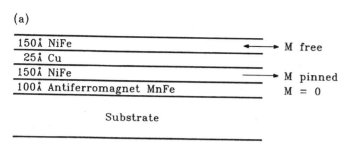

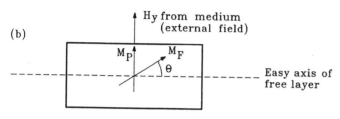

Fig. 93. Permalloy-copper spin-valve structure. (a) Layer system. (b) The effect of the external field on the free layer magnetization angle. Reprinted with permission from [234], © 1996, Academic Press.

5.4.6. Magnetoresistive RAM Arrays

The nonvolatile, magnetoresistive random access memory (MRAM) device fabricated with giant magnetoresistance (GMR) material is an excellent candidate for competing in the random access memory market. The basic structure of a memory cell is a patterned stripe of two magnetic NiFeCo layers separated by a nonmagnetic Cu interlayer. The resistance of the film is high when the magnetic moments of the two magnetic layers are antiparallel and low when they are parallel. Figure 94 shows GMR elements that form a sense line, and isolated perpendicular word lines [285]. Both reading and writing operations require simultaneous application of sense and word currents. These memory cells do not require a tran-

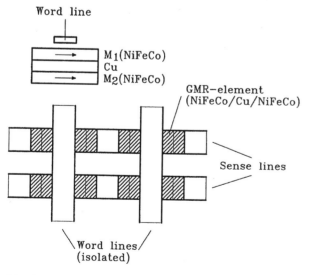

Fig. 94. Schematic illustration of a GMR bit array (magnetoresistive random access memory). Reprinted with permission from [285], © 1997, American Institute of Physics.

sistor per cell, which facilitates high densities. The theoretical limit of the information storage density is about 400 Gbit/in^2 with 0.5-ns read and 0.7-ns write times.

Acknowledgments

We are indebted to Michael Holzmüller for bibliography assistance and to Günther Stangl for technological contributions. Furthermore, we want to thank Thomas Zottl for artwork and drawings.

REFERENCES

1. R. Becker and W. Döring, "Ferromagnetismus." Springer-Verlag, Berlin, 1939.
2. J. A. Stratton, "Electromagnetic Theory." McGraw–Hill, New York/London, 1941.
3. R. M. Bozorth, "Ferromagnetism." Van Nostrand, New York, 1951.
4. E. Kneller, "Ferromagnetismus." Springer-Verlag, Berlin, 1962.
5. W. F. Brown, "Magnetostatic Principles in Ferromagnetism." North-Holland, Amsterdam, 1962.
6. S. Chikazumi, "Physics of Magnetism." Wiley, New York, 1964.
7. H. Hofmann, "Das elektromagnetische Feld." Springer-Verlag, Vienna/New York, 1974.
8. C. Heck, "Magnetic Materials and Their Technical Applications." Crane and Russak, New York, 1974.
9. E. P. Wohlfarth, Ed., "Ferromagnetic Materials." North-Holland, Amsterdam, 1982.
10. R. Boll, "Weichmagnetische Werkstoffe." Siemens AG, Berlin/Munich, 1990.
11. D. C. Jiles, "Introduction to Magnetism and Magnetic Materials," 2nd ed. Chapman & Hall, London, 1998.
12. G. M. Fasching, "Werkstoffe für die Elektrotechnik." Springer-Verlag, Vienna/New York, 1994.
13. R. C. O'Handley, "Modern Magnetic Materials." Wiley, New York, 2000.
14. J. C. Maxwell, "A Treatise on Electricity and Magnetism." Clarendon, Oxford, 1873.
15. F. Paschke, Rotating Electric Dipole Domains as a Loss-Free Model for the Earth's Magnetic Field, Session Reports of the Austrian Academy of Sciences, 1998, Vol. 207, pp. 213–228.
16. J. A. Osborn, Phys. Rev. 67, 351 (1945).
17. A. Hubert and R. Schäfer, "Magnetic Domains." Springer-Verlag, Berlin/Heidelberg/New York, 1998.
18. H. Lawton and K. H. Stewart, Proc. R. Soc. London, Ser. A 193, 72–88 (1948).
19. H. Hauser, J. Appl. Phys. 77, 2625 (1995).
20. G. Bertotti, "Hysteresis in Magnetism." Academic Press, San Diego/London/Boston, 1998.
21. F. Preisach, Z. Phys. 94, 277 (1935).
22. D. C. Jiles and D. L. Atherton, J. Magn. Magn. Mater. 61, 48 (1986).
23. H. Hauser, J. Appl. Phys. 75, 2584 (1994).
24. A. Inoue and A. Makino, J. Phys. IV 8, 3 (1998).
25. W. H. Meiklejohn and C. P. Bean, Phys. Rev. 102, 1413 (1956).
26. W. H. Meiklejohn and C. P. Bean, Phys. Rev. 105, 904 (1957).
27. W. H. Meiklejohn, J. Appl. Phys. Suppl. 33, 1328 (1962).
28. I. S. Jacobs and C. P. Bean, in "Magnetism" (G. T. Rado and H. Suhl, Eds.), Vol. 3, pp. 272–350. Academic Press, New York, 1963.
29. A. Yelon, in "Physics of Thin Films" (M. H. Francombe and R. W. Hoffman, Eds.), Vol. 6, pp. 213–223. Academic Press, New York, 1971.
30. A. E. Berkowitz and J. H. Greiner, J. Appl. Phys. 36, 3330 (1965).
31. J. S. Kouvel, J. Phys. Chem. Solids 24, 795 (1963).
32. T. Satoh, R. B. Goldfarb, and C. E. Patton, J. Appl. Phys. 49, 3439 (1978).

33. S. H. Charap and E. Fulcomer, *Phys. Status Solid: A* 11, 559 (1972).

34. E. C. Stoner and E. P. Wohlfarth, *Philos. Trans. R. Soc. London, Ser. A* 240, 599–642 (1948).

35. D. Givord, O. F. K. McGrath, C. Meyer, and J. Rothman, *J. Magn. Magn. Mater.* 157, 245 (1996).

36. L. Néel, *C. R. Acad. Sci.* 237, 1468 (1953).

37. U. Gradmann and J. Mueller, *Phys. Status Solidi* 27, 313 (1968).

38. C. Chappert and P. Bruno, *J. Appl. Phys.* 64, 5736 (1988).

39. F. J. A. den Broeder, W. Hoving, and P. J. H. Bloemen, *J. Magn. Magn. Mater.* 93, 562 (1991).

40. R. Jungblut, M. T. Johnson, J. A. de Stegge, and F. J. A. den Broeder, *J. Appl. Phys.* 75, 6424 (1994).

41. R. Bergholz and U. Gradmann, *Phys. Rev. Lett.* 59, 59 (1987).

42. P. Bruno, *Phys. Rev. B: Condensed Matter and Materials Physics* 39, 865 (1989).

43. D. S. Wang, R. Wu, and A. J. Freeman, *Phys. Rev. B: Condensed Matter and Materials Physics* 48, 15886 (1993).

44. E. R. Callen and H. B. Callen, *Phys. Rev. A: At., Mol., Opt. Phys.* 139, 455 (1965).

45. F. Ono, *J. Phys. Soc. Jpn.* 48, 843 (1980).

46. S. Ohnishi, A. J. Freeman, and M. Weinert, *Phys. Rev. B: Condensed Matter and Materials Physics* 28 (1983) 6741.

47. D. S. Wang, R. Wu, and A. J. Freeman, *Phys. Rev. Lett.* 70, 869 (1993).

48. D. S. Wang, R. Wu, and A. J. Freeman, *J. Appl. Phys.* 73, 6745 (1993).

49. D. S. Wang, R. Wu, and A. J. Freeman, *J. Magn. Magn. Mater.* 129, 237 (1995).

50. R. Richter, J. G. Gay, and J. R. Smith, *Phys. Rev. Lett.* 54, 2704 (1985).

51. U. Gradmann, in "Handbook of Magnetic Materials" (K. H. J. Buschow, Ed.), Vol. 7. Elsevier, Amsterdam, 1993.

52. H. Neal Bertram, "Theory of Magnetic Recording." Cambridge Univ. Press, Cambridge, UK, 1994.

53. E. Fulcomer and S. H. Charap, *J. Appl. Phys.* 43, 4184 (1972).

54. O. Bostanjoglo and P. Kreisel, *Phys. Status Solid: A* 7, 173 (1971).

55. O. Bostanjoglo and W. Giese, *Int. J. Magn.* 3, 135 (1972).

56. P. D. Battle, A. K. Cheetham, and G. A. Gehring, *J. Appl. Phys.* 50, 7578 (1979).

57. H. P. J. Wijn, Ed., "Landolt-Börnstein, Numerical Data and Functional Relationships in Science and Technology," Vol. III/27g, p. 30. Springer-Verlag, Berlin, 1991.

58. W. L. Roth, *J. Appl. Phys.* 31, 200 (1960).

59. W. L. Roth and G. A. Slack, *J. Appl. Phys. Suppl.* 31, 325S (1960).

60. G. A. Slack, *J. Appl. Phys.* 31, 1571 (1960).

61. J. Kanamori, *Prog. Theor. Phys.* 17, 197 (1957).

62. D. Mauri, H. C. Siegmann, P. S. Bagus, and E. Kay, *J. Appl. Phys.* 62, 3047 (1987).

63. N. Smith and W. C. Cain, *J. Appl. Phys.* 69, 2471 (1991).

64. A. E. Berkowitz and K. Takano, *J. Magn. Magn. Mater.* 200, 552 (1999).

65. A. P. Malozemoff, *Phys. Rev. B: Condensed Matter and Materials Physics* 35, 3679 (1987).

66. A. P. Malozemoff, *Phys. Rev. B: Condensed Matter and Materials Physics* 37, 7673 (1988).

67. L. Néel, *Ann. Phys. (Leipzig)* 2, 61 (1967).

68. N. C. Koon, *Phys. Rev. Lett.* 78, 4865 (1997).

69. T. C. Schulthess and W. H. Butler, *Phys. Rev. Lett.* 81, 4516 (1998).

70. K. Takano, R. H. Kodama, A. E. Berkowitz, W. Cao, and G. Thomas, *Phys. Rev. Lett.* 79, 1130 (1997).

71. K. Takano, R. H. Kodama, A. E. Berkowitz, W. Cao, and G. Thomas, *J. Appl. Phys.* 83, 6888 (1998).

72. D. C. Khan and R. A. Erickson, *J. Phys. Chem. Solids* 29, 2087 (1968).

73. D. Martien, K. Takano, A. E. Berkowitz, and D. J. Smith, *Appl. Phys. Lett.* 74, 1314 (1999).

74. T. J. Moran, J. M. Gallego, and I. K. Schuller, *J. Appl. Phys.* 78, 1887 (1995).

75. K. T.-Y. Kung, L. K. Louie, and G. L. Gorman, *J. Appl. Phys.* 69, 5634 (1991).

76. R. Jungblut, R. Coehoorn, M. T. Johnson, J. van de Stegge, and A. Reinders, *J. Appl. Phys.* 75, 6659 (1994).

77. C. Tsang and K. Lee, *J. Appl. Phys.* 53, 2605 (1982).

78. H. Fujiwara, K. Nishioka, C. Hou, M. R. Parker, S. Gangopadhyay, and R. Metzger, *J. Appl. Phys.* 79, 6286 (1996).

79. C. Hwang and T. Nguyen, *Mater. Res. Soc. Symp. Proc.* 232, 211 (1991).

80. B. Y. Wong, C. Mitsumata, S. Prakash, D. E. Laughlin, and T. Kobayashi, *J. Appl. Phys.* 79, 7896 (1996).

81. R. Schäfer, W. K. Ho, J. Yamasaki, A. Hubert, and F. B. Humphrey, *IEEE Trans. Magn.* 27, 3678 (1991).

82. L. Gunther and B. Barbara, *Phys. Rev. B: Condensed Matter and Materials Physics* 49, 3926 (1994).

83. S. Mangin, G. Marchal, W. Wernsdorfer, A. Sulpice, D. Mailly, and B. Barbara, *J. Magn. Magn. Mater.* 165, 13 (1997).

84. S. Mangin, C. Bellouard, G. Marchal, and B. Barbara, *J. Magn. Magn. Mater.* 165, 161 (1997).

85. W. Wernsdorfer, K. Hasselbach, A. Benoit, G. Cernicchiaro, D. Mailly, B. Barbara, and L. Thomas, *J. Magn. Magn. Mater.* 151, 38 (1995).

86. B. Dieny, D. Givord, and J. M. Ndjaka, *J. Magn. Magn. Mater.* 93, 503 (1991).

87. B. Barbara and M. Uehara, Institute of Physics Conference Series, no. 37, 1978, Chap. 8, p. 203.

88. V. Zablotskii and E. Soika, *J. Magn. Magn. Mater.* 182, 31 (1998).

89. A. H. Eschenfelder, "Magnetic Bubble Technology," p. 496. Springer-Verlag, New York, 1981.

90. H. Suguo, N. Xiang-fu, and H. Bao-Shan, *Acta Phys. Chim.* 37, 1703 (1988).

91. R. Suzuki and Y. Sugita, *IEEE Trans. Magn.* 14, 210 (1978).

92. A. P. Malozemoff and J. C. Sloncztwski, *Phys. Rev. Lett.* 29, 952 (1972).

93. I. Tomas, G. Vertesy, and N. Balasko, *J. Magn. Magn. Mater.* 43, 287 (1984).

94. M. Pardavi-Horvath and G. Vertesy, *J. Appl. Phys.* 58, 3827 (1985).

95. G. Vertesy and I. Tomas, *J. Appl. Phys.* 77, 6426 (1995).

96. A. P. Malozemoff and J. C. Sloncewski, in "Applied Solid State Science" (R. Wolfe, Ed.), Advances in Materials Research. Academic Press, New York, 1979.

97. V. G. Bar'yakhtar, M. V. Chetkin, B. A. Ivanov, and S. N. Gadetskii, "Dynamics of Topological Magnetic Solitons," Springer Tracts in Modern Physics, Vol. 129. Springer-Verlag, Berlin, 1994.

98. I. F. Lyuksyutov and H.-U. Everts, *Phys. Rev. B: Condensed Matter and Materials Physics* 57, 1957 (1998).

99. H. Leschhorn, T. Nattermann, S. Stepanow, and L.-H. Tang, *Ann. Phys. (Leipzig)* 6, 1 (1997).

100. S. Lemerle, J. Ferré, C. Chappert, V. Mathet, T. Giamarchi, and P. LeDoussal, *Phys. Rev. Lett.* 80, 849 (1998).

101. A. Kirilyuk, J. Ferré, J. Pommier, and D. Renard, *J. Magn. Magn. Mater.* 121, 536 (1993).

102. A. Hubert, *Phys. Status Solidi* 38, 699 (1970).

103. R. P. Cowburn, J. Ferré, S. J. Gray, J. A. C. Bland, *Appl. Phys. Lett.* 74, 1018 (1999).

104. R. P. Cowburn, J. Ferré, J.-P. Jamet, S. J. Gray, and J. A. C. Bland, *Phys. Rev. B: Condensed Matter and Materials Physics* 55, 11,593 (1997).

105. R. P. Cowburn, S. J. Gray, and J. A. C. Bland, *Phys. Rev. Lett.* 79, 4018 (1997).

106. R. P. Cowburn, S. J. Gray, J. Ferré, J. A. C. Bland, and J. Miltat, *J. Appl. Phys.* 78, 7210 (1995).

107. K. I. Sixtus and L. Tonks, *Phys. Rev.* 37, 930 (1931).

108. E. W. Lee and D. R. Callaby, *Nature (London)* 182, 254 (1958).

109. R. P. Cowburn, J. Ferré, S. J. Gray, and J. A. C. Bland, *Phys. Rev. B: Condensed Matter and Materials Physics* 58, 11,507 (1998).

110. Y. S. Didosyan, H. Hauser, V. Y. Barash, and P. L. Fulmek, *J. Magn. Magn. Mater.* 177, 203 (1998).

111. Y. S. Didosyan, H. Hauser, V. Y. Barash, and P. L. Fulmek, *J. Phys. IV* 8, 693 (1998).

112. R. W. DeBlois, *J. Appl. Phys.* 29, 459 (1958).

113. N. Menyuk, *J. Appl. Phys.* 26, 692 (1955).

114. R. Allenspach, M. Stampanonii, and A. Bischof, *Phys. Rev. Lett.* 65, 3344 (1990).

115. R. Allenspach, *J. Magn. Magn. Mater.* 129, 160 (1994).

116. B. Raquet, R. Mamy, and J. C. Ousset, *J. Magn. Magn. Mater.* 165, 492 (1997).

117. C. Amand, P. Peyrade, R. Marny, M. D. Ortega, M. Goiran, A. R. Fert, B. Raquet, and J. C. Ousset, *Phys. Status Solidi B* 195, 1671 (1996).

118. E. Fatuzzo, *Phys. Rev.* 127, 1999 (1962).

119. M. Labrune, S. Andrieu, F. Rio, and P. Bernstein, *J. Magn. Magn. Mater.* 80, 211 (1989).

120. B. Raquet, R. Manny, and J. C. Ousset, *Phys. Rev. B: Condensed Matter and Materials Physics* 54, 4128 (1996).

121. M. Avrami, *J. Chem. Phys.* 7, 1103 (1939).

122. M. Avrami, *J. Chem. Phys.* 8, 212 (1940).

123. M. Avrami, *J. Chem. Phys.* 9, 177 (1941).

124. E. Fatuzzo and W. J. Merz, *Phys. Rev.* 116, 61 (1959).

125. F. Rio, P. Bernstain, and M. Labrune, *IEEE Trans. Magn.* 23, 2266 (1987).

126. A. Kirilyuk, J. Ferré, J. Pommier, and D. Renard, *J. Magn. Magn. Mater.* 121, 536 (1993).

127. J. Pommier, P. Meyer, O. Pénissard, J. Ferré, R. Bruno, and D. Renard, *Phys. Rev. Lett.* 65, 2054 (1990).

128. B. Raquet, M. D. Onega, M. Goiran, A. R. Fert, L. P. Redoules, R. Marny, L. C. Ousset, A. Sdaq, and A. Kmhou, *J. Magn. Magn. Mater.* 150, 15 (1995).

129. C.-A. Chang, *J. Appl. Phys.* 68, 4873 (1990).

130. R. Naik, C. Kota, J. S. Payson, and G. L. Dunifer, *Phys. Rev. B: Condensed Matter and Materials Physics* 48, 1008 (1993).

131. W. W. Clegg, N. A. E. Heyes, E. V. Hill, and C. D. Wright, *J. Magn. Magn. Mater.* 95, 49 (1991).

132. G. Bochi, H. J. Hug, D. I. Paul, B. Stiefel, A. Moser, I. Parashikov, H. J. Güntherodt, and R. C. O'Handley, *Phys. Rev. Lett.* 75, 1839 (1995).

133. G. Bochi, C. A. Ballentine, H. E. Inglefield, C. V. Thompson, and R. C. O'Handley, *Phys. Rev. B: Condensed Matter and Materials Physics* 53, 1729 (1996).

134. P. Rosenbusch, J. Lee, G. Lauhoff, and J. A. C. Bland, *J. Magn. Magn. Mater.* 172, 19 (1997).

135. P. Kasirah, R. M. Shelby, J. S. Best, and D. E. Horne, *IEEE Trans. Magn.* 22, 837 (1986).

136. P. Büscher and L. Reimer, *Scanning* 15, 123 (1993).

137. J. Lee, G. Lauhoff, and J. A. C. Bland, *Europhys. Lett.* 35, 463 (1996).

138. L. Néel, *J. Phys. Radium* 12, 339 (1951).

139. J. Pommier, P. Meyer, G. Penissard, J. Ferre, P. Bruno, and D. Renard, *Phys. Rev. Lett.* 65, 2054 (1990).

140. M. Labrune, S. Andrieu, F. Rio, and P. Bernstein, *J. Magn. Magn. Mater.* 80, 211 (1989).

141. A. Kirilynk, J. Ferre, J. Pommier, and D. Renard, *J. Magn. Magn. Mater.* 121, 536 (1993).

142. L. Néel, *Ann. Geophys.* 5, 99 (1949).

143. I. Hashim, B. Park, and H. A. Atwater, *Appl. Phys. Lett.* 63, 2833 (1993).

144. R. Naik, M. Ahmad, G. L. Dunifer, C. Kota, A. P. Ketang, U. Rao, and J. S. Payson, *J. Magn. Magn. Mater.* 121, 60 (1993).

145. M. Cowley, "Diffraction Physics." North-Holland, Amsterdam, 1981.

146. A. S. Arrott, in "Ultrathin Magnetic Structures I" (B. Heinrich and J. A. C. Bland, Eds.), Springer-Verlag, Berlin, 1994.

147. J. Ferré, V. Grolier, P. Meyer, S. Lemerle, A. Maziewski, E. Stefanowicz, S. V. Tarasenko, V. V. Tarasenko, M. Kisielewski, and D. Renard, *Phys. Rev. B: Condensed Matter and Materials Physics* 55, 15,092 (1997).

148. V. S. Gornakov, V. I. Nikitenko, L. H. Bennett, H. J. Brown, M. J. Donahue, W. F. Egelhoff, R. D. McMichael, and A. J. Shapiro, *J. Appl. Phys.* 81, 5215 (1997).

149. J. P. C. Bernards, *J. Magn. Magn. Mater.* 123, 142 (1993).

150. H. J. Richter, *J. Magn. Magn. Mater.* 111, 201 (1992).

151. T. Bolhuis, L. Abelmann, J. C. Lodder, and E. O. Samwel, *J. Magn. Magn. Mater.* 193, 332 (1999).

152. H. Hauser, F. Haberl, J. Hochreiter, and M. Gaugitsch, *Appl. Phys. Lett.* 64, 2448 (1994).

153. F. Haberl, J. Hochreiter, M. Gaugitsch, and H. Hauser, *Jpn. J. Appl. Phys.* 33, 2752 (1994).

154. L. M. Falicov, D. T. Pierce, S. D. Bader, R. Gronsky, K. B. Hathaway, H. J. Hopster, D. N. Lambeth, S. S. P. Parkin, G. Prinz, M. Salamon, I. K. Schuller, R. H. Victora, J. Mater. Res. 5, 1299 (1990).

155. Y. R. Shen, "The Principles of Nonlinear Optics." Wiley, New York, 1984.

156. G. L. Richmond, J. M Robinson, and V. L. Shannon, *Prog. Surf. Sci.* 28, 1 (1988).

157. R.-P. Pan, H. D. Wei, and Y. R. Shen, *Phys. Rev. B: Condensed Matter and Materials Physics* 39, 1229 (1989).

158. R.-P. Pan and Y. R. Shen, *Chin. J. Phys. (Taipei)* 25, 175 (1987).

159. W. Hübner and K. H. Bennemann, *Phys. Rev. B: Condensed Matter and Materials Physics* 40, 5973 (1989).

160. J. Reif, J. C. Zink, C. M. Schneider, and J. Kirschner, *Phys. Rev. Lett.* 67, 2878 (1991).

161. G. Spierings, V. Koutsos, H. A. Wierenga, M. W. J. Prins, D. Abraham, and T. Rasing, *Surf. Sci.* 287, 747 (1993).

162. G. Spierings, V. Koutsos, H. A. Wierenga, M. W. J. Prins, D. Abraham, and T. Rasing, *J. Magn. Magn. Mater.* 121, 109 (1993).

163. H. A. Wierenga, W. de Jong, M. W. J. Prins, T. Rasing, R. Vollmer, A. Kirilyuk, H. Schwabe, and J. Kirschner, *Phys. Rev. Lett.* 74, 1462 (1995).

164. M. Weber and A. Liebsch, *Phys. Rev. B: Condensed Matter and Materials Physics* 35, 7411 (1987).

165. A. Liebsch, *Phys. Rev. B: Condensed Matter and Materials Physics* 36, 7378 (1987).

166. R. Vollmer, M. Straub, and J. Kirschner, *Surf. Sci.* 352, 684 (1996).

167. P. Guyot-Sionnest, W. Chen, and Y. R. Shen, *Phys. Rev. B: Condensed Matter and Materials Physics* 33, 8154 (1986).

168. Y. R. Shen, in "Chemistry and Structure at Interfaces: New Laser and Optical Techniques" (R. B. Hall and A. B. Ellis, Eds.), pp. 15ff. VCH, Deerfield Beach, Florida, Basel 1986.

169. S. Janz and H. M. van Driel, *J. Nonlinear Opt. Phys.* 2, 2 (1992).

170. T. Rasing, *J. Magn. Soc. Jpn.* 20, 13 (1996).

171. W. Hübner and K. H. Bennemann, *Phys. Rev. B: Condensed Matter and Materials Physics* 52, 13,411 (1995).

172. B. Koopmans, M. Groot Koerkamp, T. Rasing, and H. van den Berg, *Phys. Rev. Lett.* 74 (1995).

173. M. Groot Koerkamp and T. Rasing, *Surf. Sci.* 352, 933 (1996).

174. A. V. Petukhov, U. Lyubchanskii, and T. Rasing, *Phys. Rev. B: Condensed Matter and Materials Physics* 56, 2680 (1997).

175. W. D. Doyle, S. Stinnet, C. Dawson, and L. He, *J. Magn. Soc. Jpn.* 22, 91 (1998).

176. M. R. Freeman and J. F. Smyth, *J. Appl. Phys.* 79, 5898 (1996).

177. M. R. Freeman, W. K. Hiberi, and A. Stankiewicz *J. Appl. Phys.* 83, 6217 (1998).

178. T. J. Silva and A. B. Kos, *J. Appl. Phys.* 81, 5015 (1997).

179. M. Bonfim, K. Mackay, S. Pizzini, A. San Miguel, H. Tolentino, C. Giles, T. Neisius, M. Hagelstein, F. Baudelet, C. Malgrange, and A. Fontaine, Nanosecond-resolved XMCD on ID24 at the ESRF to investigate the element-selective dynamics of magnetization switchin of Gd-Co amorphous thin film. *J. Synchrotron Radiat.* 5, 750–752 (1998).

180. R. H. Koch, J. G. Deak, D. W. Abraham, P. L. Trouilloud, R. A. Altman, Yu Lu, W. J. Gallagher, R. E. Scheurlein, K. P. Roche, and S. S. Parkin. *Phys. Rev. Lett.* 81, 4512 (1998).

181. W. F. Brown, *Phys. Rev.* 130, 1677 (1963).

182. Y. Martin and H. K. Wickramasinghe, *Appl. Phys. Lett.* 50, 1455 (1987).

183. D. Rugar, H. J. Mamin, and P. Guenther, *J. Appl. Phys.* 68, 1169 (1990).

184. U. Hartmann, T. Goddenhenrich, H. Lemke, and C. Heiden, *IEEE Trans. Magn.* 26, 1512 (1990).

185. R. Proksch and E. D. Dahlberg, *J. Magn. Magn. Mater.* 104, 2123 (1992).

186. R. D. Gomez, A. A. Adly, I. D. Mayergoyz, and E. R. Burke, *IEEE Trans. Magn.* 29, 2494 (1993).

187. R. D. Gomez, E. R. Burke, and I. D. Mayergoyz, *J. Appl. Phys.* 79, 6441 (1996).

188. U. Hartmann, *J. Magn. Magn. Mater.* 157, 545 (1996).

189. K. L. Babcock, V. B. Elings, and J. Shi, *Appl. Phys. Lett.* 69, 705 (1996).

190. M. E. Hale, H. W. Fuller, and H. Rubinstein, *J. Appl. Phys.* 30, 789 (1959).

191. M. Mankos, M. R. Scheinfein, and J. M. Cowley, *J. Appl. Phys.* 75, 7418 (1994).

192. A. Tonomura, *Rev. Mod. Phys.* 59, 639 (1987).

193. R. E. Dunin-Borkowski, M. R. McCartney, B. Kardynal, and D. J. Smith, *J. Appl. Phys.* 84, 374 (1998).

194. S. J. Hefferman, J. N. Chapman, and S. McVitie, *J. Magn. Magn. Mater.* 83, 223 (1990).

195. S. McVitie, J. N. Chapman, L. Zhou, L. J. Heyderman, and W. A. P. Nicholson, *J. Magn. Magn. Mater.* 148, 232 (1995).

196. S. S. P. Parkin, R. Bhadra, and K. P. Roache, *Phys. Rev. Lett.* 66, 2152 (1991).

197. H. Holloway and D. J. Kubinski, *J. Appl. Phys.* 79, 7090 (1996).

198. D. J. Kubinski and H. Holloway, *J. Magn. Magn. Mater.* 165, 104 (1997).

199. L. J. Heyderman, J. N. Chapman, and S. S. P. Parkin, *J. Phys. D: Appl. Phys.* 27, 881 (1994).

200. P. R. Aitchison, J. N. Chapman, K. J. Kirk, D. B. Jardine, and J. E. Evetts, *IEEE Trans. Magn.* 34, 1012 (1998).

201. L. Reimer, Springer Series in Optical Sciences, Vol. 36. Springer-Verlag, Berlin, 1984.

202. I. R. McFadyen and J. N. Chapman, *EMSA Bull.* 22, 64 (1992).

203. K. Tsuno and T. Taoka, *Jpn. J. Appl. Phys.* 22, 1041 (1983).

204. J. N. Chapman, R. P. Ferrier, L. J. Heyderman, S. McVitie, W. A. P. Nicholson, and B. Bormans, *Electron Microsc. Anal.* 27, 1 (1993).

205. J. N. Chapman, A. B. Johnston, L. J. Heyderman, S. McVitie, W. A. P. Nicholson, and B. Bormans, *IEEE Trans. Magn.* 30, 4479 (1994).

206. M. E. Hale, H. W. Fuller, and H. Rubenstein, *J. Appl. Phys.* 30, 789 (1959).

207. J. N. Chapman, *J. Phys. D: Appl. Phys.* 17, 623 (1984).

208. J. N. Chapman, G. R. Morrison, J. P. Jakubovics, and R. A. Taylor, in "Electron Microscopy and Analysis, 1983" (P. Doig, Ed.), p. 197. IOP Conference Series no. 68, 1984.

209. A. B. Johnston and J. N. Chapman, *J. Microsc.* 179, 119 (1995).

210. N. H. Dekkers and H. de Lang, *Optik* 41, 452 (1974).

211. H. Rose, *Ultramicroscopy* 2, 251 (1977).

212. G. R. Morrison, H. Gong, J. N. Chapman, and V. Hrnciar, *J. Appl. Phys.* 64, 1338 (1988).

213. J. N. Chapman and M. R. Scheinfein, *J. Magn. Magn. Mater.* 200, 729 (1999).

214. J. N. Chapman, I. R. McFadyen, and S. McVitie, *IEEE Trans. Magn.* 26, 1506 (1990).

215. D. M. Mattox, "Handbook of Physical Vapour Deposition (PVD) Processing." Noyes, Westwood, NJ, 1998.

216. J. L. Vosse and W. Kern, Eds., "Thin Film Processes." Academic Press, Orlando, FL, 1978.

217. H.-R. Tränkler and E. Obermeier, Eds., "Sensortechnik, Handbuch für Praxis und Wissenschaft." Springer-Verlag, Berlin Heidelberg, 1998.

218. H. Frey and G. Kienel, Eds., "Dünnschichttechnologie." VDI Verlag, Düsseldorf, 1987.

219. L. I. Maissel and R. Glang, Eds., "Handbook of Thin Film Technology." McGraw–Hill, New York, 1970.

220. S. M. Sze, Ed., "VLSI Technology." McGraw–Hill, New York, 1988.

221. M. Sato, S. Ishio, and T. Miyazaki, *IEEE Transl. J. Magn. Jpn.* 9, 44 (1994).

222. S. L. Burkett, J. Yang, D. Pillai, and M. R. Parker, *J. Vac. Sci. Technol., B* 14, 3131 (1996).

223. Y. B. Xu, E. T. M. Kernohan, D. J. Freeland, M. Tselepi, J. A. C. Bland, S. Holmes, and D. A. Ritchie, *Sens. Actuators A* 81, 258 (2000).

224. A. Y. Toporov, P. I. Nikitin, M. V. Valeiko, A. A. Beloglazov, V. I. Konov, A. M. Ghorbanzadeh, A. Perrone, and A. Luches, *Sens. Actuators A* 59, 323 (1997).

225. D. A. Tulchinsky, M. H. Kelley, J. J. McClelland, R. Gupta, and R. J. Celotta, *J. Vac. Sci. Technol., A* 16, 1817 (1998).

226. A. Tomozawa, E. Fujii, H. Torii, and M. Hattori, *IEEE Transl. J. Magn. Jpn.* 9, 146 (1994).

227. J.-M. Quemper, S. Nicolas, J. P. Gilles, J. P. Grandchamp, A. Bosseboeuf, T. Bourouina, and E. Dufour-Gergam, *Sens. Actuators A* 74, 1 (1999).

228. C. Liu and Y. W. Yi, *IEEE Trans. Magn.* 35, 1976 (1999).

229. K. Attenborough, H. Boeve, J. de Boeck, G. Borghs, and J.-P. Celis, *Sens. Actuators A* 81, 9 (2000).

230. P. B. Lim, K. H. Shin, M. Inoue, K. I. Arai, M. Izaki, K. Yamada, and T. Fujii, *Sens. Actuators A* 81, 236 (2000).

231. J. Hong, J. A. Caballero, E. S. Lambers, J. R. Childress, and S. J. Pearton, *J. Vac. Sci. Technol., B* 16, 3349 (1998).

232. B. Khamsehpour, C. D. W. Wilkinson, J. N. Chapman, and A. B. Johnston, *J. Vac. Sci. Technol., B* 14, 3361 (1996).

233. K. B. Jung, E. S. Lambers, J. R. Childress, S. J. Pearton, M. Jenson, and A. T. Hurst, Jr., *J. Vac. Sci. Technol., A* 16, 1697 (1998).

234. J. C. Mallinson, "Magneto-Resistive Heads, Fundamentals and Applications." Academic Press, San Diego, 1996.

235. C. S. Roumenin, "Solid State Magnetic Sensors, Handbook of Sensors and Actuators 2" (S. Middelhoek, Ed.). Elsevier Science B. V., Amsterdam, 1994.

236. P. Ripka (Ed.), "Magnetic Sensors and Magnetometers." Artech, Boston, 2001.

237. T. J. Coutts, Ed., "Active and Passive Thin Film Devices." Academic Press, London, 1978.

238. H. Hauser, G. Stangl, and J. Hochreiter, Sensors and Actuators A 81, 27 (2000).

239. D. J. Mapps, Y. Q. Ma, and M. A. Akhter, *Sens. and Actuators A* 81, 60 (2000).

240. W. Thomson, *Proc. R. Soc. London, Ser. A* 8, (1857).

241. U. Dibbern, in "Sensors" (W. Göpel, J. Hesse, and J. N. Zemel, Eds.), Vol. 5, "Magnetic Sensors" (R. Boll and K. J. Overshott, Vol. Eds.), pp. 342–379. VCH, Weinheim, 1989.

242. D. J. Mapps, *Sens. Actuators A* 59, 9 (1997).

243. T. R. McGuire and R. I. Potter, *IEEE Trans. Magn.* 11, 1018 (1975).

244. H. Hauser, G. Stangl, J. Hochreiter, R. Chabicovsky, W. Fallmann, and K. Riedling, *J. Magn. Magn. Mater.* 216, 788 (2000).

245. J. McCord, A. Hubert, G. Schröpfer, and U. Loreit, *IEEE Trans. Magn.* 32, 4803 (1996).

246. J. S. Y. Feng, L.T. Romankiw, and D. A. Thompson, *IEEE Trans. Magn.* 13, 1466 (1977).

247. P. Aigner, G. Stangl, and H. Hauser, *J. Phys. IV* 8, 461 (1998),

248. Y. J. Song and S. K. Joo, *IEEE Trans. Magn.* 32, 5 (1996).

249. B. H. L. Hu, K. Ju, C. C. Han, D. Chabbra, Y. Guo, C. Horng, J. Chang, T. Torng, G. Yeh, B. B. Lal, S. Malhotra, Z. Jiang, M. M. Yang, M. Sullivan, and J. Chao, *IEEE Trans. Magn.* 35, 683 (1999),

250. R. E. Fontana, Jr., S. A. MacDonald, H. A. A. Santini, and C. Tsang, *IEEE Trans. Magn.* 35, 806 (1999).

251. A. Siritaratiwat, E. W. Hill, I. Stutt, J. M. Fallon, and P. J. Grundy, *Sens. Actuators A* 81, 40 (2000).

252. T. Kanda, M. Jimbo, S. Tsunashima, S. Goto, M. Kumazawa, and S. Uchiyama, *IEEE Transl. J. Magn. Jpn.* 9, 103 (1994).

253. M. Nawate, S. Ohmoto, R. Imada, and S. Honda, *IEEE Transl. J. Magn. Jpn.* 9, 38 (1994).

254. C. Christides, S. Stavroyiannis, G. Kallias, A. G. Nassiopoulou, and D. Niarchos, *Sens. Actuators A* 76 167 (1999).

255. J. A. Caballero, Y. D. Park, J. R. Childress, J. Bass, W.-C. Chiang, A. C. Reilly, W. P. Pratt, Jr., and F. Petroff, *J. Vac. Sci. Technol., A* 16, 1801 (1998).

256. M. Vieth, W. Clemens, H. van den Berg, G. Rupp, J. Wecker, and M. Kroeker, *Sens. Actuators A* 81, 44 (2000).

257. A. J. Devasahayam and M. H. Kryder, *IEEE Trans. Magn.* 35, 649 (1999).

258. A. Tanaka, Y. Shimizu, H. Kishi, K. Nagasaka, H. Kanai, and M. Oshiki, *IEEE Trans. Magn.* 35, 700 (1999).

259. N. J. Oliveira, P. P. Freitas, S. Li, and V. Gehanno, *IEEE Trans. Magn.* 35, 734 (1999).

260. Y. Hamakawa, M. Komuro, K. Watanabe, H. Hoshiya, T. Okada, K. Nakamoto, Y. Suzuki, M. Fuyama, and H. Fukui, *IEEE Trans. Magn.* 35, 677 (1999).

261. P. P. Freitas, F. Silva, N. J. Oliveira, L. V. Melo, L. Costa, and N. Almeida, *Sens. Actuators A* 81, 2 (2000).

262. J. L. Leal and M. H. Kryder, *IEEE Trans. Magn.* 35, 800 (1999).

263. J.-G. Zhu, *IEEE Trans. Magn.* 35, 655 (1999).

264. Y. Nishibe, H. Yamadera, N. Ohta, K. Tsukada, and Y. Nonomura, *Sens. Actuators A* 82, 155 (2000).

265. L. V. Panina and K. Mohri, *Sens. Actuators A* 81, 71 (2000).

266. M. Yamaguchi, M. Takezawa, H. Ohdaira, K. I. Arai, and A. Haga, *Sens. Actuators A* 81, 102 (2000).

267. W. P. Jayasekara, S. Khizroev, M. H. Kryder, W. Weresin, P. Kasiraj, and J. Fleming, *IEEE Trans. Magn.* 35, 613 (1999).

268. W. J. O'Kane, M. P. Connolly, T. K. McLaughlin, D. K. Milne, A. Al-Jibouri, C. J. O'Kane, and A. J. Gillespie, *IEEE Trans. Magn.* 35, 718 (1999).

269. H. Muraoka, K. Sato, Y. Sugita, and Y. Nakamura, *IEEE Trans. Magn.* 35, 643 (1999).

270. J. K. Howard, *J. Vac. Sci. Technol., A* 4, 1 (1986).

271. T. Hikosaka and Y. Tanaka, *IEEE Transl. J. Magn. Jpn.* 9, 4 (1994).

272. C. W. Chen, *J. Mater. Sci.* 26, 3125 (1991).

273. H. Ikekame, Y. Tosaka, K. Sato, K. Tsuzukiyama, and Y. Togami, *IEEE Transl. J. Magn. Jpn.* 9, 49 (1994).

274. Y. Takeda, T. Umezawa, K. Chiba, H. Shoji, and M. Takahashi, *IEEE Trans. Magn.* 35, 2166 (1999).

275. K. Shirakawa, H. Kurata, M. Mitera, O. Nakazima, and K. Murakami, *IEEE Transl. J. Magn. Jpn.* 9, 61 (1994).

276. M. Yamaguchi, S. Arakawa, and K. I. Arai, *IEEE Transl. J. Magn. Jpn.* 9, 52 (1994).

277. E. Quandt and A. Ludwig, *Sens. Actuators A* 81, 275 (2000).

278. N. Tiercelin, P. Pernod, V. Preobrazhensky, H. Le Gall, J. Ben Youssef, P. Mounaix, and D. Lippens, *Sens. Actuators A* 81, 162 (2000).

279. T. Honda, Y. Hayashi, M. Yamaguchi, and K. I. Arai, *IEEE Transl. J. Magn. Jpn.* 9, 129 (1994).

280. Y. Hayashi, T. Honda, M. Yamaguchi, and K. I. Arai, *IEEE Transl. J. Magn. Jpn.* 9, 124 (1994).

281. T. M. Liakopoulos and C. H. Ahn, *Sens. Actuators A* 77, 66 (1999).

282. W. J. Karl, A. L. Powell, R. Watts, M. R. J. Gibbs, and C. R. Whitehouse, *Sens. Actuators A* 81, 137 (2000).

283. H. Chiriac, M. Pletea, and E. Hristoforou, *Sens. Actuators A* 81, 166 (2000).

284. T. Meydan, P. I. Williams, A. N. Grigorenko, P. I. Nikitin, A. Perrone, and A. Zocco, *Sens. Actuators A* 81, 254 (2000).

285. K. Nordquist, S. Pendharkar, M. Durlam, D. Resnick, S. Tehrani, D. Mancini, T. Zhu, and J. Shi, *J. Vac. Sci. Technol. B*, 15, 2274 (1997).

Chapter 9

MAGNETOTRANSPORT EFFECTS IN SEMICONDUCTORS

Nicola Pinto, Roberto Murri
INFM, Dipartimento di Matematica e Fisica, Università di Camerino,
62032 Camerino, Italy

Marian Nowak
Silesian Technical University, Institute of Physics, 40-019 Katowice, Poland

Contents

Handbook of Thin Film Materials, edited by H.S. Nalwa
Volume 5: Nanomaterials and Magnetic Thin Films
Copyright © 2002 by Academic Press
All rights of reproduction in any form reserved.

ISBN 0-12-512913-0/$35.00

NOTATION

a_B	Bohr radius = 0.053 nm	n_L	number of states per Landau level, spin and per unit area
$\vec{B}$	magnetic induction field	n_S	sheet carrier density or carrier density per unit area
c_l	components of the elasticity tensor	N	density of atoms, impurities, etc.
D	optical deformation potential constant	N_A	density of acceptor impurities
D_n	electron diffusion coefficient	N_A^-	density of ionized acceptor impurities
e	electron charge	N_{An}	density of neutral acceptor impurities
$\vec{E}$	electric field	N_C	effective density of states
$\vec{E}_H$	Hall field	N_D	density of donor impurities: $N_D = N_D^+ + N_{Dn}$
E_i	electric field components	N_D^+	density of ionized donor impurities
$\vec{E}_{ion}$	electric field at interface due to ionized impurity space charge density	N_{Dn}	density of neutral donor impurities
		N_q	optical phonon distribution function
$\vec{E}_p$	planar Hall field	p	hole density
$f(\varepsilon_j)$	Fermi–Dirac distribution function	P	Ettingshausen coefficient
f_{cor}	correction function for van der Pauw technique of measure	q_l	wave vector for longitudinal propagating waves
		Q	Nernst coefficient
F_j	Fermi integrals	r	electric resistance
g	spectroscopic splitting factor of electron levels	r_c	cyclotron radius in free space
g_j	number of states of a jth level, at energy ε_j	r_H	Hall factor, defined as $r_H = \mu_H/\mu$
		R	Hall coefficient
G	planar Hall coefficient	R_a	adiabatic Hall coefficient: $R_a = R + P\Omega$
$G_\square$	sheet conductance	R_e	Hall coefficient for electron
h	Planck's constant	R_h	Hall coefficient for holes
$\hbar$	Dirac's constant, $h/2\pi = 1.0546 \times 10^{-34}$ Js	R	resistance
H_n	Hermite polynomial of order n	$R_\square$	sheet resistance
$\vec{I}$	electric current vector	R_{xy}	Hall resistance in a two-dimensional system: $R_{xy} = V_y/I_x$
I_j	current vector components along the j direction, with $j \in \{x, y, z\}$	R_{xx}	resistance in a magnetic field for a two-dimensional system: $R_{xx} = V_x/I_x$
$\vec{k}$	wave vector of a charge carrier	R_y	Rydberg constant
k_B	Boltzmann constant	S	Righi–Leduc coefficient
K^2	electromechanical coupling coefficient	t	sample thickness
l	carrier mean free path	T	semiconductor or lattice temperature
l_B	characteristic magnetic length	T_e	electron temperature
L	distance; sample length	T_D	Dingle temperature
L_D	Debye length: $L_D = \sqrt{D_n \tau_d}$	u	electrochemical potential: $u = \varepsilon_F - eV$
M	magnetoresistance coefficient: $M + 1 = [r(B)/r(B=0)]B^{-2}$	u^*	reduced electrochemical potential: $u/k_B T$
		$\vec{u}_i$	unit vectors of orthogonal cartesian axes, with $i \in \{x, y, z\}$
m	effective mass	U_G	semiconductor free energy: $U_G = U - TS$
m_0	free electron mass	ν	carrier speed
m_l	longitudinal effective mass	ν_F	carrier speed at Fermi energy
m_{lh}	light hole effective mass	ν_d	carrier drift velocity
m_{hh}	heavy hole effective mass	V	electrostatic potential
m_t	transverse effective mass	V_H	Hall potential
n_e	electron concentration	W	reduced energy: $W = \varepsilon_F/k_B T$
n_h	hole concentration	$W(i, j)$	transition rate between the states i and j
n_j	number of electron in the jth level, at energy ε_j	α	polar coupling constant
		β	warm electron transport coefficient

χ	dielectric constant
χ_H	high-frequency dielectric constant
χ_0	vacuum permittivity
ε	energy
ε_A	acceptor level energy
ε_{ac}	acoustic deformation potential constant
ε_C	conduction band edge energy
ε_D	donor level energy
ε_F	Fermi level: $\varepsilon_F = (\partial U_G \, \partial n)_{T=\cos t}$
ε_g	energy gap
ε_j	energy of an electron in a jth level
Φ_H	Hall angle: $\Phi_H = \tan^{-1}(E_H/E_x)$
$\Gamma(j)$	gamma function
η	reduced Fermi energy: $\eta = \varepsilon_F/k_B T$
λ_c	cyclotron wavelength in free space
μ	drift mobility
μ_0	zero field mobility
μ_B	Bohr magneton
μ_e	electron mobility
μ_h	hole mobility
μ_H	Hall mobility
ν	filling factor of Landau levels
θ_H	Heaviside step function
Θ	Debye temperature
ρ	mass density
$\underline{\underline{\rho}}$	resistivity tensor
ρ_B	resistivity in a magnetic induction field
ρ_{ij}	resistivity tensor components
ρ_I	remote impurity density
ρ_r	resistivity
$\underline{\underline{\sigma}}$	conductivity tensor
σ_e	electron conductivity
σ_h	hole conductivity
σ_{ij}	conductivity tensor components
τ_d	dielectric relaxation time
τ_e	mean time between collisions
τ_m	momentum relaxation time
ω	angular frequency of a wave
ω_C	cyclotron angular frequency: $\omega_C = |e|B/m$
$\omega_{C,e}$	cyclotron angular frequency for electrons
$\omega_{C,h}$	cyclotron angular frequency for holes
ω_l	angular vibration frequency for longitudinal propagating waves
ω_0	angular frequency of longitudinal optical phonon
ω_t	angular frequency of transverse acoustical phonon
Ω	thermoelectric power
Ξ	deformation potential constant

1. INTRODUCTION

It is worth noting the relevant role played by several electronic transport coefficients in the correct operation of any semiconductor device, from the simplest to the most complex.

To work, every semiconductor device must be inserted in an electric circuit, and it will be passed by charge carriers, either electrons or holes. The electrical performances of the semiconductor device and of the electrical circuit in which it is inserted will depend on, among several factors, how the carriers move, more or less freely, inside the semiconductor lattice.

Electronic transport coefficients such as the Hall coefficient, the mobility, the resistivity, impurity defects, density, etc., not only allow the exploration of the fundamental physical properties of the investigated material; they constitute the starting point for the design and fabrication of new and more reliable electronic devices.

Devices based on semiconductor heterostructures such as MODFETs (Modulation Doped Field Effect Transistors), SLs (Superlattices), lasers, etc. are now currently available thanks also to the great knowledge and experience gained, in past decades, in the study of electronic transport properties of charge carriers in semiconductors.

This chapter is an attempt to give an outlook on the main physical concepts and experimental techniques used to study the electronic transport coefficients in the presence of an applied magnetic and/or electric field.

Moreover, we made some selections in the techniques to be described, avoiding a complete derivation of every mathematical result, giving only the main and final equations.

We have attempted to provide the physical statements of the assumptions and approximations that constrain the use of a particular result and the application of techniques based upon it.

We are aware of the huge quantity of papers on theoretical calculations and experimental results published up to now and therefore of the difficulties of giving a complete citation of them.

We choose to restrict our attention to the techniques making use of well-established and accepted physical principles. In general, we have tried to describe the application of these techniques to a few selected materials, which indicated in a clear way the principles involved.

This work can be considered an introductory guide to the problematic and different techniques used to derive the electronic magnetotransport properties of semiconductors.

Throughout the chapter we provide references to excellent milestone publications that treat the subject more deeply.

2. INFLUENCE OF MAGNETIC FIELD ON EQUILIBRIUM CARRIER DENSITY IN SEMICONDUCTORS

In semiconductors, electrons and holes move as almost free carriers, but their energy distribution, among the available levels, obeys Fermi–Dirac statistics. At thermal equilibrium with the lattice vibration, the ratio n_j/g_j is the probability of occupancy of an energy level, ε_F, with degeneracy g_j [1],

$$\frac{n_j}{g_j} = \frac{1}{1 + e^{(\varepsilon_j - \varepsilon_F)/k_B T}} \qquad (2.1)$$

which is known as the Fermi–Dirac distribution function $f(\varepsilon_j)$. The quantities ε_F, k_B, and T are the Fermi energy, the Boltzmann constant, and the semiconductor temperature, respectively.

The Fermi level ε_F, defined as $\varepsilon_F = (\partial U_G / \partial n)_{T=\cos t}$ (where U_G is the free energy), is the energy level that has a probability of 0.5 of being occupied. Thermodynamics arguments show that ε_F is equal to the electron's chemical potential or average free energy per carrier. At thermodynamic equilibrium, in systems containing different groups of electrons, such as two different semiconductors in contact or a compound material, the chemical potentials of all groups of charge carriers must be the same. If an external electric field $\vec{E}$ is applied to the solid, then ε_F should be replaced by the electrochemical potential, u,

$$u = \varepsilon_F - eV \tag{2.2}$$

where V is the electrostatic potential associated with $\vec{E}$.

To compute the number dn of electrons occupying a level of energy $\varepsilon(\vec{k})$,

$$dn = f(\varepsilon)g(\vec{k})\,d\vec{k} \tag{2.3}$$

it is necessary to evaluate the density of states (DOS) $g(\vec{k})$. Cyclic boundary conditions of the wave function at the ends of the crystal give $1/(2\pi)^3$ states per unit volume of the crystal and per unit volume in $\vec{k}$ space. The number of states per unit volume of the crystal of the $\vec{k}$ space, lying between $\vec{k}$ and $\vec{k} + d\vec{k}$, i.e., $g(\vec{k})d\vec{k}$, is given by

$$g(\vec{k})d\vec{k} = 1/(2\pi)^3 d\vec{k} \tag{2.4}$$

The distribution $n(\varepsilon)$ is obtained by substituting (2.4) in (2.3) and then integrating,

$$n = 2\left(\frac{1}{2\pi}\right)^3 \int_0^\infty \frac{1}{1 + e^{(\varepsilon - \varepsilon_F)/k_B T}}\,d\vec{k} \tag{2.5}$$

where the factor 2 arises from the possibility of accommodating on the same level electrons of opposite spins.

In (2.5), the upper limit of integration, rigorously determined by the extension of the Brillouin zone, has been replaced by infinity for mathematical convenience, assuming that $f(\varepsilon)$ rapidly falls to zero for $\varepsilon > \varepsilon_F$.

The charge carrier distribution can be altered by the application of high-intensity electric fields or by generating excess electrons and holes with light sources.

In (2.3) DOS, $g(\vec{k})$, and the Fermi energy, ε_F, are the unknown quantities whose determination is needed to compute transport coefficients: $g(\vec{k})$ can be obtained from the energy dispersion relations, $\varepsilon = \varepsilon(\vec{k})$, whereas ε_F will be computed from the semiconductor band structure, its doping level, and its temperature.

In the following, the relations for the charge carrier density, n, will be derived for two main cases: (a) parabolic bands and (b) nonparabolic bands. Moreover, the expression of ε_F as a function of the doping level will be taken into account.

Finally, the effect on ε_F of a high-intensity magnetic field will be considered.

It must be noted that equations of the carrier concentration make it possible to obtain the expression for ε_F, the most relevant physical quantity, by simple algebraic manipulations. The carrier concentration, however, is most easily and directly measured by experiments.

a. Parabolic Bands. The $\varepsilon = \varepsilon(\vec{k})$ energy dispersion relation for isotropic parabolic bands is [1]

$$\varepsilon = \frac{\hbar^2 k^2}{2m} \tag{2.6}$$

After substituting in (2.5) and expressing the carrier concentration, n, as a function of energy, ε, we get [2]

$$n = 4\pi\left(\frac{2m}{h^2}\right)^{1/2} \int_0^\infty \frac{\varepsilon^{1/2}}{1 + e^{(\varepsilon - \varepsilon_F)/k_B T}}\,d\varepsilon \tag{2.7}$$

The integral in (2.7) cannot be solved analytically. Simplified expressions can be obtained by considering the cases (a_1) $\varepsilon_F < 0$ with $|\varepsilon_F| \gg k_B T$ (nondegenerate semiconductor), (a_2) $\varepsilon_F > 0$ with $|\varepsilon_F| <\sim k_B T$ (degenerate case), and (a_3) $\varepsilon_F > 0$ with $|\varepsilon_F| \ll k_B T$ (highly degenerate case).

a_1. Neglecting 1 in the denominator of (2.7), it follows that

$$n = 2\left(\frac{2\pi m k_B T}{h^2}\right)^{3/2} e^{\varepsilon_F / k_B T} = N_C e^{\varepsilon_F / k_B T} \tag{2.8}$$

where $N_C = 2(2\pi m k_B T / h^2)^{3/2}$ represents the effective DOS.

a_2. Solutions of (2.7) are given numerically. The integrals

$$F_j(n) = \frac{1}{\Gamma(j+1)} \int_0^\infty \frac{W^j}{1 + e^{(W - \eta)}}\,dW \tag{2.9}$$

are also known as Fermi integrals, where $\Gamma(j)$ denotes the gamma function, $W = \varepsilon/k_B T$ is the reduced energy, and $\eta = \varepsilon_F/k_B T$ is the reduced Fermi energy. Substituting in (2.9) the expression for W and η, the relation for n is obtained as

$$n = N_C \frac{2}{\sqrt{\pi}} \int_0^\infty \frac{W^j}{1 + e^{(W - \eta)}}\,dW = N_C F_{1/2}(\eta) \tag{2.10}$$

a_3. In the case of extremely degenerate semiconductors ($\eta \ll 1$), the relation (2.10) can be simplified by taking into account that the Fermi distribution function can be approximated to 1 for $W \leq \eta$, and to 0 for $W > \eta$. Then

$$n = N_C \frac{2}{\sqrt{\pi}} \int_0^\eta W^{1/2}\,dW = N_C \frac{4}{3\sqrt{\pi}} \eta^{3/2} \tag{2.11}$$

All of the formulas obtained above can be extended to semiconductors with an ellipsoidal constant energy surface, replacing the relation for N_C with [2]

$$N_C = 2\left(\frac{2\pi m_d k_B T}{h^2}\right)^{3/2} \tag{2.12}$$

where $m_d = (m_1 m_2 m_3)^{1/3}$ is the DOS effective mass.

It can be shown that for warped constant energy surfaces, typical of p-type materials, similar expressions hold, approximating the warped surfaces with spherical ones and properly defining the mass m_d [2].

b. Nonparabolic Bands. The isotropic energy dispersion function has the form

$$\gamma(\varepsilon) = \frac{\hbar^2 k^2}{2m} \qquad (2.13)$$

where m is the band-edge effective mass and the function $\gamma(\varepsilon)$ is usually expressed as $\gamma(\varepsilon) = \varepsilon(1 + \beta\varepsilon)$, where $\beta \cong 1/\varepsilon_g$, with the energy ε measured from the band edge. Substituting this last relation and (2.13) in (2.5), we get

$$n = N_C \left[F_{1/2}(\eta) + \frac{15}{4}\beta_n F_{3/2}(\eta) \right] \qquad (2.14)$$

with $\beta_n = \beta k_B T$. Equation (2.14) has been derived assuming that $\beta\varepsilon_F$ and β_n are small.

Equation (2.14) assumes a form similar to that of (2.10), defining a DOS effective mass, m_d, as follows:

$$m_d = m \left[1 + \frac{15}{4}\beta_n \frac{F_{3/2}(\eta)}{F_{1/2}(\eta)} \right]^{2/3} \qquad (2.15)$$

If the Fermi energy or the thermal energy is close to the value of ε_g, then (2.15) no longer holds and must be modified in

$$m_d = m \left[\frac{2}{\sqrt{\pi}} \int_0^\infty \frac{x^{1/2}(1 + \beta_n x)^{1/2}(1 + 2\beta_n x)}{e^{x-\eta} + 1} dx \right]^{2/3}$$
$$\times \left(F_{1/2}(\eta) \right)^{-2/3} \qquad (2.16)$$

The ratio m_d/m versus η is plotted in Figure 1 for both (2.15) (broken line) and (2.16) (continuous line) and for different values of β_n. As indicated in the figure, the difference between the two equations is reduced until $\beta_n \le 0.1$ and then diverges when $k_B T$ approaches the energy gap. The equations (2.15) and (2.16) hold for ellipsoidal constant energy surfaces, with m replaced by $(m_1 m_2 m_3)^{1/3}$.

2.1. Effects Caused by High-Intensity Magnetic Fields

The effects of a magnetic field on electronic band levels may be evident in transport experiments under particular conditions. In fact, for a conducting material at low temperature and in high-intensity magnetic fields, electrons may be forced to move in circular paths called cyclotron orbits, perpendicular to the direction of $\vec{B}$.

The theory was originally developed by Landau [3] to explain the diamagnetism of a quasi-free electron in a solid, which is not predicted by classical theory.

Landau's theory has shown that in the presence of a magnetic field the transverse components of the wave vector are quantized. These quantized levels (called Landau levels) are obtained by the solution of the Schrödinger equation in the presence of a magnetic field, $\vec{B} = \vec{\nabla} \wedge \vec{A}$, where $\vec{A}$ is a magnetic vector potential for the Coulomb gauge, i.e., $\vec{\nabla} \cdot \vec{A} = 0$.

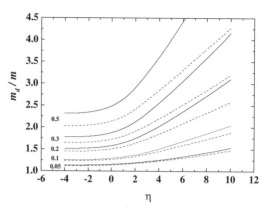

Fig. 1. Density of states effective mass ratio as a function of the reduced Fermi energy, for different values of β_n (see text). Continuous line: exact relation from Eq. (2.16). Broken line: approximation as obtained from (2.15). Reprinted with permission from [2], ©1980, Springer-Verlag.

After some substitutions the Schrödinger equation assumes the form

$$\left(-\frac{\hbar^2}{2m}\frac{d^2}{dx_1^2} + \frac{e^2 B^2}{2m}x_1^2 \right)\psi_1(x_1) = \left(\varepsilon - \frac{\hbar^2 k_z^2}{2m} \right)\psi_1 \qquad (2.17)$$

where $x_1 = x - \hbar k_y (|e|B)^{-1}$. The solution, Ψ, of (2.17) and its eigenvalues are, respectively,

$$\Psi = C \left\{ e^{\left[(|e|B/2h)(x(\hbar k_y/|e|B))^2 \right]} \right\}$$
$$\times H_n \left(x - \frac{\hbar k_y}{|e|B} \right) e^{i(k_y y + k_z z)} \qquad (2.18)$$

$$\varepsilon_l = \left(l + \frac{1}{2} \right)\hbar\omega_C \qquad (2.19)$$

where C is the normalization constant, H_n is the Hermite polynomial of order n, l is an integer, and

$$\omega_c = \frac{|e|B}{m} \qquad (2.20)$$

is the angular frequency of cyclotron resonance.

Then, in the presence of $\vec{B}$ the energy eigenvalues of a quasi-free electron are given by the relation

$$\varepsilon_{l,k_z} = \frac{\hbar^2 k_z^2}{2m} + \left(l + \frac{1}{2} \right)\hbar\omega_C \qquad (2.21)$$

The plot of this function is shown in Figure 2 for both electrons and holes, whose angular frequencies of cyclotron resonance are $\omega_{C,e}$ and $\omega_{C,h}$, respectively.

If electron spin is taken into account, the term $\pm\frac{1}{2}g\mu_B B$ must be added to (2.21), where g and μ_B are the spectroscopic splitting factor (Landè factor) and the Bohr magneton, respectively.

The previous relations have been derived, neglecting electron spin and assuming an isotropic effective mass, not

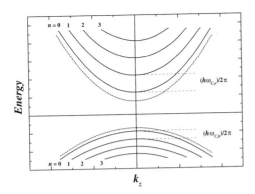

Fig. 2. Continuous line: Landau energy levels, for both electrons and holes, of an isotropic semiconductor in a quantizing magnetic field. The quantities $\omega_{C,e}$ and $\omega_{C,h}$ are the cyclotron angular frequencies for electrons and holes. Broken line: energy levels without the magnetic field. Reprinted with permission from [2], ©1980, Springer-Verlag.

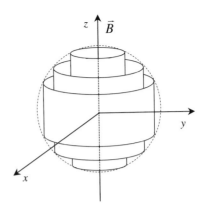

Fig. 3. Energy surfaces in a quantizing magnetic field oriented along the z axis direction (parallel to the cylinder axes).

depending on the energy. For ellipsoidal constant energy, the Landau level surfaces, considering the electron spin, assume the form

$$\varepsilon = |e|\hbar(m_1 m_2 m_3)^{-1/2}$$
$$\times (\alpha_1^2 m_1 + \alpha_2^2 m_2 + \alpha_3^2 m_3)^{1/2} B(l + \tfrac{1}{2})$$
$$\pm \tfrac{1}{2} g\mu_B B \qquad (2.22)$$

where α_1, α_2, and α_3 are the direction cosines of $\vec{B}$ with respect to the direction of the ellipsoid principal axes; m_1, m_2, and m_3 are the effective tensor mass components.

In the case of nonparabolic bands the relation (2.20) will be modified, taking into account the appropriate ε–k relation [2].

In the presence of a quantizing magnetic field, $\vec{B}$, the energy of charge carriers will be quantized only in directions transverse to that of $\vec{B}$. Moreover, the conduction band bottom will be raised by $\hbar\omega_{c,n}$, whereas the top of the valence band (V.B.) will be lowered by $\hbar\omega_{c,p}$.

From Figure 2 it must be noted that the Landau level spacing for electrons is generally different from that for holes, because the cyclotron frequency ω_C depends on the effective mass (see Eq. (2.20)).

This energy quantization causes interesting transport properties, such as Shubnikov–de Haas oscillations in magnetoresistance, de Haas–Van Alphen oscillation in magnetic susceptibility, magnetophonon oscillation in magnetoresistance and thermoelectric power, and cyclotron resonance and oscillations in the absorption coefficient of light.

For a degenerate semiconductor, which in a zero field has a spherical constant energy surface (Fermi sphere), the allowed k values in a magnetic field in the z direction lie on the surfaces of a set of concentric cylinders whose axes are along the k_z direction shown in Figure 3. As $\vec{B}$ is increased, the radii of the cylinders become larger and their number within a Fermi sphere of an energy that is equal to the Fermi energy ε_F becomes smaller. For the case $\hbar\omega_c \gg \varepsilon_F$ only the innermost

cylinder with quantum number $l = 0$ is available for accommodation of the carriers; this condition is called the quantum limit.

To be able to measure quantum effects, $\hbar^2 k_z^2/2m$ in (2.21) has to be much smaller than $\hbar\omega_c$, which is equivalent to the condition $k_B T \ll \hbar\omega_c$. Even for small effective masses, temperatures down to a few K are necessary to observe quantum effects with currently available magnetic field intensities. At such low temperatures the carrier gas is degenerate and the average energy must be about $\hbar\omega_c$.

The application of a quantizing magnetic field will alter the DOS, and then (2.5) must be modified. In these conditions, the applied magnetic field will change the DOS [2]. The DOS lying between ε and $\varepsilon + d\varepsilon$ becomes

$$g(\varepsilon)d\varepsilon = \frac{1}{4\pi^2}\left(\frac{2m}{\hbar^2}\right)^{3/2}\hbar\omega_C$$
$$\times \sum_{l=0}^{l_{max}}\left[\varepsilon - (2l+1)\frac{\hbar\omega_C}{2}\right]^{1/2} d\varepsilon \qquad (2.22)$$

In (2.23) spin degeneracy has been taken into account, whereas l_{max} is defined as

$$(2l_{max}+1)\frac{\hbar\omega_C}{2} < \varepsilon < (2l_{max}+3)\frac{\hbar\omega_C}{2} \qquad (2.23)$$

The total number of states between ε and $\varepsilon + d\varepsilon$ is obtained by summation over the various subbands. The subband corresponding to l_{max} has the property that the energy ε is defined between the top edge of this subband and the bottom edge of the next higher subband. So, when the intensity of $\vec{B}$ is large enough to produce a quantizing effect the DOS has the form shown in Figure 4 for parabolic bands. Figure 4 indicates a periodic change in the DOS with the applied field. This periodic behavior exhibited by the DOS is the main cause of oscillations observed in some transport properties at high magnetic fields.

The DOS depicted in Figure 4 refers to an ideal case of a perfect crystalline structure. In real crystals imperfections and thermal vibrations broaden the energy subbands and cause the DOS to be finite at singularity points [4].

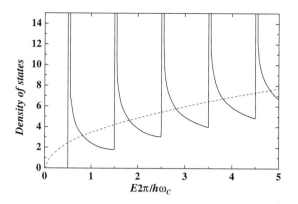

Fig. 4. Continuous line: density of states in a quantizing magnetic field for parabolic bands (Eq. (2.22)). Broken line: DOS at $\vec{B}=0$. Reprinted with permission from [2], ©1980, Springer-Verlag.

The discrete behavior of the DOS is revealed by measurements at sufficiently low temperatures, to reduce broadening due to thermal vibrations.

2.2. Fermi Level at High Magnetic Field

The Fermi level is affected by a strong magnetic field. In fact, from (2.1) and (2.23) we get

$$n = N_C \theta \sum_{l=0}^{\infty} F_{-1/2}\left[\eta - \left(l+\frac{1}{2}\right)\theta\right] \qquad (2.24)$$

where $\theta = \hbar\omega_C/k_B T$ and $F_{-1/2}$ is a Fermi integral.

Figure 5 reproduces the plot of n/N_C as a function of the quantity $\eta - \theta/2$, for different values of $l\theta$. Figure 5 clearly indicates the change in the Fermi level caused by the quantizing magnetic field $\vec{B}$. In general, ε_F is lowered when $\vec{B}$ is applied, and the material becomes less degenerate.

The relation (2.24) holds for any intensity of $\vec{B}$, and a simplified expression can be derived for nondegenerate and extremely degenerate cases [5].

The extreme quantum limit is also an interesting case that may be exploited experimentally. When the intensity of the

applied magnetic field is strong enough to cause $\hbar\omega_C > \varepsilon_F$, the Fermi level is below the bottom of the second subband, and only the first subband is filled by electrons (i.e., only the innermost cylinder will be inside the Fermi sphere). Under this condition, and for isotropic parabolic bands, the relation (2.24) simplifies to

$$n = N_C \theta F_{-1/2}\left[\eta - \frac{\theta}{2}\right] \qquad (2.25)$$

In the case of ellipsoidal constant energy surfaces (2.25) holds when N_C is replaced with (2.12) and ω_C is replaced with

$$\omega_C = (m_1 m_2 m_3)^{-1/2}\left(m_1\alpha_1^2 + m_2\alpha_2^2 + m_3\alpha_3^2\right)^{1/2}|e|B \qquad (2.26)$$

where the α values are the direction cosines of $\vec{B}$ with respect to principal axes of the ellipsoid.

For a nonparabolic band and if $\varepsilon < \varepsilon_g$ (a hypothesis satisfied in most practical cases) it can be demonstrated that a simplified relation [6]

$$n \cong N_C \theta \sum_{l=0}^{\infty}\left[\left(1+\frac{3\beta b}{2}\right)F_{-1/2}(\eta') + \frac{3}{4}\beta_n F_{1/2}(\eta')\right]a^{-1/2} \qquad (2.27)$$

applies, where $a = 1 + \beta(l+\frac{1}{2})\hbar\omega_C$, $b = (l+\frac{1}{2})\hbar\omega_C/a$, and $\eta' = (\varepsilon_F - b)/k_B T$.

Figure 6 is the plot of the ratio n/N_C vs. $\eta - \theta/2$ for several values of β_n. The curves reproduce the change in the Fermi level with the magnetic field. It can be shown [2] that a quantizing magnetic field will also shift the conduction band edge.

2.3. Fermi-Level Dependence on Impurity Concentration

At the beginning of this section we stated that the Fermi-level position can also be affected by the density of impurities, neutral and ionized.

The nature of the impurities can be two fold: they can be the result of the semiconductor doping and/or the effect of

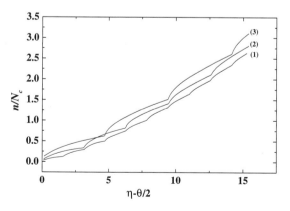

Fig. 5. The ratio n/N_C from Eq. (2.25) has been plotted as a function of $\eta - \theta/2$ (the reduced Fermi energy shifted by $\theta/2$). The numbers in parentheses, close to the curves, correspond to $\theta = 1.58$ and (1) $l = 1$; (2) $l = 2$; (3) $l = 3$. Reprinted with permission from [2], ©1980, Springer-Verlag.

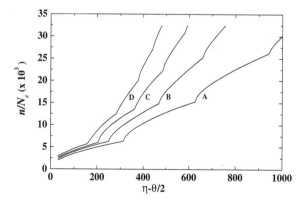

Fig. 6. Plot of the ratio n/N_C (Eq. 2.28) as a function of $\eta - \theta/2$ (the reduced Fermi energy shifted by $\theta/2$). (A) $\beta_n = 0$; (B) $\beta_n = 1$; (C) $\beta_n = 2$; (D) $\beta_n = 3$. Reprinted with permission from [2], ©1980, Springer-Verlag.

unintentional contamination of the sample due to the growth technique. It is not simple to distinguish and to separate the contributions of the two groups of impurities. A further complication arises from the different energy levels associated with each impurity because of different excitation states and degrees of degeneracy.

In any case, for some electronic transport properties to be interpreted, it is necessary to obtain an estimate of the density of neutral and ionized impurities. A general relation for the Fermi-level dependence on the impurity density is difficult to derive.

As an example we will consider a semiconductor containing an excess of donor atoms with a single energy level. We will assume that the Fermi level is always higher than the minor donor levels eventually present in the material.

Moreover, we will suppose that the semiconductor contains acceptor levels that will capture N_a electrons per unit volume. Finally, we will consider negligible the intrinsic electron generation (low temperature of the material). Let n be the total number of electrons in the conduction band,

$$n = N_D \left[1 + \beta e^{\varepsilon_F - \varepsilon_D / k_B T} \right]^{-1} - N_A \qquad (2.28)$$

where ε_D is donor level energy measured from the bottom of the conduction band and $N_D = N_D^+ + N_{Dn}$, where N_D^+ and N_{Dn} are the densities of ionized donors and neutral major donors, respectively. The variable β represents the donor level degeneracy. A typical β value is 2, taking into account spin degeneracy. The relation (2.29) allows the ε_F evaluation, provided N_D, N_A, and the band structure are known.

Figure 7 shows the temperature behavior of ε_F and n/N_D, for three different doping densities, corresponding to nondegenerate, moderate degenerate, and degenerate cases. On the same figure are plotted the behaviors for different values of the compensation ratio, K.

The effect of K on the Fermi level will be a shift from the donor level position, upward or downward, depending on whether the factor $\beta(N_D - N_A)/N_a$ is larger or smaller than unity. In any case, the detailed form of the ε_F curves from temperature will depend on the semiconductor nature due to the different values of ε_D and N_C [5].

A final consideration regards the effects caused by a high-intensity magnetic field, which, as shown in a previous section, affects the Fermi-level position. In the case of donor levels, the magnetic field will increase the ionization energy globally with a lowering of the electron density. At low temperatures this fact may cause electron freezing on the donor levels [7] (see also Section 2.2.4).

A last remark regarding the range of applicability of the relation (2.28): it is valid only for semiconductors that are not heavily doped.

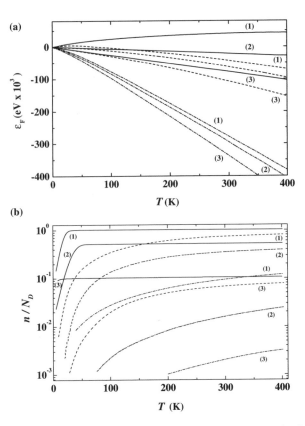

Fig. 7. (a) Plot of Fermi level ε_F as a function of the temperature for three doping levels (N_D) and three compensation ratios (K). The effective mass is $m = 0.068 m_0$, and the donor ionization level is $E_D = 5$ meV. Continuous line: $N_D = 10^{13}$ at./cm³; broken line: $N_D = 10^{17}$ at./cm³; broken and dotted line: $N_D = 10^{19}$ at./cm³. (1) $K = 0$; (2) $K = 0.5$; (3) $K = 0.9$. (b) Plot of the ratio n/N_D vs. temperature for three doping levels (N_D) and three compensation ratios (K). Same conditions and values as in (a). Reprinted with permission from [2], ©1980, Springer-Verlag.

2.4. Basic Equation of Charge Carrier Motion in Electric and Magnetic Fields

2.4.1. Electric Field

In this section we consider the fundamental equations of charge carrier motion in the presence of an electric and/or a magnetic field. This argument has been treated by several authors in excellent publications [8–10], and in this section we intend to give the main result for the simple case of a semiconductor material with an isotropic band structure.

Consider a conductor material and in a small applied external electric field, $\vec{E}$. The force caused by the field will alter the charge distribution inside the solid from that given by $f(\varepsilon)$ (see Eq. 2.1). If, in the presence of the electric field $\vec{E}$, a steady state is reached, then the carrier distribution will be described by function, $f_1(\varepsilon)$, which will take into account the charge drift velocity in the direction of $\vec{E}$. In the hypothesis of a small $\vec{E}$, the two functions will differ slightly, and if the applied field is removed, $f_1(\varepsilon)$ will revert to $f(\varepsilon)$.

Under these conditions, the rate at which equilibrium is established is governed by the equation

$$\frac{df}{dt} = \frac{f_1(\varepsilon) - f(\varepsilon)}{\tau_m} \tag{2.30}$$

where τ_m has the dimensions of a time, characteristic of the system, which is called the momentum relaxation time. It can be shown that (2.30) is valid in the majority of circumstances; any exceptions are more likely at very low temperatures [8].

To obtain an expression for the electrical conductivity it is necessary to consider (a) that electrons will acquire a drift velocity, $\delta \bar{v}$, due to the field $\vec{E}$, in a time interval, δt, between two consecutive collisions,

$$\delta \vec{v} = -\frac{e\vec{E}}{m} \delta t \tag{2.31}$$

where m is the electron effective mass.

(b) The electron motion inside the crystal will be hindered by several factors, such as the interaction with the crystal vibrations (phonons), impurities and structural imperfections of various natures (a detailed description of the main carrier scattering mechanisms is given in Section 2.3). These scattering processes will tend to restore the carrier distribution to that of equilibrium $f(\varepsilon)$.

The equation describing the rate of change of the distribution function when an electric field is applied is given by

$$\left(\frac{df(\varepsilon)}{dt}\right)_{drift} = \left(\frac{df(\varepsilon)}{dv}\right)\frac{dv}{dt} = -\frac{eE}{m}\frac{df(\varepsilon)}{dv} \tag{2.32}$$

In the condition of steady state and in the approximation of a mean relaxation time, the Boltzmann equation for transport can be used. Then

$$\left(\frac{df(\varepsilon)}{dt}\right)_{drift} + \left(\frac{df(\varepsilon)}{dt}\right)_{scattering} = 0 \tag{2.33}$$

Considering that in the absence of the field $\vec{E}$ the drift term in (2.33) will vanish and that the term due to scattering is equal to (2.30), from (2.30), (2.32), and (2.33) we obtain

$$f_1(\varepsilon) = f(\varepsilon) + \frac{eE}{m}\tau\frac{df(\varepsilon)}{dv} \tag{2.34}$$

From this last relation an expression for the current density $\bar{j}$ flowing in the x direction can be derived,

$$j_x = -e\int_0^\infty v_x f_1(\varepsilon)\phi(\varepsilon)d\varepsilon$$

$$= -\int_0^\infty \frac{e^2 E_x}{m}\tau_m\frac{df}{dv_x}v_x\phi(\varepsilon)d\varepsilon \tag{2.35}$$

where the function $\phi(\varepsilon)$ represents the energy level distribution in a simple isotropic band.

From (2.35) divided by the field E_x we can define another important parameter for the charge carrier transport, i.e., the electrical conductivity σ:

$$\sigma = \frac{j_x}{E_x} = -\int_0^\infty \frac{e^2\tau_m}{m}v_x\frac{df}{dv_x}\phi(\varepsilon)d\varepsilon \tag{2.36}$$

Assuming for $\phi(\varepsilon)$ a parabolic dependence on the energy ε, the simplified expression for the Fermi distribution function (neglecting 1 in the denominator of Eq. (2.1)), we may write

$$\sigma = \frac{4ne^2}{3\pi^{1/2}m}\int_0^\infty \tau_m W^{3/2}e^{-W}dW \tag{2.37}$$

where W is the reduced energy and in (2.37) the expression for the Fermi energy has been obtained by the relation for the carrier density, n (Eq. (2.7)).

To resolve Eq. (2.37) it is necessary to give an explicit expression for the relaxation time constant τ, which will depend on the scattering mechanisms active in the semiconductor material. It can be shown that τ depends on the energy and can be written as

$$\tau_m = \zeta\varepsilon^p = \zeta(k_B T)^p W^p \tag{2.38}$$

where ζ and p are constants whose values may be determined by experiments.

Substituting the relation (2.38) in (2.37) we obtain

$$\sigma = \frac{4ne^2}{3\pi^{1/2}m}\zeta(k_B T)^p\int_0^\infty W^{p+3/2}e^{-W}dW$$

$$= \frac{4ne^2}{3\pi^{1/2}m}\zeta(k_B T)^p\Gamma\left(p+\frac{5}{2}\right). \tag{2.39}$$

It can be shown that the conductivity expression obtained by the Drude–Lorentz theory can be derived from Eq. (2.39). In fact, if l defines the mean free path of an electron between two consecutive collisions, and it is independent of energy. Then

$$l = v\tau_m = \tau_m\left(\frac{2\varepsilon}{m}\right)^{1/2} \tag{2.40}$$

Moreover, $\zeta = l(m/2)^{1/2}$ and $p = 1/2$; the conductivity then assumes the form

$$\sigma = \frac{4}{3}\frac{ne^2 l}{(2\pi mk_B T)^{1/2}} \tag{2.41}$$

Equation (2.41) can also be expressed in the form

$$\sigma = \frac{ne^2\langle\tau_m\rangle}{m} = ne\mu \tag{2.42}$$

where $\langle\tau_m\rangle$ is the average value of the momentum relaxation time of the carrier, and μ is the mobility (drift), defined as

$$\mu = \frac{e}{m}\langle\tau_m\rangle \tag{2.43}$$

The mobility represents the average velocity gained by a carrier in an applied electric field. This transport coefficient may give important physical information about the scattering mechanisms active in the semiconductor material.

2.4.2. *Electric and Magnetic Fields*

If both electric and magnetic fields are applied to a semiconductor, the resulting force $\vec{F}$ acting on an electron is given by

$$\vec{F} = -(e\vec{E} + e\vec{\nu} \wedge \vec{B}) \tag{2.44}$$

It may be shown that in the presence of a thermal gradient and a magneto-electric field the Boltzmann equation [9] assumes the form

$$-\frac{e}{m}(\vec{E} + \vec{\nu} \wedge \vec{B})\vec{\nabla}_\nu f + \vec{\nu} \cdot \vec{\nabla}_r f = -\frac{f_1 - f}{\tau} \tag{2.45}$$

A solution of this equation requires the consideration of a perturbed function with the form [8]

$$f_1 = f - \vec{\nu} \cdot \vec{c}(\varepsilon)\mathrm{d}f/\mathrm{d}\varepsilon \tag{2.46}$$

After some calculations and supposing that the perturbed function (2.46) differs slightly from the equilibrium distribution function, we get the equation

$$-e\vec{E} \cdot \vec{\nu} + k_\mathrm{B}T\vec{\nu} \cdot \vec{\nabla}_r\left(\frac{\varepsilon - \varepsilon_\mathrm{F}}{k_\mathrm{B}T}\right)$$
$$+\frac{e}{m}\vec{\nu} \cdot (\vec{B} \wedge \vec{c}(\varepsilon)) = \frac{1}{\tau_\mathrm{m}}\vec{\nu} \cdot \vec{c}(\varepsilon) \tag{2.47}$$

To obtain an explicit relation for the function $\vec{c}(\varepsilon)$ it is sufficient to assume, without any loss of generality, that the magnetic field $\vec{B}$ is oriented along the z axis and that the electric field and the thermal gradient are in the x, y plane. Under these conditions $\vec{c}(\varepsilon)$ will be defined only by the two components in the x, y plane,

$$c_x = \frac{\xi - (\omega_\mathrm{C}\tau_\mathrm{m})s}{1 + (\omega_\mathrm{C}\tau_\mathrm{m})^2}$$
$$c_y = \frac{s - (\omega_\mathrm{C}\tau_\mathrm{m})\xi}{1 + (\omega_\mathrm{C}\tau_m)^2} \tag{2.48}$$

where

$$\xi = -\tau_\mathrm{m}\left[eE_x - k_\mathrm{B}T\frac{\mathrm{d}}{\mathrm{d}x}\left(\frac{\varepsilon_\mathrm{F}}{k_\mathrm{B}T}\right) - \frac{\varepsilon}{k_\mathrm{B}T}\frac{\mathrm{d}}{\mathrm{d}x}(k_\mathrm{B}T)\right]$$
$$s = -\tau_\mathrm{m}\left[eE_y - k_\mathrm{B}T\frac{\mathrm{d}}{\mathrm{d}y}\left(\frac{\varepsilon_\mathrm{F}}{k_\mathrm{B}T}\right) - \frac{\varepsilon}{k_\mathrm{B}T}\frac{\mathrm{d}}{\mathrm{d}y}(k_\mathrm{B}T)\right] \tag{2.49}$$

In (2.48) the quantity ω_C is the cyclotron angular frequency (see Eq. (2.20)) and represents an important parameter in the semiconductor material (see Section 2.8). The value of the product $\omega_\mathrm{c}\tau_\mathrm{m}$ defines the range of applicability of the theory given above.

If $\omega_\mathrm{c}\tau_\mathrm{m} \approx 1$, the theory breaks down and the quantum mechanical treatment must be considered. In fact, if the condition $\omega_\mathrm{c}\tau_\mathrm{m} \approx 1$ is realized, charge carriers will describe a trajectory with a radius of curvature comparable to their mean free path, with a quantization of their energy in a direction normal to that of the applied magnetic field (see Section 2.1). The analysis may be extremely complex.

For the components of the current density in the x, y plane we get

$$j_x = -e\int_0^\infty \nu_x^2 c_x \frac{\mathrm{d}f}{\mathrm{d}\varepsilon}\phi(\varepsilon)\mathrm{d}\varepsilon$$
$$j_y = -e\int_0^\infty \nu_y^2 c_y \frac{\mathrm{d}f}{\mathrm{d}\varepsilon}\phi(\varepsilon)\mathrm{d}\varepsilon \tag{2.50}$$

In the presence of an applied electric field, $\vec{E}$, the Fermi energy must be replaced by the electrochemical potential, u (see Eq. (2.2)). In our case, to derive the expression for the electrical conductivity, it is convenient to introduce the reduced electrochemical potential, u^*,

$$u^* = \frac{u}{k_\mathrm{B}T} = \frac{\varepsilon_\mathrm{F} - eV}{k_\mathrm{B}T} \tag{2.51}$$

where V is the electrostatic potential associated with the electric field $\vec{E}$.

The electrical conductivity is then given by

$$\sigma = ej_x\left[\frac{d}{dx}(k_\mathrm{B}Tu^*)\right]^{-1} \tag{2.52}$$

For the drift mobility we get

$$\mu_\mathrm{d} = \frac{j_x}{n}\left[\frac{d}{dx}(k_\mathrm{B}Tu^*)\right]^{-1} \tag{2.53}$$

2.4.3. *Mixed Conductors*

In real semiconductors, electrons, holes, or both may take part in the conduction process at the same time.

If it is assumed that the charge carriers in different bands do not interact with each other, then it is possible to write a set of equations similar to (2.52) and (2.53), for each group of electrons contributing to the conduction, in addition to the set representing the resultant behavior. These equations define the electrical properties that the solid would have if only those electrons contributed to the conduction [8]. Thus, if band 1 contains electrons while band 2 contains holes, the combination is made by applying the following rules:

(i) The electrochemical potentials u^* for the whole system, at equilibrium:

$$u^* = u_1^* = -u_2^* \tag{2.54}$$

where u_1^* and u_2^* are the electron and hole electrochemical potential, respectively.

(ii) The electric currents are the sum of the contributions from the individual bands:

$$j_x = j_{x_1} + j_{x_2} \tag{2.55}$$

General expressions can be calculated for all of the transport property coefficients valid for arbitrary magnetic fields in terms of the contributions from the individual bands under the same conditions.

The general formulae considering also the thermal effects have been computed by Putley [8], together with their approximation for the case of a weak intensity magnetic field. In this last case the electrical conductivity is independent of the magnetic field intensity, and it is expressed by

$$\sigma = \sigma_1 + \sigma_2 = ne\mu_n + pe\mu_p \qquad (2.56)$$

where n and p are the electron and hole concentrations, and μ_n and μ_p are their drift mobilities, respectively.

2.5. The Conductivity Tensor

If a weak electric field is applied to a semiconductor, with or without a magnetic field, the current density induced in the material [11–13] will obey the microscopic relationship

$$|\vec{j}|_l = \sum_m \sigma_{lm}(\vec{B})|\vec{E}|_m \qquad (2.57)$$

where σ_{lm} are the components of the electrical conductivity tensor, $\underline{\underline{\sigma}}$, depending on the magnetic field $\vec{B}$. The electrical conductivity $\underline{\underline{\sigma}}$ is a second-rank tensor and is symmetric, because its components σ_{lm} obey the Osanger relation [14]

$$\sigma_{lm}(\vec{B}) = \sigma_{lm}(-\vec{B}) \qquad (2.58)$$

For the seven crystallographic systems the components σ_{lm} of the conductivity tensor are quoted in Table I.

As is known, for nonspherical constant-energy surfaces, the effective mass is a tensor quantity, and the current density, $\vec{j}$, due to carriers in a single valley, is not collinear with the electric field $\vec{E}$, even when it acts alone. But for semiconductor solids with a cubic symmetry, and at low fields, the total current due to carriers in all of the valleys is in the direction of the electric field, as the cross terms cancel out because of symmetry. On the other hand, even in single-valley semiconductors with isotropic carrier effective mass, the current density components in the presence of a magnetic field will obey the equation (2.57), and the conductivity tensor $\underline{\underline{\sigma}}$ will assume the form

$$\underline{\underline{\sigma}} = \begin{pmatrix} \sigma_{xx} & \sigma_{xy} & 0 \\ \sigma_{yx} & \sigma_{yy} & 0 \\ 0 & 0 & \sigma_{zz} \end{pmatrix} \qquad (2.59)$$

Table I. Tensor Components σ_{ij} of the Electrical Conductivity for the Seven Main Crystallographic Systems

Crystallographic system	Tensor components
Cubic	$\sigma_{xx} = \sigma_{yy} = \sigma_{zz}$; $\sigma_{xy} = \sigma_{xz} = \sigma_{yz} = 0$
Orthorhombic	$\sigma_{xx}, \sigma_{yy}, \sigma_{zz}$; $\sigma_{xy} = \sigma_{xz} = \sigma_{yz} = 0$
Monoclinic	$\sigma_{xx}, \sigma_{yy}, \sigma_{zz}, \sigma_{xz}$; $\sigma_{xy} = \sigma_{yz} = 0$
Triclinic	$\sigma_{xx}, \sigma_{yy}, \sigma_{zz}, \sigma_{xy}, \sigma_{xz}, \sigma_{yz}$
Trigonal	
Tetragonal	$\sigma_{xx} = \sigma_{yy}, \sigma_{zz}$; $\sigma_{xy} = \sigma_{xz} = \sigma_{yz} = 0$
Hexagonal	

assuming that $\vec{B}$ has been applied in the z direction. We note that the conductivity component in the z direction is unaffected by the field $\vec{B}$, as it does not generate any force in its own direction. The other two diagonal components, σ_{xx} and σ_{yy}, are equal to each other because of symmetry. Moreover, the nondiagonal components σ_{xy} and σ_{yx} would be odd functions of the magnetic field, as the forces in the cross directions due to the magnetic field are proportional to it. Considering that σ_{xy} and σ_{yx} are odd functions of $\vec{B}$ and the relation (2.58) must hold, it follows that

$$\sigma_{xy} = -\sigma_{yx} \qquad (2.60)$$

The current density $\vec{j}$ may be written in terms of the conductivity tensor and the electric field components as

$$\begin{aligned} j_x &= \sigma_{xx}E_x + \sigma_{xy}E_y \\ j_y &= -\sigma_{xy}E_x + \sigma_{yy}E_y \\ j_z &= \sigma_{zz}E_z \end{aligned} \qquad (2.61)$$

where E_x, E_y, and E_z are the components of the electric field in the three directions. Because in an actual experiment the samples are rectangular in shape, $j_y = 0$, and from (2.61) we obtain

$$E_y = \sigma_{xy}E_x/\sigma_{yy} \qquad (2.62)$$

Experimentally the current direction is usually determined instead of the field, for example, by means of a long, thin sample with leads at the ends. Therefore, the resistivity tensor, $\underline{\underline{\rho}}$, is generally of interest. It is the inverse of the conductivity tensor $\underline{\underline{\sigma}}$,

$$\underline{\underline{\rho}} = \underline{\underline{\sigma}}^{-1} \qquad (2.63)$$

the components of which are defined by

$$E_l = \sum_m \rho_{lm}j_m \qquad (2.64)$$

An exception is the Corbino disk geometry [12], in which the components of the conductivity tensor are measured directly.

2.6. Scattering Mechanisms of Charge Carriers in Semiconductors

If charge carriers were in an ideal semiconductor, with a perfect crystalline structure, their motion would take place without any hampering. The application of an external electric field would cause a uniformly accelerated linear motion in the direction of the field.

But our experience teaches that the average drift velocity of the carriers reaches a limiting value in real crystalline materials because of different collision processes, undergone by the charge carriers during their motion, with crystalline "imperfections" [1, 2]. The collision process (also known as the scattering mechanism) is the result of an interaction between the carrier wave function (stationary Bloch functions) and the potential associated with the imperfections.

If the scattering processes are not too frequent, we may assume that the charge carrier always remains in a stationary state, but during the interaction with the imperfection. As a consequence the interaction will change the carrier wave function and will affect its momentum and energy. The new carrier state will be preserved until another collision takes place. If the number of collisions or its intensity is small, the whole process may be treated with the perturbation theory.

The nature of these collision processes can be very different, and their contribution to the overall mobility will be generally dependent, among other factors, on the temperature range considered. Figure 8 reproduces the temperature dependence of the mobility for different collision processes, as obtained by theoretical calculations for a sample of GaAs [15]. The experimental data plotted in Figure 8 demonstrate the agreement with the theoretical predictions. In the remain of this section the expressions of the Hall mobility as a function of the sample temperature will be derived.

2.6.1. Neutral Impurity Scattering

Among the scattering mechanisms, collisions with impurity atoms are the most common. The presence of impurities in the semiconductor lattice must be considered an unavoidable consequence of their growth process. Any growth technique produces imperfections in the crystal structure, such as crystallographic defects (dislocations, stacking faults, point defects, etc.) and impurities (neutral or ionized).

The presence of impurity atoms in the solid lattice, either as substitutional atoms (i.e., in a crystal position of the host lattice) or interstitial atoms (i.e., anywhere inside the lattice structure) introduces a modification of the periodic lattice potential in the form of a small perturbation. In addition, their presence in the semiconductor lattice will introduce energy levels in the energy gap of the material, that are generally placed just below the conduction band edge (donor atoms) or

just above the valence band edge (acceptor atoms), a few meV a part.

At very low temperatures (around liquid helium temperature) these impurities are neutral. The scattering process with electrons has been treated by Erginsoy, who applied the method of partial waves [16]. The result of the numerical calculation allows us to write a relation for the carrier mobility that is independent of the temperature,

$$\mu = \frac{e}{20 a_{\mathrm{B}} \hbar \chi N^{\chi}} \frac{m}{m_0} \qquad (2.65)$$

where a_{B} is the Bohr radius, χ is the dielectric constant, and N is the impurity atom density.

The mechanism of scattering by neutral impurities is difficult to observe because it is generally accompanied by other collision processes such as ionized impurity scattering and phonon scattering. To obtain information by this scattering mechanism it is necessary to use the method of cyclotron resonance (see Section 2.5) or that of ultrasonic wave attenuation.

2.6.2. Ionized Impurity Scattering

A relevant collision process in semiconductors is scattering by ionized atoms. At temperatures above liquid He the atoms of impurities have a finite probability of losing one electron (donor atoms, positively charged) or of capturing one electron (acceptor atom, negatively charged), and the density of ionized atoms will increase with the rise in temperature. An electron, approaching a charged impurity, will be deflected from its trajectory.

Calculation of the differential cross section of scattering is done with the Born approximation method. The result of this calculation is

$$\sigma(\theta) = \left(\frac{L}{2 \sin^2 \frac{\theta}{2}} \right)^2 \qquad (2.66)$$

where

$$L = \frac{Z e^2}{4 \pi \chi m v^2} \qquad (2.67)$$

is the distance for which the potential energy equals twice the kinetic energy. In (2.67) Ze is the charge of the ionized impurity.

The angle θ appearing in (2.66) is the angle of electron path deviation after the interaction with the Coulomb potential of an impurity ion, which is assumed to extend to infinity.

The calculation of the momentum relaxation time, τ_{m}, for the solution of the Boltzmann transport equation presents some difficulties due to the divergence of (2.66) for $\theta = 0$.

A possible solution to the divergence of (2.66) was found by Brooks [17], who considered a Coulomb potential of an impurity ion, screened by the charge carriers inside the solid,

$$V(r) = -\frac{Z|e|}{4 \pi \chi r} e^{-r/L_{\mathrm{D}}} \qquad (2.68)$$

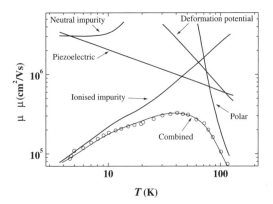

Fig. 8. Mobility as a function of the temperature for a sample of n-type GaAs. Circles: experimental data; continuous line: theoretical contributions to the overall (combined) mobility, taking into account different scattering mechanisms. Reprinted with permission from [15], ©1976, Elsevier Science.

with a typical screening distance given by the Debye length, L_D.

The results of the Brooks and Herring calculation were the expressions for the momentum relaxation time and the mobility,

$$\tau_m = \varepsilon^{3/2} \frac{16\pi\chi^2 \sqrt{2m}}{Z^2 e^4 N_I [\ln(1+\beta) - \beta/(1+\beta)]} \quad (2.69)$$

where N_I is the total ionized impurity concentration.

Assuming a nondegenerate electron gas, averaging over the energy, ε, in (2.69) gives

$$\langle \tau_m \rangle = \frac{4\sqrt{\pi}}{3} \int_0^\infty \tau_m \left(\frac{\varepsilon}{k_B T}\right)^{3/2} e^{e/k_B T} \frac{d\varepsilon}{k_B T} \quad (2.70)$$

In (2.70), replacing ε in β so that the integrand is a maximum (this happens for $\varepsilon = 3k_B T$), and from (2.43) we may write the expression for ionized impurity scattering,

$$\mu_{BH} = \frac{2^{7/2} 16\chi^2}{\pi^{1/2} Z^2 e^3 m^{1/2} N_I \left[\ln(1+\beta_{BH}^2) - \beta_{BH}^2/(1+\beta_{BH}^2)\right]} (k_B T)^{3/2} \quad (2.71)$$

where

$$\beta_{BH} = \frac{2mL_D}{\hbar} \left(\frac{2}{m} 3k_B T\right)^{1/2} \quad (2.72)$$

The solution to the divergence of the differential scattering cross section (2.66) proposed by Conwell and Weisskopf [18] was to consider a minimum scattering angle θ_{min}. The result of their calculation is quite similar to those of Brooks and Herring:

$$\mu_{CW} = \frac{2^{7/2} 16\chi^2}{\pi^{1/2} Z^2 e^3 [\log_{10}(1+\beta_{CW}^2)]} m^{-1/2} \frac{1}{N_I} (k_B T)^{3/2} \quad (2.73)$$

where

$$\beta_{CW} = \frac{1}{Z} \frac{\chi}{16} \frac{T}{100K} \left(\frac{2.35 \times 10^{19} \text{cm}^{-3}}{N_I}\right)^{1/3} \quad (2.74)$$

has been expressed in practical units. β_{CW} does not depend on the carrier concentration but on the impurity density.

Figure 9 reproduces the behavior of the mobility (Eqs. (2.71) and (2.73)) as a function of the impurity concentration at two different temperatures. As indicated in the figure, the calculations carried out by Brooks and Herring and by Conwell and Weisskopf begin to differ markedly when the impurity concentration reaches a value ranging from 10^{17} atoms/cm^3 to 10^{18} atoms/cm^3, depending on the sample temperature. However, beyond about 10^{18} atoms/cm^3 many semiconductors become degenerate, and the previous relations no longer hold.

In the case of a compensated semiconductor (i.e., a material containing both donor and acceptor impurities), a modified Brooks–Herring relation was obtained by Falicov and Cuevas [19] (in agreement with experimental data [20]). They assumed that the distribution of carriers, ionized acceptors,

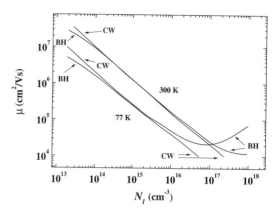

Fig. 9. The plotted curves reproduce the mobility as a function of the impurity concentration for two fixed temperatures, using the Brooks–Herring (BH) (Eq. (2.71) in the text) and Conwell–Weisskopf (CW) (see Eq. (2.73)) relations for an uncompensated semiconductor. The following parameters have been used for the curves: $m = m_0$, $\chi = 16\chi_0$, and $Z = 1$. Reprinted with permission from [1], ©1989, Springer-Verlag.

and ionized donors was as "frozen" in the configuration of minimum electrostatic energy, and they supposed that $N_A < N_D^+ < N_D$. They found the relation

$$\frac{1}{\tau_m} = \frac{mN_A}{4\pi\hbar^3} \left(\frac{4\pi e^2}{\chi}\right) \frac{1}{k^2} \left[\ln(1+\eta) + \frac{\eta}{(1+\eta)}\right] \quad (2.75)$$

In (2.75) the quantity in the square brackets depends very weakly on the energy, whereas the strong dependence on the energy is for k^2. Following a calculation procedure similar to that used by Brooks and Herring and substituting for k in η the value that maximizes the integrand (as done in the case of the Brooks–Herring relation), Falicov and Cuevas derived a relation for the mobility,

$$\mu_{FC} \cong \frac{2^{5/2}\chi^2}{\pi^{3/2}|e|^3 [\ln(1+\eta_0) + \eta_0/(1+\eta_0)]} \frac{1}{N_A} m^{-1/2} (k_B T)^{3/2} \quad (2.76)$$

where

$$\eta_0 = \frac{6mk_B T}{\pi^{2/3} \hbar^2 (N_D - N_A)^{2/3}} \quad (2.77)$$

The relation (2.76) differs from the relation (2.71) in the sign of the second term in the square brackets.

A review of the collision processes by ionized impurity scattering was published by Chattopadhyay and Queisser [21].

2.6.3. Scattering by Phonons

In all of the solid materials, however perfect they are, charge carriers undergo an unavoidable and uncontrollable scattering process due to an intrinsic physical property of the crystal: the presence at any temperature of lattice vibrations. As a result of lattice vibrations, the periodic potential of the crystal changes with time and causes variations in the electron states. As is known, from quantum mechanics applied to solid state theory [22], the lattice vibration can be described as lattice particles (called phonons), the nature of which will depend on the kind of vibration. Therefore, the phonon scattering (i.e., the interaction of electrons with the lattice vibrations) will be different, depending on the phonon type.

Phonons are quasi-particles that follow Bose–Einstein statistics, and they interact with electrons as particles obeying the principles of momentum and energy conservation.

In almost all semiconductors, there are two species of atoms or double periodicity or both. The dispersion relations (i.e., the ω versus q relation, where ω is the phonon angular frequency and q is its wave vector) comprise four branches, divided into two groups.

For one group the branches are called *acoustical*, and the phonon frequency ω increases with q, reaching a limiting value when q approaches the Brillouin zone boundary. Neighboring atoms in the lattice oscillate in phase similarly to a propagating acoustic wave. These two curves, correspond to longitudinal (LA) and transverse acoustic (TA) vibrations, and the phonons are called *acoustic phonons*.

In the second group the branches are called *optical*, and ω decreases with q. They are the longitudinal and transverse optical vibrations (LO and TO, respectively). The motion of the neighboring atoms takes place in the opposite phase, the phonons are called *optic phonons*.

In the theory of phonon scattering it is also necessary to consider the nature of the perturbing potentials caused by lattice vibrations.

Acoustic phonons generate a perturbing potential in two different ways:

a_1. In one way an acoustic wave changes the distances among the atoms, affecting both the energy gap and the edges of the conduction and valence bands. These changes from site to site of the crystal produce a potential discontinuity, called the deformation potential, whose magnitude is proportional to the lattice strain caused by the phonons. Consequently, the interaction of the charge carriers with phonons is called acoustic deformation potential scattering.

b_1. The second perturbing potential involves materials with two kinds of atoms (e.g., SiC), partially or completely ionized. In this case, the oscillations of the lattice atoms produce a piezoelectric field whose intensity may depend on the position of the atoms in the crystal and the system symmetry. In particular, systems with lower symmetry (e.g., sphalerite structure, zincblende lattice, or trigonal lattice) generate stronger piezoelectric fields.

This collision process is called piezoelectric scattering, and it is important for compound semiconductors, below room temperature in the low to medium range of temperature (see Fig. 8). Similarly to acoustic phonons, optic phonons create two distinct perturbing potentials.

a_2. Optic phonons produce a distortion of the lattice and then a perturbing potential proportional to the optic strain. The process of the collision of electrons with phonons by means of this potential strongly depends on the band structure and the symmetry of the crystal.

This kind of interaction is weak for electrons at the Γ point-minima or for minima along the $\langle 100 \rangle$ directions, but it is strong for minima along the $\langle 111 \rangle$ directions. With the exception of aluminum, gallium, and lead compounds, this kind of collision process is of little importance, assuming that for almost all of the other semiconductors, the lowest minimum is at the Γ point.

b_2. The other type of perturbing potential is characteristic of ionic compound semiconductors. Optic phonons displace the neighboring lattice ions, causing a change in the dipole moments. The perturbing potential is associated with this polarization induced by optic phonons; the electron interaction process is called polar optic phonon scattering. This scattering mechanism is one of the most relevant collision processes above liquid nitrogen temperature (see Fig. 8).

All of the phonon scattering processes considered above share a common characteristic: they take place inside the same valley; i.e., the electron remains in the same Γ point valley in ε–k space before and after the lattice scattering. Therefore, the collision interaction is called intravalley scattering. In intravalley scattering, only phonons with short wave vectors are involved.

For semiconductors with the lowest minima in the $< 100 >$ or $< 111 >$ directions, electrons could be scattered between two different valleys. This interaction, called intervalley scattering, requires phonons of long wave vector, and, in the presence of high electric fields, it may cause charge carriers scattering from a low valley to a higher, nonequivalent valley.

2.6.4. Acoustic Deformation Potential Scattering

For this kind of scattering mechanism and for charge carriers in thermal equilibrium with the lattice, understanding the relations of the momentum relaxation time and the electron mobility requires the theorem of deformation potential, developed by Bardeen and Shockley [23]. In their work, Bardeen and Shockley assumed a small displacement of the crystal atoms and considered a linear variation of the conduction and valence band edge with the lattice parameter. Under these conditions the perturbing potential is proportional through an acoustic deformation potential constant, ε_{ac}, to the amplitude of the perturbing acoustic wave.

The momentum relaxation time can be written as

$$\frac{1}{\tau_m} = \frac{\nu \varepsilon_{ac}^2}{\pi \hbar^4 c_l} m^2 k_B T \qquad (2.78)$$

where m is the density of states effective mass and c_l is the longitudinal elastic constant. In a cubic lattice and for an acoustic wave propagating along the $\langle 100 \rangle$ direction, $c_l = c_{11}$. For a $\langle 110 \rangle$ direction $c_l = (c_{11} + c_{12} + c_{44})/2$, and for a $\langle 111 \rangle$ direction $c_l = (c_{11} + 2c_{12} + 4c_{44})/3$; whereas for other directions, acoustic waves are not rigorously longitudinal. c_{11}, c_{12}, and c_{44} are the components of the elasticity tensor.

For the electron mobility we get

$$\mu = \frac{2\sqrt{2\pi}(e\hbar^4 c_l)}{3\varepsilon_{ac}^2} m^{-5/2} (k_B T)^{-3/2} \qquad (2.79)$$

The mass dependence of the mobility, expressed in (2.79), clearly shows that the highest value of the mobility is obtained by semiconductors with smaller effective mass. For example, in a n-type Ge sample at 100 K, the mobility attains a value of 3×10^4 cm^2/Vs. In any case, at higher temperatures the contribution to the collision processes by optical deformation potential scattering modifies the exponent of temperature dependence mobility. It has been found experimentally that the mobility is proportional to $T^{-1.67}$ [1].

Even though acoustic deformation potential scattering is essentially an elastic process, at high intensity electric fields the mobility becomes field strength dependent. To obtain the relations for the momentum relaxation time and for the mobility, we have to consider the energy loss by carriers that is transferred to the lattice, i.e., the so-called hot carrier transport. It is shown in Section 2.5 that in a high electric field the electrons acquire sufficient energy from the electric field that their temperature, T_e, called the electron temperature, may be much higher than that of the crystal lattice, T.

It can be shown that the average energy that charge carriers transfer to the lattice, per unit time [24], assumes the form [1]

$$\left\langle -\frac{d\varepsilon}{dt} \right\rangle_{\text{Coll}} = \frac{2m\varepsilon_{\text{ac}}^2}{\pi^{3/2}\hbar\rho} \left(\frac{2mk_BT_e}{\hbar^2} \right)^{3/2} 2\frac{T_e - T}{T} \frac{F_1(\eta)}{F_{1/2}(\eta)} \quad (2.80)$$

The relation (2.80) can be simplified by defining an average reciprocal momentum relaxation time,

$$\left\langle \tau_m^{-1} \right\rangle = \frac{2m^2\varepsilon_{\text{ac}}^2 q_1^2 k_B T}{\pi^{3/2}\hbar^4\rho\omega_1^2} \left(\frac{2k_BT_e}{m} \right)^{1/2} \frac{F_1(\eta)}{F_{1/2}(\eta)} \quad (2.81)$$

With Eq. (2.81) the quantity (2.80) assumes the following expression, valid for degeneracy and nondegeneracy:

$$\left\langle -\frac{d\varepsilon}{dt} \right\rangle_{\text{Coll}} = 4m\frac{\omega_1^2}{q_1^2}\left\langle \tau_{m1}^{-1} \right\rangle \frac{T_e - T}{T} \quad (2.82)$$

And in the case of nondegeneracy the carrier mobility is given by

$$\mu = \frac{8}{3\pi} \frac{e}{m\left\langle \tau_m^{-1} \right\rangle} \quad (2.83)$$

If μ_0 denotes the zero field mobility (i.e., the carrier mobility without an applied electric field), the relation (2.83) can be rewritten as

$$\mu = \mu_0 \left(\frac{T}{T_e} \right)^{1/2} \quad (2.84)$$

Moreover, it can be shown [1] that in the presence of an electric field of intensity E, carrier energy, gained per unit time by the field, is rapidly redistributed in all directions because of carrier–carrier collisions. In other words, the following equation holds:

$$-\left\langle \frac{\partial\varepsilon}{\partial T} \right\rangle_{\text{coll}} = \mu eE^2 \quad (2.85)$$

From (2.82) and making use of (2.83), (2.84), and (2.85), an expression for the ratio T_e/T can be obtained as a function of the electric field intensity:

$$\frac{T_e}{T} = \frac{1}{2}\left[1 + \sqrt{1 + \frac{3\pi}{8}\left(\frac{q_1\mu_0 E}{\omega_1} \right)^{\frac{1}{2}}} \right] \quad (2.86)$$

This last equation can be simplified in the case of warm or hot carriers.

For warm carriers (i.e., $T_e - T \ll T$) (2.86) reduces to

$$\frac{T_e}{T} = 1 + \left(\frac{3\pi q_1^2\mu_0^2}{32\omega_l^2} \right)E^2 \quad (2.87)$$

with the electron temperature depending on the square of the electric field; whereas, for hot carriers (i.e., $T_e \gg T$),

$$\frac{T_e}{T} = \left(\frac{3\pi}{32} \right)^{1/2}\frac{q_1}{\omega_1}\mu_0 E \quad (2.88)$$

The electron temperature depends linearly on E.

For warm carriers (i.e., when T_e is not much higher than T; see Section 2.5), the mobility dependent on the field intensity assumes the form

$$\mu = \mu_0(1 + \beta E^2) \quad (2.89)$$

where β, known as the warm electron transport coefficient, is given by

$$\beta = -\frac{3\pi}{64}\left(\frac{q_1\mu_0}{\omega_1} \right)^2 \quad (2.90)$$

For hot carriers,

$$\mu = \left(\frac{32}{3\pi} \right)^{1/4}\left(\frac{\omega_1}{q_1}\mu_0 \right)^{1/2}E^{-1/2} \quad (2.91)$$

Experimental works, carried out at low temperatures, have shown that in the presence of high electric fields acoustic phonon scattering is the main collision mechanism, and ionized impurity scattering can be neglected. Figure 10 reproduces the behavior of the mobility as a function of the electric field intensity for an n-type Ge sample, at two temperatures (77 K and 20 K) [25].

Consistent deviations from Ohm's law depend on the value of the quantities in the second set of parentheses: for $\omega_1/q_1 = 5 \times 10^5$ cm/s and $\mu_0 = 5 \times 10^3$ cm^2/Vs a critical field intensity of 100 V/cm is attained, which produces a rapid decrease in the mobility with the electric field intensity. The rapid raise of the mobility at about 15 V/cm for the curve at 20 K may be caused by ionized impurity scattering.

2.6.5. Combination of Impurity Ionized and Acoustic Deformation Potential Scattering

Among the different collision processes acting simultaneously inside the crystal lattice, ionized impurity and acoustic deformation potential scattering may have an important role in the

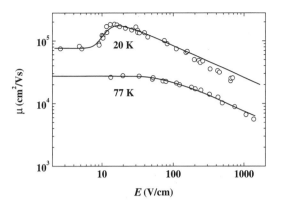

Fig. 10. Mobility of n-type Ge as a function of the electric field intensity at 20 K and 77 K, indicating the rapid decrease (see Eq. (2.91)) of the mobility values approaching a field intensity of about 100 V/cm. The bump in the 20 K curve at about 15 V/cm may be caused by ionized impurity scattering. Reprinted with permission from [25], ©1953, American Physical Society.

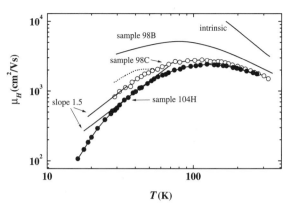

Fig. 11. Hall mobility as a function of the temperature for n-type Ge samples. Open and full circles refer to Zn doping (samples labelled 98C and 104H). Sample 98B has been doped with 5×10^{16} Sb atoms/cm^3. An intrinsic n-type Ge specimen has been shown for comparision. For all of the doped samples the continuous change from the ionized impurity scattering (below about 80 K) to lattice scattering is evident (above 200 K). Reprinted with permission from [27], ©1956, American Physical Society.

same temperature range. The overall scattering rate is obtained by summing the reciprocal momentum relaxation times,

$$\frac{1}{\tau_m} = \left(\frac{1}{\tau_m}\right)_{ac} + \left(\frac{1}{\tau_m}\right)_{ion} \tag{2.92}$$

with the ratio between the two relaxation times given by

$$\frac{(\tau_m)_{ac}}{(\tau_m)_{ion}} = q^2 \left(\frac{\varepsilon}{k_B T}\right)^{-2} \tag{2.93}$$

where the quantity q^2 has the form

$$q^2 = 6 \frac{\mu_{ac}}{\mu_{ion}} \tag{2.94}$$

where μ_{ac} and μ_{ion} are given by the relations (2.79) and (2.71), respectively.

Equation (2.92) can be rewritten by using (2.93) and (2.94):

$$\tau_m = (\tau_m)_{ac} \left(\frac{\varepsilon}{k_B T}\right)^2 \left[q^2 + \left(\frac{\varepsilon}{k_B T}\right)^2\right] \tag{2.95}$$

The mobility for a Maxwell–Boltzmann distribution, assuming an electron temperature T_e instead of the lattice temperature, T, assumes the form

$$\mu = \mu_{ac} \lambda^{1/2} \left[1 - \lambda^2 q^2 g(q')\right] \tag{2.96}$$

where $\lambda = T/T_e$, $q' = \lambda q$, and $g(q')$ is given by

$$g(q) = -Ci(q)\cos(q) - Si(q)\sin(q) = -\frac{df}{dq}$$
$$f(q) = Ci(q)\sin(q) - Si(q)\cos(q) = \frac{1}{q} + \frac{dg}{dq} \tag{2.97}$$

with $Ci(q) = \int_\infty^q (\sin(t))/t \, dt$ and $Si(q) = \int_\infty^q (\cos(t))/t \, dt$.

For a thermal carrier (i.e., $\lambda = 1$), the zero field mobility becomes [26]

$$\mu = \mu_{ac}\left[1 - q^2 g(q)\right] \tag{2.98}$$

At temperatures higher than 77 K the mobility presents a gradual change from an ionized impurity scattering to a phonon collision process [27], as indicated in Figure 11 by Hall mobility as a function of the temperature for two samples of Zn-doped Ge. Zn atoms are doubly ionized donors in Ge, resulting in a particularly efficient mechanism of ionized impurity scattering, due to the Z^2 dependence in the relations (2.71) and (2.73).

For one of the samples (104H) shown in Figure 11, the change from the $T^{3/2}$ slope, below 30 K, cannot be assigned to a change in the ion charge with the temperature, but to the impurity band conduction.

2.6.6. Piezoelectric Scattering

In a compound semiconductor whose lattice is made of two atoms of different electronegativities, charge carriers may undergo collision processes by longitudinal acoustic waves due to piezoelectric scattering, as shown by Hutson [28] and by Meyer and Polder [29]. It can be shown that the momentum relaxation time assumes the form

$$\tau_m = \frac{2^{3/2}\pi\hbar^2\chi}{e^2 K^2} \varepsilon^{1/2} m^{-1/2} (k_B T)^{-1} \tag{2.99}$$

where K^2 is a dimensionless quantity called the electromechanical coupling coefficient, defined by

$$K^2 = \frac{e_{pz}^2}{e_{pz}^2 + \chi c_1} \tag{2.100}$$

with e_{pz}^2 is a factor of proportionality on the order of 10^{-5} As/cm^2, which can be obtained from piezoelectric measurements.

Finally, the mobility for nondegenerate thermal carriers is expressed by

$$\mu = \frac{16}{3}\sqrt{2\pi} \frac{\hbar^2\chi}{eK^2} m^{-3/2} (k_B T)^{-1/2} \tag{2.101}$$

A typical mobility value at 100 K, due to piezoelectric scattering, is about 8×10^5 cm^2/Vs, assuming $\chi = 10\chi_0$, $m = 0.1m_0$, and $K^2 = 10^{-3}$. This mobility value is large compared with that due to the nonpolar acoustic collision process, which is the competing scattering mechanism in semiconductors with partly ionic bonds, in the same temperature range.

In a pure polar semiconductor, such as GaAs with a very low density of dislocation and crystallographic defects, the piezoelectric scattering may be the dominant collision process at low temperatures (about 10 K), a mobility value about five times lower than the mobility value dominated by acoustic deformation potential scattering. However, even at the very high level of material quality available with the most recent growth techniques, ionized impurity scattering cannot be excluded. Therefore, this kind of interaction mechanism of charge carriers with lattice vibration is difficult to observe and to identify in a clear way.

2.6.7. Optical Deformation Potential Scattering

In nonpolar semiconductors charge carriers may be scattered by longitudinal optical phonons. Because of the higher values of the phonon energies, which are comparable to the thermal carrier energies, this kind of collision process must be treated as inelastic. Therefore, scattering rates for transitions from a state $\vec{k}$ to a state $\vec{k} - \vec{q}$ or to a state $\vec{k} + \vec{q}$ must be considered different.

It can be shown [1] that even in this case of inelastic scattering it is possible to define a momentum relaxation time. The equation for τ_{m} derived by the calculation of Harrison [30] is

$$\frac{1}{\tau_{\mathrm{m}}} = \frac{D^2 N_{\mathrm{q}}}{2^{1/2} \pi \rho \hbar^2} m^{3/2} (k_{\mathrm{B}}\Theta)^{-1} \left\{ (\varepsilon + k_{\mathrm{B}}\Theta)^{1/2} \right. \\ \left. + e^{\Theta/T} \mathrm{Re}\left[(\varepsilon - k_{\mathrm{B}}\Theta)^{1/2} \right] \right\} \quad (2.102)$$

where D is the optical deformation potential constant and N_{q} is the optical phonon distribution function,

$$N_{\mathrm{q}} = \left(e^{\hbar\omega_0/k_{\mathrm{B}}T} - 1 \right)^{-1} = \left(e^{\Theta/T} - 1 \right)^{-1} \quad (2.103)$$

In (2.102) Re stands for "real part of," and it ensures a rapid vanishing of the term in the square brackets (i.e., of the phonon emission term) when $\varepsilon < k_{\mathrm{B}}\Theta$ (i.e., for $\varepsilon < \hbar\omega_0$).

It can be shown that τ_{m} in Eq. (2.102) rapidly decreases for values of $\varepsilon/k_{\mathrm{B}}\Theta > 1$. This decrease in τ_{m} is associated with the emission of optical phonons.

For a nondegenerate carrier gas obeying the Maxwell–Boltzmann distribution, the zero field mobility μ_0 is

$$\mu_0 = \frac{4\sqrt{2\pi}}{3} \frac{e\hbar^2\rho}{D^2} m^{-5/2} (k_{\mathrm{B}}\Theta)^{1/2} f\left(\frac{T}{\Theta}\right) \quad (2.104)$$

where the function $f(T/\Theta)$ has the form

$$f\left(\frac{T}{\Theta}\right) = (2z)^{5/2}(e^{2z} - 1)$$

$$\times \int_0^\infty \frac{y^{3/2} e^{-2zy}}{\sqrt{y+1} + e^{2z}\,\mathrm{Re}\left[\sqrt{y-1}\right]}\, dy \quad (2.105)$$

with $z = \Theta/2T$ and $y = \varepsilon/k_{\mathrm{B}}\Theta$.

For high-intensity electric fields the momentum relaxation time and the mobility may be derived by applying the momentum balance equation [1]

$$\frac{1}{\bar{\tau}_{\mathrm{m}}} = \frac{2m^{3/2}D^2}{3\pi^{3/2}\hbar^2\rho (k_{\mathrm{B}}\Theta)^{1/2}} \frac{(\lambda z)^{3/2}}{\sinh(z)} \left\{ \cosh[(1-\lambda)z]\mathrm{K}_2(\lambda z) \right. \\ \left. + \sinh[(1-\lambda)z]\mathrm{K}_1(\lambda z) \right\} \quad (2.106)$$

where $z = \Theta/2T$, $\lambda = T/T_{\mathrm{e}}$, and K_1 and K_2 are two modified Bessel functions.

For $\lambda = 1$, μ_0 is calculated by the approximation of the momentum balance equation,

$$\mu_0 = \frac{4\sqrt{2\pi}}{3} \frac{e\hbar^2\rho}{D^2} m^{-5/2} (k_{\mathrm{B}}\Theta)^{1/2} g\left(\frac{T}{\Theta}\right) \quad (2.107)$$

which differs from (2.104) for the function $g(T/\Theta)$:

$$g\left(\frac{T}{\Theta}\right) = 9\pi 2^{-7/2} z^{-3/2} \frac{\sinh(z)}{\mathrm{K}_2(z)} \quad (2.108)$$

At low temperatures ($T < 2\Theta$) the mobility values obtained with the balance equation (2.107) are lower than those obtained by the normal procedure (2.104).

2.6.8. Polar Optical Scattering

This kind of collision process between carriers and optical phonons takes place in polar semiconductors, similarly to piezoelectric scattering for acoustical phonons. As mentioned before, for optical deformation potential scattering, the inelastic nature of the electron–phonon interaction should not allow the use of a momentum relaxation time. However, using an approximation [1], it is possible to derive an expression for τ_{m},

$$\frac{1}{\tau_{\mathrm{m}}} = \alpha\omega_0 \left(\frac{\hbar\omega_0}{\varepsilon}\right)^{1/2} N_{\mathrm{q}} \left[\ln\left|\frac{a+1}{a-1}\right| + e^{\Theta/T}\ln\left|\frac{b+1}{1-b}\right| \right] \quad (2.109)$$

where α is a dimensionless quantity called the polar coupling constant,

$$\alpha = \frac{1}{137}\sqrt{\frac{mc^2}{2k_{\mathrm{B}}\Theta}}\left(\frac{1}{\chi_{\mathrm{opt}}} - \frac{1}{\chi}\right) \quad (2.110)$$

whereas a and b have the following forms:

$$a = \left(1 + \frac{\hbar\omega_0}{\varepsilon}\right)^{1/2}, \qquad b = \mathrm{Re}\left[\left(1 - \frac{\hbar\omega_0}{\varepsilon}\right)^{1/2}\right] \quad (2.111)$$

A simplified relation of (2.109) is achievable at low temperature, $T \ll \Theta$, assuming that $\varepsilon \ll \hbar\omega_0$, $b \approx 0$, and $N_{\mathrm{q}} \approx e^{-\Theta/T}$:

$$\tau_{\mathrm{m}}^{-1} \approx 2\alpha\omega_0 e^{-\Theta/T} \quad (2.112)$$

The simplified equation (2.112) for τ_m^{-1} is the result of the product of the polar constant and the probability of absorption of an optical phonon.

From the (2.112) we can derive the relation for the mobility,

$$\mu = \frac{e}{m}\tau_m = \frac{|e|}{2m\alpha\omega_0}e^{\Theta/T} \qquad (2.113)$$

A mobility value of 2.2×10^5 cm^2/Vs is obtained by (2.113) for n-type GaAs at 100 K, using the following values for the quantities appearing in (2.113): $\alpha = 0.067$; $\Theta = 417$ K, and $m/m_0 = 0.072$. The computed value of the mobility is comparable to the experimental one, even though at lower temperatures than 100 K the main scattering mechanism in GaAs and in compound materials in general is still impurity scattering. For this reason (2.113) is difficult to check.

A different kind of calculation, based on variational methods [10, 31], may be applied to find an expression of τ_m.

At low temperatures, $T \ll \Theta$, the mobilities [32] computed by two methods agree fairly well, whereas for high temperatures the relation (2.113) must be multiplied by $8\sqrt{T/9\pi\Theta}$. For intermediate temperatures it is necessary to apply numerical methods. It is found, by variational methods, that the momentum relaxation time depends on the energy by according to the relation $\tau_m(\varepsilon) \propto \varepsilon^{f(T)}$, where $f(T)$ is a function that is strongly dependent on the semiconductor temperature [32].

Further methods of calculation for polar optical scattering are the Monte Carlo procedure [33, 34], the iterative method [35], and its variant based on the concept of self-scattering [36].

In some cases where $\alpha > \approx 1$, such as in alkali halides, collision processes could be described with the concept of polarons. These are localized regions of the lattice that are polarized in the proximity of an electron as a consequence of the strong polar coupling. This lattice polarization moves together with the electron called, for this reason, a polaron. Moreover, the polaron effective mass, m_{pol}, is higher than the electron effective mass without polarization and is proportional to the α value.

For a weak coupling, $\alpha \ll 1$, the polarization can be considered a small perturbation, and the electron mass can be approximated by [22]

$$\frac{m_{pol}}{m} \approx 1 + \frac{\alpha}{6} \qquad (2.114)$$

where m is the effective electron mass without the polarization.

At high concentrations of charge carriers, n, in the polar material the interaction between electrons and phonons is screened and $\alpha = \alpha(n)$ [37]. The term *hot polaron* [38, 39] denotes hot carriers in a polar semiconductor, whereas the name *small polarons* [40] designates those polarons with a spatial extent in the lattice less than or similar to that of the lattice parameter.

Another important effect observed in polar semiconductors is ballistic transport. In particular, a Monte Carlo simulation

carried out by Ruch [41] on GaAs showed that an electric field intensity of 5 kV/cm, applied for a short time on the order of 10^{-12} s, produces a carrier speed three times higher than that produced by a field of 1 kV/cm (about 10^7 cm/s).

During this short interval of time carriers cover a distance on the order of a few hundred nanometers. In transistors with a short base (on the order of a few tens of nanometers) the transit time of the carriers is shorter than the energy relaxation time, and, then, the carriers cannot decrease their speed. Under these conditions the mobility still remains high at the zero-field value μ_0.

This ballistic transport should also occur in Si, but in polar semiconductors, such as GaAs, the transient effect is about one order of magnitude longer.

2.6.9. Carrier–Carrier Scattering

In conductors and semiconductors the scattering of an electron by another electron, known as e-e scattering, has only a little effect on the mobility, because the overall momentum of charge carriers is not affected. However, other collision processes may be enhanced by the unavoidable presence of e-e scattering. As an example we may consider an optical phonon scattering mechanism and an electron losing a photon of energy $\hbar\omega_0$. Because of e-e scattering this electron may replaced, in the k-space, by another carrier with the same energy, but the energy loss rate will increase.

Theoretical calculations of the mobility change caused by e-e scattering have been done both for nonpolar semiconductors by Appel [42] and for polar semiconductors by Bate et al. [43]. Assuming a nondegenerate material in which the ionized impurity scattering is dominant, the mobility $\mu_I(0)$ decreases to a value μ_I (e-e) in the presence of e-e scattering. The reduction factor (i.e., the mobility ratio μ_I (e-e)/ $\mu_I(0)$) is about 0.6 for no degeneracy and about 1 (i.e., no reduction) for highly degenerate semiconductors.

For electron concentrations between these two extremes Bate et al. [43] calculated the correction factor, μ_I (e-e)/

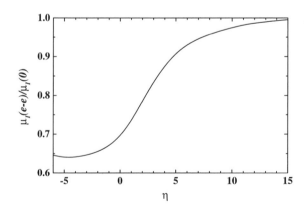

Fig. 12. Electron-electron scattering correction factor for the mobility dominated by ionized impurity scattering, plotted as a function of the Fermi reduced energy η. Reprinted with permission from [43], ©1965, Elsevier Science.

$\mu_{\mathrm{I}}(0)$, as a function of η. Figure 12 reproduces this correction function. As indicated by the curve in Figure 12, e-e scattering significantly affects the mobility values at $\eta < 7$, which, in the case of InSb at 80 K, correspond to an electron concentration below 10^{17} cm^{-3}.

At lower electron densities e-e scattering affects polar optical scattering, which is the dominant collision process. For n values lower than 10^{14} cm^{-3} e-e scattering effects tend to disappear, as well as in the case where the dominant scattering mechanism is by acoustical deformation potential.

For n-type Ge with a donor content of about 6×10^{14} cm^{-3} the highest value of the correction factor is 0.94 at 35 K [44].

2.6.10. Scattering by Dislocations

Crystal solids, even those of the highest quality that growth technologies can currently produce, contain a more or less high density of defects and imperfections. Some of these defects, as already seen in the previous section, have a direct influence on transport mechanisms of charge carriers. Among the structural defects of the crystal lattice, dislocations have a particular relevance because their control and reduction during the growth process make it possible to reach and to maintain high mobilities either in bulk materials or in heterostructure-based devices [45, 46].

One simple classification of the dislocation is according to edge type and screw type [47]. In particular, edge-type dislocations (formed by a slip of a row of atoms in a direction perpendicular to the dislocation line) are responsible for the generation of deep levels, in the energy gap of several semiconductors [48], which may act as electron and hole traps [49].

Theoretical models consider dislocations as an electrically charged line, surrounded by a cylinder of space charge. Carriers that are moving inside the space charge region are deflected, and their mobility is decreased. Pödor [50] calculated the differential cross section for this kind of scattering process and then the reciprocal of the momentum relaxation time,

$$\tau_{\mathrm{m}}^{-1} = \frac{Ne^4 L_{\mathrm{D}} f^2}{8\chi^2 a^2 m^2}\left(v_\perp^2 + \frac{\hbar^2}{4m^2 L_{\mathrm{D}}^2}\right)^{-3/2} \qquad (2.115)$$

where N is the dislocation line density per unit area, $v_\perp$ is the speed component perpendicular to the dislocation line, a is the distance between imperfection centers along the dislocation line, and f is their occupation probability [51].

In the limit of high temperatures the second term in parentheses can be neglected, and from (2.115) we obtain the expression for the mobility in a nondegenerate material:

$$\mu = \frac{30\sqrt{2\pi}\left(a^2\chi^2\right)}{Ne^3 f^2 L_{\mathrm{D}} m^{1/2}}(k_{\mathrm{B}}T)^{3/2} \qquad (2.116)$$

In Figure 13 electron mobility is plotted as a function of the temperature for a Ge sample. The figure indicates the decrease in the mobility caused by the interaction of electrons with dislocations, generated by the sample bending.

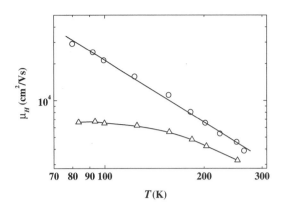

Fig. 13. Electron mobility as a function of the temperature in n-type Ge. Circles: experimental points for the undeformed sample. Triangles: bent crystal, with a bending axis [112] and a bending radius of 50 mm. The continuous line has been drawn with the use of (2.116) and experimental values for the bent sample. Reprinted with permission from [50], ©1966, John Wiley & Sons.

The straight line with the relative experimental points is the mobility for an undeformed Ge sample ($N = 4 \times 10^3$ cm^{-2}), the lower curve for the bent sample. The value of the dislocation density, N, has been computed from the bending radius.

The two curves, for the bent sample, agree qualitatively. Even though not explicitly reported in the work of Pödor, the measurements were probably carried out on the crystal as a whole. Van Weeren et al. [52] demonstrated that the method of cutting the bent crystal deeply affects the Hall mobility. Figure 14 A and B shows the experimental values of the Hall mobility in a n-type Ge crystal, for two bending directions, along [110] and [211], respectively. In both figures, the points labeled A were obtained by cutting the bent sample below or above the neutral plane and performing the Hall measurement with the current lines parallel ($A_\parallel$) and perpendicular ($A_\perp$) to the bending axis. The experimental points labeled B were obtained from a sample cut at the neutral plane.

As clearly indicated by Figure 14, not only the bending direction, but also the direction of motion of the charge carriers with respect to the dislocation line affects the mobility as well.

2.7. Hot Electron Effects

When an electric field is applied to a conductor, the charge carriers may move in the crystal lattice, causing a transfer of energy from one side of the sample to the other.

The source of the applied electric field, $\vec{E}$, continuously supplies the energy to the carriers at a rate given by $\vec{j} \cdot \vec{E}$ (where $\vec{j}$ is the current density). However, the total energy of the charge carriers cannot increase indefinitely because several collision processes (see Section 2.3) cause an energy transfer to lattice atoms. Among the different scattering mechanisms considered so far, there is also inelastic phonon scattering, in which an electron can emit or absorb a phonon. The lattice crystal will absorb energy from the electron when a phonon is emitted, and it delivers energy to the electron when a phonon

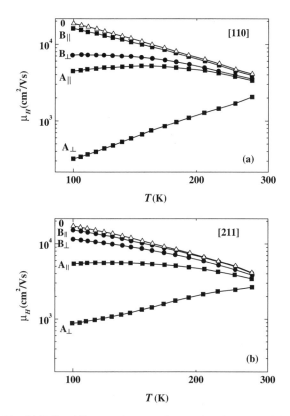

Fig. 14. (a) Hall mobility vs. T in n-type Ge crystals. Triangles: undeformed material. Circles and squares: crystal bent, at 1103 K, about the [110] direction. For the meaning of the curves labeled A and B, see the text. (b) Hall mobility vs. T in n-type Ge crystals. Triangles: undeformed material. Circles and squares: crystal bent, at 1103 K, about the [211] direction. For the meaning of the curves labeled A and B, see the text. Reprinted with permission from [52], ©1967, John Wiley & Sons.

is absorbed. In the absence of an electric field there is a balance between the emission and absorption of phonons, with no net transfer of energy from the electron system to the lattice system or vice versa.

Considered from a thermodynamic point of view, this process implies that the temperature of the electron, T_e, and that of the lattice system, T, are identical.

In contrast, the application of an external electric field increases the energy of the electron system and, then, its temperature. But as the energy increases electrons emit more phonons than they absorb, and there is a net transfer of energy from the electron system to the crystal lattice. This energy transfer process continues with time until $T_e > T$.

Instead of a continuous increase in the energy of the electrons, the whole system (electrons and lattice atoms) may reach a new equilibrium, when the difference between the electron temperature and the lattice temperature is such that the rate of gain of energy of the electrons from the electric source is balanced by the rate of loss of energy to the crystal lattice. Hence, an electron temperature higher than the lattice temperature is required to produce a steady-state condition. As long as the applied electric field is of low intensity, the required rise in T_e is sufficiently small that the difference

between the electron temperature and the lattice temperature does not produce a sizable change in the values of the transport coefficients, and the whole system can be assumed to be same temperature.

A consequence of this fact is the validity of Ohm's law, i.e., the current density varies linearly with the electric field at low electric fields. As the electric field is increased, T_e becomes much higher than T, and the current density begins to depend nonlinearly on the electric field, even if T and the electron concentration are kept constant. For intermediate values of the electric field intensity the nonlinearity may be approximated by a square law;

$$|\vec{j}| = \sigma_0 |\vec{E}|(1 + \beta |\vec{E}|^2) \qquad (2.117)$$

which may be derived from (2.87) with the use of the relation (2.42).

For these field intensities the T_e values are not too high, and such conditions are referred to as warm-electron conditions [53–64]. For very large electric fields, however, the temperature T_e may become an order of magnitude higher than the lattice temperature, and the experiments indicate a failure of Ohm's law (see Fig. 10).

In addition to the failure of Ohm's law, several new effects are observed:

1. For cubic semiconductors the electrical conductivity and Hall mobility are no longer isotropic and scalar [65–70].
2. The current density reaches a saturation value instead of linearly rising with the electric field [71–72].
3. A new phenomenon, not observed at low fields, occurs, consisting of the transfer of electrons from one valley of the energy band structure to another valley. This effect is known as the transferred electron effect [73–76]. A consequence of this effect is the production of a negative differential mobility, which has been of great technical importance in realizing solid-state microwave and millimeter wave sources.
4. In addition to the phenomena already observed, there are suggestions of the existence of new and unexpected phenomena. In fact, in some experiments cooling of the electrons [77–80] or a runaway condition at high electric fields [81] have been observed. These effects may be theoretically explained, assuming that piezoelectric and polar optic phonon scattering are predominant.

Moreover, theoretical calculations demonstrate that the electrical conductivity [82] and the diffusion constant [83, 84] may have negative values.

In recent years attention has been devoted to the study of electron transport at high fields which, because of the very high electron temperatures, has been called hot electron transport. With the novelty and peculiarity of the observed phenomena in mind, several theoretical approaches and solutions to hot-electron transport have been developed. Reviews of the main analytical theories have been given by Nag [2] and by Ferry [85].

2.8. Cyclotron Resonance

A charge carrier, either an electron or a hole, moving in a dc magnetic field B at an oblique angle, will describe a helix path around the direction of B with an angular frequency given by the cyclotron frequency (2.20). The cyclotron radius and the corresponding wavelength, in free space, are expressed by

$$r_c = \sqrt{\frac{\hbar}{eB}} \qquad (2.118)$$

$$\lambda_c = 1.07 \frac{m/m_o}{B} [\text{cm}] \qquad (2.119)$$

Assuming a magnetic field B intensity on the order of 0.1 T and an effective mass $m \approx m_o$, the wavelength falls in the microwave range. Hence, microwaves incident on the sample and polarized at an angle to $\vec{B}$ are absorbed. For $\omega = \omega_C$ the absorption shows a resonance peak; this phenomenon is called cyclotron resonance [86].

From a measurement of the resonance frequency, the value of the effective mass m is obtained (2.20). As in any resonance observation, a peak is not found for the case of strong damping, i.e., if most carriers make a collision before rotating through at least one radian. Because the number of collisions per unit time is τ_m^{-1} a condition for resonance is the condition

$$\omega_c > \tau_m^{-1} \qquad (2.120)$$

Therefore, to observe a cyclotron resonance in the microwave region it is necessary for the collision time to be at least $\tau_m > 10^{-10}$s at liquid helium temperatures. This value represents a limit that is difficult to reach in all but a few semiconductors with an extremely low concentration of impurities and defects. In these high-quality materials ionized and neutral impurity scattering processes are negligible.

Instead of microwaves, cyclotron resonance may be observed by infrared radiation and strong magnetic fields, which are available in pulsed form (up to about 100 T) or in a continuous way with superconducting magnets (up to about 25–30 T). In this case, even at room temperature the resonance condition may be fulfilled ($\tau_m = 10^{-13}$s corresponds to a wavelength of $\lambda_c = 185.9 \ \mu$m, which at about 30 T requires $m/m_o \leq 0.57$).

For practical reasons due to the production and use of large (pulsed) magnetic fields, infrared cyclotron resonance at room temperature is generally limited to carriers with effective masses of less than about 0.2 m_o.

At the large fields involved, there is a spin splitting of the levels and a selection rule for the quantum number $M (\Delta M = 0)$ for linearly polarized fields in the Faraday configuration. The Faraday configuration is realized when a wave propagates parallel to the magnetic field direction. Moreover, for nonparabolic bands and in large magnetic fields, the effective mass m depends on the magnetic field intensity B.

As already shown in section 2, to observe the Landau quantization a low temperature is needed to fulfill the condition $k_B T < \hbar \omega_C$. For temperatures below 77 K and $\omega_c \approx 10^{13}$s this

condition is satisfied. Cyclotron resonance may then be considered a transition between successive Landau levels.

A simplified formulation of cyclotron resonance can be given, assuming the equation of motion of a charge carrier in a static magnetic field $\vec{B}$ and in an ac electric field $|\vec{E}| \propto e^{i\omega t}$. Excluding the energy distribution of carriers, the equation assumes the form

$$\frac{d}{dt}(m\vec{v}_d) + \frac{m\vec{v}_d}{\tau_m} = e[\vec{E} + (\vec{v}_d \wedge \vec{B})] \qquad (2.121)$$

Assuming the Faraday configuration and choosing an appropriate Cartesian coordinate system, it can be easily demonstrated [1] that the power absorbed by carriers is given by

$$P_\pm(\omega) = \frac{1}{2}|\vec{E}|^2 \frac{\sigma_o}{1 + \tau_m^2(\omega \pm \omega_c)^2} \qquad (2.122)$$

where σ_0 is the static conductivity (see Eq. 2.42) and the signs in parentheses denote a right-handed (+) and a left-handed (−) circularly polarized electric field.

For a linear polarization that can be assumed to be composed of two circular polarizations rotating in opposite directions,

$$P(\omega) = P_+ + P_- = \sigma_o|\vec{E}|^2 \frac{1 + (\omega^2 + \omega_c^2)\tau_m^2}{[1 + (\omega^2 - \omega_c^2)\tau_m^2]^2 + 4\omega_c^2\tau_m^2} \qquad (2.123)$$

At cyclotron resonance, $\omega = \omega_C$, and assuming the condition (2.120) is valid, the previous relation (2.123) simplifies in

$$P(\omega_c) = \frac{1}{2}\sigma_o|\vec{E}|^2 \qquad (2.124)$$

Under resonance conditions, the conductivity of both electrons and holes will be equal to σ_0 (the dc conductivity), whereas for $\omega \neq \omega$ the conductivity will be lower than σ_0. At low frequencies the power, $P(0)$, absorbed by the carriers is

$$P(0) = \frac{\sigma_o|\vec{E}|^2}{1 + \omega_c^2\tau_m^2} = \frac{P_o}{1 + \omega_c^2\tau_m^2} \qquad (2.125)$$

Figure 15 shows the ratio P/P_o plotted as a function of ω_c/ω, for different values of the product $\omega\tau_m$. Assuming that usually in the experiment ω is kept constant and the intensity of $\vec{B}$ is changed, the quantity ω_c/ω becomes proportional to $\vec{B}$. The curves in Figure 15 indicate how the condition of resonance disappears when $\omega\tau_m \leq 1$.

Figure 16 reproduces the experimental results of cyclotron resonance for n-type silicon at 4.2 K and for three different orientations of the magnetic field $\vec{B}$. For the field $\vec{B}$ aligned along the $\langle 111 \rangle$ crystallographic direction there is a single resonance, and along the $\langle 001 \rangle$ and $\langle 110 \rangle$ directions two resonance peaks appear. This can be interpreted by the many-valley model of the conduction band. Because there is only one peak in the $\langle 111 \rangle$ direction, the energy ellipsoids must be located along the $\langle 100 \rangle$ directions, which are then equivalent relative to the magnetic field direction.

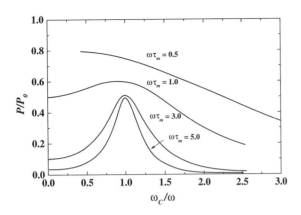

Fig. 15.   Normalized microwave power P/P_0 absorbed by electrons as a function of the ratio ω_c/ω. Reprinted with permission from [88], ©1954, Elsevier Science.

For an interpretation of Figure 16, it is necessary to solve Eq. (2.121), assuming a mass tensor $\underline{\underline{m}}$ in a coordinate system where the tensor is diagonal. For the resonance condition $\omega = \omega$ the mass assumes the form

$$m = \sqrt{\frac{m_x m_y m_z}{\alpha^2 m_x + \beta^2 m_y + \gamma^2 m_z}} \qquad (2.126)$$

where α, β, and γ denote the direction cosines of $\vec{B}$ with respect to the three coordinate axes.

If we take the valley in the $\langle 001 \rangle$ direction ($m_x = m_y = m_t = m_1/K$; $m_z = m_1$) and $\vec{B}$ in a $[1\bar{1}0]$ plane with $\alpha = \beta = (\sin\theta)/\sqrt{2}$ and $\gamma = \cos\theta$, Eq. (2.122) becomes

$$m = \frac{m_1}{\sqrt{K^2 \cos^2\theta + K \sin^2\theta}} \qquad (2.127)$$

The valleys in the $\langle 100 \rangle$ and $\langle 010 \rangle$ directions have the same effective mass because they are symmetric to the $[1\bar{1}0]$ plane:

$$m = \frac{m_1}{\sqrt{K^2 \cos^2\theta + \frac{1}{2}K(K+1)\sin^2\theta}} \qquad (2.128)$$

For the $\langle 111 \rangle$ direction $\alpha = \beta = \gamma = 1/\sqrt{3}$, the relation (2.122) yields

$$m = m_1 \sqrt{\frac{3}{K(K+2)}} \qquad (2.129)$$

Equations (2.123) and (2.124) have been used to fit the data points of Si effective mass, obtained by cyclotron resonance experiments. The fit and the data are shown in Figure 17 as a function of the angle between $\vec{B}$ and the $\langle 001 \rangle$ direction. The effective masses used are $m_1/m_0 = 0.90 \pm 0.02$ and $m_t/m_0 = m_t/(Km_0) = 0.192 \pm 0.001$. For any $\vec{B}$ direction that is not in the $[1\bar{1}0]$ plane, there should be three resonant frequencies [87].

In n-Ge, which has four ellipsoids on the $\langle 111 \rangle$ and equivalent axes, there are, in general, three frequencies of resonance if B is located in the $[1\bar{1}0]$ plane; otherwise there are four. The analysis of the data yields [88] the following effective mass values: $m_1/m_0 = 1.64 \pm 0.03$ and $m_t/m_0 = 0.0819 \pm 0.0003$.

Figure 18 reproduces the cyclotron resonance curves for p-Ge as a function of the angle between the magnetic field direction and the $\langle 001 \rangle$ direction [89]. The two curves indicate two effective masses: the light hole mass is isotropic in the $[1\bar{1}0]$ plane and has a value of $m_{lh} = 0.043m_0$, and the heavy hole mass, m_{hh}, varies between $0.28m_0$ and $0.38m_0$, depending on the angle θ.

For warped-sphere constant energy surfaces (such as for Si and Ge valence bands) a method of calculation of the

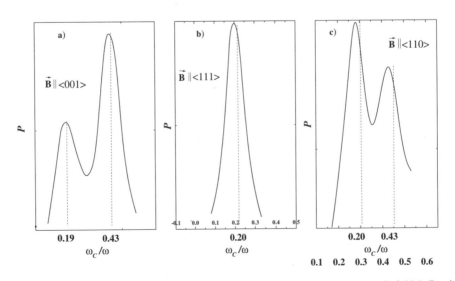

Fig. 16.   Microwave power P vs. ω_c/ω absorbed at 23 GHz by n-type Si for three different directions of the magnetic field B. Reprinted with permission from [88], ©1954, Elsevier Science.

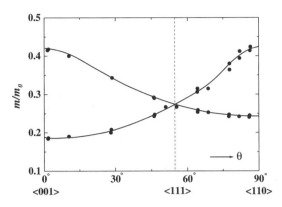

Fig. 17. Normalized electron effective mass as a function of the angle between the $\langle 001 \rangle$ direction and $\vec{B}$ lying in the $[1\bar{1}0]$ plane. Points represent data for n-Si at 4 K from cyclotron resonance measurements. Reprinted with permission from [87], ©1960, American Physical Society.

cyclotron effective mass was proposed by Shockley [90]. If $\vec{B}$ lies in the $[1\bar{1}0]$ plane the equation for the effective mass is

$$\frac{m_\pm}{m_o} = \left(A_o \pm B'_o\right)^{-1}\left[1 + \frac{1}{32}\left(1 - 3\cos^2\theta\right)^2 \Upsilon + \cdots\right] \quad (2.130)$$

where the subscript $+$ refers to light holes and $-$ refers to heavy holes and

$$B'_o = \sqrt{B_o^2 + \frac{1}{4}C_o^2}; \qquad \Upsilon = \mp\frac{C_o^2}{2B'_o(A_o \pm B'_o)} \quad (2.131)$$

where A_o, B_o and C_o are constants whose value is determined by cyclotron resonance measurements.

The cyclotron heavy-hole mass in Ge, obtained by fitting (2.131) (for the case of the minus sign) to the experimental data, is $0.3m_0$, whereas for the split-off valence band the effective mass is $0.075m_0$.

For n-type InSb, at room temperature, an effective mass of $0.013m_0$ [91, 92] was obtained from infrared cyclotron resonance measurements. This value increases to $0.0145m_0$ at

77 K. The same technique has been used for n-type doped InAs, InP, and GaAs [91, 92].

3. HALL AND GALVANOMAGNETIC EFFECTS

3.1. Hall Effect

The Hall effect in conducting materials has always played an important role since its discovery in 1879 by E. H. Hall [93]. To explain the Hall effect and to find the relation among the main physical quantities involved, to consider a slab of a conductor, for example, in the shape of a parallelepiped. In a cartesian system (see Fig. 19), if a current density $\vec{j}_x = j_x \vec{u}_x$ flows along the x axis direction and an induction magnetic field $\vec{B}_z = B_z \vec{u}_z$ is applied parallel to the z axis, orthogonal to the current flow direction, an electric field $\vec{E}_y = E_y \vec{u}_y$ (called a Hall field) is set up, mutually perpendicular to the directions of $\vec{j}_x$ and $\vec{B}_z$. The intensity of $\vec{E}_y$ is proportional to j_x and B_z by a quantity R called the Hall coefficient:

$$E_y = R j_x B_z \quad (3.1)$$

The Hall effect can be physically explained by the fact that charge carriers, forming an electrical current, $\vec{j}_x$, experience a Lorentz force $\vec{F}_L = e v \vec{u}_x \wedge \vec{B}_z$ in a direction orthogonal to the plane formed by the vectors $\vec{B}_z$ and $\vec{j}_x = nqv\vec{u}_x$, where n is the concentration of the charged particles. This force deflects the charge carriers toward the bounds of the slab and creates an electric field, $\vec{E}_H = E_y \vec{u}_y$, whose force, $e E_y \vec{u}_y$, will counterbalance the Lorentz force and, then, the further build-up of charge:

$$e\vec{v}_x \wedge \vec{B}_z = e\vec{E}_H \quad (3.2)$$

With the choices made in Figure 19 and considering the definition of the current density, (2.2) can be written as

$$v B_z = E_y \quad (3.3)$$

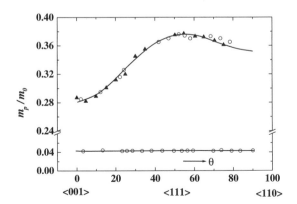

Fig. 18. Holes effective mass as a function of the angle between the $\langle 001 \rangle$ direction and $\vec{B}$ lying in the $[1\bar{1}0]$ plane. Points represent data for p-Ge at 4 K from cyclotron resonance measurements. Reprinted with permisssion from [89], ©1957, American Physical Society.

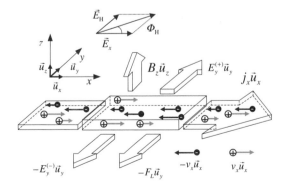

Fig. 19. The classical Hall effect. Both electrons and holes are deflected by the Lorentz force, $-F_L\vec{u}_y$, in the same direction as shown by the large arrows. The Hall field for holes, $E_H^{(+)}\vec{u}_y = E_y^{(+)}\vec{u}_y$, has the direction of the positive y axis, whereas for electrons, $-E_H^{(-)}\vec{u}_y = -E_y^{(-)}\vec{u}_y$ has the opposite direction.

Hence, the magnitude of the field $\vec{E}_y$ can be expressed as

$$E_y = \frac{1}{ne} B_z j_x \qquad (3.4)$$

From the comparison of (3.4) with (3.1) the expression for R follows:

$$R = \frac{E_y}{B_z j_x} = \frac{1}{ne} \qquad (3.5)$$

In the last relation e represents the magnitude of the electron charge. The convention about the sign of R is to take it as negative when e is negative. The unit measure of R in the S.I. system is m³/C, but cm³/C is the generally preferred unit of measure.

The Hall coefficient is a powerful means for studying the transport properties of conductors and semiconductors, because of its simple relation to the electric charge concentration n. The measurement of R can give a direct estimate of the carrier concentration contributing to the transport process.

For a simple isotropic band a more detailed treatise of the Hall effect (similar to that adopted for electrical conductivity; see Section 2.1) allows us to write a general expression for the Hall coefficient,

$$R = \frac{3m}{2e^2 B_z} \frac{\mathscr{H}_1}{\mathscr{H}_1^2 + \mathscr{K}_1^2} \qquad (3.6)$$

where $\mathscr{H}_1$ and $\mathscr{K}_1$ are defined as:

$$\begin{aligned} \mathscr{H}_i &= \int_0^\infty \frac{\varepsilon^i \tau_m}{1 + (\omega_c \tau_m)^2} \frac{df}{d\varepsilon} \phi(\varepsilon)\, d\varepsilon \\ \mathscr{K}_i &= \int_0^\infty \frac{(\omega_c \tau_m)\varepsilon^i \tau_m}{1 + (\omega_c \tau_m)^2} \frac{df}{d\varepsilon} \phi(\varepsilon)\, d\varepsilon \end{aligned} \qquad (3.7)$$

If the term $(\omega_c \tau_m)^2$ can be neglected (i.e., terms in B_z^2 are negligible) (3.6) simplifies to

$$R = \frac{3m}{2e^2 B_z} \frac{H_1}{K_1^2} \qquad (3.8)$$

where

$$\begin{aligned} H_i &= -\frac{2e B_z}{\pi^{1/2} m} n \zeta^2 (k_B T)^{2p+i-1} \Gamma\left(2p+i+\frac{3}{2}\right) \\ K_i &= -\frac{2}{\pi^{1/2}} n \zeta^2 (k_B T)^{p+i-1} \Gamma\left(p+i+\frac{3}{2}\right) \end{aligned} \qquad (3.9)$$

where ζ and p are the same constants introduced for (2.38). Therefore R assumes the form

$$R = -\frac{3\pi^{1/2}}{2} \frac{1}{ne} \frac{\Gamma(2p+5/2)}{\Gamma^2(p+5/2)} \qquad (3.10)$$

Figure 20 reproduces the behavior of R as a function of T for a p-type InSb sample at three intensities of the magnetic induction field $\vec{B}$. According to Eq. (3.6) the Hall coefficient decreases with increasing $\vec{B}$ intensity.

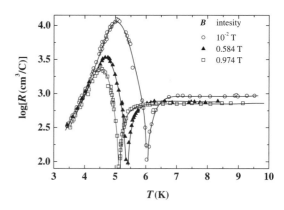

Fig. 20. Hall coefficient, R, as a function of T for a p-type InSb sample at three intensities of the magnetic induction field $\vec{B}$. According to Eq. (3.6) the Hall coefficient decreases with increasing $\vec{B}$ intensity. Reprinted with permission from [296], ©1957, Institute of Physics.

Figure 21 shows the typical behavior of R vs. the reciprocal temperature for several doping levels in both n-type and p-type InSb samples.

Combining Eqs. (3.10) and (2.38), we obtain the expression for another electronic transport coefficient, μ_H, which is known as the Hall mobility:

$$\mu_H = |R\sigma| = \frac{e}{m} \zeta (k_B T)^p \frac{\Gamma(2p+5/2)}{\Gamma(p+5/2)} \qquad (3.11)$$

And for the drift mobility we get

$$\mu = \frac{4}{3\pi^{1/2} m} \zeta (k_B T)^p \Gamma(p+5/2) \qquad (3.12)$$

We can observe that the expression for the Hall coefficient (3.10) can be rewritten as

$$R = -\frac{\mu_H}{\mu} \frac{1}{ne} = -\frac{r_H}{ne} \qquad (3.13)$$

where the ratio of the Hall mobility to the drift mobility is called the Hall factor, r_H. The Hall factor is a quantity of great

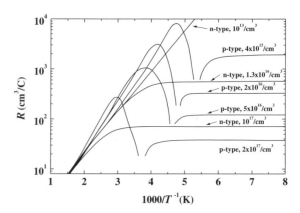

Fig. 21. Typical behavior of R vs. the reciprocal temperature for several doping levels in both n-type and p-type InSb samples. Reprinted with permission from [1], ©1989, Springer-Verlag.

interest in the discussion of the electronic transport properties of semiconductors, and its theoretical and experimental determination is extremely difficult. The r_H values are generally in the range of 1–2 for most semiconductors. Determination of the r_H value is important because from a direct measure of R the carrier concentration, n, can be computed from (3.13).

Where there is doubt about the most probable value of r_H, putting it equal to unity is not likely to introduce an error of more than 20–30% in the n value.

In the presence of both types of charge carriers, with concentration n_e and n_h, respectively, the Hall coefficient can be rearranged in the following way:

$$R = \frac{r_H}{e} \frac{n_h - n_e (\mu_e/\mu_h)^2}{(n_h + n_e(\mu_e/\mu_h))^2}. \tag{3.14}$$

Finally, another useful quantity can be derived from Figure 19. It can easily be demonstrated that in the presence of a magnetic field the directions of the electric field $\vec{E}$ ($\vec{E} = \vec{E}_x + \vec{E}_H$; see Figure 19) and the current density $\vec{j}$ differ. In fact, from (3.4), (2.36), and the definition of the current density, it follows that

$$E_y = \frac{1}{nq} B_z j_x = \frac{1}{nq} B_z \sigma E_x = R\sigma B_z E_x, \tag{3.15}$$

Then

$$R\sigma B_z = \frac{E_y}{E_x} = tg\,\Phi_H \tag{3.16}$$

The relation (3.16) is valid in the limit of small Φ_H, i.e., if the intensity of $\vec{B}$ is so low that σ does not depend on $\vec{B}$ (see Sections 2.1 and 2.2).

3.2. Galvanomagnetic Effects

3.2.1. Ettingshausen Effect

Soon after the discovery of the Hall effect, other effects were observed and explained in several experimental conditions. One of these is the Ettingshausen effect: if in a slab of a conducting material a current I_x is allowed to flow, electrons will be deflected in a different way when a magnetic field $\vec{B}_z = B_z \vec{u}_z$ is applied orthogonally to the current direction. This fact is due to the distribution of the electron velocities, which are not all exactly equal. Hence, the faster electrons will experience a different Lorentz force with respect to the slower ones. This means that more energy will be transported to one side of the solid than to the other, and therefore a transverse temperature difference will be set up. This effect is called the Ettingshausen effect and is illustrated in Figure 22.

If the specimen is placed in contact with a constant temperature bath, then this temperature difference will be eliminated by heat transfer to or from the bath. The effect will be greatest when the specimen is isolated from its surroundings. For this reason this effect and others of a similar type are called adiabatic effects.

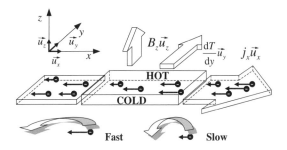

Fig. 22. The Hettingshausen effect. Slow electrons tend to suffer a greater deflection than fast ones. Because the slow electrons carry less energy than the faster ones, the side where they are deflected will tend to be colder than the opposite one. The large arrow oriented along the y direction shows the direction of the temperature gradient.

The Ettingshausen coefficient, P, is defined by the equation

$$\frac{dT}{dy} = PB_z I_x \tag{3.17}$$

and its units are cm³KJ⁻¹(practical) or m³KJ⁻¹ (S.I.)

The presence of an Ettingshausen temperature gradient modifies the Hall coefficient. If the solid has a thermoelectric power, Ω [8, 94], then there will be a transverse Ettingshausen–Seebeck field, $P\Omega$, which cannot be separated from the Hall field. Therefore what is obtained when the Hall coefficient is measured under these conditions is an adiabatic coefficient R_a related to the isothermal value:

$$R_a = R + P\Omega \tag{3.18}$$

Equation (3.18) can be verified by thermodynamic arguments. In practice, the term $P\Omega$ may be ignored for two main reasons: measurements of the main electronic transport coefficient and the Hall coefficient are generally carried out under isothermal conditions; in most circumstances the experimental error in R is higher than the magnitude of $P\Omega$, a possible exception being near the temperature at which the Hall coefficient of a mixed conductor is zero.

3.2.2. Nernst and Righi–Leduc Effects

If instead of passing an electric current along a specimen, a temperature gradient is set up, then electrons will tend to diffuse from the hot to the cold end. This gives rise to the Seebeck thermoelectric effect and, in the presence of a magnetic field, to two additional effects, the Nernst and Righi–Leduc effects, which are analogous to the Hall and Ettingshausen effects, respectively.

In the Nernst effect, electrons attempting to diffuse down a temperature gradient are deflected by a magnetic field, but a transverse electric field is set up to balance out the Lorentz force as in the Hall effect. The Nernst coefficient, Q, is defined by the equation

$$E_y = QB_z \frac{dT}{dx} \tag{3.19}$$

The units for the Nernst coefficient are $m^2 s^{-1} K^{-1}$ (S.I.) or $cm^2 s^{-1} K^{-1}$ (practical).

Like the Ettingshausen effect, the Righi–Leduc effect originates in the distribution of electron velocities, and, under adiabatic conditions, a transverse temperature gradient is set up. The Righi–Leduc coefficient, S, is defined by the relation

$$\frac{dT}{dy} = SB \frac{dT}{dx} \qquad (3.20)$$

Because of the analogy between the Hall and the Righi–Leduc effects (compare Eq. (3.20) with the (3.16)), the units of measure for the two effects are the same.

As regard the signs for the coefficients P, Q, and S, Figure 23 shows the Gerlach sign convention for the galvanomagnetic coefficients, relative to the current direction, longitudinal temperature gradient, and magnetic field when the charge carriers are negative electrons. As with the Hall coefficient, these coefficients are then given a negative sign. For the Righi–Leduc effect, however, the sign of the coefficient S depends upon that of the charge carrier, as it does for the Hall coefficient, but the other two effects are independent of this. Their signs are determined by a number of factors, such as the presence of mixed conduction, the nature of the carrier collision processes, and the presence of phonon drag.

3.3. Generalized Definition of the Hall Coefficients

The relations between the conductivity (2.36), Hall coefficient (3.5), and Hall mobility (3.11) have been obtained, assuming the validity of Ohm's law, in the presence of low-intensity magnetic induction fields.

In general, some conductors and semiconductors indicate an increase in electrical resistance when placed in a magnetic field. This effect is called magnetoresistance (see Section 4); the resistance raise depends on $|\vec{B}|^2$ for low-intensity magnetic fields.

The magnetoresistance is conveniently defined by a magnetocoefficient M,

$$M + 1 = \frac{1}{B^2} \frac{r(B)}{r(B=0)} \qquad (3.21)$$

where $r(B)$ and $r(B=0)$ are the specimen resistance with and without the application of a magnetic induction field $\vec{B}$, respectively.

In studying this effect the field $\vec{B}$ may be applied in any arbitrary direction relative to the current direction. Therefore, in the relation (3.21) the direction of $\vec{B}$ must be specified.

For the above-mentioned difficulties it is convenient to generalize the definitions of the Hall coefficient (3.5) and of the magneto-resistance coefficient (3.21) relating the electric field, the magnetic field, and the electric current. If the electric field and current density directions differ, then it is necessary to specify all three components of the electric field. The component parallel to the direction of the current gives the conductivity as in (2.36). The component perpendicular to the current and to the plane containing the current and the magnetic field is the Hall field as defined by (3.1). This leaves a third component, which is at right angles to the current but lies in the plane containing the current and the magnetic field. This component has been called the planar Hall effect, by analogy with the usual Hall effect. The most convenient way of generalizing these coefficients is to rewrite relations (3.5) and (3.21) in vector notation and to introduce a third equation for the planar Hall coefficient G:

$$R = \frac{\vec{E} \cdot (\vec{B} \wedge \vec{j})}{(\vec{B} \wedge \vec{j})^2} \qquad (3.22)$$

$$M + 1 = \frac{1}{B^2} \frac{\vec{E} \cdot \vec{j}}{(\vec{E} \cdot \vec{j})^2_{B=0}} \qquad (3.23)$$

$$G = \frac{[\vec{E} \cdot (\vec{B} \wedge \vec{j}) \wedge \vec{j}]}{(\vec{B} \wedge \vec{j})^2 (\vec{B} \cdot \vec{j})} \qquad (3.24)$$

In all of the last three relations the numerator is a scalar proportional to the desired component of the electric field. The denominator contains terms that cancel the proportionality factor in the numerator and introduce the correct scaling to make Eqs. (3.22) and (3.23) correspond to (3.5) and (3.21), respectively. Equation (3.24) shows that G is proportional to E/IB^2 and that the effect vanishes when $\vec{B}$ is parallel to $\vec{j}$ or at right angles to it, reaching a maximum value when the angle between $\vec{B}$ and $\vec{j}$ is $45°$.

4. MAGNETORESISTANCE

The electrical and magnetic properties of materials were once generally assumed to be isotropic, because early experimental works were carried out on polycrystalline material. Currently the large quantity of crystalline solids, grown by several

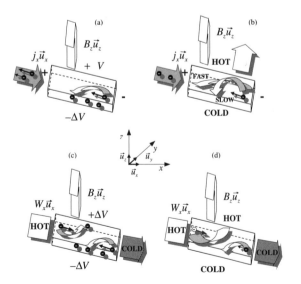

Fig. 23. Gerlach sign convention for galvanomagnetic effects. (a) Hall; (b) Ettingshausen; (c) Nernst; (d) Righi–Leduc. The drawing takes electrons into consideration. The polarity of the Nerst and Righi-Leduc effects is determined by the fact that slow electrons, diffusing from the cold to the hot side of the sample, suffer a greater deflection than fast ones, from the hot side. All coefficients are given the negative sign if they conform to the polarity shown.

techniques, may make possible the observation of anisotropic effects. The importance of this is illustrated, for example, by theoretical and experimental work on silicon and germanium. These two elements form crystals with a diamond structure, which is a type of cubic structure. It can be shown from considerations of symmetry alone that in any cubic structure the electrical and thermal conductivities and the Hall effect are isotropic, but other transport properties could not be.

One example of the anisotropic property is the magnetoresistance effect, consisting of an increase in the electrical resistance in a conducting material when a magnetic field is applied. This effect can physically be explained, assuming that the mean free path of electrons is independent of the applied electric and magnetic fields. The justification for this assumption is the fact that the perturbing effect of the applied fields upon the electronic distribution, under normal circumstances, will be small compared with the restoring forces. When a magnetic field is applied, the trajectories of electrons will be deflected between two consecutive collisions. Therefore, the component of motion in the direction of the applied electric field will be reduced by the presence of the magnetic field, causing an increase in the resistance.

For an isotropic solid containing electrons that obey Boltzmann statistics, theoretical calculations show that the magnetoresistance should be observed when a magnetic field is applied orthogonally to the electric field direction (as in the configuration used in measuring the Hall effect), but no effect should be obtained if the magnetic field is applied parallel to the electric field (longitudinal magnetoresistance), as can easily be demonstrated by applying the Lorentz force relation.

However, this conclusion is not always experimentally fulfilled, and in the case of both germanium and silicon the longitudinal and transverse effects are comparable. This fact is a clear indication that the energy bands occupied by the conduction electrons exhibit a degree of anisotropy, although they must still have cubic symmetry.

In Ge and Si the increase in the electrical resistance is considerable because of the numerous energy minima in the conduction band. In contrast, the magnetoresistance effect caused by shape and inhomogeneities can be neglected because the electron mobility is not very high. With III–V compounds just the opposite is true. The compounds with a high electron mobility, like InSb, InAs, GaAs, and InP, have a simple spherical conduction band with a minimum at $k = 0$ and a very small magnetoresistance effect for n-type conductivity. A review of the magnetoresistance effect in indium antimonide and in other III–V semiconductors was given by Weiss [95].

Experimental results obtained by Pearson and Herring [96] on n-type silicon show that the magnetoresistance effect depends on the angle between the electric and magnetic fields, for differently oriented specimens. Generally speaking, it is found that the magnitude of the effect depends on the crystalline orientation of the specimen. The interpretation of this kind of result is difficult in the absence of any other evidence or of any calculation of the energy band system.

Band structure calculations, even though they are possible in principle, are not fully self-consistent, because of approximations, and rely on experimental observations to resolve ambiguities. It has been possible to unravel the behavior of germanium and silicon with the use of two further experimental techniques: (a) the cyclotron resonance method to measure the effective masses and (b) high-resolution infrared spectroscopy to investigate the absorption spectrum associated with transitions of electrons from the valence to the conduction band.

4.1. Transverse Magnetoresistance Effect

If strong magnetic fields are applied (i.e., $\tau\omega_C = (e/m)\tau B \gg 1$) to crystalline conductors, all of the transport properties will become dependent upon the magnetic fields (see Sections 2.4, 2.5, and 3.), the most important effect being the appearance of magnetoresistance. It is found experimentally that in many cases the resistance is a function of both the transverse and longitudinal components of the magnetic field. The magnetoresistance theory for an isotropic conductor does not admit a longitudinal effect. To explain this, proper account has to be taken of the crystalline structure of actual solids.

It is interesting to consider and to discuss the results for two cases: (1) low-intensity magnetic fields (i.e., $\tau\omega_C = (e/m)\tau B \ll 1$); and (2) high-intensity magnetic fields ($\tau\omega_C = (e/m)\tau B \to \infty$).

(1) In this case, Putley showed [8] that the conductivity is expressed by the relation

$$\sigma = \sigma_0 \left\{ 1 - \mu_H^2 B^2 \left[\frac{\Gamma(p+5/2)\Gamma(3p+5/2)}{\Gamma^2(2p+5/2)} - 1 \right] \right\} \quad (4.1)$$

where $\mu_H = R_0\sigma_0$ is the Hall mobility for weak-intensity fields. If, for example, lattice scattering is considered ($p = -1/2$), (4.1) reduces to [8]

$$\sigma = \sigma_0 \left[1 - \mu_H^2 B^2 \left(\frac{8}{\pi} - 2 \right) \right] \quad (4.2)$$

and then the magnetoresistance ratio, M (see Eq. (3.21)), of the material assumes the following form [8]:

$$M = \left(\frac{8}{\pi} - 2 \right) \mu_H^2 \quad (4.3)$$

Equations (4.1) and (4.2) show that for low-intensity magnetic fields the conductivity should depend on B^2 and that the rate of variation should be a factor on the order of unity times μ_H^2. Both of these conclusions are experimentally supported in the majority of cases.

(2) In the high-intensity field limit ($\tau\omega_C \to \infty$) the conductivity is given by [8]

$$\sigma = \frac{9\pi}{16} \frac{\sigma_0}{\Gamma(p+5/2)\Gamma(5/2-p)} \quad (4.4)$$

which for lattice scattering ($p = -1/2$) simplifies to

$$\sigma = \frac{9\pi}{32} \sigma_0 \quad (4.5)$$

These last two results suggest that, at high fields, the conductivity should tend to a limit value depending on the scattering mechanism. Moreover, this result is not well substantiated by experiments, because it may be invalidated both by inhomogeneities in the samples and by quantum effects that do not lead to a conductivity tending to an upper limit.

4.2. Longitudinal Magnetoresistance Effect

As said before, anisotropy in some semiconductors can lead to a longitudinal magnetoresistance when a magnetic induction field is applied parallel (i.e., longitudinally) to the current of measure in the specimen.

Many semiconductor materials have crystal structures of cubic symmetry. These include the diamond (Si, Ge), zinc blende (InSb, InAs), and rock salt (PbS, PbSe, PbTe) structures.

Seitz [97] investigated the effect of a magnetic field on cubic solids and showed that although the conductivity in the absence of a magnetic field and the Hall effect for small fields is isotropic, the magnetoresistance effect for small fields need not be.

If terms of higher power than B^2 can be neglected, the current vector $\vec{I}$ and the electric field $\vec{E}$ are linked by the relation

$$\vec{E} = \rho_0\left[\vec{I} + a\left(\vec{I} \wedge \vec{B}\right) + bB^2\vec{I} + c\vec{B}\left(\vec{I} \cdot \vec{B}\right) + d\left(\underline{\underline{T}} \cdot \vec{I}\right)\right] \quad (4.6)$$

where σ_0 is the conductivity in a zero magnetic field and $\underline{\underline{T}}$ has the form

$$\underline{\underline{T}} = \begin{bmatrix} B_X^2 & 0 & 0 \\ 0 & B_Y^2 & 0 \\ 0 & 0 & B_Z^2 \end{bmatrix} \quad (4.7)$$

where the axes are aligned with the principal direction of the cube. In an isotropic crystal that obeys classical statistics, the quantities in (4.6) are

$$\rho_0 = \frac{1}{\sigma_0} \quad (4.8)$$

$$a = \frac{3\pi}{8}\mu \quad (4.9)$$

$$b = -\frac{9\pi}{64}(\pi - 4)\mu^2 \quad (4.10)$$

$$c = -b \quad (4.11)$$

$$d \simeq 0 \quad (4.12)$$

If $\delta\rho$ represents the change in the specimen resistance with the application of a magnetic field, then the magnetoresistance coefficient, M, can be obtained from (4.6)

$$M = \frac{\delta\rho}{B^2\rho_0} = b + c\left(\sum_{l=1}^{3} i_l m_l\right)^2 + d\sum_{l=1}^{3} i_l^2 m_l^2 \quad (4.13)$$

where i_l and m_l are the direction cosines of the current vector and magnetic induction field, respectively.

Table II. Number of Constants to be Determined to Apply the Relation (4.6), for Different Directions of Current, $\vec{I}$, and the Magnetic Induction, $\vec{B}$

$\vec{I}$ direction	$\vec{B}$ direction	M
100	100	$b + c + d$
100	010	b
110	001	b
110	110	$b + d/2$
111	111	$b + c + d/3$
111	110	$b + d/3$

Thus we see that these equations contain five constants. σ_0 is found at a zero magnetic field and a is obtained from the Hall coefficient. The remaining three constants, b, c, and d, are contained in the equation of the magnetoresistance ratio and can be determined by measuring the magnetoresistance effect for suitable orientations of the current and magnetic field. Their meaning is the following: d vanishes in an isotropic crystal, and b describes the transverse and $b + c$ the longitudinal magnetoresistance. A nonvanishing d is characteristic of anisotropy. Table II shows the constants (b, c, and d) to be determined for several directions of the vectors $\vec{I}$ and $\vec{B}$.

The three constants (b, c, and d) can be determined on one specimen, measuring all of the components of $\vec{E}$ and so determining it completely. In fact, the magnetoresistance effect measures the component of $\vec{E}$ parallel to $\vec{I}$ (see Section 3.3).

From Eq. (4.6) we can derive an expression for the intensity of the planar Hall field, $\vec{E}_p$, given by

$$E_p = GIB^2 \sin(2\Psi) \quad (4.14)$$

where Ψ is the angle between the current direction and the magnetic induction. Experimentally, G assumes a form similar to the Hall coefficient R,

$$G = \frac{Vt}{B^2I} \quad (4.15)$$

where t is the specimen thickness orthogonal to V. G can be calculated from Eq. (4.6), and Table III gives values for two useful sample orientations.

It is clear from the expressions of M and G given in Tables II and III that the three coefficients b, c, and d can be determined from measurements on one specimen only if both G and the magnetoresistance coefficient are measured. For Ge, experiments of the planar Hall effect carried out by Goldberg and Davis [98] and of the magnetoresistance effect

Table III. Relation for the Planar Hall Coefficient, G, for Two Relevant Directions of the Current and the Electric Field

$\vec{I}$ direction	$\vec{E}$ direction	Normal to plane containing $\vec{B}$	G
100	010	001	$\dfrac{c}{2\sigma_0}$
110	$\bar{1}10$	001	$\dfrac{c+d}{2\sigma_0}$

(both p-type and n-type) by Pearson and Suhl [99] agree with the theory and the relations given above. Similar results were found by Allgaier [100] for PbTe and by Glicksman [101] for Si samples.

It is worth mentioning that all of these effects are extremely sensitive to the impurity content and to the surface preparation of the specimens and the contacts with them. Moreover, shape and inhomogeneities give rise to a change in resistance in a magnetic field.

The interpretation of the experimental data indicates, in general, a dependence upon the product $m\tau_{\mathrm{m}}$. Any departure from the form expected for an isotropic solid implies some degree of anisotropy in m, τ_{m}, or both. Anisotropy in m means that the constant energy surfaces in k-space are no longer spheres. Even though m is anisotropic, τ_{m} could be a simple function of energy only, so that it need not be anisotropic also. Putley [8] carried out calculations of both the Hall coefficient and the constants b, c, and d with the use of a multiellipsoid model. In his calculation τ_{m} was assumed to be a simple function of energy or a tensor with the same symmetry as the energy surfaces, and the charge carriers were assumed to obey classical statistics.

In strong magnetic fields the theory predicts a saturation of the magnetoresistance effect, although closed expressions for the transverse effect can be obtained only for $\tau_{\mathrm{m}} \propto \varepsilon^{-1/2}$.

The limiting values of magnetoresistance for multiellipsoidal surfaces have been computed by Putley [8]. These results are valid only if the condition $\hbar\omega_{\mathrm{C}} \ll k_{\mathrm{B}}T$ is satisfied; otherwise a quantum-mechanical treatment is required.

Saturation no longer occurs in the quantum limit, and, in fact, a change in resistance proportional to B may be observed at high fields, as observed by Putley in PbTe [102]. The analysis of the magnetoresistance effect in semiconductors with a symmetry different from the cubic symmetry, such as in $\mathrm{Bi}_2\mathrm{Te}_3$, may be very complex because of the further complications that may arise from the anisotropy of conductivity even in the absence of a magnetic field.

4.3. Behavior of Typical Semiconductors

Among the more studied semiconductors, III–V materials, especially those that are homogeneous and rodlike, have been investigated in recent years, because of their very high intrinsic mobility, even at room temperature. One of the best studied III–V semiconductor has been indium antimonide, because with this compound it has been possible in many cases to separate the possible mechanisms causing the appearance of magnetoresistance, by preparation or by special measurements. In the following will be reported the main results for several III–V bulk crystals.

4.3.1. Indium Antimonide

4.3.1.1. Intrinsic and Mixed Conductivity

Longitudinal magnetoresistance was never observed in this material at room temperature, and from the results on n-type

material, Rupprecht et al. [103] concluded that the electrons in the conduction band do not produce magnetoresistance. A small change in the magnetoresistance was observed by Hilsum and Barrie [104] in the transition range from intrinsic to p-type material, with a value depending on the impurity concentration.

A different behavior was found in the range of mixed conductivity, that is, when both holes and electrons contribute to the conduction. In this case, the magnetoresistance increase tends toward a saturation value with increasing magnetic field intensity. Thus a magnetoresistance effect can occur in the range of mixed conductivity, even with no magnetoresistance contribution from the conduction or the valence bands.

A formula for the resistivity in a magnetic field for two-band conductivity has been derived by Chambers [105],

$$\rho_{\mathrm{B}} = \frac{1}{\sigma_{\mathrm{e}}(1 - R_{\mathrm{e}}/bR_{\mathrm{h}})}$$
$$\times \frac{1 + (\mu_{\mathrm{h}}B)^2(1 - bR_{\mathrm{e}}/R_{\mathrm{h}})/(1 - R_{\mathrm{e}}/bR_{\mathrm{h}})}{1 + (\mu_{\mathrm{h}}B)^2(1 - R_{\mathrm{e}}/R_{\mathrm{h}})^2/(1 - R_{\mathrm{e}}/bR_{\mathrm{h}})^2} \quad (4.16)$$

where R_{e} and R_{h} are of opposite signs and b (a positive quantity) is the ratio of electron to hole mobility. All quantities in the previous relation may still depend on the magnetic field. It is interesting to note that in (4.16) for $R_{\mathrm{e}} \neq -R_{\mathrm{h}}$ (i.e., for nonintrinsic conductivity) ρ_{B} has a saturation value if the variables $\mu_{\mathrm{e}}, \mu_{\mathrm{h}}, R_{\mathrm{e}}, R_{\mathrm{h}}$, and b have saturation values.

At temperatures higher than room temperature, a transverse magnetoresistance effect has been observed in the intrinsic range of indium antimonide. From the experimental behavior of the hole mobility, Schönwald [106] proposed a three-band model (two bands for heavy and light holes and one band for electrons) that better approaches the experimental results than does the two-band relation (4.16).

4.3.1.2. p-Type InSb

Magnetoresistance of p-type specimens decreases to very low values following the transition into the extrinsic conductivity range. This behavior was found by Weiss [107], Harman et al. [108], and Champness [109]. Champness [110] attempted to use magnetoresistance measurements at low temperatures to get information about the fast holes in p-type InSb.

Cooling the material from about 77 K to about 4 K, Fritzsche and Lark-Horovitz [111] found a reversal in the sign of the magnetoresistance near 10 K, for specimens with hole concentrations of $1-5 \times 10^{15}/\mathrm{cm}^3$, whereas a longitudinal effect was observed by Frederikse and Hosler [112] in a similar range of doping. For the investigated specimens, they found that the magnetoresistance was strongly dependent on surface treatment.

Different explanations for the negative magnetoresistance were suggested by Mackintosh [113], Stevens [114], and Toyozawa [115].

4.3.1.3. n-Type InSb

InSb has the highest electron mobility of all known semiconductors. The condition needed to observe quantistic effects (i.e., $\mu B = 10^4$ cm²T/Vs) is fulfilled at room temperature with $\mu_e = 7.6 \times 10^4$ cm²/Vs in a field of about $1.3 \times 10^{-1} T$.

For single crystals with donor impurity concentrations from 2×10^{16}/cm³ up to 10^{18}/cm³, at temperatures from 77 K to 4.2 K, the transverse magnetoresistance is practically indistinguishable from the changes caused by inhomogeneity in the samples, measured up to a field intensity of 1 T. Hence, for homogeneous specimens measured in a field of 1 T and with electron concentrations higher than 2×10^{16}/cm³, n-type InSb does not possess a magnetoresistance effect in the conduction band at all temperatures.

4.3.2. Indium Arsenide

InAs has a band structure similar to that of InSb and exhibits a very high mobility with a value, at room temperature, of 3×10^4 cm²/Vs for an electron concentration of 1.7×10^{16} cm⁻³. Because of the higher effective mass of $0.02m_0$ and the larger band gap, the shape of the conduction band deviates less from the parabolic form than it does in InSb.

The few magnetoresistance measurements available in the literature were first carried out by Weiss [116] with the use of a transverse field up to 3.3 T, on a single crystal with an electron concentration of 5.5×10^{16} cm⁻³. In fields up to 3 T, $\Delta \rho / \rho_0 \propto B^{1.65}$. The exponent 1.65 in the field dependence of $\Delta \rho / \rho_0$ for InAs has not been explained.

Zatova et al. [117] observed a negative magnetoresistance at low temperatures, which is related to a maximum in the Hall coefficient at temperatures above the zero transition of $\Delta \rho / \rho_0$. They explained the negative magnetoresistance with a conduction in an impurity band.

4.3.3. Indium Phosphide

Glicksman [118] studied a single crystal of InP ($\mu_e = 4600$ cm²/Vs for $n_e = 6 \times 10^{15}$ cm⁻³) in a transverse magnetic field of 1 T, at room temperature. The longitudinal magnetoresistance was smaller by two orders of magnitude.

The experimental results obtained by Glicksman on n-type InP with $n_e < 10^{16}$ cm⁻³ indicated an isotropic conduction band. For higher concentrations, a small anisotropy may be observed. This can be explained by the second minimum in the conduction band, the energy of which is only slightly higher than that of the main minimum at $k = 0$.

4.3.4. Gallium Antimonide

InSb and InAs have only one minimum in the conduction band, and it is near $k = 0$. Therefore, they are isotropic semiconductors. In contrast, GaSb has a second minimum in the conduction band, which has been successfully used to explain

the galvanomagnetic [119, 120], optical [120], and piezoresistance [121] effects. Magnetoresistance, which is a second-order effect in B, becomes anisotropic and is useful in providing information about the position of the second energy minimum in k-space.

It should also be noted that the magnetoresistance vanishes at 4.2 K for $|R| > 5$ cm³/As. This behavior corresponds to the increase in electron degeneracy with spherical energy surfaces in k-space. The steep increase in magnetoresistance at low temperatures with $|R| < 5$ cm³/As is caused by a mixed conductivity magnetoresistance, which is observed when the second minimum starts to be occupied.

On a p-type specimen, Becker et al. [119] found the simultaneous decrease in the Hall coefficient with increasing B. This indicates conductivity by two types of holes, with the minima of the two hole bands lying near each other.

4.3.5. Gallium Arsenide

The experimental results of the magnetoresistance effect in GaAs indicated a marked dependence on the impurity concentration. In fact, Glicksman [118] studied the anisotropy of the magnetoresistance effect of n-type single crystals of GaAs at room tempereture. The transverse magnetoresistance was negligeable up to $n_e = 4 \times 10^{17}$ cm⁻³ with a field of 1 T, and an anisotropy could not be found. In contrast, Kravchenko and Fan [122] found a remarkable longitudinal magnetoresistance as well as a distinct anisotropy on a specimen with $n_e = 8.5 \times 10^{14}$ cm⁻³ at room tempereture. Emel'yanenko and Nasledov [123, 124] observed a negative transverse magnetoresistance on n-type GaAs at temperatures below 60 K. From these results they derived an impurity band conduction that is related to a negative magnetoresistance effect like that observed in InSb.

5. QUANTUM EFFECTS IN LARGE MAGNETIC FIELDS

The magnetic induction, $\vec{B}$, applied to a semiconductor must be regarded as "large" when the product μB exceeds 10^4 cm² T/Vs. This implies that with materials having mobilities of 10^5 to 10^6 cm²/Vs the previous condition is satisfied for intensities of $B = 10^{-1}$ to 10^{-2} T.

Often, the large field condition is expressed in rationalized units as $\mu B > 1$ and, making use of the relation between μ and τ, can be rewritten as

$$\mu B = \frac{e\tau}{m} B = \omega_C \tau > 1 \qquad (5.1)$$

As already seen in Section 2.1, if condition (5.1) holds, the motion of the electron is quantized with energy levels given by the Landau levels (Eq. (2.21), for parabolic bands).

When condition (5.1) applies, an electron will pass through several orbits before undergoing a lattice collision. Under these conditions the Hall angle (3.16) is nearly $\pi/2$, and the electron motion will be made up of a rotation of frequency

ω_C about the direction of $\vec{B}$ and a free motion parallel to the Hall field $\vec{E}_H$.

If $\hbar\omega_C \ll k_B T$ the low-field theory already described will be applicable, but if $\hbar\omega_C \gg k_B T$ the quantization of the electronic distribution must be considered.

For a highly degenerate electron gas, such as is normally found in metals and in high-mobility semiconductors (e.g., InSb or InAs) at low temperatures, the Fermi level $\varepsilon_F > k_B T$. In this case the discussion of the quantization must be divided into two ranges: (a) $\hbar\omega_C < \varepsilon_F$ and (b) $\hbar\omega_C > \varepsilon_F$. In case (a) many quantum levels will be occupied, but in case (b) only the lowest level will be occupied (see Section 2.2). This condition is referred to as the "quantum limit."

Several workers carried out studies on electronic transport properties, such as the magnetoresistance, in this high-strength $\vec{B}$ regime. The results on the magnetoresistance depend upon the assumptions made concerning the degree of degeneracy and the scattering processes. Many theoretical calculations show a deviation from the B^2 variation found at low fields, and they do not indicate saturation as predicted by the quasi-free electron theory (Eq. 4.4).

When case (a) holds the magnetoresistance can show an oscillatory behavior, which is called the Shubnikov–de Haas effect. Suppose that at low magnetic field, the Fermi level is $\varepsilon_F \simeq (l + \frac{1}{2})\hbar\omega_C$, where l is an integer. As the field is increased the energy of the nth level will rise above ε_F, and then this level will rapidly become depopulated, with the electrons entering the lower levels. If the field is further increased, the electron distribution will not change until the next level approaches the Fermi level. These changes in electron distribution generate a periodic variation in the magnetic susceptibility (the de Haas–van Alphen effect) and a corresponding modulation on the resistance, Hall effect, and other transport properties. A detailed analysis shows that ε_F is affected by the field and the whole electron distribution will be raised by $\hbar\omega_C/2$ (the zero point energy).

In this section we describe some of the effects (such as the Shubnikov–de Haas effect, magnetic freeze-out, and magnetophonon oscillations) that may be observed in some semiconductors under particular conditions (detailed below) when a high-intensity magnetic field is applied.

5.1. Shubnikov–de Haas Oscillation

In a high-intensity magnetic field, among the properties exhibiting an oscillatory behavior there is the magnetoresistance. Its oscillations are called the Shubnikov–de Haas (SdH) effect [125].

In view of the complexity of the general expressions, we will give results of theoretical calculation of the SdH effect for the simple parabolic band structure. A further simplification is to consider only elastic and isotropic scattering, such as those by impurities (at low temperatures) and by acoustic phonons (below room temperature, at medium-high temperatures). As said in Section 4, the magnetoresistance may either longitudinal or transverse. In both cases SdH oscillations have been observed.

Adams and Holstein [126] calculated $\Delta\rho/\rho$ for the longitudinal case. Neglecting the nonoscillatory contribution, their results yield

$$\frac{\Delta\rho_\parallel}{\rho} = \sum_{r=1}^{\infty} b_r \cos\left(\frac{2\pi\varepsilon_F}{\hbar\omega_C} r - \frac{\pi}{4}\right) \tag{5.2}$$

where

$$b_r = (-1)^r \sqrt{\frac{\hbar\omega_C}{2\varepsilon_F r}} \cos\left(\frac{\pi g m}{2m_o} r\right)$$
$$\times r \frac{2\pi^2 k_B T/\hbar\omega_C}{\sinh(r 2\pi^2 k_B T/\hbar\omega_C)} e^{-(r 2\pi^2 k_B T_D/\hbar\omega_C)} \tag{5.3}$$

where g is the Landé factor of the carrier [127, 128] and T_D is the Dingle temperature [129], which is a measure of the natural line width of the transitions between adjacent Landau levels.

In a plot of $\Delta\rho/\rho$ vs. $1/B$, the period of the oscillation is independent of B, but it will depend on the carrier concentration (2.24). This last dependence may be used to compute the charge carrier density in the material. The coefficients b_r rapidly decrease with the increase in r, and in practice only the $r = 1$ term will prevail in the sum (5.2).

A simplification of (5.3) may be obtained, approximating the hyperbolic sine at the denominator in the (5.3) by an exponential:

$$b_r = (-1)^r \sqrt{\frac{\hbar\omega_C}{2\varepsilon_F r}} \cos\left(\frac{\pi g m}{2m_o} r\right)$$
$$\times r \frac{2\pi^2 k_B T}{\hbar\omega_C} e^{-[2r\pi^2 k_B (T+T_D)/\hbar\omega_C]} \tag{5.4}$$

In Eq. (5.4) the temperature T represents, essentially, the carrier temperature T_e. The Shubnikov–de Haas effect can be used to determine T_e [130] for a degenerate electron gas, which is heated by a strong electric field $\vec{E}$.

For the transverse magnetoresistance and assuming isotropic scattering [4], the expression for the SdH effect becomes

$$\frac{\Delta\rho_\perp}{\rho} = \frac{5}{2} \sum_{r=1}^{\infty} b_r \cos\left(\frac{2\pi\varepsilon_F}{\hbar\omega_C} r - \frac{\pi}{4}\right) + \Upsilon \tag{5.5}$$

where b_r is given by the relation (5.3) and the other quantities in (5.5) have the form

$$\Upsilon = \frac{3}{4} \frac{\hbar\omega_c}{2\varepsilon_F} \left\{ \sum_{r=1}^{\infty} b_r \left[\alpha_r \cos\left(\frac{2\pi\varepsilon_F}{\hbar\omega_c} r\right) + \beta_r \sin\left(\frac{2\pi\varepsilon_F}{\hbar\omega_c} r\right) \right. \right.$$
$$\left. \left. - \ln\left(1 - e^{4\pi k_B T_D/\hbar\omega_c}\right) \right] \right\} \tag{5.6}$$

$$\alpha_r = 2r^{1/2} \sum_{s=1}^{\infty} \frac{1}{[s(r+s)]^{1/2}} e^{-4\pi s k_B T_D/\hbar\omega_c} \tag{5.7}$$

$$\beta_r = r^{1/2} \sum_{s=1}^{r-1} \frac{1}{[s(r-s)]^{1/2}} \qquad (5.8)$$

The term Υ in Eq. (5.5) is the contribution to the oscillatory part of the transverse resistivity due to transitions in which the quantum number n changes, and it diverges if the level width $k_B T_D$ goes to zero. Its relative importance depends on the ratio $k_B T_D / \varepsilon_F$. In practice Υ would appear to be unimportant when collisions are frequent enough to damp out $r > 1$ harmonics in the oscillations [126].

The oscillatory magnetoresistance effects in III–V compounds have been studied in InSb by Frederikse and Hosler [112], Amirkhanov et al. [131]; in InAs by Sladek [132], Frederikse and Hosler [133], Shalyt and Éfros [134], and Amirkhanov et al. [135]; and in GaSb by Becker and Fan [136] and Becker and Yep [137].

In all of the semiconductors listed above, the oscillations have been due to electrons, and the equations (5.2) and (5.5) can be applied, with some little change, such as the replacement of the constant $-\pi/4$ by a phase φ. In GaSb, oscillations due to the subsidiary maxima were not observed, but effects of their occupation occur as discussed below.

The existence of degenerate free carriers at low temperatures can be attributed to screening out of the shallow impurity potential due to the high concentration of carriers. We shall be concerned here only with situations in which the carrier concentration remains constant; that is, freeze-out effects (see Section 5.2) will be neglected. Various features of the results will be discussed separately.

(1) Periods. The period and phase of the oscillations can be obtained by plotting nodal or extremal values of B^{-1} versus integers. From the period of the oscillations the electron density, for spherical energy surfaces, should be directly computed. The obtained value should be independent of the effective mass or deviations from parabolic behavior. For GaSb the agreement was with the low-temperature Hall coefficient, which reflects the high mobility carrier concentration. For InAs the results of experiments [112, 132, 134, 135] made it possible to clearly distinguish and determine the periods due to a strong increase in the ordinary magnetoresistance effect. The agreement between the calculations and the measured values was very good.

(2) Phase. The phase φ, for the model proposed above, should theoretically be given by $-\pi/4$ for both the longitudinal and transverse cases (see equations 5.2 and 5.5). A phase reversal due to the g-factor is not expected, although a reduction in amplitude should occur. Unfortunately, the phase is difficult to obtain accurately because of the extrapolation. The best agreement is found in the lowest temperature measurements in InAs [132].

(3) Amplitude. From the expression of b_r (Eq. (5.3)) it comes out that the amplitude of the oscillatory magnetoresistance, both longitudinal and transverse, is dependent on the temperature and the magnetic field. The temperature dependence of the amplitude gives a value of the cyclotron frequency and hence the effective mass. At the lowest temperatures [132, 136] a dependence of $\chi/\sinh \chi$, where $\chi = 2\pi^2 k_B T / \hbar \omega_C$, is found.

The magnetic field dependence of the amplitude involves both the temperature and the collision broadening. The effects can be separated by dividing the amplitude by $(\hbar \omega_C / \varepsilon_F)^{1/2} \chi / \sinh \chi$ and plotting the logarithm of the result versus B^{-1}. The slope of the resulting straight line gives the Dingle temperature, T_D. Excellent fit with the form of the theoretical expression was obtained for GaSb [136]. For InAs, Sladek [132] discussed the amplitude of the oscillations as a function of B and T and showed that the nature of the oscillations for monocrystalline and polycrystalline material were practically coincident. This fact was a confirmation of the isotropic spherical conduction band.

For the transverse case, if Υ (Eq. (3.6)) can be neglected, as it apparently can here [138], the amplitude should theoretically be 2.5 times higher than the longitudinal one. Experimentally the transverse amplitude is greater than the longitudinal amplitude, but by a slightly smaller factor. The theoretical value of 2.5 is dependent on the isotropic scattering mechanism assumed. A calculation based on ionized impurity scattering, which is thought to be the mechanism involved at least in GaSb [136, 139], would be desirable. The order of magnitude of the extrapolated amplitudes is in agreement with the theoretical result [4].

(4) Effective mass. For III–V compounds, the effective masses computed from the Shubnikov–de Haas effect are in good agreement with that obtained by other experiments. For GaSb, Becker and Yep [137] report a concentration dependence of the effective mass, which is expected from the nonparabolicity of the bands.

(5) Level broadening. The Dingle temperature T_D can be compared with a "mobility temperature," $T_m = \hbar/2\pi k \tau_m$, and for the III–V compounds the two temperatures correlate well. In fact, for collision processes due to ionized impurity it is expected that T_D can be slightly greater than T_m. The former depends on the carrier lifetime, and the latter depends on the momentum relaxation time, and the two are equal only for isotropic scattering. For GaSb [136], as we discuss below, more than one band is occupied, but T_D and T_m are due to the same band. The results of Becker and Fan [136] on tellurium-doped GaSb are particularly interesting in that the collision broadening effect decreases as carrier concentration is increased, resulting in a striking change in magnetoresistance amplitude with carrier concentration. GaSb has interesting galvanomagnetic properties [119] because of the existence of subsidiary $\langle 111 \rangle$ minima with a high density of states located 0.08 eV above the low-mass $k = 0$ conduction band minimum. The transport properties therefore change considerably when, because of high carrier concentration or high temperature, the subsidiary minima are occupied. The decrease in Dingle temperature occurs in just the region of concentration in which the subsidiary minima are becoming populated. A calculation carried out by Robinson and Rodriguez [139] showed a change in collision rate due to ionized impurity scattering within the $k = 0$ minimum. Assuming an ionized impurity potential, modified by electronic screening, they showed

that when the carrier density becomes large enough to populate the subsidiary bands, the increased density of states results in a greatly enhanced screening, so that the impurity scattering decreases. Their theoretical prediction of a greater decrease in T_D and T_m is in agreement with experiment.

5.2. Freeze-Out Effects

An impurity center in a semiconductor, at temperatures above about 1.8–2 K, may show an ionization energy depending on the intensity of an applied magnetic field. This effect, known as magnetic freeze-out, was discovered during experiments suggested by the theoretical work of Yafet, Keyes, and Adams (YKA) [7], which was concerned with the behavior of the ionization energy of a hydrogen atom in a magnetic field. The experiments demonstrated that when a H atom is placed in an intense magnetic field, the energy difference between the Landau levels of a free electron becomes comparable to Rydberg ionization energy for a H atom in the absence of a magnetic field. Under these conditions the ionization energy of the atom is raised, and the effect can be described in terms of a parameter,

$$\gamma = \frac{\hbar \omega_C}{2 R_y} \qquad (5.9)$$

where R_y is the Rydberg energy.

The increase in the ionization energy is due to a compression in all directions of the atom wave function by the applied magnetic field. In the case of semiconductors this affect will affect hydrogenic impurities in the presence of a high-intensity magnetic field. In fact, because of the low effective mass and high dielectric constant the Bohr radius, a_B, of impurity centers will be generally greater than the distance between nearest neighbors, even for very low concentrations (see below). If, however, an intense magnetic field is applied, the radius of the cyclotron orbit ($r = mv/eB$) may become comparable to the distance between nearest neighbors, and then the impurity centers will be localized and their ionization energies will be changed. In fact, electrons occupying the localized impurity centers may still be mobile in the impurity band, and by application of an electric field of about 1 V/cm they may be excited by impact ionization to the conduction band.

To observe the freeze-out effect, values of $\gamma \geq 1$ are necessary, and from (5.9) we can write an expression for the intensity (in Tesla) of the magnetic induction, B, making $\gamma = 1$,

$$B = 2.35 \times 10^5 \left(\frac{m}{m_0 \chi_H} \right)^2 \qquad (5.10)$$

where χ_H is the high-frequency dielectric constant. For an isolated hydrogen atom, the intensity of B required to satisfy condition (5.10) is 20,000 T, which is practically impossible to produce. In semiconductors containing hydrogenic impurities, the combination of large dielectric constants and small effective masses can bring the field required into the experimentally accessible region. For example, the value of B (see Eq. (5.10))

for Ge is 37 T [140], which is still too large to allow a reasonable increase in γ from $\gamma = 1$. For InSb, B is about 0.2 T [140], which makes it possible to reach a value of γ of about 50 with a field intensity of 10 T, which is easily available with commercial superconducting magnets. Even though for other semiconductors such as PbTe or n-type CdHgTe alloy [14.2] the field intensity required is smaller than that for the previous cases, for other reasons difficult to experimentally fulfill (detailed below), the freeze-out effect cannot be experimentally verified.

In principle, the YKA theory, even though it was developed for an isolated H atom, it could be check for some of the above mentioned semiconductors, applying strong enough magnetic field. However, the hydrogenic impurities present in a semiconductor behave like isolated atoms if the spacing between nearest neighbors is at least 10 times the Bohr radius, a_B. Moreover, the factors that reduce the ionization energy increase the Bohr radius:

$$\varepsilon_{ion} = 13.6 \left(\frac{m}{m_0 \chi_H^2} \right) \text{ [eV]} \qquad (5.11)$$

$$a_B = 5.29 \times 10^{-11} \left(\frac{m_0}{m} \right) \chi_H \text{ [m]} \qquad (5.12)$$

From relation (5.12) we can calculate the impurity density at which the average spacing between neighboring impurities is about $10 a_B$. For Ge, we obtain a value of about 3.4×10^{15} at./cm^3, corresponding to the maximum concentration at which impurities behave as isolated centers. For InSb, this value falls to 1.3×10^{12} at./cm^3, which is much smaller than the concentration in the best available material.

In the case of PbTe and HgCdTe alloys the situation is even worse, because the maximum impurity concentration should be about 2×10^{11} and 4×10^7 at./cm^3, respectively, and the minimum obtainable impurity concentration is about 10^{16} to 10^{17} cm^{-3}.

Among the possible candidates, InSb comes closest to the requirements for the observation of discrete behavior of the freeze-out effect in semiconductors. In the purest n-type InSb the interaction between the impurities is so strong that it behaves in a metallic fashion down to about 1 K, whereas, in p-type InSb, some evidence of discrete behavior might be expected for concentrations of about 10^{13} at./cm^3, which is one order of magnitude greater than the above-reported value.

Even though InSb is not suitable for a direct test of the YKA theory, an indirect proof may be obtained by a measure of the Hall coefficient, as done by Keyes and Sladek [141–143], Frederikse and Hosier [144], and Putley [145, 146].

In the experimental work carried out by Putley [140], the Hall coefficient of an n-type InSb sample was measured as a function of the temperature, for several values of the magnetic field intensity. The results indicate an almost exponential growth of the Hall coefficient with the intensity of B in the low-temperature region, around 2 K. This behavior is to be expected if the carriers are falling out of the conduction band into a shallow impurity level. With decreasing temperature from about 1.8 K down to about 1 K, the Hall coefficient falls slightly or becomes constant. As suggested by Putley [140], this kind of behavior could be the result of interactions between neighboring pairs of impurities and/or degeneracy effects of electron gas at low temperature.

Figure 24 reproduces the behavior of the quantity $(R_He)^{-1}$ as a function of the B intensity, for two samples of n-type InSb, containing different carrier concentrations (at $B = 0$). Figure 24 shows how, for the sample with the higher carrier density (about 2×10^{16} electrons/cm³) and the higher impurity concentration, the overlap of the impurity wave functions lasts for much higher magnetic field intensities than for the other specimen.

With regard to behavior above 1.8 K in a magnetic field, we need to consider the conduction band splitting into the Landau subbands induced by the intense magnetic field (see Section 2). Furthermore, we will assume that the extreme quantum limit is satisfied ($\hbar\omega_C \gg k_BT$), and we will make use of the classical statistics. Under these conditions the total number of electrons in the conduction band can be derived from the relation (2.25)

$$n = \frac{1}{2}N_C\theta F_{-1/2}\left[\eta - \frac{\theta}{2}\right] = N_B e^{(\eta - \theta/2)} \quad (5.13)$$

where the factor $\frac{1}{2}$ has been introduced because in InSb the large g-factor splits the Landau levels further, into two bands with opposite spins, and the splitting is comparable to $\hbar\omega_C/2$. In the current hypothesis of quantum extreme limit the lowest level will contain electrons of only one spin direction.

Using relation (5.13) for the Fermi energy and assuming, in the presence of the applied field $\vec{B}$, that the donor density N_D has a ground state energy ε below the lowest Landau level and that acceptors are present at a concentration N_A, the following equation is found for the temperature dependence of n:

$$\frac{(N_A + n)n}{N_D - N_A - n} = N_B e^{-\varepsilon/k_BT} \quad (5.14)$$

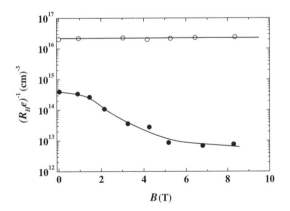

Fig. 24. Filled points: magnetic freeze-out effect of carriers in n-type InSb at 4.2 K having a carrier concentration of 4×10^{14} electrons/cm³; open points: absence of any freeze-out effect in the same material, but with a carrier density of 2×10^{16} electrons/cm³. Reprinted with permission from [141], ©1956, Elsevier Science.

This relation differs from that for a compensated impurity level in a semiconductor, only for the expression of N_B ($N_B = |e|B(2\pi mk_BT)^{1/2}/h^2$), which, in this case, represents the effective density of states in the Landau level.

By fitting Eq. (5.14) to the experimental data between 4 K and 2 K, we can compute the values of ε, N_D and N_A. The value of $N_D - N_A$ has been obtained from measurements at 77 K. Within the limits of 20%, N_D and N_A are independent of B, whereas the result of the fitting shows that only ε depends on B.

A comparison of the ionization energy values with those of the YKA theory shows that although the existence of magnetically dependent energy levels is confirmed, the behavior differs markedly from that predicted by the theory, which has been developed for an isolated impurity (H atom).

In low-intensity magnetic fields the theory predicts a metallic behavior (experimentally confirmed) if the impurity concentration is higher than the limit given in the above-mentioned examples. However, above a certain critical field (depending on the purity of the sample) discrete levels appear. The limited results available suggest that above the critical field the ionization energy, ε, increases linearly. Measurements have not been extended to sufficiently high fields to show whether at high enough fields the behavior tends to the YKA result.

Although the simple model of an impurity center with a magnetic field-dependent ionization energy accounts for the behavior in high fields and at temperatures above about 1.8–2 K, the behavior at lower temperatures does not accord with this model. In this case, as mentioned above, we should suppose the existence of other effects, such as multiple interaction or degeneracy effects [140].

5.3. Magnetophonon Effect

In a semiconductor material immersed in a quantizing magnetic field and below room temperature it may be possible to observe oscillations in the magnetoresistance, which are due to longitudinal optical phonons of energy $\hbar\omega_0$ and which do not require degeneracy. These oscillations occur if the optical-phonon energy is an integer multiple n of the Landau level spacing $\hbar\omega_C$:

$$\hbar\omega_0 = n\hbar\omega_C \quad (5.15)$$

By absorption of a phonon, an electron is transferred from a given Landau level i to a level $i + n$. This effect is known as the magnetophonon effect and was predicted independently by Gurevich and Firsov [147] and by Klinger [148] and observed by Firsov et al. [149] on n-type InSb at 90 K. Their experimental data for longitudinal and transverse magnetoresistance are displayed in Figure 25. The inset indicates the oscillatory behavior. These curves have been obtained by replotting the data as a function of B^{-1}. The period of the oscillation is constant and is given by

$$\Delta\left(\frac{1}{B}\right) = \frac{e}{m\omega_0} \quad (5.16)$$

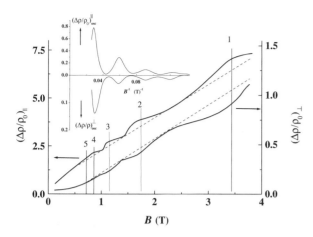

Fig. 25. Transverse ($\perp$) and longitudinal ($\parallel$) magnetoresistance for n-type InSb at 90 K. The inset shows the evidence for the oscillations of the two curves vs. B^{-1}. The numbers at the vertical lines are values of the ratio phonon frequency and cyclotron resonance frequency. Reprinted with permission from [149], ©1964, American Physical Society.

in agreement with (5.15). In n-type InSb at 3.4 T, $m/m_0 = 0.016$ and $\omega_0 = 3.7 \times 10^{13}$ s, which yields $\Delta(B^{-1}) = 3 \times 10^{-3}$/T. At resonance, the transverse magnetoresistance shows maxima, whereas the longitudinal magnetoresistance shows minima. The period is independent of the carrier density, in contrast to the Shubnikov–de Haas period (see Section 3.1). Because for large carrier densities [150] and, consequently, large impurity densities ionized impurity scattering dominates over optical-phonon scattering, no oscillations have been observed for densities of more than 5×10^{15} cm^{-3}.

Moreover, at very low temperatures, the density of optical phonons decreases and the magnetophonon effect is difficult to observe, whereas at too high temperatures the condition $\omega_c \tau_m \gg 1$ necessary for the Landau quantization is not fulfilled, even for very high-intensity magnetic fields. Because of these limitations the effect is quite small and difficult to measure, requiring sophisticated electronic techniques such as double differentiation [151].

The longitudinal magnetophonon effect occurs when two scattering processes such as inelastic scattering by an optical phonon and an elastic transition at an impurity site are operative [152].

Figure 26 [153] shows the amplitudes $\Delta\rho/\rho_0$ observed in the transverse configuration plotted as a function of the harmonic number, for n-type GaAs. The straight lines in the semilog plot suggest a dependence for $\Delta\rho$ of the form

$$\Delta\rho \propto e^{-\gamma\omega_0\omega_C} \quad (5.17)$$

where γ is a constant. For GaAs $\gamma = 0.77$, in agreement with observations.

The magnetophonon effect may be used to measure the effective mass m of charge carriers, provided the optical-phonon frequency ω_0 is known. However, whereas the effective longitudinal electron mass is generally in agreement with the cyclotron resonance value, those of both holes and transverse electrons may be significantly higher. These differences

in the effective mass have been observed in Ge [151]. The discrepancies are explained by the nonparabolicities of the bands: for magnetophonon measurements carried out at high temperatures (120 K in the case of Ge), higher parts of the bands are occupied than at 4 K (cyclotron resonance).

In polar semiconductors, because of the polaron effect (Section 2.6.8), the mass derived from magnetophonon oscillations is $1 + \alpha/3$ times the low-frequency mass and $1 + \alpha/2$ times the high-frequency or bare mass, where α is the polar coupling constant given by Eq. (2.110). For n-type GaAs a mass value of 0.06 is obtained [154].

Figure 27 presents the temperature dependence of ν_0 (panel a) and of the magnetophonon mass m (panel b), for a n-type GaAs sample [153].

The values of m obtained after a correction for the band nonparabolicity are in agreement with those obtained by other techniques such as cyclotron resonance, Faraday rotation, and interband magneto-optic absorption.

Although at temperatures below 60 K the magnetophonon effect in n-type GaAs is undetectable, Reynolds [155] observed oscillation due to electron heating by electric fields between 1 and 10 V/cm at lattice temperatures between 20 K and 50 K.

Eaves et al. [151] reported the almost linear dependence of the B^{-1} values of the peaks vs. harmonic numbers, n, for the series involving the emission of LO phonons with GaAs and CdTe. The measurements were carried out in high electric fields at 20 K (for GaAs) and 14 K (for CdTe). For a comparison the high-temperature thermal carrier peaks were reported in the same plot. The extrapolation to $B^{-1} = 0$ of the warm-carrier peaks resulted in a negative value of n, which made it possible to discover a dominant energy loss mechanism associated with the emission of an optical phonon and the capture of the warm electron by a donor atom.

For CdTe at $E > 7$ V/cm, additional peaks have been observed and identified as harmonics at higher temperatures, under zero-field conditions. These peaks obey the relation (5.15).

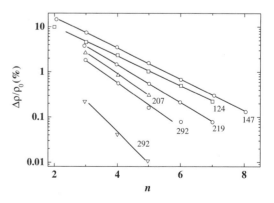

Fig. 26. Experimental values of the magnetophonon amplitude vs. the harmonic number for n-type GaAs. The numbers at the curves are the temperatures expressed in Kelvin. Reprinted with permission from [153], ©1968, EDP Sciences.

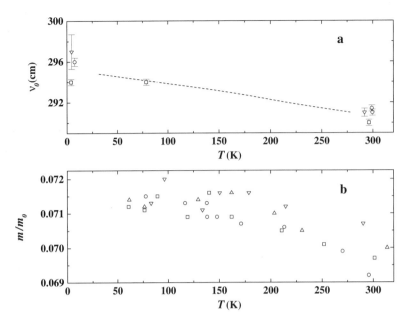

Fig. 27.   (a) Temperature dependence of the longitudinal optical (Lo) phonon frequency for GaAs, at $q = 0$. (b) Temperature dependence of the magnetophonon mass for n-type GaAs determined from the Lo phonon frequency and the magnetophonon peaks. Reprinted with permission from [153], ©1968, EDP Sciences.

For n-type InSb and GaAs at low temperatures (e.g., 11 K) and threshold fields of 0.07 V/cm and 0.8 V/cm, respectively, several peaks are observed that have been attributed to a simultaneous emission of two oppositely directed transverse acoustic phonons $\hbar\omega_t$, by electrons in the high-energy tail of the distribution:

$$2\hbar\omega_t = n\hbar\omega_C \qquad (5.18)$$

The phonons are thought to be at the X point in the phonon Brillouin zone and have energies of 5.15 meV in InSb and 9.7 meV in GaAs.

Magnetophonon effects have been observed in other electronic transport coefficients, such as in the Hall effect [156] and in thermoelectric power of n-type InSb at 120 K [157].

6. MAGNETOTRANSPORT IN LOW-DIMENSIONAL SYSTEMS AND IN HETEROSTRUCTURES

The electronic transport properties of semiconductors and solid materials depend on, among several factors, the dimensionality of the investigated system. Materials exhibiting two, one, or even zero dimensions will behave differently from the corresponding three-dimensional ones.

From the point of view of electronic transport, a low-dimensional system is that in which the charge carriers are constrained by potential barriers, so that they lose one or more degrees of freedom for motion. These systems become two-, one-, or zero-dimensional if the potential barriers confine the electrons in one, two, or three dimensions, respectively.

It must be noted that dimensionality is not an absolute property, but is related to the length scales, which determine the physical properties that have to be investigated. Several length

scales, such as specimen dimensions, de Broglie wavelength, effective Bohr radius, magnetic length, etc., can be important for transport.

The effective dimensionality of a system may be altered by changing some macroscopic physical quantities such as the sample dimensions, the carrier concentration, or the magnetic field.

Semiconductor heterostructures, such as single quantum wells (QWs), multi-quantum wells (MQWs), and superlattices (SLs), are the best (even if not unique) examples of two-dimensional systems, whose structural, optical, and electrical properties have been studied since the 1970s, when growth techniques made it possible to realize heterostructures and SLs of very high quality [158–161].

Many novel physical phenomena, including the quantized Hall effect [162], have been observed in this kind of structure, and an excellent review has been given by Ando et al. [163]. In the last two decades, the improved quality of semiconductor heterojunctions, in particular the GaAs/Al$_x$Ga$_{1-x}$As system, has allowed further new effects (e.g., fractional quantized Hall effect, ballistic motion, etc.) to be observed.

Even though semiconductor heterostructures exploit new and technologically relevant optical features, the electrical transport properties appear to be particularly interesting. In fact, heterostructures may exhibit carrier mobilities much higher than those of the corresponding bulk material because of a technique known as modulation doping, consisting of spatial separation of impurity ion dopers from the charge carriers [46, 164, 165]. This separation greatly raises the carrier mobility (especially at low temperatures) because of a significant reduction of the ionized impurity scattering.

For heterostructures of III–V group elements such as GaAs/Al$_x$Ga$_{1-x}$As, n-type doping is achieved by impurities of

Si introduced in wide band-gap material (i.e., $Al_xGa_{1-x}As$) at a suitable distance from the interface, whereas narrow band gap material (i.e., GaAs) is free from intentional doping. Some of the electrons coming from the Si ionized atoms pass from the conduction band of $Al_xGa_{1-x}As$ (working as a barrier) into the lower-lying Γ conduction band of GaAs. The transferred electrons are confined in an approximately triangular well produced by the conduction bond discontinuity and the Coulomb potential of the Si ionized atoms.

Similarly, a p-type MOD heterostructure (MODH) can be realized by replacing the Si dopant with a p-type element such as beryllium [166].

Another example of MODH is that made with elements of the IV group, Si and Ge. The SiGe system, which has been extensively studied in the last 25 years, is a promising candidate for the fabrication of fast devices [167, 168] directly integrated on the Si wafers.

For SiGe MODH, high carrier mobilities were first demonstrated by Abstreiter et al. [160] by adjusting the strain in the active layer and by modulation doping [45, 46]. Figure 28 shows [169] typical n-type heterostructures, doped with Sb, with a Si thin layer as an active channel in which electrons can move freely. Hall mobility values of $500,000$ cm^2/Vs at 0.4 K have been reported by Ismail et al. [170]. Recently Sugii et al. [171] prepared a MODH with a record electron mobility of $800,000$ cm^2/Vs at 15 K.

For p-type SiGe MODH, the active layer is a $Si_{1-x}Ge_x$ channel ($x \simeq 0.3$) cladded between two Si layers. Hole mobilities in these structures are lower than those in the n-type heterostructures.

Modulation doping is not the only way to grow a 2D system. A bidimensional electron gas can be obtained with a QW consisting of a thin layer (well) cladded between two layers with a wider band gap. Band discontinuity between the

two semiconductors along the growth direction confines the charge carriers in the well, where they will behave like a two-dimensional electron gas.

Depending on the barrier thickness, the repetition of the layers sequence, of a single QW, can create MQWs or SLs. When the barrier thickness is high enough that the electronic wave function in the well does not overlap with the adiacent ones, we have a MQW. In contrast, SLs will be realized in which the isolated energy levels of a QW separate to form minibands of energy.

6.1. Magnetotransport in Two-Dimensional Systems at Low Fields

The equations that will be given in this section hold for electrical transport properties in low fields (i.e., nonquantizing magnetic fields will be considered: $\tau\omega_C \ll 1$; see Eq. (5.1).

Magnetotransport measurements are generally carried out with the use of a standard Hall bar geometry, as shown in Figure 29.

The electrical transport is influenced directly only by the $\vec{B}$ component orthogonal to the heterointerfaces. The component of $\vec{B}$ parallel to the film surface modifies the quantization energy of the carrier motion [172], whereas the total magnetic field influences the spin splitting (Zeeman splitting).

If $\vec{E}$ is the local electric field and $\vec{j}$ is the local current density, then Ohm's law assumes the same form as in a bulk (i.e., three-dimensional) conductor (see Section 2.5),

$$\vec{E} = \underline{\underline{\rho}}\,\vec{j} \qquad (6.1)$$

where $\underline{\underline{\rho}}$ is the resistivity tensor in two dimensions. For an isotropic material the relations (see section 2.5)

$$\rho_{xx} = \rho_{yy} \qquad \text{and} \qquad \rho_{xy} = -\rho_{yx} \qquad (6.2)$$

hold. With reference to Figure 29, the total current flowing through the sample in the x direction, I_x, is given by

$$I_x = \int j_x(x, y)\mathrm{d}y \qquad (6.3)$$

which is independent of x, because of current conservation. $j_x(x, y)$ and $j_y(x, y)$ are the components of the current density vector, $\vec{j}$, whose distribution over the bar width is determined by current conservation and by the continuity of the electric field.

The coefficients of magnetotransport are defined as the ratio of longitudinal or transverse voltages to the total current I_x, and they are determined by integration over the local resistivity tensor. The Hall resistance is defined as

$$R_{xy}\left|\frac{V_y}{I_x}\right| = \frac{1}{I_x}\int E_y(x, y)\mathrm{d}y$$

$$= \frac{1}{I_x}\int[-\rho_{xy}(x, y)j_x(x, y) + \rho_{xx}(x, y)j_y(x, y)]\mathrm{d}y$$

$$(6.4)$$

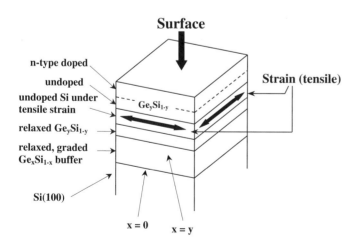

Surface

n-type doped

undoped

undoped Si under tensile strain

relaxed Ge_ySi_{1-y}

relaxed, graded Ge_xSi_{1-x} buffer

Si(100)

Strain (tensile)

Ge_ySi_{1-y}

x = 0   x = y

Fig. 28. Schematic diagram of an n-type $Si/Si_{1-x}Ge_x$ MODH. The two-dimensional electron gas is formed in the strained Si layer (well) enclosed by $Si_{1-x}Ge_x$ alloy layers (barriers). Because of the difference in the lattice parameters of Si and $Si_{1-x}Ge_x$ layers ($a_{si} \geq a_{si_{1-x}}Ge_x$) this system presents a strain-induced type II band alignment. Reprinted with permission from [169], ©1993, American Institute of Physics.

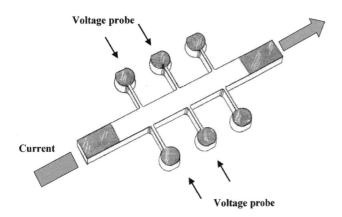

Fig. 29. Schematic of Hall bar measurements.

The resistance in a magnetic field, R_{xx}, is defined as

$$R_{xx} = \left| \frac{V_x}{I_x} \right| = \frac{1}{I_x} \int [\rho_{xx}(x,y)j_x(x,y) + \rho_{xy}(x,y)j_y(x,y)]\mathrm{d}x \tag{6.5}$$

For a homogeneous conductor the tensor $\underline{\underline{\rho}}$ is independent of x and y, and the previous relations simplify as [173, 174]

$$R_{xy} = \rho_{xy}\left(1 - \cot \Phi_H \frac{\int j_y(x,y)\mathrm{d}y}{\int j_x(x,y)\mathrm{d}y}\right) \tag{6.6}$$

$$R_{xx} = \rho_{xx}\left(\frac{\int j_x(x,y)\mathrm{d}x + \tan \Phi_H \int j_y(x,y)\mathrm{d}x}{\int j_x(x,y)\mathrm{d}y}\right) \tag{6.7}$$

The Hall mobility is defined as

$$\mu_H = \frac{\rho_{xy}}{B\rho_{xx}} \tag{6.8}$$

The electron density, n_S, in a bidimensional system can be evaluated with a Hall bar geometry [1]:

$$n_S = \frac{r_H B}{e\rho_{xy}} \tag{6.9}$$

A correct interpretation of the electrical transport properties in low-dimensional systems requires the description of the effects of electric and magnetic fields on the energy band structure of these systems. A general review of these effects has been given by Harris et al. [175], and in this section only the main relations obtained for a heterojunction interface between two semiconductors with a band offset in the conduction band will be given.

In this kind of system, energy levels form in the almost triangular potential well, ε_i, which will act as the bottom of a two-dimensional subband, with a constant density of states per unit area and energy, $D_i(\varepsilon)$, characteristic of a two-dimensional electron gas. The total two-dimensional density of states, for all of the subbands and considering the spin degeneracy, has the form

$$D_{2D}(\varepsilon) = \sum_i D_i(\varepsilon) = \sum_i \frac{m}{\pi\hbar^2}\theta_H(\varepsilon - \varepsilon_i) \tag{6.10}$$

where θ_H is the Heaviside step function. From (6.10) we can write the expression for the total electron density, n_i, of each band,

$$n_i = \frac{mk_B T}{\pi\hbar^2}\ln\left[1 + e^{(\varepsilon_F - \varepsilon_i)/k_B T}\right] \tag{6.11}$$

which for a degenerate electron gas ($T \to 0$) simplifies to

$$n_i = \frac{m}{\pi\hbar^2}(\varepsilon_F - \varepsilon_i) \tag{6.12}$$

Assuming a triangular potential well for the effective confining potential energy and a low density of charge carriers (i.e., $n_S < \frac{2\chi E_{ion}}{e}$, where E_{ion} is the electric field at the interface due to the ionized impurity space charge density), the energy levels in the well are given by [175]

$$\varepsilon_i = \left(\frac{\hbar^2}{2m}\right)^{1/3}\left[\frac{3\pi}{2}\left(i + \frac{3}{4}eE\right)\right]^{2/3} \tag{6.13}$$

6.2. Magnetotransport in One-Dimensional Systems at Low Fields

By several techniques [176–180] it is possible to further constrict a two-dimensional electron gas to create a one-dimensional system. For one-dimensional systems the density of states changes and assumes the form

$$D_{1D}(\varepsilon) = \sum_j \frac{1}{\pi\hbar}\left[\frac{2m}{\varepsilon - \varepsilon_j}\right]^{1/2}\theta_H(\varepsilon - \varepsilon_j) \tag{6.14}$$

where ε_j denotes the quantized energy level produced by electron confinement along two orthogonal directions [181, 182].

In one-dimensional systems the lateral confinement potential may be approximated by a parabolic one [183] for very narrow channels or by a square well for wider channels and higher electron densities.

6.3. Low-Dimensional Systems in High Magnetic Fields

As already seen in Section 2.8, a magnetic field $\vec{B}$, under suitable conditions, may induce cyclotron resonance of charge carriers and the formation of Landau levels (see Section 2.1). Similarly, in a two-dimensional electron gas the component of $\vec{B}$ normal to the interface, B_z, will force the electrons to move in circular orbits parallel to the interface, with the energy levels given by an equation similar to (2.21),

$$\varepsilon_{i,n,s} = \varepsilon_i + \left(n + \tfrac{1}{2}\right)\hbar\omega_C + sg\mu_B B \tag{6.15}$$

where ε_i are the energy levels due to the electric confinement of electrons in a two-dimensional plane, the second term gives the value of the quantized Landau levels, and the last term is due to spin splitting (see Section 2.1).

The two-dimensional density of states function (Eq. (6.10)) is modified and assumes the form

$$D_{2D}(\varepsilon) = \sum_{i,n,s} n_L \delta(\varepsilon - \varepsilon_{i,n,s}) \tag{6.16}$$

where n_L represents the number of states per Landau level, per spin, and per unit area, given by

$$n_L = \left(\frac{m}{2\pi\hbar^2}\right)\hbar\omega_C = \frac{eB_Z}{h} \qquad (6.17)$$

Other useful quantities are the filling factor, ν, and the characteristic magnetic length, l_B, given by

$$\nu = \frac{n_S}{n_L} = \frac{hn_S}{eB_Z} \qquad (6.18)$$

$$l_B = \sqrt{\frac{\hbar}{eB_Z}} \qquad (6.19)$$

It should be noted that electrons moving in the plane of the two-dimensional gas undergo a scattering process with a characteristic mean free path, l, given by

$$l = v_F\tau_e = \pi n_S\hbar^2/m \qquad (6.20)$$

where τ_e is the mean time between collisions. Therefore, the condition for Landau quantization requires that $l > l_B$ (more precisely, that $\omega_C\tau_m > 1$; see Eq. (5.1)) and that $\hbar\omega_C > k_BT$ to resolve the Landau levels.

The scattering processes undergone by electrons of the two-dimensional gas cause a broadening of the Landau levels. This broadening effect has been theoretically investigated by Ando [184–186].

A completely different effect is caused by a magnetic field applied parallel to the two-dimensional electron gas as discussed by Ando [172] and Beinvogl et al. [187]. In this case the energy levels are shifted upward proportionally to the increase in the extension of the wave function in the direction orthogonal to the two-dimensional gas plane. The energy separation among the levels also increases, with a possible magnetic emptying of the levels in the subband lying above the ground-state subband. Furthermore, the minimum of the energy is no longer placed at $k_x = 0$.

In the case of parallel magnetic fields with a very high intensity, or for low-intensity electric fields, the electrons with $k_x < 0$ are confined at the interface (for heterostructure systems) mainly by the magnetic field rather than by the electric field.

For one-dimensional structures, the effects of a magnetic field will depend on the width, w, of the channel compared with the length, l:

(a) If $\vec{B}$ is normal to the plane of a two-dimensional electron gas and $l_B < w$, the system will behave like a two-dimensional electron gas and therefore the relations found there will hold.

(b) If $l_B > w$, in the hypothesis of a parabolic lateral confinement potential, the energy levels become

$$\varepsilon_i = \left(i + \tfrac{1}{2}\right)\hbar\omega_T = \left(i + \tfrac{1}{2}\right)\hbar\sqrt{\omega_0^2 + \omega_C^2} \qquad (6.21)$$

where $\hbar\omega_0$ is the energy level separation at $B = 0$.

Berggren and Newson [188] extended calculations to more general cases of confinement potentials and magnetic fields.

Finally, a magnetic field applied parallel to the semiconductor interface will produce effects similar to those of the two-dimensional case, i.e., an increase in the energy level separation, which may magnetically depopulate higher subbands.

6.4. Mobility and Scattering Mechanisms in Two-Dimensional Systems

Theoretical study and experimental investigation of the carrier mobility in semiconductors have a fundamental relevance in electronic device design and fabrication. Faster operating speeds in devices require extremely high carrier mobility.

Even though extraordinary progress has been made [167, 168] in device design and fabrication, it has been quickly recognized that hot electron effects (Section 2.7) would limit device performance [189].

Theoretical models of electronic transport in two-dimensions have to explain the experimental results on the mobility dependence on temperature and carrier density. Moreover, in MODH the carrier mobility will also depend on the thickness of the undoped spacer layer.

Figure 30 shows a typical temperature dependence of the mobility in a MODH of GaAs/Al$_x$Ga$_{1-x}$As, for two spacer thickness (curves C and D). For comparison, in the same figure are plotted the mobility curves for two bulk GaAs crystals (curves A and B refer to a doped and an intrinsic semiconductor, respectively).

Figure 30 clearly indicates that far below 100 K curves B, C, and D begin to differ, and for the two-dimensional electron gas the mobility increases, lowering the temperature. Furthermore, the mobility enhancement is stronger in the MODH with the thicker spacer layer (curve D).

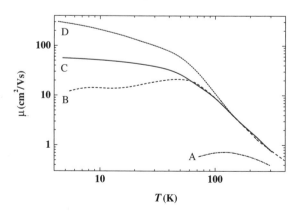

Fig. 30. Mobility as a function of the temperature in bulk GaAs and in a GaAs/Al$_x$Ga$_{1-x}$As MODH. (A) Doped bulk GaAs, $n = 10^{17}$ cm^{-3}. (B) High purity GaAs $n = 10^{14}$ cm^{-3}. (C) Two-dimensional electron gas with a 20-nm spacer. (D) Two-dimensional electron gas with a 40-nm spacer. Reprinted with permission from [175], ©1989, Institute of Physics.

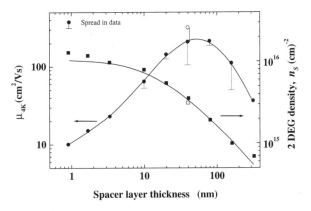

Fig. 31.   Mobility (left scale) and two-dimensional carrier density as a function of the spacer thickness of a GaAs/Al$_x$Ga$_{1-x}$As two-dimensional electron gas structure. The mobility measurement has been carried out at a constant temperature of 4.2 K. Reprinted with permission from [213], ©1986, Academic Press.

This last result may suggest that with increasing spacer layer thickness of the MODH, the mobility may be continuously enhanced. However, as shown by Figure 31, the mobility reaches a peak and then decreases, increasing the spacer layer thickness. In addition, Figure 31 indicates that the two-dimensional electron gas density continuously decreases (in the active conducting channel of the heterostructure) with the increase in the spacer thickness. This last fact represent an inconvenience, because it limits the maximum operating current in a MODH-based device.

The main features of the curves illustrated in Figure 31 have been observed by other workers [190, 191] and for other semiconductors [192].

To explain the experimental results illustrated in Figures 30 and 31, it is necessary to consider the different collision mechanisms that may be active in a bidimensional structure. Some of them are also present in the same bulk (i.e., three-dimensional) material, even if they could be modified by the specific two-dimensional nature of the carrier distribution, whereas other processes are peculiar to systems of reduced dimensionality.

For the well-known and well-investigated GaAs/Al$_x$Ga$_{1-x}$As system, several scattering mechanisms have been suggested [193–196], but analysis of the experimental data has shown that dominant processes are exerted by [195, 197–201] (1) optical and acoustic phonons, (2) remote impurities, and (3) neutral impurities. With the exception of remote impurity scattering, the theory of which has been developing, the above-mentioned mechanisms give a good explanation of the experimental results.

6.4.1. Theory of Two-Dimensional Scattering

In a bidimensional system the charge carriers are free to move in a plane, but they are confined in the direction perpendicular to that plane. Only in the plane of motion will electron momentum be conserved. If $\vec{k}_1$ and $\vec{k}_2$ are the initial and the final electron momentum, the transition rate between these

states, $W(1, 2)$, may be computed by standard techniques, if the form of the two-dimensional matrix element for the different collision mechanisms is known.

For quasi-elastic scattering, the momentum relaxation time of state 1 has the form

$$\tau^{-1}(k_1) = \int W(1, 2)(1 - \cos \alpha) \mathrm{d}^2 k_2 \qquad (6.22)$$

where α is the angle between $\vec{k}_1$ and $\vec{k}_2$.

For several scattering mechanisms acting simultaneously, the total momentum relaxation time is given by

$$\frac{1}{\tau_{\mathrm{m, Tot}}} = \sum_i \frac{1}{\tau_{\mathrm{m},i}} \qquad (6.23)$$

where $\tau_{\mathrm{m},i}$ is the momentum relaxation time of the collision process i. To find the expression for the mobility μ (see Eq. (2.43)), the expression (2.22) must be averaged over the energy:

$$\langle \tau_{\mathrm{Tot}} \rangle = \left(\int \tau_{\mathrm{Tot}} \varepsilon \left(\frac{\partial f(\varepsilon)}{\partial \varepsilon} \right) \mathrm{d}\varepsilon \right) \left(\int \varepsilon \left(\frac{\partial f(\varepsilon)}{\partial \varepsilon} \right) \mathrm{d}\varepsilon \right)^{-1} \qquad (6.24)$$

In practice, the calculation of (6.24) may be difficult, and several approaches have been suggested, based on the Matthiessen rule [195, 201], the iterative solution of Boltzmann equation [202, 203], and the memory function [196, 204, 205].

In the following we consider, as a two-dimensional prototype system a modulation doped heterostructure based on a GaAs/Al$_x$Ga$_{1-x}$As structure, which represents a system that is well known both theoretically and experimentally. This fact does not change the general conclusions that will be derived.

6.4.2. Optical Phonon Scattering

A main scattering mechanism in polar semiconductors, below room temperature, is by polar optical phonons. Figure 32 reproduces the temperature dependence of the electron mobility in GaAs/Al$_x$Ga$_{1-x}$As MODH, indicating the good agreement between theory and experiment [201]. Figure 32 also shows that deformation potential (see Sections 2.6.4 and 2.6.7) and piezoelectric scattering (see Sections 2.6.6) are negligible.

Similar results have been reported so far for a n-type MODH made of a Si strained layer cladded by Si$_{1-x}$Ge$_x$ alloy layers [46], the temperature dependence mobility of which is shown in Figure 33.

Even though the relaxation time approximation, reported before, is not strictly applicable for optical phonon scattering, different calculation methods [199, 206–208] produce slightly different results. Moreover, Lee et al. [195] and Walukiewicz et al. [201] made an approximation of the collision process, using the calculations for high-purity GaAs. In particular, they make the approximation of considering bulk GaAs phonons for Al$_x$Ga$_{1-x}$As, because the two materials have very similar dielectric and elastic constants.

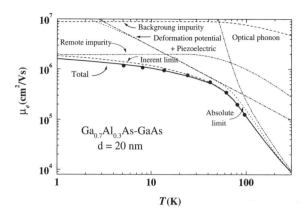

Fig. 32. Electron mobility as a function of the temperature for a GaAs/Al$_x$Ga$_{1-x}$As heterostructure. Points: experimental data for a carrier concentration of 3×10^{11} cm^{-2}. Curves: calculated mobilities, assuming a concentration of ionized impurities of 8.6×10^{11} cm^{-3}. Reprinted with permission from [201], ©1984, American Physical Society.

6.4.3. Acoustic Phonon Scattering

Acoustic deformation potential and acoustic piezoelectric scattering (see Sections 2.6.4 and 2.6.6) mechanisms have similar strengths in polar semiconductors. Theoretical and experimental work has been done to separate the two acoustic modes. The piezoelectric contribution is dominant at low carrier densities [209], whereas the opposite holds for the deformation potential scattering. The deformation potential constant, Ξ, has been calculated by several authors [201, 209–212], with a still uncertain value ranging from about 7 eV to 14 eV.

6.4.4. Remote Impurity Scattering

As already seen in Section 2.6.2, ionized impurity scattering is the main collision process at low temperatures. An expression of low-temperature mobility, limited by remote ionized impu-

rity scattering, was obtained by Lee et al. [195] for a single heterojunction two-dimensional electron gas,

$$\mu = 12.5 \left(\frac{n_S}{10^{20}} \right)^{3/2}$$

$$\times \left\{ \left(\frac{\rho_I}{10^{24}} \right) \left[\left(\frac{10^{-8}}{L_0} \right)^2 - \left(\frac{10^{-8}}{L_1} \right)^2 \right] \right\}^{-1} \quad (6.25)$$

where μ is expressed in cm^2/Vs, n_S in cm^{-2}, and ρ_I in cm^{-3}. In the previous equation (6.25) L_0 and L_1 are the distances (in cm) from the electron plane to the near and far edges of the depletion region in the doped Al$_x$Ga$_{1-x}$As, respectively.

Equation (6.25) underestimates the mobility in high-quality heterostructures. The nature of the terms appearing in (6.25) can be explained as follows:

(1) $n_S^{3/2}$ dependence of the mobility. In a two-dimensional degenerate system ($T \to 0$) the carrier density is proportional to ε_F (6.13) and the relaxation time for ionized impurity scattering varies as $\varepsilon^{3/2}$ (2.69).

(2) As already seen for impurity scattering (both neutral and ionized) in bulk (i.e., three-dimensional) material, the mobility is inversely proportional to the concentration of the scatterers (see eqs. (2.65) for neutral and (2.71 and 2.73) for ionized impurities, respectively).

(3) The last terms in the square brackets are a consequence of the Coulomb interaction ($F_e \propto r^{-2}$) and of the uniform distribution of impurity ions in the doped region (Al$_x$Ga$_{1-x}$As).

Figure 34 reproduces the experimental power law exponent, γ, for the dependence of μ on n. The typical values of γ range from 1 to 1.7, so that the term (1) is approximately correct, assuming that γ may increase slightly with spacer layer thickness [199].

It must be observed that several factors can affect remote impurity scattering, such as impurities in the surface depletion region [213–215] and the presence of shallow and deep impurity levels (DX centers) [213].

Finally, a comment on point (2): some authors believe that the hypothesis of independent remote scatterers cannot hold when the spacer layer thickness is higher than the average

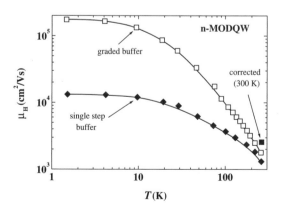

Fig. 33. Temperature dependence of Hall mobilities in n-type Si/Si$_{1-x}$Ge$_x$ MODH with the structure shown in Figure 28. The two curves indicate the effects on the mobility of the growth and the choice of the Si$_{1-x}$Ge$_x$ buffer layer (see Fig. 28). Top curve: Si$_{1-x}$Ge$_x$ buffer grown by continuously increasing the Ge content in the SiGe alloy buffer (graded composition); bottom curve: Si$_{1-x}$Ge$_x$ buffer layer grown at a constant composition. Reprinted with permission from [46], ©1997, Institute of Physics.

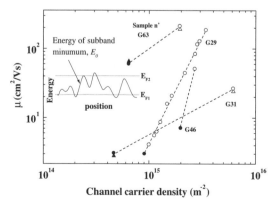

Fig. 34. Mobility as a function of the charge carrier concentration for several GaAs/Al$_x$Ga$_{1-x}$As heterostructures. Reprinted with permission from [251], ©1986, American Institute of Physics.

distance among the dopant ions. However, the different techniques adopted [216–219] predict mobility values higher than those experimentally observed.

6.4.5. Neutral Impurity Scattering

Background or neutral impurities present in a heterostructure are the result of an avoidable contamination of the material due to several factors. The density of background impurities is normally lower than 10^{14} to 10^{15} cm^{-3}, but their presence near to the two-dimensional electron gas may not be negligible, especially when the spacer layer is wide. In fact, when the thickness of the spacer is increased, the contribution of remote ionized impurities to the collision process tends to vanish, whereas that of background impurities becomes dominant. Because the two kinds of impurities have the same energy dependence of the momentum relaxation time [206], a decrease in n, which increases the spacer thickness, will result in a reduction of the mobility due to neutral impurities. This fact explains the behavior of the mobility as a function of spacer thickness (Fig. 31) and the observation of a peak.

The density of neutral impurities at which they are the dominant collision process is inversely proportional to the spacer layer thickness: the higher the density is, the narrower the spacer will be.

6.4.6. Other Scattering Mechanisms

Under particular physical conditions several other collision processes may become relevant, for example, at high carrier concentrations or in inverted heterostructures (i.e., GaAs on Al$_x$Ga$_{1-x}$As). Some of these scattering processes are caused by interface roughness, alloy, impurity at interfaces, random change in alloy composition, etc. A brief description has been given by Harris et al. [175].

6.5. Quantized Hall Effect

In 1985, the Nobel Prize in physics was awarded to Klaus von Klitzing for the discovery of the quantum Hall effect (QHE) [162], one of the most remarkable physical phenomena to be discovered in solid-state physics in recent years. This effect consists of the appearance of discontinuities in the Hall resistance, R_{xy}, of a two-dimensional electron gas when a magnetic field is applied normally to the two-dimensional electron gas plane. The R_{xy} constant values (Hall plateaus) are submultiples of the fundamental quantity h/e^2.

The QHE may be observed in the quantum limit, at very low temperatures and with the further condition of a sufficiently high concentration of two-dimensional charge carriers.

What we have above called QHE is the so-called integer quantum Hall effect (IQHE), to be distinguished from the fractional quantum Hall effect (FQHE) discovered in 1982, soon after the IQHE, by Tsui et al. [220].

The FQHE shares very similar underlying physical characteristics and concepts with the IQHE, for instance, the two-dimensionality of the system, the quantization of the Hall

resistance in units of h/e^2 with a simultaneous vanishing of the longitudinal resistance, and the interplay between disorder and the magnetic field, giving rise to the existence of extended states.

In other respects, they encompass entirely different physical principles and ideas. In particular, whereas the IQHE is believed to be a manifestation of the transport properties of a noninteracting charged particle system in the presence of a strong magnetic field, the FQHE results from a repulsive interaction between particles, giving rise to a novel form of a many-body ground state. Moreover, the mobile nonlocalized elementary excitations must carry a fractional charge, which is thought to obey unusual statistics that are neither Fermi nor Bose–Einstein.

In the next two sections, the two effects are briefly introduced, together with a description of the theoretical model suggested by Laughlin to explain the IQHE [221].

6.5.1. Integer Quantum Hall Effect

To explain the QHE it is necessary to consider the electron gas system in the quantum limit state ($\hbar\omega_c \gg k_B T$) (see Sections 2.2 and 5). Under these conditions the thermal energy and the other factors causing the broadening of the energy levels are strongly reduced. The levels are progressively filled by electrons, following the Pauli principle.

Depending on the $\vec{B}$ intensity, it may be possible that some degenerate states of the Landau level, l, are completely occupied, whereas levels at higher energies are completely empty. Under these conditions the Hall resistance is independent of the material parameters.

Recalling relation (3.4) for the Hall field, $E_y = V_y/y$, and considering that in a two-dimensional system the volume density, n, of charge carriers reduces to a surface density, n, and the current density, j_x, to a current for unit length, I_x/y, relation (3.4) for a two-dimensional electron gas can be rewritten as

$$V_y = I_x \frac{B_z}{n_S e} = I_x R_{xy} \tag{6.26}$$

In the previous relation, if n_S is replaced by a multiple integer, p, of n_L (Eq. (6.17)), then Eq. (6.26) becomes

$$R_{xy} = \frac{B_z}{n_S e} = \frac{B_z}{p n_L e} = \frac{h}{p e^2} \tag{6.27}$$

where p is the integer number of Landau levels completely filled by electrons. The previous relation is called the integer quantum Hall effect (IQHE), to distinguish it from the fractional QHE (FQHE), in which the p values may be fractions with an odd value in the denominator. However, a few considerations are necessary:

(a) Equation (2.25) is rigorously satisfied only by discrete values of the field $\vec{B}$ or of n_S and, therefore, they could not be measured.

(b) In the hypothesis of a complete filling of a Landau level n (quantum limit), the electrons cannot be scattered, and, therefore, they cannot change their state. Hence, the resistivity vanishes along the direction of charge motion.

If Eq. (6.27) is satisfied and in the quantum limit regime, it follows (point b) that

$$\rho_{xx} = 0 \qquad (6.28)$$

which means that the potential drop along the current direction (x) will be zero. Once again this condition is fulfilled only by discrete values of $\vec{B}$ or of n, which could not be experimentally measured. Despite these theoretical hindrances, the QHE has been observed experimentally with a precision in the Hall resistance values of better than 1 part in 10^7.

The QHE was first observed in an inversion layer of a Si-MOSFET (metal oxide semiconductor field effect transistor) by von Klitzing et al. [162]. In this system an electric potential, V_g, applied to the metal contact of the gate of a MOS structure, lowers the energy bands in the proximity of the oxide layer. Depending on the V_g value, the conduction band minimum of the p-type Si can be lowered below the Fermi level, in a thin layer. This bidimensional layer (called the inversion layer) will behave like a narrow potential well for electrons. Charge carriers are only free to move in this two-dimensional layer, parallel to the interface with the SiO_2 film. If a magnetic field is applied perpendicular to the two-dimensional electron gas and the temperature is lowered below 1–2 K, it is possible to observe the curves shown in Figure 35.

The two curves in Figure 35 depict the behavior of the Hall potential, U_H, and of the potential drop, U_{pp}, along the current direction as a function of the gate voltage. The other experimental parameters, such as the intensity of the magnetic field, the sample temperature, and the measurement current, were kept constant at 18 T, 1.5 K, and 10^{-6} A, respectively. The U_H curve indicates the presence of plateaus, and correspondingly, U_{pp} goes to zero (Eq. (6.28)). The plateaus are caused by the progressive filling of the Landau levels by electrons, as the potential V_g is increased.

A detailed analysis of the results obtained by von Klitzing et al. [162] indicated the excellent accuracy between theoretical predictions (Eq. (6.27)) and experimental values for different samples and geometries. This is a consequence of the fact that the equation describing the QHE holds for the sample resistance and not for the sample resistivity, which is a material-dependent parameter. If it were the resistivity rather than the resistance that was quantized, then precision measurements would be impossible because one would have to invoke assumptions of a homogeneous medium with a well-defined geometry to infer the microscopic resistivity from the macroscopic resistance. It is a remarkable feature of the QHE that this is not necessary.

The value of $R_{xy} = (6453.3 \pm 0.1)\,\Omega$ corresponds to the total occupation of the Landau level $l = 0$ ($p = 4$, for spin degeneracy and for the further degeneracy due to two equivalent minima in Si). The value $R_{xy} = (3226.2 \pm 0.1)\,\Omega$ is determined by the complete filling of the level $l = 1$ ($p = 8$). As a further check of Eq. (6.27), a value of $R_{xy} = (12906 \pm 1)\,\Omega$ is found experimentally and is assigned to the filling of the level $l = 0$ with only one spin ($p = 2$); it demonstrates the spin splitting of the Landau levels.

Since its discovery the QHE has being observed in two-dimensional electron and hole gases of several semiconductor heterostructures: in a Si inversion layer of a Si-MOSFET [162, 222–226]; in Si/SiGe MODH [170, 227–229]; in several III–V group heterojunctions made of binary, ternary, and quaternary alloys such as $GaAs/Al_xGa_{1-x}As$ [164, 165, 230–232], $In_xGa_{1-x}As/InP$ [233–234], InAs/GaSb [235–237], $InP/Al_xIn_{1-x}As$ [238, 239], $In_xGa_{1-x}As_yP_{1-y}/InP$ [240], and $In_xGa_{1-x}As/In_yAl_{1-y}As$ [241]; and in an inversion layer of a $Hg_xCd_{1-x}Te$ MISFET (metal insulator semiconductor field effect transistor) [242, 243].

However, even though the features of the QHE are practically the same in all of the materials and structures, there are differences between the experiments carried out on an inversion layer in a MOS or in a MIS structure and those on a heterostructure (i.e., $GaAs/Al_xGa_{1-x}As$). In particular, in heterostructures the two-dimensional charge carrier concentration, n, is determined by the dopant concentration and by the temperature. In a MOS or in a MIS device, as said before, the carrier density can be changed by the gate voltage. Hence to observe the QHE in semiconductor heterojunctions (to check the relations (6.27) and (6.28)) it is necessary to measure R_{xy} and ρ_{xx} as a function of the B intensity, whose variation makes it possible to fill the Landau levels.

The best results for the QHE have been obtained in $GaAs/Al_xGa_{1-x}As$ heterostructure [164, 165, 230–232], where the two-dimensional electron gas is formed in the GaAs, which does not suffer from the conduction band degeneracy, and it has an effective mass about 3 times lower than that of Si [244].

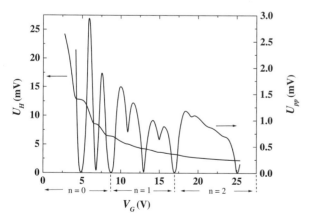

Fig. 35. Integer quantum Hall effect measured in MOSFET structure. U_H and U_{PP} represent the Hall voltage and the voltage drop, respectively, measured as a function of the gate voltage V_g. The measurements were carried out at $T = 1.5$ K with a constant source-drain current of $I = 10^{-6}$ A. The magnetic field was fixed at $B = 18T$. Reprinted with permission from [162], ©1980, American Physical Society.

Consequently, the QHE will be observed at a lower intensity of the magnetic field, and it will persist at higher temperatures.

A theoretical interpretation of the experimental results from a microscopic point of view is still difficult. Another relevant problem is the experimental observation of the existence of a continuous range of values for B and for n (instead of discrete ones) for which the relations (6.27) and (6.28) are satisfied.

The formation of a staircase in the Hall resistance, when B is decreased (the filling factor $\nu \propto B^{-1}$; see Eq. (6.18)), implies a sort of delay in the filling of the $(p+1)$th level when the pth Landau level has been completely occupied. This fact implies that some electrons will be localized in states with energy levels between two Landau levels. These electrons will not take part in the conduction process or in the Hall effect. In this way the Fermi level is pinned between two Landau levels as the B intensity is decreased, causing the appearance of the Hall plateaus.

An accurate description of the QHE theory and a review of the theoretical models proposed to explain experimental results has been given by Prange and Girvin [245].

Many attempts have been made to explain theoretically the main experimental features of the QHE. Ando et al. [163] made theoretical calculations about the influence on magnetotransport, at low temperature, of noninteracting spin-split Landau levels. Their work failed to reproduce the Hall resistance plateaus, whereas the calculated magnetoresistance, ρ_{xx}, exhibited very sharp minima instead of broad ones, as is observed in QHE experiments.

To overcome these difficulties Aoki and Ando [246] assumed a localization mechanism of electrons based on the presence of localized states surrounding a Landau level (broadened Landau level). Their working hypothesis was that in a strong magnetic field only the center of a broadened Landau level contains mobile states, whereas the states located farther away (in the wings of the distribution) are localized. In this way they reproduced the Hall resistance plateaus as long as the Fermi level remained pinned in the wings.

One of several localization mechanisms is the Anderson localization (semiclassical localization) due to disorder. In the extreme quantum limit ($B \to \infty$) electrons move along equipotential lines, provided that inelastic scattering is absent and the electric potential changes on length scales longer than that of l_B (Eq. (6.19)). In these conditions, electrons moving in cyclotron orbits will be made to drift, by the Lorentz force, in a direction normal to B, if their energy falls in the central region of the broadened Landau levels. These mobile carriers will drift along equipotential lines, which percolate through the sample. In contrast, electrons occupying energy states in the wings of the Landau level will be spatially localized.

In their work Woltjer et al. [247] explained the QHE by resistivity inhomogeneity, without invoking any localization effect. In particular, in the approach proposed by Ando et al. [163] they used a resistivity tensor, depending on the local filling factor, to describe the magnetotransport properties. Inhomogeneity of carrier concentration of a few percent causes inhomogeneity in the filling factor, which produces large resistivity changes near integer filling factors.

Considering that the Kirchoff laws govern the current distribution, quantization effects will be observed in the range of magnetic field intensities where a path with low resistance (integer filling factor) exists between the current contacts. In other words, the current will preferentially flow along these low-resistance paths. The plateaus in the Hall resistance are determined by ρ_{xy} at the integer filling factors where the minima in ρ occur.

Among the theoretical models proposed, that of Laughlin [221] is considered a fundamental contribution to the theory of the integer quantum Hall effect. In this theory, Laughlin [221] considered a closed ribbon in which a current can flow. The change in the ribbon current, i, is linked to the variation in the electronic energy, $\Delta\varepsilon$, produced by a change in the magnetic flux, $\Delta\Phi$, through the ribbon:

$$i = c \frac{\Delta\varepsilon}{\Delta\Phi} \tag{6.29}$$

If any electron determines an elementary variation in the flux $\Delta\Phi = h/e$ and if the energy change is caused by the passage of n electron from side to side in the ribbon (i.e., $\Delta\varepsilon = neV_H$), then relation (6.29) becomes

$$i = neV \frac{e}{h} = \frac{ne^2 V_H}{h} \tag{6.30}$$

The quantization of the Hall resistance will hold provided the Fermi level is pinned in the region of localized states and the energy of the Landau levels remains constant. This model has been slightly modified by Giuliani et al. [248] and by Avron et al. [249], who introduced a disordered potential produced by impurities.

6.5.2. Fractional Quantum Hall Effect

The basic phenomenon of the fractional quantum Hall effect is similar to the IQHE. It was discovered in 1982 by Tsui et al. [220] in high-quality GaAs/Al$_x$Ga$_{1-x}$As heterostructures. Experimentally, it is found that at very low temperatures and in a high-quality film (with a high carrier mobility at $B = 0$), the application of a very intense magnetic field produces Hall plateaus and deep minima structures in ρ at fractional fillings $\nu = 1/3$ [220] and $\nu = 2/3$ [165] of the lowest Landau level, in a manner analogous to that of the development of the QHE at integral fillings, as shown in Figure 36.

The existence of the QHE at fractional fillings could not be understood within the framework of the IQHE theory. The possibility of fractionally charged excitations was immediately suggested.

To observe the QHE it is necessary to define criteria for the delineation of the quantum regime, separating it from the classical regime. These criteria are listed below.

(a) Using a semiclassical picture, where the effect of disorder is to give rise to a scattering time, τ_0, the quantum condition is

$$\omega_C \tau_0 \gg 1 \tag{6.31}$$

(b) The second condition regards the minimum temperature for the onset of the FQHE, given by

$$\Delta \geq k_B T \tag{6.32}$$

$$\Delta \simeq \left(\frac{Ce^2}{2\chi}\right)\sqrt{L} \tag{6.33}$$

where Δ arises from the Coulomb repulsion between particles and represents an energy gap, assumed to exist, separating the ground state and its current-carrying excitations; C is a numerical constant, and $\sqrt{L}$ is the average interparticle spacing.

(c) A third condition is needed. It is crucial that the interaction between particles dominates over the effect of disorder. A reasonable estimate is given by the requirement that the energy gap Δ be larger than the disorder potential fluctuation $\langle(\Delta V)^2\rangle^{1/2}$:

$$\Delta \geq \langle(\Delta V)^2\rangle^{1/2} \approx \hbar\tau_0 \tag{6.34}$$

where $\hbar\tau_0$ is a rough estimate of the mean potential fluctuation. For both currently available semiconductor heterostructures and magnetic fields, Δ is about 30 times smaller than $\hbar\omega_C$.

With respect to the requirements (a), (b), and (c), the FQHE has been observed in a reduced number of semiconductor systems, when compared with the IQHE. Some of these systems are the inversion layers of Si-MOSFETs [250], GaAs/Al$_x$Ga$_{1-x}$As [165, 220, 251, 252], In$_x$Ga$_{1-x}$As/InP [253], and SiGe/Si [229, 254]. A detailed list of experimental works has been given by Prange and Girvin [245].

Let us examine the suitability for studying the FQHE, based on conditions illustrated before.

First, we examine the temperature condition. For the FQHE, the carrier mass plays no role, except when Landau level mixing is included. The dielectric constant of different materials does not change by more than about 30%.

The typical Coulomb energy scale is on the order of 4 K (0.3 meV) at intensities of the magnetic fields of about 20 T, corresponding to typical carrier densities of about 2×10^{11} cm^{-2}.

A fundamental requirement to consider is the exceedingly stringent sample quality (i.e., very high mobility at $\vec{B}=0$). For example, it appears from experiments that an electron mobility on the order of 10^5 cm^2/Vs is the minimum quality necessary in n-type GaAs/Al$_x$Ga$_{1-x}$As samples with a charge density of about 1.5×10^{11} cm^{-2}. Moreover, because the scattering time, τ_0, is the relevant parameter for the FQHE (rather than the mobility, μ, which involves an extra factor of $1/m$ for p-type materials, the mobility required is around five times smaller than in n-type materials.

The FQHE is a manifestation of the peculiar physics associated with a two-dimensional charged system in a magnetic field and the formation of a new type of correlated ground state in such a system. The phenomenon occurs as follows: a two-dimensional gas of charged particles (electrons or holes) of density n is embedded in a nearly impurity-free, static, neutralizing background. At low temperatures, with increasing magnetic field, the Hall resistance plateaus appear at quantized values of h/fe^2, when the field $B = nh/fe$ or for a filling factor $\nu \approx p$, where f is a rational fraction with an odd denominator. When these conditions are observed the ρ_{xx} vanishes. Depending on the amount of disorder and the temperature, a fractional effect corresponding to a given value of f may or may not be observed. In general, the experiments have shown that the FQHE occurs in multiples of $f = p/q$, where p and q are integers and q is odd. Some of the observed fractions are 1/3, 2/3, 4/3, 5/3, 7/3, 8/3, 1/5, 2/5, 3/5, 4/5, 7/5, 8/5, 2/7, etc. [245], which have been measured with a very high accuracy. The accuracy in the f values ranges from 3 parts in 10^5 (for $\nu = 1/3$ and $\nu = 2/3$) to 2.3 parts in 10^4 (for $\nu = 2/5$).

Several other properties have been observed in experiments on the FQHE, such as the activated behavior in certain temperature ranges of $f = 1/3$, 2/3, 4/3, and 5/3 or the linear current–voltage (I–V) characteristics at 1/3 and 2/3 down to an electric field of about 10^{-5} V/cm in the sample. A detailed description of the other results is given by Prange and Girvin [245].

It is apparent that the FQHE occurs in multiple series, involving fractions of odd denominator only. Störmer et al. [165] first noted the emergence of multiple series of p/q. The current understanding of the p/q series is based on a hierarchical model, consisting of a sequence of successive ground states that are derived from a parent state. Different hierarchies based on differing parent states have been proposed by several authors [245]. The prevailing theory is based on a parent ground state proposed by Laughlin [255] for $\nu = 1/q$, originally intended to explain the $\nu = 1/3$ effect.

The explanation of the FQHE is beyond the scope of this review chapter. A first and satisfactory theory has been given by Laughlin [245, 255]. He calculated that the state with a minimum energy for a system of interacting electrons

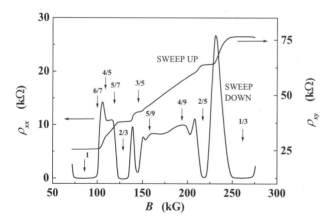

Fig. 36. Example of the fractional quantum Hall effect measured at 90 mK in a GaAs/Al$_x$Ga$_{1-x}$As heterostructure. RIGHT SCALE: Hall resistivity, ρ_{xy}; LEFT SCALE: resistivity along the Hall bar length, ρ_{xx}. The sample is gated at 700 V to a density of 2.13×10^{11} cm^{-2}. Several fractional filling factors are visible (shown at the top of the frame). Weak structures near 4/9 and 5/9 also appear in the ρ_{xx} curve. Reprinted with permission from [297], ©1984, American Physical Society.

in a magnetic field is an incompressible quantum fluid that has quasi-particles with fractional charges as excitations. The fermion character of the wavefunction of these spin-polarized Laughlin states prohibits the existence of even-denominator fractions. However, recent experiments have shown a quantization filling factor of 5/2 [256, 257]. This effect could be explained as a mixed-spin Laughlin state that can occur in the higher Landau states. Theoretical calculation and experimental investigations are still being conducted to understand this result.

7. EXPERIMENTAL TECHNIQUES

Good reviews of the electrical characterization of semiconductors for majority and minority carriers can be found in [8, 12, 258, 259]. A complete characterization of electrical conduction in a semiconductor needs to measure the free carrier mobility as well as its density. The mobility is limited by two groups of different scattering processes: intrinsic ones due to lattice vibrations (phonons) and processes associated with defects and impurities in the crystal lattice (see Section 2.6). They often can be separated by analyzing the temperature dependence of the mobility. Techniques for measuring electrical properties can be used, covering a range of methods involving specially prepared samples with ohmic contacts, probe methods, and methods requiring no electrical contacts at all. This is a possible order of accuracy and reliability in the inverse order of ease and convenience. In general, ohmic contacts and probe methods render the sample unsuitable for further use, whereas the contactless methods have no deleterious effects.

Of the probe methods, spreading resistance has a much higher spatial resolution ($\sim 10 \ \mu$m) than the four-point probe techniques (~ 1 mm) but is not an absolute method (i.e., it requires calibration using standard samples). The combination of Hall effect and resistivity measurements made on the same sample configuration probably represents the most commonly employed characterization method. By extending the measurements over a suitably wide range of temperatures, it is possible to derive values for N_D, N_A, and the ionization energies, ε_D and ε_A. A common assumption concerns the spatial uniformity of the electrical properties. In practice, considerable attention must be devoted to the methods of measuring the spatial variation of carrier density and mobility, establishing their "profiles."

We can classify measurement techniques into several groups. One uses bar (see Fig. 29) or bridge samples and van der Pauw geometry [260], another uses the four-point probe and spreading resistance methods, and a third involves methods using power dissipation from radio frequency or microwave sources. The experimental aspects are simple, and interest usually centers on interpreting the data to obtain the desired accuracy in the measured resistivity. This involves detailed consideration of sample and contact geometry. For example, measurements made on this epitaxial films may need correction for conduction in the substrate, and the interpretation of the probe measurements requires a reconstruction of the current distribution.

7.1. Resistivity of Samples with Ohmic Contacts

The need for "ohmic" contacts is stringent when one is using two terminals, but even with four terminals, it is advisable to use good ohmic contacts wherever possible. A broad account of ohmic contacts is given in [261].

7.1.1. Bar Samples

The most direct method of meausring semiconductor resistivity makes use of a bar sample (Fig. 29) with a uniform cross section along its length. Current passes through the bar by way of ohmic contacts covering the two ends of the bar, and, under the assumption of uniform current distribution, the voltage drop is measured between two-faced side arms. This geometry avoids arithmetical corrections in converting measured resistance into resistivity by Ohm's law.

7.1.2. The van der Pauw Method

The use of bar-shaped samples requires complicated sample preparation. The theorem published by van der Pauw [260] allows measurement of the resistivity in a uniform, plane sample with four line contacts placed on its edge. It should be uniform in thickness and doping level. However, it may be of arbitrary shape, but it must contain no holes. The resistivity is given by

$$\rho_r = (\pi t/\ln 2)[(R_{ABCD} + R_{BCDA})/2]f_{cor}(R_{ABCD}/R_{BCDA}) \tag{7.1}$$

A, B, C, and D, are the four line edge contacts; R_{ABCD} is the resistance measured with the current flowing through the first two labeled contacts (A and B) and the voltage measured between the other two. The function f_{cor} is a correction term depending only on the ratio between the two resistances. Murri and co-workers used the van der Pauw configuration extensively for electrical characterization of III–VI layered semiconductors, amorphous silicon, and amorphous gallium arsenide [262–267].

Equation (7.1) is concerned essentially with the sheet resistance of the sample, defined as the resistance between opposite edges of a square of arbitrary dimension, L. It is independent of L and is therefore a property of the sheet. Its dimensions are simply ohms; if only to avoid confusion, one commonly refers to it as ohms per square. The concept of sheet resistance is important for thin films, where typical lateral dimensions are much greater than the film thickness.

To make a van der Pauw measurement, bulk samples can be rapidly contacted by spring-loaded tips. However, in many cases this form of contact may show too high a resistance. It then becomes necessary to use evaporated or sputtered planar contacts, but, in these cases, the ideal van der Pauw geometry cannot be achieved. In these cases it becomes important to consider alternative contact geometries.

Errors can also arise from a number of causes, in addition to the sample nonuniformity and contact problems: Joule heating, photo effects, current leakage through the substrate and surface, and interface depletion effects (for epitaxial films). The first two can be eliminated by making measurements in dark as a function of sample current. The other two are more difficult to control. For example, a semi-insulating substrate may conduct significantly at elevated temperatures, showing activation energy of half the band gap. Possible substrated p-n junction breakdown can be eliminated by reducing the measuring voltage. Surface leakage can introduce serious errors into these measurements and must be closely controlled.

7.1.3. The Four-Point Probe

The four-point probe technique represents a development of the "bar" method described in Section 7.1.1, where voltage measuring arms are replaced by point-contact probes. As no current flows in the voltage arms, the need for ohmic contacts is no longer stringent. Resistivities between 0.1 and 10^3 Ωcm can be measured to an accuracy of about 1%. The standard four-point probe linear configuration uses the two outer probes for flowing current, and the open-circuit voltage is measured between the inner pair of probes. The resistivity is computed from Ohm's law, introducing a right proportionality constant between measured resistance and material resistivity for nonuniform current flow.

Practical problems concern the type of probe, its spring loading, the accuracy of probe spacing, contact between probes and semiconductor, preparation of the semiconductor surface, magnitude, and control of current and DC or AC signals. Possible sources of error and factors affecting the ultimate accuracy of the measured resistivity are sample geometry, heating by the measurement current, minority carrier injection, current leakage in the voltmeter circuit, and probe mechanical configuration.

Experimentally, four-point probe measurements on thin films are not different from those on bulk crystals. Attention must be paid to the fact that the required condition thickness, t, much greater than the probe spacing, s, can no longer be satisfied. In reality, t can be comparable with s. An additional complication arises when the film is supported by a substrate of high resistivity but, being thicker, with a sheet resistance similar to that of the layer. A correction factor depending on both the ratio (t/s) and the substrate resistivity must be used to calculate the actual resistivity of the layer. In practice, there are many possible combinations of resistivities and layer thicknesses.

7.2. Galvanomagnetic Effects

Experimental data on the Hall effect and resistivity over the widest temperature range and the electron or hole mobility provide an immediate indication of material quality, giving information concerning impurities, imperfections, uniformity, scattering mechanisms, etc. [8, 12, 258, 259].

7.2.1. The Hall Effect

Assuming a bar sample of thickness t, length l, and width h (see Fig. 29), Eqs. (2.36), (2.43), (3.5), (3.11), (3.15), and (3.16) define the fundamental material parameters in terms of experimentally measurable quantities, all under the assumptions of defined geometry of the sample, uniformity of electrical properties, and temperature. Moreover, side arms for the four contacts must be narrow enough not to perturb the uniform voltage drop along the bar. The length-to-width ratio must be at least 4, to avoid shorting the Hall voltage by the end contacts. These must also be either noninjecting or sufficiently far from the other side arms, to prevent injected minority carriers from recombining before reaching them. Experimentally, it is the usual practice to take averages of the voltage differences obtained for both directions of current and, when measuring R, in a magnetic field. This minimizes errors arising from misalignment of the facing Hall side arms and from thermoelectric effects arising from temperature gradients within the bar.

A uniform current distribution and a well-defined electric field are assured by the geometrical regular bar sample, but today's Hall effect measurements are made in the van der Pauw [260] geometry. Equation (7.1) makes it possible to compute ρ with a total of eight values, doing the four permutations of the contacts and reversing the current for each set. Their average is taken as the best estimate of ρ_r. Care must be used if the individual measurements differ by more than a factor of 2. The Hall coefficient is measured with the use of diagonal pairs of contacts rather than adjacent pairs,

$$R = (t/B_z)\Delta R_{ABCD} \tag{7.2}$$

where ΔR_{ABCD} is the change in R_{ABCD} produced by the applied magnetic field B_z.

The Hall mobility is then given by Eq. (3.11). Again, it is standard practice to reverse the current and magnetic field and interchange current and voltage arms, obtaining R from an average of eight readings. Murri and co-workers extensively used the Hall effect for mobility and carrier density determination of III–VI layered semiconductors, amorphous silicon, and amorphous gallium arsenide [268–272].

Ultrasonic cutting for bulk material or "sandblasting" with alumina powder for epitaxial films is commonly used to prepare samples. Epitaxial films must be isolated either by growth on semi-insulating substrates or with the use of a p-n junction. In this last case, much attention must be paid to the effect of depletion regions on the effective sample thickness. An accurate knowledge of sample thickness is necessary in all cases when carrier density is measured, and the value derived for the Hall mobility is independent of sample thickness if the electric field inside the material is uniform. The importance of the Hall effect as a characterization method lies in its ability to measure the free carrier density (see Eq. (3.13)), when current is carried by a single type of carrier. It is not true for an intrinsic semiconductor, and it may not be true for the case of mixed conduction, where, for example, $p \gg n$ but $\mu_e \gg \mu_h$.

7.2.2. Measurement of Donor and Acceptor Densities

1. Analysis of R. A plot of $\log n$ vs. T^{-1} and a careful analysis of temperature-dependent Hall effect data give information on the total impurity density (in several situations considerably different from the measured carrier density) and on the identity of the dominant shallow species (according to its ionization energy). Considering for simplicity the case of an n-type semiconductor, at high temperatures, when $k_B T \gg (\varepsilon_c - \varepsilon_D)$, all of the donors are ionized and n tends toward the exhaustion regime,

$$n = N_D - N_A \qquad (7.3)$$

so giving directly $N_D - N_A$. At lower temperatures, where $N_a < n < N_a$,

$$n \approx (\beta N_C N_D)^{1/2} \exp[-(\varepsilon_C - \varepsilon_D)/2k_B T] \qquad (7.4)$$

and the slope of $\log n$ vs. T^{-1} gives approximately $(\varepsilon_C - \varepsilon_D)/2k_B$. At still lower temperatures, $n < N_A < N_D$,

$$n \approx [\beta N_C (N_D - N_A)] \exp[-(\varepsilon_C - \varepsilon_D)/k_B T] \qquad (7.5)$$

with a slope of $(\varepsilon_C - \varepsilon_D)/k_B$.

2. Analysis of mobility measurements. The total impurity density $(N_A + N_D)$ can also be obtained from measurements of the carrier mobility, because different scattering mechanisms show different temperature dependencies and therefore limit the mobility over different temperature ranges (Section 2.6 and Fig. 8).

The method is capable of good accuracy but is time-consuming, because of the relevant quantity of Hall effect and resistivity measurements over a wide range of temperatures and the need to obtain the appropriate theoretical fit. Another relevant and critical point is the necessity to evaluate most important parameters to be use in the equations for the different semiconductors.

3. Far-infrared magneto-absorption. The methods of measuring compensation described in the two previous points are commonly used. They can be criticized as being indirect estimates and depending on fitting data to fairly complicated equations. A much more direct method exists. Neutral donors (i.e., donors that have captured free electrons from the conduction band) may absorb FIR radiation in a resonant process. The electron is excited from its lowest bound state into an excited, still bound state, and the absorption is directly proportional to the neutral donor density. If an absorption measurement is made on an n-type sample at low temperature, with effective carrier freeze-out, the absorption is proportional to $(N_D - N_A)$. Doing the experiment under conditions where the absorption coefficient, α, is proportional to the density of neutral donors, we do not need to describe the details of the absorption process. The transmitted intensity, I_t, is given by [273]

$$(I_t/I_o) = [(1 - R^2)e^{-\alpha t}(1 + k^2/n^2)][1 - R^2 e^{-2\alpha t}] \qquad (7.6)$$

where $\alpha = 4\pi k/\lambda$ R is the surface reflection for the FIR radiation with free space wavelength $\lambda(\lambda - 100\ \mu m)$, and n and k are the real and imaginary parts of the refractive index. A typical value is $\alpha \sim 100\ cm^{-1}$, and, putting $t - 5\ \mu m$, Eq. (7.6) simplifies to

$$(I_t/I_o) = [(1 - R^2)e^{-\alpha t}] \approx (1 - R^2)(1 - \alpha t) \qquad (7.7)$$

It follows that

$$(1 - \theta) = (\alpha_d - \alpha_l)$$
$$= [(1 - R^2) - (I_{t_d}/I_o)]/[(1 - R^2) - (I_{t_d}/I_o)] \quad (7.8)$$

where the subscripts d and l for α refer to dark and under band gap illumination, respectively. Knowing R through the refractive index n, $\theta = (N_A/N_D)$ can be obtained from the measured intensity ratios. The principle of the measurement is simple. However, the signal-to-noise ratio requires a laser as source of FIR radiation, but lasers have a limited range of fixed wavelengths. Moreover, a screening effect due to the high density of free carriers changes the donor ionization energy and perturbs the FIR absorption. A pulsed light may be used and the FIR absorption measured when all surplus free carriers have recombined. This time delay is short compared with the time needed for electrons trapped on donor atoms to recombine, so the absorption measures N_D.

7.3. Inhomogeneity and Effective Sample Thickness

7.3.1. Nonuniform Current Distribution

Departures from the assumed uniform current distribution can arise in various ways that are obvious or subtle. An obvious one is the band bending at a surface or, in the case of epitaxial films, an interface resulting from a charge lying in a surface or in deep states in the substrate. The current still flows parallel to the surface, and it is possible to derive general expressions for the effective Hall coefficient and Hall mobility in terms of two parallel conducting slabs, within each of which current flows uniformly (two-layer model).

We obtain as final equations for apparent conductivity and Hall mobility,

$$\sigma = (n_s + abn_B)e\mu_s/(1 + a) \qquad (7.9)$$

and

$$\mu_H = \mu_s(n_s + ab^2 n_B)/(n_s + abn_B) \qquad (7.10)$$

where $a = t_B/t_s$, t_s and t_B are the thickness of the layer and substrate, respectively, and $b = \mu_B/\mu_s$ and n_s and n_B are the doping level of layer and substrate, respectively. Equations (7.9) and (7.10) give the conditions for accurate carrier density measurements on thin epitaxial films. Taking $b = 1$, Eq. (7.10) may be written as $n_B \gg an_s$ or $\rho_B \gg a\rho_s$. Typically, $a = 10^2$ to 10^3, so we require $\rho_s \geq 10^5 \rho_L$, for a percentage error less than 1%.

7.3.2. Band Bending

Band bending arises from Fermi level pinning at a surface (surface states) or at an interface (interface states or depp trapping levels in the material on one side of the interface). We can consider an epitaxial film (e.g., n type) of thickness t_L on a semi-insulating substrate containing N_t traps at energy ε_t. Depletion layers exist at the surface and interface (x_s and x_I thick, respectively), and the effective thickness for the epitaxial layer is given by $t_{eff} = t_L - x_s - x_I$. This value must be used in evaluating Hall and resistivity data. Errors can be important for thin epilayers, where it is difficult to make the correction, because the surface barrier height is rarely known accurately.

7.3.3. Bulk Inhomogeneities

These include random fluctuations of doping density, deep impurity or damage centers, dislocations, voids, metallic conducting regions, and grain boundaries in polycrystalline material. Hall effect measurements are widely used for characterizing all of these materials, and Orton and Powell [274] have reviewed the results. Intergrain potential barriers at the boundary region generally dominate transport properties, with the Hall mobility thermally activated,

$$\mu_H = \mu_{gb} \exp(-\phi_g / k_B T) \qquad (7.11)$$

where ϕ_g is the intergrain potential barrier (i.e., the band bending), which depends on doping level, and μ_{gb} is a mobility limited by grain boundary scattering (i.e., carrier mean free path $\approx$ grain diameter).

7.4. The Hall Scattering Factor

Of more immediate interest is the Hall scattering factor, r_H, defined by Eq. (3.13). $r_H = 1$ for a degenerate electron gas originates in the fact that only carriers with the same energy close to E_F contribute to conduction. The commonly made assumption $r_H = 1$ often leads to an error of 25% or more in carrier density. Using the right values of r_H computed for different scattering processes, this error can be reduced. Assuming the relaxation time approximation, the following values are computed:

Ionized impurities: $r_H = 1.93$
Neutral impurities: $r_H = 1$
Acoustic phonons (derormation potential): $r_H = 1.18$
Optical phonons (polar materials): low temparatures: $r_H = 1$; high temperatures: $r_H = 1.1$
Piezoelectric (by acoustic phonons in polar materials): $r_H = 1.1$

Considering carriers (electrons or holes) in direct or in indirect gap materials, we can find calculation and experimental data indicating that r_H changes according to the temperature interval for the predominant scattering process, so improving the temperature interval for the predominant scattering process, so improving the analysis of the experimental data [259].

In Section 7.3.2 we discussed the fitting procedure for $n(T)$ to the experimental Hall coefficient data over a wide temperature range to compute N_D, N_A, and $(\varepsilon_C - \varepsilon_D)$.

A problem that has been widely ignored concerns the Hall scattering factor r_H, which may deviate significantly from unity and is likely to be temperature dependent. All of this leads to serious difficulty in fitting the theoretical expression for $n(T)$ to experimental $R(T)$ data (particularly if r_H itself is unknown) [259].

7.5. Magnetoresistance and the Measure of Carrier Censity

The Hall effect is the most useful for characterization purposes, but in some circumstances the magnetoresistance (see Section 4) is used to measure carrier mobility. The magnetoresistance effect is described by the magneto coefficient M (see Eq. (3.21)). The application of a magnetic field normal to the direction of current flow generates an increase in the sample resistivity quadratic in B_z,

$$\Delta\rho_r / \rho_r = [\rho_r(B_z) - \rho_{ro}] / \rho_{ro} = \kappa \mu_H^2 B_z^2 \qquad (7.12)$$

where κ is a dimensionless constant, which depends on material parameters and on sample geometry. The magnetoresistance effect is referred to as "physical" (PMR) when studied in the normal Hall bar geometry under the boundary condition $J_y = 0$. It arises purely from material properties and, in particular, from the energy dependence of carrier relaxation time. A sample geometry in which the Hall field E_y is short circuited, so that the boundary condition becomes $E_y = 0$ and J_y is nonzero, generates the "geometrical" magnetoresistance (GMR) because it occurs as a direct consequence of sample geometry, regardless of the details of relaxation time behavior. There are too many uncertainties in the precise value of κ for the PMR effect, so only the GMR effect, which is also significantly larger, can be usefully applied to material characterization [259]. Important applications are to the measurement of electron mobility in Gunn diodes and field effect transistors, where the electrode geometry short-circuits the Hall field.

7.5.1. The Geometrical Magnetoresistance Effect

The geometrical magnetoresistance effect does not suffer from the disadvantages of PMR (sample inhomogeneity, dependence of relaxation time on energy, etc.). The two commonly used geometries are the Corbino disc, where, because of cylindrical symmetry, E_y is identically zero, and the parallel plate geometry, where E_y is only zero when the electrodes are infinite, but corrections are small, provided the ratio of lateral dimension to sample length is large [259].

7.5.2. Experimental Notes

The measurements made on epitaxial structures employed a sandwich geometry formed by an ohmic contact of large area on one side and a conducting substrate on the other. The mounting system must be free from magnetic material. The magnetoresistance mobility is found by plotting the change $(\Delta R/R_0)$ against B^2 and computing the slope of the straight line that should result. The resistivity ρ_r is given by $(R_0 A/t)$, where A is the electrode area. The most common source of error in measurement on sandwich structures arises from the unknown contact resistance R_c. If this is independent of magnetic field, the measured zero-field resistance is $R_0^* = R_0 + R_c$. It follows that both the apparent mobility and the true carrier density are less than the true value. A further source of error is the existence of inhomogeneities in the sample, like the variation of the doping level in an epitaxial layer in the direction perpendicular to the plane of the layer [259].

7.6. The Characterization of High-Resistivity Materials

A working example of high-resistivity materials is the semi-insulating III–V semiconductors. For example, when the Fermi level is located near the middle of the band gap, GaAs has a resistivity $\rho_r \sim 10^9$ Ωcm, and n and p are very small ($\sim 10^6$ cm^{-3}). This high resistivity is not achieved by compensation, as is usually done, because the donor and acceptor concentrations would have to match within about 1 part in 10^8, at a level of 10^{14} cm^{-3}.

It is difficult to characterize these materials by measuring the Hall coefficient as a function of temperature, as is usually done for conducting samples. For example, the analysis is complicated by mixed conduction effects ($n \sim p$), and it would be necessary to make measurements at temperatures of up to about 1000°C to reach the exhaustion region. Look [275] gives an excellent review of all of these problems.

7.6.1. Analysis of the Experimental Data

There are three basic equations governing the electrical behavior in the mixed conduction regime (see Section 2.6.1) that is usually active in semi-insulating materials. Two of these are Eqs. (2.56) and (3.14). The latter is in the limit of a zero magnetic field. The third indicates that electron and hole densities are always related in thermal equilibrium by

$$np = n_i^2 \qquad (7.13)$$

Experimental data give values of $R_{B=0}$ and σ. The unknown quantities n, p, μ_e, and μ_h need some other theoretical or empirical relations or experimental measurements, like thermopower.

Look [275] has also provided a set of curves relating the apparent mobility and the resistivity to the true electron mobility for negative values of R_H, as for most real samples. These curves indicate that if $\rho < 10^8$ Ωcm a single carrier analysis is adequate, and while in the mixed conduction region the single carrier solutions may be double valued. The ambiguity can only be resolved by other measurements, such as the sign of the thermoelectric power. Look [275] presented an alternative way to obtain information about single carrier densities and mobilities, by measuring the magnetic field dependence of the

sample conductivity and the Hall coefficient, under low field conditions. Expressions for $\Delta\sigma$ and ΔR in the mixed conduction regime are complicated. It is experimentally found that their quadratic dependence on the magnetic field is observed to a good approximation, though the best-fitting straight lines do not always go through the origin. The more serious problem of the method lies in the uncertainty about the values of the scattering factors.

7.6.2. Practical Considerations

Galvanomagnetic measurements on high-resistivity materials present a number of practical difficulties, arising from the high resistivity ($>10^{10}$ Ωcm) of the sample at room temperature [259]. The effective impedance can be reduced with the use of high-input impedance electrometers as unity gain amplifiers in each voltage lead. All leads must be screened and be as short as possible to minimize electrical pick-up. High-impedance circuits may also have long response times, making it time consuming to record measurements with the required reversals of magnetic field and sample current direction, and requiring long-term temperature stabilization. Electrometer configurations in real circuits can be found in [276]. Hall bar samples have the advantage that the same leads and contacts directly wired to the electrometers are always used for voltage measurements, whereas switching is needed for van der Pauw measurements. Corrections for surface depletion may be significant in these high-resistance materials with deep states located near mid-gap, but terms with the dimensions of mobility do not depend on the sample thickness. In high-resistance samples parasitic conduction paths can have a significant effect on the experimental results. These conduction paths are usually associated with the preparation of the surface of the sample.

7.7. Nonuniform Material

Many situations require a measure of the possible spatial variations for a proper understanding of device behavior. Device structures may contain deliberately introduced nonuniform doping as a result of diffusion, ion implantation, or controlled epitaxial growth, or nonuniformity may arise accidentally, for example, by out-diffusion of an impurity from the substrate during epitaxy.

Moreover, carrier mobility may change considerably from the value corresponding to a given doping level in bulk material. Many situations require that both resistivity and Hall effect are measured as a function of position. The need for accurate profiling is strong: the trend toward ever smaller device dimensions does not accept departures from design profiles influencing performances significantly. Complete transport characterization requires resistivity and that either the Hall effect or magnetoresistance be measured as well.

In many cases, it is necessary to remove the surface layer, as in implanted material. Removal can be obtained by chemical etching, by ion beam etching, or by anodic oxidation, in two different ways. A sequence of thin layers is removed parallel to the semiconductor surface, with measurements made after each layer, or the sample may be angle-lapped and a series of measurements performed along the direction of the lap.

Profiling is applied to the analysis of ion-implanted or diffused regions, the depths is of which range between 1 and 10 μm. Single steps of 100–1000 Å may be needed to achieve adequate resolution.

7.7.1. Resistivity Profiling with the Four-Point Probe

The conventional in-line four-point probe measures the sheet resistivity of a thin layer with a lateral spatial resolution of about 1 mm, and for this it cannot be used with an angle-lapped sample. Other limitations are for measurements of thin films on highly conducting substrates, where most of the current flows in the substrate, or with III–V semiconductors because of high contact resistance. A typical application is the analysis of a P implantation profile in p-type Si. The total layer thickness is very much less than the probe spacing, and the measure gives an effective sheet resistance $R_\square$ averaged through the nonuniform implanted layer. The sheet resistance is given by

$$R_\square = (\pi/\ln 2)V/I \qquad (7.14)$$

measured after each removal. Resistance is then converted to the conductivity profile, $\sigma(x)$. It is important to perform the measurements in a dark box to avoid photoconductivity and with high stabilization of the temperature to minimize the effect of resistivity variation with temperature. Much care must be taken with the probe penetration, because the depth resolution is 100 Å. Additional errors may also be introduced (i) by majority carrier injection when the measurement is made on very thin layers and (ii) from the surface preparation.

7.7.2. Hall Effect Profiling

When it is necessary to obtain reliable profiles of both carrier density and mobility, the measurements of the sheet resistance and Hall effect must be combined. The analysis of the data is performed as usual.

The van der Pauw geometry is the commonly used method for performing step-by-step profile measurements. The implanted or diffused region is isolated by mesa etching, and only the central part is exposed to the etch or the anodizing solution, keeping the contacts clear.

The experiment needs to measure the sheep conductance, $G_{\square \text{avg}} = \sigma t$, and the sheet Hall coefficient, $R_{\square \text{avg}} = R_{\square \text{avg}}/t$, at each step of the layer stripping procedure t is the overall sample thickness for each particular measurement, and avg indicates that the measured quantity represents an average over the thickness of the nonuniform sample. Errors due to light exposure and temperature fluctuations must be controlled. Another important error is that resulting from surface and interface depletion due to band bending. Care should be taken

to include the effect of the Hall scattering factor r_H if accurate values of $n(x)$ and $\mu(x)$ are required.

The GMR effect may be used to derive a carrier mobility, which is related to the Hall mobility. It is therefore possible to use this as an alternative to the Hall effect in obtaining mobility profiles on nonuniform material.

7.8. Experimental Configurations

7.8.1. Hall Effect

Resistivity configurations have been described (see Section 7.1), so we will give a more detailed review of the methods for Hall effect measurement. High sensitivity requires that electrode contact resistance is very small, and rectifying barriers at the Hall electrodes may introduce error due to floating potentials. Magnetocaloric effects (Nernst, Righi–Leduc, and Ettingshausen) (see Section 3.2) can be produced when the current electrodes are of a material different from the sample. Thermoelectric voltage is produced if the Hall probes are of a material different from the specimen. When a temperature gradient is involved in the effect, reversing B or I rapidly can eliminate the error in Hall data. Thermoelectric voltages do not reverse with the current and are eliminated by reversing I and averaging the Hall voltages.

The smallest measurable Hall mobility can be estimated from

$$(R\sigma)_{\min} = (V_H)_{\min}/(E_{\max} B_{\max} t) \qquad (7.15)$$

Contact resistance, noise level, and sensitivity of the measuring circuit determine $(V_H)_{\min}$; $E_{\max}$ is limited by the allowed power dissipation and the ohmic limit; and $B_{\max}$ is determined by the magnet employed.

The Hall coefficient can be measured by the use of either DC or AC methods. DC methods require high-sensitivity electrometers with a shielded circuit. A five-arm bar sample needs a circuit to compensate for the misalignment potential difference between opposite Hall probes, particularly if the Hall mobility is small. A van der Pauw sample does not require this circuit. AC Hall voltages can be obtained by a modulation of the current and with a DC magnetic field and by a modulation of the magnetic field with a DC current or of both, with different frequencies. With a lock-in amplifier, selecting the right frequency in the first two situations or the difference (sum) of frequencies in the third, one can measure the AC Hall voltage. The sign of the Hall coefficient can be found from the phase relation between the sample current and the Hall voltage, with a reference calibration. The absolute accuracy attainable is usually determined by the homogeneity of the material and the accuracy of measurements of the physical parameters, taking into account that the van der Pauw method makes it possible to avoid the geometrical requirements for samples. In regular samples, to eliminate the shorting of the Hall field, the length-to-width ratio must be greater than 3 or 4. A pure magnetoresistance effect can be measured in specimens prepared in the shape of the so-called Corbino disc. The sample setup is with one current electrode at the center and the other around the circumference, so that current flows radially; then no Hall field can originate and magnetoresistance is a maximum. The following rapid review of the method for Hall measurements provides a list of the experimental possibilities.

1. Current, magnetic field, and Hall voltage: all three DC. The principal advantages are the simplicity of the equipment and the availability of large DC magnetic fields. Misalignment voltages due to positioning of the Hall contacts are usually large compared with Hall tension. It can be eliminated from the final result by reversal of the magnetic field or, better, with the use of a three-arms geometry: one contact on one side of the sample and the other two opposite and symmetrically placed with respect to the first one. A potentiometer connected across these two probes makes it possible to achieve a zero misalignment voltage reading when the magnetic field is zero. The stability of all readings requires a high stabilization of the temperature of the sample, also minimizing thermoelectric voltages. Moreover, the reversal of either current or magnetic field reverses V_H but not the thermoelectric voltage and hence permits a separation of the two. The measured V_H can contain a magnetoresistive term due to misalignment voltage (i.e., this changes when the magnetic field is applied). These errors can be corrected for by reversing the magnetic field: the magnetoresistance, to first order, depends on B^2 and hence is the same for B and $-B$. The Ettingshausen effect (see Section 3.2.1 and Fig. 22) and the Nernst effect (see Section 3.2.2 and Fig. 23) can both introduce errors into the Hall voltage. The contribution of the Nernst effect can be avoided by reversals in B. The Ettingshausen contribution depends on (IB) and can be avoided only by a sufficiently rapid reversal of I or B (one or both being AC signal). Because both effects are associated with the development of temperature gradients, which are slowly varying, AC methods of measure are preferable for their elimination.

2. Current: AC at frequency f_1: magnetic field: DC; Hall voltage: AC at frequency f_1. The advantages are that V_H can be amplified by regular AC methods and that thermoelectric voltages do not interfere with the measurement. Moreover, the use of large magnetic fields can be avoided. Disadvantages rise from the fact that V_H has the same frequency as I and hence misalignment voltages interfere in the measurements, being in phase with V_H. The misalignment voltages must be eliminated with a compensation circuit at $B = 0$.

3. Current: DC; magnetic field: AC at frequency; Hall voltage: AC at frequency f_2. This configuration eliminates the misalignment DC voltages from the AC Hall signal, thus avoiding the initial compensation, but limits the intensity of the magnetic field and hence the amplitude of the Hall voltage. High-sensitivity lock-in amplifiers must be used to measure the small Hall voltages and, moreover, to separate the possible pick-up voltages from V_H, with the use of the phase difference $(\pi/2)$ between the two signals. In any case, adjusting the lead wire configuration must minimize pick-up. The Nernst, Seebeck, and Ettingshausen effects do not interfere with the Hall voltage because they are more slowly varying. Nor does the

magnetoresistance interfere because, due to its dependence on B^2, voltages related to it are at $2f_2$ and DC. Care must be taken with the depth of penetration of the field into the sample, compared with its thickness.

4. Current: AC at frequency f_1: magnetic field: AC at frequency f_1; Hall voltage: DC (so-called as des Coudres method). The Hall voltage is detected at the difference frequency (really DC) between that of I and that of B. A great advantage is that the electrode misalignment voltages are at f_1. Hall voltage can be detected with the usual DC methods. The Nernst, Seebeck, and Ettigshausen effects interfere with the Hall voltage, but correction can be operated by an appropriate phase sifting either of B or I and thus reversing the Hall voltage. The magnetoresistance does not interfere because it gives rise to voltages at f_1 and $3f_1$.

5. Current: AC at frequency f_1: magnetic field: AC at frequency f_2: Hall voltage: AC at frequency $(f_1 + f_2)$ or $(f_1 - f_2)$ (so-called cross modulation). The great advantage is that the Hall voltage is AC at a frequency different from that of either the current or the magnetic field. This fact implies the noninterference of misalignment voltages and the Nernst, Seebeck, Ettigshausen, and magnetoresistive effects. Disadvantages are the relatively small magnetic fields, rather complex equipment, and the requirement for linear components and contacts in the input circuit, to avoid intermodulation of the two frequencies.

External compensation circuits for eliminating misalignment potential between Hall contacts can be designed in several configurations [277]. The circuits give a temperature-independent compensation, provided that the input current i_i is constant and that the compensating current $i_c << i_i$. As far as the systematic errors are concerned, permutation of the current polarity and magnetic field orientation makes it possible to eliminate their contribution. Table IV resumes the required permutations.

7.8.2. Review of the Literature

Boerger et al. [278] derived equations under mirror symmetry conditions, to generalize the allowed shapes for the Hall effect from rectangular parallelepiped samples and to discuss limitations and errors in the van der Pauw methods. Ryan [279] allowed the sample to be measured as it was rotated in an external DC magnetic field. The sample thus "sees" an AC

Table IV. Permutations of Current and Magnetic Field to Eliminate Spurious Effects from Hall Voltage

Mode	Potential measured between Hall contacts
$+B; +i$	$V_1 = V_H + V_M + V_E + V_T + V_N + V_R$
$-B; +i$	$V_2 = -V_H + V_M - V_E + V_T - V_N - V_R$
$+B; -i$	$V_3 = -V_H - V_M - V_E + V_T + V_N + V_R$
$-B; -i$	$V_4 = V_H - V_M + V_E + V_T - V_N - V_R$

V_H, true Hall voltage; V_M, misalignment voltage; V_E, Ettingshausen voltage; V_R, Right–Leduc voltage; V_N, Nernst voltage; V_T, thermoelectric voltage. $V_{avg} = (V_1 - V_2 - V_3 + V_4)/4 = V_H + V_E$.

magnetic field, and a pure AC Hall technique may be used. Blanc et al. [280] studied the Hall and photo-Hall effects in high-resistivity GaAs:Cu samples and the effects of deeply-lying imperfections on the properties of the initial n-type GaAs. Lavine [281] employed an alternating electric field at 10^3 Hz with a static magnetic field. The spurious voltage due to mechanical vibrations and relative movements of lead wires are discussed. Pell and Sproull [282] used an AC system for measurements on relatively noisy, low-mobility crystals. They used an AC voltage at 24 Hz for both sample and reference signal to the lock-in amplifier, and a DC magnetic field reversed every 6 s, reversing the sign of the Hall voltage and producing a square wave detected by the electronic chain. Levy [283] built a cross-modulation system with AC fields (magnetic at 60 Hz and electric at 85 Hz) and detected the Hall signal at the difference frequency of 25 Hz. Lupu et al. [284] measured the Hall effect of low-mobility samples at high temperatures. They used a double AC system, modulating the electric field at 510 Hz and the magnetic field at 2 Hz. This value is a compromise between two opposite requirements: temperature stability of the furnace, which needs a frequency as high as possible, and high intensity of the magnetic field, which needs a frequency as low as possible. Hemenger [285] describes a DC apparatus suitable for measuring high resistance with the van der Pauw method. Murri et al. [286] used double modulation (AC electric and magnetic field) to measure Hall mobility in amorphous gallium arsenide.

Carver [287] used the Corbino disk to measure Hall mobility in amorphous and crystalline semiconductors, making it possible to obtain data with an error of less than 5%. Eberle et al. [288], using n-GaAs samples with Corbino disk contacts, demonstrated symmetry breaking and self-organization by the current filaments induced by hot electrons. Taylor et al. [289, 290] explored the possibility with a Corbino disk that metallic spikes from a contact deposited on the surface of an AlGaAs/GaAs heterostructure penetrates the two-dimensional gas defining a quantum system. Yokoi et al. [291] studied the breakdown of the integer quantum Hall effect by Corbino disk in samples fabricated from GaAs/Al$_{0.3}$Ga$_{0.7}$ As heterostructures. Shikin et al. [292] proposed a theory of current voltage characteristics for a gated Corbino disk able to predict the local details in the distribution of electron density and electric potential in a current-carrying disk. Rycroft et al. [293] performed transport measurements on Bi$_2$Sr$_2$CaCu$_2$O$_8$ single crystals in both Corbino disk and strip geometry. The results indicate that surface barriers affect the behavior of the strip and that values measured with a Corbino disk are much more affordable. Aoki et al. [294] measured the spatial and spatiotemporal patterns of the current–density filaments in Corbino disks at 4.2 K with the magnetic fields parallel to the sample surface. They found the complex structure of the current filaments. Mirkovic et al. [295] found the strong suppression of the nonlinear resistivity in the vortex liquid phase in the sample with Corbino disk geometry, which excludes the role of the surface barriers.

REFERENCES

1. K. Seeger, "Semiconductor Physics," 4th ed. Springer-Verlag, Berlin, 1989.
2. B. R. Nag, "Electron Transport in Compound Semiconductors" (H.-J. Queisser, Ed.). Springer-Verlag, Berlin/Heidelberg/New York, 1980.
3. L. Landau, *Z. Phys.* 64, 629 (1930).
4. L. M. Roth and P. N. Argyress, in "Semiconductors and Semimetals" (R. K. Willardson and A. C. Beer, Eds.), Vol. 1, p. 167. Academic Press, New York, 1966.
5. J. S. Blakemore, "Semiconductor Statistics," p. 89. Pergamon, Oxford, 1962.
6. B. R. Nag and A. N. Chakrabarti, *Int. J. Electron.* 36, 275 (1974).
7. Y. Yafet, R. W. Keyes, and E. N. Adams, *J. Phys. Chem. Solids* 1, 137 (1956).
8. E. H. Putley, "The Hall Effect and Related Phenomena." Butterworths, London, 1960.
9. F. J. Blatt, "Physics of Electronic Conduction in Solids," Chap. 5, p. 109. McGraw-Hill, New York, 1968.
10. L. M. Roth, in "Basic Properties of Semiconductors" (T. S. Moss, Ed.), Vol. 1, p. 489. North-Holland, Amsterdam, 1992.
11. J. Ziman, "Electrons and Phonons." Oxford Univ. Press, Oxford, 1960.
12. A. C. Beer, "Galvanomagnetic Effects in Semiconductors, Solid State Physics," Suppl. 4. Academic Press, New York, 1963.
13. P. N. Butcher, in "Crystalline Semiconducting Materials and Devices," p. 131. Plenum, New York, 1986.
14. H. B. Callen, *Phys. Rev.* 73, 1340 (1948).
15. G. E. Stillman and C. M. Wolfe, *Thin Solid Films* 31, 69 (1976).
16. C. Erginsoy, *Phys. Rev.* 79, 1013 (1950).
17. H. Brooks, *Phys. Rev.* 83, 879 (1951).
18. E. Conwell and V. F. Weisskopf, *Phys. Rev.* 77, 388 (1950).
19. L. M. Falicov and M. Cuevas, *Phys. Rev.* 164, 1025 (1967).
20. D. Kranzer and E. Gornik, *Solid State Commun.* 9, 1541 (1971).
21. D. Chattopadhyay and H. J. Queisser, *Rev. Mod. Phys.* 53, 745 (1981).
22. O. Madelung, "Introduction to Solid State Theory" (M. Cardona, P. Fulde, and H. J. Queisser, Eds.), p. 129. Springer-Verlag, Berlin, 1981.
23. J. Bardeen and W. Shockley, *Phys. Rev.* 80, 72 (1950).
24. R. F. Greene, *J. Electron. Control* 3, 387 (1957).
25. E. M. Conwell, *Phys. Rev.* 90, 769 (1953).
26. P. P. Debye and E. M. Conwell, *Phys. Rev.* 93, 693 (1954).
27. W. W. Tyler and H. H. Woodbury, *Phys. Rev.* 102, 647 (1956).
28. A. R. Hutson, *J. Appl. Phys. Suppl.* 32, 2287 (1961).
29. H. J. G. Meyer and D. Polder, *Physica* 19, 255 (1953).
30. W. A. Harrison, *Phys. Rev.* 104, 1281 (1956).
31. D. J. Howarth and E. H. Sondheimer, *Proc. R. Soc. London, Ser. A* 219, 53 (1953).
32. H. Ehrenreich, *J. Appl. Phys. Suppl.* 32, 2155 (1961).
33. T. Kurosawa, "Proceedings of the International Conference on the Physics of Semiconductors," Kyoto, 1966, *J. Phys. Soc. Jpn. Suppl.* 21, 424 (1966).
34. W. Fawcett, A. D. Boardmann, and S. Swain, *J. Phys. Chem. Solids* 31, 1963 (1970).
35. H. Budd, "Proceedings of the International Conference on the Physics of Semiconductors," Kyoto, 1966, *J. Phys. Soc. Jpn. Suppl.* 21, 420 (1966).
36. H. D. Rees, *J. Phys. Chem. Solids* 30, 643 (1969).
37. H. Ehrenreich, *J. Phys. Chem. Solids* 8, 130 (1959).
38. M. Mikkor and F. C. Brown, *Phys. Rev.* 162, 848 (1967).
39. J. W. Hodby, J. A. Borders, and F. C. Brown, *J. Phys. C: Solid State Phys.* 3, 335 (1970).
40. T. Holstein, *Ann. Phys. NY* 8, 325, 343 (1959).
41. J. G. Ruch, *IEEE Trans.* ED-19, 652 (1972).
42. J. Appel, *Phys. Rev.* 122, 1760 (1961).
43. R. T. Bate, R. D. Baxter, F. J. Reid, and A. C. Beer, *J. Phys. Chem. Solids* 26, 1205 (1965).
44. T. P. Mc Lean and E. G. S. Paige, *J. Phys. Chem. Solids* 16, 220 (1960).
45. S. C. Jain and W. Hayes, *Semicond. Sci. Technol.* 6, 547 (1991).

46. F. Schäffler, *Semicond. Sci. Technol.* 12, 1515 (1997).

47. H. F. Mataré, "Defect Electronics in Semiconductors," John Wiley & Sons, Inc., New York, 1971.

48. G. L. Pearson, W. T. Read, Jr., and F. J. Morin, *Phys. Rev.* 93, 93 (1954).

49. W. Schröter, *Phys. Status Solidi* 21, 211 (1967).

50. B. Pödor, *Phys. Status Solidi* 16, K167 (1966).

51. F. Düster and R. Labusch, *Phys. Status Solidi (B)* 60, 161 (1973).

52. J. H. P. Van Weeren, R. Struikmans, and J. Blok, *Phys. Status Solidi* 19, K107 (1967).

53. J. B. Gunn, "Progress in Semiconductors," A. F. Gibson, R. E. Burgess Eds., Vol. 2, p. 213 (1987).

54. M. Glicksman and W. A. Hicinbothem Jr., *Phys. Rev.* 129, 1572 (1963).

55. R. J. Sladek, *Phys. Rev.* 120, 1589 (1960).

56. K. Seeger, *Z. Phys.* 172, 68 (1963).

57. P. Kastner, E. P. Roth, and K. Seeger, *Z. Phys.* 187, 359 (1965).

58. Y. Kanai, *J. Phys. Soc. Jpn.* 15, 830 (1960).

59. G. Bauer, F. Kuchar, D. Kranzer, and E. Bonek, *Acta Phys. Austria* 33, 8 (1971).

60. T. Shirakawa and C. Hamaguch, *Phys. Lett. A* 49A, 231 (1974).

61. C. T. Elliott and I. L. Spain, *J. Phys. C: Solid State Phys.* 7, 727 (1974).

62. F. D. Hughes and R. J. Tree, *J. Phys. C: Solid State Phys.* 3, 1943 (1970).

63. C. Hamaguchi and Y. Inuishi, *J. Phys. Soc. Jpn.* 18, 1755 (1963).

64. F. Kuchar, A. Phillipp, and K. Seeger, *Solid State Commun.* 11, 965 (1972).

65. W. Sasaki, M. Shibuya, and K. Mizuguchi, *J. Phys. Soc. Jpn.* 13, 456 (1958).

66. W. Sasaki, M. Shibuya, K. Mizuguchi, and G. M. Hatoyama, *J. Phys. Chem. Solids* 8, 250 (1959).

67. M. Shibuya, *Phys. Rev.* 99, 1189 (1955).

68. M. I. Nathan, *Phys. Rev.* 130, 2201 (1963).

69. B. R. Nag, H. Paria, and S. Guha, *Phys. Lett. A* 26A, 172 (1968).

70. B. R. Nag, H. Paria, and P. K. Basu, *Phys. Lett. A* 28A, 202 (1968).

71. E. J. Ryder and W. Shockley, *Phys. Rev.* 81, 139 (1951).

72. E. J. Ryder, *Phys. Rev.* 90, 766 (1953).

73. B. K. Ridley and T. B. Watkins, *Proc. Phys. Soc., London* 78, 293 (1961).

74. B. K. Ridley, *Proc. Phys. Soc., London* 82, 954 (1963).

75. C. Hilsum, *Proc. IRE* 50, 185 (1962).

76. J. B. Gunn, *Solid State Commun.* 1, 88 (1963).

77. V. V. Paranjape and T. P. Ambrose, *Phys. Lett.* 8, 223 (1964).

78. V. V. Paranjape and E. De Alba, *Proc. Phys. Soc., London* 85, 945 (1965).

79. A. C. Baynham, P. N. Butcher, W. Fawcett, and J. M. Loveluck, *Proc. Phys. Soc., London* 92, 783 (1967).

80. E. J. Aas and K. Bløtekjaer, *J. Phys. Chem. Solids* 35, 1053 (1974).

81. R. Stratton, *Proc. R. Soc. London, Ser. A* 246, 406 (1958).

82. E. Erlbach, *Phys. Rev.* 132, 1976 (1963).

83. S. V. Gantsevich, V. L. Gurevich, and R. Katilius, *Phys. Condens. Matter* 18, 165 (1974).

84. B. R. Nag and D. Chattopadhyay, *Solid State Electron.* 21, 303 (1978).

85. D. F. Ferry, in "Basic Properties Semiconductors" (T. S. Moss, Ed.), Vol. 1, p. 1039. North-Holland, Amsterdam, 1992.

86. B. Lax, "Proceedings of the International School of Physics" (R. A. Smith, Ed.), Vol. 22, p. 240. Academic Press, New York, 1963.

87. C. J. Rauch, J. J. Stickler, H. J. Zeiger, and G. Heller, *Phys. Rev. Lett.* 4, 64 (1960).

88. B. Lax, H. J. Zeiger, and R. N. Dexter, *Physica* 20, 818 (1954).

89. G. Dresselhaus, A. F. Kip, and C. Kittel, *Phys. Rev.* 98, 368 (1955).

90. W. Shockley, *Phys. Rev.* 79, 191 (1950).

91. E. D. Palik, G. S. Picus, S. Teitler, and R. F. Wallis, *Phys. Rev.* 122, 475 (1961).

92. E. D. Palik and G. B. Wright, "Semiconductors and Semimetals," Vol. 3, p. 421. Academic, New York, 1967.

93. E. H. Hall, *Am. J. Math.* 2, 287 (1879).

94. B. L. Gallangher and P. N. Butcher, in "Low-Dimensional and Mesoscopic Semiconductor Structures," p. 721. North-Holland, Amsterdam, 1992.

95. H. Weiss, in "Semiconductors and Semimetals" (R. K. Willardson and A. C. Beer, Eds.), Vol. 1, p. 315. Academic Press, New York, 1966.

96. G. L. Pearson and C. Herring, *Physica's Grav.* 20, 975 (1954).

97. F. Seitz, *Phys. Rev.* 79, 372 (1950).

98. C. Goldberg and R. E. Davis, *Phys. Rev.* 94, 1121 (1954).

99. G. L. Pearson and H. Suhl, *Phys. Rev.* 83, 768 (1951).

100. R. S. Allgaier, *Phys. Rev.* 112, 828 (1958).

101. M. Glicksman, *Prog. Semicond.* 3, 2 (1958).

102. E. H. Putley, *Proc. Phys. Soc. London, Ser. B* 68, 22 (1955).

103. H. Rupprecht, R. Weber, and H. Weiss, *Z. Naturforsch.* 15a, 783 (1960).

104. C. Hilsum and R. Barrie, *Proc. Phys. Soc., London* 71, 676 (1958).

105. R. G. Chambers, *Proc. Phys. Soc., London* A65, 903 (1952).

106. H. Schönwald, *Z. Naturforsch.* 19a, 1276 (1964).

107. H. Weiss, *Z. Naturforsch.* 8a, 463 (1953).

108. T. C. Harman, R. K. Willardson, and A. C. Beer, *Phys. Rev.* 95, 699 (1954).

109. C. H. Champness, *J. Electron. Control* 4, 201 (1958).

110. C. H. Champness, *Phys. Rev. Lett.* 1, 439 (1958).

111. H. Fritzsche and K. Lark-Horovitz, *Phys. Rev.* 99, 400 (1955).

112. H. P. R. Frederikse and W. R. Hosler, *Phys. Rev.* 108, 1136 (1957).

113. I. M. Mackintosh, *Proc. Phys. Soc., London* B69, 403 (1956).

114. K. W. H. Stevens, *Proc. Phys. Soc., London* B69, 406 (1956).

115. Y. Toyozawa, "Proceedings of the International Conference on the Physics of Semiconductors," Exeter, 1962, p. 104. Institute of Physics and the Physical Society, London, 1962.

116. H. Weiss, *Z. Naturforsch.* 12a, 80 (1957).

117. N. V. Zatova, T. S. Lagunova, and D. N. Nasledov, *Sov. Phys. Solid State* 5, 2439 (1964).

118. M. Glicksman, *J. Phys. Chem. Solids* 8, 511 (1959).

119. W. M. Becker, A. K. Ramdas, and H. Y. Fan, *J. Appl. Phys.* 32, 2094 (1961).

120. A. J. Strauss, *Phys. Rev.* 121, 1087 (1961).

121. A. Sagar, *Phys. Rev.* 117, 93 (1960).

122. A. F. Kravchenko and H. Y. Fan, in "Proceedings of the International Conference on the Physics of Semiconductors," Exeter, 1962, p. 737. Institute of Physics and the Physical Society, London, 1962.

123. O. V. Emel'yanenko and D. N. Nasledov, *Sov. Phys. Tech. Phys.* 3, 1094 (1958).

124. D. N. Nasledov and O. V. Emel'yanenko, in "Proceedings of the International Conference on the Physics of Semiconductors," Exeter, 1962, p. 163. Institute of Physics and the Physical Society, London, 1962.

125. L. Shubnikov and W. J. de Haas, *Leiden Commun.* 207a, 207c, 207d, 210A (1930).

126. E. N. Adams and T. D. Holstein, *J. Phys. Chem. Solids* 10, 254 (1959).

127. L. I. Schiff, "Quantum Mechanics," p. 441. McGraw-Hill, New York, 1968.

128. M. Cardona, *J. Phys. Chem. Solids* 24, 1543 (1963).

129. R. B. Dingle, *Proc. R. Soc. London, Ser. A* 211, 517 (1952).

130. R. A. Isaacson and F. Bridges, *Solid State Commun.* 4, 635 (1966).

131. Kh. I. Amirkhanov, R. I. Bashirov, and Yu. Zakiev, *Sov. Phys. Dokl.* 8, 182 (1963).

132. R. J. Sladek, *Phys. Rev.* 110, 817 (1958).

133. H. P. R. Frederikse and W. R. Hosler, *Phys. Rev.* 110, 880 (1958).

134. S. S. Shalyt and A. L. Éfros, *Sov. Phys. Solid State* 4, 903 (1962).

135. Kh. I. Amirkhanov, R. I. Bashirov, and Yu. É. Zakiev, *Sov. Phys. Solid State* 5, 340 (1963).

136. W. M. Becker and H. Y. Fan, Physics of Semiconductors, "Proceedings of the 7th International Conference," p. 663. Dunod, Paris, and Academic Press, New York, 1964.

137. W. M. Becker and T. O. Yep, *Bull. Am. Phys. Soc.* 10, 106 (1965).

138. S. J. Miyake, *J. Phys. Soc. Jpn.* 20, 412 (1965).

139. J. E. Robinson and S. Rodriguez, *Phys. Rev. A: At., Mol. Opt. Phys.* 135, 779 (1964).

140. E. H. Putley, in "Semiconductors and Semimetals" (R. K. Willardson and A. C. Beer, Eds.), Vol. 1, p. 289. Academic Press, New York, 1966.

141. R. W. Keyes and R. J. Sladek, *J. Phys. Chem. Solids* 1, 143 (1956).

142. R. J. Sladek, *J. Phys. Chem. Solids* 5, 157 (1958).

143. R. J. Sladek, *J. Phys. Chem. Solids* 8, 515 (1959).

144. H. P. R. Frederikse and W. R. Hosler, *Phys. Rev.* 108, 1136 (1957).

145. E. H. Putley, *Proc. Phys. Soc., London* 76, 802 (1960).

146. E. H. Putley, *J. Phys. Chem. Solids* 22, 241 (1961).

147. V. L. Gurevich and Yu. A. Firsov, *Sov. Phys. JETP* 13, 137 (1961).

148. M. I. Klinger, *Sov. Phys. Solid State* 3, 974 (1961).

149. Yu. A. Firsov, V. L. Gurevich, R. V. Parfeniev, and S. S. Shalyt, *Phys. Rev. Lett.* 12, 660 (1964).

150. A. L. Efros, *Sov. Phys. Solid State* 3, 2079 (1962).

151. L. Eaves, R. A. Stradling, and R. A. Wood, in "Proceedings of the International Conference on the Physics of Semiconductors," Cambridge MA, 1970 (S. P. Keller, J. C. Hensel, and F. Stern, Eds.), p. 816. LUSAEC, Oak Ridge, TN, 1970.

152. V. L. Gurevich and Yu. A. Firsov, *Sov. Phys. JETP* 20, 489 (1964).

153. R. A. Stradling and R. A. Wood, *J. Physique C* 1, 1711 (1968).

154. A. L. Mears, R. A. Stradling, and E. K. Inall, *J. Physique C* 1, 821 (1968).

155. R. A. Reynolds, *Solid State Electron.* 11, 385 (1968).

156. R. A. Wood, R. A. Stradling, and I. P. Molodyan, *J. Phys. (Paris) C* 3, L154 (1970).

157. S. M. Puri and T. H. Geballe, "Semiconductors and Semimetals," Vol. 1, p. 203. Academic Press, New York, (1966).

158. L. L. Chang, L. Esaki, W. E. Howard, and R. Ludeke, *J. Vac. Sci. Technol.* 10, 11 (1973).

159. H. Sakaki, L. L. Chang, R. Ludeke, C. A. Chang, G. A. Sai-Halasz, and L. Esaki, *Appl. Phys. Lett.* 31, 211 (1977).

160. G. Abstreiter, H. Brugger, T. Wolf, H. Jorke, and H. J. Herzog, *Phys. Rev. Lett.* 54, 2441 (1985).

161. J. P. Faurie, A. Million, R. Boch, and J. L. Tissot, *J. Vac. Sci. Technol., A* 1, 1593 (1983).

162. K. von Klitzing, G. Dorda, and M. Pepper, *Phys. Rev. Lett.* 45, 494 (1980).

163. T. Ando, A. B. Fowler, and F. Stern, *Rev. Mod. Phys.* 54, 437 (1982).

164. H. L. Störmer, R. Dingle, A. C. Gossard, W. Weigmann, and M. D. Sturge, *Solid State Commun.* 29, 705 (1979).

165. H. L. Störmer, A. M. Chang, D. C. Tsui, J. C. M. Hwang, A. C. Gossard, and W. Weigmann, *Phys. Rev. Lett.* 50, 1953 (1983).

166. W. I. Wang, E. E. Mendez, and F. Stern, *Appl. Phys. Lett.* 45, 639 (1984).

167. E. Kasper, *Curr. Opin. Solid State Mater. Sci.* 2, 48 (1997).

168. Y. H. Xie, *Mater. Sci. Eng.* 25, 89 (1999).

169. Y. H. Xie, E. A. Fitzgerald, D. Monroe, P. J. Silverman, and G. P. Watson, *J. Appl. Phys.* 73, 8364 (1993).

170. K. Ismail, M. Arafa, K. L. Saenger, J. O. Chu, and B. S. Meyerson, *Appl. Phys. Lett.* 66, 1077 (1995).

171. N. Sugii, K. Nakagawa, Y. Kimura, S. Yamaguchi, and M. Miyao, *Semicond. Sci. Technol.* 13, A140 (1998).

172. T. Ando, *J. Phys. Soc. Jpn.* 39, 411 (1975).

173. H. J. Lippmann and F. Kurt, *Z. Naturforsch.* 13, 462 (1958).

174. R. F. Wick, *J. Appl. Phys.* 25, 741 (1954).

175. J. J. Harris, J. A. Pals, and R. Woltjer, *Rep. Prog. Phys.* 52, 1217 (1989).

176. W. J. Skocpol, L. D. Jackel, E. L. Hu, R. F. Howard, and L. A. Fetter, *Phys. Rev. Lett.* 49, 951 (1982).

177. A. B. Fowler, A. Hartstein, and R. A. Webb, *Phys. Rev. Lett.* 48, 196 (1982).

178. H. van Houten, B. J. van Wees, M. G. J. Heyman, and J. P. Andre, *Appl. Phys. Lett.* 49, 1781 (1986).

179. T. J. Thornton, M. Pepper, H. Ahmed, D. Andrews, and G. J. Davies, *Phys. Rev. Lett.* 56, 1198 (1986).

180. H. Z. Zheng, H. P. Wei, D. C. Tsui, and G. Weimann, *Phys. Rev. B: Condensed Matter* 34, 5635 (1986).

181. K.-F. Berggren, T. J. Thornton, D. J. Newson, and M. Pepper, *Phys. Rev. Lett.* 57, 1769 (1986).

182. S. B. Kaplan and A. C. Warren, *Phys. Rev. B: Condensed Matter* 34, 1346 (1986).

183. S. E. Laux, D. J. Frank, and F. Stern, *Surf. Sci.* 196, 101 (1988).

184. T. Ando, *J. Phys. Soc. Jpn.* 36, 1512 (1974).

185. T. Ando, *J. Phys. Soc. Jpn.* 37, 622 (1974).

186. T. Ando, *J. Phys. Soc. Jpn.* 37, 1233 (1974).

187. W. Beinvogl, A. Kamgar, and J. F. Koch, *Phys. Rev. B* 14, 4274 (1976).

188. K.-F. Berggren and D. J. Newson, *Semicond. Sci. Technol.* 1, 327 (1986).

189. H. Morkoç, *IEEE Electron Device Lett.* EDL-2, 260 (1981).

190. M. Heiblum, E. E. Mendez, and F. Stern, *Appl. Phys. Lett.* 44, 1064 (1984).

191. G. Weimann, *Festkorperprobleme* 36, 231 (1986).

192. A. Schüppen, A. Gruhle, H. Kibbel, U. Erben, and U. König, *IEDM Tech. Digest* 377 (1994).

193. H. L. Störmer, *Surf. Sci.* 132, 519 (1983).

194. F. Stern, *Appl. Phys. Lett.* 43, 974 (1983).

195. K. Lee, M. S. Schur, T. J. Drummond, and H. Morkoç, *J. Appl. Phys.* 54, 6432 (1983).

196. N. Cho, S. B. Ogale, and A. Madhukar, *Appl. Phys. Lett.* 51, 1016 (1987).

197. T. J. Drummond, H. Morkoç, K. Hess, and A. Y. Cho, *J. Appl. Phys.* 52, 5231 (1981).

198. B. J. F. Lin, D. C. Tsui, M. A. Paalanen, and A. C. Gossard, *Appl. Phys. Lett.* 45, 695 (1984).

199. K. Hirakawa and H. Sakaki, *Phys. Rev. B: Condensed Matter* 33, 8291 (1986).

200. E. E. Mendez, *IEEE J. Quantum Electron.* QE-22, 1720 (1986).

201. W. Walukiewicz, H. E. Ruda, J. Lagowski, and H. C. Gatos, *Phys. Rev. B: Condensed Matter* 30, 4571 (1984).

202. M. O. Vassell and J. Lee, *Semicond. Sci. Technol.* 2, 340 (1987).

203. M. O. Vassell and J. Lee, *Semicond. Sci. Technol.* 2, 350 (1987).

204. A. Gold, *Z. Phys. B* 63, 1 (1986).

205. N. Cho, S. B. Ogale, and A. Madhukar, *Phys. Rev. B: Condensed Matter* 36, 6472 (1987).

206. K. Hess, *Appl. Phys. Lett.* 35, 484 (1979).

207. P. J. Price, *Ann. Phys. NY* 133, 217 (1981).

208. B. Vinter, *Appl. Phys. Lett.* 45, 581 (1984).

209. P. J. Price, *Surf. Sci.* 143, 145 (1984).

210. E. E. Mendez, P. J. Price, and M. Heiblum, *Appl. Phys. Lett.* 45, 294 (1984).

211. P. J. Price, *Phys. Rev. B: Condensed Matter* 32, 2643 (1985).

212. B. Vinter, *Phys. Rev. B: Condensed Matter* 33, 5904 (1986).

213. J. J. Harris, C. T. Foxon, D. E. Lacklison, and K. W. J. Barnham, *Superlattices Microstruct.* 2, 563 (1986).

214. Y. Takeda, H. Kamei, and H. Sakaki, *Electron. Lett.* 18, 309 (1982).

215. B. Etienne and E. Paris, *J. Physique* 48, 2049 (1987).

216. R. Lassnig, *Solid State Commun.* 65, 765 (1988).

217. B. K. Ridley, *Semicond. Sci. Technol.* 3, 111 (1988).

218. E. F. Schubert, L. Pfeiffer, K. W. West, and A. Izabelle, *Appl. Phys. Lett.* 54, 1350 (1989).

219. P. J. van Hall, T. Klaver, and J. H. Wolter, *Semicond. Sci. Technol.* 3, 120 (1988).

220. D. C. Tsui, H. L. Störmer, and A. C. Gossard, *Phys. Rev. Lett.* 48, 1559 (1982).

221. R. B. Laughlin, *Phys. Rev. B: Condensed Matter* 23, 5632 (1981).

222. R. J. Wagner, T. A. Kennedy, B. D. McCombe, and D. C. Tsui, *Phys. Rev. B: Condensed Matter* 22, 945 (1980).

223. G. M. Gusev, Z. D. Kvon, I. G. Neizvestnyi, V. N. Ovsyuk, and P. A. Cheremnykh, *JETP Lett.* 39, 541 (1984).

224. V. M. Pudalov and S. G. Semenchinskii, *JETP Lett.* 39, 170 (1984).

225. V. M. Pudalov and S. G. Semenchinskii, *Sov. Phys. JETP* 59, 838 (1984).

226. V. M. Pudalov and S. G. Semenchinskii, *Solid State Commun.* 51, 19 (1984).

227. H.-J Herzog, H. Jorke, and F. Schäffler, *Thin Solid Films* 184, 237 (1990).

228. D. Többen, F. Schäffler, A. Zrenner, and G. Abstreiter, *Phys. Rev. B: Condensed Matter* 46, 4344 (1992).

229. S. F. Nelson, *Appl. Phys. Lett.* 61, 64 (1992).

230. D. C. Tsui and A. C. Gossard, *Appl. Phys. Lett.* 38, 550 (1981).

231. H. L. Störmer and W. T. Tsang, *Appl. Phys. Lett.* 36, 685 (1980).

232. H. L. Störmer, Z. Schlesinger, A. Chang, D. C. Tsui, A. C. Gossard, and W. Weigmann, *Phys. Rev. Lett.* 51, 126 (1983).

233. Y. Guldner, J. P. Hirtz, J. P. Vieren, P. Voisin, M. Voos, and M. Razeghi, *J. Physique* 43, L613 (1982).

234. A. Briggs, Y. Guldner, J. P. Vieren, M. Voos, J. P. Hirtz, and M. Razeghi, *Phys. Rev. B: Condensed Matter* 27, 6549 (1983).

235. L. L. Chang and L. Esaki, *Surf. Sci.* 98, 70 (1980).

236. S. Washburn, R. A. Webb, E. E. Mendez, L. L. Chang, and L. Esaki, *Phys. Rev. B: Condensed Matter* 31, 1198 (1985).

237. E. E. Mendez, S. Washburn, L. Esaki, L. L. Chang, and R. A. Webb, in "Proceedings of the 17th International Conference on the Physics of Semiconductors" (D. J. Chadi and W. A. Harrison, Eds.), p. 397. Springer-Verlag, Berlin/Heidelberg/New York/Tokyo, 1985.

238. S. Nakajima, T. Tanahashi, and K. Akita, *Appl. Phys. Lett.* 41, 194 (1982).

239. M. Inoue and S. Nakajima, *Solid State Commun.* 50, 1023 (1984).

240. M. Razeghi, J. P. Duchemin, and J. C. Portal, *Appl. Phys. Lett.* 46, 46 (1985).

241. A. Kastalski, R. Dingle, K. Y. Cheng, and A. Y. Cho, *Appl. Phys. Lett.* 41, 274 (1982).

242. R. A. Schiebel, *Proc. IEEE Int. Conf. Electron Devices* 711 (1983).

243. W. P. Kirk, P. S. Kobiela, R. A. Schiebel, and M. A. Reed, *J. Vac. Sci. Technol., A* 4, 2132 (1986).

244. O. Madelung, "Semiconductors: Basic Data," 2nd ed. Springer-Verlag, Berlin/New York, 1996.

245. R. E. Prange and S. M. Girvin, "The Quantum Hall Effect," 2nd ed. Springer-Verlag, Berlin/New York, 1990.

246. H. Aoki and T. Ando, *Solid State Commun.* 38, 1079 (1981).

247. R. Woltjer, R. Eppenga, and M. F. H. Schuurmans, in "High Magnetic Fields in Semiconductor Physics" (G. Landwehr, Ed.), p. 104. Springer-Verlag, Berlin/New York, 1987.

248. G. Giuliani, J. J. Quinn, and S. C. Ying, *Phys. Rev. B: Condensed Matter* 28, 2969 (1983).

249. J. E. Avron and R. Seiler, *Phys. Rev. Lett.* 54, 259 (1985).

250. I. V. Kukushkin and V. B. Timofeev, "Proceedings of the International Conference on Electronic Properties of Two-Dimensional Systems," VI, 1985 San Francisco, 1985.

251. C. T. Foxon, J. J. Harris, R. G. Wheeler, and D. E. Lacklison, *J. Vac. Sci. Technol., B* 4, 511 (1986).

252. E. E. Mendez, L. L. Chang, M. Heiblum, L. Esaki, M. Naughton, K. Martin, and J. Brooks, *Phys. Rev. B: Condensed Matter* 30, 7310 (1984).

253. H. P. Wei, A. M. Chang, D. C. Tsui, and M. Razeghi, *Phys. Rev. B: Condensed Matter* 32, 7016 (1985).

254. D. Monroe, Y. H. Xie, E. A. Fitzgerald, and P. J. Silvermann, *Phys. Rev. B: Condensed Matter* 46, 7935 (1992).

255. R. B. Laughlin, *Phys. Rev. Lett.* 50, 1395 (1983).

256. R. G. Clark, R. J. Nicholas, A. Usher, C. T. Foxon, and J. J. Harris, *Surf. Sci.* 170, 141 (1986).

257. R. Willett, J. P. Eisenstein, H. L. Störmer, D. C. Tsui, and A. C. Gossard, *Phys. Rev. Lett.* 59, 1776 (1987).

258. J. W. Orton and P. Blood, "The Electrical Characterization of Semiconductors: Measurement of Minority Carriers Properties." Academic Press, London, 1990.

259. P. Blood and J. W. Orton, "The Electrical Characterization of Semiconductors: Majority Carriers and Electron States." Academic Press, London, 1992.

260. L. J. van der Pauw, *Philips Res. Rep.* 13, 1 (1958).

261. B. L. Sharma, in "Semiconductors and Semimetals" (R. K. Willardson and A. C. Beer, Ed.), Vol. 15, Chap. 1. Academic Press, New York, 1981.

262. C. Manfredotti, R. Murri, A. Rizzo, L.Vasanelli, and G. Micocci, *Phys. Status Solid: A* 29, 475 (1975).

263. V. Augelli, C. Manfredotti, R. Murri, R. Piccolo, and L.Vasanelli, *Nuovo Cimento, B* 38, 327 (1977).

264. C. Manfredotti, A. M. Mancini, A. Rizzo, R. Murri, and L. Vasanelli, *Phys. Status Solid: A* 48, 293 (1978).

265. V. Augelli and R. Murri, *J. Non-Cryst. Solids* 57, 225 (1983).

266. V. Augelli, T. Ligonzo, R. Murri, and L. Schiavulli, *Thin Solid Films* 125, 9 (1985).

267. R. Bilenchi, I. Gianinoni, M. Musci, R. Murri, and S. Tacchetti, *Appl. Phys. Lett.* 47, 279 (1985).

268. C. Manfredotti, R. Murri, A. Rizzo, and L. Vasanelli, *Solid State Commun.* 19, 339 (1976).

269. V. Augelli, C. Manfredotti, R. Murri, R. Piccolo, A. Rizzo, and L. Vasanelli, *Solid State Commun.* 21, 575 (1977).

270. V. Augelli, C. Manfredotti, R. Murri, and L. Vasanelli, *Phys. Rev. B: Solid State* 17, 3221 (1978).

271. V. Augelli, R. Murri, and T. Ligonzo, *Appl. Phys. Lett.* 43, 266 (1983).

272. V. Augelli, R. Murri, and T. Ligonzo, *Thin Solid Films* 116, 311 (1984).

273. T. Ohyama, E. Otsuka, M. Osamu, M. Yoshifumi, and K. Kunio, *Jpn. J. Appl. Phys.* 21, L583 (1982).

274. J. W. Orton and M. J. Powell, *Rep. Prog. Phys.* 43, 1274 (1980).

275. D. C. Look, "Electrical Characterization of GaAs Marterials and Devices." Wiley, New York, 1989.

276. J. Yeager and M. A. Hrusch-Tupta, Eds., "Low Level Measurements." Keithely Instruments, Cleveland, 1998.

277. H. H. Weider, in "Nondestructive Evaluation of Semiconductor Materials and Devices" (J. N. Zemel, Ed.), Chap. 2. Plenum, New York, 1979.

278. D. M.Boerger, J. J. Kramer, and L. D. Partain, *J. Appl. Phys.* 52, 269 (1981).

279. F. M. Ryan, *Rev. Sci. Instrum.* 33, 76 (1962).

280. J. B. Blanc, R. H. Bube, and H. E. MacDonald, *J. Appl. Phys.* 32, 1666 (1961).

281. J. M. Lavine, *Rev. Sci. Instrum.* 29, 970 (1958).

282. E. M. Pell and R. L. Sproull, *Rev. Sci. Instrum.* 23, 548 (1952).

283. J. L. Levy, *Phys. Rev.* 92, 215 (1953).

284. N. Z. Lupu, N. M. Tallan, and D. S. Tannhauser, *Rev. Sci. Instrum.* 38, 1658 (1967).

285. P. M. Hemenger, *Rev. Sci. Instrum.* 44, 698 (1973).

286. R. Murri, N. Pinto, and L. Schiavulli, *Nuovo Cimento, D* 15, 785 (1993).

287. G. P. Carver, *Rev. Sci. Instrum.* 43, 1257 (1972).

288. W. Eberle, J. Hirschinger, U. Margull, W. Prettl, V. Novak, and H. Kostial, *Appl. Phys. Lett.* 68, 3329 (1996).

289. R. P. Taylor, R. Newbury, A. S. Sachrajda, Y. Feng, P. T. Coleridge, J. P. Mecaffrey, M. Davies, and J. P. Bird, *Jpn. J. Appl. Phys.* 36, 3964 (1997).

290. R. P. Taylor, R. Newbury, A. S. Sachrajda, Y. Feng, P. T. Coleridge, M. Davies, and J. P. Mecaffrey, *Superlattices Microstruct.* 24, 337 (1998).

291. M. Yokoi, T. Okamoto, S. Kawaji, T. Goto, and T. Fukase, *Physica B* 251, 93 (1998).

292. V. B. Shikin and Y. V. Shikina, *Low Temperature Phys.* 25, 137 (1999).

293. S. F. W. R. Rycroft, R. A. Doyle, D. T. Fuchs, E. Zeldov, R. J. Drost, P. H. Kes, T. Tamegai, S. Ooi, A. M. Campbell, W. Y. Liang, and D. T. Foord, *Superconductor Sci. Technol.* 12, 1067 (1999).

294. K. Aoki and S. Fukui, *Physica B* 272, 274 (1999).

295. J. Mircovic and K. Kadowaki, *Physica B* 284, 759 (2000).

296. D. J. Howarth, R. H. Jones, and E. H. Puetly, *Proc. Phys. Soc. London B* 70, 124 (1957).

297. A. M. Chang, P. Berglund, D. C. Tsui, H. L. Stormer, and J. C. M. Hwang, *Phys. Rev. Lett.* 53, 997 (1984).

Chapter 10

THIN FILMS FOR HIGH-DENSITY MAGNETIC RECORDING

Genhua Pan

Centre for Research in Information Storage Technology, Department of Communication and Electronic Engineering, University of Plymouth, Plymouth, Devon PL4 8AA, United Kingdom

Contents

1. INSTRUMENTS FOR MAGNETIC MEASUREMENT

1.1. BH and MH Loop Measurement

The most commonly used instruments for MH & BH loop measurement are the BH looper, the vibrating sample magnetometer (VSM), and the alternating gradient force magnetometer (AGFM). These instruments can be divided into two categories by the techniques used: (1) inductive techniques that measure the voltage induced by a changing flux, and (2) force techniques that measure the force exerted on a magnetized sample in a magnetic field gradient. BH looper and VSM are category (1) instruments, and AGFM belongs to category (2).

1.1.1. BH Looper

A BH looper usually employs a pair of Helmholtz coils driven by an ac field of frequency from 1 to 100 Hz. The test sam-

Handbook of Thin Film Materials, edited by H.S. Nalwa
Volume 5: Nanomaterials and Magnetic Thin Films
Copyright © 2002 by Academic Press
All rights of reproduction in any form reserved.

ISBN 0-12-512913-0/$35.00

ple is placed in a pair of pickup coils, which are connected differentially so that the voltages induced by the flux change due to the applied ac field are canceled out and the voltages induced by the flux change of the test sample are enhanced. The induced voltage in the pick-up coil is given by Faraday's law, $v = N\frac{d\Phi}{dt} = NA\frac{dB}{dt}$, where N is the number of turns of the pick-up coil, and A is the cross-sectional area. The integration of the induced voltage v over time t gives the value of flux density B, which is fed to the vertical axis of a BH loop plotter (usually an oscilloscope). The H field value can be measured from the voltage across a resistor (usually 1 Ω) in series with the Helmholtz coil, or by a Hall probe next to the pick-up coil inside the Helmholtz coil.

A BH looper is normally used for a quick measurement of BH loops of soft magnetic films. It is a useful tool for determining the H_c of soft magnetic films. But it is not suitable for measuring BH loops of films with large coercivities because of the limited field value a Helmholtz coil can produce (usually smaller than 100 Oe). BH loopers are also very limited in sensitivity and can only be used with relatively large samples. Because of the difficulties in calibrating the magnetic moment, usually they are not usually used for magnetic moment measurement.

1.1.2. VSM

VSM is an instrument based on the principle of electromagnetic induction. Its basic structure is schematically shown in Fig. 1, which consists of an electromagnet, a vibration unit with a sample holder, a pair of pick-up coils, and a field sensor. A maximum field of 30 kOe is obtainable from an electromagnet field system. If a field higher than this is required, a superconducting magnet can be used, which is capable of producing a field as high as 160 kOe. The test sample (usually less than 10 mm square in size) in a uniform magnetic field can be treated as a magnetic dipole. The magnetic fluxes from the sample, which vibrates in the direction perpendicular to the applied field, induces an emf in the pick-up coil positioned near by. This emf is proportional to the magnetic moment of

the sample and is given by

$$e = \frac{3}{4\pi}\mu_0 M\omega g_1 \tag{1.1}$$

where M is the magnetic moment of the test sample, ω is the angular velocity of the vibration, and g_1 is the instrument factor. The value of g_1 is determined by measuring a sample with a known magnetic moment (calibration sample). After calibration with a standard sample, the output from the pick-up coil represents the value of the magnetic moment of the sample, and the output from the field sensor gives the applied field value. The output of the instrument is therefore a MH loop or, more accurately, a curve showing the field dependence of the magnetic moment of the test sample, from which many magnetic parameters can be obtained. VSM is one of the most commonly used magnetometers for the characterization of magnetic materials in both the research laboratory and the production factory. The ultimate sensitivity of a modern VSM is as high as 10^{-6} emu. And the field resolution can be as small as 1 mOe. It is widely used for measurements of soft magnetic films (H_c and M_S), thin film recording media (H_c, M_S, Mrt, S^*, ΔM curve), and gaint magnetoresistance (GMR) spin-valve films (MH loops, H_{ex}, H_f). When equipped with a cryostat or a heating module, VSM can be used to measure magnetic properties of films at various temperatures ranging from 4 K (with a liquid He cryostat) to 1000 K (with a heating module). This extended capability of VSM is particularly useful for the measurement of GMR, spin-valve, and tunnelling magnetoresistive (TMR) films and antiferromagnet/ferromagnet exchange systems. Details of these measurements are discussed in the corresponding sections.

1.1.3. AGFM

The alternating gradient force magnetometer (AGFM) was invented by Flanders in 1988 [1]. Its major advantage is that it has a sensitivity of 10^{-8} emu, which is 100 time more sensitive than a VSM. Its basic structure is schematically shown in Fig. 2, consisting of a field system that is the same as that of a VSM, two pairs of field gradient coils, and a piezo bimorph unit with an extension cantilever, on one end of which the

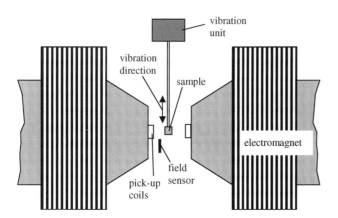

Fig. 1.   Schematic diagram of an VSM.

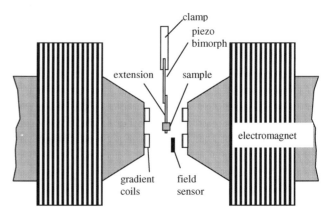

Fig. 2.   Schematic diagram of an AGFM.

magnetic sample is mounted. The test sample is magnetized by the field from the electromagnet and is at the same time subject to a small alternating field gradient, on the order of a few Oe to a few tens of Oe/mm. The alternating gradient field exerts an alternating force f_x on the sample, which is proportional to the magnitude of the field gradient dh_x/dx and the magnetic moment of the sample m_x, and is given by

$$f_x = m_x \frac{dh_x}{dx} \qquad (1.2)$$

The alternating force exerted on the sample produces a displacement of the cantilever, which is converted to the voltage output of the piezo bimorph. By operating at or near the mechanical resonance frequency of the cantilever, the sensitivity of the system is greatly enhanced. For a detailed quantitative correlation between the voltage output and the sample magnetic moment, refer to the original paper by Flanders [1].

1.1.4. Comparisons between VSM and AGFM

Although the AGFM is 100 times more sensitive than the VSM, the real advantage of using an AGFM in terms of sensitivity is not that different. This is because a VSM can accept much larger sample sizes than an AGFM. The lower sensitivity of a VSM can be compensated for by its larger sample size [2].

The disadvantages of an AGFM are associated with its alternating gradient field, which makes it unsuitable for certain measurements, such as low coercivity, magnetization decay, and remanence measurements [3]. Table I gives a brief comparison of VSM and AGFM. For a detailed technical and application review of the two types of magnetometers, refer to the article by Speliotis [2].

1.2. Magnetoresistance Measurement: Four-Point Probe Method

1.2.1. Linear Four-Point Probe

The four-point probe is a widely used method for non-destructive measurement of sheet film resistance or magnetoresistance. Figure 3 is a schematic setup of a four-point

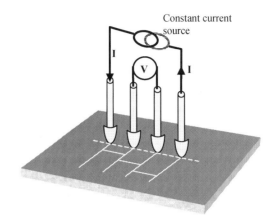

Fig. 3. Schematic diagram of a linear four-point probe.

probe method. A set of four linear microprobes is placed on a flat surface of a material with their separation distances marked in the figure as S_1, S_2, and S_3, respectively. A constant current source is connected to the two outer electrodes, and the floating potential is measured across the inner pair. This configuration has been treated in detail by Valdes [4], who showed that when the probes are placed on a material of semi-infinite volume, the resistivity is given by

$$\rho = \frac{V}{I} = \frac{2\pi}{1/S_1 + 1/S_3 - 1/(S_1 + 1/S_2) - 1/(S_2 + S_3)} \qquad (1.3)$$

When $S_1 = S_2 = S_3 = S$, Eq. (1.3) reduces to

$$\rho = \frac{V}{I} 2\pi S \qquad (1.4)$$

If the material to be measured is a thin film with a thickness of δ on an insulating substrate and $\delta \ll S$, it can be shown that [4]

$$\rho = \frac{V}{I} \frac{\delta \pi}{\ln 2} \qquad (1.5)$$

or sheet resistance R_s is

$$R_s = \frac{\rho}{\delta} = \frac{V}{I} \frac{\pi}{\ln 2} = 4.532 \frac{V}{I} \qquad (1.6)$$

Equation (1.6) implies that the value of the measured sheet resistance is independent of the probe spacing.

1.2.2. Square Four-Point Probe

A square four-point probe as shown in Fig. 4 may also used for sheet resistance measurements. In the square probe configuration, the current is fed in through any adjacent probes, and the voltage is measured from the other two. For an equally spaced square probe configuration, the sheet resistance is given by [5]

$$R_s = \frac{V}{I} \frac{2\pi}{\ln 2} = 9.06 \frac{V}{I} \qquad (1.7)$$

or the resistivity ρ of the film is

$$\rho = \delta R_s = \frac{V\delta}{I} \frac{2\pi}{\ln 2} = 9.06 \frac{V\delta}{I} \qquad (1.8)$$

Table I. Comparisons of VSM and AGFM

	VSM	AGFM
Typical parameters		
Noise floor	10^{-6} emu	10^{-8} emu
Typical sample size	6 to 10 mm	1 to 3 mm
Range of measurement	Up to a few thousand emu	Up to 1 emu
Optimal field resolution	0.001 Oe	Affected by AFG
Typical measurements		
Low-coercivity measurement	Fine	Not suitable
Magnetization decay measurement	Fine	Not suitable
Remanence measurement	Fine	Not suitable
Very thin films with high M_s	Fine	Fine
Very thin films with very low M_s	Not suitable	Fine

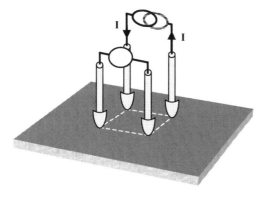

Fig. 4. Schematic diagram of the square four-point probe.

The advantage of using a square probe configuration for magnetoresistance measurement is that the current direction during measurement can be changed by a simple switch in the circuit without the need to touch the sample or change the direction of the applied field. A particular use of this is in the measurement of anisotropic properties of GMR [7].

1.3. Schematic Frequency Permeameter

Knowledge of the permeability of soft magnetic films over a wide frequency range is important for recording head applications. However, conventional permeability measurement methods, such as the toroidal method or the impedance bridge method, are not suitable for measurement of thin film head materials because thin films for heads are usually anisotropic, whereas these methods can only be used for isotropic materials. There has been considerable interest in the past two decades in the development of permeameters for the high-frequency permeability measurement of anisotropic thin film materials. One of most successful instruments of this type was developed by Grimes et al. [6], who used a shorted transmission line as a driving source and a "figure-eight" coil to sense the flux change in the sample. The output signal from one of the figure-eight coil loops is measured as the S_{21} forward transmission parameter with a HP network analyzer and is converted to the complex permeability of the sample. The instrument can operate in the frequency range between 0.1 and 200 MHz.

Figure 5 is a schematic diagram of the swept frequency permeameter. The high-frequency driving signal supplied by the network analyzer is fed into a shorted transmission line jig with a characteristic impedance Z_0 of 50 Ω. In the operation frequency range, the driving signal is considered a traveling wave with its E field perpendicular to the drive sheets and H field transverse to the drive sheets. The sample is inserted into one of the loops of the figure-eight coil with its easy axis perpendicular to the field direction for hard axis permeability measurement. The network analyzer is capable of storing in its registers the complex phasors measured by the transmission line jig at different frequencies, which is critically important for producing a permeability spectrum. It can also be interfaced, via its HPIB (IEEE 488.2) interface, for control, data logging, and data processing to a PC with HP VEE visual programming software.

According to Grimes [6], the forward transmission parameter S_{21} is given by

$$S_{21} = V_{\text{out}}/V_{\text{in}} = -i\omega\mu_0(\mu_r - 1)A_f\beta + \Gamma \qquad (1.9)$$

where V_{out} and V_{in} are the output and input voltages of the transmission line jig, i is the complex operator, ω is the angular frequency of the driving signal, A_f is the cross-sectional area of the magnetic film, $\beta(= 1/2WZ_0, W$ is the width of the drive sheets) is a geometry constant related to the shorted transmission line driving jig, and Γ represents a signal arising from a circuit response and possible loop imbalance when the permeability of the sample is unity (a magnetic saturated sample or a plain substrate sample).

In the permeability measurement, the S_{21} parameters of a sample are measured with and without a dc saturation field and are stored in the network analyzer as $S_{21}|_0$ and $S_{21}/_{\text{sat}}$, respectively. The value of $S_{21}/_{\text{sat}}$ is equal to Γ by Eq. (1.1) and is numerically subtracted by

$$\Delta S_{\text{sample}} = S_{21}|_0 - S_{21}/_{\text{sat}} = -i\omega\mu_0(\mu_r - 1)A_f\beta \qquad (1.10)$$

A reference sample with known permeance is required to determine the permeability of a test sample. A similar operation for ΔS_{ref} is carried out for the reference sample, and the permeability of the sample is calculated from the formulae

$$R = \Delta S_{\text{sample}}/\Delta S_{\text{ref}} \qquad (1.11)$$

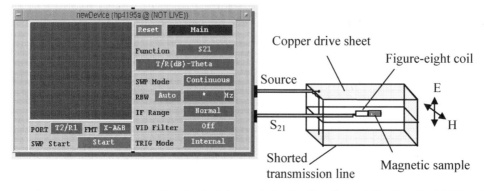

Fig. 5. A swept frequency permeameter consisting of a shorted transmission line jig, a figure-eight coil, and a HP Network analyzer.

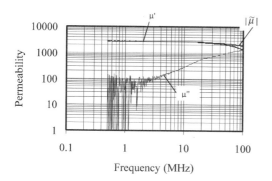

Fig. 6.   A typical hard axis permeability spectrum for thin film samples.

and

$$R = [(\mu_r - 1)A_f]_{\text{sample}}/[\mu_r - 1)A_f]_{\text{ref}}$$
$$\equiv [\mu_r t]_{\text{sample}}/[\mu_r t]_{\text{ref}} \qquad (1.12)$$

for samples with $\mu_r \gg 1$ and with the same sample size as the reference. Therefore

$$\mu_r = R[\mu_r t]_{\text{ref}}/t_{\text{sample}} \qquad (1.13)$$

where t is the thickness of the magnetic film.

The measured permeability is a complex number and is given by

$$\tilde{\mu} = \mu' - i\mu'' \qquad (1.14)$$

where μ' and μ'' are the real and imaginary parts of the complex permeability, respectively. Figure 6 is a typical permeability spectrum measured by the swept frequency permeameter.

2. BASIC PRINCIPLES OF MAGNETIC RECORDING

2.1. The Write/Read Process

The basic principle of an inductive writing process in a computer disk drive recording system is depicted in Fig. 7. The recording system is composed of a thin film recording head and a thin film recording medium. The thin film head consists of three major parts, a yoke-type magnetic core, a conducting Cu coil, and an air gap. The magnetic core is composed of a bottom yoke (P1) and a top yoke (P2), which is usually made of NiFe Permalloy or other soft magnetic films, such as CoNbTa, FeAlN, or FeTaN films. The recording medium is a thin layer (10–30 nm thick) of CoCr-based film coated with other nonmagnetic layers on a glass or Al disk substrate. When in operation (write or read), the disk spins at a speed of 3600–10,000 rpm, and the head flies on the disk surface at a typical flying height of 10–100 nm, which is also known as the head-medium spacing.

The information to be stored in a medium is first encoded and converted by signal processing into write current (write pulses) of the recording head. The write current, I, in the Cu coil produces a magnetic flux circulating in the core. The

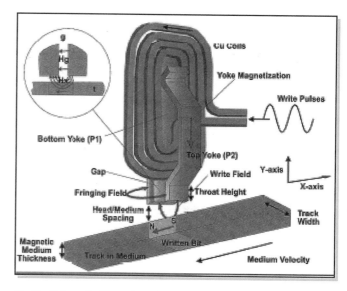

Fig. 7.   Schematic illustration of the inductive write process. Reproduced with permission from [8], copyright 1999, ICG Publishing: *Datatech*.

direction of the magnetic flux depends on the polarity of the writing current, which is usually toggled from one polarity to the other to write digital information into the recording medium. The gap permits part of the magnetic flux in the core to fringe out, intercept, and magnetize the recording medium in close proximity to the air-bearing surface of the head, forming recorded magnetization patterns in the medium (as also shown in the inset of the figure). The recording density is determined by the length of the written bit and the track width. The information thus stored in the medium can be kept for a long period (usually over 10 years) if the medium is not exposed to magnetic fields strong enough to magnetize the medium.

The recorded information can be reproduced by a replay head (an inductive replay head, a MR head, or a GMR spin-valve head). When an inductive replay head, for example, passes over a recorded magnetization pattern in the disk, the surface flux from the magnetized pattern is intercepted by the head core, and a voltage is induced in the coil, which, by Faraday's law, is proportional to the rate of change of this flux. The induced voltage is then reconstructed by signal processing into the original signal.

The magnetization pattern in the medium can be recorded longitudinally or perpendicular to the film plane, depending on the recording mode used or, in other words, depending on the type of head and medium used. The combination of a ring head and a medium with in-plane anisotropy tends to produce a recorded magnetization that is predominantly longitudinal. This is known as longitudinal recording and is the predominant recording mode used so far. An alternative to this is perpendicular recording [11], by which the magnetization is perpendicularly recorded in the medium by with the combination of a single pole head and a double-layered medium [14]. A double-layered medium consists of a perpendicular anisotropy medium and a soft magnetic back layer.

2.2. Write Field of Recording Heads

2.2.1. Write Field of Longitudinal Heads

Various magnetic heads have been used for magnetic recording. The magnetic heads for longitudinal recording are designed to produce a greater longitudinal field component H_x and, for perpendicular recording, a greater perpendicular field component H_y. The precise calculation of the magnetic field produced by a real head requires numerical computation. The most commonly accepted analytical solution for the field in front of the write gap g of a ring head with infinite size is the Karlquist approximation [12]. The longitudinal component of the writing field H_x at a point P(x, y) beneath the gap is given by

$$H_x = \frac{H_g}{\pi}\left[\arctan\left(\frac{g/2 - x}{y}\right) + \arctan\left(\frac{g/2 + x}{y}\right)\right] \quad (2.1)$$

where x and y are the horizontal and vertical coordinates of point P, and H_g is the deep gap field of the recording head. It can be seen from the above equation that the write field of a head is proportional to the deep gap field H_g. The inset of Fig. 7 gives a schematic view of the head–medium interface and the location of H_g and H_x. According to Mallinson [9], the deep gap field H_g in Oersteds for a head with write efficiency of η_w, a gap length of g (cm), a coil of N turns, and a write current of I (amperes) is given by

$$H_g = 0.4\pi\eta_w\frac{NI}{g} \quad (2.2)$$

The maximum deep gap field a head can produce is limited by the saturation magnetization M_s (or saturation flux density B_s) of the head material. Pole tip saturation occurs when

$$H_g \geq 0.6B_s \quad (2.3)$$

for thin films heads [9],

$$H_g \geq 0.5B_s \quad (2.4)$$

for ferrite heads, and

$$H_g \geq 0.8B_s \quad (2.5)$$

for metal-in-gap (MIG) heads [13].

In high-density recording, the required deep gap field for effective writing in a recording medium with a coercivity of H_c is given by [9]

$$H_g \geq 3H_c \quad (2.6)$$

To achieve effective writing, the saturation flux density of the head material for a thin film head must satisfy

$$B_S \geq 5H_c \quad (2.7)$$

or the maximum medium coercivity,

$$H_c \leq 0.2B_S \quad (2.8)$$

These criteria provide a simple guideline for the design of recording heads and the selection of head materials and medium parameters. For example, if the coercivity H_c of a recording medium for an areal density of 20 Gbit/in^2 is 3000 Oe, the required deep gap field of the writer will be 9000 Oe, and the saturation flux density of the head material for the thin film head will be 15 kG.

2.2.2. Write Field of a Perpendicular Single Pole Head

The combination of a single pole head with double-layered perpendicular media [14] effectively places the media in the write gap of the head. The deep gap field is thus given by Eq. (2.3). And the maximum writable coercivity of perpendicular recording media is equal to the deep gap field of the single pole head, which is given by

$$H_c = H_g \leq 0.6B_S \quad (2.9)$$

It can be seen from Eq. (2.9) that the maximum writable medium coercivity in perpendicular recording with a single pole head and double-layered media combination is almost three times the maximum writable coercivity in longitudinal media. High medium coercivity is essential to overcome the superparamagnetic problem at increasingly higher recording densities, which is discussed in Section 3.1.3. Perpendicular recording has a major advantage in this over its longitudinal counterpart.

2.3. Written Magnetization Transition in a Recording Medium

2.3.1. Longitudinal Transition

In digital recording, a bit of information is stored in the magnetic medium as the write current drives the writer to magnetize the medium right beneath the head gap along one of the two possible directions. An ideal written magnetization pattern representing di-bit information is shown in Fig. 8a, and the variation of magnetization along the x direction in the di-bit

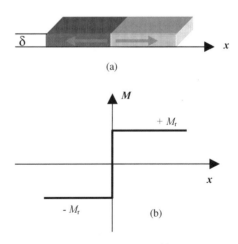

Fig. 8. A di-bit written magnetization pattern (a), and its ideal step function magnetization variation across the transition (b).

pattern is given in Fig. 8b. In such an ideal case, the magnetization written in the medium is at one of the remanent levels. The transition between the two remanent levels is a step function with zero transition length.

A real transition is not a step function. The variation of recorded magnetization in the x direction of the transition for longitudinal recording is best described by the arctangent model in which the magnetization distribution in the transition is given by

$$M(x) = \frac{\pi}{2} M_r \arctan\left(\frac{x}{a}\right) \qquad (2.10)$$

where a is the so-called transition parameter, by which the transition length l is defined as $l = \pi a$.

Fig. 9 illustrates the arctangent transition model.

When a recording bit pattern is written by the magnetic field from a recording head, the magnetized bit pattern will generate an internal field opposing the magnetization of the written bit because of the magnetic charges arising from the divergence of the written magnetization. This field is termed the transition self-demagnetizing field $\mathbf{H}_d$ and is given by

$$\mathbf{H}_d = -\int_V \frac{\rho \mathbf{r}}{|r|^3} dV \qquad (2.11)$$

where $\mathbf{r}$ is a position unit vector, V is the volume of the written magnetization pattern, and ρ is the magnetic pole density, which is defined as the divergence of magnetization $\mathbf{M}$,

$$\rho = -\nabla \cdot \mathbf{M} \qquad (2.12)$$

2.3.2. Perpendicular Transition

The perpendicular recording was first proposed by Iwasaki and Nakamura [14] from their study on the transition self-demagnetization phenomenon in longitudinal recording media. One of the fundamental differences between perpendicular recording and its longitudinal counterpart is the distribution of the demagnetizing field in the written transition. The transition self-demagnetizing field for longitudinal and perpendicular transition can be calculated with Eqs. (2.11) and (2.12) if the transition magnetization distribution is known. Figure 10 shows the transition self-demagnetizing field distribution for a

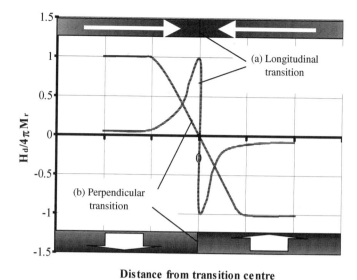

Distance from transition centre

Fig. 10. Schematic illustration of demagnetizing field distribution along a longitudinal steplike transition (a) and a perpendicular steplike transition (b).

longitudinal step-like transition (a) and a perpendicular step-like transition (b). It can be seen that the demagnetizing field in a perpendicular steplike transition is zero in the center of the transition and $4\pi M_r$ away from the center, whereas in the case of longitudinal transition, the demagnetizing field is at maximum $(4\pi M_r)$ in the center of the transition and almost zero away from the transition. This implies that it is possible to realize steplike transition (zero transition length) in perpendicular recording from the point of view of transition self-demagnetization. The ultimate recording density for perpendicular recording is therefore not limited by the transition self-demagnetizing effect.

The demagnetizing field away from the perpendicular transition $(H_d = 4\pi M_s)$, which is inherent from its thin film geometry, shears the perpendicular MH loop. For low-coercivity media, the shearing of the MH loop results in a considerably reduced remanence squareness ratio $(S \ll 1)$ and wide switching field distribution (SFD). This will cause the time decay of recorded magnetization away from transition and high media noise [16]. Figure 11 shows the sheared MH loops of high-coercivity media (solid line) and low-coercivity media (dashed line), in comparison with the unsheared loop (dotted line loop) for films without the effect of H_d. A strong anisotropy field, $[H_{k\perp} - 4\pi M_s] > 5$–$10$ kOe [15], is required to overcome this demagnetizing field and to give a stable remanent magnetization state of the medium (unity remanence squareness + a relatively large negative nucleation field H_n, as shown in the figure) [148]. A soft magnetic back layer reduces the surface magnetic pole density and therefore reduces the effect of the demagnetizing field [129].

Middleton and Wright [128] have used the Williams–Comstock model [18] to compare the transition length for perpendicular and longitudinal recording and concluded that a very narrow transition length could be realized in perpendicular recording.

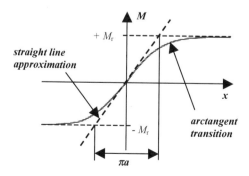

Fig. 9. The arctangent transition. Adapted from [9], copyright 1996, Academic Press.

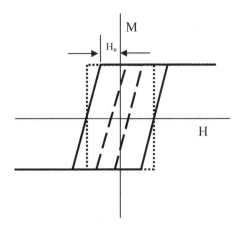

Fig. 11. MH loops of perpendicular media. Dashed line, lower coercivity MH loop sheared because of H_d and $S \ll 1$; solid line, high-coercivity MH loop sheared because of H_d and with a negative nucleation field and unity squareness; dotted line, MH loop without shearing.

3. THIN FILM RECORDING MEDIA

3.1. Physical Limits of High-Density Recording Media

3.1.1. Transition Self-Demagnetization Limit

One of the fundamental physical limits of longitudinal recording is the transition self-demagnetization limit. The self-demagnetizing field for an arctangent transition is given by [9]

$$H_d = \frac{M_r \delta x}{\pi(x^2 + a^2)} \tag{3.1}$$

where $M_r \delta$ is the product of remanent magnetization M_r and thickness δ of a medium.

The maximum value of a self-demagnetizing field $H_{d,\,max}$ occurs at $x = a$,

$$H_{d,\,max} = \frac{M_r \delta}{2\pi a} \tag{3.2}$$

Equation (1.6) implies that the maximum self-demagnetizing field in a longitudinal transition increases with the increase in linear recording density (decrease in transition parameter a). It is possible that at higher density the self-demagnetization field associated with a recorded transition might be high enough to demagnetize the recorded magnetization pattern and make the stored information disappear.

If the recording transition is self-demagnetization limited, the minimum transition length l_{min}, a recording medium can sustain is obtained from Eq. (1.6) by equating $H_{d,\,max} = H_c$,

$$l_{min} = \pi a = \frac{2\pi M_r \delta}{H_c} \tag{3.3}$$

A higher density (smaller l_{min}) can be achieved by using a smaller $M_r \delta$ or a higher H_c medium. A modern recording medium, for an areal density of 20 Gbit/in^2, for example, has a typical $M_r \delta$ value of 0.45 memu/cm^2 and a H_c of 2900 Oe [17].

The transition parameter a derived from the Williams–Comstock model [18], is given by

$$a = 2\sqrt{\frac{M_r \delta}{H_c}\left(d + \frac{\delta}{2}\right)} \tag{3.4}$$

where d is the head/medium spacing. The significance of the Williams–Comstock model is that it revealed the correlation of the transition parameter with the head and medium parameters. It confirms the self-demagnetization limit and reveals the effect of head/medium spacing on the recording density. A higher density (smaller a) is achievable by using a medium with smaller $M_r \delta$ and higher H_c, and a smaller head/medium spacing.

3.1.2. Signal-to-Noise Ratio Limit

The overall reliability of a digital recording system is determined by its bit error rate. Noise causes bit shift [19] and is one of the major error sources in a digital recording system. The minimum required signal-to-noise ratio (SNR) of a current hard disk drive is about 20 dB. Noise in a recording system arises from three predominant sources: the playback electronic circuits, the playback heads, and the recording medium [40]. In a modern recording system, the percentage of noise from each of the above three sources is typically 10%, 20%, and 70%, respectively. Therefore, the ultimate SNR limit in magnetic storage is perceived as being related to the inherent signal-to-noise capability of the media [19].

Medium noise can be separated into three somewhat distinctive sources [40]: amplitude modulation noise, particulate noise, and transition noise. Detailed physics and mathematical analysis of the three types of noise sources can be found in [40]. Thin film recording media possess virtually no amplitude modulation or particulate noise. The predominant noise source in thin film recording media is the transition noise.

Transition noise refers to the phase and amplitude fluctuations of playback voltage, which are related to the transition center of recorded bits. It increases with the increase in recording density for longitudinal media.

The medium transition noise is normally described in terms of the total noise power (NP) and the noise power spectrum (NPS). The noise power spectrum can be compared with the signal power spectrum leading to the narrow-band signal-to-noise ratio $(SNR)_n$ and the wide-band signal-to-noise ratio $(SNR)_w$ [10].

Noise in thin film media has been intensively studied in the past decade, particularly for longitudinal media [130–134]. It has been concluded that intergranular exchange coupling, which significantly enhances the size of magnetic structures and results in large-sized domains away from saturation, is the main origin of thin film medium noise. According to Bertram [135], the NP reduced by the $M_r \delta$ product of a medium is given by

$$\frac{NP}{M_r \delta} = D^2 \cdot f\left[C^*, \frac{M_r}{H_K} \cdot g\left(\frac{\delta}{D}, \frac{d}{D}\right)\right] \tag{3.5}$$

where D is the grain diameter,

$$f\left[C^*, \frac{M_r}{H_K} \cdot g\left(\frac{\delta}{D}, \frac{d}{D}\right)\right]$$

is the intergranular interaction energy function, $g\left(\frac{\delta}{D}, \frac{d}{D}\right)$ is the intergranular magnetostatic interaction energy function, d is the intergranular nonmagnetic separation, and C^* is the intergranular exchange coupling constant, which is defined by [135]

$$C^* = \frac{A^*}{K_u D}, \qquad (3.6)$$

where A^* is the effective intergranular exchange coupling energy constant and K_u is the crystalline anisotropy constant.

Equation (3.5) implies that, for a medium with a given grain diameter, the reduction of the intergranular exchange coupling constant C^* will lead to the reduction of medium noise power. For a medium with a given thickness, zero intergranular exchange coupling, and large crystalline anisotropy (large H_K), the noise power decreases as the grain diameter decreases. However, if the intergranular exchange coupling is not zero, or the crystalline anisotropy of the medium is not strong, the reduction of grain diameter will lead to very complicated effects on the noise power because the reduction of grain diameter also leads to an increase in intergranular exchange coupling (Eq. (3.6)) and intergranular magnetostatic coupling $g\left(\frac{\delta}{D}, \frac{d}{D}\right)$.

There are distinct differences in noise characteristics between longitudinal and perpendicular media. For longitudinal media, noise occurs mainly at the transition centers and increases with increasing recording density. In contrast to this, medium noise in perpendicular recording occurs away from the transition centers and decreases with increasing recording density [132].

Low-noise thin film media can usually be considered as media with very low or zero intergranular exchange coupling. In this case, the physical interpretation of media noise developed by Mallinson for particulate media also applies to thin film media. According to Mallinson, the $(SNR)_w$ at high bit density is given by [10]

$$(\text{SNR})_w = \frac{nW\lambda_{\min}^2}{2\pi} \qquad (3.7)$$

where n is the number of particles (grains) per unit volume, W is the track width, and $\lambda_{\min}$ is the minimum recording wave length. Equation (3.7) simply states that the $(SNR)_w$ for a particulate medium is equal to the number of particles in a volume of size $W \cdot (\lambda_{\min}/\pi) \cdot (\lambda_{\min}/2)$. This volume is the effective medium volume sensed by a replay head at any instant. The SNR of a medium is proportional to the number of grains per unit volume. The SNR decreases with increasing recording density (decreasing effective medium volume).

3.1.3. Gyromagnetic Switching and Superparamagnetic Limit

3.1.3.1. Gyromagnetic Switching of Magnetization Under a Reversal Field

When a reversal magnetic field H (say a field from a writer) is applied to a recording medium at $t = 0$, the magnetization vectors in the medium respond to the applied field in the form of gyromagnetic precession around the field axis at an angular velocity of about 1.76×10^7 rad/Oe-s before reaching their final equilibrium state. This is schematically shown in Fig. 12, where $\mathbf{M}(0)$ and $\mathbf{M}(\infty)$ are the two equilibrium magnetization states before and after the reversal, respectively. $\mathbf{M}(t)$ is the magnetization state at observation time t. This magnetization precession process is governed by the Landau–Lifshitz–Gilbert equation of motion in the absence of thermal perturbation (i.e., $T = 0$ K) [27],

$$\frac{d\mathbf{M}}{dt} = -\gamma(\mathbf{M} \times \mathbf{H}_{\text{eff}}) + \frac{\alpha}{M_s}\mathbf{M} \times \frac{d\mathbf{M}}{dt}, \qquad (3.8)$$

where $\mathbf{M}$ is the magnetization vector, and $\mathbf{H}_{\text{eff}}$ is the sum of all fields acting on the magnetization vector, including the external field, the anisotropy field, the demagnetizing field, and the exchange interation field. α is the Gilbert damping constant, which is material dependent, and γ is the gyromagnetic ratio ($\gamma = (|e|/2m_e c)g = 1.76 \times 10^7$ rad/Oe-s $= 2.8$ MHz/Oe, where e is the electron charge, m_e is the electron mass, c is the speed of light, and g is the Landau factor, $g = 2$).

The time required to complete the gyromagnetic switching process (the switching time) depends on the material properties (mainly anisotropy field and grain volume) and the strength of the effective field. In general, numerical computation with Eq. (3.8) is used to study the magnetization reversal dynamics. The typical switching time in thin film recording media is less than 1 ns for head fields as high as or greater than the anisotropy field of the media. This implies that thin film recording media are capable of supporting a data rate up to 1 Gbits/s. For a data rate faster than this, the dynamic gyromagnetic switching process will start to play a role [174].

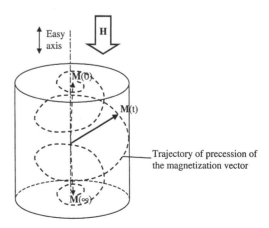

Fig. 12. Schematic illustration of gyromagnetic precession in a reversal field H applied parallel to the easy axis of the magnetic material.

Thermal fluctuation terms can also be introduced into Eq. (3.8) to study thermally assisted switching [25, 26]. In special cases where the switching time is at a moderate scale (no less than a few nanoseconds) and the material consists of uniaxial single-domain particles, semiempirical Neel–Arrhenius formalism is applicable, which is further dicussed in the next section.

3.1.3.2. Thermal Fluctuation After-Effect and Superparamagnetic Phenomenon

With the assistance of thermal energy, the magnetization of a particle can overcome an energy barrier and switch from one stable state to another. This phenomenon is known as the thermal fluctuation after-effect, which is one of the magnetic after-effects [20] and is inherent to ferromagnets.

Thermally assisted switching will take a certain time compared with the quick switching of particle magnetization in a large external field. The basic concept of thermally activated magnetization switching was first studied by Street and Woolley [21]. For uniaxial single-domain Stoner–Wohlfarth particles, or grains with coherent rotation only, the thermally assisted switching rate f (the probability per unit time of successfully crossing the barrier energy E_b) can be described by Neel–Arrhenius formalism [22].

$$f = f_0 e^{E_b/k_b T} \tag{3.9}$$

where the barrier energy E_b is the energy needed to keep the magnetization of the particle in its original state or the energy required to switch the original magnetization of the particle, and f_0 is the attempt frequency, whose dependence on field and temperature is negligible compared with that of the exponential factor, and it is typically taken to be on of the order of 10^9 s^{-1}. k_b is the Boltzmann constant and T is the absolute temperature. At temperature T, the electron spins in the magnetization of a single-domain particle are subject to thermal agitation, the energy of which is represented by $k_b T$.

The time required for the magnetization of a particle to switch, known as the relaxation time or time constant τ, is given by

$$\tau = \frac{1}{f} = \frac{1}{f} e^{E_b/k_b T} \tag{3.10}$$

At temperature T, the time constant τ depends on the value of the barrier energy E_b. The limitation of Eq. (3.10) is that the minimum time constant obtainable by the equation is 1 ns when $E_b = 0$. Therefore Eqs. (3.9) and (3.10) are only applicable to moderate time scales and to high-energy-barrier cases [30]. The energy barrier E_b depends on the anisotropy constant and volume of the particle, the applied field, and the relative orientation of the magnetization with respect to the magnetic easy axis. For a uniaxial anisotropy Stoner–Wohlfarth particle with magnetization aligned parallel to its magnetic easy axis, the energy barrier E_b in a zero external field is given by

$$E_b = K_u V \tag{3.11}$$

where K_u is the uniaxial anisotropy constant of the particle, and V is the particle volume, which is also the smallest switching volume in the thermally activated switching process (also termed activation volume). A smaller anisotropy constant K_u or a smaller particle volume V will result in a smaller energy barrier E_b which makes the particle more vulnerable to thermally assisted switching. If the energy barrier $K_u V$ at temperature T is on the same order of magnitude as $k_b T$, the magnetization of the particle is expected to be activated to overcome the energy barrier [21] under a zero external field.

The ratio of the barrier energy E_b to the thermal energy $k_b T$, $E_b/k_b T$, is known as the thermal stability factor. For non-interaction and single-domain particles, $E_b = K_u V$, the thermal stability factor becomes $K_u V/k_b T$.

According to Charap et al. [25], the behavior of particles can be classified essentially into the following three categories, according to their thermal stability factors:

1. Particles with a thermal stability factor much smaller than unity. These particles are superparamagnetic (i.e., they are internally magnetically ordered, but they lose hysteresis).
2. Particles with a thermal stability factor close to unity. The magnetic behavior of these particles is time dependent. Their response to small applied fields is linear.
3. Particles with a thermal stability factor much greater than unity. These particles respond to applied fields according to their switching characteristics, quickly acquiring an appropriate stable magnetization.

The time dependence of magnetization $M(t)$ and coercivity $H_c(t)$ for a particle of category 2 are given by [22] [23].

$$M(t) = M_0 e^{-ft} + M_\infty (1 - e^{-ft}) \tag{3.12}$$

where M_0 is the initial magnetization value and M_∞ is the ultimate value, and

$$H_c(t) = H_k \left\{ 1 - \left[\frac{k_b T}{K_u V} \ln(f_0 t) \right]^{1/2} \right\} \tag{3.13}$$

where $H_k = 2K_u/M_s$ is the effective anisotropy field of the grain (M_s is the saturation magnetization) and t is the characteristic time derived for a changing field measurement (an MH loop, for example).

The time dependence of magnetization is termed magnetization decay, which is the origin of signal amplitude decay in high-density magnetic recording media.

Time-dependent coercivity is known as dynamic coercivity. As can be seen from Eq. (5.9), dynamic coercivity is a function of the thermal stability factor and the characteristic time. "Characteristic time" refers to the field duration or pulse width if a constant field is used. In the case of a smoothly swept field, the characteristic time is derived from the rate of change of the field [24], which is discussed further in Section 3.4.3.1. Figure 13 is the dependence of media coercivity on the characteristic time and the thermal stability factor of a medium,

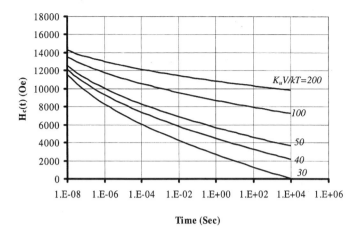

Fig. 13. Media coercivity as a function of characteristic time and thermal stability factor $K_u V/kT$, plotted with Eq. (3.13) and assuming $H_k = 16$ kOe.

plotted with Eq. (3.13). As shown in the figure, for particles with smaller thermal stability factors, the dynamic coercivity is more sensitive to the characteristic measurement time. When the measurement time scale is reduced, the dynamic coercivity of the media increases considerably. For a recording medium with a thermal stability factor of 40, the measured coercivity could vary from 2100 Oe on a time scale of 10,000 s to 12,000 Oe at 10 ns. This has two major implications for magnetic recording: the writability of recording medium at high frequency as the field duration decreases to the nanosecond regime and the long-term stability of the written data.

The time dependence of medium coercivity is the consequence of the natural response (gyromagnetic switching) of the magnetization of a material to an applied magnetic field. When a magnetic field is applied to a sample, the magnetization in the sample responds to the field in the form of precession around the field axis at a relatively high frequency of $\gamma H (= 2.8H$ MHz) in the case of free precession, where γ is the gyromagnetic ratio and H is the applied field. It will take some time to reach a static equilibrium state.

The coercivity of a magnetic material is a measure of the field required to switch the magnetization from one state to another (strictly speaking, the field required to reduce the magnetization of the sample to zero). The role of the applied field is to provide the magnetostatic energy (or Zeeman energy = $-M_s H \cos\theta$) required to overcome the energy barrier and to make the switching happen. In that sense, the applied Zeeman energy is equivalent to the thermal energy (= $k_b T$) in terms of making the magnetization switching happen. As a result of the applied field, the energy barrier term in Eq. (3.10) is reduced by the amount of the equivalent Zeeman energy [168]. For particles of uniaxial anisotropy, with the preferred axis aligned with the field direction ($\theta = 0$), the new energy barrier at field H is given by

$$E_b = K_u V \left[1 - \frac{HM_s}{2K_u}\right]^2 = K_u V \left[1 - \frac{H}{H_k}\right]^2 \qquad (3.14)$$

Substituting Eq. (3.14) into Eq. (3.10), we have

$$\tau = \frac{1}{f_0} e^{E_b/k_b T} = \frac{1}{f_0} e^{(K_u V/k_b T)[1 - H/H_k]^2} \qquad (3.15)$$

Equation (3.15) is valid for $H/H_k < 1$. The figure shows the switching time as a function of H/H_k plotted by Eq. (3.15) in the field range $H/H_k = 0.55$–0.95. We can see that the time required for a magnetization to switch depends on the strength of the applied field. For a very short measurement time scale, the field required to switch the magnetization could be very large, resulting in large coercivity. On the other hand, if the time scale of measurement is very long, the field required to switch the magnetization is reduced, resulting in smaller coercivity.

In low-frequency applications, the magnetization switching time is negligible in comparison with the duration of the applied field. However, as the recording density increases, the duration of the field from the writing head is only a few nanoseconds, and the dynamic process of magnetization reversal in such cases is no longer negligible. When the field duration is shortened, a significantly larger field is required to realize the switching within that duration (Fig. 14).

The dynamic coercivity of a recording medium is the true coercivity in the medium the writing head has to overcome during the writing process. It is therefore also called writing coercivity. As the linear recording density increases (the field pulse width decreases), the writing coercivity of a medium will increase significantly because of the time-dependent nature of the coercivity. In addition, a higher density medium has smaller grain volumes dictated by the low noise and small transition parameter requirements. This implies that the writing coercivity of the medium becomes even more sensitive to the write field duration.

3.1.3.3. Superparamagnetic Phenomenon in Thin Film Recording Media

Current thin film recording medium consists of hcp Co grains of various grain sizes with an in-plane magnetic anisotropy.

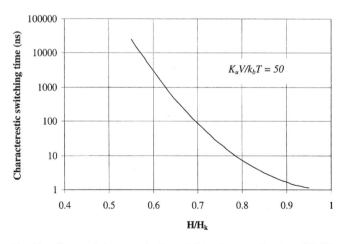

Fig. 14. Characteristic magnetization switching time as a function of H/H_k for particles with thermal stability factor $K_u V/k_b T = 50$.

Because the grains in modern thin film recording media are isolated, to a large extent, by the segregation of nonmagnetic Cr in the grain boundaries (dictated by low medium noise), a thin film medium can be assumed to be composed of uniaxial single–domain Stoner–Wohlfarth grains capable of coherent switching. Therefore, the general principle of superparamagnetism and the formulae developed in Section 3.1.3.2 apply to thin film recording media. However, some corrections are required to use these formulae for quantitative evaluation of the superparamagnetic effect in recording medium because recording medium differs from a Stoner–Wohlfarth particle in the following:

1. *Easy axis distribution.* The easy axes of hcp Co grains in the medium are randomly oriented in the film plane, and, therefore, it is unlikely that in the recorded data pattern the magnetization of each grain has a parallel alignment with its magnetic easy axis. It has been shown by Bertram and Richter [29] that for thin film media consisting of noninteracting grains with their easy axes oriented at random in the plane, the time-dependent coercivity is given by

$$H_c(t) = 0.566 H_k \left\{ 1 - 0.977 \left[\frac{k_b T}{k_u V} \ln \left(\frac{f_0 t}{\ln 2} \right) \right]^{2/3} \right\} \quad (3.16)$$

2. *Activation volume.* Grains in thin film media may not be completely free from intergranular exchange coupling. The grain volume V in the formulas discussed in this section is the smallest volume of material that reverses coherently in the thermally activated switching process. Such a volume is also known as the activation volume. For thin film media, if the grains are completely decoupled, the activation volume is equal to the grain volume. However, intergranular exchange coupling leads to activation volumes greater than a single grain volume. The activation volume in exchange-coupled medium may represent a significant number of grains. Bertram and Richter have pointed out [29] that exchange interaction between the grains reduces the pre-factor (0.566) in Eq. (5.12). A reduced pre-factor implies that the coercivity of a medium with intergranular interaction is less time dependent, or a medium is thermally more stable. For a typical medium, the pre-factor was estimated to be 0.474 [29]. However, large activation volume is the main source of medium noise [40].

3. *Grain size distribution.* The grain volume in thin film media is not the same for every grain, therefore a grain volume distribution must be considered in determining V. For typical media, grain size distribution function is a log-normal function [28]. The grain volume V in the formulas shown above should be the mean grain volume of the entire medium if they are used to assess the thermal stability of a medium. As shown in Fig. 15, a preferred grain size distribution should have a very narrow peak (solid curve) with its maximum probability density above and away from the thermally unstable grains. If the distribution peak is wide and shifted toward the superparamagnetic grain size region, the medium is thermally unstable.

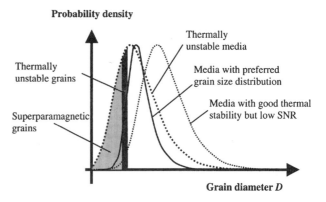

Fig. 15.   Schematic illustration of grain size distributions in recording media.

On the other hand, if the distribution curve is wide and shifted toward the larger grain region, the medium is thermally very stable but with low SNR, because the number of grains in each recorded bit is considerably reduced. There is always a trade-off in media thermal stability and SNR. Obtaining a narrow grain size distribution is one of the key techniques used in the modern thin film medium preparation process to improve the thermal stability of a medium. A more detailed mathematical treatment of the effect of grain volume distribution on time-dependent magnetic behavior can be found in [37].

4. *Transition self-demagnetization field.* In recorded data patterns of longitudinal media, as discussed in Section 2.3, a self-demagnetization field exists in each transition. The energy barrier term E_b needs to be modified to include the effect of the demagnetization field from the recorded magnetization pattern, with the from [24]

$$E_b = K_u V \left[1 - \frac{H_d}{H_k} \right]^2 \quad (3.17)$$

where H_d is the spatially varying demagnetization field from the recorded magnetization pattern. A formula for H_d of arctangent transition was given by Eq. (3.1). The track width averaged di-bit demagnetizing field at the di-bit center is given by [39]

$$H_d = \frac{8 M_r(t) \delta}{W} \left[\sqrt{1 + \left(\frac{2W}{B} \right)^2} - 1 \right] \quad (3.18)$$

where W is the recorded trackwidth and B is the bit length. It can be seen that H_d increases with increasing recording density (decreasing B and W). At high densities, the high demagnetizing field of recorded transitions in longitudinal recording reduces the stability factor from its "bulk" value, leading to a rapid decay of playback signal [36].

One of the fundamental merits of perpendicular recording over its longitudinal counterpart is its lower transition self-demagnetizing field, H_d, particularly at higher densities. This makes the perpendicular recording thermally more stable than the longitudinal recording. However, because of the inherent perpendicular demagnetizing field ($H_d = 4\pi M_s$) in the perpendicular media as discussed in Section 2.3.2, perpendicularly recorded bits tend to be unstable at lower recording densities. Special requirements for perpendicular media, such as a large anisotropy field, large coercivity, unity squareness, and a negative nucleation field, are required to overcome this inherent demagnetizing field.

3.2. Considerations of Medium Design

In summary, from the viewpoint of medium thermal stability, we require a thin film medium with large grain volume or activation volume V, a large anisotropy constant K_u, and small transition demagnetization field H_d. To maintain a high SNR (minimum 20 dB), fine grains with large K_u and minimum intergranular interaction are required. It is also known from Section 3.1 that to overcome the transition self-demagnetization limit, a medium with small $M_r\delta$ and high H_c is required. In the design of high-density recording media, the limit of grain size is set by the thermal stability requirement, typically a medium thermal stability factor of 40, i.e., $E_b = 40k_bT$ at its working temperature T. The lifetime of the stored data estimated by Eq. (3.10) is about 10 years [25]. By the definition of the thermal stability factor, we have two options for achieving the thermal stability factor of 40: using larger grain volumes or using media with a larger anisotropy constant K_u. Because a larger grain volume requires either large grain diameter or thicker films, which conflict with the SNR and transition self-demagnetization limits, media with large K_u become a preferred option.

To meet the transition self-demagnetization requirement, there has been a trend to use media with increasingly smaller $M_r\delta$ and increasingly higher H_c. The typical values of $M_r\delta$ are smaller than 1 memu/cm^2 for MR media [9] and smaller than 0.5 memu/cm^2 for spin-valve GMR media [17]. The lower limit of the $M_r\delta$ product is set by the increasingly smaller sensitivity of the read heads, i.e., by the increasingly smaller GMR ratio for the case of spin-valve replay heads due to the scaling effect, which is discussed in Section 4.4.3.

The limit of higher media coercivity is set by two factors: the available medium coercivity of the existing recording media and the available writing field of the writer due to the limitations of saturation magnetization of the head material.

The fundamental origin of high coercivity in thin film recording media is the combination of high magnetocrystalline anisotropy of the grains, smaller grain sizes, and lower intergranular interaction. The highest coercivities are obtained from films composed of the so-called Stoner–Wohlfarth single-domain grains. For media with exchange isolated grains, the maximum coercivity is given by $H_c = H_k/2$, where $H_k(= 2K_u/M_s)$ is the anisotropy field of the media. For Co-based media with an H_k of 8–10 kOe, the limiting coercivity is about 4–5 kOe. Coercivities higher than 5 kOe are obtainable from FePt thin media because of their extremely large magnetocrystalline anisotropy of the long-range ordered FePt face-centered tetragonal (fct) L1(0) structure [117–126]. However, the application of such films as longitudinal media is

inadequate because of the limited writing field a longitudinal thin film head can produce.

As discussed in Section 2.2, the maximum writable media coercivity by a longitudinal thin film head is given by $H_c \leq 0.2 \times 4\pi M_s$, where M_s is the saturation magnetization of the thin film head material. For a medium with a coercivity of 4000 Oe, the required $4\pi M_s$ of the head material will be 20 kGauss, which is almost the highest $4\pi M_s$ of the currently available head material. This imposes fundamental restrictions on the maximum writable medium coercivity in longitudinal recording and will eventually limit the maximum recording density of longitudinal recording due to the superparamagnetic effect.

On the other hand, the combination of a single-pole head with double-layered media in perpendicular recording is much more advantageous than its longitudinal counterpart in terms of maximum writable medium coercivity. Because the use of high-coercivity media is essential to overcoming the superparamagnetic problem in high-density recording, perpendicular recording has recently been considered as the technology to take the recording density beyond the superparamagnetic limit of longitudinal recording.

3.3. Preparations of Recording Media

3.3.1. CoCr-Based Longitudinal Recording Media

Media Layer Structure. The magnetic layer is the essential layer for providing the storage mechanism for a recording medium. However, a modern thin film medium usually consists of multiple layers with layer structures as shown in Fig. 21 for a typical longitudinal medium. The purpose of using the multiple-layer structure is to obtain a controlled grain size, grain isolation (Cr segregation), low intergranular exchange coupling, high coercivity, and preferred crystallographic orientation of the magnetic layer. Chemically toughened glass or ceramic disks are usually used as substrates. Al-Mg alloy substrates coated with a very thick electroless-plated NiP layer have also been used in the past as substrate materials. The surface of these substrates can be assumed to be perfectly smooth, although they are polished and textured before the deposition of films. These various layers are usually deposited on the substrates by vacuum sputtering. The role of the seed layer is to provide an initial texture and grain growth templates, while still providing a smooth surface to the underlayer [42]. It may also provide an epitaxial growth surface condition for obtaining the desired crystallographic orientation of the underlayer. This crystallographic orientation is transferred to the magnetic layer via epitaxial growth in the growth process of the magnetic layer. The grain size and grain size distribution of the magnetic layer are mainly determined by the grain sizes of the underlayer. The underlayer also has a function to promote the Cr segregation and grain isolation of the magnetic layer, which is essential for achieving high coercivity and low noise. The overcoat and lubricant layers are added to provide their protective functions.

Seed Layer. Numerous seed layer materials have been used, which include Ni-Al [32, 33], MgO [35], ZnO [35], and Co-Ti [34]. The typical thickness of these seed layers is between 50 and 100 nm. The primary functions of these seed layers are to provide the necessary conditions for the desired crystallographic texture of the underlayers and to promote fine grains.

Underlayer. The underlayer material is usually Cr or Cr-based alloys, such as CrV, CrW, CrMn, CrTi, CrNb, and CrMo. Cr, which has a bcc structure and grows either with a [001] or [110] texture normal to the film plane, is a preferred underlayer for longitudinal media because it can provide the crystallographic texture for the epitaxial growth of the CoCr-based magnetic layer with in-plane c axis orientation (Fig. 16). The use of various Cr alloys allows a more precise lattice match with the hcp Co grains in the magnetic layer and hence better crystallographic orientation of the magnetic layer. They usually follow the epitaxial growth patterns of [35] Al-Mg/NiP substrates/seed layer/Cr(001)/Co(11$\bar{2}$0) or glass (ceramic) substrates/seed layer/Cr(110)/Co alloy(10$\bar{1}$0). The Cr underlayer also promotes Cr segregation into the grain boundaries of the hcp Co grains of the magnetic layer. The additives in the underlayers, such as V, Mn, Nb, or Ti, have a second function of promoting fine grains.

Magnetic Layer. Co-based alloys, such as $Co_{86}Cr_{10}Ta_4$, $Co_{75}Cr_{13}Pt_{12}$, or $Co_{77}Cr_{13}Pt_6Ta_4$, are used as the magnetic layer. The Co-based alloy has a hcp crystal structure, which has a magnetic easy axis in the c axis direction. The longitudinal anisotropy of the medium is obtained by growing the Co-based hcp grains with their c axis lying in the film plane. This is achieved by epitaxial growth of hcp Co grains on Cr-based alloy underlayers into [11$\bar{2}$0] or [10$\bar{1}$0] textures. In both cases, the c axis of the Co grains is parallel to the film plane. The primary role of the Cr in Co-based media is to separate the Co grains through Cr segregation into the hcp Co grain boundaries. Such segregation can be further enhanced by the addition of a small amount of Ta, typically 4–6%. The addition of Pt will elongate the c axis of the hcp Co and therefore give rise to larger magnetocrystalline anisotropy [42] and hence larger coercivity.

Effect of Deposition Parameters. The goal of sputter deposition of thin film media is to obtain a medium with fine grains, good grain size distribution, excellent crystallographic orientation, zero intergranular exchange coupling, and high coercivity. In addition to the use of a seed layer, an underlayer, and the right composition of magnetic layers, sputtering parameters, such as substrate temperature, substrate bias, and Ar pressure, also play important roles. Deposition at elevated substrate temperature is one of the key process parameters for obtaining high coercivity. The optimal substrate temperature range for deposition Co-based alloys is between 200°C and 300°C. Figure 17 shows a typical substrate temperature dependence curve of media coercivity for CoCrPtTa films. H_c increases sharply with the increase in substrate temperature, reaching a maximum at about 240°C. The increased grain size due to high temperature and enhanced Cr segregation due to higher atomic mobility are mainly responsible for the increased coercivity [43]. Both dc and RF substrate bias in the range from −100 V to 400 V have been used for the deposition of thin film media to increase the coercivity. The optimal bias voltages vary with the medium layer structure, film composition, and sputtering systems. The negative substrate bias, which provides a moderate ion bombardment to the growing film surface, has the following two major effects on the film growth process: increased atomic mobility and improved film quality. The atomic mobility of the adatoms is increased because of the extra energy provided by the ion bombardment. This effect is similar to the effect of high substrate temperature. The film quality is improved because the moderate ion bombardment may effectively remove the gas impurities in the film. The effect of sputtering gas pressure on film microstructure can be explained by the classical structure zone model [115]. In the deposition of thin film recording media, higher Ar pressure is effective in promoting columnar growth, Cr segregation, and grain isolation, but at the expense of a rough medium surface.

Optimization of Medium Coercivity. The fundamental origin of high coercivity in thin film recording media is the combination of high magnetocrystalline anisotropy of the

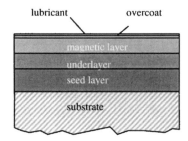

Fig. 16. Schematic diagram of the layer structure of a typical thin film longitudinal medium.

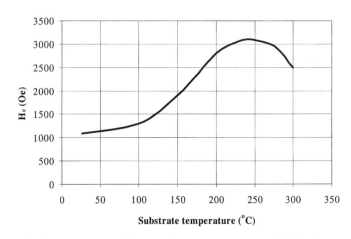

Fig. 17. Dependence of H_c on substrate temperature for CoCrPtTa films.

grains, smaller grain sizes, and lower intergranular interaction. The highest coercivities are obtained from films composed of the so-called Stoner–Wohlfarth single-domain grains. In thin film media, maximum coercivity occurs at the transition from a multidomain grain to a single-domain grain structure [42]. Therefore, the coercivity of a thin film medium usually increases with decreasing grain size because of the formation of more single-domain grains in the film. But the smallest allowable grain size (grain volume) is constrained by the superparamagnetic limit. When the grain size (volume) is smaller than a critical value, the media coercivity will decrease with the reduction in grain size, or the apparent medium coercivity will increase with the reduction in measurement time scale, as discussed in Section 3.1.3.2. This region is termed the superparamagnetic transition region.

Figure 18 shows the $M_r\delta$ dependence of coercivity of $Co_{84}Cr_{10}Ta_6$ films deposited on different underlayers with the Circulus sputtering system from BPS [117]. The general trend of the $M_r\delta$ dependence of medium coercivity is that, at very small $M_r\delta$ (or very small film thickness), the coercivity of the films increases with the increase in film thickness before reaching the maximum coercivity. This is due to the fact that the grain size in the very thin film is very small and increases with the film thickness. At the maximum coercivity point, the film is dominated by single-domain grains with the maximum allowable grain volume for sustaining the single domain state. When the film thickness (or $M_r\delta$) is greater than the critical thickness (at an $M_r\delta$ value of about 0.6), multidomain particles are formed, and the coercivity of the medium decreases with the increase in $M_r\delta$. As one of the requirements of high-density recording media is to use increasingly smaller $M_r\delta$ media ($M_r\delta < 1$ memu/cm^2 for MR media [9] and $M_r\delta < 0.5$ memu/cm^2 for GMR media [17]), it is important to prepare a recording medium with its superparamagnetic transition region below the $M_r\delta$ of the actual medium and with a slow transition into the superparamagnetic region. As can be seen from Fig. 18, the use of different Cr alloy underlayers also plays a part in determining the medium coercivities

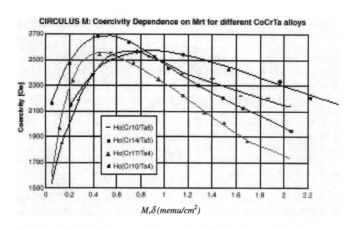

Fig. 19. Effect of film composition on the $M_r\delta$ dependence of medium coercivity. Reproduced with permission from [117], copyright 1999, ICG Publishing Ltd: *Datatech*.

below the superparamagnetic transition. Among the three Cr alloys ($Cr_{96}Ti_4$, $Cr_{80}V_{20}$, and Cr), the $Cr_{96}Ti_4$ underlayer gives the highest maximum coercivity at superparamagnetic transition and a much slower transition into the superparamagnetic region.

The $M_r\delta$ dependence of medium coercivity also varies with the composition or constituent materials of the magnetic layer. This is shown in Figs. 19 and 20, respectively. Figure 19 shows the results of CoCrTa films with various compositions deposited on Cr underlayers. These curves show that the maximum coercivity points occurs at much lower $M_r\delta$ values for magnetic layers with higher Cr concentration. The highest coercivity of just below 2700 Oe is obtained for the $Co_{81}Cr_{14}Ta_5$ alloy at an $M_r\delta$ of 0.5. The addition of the Pt to the CoCrTa ternary system, as shown in Fig. 20, can significantly enhance the value of coercivity and improve the superparamagnetic transition. A medium coercivity of over 3000 Oe is obtainable for the quaternary alloy. The superparamagnetic transition occurs at a much smaller $M_r\delta$ value (0.35 memu/cm^2), and the transition into the superparamagnetic region is much slower than the ternary systems. As

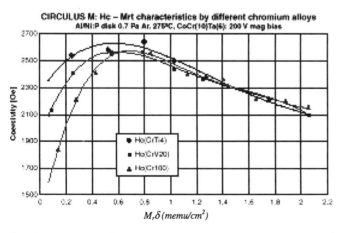

Fig. 18. Dependence of the medium coercivity of $Co_{84}Cr_{10}Ta_6$ films on $M_r\delta$ and underlayer materials. Reproduced with permission from [117], copyright 1999, ICG publishing Ltd: *Datatech*.

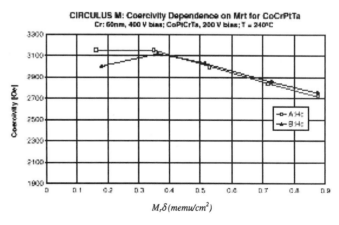

Fig. 20. Dependence of coercivity on $M_r\delta$ for CoCrPtTa alloy. A & B represent the two sides of the disk. Reproduced with permission from [117], copyright 1999, ICG publishing Ltd: *Datatech*.

discussed in Section 3.3, this is due to the fact that the addition of Pt to the CoCrTa ternary alloy will elongate the *c* axis of the hcp Co and therefore give rise to larger magnetocrystalline anisotropy [42] and larger coercivity. The increase in magnetocrystalline anisotropy K_u is mainly responsible for the improved superparamagnetic behavior because $K_u V/kT$ is also increased.

3.3.2. *CoCr-Based Perpendicular Recording Media*

Perpendicular recording media have been intensively investigated since the discovery of perpendicular recording technology by Iwasaki and Nakamura [14]. The first generation perpendicular media are CoCr-based alloys, such as CoCrTa and CoCrPtTa, which are also used as the magnetic layers of longitudinal recording media. The major difference in the preparation of perpendicular and longitudinal media is the underlayers used to control the crystallographic orientation of the hcp Co grains. As discussed in the previous section, the magnetic easy axis of the hcp Co is along the *c* axis direction. In longitudinal media, Cr or Cr alloys are used as underlayers, so that the hcp Co grains grown on them have their *c* axis lying in the film plane. In perpendicular media, underlayers such as Ti [137], TiCr [136], or Pt [138, 139] are used to provide the epitaxial growth conditions for the hcp Co grains growing with their *c* axis perpendicular to the film plane. The Ti or TiCr alloy has a hcp structure that can grow on various substrates with their *c* axis perpendicular to the film plane and therefore provide the basis of epitaxial growth for the perpendicularly oriented hcp Co grains. The Pt underlayer has a fcc structure that tends to grow on various substrates with a [111] texture, and the hcp Co grains grown on the Pt fcc (111) plane exhibit excellent [0001] texture [139]. The quality of the *c* axis orientation of perpendicular media can be examined by X-ray diffraction patterns and the rocking curve method.

Figure 21 is a schematic diagram showing the layer structures of double-layered perpendicular media. The functions of seed layer and underlayer are very similar to those of longitudinal media, providing grain size and crystallographic orientation control of the magnetic layer. The soft magnetic backlayer, usually Permalloy or Co-based amorphous alloys [140], is deposited on the underlayer to form the essential

part of the double-layered media used in combination with the single-pole head. An interlayer between the soft backlayer and the magnetic layer may be required in some cases to provide the ultimate grain size and crystallographic orientation control of the magnetic layer. The usual additives in longitudinal media, such as Ta or Pt, are used for the CoCr-based perpendicular magnetic layer to achieve the same purposes, such as promoting Cr segregation or enhancing the magnetocrystalline anisotropy.

One of the major disadvantages of CoCr-based perpendicular media is their relatively weak magnetocrystalline anisotropy with typical values of $H_{k\perp}$ between 5 and 10 kOe ($H_{k\perp} = 2K_u/M_s$). The $H_{k\perp}$ values of the CoCr-based media are almost equal to the $4\pi M_s$ values of the media, which does not meet the conditions of $[H_{k\perp} - 4\pi M_s] > 5\text{–}10$ kOe [15] and therefore are not sufficient to overcome the perpendicular demagnetizing field, $H_d = 4\pi M_s$, inherent to the thin film geometry. Because of this, the shearing of the MH loop due to the perpendicular demagnetizing field makes the squareness ratio S of a perpendicular medium substantially smaller than unity. Figure 22 shows a typical MH loop of a CoCrTa/Ti perpendicular medium, the squareness ratio of which is only 0.38. The small S causes time decay of recorded magnetization away from transition and high media noise [16]. The CoCr-based perpendicular media also exhibit a relatively small coercivity, limited by its small magnetocrystalline anisotropy, and a magnetically "dead" initial layer. All of these make the CoCr-based media unsuitable for high-density recording.

3.3.3. *High-Coercivity Perpendicular Media*

So far only two major categories of high-coercivity perpendicular media have been studied: the Co-based multilayers [142–148] and the FePt thin films [118–126]. Both types exhibit a $H_{k\perp}$ much higher than the $4\pi M_s$ values of the

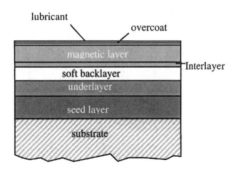

Fig. 21. Schematic diagram of the layer structure of double-layered perpendicular media.

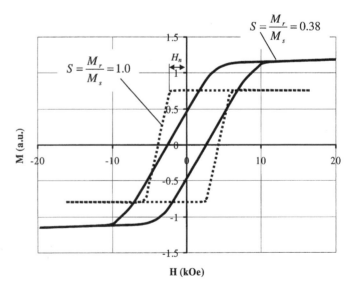

Fig. 22. Typical perpendicular MH loops of CoCrTa/Ti films (solid curves) and Co/Pd multilayers (dotted curves).

media and can easily be made to meet the condition of $[H_{k\perp} - 4\pi M_s] > 5$–$10$ kOe; i.e., they have a sufficiently high magnetocrystalline anisotropy to overcome the inherent perpendicular demagnetizing field. The high $H_{k\perp}$ value of these films made it possible to obtain high medium coercivity. Because high medium coercivity is essential to combating the medium thermal stability problem, and the combination of a single-pole head with perpendicular media has the capability to write media with much higher coercivities than in longitudinal recording, the development of high-coercivity perpendicular media has been receiving increasing attention in recent years.

Co/Pt and Co/Pd Multilayer Perpendicular Media. Co/Pt and Co/Pd multilayers were originally developed as magneto-optic recording media because of their high perpendicular anisotropy [141]. Considerable work has been done in recent years on Co/Pd or Co/Pt multilayers as perpendicular recording media. It has been found that Co/Pd multilayer perpendicular media exhibit much higher perpendicular anisotropy than CoCr-based films [150]. Their perpendicular anisotropy field ($H_{k\perp}$) typically ranges from 15 to 30 kOe, and their perpendicular coercivity ($H_{c\perp}$) from 2 to 10 kOe. The MH loops of these multilayers exhibit a unity squareness ratio because the perpendicular anisotropy energy of the multilayers is sufficiently large to overcome the perpendicular demagnetizing field. These multilayers also show little "dead" layer. Overall, they are much better perpendicular media than CoCr-based films. The only drawback found in the early work was the high medium noise associated with these multilayers [145], even when the squareness ratio of these media is unity. The high medium noise is associated with the locally reversed magnetic domains away from transition [149], which has the same origin as the medium noise in CoCr-based perpendicular media. Wu et al. [148] have found that low medium noise can be achieved in Co/Pd multilayered perpendicular media if the medium is prepared to exhibit a MH loop with a unity squareness ratio, a significantly large negative nucleation field (typically 1.5 kOe), and a small slope at coercivity. High Ar pressure and an optimized Co and Pd thickness ratio are the two key parameters for the preparation of the low-noise multilayer media. A desirable MH loop of a Co/Pd multilayer perpendicular medium is sketched in Fig. 22 (dotted curves). Wu et al. [148] also demonstrated by magnetic force microscope images that a Co/Pd multilayer with a unity squareness ratio but a zero nucleation field H_n exhibits a large number of reversed domains between the recorded transitions, whereas a Co/Pd multilayer with unity squareness and a negative nucleation field H_n exhibits no reversed domains between recorded transitions. The Co/Pt multilayers exhibit large perpendicular anisotropy, however, the media tend to be very noisy because of the large domain size originating from the strong intergranular exchange coupling in the multilayers [146]. Reduction of media noise in the Co/Pt multilayer media was achieved by using CoB/Pt or CoCrTa/Pt multilayers [146].

FePt Perpendicular Media. The long range ordered equiatomic FePt film (among the FePt, CoPt, and (CoFe)Pt family) with face-centered tetragonal (fct) L1(0) structure [117–126] are of particular interests because of their high magnetocrystalline anisotropy energy (7×10^7 erg/cm^3 [121]), high coercivity, and good corrosion resistance. The perpendicular anisotropy field of these films is between 30 and 40 kOe, and coercivities up to 5–15 kOe are obtainable. The long-range ordering of the L1(0) structure in these films is the essential condition for high coercivity.

The magnetic easy axis of the FePt fct L1(0) grain is long the c axis or the [001] direction. The FePt films can be made into longitudinal as well as perpendicular media, depending on the underlayers or substrates used during film deposition. Cr(100)/MgO, Pt, Au, and Pt seeded MgO(001) are the commonly used underlayers for obtaining perpendicular anisotropy [118–126], and FePt films deposited directly on a ZrO$_2$ substrate exhibit in-plane anisotropy [127]. The as-deposited FePt films usually exhibit the fcc phase. High-temperature annealing is required to obtain the phase transformation from fcc to fct L1(0). The required annealing temperature, depending on the underlayers, sputtering Ar pressure, and annealing gas atmosphere, ranges typically from 400°C to 600°C. Suzuki et al. found that the ordering temperature of FePt can be reduced to 450°C if a sputtering Ar gas pressure of 100 Pa is used [121]. Park et al. obtained the long-range ordered FePt fct L1(0) phase at a temperature of 350°C with the in-air annealing technique [124].

FePt films exhibit very fine magnetic domains (<80 nm) [120], and the domain size appears to be inversely proportional medium coercivity. Because of the huge perpendicular anisotropy, the MH loops of FePt films have a steep slope, a unity squareness ratio, and a large negative nucleation field, which are essential properties for low-noise and high thermal stability perpendicular media. Excellent recording performances at high densities for both single-layered and double-layered FePt media have been demonstrated [120].

3.4. Characterization of Recording Media

3.4.1. MH Loop

The magnetization versus applied magnetic field (MH loop) of a recording medium can be measured by VSM or AGFM, from which key magnetic parameters for recording media, such as coercivity H_c, remanence coercivity $H_{c,r}$, remanent magnetization M_r, $M_r\delta$, saturation magnetization M_s, and squareness ratio S, are directly obtained at various points of the loop, as shown in Fig. 23. For high-density digital recording media, it is always desirable to obtain a sharp switching between the two discrete magnetization states. The switching behavior of a medium is usually described in terms of SFD or coercive squareness S^*. The SFD measures the narrowness of the range over which the medium magnetization reverses or switches as a function of the applied field H. It is defined as the half-amplitude width ΔH_{50} normalized by H_c of the derivative dM/dH curve of an MH loop centered about H_c, as

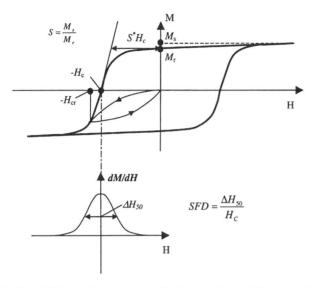

Fig. 23. MH loop of recording media showing the definition of media parameters and derivative dM/dH of a MH loop showing the definition of SFD.

also shown in Fig 23. As can be seen from the dM/dH curve in the figure, the maximum switching rate (dM/dH) of a medium in an applied field H happens around $H = H_c$. The coercive squareness S^* is defined by the equation

$$\frac{dM}{dH}\Big|_{H=-H_C} = \frac{M_r}{(1-S^*)H_C} \qquad (3.19)$$

A value of S^* close to unity is desirable for sharp switching.

3.4.2. ΔM Curve Measurements

ΔM curve is usually used to characterize the intergranular interaction of recording media. It is derived from the measurement of the isothermal remanent magnetization curve $M_r(H)$ and the dc demagnetization remanent magnetization curve $M_d(H)$. As shown in Fig 24a, when a positive field H_1 is applied to an initially ac demagnetized thin film medium and then removed, we obtain a remanence value of $M_r(H_1)$. For a larger applied field H_i, we get $M_r(H_i)$. If the process is progressively repeated until saturation is reached, we can plot a curve of $M_r(H)$ as a function of the applied field H. such a curve is called the isothermal remanence curve.

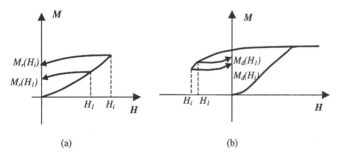

Fig. 24. Schematic illustration for the measurement of $M_r(H)$ and $M_d(H)$ curves.

Similarly, we get a dc demagnetization remanence curve $M_d(H)$, as shown in Fig 24b, by reversing a field to $-H_i$ from the positive dc saturation state, with the remanent magnetization $M_d(H_i)$ recorded as a function of reversing field until the negative saturation has been reached.

According to Wohlfarth [45], the two remanence curves of noninteraction particles obey the Wohlfarth relation

$$M_d(H) = 1-2M_r(H) \qquad (3.20)$$

Henkel [46] investigated the nature of particle interactions by plotting $M_d(H)$ versus $M_r(H)$ and suggested that the particle interaction can be characterized by a parameter ΔM, which is the deviation from the Wohlfarth relation and is given by

$$\Delta M = M_d(H) - (1-2M_r(H)) \qquad (3.21)$$

Positive values of ΔM are interpreted as a result of stabilizing interactions. Conversely, negative values of ΔM are due to destabilizing interactions [46]. Specifically, a positive value of ΔM indicates the existence of intergranular exchange interaction, and a negative value of ΔM indicates weak or no intergranular exchange interaction.

3.4.3. Thermal Stability Measurements of Recording Media

3.4.3.1. Dynamic Coercivity Measurement

The dynamic coercivity of recording media can be measured by varying the sweep rate of the applied field (sweep rate method) or by varying the field duration of a constant applied field (constant field method). In the early work for the experimental observation of time-dependent magnetic effects, the dynamic coercivity was mainly measured by the sweep field-method because of the difficulties of applying a constant field of a very short duration. The drawback of the sweep field measurement is its relatively large characteristic time (a minimum of 10^{-5} s for a 60-Hz sweep field BH loop measurement [24]) due to the available slow sweep rate, which is not comparable to the field pulse width of a writing head (typically 10^{-8} s to 10^{-9} s). Stinnett et al. [153] developed a pulse field system with microstrips capable of measuring dynamic coercivities on a time scale of 10^{-8} s. It was only recently that the constant field method has shown its advantages over the sweep field method due to the application of the spin-stand test in dynamic coercivity measurements [154–158].

VSM and AGFM Methods. The Measurement of the time dependence of media coercivity by VSM or AGFM is the most straightforward method. However, the limitation of this method is that it cannot go down very far in terms of the characteristic time due to the limit on the field sweep rate. Sharrock et al. [167] and Oseroff et al. [165] have developed a method to determine the characteristic time from the sweep field measurement. It has been found that the characteristic time (τ) depends not only on the field sweep rate (R), but also on the dynamic switching process of a specific material [167]. A series of measurements is required to determine the relationship between the field sweep rate and the characteristic time. The experimental procedure is as follows:

1. The sample is first magnetized to saturation by a field H_s.
2. A reverse field $-H_0$ is applied at a sweep rate R_0, and the time t_0 is recorded when the magnetization of the sample reaches zero at field $-H_0$.
3. Steps 1 and 2 are repeated with a reverse field $-(H_0 + \delta H)$ at the same sweep rate R_0, and the time t is recorded when M = 0.
4. Step 3 is repeated for several field values.
5. $\lg(t/t_0)$ is plotted as a function of δH. The relationship between $\lg(t/t_0)$ and δH is given by [165],

$$\ln \frac{T}{T_0} = -\alpha \delta H \qquad (3.22)$$

6. Parameter α is determined from the slope of the $\lg(t/t_0)$–δH line. And the equivalent characteristic time τ at field sweep rate R for that sample is given by [165]

$$\tau = \frac{1}{R\alpha[1 - e^{-\alpha(H_s - H_f)}]} \cong \frac{1}{R\alpha}$$

where H_s and H_f are the saturation and final fields, respectively, and α is the parameter associated with the dynamic switching process of a particular material and is determined by steps 5 and 6.

In a modern VSM or AGFM with a stepped field system, the field between measurement points can be changed at a rate as high as 2000 Oe/s without causing any significant measurement errors [1]. The characteristic time for that sweep rate with a medium of $\alpha = 0.5$ will be approximately 1 ms.

Spin-Stand Method. This is an indirect measurement of medium coercivity as a function of measurement time scale–field pulse width or pulse duration, also known as rotating disk magnetometry [156]. Several experimental techniques have been developed with this method [154–158]. Robin et al. [154] and Moser et al. [156] used the method of erasing di-bits with pulsed fields from a recording head, and the medium coercivity corresponding to a field pulse width is determined from the write current needed to erase 50% of the di-bit signal. Corrections for the dynamic behavior of the head and for the demagnetizing field associated with the di-bit transition are required in the interpretation of the measured data. The method developed by Richter et al. [155] is based on the idea of using the reverse dc eraser noise as an indicator for coercivity [151, 152]. A band of dc saturated track is recorded in one direction, then a dc writing field in the reverse direction is applied along the track with a writing head and the broadband medium noise is measured. It has been confirmed in their experiment that the peak of the noise occurs exactly at the coercivity. A demagnetizing field correction is also required. The pulse duration for every portion of the written track is approximated as $t = g/v$,

where g is the gap length of the writing head and v is the linear velocity. Because only dc currents are employed, the rise time effect is disregarded. It was argued that the effective pulse duration time is equal to the number of revolutions (also known as reptation) multiplied by the pulse duration (t) of a single revolution. In this way, the pulse duration can easily be varied from nanoseconds to 100 s (10–12 decade seconds). It was observed that the noise peaks shift toward smaller writing currents with the increase in the number of revolutions, indicating that the medium coercivity decreases with the measurement time scale (field pulse duration).

3.4.3.2. Magnetization Decay Measurement

The magnetization decay can be measured by either VSM or AGFM. However, as discussed in Section 1.1.4, VSM is a more appropriate instrument for magnetization time decay measurement because the alternating gradient field (on the order of a few Oe/mm) from the gradient field coil of an AGFM may affect the stability of the magnetization of the sample, resulting in false readings of the measurement results.

The magnetization decay for a longitudinal medium is measured by first applying a saturation field to the medium, followed by a reverse field just a few Oe smaller than the medium coercivity. While that field is held constant, the change in magnetization of the sample is measured. The so measured time decay of the magnetization will emulate the magnetization stability of the recorded bits. The reversing field applied during the time decay measurement is an equivalent to the transition self-demagnetizing field in the recording transitions. The measured time dependence of magnetization $M(t)$ is usually represented as a logarithmic function of time t[25],

$$M(t) = C + S \ln t \qquad (3.23)$$

where C and S are constants and $S(= dM/d \ln t)$ is known as the magnetic viscosity coefficient.

The thermal stability of a medium can also be represented by an equivalent field known as the fluctuation field H_f, which is derived from the measurement of S and the irreversible susceptibility χ_{irr}, and is then given by [30]

$$H_f = \frac{S}{\chi_{irr}} \qquad (3.24)$$

For a modern medium, H_f is typically a few Oe. The irreversible susceptibility χ_{irr} is usually determined from the differential of the appropriate remanence curve [159]. H_f can also be determined from the measurement of time dependence of remanence coercivity $H_{cr}(t)$ [160, 161].

The magnetization decay for perpendicular recording media is measured by applying a saturation field to the sample and measuring the change in magnetization of the sample after the field is removed (at zero field). A reversing field is not required because a perpendicular demagnetizing field already exists in the medium and a perpendicular transition is free from demagnetizing field at high densities.

3.4.3.3. Signal Decay Measurement

The time decay of recorded magnetization causes a decease in the playback signal output of recorded information over time, which can be measured by read/write tests [38, 161–163], known as the signal decay measurement. In the signal decay measurement, a MR-read/inductive-write composite head is usually used. To eliminate the effect of changes in the sensitivity of the MR head on the head signal output, the MR head signal should be calibrated with a freshly recorded reference track of the same medium and same recording density just before each measurement [38, 163]. It is also mandatory to scan the MR head over the written track to accommodate the inevitable position changes of the head while data are taken[28].

4. THIN FILMS FOR REPLAY HEADS

The history of replay heads is closely associated with the advances of magnetic recording density. Figure 25 shows the evolution history of magnetic recording heads. Early generations of heads were ring shaped heads made of laminated mu-mentals (NiFeMoCu) or ferrites (Ni, Zn-Fe$_2$O$_3$). Thin film inductive heads were first introduced to the hard disk drive market in 1979 by IBM (IBM 3370) for an areal recording density of 7.7 Mbits/in^2[175]. As the recording density increased, thinner media with higher coercivity were introduced, which required heads with high writing fields for effective overwrite and heads with higher output for acceptable signal amplitude and signal-to-noise ratio. The inductive-write/inductive-read technology, which dominated the disk drive industry up to the early 1990s, could meet both the write field and the signal-to-noise ratio requirements for the areal densities up to 500 Mbits/in^2. Although the anisotropic magnetoresistance (AMR) phenomenon was discovered as early as in 1857 and the first AMR head was invented in 1970

by Hunt [49], the first-generation disk drive product using integrated inductive-write/MR-read heads did not appear until 1991. AMR head technology allowed the extension of areal recording density to 4–5 Gbits/in^2. The discovery of the giant magnetoresistance (GMR) phenomenon [50, 51], particularly as applied to spin-valve head technology [7], has provided the technology for current disk drive replay heads, which is capable of taking the areal density up to or beyond 100 Gbits/in^2.

The focus of this section is a review of GMR multilayers and spin-valve head technology. A brief review of AMR films and AMR heads is also presented.

4.1. Anisotropic Magnetoresistance Films

Early investigations of electron transport properties of non-magnetic conducting wires revealed that there was a small difference in resistance when a conducting wire was placed parallel or perpendicular to the magnetic field direction. If $\rho_{//}$ and $\rho_\perp$ represent the resistivities when the current is parallel and perpendicular to the magnetic field, respectively, we have $\rho_\perp > \rho_{//}$ and $\Delta\rho = \rho_\perp - \rho_\parallel$. $\Delta\rho$ in general proportional to H^2, except at high magnetic fields and low temperatures [44]. This phenomenon is known as the ordinary magnetoresistance (OMR), which can be explained by the effect of Lorentz force experienced by the conduction electrons in a magnetic field.

On the other hand, in ferromagnetic materials (Fe, Co, Ni) and their alloys, the electrical resistance depends on the current direction relative to the direction of the spontaneous magnetization in the sample. In contrast to OMR, in most cases we have $\rho_\perp < \rho_\parallel$ and $\Delta\rho = \rho_\parallel - \rho_\perp > 0$ for ferromagnetic materials. This is known as the anisotropic magnetoresistance (AMR). The normalized quantity, the magnetoresistivity ratio, is given by

$$\frac{\Delta\rho}{\rho_{av}} = \frac{\rho_\parallel - \rho_\perp}{(1/3)\rho_\parallel + (2/3)\rho_\perp} \qquad (4.1)$$

where $\rho_{av} = \frac{1}{3}\rho_\parallel + \frac{2}{3}\rho_\perp$. Experimentally, ρ_{av} is usually replaced with ρ_0, which is the zero field resistivity. The difference in ρ_{av} and ρ_0 is insignificant with regard to the magnetoresistivity ratio [44].

The AMR phenomenon can be explained by spin-orbit coupling, which tends to induce an anisotropic scattering of the conduction electrons in the exchange split spin ↑ and spin ↓ 3d-subbands [47]. Comprehensive reviews of AMR in ferromagnetic 3d alloys and its basic physics can be found in [44] and [47].

4.1.1. Basic Parameters of AMR Films

4.1.1.1. MR Ratio and MR Amplitude

Because, in practice, the magnetoresistivity ratio given by Eq. (4.1) can be directly obtained from $\Delta R/R_0$ without the need to know the dimensions of the specimen ($\Delta\rho/\rho_0 = \Delta R/R_0$), Eq. (4.1) can be rewritten as

$$\frac{\Delta R}{R_0} = \frac{R(H) - R_0}{R_0} \qquad (4.2)$$

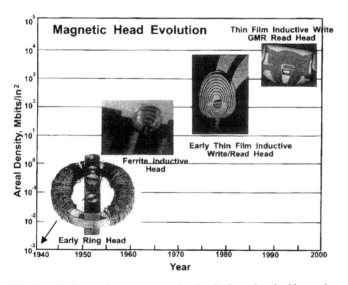

Fig. 25. Evolution of magnetic recording heads. Reproduced with permission from [8], copyright 1999, ICG Publishing: *Datatech*.

where R_0 is the isotropic magnetoresistance of the sample, which is equivalent to the magnetoresistance measured when the current is perpendicular to the internal magnetization of the film, i.e., $R_0 = R_\perp$. $R(H)$ is the resistance measured at field H. The zero level is defined as $R(H) = R_0$. The maximum MR ratio is known as the MR amplitude and is given by

$$\left(\frac{\Delta R}{R_0}\right)_{max} = \frac{R_\parallel - R_0}{R_0} \qquad (4.3)$$

i.e., the difference in resistance, $\Delta R = R_\parallel - R_\perp$, for currents flowing parallel ($R_\parallel$) and perpendicular ($R_\perp$) to the internal magnetization of the material, normalized by the perpendicular resistance (R_0).

4.1.1.2. Field Sensitivity

The differential form of the field sensitivity of an AMR film is defined by

$$S_H = \frac{\delta\left(\Delta R/R_0\right)}{\delta H} \qquad (4.4)$$

The averaged field sensitivity of an AMR film is given by the ratio of the MR amplitude and the saturation field, i.e.,

$$\overline{S}_H = \frac{\left(\Delta R/R_0\right)_{max}}{H_s} \qquad (4.5)$$

At a saturation field, the AMR film is magnetized to magnetic saturation. The averaged field sensitivity is obviously proportional to the AMR amplitude and inversely proportional to H_s. For a uniaxial anisotropy Permalloy film with infinite size (zero demagnetizing field), the saturation field H_s is equivalent to the anisotropy field H_k, which is typically 5–10 Oe. If the Permalloy film has a MR amplitude of 2.5%, the averaged field sensitivity is 0.5–0.25%/Oe.

4.1.2. AMR Measurement

Because the MR change ΔR depends on the relative orientation of the measuring current and the internal magnetization in the film, the simplest measurement of ΔR of a uniaxial anisotropy film is to measure the difference in resistance along the magnetic easy axis and hard axis, i.e., $\Delta R = R_{h.a.} - R_{e.a}$, where $R_{e.a.}$ and $R_{h.a.}$ are the resistance measured when the current is along the easy axis and hard axis, respectively.

The dependence of the magnetoresistance on the magnetic field can be measured by starting either with the current parallel or perpendicular to the easy axis of the film, but the external field must be perpendicular to the easy axis, which will give a 90° rotation of magnetization with the application of a saturation field. Figure 26 illustrates a typical four-point probe configuration, in which the measuring current is parallel to the easy axis (the **M** direction represents the easy axis of the film and the **H** direction is the applied field direction). At zero external field, the measuring current (the linear four-point probe) is aligned parallel to the internal magnetization of the film. At this point, the measured R value represents the value of $R_\parallel$, i.e., $R = R_\parallel$. As the perpendicular external field increases, the magnetization vectors in the film rotate progressively toward the field direction until saturation occurs. At saturation, the magnetization is perpendicular to the measuring current and this gives $R = R_\perp$. A typical $\Delta R/R_0 \sim H$ curve measured by the configuration shown in Fig 26a is given in Fig. 26b.

The AMR amplitude of Permalloy film is thickness dependent, ranging from 4% (1000-Å-thick film) to 1.5% (150-Å-thick film). The decrease in AMR amplitude with the decrease in film thickness is attributed to the increasing importance of surface scattering of the conduction electrons, which causes the increase in R_0, the isotropic resistance and the denominator of Eq. (4.3).

4.1.3. AMR Replay Heads

The sensing element of an AMR head (MRE) is a 150–250 Å- thick sputtered Permalloy (81Ni/19Fe, by weight) film,

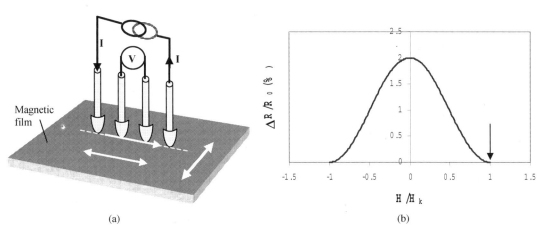

(a)

(b)

Fig. 26.   Schematic diagram showing a typical set-up for AMR measurement (a), and its corresponding MR curve (b).

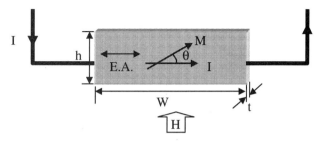

Fig. 27.   A single-domain MRE placed in a magnetic field.

patterned into strips by a photolithographic fabrication process. Because the NiFe alloy film exhibits almost zero magnetocrystalline anisotropy and magnetostriction with this composition and is also very easy to make into a uniaxial anisotropy film by applying an orientation magnetic field during film deposition, it has become the unique choice for AMR and even GMR head material [9]. Figure 27 is a schematic diagram of a MRE with width W, height h, and thickness t, placed in a magnetic field H. The induced magnetic easy axis of the Permalloy film is usually aligned parallel to the ABS during head fabrication. The field (or flux) from the recording medium is perpendicular to the easy axis of the film.

If we assume that the MRE is a single domain element, its anisotropic magnetoresistance can be expressed as a function of the angle θ between the magnetization of the MRE and the sensing current and is given by

$$R = R_0 + \Delta R \cos^2 \theta \qquad (4.6)$$

where R_0 is the isotropic resistance of the MRE (independent of θ), the value of which is equal to $R_\perp$.

ΔR is the maximum resistance change, $\Delta R = R_\parallel - R_\perp$. The θ dependence of the magnetoresistance of a single-domain MRE can be sketched as in Fig. 28, which is similar in shape (but with different axis notations) to the ΔR–H curve in Fig. 26b. If the MRE has an infinite size, then the effect of the demagnetizing field is negligible. Simple energy arguments show that, for $H \leq H_k$ [48],

$$\sin \theta = \frac{H}{H_k} \qquad (4.7)$$

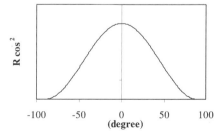

Fig. 28.   Magnetoresistance change in a MRE as a function of the angle between magnetization and current.

Substituting Eq. (4.7) into Eq. (4.6) gives a quadratic field dependence of magnetoresistance of a single-domain MRE [48],

$$R = R_0 + \Delta R \left[1 - \left(\frac{H}{H_k} \right)^2 \right] \qquad (4.8)$$

The magnitude of H_k for Permalloy is typically 5 Oe.

A real MRE is usually patterned into a rectangular shape with a finite width W and height h, as shown in Fig. 27. Because of the finite width and height of a real MRE, there are demagnetizing fields in both the vertical and horizontal directions. The values of two demagnetizing fields depend on the MRE dimensions and the magnetization angle θ, and they vary spatially. The average values of the two demagnetizing fields are given by [48]

$$H_{d,y} = 4\pi M_s \cdot \frac{t}{h} \cdot \sin \theta \qquad (4.9)$$

for a vertical demagnetizing field and

$$H_{d,x} = 4\pi M_s \cdot \frac{t}{W} \cdot \cos \theta \qquad (4.10)$$

for a horizontal demagnetizing field, where M_s is the saturation magnetization of the MRE film. It can be seen from the above equations that the average demagnetizing field is proportional to the film thickness and inversely proportional to the size of the MRE. For a saturation flux density $4\pi M_s = 10$ kG, a MRE height of 2 μm, and a thickness of 20 nm, the estimated average vertical demagnetizing field $H_{d,y}$ by Eq. (4.9) is about 100 Oe, which is 20 times the anisotropy field (H_k) of the Permalloy film.

According to Markham and Jeffers [48], because of the effect of a demagnetizing field the quadratic field dependence of Eq. (4.8) is retained over two-thirds of the resistance change, but H_k is replaced by $4(H_k + H_{d,y})/3$, resulting in

$$R = R_0 + \Delta R \left[1 - \left(\frac{H}{(4/3)(H_k + H_{d,y})} \right)^2 \right] \qquad (4.11)$$

where $H_{d,y}$ is the average vertical demagnetizing field. The resultant MR response curve for a sensor with finite size, as shown in Fig. 29, is a much broader one in comparison with that of a single-domain film with infinite size. The demagnetizing fields significantly increase the saturation field and reduce the field sensitivity of the MRE. For this reason, the thickness of the MR film for high-density magnetoresistive heads must be as thin as possible, usually less than 15 nm.

The demagnetizing field is nonuniform across the MRE and is much larger at the sensor edge than in the center. This causes the center of the sensor to saturate first and gives rise to the inflection points and skirts on the $R(H)$ curve [48], which is also shown in Fig. 29. The inflection points occur at the average angle $\theta \cong 55°$, or $H = H_k + H_{d,y}$.

Because of the quadratic field dependence of the MR response of a MRE, in magnetic recording head applications it is necessary to give an appropriate vertical bias field so that

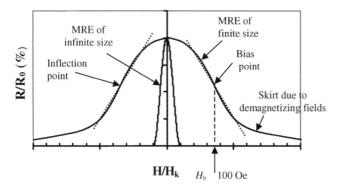

Fig. 29. MR-H curves showing the effect of a demagnetizing field due to the finite size of a MRE.

the MRE will operate in the linear region. The ideal vertical bias condition is to set the vertical bias field at the inflection point. This will give an almost linear response of the MRE to vertical magnetic flux. A typical value for a vertical bias field is around 100 Oe, or $H_b = H_k + H_{d,y}$. Various vertical bias schemes were employed to obtain the bias fields for MR heads, which include permanent magnet bias, shunt current bias, soft adjacent layer (SAL) bias, exchange bias, self-bias, double-element bias, split-element bias, barber-pole bias, and servo-bias schemes. It is also necessary to have small a horizontal bias field (typically ~5 Oe) along the easy axis of the MR film to eliminate the Barkhauson noise. For a comprehensive review of the various bias schemes and other topics related to MR heads, readers are encouraged to refer to [48] and [9].

4.2. Giant Magnetoresistance Films

The giant magnetoresistance (GMR) phenomenon was first discovered by Baibich et al. [50] and Binasch et al. [51] in 1988 in Fe/Cr multilayers grown by molecular beam epitaxy (MBE). The significance of the discovery is that the value of the observed magnetoresistance change is as high as 200% [50] in comparison with the 2–3% in AMR films. In 1990, Parkin et al. [52] demonstrated that GMR could be obtained in sputter-deposited Fe/Cr, Co/Cr, and Co/Ru multilayers and that the GMR amplitude could be even larger in sputtered multilayers than in MBE-grown samples. Since then, the magnetotransport properties of magnetic multilayers were under intensive investigation by many research groups, and many multilayer systems, such as Co/Cu [54], Fe/Cu [55], CoFe/Cu, CoNiFe/Cu [58], NiFe/Ag [56], etc. have been found to exhibit GMR.

4.2.1. Basic Parameters of GMR Films

4.2.1.1. GMR Ratio and GMR Amplitude

Giant magnetoresistance refers to the difference in resistance, $\Delta R = R_{\uparrow\downarrow} - R_{\uparrow\uparrow}$, or resistivity $\Delta\rho = \rho_{\uparrow\downarrow} - \rho_{\uparrow\uparrow}$, for a multilayer with its magnetization vectors in the neighboring layers aligned antiparallel ($R_{\uparrow\downarrow}$) and parallel ($R_{\uparrow\uparrow}$). A

GMR multilayer exhibits maximum resistance in antiparallel magnetization configuration and minimum resistance in parallel configuration. The GMR ratio, which is field dependent, is defined as [51]

$$\frac{\Delta R}{R_{\text{sat}}} = \frac{R(H) - R_{\text{sat}}}{R_{\text{sat}}} \quad (4.12)$$

where $R(H)$ is the resistance at external field H. R_{sat} is the resistance for a saturation field, i.e., when the magnetization vectors of the adjacent layers have a parallel configuration, here $R_{\text{sat}} = R_{\uparrow\uparrow} = R_{\downarrow\downarrow}$. The zero level is defined by $R(H) = R_{\text{sat}}$. The maximum GMR ratio is known as the GMR amplitude and is given by

$$\left(\frac{\Delta R}{R_{\text{sat}}}\right)_{\text{max}} = \frac{R_{\uparrow\downarrow} - R_{\text{sat}}}{R_{\text{sat}}} \quad (4.13)$$

where $R_{\uparrow\downarrow}$ is the resistance measured with antiparallel magnetization configuration, which is usually equal to the resistance at zero field, R_0.

A typical GMR ratio ($\Delta R/R_{\text{sat}}$) versus H curve is sketched in Fig. 30. The relative magnetization configurations of the adjacent layers at zero field and at saturation fields are also shown in the figure. The GMR ratio, $\Delta R/R_{\text{sat}}$, is usually measured by the four-point probe method as described in Section 5.2.2. The definition of the GMR ratio is given by Eq. (4.12).

Another commonly used definition of the GMR ratio is the one normalized by the resistance at zero field, R_0, and is given by

$$\frac{\Delta R}{R_0} = \frac{R(H) - R_0}{R_0} \quad (4.14)$$

The two different definitions of the GMR ratio given by Eqs. (4.12) and (4.14) are related to each other by [66]

$$\left(\frac{\Delta R}{R_{\text{sat}}}\right) = \frac{(\Delta R/R_0)}{(\Delta R/R_0) - 1} \quad (4.15)$$

However, the definition given by Eq. (4.12) is a more commonly used formula. This is due to the fact that the zero field

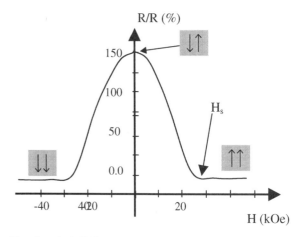

Fig. 30. A typical $\Delta R/R \sim H$ curve of GMR multilayers, also shown the magnetization orientations in the adjacent layers.

magnetic configuration is often poorly defined. If the antiferromagnetic coupling is weak or if the hysteresis of the magnetic layers is large, R_0 can depend on the magnetic history of the sample [66].

4.2.1.2. Field Sensitivity

Similar to the field sensitivity for AMR film, the averaged field sensitivity of a GMR film is defined by the ratio of the GMR amplitude and the saturation field, i.e., $(\Delta R/R_{sat})/H_s$. At saturation field, the GMR multilayer is magnetized to magnetic saturation and therefore has the lowest resistance. The averaged field sensitivity is obviously proportional to the GMR amplitude and inversely proportional to H_s.

For a magnetic sensor application, a GMR film with high field sensitivity is required, i.e., a GMR film with high GMR amplitude and a small saturation field. The differential form of field sensitivity is given by Eq. (4.4).

4.2.2. Multilayer GMR Measurement

Figure 31 shows the two probe configurations for GMR measurement. In general, GMR multilayers (superlattice) are magnetically isotropic. Therefore, the four-point probes can be placed along any direction on the film. The difference in the two configurations is as follows. In Fig. 31a the applied field is parallel to the current, whereas in Fig. 31b the applied field is perpendicular to the current. It was initially thought that GMR was isotropic [50], i.e., the GMR amplitude is independent of the angle between the applied field and the sensing current. Therefore it can be measured by either of the two configurations shown in Fig. 31. However, Dieny et al. [7] have demonstrated that GMR is also anisotropic. A difference in GMR amplitude on the order of 7% was obtained by using a square probe configuration for a $(Co/Cu/NiFe/Cu)_{10}$ multilayer. The GMR amplitude measured with current perpendicular to the applied field, as in Fig. 31b, is 7% higher than that obtained with current parallel to the applied field, as in Fig. 31a.

Some GMR multilayers (typically sandwiched films) are magnetically anisotropic. In such cases, the applied field for GMR measurement must be along the easy axis of the film to exclude the AMR contribution. If the applied field is perpendicular to the magnetic easy axis of the film (i.e., using the same probe configuration for AMR measurement as shown in Fig. 26), the GMR-H curve will be a hard axis response curve and the GMR amplitude will also include the contribution of AMR. A typical example of this can be found in [51].

4.2.3. Types of GMR Multilayer Structures

4.2.3.1. Anti-Ferromagnetically Coupled Multilayers (Superlattice)

The GMR phenomenon was first discovered in antiferromagnetically (AF) coupled multilayers. A GMR multilayer consists of a number of very thin ferromagnetic layers separated by thin layers of nonmagnetic material, normally expressed in the form of $(F/NM)_n$, in which F designates a ferromagnetic layer (transition metal Fe, Co, Ni, or their alloys), NM is a nonferromagnetic transition metal or a noble metal (V, Cr, Mn, Nb, Mo, Ru, Re, Os, Ir, Cu, Ag, or Au), and n is the number of bilayers. The archetypal system in this category is $(Fe/Cr)_n$ multilayers.

At zero magnetic field, the magnetization vectors in the adjacent ferromagnetic layers of a multilayer system are ordered antiparallel (antiferromagnetic coupling), and the film exhibits the highest resistance. With the application of a sufficiently large magnetic field, the magnetization vectors in the multilayer become aligned in the same direction as the applied field (parallel magnetization configuration) and the film exhibits the lowest resistance. The GMR amplitude is determined by the maximum and minimum resistance, as defined by Eq. (4.13).

The thickness of the magnetic layer is between 10 and 100 Å, with an optimum thickness (maximum GMR amplitude) between 10 and 30 Å. And the thickness of the nonmagnetic spacer can vary from 5 to 100 Å. A higher GMR amplitude is usually obtained at smaller spacer thicknesses. But the GMR amplitude does not increase monotonically with the reduction of spacer thickness. The oscillatory dependence of GMR on spacer thicknesses is usually observed.

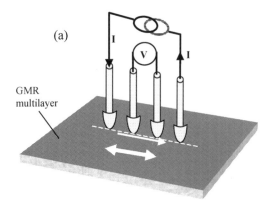
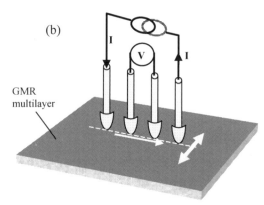

Fig. 31. Schematic diagrams of the four-point probe configurations for GMR measurement (the H field is parallel to the magnetic easy axis). (a) Current parallel to magnetic field. (b) Current perpendicular to magnetic field.

Table II. GMR Characteristics of Some Antiferromagnetically Coupled Multilayers

Multilayer system	GMR amp. (%)	Measuring temp. (k)	Saturation field(kOe)	Preparation method
(Fe 120Å/Cr 10Å/Fe 120Å)	1.5		1.5	MBE [51]
(Fe 30Å/Cr 9Å)$_{60}$	50	4.2	20	MBE [50]
Cr 100Å/(Fe 14Å/Cr 8Å)$_{50}$	150	4.2	20	Sputter [53]
Cr 100Å/(Fe 14Å/Cr 8Å)$_{50}$	30	RT		Sputter [53]
Fe 50Å/(Co 8Å/Cu 9Å)$_{60}$/Fe	115	4.2	13	Sputter [54]
(Co/Cu 25Å)$_{10}$	18.5	RT	0.5	Sputter [58]
(Co 18Å/Ru 8Å)$_{20}$	2.5	4.5	10	Sputter [52]
(Co 15Å/Cr 4Å)$_{30}$	2.5	4.5	6	Sputter [52]
(Co$_{90}$Fe$_{10}$//Cu 12Å)$_{10}$	23	RT	2	Sputter [58]
(Ni$_{76}$Fe$_{10}$Co$_{14}$/Cu)$_{10}$	11	RT	0.05	Sputter [58]
Co$_{70}$Fe$_{30}$ 4Å/Ag 15Å)	100	4.2	3	Sputter, 77K [66]
Fe 50Å/(NiFe 10Å/Cu 10Å)$_{20}$	19	RT	3	IBD [59]
(NiFe 20Å/Ag 10Å)	50	4.2	1	Sputter, 77K [62]
(Ni$_{81}$Fe$_{19}$12Å/Ag 11Å)$_{n}$	17	RT	0.3	Sputter/ann. [61]
Ni$_{80}$Fe$_{20}$/Ag	6	RT	0.01	Sputter/ann. [57]

Adapted with permission from [66], copyright 1994, Elsevier Science.

Typical GMR multilayer systems of this category, together with their GMR amplitude, saturation field, and preparation methods, are listed in Table II. It is worth noting that the GMR amplitudes and saturation fields vary considerably from one system to another. The reasons for these variations can be attributed to the influences of the combinations of magnetic/spacer materials, the interlayer coupling strength, the characteristics of the so-formed interfaces, and the effect of thicknesses of both materials on the transport properties of a particular multilayer system. Detailed discussions on the factors influencing the GMR amplitudes of GMR multilayers are given in Section 4.2.5.

The saturation fields in early GMR superlattices such as Fe/Cr or Co/Cr are very large, typically ranging from a few kOe to a few tens of kOe. Although the multilayer has a much higher GMR amplitude than does an AMR film, the field sensitivity of a GMR multilayer is much smaller than that of an AMR film because of the high H_s. The high H_s is due to two factors: the strong interlayer antiferromagnetic exchange coupling in the Fe/Cr and Co/Cr multilayer systems and the high anisotropy field of the Fe or Co films. A high external field is therefore required to magnetize the sample to saturation because it has to overcome both the large anisotropy field of the magnetic material and the antiferromagnetic exchange coupling field. It was later found that the interlayer exchange coupling exhibits an oscillatory behavior [54]. Thicker spacer layers can be used to reduce the strength of interlayer antiferromagnetic coupling and hence to obtain low saturation fields, but at the expense of lower GMR amplitudes. The use of low-anisotropy materials such as Permalloy, CoFe, or CoNiFe as the magnetic material can also reduce the saturation field. The combination of these techniques has produced GMR multilayers with quite small saturation fields, for example, (Ni$_{76}$Fe$_{10}$Co$_{14}$/Cu)$_{10}$ [58] and Ni$_{80}$Fe$_{20}$/Ag multilayers [57] with H_s on the order of 10–50 Oe.

4.2.3.2. Uncoupled GMR Multilayers

In this type of multilayer, the GMR effect is obtained in the absence of interlayer antiferromagnetic coupling. The antiparallel orientation of the magnetic moment in the successive layers is achieved by using alternate layers with different coercivities (e.g., Permalloy (low H_c) and Co (higher H_c) in a multilayer of (Permalloy/Cu/Co/Cu)$_n$). When the external magnetic field sweeps between the positive and negative saturation of the multilayer, the magnetic configuration of the successive layers changes from parallel at large field to antiparallel at fields between the two coercivities of the two layers. The uncoupled GMR multilayers have the advantages of lower saturation field and higher GMR amplitude at room temperature than antiferromagnetically coupled systems [67].

4.2.3.3. Exchange Biased Spin-Valve Sandwiches

This is the most successful type of GMR structure so far for recording head applications. An exchange-biased spin-valve sandwich is composed of essentially four films: a free layer (sensing layer), a conducting spacer, a pinned layer, and an AF exchange pinning layer. Seed and capping layers are also used in the deposition of the sandwich structure. A typical spin-valve with a layer structure of substrate/seed/F$_1$/NM/F$_2$/AF/cap is schematically shown in Fig. 32. The pinned layer F$_2$, free layer F$_1$, and Cu conducting layer are very thin, allowing conduction electrons to frequently move back and forth between the sensing and pinned layers via the conducting spacer. The magnetic orientation of the pinned layer is fixed and held in place by the adjacent antiferromagnetic exchange pinning layer, and the magnetic orientation of the sensing layer changes in response to the magnetic field, H, from the disk. A change in the magnetic orientation of the sensing layer will cause a change in the resistance of the combined sensing and pinned layers. When the magnetization of the free layer is parallel to the magnetization of the pinned layer, the spin-valve

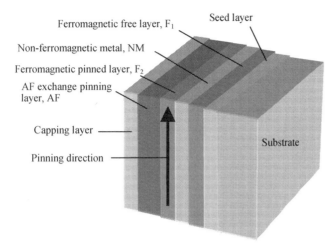

Ferromagnetic free layer, F₁
Seed layer
Non-ferromagnetic metal, NM
Ferromagnetic pinned layer, F₂
AF exchange pinning layer, AF
Capping layer
Substrate
Pinning direction

Fig. 32. Schematic diagram of a typical spin-valve layer structure.

film exhibits minimum resistance, and when the magnetization of the free layer is antiparallel to the pinned layer, it exhibits maximum resistance. The GMR amplitude is defined as the maximum resistance change normalized by the saturation resistance (minimum resistance of the film). A more detailed discussion on exchange-biased spinvalves is given in Section 4.3.

4.2.4. Basic Physics of GMR Phenomenon

The interpretation for the origin of GMR, as first proposed by Baibich et al., was based on spin-dependent scattering of conduction electrons [51]. Usually it is assumed that the conductivity of a ferromagnet is due to the 4s band electrons and 3d band holes (empty band states). But the 4s electrons are the primary electric current carriers because the 4s bands are broad and the 4s electrons have low effective mass, in contrast to the narrow 3d bands and high effective mass of the 3d holes. When a conduction electron passes through a ferromagnetic metal, scattering will occur because of impurities, structural defects, phonons, and magnons, etc. The probability of occurrence of a scattering event will depend both on the effectiveness of the scattering centre and on the availability

of a terminal state into which an electron can scatter [65]. The 3d bands in ferromagnetic materials play a very important role in providing the terminal states (holes) into which the 4s electrons can be scattered. The relaxation time of the scattered electrons in the empty states contributes to the different conductivities (resistivities) of magnetic metals.

Fig. 33 is a schematic illustration of the electronic band diagrams for the 3d ferromagnetic transition metals. The density of states function, $N(E)$, for a 3d transition metal can be divided into two components: $N(E) = N\uparrow(E) + N\downarrow(E)$, where $N\uparrow(E)$ and $N\downarrow(E)$ represent the density of states of the spin↑ and spin↓ electrons, respectively. The spin↑ direction usually refers to the direction of the magnetic moment (or spontaneous magnetization), and the spin↓ direction is opposite to the magnetic moment. For a nonmagnetic 3d transition metal, $N\uparrow(E)$ and $N\downarrow(E)$ are equal within any band. This is shown in Fig. 33a, where the 3d↑ and 3d↓ subbands have equal numbers of occupied states as well as empty states (holes). So do the 4s↑ and 4s↓ subbands. In ferromagnetic metals, as shown in Fig. 33b, because of the exchange interaction, the two 3d subbands of a ferromagnetic transition metal are split, resulting in more occupied states in 3d↑ than in 3d↓ bands and more empty states in 3d↓ than in 3d↑. The number of occupied and empty states in each 3d subbands varies with different metals.

Table III is a list of the electron states for the three ferromagnetic elements (Fe, Co, and Ni) calculated by the electronic band theory. The value of the saturation magnetic moment of a ferromagnetic material depends on the difference in the densities of the occupied states in the 3d↑ and 3d↓ subbands and is given by $m = (N_\uparrow - N_\downarrow)\mu_B$, where μ_B is the Bohr magneton. The band theory predicts that $(N_\uparrow - N_\downarrow)$ is usually a nonintegral number, which agrees with experimental data.

The 4s conduction electrons are distinguished into two families according to their spins (two-current model), the spin↑ ($S = +\frac{1}{2}$) and spin↓ ($S = -\frac{1}{2}$) electrons. It is commonly assumed that the two families have equal numbers of electrons and that at temperatures comparatively lower than the ferromagnetic ordering temperature the two families of electrons are quite independent, i.e., there is no spin mixing between

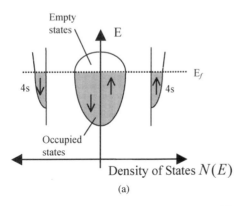

(a)

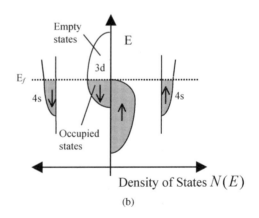

(b)

Fig. 33. Schematic illustration of the electronic band diagrams, (a) before exchange interaction and (b) after exchange interaction. As a result of exchange interaction, the 3d band of transition-metal ferromagnetic materials (Co, Ni, Fe, and their alloys) is split into two subbands of opposite spins and with different densities of empty states at the Fermi level.

Table III. Electron States of 3d Ferromagnetic Transition Metals [73]

Element	Electron configuration	Occupied states				Empty states		$N\uparrow - N\downarrow$
		3d↑	3d↓	4s↑	4s↓	3d↑	3d↓	
Fe	$3d^6 4s^2$	4.8	2.6	0.3	0.3	0.2	2.4	2.2
Co	$3d^7 4s^2$	5.0	3.3	0.35	0.35	0	1.7	1.7
Ni	$3d^8 4s^2$	5.0	4.4	0.3	0.3	0	0.6	0.6

the two spin channels. When the conduction electrons pass through a 3d ferromagnetic transition metal, they are likely to be scattered into empty energy bands of the same spins at the Fermi level. As can be seen from Fig. 33, because of the exchange interaction in 3d ferromagnets, the densities of empty states of the two 3d subbands are different at the Fermi level. The 3d↓ subband has more empty states than the 3d↑ band. This results in different scattering rates for spin↑ and spin↓ conduction electrons when they pass through the ferromagnetic metal; i.e., the spin↑ electrons will be less likely to be scattered (they can only be scattered into the empty 4s↑ band), whereas the spin↓ conduction electrons are more likely to be scattered (they can be scattered into both 3d↓ and 4s↓ empty bands) at the Fermi level. The difference in scattering rates results in different mean free paths (MFPs, or λ) of the spin↑ and spin↓ conduction electrons in ferromagnetic materials. For example, in Permalloy ($Ni_{80}Fe_{20}$), the mean free path of the spin-up electrons, $\lambda_\uparrow = 50–100$Å, is about five times longer than that of spin-down electrons, $\lambda_\downarrow = 10–20$ Å [66]. The resistivity of each channel is inversely proportional to the MFP of electrons, therefore we have $\rho_\uparrow << \rho_\downarrow$.

The electron transport behaviors in a GMR multilayer can be explained by the phenomenological model shown in Fig. 34 [65] when the electron mean free path is much larger than the thickness of the layers. As shown in Fig. 34a, at zero magnetic field, a GMR superlattice has an antiparallel magnetization configuration due to the interlayer antiferromagnetic coupling, which means that the two channels of electrons are alternately strongly and weakly scattered as they cross the successive ferromagnetic layers or interfaces. The mean free paths for both species of electrons are very short. The resistance of the multilayer in such an antiparallel magnetization configuration is a maximum. If an external field $H > H_s$ is applied, as shown in Fig. 34b, the superlattice will be in a magnetic saturation state, i.e., the internal magnetization vectors are aligned parallel. The spin↑ electrons (spins parallel to the magnetization vectors) are then weakly scattered in all layers as they pass through the multilayer and therefore have very long mean free paths (fast electrons) and low resistivity, whereas the spin↓ electrons (electron spins antiparallel to the magnetization vectors) are strongly scattered in every layer and therefore have very short mean free paths (slow electrons) and high resistivity.

The electron transport properties in multilayers can be further simplified with a parallel resistive network of the two conduction electron channels [65, 67], as also shown in Fig. 34. In the case of an antiparallel configuration (Fig. 34a), the two channels have both low and high resistivity in alternate layers.

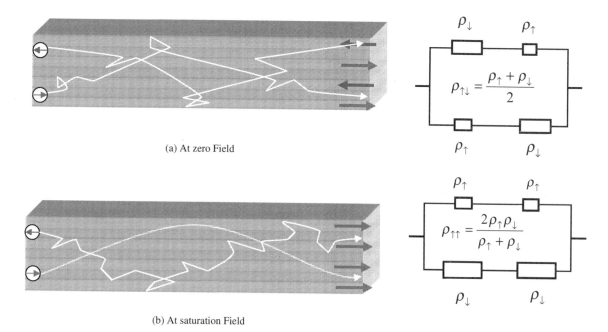

(a) At zero Field

$$\rho_{\uparrow\downarrow} = \frac{\rho_\uparrow + \rho_\downarrow}{2}$$

(b) At saturation Field

$$\rho_{\uparrow\uparrow} = \frac{2\rho_\uparrow \rho_\downarrow}{\rho_\uparrow + \rho_\downarrow}$$

Fig. 34. A phenomenological model of the spin-dependent scattering for conduction electrons in GMR multilayers. (a) Antiparallel magnetization configuration and its corresponding resistance model. (b) Parallel magnetization configuration and its corresponding resistance model.

If we designate $\rho_\uparrow$ and $\rho_\downarrow$ as the low resistivity (when the spin of electrons is parallel to the magnetic moment, i.e., spin$\uparrow$) and high resistivity (when the spin of electrons is antiparallel to the magnetic moment, i.e., spin$\downarrow$), respectively, the total resistivity of the multilayer in an antiparallel magnetic configuration can be estimated as [66]

$$\rho_{\uparrow\downarrow} = \frac{\rho_\downarrow + \rho_\uparrow}{2} \qquad (4.16)$$

In the case of a parallel magnetic configuration, the spin$\uparrow$ channel will exhibit very small resistivity and the spin$\downarrow$ down channel will exhibit very large resistivity, as schematically shown in the equivalent circuit diagram in Fig. 34b. The resulting total resistivity of the multilayer is given by

$$\rho_{\uparrow\uparrow} = \frac{2\rho_\downarrow\rho_\uparrow}{\rho_\downarrow + \rho_\uparrow} \qquad (4.17)$$

and the GMR amplitude is given by

$$\frac{\Delta\rho}{\rho_{\uparrow\downarrow}} = \frac{\rho_{\uparrow\downarrow} - \rho_{\uparrow\uparrow}}{\rho_{\uparrow\downarrow}} = \left(\frac{\rho_\downarrow - \rho_\uparrow}{\rho_\downarrow + \rho_\uparrow}\right)^2 = \left(\frac{\alpha - 1}{\alpha + 1}\right)^2 \qquad (4.18)$$

where $\alpha = \rho_\downarrow/\rho_\uparrow$ represents the contrast of the spin-dependent scattering of the conduction electrons when their spins are parallel or antiparallel to the magnetic moment.

4.2.5. Parameters Influencing the GMR Amplitude

Excellent reviews of the influence of various parameters on the GMR characteristics can be found in [66–68]. The following discussions are adapted mainly from [66].

4.2.5.1. Volume versus Interfacial Spin-Dependent Scattering

It is widely accepted that GMR originates from spin-dependent scattering (SDS). However, SDS in GMR multilayers may occur at the interfaces of F/NM layers, in bulk ferromagnetic films (volume), or in both. A question under contention has been the whereabouts of the conduction electron scattering, whether within the bulk of the magnetic layers or at the magnetic/nonmagnetic interfaces. The nature and location of the scattering centers that give rise to the SDS are important for GMR optimization. It is possible to experimentally distinguish these two types of scattering centres by introducing very thin layers of transition metal impurities at the F/NM interfaces or in the bulk of the ferromagnetic layers at various distances from the interfaces and looking at the influence of these impurities on the amplitude of the GMR [66]. Alternatively, they can be separated by measuring the transport properties of the multilayers with current perpendicular to the film plane (CPP) [76]. The variation of GMR amplitude with the thickness of magnetic layers also contains the potential information of the proportion of bulk and interfacial SDS [67]. The major findings on this topic are summarized as follows.

Sputtered Co layers seem to give rise to significant volume SDS, whereas Fe layers have very weak volume SDS [71].

However, the interfacial SDS is predominant in Co/Cu, Co/Ag, and Fe/Cr GMR multilayers. It is believed that in Co/Cu multilayers, the volume SDS is masked by the predominant interfacial SDS. When the ferromagnetic layers in the multilayer are alloy instead of pure metal, in particular with Permalloy, the volume SDS seems to be rather large and to dominate the SDS. For bulk scattering dominated systems, much thicker magnetic layers are required to alter the GMR amplitude. In contrast, in interfacial SDS-dominated systems, GMR amplitude can be significantly increased by the insertion of a very thin interfacial layer.

Typical examples of the use of interfacial SDS to improve GMR amplitude is the insertion of two very thin Co layers (5–10 Å), known as Co dusting, into NiFe/Cu multilayers [68] or into a spin-valve sandwich with a layer structure of NiFe(60 Å)/Cu(25 Å)/NiFe(40 Å)/FeMn(100 Å). The GMR amplitude of the spin valve is almost doubled as a result of Co dusting [64]. For interfacial SDS-dominated systems, the quality of the interfaces (no interdiffusion) is vitally important for high GMR amplitude. The interfacial SDS characteristic (maximum increase in GMR amplitude) is usually obtained within a minimum thickness of two to three monolayers of dusting materials. The insolubility of the interfacial dusting material with the host magnetic and nonmagnetic layers is a prime requirement.

4.2.5.2. Interfacial Roughness and Film Quality

Interfacial roughness plays various roles in influencing the GMR amplitude. In Co/Cu or Fe/Cr multilayers for which the interfacial SDS is mainly responsible for their GMR amplitudes, some interfacial roughness may increase the density of scattering centers at interfaces and may therefore lead to a larger GMR amplitude. In contrast, in multilayers for which the interfacial SDS is not as important as the volume SDS (NiFe/Cu multilayers, for example), an increase in the interfacial roughness leads to a decreased GMR amplitude [66]. In the limiting case, the interfaces of GMR multilayers cannot be too rough or too smooth. If the interfaces are too rough, the mean free path of the conduction electrons becomes too small compared with the period of the multilayer. Consequently, the coherent interplay of SDS between the successive magnetic layers cannot occur efficiently. In the opposite limit of very smooth interfaces, the specular reflection of the conduction electrons on the interfaces may be increased, leading to a channeling of electrons within each layer. Such a channeling effect will also prevent the SDS from occurring [66].

The quality of the films also influences the GMR amplitude. In particular, the existence of pinholes through the nonmagnetic spacer may significantly reduce the GMR by introducing localized ferromagnetic coupling between magnetic layers [66]. However, the quality of the crystal growth does not appear to be a necessary requirement for obtaining GMR because the largest GMR amplitudes (in Co/Cu or Fe/Cr [54]) are obtained in multilayers grown by sputtering and not by MBE.

4.2.5.3. Thickness of Magnetic Layer

The dependence of GMR amplitude on the thickness of magnetic layers was extensively studied by many research groups [66]. The general observation is that the GMR amplitude always shows a maximum when the thickness of the magnetic layers (t_F) is varied from the thickness of a monolayer to a few tens of monolayers. Typical experimental results for this are shown in Fig. 35. The optimal magnetic layer thickness l_F, at which maximum GMR amplitude occurs, depends mainly on the type of GMR structures, the location of the SDS centers, the total number of periods (n) in the case of multilayers, and the thickness of the nonmagnetic spacer and even of the buffer and capping layers. For thicker multilayers ($n > 10$), as shown in Fig. 35a, the optimal magnetic layer thickness is typically 10–30 Å, in contrast to 40–100 Å for spin-valve sandwiches, as shown in Fig.35b. For interfacial SDS-dominated systems, maximum GMR amplitudes were usually obtained at thinner magnetic layers. Dieny [66] has found from the study of spin-valve sandwiches in the form of Ft_F/Cu 25 Å/NiFe 50 Å/FeMn 100 Å, with F = NiFe, Fe, or Co, that the maximum GMR occurs at a Fe layer thickness of 50 Å for spin valves with Fe free layers in contrast to 80 Å for spin valves with NiFe free layers.

A phenomenological expression describing the dependence of GMR amplitude of multilayers or spin-valve sandwiches on the thickness of the ferromagnetic layers is given by [66]

$$\frac{\Delta R}{R}(t_F) = \left(\frac{\Delta R}{R}\right)_0 \left(\frac{1 - e^{-t_F/l_F}}{1 + t_F/t_0}\right) \qquad (4.19)$$

where t_F is the thickness of the ferromagnetic layer, l_F is the critical thickness at which the GRM amplitude is at maximum, t_0 is an effective thickness that represents the shunting of the current in the rest of the structure (i.e., in all layers except the ferromagnetic layer, whose thickness t_F is varied), and $(\Delta R/R)_0$ is a normalization coefficient that depends on the combination of F and NM materials and on the thickness of the spacer layer. The fits of experimental data with the phenomenological expression are shown in Fig. 35 by the solid lines. As pointed out by Dieny [66], a limitation of Eq. (4.19) is that it does not properly account for the various incident angles θ of the conduction electrons with respect to the plane of the layers. The effective thickness of the ferromagnetic layers seen by the conduction electrons varies as $t_F/\cos\theta$.

The decrease in GMR amplitude above the optimal thickness, as summarized by Dieny [66], is mainly due to the increased shunting current in the inner part of the magnetic layers. On the other hand, the decrease in GMR amplitude below the optimal thickness has two possible origins, depending on whether the multilayer structure has a large number of periods or is just a sandwich.

In a multilayer with a large number of periods, where the scattering on the outer surfaces (substrate, buffer, or capping layers) can be ignored, the decrease in GMR at small t_F is due to insufficient scattering of the normally strongly scattered species of electrons. As shown by Eq. (4.18), large GMR amplitude originates from the large contrast in SDS of the two species of electrons in the antiparallel and parallel magnetic configurations. Strong scattering of electrons in magnetic layers whose magnetization is antiparallel to the electron spins (spin↓) leads to a short MFP (mean free path, $\lambda \downarrow$) and high effective resistivity (ρ_H), whereas weak scattering of electrons in magnetic layers whose magnetic moment is parallel to the electron spins (spin↑) leads to a long MFP ($\lambda \uparrow$) and low effective resistivity (ρ_L). The effectiveness of SDS of electrons in the antiparallel magnetic configuration is an important factor in determining the GMR amplitude. In the case of bulk SDS, the critical thickness below which insufficient SDS will occur is determined by the shorter of the two MFPs ($\lambda \downarrow$ and $\lambda \uparrow$), which is on the order of 10–20 Å.

In the case of spin-valve sandwich structures, the decrease of GMR amplitude at small t_F is not due to the insufficient SDS of the spin↓ electrons since the optimal thickness for spin-valve sandwich is always significantly larger than $\lambda \downarrow$. However, if the magnetic layer thickness is smaller than $\lambda \uparrow$ (the longer of the two MFPs), the spin↑ electrons penetrate the magnetic layer and are scattered by the outer surfaces (substrate, AF layer, buffer layer or capping layer), which results in small $\lambda \uparrow$, increased low effective resistivity (ρ_L), or reduced SDS contrast and hence reduced GMR amplitude.

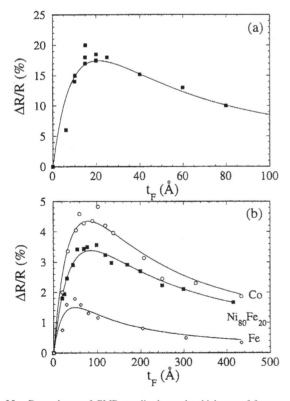

Fig. 35. Dependence of GMR amplitude on the thickness of ferromagnetic layers at room temperature in structures of (a) a multilayer of glass/($Ni_{80}Fe_{20}$ 50 Å/Cu 10 Å/$Ni_{80}Fe_{20}t_F$/Cu 10 Å)30, and (b) a spin-value sandwich of the form Ft_F/Cu 25 Å/NiFe 50 Å/FeMn 100 Å, with F = $Ni_{80}Fe_{20}$, Co, or Fe. The solid lines in both figures are fits of experimental data according to the phenomenological expression of Eq. (4.19). Reproduced with permission from [66], copyright 1994, Elsevier Science.

4.2.5.4. Thickness of Nonmagnetic Layers

The spacer thickness dependence of GMR amplitude should be discussed separately with respect to multilayer GMR and exchange-biased spin-valve sandwich-type GMR because of the different mechanisms used to obtain GMR in the two systems. Although the GMR effect in the two types of systems originates from the spin-dependent scattering and it is commonly observed that the GMR amplitudes in the two systems vary with the spacer thickness, their mechanisms for obtaining the GMR are different, however. In multilayer films, the GMR phenomenon is closely related to the interlayer antiferromagnetic coupling, and the interlayer antiferromagnetic coupling is the essential requirement for obtaining GMR, whereas in spin-valve sandwiches, the interlayer coupling between the free and pinned layers is not related to GMR, and in sensor applications the interlayer coupling is not desirable.

Effect of Spacer Thickness on Multilayer GMR–Oscillatory Interlayer Coupling. The general observation is that the multilayer GMR amplitudes and the saturation fields decrease with an increase in spacer thickness. However, they do not decrease monotonically with increasing spacer thicknesses. Instead, they oscillate in magnitude as a function of the spacer thickness with a period ranging from 12 Å in the Co/Ru system to 10–21 Å in the Fe/Cr and Co/Cr systems [52]. The GMR amplitude oscillation reflects the oscillations of the interlayer antiferromagnetic coupling through the spacer layer as its thickness varies. Figure 36 [68] shows the typical MH loops of a series of multilayers of the structure 100 Å Ru/[30 Å $Ni_{81}Fe_{19}$/Ru(t_{Ru})]$_{20}$ as the Ru spacer thickness varies from 4 Å to 44 Å. The MH loops of these samples clearly show an oscillatory variation of saturation field with Ru thickness. For the Ru spacer layer thicknesses of 4, 12, 24, and 37 Å shown in Fig. 36, the magnetization of the multilayer is saturated in

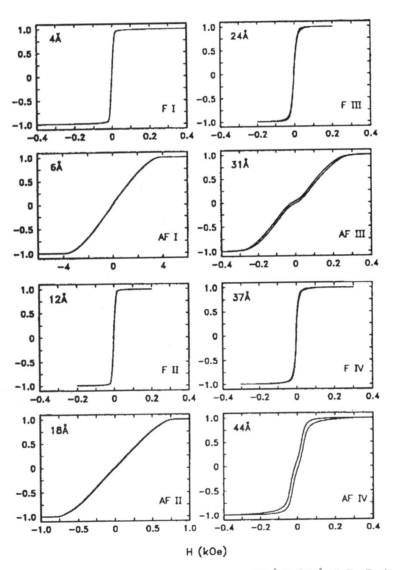

Fig. 36.   Typical MH loops at room temperature for several multilayers of the structure 100 Å Ru/[30 Å $Ni_{81}Fe_{19}$/Ru (t_{Ru})]$_{20}$ as a function of Ru spacer thickness. Reproduced with permission from [68], copyright 1994, Springer-Verlag.

a very low field of about 10 Oe. For intermediate Ru thickness the saturation fields are larger, although they decay with increasing Ru thickness. A detailed dependence of saturation field on Ru thickness is shown in Fig. 37 for multilayers with the same structures as in Fig. 36. Five oscillations in the saturation field are shown in Fig. 37 with an oscillation period of about 11 Å. In the limit of very thin Ru thickness, the coupling is antiferromagnetic. Even for a Ru layer that is just 3 Å thick, strong AF coupling is observed. This is in contrast to the Fe/Cr multilayer system, where the interlayer coupling becomes very weak as the Cr thickness is decreased below 8 Å [52].

The GMR amplitude in multilayer systems is closely related to the interlayer coupling and hence to the saturation field. The highest GMR amplitude is always obtained at the first peak, which also gives the highest saturation field. However, the optimum thickness of a nonmagnetic spacer for obtaining the largest field sensitivity $(\Delta R/R)/H_s$ is normally not at the first GMR peak. There is always a compromise between the highest GMR amplitude and the highest field sensitivity. But t_{NM} must be chosen in a range for which the coupling is antiferromagnetic for any observable GMR.

Interlayer antiferromagnetic coupling and oscillations in the coupling have been found in numerous multilayer systems, including ferromagnet/antiferromagnet (Cr, Mn) multilayers, ferromagnet/transition metal (Ru, Mo) multilayers, and ferromagnet/noble metal (Cu, Ag, Au) multilayers. A systematic study of the interlayer coupling in sputtered Co-based multilayers with 3d, 4d, and 5d transition metal spacers was carried out by Parkin [63]. The results are shown in Fig. 38. It was found from Parkin's work (see Fig. 38) that the period of oscillatory coupling is similar for most metals (typically 5–6 ML, where ML refers to monolayer), with the exception of Cr, for which the period is significantly longer (12.5 ML). Among all spacer metals, the Co/Ru system has the strongest antiferromagnetic coupling strength at its first peak (5 erg/cm^2, at 3 Å).

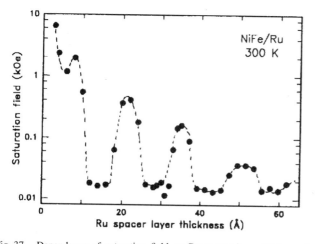

Fig. 37. Dependence of saturation field on Ru spacer layer thickness for several series of $Ni_{81}Fe_{19}$/Ru multilayers with structure 100 Å Ru/[30 Å $Ni_{81}Fe_{19}$/Ru(t_{Ru})]$_{20}$, where the topmost Ru layer thickness is adjusted to be about 25 Å for all samples. Reproduced with permission from [68], copyright 1994, Springer-Verlag.

This unique system has recently found application in synthetic spin-valve structures, which are discussed in Section 4.3.3.

Note the following:

- There are four cells beneath each element in the periodic table, marked A_1, J_1, A_1, ΔA_1 and P. A_1 (Å) is the spacer layer thickness corresponding to the position of the first peak in antiferromagnetic exchange coupling strength as the spacer layer thickness is increased; J_1 (erg/cm^2) is the magnitude of the antiferromagnetic exchange coupling strength at this first peak; ΔA_1 (Å) is the approximate range of spacer layer thickness of the first antiferromagnetic region; and P (Å) is the oscillation period.

- The most stable crystal structures of the various elements are included for reference. Note that no dependence of the coupling strength on crystal structure or any correlation with electron density is found.

- For the elements Nb, Ta, and W, only one AF coupled spacer layer thickness region was observed, so it was not possible to directly determine P. For Ag and Au no oscillatory coupling was observed in Co-based multilayers. Pd and Pt show strong ferromagnetic coupling, with no evidence of oscillatory coupling.

The oscillatory behaviour in interlayer exchange coupling has been experimentally confirmed by magnetic domain imaging of a Fe/Cr/Fe superlattice sample with a wedge-shaped Cr spacer by scanning electron microscopy with polarization analysis (SEMPA) [60]. For a detailed review of the SEMPA technique, readers are encouraged to read reference [69] by Pierce et al.

The interlayer antiferromagnetic coupling in GMR multilayers can be explained by the RKKY-like coupling, which is a type of indirect coupling between the successive magnetic layers through the spacer and is characterized by oscillations of coupling strength and phase. For a complete review of the RKKY coupling theory of GMR multilayers, please refer to reference [70] by Hathaway.

Effect of Spacer Thickness on Exchange-Biased Spin-Valve GMR. In contrast to multilayer GMR, the GMR amplitudes in exchange-biased spin valves decrease monotonically with increasing conductor spacer thickness. Typical examples of this are shown in Fig. 39 for Co/NiFe spin-valve sandwiches with Cu and Au conductor spacers [66]. As pointed out by Dieny [66], qualitatively this decrease is due to two factors: (i) the increasing scattering of the conduction electrons as they traverse the spacer layer (which reduces the flow of electrons crossing the spacer from one ferromagnetic layer to the next and therefore reduces the efficiency of the spin-valve mechanism) and (ii) the increased shunting current in the thicker spacer layers.

The variation of the GMR amplitude with spacer thickness can be represented by the following phenomenological

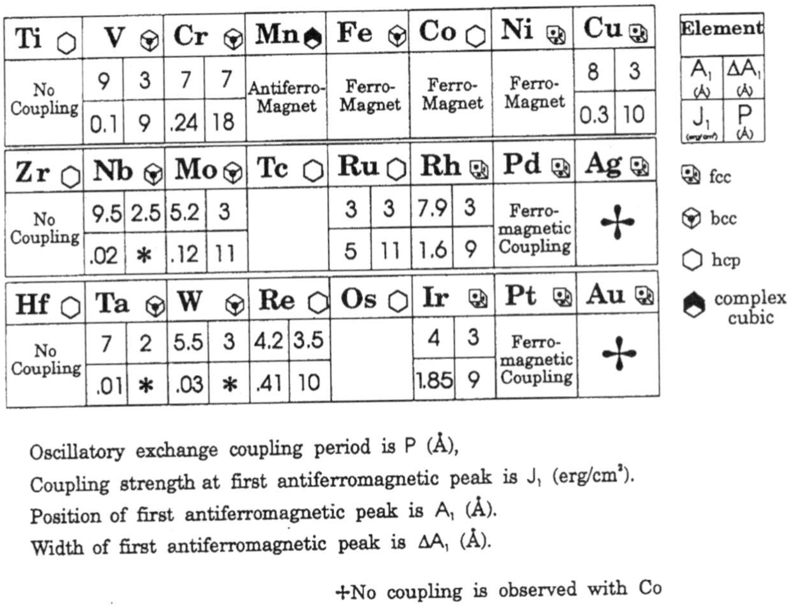

Element	Structure	A_1 (Å)	ΔA_1 (Å)	J_1 (erg/cm²)	P (Å)
Ti	hcp	No Coupling			
V	bcc	9	3	0.1	9
Cr	bcc	7	7	.24	18
Mn	complex cubic	Antiferro-Magnet			
Fe	bcc	Ferro-Magnet			
Co	hcp	Ferro-Magnet			
Ni	fcc	Ferro-Magnet			
Cu	fcc	8	3	0.3	10
Zr	hcp	No Coupling			
Nb	bcc	9.5	2.5	.02	*
Mo	bcc	5.2	3	.12	11
Tc	hcp				
Ru	hcp	3	3	5	11
Rh	fcc	7.9	3	1.6	9
Pd	fcc	Ferromagnetic Coupling			
Ag	fcc	✛			
Hf	hcp	No Coupling			
Ta	bcc	7	2	.01	*
W	bcc	5.5	3	.03	*
Re	hcp	4.2	3.5	.41	10
Os	hcp				
Ir	fcc	4	3	1.85	9
Pt	fcc	Ferromagnetic Coupling			
Au	fcc	✛			

Legend (Element box): A_1 (Å), ΔA_1 (Å); J_1 (erg/cm²), P (Å). Symbols: ▦ fcc, ⊗ bcc, ○ hcp, ⬡ complex cubic.

Oscillatory exchange coupling period is P (Å),
Coupling strength at first antiferromagnetic peak is J_1 (erg/cm²).
Position of first antiferromagnetic peak is A_1 (Å).
Width of first antiferromagnetic peak is ΔA_1 (Å).

✛No coupling is observed with Co

Fig. 38. Compilation of data on various polycrystalline Co/TM multilayers with magnetic layers composed of Co and spacer layers of the transition and noble metals. Reproduced with permission from [68], copyright 1994, Springer-Verlag.

expression, similar to Eq. (4.19), which is derived for the ferromagnetic layer thickness dependence of GMR amplitude.

$$\frac{\Delta R}{R}(t_{NM}) = \left(\frac{\Delta R}{R}\right)_1 \left(\frac{e^{-t_{NM}/l_{NM}}}{1 + t_{NM}/t_0}\right) \qquad (4.20)$$

where t_{NM} is the thickness of the nonmagnetic spacer, l_{NM} is related to the MFP of the conduction electrons in the spacer layer, t_0 is an effective thickness that represents the shunting of the current in the rest of the structure (i.e., in all layers except the nonmagnetic spacer, whose thickness, t_{NM} is varied), and $(\Delta R/R)_1$ is a normalization coefficient that depends on the combination of F and NM materials and on the thickness of the spacer layer. The fits of experimental data with the phenomenological expression are shown in Fig. 35 by the solid lines.

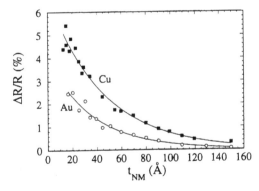

Fig. 39. Dependence of GMR amplitude at room temperature on the thickness of the nonmagnetic spacer for a spin-value sandwich of the form Si/Co 70 Å/NMt_{NM}/NiFe 50 Å/FeMn 80 Å with NM = Cu, Au. The solid lines are fits of experimental data to Eq. (4.20). Reproduced with permission from [66], copyright 1994, Elsevier Science.

It is worth noting that in exchange-biased spin-valve films there also exists the oscillatory interlayer exchange coupling of the two magnetic layers through the conductor spacer as the thickness of the spacer varies. However, the existence of this oscillatory interlayer coupling does not affect the spin-valve GMR amplitude, as the GMR spin valve is obtained through a different mechanism. This has been experimentally demonstrated by Speriosu [72] for a series of spin-valve films of (Co 30 Å/Cu t_{Cu}/Co 30 Å/FeMn); the result is shown in Fig. 40. As shown in Fig. 40a, for t_{Cu} ranging from 18 to 40 Å, the coupling between the Co layers oscillates between parallel (ferromagnetic) and antiparallel (antiferromagnetic) coupling with a period of 9 Å, which is consistent with that obtained in GMR multilayers [68]. However, the GMR amplitude of these spin valves, as shown in Fig. 40b, decreases with increasing spacer thickness, which is independent of the oscillatory coupling of the two Co layers.

4.3. Properties of Exchange-Biased Spin-Valve Films

4.3.1. Basic Principle of Operation of Spin Valves

In Section 4.2.3.3, we briefly introduced the basic structures of a spin-valve sandwich. The principle of operation of a spin-valve is again based on the spin-dependent scattering (SDS) of conduction electrons. Similar to the antiferromagnetically coupled GMR multilayers, spin-valve sandwiched films exploit the quantum nature of the conduction electrons, which have two spins, ↑and ↓. The sensing current, I, is composed of these two families of electrons. As shown in Fig. 41, when the magnetization of the free layer is parallel to the magnetization in the pinning layer; i.e., the electron spins in the magnetization of both layers are all spin↑, the spin↑ electrons

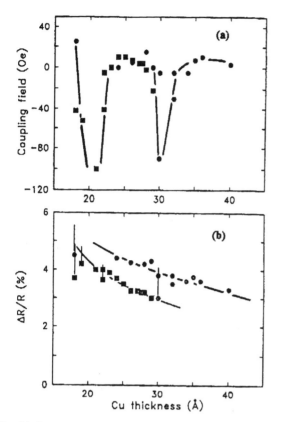

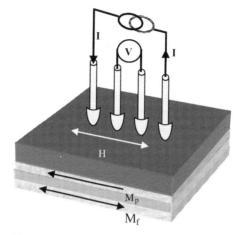

Fig. 42.   Four-point probe measurement of spin-valve GMR.

Fig. 40.   (a) Coupling field and (b) magnetoresistance vs. Cu thickness for spin valves of underlayer/Co 30 Å/Cut_{cu}/Co 30 Å/FeMn. The two sets of points correspond to different underlayer thickness, giving different shunting. Reproduced with permission from [72], copyright 1991, American Physical Society.

entation of the free layer. When the magnetic moments in the free layer and the pinned layer are antiparallel, the two channels of conduction electrons will be alternately strongly and weakly scattered as they pass across the free layer and pinned layer. This is equivalent to a two-resistor network in which both resistors have high resistance. The total resistance of the spin valve in this case is high.

4.3.2. Spin-Valve GMR Measurement

Four-point probe method.   Spin-valve GMR is characterized by low-field and high-field $R(H)$ loops together with a series of characteristic fields. The low-field and high-field $R(H)$ loops can be measured by the four-point probe method at low or high applied fields, respectively, for sheet films, or measured directly from patterned stripes for sensors. The low field is usually created with a pair of Helmholtz coils, and the high field is created with a pair of electromagnets. Figure 42 is a schematic diagram of the four-point probe setup for $R(H)$ loop measurement, in which the four-point probe (or current) is in line with the magnetic moments. The four-point probes can be arranged either parallel or perpendicular to the magnetic moments. However, as pointed out by Dieny [7], the

in the sensing current will pass through the multilayer structure without any scattering, whereas the spin↓ electrons in the sensing current are subject to heavy scattering by the spin↑ electrons in the magnetization vectors of the pinned and free layers. This is equivalent to a two-resistor parallel network in which one of the resistors has a very small resistance. The total resistance of the spin valve is therefore very low. A change in the external field will cause a change in the magnetic ori-

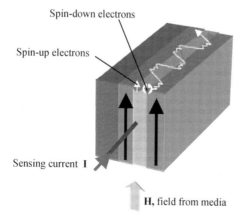

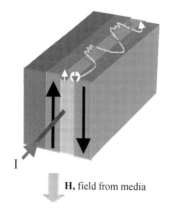

Fig. 41.

spin-valve GMR is intrinsically anisotropic, i.e., the two configurations (parallel or perpendicular) will give different GMR values. It has been observed by Dieny et al. that the GMR amplitude is larger when the magnetic moments are perpendicular to the current in comparison with the situation where the magnetic moments are parallel to the current. They attribute this to the anisotropy in the $\alpha = \rho_\uparrow / \rho_\downarrow$ ratio caused by spin-orbit coupling [7]. In their measurement, the contribution of AMR to the GMR ratio was excluded.

Low-Field R(H) Loop and Coupling Field H_f. Typical low- and high-field $R(H)$ loops of spin-valve films together with their field parameters are shown in Figs. 43 and Fig. 44, respectively. At a low applied field (typically $H < 100$ Oe), only the free layer of the spin-valve film responds to the applied field. The spin-valve film exhibits the lowest resistance when the magnetic moment in the free layer is parallel to the magnetic moment of the pinned layer, and it exhibits the highest resistance when the two magnetic moments are antiparallel. The low-field $R(H)$ loop is offset from a zero field. This is due to the presence of an interlayer exchange coupling field (H_f) between the free and the pinned layers. As discussed in 4.2.5.4, the coupling field is spacer thickness dependent and is oscillatory in nature, i.e., it may be ferromagnetic (positive) or antiferromagnetic (negative), depending on the spacer thickness, but it has no effect on the GMR amplitude of the spin valves. The ferromagnetic coupling field H_f and the coercivity of the free layer H_c(free) can be obtained from the low field $R(H)$ loop, as shown in Fig. 43. However, one must note that the offset field of the $R(H)$ loop measured by the four-point probe method represents roughly the ferromagnetic coupling field between the pinned and free layers (neglecting the field induced by measuring current in the free layer). If the $R(H)$

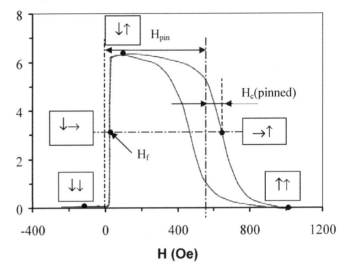

Fig. 44. Schematic illustration of a high field $R(H)$ loop of spin-valve films, field parameters and the orientation of magnetic moments in the free layer (black arrows) and pinned layer (red arrows) at various points of the loops.

loop is measured from a patterned sensor, the offset field is no longer equal to the ferromagnetic coupling field. In fact, it is equal to the sum of the ferromagnetic coupling field, the magnetostatic stray field from the pinned layer, and the field induced by the measuring current.

High-Field R(H) Loop and Pinning Field H_{pin}. When a field higher than the exchange pinning field (H_{pin}) between the AF and the pinned layer is applied, the pinned layer will also respond to the applied field. The reversal of magnetic moment of the pinned layer (red arrows in Fig. 44) results in a second $R(H)$ loop, which is offset from the zero field position by the unidirectional exchange pinning field, H_{pin}. Before the reversal of the pinned layer, as shown in Fig. 44, the spin valve exhibits the highest resistance because the free and pinned layers have an antiparallel magnetic configuration. When the magnetic moment of the pinned layer is completely reversed, the spin valve returns to the lowest resistance state. The high-field $R(H)$ loop gives a measurement of the exchange pinning field, H_{pin}, and the coercivity of the pinned layer, H_c(pinned).

VSM Measurement. The characteristic fields of spin valves can also be obtained from the MH loops measured by VSM. Figure 45 is a schematic illustration of the MH loops for spin-valve films, their field parameters, and the magnetization orientation of the pinned and free layers. Assuming that the saturation magnetizations of the free and pinned layers are equal, and in zero applied field the magnetic moments of the free and pinned layers have a parallel configuration, the spin-valve films are then at negative saturation state, as shown in Fig. 45. With the application of a positive field, the magnetization of the free layer will reverse first. When the free layer is completely reversed, the magnetization of the free layer is antiparallel to that of the pinned layer, and the spin-valve film

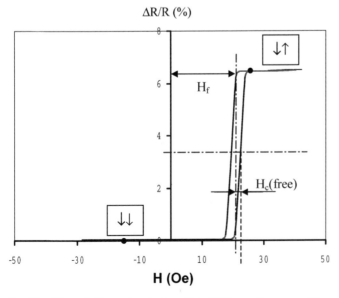

Fig. 43. Schematic illustration of a low-field $R(H)$ loop of spin-valve films, field parameters, and magnetization orientation of the pinned layer (red arrows) and the free layer (black arrows) at various points in the loop.

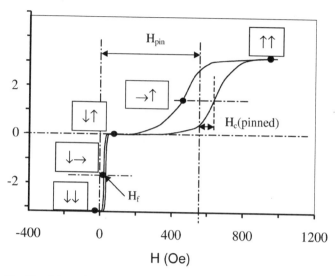

Fig. 45. Schematic illustration of MH loops of spin-valve films, their characteristic fields, and the relative orientation of the magnetic moment of the pinned layer (red arrows) and the free layer (black arrows) at various points in the loops.

will exhibit zero magnetization. As the applied field increases further, the magnetization of the pinned layer will start to reverse, giving rise to the top curve of the second MH loop. In the maximum applied field, the pinned layer is magnetized to saturation, and the spin-valve film is saturated by the applied field and has a parallel magnetic configuration in the direction opposite that of the original.

4.3.3. Types of Spin Valves

Since the discovery of the first spin-valve structure [71], many different types of spin valves have been introduced by various researchers. Here is a list of some of the spin-valve layer structures reported so far.

1. *Basic Spin Valves.* The basic layer structure of a spin valve consists of four function layers grown on a substrate coated with a Ta seed layer. One usually finishes the deposition of spin-valve multilayers with a Ta protection layer. The structure is either in the sequence of substrate/Ta/F_1/NM/F_2/AF/Ta or substrate/Ta/AF/F_2/NM/F_1/Ta, where F_1 refers to a free layer, NM to a noble metal conductor spacer, F_2 to a pinned layer, and AF to an antiferromagnetic layer. The first sequence is known as a top spin valve and the second sequence is called a bottom spin valve. Because of the deposition sequence dependence of exchange coupling between the F_2/AF layers, not all of the spin valves can be directly made into top and bottom spin-valve structures by the sputtering process. For example, if FeMn AF material is used, only top spin valves can be made because no exchange coupling between F_2/AF can be obtained if the FeMn AF layer is deposited before the ferromagnetic layer. The magnetic materials for F_1 and F_2 can either be the same or be chosen differently. The advantage of using different materials for F_1 and F_2 is that the material of F_1 can be chosen to obtain an excellent soft magnetic behavior, and F_2 can be chosen to obtain the highest exchange pinning field between the F_2 and the AF layers and maximum spin-dependent scattering. $Ni_{81}Fe_{19}$ (Permalloy) is one of the most frequently used free-layer materials because of its magnetic softness at very small thicknesses. Co, FeCo films are also used as free layers, but with a slightly larger saturation field. The largest spin-valve GMR values are obtained with Co and Permalloy layers separated by a noble metal (Cu, Ag, or Au) [66]. Cu is the most commonly used conductor spacer. Typical AF materials used in spin valves include FeMn, NiMn, IrMn, PtMn, RuRhMn, PtPdMn, NiO, α-Fe_2O_2, etc. Properties of AF exchange-coupled systems are discussed further in Section 4.4.4.

2. *Interface Engineered Spin Valves.* This type of spin valve has a layer structure of substrate/Ta/F_1/Co/NM/Co/F_2/AF/Ta, i.e., with two extra very thin Co layers (about 5–20 Å) inserted into the interfaces of free layer/Cu and Cu/pinned layer [78]. The very thin Co layers are known as "dust layers." FeCo may also be used as an interface-dust layer. As discussed in Section 4.2.5.1, dust layers are used to enhance the interfacial spin-dependent scattering and hence to enhance the GMR amplitude of the spin valves.

3. *Symmetric Spin Valves.* A symmetric spin valve [88, 89] has a layer structure of substrate/Ta/AF/F_2/NM/F_1/NM/F_2/AF/Ta; i.e., it is a combination of a top and a bottom spin valve, but with a shared free layer. The two pinned layers of a symmetric spin valve are biased in the same direction (parallel magnetization configuration). As shown in Fig. 46a, when the magnetic moments in the free and pinned layers are parallel, the symmetric spin valve exhibits a low resistance state. When the magnetic moments of the free and pinned layers are antiparallel (Fig. 4.6b), the symmetric spin valve is in a high resistance state. Because the symmetric spin valve allows double spin-dependent scattering, as shown in Fig. 4.6b, its GMR amplitude can be increased to about 130% of that of a simple spin valve, but at the expense of an almost doubled sensor thickness. Another minor drawback of this structure is that, as the free layer is between two pinned layers, it feels a double coupling strength [79].

4. *Synthetically Pinned Spin Valves.* The synthetically pinned spin valve [91, 93, 110] uses an alternative pinning mechanism to replace the AF/F exchange pinning system in conventional spin valves. Its layer structure is schematically shown in Fig. 47. It consists of an AF pinning layer, a synthetic ferrimagnet (or antiferromagnet) pinned layer (SF), a conductor spacer, and a free layer (F_1). The synthetically pinned layer has a structure of F_a/Ru/F_b, where F_a and F_b are ferromagnetic materials (Co or FeCo). The Ru layer usually has a thickness of 3–5 Å. The name "synthetic ferrimagnet" comes from the fact that the two ferromagnetic layers have different thicknesses and therefore different saturation magnetic moments. As discussed in Section 4.2.5.4, the Co/Ru/Co

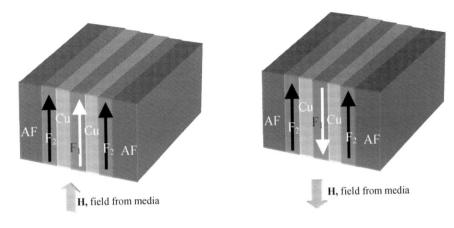

(a) Low resistance state.   (b) High resistance state.

Fig. 46.   Schematic diagram of a symmetric spin valve. (a) Low-resistance state. (b) High-resistance state.

or FeCo/Ru/FeCo multilayers are unique in that they exhibit very strong antiferromagnetic coupling ($H_{coupling}$ is up to a few thousand Oe) when the Ru thickness is in the range of 3–5 Å. Synthetically pinned spin valves exhibit the following three major advantages over conventional spin valves: (i) the antiferromagnetic pinning system becomes very rigid because of the strong RKKY indirect coupling of the two ferromagnetic layers and the reduction of the self-demagnetizing field within the pinned layer; (ii) the thermal stability of the AF exchange pinning system is much inproved [94]; (iii) there is a significant reduction of the demagnetizing field in the free layer arising from the pinned layer stray field, which allows for the reduction of the free layer thickness. All of these features are important for very small (submicron-sized) spin-valve heads, which are further discussed in Section 4.4.3.2.

5. *Synthetic Free-Layer Spin Valves.* In this type of spin valves [90], the free layer consists of a synthetic ferrimagnet (or antiferromagnet) in the form of $F_a/Ru/F_b$, similar to the ones used in synthetic pinned layers, where F_a and F_b are ferromagnetic materials (Permalloy, Co, CoFe). The advantage of such a free-layer structure is the reduced self-demagnetizing field due to flux closure within the free layer and hence

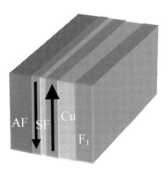

Fig. 47.   Schematic diagram of a symmetric spin valve. The pinned layer SF has a layer structure of F/Ru/F. The two F layers are antiferromagnetically coupled through the very thin Ru layer.

reduced magnetostatic interaction between the free layer and the pinned layer. The reduced self-demagnetizing field of the free layer is important to maintaining the field sensitivity of submicron-sized spin-valve sensors, which are discussed in Section 4.4.3.2. It was claimed [90, 170] that the used of a synthetic free layer can reduce the effective magnetic thickness of the free layer ($t_{eff} = (M_a t_a + M_b t_b)/M_{NiFe}$) while maintaining the large physical thickness.

6. *Specular Spin Valves.* A specular spin valve may have a layer structure similar to that of any of the above, but with an insulation antiferromagnetic layer as a specular reflection layer. In spin valves with metallic AF and capping layers, some of the conduction electrons are lost because of the shunting effect via the AF and capping layers. As a result, the number of electrons participating in the spin-dependent scattering (contributing to the GMR effect) is reduced. An insulation AF layer functions as an electron mirror, reflecting (scattering) the conduction electrons reaching the F_2/AF interface back to the multilayer and preventing them from escaping the spin-valve multilayers. Specular reflection can be further enhanced by the use of an insulation capping layer [82] or a very thin (two monolayers) Au capping layer [83]. However, the specular reflection of electrons at the interfaces must not affect the electron spins. Otherwise the GMR ratio will not be enhanced. Egelhoff et al. have obtained a GMR amplitude of 15% at room temperature for NiO bottom spin valves with a 4-Å Au capping layer [83] and 24.8% for specular symmetric spin valves [81]. Sugita et al. [84, 85] have reported that a GMR amplitude of 28% was obtainable for specular symmetric spin valves with metal/insulator (α-Fe$_2$O$_3$/Co and Co/α-Fe$_2$O$_3$) interfaces.

7. *Spin-Filter Spin Valves.* Spin-filter spin valves [171–173] allow a reduction of the free layer thickness without a significant reduction of GRM amplitude. Therefore, ultrathin free layers can be used in such a spin-valve structure. As discussed in Section 4.2.5.3, if the free layer thickness is smaller than

$\lambda \uparrow$ (the longer of the two MFPs), the spin $\uparrow$ electrons will penetrate the free layer and be scattered by the outer surfaces (substrate, AF layer, buffer layer, or capping layer), which results in smaller effective $\lambda \uparrow$, increased low effective resistivity (ρ_L), or reduced SDS contrast and hence reduced GMR amplitude. The spin-filter spin valve uses an extra high conductance layer on the other side of the free layer (the free layer is sandwiched by the Cu conducting layer and the high-conductance layer). Because of the low scattering rate of the extra high conductance layer, the $\lambda \uparrow$ of the spin $\uparrow$ electrons is no longer restricted by the free layer thickness. As a result of this, the GMR amplitude of the spin-filter spin valve does not decrease significantly as the free layer thickness is reduced to below the value of $\lambda \uparrow$.

8. *Dual Spin Valves.* As in to the dual-element MR heads [9], various types of dual spin-valve structure can be formed, depending on the combinations of magnetization pinning directions, sensing current directions, types of spin valves, and the voltage-sensing mode. Figure 48 shows two typical types of dual spin-valve structures, namely the antiparallel and the parallel dual spin valves [92]. According to [92], both types of dual spin-valve heads are thermal noise immune because of differential voltage sensing. The thermal noise immunity of heads is accomplished by connecting a differential amplifier to the output terminal of each spin valve. The antiparallel dual

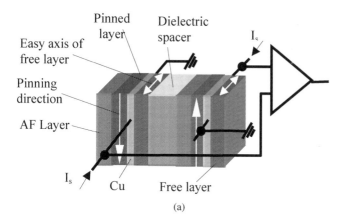

(a)

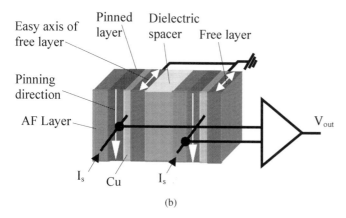

(b)

Fig. 48. Schematic diagram of the antiparallel (a) and parallel (b) dual spin valve structure. Reproduced with permission from [92], copyright 1999, IEEE.

spin-valve head exhibits greater linear dynamic range than a single spin-valve head and an almost doubled sensitivity of a single spin-valve head. It also has an ideal transfer curve passing through the origin that is independent of the bias state of its single spin valves. However, the drawback of dual spin-valve heads is that they are three terminal devices and are therefore more expensive to fabricate.

In spin-valve design, it is possible to combine various types of spin-valve structures to form a new spin valve to meet the design requirements. For example, a GMR amplitude of 15.4% was obtained for synthetically pinned symmetric spin valves with the composition Ta/Seed/IrMn/CoFe/Ru/CoFe/Cu/CoFe/NiFe/CoFe/Cu/CoFe/Ru/CoFe/IrMn/Ta, which was used in the demonstration of a 20 Gb/in^2 recording system [80].

4.4. Spin-Valve Head Engineering

4.4.1. Considerations of Spin-Valve Head Design

The major considerations for the design of high recording density read heads are as follows [86, 87].

1. *Pulse Amplitude.* The desired pulse amplitude of the reproduced signal from the increasingly smaller recorded bits must be maintained as the recording density increases. Pulse amplitude is normally measured as the track averaged amplitude (TAA), which is simply given by Ohm's law [87],

$$\Delta V = I_s \Delta R \qquad (4.21)$$

where I_s is the sensing current and ΔR is the sensor magnetoresistance change. To achieve high pulse amplitude from the increasingly smaller flux from the media, spin valves with a higher GMR ratio and higher field sensitivity are required. Unfortunately, the spin-valve field sensitivity decreases with the reduction of sensor size (or increasing t/h or t/w ratio) because of the enhanced magnetostatic interactions; this puts further demands on the increase in GMR ratio of spin-valve films.

2. *Pulse Amplitude Symmetry.* It is important to have equal amplitude of the positive and negative output pulses. To achieve this, the read sensor must be properly biased to have a linear transfer curve with a bias point at the origin of the coordinator system, as shown in Fig. 50.

3. *Stability.* Stability issues for spin-valve heads include magnetic domain stability and thermal stability. Magnetic domain stability is achieved by hard magnet bias, and thermal stability is optimized by the choice of antiferromagnetic exchange materials, which requires AF materials of high pinning field, high blocking temperature, and high temperature stability.

4. *Lifetime Reliability.* The lifetime reliability of a spin-valve head is determined by the operating temperature of the heads, which in turn depends on the joule heating effect due to $I_s^2 R$. Low sensor resistance and sensing current are therefore preferred options.

5. *ESD (Electrostatic Damage) Immunity.* The ESD immunity of a spin-valve head decreases with decreasing sensor size. For example, the ESD threshold voltage for an areal density of 1 Gb/in^2 is about 15 V, but it is reduced to about 1.5 V for 10 Gb/in^2 heads [87]. For submicron-sized sensors, ESD becomes a very serious problem. ESD can cause pinned layer reversal, sensor interdiffusion, amplitude loss, and physical damage of the sensor [86].

4.4.2. Basics of Spin-Valve Replay Heads: Sensor Bias Point

The magnetization orientations of free and pinned layers of a spin-valve read sensor in an external field H are schematically shown in Fig. 49. In spin-valve replay heads, the free layer is usually a uniaxial anisotropy film with easy axis aligned longitudinally (parallel to the ABS). The magnetic moment of the pinned layer is pinned along the vertical axis, which is ideally at a right angle with the easy axis of the free layer. This initial easy axis alignment is achieved during film deposition or post-deposition annealing. For such a magnetic configuration, the sensor magnetoresistance is proportional to the cosine of the relative angle between the magnetization vectors in the free and pinned layers and is given by [205]

$$\Delta R \approx \frac{\Delta R}{R} R_{\text{Sh}\uparrow\uparrow} \frac{W}{h} \frac{\langle \cos(\theta_1 - \theta_2) \rangle}{2} \quad (4.22)$$

where $\Delta R/R = (R_{\uparrow\downarrow} - R_{\text{sat}})/R_{\text{sat}}$ is the intrinsic magnetoresistance of the spin valve as measured on infinite samples, and $R_{\text{Sh}\uparrow\uparrow}$ is the sheet resistance measured in the parallel magnetic state. W is the length of the sensor active region (track width), and the notation $\langle \ldots \rangle$ denotes averaging over the sensor height, h.

An ideal transfer curve of a spin valve with infinite size is shown in Fig. 50. Because the uniform external field acts on the magnetic hard axis of the sensor, the response of the sensor to the field is linear. Here we only consider the ideal spin-valve, i.e, the magnetic moment of the free layer is at a right angle with the pinned layer when the external field is zero, and the ferromagnetic coupling field and the demagnetizing field are both zero. In such a case, the linear transfer curve will pass through the origin of the coordinate system and is equidistant from saturation in both positive and negative fields. The sensor exhibits minimum resistance (parallel magnetic configuration)

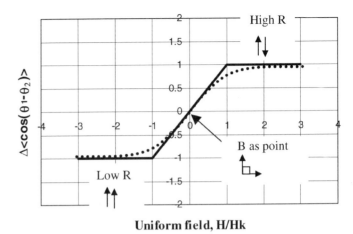

Fig. 50. Ideal spin-valve transfer curve for an infinite sensor in a uniform hard axis field (solid line) [75], and transfer curve due to the effect of non-uniform stray field from the pinned layer (dotted line).

when saturated by a negative field and maximum resistance (antiparallel magnetization configuration) when saturated by a positive field. Such a sensor is intrinsically linear, and the linear dynamic range of such an ideal sensor is $\langle \cos(\theta_1 - \theta_2) \rangle = \pm 1$, or $\Delta \langle \cos(\theta_1 - \theta_2) \rangle = 2$ [75].

In a real spin-valve film, the free layer is acted on by the ferromagnetic coupling field from the pinned layer $\mathbf{H}_f$, the stray field from the transversely saturated pinned layer $\mathbf{H}_s$, and the field generated by the sensing current $\mathbf{H}_I$ [205]. Fig. 51 is a schematic illustration of various fields acting on the free layer for a spin-valve replay head. Both $\mathbf{H}_f$ and $\mathbf{H}_s$ originate from the pinned layer. These two fields are always opposite in direction and therefore are partially canceled out by each other. As discussed in Section 4.2.5.4, $\mathbf{H}_f$ originates from the RKKY interlayer coupling, is nonmagnetic spacer thickness dependent, and may exhibit oscillatory behavior. $\mathbf{H}_f$ must be optimized to an acceptable value (typically $\mathbf{H}_f < 20$ Oe) during the spin-valve film deposition process, so that the free layer is

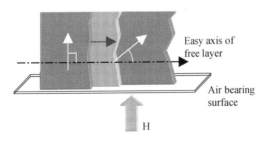

Fig. 49. Schematic diagram showing the magnetization orientations in the free and pinned layers of a spin-valve sensor.

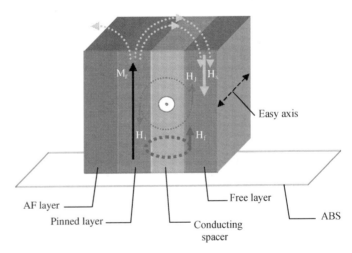

Fig. 51. Schematic illustration of the magnetic fields acting on the free layer of a spin-valve head.

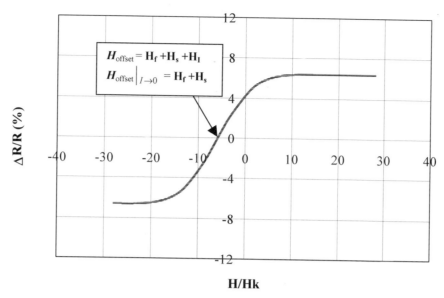

Fig. 52. R(H) curve of a patterned spin-valve sensor. The offset field $\mathbf{H}_{\text{offset}} = \mathbf{H}_f + \mathbf{H}_s + \mathbf{H}_I$ and $\mathbf{H}_{\text{offset}}\big|_{I \to 0} = \mathbf{H}_f + \mathbf{H}_s$.

reasonably decoupled from the pinned layer. $\mathbf{H}_s$ is generated by the stray flux from the transversely saturated pinned layer. Its value depends on sensor height, thickness, and saturation magnetization of the pinned layer and thickness of the free layer. The value of $\mathbf{H}_I$ depends on the magnitude of the sensing current and is given by Ampere's circuital law, $\mathbf{H}_I = \oint I \, dl$. The direction of the field $\mathbf{H}_I$, which depends on the direction of the current, can be in the direction of $\mathbf{H}_f$ or $\mathbf{H}_s$.

Because of the effect of these fields on the magnetization orientation of the free layer, the $R(H)$ transfer curve of a real spin-valve film is usually offset from the origin. If the sensor size is reduced to the submicron scale, the effect of these fields on the offset of the transfer curve of the free layer will be pronounced. Figure 52 shows a typical $R(H)$ curve of a spin-valve sensor. The offset field $\mathbf{H}_{\text{offset}}$ of the $R(H)$ curve represents the vector summation of the three fields. The value of the offset depends not only on the magnitude of the sensing current during measurement, but also on the direction of the current. If the measuring sensing current is negligibly small, the offset field is equal to the vector sum of $\mathbf{H}_f$ and $\mathbf{H}_s$.

In spin-valve head design, sensing currents with different magnitudes and in different directions are employed to balance out coupling fields from the pinned layers to obtain transfer curves with zero offset, i.e., to make $\mathbf{H}_f + \mathbf{H}_I + \mathbf{H}_s = 0$. However, this is only feasible if the sum of $\mathbf{H}_f$ and $\mathbf{H}_s$ is not so large, because the maximum sensing current density that can be used in a spin-valve head is about 6×10^7 A/cm² [98]. Too great a current leads to overheating of the sensor, which causes reduction of the GMR ratio and destabilization of the pinned layer magnetic configuration. Too large a current will also produce a strong $\mathbf{H}_I$ in the pinned layer opposing the pinned magnetization, as shown in Fig. 51, leading to a reduced pinning effect [110].

The stray field $\mathbf{H}_s$ from the pinned layer is highly nonuniform [205]; it is usually not possible to achieve $\theta_1 = 0$

everywhere over the entire free layer. θ_1 will be significantly nonzero at some points of the sensor [205]. This will cause a reduction of the linear dynamic range of the sensor [75]. However, within a considerable dynamic range, the spin-valve sensor is still linear. This is shown in Fig 50 by the dotted curve.

4.4.3. Effect of Finite Sensor Size

4.4.3.1. Effect of Self-Demagnetizing Fields

Similar to the finite-sized AMR sensors discussed in Section 4.1.3, self-demagnetizing fields also exist in spin-valve sensors in both the vertical and horizontal directions of pinned and free layers. Equations (4.8) and (4.9) can be used for estimation of the average values of the vertical and horizontal self-demagnetizing fields of spin valves. But the self-demagnetizing fields are highly nonuniform across a spin-valve sensor and are much larger at the edge than at the center [9]. The self-demagnetizing field in the pinned layer tends to rotate the pinned magnetization from its original direction (perpendicular to the ABS) to the longitudinal direction [95]. This effect is pronounced when the sensor t/w or t/h ratio increases (sensor size decreases), which will lead to amplitude asymmetry of the replay waveform. Therefore, for smaller sensor sizes, AF materials with higher exchange pinning fields are required. The use of a synthetic pinned layer is another effective solution to this. As discussed in Section 4.3.3, the self-demagnetizing field of the pinned layer can be significantly reduced because of the flux closure within the two antiferromagnetically coupled layers.

The major effect of the self-demagnetizing field in the free layer is on its transfer curve. As can be seen from Eqs. (4.8) and (4.9), the average self-demagnetizing fields are proportional to the sensor thickness and inversely proportional to the sensor width (for a horizontal demagnetizing field) or sensor height (for a vertical demagnetizing field). Because the thickness of the free layer of a spin-valve head is much smaller than an AMR film (about 1/5 to 1/10 of the thickness of an AMR film), for large-sized spin-valve sensors, the self-demagnetizing field is not significant. However, as the recording density increases, the spin-valve sensor size becomes increasingly smaller, and the effect of the self-demagnetizing field will become serious if the ratios t/h and t/w are not kept constant. For example, if a spin-valve sensor has a free layer thickness of 4 nm and a $4\pi M_s$ of the free layer of 10 kG, the estimated average vertical demagnetizing field of the sensor from Eq. (4.8) is about 20 Oe and 100 Oe, for sensor heights of 2 μm and 0.4 μm, respectively. The self-demagnetizing fields of the free layer will have two major effects on the transfer curves of a spin-valve sensor, the linear dynamic range and the field sensitivity. This is schematically illustrated in Fig. 53. Because of the self-demagnetizing field, more external field is required to rotate the magnetic moment of the sensor to the same angle, and a large saturation field is required to magnetize the sensor to saturation, which means that the field sensitivity is considerably reduced as the sensor size decreases. The self-demagnetizing field has a nonuniform spatial distribution and is much larger at the sensor edge than at the center. This causes the center of the sensor to saturate first. At lower fields, the sensor linearity is still preserved. However, because the edge magnetic moments are more difficult to saturate, the transfer curve is not linear when approaching saturation. This causes a reduction in the linear dynamic range of the sensor, i.e., $\Delta\langle\cos(\theta_1 - \theta_2)\rangle \ll 2$, or the maximum linear $\Delta R/R$ is less than the value obtained from a sheet spin-valve film. To compensate for the loss of field sensitivity and maximum linear $\Delta R/R$, the required GMR amplitude of spin-valve films increases with the reduction in sensor size (increase in the t/h ratio).

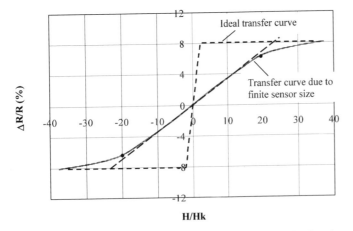

Fig. 53. Schematic illustration of spin-valve transfer curves showing the effect of a self-demagnetizing field due to the finite size of the sensor. The maximum linear dynamic range of $\Delta R/R$ is reduced as marked by the two dots on the curve (1/2 slope points [75]). Perfect bias is assumed.

4.4.3.2. The Use of Thinner Free Layers

Equations (4.8) and (4.9) imply that one of the effective ways of reducing the average self-demagnetizing field is to keep t/W or t/h down or at a constant value. This means that as the sensor dimension (W, h) decreases, the thickness of the free layer must also scale down. However, the reduction of free layer thickness will also cause lower GMR amplitude and higher $\mathbf{H_s}$ arising from the pinned layer.

As discussed in Section 4.2.5.3, the GMR amplitude of spin-valve sandwiches decreases with the reduction in free layer thickness (as shown in Fig. 39). This is due to the fact that when the free layer thickness is smaller than $\lambda\uparrow$ (the longer of the two MFPs, typically ~ 50Å [96]), the spin $\uparrow$ electrons will penetrate the free layer and be scattered by the outer surfaces (substrate, AF layer, buffer layer, or capping layer), which results in increased low effective resistivity (ρ_L) or reduced SDS contrast ($\alpha = \lambda_\uparrow/\lambda_\downarrow = \rho_\downarrow/\rho_\uparrow$) and hence reduced GMR amplitude (see Eq. (4.18)). Various new spin-valve structures, such as spin-filter spin valves, spin valves with ferrimagnet free layers, and specular spin valves, have been employed to overcome this problem.

The second problem of using a thinner free layer is the increased $\mathbf{H_s}$ arising from the pinned layer. For a given pinned layer, the magnetostatic stray field in the free layer from the pinned layer increases with the decrease in free layer thickness (the same flux passing through a smaller cross-sectional area). The increased $\mathbf{H_s}$ makes it difficult to obtain an optimum bias point for submicron-sized sensors in conventional spin valves.

Spin valves with synthetically pinned layers have been successfully used to solve this problem [93, 110]. With the use of a synthetic pinned layer, the $\mathbf{H_s}$ from the pinned layer can be significantly reduced. Leal and Kryder [110] have experimentally measured the offset field of synthetic spin-valve sensors with track widths of 10 and 1 μm, as a function of sensor height h, and compared the results with conventional spin-valve sensors. As shown in Fig. 54, for sensors with relatively large sizes ($h > 4\mu$m), the effective coupling fields of the two types of spin valves are quite similar and are on an order of magnitude of smaller than 20 Oe. As the sensor height decreases, the effective coupling field of spin-valve sensors with a conventional AF exchange bias system increases considerably, reaching 125 Oe at a sensor height of 1 μm. It is not possible to obtain a zero offset bias point for such spin-valve sensors if only the field from the sensing current is used to cancel out the offset field, because the maximum sensing current is limited by the joule heating effect and by the ESD immunity requirement. On the other hand, the effective coupling field of spin-valve sensors with synthetic pinned layers remains almost constantly below 20 Oe, even when the sensor height is reduced to 0.5 μm. The advantages of synthetic spin valves over conventional spin valves in the application of submicron-sized sensors can be clearly seen from this work. A zero offset bias point for such a sensor can easily be obtained by varying the sensing current values. This is shown in Fig. 55 by the two $R(H)$ loops of a spin-valve sensor with a 10 μm

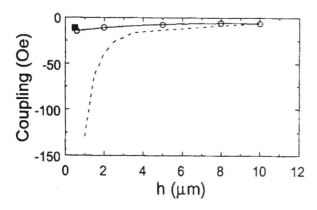

Fig. 54. Effective coupling field measured as a function of stripe, height for a synthetic spin-valve head with $W = 10$ mm (open dots) and $W = 1$ mm (solid square). The dashed line shows the expected effective coupling field as a function of stripe height for simple spin valves. Negative values are associated with antiferromagnetic coupling. Reproduced with permission from [110], copyright 1999, IEEE.

track width and a 0.5-μm sensor height, measured at two different sensing current values. The $R(H)$ loop with a current density of 3.3×10^7 A/cm^2 shows a zero offset field and therefore is a properly biased transfer curve. The slight reduction in the GMR ratio at higher current density is due to the heating effect of the current, which leads to an increase in sensor temperature and hence an increase in film resistivity and a decrease in the GMR ratio [110].

4.4.4. Optimization of AF Exchange-Biasing Materials

4.4.4.1. Basic Parameters of AF Exchange-Biased Systems

An antiferromagnet/ferromagnet exchange bias material is characterized by the following parameters: a unidirectional exchange anisotropy field (H_{ua}), interfacial coupling energy (J_k), a blocking temperature (T_B), and a critical thickness (δ_c).

Unidirectional Exchange Anisotropy Field (H_{ua}). The unidirectional exchange anisotropy field H_{ua} (Oe) is a measure of the exchange coupling strength of the antiferromagnet/ferromagnet interface. It is measured by the unidirectional

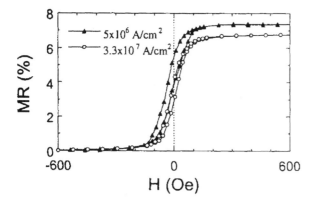

Fig. 55. $R(H)$ loops for an unshielded spin-valve head with a sensor width of 10 mm and a height of 0.5 mm, measured at two different current densities. Reproduced with permission from [110], copyright 1999, IEEE.

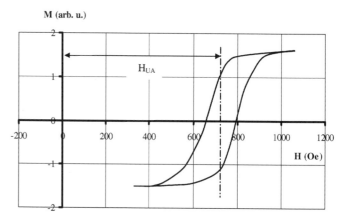

Fig. 56. Schematic illustration of the MH loop of an antiferromagnet/ferromagnet exchange-biased system, together with the definition of H_{UA}.

offset from the origin of the MH loop of the exchange-biased system, as shown in Fig. 56 for an antiferromagnet/ferromagnet bilayer system. This field pins the magnetic moment of the ferromagnet to saturation in one direction. It is therefore also known as a pinning field, H_{pin}. Other names used for the same field include bias field, exchange field, or exchange pinning field. H_{ua} is a convenient term to use because it describes the interface exchange coupling strength in terms of an effective magnetic field strength. However, H_{ua} is ferromagnetic thickness dependent., which decreases with increasing ferromagnetic thickness.

Interfacial Coupling Energy (J_k). An alternative way of describing the interface exchange coupling strength is to use the interfacial coupling energy J_k (erg/cm^2). Because J_k is a term of surface energy density, it is simply equal to the product of H_{ua}, M_s, and t_F and is given by [109]

$$J_k = H_{ua} M_s t_F \qquad (4.23)$$

where M_s and t_F are the saturation magnetization and thickness of the ferromagnetic layer, respectively. The value of J_k does not depend on the ferromagnetic thickness.

Blocking Temperature (T_B) and Thermal Stability. The exchange anisotropy pinning field of an antiferromagnet/ferromagnet system usually decreases with increasing temperature and can be plotted as a temperature dependence curve as shown in Fig. 57. The blocking temperature is the temperature at which the exchange pinning field becomes zero, which mainly depends on the crystal phase structure of the material. The blocking temperature (T_B) is one of the measures of the thermal stability of an exchange-coupled system. However, a high blocking temperature does not necessarily mean a good thermal stability. The blocking temperature distribution is more closely related to the thermal stability of the exchange-coupled system. There are various ways to characterise the blocking temperature distribution of an AF/F system [113, 114]. Here we introduce the method used by Nagai et al

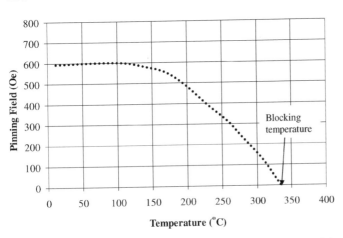

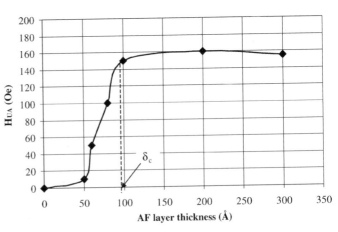

Fig. 57. Temperature dependence of the exchange pinning field and definition of the blocking temperature T_b.

Fig. 59. Schematic illustration of the definition of critical thickness δ_c of AF films.

[113]. Samples are heated to a pre-set temperature in a magnetic field (typically 3 ~ 10 k Oe) applied against the original pinning direction, soaked at that temperature for 10~15 minutes, and then cooled down to room temperature for magnetic measurement. The experiment is repeated at different soak temperatures up to the blocking temperature. The measured exchange pinning field is normalised by the original value of H_{ua} and plotted versus the soak temperature, as shown in Fig. 58. The differential of the curve represents the blocking temperature distribution. Because there is a large field applied to the sample against its pinning direction during annealing, part or all of the pinning field will be reversed as the temperature increases. As shown in Fig. 58, if the curve stays at the value of the normalized pinning field of 1.0 at temperature T, the original pinning field is stable at that temperature. As the temperature increases, the curve will start to fall off, changing sign, and, finally, at the blocking temperature, reaching -1.0 of the normalized field value, which means a complete reversal of the pinning field. The thermal stability of the pinning field can therefore be characterized by the peak temperature T_p of the blocking temperature distribution curve together with

its half peak width W_{50}. For good thermal stability, an AF/F system must have a small W_{50} and a high T_p.

Critical AF Layer Thickness (δ_c). The exchange pinning field is AF layer thickness dependent [115]. The thickness dependence usually has the general tendency shown in Fig. 59. For a particular AF/F exchange coupling system, there is always a minimum AF layer thickness, below which the exchange pinning field will start to fall off. This thickness is termed the critical thickness (δ_c) of the AF material.

4.4.4.2. Requirements for AF Materials for Spin-Valve Heads

Antiferromagnetic films play a vital role in spin valves. The basic requirements of AF materials for spin-valve head applications are high H_{ua} (or high J_k), high T_B, high thermal stability, small δ_c, high resistivity, high corrosion resistance, and ease of fabrication. The required H_{ua} for the desired output and linearity of spin-valve heads is sensor size dependent. For a sensor size of 1 μm, the required minimum H_{ua} is about 300 Oe [98] at sensor operation temperature. Smaller exchange pinning fields cause read-back waveform distortion [98], and make the output smaller [99]. High T_B and high thermal stability are important for ensuring the temperature stability of the spin-valve heads at operating temperatures (typically 150°C). A good thermal stability is also essential for improved ESD immunity. A small critical thickness δ_c of the AF layer is required to fit the spin-valve element into the increasingly smaller read gap at high densities. For example, the projected read gap length for an areal density of 40 Gb/in^2 is 50 nm if the track width of the head is 0.2 μm [97], and the required δ_c of the AF material must be smaller than 100 Å.

4.4.4.3. Properties of AF Materials

Various AF materials have been studied [97–108]. Figure 60 presents the temperature dependence curves of the exchange pinning fields of some typical AF materials [111]. Typical

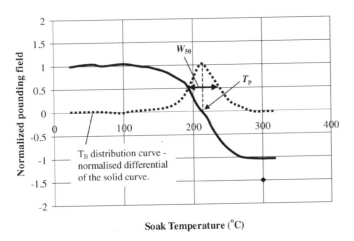

Fig. 58. Soak temperature dependence of the normalized exchange pinning field (solid curve) and its T_b distribution (dotted curve).

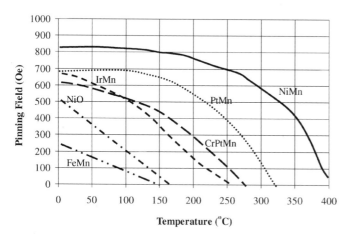

Fig. 60. Temperature dependence of exchange pinning field for some typical AF materials. Sketched base mainly on the data from [111], copyright 2000, American Institute of Physics.

properties of AF materials commonly used in spin-valve heads are also listed in Table IV. For a modern disk drive, the operating temperature of the spin-valve heads is typically around 150°C and the minimum required exchange pinning field is about 200–300 Oe [87]. It can be seen that NiMn is the best AF material discovered so far in terms of J_K, T_B, and thermal stability. However, the antiferromagnetic θ-NiMn phase is obtainable only after postdeposition annealing [104, 105] at high temperatures and over a long period, which is not a preferred process for spin valves, because of its detrimental effect on multilayer interfaces and $\Delta R/R$ ratio [79]. PtMn is the second best material but is easier to fabricate than NiMn systems. Both AF materials are suitable for use in current spin-valve heads with areal densities up to 10 Gb/in^2. But their δ_c must be considerably reduced to be used in future generation spin-valve heads. Oxide AF films such as NiO and CoO have the advantage of high corrosion resistance, zero shunting loss, and enhanced $\Delta R/R$ due to specular reflection [100]. However, the J_K and T_B of the existing oxide AF materials are too

low. The properties of AF materials depend on material, structure, underlayer, grain size, AF layer thickness, fabrication process, etc. Elemental AF materials have been well studied. Considerable work has been done for binaries. However, only very limited cases for ternaries and beyond have been studied [105]. The major technical challenge for future AF material will be to obtain high J_K and T_B and other desired properties at very small δ_c. As discussed in Section 4.3.3, exchange-bias systems with synthetically pinned layers exhibit much better properties than AF/F bilayer systems.

4.4.5. Hard Bias

As in to AMR heads, horizontal biasing is necessary for spin-valve heads, so that a single domain state is obtained in the free layer of the sensor and the single-domain state is stable against all reasonable perturbations [9]. The most commonly used horizontal biasing technique for spin-valve heads is permanent magnet hard biasing, as shown in Fig. 61. Patterned high-coercivity magnetic strips (CoCrPt, for example) are abutted against the ends of the spin-valve sensor, which provides a magnetic field that magnetizes the sensor along the horizontal direction to a single-domain state. The magnitude of the field produced by the hard magnet bias film depends on the $M_r t$ of the film. Figure 62 shows the effect of relative layer thickness of hard bias films on spin-valve transfer curves, where the relative layer thickness RL is defined as

$$RL = \frac{M_r t}{M_{s,F} t_F} \qquad (4.24)$$

Fig. 61. Schematic illustration of the permanent magnet hard bias technique.

Table IV. Typical Exchange Systems Used in Spin Valves and Their Properties

Exchange systems	J_K (erg/cm^2)	H_{ua} (Oe)	T_B (°C)	δ_c (nm)	Corrosion resistance	Thermal stability	Annealing
Ta/NiFe/FeMn		420					
[64, 106]	0.13	($t_{NiFe} = 4$ nm)	150	7	Poor	Poor	No
NiO/NiFe/Co		360					
[100, 106]	0.09	($t_{NiFe} = 4$ nm)	200	40	Very good	Poor	No
Ta/NiFe/IrMn		400					
[97, 107]	0.15	($t_{NiFe} = 2$ nm)	280	8	Good	Acceptable	Yes
Co/CrMnPt		320					
[97, 101]	0.19	($t_{Co} = 3$ nm)	320	30	Very good	Acceptable	Yes
NiFe/RuRhMn		350					
[103]	0.17	($t_{NiFe} = 2.5$ nm)	250	10	Good	Good	No
Ta/NiFe/PtMn		600					
[97]	0.2	($t_{NiFe} = 3$ nm)	340	30	Good	Very good	Yes
Ta/NiFe/NiMn		650					
[104, 106]	0.24	($t_{NiFe} = 3$ nm)	380	30	Good	Very good	Yes

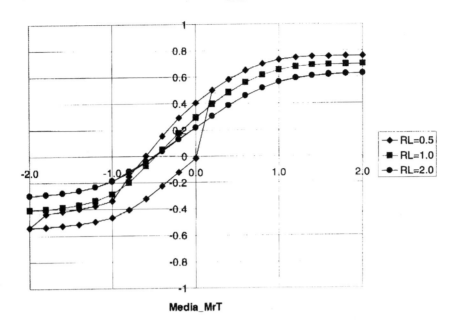

Fig. 62. Effect of hard bias layer thickness on the transfer curves of spin-value sensors. Reproduced with permission from [86], copyright 1999, IDEMA.

where $M_{s,F}$ and t_F are the saturation magnetization and thickness of the free layer, respectively. $RL = 1$ when $M_r t$ of the hard biasing film is equal to $M_{s,F} t_F$ of the free layer. As can been seen from Fig. 62, when $RL = 0.5$, the single-domain state is not formed in the free layer, and the transfer curve is affected by the moving domain walls in the free layer. When $RL = 1$, a smooth transfer curve is obtained, which indicates that the free layer is in a single-domain state. However, the bias field also causes a reduction of amplitude of the transfer curve. A further increase in the thickness of the hard bias layer ($RL = 2$) causes further reductions in the amplitude. This is due to the fact that the excessive bias field will lock the free layer in the bias field direction and the free layer will lose its sensitivity to the field from the recording media.

5. FILMS FOR WRITE HEADS

5.1. Basics of Soft Magnetic Films for Writers

5.1.1. Hysteresis Loop, Coercivity H_c, and Anisotropy Field H_k

Soft magnetic films used for film heads are usually required to have an in-plane uniaxial anisotropy. The typical easy and hard axis BH loops of a uniaxial anisotropy film are shown in Fig. 63. The BH loop along the easy axis is a nearly rectangular-shaped open loop, which reflects the fact that the magnetization reversal process along the easy axis is predominantly domain wall motion. The area S ($= \oint H \, dB$) enclosed by the loop represents the hysteresis loss per cycle in unit volume material (joules/m³ cycle). For most magnetic applications it is always desirable to make the hysteresis loss as small as possible. The coercivity H_c is a measure of the openness of the loop and therefore is the representation of the softness of

a magnetic material. The hard axis usually exhibits a straight-line loop, reflecting the fact that the magnetization reversal process along the hard axis is predominantly coherent rotation rather than domain wall motion. The effective anisotropy field (H_k) of a soft magnetic film is defined from the hard axis loop, as also shown in the figure. In thin film heads, the magnetic film is so oriented that the magnetic flux path is always along the hard axis.

The uniaxial anisotropy of a soft magnetic film can be obtained by applying a magnetic field during the film deposition or annealing process. The origin of the induced anisotropy is the short-range directional ordering, in which atomic pairs in a film tend to align with the local magnetization. A magnetic field that is strong enough to saturate the magnetization of the film (typically 50–100 Oe) is required to induce uniaxial anisotropy.

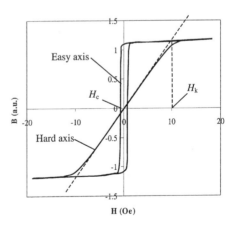

Fig. 63.   Typical BH loops of soft magnetic films.

5.1.2. Permeability and Its Frequency Dependence

The permeability μ of a soft magnetic material is defined as

$$\mu = \mu_0 \mu_r = \frac{B}{H} \qquad (5.1)$$

where μ_0 is the permeability of free space ($\mu_0 = 1$ in CGS units), μ_r is the relative permeability, B is the flux density (Gauss), and H is the magnetic field (Oe). It can be seen that $\mu = \mu_r$ in CGS units. Because of the nonlinear nature of the hysteresis loop, the permeability of a magnetic film varies with the magnitude of the applied field. Terms like initial permeability ($\mu_i = dB/dH|_{H=0, B=0}$), maximum permeability ($\mu_{max} = (B/H)_{max}$), and differential permeability ($\mu_d = dB/dH$) are frequently used to describe the field dependence of permeability.

In a high-frequency magnetic field, the permeability of a magnetic thin film is a complex number. If the high frequency field, given by

$$H = H_m \cos \omega t = H_m e^{i\omega t} \qquad (5.2)$$

is applied to a magnetic thin film, the flux density B of the magnetic film will lag behind the field H by a phase of δ because of the effect of hysteresis, eddy current, magnetic viscosity, etc. and can be represented by

$$B = B_m \cos(\omega t - \delta) = B_m e^{i(\omega t - \delta)} \qquad (5.3)$$

According to the definition of permeability of Eq. (5.1), we have

$$\tilde{\mu} = \frac{B}{H} = \frac{B_m}{H_m} e^{-i\delta} = \mu' - i\mu'' \qquad (5.4)$$

where μ' and μ'' are the real and imaginary parts of the permeability, respectively, and are given by

$$\mu' = \frac{B_m}{H_m} \cos \delta \qquad (5.5)$$

$$\mu'' = \frac{B_m}{H_m} \sin \delta \qquad (5.6)$$

and

$$|\tilde{\mu}| = \sqrt{\mu'^2 + \mu''^2} \qquad (5.7)$$

Figure 64 is a schematic representation of the phase relationship between **B** and **H**. As can be seen from the figure, the real part of the complex permeability μ' is the ratio of the in-phase component of the amplitude of the flux density B_m with the applied field and the amplitude of the applied field H_m. The imaginary part of the complex permeability μ'' is the ratio of the quadrature components of the amplitude of the flux density B_m and the amplitude of the applied field H_m. In fact, μ' is associated with the energy storage rate $w_{storage}$ of a magnetic material, and μ'' is associated with the power loss rate P_{loss} of the material. These are given by

$$w_{storage} = \frac{1}{T} \int_0^T \frac{1}{2} \mathbf{H} \cdot \mathbf{B} \, dt = \frac{1}{2} \mu' H_m^2 \qquad (5.8)$$

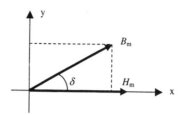

Fig. 64. Schematic illustration of the phase relationship between B_m and H_m at high frequency.

and

$$P_{loss} = \frac{1}{T} \int_0^T \mathbf{H} \cdot d\mathbf{B} = \pi f \mu'' H_m^2 \qquad (5.9)$$

The ratio of μ' and μ'' is known as the quality factor of the soft magnetic material.

Figure 65 is a schematic illustration of a typical frequency spectrum of the hard-axis complex permeability of a soft magnetic film [74]. In considering the effect of frequency on permeability, it has been assumed that the true permeability μ is independent of the frequency of the field and that only the apparent permeabilities (μ' and μ'') vary with frequency [216]. The permeability spectrum in the frequency domain can be divided into five zones. In zones I and II, the change in μ' and μ'' with frequency is insignificant. In zone III, the real part of the permeability starts to drop and the imaginary part starts to rise. The primary reasons for this could be the effects of hysteresis and eddy current damping. The resonance-type curve in zone III may be caused by the magnetic after-effect (or magnetic lag or magnetic viscosity). As the frequency further increases (zone IV), magnetic resonances (ferromagnetic resonance, domain wall resonance, and dimensional resonance) may occur, which result in a resonance-type spectrum. The corresponding frequencies for each zone may vary with different materials. The challenge of magnetic materials science and engineering is to extend zones I and II to frequencies that are as high as possible.

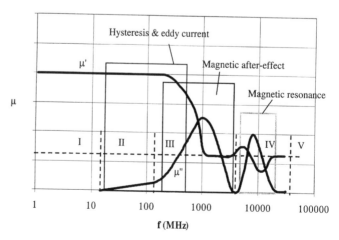

Fig. 65. Typical curves of the frequency dependence of complex permeability.

5.1.3. Magnetostriction

Magnetostriction is defined as the fractional change in length of a sample induced by the change in magnetization state of the sample. In the one-dimensional case, the magnetostriction coefficient λ is given by

$$\lambda = \frac{\Delta l}{l} \qquad (5.10)$$

where Δl is the change in length of a sample and l is the total length of the sample. λ can be positive or negative and is usually a very small number (typically on the order of magnitude of 10^{-6}). The value of λ depends on the applied magnetic field (magnetization states). The saturation magnetoresistance λ_s applies if the sample is magnetized from a demagnetized state to saturation.

For single-crystal materials, the magnetostriction is direction dependent, and their magnetostriction constants usually refer to the corresponding crystal axis. For example, λ_{100} and λ_{111} refer to the magnetostriction constants of a cubic single crystal along the (100) and (111) axes, respectively. If a material is polycrystalline, the magnetostriction of the sample along any direction is the averaged value of magnetostriction of the single crystals along that direction and is given by

$$\bar{\lambda} = \frac{\overline{\Delta l}}{l} = \frac{3}{2}\bar{\lambda}_0\left(\cos^2\theta - \frac{1}{3}\right) \qquad (5.11)$$

where θ is the angle between the field direction and the measurement direction, and $\bar{\lambda}_0$ is the magnetostriction constant when θ is zero. For a polycrystalline material, $\bar{\lambda}_0$ is isotropic and is given by

$$\bar{\lambda}_0 = \frac{2\lambda_{100} + 3\lambda_{111}}{5} \qquad (5.12)$$

5.2. Basics of Thin Film Writers

5.2.1. Reluctance and Permeance

A recording head can be analyzed with a magnetic circuit model in a way similar to that in which an electrical circuit is analyzed. The reluctance of a magnetic circuit R_m is the analogue of resistance R in an electrical circuit. It is therefore also known as magnetic resistance. Reluctance is defined as

$$R_m = \frac{l}{\mu A} \qquad (5.13)$$

where l is the length of the magnetic circuit or magnetic material, μ is the permeability, and A is the cross-sectional area of the magnetic material or magnetic circuit. The unit of reluctance, as can be seen from Eq. (5.13), is simply cm^{-1} in the CGS unit system. Like electric resistance, two or more reluctances in a series magnetic circuit add together.

In analogy to conductance in an electrical circuit, the inverse of reluctance is known as permeance P and is given by

$$P = \frac{\mu A}{l} \qquad (5.14)$$

The unit for permeance in the CGS system is cm. Like conductance, two or more permeances in a parallel magnetic circuit add together.

5.2.2. Write Head Efficiency

A magnetic recording head can be represented by a magnetic circuit model, with reluctances or permeances. In the simplest case, the head reluctance can be reduced to three reluctances: yoke reluctance R_{my}, gap reluctance R_{mg}, and leakage reluctance R_{ml}. And the magnetic head can be represented with the equivalent circuit, as shown in Fig. 66. The write head efficiency η_w is defined as

$$\eta_w = \frac{\Phi_g}{\Phi_y} \qquad (5.15)$$

Using Kirchhoff's law around the bottom loop gives

$$(\Phi_y - \Phi_g)R_{ml} - \Phi_g R_{mg} = 0 \qquad (5.16)$$

From Eqs. (5.15) and (5.16),

$$\eta_w = \frac{R_{ml}}{R_{ml} + R_{mg}} = \frac{1}{1 + R_{mg}/R_{ml}} \qquad (5.17)$$

It can be seen from Eq. (5.17) that a smaller gap reluctance results in a high write efficiency. It is not obvious from Eq. (5.17) that the write head efficiency actually depends on the yoke reluctance of the permeance of the head. In fact, the high permeance or the low reluctance of the yoke is critically important for high write efficiency, because if the permeance of the yoke is low, there will be little magnetic flux reaching the gap. Most of the flux will be lost in air before reaching the gap. If Φ_g is small, by Eq. (5.15) η_w will be small.

5.2.3. Head Inductance

The expression of inductance of a coil with N turns can be derived from Faraday's law ($e = N\frac{d\phi}{dt}$) and Lenz's law ($e = L\frac{di}{dt}$) and is given by

$$L = N\frac{d\phi}{di} \qquad (5.18)$$

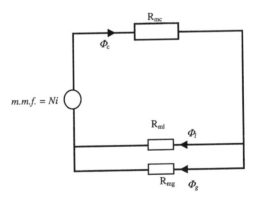

Fig. 66. A simple magnetic circuit model for a recording head.

For a circuit with constant reluctance (or constant permeability), Equation (5.18) can be written as

$$L = N \frac{\phi}{i} \qquad (5.19)$$

Equation (5.19) implies that the inductance of a coil is a measure of its flux linkages per ampere.

From the simple magnetic circuit model (as shown in Fig. 66) the current circulating in the coil of a head is given by

$$Ni = \phi \sum R_m \qquad (5.20)$$

where $\sum R_m$ is the total reluctance of a magnetic circuit. The inductance of a recording head is then given by

$$L = \frac{N^2}{\sum R_m} \qquad (5.21)$$

Head inductance can be divided into two parts [8]: the coil inductance in the absence of soft magnetic material in the yoke, and the magnetic yoke inductance contributed by the flux linkages due to the presence of the soft magnetic yoke materials. The coil reluctance varies with the coil diameter, and the magnetic yoke reluctance is proportional to the volume of the yoke material and inversely proportional to the separation of the two poles [8].

At high frequencies (high data rate recording), a high head inductance results in a high head impedance, which is not desirable because it requires a high power voltage supply.

5.2.4. Structure and Microfabrication Process of Film Heads

As has already been mentioned in Section 2, a thin film head consists of three principal functional parts, the pancake-shaped copper coil, the yoke (poles), and the front gap. They are all made of thin films. The function of the coil, which is sandwiched by the upper and lower poles, is to convert the electrical signals (write current) into magnetic fields. The yoke, which consists of two poles (P_1 and P_2) partially separated by the gap layer, the coil, and insulation layers, is used to form a low-reluctance (high permeance) magnetic circuit and to deliver the magnetic flux generated by the coil to the front gap. Only the stray field from the air-bearing surface (ABS) of the front gap is used for writing. The writing track width is defined by the strip width of the trailing pole (usually P_2), and the writing bit length is determined by the front gap length, the linear speed of the media, and the write current pulse duration.

A schematic drawing of an IBM 3370 thin film head is shown in Fig. 67 [175]. Figure 67a is an expanded cross-sectional view of the lower half of the head (from the ABS to the center of the copper coil), showing the two poles (A), insulation layers (E), the eight-turn copper coil (C), and the gap layer (D). Both the copper coil layer and the Permalloy yoke are electroplated. Insulation layers between the coil and the poles are usually made of hard-cured photoresist. The most commonly used material for the gap layer (D) is sputter-deposited Al_2O_3 film [175]. Figure 67b is a planar view of the

whole head, showing the pancake-shaped coil structure (C), the electrical contact leads of the coil, and the top magnetic pole. The pole tip structure on the ABS surface of the head is shown in Fig. 67c. The top pole or trailing pole (P_2) is slightly narrower than the bottom pole (P_1). The writing track width is defined by the width of P_2, which was 38 μm for the IBM 3370 heads. The gap length of the head was 0.6 μm.

The IBM 3370 film heads were fabricated using the so-called wet process; i.e., metal films (Permalloy and copper) were deposited by electroplating. This wet process has since been used by most thin film head manufacturers because of its economical viability. A ceramic wafer of alumina and titanium carbide [218] was used as the head substrate, which was chosen because of its high yields of chip-free rails and good durability in start–stop operation [175]. The first layer deposited on the substrates was a thin film of alumina, which was used to provide a smooth surface for further layer processing and to insulate the heads from the slightly conductive ceramic substrate. The bottom Permalloy pole (P_1) was then photolithographically patterned and electroplated, followed by the sputter deposition and patterning of the gap layer (D) (alumina) and a hard-cured photoresist insulation/planarization layer (E). The copper coil layer (C), together with the first coil lead, was then patterned and formed by electroplating, followed by another hard-cured photoresist insulation/planarization layer (E). Next, the top pole (P_2), together with the second coil lead, was patterned and electroplated on top of the hard-cured photoresist insulation layer. The same

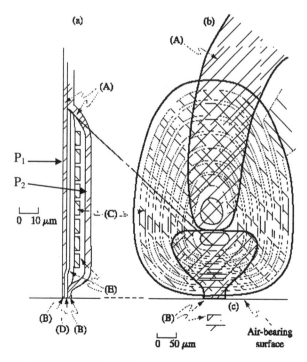

Fig. 67. IBM 3370 film head. (a) Schematic cross section showing the magnetic layers (A), pole tips (B), conductor turns (C), gap layers (D), and insulation layers (E). (b) Planar view of the film head. (c) Pole tip structure at the air-bearing surface. Reproduced with permission from [175], copyright 1996, IBM.

Table V.   IBM Inductive Film Head Products

	GA	TPI	KBPI	AD (Mb/ln.)	P_1 (μm)	P_2 (μm)	G (μm)
3370	1979	635	12.1	7.3	1.6	1.9	0.60
3375	1980	800	12.1	9.7	2.0	1.9	0.70
3380	1981	801	15.2	12.2	1.7	2.0	0.60
3380E	1985	1,386	16.2	22.5	1.6	1.9	0.60
9335	1986	1,600	16.2	25.9	1.6	1.9	0.60
3380K	1987	1,600	16.2	25.9	1.6	1.9	0.60
3380K	1987	2,089	15.2	31.7	1.6	1.9	0.60
9332	1988	2,017	23.6	47.5	3.0	3.0	0.55
3390*	1989	2,242	27.9	62.6	0.9	1.0	0.55
Aptos*	1990	2,242	27.9	62.6	0.9	1.2	0.55
3390-3*	1991	2,984	30.0	89.4	0.9	1.2	0.55
Tanba Turbo*	1992	2,436	59.8	146	3.0	3.1	0.35
Tanba-3*	1993	3,041	61.0	186	3.0	3.1	0.32
Ritz-1	1994	4,000	80.0	320	3.5	3.5	0.25
Ritz-2	1994	4,000	80.0	320	3.5	3.5	0.25

	P_1w (μm)	P_2W (μm)	Turns	Layers	R (11)	L (nH)	Efficiency (LN³)	Slider length (mm)	Slider width (mm)	Slider height (mm)
3370	38.0	34.0	8	1	7	80	1.25	4.0	3.2	0.850
3375	27.5	24.5	8	1	7	80	1.25	4.0	3.2	0.850
3380	29.5	26.5	8	1	7	80	1.25	4.0	3.2	0.850
3380E	15.0	12.0	18	1	15	350	1.55	4.0	3.2	0.850
9335	13.5	11.0	18	1	15	350	1.55	4.0	3.2	0.850
3380K	8.5	6.5	31	2	24	650	0.67	4.0	3.2	0.850
9332	10.5	8.5	31	2	24	800	0.83	4.0	3.2	0.850
3390	8.0	8.0	31	2	24	800	0.83	4.0	3.2	0.850
Aptos	8.0	8.0	31	2	24	800	0.83	4.0	3.2	0.850
3390-3	5.6	5.6	37	2	31	950	0.69	4.0	3.2	0.850
Tanba Turbo	8.4	8.4	44	2	39	1,400	0.72	2.5	1.6	0.425
Tanba-3	6.4	6.4	44	2	39	1,200	0.61	2.5	1.6	0.425
Ritz-1	6.3	4.8	45	3	29	800	0.39	2.0	1.6	0.425
Ritz-2	6.5	5.0	36	3	20	500	0.39	2.0	1.6	0.425

*Ion milled.
Reproduced with permission from [175], copyright 1996, IBM.

Permalloy layer as the top pole layer was used as the second coil lead just for process simplicity. The top pole has a physical contact with the bottom pole through a pre-patterned hole in the center of the head to ensure low magnetic reluctance of the magnetic circuit. The second coil lead was also made to have a good electrical contact with the central contact pad of the coil layer. The photoresist insulation layer (E) was limited to the region of the copper coil as shown in Fig. 67a, and consequently was not exposed, as the head was lapped to final throat height [175]. Thick ($> 25\mu$m) and large copper connection studs were electroplated through photoresist masks to provide electrical connections for the head. The last layer of the film head is a very thick (~25 μm) alumina overcoat, which is necessary for chemical and mechanical protection of the heads during further wafer processing and use. The alumina overcoat over the studs was mechanically lapped away after deposition, exposing the copper for electrical connections [175]. The performance of thin film heads depends on various head parameters and head materials. A higher areal density and a high date rate of magnetic recording are achieved with narrower tracks and a smaller gap length. Techniques employed to achieve

higher recording densities and higher data rates with inductive head technology included the used of multilayer coil structures, thick poles, separate designs for reader and writer, FIB pole trimming, and novel yoke structures [175, 186]. Typical head parameters of IBM inductive head products from 1979 to 1994 are given in Table V. Detailed reviews of the technical evolution of film head technology can be found in [175, 186].

5.2.5. Factors Determining the Yoke Permeance of Film Heads

It can be seen from Eq. (5.14) that the permeance of a film head is proportional to the permeability of the poles and their cross-sectional area and inversely proportional to the length of the magnetic circuit. Therefore there are three main factors determining the head permeance at high frequencies: the domain configuration, eddy current, and yoke length.

Domain Configuration and Control. One of the key factors in realizing high yoke permeance, particularly at high frequencies, is the control of magnetic domains of the yoke, particularly in the pole tip of the head. In film heads, the induced easy axis of the magnetic films of the yoke is always perpendicular to the flux path, so that the magnetic domains in the poles are aligned perpendicular to the flux propagation direction (easy axis domains), and the magnetization reversal process in the yoke is predominately coherent rotation rather than domain wall motion. This ensures that the permeability of the yoke is high at high frequencies and is essential for obtaining a linear transfer curve for the head. Details about domain configuration and control in thin film heads including the use of lamination to obtain an easy axis single-domain state and the effect of magnetostriction on head domain structures are discussed in Section 5.3.

Eddy Current. At high frequencies, the time-varying flux in magnetic materials will induce eddy currents in the yoke, which in turn induce magnetic fields opposing the original flux. The eddy currents act as a screen to prevent the magnetic flux from penetrating the magnetic material and consequently reduce the effective cross-sectional area of the yoke for flux carrying. The depth of a magnetic material at which the high-frequency magnetic field can penetrate is known as the skin depth (δ), which depends on the frequency of the field (ω), and the permeability (μ) and conductivity (σ) of the magnetic material, and is given by

$$\delta = \sqrt{\frac{2}{\omega\mu\sigma}} \qquad (5.22)$$

As a result of the skin depth effect, the permeance of the yoke will decrease from the outer surface to inner depths reaching its minimum at the skin depth. To overcome the eddy current effect, soft magnetic films with low conductance are desirable. For very high-frequency heads, laminated yoke materials with insulation spacers are usually used. As discussed in Section 5.3, another major characteristic of lamination is that an easy axis single-domain state of the yoke can be obtained with proper lamination, which can further improve the high-frequency response of film heads.

Yoke Length. Film heads with two or more coil layers are typical examples in which the length of magnetic circuit is reduced to achieve high permeance. These heads are more efficient than the ones with a single coil layer, but at the expense of process complexity [175].

5.2.6. Required Head Field Properties for a High-Density Inductive Writer

The primary function of a writer is to produce a writing field strong enough to reverse the magnetization in the medium. Figure 68 is a typical spatial distribution of the longitudinal field component of a film head simulated by computer. The magnitude of the field, which depends on the deep gap field H_g of the head, must be sufficiently high to achieve effective writing on recording media. As discussed in Section 2.2, the deep gap field H_g of a film head must be greater than three times the medium coercivity H_c for effective writing without

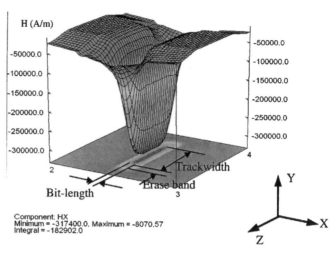

Fig. 68. A typical spatial distribution of the longitudinal write field of a film head, simulated by computer, its field mapping on a recording medium, and the resulting bit length, track width, and erase band.

pole-tip saturation. The continued increase in recording density can have two major impacts on head materials. One is the demand for higher medium coercivity to overcome the transition self-demagnetizing field (see Section 3.1) and to maintain the thermal stability of the recorded information (see Section 3.1.3). The other is that the increase in the data rate will substantially reduce the write head field duration, which makes the dynamic writing coercivity of the media even higher. Both require an increasingly high writing field, which is primarily limited by the saturation magnetization M_s or the saturation flux density B_s of the head material. High B_s head materials are therefore indispensable for meeting the requirements of high medium coercivity and high data rate.

Sharp spatial distribution of the write field is also important for achieving high track and linear densities. The sharpness of the field distribution is improved by maximizing the derivatives of the head field H_x with respect to x (down-track direction) and z (cross-track direction) at the point where $H_x = H_c$. A maximum $\delta H_x/\delta x|_{H_x=H_c}$ will ensure a sharp written transition, and a maximum $\delta H_x/\delta z|_{H_x=H_c}$ will ensure a narrow erase band, which are of critical importance to high linear and track densities [8].

5.3. Magnetic Domain Configurations in Film Heads

One of the key factors in realizing high yoke permeance, particularly at high frequencies, is the control of magnetic domains of the yoke, particularly in the pole tip of the head. In film heads, the induced easy axis of the magnetic films of the yoke is always perpendicular to the flux path, so that the magnetic domains in the yoke are aligned perpendicular to the flux propagation direction. This is essential for obtaining a linear transfer curve for the head and high permeability and hence high permeance at high frequencies. However, as a consequence of energy minimization of the magnetic system, as discussed below, typical domain patterns in a film head yoke may look like those shown in Fig. 69, where the majority of

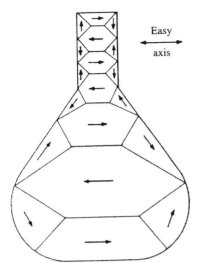

Easy
axis

Fig. 69. Typical domain configurations in a film head. Reproduced with permission from [186], copyright 1994, IEEE.

the domains are aligned parallel to the easy axis of the film, but domains along the edge of the strips are aligned parallel to the edges. The possible domain configurations in film heads will be analyzed separately for the following two cases: single-layered poles and laminated poles.

5.3.1. Domain Configurations in Single-Layered Films

In a ferromagnetic thin film, the atomic magnetic moment tends to have a perfect parallel alignment to minimize the exchange energy E_{ex}, which is given by

$$E_{ex} = -2AS^2 \cos\phi \qquad (5.23)$$

where A is the exchange integral and $A > 0$ for a ferromagnet, S is the atomic spin angular momentum, and ϕ is the angle between the neighboring atomic moments. The condition for minimum exchange energy is obviously $\phi = 0$, i.e., perfect parallel alignment of all atomic moments in the ferromagnet (single-domain state). However, as is well known, in a ferromagnetic material magnetic moments are broken into localized small-volume regions, known as magnetic domains. The domains are separated by domain walls. Only the atomic moments in each domain have a perfect parallel alignment. The formation of magnetic domains in a magnet is due to the energy minimization of the magnetic system. In addition to the exchange energy E_{ex}, other energy terms also have to be considered, such as the magnetostatic energy E_{mag}, the domain wall energy E_w, and the anisotropy energy E_a, which include the uniaxial anisotropy energy E_k and the stress-induced anisotropy energy or magnetoelastic energy E_{ela}.

The magnetostatic energy includes the external field energy E_H (or Zeeman energy) and demagnetization energy E_d (or shape anisotropy energy). For a given ferromagnet with a magnetization of M_s in an applied field $\mathbf{H}$, the Zeeman energy is given by

$$E_H = -\int \mathbf{H} \cdot d\mathbf{M} = -M_s H \cos\theta \qquad (5.24)$$

where θ is the angle between $\mathbf{M}$ and $\mathbf{H}$. E_H is at minimum when θ is 0° and maximum when θ is 180°.

The demagnetization energy E_d measures the interaction between the magnetic film and the demagnetizing field $\mathbf{H}_d$ ($= N_d \mathbf{M}$, where N_d is the demagnetization factor), which originates from the magnetic charges arising from the divergence of the magnetization given by Eq. (2.12). The general formula for E_d is given by

$$E_d = -\int \mathbf{H}_d \cdot d\mathbf{M} = -N_d \int \mathbf{M} \cdot d\mathbf{M} \qquad (5.25)$$

E_d is sample shape, size, thickness, and magnetic domain configuration dependent, and it is not always easy to calculate, particularly for multidomain samples. However, in most cases, it is one of the dominant factors determining the domain configuration of a magnetic film, which will be further discussed in Fig. 70.

In film heads, the magnetic films are usually very thick, and therefore all of the domain walls are treated as Bloch walls, the wall energy of which is given by

$$E_w = S_w \gamma_{Bloch} \qquad (5.26)$$

where S_w is the total area of walls per unit of magnetic film area, and γ_{Bloch} is the total wall energy per unit wall area (wall energy density), given by [187]

$$\gamma_{Bloch} = \frac{\gamma_0}{2}\left(\frac{\delta}{\delta_0} + \frac{\delta_0}{\delta}\right) + \frac{2\pi\delta^2 M_s^2}{\delta + t} \qquad (5.27)$$

where γ_0 is the wall energy density of the bulk material ($\gamma_0 = 0.1$ erg/cm^2 for Permalloy), δ is the total wall thickness, δ_0 is the bulk wall width ($\delta_0 = 2 \times 10^{-4}$ cm for Permalloy), t is the film thickness, and M_s is the saturation magnetization of the film.

The uniaxial anisotropy energy E_k is given by

$$E_k = K_u \sin^2\theta \qquad (5.28)$$

where K_u is the uniaxial anisotropy constant ($K_u = 1.5 \times 10^3$ ergs/cm^3 for Permalloy [187]) and θ is the angle between the magnetization vector and the easy axis. The minimum E_k is obtained when θ is zero or 180°, i.e., when the domains are oriented along the easy axis.

The stress-induced magnetoelastic energy E_{ela} is given by the following equation for polycrystalline films with isotropic magnetostriction constant λ_s, and with positive (tensile) stresses σ_x and σ_y in the x–y film plane and induced easy axis along the y axis,

$$E_{ela} = -\frac{3}{2}\lambda_s(\sigma_x \cos^2\theta + \sigma_y \sin^2\theta) \qquad (5.29)$$

The angle θ is the angle between the magnetization vector and the easy axis.

The combined anisotropy energy term is

$$E_k = \left[K_u - \frac{3}{2}\lambda_s(\sigma_x - \sigma_y)\right]\cos^2\theta \qquad (5.30)$$

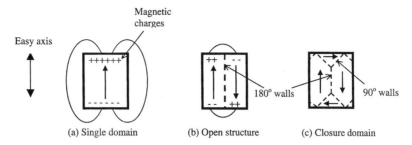

Fig. 70. Typical magnetic domain configurations of a uniaxial anisotropy film.

Summation of Eqs. (5.24) to (5.30) gives the total system energy, and the minimum total energy condition gives the most stable domain configurations. However, because the values of each energy term are interrelated, such a minimum energy condition usually can only be found with the aid of computer simulation—a subject of micromagnetics. Here we give only a rough analysis of the possible roles each energy term may play in the formation of a domain configuration.

Figure 70 shows the possible domain configurations in a uniaxial anisotropy soft magnetic film. Figure 70a is a single-domain state, in which the atomic moment of the film is aligned perfectly parallel to meet the minimum exchange energy requirement, and the domain magnetization is aligned along the easy axis to meet the minimum anisotropy energy requirement (assuming zero external field and zero magnetostriction). However, when a ferromagnet is in a single-domain state, the demagnetization energy could be very high because of the high magnetic charges in such a domain configuration, which makes the total energy of the system high.

In Fig. 70b, the single-domain state of Fig. 70a is broken up into two subdomains, with their magnetizations aligned antiparallel but still along the easy axis. The walls between the two domains are 180° Bloch walls. In such a case, the anisotropy energy term is unchanged. The demagnetization energy is considerably reduced because fewer charges are formed, because of partial flux closure in the two antiparallel domains. The exchange energy is increased slightly, and an extra energy term, the domain wall energy, is introduced into the system.

Figure 70c shows another typical type of domain configuration, the closure domain structure. Domains with 90° domain walls are formed along the strip edge. In this case, the demagnetization energy term almost vanished because of the complete flux closure within the magnetic domains. There is a slight increase in exchange energy and domain wall energy because more domains are formed. The anisotropy energy term is no longer zero because the edge domains are aligned at a 90° angle with respect to the easy axis. If the uniaxial anisotropy of the film is not sufficiently strong, the edge domains will expand further to form the so-called diamond structure. In the worst case, the edge domains will occupy the whole strip to form the hard-axis aligned domains, which are further discussed in Section 5.3.2.

The effect of the signs of magnetostriction on domain configurations can be understood from Eq. (5.30). In the case of $\sigma_x > \sigma_y$, a positive λ_s will reduce the effect of K_u, where as a negative λ_s will enhance the effect of K_u. A typical example of this was demonstrated by Narishige et al. [217], as shown in Fig. 71, for domain patterns in the pole and yoke of thin film heads with positive and negative magnetostrictive Permalloy films. Because the y dimension (easy axis) is considerably smaller than the x dimension (hard axis) in a thin film head, the tensile stresses follow $\sigma_x > \sigma_y$. Therefore, the Permalloy yoke with negative λ_s exhibits more stable easy axis aligned domains than the one with positive λ_s.

5.3.2. Domain Configurations in Multilayered Films

Multilayered films or laminated films are frequently used in film heads to improve high-frequency performance. Lamination of soft magnetic layers with an insulation spacer brings two major advantages [176]. One is that lamination can suppress eddy currents. If the thickness of each magnetic sublayer is made smaller than the eddy current skin depth, the high-frequency magnetic flux is carried by the whole thickness of the yoke, and the flux-carrying efficiency of the yoke at high frequency is greatly improved. The other is that proper lamination can eliminate domain walls and achieve a single-domain yoke, in which the contribution to the flux reversal is mainly from magnetization rotation rather than a domain wall motion mechanism. Hence, the high-frequency permeability of the

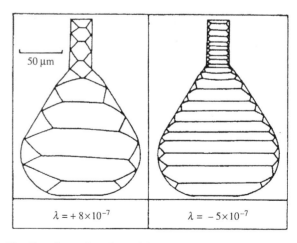

Fig. 71. Domain configurations of film heads with positive and negative Permalloy films, Reproduced with permission from [217], copyright 1984, IEEE.

head is maintained. The possible domain structures and micromagnetic analysis of laminated film heads were well studied by Slonczewski et al. [176]. Here a brief summary of the work is given.

According to Slonczewski et al. [176], a two-layer laminated Permalloy strip may have one of the three domain configurations shown in Fig. 72. Figure 72a is the periodic stripe domain pattern with hard-axis closure domains along the edge, which is similar to the domain configuration in a single-layered film, as discussed in the previous section. The formation of the edge domains is due to the strong demagnetizing field (or shape anisotropy field) along the strip edge, which overcomes the uniaxial anisotropy field and forces the magnetization aligned parallel to the edge (parallel to the shape anisotropy field). Such a shape anisotropy field is strip width and film thickness dependent and highly nonuniform in nature. When the strip width decreases, the shape anisotropy field increases and the edge closure domains expand to form the diamond pattern, as shown on the right of Fig. 72a, in which vertical 180° walls no longer exist. This is a typical problem in single-layered narrow track film heads, where diamond patterns are formed in the narrow pole tip, and the hard-axis edge domains occupy a large proportion of the pole tip but play no part in the flux reversal process.

The energy per unit strip length of each magnetic sublayer for a closure domain pattern is [176]

$$E_{\mathrm{CP}} = D \cdot \sqrt{2K_{\mathrm{u}}\gamma_{\mathrm{Bloch}}W} \qquad (5.31)$$

where D is the thickness of the magnetic sublayer, K_{u} is the uniaxial anisotropy constant, W is the width of the strip, and

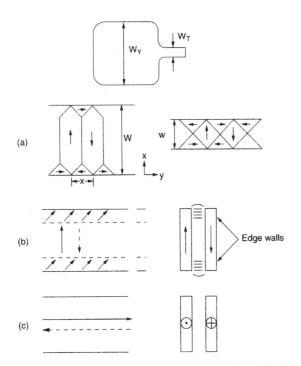

Fig. 72. Domain pattern classification of laminated strips. (a) Closure-domain pattern. (b) Easy-axis state. (c) Hard-axis state. Reproduced with permission from [176], copyright 1988, IEEE.

γ_{Bloch} is the Bloch wall energy density given by Eq. (5.27) or, alternatively, by a simple numerical expression [176],

$$\gamma_{\mathrm{Bloch}} = \frac{21.43A}{D} + 0.581K_{\mathrm{u}}D + \cdots \qquad (5.32)$$

where A is the exchange constant of the magnetic material ($A = 10^{-6}$ erg/cm for Permalloy).

The energy per unit length of each magnetic sublayer for a diamond pattern is [176]

$$E_{\mathrm{DP}} = D \cdot \left(\frac{K_{\mathrm{u}}W}{2} + \gamma_{\mathrm{Bloch}}\right) \qquad (5.33)$$

Figure 72b shows the easy axis state domain pattern of a two-layer laminate, where each layer exhibits a single domain state with domain magnetizations aligned antiparallel with each other along the easy axis direction and with two edge walls (known as edge curling walls or ECW) to form a closed magnetic circuit. In such a domain configuration, the energy per unit strip length of one magnetic sublayer is [176]

$$E_{\mathrm{EA}} = \pi K_{\mathrm{u}}D\Delta \qquad (5.34)$$

where Δ is termed the wall shape energy and is given by [176]

$$\Delta = M_{\mathrm{S}}\sqrt{\frac{\pi bD}{K_{\mathrm{u}}}} \qquad (5.35)$$

where b is the thickness of the nonmagnetic spacer.

Figure 72c shows the hard-axis state domain pattern of a two-layer laminate, in which each layer also exhibits a single-domain state, but with domain magnetizations aligned antiparallel with each other along the hard axis. This situation occurs when the strip width is very small and the energy density required to maintain the easy-axis state becomes too high in comparison with the energy density required for a hard-axis state domain, which is given by [176]

$$E_{\mathrm{HA}} = K_{\mathrm{u}}DW \qquad (5.36)$$

"Phase diagrams" showing the conditions under which the three domain configuration are stable can be constructed by using the above energy density equations for each domain pattern and using the basic material and geometry parameters such as $H_{\mathrm{k}}, D, b,$ and W. Some typical results are shown in Fig. 73 for two-layer laminated narrow Permalloy strips. These phase diagrams give a simple and clear indication of the effect of the strip width W, thicknesses (D and b) of magnetic and nonmagnetic layers, and the anisotropy field (H_{k}) on the possible domain configuration in film heads, which can be used as a rough guide in the design of thin film heads.

This simple model can be extended to multilayer systems [176].

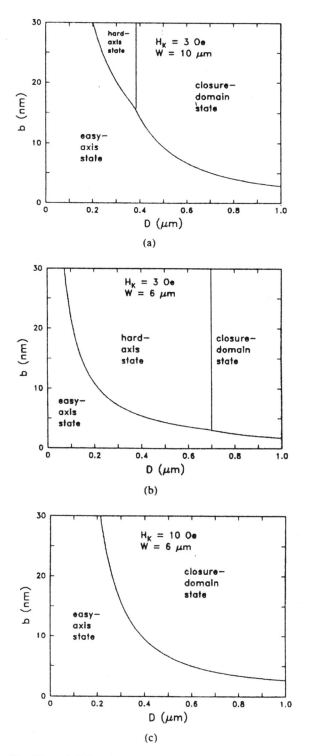

Fig. 73. Theoretical phase boundaries for narrow stips, assuming three combinations of W and H_k. Reproduced with permission from [176], copyright 1988, IEEE.

5.4. Soft Magnetic Films for Writers

In addition to the high B_s requirement, film head materials usually have a well-defined uniaxial anisotropy (for domain structure control) with small coercivity (for low hysteresis loss), reasonably high permeability, particularly at high frequencies (for high efficiency), zero or near-zero magnetostriction (for reduced domain noise/head instability), high resistivity (for reduced eddy current damping), and good thermal stability and corrosion resistance (to survive the head fabrication process and to allow for long life). Table VI lists the properties of some typical soft magnetic materials used in film heads. Soft magnetic thin films for recording head applications can be roughly classified into the following four categories: Ni-Fe alloys, FeSi-based alloys, cobalt-transition metal (Co-TM) amorphous alloys, and FeN-based nanocrystalline films. However, there are still a large number of soft magnetic materials that are not included in these four categories.

5.4.1. Ni-Fe Alloys

$Ni_{81}Fe_{19}$ (wt %), known as Permalloy, was the most popular head material for the early generation of film heads because of its excellent soft magnetic properties. The uniaxial anisotropy of Permalloy films can be obtained by applying an external magnetic field during film deposition or by postdeposition annealing [187]. One of the major advantages of using Permalloy in film heads is that it can be deposited by electroplating (wet process), which is a fast and cheap deposition process particularly suitable for mass production. Detailed descriptions of the electrodeposition process of Permalloy for film heads can be found in [187, 189]. In comparison with the vacuum deposition process (dry process), electrodeposition is an additive process for feature definition, which allows for much easier lithographic definition and control of small features [184]. This can considerably reduce the cost of fabrication of film heads. The major disadvantages of Permalloy are its low B_s and low resistivity, which is not suitable for heads used in high-density disk drives with a high data rate and high H_c media. The electroplating process also has its limitations. For example, it is difficult to make laminated films with insulation spacers, which is essential for domain control of narrow track heads and for eddy current suppression in high data rate recording.

An alternative material in the Ni-Fe alloy family is the $Ni_{45}Fe_{55}$ alloy [184, 185], which exhibits higher B_s (17 kG) and higher resistivity than Permalloy, and which can also be deposited by electroplating. Its major disadvantage is its relatively high saturation magnetostriction ($\sim 10^{-5}$).

5.4.2. FeSi-Based Alloys

Sendust (Fe-Si-Al alloy) is one of the popular materials in the FeSi-X family, which is mainly used in metal-in-gap (MIG) heads because of its superior thermal stability. To obtain desirable soft magnetic properties, the composition of Fe-Si-Al alloy must be within the so-called Sendust composition, which is typically $Fe_{73.6}Al_{9.9}Si_{16.5}$. Such a film exhibits higher B_s (11 kG), higher resistivity (84 $\mu\Omega$-cm), and higher permeability than Permalloy films.

The addition of N Sendust can improve its soft magnetic properties. For example, films with a composition of $Fe_{76.6}Al_{1.3}Si_{16.33}N_{5.8}$ prepared in Ar/N_2 ambient have been

Table VI. Properties of Some Typical Soft Magnetic Films in Recording Heads

Properties	Ni-Fe alloys		FeSi-X alloys	Co-TM amorphous alloys [177]			FeN-X nanocrystalline	
	Permalloy ($Ni_{81}Fe_{19}$) [186]	$Ni_{45}Fe_{55}$ [184, 185]	Sendust (Fe-Si-Al) [188]	$Co_{96}Zr_4$ [197]	$Co_{91}Nb_6Zr_3$ [194]	$Co_{95.2}Ta_{3.6}Zr_{1.2}$ [193]	FeTaN [178, 182]	FeAlN [179 − −181, 183]
B_s (KG)	10	17	11	~ 16	~ 13	~ 15	~ 20	~ 20
H_k (Oe)	2.5	9.5	N/A	~ 24	3 ~ 16	3 ~ 16	~ 10	~ 5
H_c (Oe)	0.1 ~ 0.5	0.4	~ 0.2	0.5	0.01	0.01	1.5	0.6
μ_r	~ 1000	~ 1700	~ 2000	~ 1500	1500 ~ 5000	1500 ~ 6000	~ 2500	~ 2500
λ_s	-1×10^{-6}	2×10^{-5}	$~ 10^{-6}$	2×10^{-6}	$~ 10^{-8}$	$~ 10^{-8}$	$~ 10^{-6}$	$~ 10^{-6}$
ρ ($\mu\Omega-cm$)	24	48	85	>100	125	>100	30 ~ 120	20 ~ 130
Recrystalization temperature (°C)	~ 425	N/A	~ 700	~ 450	~ 450	~ 450	300	500
Thermal stability	Excellent	Excellent	Excellent	~ 270°C	~ 270°C	~ 270°C	200°C	Good
Corrosion resistance	Poor	N/A	Fair	Good	Good	Good	N/A	N/A

shown to have a B_s of 13.6 kG, an H_c of 0.3 Oe, a μ_r of 2500, and a λ_s of 5×10^{-7} [199].

FeRuGaSi alloy (SOFMAX) is another type of soft magnetic material that belongs to the FeSi-X family [190]. It exhibits higher B_s (up to 13.6 kG) with improved soft magnetic properties, thermal stability, and wear and corrosion resistance compared with Sendust alloys.

5.4.3. Co-TM Amorphous Soft Magnetic Films

Co-TM (TM = Ti, Zr, Nb, Mo, Hf, W, or Ta) binary or ternary amorphous films are found to exhibit higher saturation magnetization and better soft magnetic properties than Permalloy and Sendust films. An excellent review of these films can be found in [177]. These films can be deposited by sputtering.

One of the advantages of these films is that zero or near-saturation magnetostriction can be obtained by using different glass-forming elements or by adjusting the film composition. As shown in Fig. 74, the saturation magnetostriction of Co-rich binary amorphous alloys can be either positive or negative, depending on the type of glass-forming element used. λ_s also varies with film composition. This provides us with a large number of possibilities for new ternary alloys with zero or near-zero magnetostriction. Examples of these films include $Co_{87}Nb_5Zr_8$ [192], $Co_{85}Nb_{7.5}Ti_{8.5}$ [191], $Co_{92}Ta_4Zr_4$ [193], and $Co_{91}Nb_6Zr_3$ [194].

It is also possible to add a ferromagnetic element such as Fe Co-TM binary systems to form zero or near-zero magnetostriction amorphous films. A typical example of this is $Co_{89.8}Nb_8Fe_{2.2}$ [195, 196], which exhibits zero magnetostriction and a high B_s of 14.3 kG. The Co-Nb binary system exhibits a negative λ_s. The addition of Fe atoms changes λ_s toward positive passing through zero at an Fe concentration of 2.2 at % [177].

The dependence of the saturation magnetic flux density B_s of Co-TM amorphous alloys on the concentration (y) of the glass-forming element is shown in Fig. 75 [177]. Almost all

of the values of B_s follow a common straight line increasing with the decrease in y, except for the Co-Ta alloys. The arrows in the figure indicate the critical concentration (y_c) of the glass-forming element, below which the alloy becomes crystalline. The highest B_s for each Co-TM amorphous alloy is determined by the value of y_c. Among them, the $Co_{96}Zr_4$ amorphous alloy exhibits the highest B_s (16 kG), but with nonzero magnetostriction. The values of B_s for nonmagnetostrictive amorphous alloys are always smaller than this because additional glass-forming elements other than Zr are required to achieve the zero magnetostriction. The nonmagnetostrictive condition in these amorphous alloys is achieved at the sacrifice of B_s. For example, an optimized composition of Co-Ta-Zr amorphous alloy to make B_s as high as possible within the condition of zero magnetostriction is found in $Co_{95.2}Ta_{3.6}Zr_{1.2}$ [193]. The B_s of such an alloy is about 15 kG [193].

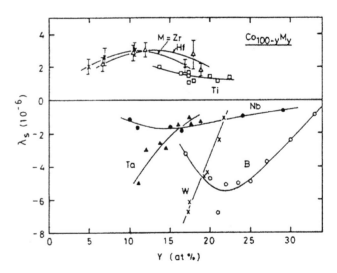

Fig. 74. Saturation magnetostriction (λ_s) of Co-TM amorphous alloys. TM is Zr, Hf, Ti, Nb, Ta, W, or B. Reproduced with permission from [177], copyright 1984, American Institute of Physics.

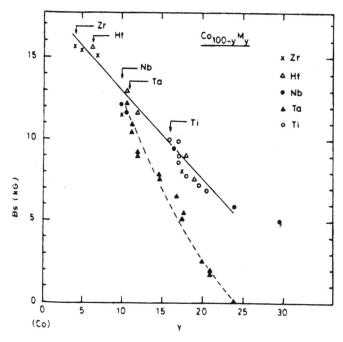

Fig. 75. Saturation magnetic flux density (B_s) of Co-TM amorphous alloys. TM is Zr, Hf, Nb, Ta, or Ti. Reproduced with permission from [177], copyright 1984, American Institute of Physics.

The amorphous state of the Co-TM alloys is metastable. The recrystallization temperature of these amorphous alloys is between 400 and 500°C, varying with the type and concentration of the constituent glass-forming element. The soft magnetic properties of these films disappear completely after recrystallization. Hayashi et al. [190] have formulated an expression based on the available data for the estimation of recrystallization temperatures (T_x) for Co-TM (TM = Zr, Nb, or Ta) amorphous alloys

$$T_x(°C) = (3.43 \pm 0.35)X_{Co} + (16.20 \pm 2.13)X_{Ta}$$
$$+ (13.44 \pm 2.07)X_{Nb} \qquad (5.37)$$
$$+ (19.79 \pm 2.59)X_{Zr}$$

where X_{Co}, X_{Ta}, X_{Nb}, and X_{Zr} are the at % of the Co, Ta, Nb, and Zr elements in the alloy, respectively. Equation (5.37) is valid for $75 \le X_{Co} \le 95, 5 \le X_{Ta}$ and X_{Nb} and $2 \le X_{Zr}$. Obviously, in terms of the contribution to the higher recrystallization temperatures, the three TM elements are in the order Zr, Ta, and Nb. The highest recrystallization temperatures are obtainable in the Co-Zr amorphous alloys with higher Zr concentrations.

The uniaxial anisotropy of Co-TM amorphous films can be obtained by using a magnetic substrate holder with a field strength of 50–100 Oe during film deposition, or by subsequent magnetic annealing after deposition. The temperature for magnetic annealing is in the range of 200–350°C. Typical values of the anisotropy field H_k of as-deposited Co-TM amorphous films is between 13 Oe and 20 Oe. The hard-axis initial relative permeability of Co-TM amorphous films is

about 1500. The anisotropy field is reduced to about 3 Oe, and permeability is increased to 5000 by rotational field annealing (RFA) [194].

The major drawback of amorphous soft magnetic films is the thermal stability of the uniaxial anisotropy. The easy axis of most Co-TM amorphous alloys becomes unstable above 200°C, which makes it difficult for them to survive the head fabrication process without degradation of easy axis orientation. However, the easy axis thermal stability of these amorphous films can be improved by an initial magnetic annealing after deposition. Ishi et al. [198] have found that CoZrTa amorphous films exhibit stable easy-axis orientation up to 270°C if the film was initially annealed in a magnetic field at 350°C.

5.4.4. FeN-Based Nanocrystalline Films

The FeN-X (X = Al, Ta, or Ti) nanocrystalline films are the only category of soft magnetic films developed so far with the highest saturation magnetization ($4\pi M_s$ up to 20 kG, only slightly lower than that of pure Fe), good soft magnetic properties (well-defined uniaxial anisotropy, near-zero magnetostriction, and small H_c), and good thermal stability for recording head applications. Thin films heads with FeAlN [183] poles have been fabricated and used in high-density recording.

The soft magnetism of FeN-X films originates from their nanocrystalline structure and can be explained by Hoffmann's ripple theory [203, 204]. These films are usually prepared by sputter deposition techniques. The key factors determining the soft magnetism of these films are grain size, their morphology, and the nature of grain boundary materials [201]. The addition of a small amount (2–3 at %) of a third element (Al, Ta, Ti, or Si) to the Fe-N binary system shows a marked effect in preventing the grain growth and in improving the soft magnetism and thermal stability of the films [179]. It was found that when the grain size of these films is reduced to below 15 nm, the magneto-crystalline anisotropy vanishes and the coercivity is considerably reduced to below 1 Oe. However, it is necessary that the grain boundary be ferromagnetic and very thin [188], so that strong intergranular exchange coupling is maintained and the nanosized grains are not magnetically isolated from each other. Grain isolation will result in high coercivity because of the single-domain particle behavior of the isolated grains. Columnar structure is not desirable because it plays the key role in promoting the perpendicular anisotropy component ($K_\perp$). Under noncolumnar growth conditions, the incorporation of nitrogen into the film first decreases $K_\perp$ to a critical value, then the N acts as a "grain refiner," and excellent soft magnetic properties can be obtained [201]. Nitrogen incorporation in FeTa films is much higher than in Fe films [201]. However, it was found that the saturation magnetization of the films did not change with nitrogen addition up to 2% lattice dilation [200]. A reversible degradation of soft magnetic properties above a critical temperature ($T_{0,c}$) was reported in FeAlN films, and $T_{0,c}$ decreases with the increase of N contents in the films [202]. The weakened intergranular exchange coupling due to the change of ferromagnetic phase to the paramagnetic phase of the grain boundary materials at temperatures above $T_{0,c}$ was believed to be responsible for the degradation of soft magnetic properties [202].

The effect of magnetic annealing on the behavior of soft magnetic properties of FeTaN films was studied by Viala et al. [206]. It was found that the soft magnetic properties of FeTaN films annealed in a field applied parallel to the original easy axis were stable up to annealing temperatures of 300°C. Higher annealing temperatures lead to grain growth and TaN precipitation, along with substantial reduction of saturation magnetization and poorer soft magnetism. Changes in stress and magnetostriction were also observed. When the FeTaN films were annealed in a transverse field at 150°C as a function of annealing time, change and switching of H_k were observed. H_k decreases with increasing of annealing time, switching at 60 min and finally stabilizing at 120 min, aligning with the annealing field. Minor and Barbard [205] have studied the thermal stability of FeTaN films and found that the thermal stability of FeTaN films was improved because of the Ta. The Ta content for optimum thermal stability was about 10 wt%.

5.4.5. Giant Moment Soft Magnetic Films

The giant magnetic moment phenomenon in Fe-N films was first reported by Kim and Takahashi in 1972 [208]. They discovered that the saturation magnetization (M_s) of polycrystalline Fe-N films deposited by evaporation in low-pressure nitrogen was as high as 2050 emu/cc ($4\pi M_s = 25.8$ kG). Fig. 76 shows the dependence of saturation magnetization of the as-deposited Fe-N films on the nitrogen pressure during deposition. The giant moment of the Fe-N films was attributed to the α''-martensite phase (α''-$Fe_{16}N_2$) in the Fe-N films, which has a bct structure, and the atomic magnetic moment associated with Fe atoms of α''-$Fe_{16}N_2$ was deduced to be

3.0 μ_B, compared with 2.4 μ_B per atom for pure Fe. The α''-martensite phase was discovered in 1951 by Jack [207]. However, its magnetic properties were not investigated until 1972.

There has been increased interest in this material over the past decade because of its potential applications in film heads. Giant magnetic moment α''-$Fe_{16}N_2$ (001) single–crystal phase ($4\pi M_s = 29$ kG) was obtained by Komuro et al. by molecular beam epitaxy (MBE) on InGaAs (001) substrates [209]. The atomic moment measured by VSM and Rutherford backscattering was 3.5 μ_B [215]. The resistivity of the pure α''-$Fe_{16}N_2$ phase is 30 $\mu\Omega$-cm [215]. Giant magnetic moment (up to 25 kG) multiphased Fe-N films have also been obtained by ion beam deposition on Si(111) substrates [210], and dc and rf magnetron sputtering on Si(001) and glass substrates [211, 212]. Most recently, a giant moment α''-$(Fe,Co)_{16}N_2$ phase was also reported by Wang and Jiang, who used facing target sputtering on Si(001) substrates [213]. Although results published to date have not been so consistent and are incompatible with theoretical predictions [214], and the reproducibility of the film was also in question, they appeared to indicate that the formation of the giant magnetic moment α'' phase depends upon the substrate material, nitrogen partial pressure, film thickness, deposition process, and in situ and post deposition heat treatment. Questions regarding the mechanism of giant magnetic moment have yet to be answered. With regard to applications, reproducibility, soft magnetism, and thermal stability of the α'' phase have to be further investigated.

Acknowledgment

The author is grateful to John Mallinson for the review of the GMR spin valve chapter.

REFERENCES

1. P. J. Flanders, *J. Appl. Phys.* 63, 3940 (1988).
2. D. Speliotis, *Data Tech* 3 (2000).
3. K. O'Grady, V. G. Lewis, and D. P. E. Dickson, *J. Appl. Phys.* 73, 5608 (1993).
4. L. B. Valdes, *Proc. I.R.E.* 42, 420 (1954).
5. D. R. Zrudsky, H. D. Bush, and J. R. Fassett, *Rev. Sci. Instrum.* 37, 885 (1966).
6. C. A. Grimes, P. L. Trouilloud, and R. M. Walser, *IEEE. Trans. Magn.* 24, 603 (1988).
7. B. Dieny, A. Granovsky, A. Vedyaev, N. Ryzhanova, C. Cowache, and L. G. Pereira, *J. Magn. Magn. Mater.* 151, 378 (1995).
8. M. L. Williams and E. Grochowski, *Data Tech* 2, 54 (1999).
9. J. C. Mallinson, "Magneto-Resistive Heads— Fundamentals and Applications," Vol. 15. Academic Press, San Diego, 1996.
10. J. C. Mallinson, "The Foundations of Magnetic Recording." Academic Press, San Diego, 1987.
11. S. Iwasaki and Y. Nakamura, *IEEE Trans. Magn.* 13, 1272 (1977).
12. O. Karlquist, *Trans. R. Inst. Technol. Stockholm* 86 (1954).
13. M. H. Kryder, "Proceedings of the Symposium on Magnetic Materials, Process and Devices," 1990, Vol. 90, p. 25.
14. S. Iwasaki and Y. Nakamura, *IEEE Trans. Magn.* 13, 1272 (1977).
15. Y. Sonobe, Y. Ikeda, and Y. Tagashira, *IEEE Trans. Magn.* 35, 2769 (1999).
16. N. Honda and K. Ouchi, *IEICE Trans. Electron.* E80-C(9), 1180 (1997).

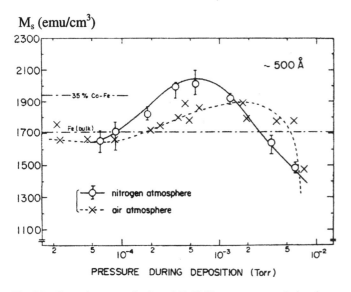

Fig. 76. Saturation magnetization of Fe-N films vs. pressure during deposition. $\bigcirc$, deposited in nitrogen atmosphere; $\times$, deposited in air (usually vacuum). Reproduced with permission from [208], copyright 1972, American Institute of Physics.

17. S. Shi, F. Liu, H. C. Tong, S. Dey, J. Kimmal, S. Malhotra, B. Lal, and M. Schultz, *IEEE Trans. Magn.* 35, 2634 (1977).

18. M. L. Williams and R. L. Comstock, "17th Annual AIP Conference Proceedings," 1971, Part 1, p. 738.

19. A Hoagland and J. Monson, "Digital Magnetic Recording." Wiley–Interscience, New York, 1991.

20. S. Chikazumi and S. Charap, "Physics of Magnetism," p. 313. Wiley, New York, 1964.

21. R. Street and J. C. Woolley, Proc. R. *London, Ser. A* John Wiley & Sons, Inc., New York/London. 62, 562 (1949).

22. M. P. Sharrock and J. T. McKinney, *IEEE Trans. Magn.* 17, 3020 (1981).

23. A. Aharoni, "Introduction to the Theory of Feromagnetism." Clarendon, Oxford, 1996.

24. M. P. Sharrock, *IEEE Trans. Magn.* 26, 193 (1990).

25. S. H. Charap, P. L. Lu, and Y. He, *IEEE Trans. Magn.* 33, 978 (1997).

26. R. W. Chantrell, J. D. Hannay, and M. Wongsam, *IEEE Trans. Magn.* 34, 1839 (1998).

27. T. L. Gilbert, *Phys. Rev.* 100, 1243 (1955).

28. H. J. Richter, *IEEE Trans. Magn.* 35, 2790 (1999).

29. H. N. Bertram and H. J. Richter, *J. Appl. Phys.* 85, 4991 (1999).

30. K. O'Grady and H. Laidler, *J. Magn Magn. Mater.* 200, 616 (1999).

31. M. El-Hilo, A. M. de Writte, K. O'Grady, and R. W. Chantrell, *J. Magn. Magn. Mater.* 117, 307 (1992).

32. J. Zou, B. Lu, D. E. Laughlin, and D. N. Lambeth, *IEEE Trans. Magn.* 35, 1661 (1999).

33. L. L. Lee, D. E. Laughlin, and D. N. Lambeth, *IEEE Trans. Magn.* 30, 3951 (1994).

34. S. Y. Hong, T. D. Lee, and K. H. Shin, *J. Appl. Phys.* 73, 5563 (1993).

35. B. Heinz and B. L. Gehman, *Data Tech* 2, 79 (1999).

36. M. Alex and D. Wachenschwanz, *IEEE Trans. Magn.* 35, 2796 (1999).

37. A. Moser and D. Weller, *IEEE Trans. Magn.* 35, 2808 (1999).

38. Y. Zhang and H. Neal Bertram, *IEEE Trans. Magn.* 34, 3786 (1998).

39. H. Neal Bertram, H. Zhou, and R. Gustafson, *IEEE Trans. Magn.* 34, 1845 (1998).

40. H. Neal Bertram, "Theory of Magnetic Recording." Cambridge Univ. Press, Cambridge, UK, 1994.

41. B. Cord, M. Geisler, E. Koparal, O. Keitel, and J. Scherer, *Data Tech* 2, 97 (1999).

42. D. N. Lambeth, E. M. T. Velu, G. H. Bellesis, L. L. Lee, and D. E. Laughlin, *J. Appl. Phys.* 79(8) 4496 (1996).

43. S. Guruswarmy, M. R. Kim, and K. E. Johnson, *Metals Mater. Processes* 8, 167 (1996).

44. T. R. McGuire and R. I. Potter, *IEEE Trans. Magn.* 11, 1019 (1975).

45. E. P. Wohlfarth, *J. Appl. Phys.* 29, 595 (1958).

46. O. Henkel, *Phys. Status Solidi* 7, 919 (1964).

47. I. A. Campbell and A. Fert, in "Ferromagnetic Material" (E. P. Wohfarth, Ed.), Vol. 3, p. 747. North-Holland, Amsterdam, 1982.

48. D. Markham and F. Jeffers, "Proceedings of the Symposium on Magnetic Materials, Processes and Devices," 1990, Vol. 90-8, p. 185.

49. R. P. Hunt, *IEEE Trans. Magn.* 6, 3 (1970).

50. M. N. Baibich, J. M. Broto, A. Fert, F. Nguyen Van Dau, F. Petroff, P. Etienne, G. Creuzet, A. Friederich, and J. Chazelas, *Phys. Rev. Lett.* 61, 2472 (1988).

51. G. Binasch, P. Grunberg, F. Saurenbach, and W. Zinn, *Phys. Rev. B: Solid State* 39, 4828 (1989).

52. S. S. P. Parkin, N. More, and K. P. Roche, *Phys. Rev. Lett.* 64, 2304 (1990).

53. E. Fullerton, M. J. Conover, J. E. Mattson, C. H. Sowers, and S. D. Bader, *Appl. Phys. Lett.* 63, 1699 (1993).

54. S. S. P. Parkin, R. Bhadra, and K. P. Roche, *Phys. Rev. Lett.* 66, 2152 (1991).

55. W. R. Bennett, W. Schwarzacher, and W. F. Egelhoff, Jr., *Phys. Rev. Lett.* 65, 3169 (1990).

56. S. F. Lee, W. P. Pratt, Jr., R. Loloee, P. A. Schroder, and J. Bass, *Phys. Rev. B: Solid State* 548 (1992).

57. T. L. Hylton, K. R. Coffey, M. A. Parker, and J. K. Howard, *Science* 261, 1021 (1999).

58. M. R. Parker, S. Hossain, and D. Seale, *IEEE Trans. Magn.* 30, 358 (1994).

59. R. Nakatani, T. Dei, T. Kobayashi, and Y. Sugita, *IEEE Trans. Magn.* 28, 2668 (1992).

60. J. Unguris, R. J. Celotta, and D. T. Pierce, *Phys. Rev. Lett.* 67, 140 (1991).

61. J. Mouchot, P. Gerard, and B. Rodmacq, *IEEE Trans. Magn.* 29, 2732 (1993).

62. B. Rodmacq, G. Palumbo, and P. Gerard, *J. Magn. Magn. Mater.* 118, L11 (1993).

63. S. S. P. Parkin, *Phys. Rev. Lett.* 67, 3598 (1991).

64. B. Dieny, V. S. Speriosu, S. Metin, S. S. Parkin, B. A. Gurney, P. Baumgart, and D. R. Wilhoit, *J. Appl. Phys.* 69, 4774 (1991).

65. R. L. White, *IEEE Trans. Magn.* 28, 2482 (1992).

66. B. Dieny, *J. Magn. Magn. Mater.* 136, 335 (1994).

67. A. Fert and P. Bruno, in "Ultrathin Magnetic Structures II" (B. Heirich and A. Bland, Eds.), pp. 82–106. Springer-Verlag, Berlin, 1992.

68. S. S. P. Parkin, in "Ultrathin Magnetic Structures II" (B. Heirich and A. Bland, Eds.), pp. 148–185. Springer-Verlag, Berlin, 1994.

69. D. T. Pierce, J. Unguris, and R. J. Celotta, in "Ultrathin Magnetic Structures II" (B. Heirich and A. Bland, Eds.), pp. 117–147. Springer-Verlag, Berlin, 1994.

70. K. B. Hathaway, in "Ultrathin Magnetic Structures II" (B. Heirich and A. Bland, Eds.), pp. 45–81. Springer-Verlag, Berlin, 1994.

71. B. Dieny, V. S. Speriosu, J. P. Nozeieres, B. A. Gurney, A. Vedeayev, and N. Ryzhanova, in "Magnetism and Structure in Systems of Reduced Dimension" (R. Farrow et al., Eds.), NATO ASI Ser. B: Physics, Vol. 309. Plenum, New York, 1993.

72. V. S. Speriosu, B. Dieny, P. Humbert, B. A. Gurney, and H. Lefakis, *Phys. Rev. B: Solid State* 44, 5358 (1991).

73. D. Dai and K. Qian, "Ferromagnetism," Vol. 1, p. 326. Science Press, Beijing, 1987.

74. S. Liao, "Ferromagnetism," Vol. 3, p. 62, Science Press, Beijing, 1987.

75. D. E. Heim, R. E. Fontana, Jr., C. Tsang, V. S. Speriosu, B. A. Gurney, and M. L. Williams, *IEEE Trans. Magn.* 30, 316 (1994).

76. T. Valet and A. Fert, *Phys. Rev. B: Solid State* 48, 7099 (1993).

77. W. P. Pratt, Jr., S. F. Lee, J. M. Slaughter, R. Loloee, P. A. Schroeder, and J. Bass, *Phys. Rev. Lett.* 66, 1991.

78. S. S. P. Parkin, *Phys. Rev. Lett.* 71, 1641 (1993).

79. J. C. S. Kools, *IEEE Trans. Magn.* 32, 3165 (1996).

80. H. C. Tong, X. Shi, F. Liu, C. Qian, Z. W. Dong, X. Yan, R. Barr, L. Miloslavsky, S. Zhou, J. Perlas, P. Prabhu, M. Mao, S. Funada, M. Gibbons, Q. Leng, J. G. Zhu, and S. Dey, *IEEE Trans. Magn.* 35, 2574 (1999).

81. W. F. Egelhoff, P. J. Chen, C. J. Powell, M. D. Stiles, R. D. McMichael, J. H. Judy, K. Takano, and A. E. Berkowitz, *J. Appl. Phys.* 82, 6142 (1997).

82. H. J. M. Swagten, G. J. Strijkers, R. H. J. N. Bitter, W. J. M. de Jonge, and J. C. S. Kools, *IEEE Trans. Magn.* 34, 948 (1998).

83. W. F. Egelhoff, P. J. Chen, C. J. Powell, M. D. Stiles, R. D. McMichael, J. H. Judy, K. Takano, A. E. Berkowitz, and J. M. Daughton, *IEEE Trans. Magn.* 33, 3580 (1997).

84. Y. Sugita, Y. Kawawake, N. Satomi, H. Sakakima, *Jpn. J. Appl. Phys., Part 1* 37, 5984 (1998).

85. H. Sakakima, Y. Sugita, and M. Satomi, and Y. Kawawake, *J. Magn. Magn. Mater.* 199, 9 (1999).

86. M. M. Dovek, "GMR Recording Head Engineering," presented at GMR Heads and Media Symposium, IDEMA, February 1999.

87. S. Gangopadhyay, "Current and Future Challenges for GMR Heads," presented at GMR Heads and Media Symposium, IDEMA, February 1999.

88. P. M. Baumgart, B. Dieny, B. A. Gurney, J. P. Nozieres, V. S. Speriosu, and D. R. Wilhoit, U.S. Patent 5,287,238 (1994).

89. T. C. Anthony, J. A. Brug, and S. Zhang, *IEEE Trans. Magn.* 30, 3819 (1994).

90. V. S. Speriosu, B. A. Gurney, D. R. Wilhoit, and L.B. Brown, *Intermag '96 Digest* AA-04 (1996).

91. D. Heim and S. S. P. Parkin, U.S. Patent 5,465,185 (1995).

92. G. Pan, S. Huo, D. J. Mapps, and W.W. Clegg, *IEEE Trans. Magn.* 35, 2556 (1999).

93. K. R. Coffey, B. A. Gurney, D. E. Heim, H. Lefakis, D. Mauri, V. S. Speriosu, and D. R. Wilhoit, U.S. Patent 5,583,725, 1996.

94. J. L. Leal and M. H. Kryder, *J. Appl. Phys.* 83, 3720 (1998).

95. S. W. Yuan and H. N. Bertram, *J. Appl. Phys.* 75, 6385 (1994).

96. B. A. Gurney, V. S. Speriosu, J. P. Nozieres, D. R. Wilhoit, and O. U. Need, *Phys. Rev. Lett.* 71, 4023 (1993).

97. R. Simmons, in "IDEMA GMR Heads and Media Symposium," Santa Clara, CA, February 10, 1999.

98. S. W. Yuan and H. N. Bertram, *J. Appl. Phys.* 75, 6385 (1994).

99. D. Liu and J.-G. Zhu, *IEEE Trans. Magn.* 31, 2615 (1995).

100. K. Nakamoto, Y. Kawato, Y. Susuki, Y. Hamakawa, and T. Kawabe, *IEEE Trans. Magn.* 32, 3374 (1996).

101. C. Cowache, B. Dieny, S. Auffret, and M. Cartier, *IEEE Trans. Magn.* 34, 843 (1998).

102. H. Hoshiya, S. Soeya, Y. Hamakawa, R. Nakatani, H. Fuyama, H. Fukui, and Y. Sugita, *IEEE Trans. Magn.* 33, 2878 (1997).

103. M. Saito, N. Hasegawa, T. Watanabe, Y. Kakihara, K. Sato, H. Seki, Y. Nakazawa, A. Makino, and T. Kuriyama, *Intermag'97 Digest* HA-06 (1997).

104. S. Araki, E. Omata, and M. Sano, *IEEE Trans. Magn.* 34, 387 (1998).

105. S. Mao, S. Gangopadhyay, N. Amin, and E. Murdock, *Appl. Phys. Lett.* 69, 3593 (1996).

106. A. J. Devasahayama and M. H. Kryder, *IEEE Trans. Magn.* 32, 4654 (1996).

107. T. Lin, C. Tsang, R. E. Fontana, and J. K. Howard, *IEEE Trans. Magn.* 31, 2585 (1991).

108. H. Hagde, J. Wang, and A. J. Devasahayam, in "IDEMA GMR Heads and Media Symposium," Santa Clara, CA, February 10, 1999.

109. B. Dieny, V. S. Speriosu, S. S. P. Parkin, B. A. Gurney, D. R. Wilhoit, and D. Mauri, *Phys. Rev. B: Solid State* 43, 1297 (1991).

110. J. L. Leal and M. H. Kryder, *IEEE Trans. Magn.* 35, 800 (1999).

111. J. P. Nozieres, S. Jaren, Y. B. Zhang, A. Zeltser, K. Pentek, and V. S. Speriosu, *J. Appl. Phys.* 87, 3920 (2000).

112. B. Dieny, M. Li, S. H. Liao, C. Horng, and K. Ju, *J. Appl. Phys.* 87, 3415 (2000).

113. H. Nagai, M. Ueno, and F. Hikami, *IEEE Trans. Magn.* 35, 2964 (1999).

114. S. Araki, M. Sano, M. Ohta, Y. Tsuchiya, K. Noguchi, H. Morita, and M. Matsuzaki, *IEEE Trans. Magn.* 34, 1426 (1998).

115. O. Allegranza and M. Chen, *J. Appl. Phys.* 73, 6218 (1993).

116. J. A. Thornton, "Evaporation and Sputtering." Noyes, Park Ridge, NJ, 1984.

117. B. Cord, M. Geisler, E. Koparal, O. Keitel, and J. Scherer, *DataTech* 2, 97 (1999).

118. B. M. Lairson, M. R. Visokay, R. Sinclair, and B. M. Clements, *Appl. Phys. Lett.* 62, 639 (1993).

119. R. F. C. Farrow, D. Weller, R. F. Marks, M. F. Toney, A. Cebollada, G. R. Harp, *J. Appl. Phys.* 79, 5967 (1996).

120. T. Suzuki, N. Honda, and K. Ouchi, *IEEE Trans. Magn.* 35, 2748 (1999).

121. T. Suzuki, N. Honda, and K. Ouchi, *J. Magn. Magn. Mater.* 193, 85 (1999).

122. J. U. Thiele, L. Folks, M. F. Toney, and D. K. Weller, *J. Appl. Phys.* 84, 5686 (1998).

123. Y. W. Lee, W. S. Cho, and C. O. Kim, *J. Korean Phys. Soc.* 35, S472 (1999).

124. C. H. Park, J. G. Na, P. W. Jang, and S. R. Lee, *IEEE Trans. Magn.* 35, 3034 (1999).

125. Y. N. Hsu, S. Jeong, D. N. Lamberth, and D. Laughlin, Session AQ-08, Intermag 2000, Toronto, Canada.

126. H. Kanazawa, T. Suzuki, G. Lauhoff, and R. Sbiaa, Session AQ-07, Intermag 2000, Toronto, Canada.

127. N. Li and B. M. Lairson, *IEEE Trans. Magn.* 35, 1077 (1999).

128. B. K. Middleton and C. D. Wright, *IERE Conf. Proc.* 54, 181 (1982).

129. S. Iwasaki, *J. Appl. Phys.* 69, 4739 (1991).

130. J. G. Zhu and H. N. Bertram, *IEEE Trans. Magn.* 26, 2149 (1990).

131. T. Chen and T. Yamashita, *IEEE Trans. Magn.* 24, 2700 (1988).

132. I. A. Bearsley and J. G. Zhu, *IEEE Trans. Magn.* 27, 5037 (1991).

133. J. G. Zhu and H. N. Bertram, *IEEE Trans. Magn.* 24, 2706 (1988).

134. R. P. Ferrier, *IEEE Trans. Magn.* 25, 3387 (1989).

135. H. N. Bertram, NATO-ASI Summer School on Applied Magnetism, July 1992.

136. M. Futamoto, Y. Hirayama, N. Inada, Y. Honda, K. Ito, A. Kikugawa, and T. Takeuchi, *IEEE Trans. Magn.* 35, 2802 (1999).

137. G. Pan, D. J. Mapps, M. A. Akhter, J. C. Lodder, P. ten Berge, H. Y. Wong, and J. N. Chapman, *J. Magn. Magn Mater.* 113, 21 (1992).

138. G. Pan, A. Okabe, and D. J. Mapps, Japanese Patent S91098992, 1992.

139. D. J. Mapps, G. Pan, M. A. Akhter, S. Onodera, and A. Okabe, *J. Magn. Magn. Mater.* 120, 305 (1993).

140. D. J. Mapps, G. Pan, and M. A. Akhter, *J. Appl. Phys.* 69, 5178 (1991).

141. Y. Maeno, H. Yamane, K. Sato, and M. Kobayashi, *Appl. Phys. Lett.* 60, 510 (1992).

142. B. M. Lairson, J. Perez, and C. Baldwin, *IEEE Trans. Magn.* 30, 4014 (1994).

143. R. Yoshino, T. Nagaoka, R. Terasaki, and C. Baldwin, *J. Magn. Soc. Jpn.* 18, 103 (1994).

144. M. Sizuki, H. Awano, N. Inaba, Y. Honda, and M. Futamoto, *J. Magn. Soc. Jpn.* 18, 451 (1994).

145. L. Wu, S. Yanase, N. Honda, and K. Ouchi, *IEEE Trans. Magn.* 33, 3094 (1997).

146. K. Ho, B. M. Laison, Y. K. Kim, G. I. Noyes, and S. Sun, *IEEE Trans. Magn.* 34, 1854 (1998).

147. L. Wu, T. Kiya, H. Honda, and K. Ouchi, *J. Magn. Magn. Mater.* 193, 89 (1999).

148. L. Wu, N. Honda, and K. Ouchi, *IEEE Trans. Magn.* 35, 2775 (1999).

149. H. Muraoka, S. Yamamoto, and Y. Nakamura, *J. Magn. Magn. Mater.* 120, 323 (1993).

150. K. Ouchi and N. Honda, TMRC99, A3 (1999).

151. H. Aoi, M. Saitoh, N. Nishiyama, R. Tsuchiya, and T. Tamura, *IEEE Trans. Magn.* 22, 895 (1986).

152. H. N. Bertram, K. Hallamasek, and M. Madrid, *IEEE Trans. Magn.* 22, 247 (1986).

153. S. M. Stinnett, W. D. Doyle, P. J. Flanders, and C. Dawson, *IEEE Trans. Magn.* 34, 1828 (1998).

154. K. Rubin, J. S. Goldberg, H. Rosen, E. Marinero, M. Doerner, and M. Schabes, *Mater. Res. Soc. Proc.* 1517, 261 (1998).

155. H. J. Richter, S. Z. Wu, and R. K. Malmhall, *IEEE Trans. Magn.* 34, 1540 (1998).

156. A. Moser, D. Weller, and M. E. Best, *J. Appl. Phys.* 85, 5018 (1999).

157. P. Dhagat, R. S. Indeck, and M. W. Muller, *J. Appl. Phys.* 85, 4994 (1999).

158. D. Wachenschwanz and M. Alex, *J. Appl. Phys.* 85, 5312 (1999).

159. S. Uren, M. Walker, K. O'Grady, and R. W. Chantrell, *IEEE Trans. Magn.* 24, 1808 (1988).

160. A. M. de Witte, K. O'Grady, and R. W. Chantrell, *J. Magn. Magn. Mater.* 120, 187 (1993).

161. H. Uwazumi, J. Chen, and J. H. Judy, *IEEE Trans. Magn.* 33, 3031 (1997).

162. Y. Hosoe, I. Tamai, K. Tanahashi, Y. Takahashi, T. Yamamoto, T. Kanbe, and Y. Yajima, *IEEE Trans. Magn.* 33, 3028 (1997).

163. Y. Hosoe, T. Kanbe, K. Tanahashi, I. Tamai, S. Matsunuma, and Y. Takahashi, *IEEE Trans. Magn.* 34, 1528 (1998).

164. P. J. Flanders and M. P. Sharrock, *J. Appl. Phys.* 62, 2918 (1987).

165. S. B. Oseroff, D. Franks, V. M. Tobin, and S. Schultz, *IEEE Trans. Magn.* 23, 2871 (1987).

166. M. P. Sharrock and J. T. Mckinney, *IEEE Trans. Magn.* 17, 3020 (1981).

167. M. P. Sharrock, *IEEE Trans. Magn.* 20, 754 (1984).

168. M. P. Sharrock and L. Josephson, *IEEE Trans. Magn.* 22, 723 (1986).

169. K. Ohashi, N. Morita, T. Tsuda, and Y. Nonaka, *IEEE Trans. Magn.* 35, 2538 (1999).

170. A. Veloso and P. Freitas, *IEEE Trans. Magn.* 35, 2568 (1999).

171. B. A. Gurney, V. S. Speriosu, J. P. Nozieres, H. Lefakis, D. R. Wilhoit, and O. U. Need, *Phys. Rev. Lett.* 71, 4023 (1993).

172. B. A. Gurney, D. Heim, H. Lefakis, O. U. Need, V. S. Speriosu, and D. R. Wilhoit, U.S. Patent 5,422,571, June 1995.
173. H. Iwasaki, H. Fukuzawa, Y. Kamiguchi, H. N. Fuke, K. Saito, K. Koi, and M. Sahashi, *Intermag'99 Digest* BA-04 (1999).
174. K. B. Klaassen and J. C. L. van Peppen, *Insight*, May/June, p. 42 (2000).
175. A. Chiu, I. Croll, D. E. Heim, R. E. Jones, Jr., P. Kasiraj, K. B. Klaassen, C. D. Mee, and R. G. Simmons, *IBM J. Res. Dev.* 40, 283 (1996).
176. J. C. Slonczewski, B. Petek, and B. E. Argyle, *IEEE Trans. Magn.* 24, 2045 (1988).
177. H. Fujimori, N. S. Kazama, K. Hirose, J. Zhang, H. Morita, I. Sato, and H. Sugawara, *J. Appl. Phys.* 55, 1769 (1984).
178. J. Hong, A. Furukawa, N. Sunn, and S. X. Wang, *IEEE Trans. Magn.* 35, 2502 (1999).
179. K. Katori, K. Hayashi, M. Hayakawa, and K. Aso, *Appl. Phys. Lett.* 54, 1181 (1989).
180. W. Maass and H. Rohrmann, *IEEE Trans. Magn.* 34, 1435 (1998).
181. W. P. Jayasekara, J. A. Bain, and M. H. Kryder, *IEEE Trans. Magn.* 34, 1438 (1998).
182. L. Varga, H. Jiang, T. J. Klemmer, and W. D. Doyle, *IEEE Trans. Magn.* 34, 1441 (1998).
183. W. P. Jayasekara, S. Khizroev, M. H. Kryder, W. Weresin, P. Kasiraj, and J. Fleming, *IEEE Trans. Magn.* 35, 613 (1999).
184. N. Robertson, H. L. Hu, and C. Tsang, *IEEE Trans. Magn.* 33, 2818 (1997).
185. J. Jury, P. George, and J. H. Judy, *IEEE Trans. Magn.* 35, 2508 (1999).
186. K. G. Ashar, "Magnetic Disk Drive Technology—Heads, Media, Channel, Interfaces, and Integration." IEEE Press, New York, 1994.
187. R. L. Comstock, "Introduction to Magnetism and Magnetic Recording." Wiley, New York, 1999.
188. O. Kohmoto, *IEEE Trans. Magn.* 27, 3640 (1991).
189. S. H. Liao, *IEEE Trans. Magn.* 26, 328 (1990).
190. K. Hayashi, M. Hayakawa, W. Ishikawa, Y. Ochiai, Y. Iwasaki, and K. Aso, *J. Appl. Phys.* 64, 772 (1988).
191. N. S. Kazama, H. Fujimori, and K. Hirose, *IEEE Trans. Magn.* 19, 1488 (1993).
192. Y. Shimada and K. Kojima, "Symposium on Perpendicular Magnetic Recording," Tohoku University, Sendai, 1982, p. 111.
193. N. Terada, Y. Hoshi, M. Naoe, and S. Yamanaka, "Digests of the 6th Annual Conference on Magnetism," Japan, 1982, p. 93.
194. G. Pan, D. J. Mapps, K. J. Kirk, and J. N. Chapman, *IEEE Trans. Magn.* 33, 2587 (1997).
195. H. Sakakima, *IEEE Trans. Magn.* 19, 131 (1983).
196. G. Pan, A. G. Spencer, and R. P. Howson, *IEEE Trans. Magn.* 27, 664 (1991).
197. Y. Shimada and H. Kojima, *J. Appl. Phys.* 53, 3156 (1982).
198. T. Ishi, T. Suzuki, N. Ishiwata, M. Nakada, K. Yamada, K. Shimabayashi, and H. Urai, *IEEE Trans. Magn.* 34, 1411 (1998).
199. N. Hayashi, S. Yosjimura, and J. Mumazawa, *J. Magn. Soc. Jpn. Suppl.* 13, 545 (1989).
200. L. Varga, H. Jiang, T. J. Klemmer, W. D. Doyle, and E. A. Payzant, *J. Appl. Phys.* 83, 5955 (1998).
201. B. Viala, M. K. Minor, and J. A. Barnard, *J. Appl. Phys.* 80, 3941 (1996).
202. P. Zheng, J. A. Bain, and M. H. Kryder, *J. Appl. Phys.* 81, 4495 (1997).
203. H. Hoffmann, *IEEE Trans. Magn.* 14, 32 (1968).
204. H. Hoffmann, *IEEE Trans. Magn.* 15, 1215 (1979).
205. H. Minor and J. A. Barbard, *IEEE Trans. Magn.* 33, 3808 (1997).
206. B. Viala, V. R. Inturi, and J. A. Barnard, *J. Appl. Phys.* 81, 4498 (1997).
207. K. H. Jack, *Proc. R. Soc. London, Ser. A* 208, 200 (1951).
208. T. K. Kim and M. Takahashi, *Appl. Phys. Lett.* 20, 492 (1972).
209. M. Komuro, Y. Kozono, M. Hanazono, and Y. Sugita, *J. Appl. Phys.* 67, 5127 (1990).
210. N. Terada, Y. Hoshi, M. Naoe, and S. Yamanaka, *IEEE Trans. Magn.* 20, 1451 (1984).
211. M. A. Russak, C. V. Jahnes, E. Klokholm, J. W. Lee, M. E. Re, and B. C. Webb, *J. Appl. Phys.* 70, 6427 (1991).
212. C. Gao, W. D. Doyle, and M. Shamsuzzoha, *J. Appl. Phys.* 73, 6579 (1993).
213. H. Y. Wang and E. Y. Jiang, *J. Phys: Condens. Matter* 9, 8547 (1997).
214. R. Coehoorn, G. H. O. Daalderop, and H. J. F. Jansen, *Phys. Rev. B: Condens. Matter* 48, 3830 (1993).
215. Y. Sugita, H. Takahashi, M. Komuro, and M. Igarashi, *J. Appl. Phys.* 79, 5576 (1996).
216. R. Bozorth, "Ferromagnetism." IEEE Press, New York, 1993.
217. S. Narishige, M. Hanazono, M. Takagi, and S. Kuwatsuka, *IEEE Trans. Magn.* 20, 848 (1984).
218. W. G. Jacobs, U. S. Patent 4,251,841, 1981.

Chapter 11

NUCLEAR RESONANCE IN MAGNETIC THIN FILMS AND MULTILAYERS

Mircea Serban Rogalski

Faculdade de Ciências e Tecnologia, Universidade do Algarve, Gambelas, 8000 Faro, Portugal

Contents

1. INTRODUCTION

The study of thin film, interface, and multilayer magnetism is, in recent years, one of the most active areas of both fundamental and more applied research. A number of differences, with potential technical importance, have been observed in the transport and magnetic properties of thin films and multilayers, as compared with bulk materials, and have been assigned to structural changes specific for systems with reduced dimensionality. Here we describe the usefulness of nuclear resonance techniques, which provide valuable microscopic information about thin film systems. Both nuclear magnetic resonance (NMR) and nuclear gamma resonance (NGR) or Mössbauer spectroscopy are very sensitive probes of the local structure and properties of materials, as the measurements are concerned with properties and energy states of single nuclei, which are in turn influenced by their own atomic electrons and the state of the solid. Nuclear resonance spectra are shaped directly by nuclear–atom interactions and indirectly by solid–atom interactions, which are themselves responsible for the main bulk properties of a given material, and, hence, the nuclear, atomic, and solid-state parameters obtained from these measurements are complementary to the bulk parameters.

The principles of nuclear resonance that are observed by element-specific techniques are introduced by considering the ^{57}Fe nucleus, which is the most convenient isotope for Mössbauer measurements, and the ^{59}Co nucleus, used in the great majority of NMR work on thin films. We then provide a brief formulation of hyperfine interactions, to show how useful information can be extracted from the nuclear resonance spectra. This makes it possible to discuss the basic hyperfine parameters determined by the internal interactions that are present in the material. By observing the magnetic hyperfine fields, which are induced by the ordered electronic magnetism, the magnitude of the magnetic moments, the orientation of magnetization, and the magnetic transition temperature can be determined. A later section describes the experimental techniques mainly used for the thin film and multilayer investigation, namely spin-echo NMR and conversion electron Mössbauer spectroscopy (CEMS). We survey the experimental work on metallic multilayers, showing that information from both NMR and NGR methods is able to confirm whether the

Handbook of Thin Film Materials, edited by H.S. Nalwa
Volume 5: Nanomaterials and Magnetic Thin Films

ISBN 0-12-512913-0/$35.00

multilayer samples and their interfaces are composed of different metallurgical or magnetic phases. Finally, the last section is devoted to amorphous, nanostructured, and granular thin films, with soft or hard magnetic properties that are of technical interest.

2. PRINCIPLES OF NUCLEAR RESONANCE SPECTROSCOPY

It has been established by experiment that some nuclei possess a total angular momentum, called the nuclear spin $\vec{I}$, whose magnitude $\sqrt{I(I+1)}\,\hbar$ and observable components $m_I\hbar$ are defined in direct analogy with the electron spin. The nuclear spin quantum number I is not restricted to $\frac{1}{2}$, but is found to take the values $0, \frac{1}{2}, 1, \frac{3}{2}, 2, \frac{5}{2}, 3, \frac{7}{2}, \ldots$, and hence m_I takes $2I + 1$ values. The same basic idea of a magnetic moment associated with a spinning electric charge, applied for the electron, can also be used in the case of a nucleus, where the magnetic moment is defined in terms of the nuclear magneton $\mu_N = e\hbar/2m_p$, analogous to the Bohr magneton, in the form

$$\vec{\mu}_I = \left(\frac{g_I \mu_N}{\hbar}\right)\vec{I} = \gamma_I \vec{I} \tag{1}$$

where g_I is the nuclear g-factor and γ_I is the nuclear gyromagnetic ratio. Our knowledge of nuclear spin number mainly relies on experiment, although many nuclear spin numbers have been established by the nuclear shell model [1]. Inasmuch as nuclei that contain a certain number of nucleons (namely $A = 2, 8, 20, 28, 50, 82,$ and 126) are very stable, an analogy with the electronic structure of inert gases (with $Z = 2, 10, 18, 36, 54,$ and 86) was suggested, by postulating that the nucleus contains a series of stationary energy levels. These are calculated in terms of three quantum numbers, n, l, and j, which characterize the radial part of the nucleon wave function, the orbital angular momentum of the nucleon within the nucleus, and the spin-orbit coupling between the spin ($\frac{1}{2}$) and orbital angular momenta of protons and neutrons. For each value of j there are $2j + 1$ degenerate levels, and, in a particular level, according to the exclusion principle, the protons pair with antiparallel spins, as do the neutrons. It follows that nuclei with even mass and charge number have $I = 0$ and are nonmagnetic; hence they do not display NMR spectra (common isotopes are ^{12}C and ^{16}O). Nuclei with even mass number and odd charge number have integral spin ($I = 1$ for ^{2}H, $I = 3$ for ^{14}N) and are suitable NMR isotopes, also exhibiting a nuclear electric quadrupole moment Q, because $I > \frac{1}{2}$. The quadrupole moment interacts with the local electric field gradients in a solid, resulting in a set of quadrupole energy levels, which is indicated by the spectra of Mössbauer isotopes. Nuclei with odd mass numbers have a half-integral spin, which is due entirely to the angular momentum of their last unpaired nucleon, $j = l \pm \frac{1}{2}$. These include ^{57}Fe, which has $I = \frac{1}{2}$, or ^{59}Co, with $I = \frac{7}{2}$. The ground and first excited nuclear levels of ^{57}Fe are illustrated in Figure 1. The energy splitting of the two ground Zeeman states by an external field

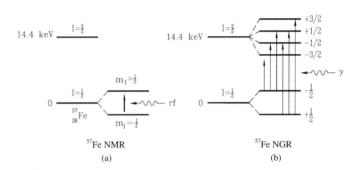

Fig. 1. The Zeeman levels in $^{57}_{26}$Fe involved in (a) NMR and (b) MS experiments.

B, shown in Figure 1a can be used to obtain the NMR signal by absorbing normal radio frequency (rf) power in transitions between the states. As $I = \frac{1}{2}$, we can preserve a close analogy with the case of the electron. At the same time, ^{57}Fe is the most convenient Mössbauer isotope that allows resonant transitions between the Zeeman states of the ground and first excited states (see Fig. 1b), when absorbing the recoil-free gamma rays emitted by a suitable radioactive source containing ^{57}Co.

The nuclear magnetic moment $\vec{\mu}_I$ interacts with the magnetic induction B at the position of the nucleus, acquiring energy,

$$E = -\vec{\mu}_I \cdot \vec{B} = -\gamma_I \hbar B_z m_I \tag{2}$$

such that the magnetic moments aligned against the field have a higher energy than those lined up with the field. As in any branch of absorption spectroscopy, we cause transitions between the ground Zeeman levels of the half-spin nucleus, as illustrated in Figure 1a, if the resonance condition is satisfied, namely:

$$\hbar\omega = E(m_I = -\tfrac{1}{2}) - E(m_I = \tfrac{1}{2}) = \gamma_I \hbar B_z = g_I \mu_N B_z \tag{3}$$

The corresponding resonance line is shown in Figure 2a for the case of a fixed frequency ω, so that by sweeping the magnetic field through the resonance value one obtains a spectrum in which the absorption of energy is plotted as a function of magnetic field strength. The alternative of sweeping the frequency at a given static field is illustrated in Figure 2b by the single-line NMR spectrum of ^{63}Cu nuclei in a 400-nm Cu film [2], where Eq. (3) yields a conversion factor of $\gamma_I = 11.29$ MHz T^{-1}.

In thin magnetic films, it is possible to perform NMR in the static component of the internal field at each nucleus, called the hyperfine field, which is induced by the electronic

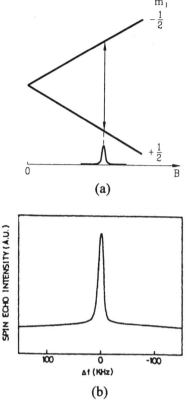

Fig. 2. Single-line NMR spectrum corresponding to the resonant transition between the Zeeman levels in $^{57}_{26}$Fe (a) and ^{63}Cu (b) NMR absorption lines produced by frequency sweeping at a static field of 7 T in a single 400-nm Cu film [2].

magnetization of the material. Sweeping the frequency without applying an external magnetic field to the sample produces most of these NMR spectra. The conversion factors between frequency and the internal field for resonance are $\gamma_I = 1.4$ MHz T^{-1} for ^{57}Fe and $\gamma_I = 10.1$ MHz T^{-1} for ^{59}Co, which is the nucleus used in the great majority of the NMR work on thin films, and, hence, the measurements are usually performed in the range of either 10–50 MHz or 180–250 MHz, respectively, which are common radio frequencies. We note that in the case of electron resonance (EPR), a resonance condition similar to Eq. (3) holds, where μ_N is replaced by the Bohr magneton $\mu_B = 1837\,\mu_N$, such that the required resonance frequency is in the microwave region. However, in both the radio and microwave frequency ranges, an extremely high accuracy in measuring frequencies and other radiation properties is actually available, and this enables the NMR (and EPR) technique to probe the energy levels of nuclei (and electrons) in the finest detail [3].

The set-up of an NMR experiment is essentially based on the time dependence of the nuclear spin eigenstates Ψ_I, induced by the magnetic interaction (Eq. (2)),

$$\Psi_I(t) = e^{-i\widehat{H}t/\hbar}\Psi_I(0) = e^{-i(-\gamma_I B_z \widehat{I}_z)t/\hbar}\Psi_I(0)$$
$$= e^{-i\widehat{I}_z\,\varphi/\hbar}\Psi_I(0) = \widehat{R}(\varphi)\psi_I(0) \quad (4)$$

where $\widehat{R}(\varphi)$ has the standard form of the rotation operator [1], written in terms of the nuclear spin operator I_z and the rotation angle φ as

$$\varphi = \omega_L t = -\gamma_I B_z t \quad \text{or} \quad \omega_L = -\gamma_I B_z = -\left(\frac{g_I \mu_N}{\hbar}\right)B_z \quad (5)$$

In other words, the angular momentum $\vec{I}$ in each of its eigenstates $\Psi_I(t)$ precesses around $\vec{B}$ with the classical Larmor frequency ω_L given by Eq. (5). The physical situation is described by a vector model of magnetic resonance, illustrated for $I = \frac{1}{2}$ in Figure 3a, and is classically described by the free motion equation,

$$\frac{d\vec{\mu}_I}{dt} = \gamma_I(\vec{\mu}_I \times \vec{B}) \quad (6)$$

The NMR transition corresponds to $\vec{\mu}_I$ passing from one orientation to the other. This requires a torque, which is obtained by applying a magnetic field $\vec{B}_1$ in the x-y plane. It is clear that no effective deflection of $\vec{\mu}_I$ is obtained if $\vec{B}_1$ is stationary in a given direction, because when $\vec{\mu}_I$ has a component along $\vec{B}_1$ it is tipped away from its cone of rotation, but then it is restored to the same cone when its component becomes antiparallel to $\vec{B}_1$. A transition occurs only if $\vec{B}_1$ rotates in the x-y plane in phase with the precession of $\vec{\mu}_I$, that is, with the Larmor frequency ω_L.

This can be achieved, as shown in Figure 3b, by applying perpendicular to the static field $\vec{B}$ a time-dependent radiofrequency field $\vec{B}_{rf}(t) = 2\vec{B}_1 \cos \omega t$ such that the total field will be composed of the static field in the z direction and a rotating field in the x-y plane:

$$B_x = B_1 \cos \omega t, \qquad B_y = B_1 \sin \omega t, \qquad B_z = B \quad (7)$$

Resonant absorption is caused by the rf frequency $\omega = \omega_L$ (Eq. (3) or (5)), which is reached by varying either ω or ω_L (i.e., the static field B), in the so-called continuous-wave

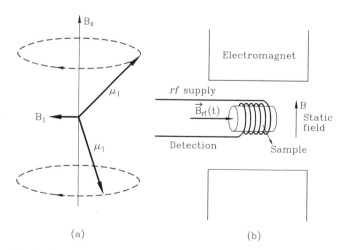

Fig. 3. Precession of the nuclear magnetic moments μ_I in the two-spin states and a rotating field B_1 in the x-y plane (a) and a schematic NMR spectrometer (b).

experiments. Assuming that the oscillating magnetic field (Eq. (7)) acts on the nucleus, across the static field $\vec{B}$, in the x direction, the resulting time-dependent perturbation is [1]

$$\widehat{V}(t) = 2\mu_N g_I B_1 \widehat{I}_x \cos \omega t = \mu_N g_I B_1 \widehat{I}_x (e^{i\omega t} + e^{-i\omega t})$$

If we take the matrix elements between two Zeeman angular momentum eigenstates, it follows that the transition rate from state ψ_{jm_j} to state $\psi_{jm'_j}$ can be derived, with Fermi's golden rule, in the form

$$P_{m_I \to m'_I} = \frac{2\pi}{\hbar^2} \mu_N^2 g_I^2 B_1^2 \left| \langle \psi_{Im'_I} | \widehat{I}_x | \psi_{Im_I} \rangle \right|^2 \delta(\omega_{m_I m'_I} - \omega)$$

where we may further use the nonvanishing matrix elements of $\widehat{I}_x$ [1] and $\gamma_I = \mu_N g_I / \hbar$ to obtain

$$P_{m_I \to m'_I} = \frac{\pi}{2} \gamma_N^2 B_1^2 [I(I+1) - m_I(m_I \pm 1)] \delta(\omega_{m_I, m_I \pm 1} - \omega) \tag{8}$$

Because μ_N is much smaller than μ_B, this transition rate is a factor of about 10^{-5} smaller for NMR transitions as compared with that of electron spin resonance (ESR), which is calculated in precisely the same way.

It might be expected that nuclear resonant absorption should also occur for γ-radiation emitted when nuclei in excited states lose their energy by radiation. However, the effect of the recoil momentum, which can be neglected for atomic radiation, becomes dominant for γ-radiation, because of its much higher energy. For emission of radiation from free atoms we simply have the recoil energy $R = p_R^2/2m_N = p_\gamma^2/2m_N = E_\gamma^2/2m_N c^2$, where m_N is the nuclear mass and the recoil momentum is equal to that of the emitted photon, $p_\gamma = E_\gamma/c$. This recoil energy prevents the observation of nuclear resonance absorption by free atoms, as suggested in Figure 4a. The Mössbauer effect arises when we consider emitting and absorbing atoms that are bound in a solid, such that they are no longer able to recoil individually [4]. In this case, for the system that consists of the atom and the entire solid in which it is embedded, we may assume that the wave function $\Phi = \Psi_N \psi$ is a product, to a first approximation, of the nuclear wave function Ψ_N and the wave function ψ of the solid, which are not affected by each other.

Before the γ-ray emission, the solid is assumed to be at rest, in a stationary state ψ_i, such that

$$\widehat{H}\psi_i = E_i \psi_i, \qquad -i\hbar \nabla \psi_i = 0 \tag{9}$$

After emitting a photon of momentum $\vec{p}_0 = \hbar \vec{k}_0$, the solid is in a state ψ_f, which also must be an eigenstate of the momentum operator, with eigenvalue $-\vec{p}_0$,

$$-i\hbar \nabla \psi_f = -\vec{p}_0 \psi_f$$

because the recoil is to small too eject the atom from the solid, and, hence, the entire solid must take up the recoil momentum. The state functions of the solid can be written in the Bloch form:

$$\psi_i = \sum_{\vec{k}} a_{\vec{k}i} e^{i\vec{k}\cdot\vec{r}}, \qquad \psi_f = \sum_{\vec{k}} a_{\vec{k}i} e^{i(\vec{k}-\vec{k}_0)\cdot\vec{r}} = -e^{-i\vec{k}_0\cdot\vec{r}} \psi_i \tag{10}$$

Then we may expand ψ_f, which is an eigenfunction of momentum, in terms of the complete system of energy eigenfunctions defined by Eq. (9), $\psi_f = \sum_n c_n \psi_n$, such that the probability of finding the solid, after the γ-radiation emission, in a stationary state E_n is

$$|c_n|^2 = \left| \int \psi_n^* \psi_f \, dV \right|^2 = \left| \int \psi_n^* e^{-i\vec{k}_0\cdot\vec{r}} \psi_i \, dV \right|^2$$

For $n = i$ the solid will remain in the same state as before, after the γ-ray emission, which means that the photon has carried away the full transition energy, and thus it was emitted without recoil energy loss, with a probability

$$f = |c_i|^2 \left| \int \psi_i^* e^{-i\vec{k}_0\cdot\vec{r}} \psi_i \, dV \right| \tag{11}$$

called the recoilless fraction. A simple expression for f can be obtained by representing the solid as a harmonic oscillator in one dimension (Einstein solid), with a ground-state energy eigenfunction and eigenvalue [1],

$$\psi_0(x) = \left(\frac{M\omega}{\pi\hbar} \right)^{1/4} e^{-M\omega x^2/2\hbar}, \qquad E_0 = \frac{1}{2}\hbar\omega$$

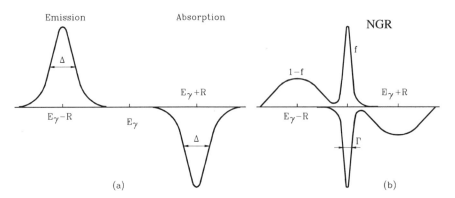

Fig. 4. Emission and absorption line shift by recoil (a) and resonant absorption in solids (b).

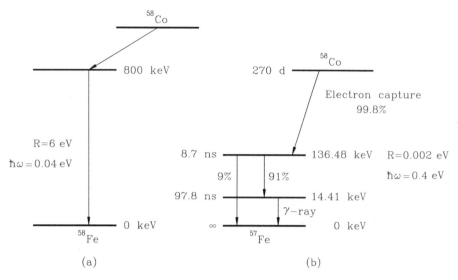

Fig. 5. Comparative decay schemes for ^{57}Fe and ^{58}Fe.

Substituting $\psi_0(x)$ and $k_0 = E_0/\hbar c$ into Eq. (11) yields

$$f_0 = \left| \int_{-\infty}^{\infty} \psi_0^*(x)\, e^{-iE_0 x/\hbar c}\, \psi_0(x)\, dx \right|^2 = e^{-E_0^2/2Mc^2\hbar\omega} = e^{-R/\hbar\omega} \tag{12}$$

If the recoil energy R is much smaller than $\hbar\omega$, which is the minimum amount of energy that the solid can accept, a significant fraction f_0 of γ-rays will escape with the full energy E_0 of the nuclear transition and without changing the internal energy of the solid. This implies that the Mössbauer γ-radiation has a Lorentzian lineshape, centered at E_γ, with the natural linewidth of the transition. There is no Doppler broadening, as this only comes from thermal excitations of the solid, such that the theoretical resolution of Mössbauer experiments is only determined by the lifetime of the excited state of the nuclear transition. There are a few isotopes where the NGR may occur, as they possess a suitable Mössbauer transition, with $R \ll \hbar\omega$. A crude estimation of the energy $\hbar\omega$ for a solid is obtained by taking $\omega = 2\pi v/\lambda$, where v

is the sound velocity and $\lambda = 2a$ (a is the lattice constant). For metallic iron, where $v = 5690$ m/s and $a = 2.9$ Å, we obtain $\hbar\omega = 0.04$ eV, and we may compare the decay schemes of two different isotopes, ^{57}Fe and ^{58}Fe, given in Figure 5. It follows that ^{57}Fe is a Mössbauer isotope, actually the best known, whereas NGR cannot be obtained with ^{58}Fe. Although the abundance of ^{57}Fe in natural iron is only 2.17%, its large recoilless fraction f provides satisfactory experimental results at room temperature. The fact that the Mössbauer resonance is specific to atoms of a particular isotope in the material results in a highly selective form of spectroscopy.

The Mössbauer spectrum is commonly obtained in the transmission geometry illustrated in Figure 6a. The γ-radiation is emitted from a single line source, which consists of a non-magnetic matrix containing radioactive ^{57}Co, which decays to ^{57}Fe, as shown in Figure 5b. The intensity transmitted through an absorber containing iron (hence ^{57}Fe) atoms in their ground state is measured as a function of E_γ, which is varied by an amount $E_\gamma v/c$ through Doppler shifting of

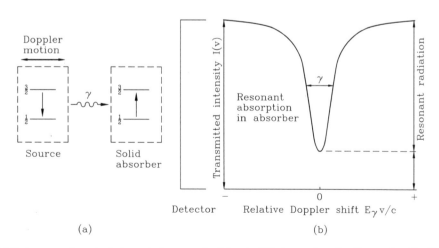

Fig. 6. Schematic Mössbauer spectrometer in transmission geometry (a) and a Mössbauer absorption line produced by Doppler scanning (b).

the source relative to the absorber. An intensity minimum (or maximum absorption) occurs at particular velocities where the source and absorber levels differ by $E_\gamma v/c$. Any increase or decrease in v from the resonance values results in a decrease in the absorption (see Fig. 6b). The natural linewidth of the 14.4 keV of ^{57}Fe is 0.095 mm/s, and the recorded spectrum is obtained as the convolution of the emission and absorption lines, such that the smallest observable Γ is 0.19 mm/s, which corresponds to an energy on the order of 10^{-8} eV. This is the order of magnitude of the nuclear–atom hyperfine interactions, which thus can be investigated by this technique.

3. HYPERFINE INTERACTIONS AND NUCLEAR RESONANCE SPECTRA IN THIN SOLID FILMS

The NMR spectrum in a solid is a measure of the response from a very large number of nuclei, which can be assumed to be at thermal equilibrium with their atomic environment, such that the m_I states are populated according to the Boltzmann distribution. This implies that the resonance condition will involve the bulk magnetization of the nuclei in the sample $\vec{M} = \sum \vec{\mu}_I / V$ rather than single nuclear magnetic moments. The expectation value of the component of $\vec{M}$ along the static field $\vec{B}$ can be written, using Eq. (1) and the density N of nuclear spins, in the form

$$M = N\gamma_I \langle I_z \rangle \tag{13}$$

where the thermal average of the nuclear spin (or nuclear spin polarization) for a two-level system, as pictured in Figure 1a, is

$$\langle I_z \rangle = \frac{\sum_{m_I} m_I \hbar e^{-E(m_I)/k_B T}}{\sum_{m_I} e^{-E(m_I)/k_B t}} \approx \frac{\sum_{m_I} m_I \hbar (1 + \gamma_I m_I \hbar B/k_B T)}{\sum_{m_I} (1 + \gamma_I m_I \hbar B/k_B T)}$$

$$= \frac{\frac{\hbar}{2}(1 + \gamma_I \hbar B/2k_B T) - \frac{\hbar}{2}(1 + \gamma_I \hbar B/2k_B T)}{1 + \gamma_I \hbar B/2k_B T + 1 - \gamma_I \hbar B/2k_B T}$$

$$= \left(\frac{\gamma_I \hbar B}{8k_B T}\right)\hbar = \left(\frac{g_I \mu_N B}{8k_B T}\right)\hbar \tag{14}$$

The magnitude of $(g_I \mu_N B/8k_B T)$ is very small at room temperature ($\sim 10^{-6}$), such that a satisfactory NMR signal from the rather small number of nuclei in a thin film can only be obtained in the low temperature range. It is convenient to describe the free precession motion of magnetization, in a coordinate system that rotates about the static field $\vec{B}$ with the same frequency ω as the radiofrequency field $\vec{B}_{rf}(t)$, in the form

$$\left(\frac{d\vec{M}}{dt}\right)_{rot} = \left(\frac{d\vec{M}}{dt}\right)_{lab} - (\vec{\omega} \times \vec{M})$$

$$= \gamma_I [\vec{M} \times (\vec{B} + \vec{B}_1)] - (\vec{\omega} \times \vec{M})$$

$$= \gamma_I (\vec{M} \times \vec{B}_{rot}) \tag{15}$$

where $d\vec{M}/dt$ in the fixed coordinate system was written in the form given by Eq. (6) for the field $\vec{B} + \vec{B}_1$, which has its static component along the z direction and a radiofrequency of magnitude B_1 in the x-y plane. The total field in the rotating frame was defined as $\vec{B}_{rot} = \vec{B} + \vec{B}_1 + \vec{\omega}/\gamma_I$. When the radiofrequency is $\omega = \omega_L = -\gamma_I B$, as given by Eq. (5), one obtains $\vec{B}_{rot} = \vec{B}_1$. Because Eq. (15) has the same form as Eq. (6), the motion in the rotating coordinate system consists of a precession of the magnetization around $\vec{B}_{rot}$, that is, around $\vec{B}_1$ at resonance ($\omega = \omega_L$). The physical situation is illustrated in Figure 7. In the absence of the $\vec{B}_{rf}(t)$ radiofrequency field, the individual magnetic moments $\vec{\mu}_I$ precess with random phase about the z direction, and the nuclear spin polarization gives rise to an initial magnetization M_z, according to Eqs. (13) and (14). If the field $\vec{B}_1$ is applied at the resonance Larmor frequency, $\vec{B}_1$ is stationary in the coordinate system that rotates with ω_L. Hence, the result drawn in the rotating frame consists of the deflection of all individual moments in the $\vec{B}_1$ direction, such that M_z decreases because of transitions and M_x, M_y become nonzero as the moments attain phase coherence (see Fig. 7b).

If the radiofrequency field is removed after a time t_w, which is the duration (width) of a rf pulse, the magnetization returns to its thermal equilibrium values, namely $M_z = M_0$ and $M_x =$

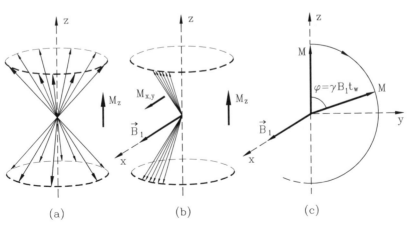

Fig. 7. Random phase precession in a static field $\vec{B}$ (a), phase coherence of individual magnetic moments represented in the rotating frame with $\omega = \omega_L$ (b), and precession of magnetization about $\vec{B}_1$ at resonance (c).

$M_y = 0$, by the so-called relaxation processes. A first mechanism consists of transferring to the lattice the energy acquired by the nuclear spin system, in a time T_1, called the spin-lattice relaxation time or longitudinal relaxation time because it is associated with M_z. If N_p and N_a are the number of atoms in the upper and lower levels and the probabilities per unit time of upward and downward relaxation transitions are w_u and w_d, we have $dN_a/dt = N_p w_d - N_a w_u$ and $dN_p/dt = N_a w_u - N_p w_d$. If we introduce the magnetization $M_z = (N_p - N_a)\mu_N$ along the z direction, and the total number of spins $N = N_p + N_a$, one has

$$\frac{dM_z}{dt} = \frac{d}{dt}(N_a - N_p)\mu_N = 2(N_p w_d - N_a w_u)\mu_N$$
$$= N\mu_N(w_d - w_u) - M_z(w_d + w_u)$$

It is convenient to set

$$w_d + w_u = \frac{1}{T_1}, \qquad N\mu_N \frac{w_d - w_u}{w_d + w_u} = M_0$$

such that we obtain the so-called longitudinal Bloch equation,

$$\frac{dM_z}{dt} = \frac{M_0 - M_z}{T_1} \tag{16}$$

The solution shows that the bulk magnetization induced by the radiofrequency field M_z returns to its equilibrium value M_0 exponentially, with the characteristic time T_1:

$$M_z - M_0 = (M_z - M_0)_{t=0} \, e^{-t/T_1} \tag{17}$$

The second mechanism consists of removing the phase coherence of the individual magnetic moments and hence decreasing the M_x and M_y values exponentially to zero through spin-spin relaxation, with a characteristic time T_2. In this case, the dephasing process is due to slight differences in the individual precessing frequencies, caused by interactions of a given spin with its neighbors. The phenomenological Bloch equations for this process are

$$\frac{dM_x}{dt} = -\frac{M_x}{T_2}, \qquad \frac{dM_y}{dt} = -\frac{M_y}{T_2} \tag{18}$$

where T_2 is called the transverse relaxation time, because it characterizes the return of magnetization to zero in the directions perpendicular to the static field.

Figure 7c suggests that the precession angle $\varphi = \gamma_I B_1 t_w$ of the magnetization out of the z direction can be chosen by selecting the duration t_w of the rf pulse and the rf amplitude B_1. Pulse NMR experiments rely upon this choice to produce particular pulse phase lengths, such as 90° and 180° pulses, corresponding to

$$\gamma_I B_1 t_w = \frac{\pi}{2} \qquad \text{or} \qquad \gamma_I B_1 (2t_w) = \pi$$

which respectively bends the magnetization from the z direction to the x-y plane or inverts its direction. Following the rf field switch off, after $t_w = \pi/2\gamma_I B_1$ (90° pulse), the free-induction decay of the magnetization is experimentally

observed, consisting of a free precession in the x-y plane and a return to the thermal equilibrium position through T_2 relaxation. For the ideal case of equivalent nuclei and $\omega = \omega_L$, the physical situation is depicted in Figure 8a. The time-dependent signal from the time space is Fourier transformed in the frequency space, which allows observation of the underlying frequencies and their linewidths. If $\omega \neq \omega_L$, the free-induction decay will appear to be modulated as a function of $|\omega - \omega_L|$, as shown in Figure 8b. As the static field is not the same for all nuclei, because of the hyperfine interactions in the solid, the effective field in the rotating frame is also different, and only a fraction of the nuclei correspond to $\vec{B}_{rot} = \vec{B}_1$ (and, hence, to a 90° rotation) during the pulse duration t_w. In a solid with different resonance frequencies, the free-induction decay signal is the result of the interference of individual signals (Figure 8c).

Therefore, there are two experimental NMR parameters of interest for the investigation of thin solid films:

(i) Resonance line position. It is clear from Figure 8 that the NMR signal allows a precise measurement of the resonance frequency. Because the nucleus interacts by either its magnetic dipole moment (with other nuclear magnetic moments, with electron spin magnetic moments, and with magnetic fields arising from the orbital electron motion) or by its electric quadrupole moment, if $I > \frac{1}{2}$ (with electric field gradients due to atomic electrons), the resonance frequency is shifted from the Larmor frequency ω_L caused by the static field $\vec{B}$ to a frequency determined by $\vec{B} + \vec{B}_{hf}$ where $\vec{B}_{hf}$, is the internal hyperfine field at the nucleus. This shift provides information about the surroundings of the nucleus. Figure 9 shows the room temperature ^{57}Fe pulse NMR spectrum of $Y_2 Fe_{17} N_{2.8}$ [5], a compound with four crystallographically unequivalent iron sites, labeled 6c, 9d, 18f, and 18h, where one would expect four values of the static internal (hyperfine) field, and hence four resonance frequencies. The assignment of the resonance peaks, centered at 44.8, 45.8, 48.2, 50.1, 52.3, and 54.5 MHz, was made in view of the crystallographic point symmetry, which causes the 9d, 18f (and 18h) sites to be divided into two groups of magnetically unequivalent sites, with relative populations of one to two, whereas the 6c sites remain magnetically unique. Therefore, each of the six Gaussian NMR lines correspond to a given Fe site in a specific environment. Also indicated in Figure 9 are the 6c and 9d peak locations for the parent $Y_2 Fe_{17}$ compound. It is known that insertion of nitrogen results in lattice expansion [5], and consequently a large Fe hyperfine field increase was observed for the 9d, 18f, and 18h sites in the nitrogenated compounds. Such behavior is generally attributed to the well-known Néel–Slater dependence of the Fe-Fe exchange interactions on

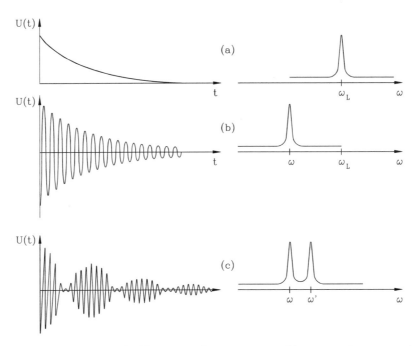

Fig. 8. NMR spectrum in the time and frequency space for (a) resonance frequency $\omega = \omega_L$, (b) resonance frequency $\omega \neq \omega_L$, and (c) two free-induction decays in the sample.

the interatomic distances. This is confirmed by the Fe (6c) resonance, whose location and linewidth are practically unchanged by nitrogenation (NMR frequency is 50.00 MHz in $Y_2 Fe_{17}$ and 50.1 MHz in $Y_2 Fe_{17} N_{2.8}$). It has been found by neutron diffraction that the interatomic distance between the two dumbbell Fe (6c) atoms and the surrounding Fe atoms does not change, whereas other Fe-Fe distances expand upon nitrogenation [6].

(ii) Resonance line shape and width. A Lorentzian shape is observed in the NMR spectra from solutions, whereas in solids it is Gaussian. This is due to a

distribution of spin states induced by the local fields, which results in a line broadening with a Gaussian envelope, as in Figure 10a. The principal mechanism of homogeneous broadening in many solids is the coupling of each nuclear magnetic moment with all neighboring nuclear moments. It can be shown that the linewidth of a Gaussian broadened line is inversely proportional to the spin-spin relaxation time T_2. For instance, the ^{59}Co NMR spectrum from multidomain Co particles consists of a single line with a linewidth of 0.7 MHz, whereas in Co films the linewidth is found to be about 10 MHz [7]. Line broadening increases as the layer thickness is decreased, as illustrated in Figure 10b for Co/Sn multilayers [8], where the low-frequency tail is associated with Co atoms in the interfacial regions. On the other hand, an anisotropic broadening has been found as the effect of varying the angle between the applied static field direction and the film surface [9]. Figure 11 shows the angular dependence of the ^{63}Cu resonance in a $[Ni(30 \text{ Å})/Cu(70 \text{ Å})]_{\times 29}$ multilayer, with a pronounced anisotropy in the linewidth associated with the external field rotation from the in-plane to the out-of-plane orientation.

The source of the magnetic hyperfine field in transition metals consists of three interaction mechanisms, namely

(i) the contact interaction, where the partly filled 3d atomic shell polarizes the s electrons, which have a finite density at the nucleus, thus creating a net spin density $\langle \vec{S} \rangle$, which gives rise to a contact field $\vec{B}_c$

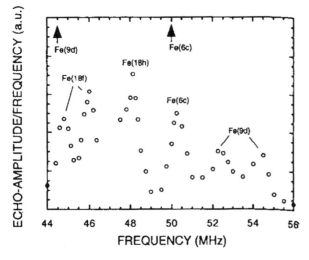

Fig. 9. The ^{57}Fe pulse NMR spectrum of $Y_2 Fe_{17} N_{2.8}$. Reproduced with permission from [5], copyright 1995, Elsevier Science.

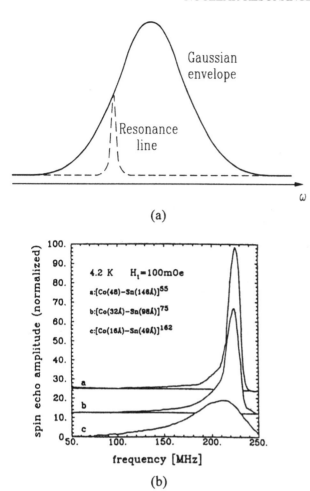

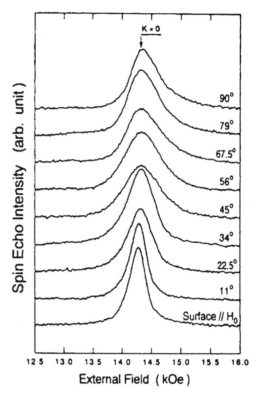

Fig. 11. Angular dependence of the ^{63}Cu resonance line in [Ni(30 Å)/Cu(70 Å)]$_{\times 29}$ multilayers. Reproduced with permission from [9], copyright 1993, Physical Society of Japan.

Fig. 10. Gaussian broadening of the resonance line in solids (a) and ^{59}Co NMR spectra in three Co/Sn multilayer systems (b) [8].

(ii) the interaction with the magnetic field $\vec{B}_{\text{orb}}$ produced at the nucleus by the orbital motion of 3d electrons, with $\langle \vec{L} \rangle = (g-2)\langle \vec{S} \rangle$, because the orbital angular momentum is normally quenched by the crystal field and partially restored by spin-orbit interaction

(iii) the dipolar interaction with the field $\vec{B}_{\text{dip}}$ arising from the 3d electron spins

These contributions account for the local hyperfine field at the transition metal nuclei, which has the form [10]

$$\vec{B}_{\text{hf}} = 2\frac{\mu_0 \mu_B}{4\pi}\left\langle \frac{1}{r^3} \right\rangle \left[-k\langle \vec{S} \rangle + \langle \vec{L} \rangle + \left(3\vec{r}\frac{\vec{S}\cdot\vec{r}}{r^2} - \vec{S} \right) \right]$$

$$= 2\frac{\mu_0 \mu_B}{4\pi}\left\langle \frac{1}{r^3} \right\rangle [-k + (g-2) + \langle 3\cos^2\theta - 1 \rangle]\langle \vec{S} \rangle \quad (19)$$

and have recently been calculated for Fe and Co in bulk material [11], as given in Table I. It is apparent that total hyperfine fields at bcc Fe and fcc Co nuclei are both antiparallel to the direction of the magnetization, as indicated by their negative signs. A large reduction in the magnitude of the surface hyperfine field as compared with the bulk value is found for both Fe and Co, because the contribution of valence s electrons

changes by about 20 T or, respectively, 10 T and becomes parallel to the magnetization.

If the ^{57}Fe Mössbauer spectrum is shaped by magnetic interactions only, there are six lines, corresponding to the transitions allowed by the $\Delta m_I = 0, \pm 1$ selection rule between the ground state ($I = \frac{1}{2}$) and the first excited state ($I = \frac{3}{2}$), which are split into two or, respectively, four Zeeman levels, as in Figure 1b. For a magnetic thin film, the relative area of these six lines is in the ratio 3:x:1:1:x:3, where x is determined by

Table I. Calculated Values of the Hyperfine Fields (T) for bcc Fe and fcc Co

Shell	bcc Fe	fcc Co
1s	−2.05	−1.84
2s	−52.18	−43.69
2p	0.17	−0.16
2s	29.1	26.09
3p	−0.07	0.07
Core sum	−25.02	−19.36
s	−7.28	−7.54
p	0.09	0.17
d	1.46	4.46
f	−0.03	−0.01
Valence sum	−5.77	−2.92
Total HF	−30.79	−22.28
Experimental HF	−33.3	−21.69

From [11].

the angle θ between the γ-radiation beam and the magnetic induction $\vec{B}_{hf}$ in the form

$$x = \frac{4\sin^2\theta}{1+\cos^2\theta} \qquad (20)$$

This predicts a ratio of 3:4:1:1:4:3 when the film is magnetized in its own plane and of 3:0:1:1:0:3 for magnetization normal to the plane. The overall splitting of the sextet is proportional to the $\vec{B}_{hf}$ experienced by the nucleus. The hyperfine field for metallic compounds is proportional, to a first approximation, to the Fe atomic magnetic moment with a proportionality constant in the range of $12–15 T/\mu_B$. Because the experimental linewidth, which typically is on the order of 0.25–1.3 mm/s in crystalline solids, corresponds to a range of 1–1.2 T for the hyperfine field, it follows that resolved six-line patterns can be obtained from compounds where $B_{hf} > 10$ T. In a metallic iron foil at room temperature, normally used for spectra calibration, we have $B_{hf} = 33.3$ T. The highest values of the hyperfine field, in the range of 45–55 T, are observed in the high-spin iron (III) compounds, and this can be explained in terms of the iron spin state dependence of B_{hf} (Eq. (19)). The metallic Fe atom has eight electrons in a $3d^6 4s^2$ configuration, the ion Fe^{2+} or iron (II) has a $3d^6$ configuration, and the Fe^{3+} or iron (III) ion has a $3d^5$ configuration, and this leads to the following spin states:

(i) $S = \frac{5}{2}$ $(5d\uparrow)$ high-spin iron (III)
(ii) $S = 2$ $(5d\uparrow 1d\downarrow)$ high-spin iron (II)
(iii) $S = \frac{1}{2}$ $(3d\uparrow 2d\downarrow)$ low-spin iron (III)
(iv) $S = 0$ $(3d\uparrow 3d\downarrow)$ low-spin iron (II)

Experiment shows, for instance, that in high-spin iron (II) compounds the hyperfine fields are significantly smaller, namely 20–25 T.

In magnetic compounds, where Fe occupies different crystallographic sites, the spectrum consists of a superposition of sextets with different hyperfine fields. Figure 12 shows Mössbauer spectra recorded in the temperature range between 5 K and the Curie point (643 K) from $UFe_{10}Si_2$[12]. This material crystallizes in the body-centered tetragonal $ThMn_{12}$ structure, where Fe (and Si) atoms are distributed among the 8f, 8j, and 8i sites. In the mean-field approximation, the B_{hf} at each site is mainly determined by the Fe-Fe interatomic distances d_{FeFe} and the average number NN of Fe nearest neighbors of an Fe atom at each crystallographic position. This indicates that we expect the hyperfine fields to be in the order $B_{hf}(8f) < B_{hf}(8j) < B_{hf}(8i)$. As the recoil-free fractions of the different sites are assumed to be equal, each of the three sextets recorded below the Curie temperature has a relative absorption area that is proportional to the relative iron population of the crystallographic position, namely I(8i):I(8j):I(8f)= 38:36:26. This corresponds to the site occupation factors calculated from neutron diffraction data [13], which are 100%, 93%, and 61% for the 8i, 8j, and 8f sites, respectively, indicating that Si atoms enter substitutionally rather than interstitially into the lattice.

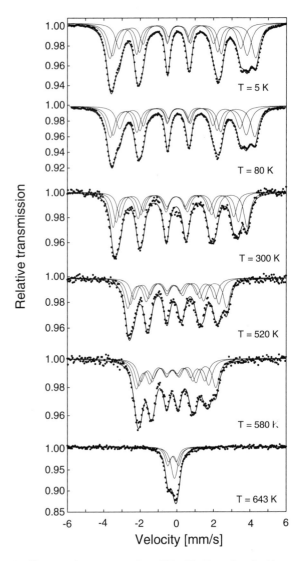

Fig. 12. ^{57}Fe Mössbauer spectra from $UFe_{10}Si_2$. Reproduced with permission from [12], copyright 1997, Elsevier Science.

The temperature dependence of the hyperfine fields, assumed to be proportional to the magnetic moments, is normally fitted by a Brillouin function, $B_{hf}(T) = B(T/T_c)B_{hf}(0)$, which determines the hyperfine field at $T = 0$ and the transition temperature T_c. It is often convenient to use the representation

$$\frac{B_{hf}(T)}{B_{hf}(0)} = 1 - D_s\left(\frac{T}{T_c}\right)^n \qquad (21)$$

where D_s is a site-dependent constant. It has been shown [14] that $n = \frac{3}{2}$ or $n = 2$ if the temperature dependence of the magnetization is governed by spin wave excitations or by single-particle excitations, respectively. Figure 13 shows a cubic temperature dependence of the site hyperfine fields, which is similar to that observed in Y_6Fe_{23}, although no appropriate excitation mechanism has been proposed yet.

Taking into account that the nucleus is a continuous distribution $\rho_N(\vec{r})$ of positive charges within a small volume rather

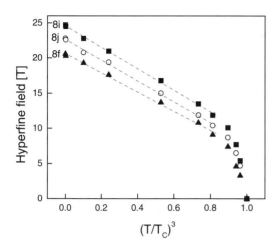

Fig. 13. $B_{hf}(T/T_{cf})$ in $UFe_{10}Si_2$. The dashed line is calculated with Eq. (21), where $n = 3$. Reproduced with permission from [12], copyright 1997, Elsevier Science.

than a point charge, there is an interaction with the external electric potential of the form

$$E = \int \rho_N(\vec{r})V(\vec{r})\,dV = V(0)\int \rho_N(\vec{r})\,dV$$

$$+ \sum_{i=x,y,z}\left(\frac{\partial V}{\partial x_i}\right)_0 \int \rho_N(\vec{r})\,x_i\,dV$$

$$+ \frac{1}{2}\sum_{i,j=x,y,z}\left(\frac{\partial^2 V}{\partial x_i \partial x_j}\right)_0 \int \rho_N(\vec{r})x_i x_j\,dV + \cdots \quad (22)$$

The first term defines the monopole interaction, which remains constant during nuclear transitions and will be disregarded. The second term gives the dipolar interaction, which vanishes because the nuclear charge distribution is an even function of coordinates such that the expectation value of the nuclear electric dipole moment is zero. The third term can be rewritten in terms of the electric field gradient components $\partial^2 V/\partial x_i \partial x_i = V_{ij}$, which form a symmetric matrix, and hence can be diagonalized by coordinate rotation:

$$E = \frac{1}{2}\sum_{i=x,y,z} V_{ii}\int \rho_N(\vec{r})x_i^2\,dV = \frac{1}{6}\sum_{i=x,y,z} V_{ii}\int \rho_N(\vec{r})r^2\,dV$$

$$+ \frac{1}{6}\sum_{i=x,y,z} V_{ii}\int \rho_N(\vec{r})(3x_i^2 - r^2)\,dV \quad (23)$$

In view of the Poisson equation $V_{xx} + V_{yy} + V_{zz} = \nabla^2 V = -\rho_e/\varepsilon_0$ it is convenient to express the electron charge

density in terms of the probability density of s-electrons at the nucleus $\rho_e = -e|\Psi_s(0)|^2$ and to introduce both the mean square nuclear radius $\langle R_N^2 \rangle = \int \rho_N(\vec{r})r^2 dV/Ze$ and the nuclear quadrupole moment of components $Q_{ii} = \int \rho_N(\vec{r})(3x_i^2 - r^2)dV$. This gives

$$E = \frac{Ze^2}{6\varepsilon_0}\langle R_N^2 \rangle |\Psi_s(0)|^2 + \frac{1}{6}\sum_{i=x,y,z} V_{ii}Q_{ii} \quad (24)$$

The first term describes the nuclear size effect, called chemical or isomer shift, which causes energy levels of two isotopes with different nuclear radii $\langle R_N^2 \rangle$, or embedded in different atomic surroundings $|\Psi_s(0)|^2$, to be slightly different, as in Figure 14a. The second term in Eq. (24) depends on the nuclear shape, which is measured by the intrinsic quadrupole moment, defined as

$$eQ' = \int \rho(\vec{r})r^2(3\cos^2\theta - 1)\,dV \quad (25)$$

We see that $Q' = 0$ for a spherical charge distribution, $Q' > 0$ for a distribution elongated at the poles (Fig. 14b) and $Q' < 0$ for a distribution flattened at the poles (see Fig. 14c). It is apparent that the quadrupole interaction will depend on the orientation of the deformed nucleus.

By the appropriate choice of axes, the electric field gradient is described by two parameters only, namely V_{zz} and the asymmetry factor $\eta = (V_{xx} - V_{yy})/V_{zz}$. If we define $\cos\theta$ in Eq. (25) in terms of the nuclear quantum numbers m_I and I in the form

$$\cos\theta = \frac{I_z}{|\vec{I}|} = \frac{m_I}{\sqrt{I(I+1)}} \quad (26)$$

the quadrupole interaction term from Eq. (24) becomes

$$E_Q = \frac{eQ'V_{zz}}{8}(3\cos^2\theta - 1)$$

$$= \frac{eQV_{zz}}{4I(2I-1)}\left[3m_I^2 - I(I+1)\right] \quad (27)$$

where Q is the effective quadrupole moment actually observed by experiment. This is always less than the intrinsic value Q', because of the nuclear spin precession caused by the quadrupole interaction. As mentioned before, $E_Q = 0$ for $I = \frac{1}{2}$, and the two-level splitting plotted in Figure 14d corresponds to the first excited state of ^{57}Fe, with $I = \frac{3}{2}$.

Because ^{57}Fe atoms in the source and absorber have different chemical environments, each Mössbauer absorption line

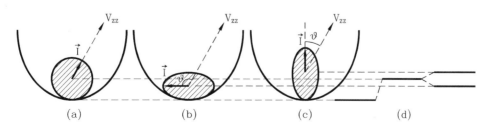

Fig. 14. Graphical interpretation of the nuclear size (a), and nuclear shape effects (b and c) on the nuclear energy levels (d).

(and hence the center of the spectrum) is shifted by a constant isomer shift δ. This is usually measured relative to an α-Fe calibration foil at room temperature, the center of which defines the zero velocity and the velocity scale. If $v_1, v_2, \ldots, v_6$ are the peak positions of a Mössbauer sextet on this scale, the isomer shift is given by $\frac{1}{6}\sum v_i$, and this can be either positive or negative, in the range -0.1 to 0.2 mm/s for metallic iron compounds and alloys. As expected from the second-order Doppler shift, the temperature dependence of the isomer shift is given by

$$\delta(T) = \delta(0) - A\left(T - \frac{B}{T}\right) \tag{28}$$

in the Debye model approximation, that is, for $T > 0.4\theta_D$, where θ_D is the Debye temperature [15]. This behavior is illustrated in Figure 15 for the three Fe sites in UFe$_{10}$Si$_2$ compounds, showing that A increases with increasing Si substitution at different sites, as the constant A is inversely proportional to the mass of the oscillators. Values of θ_D between 480 and 560 K have been obtained [12].

The effect of the quadrupole interaction can be observed in the spectrum of UFe$_{10}$Si$_2$ at the Curie point ($T_c = 643$K), plotted in Figure 12, which reduces to a superposition of three doublets, each of them characteristic for the splitting of the nuclear excited state ($I = \frac{3}{2}$) of ^{57}Fe, in the paramagnetic state, into two levels, as described by Eq. (27). The different isomer shifts at the three sites lead to an asymmetric shape of the Mössbauer spectrum. The quadrupole splitting is a measure of the asymmetry of the electron charge distribution around the iron nucleus, such that the larger the splitting of the two-line pattern, the larger is the local distortion from cubic symmetry.

Below the Curie temperature, it is expected that the Mössbauer spectrum will be shaped by both the magnetic

hyperfine and electric quadrupole interactions, such that the ^{57}Fe Zeeman levels become

$$E_e = -\gamma_e B_{hf} m_I + \frac{eQV_{zz}}{4I(2I-1)}[3m_I^2 - I(I+1)],$$

$$E_g = -\gamma_g B_{hf} m_I \tag{29}$$

In this case the separations between the excited-state Zeeman levels are no longer equal, and the absorption lines are not equally spaced in energy. Normally, the quadrupole interaction is small and the corresponding perturbation is smaller than that produced by the magnetic interaction. This results in sextets that are asymmetrically positioned relative to the center of the spectrum, with the two highest velocity lines situated closer together than the two lowest velocity lines. This quadrupole shift of each sextet is immediately expressible in terms of the six peak positions $v_1, v_2, \ldots, v_6$ by $\varepsilon = eQV_{zz}/4 = (v_1 + v_6 - v_2 - v_5)/2$. The temperature dependence of the quadrupole shift, plotted in Figure 16 for the UFe$_{10}$Si$_2$ compounds, follows, in the temperature range 80–600 K, the equation

$$\frac{\varepsilon(T)}{\varepsilon(T_c)} = 1 - C\left(\frac{T}{T_c}\right)^{3/2} \tag{30}$$

This $T^{3/2}$ dependence is common for metals and dilute binary alloys. Attempts have been made to understand this behavior on the basis of the effect of the mean square displacement of atoms on the electric field gradient, when conduction electrons are present in the lattice [16].

The competing magnetic and electric interactions can be clearly separated at the spin reorientation transitions. ^{57}Fe Mössbauer spectra of Er$_2$Fe$_{14}$B, recorded at different temperatures from 295 K to 340 K, have been used to obtain direct information on the behavior of each Fe site individually through the magnetic transition [17]. As seen in Figure 17a, Mössbauer spectra indicate that an abrupt change occurs at $T_{SR} = 323$ K. Below the spin reorientation temperature T_{SR} the

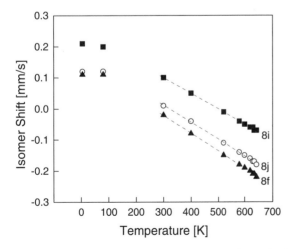

Fig. 15. Temperature dependence of the isomer shift for each site in UFe$_{10}$Si$_2$, calculated from the spectra plotted in Figure 12. Reproduced with permission from [12], copyright 1997, Elsevier Science.

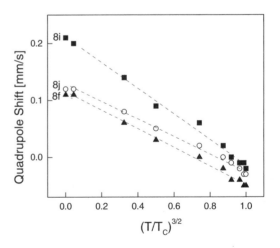

Fig. 16. Temperature dependence of site quadrupole shifts in UFe$_{10}$Si$_2$. Reproduced with permission from [12], copyright 1997, Elsevier Science.

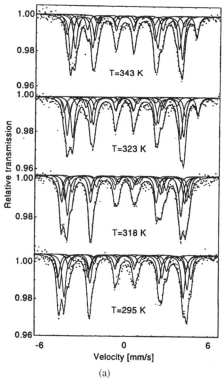

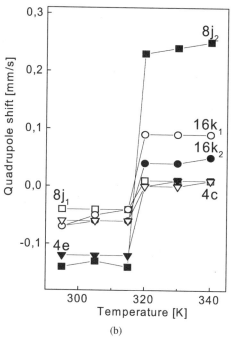

Fig. 17. Mössbauer spectra of $Er_2Fe_{14}B$ between room temperature and 343 K (a), and the temperature dependence of the quadrupole shifts at various Fe sites (b). Reproduced with permission from [17], copyright 1997, American Institute of Physics.

crystals with magnetization parallel to the c axis ($\theta = 0$). The hyperfine fields at the six Fe sites decrease in the order $8j_2 > 16k_2 > 8j_1 > 4e > 16k_1 > 4c$, in agreement with the decreasing number of iron near neighbors for each site and the number of adjacent boron atoms that behave like electron donors to the iron atoms, reducing the hyperfine field [18, 19]. There are abrupt changes in all of the hyperfine fields that decrease with increasing temperature, as expected in the presence of a magnetic transition at 323 K. The presence of the high-energy satellite peak at 5 mm/s in the 323 K spectrum only can be explained in terms of the temperature dependence of the six quadrupole shifts, represented in Figure 17b. It is apparent that, at 323 K, there is a large increase, by a factor of -2, in the quadrupole shift at the $8j_2$ site, and there are abrupt changes in all of the remaining quadrupole shifts. This is predicted by the quadrupole shift formula, Eq. (27), for the case of an axially symmetric electric field gradient, if the magnetic field is rotated from $\theta = 90°$ to $\theta = 0°$. Because the $8j_2$ site is known to be almost axially symmetric about the c axis [20], its quadrupole energy shift is indicative of a rotation of the moments from the basal plane (at 320 K) to the c axis (at 323 K). The quadrupole shifts of the remaining sites, where the symmetries of the electric field gradients are not known, also change signs between 320 and 323 K, and this is consistent with a rotation of the magnetic moments having occurred at 323 K.

The orientation of spontaneous magnetization is valuable information with respect to both basic and applied thin film magnetism. Preferential in-plane orientation of magnetization is associated with a dominant shape anisotropy, and thin films and superlattices with perpendicular spin structure are of interest for technical applications.

To summarize, the Mössbauer parameters of interest for the investigation of solids are

(i) The magnetic and/or quadrupole splitting of the Mössbauer lines. The larger the number of iron near neighbors, the larger is the magnetic splitting, and the larger the deviation from the electric cubic symmetry, the larger is the quadrupole splitting/shift. Information is obtained from the line splitting on both the magnetic dipole and electric quadrupole interactions, in both the ground and excited nuclear states. This should be compared with the NMR experiments, where only the nuclear ground state is involved.

(ii) The isomer shift of the sextet and doublet subspectra, which is proportional to the s-electron density at the nucleus. This provides information about the Wigner–Seitz cell volume, in the sense that the larger the Wigner–Seitz cell volume of an Fe site, the larger is the isomer shift.

(iii) The width and shape of Mössbauer lines. Line broadening accounts for relaxation processes, as in the case of NMR experiments, through spin-spin and spin-lattice interactions.

spectra consist of a broad and almost symmetrical hyperfine sextet, associated with the magnetization direction perpendicular to the c axis ($\theta = 90°$) whereas above T_{SR} the spectra are more structured and include a well-resolved satellite peak at about 5 mm/s that is characteristic for the $R_2Fe_{14}B$ poly-

4. EXPERIMENTAL TECHNIQUES FOR THIN FILM CHARACTERIZATION

In pulse NMR experiments, the free-induction decay signal is often caused by an effective spin-spin relaxation time T_m, which includes the intrinsic spin-spin relaxation time T_2 and a contribution T_2^* due to the magnetic field inhomogeneities ($1/T_m = 1/T_2 + 1/T_2^*$), resulting in a linewidth inversely proportional to T_m. The inhomogeneous broadening can be eliminated by the spin-echo experiments, which consist of using, as illustrated in Figure 18a, two short pulses of duration t_w (90° pulse) and $2t_w$ (180° pulse), separated by a time interval τ, subject to the condition

$$\frac{2\pi}{\omega} < t_w < T_m < T_2^* < \tau \ll T_2 \tag{31}$$

In other words, t_w should exceed several rf periods $2\pi/\omega$ but should be short compared with the reciprocal of the linewidth in frequency units. The 90° pulse bends the magnetization into the x-y plane, where the field inhomogeneities cause the nuclear spins to precess with slightly different frequencies and hence to dephase with time. A subsequent 180° pulse at time τ will invert the directions of all spins (and hence their relative positions), and because each spin continues to precess with its former frequency, they will be perfectly reclustered at $t = 2\tau$. Therefore, a maximum signal, or spin echo, is observed at $t = 2\tau$, and this begins to decay as the nuclear spins are again dephased for $t > 2\tau$ (Fig. 18b):

$$S(t) = S(0)e^{-t/T_2} \tag{32}$$

The width of the echo is on the order of T_2^*, because it is equivalent to two free-induction decays back to back, as in Figure 18a. The strength of the spin echo, $S(0)$, is attenuated for materials with short relaxation times T_2, on the order of a few microseconds, where the free-induction decay might become undetectable as the effect of saturation. T_2 may be increased by decreasing the temperature. The magnitude of the NMR signal is also diminished because of time-dependent field inhomogeneities, which cause less than all spins to be reclustered after 2τ. Assuming that all of the nuclear spins

have been rotated by the rf pulse of width t_w, $S(0)$ will be proportional to both the magnetization (Eq. (13)) and the NMR frequency, and, hence, for a two-level system, it takes the form

$$S(0) = \text{const.} \frac{N\gamma_I^2\hbar^2 B\omega}{8k_B T} \tag{33}$$

Because $S(0)$ is proportional to the reciprocal of the absolute temperature, the NMR investigation of thin magnetic films is usually limited to the helium range temperature.

Most NMR spin-echo experiments on thin magnetic films are performed on ^{59}Co, where $I = \frac{7}{2}$ and one may expect the combined effect of magnetic and quadrupole interactions. However, reasons are given for observing, in metallic compounds, only the $\frac{1}{2} \rightarrow -\frac{1}{2}$ transition [21]. The position of the NMR peaks is quite different for the three possible metallic structures: bcc Co (198 MHz), fcc Co (217 MHz), and hcp Co (228 MHz), as shown in Figure 19 [22]. The pure Co phases have established crystallographic and magnetic properties [23]. The fcc and hcp spectra can both be identified in bulk samples, whereas cobalt films have been prepared, up to some critical thickness, in all three phases, bcc, fcc, and hcp, with appropriately chosen substrates and deposition conditions.

It has been shown that the ferromagnetic bcc Co may be associated with a minimum energy of the cubic lattice, with a lattice constant of 2.827 Å, and a magnetization of 1.53 μ_B/Co atom. A long transverse relaxation time $T_2 = 180$ μs, indicative of a nanocrystalline material, has been determined for a 357-Å-thick film grown by molecular beam epitaxy [24, 25]. However, the bcc phase is found to be metastable [26], and in practice bcc Co deposited on a suitable GaAs substrate tends to transform into hcp Co. The influence of the layer thickness on this process is illustrated in Figure 20 for a series of Co$_x$Fe$_{24}$Å superlattices [27], grown on GaAs (110). The intensity of the strong narrow line at 198 MHz from bcc Co increases with respect to that of a broad distribution at about 212 MHz, assigned to the Co/Fe interface, up to a Co thickness of 21 Å (where the hcp Co signal at 227 MHz begins to appear). It is still observed in the 42-Å-thick layers.

The fcc Co phase, with a lattice constant of 3.5445 Å, a magnetization of 1.751 μ_B/Co atom, and a transverse relax-

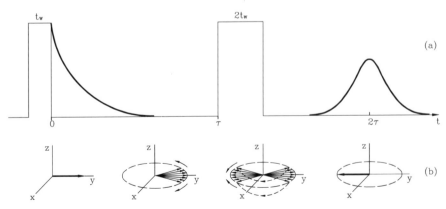

Fig. 18.   Schematic representation of the spin-echo method (a) and the formation of an echo (b).

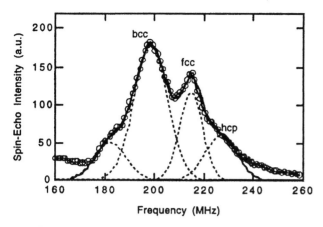

Fig. 19. NMR spectrum and Gaussian fits of $[\tau MnAl(44\ \text{Å})/Co(17\ \text{Å})]_{\times 8}$ multilayers [22]. Reproduced with permission from [22], copyright 1998, IEEE.

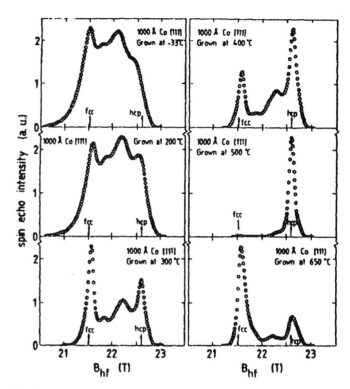

Fig. 21. NMR spectra of Co films on mica substrate. Reproduced with permission from [28], copyright 1994, American Physical Society.

ation time $T_2 = 50\ \mu s$, is found to be stable in fine particles. Its NMR spectrum consists of a single line with a peak value at 217 MHz. Above a critical film thickness, which is dependent on the material used as the substrate and on its temperature, the fcc phase begins to transform to the equilibrium form of Co at low temperature, with hcp structure. This has lattice constants $a = 2.5070$ Å and $c = 4.0698$ Å, a transverse relaxation time $T_2 = 40\ \mu s$, and an anisotropic hyperfine field (HF). A resonance line of 220 MHz is associated with spins in the basal plane, and a value of 228 MHz with those along the c axis. As a result, the resonance line of hcp Co is usually observed as a broad line centered at about 225 MHz. The magnetization of $1.72\ \mu_B$ per Co atom lies in the film plane in a zero applied field. Figure 21 shows, for a 1000-Å Co film grown on mica [28], that the fcc phase formation is favored by deposition at low and high substrate temperatures,

whereas a pure hcp phase, which is the stable structure for the bulk material, only forms in a narrow temperature interval near about 500°C. The intermediate NMR signals are assigned to a stacking fault structure.

Although most Mössbauer measurements made on thin magnetic films are performed in transmission geometry, it is possible to detect the NGR from backscattered conversion X-rays and conversion electrons arising in many Mössbauer isotopes from internal conversion of γ-radiation. The resonant and nonresonant interactions of γ-radiation are illustrated in Figure 22 [29]. After resonant absorption of the 14.4-keV photon emitted by the source, the ^{57}Fe nucleus in the excited state $I = \frac{3}{2}$ releases the energy either by emitting a photon or by transferring to an outer electron. Those electrons that are ejected in this manner (predominantly from the K shell but also from the L and M shells), are called K (or L, or M) conversion electrons. As seen in Figure 22, the inner electron conversion may be followed by the emission of a KLL or KLM Auger electron. A KLM electron is emitted from the M shell, because of the energy transferred by an L electron falling into the hole created by the K conversion electron. Alternative to the Auger process is the emission of characteristic X-rays, when a hole in the K shell is filled. The probability of this process is in the ratio 27:63 to that of the KLL Auger electron emission. In addition to the resonant electrons, a strong background is created by secondary electrons through Compton and photoelectric effects. We observe that the inner-electron conversion process occurs in ^{57}Fe far more frequently than the γ-photon emission, and this makes this isotope particularly

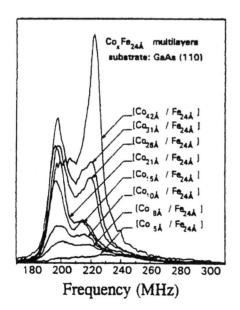

Fig. 20. ^{59}Co NMR spin-echo spectra of $Co_xFe_{24\text{Å}}$ multilayers deposited on Ga (110). Reproduced with permission from [27], copyright 1996, Kluwer.

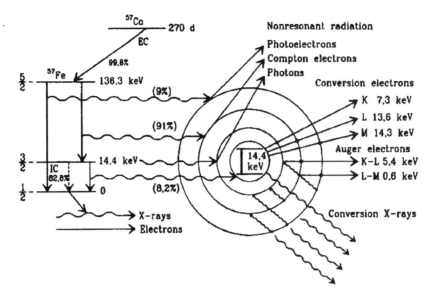

Fig. 22. Conversion electron and conversion X-ray emission associated with nuclear gamma resonance.

suitable for the CEMS technique. For every 100 Mössbauer γ-photon absorptions, only nine lead to reemission of γ-photons, and 81 result in K (7.3 keV) conversion electrons, and the rest in L (13.6 keV) conversion electrons.

Conversion electrons lose their kinetic energy through elastic and inelastic scattering processes, when traveling through material, such that only those electrons that are produced near the surface are able to escape the sample. The escape depths for K electrons, X-rays, and γ-rays are about 1000 Å, 15 μm, and 20 μm, respectively. In the integral scattering measurements, the electrons leaving the surface with energy between near zero and 13.6 keV are counted, giving the experimental limits of investigation by CEMS at 50 Å and 3000 Å. The same limits for the conversion X-ray Mössbauer spectroscopy are found to be 0.4 μm and 4 μm. To obtain good resolution with these techniques, gas flow proportional counters have been designed that could be used to detect either X-rays or electrons with the choice of the proper gas mixture: argon/methane for X-rays and helium/methane for electrons. The sample is placed inside the proportional counter so that the sample surface is exposed directly to the gas flow. In the case of electron detection this feature is required to avoid the attenuation that would occur if there were a detector window separating the sample surface from the counter gas. It is of particular interest to measure backscattered radiation and γ-ray transmission simultaneously, because of the different ranges in penetrating materials for conversion electrons, conversion X-rays, and resonant γ-rays, so that one might scan the sample by varying the radiation and thus give a good assessment of the different compositions on the surface and on the whole volume. This can be particularly useful if a sample consists of layers of different phases. A cylindrical electron counter can be combined with a toroidal X-ray counter to obtain a suitable detection assembly for this purpose, which is shown in Figure 23 [30].

Application of CEMS experiments to the study of Fe/Ti multilayers deposited by rf sputtering [31] shows that modulated atomic (crystalline/amorphous) and magnetic (ferro/paramagnetic) structures depend on the layer thickness. Figure 24 shows a series of CEMS spectra recorded for the indicated d_{Fe}/d_{Ti} thickness ratio, in the range 30 Å/30 Å to 2.5 Å/2.5 Å. It is clear the magnetic structure changes, as the

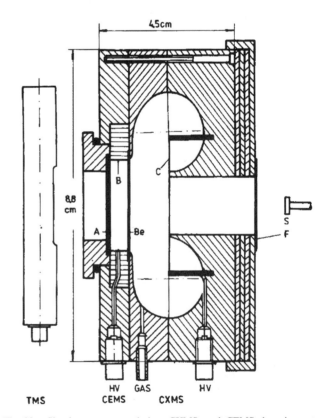

Fig. 23. Simultaneous transmission, CXMS, and CEMS detection set-up. Reproduced with permission from [30], copyright 1996, Institute of Physics.

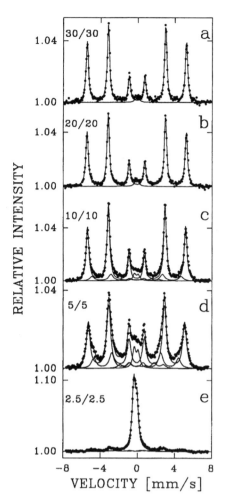

Fig. 24. CEMS spectra from Fe/Ti multilayers of different d_{Fe}/d_{Ti} thicknesses. Reproduced with permission from [31], copyright 1997, Institute of Physics.

as well as on topology and strains, and this is complementary to X-ray diffraction studies. Moreover, NMR and CEMS appear to be powerful tools for metallic multilayer investigation because of their capacity to be used as a probe for both structural and the magnetic local properties, which often can be directly related to the macroscopic magnetic and transport properties. Major problems to solve in metallic multilayers are

(i) Phase analysis and information on the topology for interfaces and individual layers embedded in the structure. Both NMR and CEMS provide the possibility of probing the nanostructure of multilayers and their real interfaces, which are not ideally sharp, as taken into account in the band structure calculations [32], providing information on the degree of intermixing or interface roughness at the atomic scale.

(ii) Influence of the substrate and the procedure of deposition on structure and the extent of mixing at an interface, given various combinations of elements on which the interface reactivity depends. This can be estimated by analyzing the hyperfine field distribution observed by both NMR and CEMS.

(iii) Modification of magnetic properties (that presumably occurs because of changes in the structure of the multilayer) through thermal annealing, which still provides the most effective way of controlling material parameters. The sensitivity of both NMR and CEMS to the nearest-neighbor environment permits structural changes on a scale of a few angstroms to be examined.

thickness decreases, from almost crystalline α-Fe (a), as indicated by the sextet with HF = 32.8 T, to the almost amorphous Fe-Ti phase (e), which corresponds to the QS doublet. The amorphous FeTi phase is formed during deposition at the interfaces between Fe and Ti layers. From the relative line intensity ratio in the sextet (Eq. (20)), it is derived that the spins are aligned predominantly in the plane of the film, although the spectral paramagnetic component corresponding to the FeTi interfacial regions suggests a random spin orientation at the interfaces.

5. NUCLEAR RESONANCE STUDY OF METALLIC MULTILAYERS

In recent years there has been a growing interest in using ^{59}Co NMR spin-echo measurements and ^{57}Fe CEMS experiments for the study of metallic multilayers, either as local probes of transition metal magnetism or as nondestructive methods for new material testing. As specified in the previous section, these methods provide information on the atomic structure and hence on the structural perfection and lattice parameters,

Application of the ^{59}Co NRM spin-echo technique is based on the sensitivity of the hyperfine field B_{hf} (Eq. (19)) to the nearest neighbors of the Co ion. Replacement of one near-neighbor Co ion reduces the s-electron polarization and results in a discrete shift of B_{hf}. For instance it was established that the resonance frequency decreases by about 18 MHz per Co atom replaced by Cu in fcc cobalt [33]. This feature enables one to probe the topology of the interface in Co/Cu multilayers. On the other hand, B_{hf} depends on the atomic distance, such that $\Delta B_{hf}/B_{hf} = -1.16 \Delta V/V$, where V is the atomic volume. Hence the shift of the NMR spectrum might be related to strain in the lattice. Finally, B_{hf} depends on local symmetry, namely $B_{hf} = 21.6$ T in fcc and $B_{hf} = 22.5$ T in hcp. A typical NMR spectrum from a $[Co(12.3 \text{ Å})/Cu(42 \text{ Å})]_{\times 40}$ multilayer system, prepared by electron-beam evaporation in ultrahigh vacuum [34], is shown in Figure 25. Spectra for different thicknesses of Co (or Cu) are qualitatively similar but shifted in frequency.

It is apparent that the main line in the spectrum is situated close to the resonance frequency of 217 MHz of the fcc bulk-surrounded Co. There is no higher frequency; hence the amount of hcp Co (which is expected at about 225 MHz) is very small. The main line is thus assigned to the bulk Co layers. Experiment shows that the intensity ratio between this line

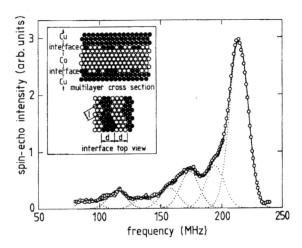

Fig. 25. ^{59}Co NRM spin-echo spectrum for a $Co_{12}Cu_{42}$ multilayer. Reproduced with permission from [34], copyright 1991, American Physical Society.

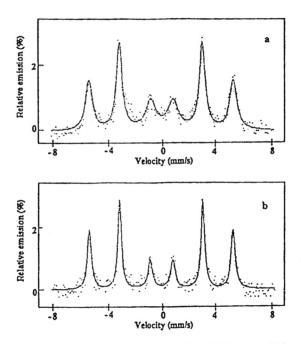

Fig. 26. CEMS spectra of an Fe/Au superlattice with Fe layer thicknesses of (a) 18 Å and (b) 48 Å. Reproduced with permission from [35], copyright 1997, IEEE.

and the lower frequency part of the spectrum linearly increases with the Co layer thickness d_{Co}, which implies that the lower resonance frequencies can be assigned to Co atoms at the interfaces, where one or more nearest-neighbor Co atoms are replaced with Cu. This assignment is supported by the observation that the linewidths of the satellite lines have half the value for the main peak. In other words, the spin-spin relaxation time T_2 at low frequencies is twice the value measured for the bulk Co layers, indicating a different origin of the NMR signals. The spectrum is well fitted by Gaussian lines with a constant separation of about 19 MHz, which is close to the previously mentioned shift of 18 MHz, and hence the satellite lines can be assigned in turn to interface Co atoms having 11 to 6 near neighbors. For a flat interface one would expect a two-line spectrum, with a main line for the bulk where Co atoms have 12 near neighbors and a single interface line, corresponding to 9 Co neighbors, as the growth was [111] oriented. Spectra fitting with a model shown in the inset of Figure 25, involving monoatomic steps at an average separation d and average length l, both on the order of two atomic distances, gives a better agreement than the complete random distribution, which would imply a flat but diffused interface layer. This means that in the interface layer the Co atoms are mainly surrounded by Co, and the Cu atoms by Cu.

CEMS spectra of Fe/Au prepared by rf diode sputtering have been recorded to obtain information about Fe atoms present both in the bulk of the layer and at the interface [35]. The hyperfine field for all of the samples studied has been found to be close to that of bulk Fe. In agreement with the M-H loop measurements, the hyperfine field was found to be parallel to the film plane, as indicated by the second and fifth peaks in all CEMS spectra, which are the most intense. However, as the Fe layer thickness decreases below 20 Å, there is an increase in both linewidth and the isomer shift, which is apparent from Figure 26, and is attributed to the presence of two types of Fe sites in the thinner layer. One type was assigned to Fe atoms in the bulk of the Fe layer, whereas

a second type would correspond to those close to the interface, having some Au atoms as first neighbors. The effect of this second type is more pronounced as the Fe layer thickness decreases.

The perpendicular magnetic anisotropy in Tb/Fe multilayers has been suggested to originate from amorphous alloyed interfaces [36], with a stronger intermixing occurring at the bottom (Fe deposited on Tb) than at the top (Tb deposited on Fe) interface [37]. The effect of each type of interface has been investigated separately by placing 50-Å-thick Ag layers at the top or bottom interface only. CEMS spectra recorded at 80 K [38] show that the relative spectral areas of the interface component are 25%, 28%, and 33% for the top, bottom, and both interfaces, respectively, (see Fig. 27). Hence, the average Fe thickness per interface that is affected has been derived to be 4.4 Å, 4.9 Å, and 5.3 Å, respectively. A canted average Fe spin orientation of 35°, 32°, and 30°, with perpendicular Fe spin components, was calculated from Eq. (20) for the S2 (top) ([Tb(14 Å)/Ag(50 Å)/Fe(35 Å)]$_{\times 10}$), S1 (bottom) ([Tb(14 Å)/Fe(35 Å)/Ag(50 Å)]$_{\times 10}$), and S3 (both) ([Tb(14 Å)/Fe(35 Å)]$_{\times 10}$) samples, respectively. The conclusion is that the larger interfacial roughness, due to a larger Fe-Tb intermixing, of the bottom interface S1 induces a smaller canting angle and, hence, a stronger perpendicular magnetic anisotropy than the top interface S2.

The effects of both the buffer layer and the growth techniques on the interfacial quality have been demonstrated by comparing the NMR spectra of the Co/Cu interface from multilayers deposited on a Si substrate [39]. One fcc or hcp ^{59}Co bulk line and a single interface line would be expected from ideal planar interfaces. The NMR spectrum obtained from

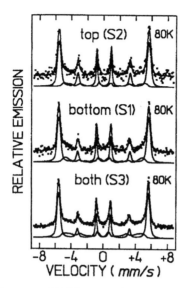

Fig. 27. CEMS spectra at 80 K from samples S2, S1, and S3 of Tb-Fe-Ag multilayers. Reproduced with permission from [38], copyright 1998, IEEE.

multilayers evaporated on a single crystal Cu buffer (Fig. 28a) is close to this ideal interfacial spectrum, with a broad line centered at approximately 222 MHz, which is an envelope for the fcc Co line at 217 MHz and the hcp Co line at about 228 MHz, where Co is surrounded by 12 Co nearest neighbors. The third satellite line at ~168 MHz is assigned to an abrupt interface, corresponding to the atomic environment where three Co atoms have been replaced by Cu. The spec-

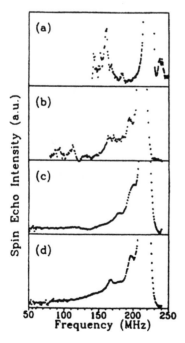

Fig. 28. ^{59}CoNMR spectra of Co/Cu multilayers grown on Si substrates by (a) UHV evaporation with a single-crystal Cu buffer, (111) texture, (b) UHV evaporation with a Cu buffer, (c) rf diode sputtering with an Fe buffer, and (d) dc magnetron sputtering with a Ta buffer, (111) texture. Reproduced with permission from [39], copyright 1993, Elsevier Science.

trum from Figure 28b follows the same separation regularities discussed in Figure 25. A Co/Cu multilayer with extensive interfacial mixing is obtained by rf diode sputtering (Fig. 28c), as indicated by a strong first satellite line (11 Co NN), a smaller second satellite, and an extended tail. The presence of the third satellite line when the multilayer is deposited by dc magnetron sputtering, as in Figure 28d, suggests the presence of islands of ideal planar interface separated by alloy-like regions.

The main purpose of studying metallic multilayers is to examine the link between the structure of the interface and the magnetic/magnetotransport properties. NMR studies have established an unambiguous correlation between magnetoresistance and structure, showing that interfacial changes at the atomic scale, depending on the deposition conditions, modify the spin-dependent scattering and therefore the magnetoresistance. In sputtered Co/Cu multilayers the interface structure derived from ^{59}Co NMR spectra has been correlated with the magnetoresistance ratio $\Delta\rho/\rho$, as determined by the underlying deposition conditions, namely the sputtering bias voltage. The spectra recorded from $[Co(10 \text{ Å})/Cu(10 \text{ Å})]_{\times 100}$ superlattices [40], prepared by ion-beam sputtering from Co and Cu targets on MgO (110) single-crystal substrates, at four typical values of the acceleration voltage V_B (400 eV ($\Delta\rho/\rho = 5.5\%$), 600 eV ($\Delta\rho/\rho = 50.6\%$), 1 keV ($\Delta\rho/\rho = 33.15\%$), and 1.4 keV ($\Delta\rho/\rho = 1.73\%$)), are shown in Figure 29. It can be seen that the maximum magnetoresistance ratio corresponds to the smallest layers of interface alloy regions. This indicates that a small degree of interface alloy region is desirable, because in these regions the magnetic properties of the Co atoms and the interlayer coupling are changed to decrease the spin-dependent scattering, and this is considered to be essential for causing a giant magnetoresistance effect.

The influence of the sputtering parameters on the phase structure and magnetic texture of $Nd_2Fe_{14}B$ thin films, grown by dc magnetron sputtering, has been investigated by CEMS [41] to optimize the magnetic characteristics of a multilayered two-phase magnet, consisting of thin alternating $Nd_2Fe_{14}B/Fe$ hard and soft magnetic layers. The CEMS spectra of 700-nm-thick $Nd_2Fe_{14}B$ films deposited under variable argon pressure on boron-silicate-glass substrates, heated at 825 K, all consist of a magnetic contribution from the $Nd_2Fe_{14}B$ phase and a nonmagnetic contribution assigned to Fe atoms in Nd-rich regions. At low pressure, the thin film contains almost only the $Nd_2Fe_{14}B$ phase, which is highly textured with magnetic moments close to the normal of the sample, whereas for deposition pressures above 0.1 mbar, the magnetic texture is lost, as the CEMS spectra are fitted by $\theta = 54°$, according to Eq. (20), which corresponds to a random distribution of the directions of the magnetic moments. Here the central paramagnetic contribution is higher, suggesting the decoupling of Fe grains by Nd-rich regions. Among the spectra of the trilayers $1.15d$ [$Nd_2Fe_{14}B$]/dFe/$1.15d$ [$Nd_2Fe_{14}B$] deposited with $d = 180$ nm, 30 nm, and 7 nm, given in Figure 30a, b, and c, respectively, it is interesting to compare the last two spectra, where CEMS information on both the $Nd_2Fe_{14}B$ and

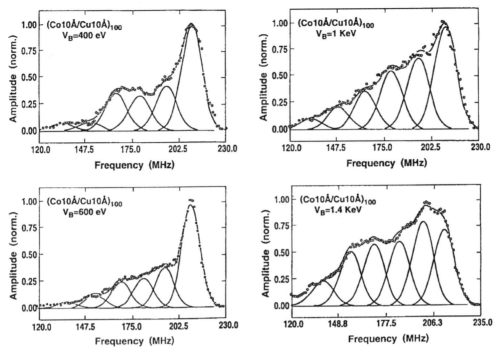

Fig. 29. ^{59}Co NMR spin-echo spectra of superlattices with 10 Å Co and 10 Å Cu thicknesses for different V_B values. Reproduced with permission from [40], copyright 1993, Elsevier Science.

Fe layers is available. The Fe moments lie in the plane of the 30-nm-thick layer, as also observed in Fe/Tb multilayers [42], and there is a pronounced magnetic texture normal to the $Nd_2Fe_{14}B$ layers. The α-Fe phase becomes less crystallized as the layer thickness decreases down to 7 nm, and a related

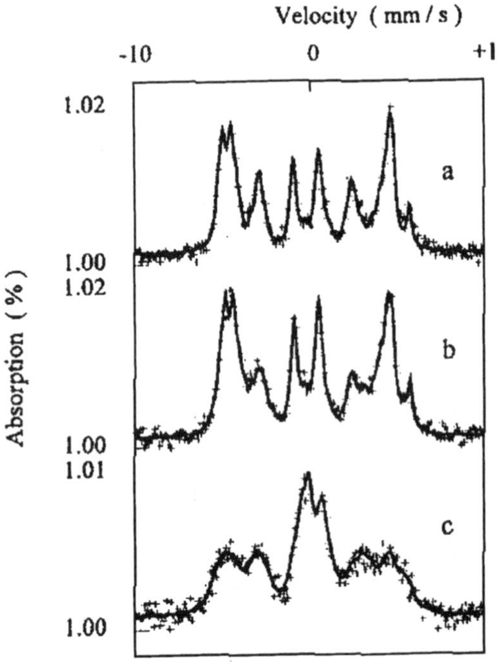

loss of texture of $Nd_2Fe_{14}B$ is apparent, because of the poor crystallization.

Thermal annealing of metallic multilayers at moderate temperatures (200–400°C) may result in considerable modification of their magnetic and magnetotransport properties, due to some structural changes. A systematic NMR study of Co/Ag multilayers, with Co thicknesses of 8 Å and 15 Å, illustrated in Figure 31, indicates that the main annealing effects are strain relaxation, phase separation, and a smoother interface promotion [43]. All of the spectra exhibit a bulk (>20 T) and an interface (<20 T) line, both much broader than similar lines found in Co/Cu multilayers, and this is assigned to the presence of a mixture of fcc Co, hcp Co, and stacking faults. Strain relaxation causes a visible shift in the main line to higher frequencies, due to a growth of Co clusters, and improves the interface quality, as seen from the flatter interface line at 17.5'T. However, the mechanism for the observed change in magnetoresistance has been related to the change in the cluster size and separation, rather than to strain relaxation.

The effect of thermal treatment on interfacial phenomena has been studied by CEMS in FeAl multilayers, deposited by electron-beam evaporation [44]. A comparison between the spectra relative to (a) Fe(30 Å)/Al(20 Å) and (b) Fe(15 Å)/Al(20 Å) multilayers is illustrated in Figure 32. The multilayers were thermally treated in vacuum at 420 K for 24 h. It is apparent that the paramagnetic contribution increases, and the ferromagnetic sextet is replaced by a superparamagnetic central line, assigned to fine particles, as the Fe layer thickness decreases. In other words, a higher hyperfine field, associated with a Fe-Al solid solution, prevails in the thicker Fe layer. This indicates that, with a decreasing in the thickness of the Fe layer to 15 Å, there is a reduction of the iron

Fig. 30. CEMS spectra of the trilayer 1.15d [$Nd_2Fe_{14}B$]/dFe/1.15d [$Nd_2Fe_{14}B$] deposited by sputtering at 0.1 mbar, with an Fe layer thickness d of (a) 108 nm, (b) 30 nm, and (c) 7 nm. Reproduced with permission from [41], copyright 1999, Elsevier Science.

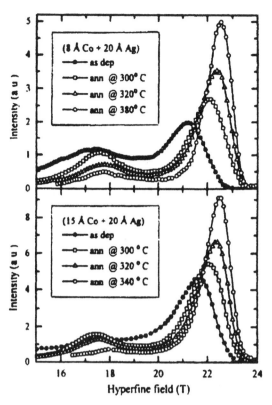

Fig. 31. NMR spectra of Co/Ag multilayers as a function of annealing temperature. Reproduced with permission from [43], copyright 1995, Elsevier Science.

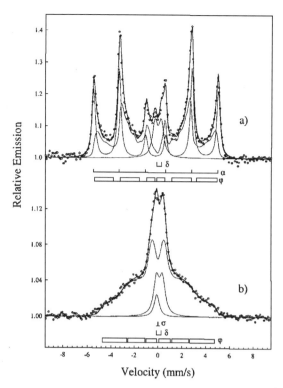

Fig. 32. CEMS spectra for thermally treated (a) Fe(30 Å)/Al(20 Å) and (b) Fe(15 Å)/Al(20 Å) multilayers. Reproduced with permission from [44], copyright 1999, Elsevier Science.

grain core, presumably caused by Al diffusion. The presence of the superparamagnetic contribution supports the suggestion that the particle size distribution shifts toward smaller values, and this is consistent with the magnetic results, which show a decrease in the saturation magnetization from 1.7 MA/m for Fe(30 Å)/Al(20 Å) to 1 MA/m for Fe(15 Å)/Al(20 Å).

Considerable research has been done in recent years on the structural and magnetic properties of metallic multilayers with technical applications. For instance, FeXN (X = Ta, Al) films and multilayer systems like FeN/SiO$_2$, FeTaN/SiO$_2$, and FeAlN/SiO$_2$(Al$_2$O$_3$) have been found to be potential pole piece materials for magnetic write heads. It was observed that FeTaN films exhibit a combination of high saturation magnetization and good soft magnetic behavior, which is suitable for high-density recording heads. As storage density increases, any new pole piece material must be characterized by low coercivity, high permeability over a wide frequency range, small magnetostriction, and appropriate thermal stability. Improved magnetic properties over single-layer films have been found for multilayers, where FeTaN films are separated by high-resistivity spacer layers. Recently, a new multilayer [FeTaN/TaN]$_n$, deposited by dc magnetron sputtering, where high-resistivity TaN spacer layers of 30–40 Å provide a better electrical insulation than SiO$_2$ or Al$_2$O$_3$, was reported to be a good candidate for high-magnetization pole piece fabrication, provided magnetostriction is lowered [45]. Phase analysis by transmission spectroscopy and CEMS was used as a local

probe for the optimization of the deposition conditions, magnetic structure, and thermal stability of these multilayers.

The [FeTaN(3200 Å)/TaN(50 Å)]$_n$ multilayers were deposited up to a total thickness of 1.62 μm on water-cooled mica substrates. For FeTaN film deposition, 3-inch Fe-10 wt% Ta alloy targets were used, in a 3-mTorr Ar + N$_2$ atmosphere, at variable nitrogen partial pressure p_{N_2} from 0 to 0.12 mTorr. An aligning field of 20 Oe was applied during deposition. High-resistivity TaN films were deposited from 3-inch Ta targets in a 5-mTorr Ar + N$_2$ atmosphere, at constant $p_{N_2} = 0.19$ mTorr. The nitrogen content of FeTaN films was determined to be in the atomic ratio Fe:Ta:N of 93.8:2.4:3.8 for a partial pressure $p_{N_2} = 0.12$ mTorr used during deposition [45]. Transmission ^{57}Fe Mössbauer spectra were obtained at room temperature and high temperature on a constant acceleration spectrometer that utilized a ^{57}Co/Rh source. The CEMS measurements were performed at room temperature with a He-4% CH$_4$ gas-flow electron counter, depicted in Figure 23a. The resulting spectra were calibrated with α-Fe foil at room temperature and fitted with the integrated least-squares computer program Mosswin [46].

All of the transmission Mössbauer spectra, recorded at room temperature, show a dominant contribution from the magnetic sublattice with a HF of 33.4 T (α-Fe phase), irrespective of the deposition conditions (Fig. 33). However, larger linewidths than those expected for metallic Fe fit this subspectrum, indicating a spreading in magnitude of the HF values, and this is consistent with the fine grain size calculated from

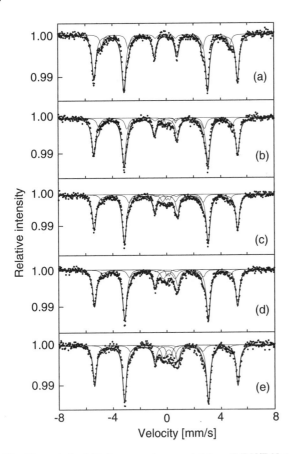

Fig. 33. Transmission Mössbauer spectra recorded from FeTaN/TaN multi-layers deposited at variable nitrogen partial pressure: (a) p_{N_2} = 0, (b) p_{N_2} = 0.05 mTorr, (c) p_{N_2} = 0.07 mTorr, (d) p_{N_2} = 0.10 mTorr, (e) p_{N_2} = 0.12 mTorr.

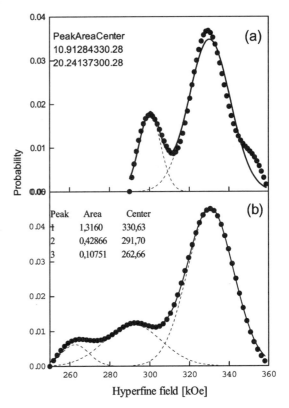

Fig. 34. Hyperfine field distribution in the transmission spectra from FeTaN layers.

the X-ray diffraction patterns [15]. Additional magnetic phases might be assigned to Fe nitrides, as it is known that iron and nitrogen may form three compounds that are ferromagnetic at room temperature: Fe_8N, Fe_4N, and ε-$Fe_{3.2}N$. Ferromagnetic iron nitride compounds Fe_xN ($x > 3$) have high mechanical hardness, large saturation magnetization, and chemical stability superior to those of pure Fe, in agreement with the magnetic behavior of the multilayer system, either as cast or after annealing, as determined by vibrating sample magnetometry [45]. To identify the Fe_xN magnetic phases, the HF distribution of the magnetic sextets has been calculated with the Mosswin fitting program, and two kinds of distribution have been observed, as shown in Figure 34. The HF distribution curves consist of two Gaussian contributions if the nitrogen partial pressure during FeTaN layer deposition does not exceed $p_{N_2} = 0.5$ mTorr, corresponding to the transmission spectra given in Figure 33a and b and exhibits a third contribution if p_{N_2} increases to 0.12 mTorr, as in Figure 33c–e. The dominant contribution is distributed in both cases about an average HF of 33 T, with the same standard deviation, indicating the α-Fe phase origin. A second magnetic phase, with a narrow distribution about an average HF of 30 T, can be identified in Figure 34a as iron nitride (Fe_xN, with $x > 3$), with one interstice in the Fe nearest neighborhood occupied by N. As

the nitrogen partial pressure increases, this phase is replaced by two components, corresponding to broad HF distributions about 30 T and 26.2 T average field values. These are likely to be due to disorder in the occupation of interstitial sites by nitrogen atoms in the Fe_xN structure, which approaches an iron-nitrogen solid solution where Fe occupies two distinct positions, with one and two interstitial N near neighbors. Hence, the calculated HF distributions indicate the formation of ferromagnetic Fe_xN nitrides in the FeTaN layers, the relative concentration of which was determined by the standard fitting procedure for the alloy spectra. The line broadening found for the Fe_xN subspectra is consistent with some disorder in the interstitial distribution of nitrogen atoms. The dependence of the calculated relative intensities, which give the ferromagnetic phase content, on the partial pressure of nitrogen during deposition is represented in Figure 35. In these Mössbauer measurements, the γ-rays were perpendicular to the plane of the sample, and the spin orientation in the FeTaN layers with respect to the γ-ray direction was determined from the line intensity ratio in the Zeeman sextets 3:x:1:1:x:3 (Eq. (20)). For all of the magnetic phases a 3:4:1:1:4:3 relative intensity ratio was found, indicating in-plane magnetic anisotropy. However, Mössbauer spectra indicate some structural disorder associated with Fe_xN, and this might account for the relatively high easy axis dispersion ($\alpha_{50} \sim 20°$) [45], which is a drawback for head fabrication.

Transmission Mössbauer spectra shown in Figure 33 also indicate the presence of paramagnetic Fe, which can be

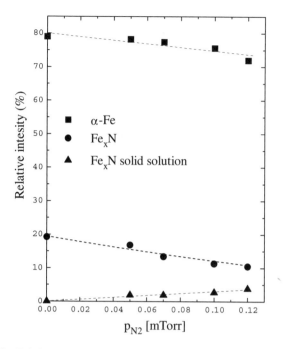

Fig. 35. Relative concentration of the magnetic phases versus nitrogen partial pressure.

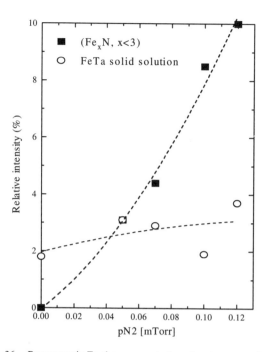

Fig. 36. Paramagnetic Fe phase concentration at various p_{N_2} values.

understood as well in terms of nitride formation. Because the Curie temperature of Fe nitrides decreases with increasing nitrogen content below room temperature, a broad central doublet (QS = 0.57 mm/s and IS = 0.3 mm/s) at a concentration that increases with increasing p_{N_2} during deposition might be assigned to ε-Fe$_x$N with $2 < x < 3$ [47]. There is also a second Fe phase (QS = 0.22 mm/s and IS = 0.2 mm/s) with a 2% contribution, which seems to be unrelated to the deposition conditions. This last contribution is assumed to be due to FeTa solid solution, as suggested by a 2.4% concentration determined for Ta. The relative content of the paramagnetic phases is represented in Figure 36 as a function of nitrogen partial pressure.

The information provided by the transmission Mössbauer is integrated over the five FeTaN layers, including the interfacial regions between the magnetic and spacer layers, whereas that derived from the CEMS spectra is limited to a maximum depth of 3000 Å and hence is relevant for the upper FeTaN layer structure only. All of the CEMS spectra recorded from the same systems are similar, as seen in Figure 37, showing a higher relative contribution from the ferromagnetic Fe nitride phase than the average derived from the transmission spectra. There is also a striking feature with the virtual absence of any paramagnetic contribution in the upper FeTaN layer, up to the relevant depth of about 3000 Å. Phase analysis results on α-Fe and Fe$_x$N concentration, given by CEMS spectra, are comparatively represented in Figure 38, in terms of p_{N_2}. It can be seen that the relative content in metallic Fe and ferromagnetic Fe nitride derived from the CEMS spectra is almost independent of the deposition conditions, whereas this content decreases with increasing p_{N_2} in the transmission spectra

(see Fig. 35). This is due to an increase in the relative ε-Fe$_x$N ($2 < x < 3$) fraction (Fig. 36), which is proportional to the square of p_{N_2}. In other words, the paramagnetic ε-Fe$_x$N nitride is probably formed at the interface region between the magnetic layer FeTaN and the TaN spacer layer. This result is

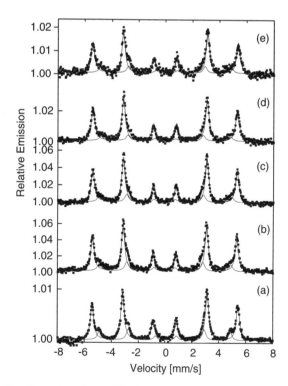

Fig. 37. Conversion electron Mössbauer spectra obtained from FeTaN/TaN multilayers deposited at variable nitrogen partial pressure: (a) $p_{N_2} = 0$, (b) $p_{N_2} = 0.05$ mTorr, (c) $p_{N_2} = 0.07$ mTorr, (d) $p_{N_2} = 0.10$ mTorr, (e) $p_{N_2} = 0.12$ mTorr.

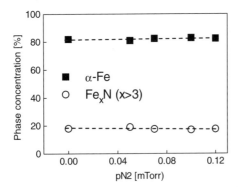

Fig. 38. Fe phase concentration in FeTaN layers derived from CEMS Mössbauer spectra.

likely to be related to the positive magnetostriction of the system, which is on the order of 5×10^{-6} for $p_{N_2} = 0.12$ mTorr [45] and has to be minimized to improve the magnetic sensor response. It has been shown by experiment that magnetostriction increases with increasing nitrogen content and the formation of nitride phases. It is reasonable to assume that the interfacial ε-Fe$_x$N $(2 < x < 3)$ phase, revealed by the Mössbauer spectra, may account for this relatively high magnetostriction. Consequently, controlling the deposition conditions for the soft FeTaN layers might lower it.

Transmission spectra have also been recorded above room temperature, up to the Curie point of the material, from the multilayer system deposited at $p_{N_2} = 0.12$ mTorr, where the paramagnetic phase concentration is maximized. The temperature dependence of the relevant phase content is plotted in Figure 39. Some of the transmission spectra recorded at various temperatures up to 733°C, which is a value close to the Curie point of metallic Fe, are reproduced in Figure 40. The measurement cycle above room temperature corresponds to a thermal treatment of the system, whose irreversible effects are apparent from the postannealing recorded spectrum, where the phase assigned to a FeTa solid solution is still present. These results demonstrate a good compositional stability of the multilayer system up to 430°C, which is consistent with the fact that hexagonal iron nitrides will only decompose to

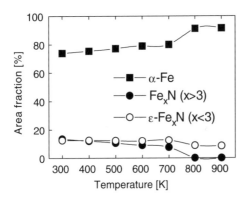

Fig. 39. Temperature dependence of phase concentration derived from transmission Mössbauer spectra.

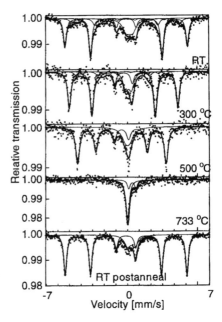

Fig. 40. Transmission Mössbauer spectra above room temperature.

Fe and nitrogen at 452°C. The temperature dependence of the magnetic splitting of the dominant phase, which for Fe and Fe alloys is proportional to magnetization, follows the expected Brillouin type dependence, typical for metallic Fe. This accounts for the stability of the magnetic behavior at temperatures below 430°C, which was also indicated by vibrating sample magnetometer measurements after sample annealing at 350°C.

6. NUCLEAR RESONANCE SPECTROSCOPY IN AMORPHOUS, NANOSTRUCTURED, AND GRANULAR THIN FILMS

There is a large class of magnetic materials with a common feature in their ^{59}Co NMR spectra and ^{57}Fe Mössbauer spectra that consists of broadening of the spectral lines, and this is caused by a spread in the values of the hyperfine fields. This distribution is determined in turn by the range of local surroundings of the Co and Fe sites, respectively, so that the magnetic detail of these solids may be probed by analysis of broadening.

Figure 41 shows the NMR spin echo spectrum of ^{59}Co in the amorphous (open circles) and recrystallized (closed circles) Co$_{78}$Zr$_{22}$ thin film alloy [48]. It can be seen that recrystallization results in a narrower spectrum, with a well-pronounced structure associated with the particular crystallographic Co sites, which belong to the fcc Co and the crystalline Co$_{23}$Zr$_6$ phases. The broad signal, with a linewidth of about 33 MHz, is typical for an amorphous phase, where a ^{57}Co hyperfine field distribution is assumed. This experimental spectrum was fitted with equally spaced Gaussian lines, each corresponding to Co with 1, 2, 3, ... Zr neighbors, respectively. The component line intensities follow the binomial dis-

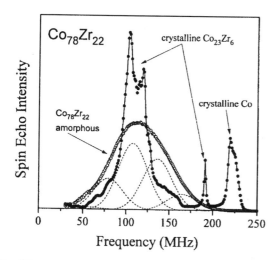

Fig. 41. ^{59}CoNMR spin-echo spectrum of $Co_{78}Zr_{22}$. Reproduced with permission from [48], copyright 1996, Elsevier Science.

tribution law for 12 near neighbors of Co, proving a random, densely packed structure of the film.

The form of the hyperfine field distribution can be extracted by analyzing the broadening of Mössbauer lines with standard fitting procedures [46, 49, 50], all based on trigonometric series expansions of the distribution probability $P(H)$. The broadened Mössbauer lines of a series of spectra recorded from $Fe_{81}B_{13.5}Si_{3.5}C_2$ amorphous ribbons between room temperature and 410°C (Fig. 42a) indicate that below the ordering temperature of 405°C, there is a spatial distribution of the exchange interactions, plotted in Figure 42b, with an almost linearly decreasing linewidth as temperature increases [51].

A similar behavior has been found for granular materials, as shown in Figure 43 (see also the inset), where the hyperfine field distributions were derived from the room temperature Mössbauer spectra of the as-cast and annealed samples of $Fe_{25}Cu_{75}(\%)$ granular alloy, prepared by splat cooling [52]. The distribution mainly consists of a Gaussian, centered about an average hyperfine field (which increases upon annealing toward the value of 33.3 T, characteristic of bulk Fe ferromagnetism). The line broadening of the Mössbauer ferromagnetic subspectrum can be discussed, below the blocking temperature, which is defined in Figure 44, in terms of the so-called collective magnetic excitations, where the magnetization vector of each Fe grain may fluctuate in directions close to its own easy axis faster than the time scale of Mössbauer spectroscopy, so obtaining an observed magnetic splitting H_{obs} distributed about the average HF value $\langle H \rangle$. When the thermal energy $k_B T$ (where T is room temperature in our case) is much smaller than the anisotropy energy KV, where K is the effective anisotropy constant and V is the mean volume of the particle, the magnitude of the magnetic splitting becomes [53]

$$\frac{H_{obs}}{H_0} = 1 - \frac{k_B T}{2KV} \qquad (34)$$

This linear temperature dependence is a reasonable approximation of the Brillouin-type dependence of magnetization for

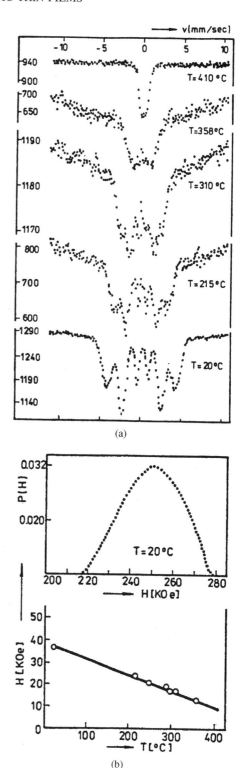

(a)

(b)

Fig. 42. Mössbauer spectra of a $Fe_{81}B_{13.5}Si_{3.5}C_2$ ribbon sample at variable temperatures (a) and their hyperfine field distribution (b). Reproduced with permission from [51], copyright 1990, Kluwer.

T below the blocking temperature (see Fig. 44). Here H_0 is the saturation field, given by the magnetic splitting in a bulk Fe ferromagnet at very low temperature.

Because H_{obs} is dependent on the grain size, the broadening of the Mössbauer lines due to a HF distribution also implies

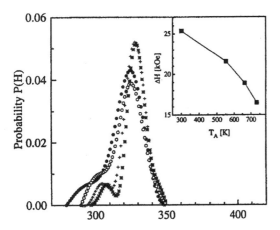

Fig. 43. Hyperfine field distribution $P(H)$ for a $Fe_{25}Cu_{75}$ granular alloy. Reproduced with permission from [52], copyright 1996, Elsevier Science.

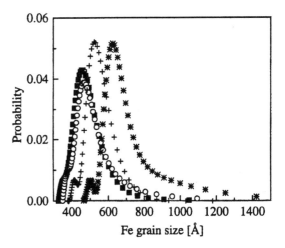

Fig. 45. Grain size distribution evaluated for $Fe_{25}Cu_{75}$ alloy. Reproduced with permission from [52], copyright 1996, Elsevier Science.

a broad size distribution of the ferromagnetic grains. Equation (34) shows that the mean value of the Fe grain diameters $\langle D \rangle$ might be related to the average HF value $\langle H \rangle$. Because each hyperfine field H_{obs}, measured at room temperature T, can be associated with an average volume V of the Fe grains, the hyperfine field distribution represented in Figure 43 can be easily transformed into a distribution of grain diameters, described by

$$D = \left[\frac{3k_B T}{\pi K} \frac{H_0}{H_0 - H_{obs}} \right]^{1/3} = \langle D(T_A) \rangle \left[\frac{H_0 - \langle H(T_A) \rangle}{H_0 - H_{obs}} \right]^{1/3} \tag{35}$$

where, for simplicity, it is assumed that the Fe ferromagnetic grains are of spherical shape and that the anisotropy constant K is not affected by annealing. It is necessary to calibrate the graphical representation of Eq. (35), using, for instance, the $\langle D \rangle$ values estimated by the Scherrer analysis of the Fe line profile from the X-ray diffraction patterns at different annealing temperatures T_A. The result, plotted in Figure 45, is similar to that predicted by the log-normal distribution,

$$f(D) = \frac{1}{\sqrt{2\pi}\sigma D} \exp\left[\frac{\ln^2(D/\langle D \rangle)}{2\sigma^2} \right] \tag{36}$$

Because the exchange interactions do not extend much beyond the nearest neighbors, it is also interesting to discuss the structural characteristics of amorphous materials in terms of the

grain size, which in this case is defined by the range of the local magnetic order. This concept has been used, for instance, to compare the coercivity data from $Fe_{81}B_{13.5}Si_{3.5}C_2$ amorphous ribbons and laser-ablated films with the same nominal composition [54], on the grounds of the grain refinement promoted by laser ablation of the ribbon target material. The SQUID data are reproduced in Figure 46. The CEMS spectra from both the film and the ribbon target exhibit the broadened six-line pattern, as seen in the insets of Figure 47 and, hence, correspond to a distribution of the hyperfine fields at the Fe nuclei. The result plotted in Figure 47 indicates a broad distribution of ferromagnetic grain diameters and hence of the range of local magnetic order, with a shape similar to that of the log normal distribution, for both the meltspun ribbon material and the laser-ablated films. It is clear, however, that there is a shorter range for the exchange interactions in the laser-ablated films than in the meltspun material of the ribbon target, and this is consistent with the conclusion derived from the SQUID data, namely a reduction of the saturation magnetization, associated with an increase in the coercive field for the thin films.

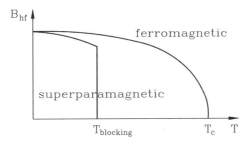

Fig. 44. Superparamagnetic relaxation of small ferromagnetic particles.

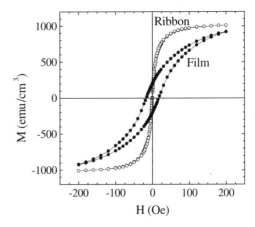

Fig. 46. SQUID data at 4 K from $Fe_{81}B_{13.5}Si_{3.5}C_2$ ribbon and film samples [54].

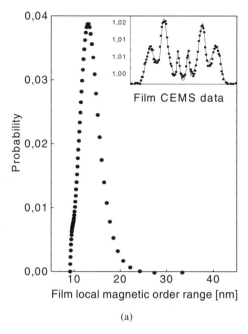

(a)

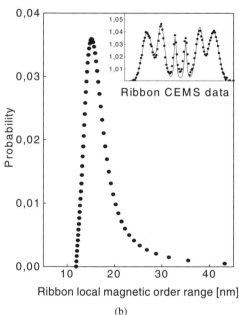

(b)

Fig. 47. Hyperfine field distribution (a) and short-range order distribution (b) in $Fe_{81}B_{13.5}Si_{3.5}C_2$ ribbon and thin film, as derived from CEMS spectra [54].

Ferromagnetic nanocrystalline materials, prepared by annealing from the amorphous state, have been found to exhibit enhanced soft magnetic properties at high frequency (high effective permeability, low coercivity, and small magnetic core loss) compared with those in amorphous form. CEMS spectra from ultrathin $Fe_{83}B_9Nb_7Cu_1$ alloy (Fig. 48) show how the amorphous component has been gradually transformed, by flash annealing [55], into the microstructure of a nanocrystalline phase. Up to the annealing temperature of 623 K (Fig. 48a and b), a typical broadened six-line pattern indi-

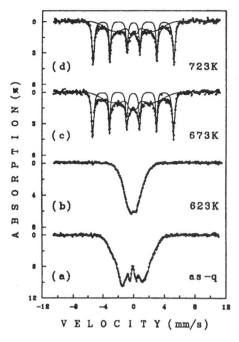

Fig. 48. Room temperature spectra of ultrathin $Fe_{83}B_9Nb_7Cu_1$ alloys, ascast (a) and after flash annealing at 623 K (b), 673 K (c), and 723 K (d). Reproduced with permission from [55], copyright 1997, IEEE.

cates disordered atomic arrangements, in which the strength of the hyperfine field changes from one site to another because of a structurally unequivalent Fe environment. Above 673 K the amorphous component coexists with α-Fe grains about 6 nm in diameter. The study of the soft magnetic characteristics reveals that the optimum annealing temperature, leading to a maximum value of the effective permeability at 1 MHz, is 723 K.

Desirable soft magnetic properties can be achieved by reducing the grain size in Fe-based alloys, through the addition of a B, N, or C element. As-cast Fe-Zr-N thin films, prepared by rf magnetron sputtering [56], exhibit soft magnetic properties with an effective permeability of 1400 or more at 1 MHz. The room temperature CEMS spectrum of the as-deposited $Fe_{76}Zr_8N_{16}$ thin film, given in Figure 49, provides a probe for the local structure: three kinds of magnetic Fe sites, with fairly good crystalline quality and almost the same concentration, are assigned to α-Fe (HF = 33.1 T), to Fe atoms surrounded by interstitial nitrogen atoms (HF = 32.1 T), and to Fe atoms surrounded by Zr atoms (HF = 27.6 T), respectively. It has been assumed that substitutional Zr atoms interact chemically with iron, and this leads to a decrease in the saturation magnetic flux density of α-Fe.

However, an increase in coercivity has been observed as the grain size is reduced by thin film deposition from bulk targets with good soft magnetic properties. Soft NiZn ferrite films, deposited by laser ablation with stoichiometries close to that of the target, have been studied by comparison to the bulk material, by CEMS and VSM [57]. From the magnetization curves of NiZn ferrite, shown in Figure 50, the film saturation magnetization was found to be 660×10^5 J/T/m³,

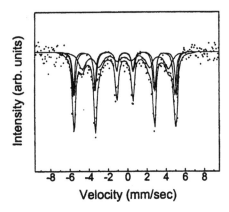

Fig. 49. Room temperature Mössbauer spectrum for an as-cast Fe$_{76}$Zr$_8$N$_{16}$ nanostructured thin film. Reproduced with permission from [56], copyright 1998, American Institute of Physics.

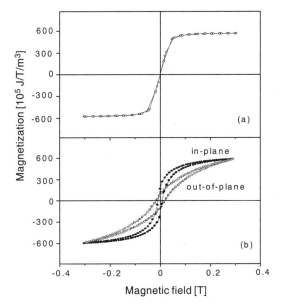

which is close to the value for the bulk material, as expected from stoichiometric NiZn ferrite. The coercive field of about 25.5 kA/m is larger than the values found for the target, and this can be attributed to a decrease in the crystallite size. This conclusion is supported by the Mössbauer data analysis. The transmission spectrum from the NiZn target material (Fig. 50c) shows typical relaxation features, which appear as two sextets, $H(A) = 36.4$ T and $H(B) = 30.6$ T, and a quadrupole split doublet ($\Delta E_Q = 1.31$ mm/s, IS $= 0.35$ mm/s). The corresponding CEMS parameters of the film, $H(A) = 35.7$ T and $H(B) = 30.3$ T, are consistent with the saturation magnetization data obtained by VSM. There is a linewidth increase in the CEMS spectra, with respect to the transmission data, which makes it possible to estimate the crystallite size distribution, as described above. The average crystallite size is found to be in the range of 300–600 nm, in agreement with the atomic force microscopy image analysis [58]. This crystallite size is smaller than in the bulk material and results in an increased number of low-angle grain boundaries, which act as pinning sites for domain walls and are responsible for larger H_c values [59].

NMR has been found to be appropriate for the study of microstructure, with the reduction of lateral dimensions of thin films and multilayers and the advent of magnetic nanowires, obtained by electrodeposition in the pores of polycarbonate membranes. Typical nanowires have a length on the order of 10 μm and consist of alternating metallic layers with diameters in the range of 30–500 nm and a thickness in the nanometer range, as illustrated in Figure 51a [60]. The NMR spectra shown in Figure 51b indicate that both fcc and hcp Co are present in the Co/Cu nanostructured nanowires, whereas the Co(100 Å)/Cu(100 Å) multilayers only exhibit the fcc Co signal [61]. From the relative intensity of the low-frequency satellites, arising from Co atoms with fewer than 12 neighbors, it has been possible to determine that Cu is present in the form of clusters of about 30 atoms.

The nuclear resonance study of thin granular films has intensified in recent years, based the intimate relation between structure and the magnetic response of the material, given the

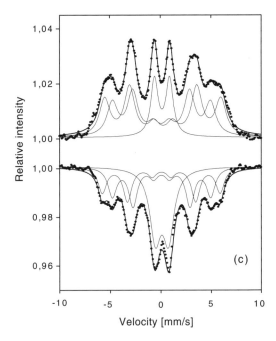

Fig. 50. VSM magnetization loops of NiZn ferrite target (a) and film (b), and the corresponding transmission Mössbauer spectrum from NiZn target and CEMS spectrum from laser-ablated film (c). Reproduced with permission from [57], copyright 1998, American Institute of Physics.

potential technical importance of giant (GMR) and tunneling (TMR) magnetoresistance effects in granular materials. It has been found [62] that the TMR effect, which can be used in magnetic field sensors and potentially in magnetic random access memories, is not limited to the trilayer junction structure where it was first observed [63]; it can also be obtained in granular solids consisting of ultrafine magnetic particles in an insulating matrix. Moreover, granular films are easier to prepare than trilayer junctions. Figure 52 shows room temperature Mössbauer spectra from Fe-Pb-O granular films, prepared by rf magnetron sputtering and then thermally treated at

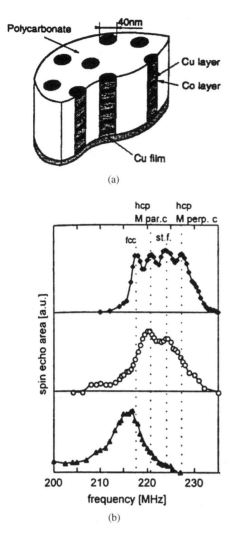

(a)

(b)

Fig. 51. Schematic representation of an array of Co/Cu nanowires in the cylindrical pores of a membrane (a). Reproduced with permission from [60], copyright 1999, Elsevier Science. ^{59}CoNMR spectra of Co/Cu nanowires under normal deposition conditions (top), Co/Cu nanowires deposited with highly diluted electrolyte (middle), and Co/cu multilayers (bottom) (b). Reproduced with permission from [61], copyright 1998, IEEE.

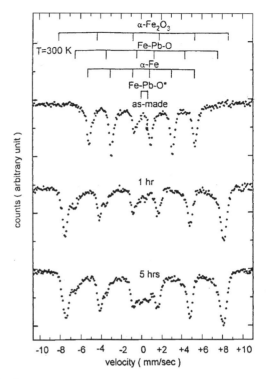

Fig. 52. Room temperature Mössbauer spectra from Fe-Pb-O granular films, ascast and annealed at 400° C for 1 and 5 h. Reproduced with permission from [64], copyright 1998, IEEE.

400° C, in an Ar atmosphere, for various annealing times [64]. The as-deposited film contains Fe grains in a PbO insulating matrix and exhibits a negligible TMR effect. After annealing, the spectrum mainly consists of two sextuplets, one assigned to Fe_2O_3 (HF = 51 T) and the other to a Fe-Pb-O magnetic phase (HF = 43.2 T), which preserve the same relative intensity irrespective of the annealing time. The magnetic tunneling barrier formed between PbO and Fe_2O_3 is assumed to enhance the TMR effect, which exhibits a maximum of about 10% for an annealing time of 1 h.

A Mössbauer spectroscopy study has been carried out in $(Fe_{50}Co_{50})_xAg_{1-x}$ (x is the volume fraction) granular films of about 400-nm thickness, deposited by electron beam coevaporation, which have been noticed by their isotropic GMR [65]. Figure 53a shows several room temperature Mössbauer spectra recorded from films with different volume concentrations x of $Fe_{50}Co_{50}$ ($x \geq 0.26$). All of the spectra can be fitted with three subspectra: a sextet, a broad sextet, and a doublet, respectively

assigned to a metallic phase consisting of clusters with hyperfine fields of about 35 T, a disordered ferromagnetic phase (disordered solid solution with distributed hyperfine field values), and a nonferromagnetic phase (attributed to Fe atoms in silver-rich regions). The variations with x of the relative resonance areas and hyperfine fields are represented in Figure 53b. The main effect of increasing the volume concentration x of $Fe_{50}Co_{50}$ is to increase the fractional volume of the ferromagnetic phase at the expense of the disordered ferromagnetic phase. Note that the resonant area of the disordered ferromagnetic phase has a sharp decrease from 50% to 36% followed by an almost steady behavior toward an average volume fraction of about 30%. The hyperfine field of the ferromagnetic clusters takes values above the 33.3 T value, typical for α-Fe, and this suggests the onset of a collective interaction between clusters that might be promoted by the decrease in their separation, as the volume fraction x increases. The subspectrum corresponding to the disordered ferromagnetic phase can be interpreted in terms of a hyperfine magnetic field distribution, with the average hyperfine field increasing, as x increases, from 25.4 T to 30.4 T. This indicates the presence of Ag atoms dissolved into the disordered phase to form a solid solution. In other words, a decrease in x favors a diminution of the average hyperfine field through the incorporation of nonmagnetic atoms, a process that exhibits saturation, and this appears to be almost obtained for $x = 0.26$. These results are consistent with the typical GMR effect, plotted in Figure 54 in the form of the dependence of magnetoresistance on the volume concentration x. At low magnetic metal concentrations the value of the negative magnetoresistance is small. It is caused by the low number

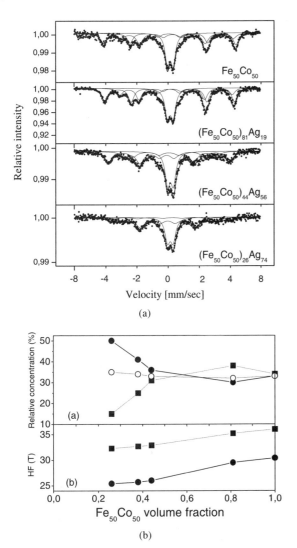

Fig. 53. Room temperature transmission Mössbauer spectra from $(Fe_{0.5}Co_{0.5})_x Ag_{1-x}$ granular films with various x values (a), and the dependence of the relative areas of the metallic (■) and the disordered (●) ferromagnetic phases on the volume fraction x (b) [65].

and small size of magnetic clusters. The maximum GMR of 9.7% is found for $x = 0.19$. A comparison between the results plotted in Figures 53b and 54 shows that the GMR enhancement at low x values can be associated with a high fractional

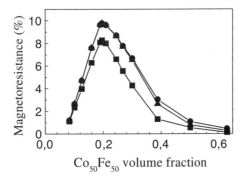

Fig. 54. Magnetoresistance $\Delta R/R\%$ as a function of $Co_{50}Fe_{50}$ volume fraction for granular films with Ag matrix in three different geometries, P (■), II (●), and T (▲) [65].

volume of the disordered ferromagnetic phase, where the two sets of results show almost the same dependence on x.

^{57}Fe CEMS spectra were also used as a local probe for studying the Fe clusters in ^{57}Fe-implanted silver films previously deposited by laser ablation. The films were implanted with ^{57}Fe or ^{56}Fe ions at doses of 1 to 7.5×10^{16} at/cm^2 and an ion energy of 50 to 180 keV. The mass resolution made possible the implantation of ^{57}Fe with only a ^{56}Fe contamination of less than 5% [66]. The calculated projected range of Fe in the Ag films is close to 530 Å, with a standard deviation of 315 Å. The local structure of the implanted iron was revealed by CEMS spectra analysis [67]. Figure 55a shows the CEMS spectra recorded at room temperature, where we may distinguish two different (structural/aggregation) states of the implanted Fe. The singlet with isomer shifts IS = 0.5 mm/s in the FeAg1 spectrum (see Fig. 55a and Table II) is likely to be due to individual Fe atoms dissolved in the fcc Ag matrix and, hence, surrounded by 12 nearest-neighbor Ag atoms. The large width suggests that Fe atoms are situated at damaged regions of the lattice. For the Fe/Ag film this phase contributes only 46% of the total Fe amount. We observe an additional doublet in the CEMS spectrum, here assigned to Fe atoms dissolved in Ag, but with one or more Fe atoms among their 12 near neighbors. The presence of both Fe and Ag atoms in this shell destroys the cubic symmetry at the central Fe site, resulting in a large quadrupole splitting, QS = 1.1 mm/s. From the relative intensity of the QS doublet we estimate that 37% of the Fe ions in Fe/Ag films are in this nonsymmetric phase. Thus, we obtain a relative content of 83% of Fe atoms dissolved in the Ag matrix (therefore in a nonferromagnetic state), as shown in Table II. The presence of the quadrupole interaction also indicates a less packed structure that should be related to the high ablation rate obtained from the Ag target. Finally, the singlet at IS = 0.08 mm/s has isomer shifts close to that of metallic iron and hence is assigned to sufficiently large Fe clusters, where each Fe atom has 12 Fe nearest neighbors, as in the bcc structure of normal α-Fe. It is reasonable to assume that the six-line pattern expected from metallic Fe is not seen at room temperature because the Fe clusters exhibit superparamagnetic relaxation, and the six-line pattern then collapses into a singlet. This metallic Fe phase has a relative intensity of 17% in the FeAg1 film.

The Mössbauer spectra from the FeAg2 and FeAg3 samples (see Fig. 55a) can be discussed in terms of the same three components, which have similar fitting parameters. The presence of isolated Fe atoms is only slightly diminished by an increase in the implantation dose, as indicated by the relative intensity of 38% coming from the singlet situated at IS = 0.43 or 0.44 mm/s with respect to α-Fe. However, the dominant contribution is now given by singlets assigned to large Fe clusters, with superparamagnetic behavior at room temperature, and the relative contribution of small clusters decreases with increasing dose. This indicates an enhanced diffusion due to the high concentration of vacancies, leading to clustering and precipitation. An increase of the GMR response for the samples FeAg2, FeAg3, and FeAg4, as compared with the

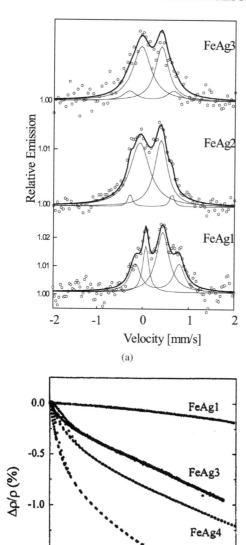

Table II. Mössbauer Parameters of ^{57}Fe-Implanted Ag Films

Sample	IS (mm/s)	QS (mm/s)	Area (%)	Phase
FeAg1	0.5	—	46	Isolated atoms
	0.36	1.1	37	Small clusters
	0.08	—	17	Large clusters
FeAg2	0.43	—	38	Isolated atoms
	0.2	1.04	5	Small clusters
	−0.09	—	55	Large clusters
FeAg3	0.44	—	38	Isolated atoms
	0.2	1.09	13	Small clusters
	−0.06	—	49	Large clusters

From [68].

Fig. 55. Room temperature conversion electron Mössbauer spectra for FeAg1, FeAg2, and FeAg3 films (a), and magnetoresistance of the FeAg films taken at 10 K for magnetic fields up to 1 Tesla (b). Reproduced with permission from [67], copyright 1998, Elsevier Science.

sample FeAg1, was observed, as seen in Figure 55b, and this can be attributed mainly to the enhancement of the implantation doses from 1×10^{16} to 6×10^{16} at/cm^2, which increases the concentration in the Ag films. This change in the concentration of Fe clusters (large and small), clearly indicated by the CEMS spectra, reduces the distance between clusters, with direct consequences for the GMR effect, which mainly depends on the ratio between the electron mean free path and the distance between clusters.

7. CONCLUDING REMARKS

This review was largely devoted to an analysis of experimental work on thin films, interfaces, and multilayers, rather than attempting to survey the vast literature on the preparation of systems with reduced dimensions or the calculation of their magnetic properties. The reader is referred to some relevant publications for the general problems of growth and characterization of thin films and multilayers [69–71] or for computational work [72, 73].

Several NMR and NGR spectroscopy results on thin films and multilayer systems have been described, demonstrating that ^{57}Fe and ^{59}Co are useful probes for the study of structural and magnetic properties at the atomic scale in thin metallic layers. If the NMR and Mössbauer isotope probes are located at interfaces, the structure of the interface can be determined from site-selective hyperfine field observation. The unique information that the nuclear resonance techniques may provide, on phase analysis, the influence of the deposition conditions on the layer structure, or the modification of magnetic behavior through thermal annealing, has been summarized. The properties of amorphous, nanostructured, and granular thin films have been discussed on the grounds of the hyperfine field distribution that shapes the nuclear resonance spectra.

Some other important applications of ^{59}Co NMR and ^{57}Fe NGR, such as the magnetism of ultrathin films, magnetic relaxation effects, or nuclear resonant synchrotron experiments, as well as relevant results obtained with other nuclei like ^{51}V or ^{55}Mn, which are useful NMR isotopes, or the Mössbauer probes ^{119}Sn and ^{197}Au, have not been discussed here, and the reader is referred to recent extensive reviews on this matter [7, 74–76].

Acknowledgments

The author thanks the following people for their collaboration: J. B. Sousa, J. C. Soares, P. P. Freitas, and M. C. Abreu. Financial support from the Fundação para a Ciência e a Tecnologia (Programa PRAXIS XXI) is gratefully acknowledged.

REFERENCES

1. M. S. Rogalski and S. B. Palmer, "Quantum Physics." Gordon & Breach, London, 1999.
2. Q. I. Jin, Y. B. Xu, H. R. Zhai, C. Hu, M. Lu, Q. S. Bie, Y. Zhai, G. L. Dunifer, H. M. Naik, and M. Ahmad, *Phys. Rev. Lett.* 72, 768 (1994).
3. G. Schatz and A. Weilinger, "Nuclear Condensed Matter Physics." Wiley, Chichester, 1996.
4. T. E. Cranshaw, B. W. Dale, G. O. Longworth, and C. E. Johnson, "Mössbauer Spectroscopy and Its Applications." Cambridge Univ. Press, Cambridge, U.K., 1985.
5. Y. D. Zhang, J. I. Budnick, N. X. Shen, W. A. Hines, G. W. Fernando, and T. Manzur, *J. Magn. Magn. Mater.* 140-44, 987 (1995).
6. T. Kajitani, Y. Morii, S. Funahashi, T. Iriyama, K. Kobayashi, H. Kato, Y. Nakagawa, and K. Hiraya, *J. Appl. Phys.* 73, 6032 (1993).
7. P. C. Riedi, T. Thomson, and G. J. Tomka, in "Handbook of Magnetic Materials" (K. H. J. Buschow Ed.), Vol. 12, pp. 97–258. Elsevier Science, Amsterdam, 1999.
8. E. Jedryka, P. Panissod, P. Guilmin, and G. Marchal, *Hyperfine Interact.* 51, 1103 (1989).
9. A. Goto, H. Yasuoka, H. Yamamoto, and T. Shinjo, *J. Phys. Soc. Jpn.* 62, 2129 (1993).
10. D. P. E. Dickson and F. J. Berry, Eds., Mössbauer Spectroscopy." Cambridge Univ. Press, Cambridge, U.K., 1986.
11. G. Y. Guo and H. Ebert, *Phys. Rev. B: Condens. Matter* 53, 2492 (1996).
12. J. C. Waerenborgh, M. S. Rogalski, M. Almeida, and J. B. Sousa, *Solid State Commun.* 104, 271 (1997).
13. J. A. Paixão, J. C. Waerenborgh, M. S. Rogalski, M. Almeida, J. B. Sousa, A. P. Gonçalves, J. C. Spirlet, and M. Bonnet, *J. Phys: Condens. Matter* 10, 4071 (1998).
14. P. C. M. Gubbens, J. H. F. van Apeldoorn, A. M. van der Kraan, and K. H. J. Buschow, *J. Phys. F* 4, 921 (1974).
15. C. K. Shepard, M. A. Park, P. A. Polstra, and J. G. Mullen, *Hyperfine Interact.* 92, 1227 (1994).
16. J. K. Shrivastava, S. C. Birghava, P. K. Iyengar, and B. V. Thosar, "Advances in Mössbauer Spectroscopy." Elsevier, Amsterdam, 1983.
17. M. M. Amado, R. P. Pinto, M. E. Braga, M. S. Rogalski, and J. B. Sousa, *J. Appl. Phys.* 81, 5784 (1997).
18. H. M. van Voort, D. B. de Mooij, and K. H. J. Buschow, *J. Appl. Phys.* 57, 5414 (1985).
19. G. J. Long, R. Kulasekere, O. A. Pringle, F. Grandjean, and K. H. J. Buschow, *J. Magn. Magn. Mater.* 117, 239 (1992).
20. J. F. Herbst and W. B. Yelon, *J. Appl. Phys.* 57, 2343 (1985).
21. C. P. Slichter, in "NMR in Solids" (L. van Gerven, Ed.), NATO-ASI. Plenum, pp. 3–16. Plenum, New York, (1977).
22. C. Bruynseraede, G. Lauhoff, J. A. C. Bland, G. Strijkers, J. De Boeck, and G. Borghs, *IEEE Trans. Magn.* 34, 861 (1998).
23. P. C. Riedi, T. Dumelow, M. Rubinstein, G. A. Prinz, and S. B. Quadri, *Phys. Rev.* 36, 4595 (1987).
24. G. A. Prinz, *J. Magn. Magn. Mater.* 100, 469 (1991).
25. Y. U. Idzerda, W. T. Elam, B. T. Jonker, and G. A. Prinz, *Phys. Rev. Lett.* 62, 2480 (1989).
26. A. Y. Liu and D. J. Singh, *Phys. Rev. B: Solid State* 47, 8515 (1993).
27. P. Panissod, J. P. Jay, C. Mény, M. Wojcik, and E. Jedryka, *Hyperfine Interact.* 97–98, 75 (1996).
28. H. A. M. de Gronckel, P. J. H. Bloemel, E. A. M. van Alphen, and W. J. M. de Jonge, *Phys. Rev. B: Condens. Matter* 49, 11327 (1994).
29. M. S. Rogalski, *Nondestr. Test. Eval.* 15, 15 (1998).
30. I. Bibicu, M. S. Rogalski, and G. Nicolescu, *Meas. Sci. Technol.* 7, 113 (1996).
31. M. Kopcewicz, T. Stobiecki, M. Czapkiewicz, and A. Grabias, *J. Phys: Condens. Matter* 9, 103 (1997).
32. S. Ohnishi, M. Weinart, and A. J. Freeman, *Phys. Rev. B: Solid State* 30, 36 (1984).
33. S. Nasu, H. Yasuoka, Y. Nakamura, and Y. Murakami, *Acta Metall.* 22, 1057 (1974).

34. H. A. M. de Gronckel, K. Kopinga, W. J. M. de Jonge, P. Panissod, J. P. Schillé, and F. J. A. den Broeder, *Phys. Rev B: Condens. Matter* 44, 9100 (1991).
35. R. Krishnan, A. Das, J. P. Eymery, M. Porte, and M. Tessier, *IEEE Trans. Magn.* 33, 3697 (1997).
36. F. Badia, A. A. Badry, X. X. Zhang, J. Tejada, R. A. Brand, B. Scholz, and W. Keune, *J. Appl. Phys.* 70, 6206 (1991).
37. F. Richomme, B. Scholz, R. A. Brand, W. Keune, and *J. Teillet, J. Magn. Magn. Mater* 156, 195 (1996).
38. O. Marks, T. Ruckert, J. Tappert, W. Keune, W.-S. Kim, and W. Kleemann, *IEEE Trans. Magn.* 34, 834 (1998).
39. P. Panissod and C. Mény, *J. Magn. Magn. Mater.* 126, 16 (1993).
40. Y. Saito, K.Ynomata, A. Goto, and H. Yasuoka, *J. Magn. Magn. Mater.* 126, 466 (1993).
41. S. Steyaert, J. M. Le Breton, S. Parhofer, C. Kuhrt, and J. Teillet, *J. Magn. Magn. Mater.* 196–197, 48 (1999).
42. F. Richomme, J. Teillet, A. Fnidiki, P. Auric, and Ph. Houdy, *Phys. Rev. B: Condens. Matter* 54, 416 (1996).
43. E. A. M. van Alphen, P. A. A. van de Heijden, and W. J. M. de Jonge, *J. Magn. Magn. Mater.* 144, 609 (1995).
44. M. Carbucicchio, M. Rateo, G. Ruggiero, M. Solzi, and G. Turilli, *J. Magn. Magn. Mater.* 196–197, 33 (1999).
45. S. X. Li, P. P. Freitas, M. S. Rogalski, M. M. P. Azevedo, J. B. Sousa, Z. N. Dai, J. C. Soares, N. Matsakawa, and H. Sakakima, *J. Appl. Phys.* 81, 4501 (1997).
46. M. M. P. Azevedo, M. S. Rogalski, and J. B. Sousa, *Meas. Sci. Technol.* 8, 1 (1997).
47. D. H. Mosca, P. H. Dionisio, W. H. Schreiner, I. J. R. Baumvol, and C. Achete, *J. Appl. Phys.* 67, 7514 (1990).
48. M. Wojcik, E. Jedryka, S. Nadolski, T. Stobiecki, and M. Czapkiewicz, *J. Magn. Magn. Mater.* 157–58, 220 (1996).
49. J. Hesse and H. Rübartch, *J. Phys. E: Sci. Instrum.* 7, 526 (1974).
50. G. LeCaër and J. M. Dubois, *J. Phys. E: Sci. Instrum.* 12, 1083 (1979).
51. D. Barb, M. S. Rogalski, M. Morariu, H. Chiriac, and V. Frunza, *Hyperfine Interact.* 55, 897 (1990).
52. M. S. Rogalski, M. M. P. Azevedo, and J. B. Sousa, *J. Magn. Magn. Mater.* 163, L257 (1996).
53. S. Morup, *J. Magn. Magn. Mater.* 37, 39 (1983).
54. J. A. Mendes, M. S. Rogalski, and J. B. Sousa, to be published.
55. C. S. Kim, S. B. Kim, H. M. Lee, Y. R. Uhm, K. Y. Kim, T. H. Noh, and H. N. Ok, *IEEE Trans. Magn.* 33, 3790 (1997).
56. J. S. Baek, S. C. Yu, W. Y. Lim, C. S. Kim, T. S. Kim, and C. O. Kim, *J. Appl. Phys.* 83, 6646 (1998).
57. M. M. Amado, M. S. Rogalski, L. Guimaraes, J. B. Sousa, I. Bibicu, R. G. Welch, and S. B. Palmer, *J. Appl. Phys.* 83, 6852 (1998).
58. R. G. Welch, J. Neamtu, M. S. Rogalski, and S. B. Palmer, *Solid State Commun.* 97, 355 (1996).
59. C. M. Williams, D. B. Chrisey, P. Lubitz, K. S. Grabowski, and C. M. Cotell, *J. Appl. Phys.* 75, 1676 (1994).
60. A. Fert and L. Piraux, *J. Magn. Magn. Mater.* 200, 338 (1999).
61. B. Doudin, J. E. Wegrowe, S. E. Gilbert, V. Scarani, D. Kelly, J. P. Mayer, and J.-Ph. Ansermet, *IEEE Trans. Magn.* 34, 968 (1998).
62. T. Furubayashi and I. Nakatani, *J. Appl. Phys.* 79, 6258 (1995).
63. M. Julliere, *Phys. Lett. A* 54, 225 (1975).
64. J. H. Hsu, Y. H. Huang, P. K. Tseng, and D. E. Chen, *IEEE Trans. Magn.* 34, 909 (1998).
65. M. M. Amado, M. S. Rogalski, A. M. L. Lopes, J. P. Araujo, M. M. P. de Azevedo, M. A. Feio, A. F. Kravets, A. Ya. Vovk, G. N. Kakazei, and J. B. Sousa, to be published.
66. J. C. Soares, L. M. Redondo, C. M. de Jesus, J. G. Marques, M. F. da Silva, M. M. P. de Azevedo, J. A. Mendes, M. S. Rogalski, and J. B. Sousa, *J. Vac. Sci. Technol. A* 16, 1812 (1998).
67. L. M. Redondo, C. M. de Jesus, J. G. Marques, M. F. da Silva, J. C. Soares, M. M. P. de Azevedo, J. A. Mendes, M. S. Rogalski, and J. B. Sousa, *Nucl. Instrum. Methods* 139, 350 (1998).
68. J. B. Sousa, M. M. P. Azevedo, M. S. Rogalski, Yu. G. Pogorelov, L. M. Redondo, C. M. de Jesus, J. G. Marques, M. F. da Silva,

J. C. Soares, J. C. Ousset, and E. Snoeck, *J. Magn. Magn. Mater.* 196–197, 13 (1999).

69. A. Fert and P. Bruno, "Ultrathin Magnetic Structures." Springer-Verlag, Berlin, 1992.

70. B. Heinrich and J. A. C. Bland, Eds., "Ultrathin Magnetic Structures I and II." Springer-Verlag, Berlin, 1994.

71. D. B. Chrisey and G. K. Hubler, Eds., "Pulsed Laser Deposition of Thin Films." Wiley, New York, 1994.

72. V. L. Moruzzi and P. M. Marcus in "Ferromagnetic Materials" (K. H. J. Buschow, Ed.), Vol. 7, p. 97 North-Holland, Amsterdam, 1993.

73. M. S. S. Brooks and B. Johansson, in "Ferromagnetic Materials" (K. H. J. Buschow, Ed.), Vol. 7, p. 139. North-Holland, Amsterdam, 1993.

74. T. Shinjo, *Surf. Sci. Rep.* 12, 49 (1991).

75. M. Przybylski, *Hyperfine Interact.* 113, 135 (1998).

76. T. Shinjo and W. Keune, *J. Magn. Magn. Mater.* 200, 598 (1999).

Chapter 12

MAGNETIC CHARACTERIZATION OF SUPERCONDUCTING THIN FILMS

M. R. Koblischka

Experimentalphysik, Universität des Saarlandes, D-66041 Saarbrücken, Germany

Contents

Handbook of Thin Film Materials, edited by H.S. Nalwa
Volume 5: Nanomaterials and Magnetic Thin Films

ISBN 0-12-519213-0/$35.00

1. INTRODUCTION

Superconducting thin films play an important role in a variety of applications like microwave antennas and mixers, superconducting electronics, magnetic superconducting quantum interference device (SQUID) sensors, and fault current limiters (for recent developments in this field, see, e.g., the proceedings of the Applied Superconductivity Conference [1], the European Applied Superconductivity Conference [2], or the International Superconductivity Conference in Japan [3]). With the development of high-T_c superconductors, the research in this field has increased considerably. An important task for present research is to produce large-scale superconducting films (mostly consisting of $YBa_2Cu_3O_{7-\delta}$, YBCO) with diameters of up to 10 inches [4]. Furthermore, deposition methods are developed into continuous techniques in the production of so-called second-generation tapes [5].

In thin film samples, the largest critical current densities, j_c, are reached as compared with other types of superconductors (bulk, wires, or tapes); the critical current densities are coming close to the depairing current density (i.e., the Cooper pairs are broken up), which defines the maximum possible current density of a superconductor. Therefore, it is a very important task to characterize these superconducting thin films to understand their magnetic properties (e.g., the origin of the strong flux pinning), but also to perform quality analysis of commercially produced superconducting thin films.

In this chapter, the pecularities of the current flow in superconducting thin films are discussed in detail. For thin films with the magnetic field applied perpendicular to their surface (now called "perpendicular geometry"), the classical Bean model of the critical state is no longer valid, as in this model a long cylinder with the field applied along the axis is considered (longitudinal geometry). Interestingly enough, the theory describing the perpendicular geometry was only developed very recently [6], even though superconducting thin film samples were also available with conventional superconductors. The large demagnetization factors of the thin film samples play an important role in the understanding of magnetic measurements. As an example, grain boundaries are found to act as weak links as the flux penetration along them is faciliated by the large demagnetization factor. The results of this effect are discussed in detail in Section 3.

Magneto-optic (MO) imaging [7,8] of flux patterns played a considerable role in establishing the differences between the perpendicular and longitudinal geometries; therefore, the main part of Section 2 is devoted to a description of this technique. Some different experimental techniques, which are also employed for the study of flux distributions, are described as well. Another important tool for characterizing the magnetic properties of superconducting thin films are integral magnetization, AC susceptibility, and magnetotransport measurements, which are discussed in Section 2.

The consequences of perpendicular geometry for flux distributions in superconducting thin films is outlined in Section 3. Examples of flux patterns of various sample geometries, field-cooled states, and current-induced flux patterns are presented.

By means of MO imaging, the field distributions in and around superconducting thin film samples can be directly visualized; so several types of experiments are presented in Section 4. A main issue for the present research is the understanding of flux patterns of samples with structural defects (e.g., for quality assessment); the basics of the current flow around defects are also described in Section 4. Grain boundaries play an important role as current-limiting defects within superconducting samples (thin films, bulks, and especially in wires and tapes of high-T_c superconductors). Thin films can serve as model samples, as grain boundaries can be generated in a controlled way by the use of bicrystalline substrates; thus their effect on the flux distributions and current flow is discussed in Section 4 as well. Furthermore, patterned thin films may be used to model a layer of grains of a superconducting tape; an example of this is also given in Section 4. Thick films (i.e., superconducting films with a thickness $d > 1\mu m$) are currently receiving more and more attention, as the fabrication develops toward so-called coated conductors (see, e.g., [2, 3]). Flux patterns of such thick films are presented as well. To conclude this section, MO experiments on YBCO thin films with a high time resolution are discussed.

Finally, the flux pinning properties of superconducting thin films are discussed in Section 5, and differences from bulk superconductors are given. Another effect of the large demagnetization factor of superconducting thin films leads to peculiar problems when superconducting thin films are measured in magnetometers; this topic is also discussed in Section 5. Finally, a summary is given in Section 6.

2. LOCAL AND INTEGRAL MAGNETIZATION MEASUREMENTS

To characterize the magnetic properties of superconducting thin films, there are two possibilities: one is the measurement of the integral properties of a sample, i.e., the magnetization m or torque $\tau = m \times B$; the other one is to employ a technique allowing measurement of the field distribution around a thin film sample locally. To achieve this goal, several techniques are described in the literature [9,10]; a recent review of local magnetic probes in superconductors was given by Bending [11] and de Lozanne [12]. The most important of all of these techniques are the micro Hall probes and the MO flux visualization, which is based on the Faraday effect. The MO technique has the advantage of visualizing the flux patterns in a direct way, and the observation window can be varied from details in the micrometer range to whole samples; therefore, this technique presents an ideal tool for the study of the general properties of flux distributions. Thus, the MO imaging technique is described in detail in this chapter, and MO images are used to illustrate the pecularities found in the flux distributions of supercoducting thin films. Another section is devoted to the presentation of different local investigation techniques, which offer some additional experimental possibilities.

In this article, the general properties of flux distributions in superconducting thin films are discussed, as these are fundamentally different from those obtained from bulk samples. The flux distributions presented are obtained mainly by the MO technique. Furthermore, the properties of magnetization hysteresis loops on superconducting thin film samples are described in detail.

2.1. Other Local Techniques

2.1.1. Micro Hall Probes and Hall Probe Arrays

The recent development of micro Hall probes makes possible the study of flux density profiles with a reasonably good spatial resolution [9]. Sensor sizes down to $10 \times 10 \ \mu m^2$ have been achieved. However, for most investigations, the micro Hall probes were superseded by Hall probe arrays, which combine up to 20 Hall probes on a single chip [13]. With such Hall probe arrays, the flux density profiles and local magnetization loops can be scanned with much better precision. Studies of this type have mainly been carried out on single crystalline samples, but have also been done very recently on superconducting thin films [14, 15]. McElfresh et al. investigated the local time-dependent magnetization of superconducting $YBa_2Cu_3O_{7-\delta}$ films in the presence of a transport current [16].

2.1.2. Scanning Hall Probes

The scanning Hall probe technique provides a mechanical scan over the superconducting sample [17,18]. For a recent review of this field, see the review article by Bending [11]. This technique has been employed on high-T_c thin films, as well as on Nb thin films recently [18]. The scan range of such a scanning Hall probe microscope is about $150 \times 150 \ \mu m^2$, with a maximum spatial resolution of about 200 nm [11], where the scanning is performed by means of a STM tip (STM tracking mode).

2.1.3. Low-Temperature Scanning Electron Microscopy

Another development of a low-temperature imaging technique is the scanning electron microscope (SEM) equipped with a low-temperature stage. The development of this technique was reviewed by Hübener [19] and Koelle and Gross [20]. Figure 1 shows a SQUID washer, fabricated from a YBCO thin film, which was field-cooled in a field of 2.5 μT.

The temperature during the measurement was 77 K [21]. The voltage of the microscope was 10 KV, and the current was 12 nA. This experiment was carried out at the University of Tübingen, Germany. In this group, several such experiments were carried out on superconducting YBCO thin films and Josephson junctions.

2.1.4. Scanning SQUID Technique and Magnetic Force Microscopy

Recent developments of the scanning SQUID technique have been based exclusively around DC SQUIDs in flux-locked

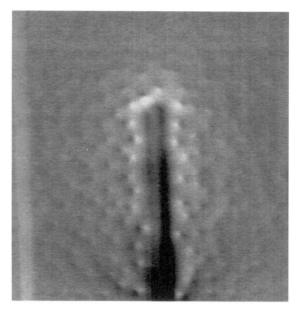

Fig. 1. LT-SEM image of a SQUID washer. The sample was field-cooled to 77 K in a magnetic field of 2.5 μT. The vortices are imaged as bright spots around the split. Image provided by R. Straub, University of Tübingen.

loops. The scanning SQUID technique currently achieves the highest field resolution, so that vortices can easily be visualized [22]. The drawback of the technique is that in most setups the SQUID (fabricated from a low-T_c thin film) and the sample to be investigated have to be at the same temperature. Therefore, many interesting experiments in the field of high-T_c superconductivity cannot be carried out.

Magnetic force microscopy (MFM) was also successfully applied to YBCO thin films [23, 24]. This technique also makes it possible to readily measure the forces between the vortices. For details of this technique, see the review by Bending [11].

2.2. Samples Used in This Study

A variety of superconducting thin films are used in this study; mainly of the high-T_c superconductor $YBa_2Cu_3O_{7-\delta}$ (YBCO), which can now be produced with a very high quality. The thin films were prepared by means of the laser ablation technique [25], using a deposition pressure of 0.33 mbar and a substrate temperature of 775°C. The film thickness was $d = 250–300$ nm with the c axis oriented perpendicular to the film plane. Finally, patterns were formed by means of Ar ion milling. Mainly $SrTiO_3$ was used as the substrate material. More complicated structures were prepared by means of electron beam lithography (see, e.g., the model sample presented in Section 4) and laser cutting. Many features of the flux patterns are presented on samples of a batch of YBCO thin films, patterned in a rectangular shape (thickness of the film $d = 250$ nm, lateral dimensions $1600 \times 600 \ \mu m^2$); these samples were prepared at the NKT Research Center by Vase et al. [25]. These films were found to exhibit excellent superconducting properties, including some samples that did not show any microstructural defects.

2.3. Magneto-Optic Flux Visualization

An extended review of magneto-optic flux visualization on high-T_c superconductors was published recently by Koblischka and Wijngaarden [7]; early MO experiments including experiments on heavy-ion irradiated superconducting thin films were summarized by Schuster and Koblischka [8]. MO experiments on low-T_c samples, including Nb and Pb thin films, were summarized by Hübener [26].

Field distributions are obtained by the rotation of the polarization plane of linearly polarized light, which passes a magneto-optically active layer exposed to the magnetic field of the underlying superconductor. From flux-free regions the light is reflected without rotation and thus cannot pass the analyzer, which is set in a crossed position with respect to the polarizer. The images presented here are, therefore, maps of the z component of the local magnetic field, $B_{i,z}$. As the MO active layer, we use mainly a Bi-doped yttrium-iron-garnet (YIG) film with in-plane anisotropy. These films provide a high Faraday rotation in a wide temperature range up to the superconducting transition temperature of the superconducting thin films (Fig. 2). Another possibility, which provides the maximum spatial resolution of the technique ($\approx 0.5\ \mu$m), but works only at low temperatures (2–20 K), would be the use of Eu-based chalcogenide thin films, as outlined in [7]. As shown in Figure 3, two arrangements for the magneto-optic active layer are possible. The MO layer (and a reflective coating) can be evaporated directly onto the surface of the superconducting sample. This arrangement achieves the highest possible spatial resolution of the MO technique; however, the sample may be damaged during the evaporation procedure. Furthermore, the sample cannot be used otherwise. In the other arrangement, there is a substrate holding the MO layer and the reflective coating. The substrate with the MO layer ("MO indicator")

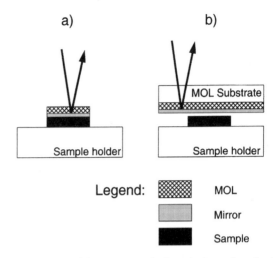

Fig. 3. Arrangement of the magneto-optically active layer, the reflective mirror layer (mostly Al), and the sample. (a) Arrangement used in the HRF technique in the early stages of MO imaging, to obtain the highest possible spatial resolution. (b) Arrangement permitting a nondestructive observation technique; this is the principle for the YIG garnet films.

is then simply laid on top of the sample. The YIG films are made following this principle, and the Eu-chalcogenides may be arranged in this way to obtain a nondestructive flux imaging technique, as suggested by Koblischka [27].

The microscope provides a uniform light field and a heat filter, which ensures that the heat load on the sample is negligible. The images are recorded with a charge-coupled device (CCD) video or digital camera (each image typically consists of 1536×1024 pixels per frame) and are subsequently transferred to a computer for processing and storage. In the magneto-optic apparatus the sample was mounted on the cold finger of an optical helium flow cryostat [28]. Samples are glued to a copper holder with conductive carbon cement [29] to ensure a good thermal contact, and the indicator film is laid on the sample surface and centered.

The magnetic field is normally applied perpendicular to the sample surface (i.e., along the c axis of the YBCO) with a copper solenoid with $B_{e,\max} = \pm 120$ mT. Other apparatuses for MO imaging can reach magnetic fields around 500 mT, and one setup was constructed that provided fields of up to 7 T with the use of a superconducting coil [7].

A variety of results obtained by means of the MO flux visualization technique will be described in detail in Sections 3 and 4.

2.4. Magnetization Measurements

Measurement of induced magnetic moments of superconductors is a widely used method for the study of critical currents [30]. Magnetic measurements are mostly made with SQUID (superconducting quantum interference device) magnetometers or vibrating sample magnetometers (VSMs). The latter make possible a rapid measurement of magnetic moments in the regimes of both constant and sweeping fields. In the study of superconductors, the former regime corresponds to a conventional relaxation experiment, the latter

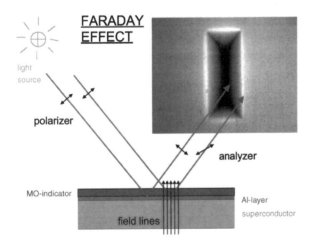

Fig. 2. Schematical drawing of a Faraday effect. Polarizer and analyzer are set in a crossed position with respect to each other. The superconductor is covered by an MO indicator film, which carries at the bottom an Al layer to enhance the reflectivity. On the right side, an MO image is shown of a YBCO thin film as an example. Flux is imaged as bright areas; the Meissner area (no field lines) stays dark. The incident light beam in the real experiment is perpendicular to the sample surface; the drawn angle is only for clarity.

to measurements of magnetic hysteresis loops (MHLs). One problem for both measurement techniques is the small available sample space, which makes it difficult to mount thin films with a typical size of 10×10 mm^2. Another technique that can be used to measure superconducting thin films is torque magnetometry, where the torque $\tau = m \times B$ is recorded during field sweeps. This technique is especially well suited to dynamic relaxation experiments on large superconducting thin film samples, as shown by van Dalen et al. [31].

Some details of measurements performed with the integral magnetization technique are described in Section 5, including a description of flux pinning. Here, a summary of the different types of experiments performed on superconducting thin films is given. In the beginning of the research on high-T_c superconductors, many magnetization experiments on thin films were performed, often together with a comparison of the results to transport measurements [32]. Another important topic is the study of the angular dependence of magnetization [33]. As the high-T_c superconductors have a layered crystal structure, the vortices do exist only in the Cu-O-planes as so-called pancake vortices. Therefore, the angular dependence of the magnetization of a thin epitaxial film should follow a 2D scaling behavior. The early measurements showed that such a 2D scaling is largely followed, but newer measurements revealed that there are deviations.

In Figure 4, MHLs obtained on a YBCO thin film by means of a SQUID magnetometer are presented for temperatures between 20 K and 85 K. The YBCO thin film sample used in this measurement is the one shown in Figure 2, without structural defects. All MHLs show a pronounced central peak at $\mu_0 H_a = 0$ T. Details about this position of the central peak are discussed in Section 5. The magnetization, m, is monotonously decaying from this peak position, in contrast to bulk superconducting samples, where often a secondary maximum ("fishtail peak") can be observed. In superconducting thin film samples no such anomalous peak in the magnetization was ever observed, which is a consequence of the larger critical current density achieved in superconducting thin films. According to the recent understanding of the secondary peak

effect, the higher critical current density due to flux pinning of the δl type does not allow a peak effect to appear, which is provided by the considerably weaker δT_c pinning [34].

An important piece of information comes from flux creep experiments, where the pinning potential, U_0, can be determined. The so-called vortex matter and flux creep experiments were reviewed by Blatter et al. [35] and Yeshurun et al. [36]. Here it is only important to mention that conventional ($H_a =$ const.) and dynamic relaxation [37] experiments can be performed. For superconducting thin films, dynamic relaxation (DR) is especially advantageous, as the full penetration field of thin film samples is typically very small. Examples of these types of measurements can be found in [31].

Superconducting thin film samples are ideally suited to the dynamic relaxation technique because of the small values of the full penetration field, H^*, of superconducting thin films [31, 38, 39]. It is therefore possible to measure a set of minor loops around a given target field with the use of different sweep rates of the external magnetic field. An example of such a measurement is presented in Figure 5; here a YBCO thin film is measured at magnetic field sweep rates between 0.6 mT/s and 40 mT/s.

Another reason why dynamic relaxation experiments on superconducting thin film samples are better suited as conventional ones is the fact that the flux distribution in superconducting thin films is very sensitive to small changes of the applied magnetic field (Figs. 6 and 7). Minor field over- and undershoots (which are quite common during the approach of a target field by a superconducting laboratory magnet) may alter the flux creep behavior completely. One such example is given by Griessen et al. [40], where the flux creep rate can be positive or negative or even be completely canceled out, depending on the magnetic history.

2.5. AC Susceptibility Measurements

A widely used tool for determining the superconducting transition temperature, T_c, of superconducting samples is the AC susceptibility measurement [41]. A variety of measurements

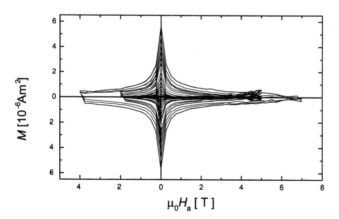

Fig. 4. Magnetization hysteresis loops (MHLs) measured on a YBCO thin film (the sample is the same sample as shown in Fig. 2) in the temperature range 20–85 K.

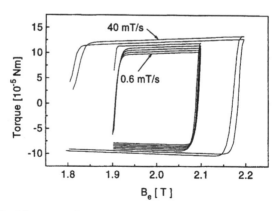

Fig. 5. Minor torque hysteresis loops measured at $T = 15$ K and $B_e = 2.0$ T on a YBCO thin film with an oxygen content of 6.85. The magnetic field sweep rates used here are $dB_e dt = 40$ (outer curve), 20, 10, 5, 2.5, 1.25, and 0.6 (inner curve) mT/s, respectively. Reprinted with permission from [38].

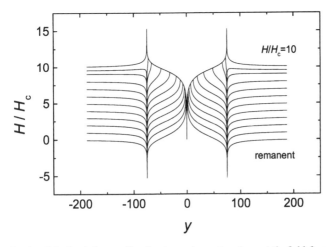

Fig. 6. Calculated flux profiles for decreasing external magnetic field from $H/H_c = 10$ toward the remanent state. The strip width is between $y \pm 75$ (arbitrary units).

on superconducting samples is reviewed in [42]. Typically, measurements of AC susceptibility are performed on polycrystalline superconducting samples, where the imaginary part of the AC susceptibility enables one to distinguish between the intragrain and the intergrain contributions. Therefore, this type of measurement currently plays a large role in the study of wires and tapes of high-T_c superconductors. However, with specially shaped coils according to the geometry of the thin film samples, this technique also provides a fast and reliable determination of superconducting properties of thin film samples, mainly concerning the superconducting transition temperature and the details of the superconducting phase transition, which yield important information on the quality of thin film samples [25].

Moreover, detailed experiments studying the AC response of different sample geometries (here the superconducting thin films permit the patterning of various sample geometries) have been performed [43–45] to verify the theoretical predictions by Brandt et al. [46], Mints [47], and others [48].

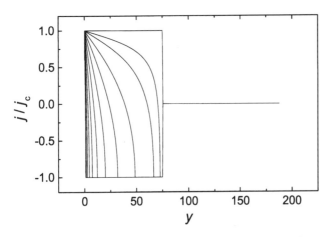

Fig. 7. Calculated current distribution in one half-width of the sample, corresponding to the profiles of Figure 6.

2.6. Magnetotransport

Properties of superconducting thin films are widely characterized by direct transport measurements. The geometry of thin films makes it possible to employ small currents and to achieve a large critical current density. Therefore, thin films are often patterned into narrow bridges with typical dimensions of some 100 μm in length and a width of about 10 to 100 μm. Here, I mention only two important types of experiments from the vast literature on this topic (see, e.g., a review by Wördenweber [49]).

A very important measurement was carried out to study the effect of anisotropy in magnetic fields [50,51] on both YBCO and $Bi_2Sr_2CaCu_2O_{8+\delta}$ thin films, revealing the much larger anisotropy of the latter compound. Many more investigations considered the effects of flux pinning in the thin film samples, mostly on the YBCO system [52–58], but also in some anisotropic layered low-T_c systems [59].

Measurement of I–V characteristics also makes possible the study of flux creep phenomena with the transport measurement technique. Moreover, comparison experiments have been performed, as magnetization techniques and I–V-measurements cover different time windows of the vortex relaxation [60,61]. A scaling approach of the resistivity data makes possible the study of transitions in the vortex system; this is especially interesting for the case of the less-common high-T_c systems, which mostly have a larger anisotropy as compared with the YBCO system [62], but there are also still many open questions concerning the vortex matter in YBCO [63].

3. THEORETICAL DESCRIPTION OF THE IDEAL FLUX PATTERNS

MO imaging has contributed a great deal to the development of the theoretical description of flux and current distributions in perpendicular geometry (i.e., the magnetic field is applied perpendicular to the sample surface; thus large demagnetization factors prevail). This situation is in stark contrast to the longitudinal geometry, where the field is applied along the long axis of a superconducting cylinder [64]. In an early stage, numerical simulation models were developed [65–68], until Brandt et al. [69], Brandt and Indenbom [70], and, at the same time, Zeldov et al. [71] presented analytical solutions of the problem.

These newly developed analytical models for calculating field and current distributions in thin samples [69–75, 84] with a thickness d much smaller than the London penetration depth, λ_L, and the external magnetic field applied perpendicular to the sample surface (perpendicular geometry) can be used as a starting point. Calculations are possible for various sample geometries, for the infinitely long stripe, for circles, and for rectangles [75]. The most complete description with a field-dependent current density was given recently by Shantsev et al. [76].

Measurements of flux distributions on superconducting thin films are ideally suited to verification of these theoretical calculations [77–83], as thin film samples can now be prepared in excellent quality.

3.1. Infinitely Long Strip

For our purpose, it is necessary to calculate the current distributions for general magnetic history. As an example, I have chosen an infinitely long strip (along z) with width $2a$ ($-a \leq y \leq a$). The current is taken to be independent of the magnetic field (Bean model). Defining a critical sheet current $J_c = j_c d$ and a critical field $H_c = J_c / \pi$, one may write the sheet current $J(y)$ and the perpendicular magnetic field in an external field $H_a(t)$, which is slowly increased from zero as [70]

$$
J(y) = \begin{cases} 2 \dfrac{J_c}{\pi} \arctan \dfrac{cy}{(b^2 - y^2)^{1/2}}, & |y| < b \\ J_c y / |y|, & b < |y| < a \end{cases} \tag{1}
$$

$$
H(y) = \begin{cases} 0, & |y| < b \\ H_c \operatorname{arctanh} \dfrac{(y^2 - b^2)^{1/2}}{c|y|}, & b < |y| < a \\ H_c \operatorname{arctanh} \dfrac{c|y|}{(y^2 - b^2)^{1/2}}, & |y| > a \end{cases} \tag{2}
$$

where

$$
b = a / \cosh(H_a / H_c) \tag{3}
$$

$$
c = (a^2 - b^2)^{1/2} / a = \tanh(H_a / H_c). \tag{4}
$$

Here b denotes the position of the flux front, and $a - b$ is the penetration depth of the magnetic flux. The sheet current $J(r)$ in a disk of radius a is also given by Eq. (1), but with y replaced by r and with $H_c = J_c / \pi$ replaced by $H_c = J_c / 2$ in Eqs. (3) and (4).

Equations (1) and (2) represent the virgin current and field distributions, i.e., when the external magnetic field is raised from zero. From these solutions one may obtain the general solutions for arbitrary magnetic history $H_a(t)$. If the external magnetic field $H_a(t)$ is cycled between the values $+H_0$ and $-H_0$ one obtains in the half-period with decreasing H_a

$$
J \downarrow (y, H_a, J_c) = J(y, H_0, J_c) - J(y, H_0 - H_a, 2J_c) \tag{5}
$$

and

$$
H \downarrow (y, H_a, J_c) = H(y, H_0, J_c) - H(y, H_0 - H_a, 2J_c) \tag{6}
$$

Similar expressions are found for other situations when H_a is cycled from $-H_a$ to H_a [70].

This model replaces the conventional Bean model in perpendicular geometry. Several extensions have been published, e.g., allowing for a field dependence of the critical currents [85, 86] and flux creep [87].

The important difference in the longitudinal geometry is the field "overshoot" along the sample edges due to the large demagnetization factor. Furthermore, during flux creep, an interesting observation can be made: along the line $B_{i,z} = B_a$, there is no flux creep, and the direction of creep is different inside and outside this line. The strong field overshoot along the sample edges decays, and flux penetrates deeper into the sample inside this line. This line, now called the "neutral line," was predicted in calculations by Gurevich and Brandt [88] and was first observed by Koblischka et al. [89] in very thin $DyBa_2Cu_3O_{7-\delta}$ single crystals before the modeling was developed, and later by Schuster et al. [90] in YBCO thin films.

3.2. Other Sample Geometries

Up to now, only an infinitely long superconducting strip was considered. By some extensions, the flux and current patterns of squares, rectangles, and circles can also be calculated (see, e.g., [75]).

As the concept of d lines [91] is important for the interpretation of our observations, we just recall in the following some characteristic properties of the d lines:

(i) The so-called d^+ lines are formed where the currents are forced to make a sharp turn to flow parallel to the sample edges and appear in the MO images as dark lines. The d^- lines are generated at the sample edges (= field overshoot due to demagnetization effects) and along structural defects that facilitate the flux penetration process [92]; these lines appear as bright lines.

(ii) The location of the d^+ lines is determined by the whole sample geometry. In particular, differences in the critical current densities in different regions of the sample contribute strongly to the appearance of d^+ lines. So, the d^+ lines reflect the magnetic behavior of the entire sample.

(iii) The flux lines are not able to cross the d^+ lines. This follows from the flux motion toward and away from these lines when the external magnetic field is, respectively, enhanced or reduced. Thus, the d^+ lines divide the superconductor into independent areas of flux motion and determine a skeleton of the flux behavior.

(iv) The d^+ and d^- lines do not change their position or intensity when the external magnetic field is lowered or reversed, although the magneto-optically detected intensities of the d^+ and d^- lines are reversed in the remanent state. Furthermore, the d^+ lines depend on the d^- lines, thus allowing the determination of any differences in current density inside a superconducting sample.

Figure 8 presents the ideal geometric pattern of the d lines, obtained on a YBCO thin film sample without any microstructural defects. In the presence of an applied magnetic field, the d^+ lines are always represented as dark lines, whereas in the remanent state, the d^+ lines appear as bright lines. The d^- lines along the sample edges are bright during increasing field and represent the large demagnetization factor of superconducting thin films in perpendicular geometry. In the remanent

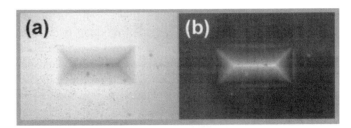

Fig. 8. Ideal flux patterns of a superconducting thin film, patterned into a rectangular shape. In (a), the applied field is 110 mT, and (b) presents the corresponding remanent state ($\mu_0 H_a = 0$ T) after a field of 110 mT was applied. The observation temperature is $T = 18$ K. The dark (resp. bright) spots visible in the image stem from imperfections in the MO indicator film.

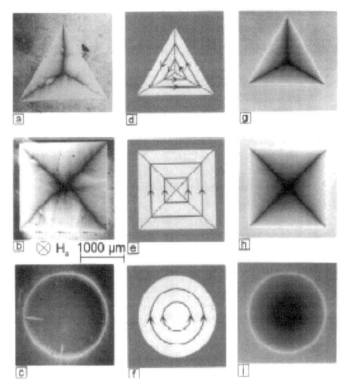

Fig. 10. Experimentally determined magnetic flux density distribution pattern of a triangular, square, and circular $YBa_2Cu_3O_{7-\delta}$ film after zero-field cooling and application of an external magnetic field of (a) 13.5 mT (b) 100.3 mT, and (c) 150.7 mT. Images (d) to (f) show the current distribution in the high-field region and accordingly, (g)–(i) the calculated flux density distribution. Reprinted with permission from [93], © 1997, Elsevier Science.

state, vortices of opposite polarity enter the sample, so the d^- lines appear again as bright lines (whereas they are dark during reduction of the applied field, see Figure 12. The angle α between the sample edges and the d lines is 45°, and the flux penetration distance from the sample edge to the respective d line is equal; corresponding to an isotropic situation. This ideal geometric shape of the d^+ lines may be disturbed by the presence of defects, thus the determination of α may be nontrivial. Therefore, the concept of d lines plays an important role in the understanding if flux patterns observed on samples with structural defects or samples containing areas of different current density, as is shown in Section 4.

Figure 9 presents an ideal field distribution of a square-shaped YBCO thin film. Moreover, this sample is practically free of defects. The angle α between the d lines and the sample edges is 45°, and the flux penetration in all directions is equal. This situation corresponds to $j_{c,a} = j_{c,b}$; i.e., the critical currents in the a and b directions are equal in size. The determination of α can therefore be used to determine the anisotropy parameters directly from the MO images.

Figure 10 presents MO flux patterns of a triangular, square, and circular $YBa_2Cu_3O_{7-\delta}$ thin films after zero-field cooling and application of (a) 13.5 mT, (b) 100.3 mT, and (c) 150.7 mT. The images (d) to (f) show the current distribution in the high-field-region and accordingly, images (g) to (i) show the calculated flux density distribution. The samples are not fully defect-free, but a comparison with the calculated flux density distributions (g–i) is nevertheless possible.

The self-field of such regular polygons with thickness D can be calculated by dividing the sample into N isosceles triangles in which the current flows homogeneously in one direction ($N = 2, 3, 4$, and ∞ belongs to strip, triangle, square, and circle) [93]. The self-field of such a basic bridge is known analytically [94], and thus the self-field of the whole sample is obtained by a superposition.

3.3. Current-Induced Flux Patterns

Flux patterns can be induced not only by a magnetic field, but also by an applied transport current. The model calculations [70, 71] also make possible a treatment of this situation. Here, a typical magneto-optical experiment is presented.

In Figure 11, MO images of a current-carrying state are shown; the experiment was performed at the University of Oslo [95]. The sample is in this case a YBCO thin film patterned into the shape of a long bridge ($500 \times 110 \mu m^2$). In

Fig. 9. Ideal flux patterns of a superconducting thin film, patterned into a square shape. The applied field is 80 mT, and the observation temperature is 18 K.

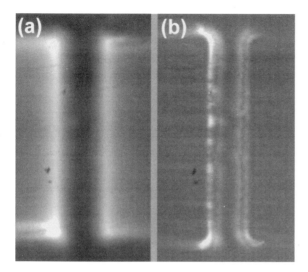

Fig. 11. (a) MO image of the flux distribution with the superconducting bridge in the current-carrying state with $I = 4.16$ A. (b) MO image of the flux distribution with the bridge in the remanent state after the current application presented in (a). Tow inner bright lines near the sample edges represent the original trapped flux, the outer bright lines represent the penetration of oppositely directed return field, and the dark lines between them correspond to regions of vortex–antivortex annihilation. To observe this current-induced remanent state, strong contrast enhancement is necessary. Observation at $T = 18$ K. Image provided by M. Gaevski and D. V. Shantsev [95].

(a), the bridge carries a current I of 4.16 A. The image (b) shows the remanent flux after the transport current is switched off. Two inner bright lines near the sample edges represent the original trapped flux, the outer bright lines represent the penetration of an oppositely directed return field, and the dark lines between them correspond to regions of vortex–antivortex annihilation. To observe this current-induced remanent state, strong contrast enhancement is necessary. This situation cor-

responds well to the calculated flux density profiles for this situation.

3.4. Magnetization Loop of a YBCO Thin Film

By means of magneto-optic imaging, a complete magnetization loop of a superconducting thin film can be visualized. Figure 12 shows such a loop together with a magnetization curve measured with a SQUID magnetometer at the same temperature.

The YBCO thin film shown here comes from the same batch as the one presented in Figure 2, but has one tiny defect of type (ii) (see Section 4). During initial flux penetration (virgin curve), one observes first only the stray fields around the sample, before the flux penetration sets in. Just before reaching the main loop, the flux penetration at the defect becomes visible. With a further increase of the field, the typical d-line pattern for isotropic rectangular samples is formed, slightly disturbed by the presence of the defect. At 150 mT, the full penetration field is reached, with a narrow dark d-line in the sample center. On field reversal, the flux density along the outer rim of the sample decreases, and this rim becomes dark. In the middle of the so-called reverse leg of the MHL, the outer rim is completely dark, and the defect is now a dark spot. In the remanent state, vortices of opposite polarity have entered the sample and form a bright rim along the sample edge. In an increasing negative field, these vortices of opposite polarity continue to enter the sample, thereby annihilating the pinned vortices. This annihilation process is complete at about -75 mT. From there on, only the vortices of opposite polarity are residing in the sample. Finally the flux pattern at -150 mT corresponds exactly to the one at $+150$ mT.

Figures 13–17 present flux density profiles for the entire MO loop. Note that the flux profiles shown correspond to the

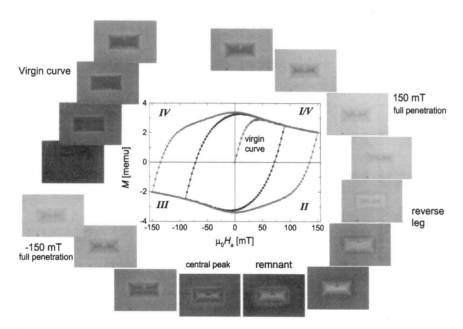

Fig. 12. MO visualization of a magnetization loop of a superconducting thin film, together with a magnetization loop measured with a SQUID magnetometer. The MO images follow the MHL up to ±150 mT; a minor MHL to ±100 mT is also shown. The MO images represent branches I, II, and III.

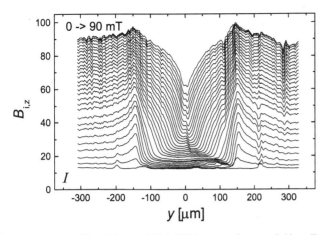

Fig. 13. Flux profiles during an initial field increase after zero-field cooling of the sample: MHL branch I.

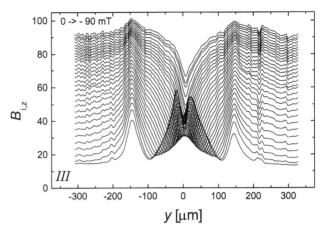

Fig. 15. Flux profiles obtained in increasing magnetic field from 0 mT to 90 mT, starting from the remanent state: MHL branch III.

smaller loop ±90 mT as shown in Figure 12. The flux profiles are taken on a line through the middle of the sample, in steps of 3 mT.

The flux profiles reveal the differences from the idealized case of the calculations (see, e.g., Figs. 6 and 14, which correspond to the same situation). The current density of real thin films depends on the applied field, and the thickness of the MO indicator film has some effect on the extrema. Improved calculation models permit a more precise calculation, taking into account both the field dependence of j_c and the real thickness and distance of the MO indicator film [96]. However, for most experimental situations the simplest case delivers reasonable results.

3.5. Flux Patterns after Field Cooling

Up to now, most of the MO images presented here were taken in applied magnetic fields after the sample was zero-field cooled (ZFC). In this section, images after field cooling (FC) in various applied fields are presented, which yield different types of flux patterns, as in the ZFC case. In Figure 18, FC experiments are shown, with both a field applied and a FC remanent state. The sample is again the rectangular YBCO thin film without any defects, so the observations have a prototypical character.

During FC of YBCO thin films with a magnetic field applied, the sample position cannot be distinguished from the background in (a) and (c). The field distribution inside and outside the sample is completely homogeneous. In both images (a) and (b) with a field of 1 mT, a contrast enhancement is performed, otherwise the images would be completely black. On removal of the applied field (b), the sample is completely penetrated by flux (the location of the individual vortices, however, can not currently be resolved by the MO technique), and the stray field caused by the pinned vortices is clearly visible. The images (c) and (d) present the same experiment, but the sample was cooled in a field of 20 mT. Now, the remanent state in (d) clearly reveals a fully penetrated state. As the applied field during FC was larger than the self-field, vortices of opposite polarity can penetrate the sample, and between these and the pinned vortices, a dark annihilation zone is formed. The absence of large flux density gradients

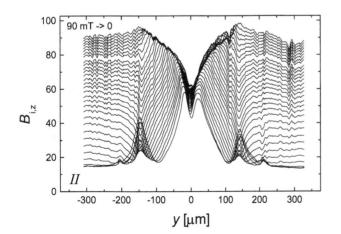

Fig. 14. Flux profiles obtained in decreasing magnetic field from 90 mT to 0 mT (remanent state): MHL branch II.

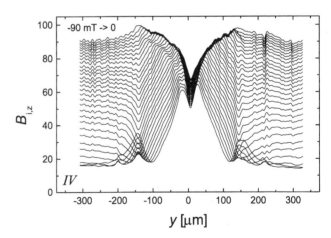

Fig. 16. Flux profiles obtained in decreasing magnetic field from 90 mT to 0 mT (remanent state): MHL branch IV.

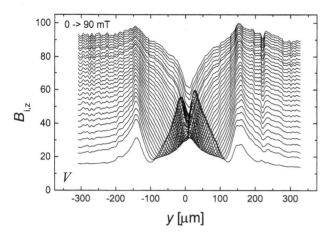

Fig. 17. Flux profiles obtained in increasing magnetic field from 0 mT to 90 mT: MHL branch V.

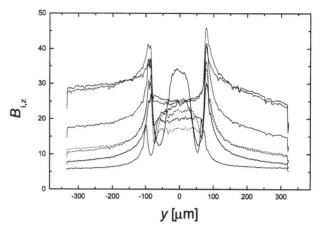

Fig. 19. Flux density profiles for FC remanent states, obtained after field-cooling in fields of 100 mT, 90 mT, 50 mT, 40 mT, 30 mT, 20 mT, and 10 mT. In small applied fields, there is a broad maximum in the center with a homogeneous field distribution. The larger the applied field during cooling, the more vortices of opposite polarity are generated when the applied field is removed.

and of the d-lines indicates that there is only a small current flow in the FC states.

Figure 19 presents flux profiles obtained in the FC remanent state after cooling in various applied fields (100 mT, 90 mT, 50 mT, 40 mT, 30 mT, 20 mT, and 10 mT) down to $T = 18$ K. The flux profiles reveal that a nearly homogeneous flux distribution is achieved by FC in small applied fields. In larger applied fields, vortices of opposite polarity can eventually enter the sample, and an annihilation zone is observed inside the sample.

Figure 20 presents the changes of the flux patterns when such a FC remanent state is warmed to the T_c of the superconductor. Figure 20a shows the starting pattern, created by

FC in a field of 20 mT and subsequently removing the field at $T = 18$ K. With decreasing current density (flux pinning) with increasing temperature, the annihilation zone moves inward. Due to this movement, flux density gradients develop, and the d-line pattern becomes clearer. Finally, at $T = 60$ K (e), the pattern practically corresponds to a remanent state obtained after ZFC at this temperature. Here, it should be noted that in this type of experiment no traces of "flux turbulence" as in YBCO or NdBCO single crystals were ever observed [97], which is again a consequence of the larger critical current densities in thin film samples.

Figure 21 shows the flux density profiles corresponding to Figure 20. These profiles clearly reveal the appearance of the annihilation zone ($B_{i,z} = 0$ T) into the sample during warming. This is a direct manifestation of the weakening of the flux pinning forces in the sample. At the higher temperatures, the remaining pinned flux is practically compressed on the d lines. Also in the flux profiles, no trace of flux turbulence is found, and the flux fronts are completely stable.

3.6. Current Flow Reconstruction

First attempts to solve this problem incorporated a numerical procedure that made it possible to determine the flux density profiles from a given current distribution in the disk geometry [65–67]. Furthermore, the aspect ratio of the sample could be varied easily, requiring a self-consistent iteration procedure. Based on this approach, but only for an infinitely thin current layer, Theuss et al. [68,98] described a fitting procedure of flux profiles obtained by MO imaging, which allowed the determination of the unknown field component, B_r, and the critical current density, j_c, from the map of B_z in a fully penetrated state. Such a treatment is of great importance to obtaining quantitative results from the MO images or Hall probe measurements. This approach is restricted, however, to the disk

Fig. 18. FC experiment on the rectangular YBCO thin film. 1, applied field 1 mT; 2, applied field 20 mT. Note that the gray level was adjusted in (1) to show a pattern. In an applied magnetic field no trace of the sample position can be observed; the field is homogeneously distributed. After removal of the external magnetic field, a FC remanent state is formed. In 1, the stray field can be clearly observed; the sample center is penetrated by flux, which is close to the detection limit. In case 2, the fully penetrated state can be clearly observed; the applied field during the field-cooling was larger than the self-field of the sample, so negative vortices are generated, which can penetrate the sample.

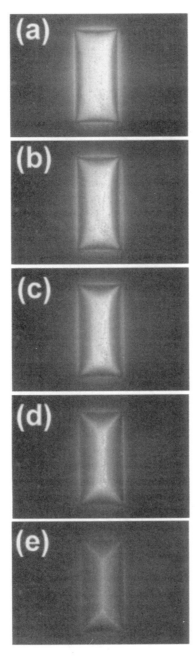

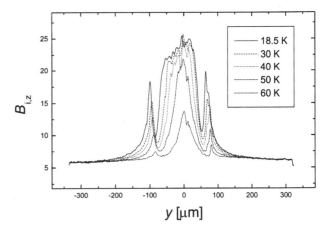

Fig. 21. Flux profiles recorded during warmup of a FC remanent state. Profiles are taken at temperatures 18 K, 30 K, 40 K, 50 K, and 60 K, corresponding to the images of Figure 20.

Fig. 20. Warming-up experiment of an FC remanent state. Images are taken at temperatures $T = 18.5$ K, 30 K, 40 K, 50 K, and 60 K. The corresponding flux profiles are presented in Figure 21.

geometry and cannot treat an abnormal current flow caused by structural defects or weak links.

Further development has made it possible to calculate analytically flux profiles in other geometries like strips [70, 71, 75, 84], disks [99], and rectangles [100] of thin samples with $d \leq \xi$, which implies that the flowing currents are confined to an infinitesimally thin current sheet only. This treatment also makes it possible to incorporate the effects of flux creep [87, 101] and of field-dependent current densities [85].

Quite recently, approaches were made to determine not only the current density from the maps of B_z, but also the

underlying current distribution. Several workers have investigated this inversion problem for the special case of the thin current sheet. Pashitski et al. [102] developed an inversion scheme that allows the study of inhomogeneous current distributions in the strip geometry, e.g., as in an Ag-sheathed $(Pb,Bi)_2Sr_2Ca_2Cu_3O_{10+\delta}$ (Bi-2223) tape or an $YBa_2Cu_3O_{7-\delta}$ (YBCO) thick film. This method assumes that the field distribution outside the superconducting strip with defects corresponds directly to that of a homogeneous superconducting strip of the same dimensions. Their analysis is, therefore, restricted to the strip geometry.

Very recently, Wijngaarden et al. [103] discussed the important generalization to bulk current flow, i.e., when the current sheet is of finite thickness. This extension makes it possible to derive local currents in a flat and arbitrarily thick superconducting sample. This is very important for most high-T_c samples, as the approach using an infinitesimal thin current layer is, strictly speaking, only valid for thin films with a thickness of about 100 nm (thus being less than the London penetration depth, λ_L). Most single crystals and the Bi-2223 tapes have a thickness ranging between several micrometers up to millimeters, which is considerably larger than λ_L. Their analysis leads to an inversion procedure based on an iterative solution of a matrix equation. This kind of analysis requires a great deal of computer power.

This problem was overcome by Johansen et al., who found an analytical solution of the inversion problem as described in [104]. The exact solution allows a simple and efficient inversion scheme to be implemented. Our method allows the treatment of any sample geometry, provided that the MO image used as a starting point provides a B_z map of the entire sample plus the outside field distribution. Furthermore, it is possible to account for the distance between the MO layer and the sample (or the Hall probes and the sample), and even for the finite thickness of the MO layer itself. The importance of this feature is illustrated in Figure 26.

In general, the inverse problem does not have a unique solution. Hence, we restrict the discussion to cases where

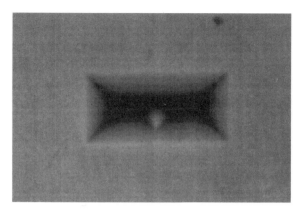

Fig. 22.   The starting point: flux distribution of a rectangular superconducting thin film with one structural defect. This is the same sample as shown in Figure 12.

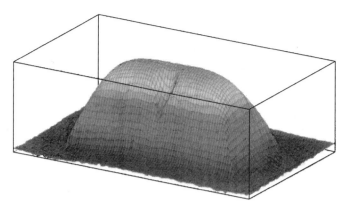

Fig. 24.   3D plot of the local magnetization.

the current is two-dimensional (2D). Moreover, we assume according to the MO observations that the plane of observation is coplanar with the current flow, and that the perpendicular component of the field is measured.

One such current flow reconstruction is given as an example in Figures 22–25. Figure 22 shows the original MO image, the rectangular YBCO thin film with one defect, and in Figure 23 the field distribution is shown as a two-dimensional plot. Figure 24 presents the conversion of the field map into the local magnetization. The derived current streamline pattern is shown in Figure 25, where one sees that the inversion correctly reproduces the rectangular flow of the critical state.

The calculations presented here were all made on a PC with the program Mathcad 6.0.

The analysis of the inverse problem can be extended even further for field maps obtained with the MO imaging method. In the Faraday-active indicator each infinitesimal layer, dz, contributes to the Faraday rotation by

$$d\theta_F(x, y) = VB_z(x, y; z)\, dz \qquad (7)$$

where V is the effective Verdet constant of the MO material (a factor 2 due to the double pass configuration is absorbed). After the light has exited the MO layer of thickness t', the plane of polarization has acquired a total rotation, which in Fourier space becomes

$$\theta_F(\vec{k}) = V \int_h^{h+t'} e^{-kz}\, B_z(\vec{k}; 0)\, dz$$

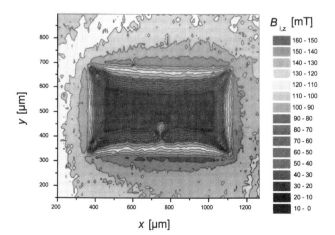

Fig. 23.   2D graph of a field distribution, corresponding to the image of Figure 22.

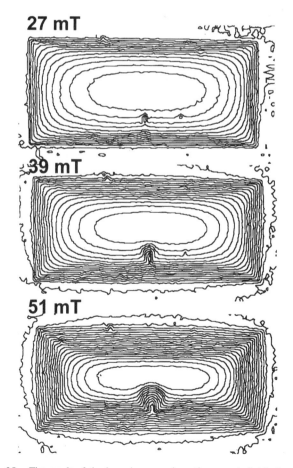

Fig. 25.   The result of the inversion procedure: the current distribution for three different applied magnetic fields.

$$= Vt' e^{-kh} \frac{1 - e^{-kt'}}{kt'} B_z(\vec{k}; 0) \qquad (8)$$

Thus, in addition to the smearing of the field across the gap distance h, as given by

$$B_z(\vec{k}; h) = e^{-kh} B_z(\vec{k}; 0), \qquad (9)$$

the finite thickness of the MO indicator will also contribute by a similar filtering. The role of the latter effect depends on the dimension t' compared with h, as described by Eq. (8).

The frequently used indicators made from Bi-doped yttrium-iron-garnet (YIG) with in-plane anisotropy have a thickness of 4–5 μm [105]. Thus, being of the same order of magnitude as the sample-indicator gap, the additional smearing becomes significant, and proper recalculations from such MO images should take the effect into account. Note then that the additional factor, $(1 - e^{-kt'})/kt'$, representing the MO indicator response, can immediately be included in the inversion scheme presented by Johansen [106].

Figure 26 presents two flux distributions obtained on the same sample. The two images are taken at the same external applied field of 70 mT, but the distance between the MO indicator and the sample is increased in (b) to 50 μm by means

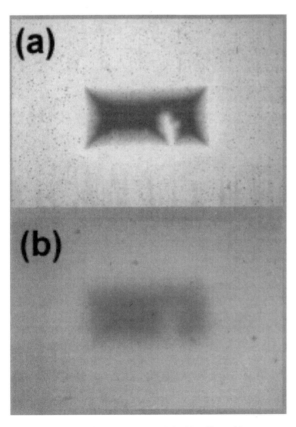

Fig. 26. Distance experiment on a YBCO thin film with one structural defect. Both images are taken at an applied field of 70 mT. (a) The "normal" configuration; i.e., the MO indicator is laid directly on the substrate of the sample. (b) The flux pattern with the MO indicator laying on two copper tapes with a thickness of 50 μm. The differences in the flux distributions are very clear and illustrate the importance of correcting for the distance between the MO indicator and the sample when the current reconstruction is performed.

of a thin copper tape, which is placed on the substrate of the thin film sample.

The flux pattern of (b) is clearly much more diffused; the fine structures of the d line pattern and of the flux distribution around the defect are washed away. However, the basic distance of flux penetration into the sample remains the same. This experiment illustrates nicely the importance of correcting for the distance between the sample and the MO indicator in the current reconstruction. To achieve the highest possible resolution, the finite thickness of the MO layer should also be considered.

In summary, this scheme provides an excellent tool for deriving the current flow pattern directly from the MO images or Hall probe scans.

4. FLUX PATTERNS AROUND DEFECTS AND SPECIAL MAGNETO-OPTIC EXPERIMENTS

4.1. Flux Creep Experiments

Not only can flux creep experiments be performed in magnetometers (as discussed in Section 2.4), they can also be done with the MO technique, which permits a local study of flux creep phenomena.

In Figure 27 we present a dynamic relaxation experiment at $T = 18$ K, and the flux structure is always recorded at $\mu_0 = 50$ mT. The sweep rates range from 10 mT/s (a) and 1.6 mT/s (b) to 0.2 mT/s (c).

The rectangular thin film used in this experiment was shown on to be of excellent quality, without any defects disturbing the flux patterns [92]. This permits a direct comparison of the MO images [107] to the theoretical calculations of flux profiles in perpendicular geometry [71, 75, 85]. At this relatively low temperature and small fields, the influence of the sweep rate on the flux patterns can already been seen quite clearly with respect to the amount of Meissner phase residing in the center of the sample. The slower the sweep rate during the sweep up to the target field, the smaller the flux gradients measured in the sample. If the pinning potential, U, of the sample is quite small, then very homogeneous flux patterns may be formed when is the magnetic field is slowly swept. It is important to point out that all of these subsequent experiments are started from a completely virgin state; e.g., after each experimental run the sample is warmed up to a temperature $T > T_c$ while all optical settings are kept in a fixed position. This was done to compare these results with results obtained by the procedure used in magnetometry, where the so-called minor hysteresis loops (hysteresis loops measured around a given central field value) are measured starting with the fastest sweep rate, and subsequently the sweep rate is reduced after each loop (see, e.g., the measurements presented in [31, 37, 108]). The MO flux distributions can prove that these two procedures yield identical results.

4.2. Visualization of Meissner Currents

Here we show that it is possible to perturb the Meissner sheet current flow, generating new magneto-optically visible flux

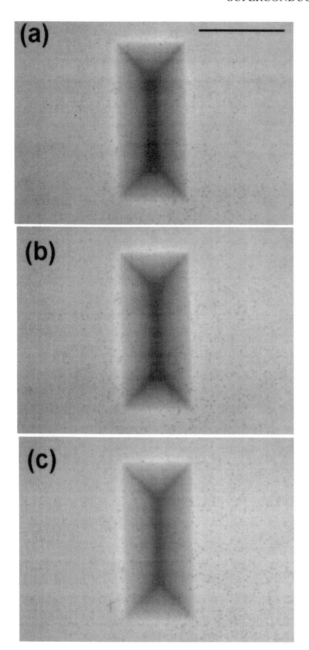

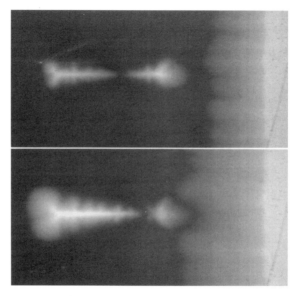

Fig. 28. Visualization of Meissner currents in a YBCO thin film. The sample is a YBCO thin film with 250-nm thickness at an observation temperature $T = 15$ K. The slit is 500 μm long. (a) Applied field of 30 mT. The flux front enters the sample from the right side; however, there is flux penetration into the sample starting from the slit. The right lobe at the slit has an opposite polarity as compared with the penetrating flux. (b) Applied field 40 mT. As the "regular" flux front approaches the right flux lobe, annihilation of reverse flux takes place, indicated by the dark border ("annihilation zone") between the flux front and the lobe. Photo provided by M. Baziljevich, University of Oslo.

Fig. 27. Flux distributions on a rectangular patterned YBCO thin film, obtained at $T = 18$ K. The marker is 1 mm long. Note the excellent quality of the sample, which is entirely free of structural defects, thus allowing a direct comparison with the theoretical calculations of flux distributions. The external magnetic field is swept in all images to $\mu_0 H_{\text{ext}} = 50$ mT, but using different sweep rates dH_{ext}/dt, (a) -10 mT/s, (b) -1.6 mT/s, and (c) -0.2 mT/s. It is clearly visible that a slow sweep rate leads to a deeper penetration of flux. All of these experiments are started from a zero-field cooled virgin state.

patterns inside the Meissner visible flux patterns inside the Meissner region [109].

The YBCO film was laterally shaped by Ar ion milling resulting in a 5-mm-long rectangular strip with a half-width of $a = 1.6$ mm. Three slits were carefully cut out of the film with a laser. Two of the slits had a length of 0.5 mm, and the third a length of 0.7 mm. The MO images of the film shown in Figure 28 were obtained after zero field cooling to 18 K

followed by a slow increase in the applied field. At 10 mT, visualization shows that the sample screens most of the field effectively, as the vortex penetration front is located less then 100 mm from the sample edge. The dark central area shows the Meissner region. The image in Figure 28 contains two distinct twin lobes of flux located deep within the Meissner region. The twin lobes obviously coincide with the positions of two of the slits. In contrast to this, a slit running parallel to the sample edge has no flux structure associated with it.

By changing the polarizers to employ the Faraday dispersion as a color discriminator of different flux polarities, we find that the lobe in a pair located near the sample edge always has reverse polarity relative to the applied field. The inner lobe has the expected polarity. The twin lobes are therefore bipolar flux structures. At an applied field of 50 mT, shown in Figure 28, the twin lobes become asymmetric as the vortex penetration front, seen as the cushionlike contour, reaches the outer lobes. When this occurs, the flux and the antiflux are brought together, and substantial annihilation occurs.

Figure 29 presents a YBCO thin film with a series of slits with increasing spacing. The applied magnetic field is 5 mT at $T = 15$ K. At this low applied field after zero-field cooling, the visible flux is entirely due to the Meissner currents, flowing in the sample.

4.3. Flux Patterns of a Superconducting Bend

Here we look at the flux patterns obtained on bends, patterned into superconducting thin films. There are, in general, two

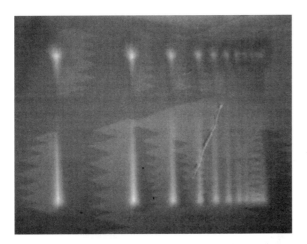

Fig. 29. YBCO thin film with a row of slits in a low applied magnetic field after zero-field cooling. $B_a = 5$ mT, $T = 15$ K. Photo provided by M. Baziljevich, University of Oslo.

possibilities for creating such a structure: a rounded bend with a certain radius, and an edge-type bend.

Bends of such types may occur in superconducting circuits used for superconducting electronics. Here it should be noted that a typical design is of the rectangular type, similar to semiconducting circuits. However, from the two MO images presented in Figure 30 it is clear that an unhindered current flow is only achieved in a rounded bend, as the current lines are running continuously through the bend. In contrast to this behavior, the inward-curved part of the rectangular bend is a source for additional flux penetration, thus disturbing the current flow. Therefore, the design of superconducting electronics has to consider this problem seriously. However, it should be noted that a difference in transport current between the two bends will only be visible at large currents.

4.4. Analysis of Flux Patterns in the Presence of Defects

Structural defects play an important role in superconducting thin films because of the large demagnetization factor in the perpendicular geometry. As a consequence, even small obstacles may alter the current flow considerably.

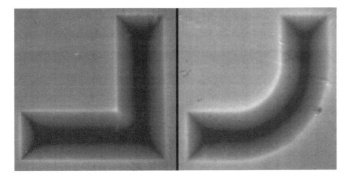

Fig. 30. YBCO thin films, patterned into bends that may occur in superconducting circuits. Left: rectangular bend; right: rounded bend. The overall length of the sample is 5 mm; the applied field is 50 mT. The MO images were obtained at $T = 18$ K.

As shown by Koblischka [92], the defects found in superconducting thin films can be classified into two categories by their influence on the flux patterns obtained:

(i) "extended" defects formed by a weakly superconducting material (typical size between some micrometers up to millimeters) that forces the currents to flow along them thus leading to faciliation of flux penetration along the defect (these defects form channels for easy flux penetration into the sample), and

(ii) small obstacles of nonsuperconducting material (typical size of micrometers and even less) where the currents can flow around them. No enhancement of the flux penetration is found, but a parabolic discontinuity line of the currents is formed.

In Figures 31 and 32 schematical current patterns for both defect classes are shown. In both figures, the critical currents are drawn with dashed lines. The typical feature of class (i) is an extended defect line (Fig. 31), where flux penetrates the sample preferentially. This leads to the formation of a finger-like flux pattern. Flux density gradients are always pointing away from the defect line, and at the end of the flux front, the currents are forced into a U-turn [110]. At the beginning of the defect line the currents form a wide turn toward the defect line, as the currents are always flowing along the sample edges and along the defect lines. The defect line, therefore, forms a quasi-sample edge within the sample and can be described by a d^- line. If the defect line is sufficiently long, one can observe that the flux penetration is largest in the center, as in the case of a real sample edge, thus forming a "lip"-like flux pattern along such a defect. The position of the accompanying d^+ lines depends on how much flux rushes along the defect, i.e., on the weakness of the superconductor in the defect area.

Figure 32 shows the current flow around a small obstacle of class (ii), here symbolized as a round object. When the

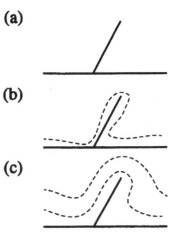

Fig. 31. A defect of linear type [class (i)], symbolized by a bold line, leads to an enhanced flux penetration along it. Consequently, the currents have to follow as indicated by the flux density gradients in the magneto-optical images. In contrast to the defect shown in Figure 32, there is no bottleneck between the regular flux front and the flux penetration along the defect line.

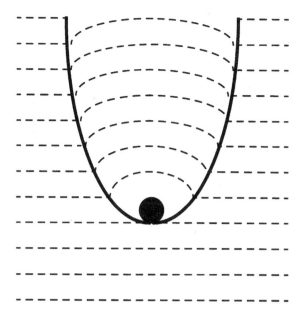

Fig. 32. A Schematical drawing of the current distribution in the vicinity of a cylindrical obstacle near the edge of a superconductor, following a description given by Campbell and Evetts [110]. The sample edge is located at the bottom of the figure. The spatially constant current density j_c is symbolized by equidistant current lines (dashed lines). When the magnetic flux penetrating from the sample edge reaches the obstacle, the current lines have to form turns to flow around the obstacle on arcs of concentric circles. These bending points lie on a parabola. Penetrating flux inside of the parabola has always to pass the defect; thus the obstacle is forming a "bottleneck" for the vortices.

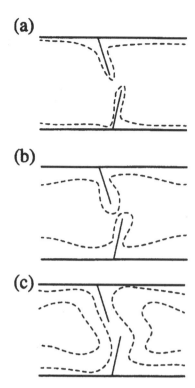

Fig. 33. In the case of a high defect density, two neighboring defects may be found in a sample (a). The flux penetration and current flow are identical to those of Figure 31 as long as the magnetic field is small (b). With raising field the vortices penetrating the sample from both sides may merge together (c). This forces the currents to flow along both defect lines, thus causing the formation of a current loop.

currents come close to this defect, they are forced into a slight turn and follow the circumference of the obstacle [110]. All of the points where the current lines turn around lie on a parabola that can be clearly observed in the measured flux distributions. This parabola is also a discontinuity line for the currents (d^+ line). In contrast to the extended defect, all vortices entering the inner area of the parabola have to move in at the defect itself, as the vortices cannot cross a d^+ line. Such a defect is therefore a "bottleneck" for the flux penetration; the presence of such defects may trigger flux jumps [91], which can indeed be observed during fast field sweeps.

If the density of defects is sufficiently large, we may encounter an area where two neighboring defects are located (see Fig. 33). At small fields, the current flow is analogous to that in Figure 31. During flux penetration in an increasing magnetic field, it may happen that the flux fronts meet each other. Now the current pattern changes completely, as the currents can now flow along both defect lines and form new, smaller current loops. Consequently, the shielding currents are now also flowing in smaller subloops. Now the length scale of the currents has changed and the sample is split up magnetically into smaller pieces. If the external magnetic field is increased further, the parts of the sample will behave completely independently from each other. Such behavior is indicated by the presence of d^+ lines in each part of the sample. This is an example of magnetically induced granularity, with the defect lines acting as weak links at higher magnetic fields.

An example of typical defects found in superconducting thin films is given in Figure 34. This figure presents four pieces of YBCO thin film patterned into rectangles of $1600 \times 800 \ \mu m^2$. One piece of this batch showed no defect at all (see Figure 8), and another one had only one small defect (see Figure 12). Figure 34 presents typical results that were found within the same batch of samples, indicating that a

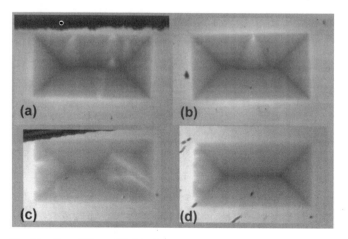

Fig. 34. Four different YBCO thin films of the same batch, patterned into rectangles with a size of $1600 \ \mu m \times 600 \ \mu m$. The applied field is 80 mT; The observation temperature is 10 K.

completely defect-free sample is quite rare. In Figure 34a three defects are visible close to the sample edge, which were introduced into the sample by means of the patterning process. The two small defects inside the sample may be due to the mask removal or defects created during the laser ablation process (flux droplets). Otherwise, the full d line pattern can be observed; practically undisturbed by the present defects. Figure 34b presents a sample with only one defect close to the sample edge; again due to the patterning. Figure 34c is an example showing extended defects, which stem from scratches in the substrate. Here the d line pattern is completely disturbed. Figure 34d shows a nearly perfect sample, where on the left side a problem must have occurred during the etching, leading to a very bright d^- line, followed by several stripes in the MO pattern. Otherwise the sample behaves very homogeneously.

Figure 35 presents the flux distribution on a YBCO thin film (size 10×10 mm^2) showing two sample edges. The sample contains numerous defects, which cause a complicated flux pattern that is entirely different from what is expected for an ideal, homogeneous sample. The "regular" flux penetration starting from the sample edges and, hence, the pattern of the d lines is completely suppressed. Most of the extended defects in this sample stem from a substrate that is not properly polished. Integral magnetic measurements (and transport current measurements) lead to wrong results, as the real size of the current loops is not known.

Further investigations of the influence of the substrate on the magnetic flux patterns in superconducting thin films were performed by Surdeanu et al. [111]. This work again demonstrated the advantages of the local investigation technique.

4.5. Grain Boundary Studies

A special application of the MO technique in the study of defects in superconducting thin films is devoted to the study of grain boundaries [112,113]. Planar crystalline defects, such as grain and twin boundaries, intergrowths, and stacking faults can impose limitations on the critical current density J_c of high-T_c superconductors because the short coherence length of these materials tends to make planar defects weakly coupled [114]. High-angle grain boundaries are particularly effective barriers to current flow. There is an extensive literature on the dependence of the intergrain critical current density J_b on the misorientation angle Θ of bicrystals [115–129]. MO studies can also be very helpful in tracing the crossover from strongly to weakly coupled grain boundaries with increasing Θ and in determining whether boundaries are uniform in their properties. These characteristics may prove to be important for further optimization of polycrystalline high-T_c materials, especially in light of recent reports of extremely high J_c values observed in YBa$_2$Cu$_3$O$_{7-\delta}$ thick-film composites.

The series of MO images given in Figure 36 illustrates the development of a weak-link behavior with an increasing misorientation angle of the grain boundary. All images are taken in an applied magnetic field of ∼40 mT. A 3° grain boundary is hardly visible in the MO image (as compared with other structural defects in the sample), and the current density across the boundary is 95% of the original J_c value. From 5° on, the weak-link behavior is fully developed and the according current discontinuity lines (d^+ lines) are easily visible. Accordingly, the current density across the grain boundary is also considerably reduced. This decrease in the current density is directly visible from the MO images as the angle of the dark d^+ line with the sample contour line. This is further illustrated in Figure 37.

Figure 37 presents the flux and current distribution around a 5° grain boundary in detail. The first image is the real MO image; the second gives the corresponding contour map of $B_z(x, y)$. The third figure illustrates the current flow around the grain boundary, with the approach sketched in Section 4.4. The direction of the current flow is indicated by arrows; the grain boundary is a d^- line. From such an analysis, it is possible to determine $J_b(\Theta)$ independently of the current contribution from the grains, which is always present in magnetization or transport current studies.

The MO imaging technique now further permits the extraction of J_c and $J_b(\Theta)$ as a function of temperature (Fig. 38). This analysis was performed in [112], with the result that the noticeable differences found at low temperatures become less pronounced at higher temperatures (the measurements were performed up to $T = 77$ K). This weaker temperature dependence of the intragrain $J_b(\Theta)$ is attributed to additional flux pinning at the grain boundaries provided by the grain boundary dislocations. Low-angle grain boundaries are not continuous interfaces, but rather a chain of edge dislocations separated by regions of comparatively undisturbed crystal lattice [116, 122, 130]. The dislocation cores are believed to suppress T_c locally; at the same time, the chain of dislocation cores could provide additional flux pinning for the intergrain vortices. This extra flux pinning may then become more significant at higher temperatures.

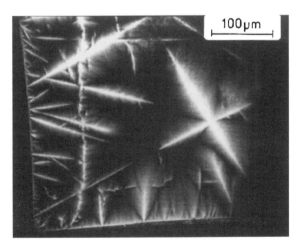

Fig. 35. Flux distribution on a YBCO thin film with a total size of 10×10 mm^2 at $T = 5$ K and an external magnetic field of 0.1 T applied perpendicular to the sample surface. Nearly the entire flux penetration into the sample goes through the various defects; the regular flux penetration starting from the sample edges is completely suppressed.

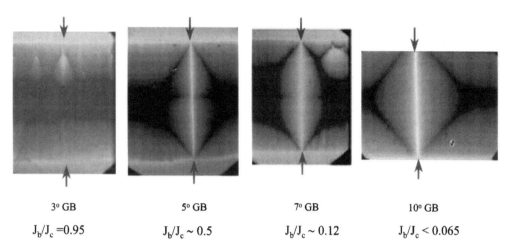

3° GB	5° GB	7° GB	10° GB
$J_b/J_c = 0.95$	$J_b/J_c \sim 0.5$	$J_b/J_c \sim 0.12$	$J_b/J_c < 0.065$

Fig. 36. Magnetic flux behavior around 3°, 5°, 7°, and 10° grain boundaries in superconducting $YBa_2Cu_3O_{7-\delta}$ thin films. MO images taken at $T = 7K$, $\mu_0H_a \approx 40$ mT, $H \parallel c$. The arrows point to the position of the d^+ lines. At an angle of 3°, the position of the d^- line is hardly visible; at all angles above 5°, the d^- line is well developed. Image provided by A. Polyanskii, University of Wisconsin-Madison.

4.6. Large Thin Film Samples

Figure 39 presents a thin film sample with a diameter of 8 mm. This sample size is about the maximum for which observations can be carried out in a regular MO apparatus. The image shown is a field-cooled remanent state; i.e., the sample was field-cooled to 18 K in a field of 30 mT, and the field was subsequently removed. In this way, a fully penetrated state is obtained, even though the available field would otherwise not be sufficient to achieve a fully penetrated state of such a large sample. The entering vortices of opposite polarity effectively "scan" the structural defects along the sample edges due to the annihilation process [131]. The superconducting thin film shown has an excellent quality and was subsequently used for levitation experiments as described in [132].

Figure 40 presents an MO image from a large thin film sample taken at the University of Augsburg, with the use of a special MO setup with an inhomogeneous magnetic field [133–136]. The use of an inhomogeneous magnetic field was suggested by Xing et al. [137]; this experimental arrangement allows the nondestructive study of extended thin film structures, but an inhomogeneous magnetic field has to be employed, which does not allow, for example, the current reconstruction procedure described in Section 3.6, but has the advantage that the field homogeneity and the size of the magnet do not limit the sample sizes.

The MO image is composed of 18 individual frames taken during an x-y scan of the entire sample. The maximum applied field is 45 mT; the observation temperature is $T = 45$ K. To

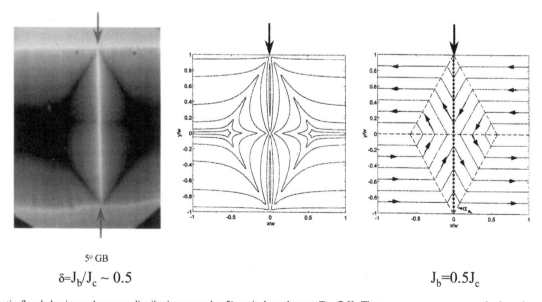

5° GB

$\delta = J_b/J_c \sim 0.5$

$J_b = 0.5J_c$

Fig. 37. Magnetic flux behavior and current distribution around a 5° grain boundary at $T = 7$ K. The transport current across the boundary is only 50% of the original value. The left image presents the original MO flux pattern; the middle image gives the corresponding contour map of $B_z(x, y)$. The third figure illustrates the current flow around the grain boundary, using the approach sketched in Section 4.4. Reprinted with permission from [112], © 1996, American Physical Society.

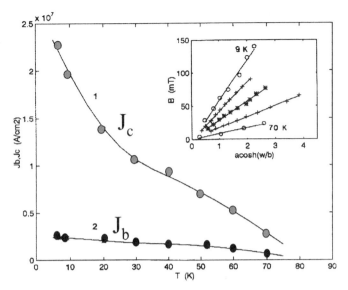

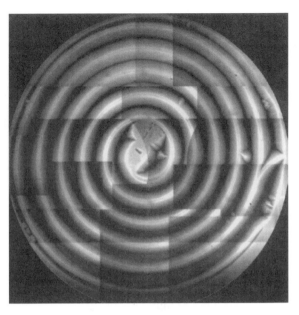

Fig. 38. Temperature dependence of the intergrain J_c (upper curve 1); the data are taken from the thin film sample with a 7° grain boundary. The lower curve (2) represents the temperature dependence of J_b. The inset shows the linear fit of the observed width of the Meissner region $2b$ for $T = 9, 20, 40, 60,$ and 70 K. Reprinted with permission from [112], © 1996, American Physical Society.

Fig. 40. Flux pattern obtained on a YBCO thin film patterned into a spiral-like structure. The MO image is composed of 18 individual frames taken during an $x - y$ scan of the entire sample. The maximum applied field is 45 mT; the observation temperature is $T = 45$ K. The field is generated by means of a permanent magnet. This image was taken at the University of Augsburg, Germany, courtesy of M. Kuhn [133].

generate the applied magnetic field, a permanent magnet is employed. The magnetic field has an axial symmetry for the homogeneous field in the x direction. In the y direction, the applied field shows a large field gradient. The maximum field is in all frames in the middle of each frame. The MO image reveals several macroscopic defects within the superconducting thin film; the large extended defect in the center stems

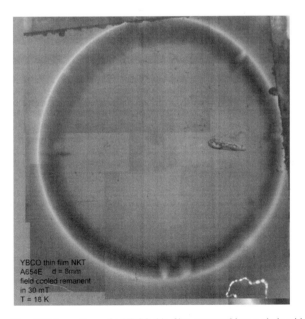

Fig. 39. MO flux pattern of a YBCO thin film, patterned into a circle with a diameter of 8 mm. The sample was field-cooled in a magnetic field of 30 mT, then the field was removed; i.e., the image presents a field-cooled remanent state.

from burning of the thin film bridge after a large current is applied.

The group at the University of Konstanz had developed a setup that allows the investigation of 3-inch wafers at liquid nitrogen temperature (77 K), where the achievable MO contrasts are considerably lower. This allows a sample testing even under conditions were cooling with liquid helium or closed-cycle refrigerators is not available. This setup again uses a homogeneous magnetic field but requires large MO indicator films of the same size as the samples to be investigated [138]. Figure 41 presents a MO image of a 3-inch double-sided YBCO wafer at $T = 10$ K, at homogeneous magnetic field of 86 mT. The image was taken with a 12-bit slow-scan CCD camera. The arrows in the image point to positions were the wafer was fixed during fabrication. It is clearly visible (note the different d line pattern as compared with, for example, the image presented in Fig. 10) that the deviation from the perfect round shape leads to a considerable change in the current flow.

These developments of MO flux visualization in large-scale samples offers new possibilities for the use of MO imaging for quality control of superconducting samples.

4.7. Current Flow in Samples with Variation of Current Density

The principles laid out with the d-line concept (see Section 3) can also be applied to much more complicated flux patterns. Here two YBCO thin films are presented, the left half of which was irradiated by 1.0-GeV lead ions, but with two different fluences. In Figure 42a, a fluence of 9.0×10^{10} ions/cm^2 was

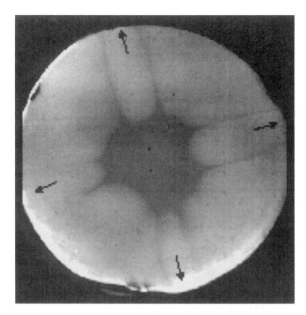

Fig. 41.   Flux distribution of a 3-inch double-sided YBCO wafer at $T = 10$ K and at a homogeneous magnetic field of 86 mT. The image was taken with a 12-bit slow-scan CCD camera. The arrows in the image point to positions were the wafer was fixed during fabrication. It is clearly visible that the deviation from the perfect round shape leads to a considerable change in the current flow. Image provided by B. -U. Runge, University of Konstanz [138].

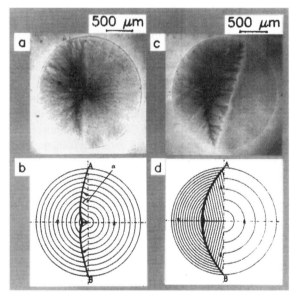

Fig. 42.   (a) Flux distribution obtained from a partly irradiated YBCO thin film at an external applied field of 273 mT. The left part of the film was irradiated with 1.0-GeV lead ions at a fluence of 9.0×10^{10} ions/cm². The asymmetric flux distribution due to the irradiation-induced enhancement of the critical current density is clearly visible. (b) Current distribution of the thin film presented in (a). The constant critical current density in each half of the sample is symbolized by equidistant concentric circles. (c) Flux distribution obtained from a YBCO thin film at an external magnetic field of 273 mT; the left half of this sample was irradiated with lead ions, but with a higher fluence of 30.0×10^{10} ions/cm². The difference in the critical current densities in both parts is apparently larger than in the case presented in (a). (d) Current distribution of the sample presented in (c).

used; in Figure 42c a fluence of 30.0×10^{10} ions/cm² was used. The asymmetry of the magnetic flux distribution indicates the differences in the critical current densities in the two halves of each sample. The difference in the two critical current densities, $j_c(0)$ (unirradiated area) and $j_c(\Phi t)$ (irradiated area) is visualized by different densities of current lines in the two areas. In the unirradiated part of the sample (right), $j_c(0)$ is flowing parallel to the edge (i.e., on concentric circles), as in an ideal film. At the diameter dividing the irradiated area from the unirradiated area (left), the current lines have to bend to the angle α according to

$$j_c(0) = j_c(\Phi t) \sin \alpha \qquad (10)$$

Thus, the concentric circles go over into a denser set of parallel current lines in the irradiated area. In this area, the currents $j_c(\Phi t)$ are also forming concentric circles. From geometrical considerations, the position of the d^+ lines can be calculated. Furthermore, the measurement of the angle α from the image allows one to obtain the relation between the two current densities, $j_c(\Phi t)/j_c(0)$.

4.8. Sample for Modeling Properties of Bi-2223 Tapes

In this section, MO flux patterns on a sample used to model a layer of grains in a superconducting tape is presented.

Properties of granular samples are important to be understand in detail, as most future applications will involve high-T_c superconductors with a certain amount of granularity, whether it is $(Pb,Bi)_2Sr_2Ca_2Cu_3O_{10+\delta}$ (Bi-2223) tapes for power applications or bulk, melt-processed $(RE)Ba_2Cu_3O_{7-\delta}$ (RE = rare earths) for levitation or permanent magnets [2]. In such granular samples, there are always two different contributions to the current flow: the intragranular currents (inside the individual grains), j_{grain} or j_{intra}, and the intergranular currents, j_{inter} or j_{trans} (between the grains, i.e., the "transport currents") [139, 140]. As these contributions are generally not independent from each other, model descriptions are required. Recently, in Bi-2223 tapes an anomalous position of the central (low field) peak in magnetization loops was observed [141, 142]. Such a sample was originally suggested by Koblischka et al. [141].

The model sample was prepared by means of laser ablation on a $LaAlO_3$ substrate. The thickness of the $YBa_2Cu_3O_{7-\delta}$ (YBCO) thin film is 150 nm. The film is then patterned into a hexagonal close-packed lattice of disks ($2r = 50$ μm) by means of electron beam lithography. The disks are touching each other at the circumferences to allow the flow of an intergranular current; this contact width is chosen to be $w = 3.5$ μm. The sample has an overall size of 4×4 mm², comprising ~8000 disks. The transition temperature, T_c, after the patterning process is about 83 K. Figure 43 shows the original drawing of the model structure, and in Figure 44, the realized structure, patterned into a YBCO thin film, is presented.

An MO flux pattern is presented in Figure 45, at a low magnification. At $B_e = 18$ mT, flux enters the structure starting from the sample edges, so a flux front can be observed, even

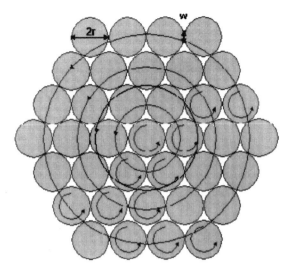

Fig. 43. Schematical drawing of the model sample.

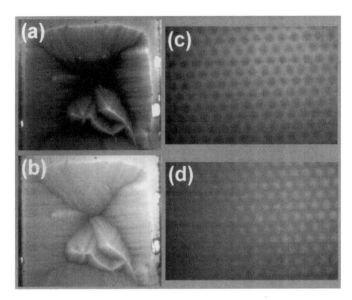

Fig. 45. (a and b) Flux patterns of the model sample at low magnification, 18 mT and 45 mT. (c and d) High magnification revealing the behavior of individual disks. (c) 18 mT. (d) Remanent state.

though the model sample is not a continuous thin film. This is a quite surprising observation, revealing that vortices exist in the intergranular area, thus forming an "effective medium." The set of current discontinuity lines [91] (dark) generated is direct proof that this flux pattern is evidently due to the intergranular currents. These observations have further direct consequences for the interpretation of flux patterns of samples with granularity. This is especially important for the observations of flux penetration carried on Bi-2223 tapes, which are a special high-T_c system in the sense that large transport currents can flow, but there is also a certain amount of granularity left in the system. Flux patterns of such superconductors are complicated to interpret properly. From our present observations, we can confirm that the generation of d lines in an effective medium is possible. This directly implies that the flux patterns revealing the appearance of d lines (which are due to the flow of currents in the entire sample) are due to the flow of intergranular currents. Effects of the individual grains are then superimposed on this pattern. In Bi-2223 tapes, flux patterns at elevated temperatures were found to be nearly homogeneous, but at low temperatures clear signs of granular-

ity were observed [143, 144]. At temperatures of $T > 50$ K, the contributions of the intragranular currents inside the Bi-2223 grains are small, and the flux patterns observed are mainly due to the intergranular currents, thus causing a nearly homogeneous flux distribution [143, 144]. From the observations on our model sample, we can now confirm that these uniform flux patterns are unambiguously due to the flow of the intergranular currents. The contribution of the Bi-2223 grains to the flux patterns can only be seen at low temperatures, as the intragranular current density increases steeply with decreasing temperature.

A discussion of the results obtained with this model sample and the comparison of the local flux distribution with the integral properties measured by a SQUID magnetometer can be found in [145]. Furthermore, temperature-dependent magnetization measurements in small applied magnetic fields showed that this model sample exhibits the paramagnetic Meissner effect [146].

4.9. Thick Superconducting Films

Finally, MO images of a thick YBCO film are presented. The YBCO film is about 3 μm thick and is prepared with an electrophoretic deposition technique [148], followed by a melt-processing step. In contrast to the laser-ablated thin films, the superconducting film does not grow homogeneously, so single superconducting grains can be observed. The connections between the grains may vary in characteristics, depending on the misorientation angle between them. Figure 46 presents magneto-optic flux patterns obtained during reduction of the external magnetic field. As already described in Section 4.6, the observation of flux patterns during field reduction results from quasi-scans of the defect structure of a superconducting sample. Here, the weak links between the strongly superconducting YBCO grains become clearly visible; in the remaining

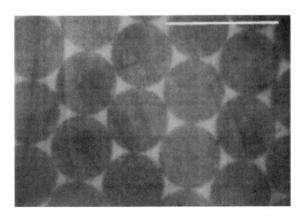

Fig. 44. Polarization image of the realized structure; the diameter of each disk is $2r = 50$ μm, and the overlap of the disks is 3.5 μm.

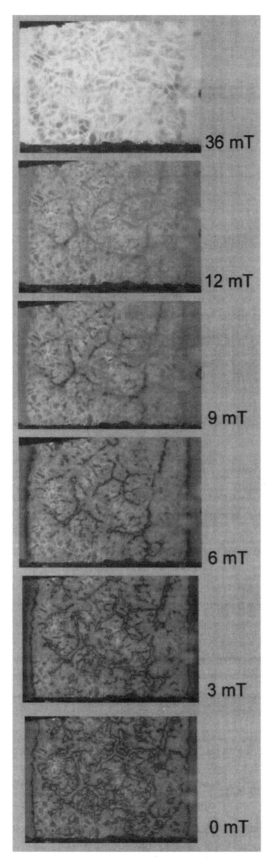

Fig. 46. Flux patterns on an YBCO thick film. Observation temperature, 18 K. The magnetic field is reduced from 90 mT to 36 mT, 12 mT, 9 mT, 6 mT, and 3 mT, and, finally, the remanent state is reached.

field of 12 mT, the flux has been annihilated completely in the weakest channels. When the external magnetic field is reduced even further, the penetration/annihilation will gradually proceeds, until in the remanent state trapped flux remains only within the YBCO grains.

These experiments on the YBCO thick films demonstrate the sensitivity of the MO technique to crack structures in superconducting samples, which makes the MO technique an ideal tool for investigating defect structures on a length scale in the millimeter to micrometer range.

4.10. Magneto-Optic Studies with High Time Resolution

The use of Eu-based chacogenide films makes it possible to reach an extremely high time resolution that cannot be achieved with any other flux visualization technique [7, 11]. Hübener [26] developed a stroboscopic technique to visualize the movement of a macro-vortex in a lead film from one pinning site to another. The time resolution of these experiments was about 10^{-6} s. Recently, a series of experiments was carried out at the University of Konstanz that reached a time resolution in the nanosecond range. The first experiments [149, 150] studied the magnetic flux patterns triggered by a magnetic instability into the Meissner state of a superconducting thin film. These experiments were then subsequently extended to a study of the flux patterns in different applied magnetic fields [151, 152]. The equilibrium current distribution was then disturbed by a laser pulse (generated with a frequency doubled Nd:YAG laser), which is focused on the thin film from the substrate side.

Figures 47 and 48 present typical flux patterns after the laser pulse in different applied magnetic fields. In contrast

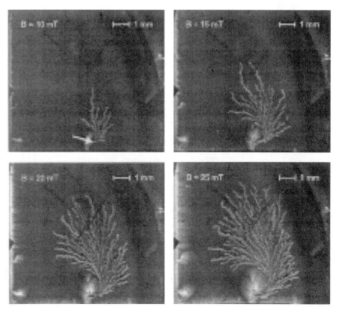

Fig. 47. Magneto-optical images of a YBCO film zero field cooled down to 10 K. The flux distribution after a 10 mJ/cm^2 laser pulse focused to a 30-μm spot close to the bottom edge of the image (arrow). The area of the pattern increases linearly with the applied external magnetic field above a threshold ($B_{ext} =7.5$ mT). Reprinted [151], © 2000, with permission from Elsevier Science.

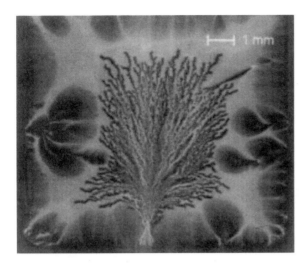

Fig. 48. MO image taken under conditions similar to those in Figure 47. Instead of the increasing B_{ext}, the external field (here 45 mT) was switched off ($\Delta B_{ext} < 0$) before the laser pulse was applied. Reprinted with permission from [151], © 2000 Elsevier Science.

to the more or less homogeneous flux fronts that propagate toward the sample center with increasing applied magnetic field, the instability develops in the form of a dentritic pattern, yielding quite spectacular MO images. A similar structure was also found in Nb crystals by means of the decoration technique [153].

5. MAGNETIZATION MEASUREMENTS

In this section, integral magnetization measurements and some peculiar observations on superconducting thin film samples are presented. Magnetization hysteresis loops (MHLs) of homogeneous thin film samples exhibit the central peak exactly at 0 T, which was very recently demonstrated both experimentally and theoretically [96]. Flux pinning forces, as defined by $F_p = j_c \times B$, are found to exhibit a peak at $h_0 = 0.33$ [154], which is a strong indication of flux pinning at normally conducting or insulating pinning sites, so-called δl-pinning [35], corresponding to the results of Dew-Hughes [155], Kramer [156], and Webb et al. [157], obtained with conventional superconductors. This result was recently confirmed by Jooss et al. [158] and Dam et al. [159] for different types of YBCO thin films.

Another phenomenon that occurs during magnetization measurements of superconducting thin films is the appearance of additive paramagnetic moments. As shown by Pûst et al. [160], this additional paramagnetic moment is a consequence of the large demagnetization factor of the samples, together with the inhomogeneity of the applied external magnetic field in the measurement apparatus.

5.1. Magnetization Measurements of Superconducting Thin Films

Magnetization loops of superconducting thin films are shown in Figure 4. In this section, I focus on some features observed when magnetization loops of thin film samples are measured.

An ever-present feature of MHLs is a peak in the magnetization located at an applied field B_{cp} near zero. The central peak is due to a field dependence of the critical current density, $J_c(B)$, monotonously decreasing at small fields. When the sample is a long cylindrical body placed in a parallel applied field, the peak position can be calculated analytically within the critical-state model for several $J_c(B)$ dependencies [147]. One always finds on the descending field branch that $B_{cp} < 0$, and similarly $B_{cp} > 0$ on the ascending branch of large-field loops. In simple terms the shift is a consequence of the local flux density, B, lagging behind the applied field. Consider a long thin superconducting strip with edges located at $x = \pm w$, the y axis pointing along the strip, and the z axis normal to the strip plane. The magnetic field, B_a, is applied along the z axis, so screening currents flow in the y direction. B is the z component of magnetic induction in the strip plane. The sheet current is defined as $J(x) = \int j(x, z)dz$, where $j(x, z)$ is the current density and the integration is performed over the strip thickness, $d \ll w$.

From the Biot–Savart law for the strip geometry, the flux density is given by [70]

$$B(x) - B_a = -\frac{\mu_0}{2\pi} \int_{-w}^{w} \frac{J(u)\,du}{x - u} \qquad (11)$$

Assume that the strip is in a fully penetrated state, i.e., the current density is everywhere equal to the critical current density, $J(x) = \text{sign}(x)J_c[B(x)]$. This state can be reached after a very large field is applied, reducing it to some much smaller value. The field distribution then satisfies the integral equation

$$B(x) - B_a = -\frac{\mu_0}{\pi} \int_0^w \frac{J_c[B(u)]}{x^2 - u^2}u\,du \qquad (12)$$

In the remanent state, $B_a = 0$, the flux density profile $B(x)$ has an interesting symmetry. This is seen by changing the integration variable in Eq. (12) from u to $v = \sqrt{w^2 - u^2}$. We then obtain

$$B(x) - B_a = \frac{\mu_0}{\pi} \int_0^w \frac{J_c[B(\sqrt{w^2 - v^2})]}{w^2 - x^2 - v^2}v\,dv \qquad (13)$$

Replacing x with $\sqrt{w^2 - x^2}$ in Eq. (12), one obtains a similar equation for $B(\sqrt{w^2 - x^2})$,

$$B(\sqrt{w^2 - x^2}) - B_a = -\frac{\mu_0}{\pi} \int_0^w \frac{J_c[B(u)]}{w^2 - x^2 - u^2}u\,du \qquad (14)$$

By comparing Eqs. (13) and (14) at $B_a = 0$, we conclude that

$$B(x) = -B(\sqrt{w^2 - x^2}) \qquad (15)$$

is generally valid if J_c depends only on the absolute value of the magnetic induction, $J_c(|B|)$. This symmetry is immediately evident for the case $J_c = \text{const.}$, i.e., the Bean model, when Eq. (12) reduces to

$$B(x) = B_a + B_c \ln \frac{\sqrt{w^2 - x^2}}{x}, \qquad B_c = \frac{\mu_0 J_c}{\pi} \qquad (16)$$

In a similar way one can prove also another symmetry relation valid at $B_a = 0$, namely

$$\mathscr{D}(x) = \mathscr{D}(\sqrt{w^2 - x^2}) \qquad (17)$$

where $\mathscr{D}(x) = \partial B(x)/\partial B_a$.

Consider now the magnetic moment per unit length of the strip, $M = 2\int_0^w J(x)x\,dx$. Differentiating M with respect to B_a and taking into account that $B(x)$ changes sign at $x = w^* = w\sqrt{2}$, one has, after splitting the integral into two parts,

$$\frac{\partial M}{\partial B_a} = 2\left(\int_0^{w^*} - \int_{w^*}^w\right)\frac{\partial J_c(|B(x)|)}{\partial |B|}\mathscr{D}(x)x\,dx \qquad (18)$$

Then, replacing x in the second integral with $\sqrt{w^2 - x^2}$ and using Eq. (15), we come to

$$\frac{\partial M}{\partial B_a} = 0 \qquad \text{at} \quad B_a = 0 \qquad (19)$$

Consequently, on a major MHL the magnetic moment has an extremum in the remanent state for any $J_c(B)$ dependence. An iterative numerical solution of Eq. 12 for various $J_c(B)$ dependences showed that a maximum for a decreasing $J_c(B)$ is obtained, and a minimum is obtained if $J_c(B)$ has a pronounced second peak, the so-called fishtail behavior. Figure 49a shows the central part of the MHLs for the uniform film. The measurements were carried out over a wide range of temperatures to probe different $J_c(B)$ dependences.

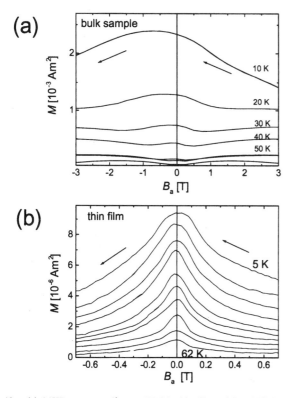

Fig. 49. (a) MHLs measured on a YBCO thin film with a VSM magnetometer. (b) MHLs measured on a bulk sample (YBCO single crystal) for comparison.

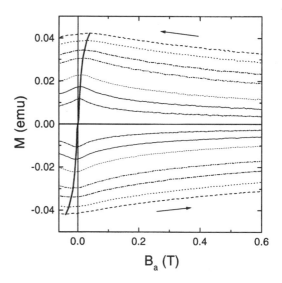

Fig. 50. MHLs of the model sample, YBCO thin film with a artificial granularity. Note the occurrence of the central peak at positive applied fields when the external magnetic field is decreased. At low temperatures, the effect becomes more pronounced.

Here there is evidently a pronounced central peak at all temperatures. Within the experimental resolution the position of the peak is located at a zero applied field. The observation that the peak remains at $B_a = 0$ over the entire temperature range is in full agreement with our general analytical result. A different result is obtained for the model sample with artifical granularity, as shown in Figure 50.

Indeed, this sample exhibits the central peak position at positive field values, as predicted in the original paper by Koblischka et al. [141], (see also Section 4.8).

Note that the result was derived for an infinite strip fully penetrated by the magnetic field. In a thin sample, in contrast to long cylinders, the flux fronts move in response to a changing applied field at an exponential rate [70]. Therefore, a fully remagnetized state is reached quickly after the direction of a field sweep is reversed, and our experimental conditions are consistent with the assumptions made in the theory. The only exception is the finite length of the strip. However, our results show that it is not a crucial factor for the peak position (Fig. 51).

5.2. Flux Pinning in Thin Films

Integral magnetization measurements of superconducting thin films are well suited to determination the volume flux pinning forces, in analogy to bulk superconductors. With the use of pinning force scaling, which was developed for conventional superconductors by Dew-Hughes [155] and Kramer [156,157], a peak is obtained at $h_0 = 0.33$ when the normalized pinning forces are plotted, $F_p/F_{p,\max}$ versus a reduced field, $h = H_a/H_{irr}$. H_{irr} denotes the irreversibility field, which provides the upper limit for strong flux pinning, and H_a is the applied field. The peak position at $h_0 = 0.33$ indicates a dominant flux pinning at insulating or normally-conducting pinning sites,

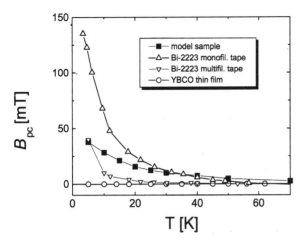

Fig. 51. Summary of the peak positions of a variety of superconducting samples as a function of temperature. Note also that above 55 K, the anomalous positions of the central peak vanish.

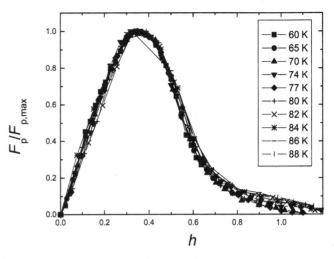

Fig. 52. Pinning force scaling of a YBCO thin film, measured at temperatures between 60 and 88 K. The sample is identical to the one presented in Figure 4.

which have a size comparable to that of the superconducting coherence length, ξ. This type of pinning is called δl-pinning [35]. Another important source of flux pinning is the so-called δT_c-pinning, where flux pinning is provided by a spatial variation of the superconducting transition temperature, T_c, throughout the sample and, hence, a variation of the material parameter, κ. Many such scaling experiments of the flux pinning forces or of the critical current densities were carried out in the literature, by means of magnetization measurements as well as by magnetotransport measurements [161].

Using different investigation techniques, Jooss et al. [158] and Dam et al. [159] concluded that δl-pinning is the dominant pinning in YBCO thin films, which confirms the result of the pinning force scaling and the older work by Griessen et al. [162] and van Dalen et al. [108], who used an inversion scheme to calculate the true critical current density (i.e., the current density unaffected by flux creep). This result for thin films is in contrast with recent measurements of pinning forces on bulk melt-processed YBCO and NdBCO samples, where δT_c-pinning plays an important role as well [34] (Figs. 52 and 53). This behavior can be explained as follows: by nature, δT_c-pinning is weaker as compared with δl-pinning (δl-pinning is practically a special case of δT_c-pinning with $T_c = 0$ K). Therefore, in bulk samples, with critical current densities much smaller than those of thin films, the effects of δT_c-pinning can still be observed. Accordingly, irradiation experiments producing a large number of δl-pinning sites [163] clearly demonstrated that flux pinning is then completely dominated by δl-pinning. Another proof for this assumption comes from irradiation experiments performed on superconducting YBCO thin films [98], where irradiation with lead ions could not produce further increase in the critical current density, even though at the same fluence a drastic increase in the j_c of single crystals was measured.

5.3. Virtual Additive Moments

In VSM measurements, the amplitude of sample vibrations is small (typically a few tenths of a millimeter), and the effect of sample translations during the measurement can usually be neglected. The finite height of the MHL ($m^+ - m^-$), where m^+ and m^- are the moment values on the descending and ascending field branches, respectively, is a result of an equilibrium between induction of critical currents by a change of the external field and the simultaneous relaxation of the induced current due to thermal activation. Therefore, the shape of a MHL depends strongly on the field sweep rate R [37]. The MHL is narrow at low R and wide at high R. In most superconducting films (if these do not contain magnetic ions like, e.g., Gd or Nd) the reversible magnetic moment, defined as $m_r = (m^+ + m^-)/2$, is negligible, and the MHL is therefore

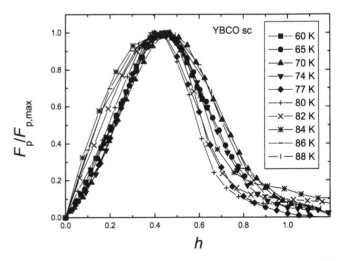

Fig. 53. Pinning forces of YBCO single crystal for comparison, measured in the same temperature range. The scaling is not nearly as good. The nonscaling even indicates some crossover of pinning regimes. See the detailed discussion in [34].

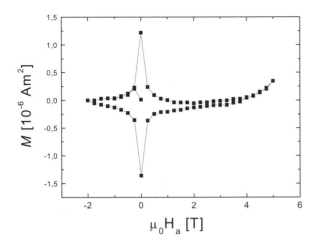

Fig. 54. MHL measured on a YBCO thin film showing a virtual additive paramagnetic moment; $T = 70$ K. The measurement was performed with a SQUID magnetometer.

nearly symmetrical with respect to the field axis, irrespective of the field sweep rate (Fig. 54).

However, not only were MHLs measured on superconducting thin films at low sweep rates found to be narrower; they also exhibited an anomalous additive reversible moment (i.e., they were shifted at high positive fields toward positive moment values and at high negative fields toward negative moment values) [160]. Measurements at larger sweep rates did not reveal such an offset. This peculiar behavior was particularly evident at intermediate temperatures (40–65 K). As shown in Figure 54, this additional paramagnetic moment is also visible in SQUID measurements.

Půst et al. [160] presented a model to explain this special feature of magnetic measurements of thin films, and the appearance of this virtual magnetic moment is ascribed to field inhomogeneities. Furthermore, such virtual additive moments are seen in general when large superconducting thin films are measured laterally on any type of magnetometer.

5.4. Flux Creep Measurements

Flux creep, i.e., the time relaxation of magnetic flux in a superconducting sample, was first described by Webb et al. for conventional superconductors. In high-T_c superconductors, effects of flux creep play a much more important role, as the higher κ values cause smaller pinning potentials. Therefore, flux creep effects are dominant in high-T_c samples at elevated temperatures above 77 K.

As already mentioned in Section 2.4, an important piece of information concerning the pinning potential can be obtained from flux creep measurements. Superconducting thin films are especially well suited for flux creep measurements with the dynamic technique employing various sweep rates of the external magnetic field. This is due to the fact that the field of full penetration is typically quite small in thin film samples, so relatively small minor loops are sufficient to obtain the critical current density as a function of the field sweep rate. Such

an experiment is shown in Figure 5. Flux creep measurements on YBCO thin films revealed that the dominant flux pinning mechanism is of the δl type [108, 162], which corroborates results obtained with other techniques.

Sheriff et al. [15] investigated the thickness dependence of the magnetic properties of laser-ablated YBCO thin films with thicknesses between 35 and 300 nm. As the thickness is decreased, domination of the surface pinning mechanism over the bulk pinning gives rise to increasing critical current density, j_c. The results presented in their paper contain significant implications related to applications of superconducting thin films. At low temperatures, thin samples with large j_c are preferable, whereas at elevated temperatures it is important to consider the time window of the application. For DC applications, thicker films are preferable because of their lower relaxation rates.

6. CONCLUSIONS

In this chapter, local (magneto-optic flux patterns) and integral magnetization measurements on superconducting thin films were presented. Because of the large demagnetization factor of such thin film samples, the magnetic behavior exhibits several peculiar features. The local investigations of flux patterns by means of the MO technique allow the study of defect structures in commercially produced thin film samples, thus the MO technique is a very sensitive tool for quality assessment. Furthermore, a variety of new physical phenomena (flux and current distributions in perpendicular geometry, Meissner currents, current-induced flux patterns) have been observed on superconducting thin films.

Acknowledgments

The MO experiments were performed at Risø National Laboratory, University of Oslo, and the MPI für Metallforschung, Stuttgart. Magnetization measurements were performed during my stay at SRL/ISTEC, Tokyo. I express my thanks to E. H. Brandt, M. Baziljevich, T. H. Johansen, D. V. Shantsev, L. Půst, M. McElfresh, T. Tamegai, N. H. Andersen, B. Stritzker, P. Leiderer, P. Kleiner, H. Kronmüller, and M. Murakami for valuable discussions. M. Baziljevich, D. V. Shantsev, R. Straub, and M. Kuhn provided unpublished results of their experiments. The samples shown in this article were prepared by Y. Shen (NKT Research Center, Brøndby, Denmark) and by B. Nilsson and T. Claeson (Chalmers University, Gothenburg, Sweden).

The work presented here was partly subsidized by the Research Council of Norway and the Danish Research Council. Furthermore, I thank the STA (Japanese Science and Technology Agency) for support during my stay at SRL/ISTEC, Tokyo.

REFERENCES

1. "Proceedings of the 1998 Applied Superconductivity Conference," Desert Springs Resort, Palm Desert, CA, September 13–18, 1998. *IEEE Trans. Appl. Supercond.* 9, No. 2 (Parts I–III) (1999).

2. H. Rogalla and D. H. A. Blank, Eds., "Proceedings of the 3rd EUCAS Conference," Veldhoven, the Netherlands, Institute of Physics, Bristol, U.K. IOP Conference Series 158.

3. "Advances in Superconductivity, Proceedings of the 12th ISS'99," October 17–19, Morioka, Vol. 12. Springer-Verlag, Tokyo, 2000.

4. M. Bauer, R. Semerad, H. Kinder, J. Wiesmann, J. Dzick, and H. C. Freyhardt, *IEEE Trans. Appl. Supercond.* 9, 2244 (1999).

4a. R. Utz, R. Semerad, M. Bauer, W. Prusseit, P. Berberich, and H. Kinder, *IEEE Trans. Appl. Supercond.* 7, 1272 (1997).

5. D. K. Finnemore, K. E. Gray, M. P. Maley, D. O. Welch, D. K. Christen, and D. M. Kroeger, *Physica C* 320, 1 (1999).

5a. R. H. Hammond, in "Advances in Superconductivity, Proceedings of the 11th ISS'98," November 16–19, Fukuoka, Vol. 11, No. 1, p. 43. Springer-Verlag, Tokyo, 1999.

6. E. H. Brandt, *Rep. Prog. Phys.* 58, 1465 (1995).

7. M. R. Koblischka and R. J. Wijngaarden, *Supercond. Sci. Technol.* 8, 199 (1995).

8. T. Schuster and M. R. Koblischka, in "Studies of High-Temperature Superconductors," (A. V. Narlikar, Ed.) Vol. 14, p. 1. Nova Publishers, Commack, NY, 1995.

9. M. Konczykowski, F. Holtzberg, and P. Lejay, *Supercond. Sci. Technol.* 4, S445 (1991).

9a. T. Tamegai, in "Advances in Superconductivity, Proceedings of the 5th ISS'92," November 16–19, Kobe, Vol. 5, p. 507. Springer-Verlag, Tokyo, 1993.

9b. M. Konczykowski, L. Burlachkov, Y. Yeshurun, and F. Holtzberg, *Physica C* 194, 155 (1994).

10. J. R. Kirtley, C. C. Tsuei, M. Rupp, J. Z. Sun, L. S. Yu-Jahnes, A. Gupta, M. B. Ketchen, K. A. Moler, and M. Bushan, *Phys. Rev. Lett.* 76, 1336 (1995).

11. S. J. Bending, *Adv. Phys.* 48, 449 (1999).

12. A. de Lozanne, *Supercond. Sci. Technol.* 12, R43 (1999).

13. Y. Abulafia, M. McElfresh, A. Shaulov, Y. Yeshurun, Y. Paltiel, D. Majer, H. Shtrikman, and E. Zeldov, *Appl. Phys. Lett.* 72, 2891 (1998).

13a. D. T. Fuchs, E. Zeldov, T. Tamegai, S. Ooi, M. Rappaport, and H. Shtrikman, *Phys. Rev. Lett.* 80, 4971 (1998).

13b. Y. Abulafia, A. Shaulov, Y. Wolfus, R. Prozorov, L. Burlachkov, Y. Yeshurun, D. Majer, E. Zeldov, H. Wuhl, V. B. Geshkenbein, and V. M. Vinokur, *Phys. Rev. Lett.* 79, 3795 (1997).

14. P. Esquinazi, A. Setzer, D. Fuchs, Y. Kopelevich, E. Zeldov, and C. Assmann, *Phys. Rev. B: Solid State* 60, 12,454 (1999).

15. E. Sheriff, R. Prozorov, Y. Yehurun, A. Shaulov, G. Koren, and C. Chabaud-Villard, *J. Appl. Phys.* 82, 4417 (1997).

16. M. W. McElfresh, E. Zeldov, J. R. Clem, M. Darwin, J. Deak, and L. Hou, *Phys. Rev. B: Solid State* 51, 9111 (1995).

17. A. Oral, S. J. Bending, and M. Henini, *Appl. Phys. Lett.* 69, 1324 (1996).

18. S. S. James, S. B. Field, J. Seigel, and H. Shtrikman, *Physica C* 332, 445 (2000).

19. R. P. Hübener, *Rep. Prog. Phys.* 47, 175 (1984).

20. D. Koelle and R. Gross, *Rep. Prog. Phys.* 57, 651 (1994).

21. S. Keil, R. Straub, R. Gerber, R. P. Hübener, D. Koelle, R. Gross, and K. Barthel, *IEEE Trans. Appl. Supercond.* 9, 2961 (1999).

22. J. R. Kirtley, M. Ketchen, K. W. Stawiasz, J. Z. Sun, W. J. Gallagher, S. H. Blanton, and S. J. Wind, *Appl. Phys. Lett.* 59, 2609 (1991).

22a. J. R. Kirtley, C. C. Tsuei, M. Rupp, J. Z. Sun, L. S. Yu-Jahnes, A. Gupta, M. B. Ketchen, K. A. Moler, and M. Bushan, *Phys. Rev. Lett.* 76, 1336 (1995);

23. A. Moser, H. J. Hug, Th. Jung, U. D. Schwarz, and H. J. Günterodt, *Meas. Sci. Technol.* 4, 769 (1993).

23a. A. Moser, H. J. Hug, I. Parashikov, B. Stiefel, O. Fritz, H. Thomas, A. Baratoff, H. J. Günterodt, and P. Chaudhari, *Phys. Rev. Lett.* 74, 1847 (1995).

24. H. J. Hug, A. Moser, I. Parashikov, B. Stiefel, O. Fritz, H. J. Günterodt, and H. Thomas, *Physica C* 235–240, 2695 (1994).

25. P. Vase, Y. Shen, and T. Freltoft, *Physica C* 180, 90 (1991).

26. R. P. Hübener, "Magnetic Flux Structures in Superconductors," Springer Verlag, New York, 1979.

27. M. R. Koblischka, in "Proceeding of the 2nd European Conference on Applied Superconductively," July 3–6, Edinburgh, IOP Conference Series, Series 148, p. 579 (1995).

27a. M. R. Koblischka, Thesis, University of Stuttgart, 1992.

28. M. Baziljevich, Thesis, University of Oslo, 1996.

29. Nelbauer Chemikalien, "Thermal Conducting Carbon Glue, Leit-C." Neubauer Chemikalien, Münster, Germany.

30. S. Senoussi, *J. Physique* 2, 1041 (1992).

31. A. J. J. van Dalen, R. Griessen, J. C. Martinez, P. Fivat, M. Triscone, and Ø. Fischer, *Phys. Rev. B: Solid State* 53, 896 (1996).

32. M. J. Scharen, A. H. Cardona, J. Z. Sun, L. C. Bourne, and J. R. Schrieffer, *Jpn. J. Appl. Phys.* 30, L15 (1991).

32a. H. Teshima, A. Oishi, H. Izumi, K. Ohata, T. Morishita, and S. Tanaka, *Appl. Phys. Lett.* 58, 2833 (1991).

32b. C. T. Blue, C. A. Blue, and P. Boolchand, *J. Appl. Phys.* 72, 1021 (1992).

32c. Chang Jun Liu, C. Schlenker, J. Schubert, and B. Stritzker, *Phys. Rev. B: Solid State* 48, 13,911 (1993).

32d. W. Xing, B. Heinrich, Hu Zhou, A. A. Fife, and A. R. Cragg, *J. Appl. Phys.* 76, 4244 (1994).

32e. K. Frikach, S. Senoussi, S. Hammond, A. Fert, and P. Manuel, *Physica C* 235–240, 2849 (1994).

32f. V. N. Trofimov and A. V. Kuznetsov, *Physica C* 235–240, 2853 (1994).

32g. A. Neminsky, J. Dumas, C. Schlenker, H. Karl, and B. Stritzker, *Physica C* 235–240, 2907 (1994).

32h. X. G. Qiu, N. Kanda, J. Nishio, M. Kawasaki, and H. Koinuma, *Physica C* 282–287, 1297 (1997).

33. G. Ravi Kumar, M. R. Koblischka, J. C. Martinez, R. Griessen, B. Dam, and J. Rector, *Physica C* 235–240, 3053 (1994).

33a. S. F. Kim, Z. Z. Li, and H. Raffy, *Physica C* 244, 78 (1995).

33b. A. V. Kuznetsov, A. A. Ivanov, D. V. Eremenko, and V. N. Trofimov, *Phys. Rev. B: Solid State* 52, 9637 (1995).

33c. E. V. Blinov, R. Laiho, E. Lähderanta, A. G. Lyublinsky, and K. B. Traito, *Phys. Rev. B: Solid State* 53, 79 (1996).

33d. R. Prozorov, A. Poddar, E. Sheriff, A. Shaulov, and Y. Yeshurun, *Physica C* 264, 27 (1996).

33e. E. P. Thrane, C. Schlenker, J. Dumas, and R. Buder, *Phys. Rev. B: Solid State* 54, 15,518 (1996).

33f. A. Wienss, G. Jakob, P. Voss-de Haan, and H. Adrian, *Physica C* 280, 158 (1997).

33g. H.-H. Wen, P. Ziemann, H. A. Radovan, and T. Herzog, *Physica C* 305, 185 (1998).

33h. M. Suzuki, K. Miyahara, S. Kubo, S. Karimoto, K. Tsuru, and K. Tanabe, in "Advances in Superconductivity, Proceedings of the 7th ISS'94," November 8–11, Kitakyushu, Vol. 7, No. 1, pp. 213–218. Springer-Verlag, Tokyo, 1995.

33i. S. K. Hasanian, I. Ahmad, and R. Semerad, *Supercond. Sci. Technol.* 12, 633 (1999).

34. M. R. Koblischka and M. Murakami, *Supercond. Sci. Technol.* 13, 738 (2000).

35. G. Blatter, M. V. Feigel'man, V. B. Geshkenbein, A. I. Larkin, and V. M. Vinokur, *Rev. Mod. Phys.* 66, 1125 (1994).

35a. M. V. Feigel'man, V. B. Geshkenbein, and V. M. Vinokur, *Phys. Rev. B: Solid State* 43, 6263 (1991).

36. Y. Yeshurun, A. P. Malozenoff, and A. Shaulov, *Rev. Mod. Phys.* 68, 911 (1996).

37. L. Půst, J. Kadlecova, M. Jirsa, and S. Durčok, *J. Low Temp. Phys.* 78, 179 (1990).

38. A. J. J. van Dalen, Thesis, Free University of Amsterdam, 1995.

39. A. Hoekstra, J. C. Martinez, and R. Griessen, *Physica C* 235–240, 2955 (1994).

40. R. Griessen, J. G. Lensink, T. A. M. Schroder, and B. Dam, *Cryogenics* 30, 563 (1990).

41. Y.-Q. Li, D. Noh, B. Gallois, G. S. Tompa, P. E. Norris, and P. A. Zawadzki, *J. Appl. Phys.* 68, 3775 (1990).

41a. R. B. Flippen, C. R. Fincher, D. W. Face, and W. L. Holstein, *Physica C* 193, 145 (1992).

41b. Q. H. Lam, C. D. Jeffries, P. Berdahl, R. E. Russo, and R. P. Reade, *Phys. Rev. B: Solid State* 46, 437 (1992).

41c. W. Xing, B. Heinrich, J. Chrzanowski, J. C. Irwin, H. Zhou, A. Cragg, and A. A. Fife, *Physica C* 205, 311 (1993).

41d. M. Ye, M. Mehbod, J. Schroeder, and R. Deltour, *Physica C* 235–240, 2999 (1994).

41e. X. Castel, M. Guilloux-Viry, A. Perrin, C. Le Paven Thivet, and J. Debuigne, *Physica C* 255, 281 (1995).

41f. M. Björnander, J. Magnusson, P. Svedlindh, P. Nordblad, D. P. Norton, and F. Wellhofer, *Physica C* 272, 326 (1996).

41g. M. Wurlitzer, M. Lorenz, K. Zimmer, and P. Espuinazi, *Phys. Rev. B: Solid State* 55, 11,816 (1997).

41h. B. J. Jonsson, K. V. Rao, S. H. Yun, and U. O. Karlsson, *Phys. Rev. B: Solid State* 58, 5862 (1998).

41i. S. A. Aruna, P. Zhang, F. Y. Lin, S. Y. Ding, and X. X. Yao, *Supercond. Sci. Technol.* 13, 356 (2000).

42. R. A. Hein, Th. I. Francavilla, and D. H. Liebenberg, Eds., "Magnetic Susceptibility of Superconductors and Other Spin Systems," Plenum Press, New York, 1991.

43. W. Xang, B. Heinrich, H. Zhou, A. A. Fife, and R. A. Cragg, *IEEE Trans. Appl. Supercond.* 5, 1420 (1995).

44. W. Xing, B. Heinrich, J. Chrzanowski, J. C. Irwin, H. Zhou, A. Cragg, and A. A. Fife, *Physica C* 205, 311 (1993).

45. Th. Herzog, A. Radovan, P. Ziemann, and E. H. Brandt, *Phys. Rev. B: Solid State* 56, 2871 (1997).

46. E. H. Brandt, *Phys. Rev. Lett.* 67, 2219 (1991).

46a. E. H. Brandt, *Phys. Rev. B: Solid State* 50, 13,833 (1994).

47. R. G. Mints and E. H. Brandt, *Phys. Rev. B: Solid State* 54, 12,421 (1996).

48. D. Gugan and O. Stoppard, *Physica C* 213, 109 (1993).

48a. J.R. Clem and A. Sanchez, *Phys. Rev. B: Solid State* 50, 9355 (1994).

48b. O. Stoppard and D. Gugan, *Physica C* 241, 375 (1995).

48c. E. Moraitakis, M. Pissas, and D. Niarchos, *Physica C* 235–240, 3211 (1994).

48d. A. Gurevich and E. H. Brandt, *Phys. Rev. B: Solid State* 55, 12,706 (1997).

49. R. Wördenweber, *Rep. Prog. Phys.* 62, 187 (1999).

50. B. Roas, L. Schultz, and G. Endres, *Appl. Phys. Lett.* 53, 1557 (1988).

51. P. Schmitt, P. Kummeth, L. Schultz, and G. Saemann-Ischenko, *Phys. Rev. Lett.* 67, 267 (1991).

52. H. Adrian, C. Tome-Rosa, G. Jakob, A. Walkenhorst, M. Maul, M. Schmitt, P. Przyslupski, G. Adrian, M. Huth, and Th. Becherer, *Supercond. Sci. Technol.* 4, S166 (1991).

53. A. Walkenhorst, C. Tome-Rosa, C. Stölzel, G. Jakob, M. Schmitt, and H. Adrian, *Physica C* 177, 165 (1991).

54. G. Jakob, M. Schmitt, Th. Kluge, C. Tome-Rosa, P. Wagner, Th. Hahn, and H. Adrian, *Phys. Rev. B: Solid State* 47, 12,099 (1993).

55. P. Fischer, H.-W. Neumüller, B. Roas, H. F. Braun, and G. Saemann-Ischenko, *Solid State Commun.* 72, 871 (1989).

56. G. Friedl, B. Roas, M. Romheld, L. Schultz, and W. Jutzi, *Appl. Phys. Lett.* 59, 2751 (1991).

56a. T. Schuster, M. R. Koblischka, H. Kuhn, H. Kronmüller, G. Friedl, B. Roas, and L. Schultz, *Appl. Phys. Lett.* 62, 768 (1993).

57. H. Yamasaki, K. Endo, S. Kosaka, M. Umeda, S. Yoshida, and K. Kajimura, *Phys. Rev. Lett.* 70, 3331 (1993).

58. R. Wördenweber and M. O. Abd-El-Hamed, *J. Appl. Phys.* 71, 808 (1992).

58a. R. Wördenweber *Phys. Rev. B: Solid State* 46, 3076 (1992).

59. M. C. Hellerqvist and A. Kapitulnik, *Phys. Rev. B: Solid State* 56, 5521 (1997).

60. H.-H. Wen, H. G. Schnack, R. Griessen, B. Dam, and J. Rector, *Physica C* 241, 353 (1995).

61. H.-H. Wen, P. Ziemann, H. A. Radovan, and S. L. Yan, *Europhys. Lett.* 42, 319 (1998).

61a. H.-H. Wen, Z.-X. Zhao, S.-L. Yan, L. Fang, and M. He, *Physica C* 312, 274 (1999).

62. J. Deak, M. W. McElfresh, D. W. Face, and L. W. Holstein, *Phys. Rev. B: Solid State* 52, R3880 (1995).

63. M. Friesen, J. Deak, L. Hou, and M. W. McElfresh, *Phys. Rev. B: Solid State* 54, 3525 (1996).

64. C. P. Bean, *Phys. Rev. Lett.* 8, 250 (1962).

64a. Y. B. Kim, C. F. Hempstead, and A. R. Strnad, *Phys. Rev. Lett.* 9, 306 (1962).

65. D. J. Frankel, *J. Appl. Phys.* 50, 5402 (1979).

66. M. Däumling and D. C. Larbalestier, *Phys. Rev. B: Solid State* 40, 9350 (1989).

67. M. Conner and A. P. Malozemoff, *Phys. Rev. B: Solid State* 43, 402 (1991).

68. H. Theuss, *Physica C* 190, 345 (1991).

69. E. H. Brandt, M. V. Indenbom, and A. Forkl, *Europhys. Lett.* 22, 735 (1993).

70. E. H. Brandt and M. V. Indenbom, *Phys. Rev. B: Solid State* 48, 12,893 (1993).

71. E. Zeldov, J. R. Clem, M. McElfresh, and M. Darwin, *Phys. Rev. B: Solid State* 49, 9802 (1994).

72. V. Kuznetsov, D. V. Eremenko, and V. N. Trofimov, *Phys. Rev. B: Solid State* 59, 1507 (1999).

73. K. V. Bhagwat and P. Chaddah, *Physica C* 224, 155 (1994).

73a. K. V. Bhagwat and P. Chaddah, *Physica C* 280, 52 (1997).

74. A. N. Artemov, *JETP Lett.* 68, 492 (1998).

75. E. H. Brandt, *Phys. Rev. B: Solid State* 54, 4246 (1996).

76. D. V. Shantsev, Y. M. Galperin, and T. H. Johansen, *Phys. Rev. B: Solid State* 61, 9699 (2000).

77. D. Glatzer, A. Forkl, H. Theuss, H.-U. Habermeier, and H. Kronmüller, *Phys. Status Solidi B* 170, 549 (1992).

78. T. Schuster, M. R. Koblischka, H. Kuhn, B. Ludescher, M. Leghissa, M. Lippert, and H. Kronmüller, *Physica C* 196, 373 (1992).

79. P. Brüll, D. Kirchgässner, and P. Leiderer, *Physica C* 182, 339 (1991).

80. P. Brüll, R. Steinke, P. Leiderer, J. Schubert, W. Zander, and B. Stritzker, *Supercond. Sci. Technol.* 5, 299 (1992).

81. D. Kirchgässner, P. Brüll, and P. Leiderer, *Physica C* 195, 157 (1992).

82. P. D. Grant, M. W. Denhoff, W. Xing, P. Brown, S. Govorkov, J. C. Irwin, B. Heinrich, H. Zhou, A. A. Fife, and A. R. Cragg, *Physica C* 229, 289 (1994).

83. S. A. Govorkov, A. F. Khapikov, B. Heinrich, J. C. Irwin, R. A. Cragg, and A. A. Fife, *Supercond. Sci. Technol.* 9, 952 (1996).

83a. S. Govorkov, A. A. Fife, G. Anderson, V. Haid, H. Zhou, B. Heinrich, and J. Chrzanowski, *IEEE Trans. Appl. Supercond.* 7, 3235 (1997).

84. Th. Schuster, H. Kuhn, E. H. Brandt, M. V. Indenbom, M. R. Koblischka, and M. Konczykowski, *Phys. Rev. B: Solid State* 50, 16,684 (1994).

85. J. McDonald and J. R. Clem, *Phys. Rev. B: Solid State* 53, 8643 (1996).

86. D. V. Shantsev, Y. M. Galperin, and T. H. Johansen, *Phys. Rev. B: Solid State* 60, 13,112 (1999).

87. E. H. Brandt, *Rep. Prog. Phys.* 58, 1465 (1995).

88. A. Gurevich and E. H. Brandt, *Phys. Rev. Lett.* 73, 178 (1994).

89. M. R. Koblischka, T. Schuster, B. Ludescher, and H. Kronmüller, *Physica C* 190, 557 (1992).

90. T. Schuster, H. Kuhn, and E. H. Brandt, *Phys. Rev. B: Solid State* 51, 697 (1995).

91. Th. Schuster, M. V. Indenbom, M. R. Koblischka, H. Kuhn, and H. Kronmüller, *Phys. Rev. B: Solid State* 49, 3443 (1994).

91a. Th. Schuster, H. Kuhn, E. H. Brandt, M. V. Indenbom, M. Kläser, G. Müller-Vogt, H. U. Habermeier, H. Kronmüller, and A. Forkl, *Phys. Rev. B: Solid State* 52, 10,375 (1995).

92. M. R. Koblischka, *Supercond. Sci. Technol.* 9, 271 (1996).

92a. M. R. Koblisehka, *Physica C* 259, 135 (1996).

93. A. Forkl, C. Jooss, R. Warthmann, H. Kronmüller, and H. U. Habermeier, *J. Alloys Comp.* 251, 146 (1997).

94. A. Forkl and H. Kronmüller, *Phys. Rev. B: Solid State* 52, 16,130 (1995).

95. M. E. Gaevski, A. V. Bobyl, D. V. Shantsev, Y. M. Galperin, T. H. Johansen, M. Baziljevich, H. Bratsberg, and S. F. Karmanenko, *Phys. Rev. B: Solid State* 59, 9655 (1999).

96. D. V. Shantsev, M. R. Koblischka, Y. Galperin, T. H. Johansen, L. Půst, and M. Jirsa, *Phys. Rev. Lett.* 82, 2947 (1999).

97. T. Frello et al., *Phys. Rev. B: Solid State* 59 R6639 (1999).

97a. M. R. Koblischka, M. Murakami, T. H. Johansen, M. Baziljevich, T. Frello, and T. Wolf, *J. Low Temp. Phys.* 117, 1483 (1999).

98. Th. Schuster, H. Kuhn, M. R. Koblischka, H. Theuss, H. Kronmüller, M. Leghissa, M. Kraus, and G. Saemann-Ischenko, *Phys. Rev. B: Solid State* 47 (1993) 373.

99. P. N. Mikheenko and Yu. E. Kuzovlev, *Physica C* 204, 229 (1993).

100. E. H. Brandt, *Phys. Rev. Lett.* 74, 3025 (1995).

101. Th. Schuster, H. Kuhn, E. H. Brandt, M. V. Indenbom, M. Kläser, G. Müller-Vogt, H.-U. Habermeier, H. Kronmüller, and A. Forkl, *Phys. Rev. B: Solid State* 52, 10,375 (1995).

102. A. E. Pashitski, A. Gurevich, A. A. Polyanskii, D. C. Larbalestier, A. Goyal, E. D. Specht, D. M. Kroeger, J. A. DeLuca, and J. E. Tkaczyk, *Science* 275, 167 (1997).

103. R. J. Wijngaarden, H. J. W. Spoelder, R. Surdeanu, and R. Griessen, *Phys. Rev. B: Solid State* 54, 6742 (1996).

104. T. H. Johansen, M. R. Koblischka, Yu. Galperin, and M. Baziljevich, in preparation.

105. T. H. Johansen, M. Baziljevich, H. Bratsberg, Y. Galperin, P. E. Lindelof, Y. Shen and P. Vase, *Phys. Rev. B: Solid State* 54, 16,264 (1996).

106. T. H. Johansen, private communication.

107. M. R. Koblischka, T. H. Johansen, H. Bratsberg, Y. Shen, and P. Vase, *J. Phys. C: Solid State Phys.* 9, 10,909 (1997).

108. A. J. J. van Dalen, R. Griessen, S. Libbrecht, Y. Bruynseraede, and E. Osquiguil, *Phys. Rev. B: Solid State* 54, 1366 (1996).

109. M. Baziljevich, T. H. Johansen, H. Bratsberg, Y. Shen, and P. Vase, *Appl. Phys. Lett.* 69, 3590 (1996).

110. A. M. Campbell and J. E. Evetts, *Adv. Phys.* 21, 199 (1972).

111. R. Surdeanu, R. Wijngaarden, B. Dam, J. Rector, R. Griessen, C. Rosseel, Z. F. Ren, and J. H. Wang, *Phys. Rev. B: Solid State* 58, 12,467 (1998).

111a. R. Surdeanu, R. J. Wijngaarden, E. Visser, J. M. Huijbregtse, J. H. Rector, B. Dam, and R. Griessen, *Phys. Rev. Lett.* 83, 2054 (1999).

112. A. A. Polyanskii, A. Gurevich, A. E. Pashitski, N. F. Heinig, R. D. Redwing, J. E. Nordman, and D. C. Larbalestier, *Phys. Rev. B: Solid State* 53, 8687 (1996).

112a. A. Polyanskii, A. Pashitski, A. Gurevich, J. A. Parrell, M. Polak, D. C. Larbalestier, S. R. Foltyn, and P. N. Arendt, in "Advances in Superconductivity, Proceedings of the 9th International Symposium on Superconductivity (ISS '96)," October 21–24, 1996, Sapporo, Vol. 9, No. 1, p. 469. Springer-Verlag, Tokyo, 1997.

113. N. F. Heinig, G. A. Daniels, M. Feldmann, A. Polyanskii, D. C. Larbalestier, P. Arendt, and S. Foltyn, *IEEE Trans. Appl. Supercond.* 9, 1614 (1999).

113a. M. B. Field, A. Pashitski, A. Polyanskii, D. C. Larbalestier, A. S. Parikh, and K. Salama, *IEEE Trans. Appl. Supercond.* 5, 1631 (1995).

114. G. Deutscher and K. A. Muller, *Phys. Rev. Lett.* 59, 1745 (1987).

115. P. Chaudhari, J. Mannhart, D. Dimos, C. C. Tsuei, J. Chi, J. Oprysko, and M. Scheuermann, *Phys. Rev. Lett.* 60, 1653 (1988).

115a. D. Dimos, P. Chaudhari, J. Mannhart, and F. K. Le-Goues, *Phys. Rev. Lett.* 61, 219 (1988).

115b. J. Mannhart, P. Chaudhari, D. Dimos, C. C. Tsuei, and T. R. McGuire, *Phys. Rev. Lett.* 61, 2476 (1988).

116. D. Dimos, P. Chaudhari, and J. Mannhart, *Phys. Rev. B: Solid State* 41, 4038 (1990).

117. Z. G. Ivanov, P. A. Nilsson, D. Winkler, J. A. Alarco, T. Claeson, E. A. Stepantsov, and A. Ya. Tzalenchuk, *Appl. Phys. Lett.* 59, 3030 (1991).

118. D. K. Lathrop, B. H. Moeckly, S. E. Russek, and R. A. Buhrman, *Appl. Phys. Lett.* 58, 1095 (1991).

119. M. Strikovsky, G. Linker, S. Gaponov, I. Mazo, and O. Mayer, *Phys. Rev. B: Solid State* 45, 12,522 (1992).

120. R. Gross and B. Mayer, *Physica C* 180, 235 (1991).

121. R. Gross, in "Interfaces in High-T_c Superconducting Systems" (S. L. Shinde and D. A. Rudman, Eds.), p. 176. Springer-Verlag, New York, 1993.

122. M. F. Chisholm, and S. J. Pennycook, *Nature (London)* 351, 47 (1991).

123. A. H. Cardona, H. Suzuki, T. Yamashita, K. H. Young, and L. C. Bourne, *Appl. Phys. Lett.* 62, 411 (1993).

124. B. Mayer, L. Alff, T. Trauble, and R. Gross, *Appl. Phys. Lett.* 63, 996 (1993).

125. M. Kawasaki, E. Sarnelli, P. Chaudhari, A. Gupta, A. Kussmaul, J. Lacey, and W. Lee, *Appl. Phys. Lett.* 62, 417 (1993).

126. E. Sarnelli, P. Chaudhari, W. Y. Lee, and E. Esposito, *Appl. Phys. Lett.* 65, 362 (1994).

127. T. Nabatame, S. Koike, O. B. Hyun, and I. Hirabayashi, *Appl. Phys. Lett.* 65, 776 (1994).

128. A. Marx, U. Fath, W. Ludwig, R. Gross, and T. Amrein, *Phys. Rev. B: Solid State* 51, 6735 (1995).

129. T. Amrein, L. Schultz, B. Kabius, and K. Urban, *Phys. Rev. B: Solid State* 51, 6792 (1995).

130. N. F. Heinig, R. D. Redwing, I. F. Tsu, A. Gurevich, J. E. Nordman, S. E. Babcock, and D. C. Larbalestier, unpublished observations.

131. M. R. Koblischka, A. Das, M. Muralidhar, N. Sakai, and M. Murakami, *Jpn. J. Appl. Phys.* 37, L1227 (1998).

132. A. B. Riise, T. H. Johansen, H. Bratsberg, M. R. Koblischka, and Y. Q. Shen, *Phys. Rev. B: Solid State* 60, 9855 (1999).

133. M. Kuhn, Thesis, University of Augsburg, 2000.

134. M. Kuhn, B. Schey, W. Biegel, B. Stritzker, J. Eisenmenger, and P. Leiderer, *Rev. Sci. Instrum.* 70, 1761 (1999).

135. M. Kuhn, B. Schey, R. Klarmann, W. Biegel, B. Stritzker, J. Eisenmenger, and P. Leiderer, *Physica C* 294, 1 (1997).

136. M. Kuhn, B. Schey, W. Biegel. B. Stritzker, J. Eisenmenger, P. Leiderer, B. Heismann, H.-P. Kramer, and H.-W. Neumüller, *IEEE Trans. Appl. Supercond.* 9, 1844 (1999).

137. W. Xing, B. Heinrich, Hu Zhou, A. A. Fife, and A. R. Cragg, *J. Appl. Phys.* 76, 4244 (1994).

138. J. Eisenmenger, J. Schiessling, U. Bolz, B.-U. Runge, P. Leiderer, M. Lorenz, H. Hochmuth, M. Wallenhorst, and H. Dötsch, *IEEE Trans. Appl. Supercond.* 9, 1840 (1999).

139. J. R. Clem, *Physica C* 153–155, 50 (1988).

140. G. Waysand, *Europhys. Lett.* 5, 73 (1988).

141. M. R. Koblischka, L. Půst, A. Galkin, and P. Nalevka, *Appl. Phys. Lett.* 70, 514 (1997).

142. K.-H. Müller, C. Andrikis, and Y. C. Guo, *Phys. Rev. B: Solid State* 55, 630 (1997).

143. M. R. Koblischka, T. H. Johansen, H. Bratsberg, and P. Vase, *Supercond. Sci. Technol.* 12, 113 (1999).

144. M. R. Koblischka, T. H. Johansen, and H. Bratsberg, *Supercond. Sci. Technol.* 10, 693 (1997).

145. M. R. Koblischka, L. Půst, A. Galkin, P. Nalevka, M. Jirsa, T. H. Johansen, H. Bratsberg, B. Nilsson, and T. Claeson, *Phys. Status Solidi A* 167, R1 (1998).

145a. M. R. Koblischka, L. Půst, A. Galkin, P. Nalevka, M. Jirsa, T. H. Johansen, H. Bratsberg, B. Nilsson, and T. Claeson, *Phys. Rev. B* 59, 12,114 (1999).

146. M. R. Koblischka, L. Půst, N. Chikumoto, M. Murakami, B. Nilsson, and T. Claeson, *Physica B* 284–288, 599 (2000).

147. D.-X. Chen and R. B. Goldfarb, *J. Appl. Phys.* 66, 2489 (1989).

147a. P. O. Hetland, T. H. Johansen, and H. Bratsberg, *Appl. Supercond.* 3, 585 (1995).

147b. T. H. Johansen and H. Bratsberg, *J. Appl. Phys.* 77, 3945 (1995).

148. T. C. Shields, J. B. Langhorn, S. C. Watcham, J. S. Abell, and T. W. Button, *IEEE Trans. Appl. Supercond.* 7, 1478 (1997).

149. P. Leiderer, J. Boneberg, P. Brüll, V. Bujok, and S. Herminghaus, *Phys. Rev. Lett.* 71, 2646 (1993).

150. V. Bujok, P. Brüll, J. Boneberg, S. Herminghaus, and P. Leiderer, *Appl. Phys. Lett.* 63, 412 (1993).

151. U. Bolz, J. Eisenmenger, J. Schiessling, B.-U. Runge, and P. Leiderer, *Physica B* 284–288, 757 (2000).

152. B.-U. Runge, U. Bolz, J. Boneberg, V. Bujok, P. Brüll, J. Eisenmenger, J. Schiessling, and P. Leiderer, *Laser Phys.* 10, 53 (2000).

153. C. A. Duran, P. L. Gammel, R. E. Miller, and D. J. Bishop *Phys. Rev. B: Solid State* 52, 75 (1995).

154. M. R. Koblischka, *Physica C* 282–287, 2193 (1997).

155. D. Dew-Hughes, *Philos. Mag.* 30, 293 (1974).

156. E. J. Kramer, *J. Appl. Phys.* 44, 1360 (1973).

157. W. A. Fietz, M. R. Beasley, J. Silcox, and W. W. Webb, *Phys. Rev.* 136, 335 (1964).

158. C. Jooss, R. Warthmann, H. Kronmüller, T. Haage, H.U. Habermeier, and J. Zegenhagen, *Phys. Rev. Lett.* 82, 632 (1999).

159. B. Dam, J. M. Hujibregtse, F. C. Klaassen, R. C. F. van der Geest, G. Doornbos, J. H. Rector, A. M. Testa, S. Freisem, J. C. Martinez, B. Stäuble-Pümpin, and R. Griessen, *Nature (London)* 399, 439 (1999).

160. L. Půst, D. Dlouhy, and M. Jirsa, *Supercond. Sci. Technol.* 9, 814 (1996).

160a. D. Dlouhy, L. Půst, M. Jirsa, and V. Gregor, "Proc. 7th IWCC," Alpbach, 24.1.–27.1 1994, Austria (H. W. Weber, Ed.), p. 121, World Scientific, Singapore, 1994.

161. L. Civale, M. W. McElfresh, A. D. Marwick, F. Holtzberg, C. Feild, J. R. Thompson, and D. K. Christen, *Phys. Rev. B: Solid State* 43, 13,732 (1991).

161a. L. Klein, E. R. Yacoby, Y. Yeshurun, A. Erb, G. Müller-Vogt, V. Breit, and H. Wühl, *Phys. Rev. B: Solid State* 49, 4403 (1994).

161b. J. S. Satchell, R. G. Humphreys, N. G. Chow, J. A. Edwards, and M. J. Kane, *Nature (London)* 334, 331 (1988).

161c. T. Nishizaki, T. Aomine, I. Fujii, K. Yamamoto, S. Yoshii, T. Terashima, and Y. Bando, *Physica C* 181, 223 (1991).

161d. H. Yamasaki, K. Endo, S. Kosaka, M. Umeda, S. Yoshida, and K. Kajimura, *Phys. Rev. Lett.* 70, 3331 (1993).

162. R. Griessen, H. H. Wen, A. J. J. van Dalen, B. Dam, J. Rector, H. G. Schnack, S. Libbrecht, E. Osquiguil, and Y. Bruynseraede, *Phys. Rev. Lett.* 72, 1910 (1994).

163. H. W. Weber, private communication.

Index